Test and Evaluation of Aircraft Avionics and Weapon Systems

Test and Evaluation of Aircraft Avionics and Weapon Systems

2nd Edition

Robert E. McShea

Edison, NJ
theiet.org

Published by SciTech Publishing, an imprint of the IET.
www.scitechpub.com
www.theiet.org

First edition 2010
Second edition 2014

Editor: Dudley R. Kay

10 9 8 7 6 5 4 3 2 1

ISBN 978-1-61353-176-1 (hardback)
ISBN 978-1-61353-180-8 (PDF)

Typeset in India by MPS Limited
Printed in the USA by Sheridan Books Inc.
Printed in the UK by CPI Group (UK) Ltd, Croydon

This Book is Dedicated to
Marlenni Belinni
Friends till the end

Contents

Preface xv
Preface to First Edition xix
List of Acronyms xxi

---

1   What is Avionics Flight Test and Why Do We Need It 1

1.0   Avionics Flight Test   1
1.1   Historical Background   2
1.2   Flight Test Organization   4
1.3   Flight Test Objectives   5
1.4   The Need for Flight Test   6
1.5   Classifying the Program   8
1.6   Contractual Requirements   9
1.7   Test Team Composition   11
1.8   Team Member Responsibilities   13
1.9   Understanding Terms   15
1.10   Contractor and Subtier Specifications   20
1.11   Customer Expectations   21
1.12   Formulating the Program   22
1.13   Summary   25
1.14   Selected Questions for Chapter 1   26

---

2   Time, Space, Position Information 29

2.0   An Overview of Time, Space, Position Information   29
2.1   Sources of TSPI   30
2.2   Ground-Based Tracking Radar   30
2.3   Radar Characteristics   32
2.4   Radar Accuracies   35
2.5   Geometric Dilution of Precision   38
2.6   Earth Models   39
2.7   Theodolites   41
2.8   Theodolite Limitations   44
2.9   Global Positioning System   44
2.10   Stand-Alone GPS Receivers   44

2.11 Inertial-Aided GPS Receivers   46

2.12 Differentially Corrected GPS   48

2.13 Sensor Fusion   48

2.14 Ranges   49

2.15 Range Assets   49

2.16 Unique Services   50

2.17 Interfacing with the Range   51

2.18 Time Alignment   53

2.19 Summary   55

2.20 Exercises   55

2.21 Answers to Exercises   55

2.22 Selected Questions for Chapter 2   58

---

3 MIL-STD-1553 and Digital Data Busses: Data Reduction and Analysis   61

3.0 Overview   61

3.1 Historical Background   62

3.2 1553 System Architecture   62

3.3 A Bit About Bits   65

3.4 1553 Word Types   67

3.5 Data Encoding   67

3.6 Word Formats   68

3.7 Command Words   68

3.8 Anomalous Command Word Conditions   70

3.9 Command Word Summary   71

3.10 Status Words   71

3.11 Data Words   73

3.12 Message Contents   74

3.13 Bus Controller Design   75

3.14 Sample Bus Configurations   76

3.15 The Flight Tester's Task   79

3.16 MIL-STD-1553 Summary   80

3.17 Other Data Busses   81

3.18 Potential Problems for the Analyst   122

3.19 Data Acquisition, Reduction, and Analysis   124

3.20 Selected Questions for Chapter 3   136

---

4 Communications Flight Test   139

4.0 Overview   139

4.1 Communications Basics   140

4.2   Aircraft Communications Equipment   142

4.3   Test Requirements   144

4.4   The Three Steps in Avionics Testing   144

4.5   Communications Test Plan Matrix   146

4.6   Executing the Matrix   148

4.7   Other Considerations in Communications Tests   149

4.8   Effects of Stores, Landing Gear, and Flaps   149

4.9   Effects of Weather   149

4.10 Logistics   150

4.11 Boredom   151

4.12 Speech Intelligibility   151

4.13 Electromagnetic Interference/Electromagnetic Compatibility   154

4.14 EMI/EMC Testing   158

4.15 EMI/EMC Test Issues   161

4.16 EMI/EMC Elimination   161

4.17 Selected Questions for Chapter 4   164

5   Navigation Systems   167

5.0   Introduction   167

5.1   History   168

5.2   Basic Navigation   168

5.3   Radio Aids to Navigation   175

5.4   Radio Aids to Navigation Testing   184

5.5   Inertial Navigation Systems   191

5.6   Doppler Navigation Systems   207

5.7   Global Navigation Satellite Systems (GNSS)   213

5.8   Identification Friend or Foe   234

5.9   Data Links   238

5.10 Selected Questions for Chapter 5   259

6   Part 23/25/27/29 Avionics Civil Certifications   267

6.0   Introduction   267

6.1   FAA Type Certification History   268

6.2   Federal Aviation Regulations in the Code of Federal Regulations   270

6.3   Other Rules and Guidance   273

6.4   FAA Type Certification Process   274

6.5   Avionics Software Considerations in the Technical Standard Order Process   282

6.6   Certification Considerations for Highly Integrated or Complex Systems   285

6.7   Important Notes for Evaluators Concerning Documentation   298

6.8   Differences between EASA and FAA Documentation   300

6.9   Cockpit Controls and Displays Evaluations   301

6.10 Weather RADAR Certification   322

6.11 Airworthiness Approval of Positioning and Navigation Systems   329

6.12 Reduced Vertical Separation Minimums   348

6.13 Proximity Warning Systems   356

6.14 Terrain Awareness and Warning System   382

6.15 Flight Guidance Systems   410

6.16 Landing Systems   459

6.17 Flight Management Systems   504

6.18 Enhanced Vision Systems (EVS)   520

6.19 Summary   527

6.20 Selected Questions for Chapter 6   528

7   Electro-optical and Infrared Systems   535

7.0   Introduction   535

7.1   Infrared History   536

7.2   IR Radiation Fundamentals   547

7.3   IR Sources   552

7.4   The Thermal Process   554

7.5   Atmospheric Propagation of Radiation   556

7.6   Target Signatures   562

7.7   EO Components and Performance Requirements   567

7.8   Passive EO Devices   581

7.9   Laser Systems   588

7.10 Passive EO Flight Test Evaluations   594

7.11 Active EO Systems   614

7.12 Selected Questions for Chapter 7   618

8   Radio Detection and Ranging – Radar   625

8.0   Introduction   625

8.1   Understanding Radar   626

8.2   Performance Considerations   629

8.3   Radar Utility   630

8.4   Radar Detections   636

8.5   Maximum Radar Detection   641

8.6   Radar Sample Applications   645

8.7   Pulse Delay Ranging Radar Modes of Operation and Testing   652

8.8   Doppler and Pulse Doppler Modes of Operation and Testing   668

8.9   Air-to-Ground Radar   694

8.10   Millimetric Wave Radar   710

8.11   Miscellaneous Modes of Radar   712

8.12   Some Final Considerations in Radar Testing   712

8.13   Selected Questions for Chapter 8   714

9   Electronic Warfare   719

9.0   Introduction   719

9.1   Electronic Warfare Overview   720

9.2   The Threat   724

9.3   Air Defense Systems   728

9.4   Electronic Attack   729

9.5   Noise Jamming   734

9.6   Deception Jamming   738

9.7   Chaff Employment   742

9.8   Flare Employment   743

9.9   Electronic Protection Measures   745

9.10   Electronic Warfare Systems Test and Evaluation   748

9.11   Finally   760

9.12   Selected Questions for Chapter 9   760

10   Air-to-Air/Air-to-Ground Weapons Integration   763

10.0   Introduction   763

10.1   Weapons Overview   764

10.2   Stores Management System   767

10.3   Air-to-Air Missiles   785

10.4   Air-to-Ground Weapons   803

10.5   MIL-HDBK-1763 Test Requirements   812

10.6   AGARD Flight Test Techniques Series, Volume 10 Requirements   820

10.7   Weapons Delivery Considerations for Helicopters   824

10.8   Selected Questions for Chapter 10   828

11   A Typical Avionics Integration Flight Test Program   831

11.0   Introduction   831

11.1   Vehicle Test Requirements   831

11.2   Avionics Test Requirements   833

11.3   Test Planning   835

11.4 Responsibilities of the Test Team   843

11.5 Analysis and Reporting   848

11.6 Selected Questions for Chapter 11   849

---

12   Unmanned Aerial Vehicles (UAV)   851

12.0 Introduction   851

12.1 UAV Types   852

12.2 Interoperability   856

12.3 The Airworthiness Certificate   857

12.4 UAS Communications Architecture   860

12.5 Navigation   886

12.6 Autopilots   889

12.7 Sense and Avoid Systems   892

12.8 Payload   899

12.9 Optionally Piloted Aircraft (OPA)   901

12.10 Summary   902

12.11 Selected Questions for Chapter 12   902

---

13   Night Vision Imaging Systems (NVIS) and Helmet Mounted Displays (HMD)   907

13.0 Introduction   907

13.1 Overview   908

13.2 Image Intensification (I^2) Technology   909

13.3 NVG Human Factors Issues   918

13.4 Lighting Specifications   920

13.5 Interior NVIS Lighting Methods   924

13.6 Exterior Lighting Methods   929

13.7 Test and Evaluation of NVIS Equipment   931

13.8 Some Final Considerations   938

13.9 Helmet Mounted Display Systems (HMD)   939

13.10 HMD Components   942

13.11 Test and Evaluation of HMD Equipment   949

13.12 Selected Questions for Chapter 13   954

---

14   Acquisition, Test Management, and Operational Test and Evaluation   957

14.0 Overview   957

14.1 Applicable Documentation   958

14.2 The Acquisition Process   958

14.3 The Operational Requirements Document (ORD)   965

14.4 The Test and Evaluation Master Plan (TEMP)   966

14.5 Operational Test and Evaluation   967

14.6 OT&E Test Plan Structure   970

14.7 Reliability, Maintainability, Logistics Supportability, and Availability (RML&A)   978

14.8 Summary   981

14.9 Selected Questions for Chapter 14   981

Index   983

Preface

Overview and Goals

The original purpose of this text was to offer the test community, especially the Avionics and Weapons Systems Test Community, a primer on the basics of Test and Evaluation. The text is a compilation and refinement of experiences and methods gleaned from a multitude of programs and test engineers over the past 30 years, and is designed to help students and readers avoid common mistakes. The text is an invaluable companion to evaluators of fighter or heavy aircraft systems, military or civilian programs both manned and unmanned.

A question which is commonly asked is "Where is the Flight Test Cookbook?" Unfortunately, there is no such book, and readers may be disappointed that this text is not a cookbook either. Each task undertaken in the real world is unique and requires some rework to the established rules. Avionics Flight Test is no different, and in fact is probably more dynamic than any other area of test. An engineer, in order to be successful, must be schooled in the fundamentals of avionics test, understand the ground-rules and possess a knowledge of similar systems which have been previously tested. This knowledge, with a little bit of common sense, engineering judgment and luck makes an Avionics Flight Test Engineer. This text will identify the ground rules and provide information on systems which were previously tested. Lessons learned (good and bad) are addressed in each chapter. Exercises to reinforce the instructional material are also included. Engineering judgment was issued to you when you graduated college (whether you know it or not) and you must supply the common sense. Luck normally comes to those who know what they're doing and is only viewed as luck by those who don't know what they're doing.

The best features of the book are its readability and usability. The author has received numerous comments and feedback that the text is being used on a daily basis throughout the industry. The text has been remarkably successful and the author is honored that it is currently being used as course reference in many short courses and schools around the world. The revision will keep the format, style and form of the original text. The intent of this revision is not to change the flavor of the original text, but rather expand and improve upon it.

Significant Changes to the Original Text

Comments received from readers indicated a desire to include additional material covering a multitude of systems. The system which generated the most interest was the unmanned aircraft program as it has gained much attention in the past five years. A new chapter (Chapter 12) is now devoted to Unmanned Aerial Vehicles (UAV), concentrating on the systems aspects of these programs. The chapter builds on previous sections of the text and then expounds on some of the difficulties encountered when evaluating these systems. A second request was a natural extension of EO/IR evaluations: Night Vision Systems and Helmet Mounted Displays found in Chapter 13. An Operational Test and Evaluation chapter was added to discuss the differences between development and operational testing in finer detail than previously covered.

Due to significant changes in civilian regulatory and advisory material, Chapter 6, Civil Certification, has been entirely re-written. Comments from rotary wing readers requested that helicopter unique certification issues be addressed and they are now included in the chapter. Other sections throughout the text have been edited and updated based on the latest advances in technologies. An example is the enhancement of Global Navigation Satellite Systems (GNSS) and Space Based Augmentation Systems (SBAS) found in Chapter 5.

Rationale

The author has been working on this book for a number of years with the express purpose of providing students a reference to fall back on when involved in a multitude of programs. There are textbooks that describe Avionics Systems, but not how to evaluate them for their usefulness. There are textbooks available that describe acceptable test techniques for evaluating an aircraft's handling and flying qualities but none devoted to test techniques for evaluating avionics and weapons systems. Many of the modules, and hence the Chapters, are unique, and to my knowledge, are not taught anywhere else but at the National Test Pilot School. The text provides these modules to the Test and Evaluation community in the hopes that it will make life a little simpler.

Audience

The audience for this book will encompass Flight Test Engineers, Program Managers, military and civilian aerospace workers, Flight Test Range and Instrumentation support. The book is appropriate for newly hired or assigned personnel at these locations as well as seasoned veterans in the Test and Evaluation areas. The text is not limited to U.S. locations as it will find practical applications in the international arena, particularly for aerospace companies and flight test centers.

Organization of Content

This book is written to accompany a six month course of instruction in Avionics and Weapons Systems Evaluation which was developed by the author and currently presented at the National Test Pilot School in Mojave, CA. The book reflects the culmination of the development of this course as accredited by ABET and approved by the State of California as a requisite for a Master of Science Degree in Flight Test Engineering. The chapters of the book are structured to each 40 hour course of curriculum taught during this six month period and are sequenced exactly as they are in the syllabus. The preliminary chapters introduce the student to the fundamentals of Test and Evaluation and their relationship to aircraft avionics and weapons systems, whereas subsequent chapters discuss specific aircraft systems such as radar, IR Imaging devices, navigation systems and weapons. Chapter 6 deals specifically with civil certification procedures and airworthiness and will prove invaluable to those individuals seeking to obtain airworthiness certification for avionic systems installed in civilian aircraft.

Each chapter covers the basic theory of operation of the systems discussed but does not attempt to re-write the many excellent textbooks available. References used in the course are cited for the reader when additional information on theory or operation is required. Formulas are not developed but are included when required to understand the subject. Each chapter concludes with the methods to successfully evaluate a multitude of aircraft avionics and weapons systems. Sample questions for each Chapter are used to reinforce key ideas and concepts.

Acknowledgments

The author would like to first and foremost thank the hundreds of students who have used draft copies of this text and provided comments, corrections and personal experiences to make it a better reference. The author owes a special debt of gratitude to Chuck Antonio, Aerospace Medicine/Night Vision Systems Pilot Instructor, National Test Pilot School, Doctor of Medicine, Medical University of South Carolina, Charleston SC, Graduate, US Navy Flight Training and Flight Surgeon School, NAS Pensacola FL, and Edward "Dick" L. Downs, Night Vision Systems Pilot Instructor, National Test Pilot School, BSc Aeronautical Engineering, Imperial College London, without whose help the chapter on Night Vision Systems would not have been possible. Also, a sincere thank you should be given to the numerous community reviewers who gave of their time to improve the draft manuscript. Finally, thanks to SciTech- IET Editor Dudley Kay and his colleagues for their work in bringing this title to fruition.

Despite all the reviewing and editing, any errors that remain are my own responsibility. I would very much appreciate being informed of errors found or suggestions for improvement to subsequent printings and editions.

Bob McShea
Deputy Director for Systems Flight Test Instruction,
Flight Test Engineer Instructor
National Test Pilot School
Mojave, CA
bmcshea@ntps.edu
February 2014

Preface to First Edition

The purpose of this text is to offer the test community, especially the avionics and weapons systems flight test community, insight into the differences in avionics testing from our brethren in the vehicle test world. Whenever flight test is mentioned in conversation, thoughts are immediately drawn to visions of those magnificent men in their flying machines, breaking the sound barrier, performing max performance turns, pressing the "q" limit, or some other glorious adventure. In reality, most of today's flight testing involves the upgrading of aircraft with state-of-the-art avionics equipment, making bombs "smart," and providing aircrew members with current, correct, and understandable information. Strain gauges, pressure transducers, and temperature probes that provide information in hundreds of samples per second have given way to MIL-STD-1553, ARINC 429, and high-speed data busses that provide information in megabits per second. Vehicle flight test has not been replaced with avionics flight test, and the differences between the two are substantial. A newer type of flight test, with its own set of rules, has evolved over the last 10 to 15 years; this text attempts to explain those changes and identify what is expected of today's avionics flight test engineer.

A question that is commonly asked is, "Where is the flight test cookbook?" Unfortunately, there is no such book, and readers may be disappointed that this text is not a cookbook either. Each task undertaken in the real world is unique and requires some rework to the established rules. Avionics flight test is no different and, in fact, is probably more dynamic than any other area of testing. An engineer, in order to be successful, must be schooled in the fundamentals of avionics testing, understand the ground rules, and possess a knowledge of similar systems that have been previously tested. This knowledge, with a little bit of common sense, engineering judgment, and luck, makes an avionics flight test engineer. This text will identify the ground rules and provide information on systems that have been tested previously. Lessons learned (good and bad) are addressed in each chapter. Exercises to reinforce the instructional material are also included. Engineering judgment was issued to you when you graduated college (whether you know it or not) and you must supply the common sense. Luck normally comes to those who know what they're doing and is only viewed as luck by those who don't know what they're doing.

This text is a compilation of experiences and methods gleaned from a multitude of programs and test engineers solicited over the past 25 years, and hopefully will prevent the reader from making some of the same mistakes that were made previously. Humor is included in the discussions, and the "war stories" are true. The text will prove to be an invaluable companion to fighter and heavy aircraft, U.S. and foreign national, and military and contractor testers. Happy testing.

List of Acronyms

A

A/A	Air-to-Air
A/C	Aircraft
A/D	Analog to Digital
A/G	Air-to-Ground
A/RFM	Airplane/Rotorcraft Flight Manual
AAA	Anti-aircraft Artillery
AAD	Assigned Altitude Deviation
ABC	Auto Brightness Control
ABF	Annular Blast Fragmentation
ABL	Airborne LASER
ABS	Advanced Brake System
ABSAA	Airborne Sense-And-Avoid
AC	Alternating Current
AC	Advisory Circular
ACAS	Active Collision and Avoidance Systems
ACC	Air Combat Command
ACK	Acknowledgment
ACK	Acknowledge
ACLS	Automatic Carrier Landing System
ACM	Air Combat Maneuvering
ACO	Aircraft Certification Office
AD	Airworthiness Directives
ADC	Air Data Computer
ADI	Attitude Direction Indicator
ADL	Armament Datum Line
ADL	Avionics Development Lab
ADS	Automatic Dependent Surveillance
ADS-B	Automatic Dependent Surveillance Broadcast

AEEC	Airlines Electronic Engineering Committee
AEG	Aircraft Evaluation Group
AEIS	Aircraft/Store Electrical Interconnection System
AEWS	Airborne Early Warning System
AFDX	Avionics Full Duplex Switched Ethernet
AFOTEC	Air Force Operational Test and Evaluation Center
AGARD	Advisory Group for Aerospace Research and Development
AGC	Automatic Gain Control
AGHME	Aircraft Geometric Height Measurement Element
AGL	Above Ground Level
AH	Absolute Humidity
AHRS	Attitude Heading Reference System
AI	Airborne Intercept
AIAA	American Institute of Aeronautics and Astronautics
AIF	Avionics Integration Facility
AIL	Avionics Integration Lab
AIM	Aeronautical Information Manual
ALARM	Air Launched Anti-radiation Missiles
ALI	Aerospace Lighting Institute
AM	Aeromechanical
AM	Amplitude Modulation
AMC	Acceptable Method of Compliance
AMLCD	Active-matrix LCD

AMRAAM	Advanced Medium Range Air-to-Air Missile		ATC	Air Traffic Control
ANL	Automatic Noise Leveling		ATCRBS	Air Traffic Control RADAR Beacon System
ANP	Actual Navigation Performance		AT-FLIR	Advanced Targeting FLIR
ANSI	American National Standards Institute		ATG	Auto-gating
ANVIS	Aviator Night Vision Imaging Systems		ATM	Air Traffic Management
			ATP	Advanced Targeting Pod
AO	Aero-optical		ATS	Automatic Thrust System
AOJ	Angle-on-Jam		AURT	All-Up-Round Test
AOR	Angle Only Ranging		AVHRR	Advanced Very High Resolution RADAR
APU	Auxiliary Power Unit			
ARAC	Aviation Rulemaking Advisory Committee		AVMUX	Avionics Bus
			AWACS	Airborne Warning and Control System
ARC	Automatic Radio Compass			
ARDP	Airborne RADAR Display Processor		**B**	
ARDS	Advanced Range Data System		Baro	Barometric Altitude
ARINC	American Radio Incorporated		BER	Bit Error Rate
ARM	Anti-Radiation Missile		BET	Best Estimate Trajectory
ARMUX	Armament Bus		BFL	Bomb Fall Line
ARP	Aerospace Recommended Practices		BHMD	Binocular Helmet Mounted Display
ARSP	Airborne RADAR Signal Processor		BIT	Built-in-Test
			BR	Break Rate
ASA	American Standards Association		B-RNAV	Basic Area Navigation
			BRU	Bomb Release Unit
ASE	Altimetry System Error		BSP	Bright Source Protection
ASG	Aperiodic Synchronization Gap		BSR	Blip-Scan Ratio
ASI	Aircraft Station Interface		BVR	Beyond Visual Range
ASPJ	Airborne Self-Protection Jammer		**C**	
ASR	Airport Surveillance RADARS		C/A	Coarse Acquisition Code
			C2	Command and Control
ASTM	American Society for Testing and Materials		C2P	Command and Control Processor
AT	Aperiodic Access Time-out		C3I	Command, Control and Communication Intelligence
ATARS	Advanced Tactical Airborne Reconnaissance System		C4I	Command, Control and Communication Intelligence and Computers

C4ISR	Command, Control, Communications, Computers, Intelligence, Surveillance and Reconnaissance
CAA	Civil Aeronautics Authority
CAA	Civil Aeronautics Administration
CAB	Civil Aeronautics Board
CADC	Central Air Data Computer
CAN	Controller Area Network
CANCO	Can't Comply
CandD	Controls and Displays
CANTPRO	Can't Process
CAS	Calibrated Airspeed
CAW	Cautions, Advisories and Warnings
CBA	Capability-Based Assessment
CBR	chemical/biological/radiological
CCA	Common Cause Analysis
CCD	Charge-Coupled Device
CCIP	Continuous Computing Impact Point
CCTV	Closed-circuit TV
CDD	Capability Development Document
CDI	Course Deviation Indicator
CDL	Common Data Link
CDRL	Contract Data Requirement List
CDV	Command Directed Vehicle
CE	Conducted Emission
CE	Clutter Elimination
CE	Concatenation Event
CEA	Circular Error Average
CEDP	Compatibility Engineering Data Package
CEP	Circular Error Probable
CFAR	Constant False-Alarm-Rate
CFIT	Controlled Flight Into Terrain

CFR	Code of Federal Regulations
CG	Center of Gravity
CGS	Chirp Gate Stealer
CLHQ	Closed Loop Handling Qualities
CMOS	Complementary Metal Oxide Semi-conductor
COA	Certificate of Authorization
COI	Critical Operational Issues
COMINT	Communications Intelligence
COS	Continued Operational Safety
COSRO	Conical Scan Receive Only
CPA	Closest Point of Approach
CPDLC	Controller Pilot Data Link Communications
CR	Combat Rate
CRC	Cyclical Redundancy Check
CRM	Crew Resource Management
CRT	Cathode Ray Tube
CRT	CDD Requirements Correlation Table
CS	Conducted Susceptibility
CS	Control Symbol
CSI	Carriage Store Interface
CSMA	Carrier Sense Multiple Access
CSMA/CR	Carrier Sense, Multiple Access with Collision Resolution
CTF	Combined Test Force
CTP	Critical Technical Parameters
CTS	Clear to Send
CV	Carrier
CW	Continuous Wave
CWS	Control Wheel Steering

D

D/A	Digital to Analog
DA	Destination Address
DAR	Designated Avionics Representative

DAS	Detector Angular Subtense		DP	Decision Point
DAS	Distributed Aperture System		DRA	Defense Research Associates
DAS IRST	Distributed Aperture System Infra-red Search and Track System		DSARC	Defense Systems Acquisition Review Council
DBS	Doppler Beam Sharpening		DSL	Digital Subscriber Line
DC	Direct Current		DSR	Data Set Ready
DCCIP	Delayed Continuous Computed Impact Point		DT&E	Development Test and Evaluation
DCD	Data Carrier Detect		DT/OT	Development/Operational Test
DCE	Data Communications Equipment		DTE	Digital Terrain
DDU	Digital Data Unit		DTE	Data Transmission Equipment
DE	Directed Energy		DTR	Data Terminal Ready
DEP	Design Eye Position			
DER	Designated Engineering Representatives		**E**	
			E3	Electromagnetic Environmental Effects
DERA	Defence Evaluation Research Agency		EA	Electronic Attack
DF	Direction Finding		EADI	Electronic Attitude Direction Indicator
DFCS	Digital Flight Control System		EASA	European Aviation Safety Administration
DID	Data Item Description		EBR	Enhanced Bit Rate
DITS	Digital Information Transfer System		ECCM	Electronic Counter-Counter Measures
DME	Distance Measuring Equipment		ECM	Electronic Counter Measures
			ECP	Engineering Change Proposal
DMIR	Designated Manufacturing Inspection Representative		ED	End Delimiter
DNS	Doppler Navigation System		EDR	Expanded Data Request
DOA	Direction of Arrival		EEC	Essential Elements of Capability
DoD	US Department of Defense		EFAbus	Eurofighter bus
DoDD	Department of Defense Directive		EFIS	Electronic Flight Instrument System
DoDI	Department of Defense Instruction		EFVS	Enhanced Flight Vision System
DOE	Design of Experiments		EGNOS	European Geostationary Navigation Overlay Service
DOP	Dilution of Precision			
DOT	Department of Transportation		EGT	Exhaust Gas Temperature
DOTMLPF	Doctrine, Organization, Training, Materiel, Leadership and Education, Personnel and Facilities		EIA	Electronic Industries Alliance
			EL	Electroluminescence
			ELINT	Electrical Intelligence

EM	Electromagnetics
EM	Electro-Magnetic
EMC	Electro-magnetic Compatibility
EMCON	Emission Control
EMI	Electro-magnetic Interference
EMS	Emergency Medical Services
EO	Electro-optics
EO-DAS	Electro-Optical Distributed Aperture System
EOF	End of Frame
EOTS	Electro-Optical Targeting System
EP	Electronic Protection
EPE	Expected Positional Error
ERP	Effective Radiated Power
ES	Electronic Warfare Support
ESA	European Space Agency
EU	Engineering Unit
EVS	Enhanced Vision Systems
EW	Electronic Warfare
EWISTL	Electronic Warfare Integrated Systems Test Laboratory
EWR	Early Warning RADAR

F

FA	False Alarm
FA	Frequency Agility
FAA	Federal Aviation Administration
FAR	Federal Aviation Regulation
FC	Fibre Channel
FC	Frame Control
FC-AE-ASM	Fibre Channel-Avionics Environment Anonymous Subscriber Messaging protocol
FCS	Flight Control System
FCS	Frame Check Sequence
FD	Flight Director
FDE	Force Development Evaluation

FEATS	Future European Air Traffic Management System
FED	Field Emission Display
FGS	Flight Guidance System
FHA	Functional Hazard Assessment
FLIR	Forward-Looking Infra-red
FLTA	Forward-Looking Terrain Avoidance
FM	Frequency Modulation
FMC	Fully Mission Capable
FMR	Frequency Modulation Ranging
FMS	Flight Management System
FOC	Full Operational Capability
FOG	Fiber-optic Gyro
FOM	Figure of Merit
FOR	Field of Regard
FOT&E	Follow-on OT&E
FOTD	Fiber-optic Towed Decoy
FOV	Field-of-View
FPA	Focal Plane Array
FRD	Functional Requirements Document
FRL	Fuselage Reference Line
FRP	Full Rate Production
FRUIT	False Replies Uncorrelated in Time
FRV	Federal Republic of Yugoslavia
FSD	Full Scale Development
FSDO	Flight Standards District Office
FSPI	Full Speed Peripheral Interface
FSS	Flight Service Station
FTA	Fault Tree Analysis
FTE	Flight Technical Error
FTJU	Fuel Tank Jettison Unit
FTT	Fixed Target Track

G

GAGAN	Indian GPS Aided Geo Augmented Navigation
GAI	GPS Aided INS
GAMA	General Aviation Manufacturers Association
GATM	Global Air Traffic Management System
GBAS	Ground-based Augmentation System
GBS	Global Broadcast Service
GCE	Ground Control Element
GCI	Ground Controlled Intercept
GCU	Gun Control Unit
GDOP	Geometric Dilution of Precision
GFE	Government Furnished Equipment
GFI	General Format Identifier
GLONASS	Russian Federation Ministry of Defense Global Navigation Satellite System
GLS	GNS Landing System
GMR	Ground Mobile Radio
GMTI	Ground Moving Target Indicator
GMTT	Ground Moving Targat Track
GMU	GPS Monitoring Unit
GNSS	Global Navigation Satellite System
GPS	Global Posining System
GPWS	Ground Proximity Warning Systems
GRAS	Australian Ground Based Regional Augmentation System
GRI	Group Repetition Interval
GTRI	Georgia Tech Research Institute

H

HARM	High Speed Anti-radiation Missile
HAT	Height Above Touchdown
HAT	Height Above Terrain
HAV	Hybrid Air Vehicle
HAVCO	Have Complied
HDD	Heads Down Display
HDDR	High Density Digital Recorder
HDOP	Horizontal Dilution of Precision
HEL	High-energy LASER
HFOV	Horizontal Field-of-View
HFR	Height Finder RADARS
HIRF	High Intensity Radiated Fields
HMD	Helmet Mounted Displays
HMI	Human Machine Interface
HMS	Helmet Mounted Sight
HOP	Head Orientation and Position
HOTAS	Hands on Stick and Throttles
HP	High Precision
HPRF	High PRF
HSFLIR	Head Steered Forward Looking Infra-red
HSI	Heading Situation Indicator
HSPI	High Speed Peripheral Interface
HTS	Helmet/Head Tracking Systems
HUD	Heads-Up-Display

I

I/S	Interference to Signal ratio
I2	Image Intensification
IADS	Integrated Air Defense System
IAG	Inertially Aided GPS
IAS	Indicated Airspeed
IBIT	Initiated BIT
ICAO	International Civil Aviation Organization
ICD	Initial Capabilities Document
ICD	Interface Control Document
ICS	Intercommunications System
IDE	Identifier Extension

IDS	Interface Design Specification
IFAST	Integrated Facility for Avionic Systems Testing
IFF	Interrogate Friend or Foe
IFOV	Instantaneous Field-of-View
IFR	Instrument Flight Rules
IGI	Integrated GPS/INS
IJMS	Inadvertent Jettison Monitoring System
ILS	Instrument Landing Systems
ILS	Integrated Logistics Support
INCITS	International Committee for Information Technology Standards
iNET	Integrated Network-Enhanced Telemetry
INFO	Information
INS	Inertial Navigation System
IOFP	Interim Operational Flight Program
IOT&E	Initial OT&E
IPF	Interference Protection Feature
IPT	Integrated Product Teams
IR	Infrared
IRCCM	Infrared Counter-Counter Measures
IRLS	IR Line Scanner
IRNSS	Indian Regional Navigation Satellite System
IR-OTIS	Infrared Optical Tracking and Identification System
IRS	Inertial Reference System
IRST	Infra-red Search and Track
ISAR	Inverse SAR
ISM	Industrial, Scientific and Medical
ITV	Inert Test Vehicles
IU	Interface Unit

J

J/S	Jamming to Signal Ratio
JAA	Joint Aviation Administration
JAUS	Joint Architecture for Unmanned Systems
JDAM	Joint Direct Attack Munition
JEM	Jet Engine Modulation
JETDS	Joint Electronics Type Designation System
JIC	Joint Intelligence Center
JPL	Jet Propulsion Lab
JSF	Joint Strike Fighter
JSTARS	Joint Surveillance Target Attack Radar System
JTA	Joint Technical Architecture
JTIDS	Joint Tactical Information Distribution System
JTRS	Joint Tactical Radio System
JTT	Joint Test Team
JU	JTIDS Unit
JWICS	Joint Worldwide Intelligence Communications System

K

KEAS	Knots Equivalent Airspeed
KPP	Key Performance Parameters
KSA	Key System Attributes
KTAS	Knots True Airspeed

L

LAAS	Local Area Augmentation System
LAN	Local Area Network
LANDSAT	Land Remote Sensing Satellite
LANTIRN	Low Altitude Navigation and Targeting Infrared for Night
LARS	Launch Acceptability Regions
LASER	Light Amplification by Stimulated Emission of Radiation

LCD	Liquid Crystal Display		MDD	Materiel Development Decision
LDU	Link Data Unit		MDT	Mean Down Time
LED	Light-emitting Diodes		MDTCI	MUX Data Transfer Complete Interrupt
LEMV	Long Endurance Multi-inteklligence Vehicle		MEL	Master Equipment List
LEO	Low Earth Orbiting		MEO	Medium Earth Orbiting
LLC	Logical Link Control		MFA	Multiple Filter Assemblies
LNAV	Lateral Navigation		MFD	Multi-function Display
LOA	Letter of Authorization		MFL	Maintenance Fault List
LORAN	Long Range Navigation		MFOV	Medium Field-of-View
LORO	Lobe-On Receive Only		MFR	Multiple Frequency Repeater
LOS	Line-of-Sight		MIAA	Multiple Intruder Autonomous Avoidance
LPI	Low Probability of Intercept		MIDO	Manufacturing Inspection District Office
LPRF	Low PRF			
LPV	Localizer Performance with Vertical guidance		MIDS	Multifunctional Information Distribution System
LRF	LASER Range Finder		Mil-Std	Military Standard
LRIP	Low Rate Initial Production		MIP	Mean Impact Point
LRR	Long Range RADAR		MISB	Motion Imagery Standards Board
LRU	Line Replaceable Unit			
LSB	Least Significant Bit		MIU	High Speed Media Interface Unit
LTE	Launch to Eject			
LTM	LASER Target Markers		MLC	Main Lobe Clutter
LWIR	Long Wave Infra-red		MLS	Microwave Landing System
			MMEL	Master Minimum Equipment List
M				
MAC	Media Access Control		MMR	Multi-mode Receiver
MagVar	Magnetic Variation		MMSI	Miniature Munitions/Stores Interface
MAJCOM	Major Command			
MAL	Maximum Aperiodic Length		MMW	Millimetric Wave
MALD	Miniature Air Launched Decoy		MNPS	Minimum Navigation Performance Standards
MASPS	Minimum Aviation System Performance Standards		MOA	Measures of Assessment
MAU	Medium Attachment Unit		MOAT	Missile-on-Aircraft-Test
MBIT	Maintenance BIT		MOE	Measures of Effectiveness
MCP	Micro Channel Plate		MOP	Measures of Performance
MDA	Minimum Descent Altitude		MOPS	Minimum Operational Performance Standards
MDB	Measurement Descriptor Blocks		MOS	Measures of Suitability

MPAD	Man-Portable Air Defense		NCS	Net Control Station
MPEG	Motion Pictures Experts Group		NCTR	Non-cooperative Target Recognition
MPRU	Missile Power Relay Unit		NDB	Non-directional Beacon
MRA	Minimum Release Altitude		NED	North, East and Down
MRC	Minimum Resolvable Contrast		NETD	Noise Equivalent Temperature Difference
MRΔT	Minimum Resolvable Temperature Difference		NFOV	Narrow Field-of-View
MSB	Most Significant Bit		NIMA	National Imagery and Mapping Agency
MSI	Mission Store Interface		NMSU	New Mexico State University
MSL	Mean Sea Level		NOTAM	Notices to Airmen
MSOT&E	Multi-service OT&E		NPG	Network Participation Groups
MSPS	Minimum System Performance Standards		NRZ	Non Return to Zero
			NRZI	Non Return to Zero Invert
MTBCF	Mean Time Between Critical Failure		NS	No Statement
			NTR	Net Time Reference
MTBF	Mean Time Between Failure		NVD	Night Vision Devices
MTF	Modulation Transfer Function		NVG	Night Vision Goggles
MTI	Moving Target Indicator		NVG	Night Vision Goggles
MTR	Moving Target Rejection		NVIS	Night Vision Imaging Systems
MTSAT	Multi-function Transport Satellite		**O**	
MTTR	Mean Time to Repair		OA	Operational Assessment
MUH	Minimum Use Height		OBEWS	On-board EW Simulator
MWIR	Mid-wave Infra-red		OBS	Omni-bearing Selector
MWS	Missile Warning System		OCRD	Operational Capability Requirements Document
N			OEM	Original Equipment Manufacturer
NACK	Non-Acknowledgments		OFP	Operational Flight Program
NAD	North American Datum		OI	Operational Instruction
NAGU	Notice Advisories to GLONASS Users		OLED	Organic LED
NANU	Notice Advisory to NAVSTAR Users		Ops Specs	Operations Specifications
			OPSEC	Operations Security
NAS	National Airspace System		OPTEVFOR	United States Navy Operational Test and Evaluation Force
NAVAIRINST	Naval Air Instruction			
NBC	Nuclear, Biological, Chemical		ORD	Operational Requirements Document
NBRN	Narrow Band Repeater Noise Jammer		ORT	Operational Readiness Test

OT	Other Traffic
OT&E	Operational Test and Evaluation
OTH	Over-the-Horizon
OUE	Operational Utility Evaluation
OUI	Organizationally Unique Identifiers

P

PA	Physical Address
PAL	Pilot Assist Lock-on
PBIT	Power-up BIT
PBIT	Periodic BIT
PC	Personal Computer
PCM	Pulse Code Modulation
PD	Pulse Doppler
PDA	Premature Descent Alert
PDOP	Positional Dilution of Positional
PF	Pilot Flying
PFD	Primary Flight Display
PFI	Primary Flight Instruments
PFL	Pilot Fault List
PIC	Pilot in Command
PID	Packet Identifier
PIRATE	Passive Infrared Airborne Track Equipment
PK	Probability of Kill
PLD	Program Load Device
PMA	Program Management Activity
PMC	Partially Mission Capable
PNF	Pilot Not Flying
PoC	Proof of Concept
POC	Point of Contact
POT	Position and Orientation Tracker
PPI	Plan Position Indicator
PPLI	Precise Participant Location and Identification
PPS	Program Performance Specification

PR	Pulse Repeater
PR	Preamble
PRF	Pulse Repetition Frequency
PRI	Pulse Repetition Interval
PRN	Pseudo Random Noise Jammer
P-RNAV	Precision Area Navigation
PSG	Periodic Synchronization Gap
PSSA	Preliminary System Safety Assessment
PSTT	Pulse Single Target Track
PUF	Probability of Unsafe Flight
PVU	Position Velocity Update
PW	Pulse Width
PWD	Pulse Width Discriminator

Q

QOT&E	Qualification OT&E
QZSS	Quazi-zenith Satellite System

R

R&M	Reliability and Maintainability
R/C	Receipt/Compliance
RA	Resolution Advisory
RA	Raid Assessment
RADAR	Radio Detection and Ranging
RAIM	Receiver Autonomous Integrity Monitor
RAM	RADAR Absorption Material
RAT	Ram Air Turbine
RCR	Raid Count Request
RCS	RADAR Cross Section
RD	Random Doppler
RE	Radiated Emission
RelNav	Relative Navigation
RF	Radio Frequency
RFI	Returned for Inventory
RGPO	Range Gate Pull-Off
RH	Relative Humidity

RHAW	RADAR Homing and Warning		**S**	
RIM	Range Interface Memorandum		SA	Source Address
RLG	Ring Laser Gyro		SAE	Society of Automotive Engineers
RML&A	Reliability, Maintainability, Logistics Supportability, and Availability		SAM	Situation Awareness Mode
			SAM	Surface-to-Air Missile
RMS	Root Mean Square		SAR	Search and Rescue
RNASP	Routing Network Aircraft System Port		SAR	Synthetic Aperture RADAR
			SAS	Stability Augmentation Systems
RNAV	Area Navigation		SATCOM	Satellite Communications
RNP	Required Navigation Performance		SAW	Selectively Aimable Warhead
ROC	Required Obstacle Clearance		SCATI	Special Use Category I GPS Landing System
ROIC	Readout Integrated Circuit		SCMP	Software Configuration Management Plan
RPV	Remotely Piloted Vehicle			
RRN	Recurrence Rate Number		SCR	Signal-to-Clutter Ratio
RS	Radiated Susceptibility		SD	Start Delimiter
R-S	Reed-Solomon Encoding		SDI	Source/Destination Identifier
RSAM	Repeater Swept Amplitude Modulation		SDU	Secure Data Unit
			SEAD	Suppression of Enemy Air Defenses
RSS	Root Sum Square			
RTC	Required Terrain Clearance		SEAM	Sidewinder Expanded Acquisition Mode
RTCA	Radio Technical Commission for Aeronautics		SFAR	Special Federal Aviation Regulation
RTO	NATO Research and Technology Organization		SFD	Start Frame Delimiter
			SG	Synchronization Gap
RTO	Responsible Test Organization		SICP	Subscriber Interface Control Program
RTP	Real-time Transport Protocol			
RTR	Remote Transmission Request		SID	Standard Instrument Departures
RTS	Request to Send			
RTT	Round-trip Timing		SIGINT	Signal Intelligence
RTZ	Return to Zero		SIL	Software Integration Lab
RVR	Runway Visibility Range		SINPO	Signal strength, Interference, Noise, Propagation disturbance, Overall rating
RVSM	Reduced Vertical Separation Minima			
RWR	RADAR Warning Receiver		SIPRNet	Secret Internet Protocol Router Network
RWS	Range While Search			
RXD	Received Data		SLB	Sidelobe Blanking

SLC	Side Lobe Clutter
SLC	Sidelobe Cancellation
SLS	Side Lobe Suppression
SMP	Stores Management Processor
SMPD	Single-carrier Modulation Photo Detector
SMS	Stores Management System
SNAS	Chinese Satellite Navigation Augmentation System
SNR	Signal-to-Noise Ratio
SOA	Service Oriented Architecture
SOF	Start of Frame
SOJ	Stand-off Jamming
SOT	Start of Transmission
SOW	Statement of Work
SP	Standard Precision
SPADATS	Space Debris and Tracking Station
SPIE	International Society for Optical Engineering
SPIE	Society of Photographic Instrumentation Engineers
SPJ	Self Protection Jamming
SQAP	Software Quality Assurance Plan
SRB	Safety Review Board
SRR	Substitute Remote Request
SSA	System Safety Assessment
SSE	Static Source Error
SSEC	Static Source Error Correction
SSI	Software Support and Integration
SSM	Signal/Sign Matrix
SSV	Standard Service Volume
STANAG	Standing NATO Agreement
STANAG	NATO Standardization Agreement
STAR	Standard Terminal Arrival Routes

STC	Supplemental Type Certificate
STC	Sensitivity Time Constant
STI	Speech Transmission Index
STT	Single Target Track
SUAV	Surrogate UAV
SUT	System Under Test
SWAG	Scientific Wild-ass Guess
SWIR	Short Wave Infra-red
T	
T&E	Test and Evaluation
T/R	Transmit/Receive
TA	Traffic Advisory
TA	Traffic Alert
TA	Terrain Avoidance
TACAN	Tactical Air Navigation
TAD	Towed Air Decoy
TADIL	Tactical Digital Information Link
TAE	Track Angle Error
TAMMAC	Tactical Area Moving Map Capability
TAS	True Airspeed
TAS	Traffic Advisory Systems
TAWS	Terrain Awareness Systems
TC	Type Certificate
TCAS	Traffic Alert and Collision Avoidance System
TCP	Transmission Control Protocol
TD&E	Tactics Development and Evaluation
TD.	Time Difference
TDI	Time Delay Integration
TDMA	Time Division Multiple Access
TDS	Technology Development Strategy
TEC	Total Electron Content
TER	Triple Ejector Rack

TERPS	U.S. Standard for Terminal Instrument Procedures	TTPF	Target Transfer Probability Function
TF	Terrain Following	TVE	Total Vertical Error
TFP	Terrain Following Processor	TWS	Track While Scan
TFT	Thin-film Transistor	TXD	Transmitted Data
TG	Terminal Gap	**U**	
TI	Thermal Imager	UAS	Unmanned Aerial System
TI	Transmit Interval	UAV	Unmanned Aerial Vehicles
TIA	Type Inspection Authorization	UCAS	Unmanned Combat Air System
TIM	Terminal Input Message		
TIP	Tetrahedral Inertial Platform	UDP	User Datagram Protocol
TIR	Type Inspection Report	UR	Utilization Rate
TIS	Traffic Information Service	USAARL	US Army Advanced Research Laboratory
TIS	Test Information Sheets		
TLAR	That Looks About Right	USAF	United States Air Force
TM	Telemetry	USB	Universal Serial Bus
TmNS	Telemetry Network System	USN	United States Navy
TOA	Time of Arrival	UTC	Universal Time Code
TOGA	Takeoff and Go-around	UV	Ultraviolet
TOM	Terminal Output Message	**V**	
TPIU	Transmission Protocol Interface Unit	V	Voltage
		V/H	Velocity to Height Ratio
TPWG	Test Planning Working Group	VA	Visual Acuity
TQ	Time Quality	VandV	Validation and Verification
TRB	Test Review Board	VCS	Visually Coupled System
TRSB	Time Reference Scanning Beam	VDOP	Vertical Dilution of Precision
		VFOV	Vertical Field-of-View
TSB	Time Slot Block	VFR	Visual Flight Rules
TSDF	Time Slot Duty Factor	VGPO	Velocity Gate Pull-Off
TSE	Total System Error	VLAN	Virtual Local Area Network
TSEC	Transmission Security	VNAV	Vertical Navigation
TSO	Technical Standard Order	VOR	VHF Omni-Directional Range
TSOA	Technical Standard Order Authorization	VORTAC	VOR/TACAN
		VOT	VOR Test Facility
TSPI	Time Space Position Information	VS	Velocity Search
		VTAS	Visual Target Acquisition System
TSS	Test Summary Sheets		
TTNT	Tactical Targeting Network Technology	VVI	Vertical Velocity Indicator

W

WAAS	Wide Area Augmentation System	WRA	Weapons Replaceable Assembly
WC	Word Count	WSAR	Weapons System Accuracy Report
WCA	Warning, Caution and Advisory	WSEP	Weapons System Evaluation Program
WFOV	Wide Field-of-View	WWI	World War I
WGS	World Geodetic System	WWII	World War II
WILCO	Will Comply	**X**	
WOW	Weight on Wheels	X-EYE	Cross Eye
WOW	Weight off Wheels	X-POLE	Cross Polarization

What is Avionics Flight Test and Why Do We Need It

Chapter Outline

1.0	Avionics Flight Test	1
1.1	Historical Background	2
1.2	Flight Test Organization	4
1.3	Flight Test Objectives	5
1.4	The Need for Flight Test	6
1.5	Classifying the Program	8
1.6	Contractual Requirements	9
1.7	Test Team Composition	11
1.8	Team Member Responsibilities	13
1.9	Understanding Terms	15
1.10	Contractor and Subtier Specifications	20
1.11	Customer Expectations	21
1.12	Formulating the Program	22
1.13	Summary	25
1.14	Selected Questions for Chapter 1	26

1.0 | AVIONICS FLIGHT TEST

There is little written on the specific subject of avionics and weapons systems testing, but there are articles and tutorials available that address the general subject of testing and experiments. One source on testing is *The Logic of Warfighting Experiments*, by Richard A. Kass (Washington, DC: CCRP Press, 2006), which is available online at the time of this writing at http://www.dodccrp.org/files/Kass_Logic.pdf. The U.S. Air Force, and in particular the 46th Test Wing at Eglin Air Force Base, Florida, has been promoting the use of "design of experiments" (DOE) as the preferred method of constructing and analyzing test programs. As a reference, the reader is encouraged to read any of the numerous papers on this subject by Gregory T. Hutto and James M. Higdon. The most recently published is American Institute of Aeronautics and Astronautics (AIAA) paper 2009-1706, "Survey of Design of Experiments (DOE) Projects in Development Test CY07-08," which is also available online at http://pdf.aiaa.org/preview/CDReadyMTE09_2104/PV2009_1706.pdf.

1.1 | HISTORICAL BACKGROUND

The flight test community as a whole has been driven to do more with less in recent years. It has been forced to become more efficient with the available dollars for testing and evaluation. In the past, flight testing was divided among three separate groups of testers:

- Contractor flight testing
- Development test and evaluation (DT&E)
- Operational test and evaluation (OT&E)

Each of these groups was assigned a charter and was responsible for the testing and reporting of their respective tasks. There was a bit of overlap of tasks and, as a result, some of the testing was repetitive; different results from similar tests sometimes led to heated confrontations. In all fairness, the test communities from each of these groups were seeking to evaluate similar tasks from different perspectives.

Contractor flight testing of avionics systems is first and foremost developmental in nature. Perhaps as much as 90% of the test program is dedicated to development, whereas the final 10% is the ultimate demonstration of the finished product. Even this demonstration is a "test to spec program," where the contractor is on the hook to show specification compliance. Completion of the program, and therefore payment, is contingent upon this successful demonstration. It should be obvious that the contractor will be reluctant to perform any testing that is above and beyond specification compliance (performance testing, for example), since this will reduce the time available for specification compliance testing. The contractor is also not being paid to demonstrate performance.

Development test and evaluation (DT&E) is chartered with ensuring that the system under test is indeed ready for operational test and evaluation (OT&E). Since OT&E is interested in the true performance and functionality of the system under real-world scenarios, the DT&E community is saddled with a rather large burden. DT&E has to validate the specification compliance testing previously done by the contractor and then accomplish sufficient performance testing to ensure that OT&E will not dismiss the system outright. Specification compliance verification is used as a baseline; additional performance testing would be moot if the system cannot perform to specifications.

Operational test and evaluation (OT&E) has no interest in specification compliance verification and is only interested in the system's capability to perform the desired mission. For example, OT&E was interested in the ability of the F-14D to perform fleet defense, not in its ability to detect a nonmaneuvering target at high altitude at some specified range. The latter is a specification compliance issue and not a performance or operational utility issue. In many cases, OT&E would determine that the specifications were so poorly worded that the system, although it met the specifications, could not perform the mission. Of course, the opposite is also true; the system may fail to meet specification compliance but is found to be acceptable in meeting mission requirements.

The existence of these three independent organizations necessitated three periods within the master schedule for flight testing. In fact, most test programs included the contractor test period interspersed with DT and OT periods. These DT and OT periods were scheduled at major milestones within the schedule coinciding with incremental capabilities of the system. These periods lasted for weeks, and sometimes months, and severely impacted contractor development testing, because during these periods software was frozen and further development work was halted. The rationale here was to not allow the contractor to fritter away all of the test and evaluation (T&E) monies without periodic progress checks. This rationale worked fine as long as the T&E schedule was not shortened and development proceeded as expected. In reality, the test schedule was always shortened and development was always lagging. These real-world developments caused considerable consternation among the three groups, since none of the test organizations had any inclination to giving up scheduled test time.

In the mid-1980s, an idea to combine some of the testing among the three groups was floated to the test community. The idea was to incorporate some or all of the DT&E testing with the contractor testing. This way the separate DT&E test periods could be absorbed by the contractor test periods and thereby shorten the total T&E effort. The contractor would simply have to include the DT&E test objectives in the contractor test plan. Since the objectives would be combined, the next logical step would be to combine the test teams. It was from this initial proposal that the Combined Test Force (CTF), Joint Test Team (JTT), or any of many other names this union was called was born. Note that these were not integrated product teams (IPTs); they will be addressed later.

The initial CTFs had some growing pains. The first attempts at participatory flight testing simply deleted the DT&E test periods and shortened the T&E effort accordingly. Unfortunately, the contractor test periods were not expanded, even though it was expected that they would wholly incorporate the DT&E effort. In the ideal world, many of the DT&E test objectives and test points would match one for one with the contractor test plan (specification compliance). In the real world, the DT&E community structured its testing to try and accomplish both specification compliance and performance with the same test objective. This created quite a dilemma for the contractor test organization, especially since, during this period of cooperation, no one changed the contracts. Contractors were still bound to satisfy a contract that was rooted in specification compliance; there were no stipulations for additional testing. Contractor test teams were routinely caught in the middle of the "customer satisfaction" and program direction battle. On the one hand, they tried to incorporate the customer's testing; on the other, they were accountable to program management to test only what was in the contract.

There were other ancillary problems which became evident during these phases of joint testing. Since testing would be combined, the teams would be combined. This idea would have been a good one if it had not been tinkered with. the premise of this idea was that consolidation would save money. The most obvious savings were reaped by eliminating the DT&E periods. A second method of savings, courtesy of the accountants, was the elimination of manpower due to consolidation. Contractor test teams were reduced by the number of DT&E personnel which were assigned to the JTT and vice versa. Once again, the charters of each of these organizations were never adjusted, so the workload remained the same, but the manpower was reduced. Each organization was still responsible for such things as data reduction, analysis, and generation of test reports. The contractor organization was still responsible for the instrumentation of the aircraft, formulation of the test plan, telemetry station setup, and engineering unit (EU) conversion of data. Additional tasks of training new DT&E members and converting data into a format usable by the DT group was added.

As time went on, it was determined that even more time and money could be saved if the OT&E community was also brought into the fold and a true CTF could be formed. Problems similar to those previously presented rose anew. One unique problem with this setup involved the specific charter of the OT&E community. Any member of the OT&E

community will tell you that they are chartered by the U.S. Congress to ensure that the taxpayers are getting what they pay for and the users are getting a system that is capable of performing the designated mission. This is an integral part of the checks and balances of the acquisition system, but it leads to serious problems when trying to consolidate the test community. How can the OT&E testers be melded with the CTF and still retain their autonomy? In reality, operational testers have everything to gain and nothing to lose in this situation. They will not give up their dedicated periods of testing and are able to interject special testing into the combined test plan. This special testing is over and above the development/specification compliance test plan.

The greatest problem with consolidation is the failure of all team members to understand that avionics testing programs, by nature, are developmental. That is, software-dependent systems are incremental. There is no such thing as 100% software (full system capability), especially with the initial deliveries. If this fact is understood, specification compliance and validation testing will never be performed early in a program. In fact, this testing should occur at the end of the program. The differences between development, demonstration, qualification, and performance are covered later in this chapter. OT&E organization and testing is covered in chapter 14.

▮ 1.2 ▮ FLIGHT TEST ORGANIZATION

Ideally flight test should be an independent organization (i.e., not reporting to the Engineering Department) tasked with impartially evaluating and reporting on a vehicle or system. the discussion here will address avionics and weapons systems. The personnel within this organization should be knowledgeable of the system under test. They should understand the basic operation of the system, the interfaces with other avionics systems aboard the aircraft, and where and how the system will be employed. In addition to basic knowledge of the system, the tester must also be knowledgeable of applicable system specifications. Specifications may be defined as the minimum standard to which the system must perform. Many of the MIL-STD specifications we have been familiar with in the past have given way to minimum operating capabilities, desired objectives, or essential elements of capability (EEC). It should be noted that the specifications for a program were probably formulated many years earlier by individuals who may no longer be involved with the program. The specifications may be poorly written or, worse yet, be seemingly unattainable. These problems lead to a more difficult task of test plan preparation for the flight test community. An example is the inputs to the cockpit design by the Pilot's Working Group. In many test programs, a Pilot's Working Group is established during the design phase to solicit input from operational users for the design of controls, displays, and the human machine interface (HMI). These inputs are used by design engineers and incorporated into the final product. Five years later, when the pilots have all moved to different assignments and the system enters flight testing, no one can fathom the reason for the display presentations.

It is with these specifications that we must formulate the foundation of our test plan. Above all, this organization of testers must have the ability to "call somebody's baby ugly." This is a whimsical way of stating that the organization has the ability to remain independent of Development Engineering and maintain objectiveness and impartiality. Some flight test organizations within corporations and defense agencies are organizationally within the Engineering Department. This is unfortunate because the Engineering

Department more often than not is responsible for designing and implementing the system. It is pretty hard to call somebody's baby ugly when you and the baby have the same father.

In the ideal world once again, the flight test organization will be a composite test team. This team will put aside all parochial interests and focus on the common goal. Members of the team would include flight test engineers, aircrew, and analysts from the contractor and customer development and operational test organizations. These members will provide the core of the test structure. In addition to flight test engineers, the team may include the design engineers, who act as consultants and provide the technical expertise. There may be some vendor or subcontractor engineers for those unique systems not designed by the primary contractor—an inertial navigation system (INS) or radar, for example. Instrumentation personnel also provide a key role on the team, ensuring that data are available in a format that is usable. Site support personnel such as maintenance, integrated logistics support (ILS), and range support provide an orderly flow to test operations.

1.3 | FLIGHT TEST OBJECTIVES

Perhaps the most important goal of the flight test team is to validate that the system being tested meets the customer's needs and requirements. We are not validating the fact that the customer knows what they want, but rather that our system will fulfill their requirements. This is accomplished by ensuring the top-level specifications are satisfied and by performing some measure of system performance testing. In an avionics program, the first step is to identify system problems and anomalies. Some readers may know this portion of the program as "fly-fix-fly," which indeed it is. Unfortunately, once again, there are some acquisition experts who believe that fly-fix-fly is a product of an age long since passed. With all of our investment in state-of-the-art simulators, some believe that all problems can be worked out in the laboratory prior to the flight test, hence there is no need for a fly-fix-fly or development program. This tenet is, of course, ridiculous to those who have been involved in the testing and evaluation of avionics systems.

Along with the identification of problems and anomalies comes providing troubleshooting assistance. The most helpful thing a flight test engineer can do in the area of troubleshooting is to provide data. It cannot be emphasized enough that the identification of problems is only half of the job. If a problem is presented to an engineer without proof (a picture of the baby), it has a very small chance of being corrected in a timely manner. Flight test engineers should live and die by data; it is the fruit of their labor. In addition to providing data, the test engineer should be ready to assist with the problem resolution. The test engineer should be the closest individual to the problem. It was the engineer who should have designed the test, then collected, reduced, and analyzed the data. Yet they are quite happy to report that a problem exists in a system without assisting in finding a solution to the problem. As mentioned previously, engineering judgment is issued to you at graduation. It is OK to act like an engineer and provide an educated guess as to where the problem lies and how it can be fixed.

In many cases, problems can be as simple as a busy bit being set continuously. An obvious solution might be to reset the bit. The flight test engineer should know this by looking at the data. The design engineer, upon hearing of the problem, but without supporting data and the flight test engineer's insight, may spend as much time researching the problem as the test engineer in order to come to the same conclusion. This may seem a trivial matter, however, if multiplied by a large number of anomalies, one can see how the system can easily be bogged down. This logjam can be broken up with good flight test engineering.

The final goal of the test team should be fair and accurate reporting of all tests conducted and the results obtained. These reports should document the system performance versus specified requirements and must be supported by the data. Once again, the independence of the organization is critical in providing an impartial report.

▌ 1.4 ▌ THE NEED FOR FLIGHT TEST

The previous section addressed the foolishness of eliminating the fly-fix-fly concept from flight test planning. Some quarters feel that there is no need for flight testing at all, and here is a good place to put that issue to rest. Simulation and laboratory development are excellent tools in placing a system into service. In fact, experience has shown that approximately two-thirds of all problems discovered in an avionics program are found in development labs and simulators. This is an excellent method for reducing flight time, and with the advent of more aircraft-representative development labs, the percentages should increase, but to what degree? Avionics systems are becoming increasingly complex, with hard-to-simulate dependencies on other avionics systems and environmental conditions.

There are two major drawbacks to labs and simulators. The first hurdle is the lab's approximation of the aircraft; space constraints, lack of available flight-qualified hardware and software, and line runs/losses between line replaceable units (LRUs) are a few examples of laboratory limitations. Real-world variables such as acceleration, vibration, acoustics, and loads cannot be adequately modeled without collecting flight data first. The second hurdle is the simulation itself (i.e., simulation is not real; it is a simulation). Simulations cannot account for all real-world dynamics. It would be impossible to evaluate the human factor implications of a system's operation in moderate turbulence in a simulator; these are things normally discovered during actual operation in a flight test environment. Attempts are made to make the simulations as real as possible, however, the number of variables can be astronomical. These considerations are responsible for the number of "lab anomalies" that are seen in all flight test programs. Even more interesting is the fact that some functions fail in lab testing and pass in flight testing, and some which are successful in the lab may fail on the aircraft.

Simulations can be compared to software development. As a rule, most simulations are not developed by the user, and the programmers attempt to keep as many variables as possible constant in the algorithms. Man-in-the-loop simulators can reduce the problems caused by nonusers developing the simulation. The second problem is a little trickier to reduce. The number of variables that can affect an algorithm can be large. A simulation attempting to replicate the real world would entail developing a program that would be quite complex. The simple bombing problem which follows demonstrates how the variables can multiply very rapidly if we attempt to replicate real-world situations.

If we attempt to drop a bomb from an aircraft with the intention of hitting an object on the ground, the pilot must consider a number of variables which directly affect the impact point. The first of these variables involves the ballistics of the bomb.

These ballistics determine the flight characteristics of the bomb. By knowing how the bomb will fly, downrange can be computed by knowing where in space the bomb was released (altitude, airspeed, g, and attitude). If the pilot deviates from any of these desired conditions, the impact point of the bomb will change. Lead computing sights and target tracking boxes on today's fighter aircraft compute a constantly changing impact point based on changes in the input variables.

If a programmer wants to simulate these conditions, he has to take into account the four variables mentioned. But are there only four variables? Altitude can be above ground level (AGL), mean sea level (MSL), or height above the ellipsoid (utilizing any one of a number of earth models). Airspeed is measured in three axes, normally North, East, and Down, or East, North, and Up (depending upon the aircraft convention). Affecting velocity is wind and wind gusts, also measured in three axes. Acceleration (g) is also measured in three axes. Attitude refers to the pitch, roll, and yaw of the platform; the true attitude of the platform may differ from the reference the platform is using for the computations (INS, global positioning system [GPS], attitude heading reference system, etc.). By taking all of theses components into account, the number of variables grows to 12 or more. If assumptions are made—the pilot will always be at the correct altitude, 1g level flight—the computations for the simulation become much simpler.

Although the preceding example may be an oversimplification of how simulations are developed, it does explain the principle. By eliminating variables, the simulation becomes easier and the coding more manageable. On the flip side, the simulation becomes further removed from the real-world aircraft. The infrared (IR) signature of an F-5 aircraft at a given aspect in clear air on a standard day is probably well known, but that same F-5 flying in desert haze over mountainous terrain in August is probably not known. It is probably not known because all of the variables have changed and cannot be accounted for in a simulation. The only way a database can be built for a simulation would be to collect a massive amount of data for varying terrain, haze, and temperature conditions. This is performance data that can only be obtained in a flight test program.

One last example: radar detection and classification of rain. If the capability of a radar to detect rain is a requirement, it may be advantageous to test this capability within the confines of a lab. After all, the size of raindrops and reflectivity based on frequency are known quantities. The problem with this task is that no attention has been given to the terrain that the rain is falling over. The performance of the radar will change if the terrain is a runway, forested, a sand dune, mountainous, or the ocean. A massive database would have to be obtained for varying rain rates over various terrains in order to quantify true radar performance or specification compliance. Once again, a flight test program would probably be more efficient than creating an additional test program for the purpose of collecting rain data.

In today's test environment, labs, simulators, and flight tests coexist and work together in order to meet program goals. Simulations and mathematical models attempt to predict system performance to the extent of their own capabilities, and are excellent tools for discovering gross anomalies in the system. Flight tests provide data on functions that cannot be evaluated in the lab, and the data that are collected can help to validate or improve the models used in the lab. A constant supply of real flight data allows the models and simulations to mature and become better predictors of system performance. It can be reasonably assumed that simulation will never supplant flight testing. It is also readily apparent that flight testing should not be undertaken without first undertaking extensive lab evaluation.

██ 1.5 █ | **CLASSIFYING THE PROGRAM**

Avionics test programs are of two types: subsystem/system development and system integration. Subsystem/system development encompasses the development and laboratory testing of a new or improved avionics system. An example of this classification would be the development of a new airborne radar or the development of an upgraded radio system. System integration is the type of program which most flight test engineers are familiar with. These programs entail the installation and integration of a proven system into an existing platform. A prime example would be the installation of GPS into an AV-8B. GPS is a proven system, more than likely provided as government-furnished equipment (GFE). A full-scale development (FSD) program may consolidate these two programs under one umbrella where there is some new development as well as system integration. FSD programs are becoming increasingly rare, as military services and corporations are electing to direct monies to upgrade programs rather than the development of new airframes. Examples of recent FSD programs are the F-35, A-400 and the F-22.

Unlike vehicle programs (when the term "vehicle programs" is used it refers to performance and flying qualities programs and analog-type systems such as hydraulics, fuel, pneumatics, etc.), avionics programs deal extensively with software. This implies that the program will be incremental in nature; software is never delivered to test with full system capabilities. In fact, there may be many deliveries, or builds, sent to the test organization. If we take the case of radar software in a subsystem/system development program, the deliveries may be scheduled as

- Build A: raw sensor development
- Build B: system integration
- Build C: weapon support
- Build D: special modes
- Build E: reliability/maintainability/qualification/sell-off

In this scenario, the first delivered build to the test community (Build A) will contain only basic capabilities for the radar. The system will not play with the other aircraft systems and may be good only for a "smoke check" (turning the system on does not cause a fire) in the aircraft. Build B allows for system integration, which may provide the first opportunity for viewing radar information on the displays, employing basic radar modes, and possibly establishing track files. The system will not support weapons nor will it feature special modes such as frequency agility. Build C is the first opportunity for the radar to support weapons delivery and a missile in flight. Build D provides all radar functions with the exception of final sell-off and qualification. Build E is the sold-off version of the radar software and will have been subjected to all environmental, mean time between failure (MTBF), and false alarm tests.

The schedule of these deliveries is normally determined at the start of the program and will drive the conduct of the flight test program. In the previous example, the deliveries may be scheduled at 6 month intervals for a 3 year test program (6 months of tests per delivery). Figure 1.1 shows the relationship of radar development to the overall schedule.

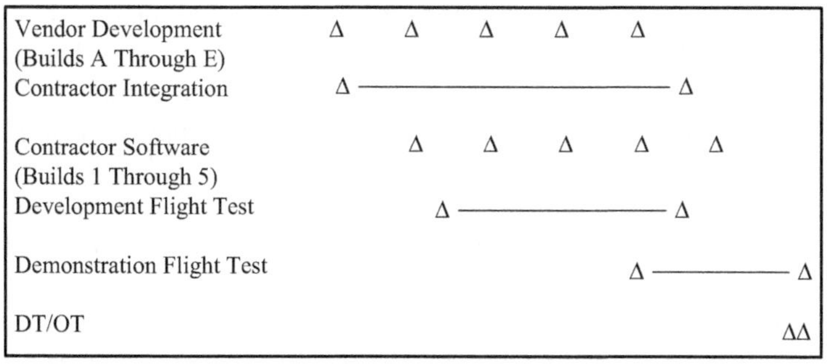

Vendor Development (Builds A Through E)	Δ	Δ	Δ	Δ	Δ	
Contractor Integration	Δ —————————————— Δ					
Contractor Software (Builds 1 Through 5)		Δ	Δ	Δ	Δ	Δ
Development Flight Test		Δ ——————— Δ				
Demonstration Flight Test				Δ ——————— Δ		
DT/OT						ΔΔ

FIGURE 1.1 ■ Software Delivery as it Affects Flight Testing

The figure also shows the interdependence of the flight test organization on the contractor and vendor software deliveries. The integration contractor must await a new radar software delivery prior to proceeding with its own integration work. This integration work will entail melding the radar software with the aircraft software. A software release for testing by the contractor will include the last version of radar software received. Should any of these milestones (Δ) slip, one of two things must happen: 1) Each successive integration or test period must move to the right; or 2) Each successive integration period or test period must be shortened by the length of the slip. For those individuals working integration, the news is not bad; the integration period merely slips to the right. For those individuals performing flight testing, the news may not be so good; the flight test period may be shortened by the length of the slip.

The rationale is not that hard to comprehend. The milestones after the flight test period (in our case, DT/OT) are the final milestones on the master schedule. These milestones indicate a few things. The first is that this is the end of the program and this is where the customer will make a final decision on the product. The second implication is that this is where the money typically runs out. Both of these scenarios are true and they dictate that the final milestones after flight testing are generally prohibited from moving to the right. It is OK to move to the left, and this is seen quite often when budgets seem to be overrunning and costs need to be trimmed. Therefore the first rule of avionics flight testing is that, for testers, the final milestone must never move to the right. As flight testing is compressed to the right, certain testing may be deleted and some functions may not be evaluated at all. This is a problem for flight testers if they do not address the problem with program management. Program management must be advised that with a reduction in testing comes a reduction in confidence level. Confidence level is your belief in the accuracy of the data you provide in the final report. Most flight test programs are estimated based on an 85% to 90% confidence level. A reduction in the amount of data collected will cause a reduction in confidence, and this fact needs to be included in the final report.

1.6 | CONTRACTUAL REQUIREMENTS

Whether you are a civilian contractor or a Department of Defense (DOD) tester, each flight test program has contractual obligations that must be met. A contractual item that your organization may be responsible for may be an interim or final report. In the civilian world, responsibilities of the test organization may be found in the contract data requirement list (CDRL) section of the contract. Each CDRL item has an associated data item description (DID) which explains in detail what the contractor is required to submit, in what format, to whom, when, and how many copies. DT&E and OT&E organizations are typically required to submit interim and final reports on the system being tested. Format and submittal of these CDRLs is governed by regulations or procedures. As each test program becomes more and more consolidated, the responsibility for meeting the CDRL submittal will fall to the CTF. Some programs have moved away from the formal specification compliance submittals common in past programs. These programs have attempted to define system performance in terms such as a statement of objectives or essential employment capability (EEC). Although the names and terminology have changed, the responsibility for reporting progress and performance to the customer is still a requirement.

Depending on the program, the first report that may be delivered to the customer is an instrumentation list. As mentioned previously, compliance and performance can only be proven when substantiated with data. The instrumentation list provides a list of all parameters that will be collected, reduced, and analyzed during the program in order to meet this goal. In addition to data parameters taken directly from the aircraft or system under test, the list must also include data parameters that will be used for "truth." This list includes parameters that are obtained from pulse code modulation (PCM) encoding systems, data busses, and other data collection systems, along with their respective data rates or sample sizes, accuracy, and scaling factors.

In addition to identifying the parameters that will be used for compliance or performance assessment, the customer is also interested in how you intend to manipulate the data. You must provide signal conditioning information, data reduction routines, filters, and processing algorithms in order to show that the data will not be over-massaged. If data are to be obtained in real time, the customer will want to know merge routines, sample rates, and if there is any data latency or senescence in the processing system. The time alignment system you have chosen, system time (GPS, satellite, manual), and clock drift should also be provided.

A newer report that has routinely been requested in recent programs is the weapons system accuracy report (WSAR). This report is delivered in two sections, one at the beginning of the program and the other at the completion of the program. The first section deals with the accuracy of the time space position information (TSPI), or truth data. As systems become more and more accurate, it is becoming increasingly difficult to provide a truth system which is at least four times as accurate as the system under test. Yes, four times as accurate (which is the second rule of avionics testing). In many cases we are forced into using some form of sensor fusion in order to obtain the accuracy we require. The customer may request that sensor fusion routines, accuracies, filters, weather corrections, and earth models used for truth be provided in this report. The second phase of this report is issued after the test program has been completed, and addresses the overall system accuracy based on the results of testing. For example, although the accuracy of a GPS is 15 m while on the bench, after being integrated into our platform the accuracy may now be 17 m. The 2 m degradation can be a result of many things: timing, latency, senescence, update rates, etc. Regardless, the total system performance is 17 m, and that is what will be reported.

The submittals that most test engineers are already familiar with are flight test plans and reports. Flight test plans are submitted prior to the start of the actual test program and are normally subject to the approval of the customer. Depending on the organization or affiliation, the formats of these reports will vary. Historically, contractors have been able to submit their test plans in "contractor format." As CTFs become more the norm, test plans must fit the format required by the customer. The U.S. Air Force (USAF), as well as the U.S. Navy (USN), require that the flight test plan conform to a format specified by regulation. The USN, for example, uses Naval Air Instruction (NAVAIR-INST) 3960.4B, "Project Test Plan Policy and Guide for Testing Air Vehicles, Air Vehicle Weapons, and Air Vehicle Installed Systems."

Reports are divided into three categories: pilot reports, flight reports, and final reports. Pilot reports are normally a requirement for all avionics programs and are submitted on a per flight basis; these reports are sometimes called dailies in some test organizations. Depending upon the program, flight reports may or may not be required. These reports may be required only for the initial flights of the program, all flights of the program, or may be in the form of status reports which are required after each block of testing. Final reports are always required, and as with the test plan, may be in either a "contractor format" or customer format. The test community is usually responsible for the latter two reports, while the test aircrew normally submits the pilot reports. The DID addresses these requirements, as well as the currency and time period allotted for submittal.

1.7 | TEST TEAM COMPOSITION

Prior to the start of the test program, or even the test planning phase, a coherent test team must be put into place. Although this sounds like common sense, in many programs it is often overlooked. As mentioned in section 1.5, flight testing is the last operation in the program master schedule. As such, funding for flight testing does not ramp up until the flight test period is imminent. Prior to this funding, a skeleton crew is normally authorized to perform planning, instrumentation development, data reduction routine preparation, and scheduling, which is no small task. It is the responsibility of the flight test manager to present the case to program management in order to obtain the required funding and manpower early in the program. Failure to do so may cause the flight test community to be behind schedule, which may eventually lead to failure.

Task teaming needs to be implemented as early as possible in the flight test cycle. It has been proven on numerous occasions that early teaming is essential to the success of the program. The first requirement of teaming is to identify all groups that are required for success. Today's jargon states that all members who either own or contribute to the "process" need to be included. Once individuals or groups are identified, the team must identify the goals of the group and team member responsibilities, and assign accountability. Most importantly, the group must be colocated. Excessive travel and distance is detrimental to a group's progress, and in many cases, actually prevents a team from carrying out its charter successfully. This situation is best demonstrated if we compare test site colocation to the children's game of telephone. In the game, a group of children is placed in a line and a little ditty is told to the first child, who then whispers it to the next child. When the last child receives the message, we ask that child to tell us what he heard. Everyone gets a big chuckle as the child tells us something that does not sound anything like what we told the first child. This is true when the test team is not colocated; communication gets garbled in transit. This is the third rule of avionics testing.

Whether test facilities are located 2 miles apart or 20 miles apart, the results are always the same. Communications become garbled when the test group is not colocated.

The task team should outline a new organizational structure at the outset. A team leader should be appointed and given the responsibility and the authority to carry out the goals of the team. All too often team leaders are given the responsibility for the team's actions, but they are not given the authority required to lead effectively. In this scenario, we have a built-in scapegoat for anything that goes wrong in the program. This scenario also weakens a leader's ability to assign accountability to individuals within the group.

Within CTFs or JTTs, some unique problems may occur with the team leadership. A CTF or JTT may combine contractor, DT&E, and OT&E testers under one umbrella. In some cases, an individual from each of the groups may be assigned as a joint team leader. As the charters of these groups are different, there will be some jockeying for control of the direction of the team. In the end, it may be the strongest personality that wins out.

The individuals and groups assigned to the flight test task team are those individuals previously mentioned in section 1.2: development and software engineering, flight test engineering, subcontractors and vendors, and DT&E and OT&E customer representatives. Each of these groups have specific responsibilities to the task team and are contributors to the overall goals of the team.

The goals are set by first establishing the priorities of the program. The team must determine if the program is one of specification compliance, development, or system performance evaluation, or some combination of the three. Once the priorities are set, schedules can be developed in order to meet these goals. In formulating these schedules, it is important for the team to exercise some judgment in realism. A careful review of similar previous programs should give an indication of how long and how difficult the program will be. The bottom line is to make schedules realistic rather than optimistic.

The foregoing description may be recognizable as an IPT, which is very much in vogue today. The premise is the same as task teaming, and in fact, IPTs are an outgrowth of the task team. The greatest difference is that IPTs are formed for all "processes," and have been used in many organizations to improve flow efficiency and, therefore, cut costs. These teams have been highly effective in areas such as assembly lines and customer assistance, but their effectiveness as a tool for the flight test effort is in doubt.

Under an IPT-based flight test program, changes must be coordinated through all affected IPTs. If, for example, an anomaly is detected during test and a possible solution is determined by data analysis, the fix and retest may take several months to accomplish. The problem and fix must first be brought before the avionics IPT, who determine if the problem and the associated correction are acceptable. Assuming they are acceptable, the engineering change proposal (ECP) IPT is next to be briefed. This IPT will determine if the fix is required (i.e., within scope), how much it will cost, the timeline involved, and if it will fit in the available schedule. If approved, the fix is presented to the software IPT, which determines if the fix can be done, how much it will cost, the timeline involved, and if it will fit in the software build schedule. Should the fix make it past this hurdle, it is then presented to the flight test IPT, who determines if the procedure requires modification to the existing test program, if additional safety concerns materialize, if additional testing is required, and whether or not the test can be accomplished within the existing testing budget for that particular system. If it is agreed that the test can be accomplished, it is forwarded to the execution IPT. This IPT determines if the test can be accomplished with available resources and test assets and in the available schedule. If approved here, the final stop is the integrated test team IPT, who determines if the test is in the best interest of the contractor, DT&E, and OT&E. If it is, then the test is scheduled. Each of these processes is on the order of weeks if the IPTs meet once a week; the entire iteration is a long and tedious one. When the schedule of software builds is taken into account, the delivery of the fix may be several months away if a build was missed while the fix was in the IPT loop. Thus, by establishing an IPT for each process, we have impeded the progress of flight testing by an order of magnitude.

There are other problems associated with IPT-based flight test programs. Because there are so many identifiable processes within the flight test community, there tend to be a like number of IPTs. This phenomenon leads to two distinct problems. The first is the sheer number of personnel required to staff these teams. The personnel problem can be handled in one of two ways: either hire more staff or allow the same staff to become team members of multiple IPTs. Most flight test organizations utilize a combination of these solutions, and the results are equally disappointing. More personnel are hired to create a process that is supposed to save money, or personnel are so busy attending to IPT business that there is no time to perform the actual job of flight testing.

Because of the number of IPTs created, there exists an overlap of responsibilities. This particular problem is amplified as individuals serve on many different teams. If there is no one responsible person who can resolve problems between teams, warfare between the fiefdoms will ensue. Unlike the autocratic flight test of a few years back, where the program manager controlled the purse strings and had absolute authority, today's IPTs are generally independent, with their own budgets. As a result, IPTs tend to become self-serving and an impediment to the process they were designed to assist.

1.8 | TEAM MEMBER RESPONSIBILITIES

Each member of the task team is assigned specific responsibilities that must be accomplished in order to successfully complete the flight test program. Some liberty has been taken here to identify each member's (group) responsibilities to the team. Some organizations may place these responsibilities in different organizations; nevertheless, all of these tasks must be completed by someone.

Development and software engineering have the primary responsibility for providing the task team with all applicable documentation. They are tasked with preparing and maintaining the design, performance, and integration specifications as well as measurement descriptor blocks (MDBs). An MDB is the description of a parameter on the data bus in relation to where the parameter comes from and where it is going to. It identifies the 1553 (MIL-STD-1553 military data bus, covered extensively in chapter 3 of this text) message number, word, start bit, and stop bit for every parameter on the bus. This information is the life blood of the test team, because this is how data are extracted from the system. For example, if we are testing the radar altimeter, we would have to know where the parameter "radar altitude" is on the data bus. The MDB provides us with this information. MDB information is contained in the integration design specification (IDS). This specification must be updated with each release, as the locations of the parameters on the data bus are likely to change with each new release (parameters are added, deleted, etc.).

The program performance specification (PPS) identifies the total system operation and is similar to the system operations section of military technical orders (Dash 1 for the USAF, NATOPS for the USN). When testing a navigation system, it is imperative that we understand the hierarchy of the systems and what can be expected as navigation systems gracefully degrade. This information is found in the PPS.

The functional requirements document (FRD) describes the operations that take place within the system line replaceable unit. This information is important to the flight tester because it determines what functions will need to be tested. In testing system compliance, we must know what is happening to the data within the system in order to determine its performance. An FRD is written for each avionics system in the aircraft which also describes the hardware interfaces and data flow between modules.

It should be evident that this documentation is mandatory information for the test team. Instrumentation tasking depends on up-to-date parameter information. In most programs, the first iteration of these documents will be the last time that the information is current. Unfortunately, documentation is probably the last priority for development and software engineering. It is for this reason that a close liaison must be developed between the flight test team lead engineer and the cognizant engineer. This liaison will ensure that changes in parameter location and identification are transmitted to the test team in a timely manner.

In addition to supplying the applicable documentation to the test team, the engineering community is also responsible for a number of other tasks. They are the responsible agency for accepting and testing government furnished equipment/contractor furnished equipment (GFE/CFE). All upfront integration testing is conducted by the engineering group. This task includes the integration of software and hardware on all GFE/CFE systems, as well as vendor unique systems. They prepare a lab test plan and conduct the actual lab tests. These tests are normally conducted in an avionics integration facility (AIF); also known as an avionics integration lab (AIL), software integration lab (SIL), avionics development lab (ADL), and software support and integration (SSI). Results of lab testing are furnished in final and interim reports. Successful integration and lab testing will result in a software release to test. This release will contain the functionality of the software, affected systems, and limitations to test. A final task of the engineering community is to provide a point of contact (POC) for each system under test.

The flight test engineer's primary task is to ensure compliance with the top-level specifications, statement of objectives, or EECs. In order to accomplish this task, they must formulate a test plan, coordinate the plan through the task team, prepare schedules to meet the plan, write test cards, brief, conduct and debrief the flight. The flight test organization is responsible for collecting flight data, reducing it to a usable format, and analyzing the data with respect to the task team goals. Data analysis provides the basis for meeting the primary goal. An accurate representation of the results of the test are provided in interim and final reports. Ancillary tasks are the documentation of anomalies discovered during the testing phase and assisting in the troubleshooting of these anomalies with the engineering group. Troubleshooting involves supplying data to the cognizant engineer as well as attempting to pinpoint the cause of the anomaly. Planning and retesting after problem resolution as part of the fly-fix-fly evolution, as well as the formulation of a regression test plan, complete the flight test tasks.

The subcontractor and vendor test representatives provide onsite technical expertise for their respective systems. These engineers are critical to the test effort, in that most of the problems experienced during the integration of these vendor systems should have been seen before in previous vendor applications. They have corporate knowledge for their systems. Another important task of these representatives is the interpretation of data collected by vendor-unique data collection systems. Many systems that are developed by subcontractors contain data collection routines, outputs, and recording instruments that can only be interpreted by vendor-unique data reduction systems. These systems are highly specialized, and purchasing one may be cost prohibitive. In these situations, the vendor may also be contracted to reduce and analyze data from this unique system. This flies in the face of the need for the flight test community to be an independent organization. In this case, we have the contractor who developed the system now providing analysis as to whether the system is performing to specifications. Although this is similar to the fox guarding the chicken coop, safeguards can be employed in order to ensure the integrity of the flight test process. The simplest way of avoiding conflict is to assign an individual from the team as a data analyst to assist in the vendor's analysis. Personalities play a key role here, as the vendor is naturally going to be somewhat protective of his system. This is human nature, as DT&E and OT&E personnel assigned to a contractor test team can attest.

The customer representatives (DT&E and OT&E) should be fully integrated with the task team. The ideal team will be comprised of analysts from all agencies, sharing the workload of collection, reduction, and analysis in support of the common goal. In an FSD program, where multiple systems are being tested, it is not uncommon to have a contractor engineer leading the radar team, a DT&E engineer leading the navigation team, an OT&E engineer leading the electronic warfare team, a vendor engineer leading the missile integration team, and so on. In addition to performing analyst functions, customer representatives normally provide operational aircrew to support the tests and may have the authority to recommend approval for specification compliance demonstrations.

1.9 | UNDERSTANDING TERMS

There is great confusion in the flight test community regarding the true meaning of terms in relation to a flight test program. One of the first tasks of the task team is to identify realistic goals. This can be a difficult endeavor if the type of program is not well defined. Aside from system development or subsystem integration, a tester needs to know if the program is either development or demonstration (sometimes identified as "qualification"). There are major differences between the two, and a failure to realize the correct type may cause an underbidding of the program by as much as an order of magnitude. A development program is the type previously described as fly-fix-fly. A demonstration program entails a proof of specification compliance, with no development work involved. It is a straightforward test program whose satisfactory completion normally is approval for production.

A development program is heavily dependent on the software delivery schedule and the maturity of each of the releases (that is not to say that hardware deliveries may also impact program schedules). In this type of program, the primary task of the test team is to identify problems in hardware and software as early as possible, isolate the cause, and retest as necessary. Figure 1.2 shows the fly-fix-fly process and its relationship to the test community. In order to successfully complete any avionics flight test program on time and within budget, the cycle time of this process must be made as short as possible.

It is readily apparent from the figure that the tester could get into an infinite loop if a discovered anomaly never gets resolved to 100%. This fact highlights the fourth and fifth rules of avionics flight testing: 4) the duration of the flight test program is directly

FIGURE 1.2 ■
The Fly-Fix-Fly
Process

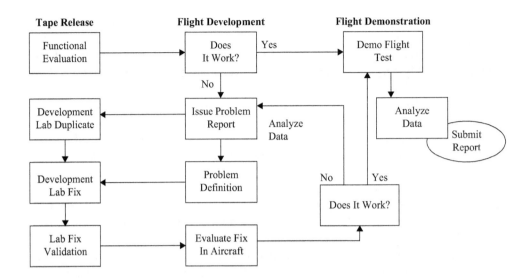

FIGURE 1.2 ■ The Fly-Fix-Fly Process

proportional to the duration of the fly-fix-fly process, and 5) no software release is ever 100% mature. In other words, testers must do everything in their power to accelerate the fly-fix-fly cycle, as well as know when to stop testing a particular function. The decision to stop the evaluation process on a particular function is one of the most difficult tasks a tester is presented with. More often than not, liaison with the customer will alleviate some of these difficulties. The customer (end user) will know better than anyone if the fleet/squadron will be able to perform its desired mission with less than optimum performance of the particular function. The tester may also face pressure from the engineering/software development community, who will be adamant that the function is fixed in the next build.

I digress a moment here in order to explain what is meant by a software build. The software build is called many things, and we frequently interchange these expressions in flight test discussions. This text may also inadvertently use different terminology to explain this concept. The software build may also be called a tactical release, operational flight program (OFP), build release, interim operational flight program (IOFP), or software release. Aside from the OFP, all of these terms relate to an incremental release of software, with each subsequent release containing more mature functionality. The OFP is normally released to the customer for operational use (i.e., the test program is complete), and therefore this term will not be used in our discussions.

The term software build conjures up visions of a floppy disk or nine-track or some other sort of electronic media used to program the aircraft mission computer. This is true to the extent that the build normally arrives at the aircraft in some sort of media format, but it contains much more than mission computer coding. In fact, a software build will contain code for all processors aboard the aircraft. Figure 1.3 shows a typical example of a hypothetical software build delivered to an aircraft for test. Avionics build 1.0 contains all of the software programs for each processor aboard the aircraft. If any of these sub-builds within Avionics build 1.0 are changed, then the overall build number must also be changed.

In the case cited in Figure 1.3, the software in Display Processor 1 was changed, perhaps to correct an anomaly found in testing, which necessitated a change to the overall software build. Upon closer reflection, it should become painfully obvious why

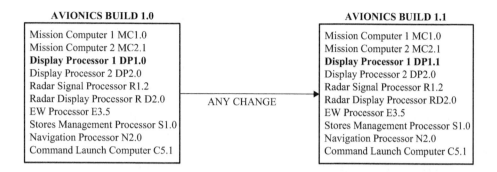

AVIONICS BUILD 1.0

Mission Computer 1 MC1.0
Mission Computer 2 MC2.1
Display Processor 1 DP1.0
Display Processor 2 DP2.0
Radar Signal Processor R1.2
Radar Display Processor R D2.0
EW Processor E3.5
Stores Management Processor S1.0
Navigation Processor N2.0
Command Launch Computer C5.1

ANY CHANGE

AVIONICS BUILD 1.1

Mission Computer 1 MC1.0
Mission Computer 2 MC2.1
Display Processor 1 DP1.1
Display Processor 2 DP2.0
Radar Signal Processor R1.2
Radar Display Processor RD2.0
EW Processor E3.5
Stores Management Processor S1.0
Navigation Processor N2.0
Command Launch Computer C5.1

FIGURE 1.3 ■
Software Build
Contents

the fly-fix-fly cycle is normally a slow, tedious process. If any small change in the subsystem software requires a major build release, it is more efficient for the software community to collect a group of changes and incorporate all of these changes at once. In this way, only one major release is required instead of the many it would take by incorporating changes as they come in. Although good for the engineering software community, this logic can be a killer for the test community. Testing often comes to a grinding halt while testers await new software.

How then can we accelerate the development cycle in order to complete the test program on time and within budget? One way is to allow the testing to continue with software patches. A patch allows a specific portion of the software to be modified without impacting the entire build. This method allows testing to continue with fixes that ordinarily would not be seen until a new software release. A problem with this philosophy is that eventually all of these patches must be incorporated into a major release. This process is known as a software recompile, and it can create serious problems for the software tester. After a recompile, serious regression testing must be accomplished on the new build to ensure two things: 1) that all of the patches were incorporated correctly in the new release, and 2) and most difficult, to determine if any other portion of the code was affected by the incorporation of these patches. Remember that patches do not affect the entire build because they are only written for a specific portion of the software; they can be thought of as an addendum to the code. The recompile, however, replaces code. Regression testing in the lab will determine the extent, if any, of the impact of patch incorporation. Experience has shown that software recompiles that incorporate large numbers of patches are functionally inferior to the software they replace, and two or three iterations may be required in order to regain that functionality. Historically the U.S. Navy has allowed patches on avionics integration programs, whereas the U.S. Air Force is currently not allowing patches.

It should also be remembered from the previous discussions that data acquisition software is also dependent on software releases. Each software release will entail changes to the software used to collect data. The data acquisition routines that are used for the current software release may only be good for that release. Subsequent releases may require updates to these routines, and a new version number must be assigned with each new release. The data acquisition routines are married with the software releases and are cataloged for future reference. This is configuration management, which will be addressed in a later chapter.

Now that software releases are understood, we can continue our discussion on the differences between development and demonstration programs. In addition to the fly-fix-fly concept of the development program, there are other objectives that the test community must address. The flight test team will validate engineering test results. That is to say, the team will determine if the systems in the aircraft work the same as in the lab. Section 1.4 touched on this issue briefly while addressing the need for flight testing. The development test program validates the lab simulations and mathematical models as well as provides flight data to enhance these simulations. Since there will be many conditions and functions that are not testable in the lab, the development test community incorporates these functions in the flight test plan. The development team also has the responsibility of uncovering any design discrepancies not found in the lab, providing data, and supporting the troubleshooting of anomalies. The final task is to incorporate the refly effort into the flight test plan and account for regression testing.

Diametrically opposed to the development test program is the demonstration, or qualification, flight test program. In this type of program, the tester's responsibility is to ensure that the system is compliant with the applicable specifications. It is often assumed, albeit wrongly, that there is no development work required for this type of program. A program is usually deemed as demonstration when proven hardware/software is to be installed in a proven aircraft or if a decision is made that all development work will be accomplished in the lab. There are fallacies in each of these assumptions; the latter was debunked in section 1.4. Although proven hardware/software is being installed in the proven test bed, there is still the small matter of integration which must be addressed. An installation of a MIL-STD-1553 compatible tactical air navigation (TACAN) system into an aircraft today is much more involved than previous tasks of ripping out an old system and widening the hole for a new system. Avionics systems of today are not stand-alone. These systems must communicate on the bus with more than one system and are required to interface with the host aircraft's software design protocols. If this task were as simple as it sounds, then GPS integration into any aircraft would not take upwards of a year to complete. If a mistake is made, and a development program is classified as demonstration only, it is likely the budget will run out a quarter of the way through the program or some serious compromises will have to be made. Historically the ratio of development to demonstration is 4:1, and this is the sixth basic rule of avionics flight testing.

Even though there is no such thing as a demonstration-only avionics flight test program, the tasks associated with a demonstration program must still be satisfied. Normally the program will be set up for development and demonstration, with the demonstration tasks being held back until development is complete. Demonstration is extremely important, because 99% of the time, final payment is based on successful completion of the qualification phase (the contractual obligations have been met). In this phase, the test community is responsible for ensuring that the system can satisfy the top-level specifications. These top-level specifications identify the minimum level of acceptable performance. In the past, top-level specifications were the degree to which the system had to be demonstrated and were used as the basis for formulating the flight test plan. In today's environment, many of the top-level specifications have been replaced with design objectives or EECs. Although the names have changed, the essential information and requirements remain the same for the test community. It is imperative that testers understand the demonstration requirements, for without them the program can not be intelligently planned.

Normally, top-level specifications are of the MIL-STD variety and are very specific as to what the avionics systems are and how they should perform. An example of a top-level specification is MIL-D-8708B(AS); the AS stands for "as supplemented." This is the specification for fighter aircraft and identifies all of the possible systems that may be incorporated into an aircraft. MIL-D-8708 has many addenda, and each of the addendum represents a particular aircraft. Addendum 100, for example, refers to the F-14D. The addenda instruct the reader as to which of the paragraphs in MIL-D-8708 are applicable to a particular type of aircraft. The addenda also refer the reader to a subsystem document if there are further specifications to be met. For example, under Radar in Addendum 100 to MIL-D-8708B(AS), the reader is referred to a radar specification, SD-4690. SD-4690 contains the specifications to be met by the aircraft-specific radar; in this case, the APG-66. SD-4690, in turn, refers the reader to another specification that contains additional specifications for the APG-66 operating in special modes. This document is classified SECRET and therefore cannot be included in SD-4690. This third document is SD-4619. This flow of documentation is typical of specifications for military and federal agency avionics systems. It is important for the tester to understand this flow and not be deceived into thinking that specification compliance is meeting the objectives of the basic specification (in this case MIL-D-8708B(AS)). The following is an example of the level of detail a tester might expect to see in these specifications:

- MIL-D-8708B(AS), Addendum 100 might state:

  "the onboard radar shall be able to detect and track targets in the pulse Doppler mode."

- Referring to the same paragraph in MIL-D-8708, SD-4690 may state:

  "the APG-XX radar shall detect and track a 2 square meter target at 35,000 ft with a total closure of 1200 knots under maneuvers to 5g at a range of 95 nautical miles in the pulse Doppler mode."

- SD-4690 may in turn refer the reader to SD-4619, which may state:

  "the APG-XX radar shall detect and track a 2 square meter target at 35,000 ft with a total closure of 1200 knots exhibiting angle deception noise swept amplitude modulation (NSAM) at a range of not less than 50 nautical miles with elevation and angle accuracy of not less than 1.5° in the pulse Doppler mode."

Each of the cited paragraphs above are bona fide requirements that must be demonstrated by the test team. If the tester elects to only demonstrate the basic specification (MIL-D-8708), the test plan is going to be very simple and very incomplete. The plan would be so simple in fact that the total test program would not be much longer than one flight, and that flight would basically be a "smoke" check of the radar. As long as the radar did not burn up after it was turned on, it would probably meet the specifications. The second paragraph is more representative of what the test plan should be built on, as it calls for some performance numbers that must be verified. The third paragraph is even more specific and requires more test planning and effort in order to meet the desired specifications. Each of these paragraphs is a mandatory criteria that must be demonstrated. Satisfaction of the basic specification does not meet specification compliance for the system.

The foregoing discussion should trigger some thinking; prior to initiating any test planning activity, an in-depth review of the specifications must be accomplished. If this review is not accomplished, and all of the test requirements are not known, serious underestimation and underbidding of the program will occur. On a related note, consider the effect of a poorly written specification on the quality of the product. If the basic specification is the only specification issued, would we expect the same product as if we had issued all three specifications? In some quarters, the answer is yes. Some agencies honestly believe that if contractors are given only the minimum guidance, they will produce the best possible product to meet the customer's needs. It is believed that there is no so such thing as a "lying, cheating contractor," and given half a chance, the contractor will work miracles. The thought process is to release the contractor of the burden of specifications and let them build the best product possible in the shortest possible time. Obviously it is the contractor who knows best how to build the best system. The problem, of course, is what happens at the end of the flight test program when the money and assets are gone and the system is not exactly what the customer needs? All the contractor needs to do is point to the contract, and the lack of detail, and say the contract has been satisfied. The customer really has no options for recourse except to deny future contracts. This points out the fact that the customer ultimately "owns" the risk in a development program. The customer must ensure that the requirements are accurate, complete, verifiable, and fully understood by the contractor. On the other hand, requirements equate to cost, so the customer should not have more requirements than is absolutely necessary.

This is not to say that this scenario will happen; however, the option is there and customers may wind up paying twice for a system they desire. Contractors, as a rule, will not offer this concept to their subcontractors, and this leads us into the next discussion on contractor and subcontractor specifications.

1.10 | CONTRACTOR AND SUBTIER SPECIFICATIONS

In addition to validation of the top-level specifications previously discussed, the test community is also tasked with the responsibility of validating contractor and subcontractor specifications. These are sometimes called subtier or C-level specifications and are used to guide the design of avionics systems and their interfaces with other avionics systems aboard the aircraft. For the sake of completeness, A-, B-, and C-level specifications can be defined as

- A-level: high-level customer requirements.
- B-level: requirements levied on the prime contractor (perhaps withholding some margin, or removing some requirements the customer might choose to delay until future variants).
- C-level: the specification documents the prime contractor levies on subcontractors.

The IDS, PPS, and FRD were addressed in section 1.8. The prime contractor is responsible for these specifications. Should the prime contractor subcontract work to other vendors, an interface specification must be written for each of the systems that have been subcontracted. These documents describe the interfaces required between subsystems and prime systems to allow functionality. A simpler way of describing these documents would be to say that the interface specifications are memorandums of agreement between the two companies.

Since one of the primary jobs of the avionics tester is to validate the travel of data from the sensor through each of the related LRUs to the final output to the aircrew, they must in fact validate the interface specifications. Testing may uncover discrepancies or incompatibilities between the IDS and the subcontractor interface specifications. Any incompatibilities will likely cause a failure of the system being tested. At first glance it may appear as though the system under test is at fault; however, under closer scrutiny we may find that the system is performing as designed. An example may clarify this point: In one particular program, the entire avionics system was to perform to class A power specifications. An electronic system can be designed to operate under three classes of power (A, B, or C). These classes may be thought of as operation under differences in power quality, transients, and recovery time. An LRU specified to operate to class A may be required to accept a larger transient of power for a longer period of time without adverse operation than a system designed to meet class B power. In our particular example, the specification given to the subcontractor stated class A power requirements; however, the chart included with the text was one for class B. The subcontractor designed the LRU to the weaker requirement (class B). Of course, the first time the aircraft transferred from external to internal power, the system dropped offline. Who is at fault? Is the system unacceptable?

To the user, the system is definitely unsatisfactory, yet to the subcontractor, the system was built as designed and performs to specifications. The prime contractor in this case is stuck in the middle with "egg on their face" and is forced to fund an engineering redesign (this particular venture cost the contractor well over a million dollars and a year of lost time). An avionics flight tester identified the problem and cause, and pinpointed the documentation error.

1.11 | CUSTOMER EXPECTATIONS

The test community's customer may be the final customer (the buyer) or it may be local program management. Each of these customers has different expectations of the avionics test program. The final customer will expect, at a minimum, that the top-level specifications have been satisfied and that all CDRLs have been fulfilled. Since the test team is more than likely a task team that includes DT&E and OT&E members who were involved in the test plan preparation, the final customer will also expect some measure of performance data.

Local program management is a completely different customer than the buyer. Local program management is tasked with maintaining fiscal accountability and adherence to the program master plan. Fiscal accountability is easy to understand; there is only x amount of money allotted to the program and there must be somewhat of a profit margin. Normally program management maintains a contract-oriented regimen that only permits testing strictly called out in the contract. Program management is also concerned with maintaining progress in the test program. Progress is measured against milestones, and the final milestone is looming ever larger on the horizon. The test community is technical by design, and communicating with an organization (i.e., program management) that is not technical in nature can be a frustrating experience.

In order to maintain program discipline and fiscal accountability (and mental sanity), a constant liaison between the program office and the technical test team must be maintained. The first consideration in this liaison is maintaining a balance between the program milestones and technical merit. This can best be illustrated by the "flight around the flagpole" (a flight accomplished for the sole purpose of demonstrating to the world that your program is alive and well), of which all test programs have been guilty. Unfortunately one of the methods used to gauge progress in a flight test program is flight rate; the more you fly, the more progress you are making, and vice versa. One of the worst things that can happen in a flight test program is for the test vehicle to be on the ground, either intended or unintended. The unintended portions include weather, maintenance, or instrumentation problems, and are nonproductive periods of time that cannot be avoided. But what happens if the test team suggests that flying be halted for a period of time so data can be analyzed, or worse, the team needs new software and is devoid of constructive tests. More often than not, this halt in flying indicates a halt in progress which can be easily alleviated by a continuance of flying. Therefore the test team is directed to resume flying (progress). This example may seem silly to some, but rest assured, it does occur in today's environment. It is the test engineer's responsibility to plead their case to management and present the rationale for standing down. The test engineer must be prepared to justify every test. Conversely, the test engineer must also be prepared to explain why a test is a waste of time/money, even while program management is demanding that testing start/resume in order to "show progress."

It must also be remembered that flight testing is dependent on the configuration of the hardware as well as software in the test bed. Asset requirements, hardware as well as software capability, must be relayed to program management in a timely manner in order to preclude flying with the wrong configurations. If the test plan calls for testing data link at a certain period of time and the hardware is not available, the applicable milestones showing these tests must be moved to the right on the program schedule. Failure to do so sets up a conflict between the program and the test team when the tests are not complete. Similarly, when the software does not support the test milestones, a similar scenario will occur. Above all, it must be remembered that scheduling, briefing, and conducting a nonproductive mission takes as much time, effort, and manpower as a productive flight.

1.12 | FORMULATING THE PROGRAM

Understanding customer requirements, test team goals and responsibilities, and the nature of the test program allows us to formulate, or estimate, the program. It is important to realize that undertaking an estimate without first determining these aspects is a recipe for failure. Each estimate should follow a tried and true methodology: determine the type of program and the required outcome, determine the assets required to perform the task in the time allotted, and estimate the task. We have discussed the factors that affect the type of program and will now address the assets required for the time allowed.

By knowing the nature of the development, demonstration, or performance requirements of a system, we can set out to identify each of the following:

- The number and type of test vehicles required
- Instrumentation
- Support facilities
- Size and composition of the test team
- Special hardware or software requirements
- Schedule of software releases

Normally the number and type of test vehicles are predetermined by program management, the customer, fiscal constraints, or some combination of these factors. In estimating a program where the final determination of assets has not been ascertained, the tester may be asked to propose a number which he feels is necessary in order to complete the program in the time allowed. The key factor here is time.

It is an excellent idea to first research past programs with similar requirements and examine the total sortie (flight) count and total time for program completion. This information is archived in many locations within test organizations and is just begging to be read prior to a new test program. These past histories probably reside next to the "lessons learned" notes from the same program. Once again, it is unfortunate that many testers would rather reinvent the wheel than learn from others' experiences. Past history and lessons learned from similar previous programs provide invaluable insight into what can be expected in the program you are estimating if the lessons are applied appropriately to the test activity at hand. There are a series of reports available for reference regarding the aforementioned discussion. The Rand Corporation has been involved with Project AIR FORCE and has issued a number of reports documenting trends and lessons learned on T&E projects. The first one that I would like to reference is Test and Evaluation Trends and Costs for Aircraft and Guided Weapons which was published in 2004 and covers many of the topics previously addressed. The second Project AIR FORCE report is entitled Lessons Learned from the F/A-22 and F/A-18E/F Development Programs, Rand, 2005. Quite an interesting read.

Consider the case where a previous radar program spanned 3 years and 300 sorties and the data time per flight would have been 1 hr. On closer examination, we see that only one test bed was used for the program, and that the initial schedule called for the test bed to fly 10 sorties per month. The actual flight rate was something less, about 8.33 sorties per month. This information tells us a great deal about a radar program. It not only tells us the total number of sorties required, but it also tells us that there should be some contingency factor of about 20% to account for loss of scheduled sorties from the plan. This loss could be due to maintenance (scheduled or unscheduled), weather, software, instrumentation, or personnel problems (sick, holidays, vacations, etc.). If we used an historical program that was accomplished at Edwards Air Force Base, where there are on average 330 days of flyable weather, and compare it to a current program being conducted at Patuxent River, MD, where weather does pose a problem, some adjustments would have to be made. A contingency of 20% is realistic and can be used in estimating any avionics or weapons system test program. The numbers also show that if flight time can be increased or if assets can be added, the total program time will decrease. If our historical program used an F/A-18C as a test bed and our test bed is an F-15E, the flight time will indeed be doubled and the program should be halved. Care must be taken in the amount of time compression we anticipate in a test program. As we compress time, we place a larger burden on the flight test analysts to turn data around more rapidly and on software engineering to compress the fly-fix-fly cycle. Either of these burdens may be eased by increasing manpower, but the incurred cost may outweigh the benefit of compressing the test program.

In order to show compliance, troubleshoot anomalies, and report performance, we must provide data. Data can only be provided if the test bed is instrumented. A second phase of formulating the program is determining what instrumentation is required, in what format data should be presented, and whether these data need to be transmitted to the ground in real time or be postflight processed. In addition, the type of recording—video and data—needs to be determined, as well as time synchronization and whether the development lab needs to be instrumented. The PCM format and whether edited data bus or full data bus will be recorded also needs to be determined.

Each of the items in the instrumentation package directly impacts the length and cost of the avionics program. Real-time data will shorten the program considerably; however, the cost may be prohibitive. Actual labor to instrument the aircraft may take longer than the program allows or equipment may not be available. Special instrumentation such as onboard simulators and stimulators have very long lead items and need to be accounted for in the estimate if they are mandatory for success. As is the case with test vehicles, the test engineer must be aware of the trade-offs between time, money, and manpower affected by the type of instrumentation package chosen.

Often overlooked when trying to formulate a test program are the seemingly benign support facilities. Test range support needs to be addressed, and the level of support that is necessary for successful program completion needs to be accurately conveyed to the range directorate. This is addressed in some detail in section 2.16 of this text.

In addition to consolidation of manpower within the test teams, the current policy in the United States is to consolidate test sites as well. Flight testing of the future may be at either an East Coast or West Coast range facility. Smaller government and contractor facilities will be forced to unify under one of these two facilities. Work in this area is already well under way, and it is yet to be seen who the big winners are.

Nothing is more distressing for a tester than to arrive at a test facility and find out that there is no place to sit, or there's no phone or computer access, or that the test article has to brave the elements because there is no hangar space. All of these scenarios are distinct possibilities if there was no upfront planning and coordination on the part of the test team. Do not be lulled into a false sense of security and into believing that since the test program has been directed to use a designated facility that all is in readiness. In fact, there are additional costs that need to be accounted for. Other than workspace areas and telephone and computer access, there are other areas that must be considered and which are often overlooked.

Clearances and certifications may not be transferable from site to site; unless pre-coordinated, program clearance at one site does not automatically grant access to another. The ability to transmit radio frequencies requires a clearance and a deconfliction. In the United States, this is obtained through the regional frequency council; radio frequencies also cover telemetry frequencies. A clearance to transmit data will also be required, and in many cases can only be done if the data is encrypted, which can lead to a cascading list of additional requirements. Encryption requires additional hardware and software, and in most cases raises the security classification. With enhanced security comes enhanced security measures in terms of access, storage, and handling. Data must be logged, controlled, and stored in special facilities and access to the data must be controlled. Crypto keys are controlled, change from day to day, and require specialized personnel to load the keys into aircraft and telemetry processors. All of these additional requirements will have an impact on the cost of the program.

Certification in a particular field, such as hydrazine handling, may not be valid at the test site, especially if you have a contractor certificate going to a government test site. Military personnel are not exempt either, because in many cases the Navy does not recognize Air Force certification, and vice versa.

Data reduction and processing facilities at the test location may not be able to handle the installed instrumentation system, or the data rate, or support rapid turnaround of data. In the worst-case scenario, a test team estimates the length of a program based on a 24 hr turnaround of data, only to find out later that data will not be available for analysis for 72 hr. In a similar vein, if data must shipped to the engineering group for confirmation of results or for troubleshooting purposes and they are located many miles away, the program will be delayed as well. This can be disastrous to the efficient flow of the program. There are two options: maintain the schedule even though the data will be delayed, or slip the schedule to accommodate the delay of data. From our previous discussions, you know that you are not going to slip the program, because that costs money. On the other hand, if you maintain the schedule, you will probably be flying without the benefit of knowing the last flight's performance. The end result of such a miscalculation is wasted or lost flights and a program coming to an end without obtaining all the necessary data. Knowing all of the limitations of a test program, not just system limitations, is imperative if you intend to be successful.

The requirements for special hardware and software drive the program schedule and cost. Special instrumentation, such as data bus editors or word grabbers, may cost some additional funds, but more importantly, they require a long lead time. Times of one year or more to design, fabricate, install, and test a special piece of instrumentation are not uncommon. You cannot hinge the success of your program on the acquisition of key pieces of data if you do not know how you are to capture that data. Never assume that all data are always available with off-the-shelf, time-tested instrumentation. This is another reason why instrumentation personnel need to be brought into the program early.

As previously mentioned, knowledge of the capabilities and limitations of the installed software is imperative to efficient testing. Similarly, you must know the schedule of builds in order to put together a rational schedule. Remember, incremental builds are the norm in avionics testing, and the length of the test program is directly proportional to the functional capabilities available in each build. The schedule will also stretch with the amount of time it takes to fix a problem in the software. How does today's tester understand the capabilities and limitations of software? They need to be in the lab during validation and verification testing long before the software gets to the test bed. This is the only way a tester can understand the maturity of a software system. Armed with this knowledge, a realistic schedule can be formulated.

1.13 | SUMMARY

Before putting pen to paper or scheduling your first test, it is important to realize that there are many variables to be considered in the world of avionics systems flight testing. The rules you may have learned in vehicle testing may not apply. The single largest difference is that you are dealing with a nasty little item called software. Do not ever believe the adage that software changes are "transparent to the user," or that "we didn't change anything in that module." Software has extremely long tentacles, and changes in one module may affect other seemingly unrelated modules. Scheduling test programs can be an interesting effort, as there are so many items that can affect them: software maturity, fix cycle, delivery schedule of builds, data turnaround time, etc. It takes a tester adequately versed in these parameters to make an avionics test program work properly. The tester that recognizes these variables, reads the "lessons learned" file, and applies them appropriately to the test activity under consideration is the one who will succeed.

Remember the six basic rules when performing avionics and weapons systems testing:

1. The final milestone (Δ) never moves to the right.
2. Truth data (TSPI) needs to be four times as accurate as the system under test.
3. Whether test facilities are located 2 miles apart or 20 miles apart, the results are always the same: communications are garbled when the test group is not colocated.
4. The duration of the flight test program is directly proportional to the duration of the fly-fix-fly process.
5. No software release is ever 100% mature.
6. Historically the ratio of development work to qualification testing is 4:1.

▌ 1.14 ▐ SELECTED QUESTIONS FOR CHAPTER 1

1. What are the major differences between vehicle testing and systems testing?
2. Name three major objectives of flight testing.
3. What is a measurement descriptor block?
4. What is the bulk of testing in today's environment?
5. What is an avionics build (software release)?
6. How come it takes so long to receive new/updated software?
7. Software is incremental. How does this fact affect systems testing?
8. What is a CDRL?
9. What is a DID?
10. What is the difference, if any, between DT&E development, demonstration, and performance testing?
11. Why are data important when identifying problems?
12. Who has the responsibility of ensuring that the system meets the specifications?
13. What is meant by the term "composite test team"?
14. What are "lessons learned" and how do they apply to flight testing?
15. What is fly-fix-fly?
16. What factors affect the length of a system's flight test program?
17. How would you describe avionics systems flight testing?
18. Name the advantages of using real-time data.
19. What is an IPT? What is its purpose?
20. Who are the key individuals that should be a part of the flight test team?
21. What is OT&E responsible for?

22. Why can simulation not be used in lieu of flight testing?

23. Give three examples of objectives that cannot be accomplished in the simulator.

24. Is combining DT&E and OT&E a good idea? Why or why not?

25. What is a top-level specification?

26. What is the program manager's purpose?

27. Will there ever be a difference of opinion between flight test and program management?

CHAPTER 2

Time, Space, Position Information

Chapter Outline

2.0	An Overview of Time, Space, Position Information	29
2.1	Sources of TSPI	30
2.2	Ground-Based Tracking Radar	30
2.3	Radar Characteristics	32
2.4	Radar Accuracies	35
2.5	Geometric Dilution of Precision	38
2.6	Earth Models	39
2.7	Theodolites	41
2.8	Theodolite Limitations	44
2.9	Global Positioning System	44
2.10	Stand-Alone GPS Receivers	44
2.11	Inertial-Aided GPS Receivers	46
2.12	Differentially Corrected GPS	48
2.13	Sensor Fusion	48
2.14	Ranges	49
2.15	Range Assets	49
2.16	Unique Services	50
2.17	Interfacing with the Range	51
2.18	Time Alignment	53
2.19	Summary	55
2.20	Exercises	55
2.21	Answers to Exercises	55
2.22	Selected Questions for Chapter 2	58

2.0 | AN OVERVIEW OF TIME, SPACE, POSITION INFORMATION

All testing of avionics and weapons systems requires that the tester prove the performance of the system under test by measuring its performance against some known quantity. This known quantity falls under the heading "truth data," or time, space, position information, affectionately known as TSPI. The concept is slightly different from the testing accomplished on the vehicle side of the house. In loads testing for example, the tester relies on instrumented parameters to determine the "true" bending and torsion on a member. The accuracy of the result depends entirely on the accuracy of the instrumentation system. The tester in this case is measuring the truth and comparing this answer to the predicted results. In avionics and weapons system testing, we still collect instrumented parameters such as latitude, longitude, and altitude, but we compare them to an outside reference, a truth. This truth is generated by calibrated reference systems such as cinetheodolites or an on-board global positioning system (GPS). It is important that our truth system be accurate enough to measure the expected performance of the system under test. Unfortunately, all too often testers use whatever TSPI system is available without considering the impact on the results of the test. In some cases, highly accurate avionics systems are being measured by TSPI systems that are not as accurate as the system under test! This chapter discusses what systems are available, how the results can be improved upon, and defines a common-sense approach for identifying required test assets.

2.1 | SOURCES OF TSPI

The first place a tester needs to visit prior to any avionics test is the range directorate. These are the people responsible for providing the user with truth data. In some organizations, the range directorate and data support and instrumentation may be the same organization. You will find these organizations extremely helpful in determining what sources are available, and more importantly, what accuracies to expect. The novice in flight testing may want to spend some time with this organization in order to become familiar with TSPI in general.

There are other sources of information available on the subject as well. Each test range will be able to provide a range user's handbook that describes the capabilities, and sometimes the prices, of the range assets. There are some very good range websites, unfortunately many of them require government access. One particularly good source of information is the Defense Test and Evaluation Professional Institute, located at Pt. Mugu, CA. They have put together a CD-ROM course on TSPI that is very beneficial and may be available to your organization.

2.1.1 Summary of TSPI Sources

The sources that will be discussed in the following sections, as well as a summary of their attributes, are contained in Table 2.1. Other subject areas that will be covered are sensor fusion, ranges, and time alignment.

2.2 | GROUND-BASED TRACKING RADAR

Until the proliferation of satellite based positioning systems, the most common form of TSPI service was provided by ground-based tracking radars. These systems are most common on U.S. military ranges and have been used for many years. The most common form of the ground-based tracking radar can trace its origins back to the U.S. Army Nike

TABLE 2.1 ▪ TSPI Sources and Attributes

TSPI Source	Positive Attributes	Negative Attributes	Relative Cost	Availability
Ground-based tracking radar	1) Long-range system 2) High x, y, z accuracy at short ranges	1) Accuracy degrades as range increases 2) Accuracy degrades as elevation angle decreases	Moderate	Good, available at most ranges
Film-read cinetheodolite	1) Highly accurate in x, y, z position	1) Time-consuming processing will cause delays in receiving data 2) Weather dependent	Very high	Low, specialized
Optical tracking cinetheodolite	1) Highly accurate in x, y, z position 2) High data rates in real time	1) Weather dependent 2) Limited range 3) Limited aircraft velocity	High	Low, except for specialized weapons delivery ranges
LASER trackers	1) Highly accurate in x, y, z position 2) High data rates in real time	1) Weather dependent 2) Limited range 3) Limited aircraft velocity	High	Low, except for specialized weapons delivery ranges
Stand-alone GPSs	1) Good x, y position accuracy if nonmaneuvering 2) Readily available	1) Low data rates 2) Limited maneuvering 3) Susceptible to signal loss	Very low	Plentiful
Inertial-aided GPSs	1) Extremely good velocity and acceleration accuracy 2) Good x, y position accuracy	1) Susceptible to signal loss 2) May coast out	Low	Medium, pods are available at most military ranges
Differential GPSs	1) Extremely good x, z positional accuracy 2) If differentially corrected and aided by barometric altitude it is the best TSPI system available	1) Susceptible to signal loss	Low	Good, most test facilities

Hercules Air Defense Systems (Figure 2.1). Many of these systems have been upgraded to solid-state technology and provide fairly accurate position information to test organizations. An advantage of the ground-based tracker is its ability to track targets at long ranges out to the limits of line of sight.

In order to understand the capabilities of this TSPI source, as well as its limitations, it is important to note the key aspects of radar operation. For any TSPI source, we are interested in determining true position. In order to determine the three-dimensional (3D) position of an object in space, we need to determine that object's latitude, longitude, and altitude. Radars provide us with a 3D location of an object in space in relation to the radar pedestal, or ownship location.

The tracking radars we will come in contact with are pulse radars. Radio energy is directed (transmitted) out of the antenna dish in distinct pulses. The radar then listens (receives) for the return of these pulses. Since we know that radio energy travels at the speed of light, we merely multiply the speed of light by the amount of listening time it

takes for the pulse to return and divide by 2 (since the energy travels a roundtrip and we only want the one-way distance. This computation gives us the range to the target (note that this is slant range, and not ground range). By knowing the pedestal orientation in relation to true north, it is an easy task to measure the bearing to the target. This measurement is known as the azimuth. Similarly, if we know the reference of level to the earth, we can compute how many degrees of tilt are used to detect the signal. This measurement is called the elevation. If the fixed position of the pedestal is known (latitude, longitude, and altitude), the measured range, azimuth, and elevation can be put into a coordinate conversion routine and the detected object's latitude, longitude, and altitude can be computed.

Converting range and bearing from a fixed point to a latitude, longitude, and altitude is a relatively simple procedure. If math is not your forte, there are a number of software programs available, many of them free on the World Wide Web. Just enter the keywords "coordinate conversion." If you are receiving data from an established range, you may request your data in the form of either range, azimuth, and elevation, or latitude, longitude, and altitude and the conversion is performed for you.

2.3 | RADAR CHARACTERISTICS

There are some considerations the tester must address prior to using a radar as a TSPI source. The first of these considerations is the ability of the radar to track the target. In all quotes of accuracy of a ground-based tracking radar, it is assumed that the object to be tracked is equipped with an onboard beacon. These beacons, which operate in either the C or X band, allow the ground-based radar to track a point source of radiated energy. The beacon acts as a transponder. The ground-based radar interrogates the beacon and the beacon, in turn, responds back to the radar after a set delay. Since the frequency of the response is the same frequency as the radar, very accurate tracking can be accomplished. The azimuth and elevation are determined by a passive detection of

FIGURE 2.2 ■
Effects of RCS on
the Reflectivity of
Radio Waves

the beacon reply and range is measured by the time of the reply. If a beacon is not available, or the beacon is "blanked" from being seen on the ground by aircraft maneuvering, the radar must "skin" track, which causes drift and loss of signal return from time to time, thus degrading the position trace. The problem will also occur if the test aircraft dispenses chaff (chaff is an expendable described in chapter 9 of this text). This problem is most severe when attempting to track dynamic targets whose aspect angle, and hence radar cross-section (RCS), is ever changing. RCS is described in detail in chapter 8 of this text.

As can be seen in Figure 2.2, the aspect of the target as seen by the radar has a significant effect on the amount of energy returned to the radar. In some cases, a relatively small change in the aspect may cause a strong return to suddenly disappear and the radar will transfer tracking to a stronger return from the aircraft, affecting the reported position. The effect on a TSPI solution is readily apparent—dropouts of the radar track, that is, a loss of consistency causing an erratic trace. Military ranges in the United States require aircraft on their ranges to have a beacon installed on the aircraft as a condition of entry. A consideration for "stealth" aircraft may require that low observable testing be accomplished on an aircraft not used for systems testing, since those aircraft will be dirty with the existence of an external beacon.

Some readers may be confused with the terminology of X and C bands. Depending on your background, these bands (which equate to a specific grouping of frequencies) may look more familiar if they were called I and G bands, which is the North Atlantic Treaty Organization (NATO) standard. Back when radars were being developed, the "microwave" frequency band was broken up into specific groups and labeled with a seemingly random nomenclature; these are the "old" radar frequency designations. In reality, the lettering scheme used does have real meaning. L band represents long wave, whereas S band represents short wave, and C band is a compromise between L and S bands. Table 2.2 provides the nomenclature and frequency bands for both the "old" and "new" radar frequency designations. Test engineers working with electronic warfare do not use the "old" designations, but rather use the NATO, or "new" standard: C band through M band (lowest to highest frequency).

Since we are dealing with radar, it is important for us to understand some of the properties of radio waves. First, radio waves cannot travel through rocks, buildings, or other manmade structures. This is important for the tester in determining if there are any blockage areas that may impede the radar's ability to track the target. A second property is that since these are radio waves, they are restricted to a line of sight (LOS) with the target. Since the earth is curved, flying at a constant altitude away from the radar will

TABLE 2.2 ■ Radar Band Designations

IEEE US (old radar designation)	Origin	Frequency Range	Wavelength	NATO, US ECM (new radar designation)
W	W follows V in the alphabet	75–111 GHz	400–270 mm	M (60–100 GHz)
V	Very short	40–75 GHz	700–400 mm	L (40–60 GHz)
K_A	Kurtz (above)	26–40 GHz	1.1–0.7 cm	K (20–40 GHz)
K	Kurtz (which is German for short)	18–26 GHz	1.6–1.1 cm	J (10–20 GHz)
K_U	Kurtz (under)	12.4–18 GHz	2.5–1.6 cm	
X	Used in World War II for fire control as an "+" for crosshairs	8–12.4 GHz	3.7–2.5 cm	I (8–10 GHz)
C	Compromise between S and X	4–8 GHz	7.5–3.7 cm	H (6–8 GHz) G (4–6 GHz)
S	Short wave	2–4 GHz	15–7.5 cm	F (3–4 GHz) E (2–3 GHz)
L	Long wave	1–2 GHz	30–15 cm	D (1–2 GHz)
UHF		0.3–1 GHz	<1 m–30 cm	C (0.5–1 GHz)

FIGURE 2.3 ■
Effects of Multipath
on Radar Reception

eventually lead you to be below the horizon; thus none of the radio energy will be reflected by your aircraft, and hence there will be no target returns. Radio waves are also refracted as they travel through the atmosphere. This bending is dependent on the frequency employed as well as the atmospherics of the day. We, as users of the data, do not have to worry about the inaccuracy of the radar pointing angle due to refraction because an "index of refraction" is applied by the pedestal operators at the site.

A fourth property is the concept of multipath (Figure 2.3). Since radio waves cannot penetrate solid objects, they are reflected by those objects. The earth happens to be a very good reflector of radio energy. As a radar attempts to track a target low on the horizon, it will receive a return from the target (direct reflection) and a second return from the earth (radio energy reflected from the target which subsequently is rereflected off of the ground). The tracking radar encounters two problems: first, the phase of the multiple returns may cancel each other out, depending on how far out of phase they are; and second, it cannot determine the true target reflection from the rereflected energy from the earth, thus destroying the tracking solution (elevation accuracy is lost). In some cases, the aircraft may appear to be below the ground elevation. For this reason, many ranges will not guarantee radar tracking accuracy at elevation angles of less than 2°.

For the tester, it is important to calculate the minimum altitude allowable for the aircraft during the flight which will ensure elevation angles to the aircraft of more than 2°. To allow for refraction, it is common for ranges to use a 4/3 earth model which approximates the effect. If the 4/3 earth model is assumed, a good rule of thumb to ensure a 2° elevation is that for every 4 nm the aircraft is away from the radar, altitude must be increased by 1000 ft. For example, if the aircraft is 100 nm from the site, then the aircraft would have to be at 25,000 ft:

$$100 \div 4 = 25 \times 1000 \text{ ft} = 25,000 \text{ ft}.$$

It can be seen that there will be a range where it will become impossible to maintain the required altitude for LOS. It should also be apparent that if the test cards specify a particular altitude to fly, that will dictate a maximum range allowable due to range asset restrictions. The good news is that the radar may have a maximum range of 300 nm. The bad news is that your crew will need pressure suits to maintain the 2° restriction.

2.4 | RADAR ACCURACIES

Perhaps the greatest concern for the tester, however, is the actual accuracy one can expect from the system. All radar vendors deliver the tracking systems with a stated vendor accuracy. This accuracy is tempered with the caveat that accuracy is based on a schedule of calibrations. That is, the accuracy can only be guaranteed if the specified calibrations have taken place. There are some important issues related to calibration. Pedestal level, surveyed location, radar center of radiation (with respect to the antenna), and a host of instrumentation errors can all affect data, whether it be a TSPI radar or testing of a radar system. Often they must be corrected postmission with simple or complex bias removal. Contact with your range directorate can give you these answers.

There are many types of calibrations, ranging from star calibrations to beacon ranging tests. Many ranges have calibration spots on the taxiways as a final sanity check prior to takeoff. The accuracies that can be provided are normally found in the range handbook. The book will state what can be provided to the test community for each TSPI system available. Figure 2.4 is a sample page from one of these handbooks.

The meanings of each of the columns are pretty obvious. The first column indicates that data being provided is TSPI. The next column provides the user with the guaranteed accuracy that will be provided as long as the conditions in the remarks column are met. Skipping over to the remarks column, we can see that a C-band beacon is required on this range. The maximum range from the pedestal is 125 nm (not really too good for a ground-based tracking radar). The next two items are really excuse statements. As mentioned previously, there will be some propagation and delay errors as well as multipath in all radar systems. Because of multipath, this range cautions you to remain at elevation angles greater than 2°. In addition to the remarks

FIGURE 2.4 ■

Sample Range
Handbook Radar
Accuracy

Data Required	Data Accuracy	Activity Area	Output Rate	Data Products	Remarks
TSPI	±0.5 mrad azimuth and elevation ±5.0 yards range	Controlled areas above 5° elevation	20 samples per second	Printouts Magnetic tape Plots	1. C-band AN-TPQ-39V-0 2. Beacon-assisted track 3. 125 nm max range 4. Exclusive of atmospheric delay and multipath 5. Exclusive of beacon delay and propagation variations 6. Data degrades significantly at elevation angles of less than 2°

section, the chart also tells you where you may operate and what data products and sample rates are available. The data rate will also affect the accuracy. How far the aircraft travels between samples and how much it can deviate from a straight-line path directly dictate the required data rate. Later we will discuss data rates and the importance of time alignment, but for the time being, try to remember that we would like the output rate of data to be as high as possible.

The units in the accuracy column may look a little strange to us, as they are not units that we really deal with in the test community. This is because the radar community is a little strange and likes to stand apart from the rest of the crowd. The range accuracy is not too difficult to understand. The chart tells us that for any range out to 125 nm the error we might expect is no greater than plus or minus 15 ft. This is pretty good accuracy, but as stated previously, with a pulse radar with or without beacon enhancement, excellent range accuracy should be expected because it is a direct measurement. Azimuth and elevation accuracies are a different bird completely.

The accuracies are given in milliradians, which pilots call mils. But just what is a milliradian? A milliradian (1 mrad) is equal to 0.001 rad. A radian is an angular measurement, and there are 2π radians in a circle (6.2832 rad = 360°). The error can be thought of as an inclusive angle radiating outward from the radar. But as testers, we are really interested in an error budget; or stated another way, for any point in time, how accurate is our TSPI? We would really like to know what our total error is, in feet. If we take an aircrew's idea of what exactly a mil is, it may enlighten us. To an aircrew, this is a mil (.). The dot inside the parenthesis, which they call a "pipper," located at the center of their aircraft's target reticle.

To the aircrew, that angular measurement can be represented by a symbol of some width relative to a fixed distance. To further explain:

- One mil subtends 1 ft at a range of 1000 ft, or as a practical example, a 20 ft wingspan will be completely covered with a dot which is 20 mils wide at 1000 ft. This is how pilots can tell the range of an object even without a sensor lock-on. It is called reticle matching or stadiametric ranging. (It should be noted that in European countries and in some ground test organizations, the relationship is given as 1 mil subtends 1 m at 1 km.)

- If we use a radar mile of 6000 ft, 1 mil would subtend 6 ft.
- The example in Figure 2.4 states the azimuth and elevation accuracy of this particular radar is 0.5 mrad. By using the preceding logic:

$$1 \text{ mil} = 1 \text{ ft @ } 1000 \text{ ft}$$
$$0.5 \text{ mil} = 0.5 \text{ ft @ } 1000 \text{ ft}$$
$$0.5 \text{ mil} = 0.5 \text{ ft} \times 6 \text{ @ } 6000 \text{ ft } (1 \text{ mile}) = 3 \text{ ft}$$
$$0.5 \text{ mil} = 3 \text{ ft} \times 50 \text{ miles} = 150 \text{ ft}$$
$$0.5 \text{ mil} = 3 \text{ ft} \times 100 \text{ miles} = 300 \text{ ft}$$

By using the previous example, a tester can determine the accuracy of this TSPI system at any point in time. If you are running a particular test and you require 300 ft accuracy, you must fly no further than 100 miles from the pedestal. This type of information can prove to be invaluable in a test situation where the prebriefed setup has to be moved due to weather, traffic, etc. Where can we go and maintain the accuracy that we need? What are the ramifications if we conduct a test in an area where the truth data are determined to be insufficient? We will need to refly the objective, which is sure to make the program manager's day. Of course, if you would really like to convert thousandths of radians to radians to degrees and then complete the triangle to determine the error, you can. The answer will be the same, just a little bit behind real time.

Unlike range accuracy, which is constant out to the maximum range of the radar, azimuth and elevation accuracy are completely range dependent. This is because the fixed angle of the pencil beam (assuming the azimuth error of the radar is equal to the elevation error) subtends more area with increasing range. This can readily be seen in Figure 2.5. The cross-sectional area of the radar beam can also be viewed as the maximum error the tester can expect. When producing an error budget for the test (maximum error in the TSPI system), the test engineer must consider the errors in three dimensions. Since a radar will produce range, azimuth, and elevation measurements, that is where we will concentrate our analysis. We know the range error is constant and the azimuth and elevation will vary based on the range from the pedestal. So the error budget can be imagined as a giant coin in the sky with the depth of the coin equal to twice the error in range and the diameter of the coin varying with respect to range.

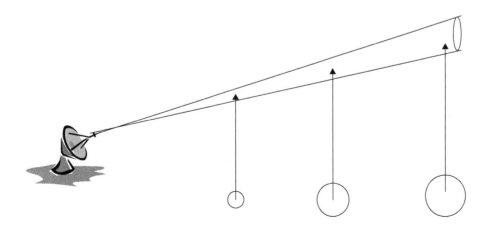

FIGURE 2.5 ▪ Beam Cross-Sectional Area in Relation to Range

FIGURE 2.6 ■
Radar Error Budget

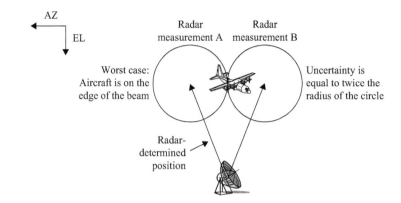

Figure 2.6 illustrates that the error budget is the same for all points within the circle. The worst case is when the target is anywhere on the edge of the error circle. Radar measurement A determines and reports the aircraft's position. In the next measurement, it again reports the aircraft's position, but to the right of the first measurement (radar measurement B). The difference between these two measurements is the uncertainty, which is equal to the distance subtended by the inclusive angle. As an example:

- What is the error for a radar tracking a target at 100 miles whose azimuth and elevation accuracies are 0.15 mrad?
- 1 mil at 1000 ft = 1 ft
- 1 mil at 6000 ft = 6 ft
- 0.15 mrad at 6000 ft = 0.9 ft
- 0.15 mrad at 100 miles = 90 ft
- Total error budget is 90 ft in azimuth and 90 ft in elevation

In the previous example, it was determined that our radar had an error of 90 ft at a range from the pedestal of 100 miles. If we were performing an inertial navigation system (INS) navigation run, and our specification stated that positional accuracy in the *x*-*y* plane must be accurate to within 300 ft, could we perform the test? The answer is not at 100 miles. Why, you say? We just proved that the radar was accurate to within 90 ft at 100 miles. The answer lies in the realm of confidence. In systems testing, the truth data must be four times as accurate as the system under test. So in our particular INS test, the truth must be accurate to 300/4 = 75 ft. This radar can be used as a truth source, but the test must be performed within 83 miles of the pedestal (75 ft ÷ 0.9 ft = 83.33 miles).

▬▬ 2.5 ▮ GEOMETRIC DILUTION OF PRECISION

One last concept with regards to accuracy is called the geometric dilution of precision (GDOP). All TSPI sources have a GDOP, and as the name implies, it is a dilution of precision based on your reference position to the TSPI source. Accuracy can degrade in certain quadrants for many reasons. A radar may have accuracy degradation due to blockage areas. The GPS suffers degradations due to jamming, galactic interference, multipath, and constellation availability. These degradations can be plotted to give the tester an idea of where the best coverage is available.

FIGURE 2.7 ■
Cinetheodolite
GDOP

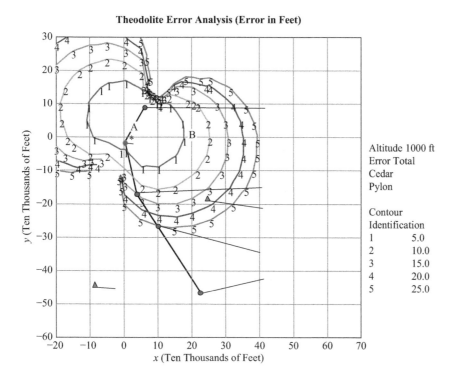

Theodolite Error Analysis (Error in Feet)

Altitude 1000 ft
Error Total
Cedar
Pylon

Contour
Identification
1	5.0
2	10.0
3	15.0
4	20.0
5	25.0

Figure 2.7 shows the GDOP for a cinetheodolite tracking system. There are a series of contour lines that are numbered in the figure. These numbers indicate that if you are within the contour, your accuracy will be the system's best accuracy, multiplied by the contour number. If you were flying at point A, you are within contour 1, so your accuracy would be 1× the stated accuracy of the system. If you were flying at point B, you are within contour 2, and your accuracy would degrade (or dilute) to 2× the stated accuracy. Put another way, if your TSPI system can provide a 50 ft accuracy, it can only do so within the area bounded by the contours for a GDOP equal to 1. As you fly into an area outside of contour 1 (best accuracy), the accuracy will degrade to a number proportional to the GDOP contour number. GDOPs for range TSPI systems are provided by the range directorate. GDOPs for a GPS can be obtained online prior to flight from a number of Websites, or real time from the GPS itself. **NOTE:** GDOP is a 3D value. Other dilutions of precision (DOPs) may be given for a system. For example, a GPS may provide the user with a horizontal (HDOP), vertical (VDOP), or positional (PDOP) dilution of precision. If the user is only interested in the horizontal degradation of the GPS (as is the case with barometric-aided systems), the user would reference the GPS HDOP.

The position of the DOP contours as depicted in Figure 2.7 is also dependent on the earth model that is being used. We will look at what an earth model is in the next section.

2.6 | EARTH MODELS

An earth model is a mathematical representation of the earth. The explanation presented here is simplistic, and my apologies to the geodesy gurus. For a more in-depth explanation, I refer readers to any one of a number of texts on the subject. I would

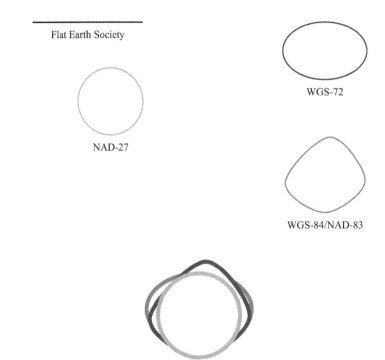

recommend *Geodesy for the Layman*, by Capt. R. K. Burkard (Washington, DC: Defense Mapping Agency, 1984; available at http://www.ngs.noaa.gov/PUBS_LIB/ Geodesy4Layman/toc.htm).

If we were to draw a picture of the earth, we would have to know the equatorial and polar radii, and we would have to know the function of the curvature (or flattening) of the earth. The picture will be a function of the parameters we choose. Mathematicians have always wanted to define the precise shape of the earth, and it has been an ongoing quest. Figure 2.8 shows some of the common depictions of the earth.

We started out believing (and some still do) that the earth is flat, and eventually saw the earth as a spheroid. The spheroid shape was the norm for quite a while, until mathematicians decided that it was not really round at all, but really an ellipsoid. This theory was also debunked, and now we believe that the earth is really an oblate spheroid, or more commonly known as a geoid. The numbers under each of the shapes in the figure refer to the datum, or mathematical description, of the earth. NAD refers to North American Datum and WGS refers to World Geodetic System, or World Geodetic Survey (depending on the document you read). The numbers refer to the year in which the survey was completed; NAD-27 is the North American Datum, 1927; WGS-72 is the World Geodetic System, 1972. These datums are by no means universal. Individual nations and groups of nations use their own unique datum, causing some confusion. Avionics systems and TSPI systems output position referenced to an earth model. In addition, charts and maps are printed with reference to an earth model. The problem is obvious if we superimpose competing earth models over one another.

You can see in Figure 2.9 that the shapes do not match up and there are some areas where gross differences exist. Attack pilots notice these differences when using GPSs that are outputting position in WGS-84 and comparing the GPS position to high-resolution charts referenced to NAD-27. They do not match. There can be differences of up to 1500 ft when comparing WGS-84 to NAD-27. The differences between WGS-72 and WGS-84 are much smaller, averaging anywhere from a few tens to 100 ft. There is a solution, of course. Either use a chart that uses WGS-84 datum or output the GPS position in NAD-27. The important thing to remember here is to not mix apples and oranges. When testing any avionics or weapons system, ensure that the system under test and the TSPI are in like earth models. Most countries and test organizations are converting to WGS-84 as a universal standard, but many maps and charts used by aviators and mariners are referenced to older earth models.

2.7 | THEODOLITES

Another device that may be used as TSPI is called a theodolite. It may be either a cinetheodolite or a laser theodolite. The position of your aircraft is determined by aerotriangulation of lines of azimuth and elevation or arcs of range. Both systems capitalize on optical systems that provide excellent attributes of azimuth, elevation, and range.

2.7.1 Cinetheodolites

There are two types of cinetheodolites: the first is a film-read theodolite that corrects radar position postflight by reading the film time history of the event; the second computes the position of an aircraft by aerotriangulation of azimuth and elevation lines from multiple trackers. In the first case, an aircraft is tracked by a ground-based radar. A camera is bore-sighted to the radar and records the position of the radar cursor relative to the tracked aircraft. Postflight, a technician reviews the recorded film and manually corrects the radar position for each frame to the actual target position. Figure 2.10 depicts the correction process.

As can be seen in Figure 2.10, the radar shows an erroneous position of the aircraft slightly right and high. The technician applies a delta azimuth (ΔAz) and delta elevation (ΔEl) to the radar position to move the radar to the desired point on the aircraft. This process is repeated frame by frame for the entire time of interest. Since this is a manual

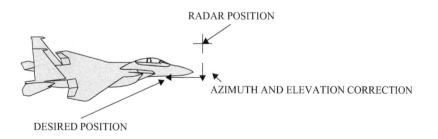

RADAR POSITION

AZIMUTH AND ELEVATION CORRECTION

DESIRED POSITION

FIGURE 2.10 ▪
Film-Read
Cinetheodolite

process, it is very time consuming. A 7- to 10-day data turnaround time is not unusual for this type of TSPI. It does provide a high degree of accuracy, however, it will cost the user more in terms of time and money. Figure 2.11 provides a range handbook sample for a film-read cinetheodolite.

A second type of cinetheodolite uses optics only to track the airborne target. Since optical systems (cameras) provide excellent azimuth and elevation, but not range (the radar in the previous example provided the range), a different method must be employed in order to determine the aircraft's position. For these systems, multiple optical trackers are placed on surveyed points and aerotriangulation is used to determine the aircraft's position. A minimum of three trackers are needed to obtain the position, and additional trackers will provide enhanced accuracy. The aircraft flies between these trackers and a coordinate conversion is performed by the tracking systems to obtain the latitude, longitude, and altitude of the target. Figure 2.12 depicts the target setup. This system may also employ postflight film reading (as described previously) to enhance system accuracy.

FIGURE 2.11 ▪
Instrumentation
Theodolites

Data Required	Data Accuracy	Activity Area	Output Rate	Data Products	Remarks
TSPI	±0.3 mrad azimuth and elevation in real time ±0.1 mrad azimuth and elevation postflight film read root mean squared (RMS) error	Within a 10 × 15-mile area to the east of the range	20 samples per second	Printouts Magnetic tape plots	Weather dependent Obstruction to the west of the shoreline

FIGURE 2.12 ▪
Aerotriangulation
Cinetheodolites

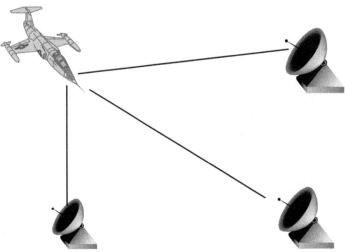

Lines of Azimuth and elevation provided by the trackers

Both types of cinetheodolites produce accuracies on the order of 0.1 mrad, about twice the accuracy of a ground-based tracking radar. Unfortunately this increase in accuracy is directly proportional to the price of the service. There are also limitations with these systems not found with radar. These limitations are addressed in section 2.8.

2.7.2 Laser Trackers

Another type of theodolite is the laser tracking system. The laser system is very similar to the second type of cinetheodolite in that it uses aerotriangulation methods from known points on the ground. Since the laser is an excellent measure of range, this system aerotriangulates with arcs of range rather than lines of azimuth and elevation. The methodology is very similar to another system in everyday use, the GPS. Figure 2.13 shows the difference between the laser tracker and the cinetheodolite.

Lasers provide accuracies comparable to cinetheodolites, on the order of 0.1 mrad. Figure 2.14 provides a range handbook excerpt for a laser tracking system. In order to utilize this system, the range directorate must install laser reflectors on the sides of the test aircraft. These are multifaceted mirrors that provide an orthogonal reflecting surface at every look angle from the aircraft.

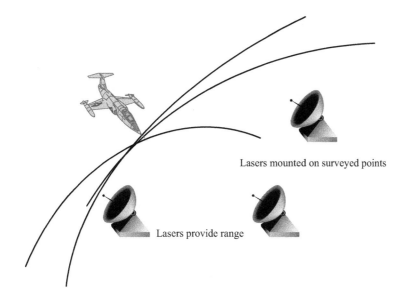

Lasers mounted on surveyed points

Lasers provide range

FIGURE 2.13 ■
Aerotriangulation by Arcs of Range

Data Required	Data Accuracy	Activity Area	Output Rate	Data Products	Remarks
TSPI	±0.1 mrad azimuth and elevation ±1–3 ft range	Runway coverage and over the station	Real time: 20 samples per second Postflight: 100 samples per second	Printouts Magnetic tape Plots	Weather dependent Mobile unit can be relocated with advance notice and funding

FIGURE 2.14 ■
Laser Tracking System

▌▌ 2.8 ▌ THEODOLITE LIMITATIONS

Theodolite systems operate in the optical spectrum and are limited in many ways. The most obvious limitation is visibility. The target cannot be tracked if it cannot be seen. The laser will be affected by the amount of moisture in the air, and may provide ranging to droplets in a rain shower rather than the aircraft. Smoke, haze, fog, and smog all affect optical systems. Obstructions such as buildings, mountains, and LOS to the target also affect the visibility. Another obvious limitation is range. Laser trackers are limited in power as to not injure the aircrew, hence their range is limited. Theodolites are operated by human operators who must be able to see the target in order to track it. The target becomes smaller with range and is therefore harder to track accurately. Aircraft speed is also a limitation, as operator slew rates must be taken into account. Figures 2.11 and 2.14 identify relatively small areas of coverage, which is typical of most theodolite systems. It is possible to combine a laser rangefinder with a cinetheodolite system that can provide the user with range, azimuth, and elevation in real time.

▌▌ 2.9 ▌ GLOBAL POSITIONING SYSTEM

The GPS has proven to be a relatively inexpensive positioning device for the test community. The basic GPS concept, accuracies, and limitations of the system are described in detail in chapter 5 and will not be repeated here. It may be advisable for the reader who is not familiar with the GPS to review the section in chapter 5 and then return to see how it is used as a TSPI system. TSPI systems that utilize the GPS as a positioning device may be stand-alone, inertial (INS) aided, differentially corrected, or a combination of any of these. It is important that testers understand the limitations of the GPS and have some method of validating the information provided is true and accurate. Many range directorates have programs in place that validate the output of GPS-based TSPI and notify the user when GPS accuracies degrade.

▌▌ 2.10 ▌ STAND-ALONE GPS RECEIVERS

A stand-alone GPS receiver is the most common and least expensive of all the GPS TSPI systems. It also has the most problems. The GPS is a navigational system that has been adopted for use as a TSPI system. In the four previous examples, the TSPI systems directly measured some or all of the parameters of range, azimuth, and elevation. These systems are *positioning* systems because they directly measure the position of an object. What is the GPS? The only parameter that the GPS can directly measure is time. From differences in the time a signal was sent to the time it was received, the GPS calculates range. The computations for position are then accomplished the same way as with a laser tracker. The GPS does not measure position, velocity, or acceleration. Is the GPS, then, a positioning system? We really need to examine the outputs of the GPS and determine where they come from. This can be accomplished by examining the nine-state vector. All navigation systems output the nine-state vector shown in Figure 2.15.

As previously described, the GPS measures time, calculates range, and uses aerotriangulation to determine an aircraft's position. That position is defined as latitude, longitude, and altitude. Position in the horizontal plane will be good only if the DOP is good.

Nine-State Vector Parameters	
Latitude	
Longitude	
Altitude	
North velocity	
East velocity	
Down velocity	
North acceleration	
East acceleration	
Down acceleration	

FIGURE 2.15 ■ The Nine-State Vector

Most test organizations require a DOP of 4 or better for the GPS to be used as TSPI. Assuming the DOP is good, latitude and longitude will be good as well. Due to the coplanar nature of the GPS constellation (all satellites are in a plane above me), there is nothing to help bound the error in the vertical (altitude). In fact, most systems, and all military systems, opt to aid (or substitute) the GPS with barometric altitude. So the answer to the original question would be "Yes, the GPS appears to be a positioning device, but only in the horizontal plane." This is a true statement with some limitations.

The first limitation applies to all GPS TSPI systems as well as all GPSs, and that is the lack of system integrity within the GPS. No GPS can be trusted unless monitored by an external integrity monitor. This monitor may be a receiver autonomous integrity monitor (RAIM), an additional navigation system used to cross-check the GPS, a space based augmentation system (SBAS) signal, or a differential GPS. A GPS may not be used as a TSPI system unless one of these monitors is in place.

A second limitation unique to stand-alone receivers is the aircraft attitude. As an aircraft banks, the visibility of the GPS antenna is impeded and may start to lose satellites. If the bank angle is increased so the aircraft is now inverted, the system will not have enough satellites for position determination. As the aircraft is rolled upright, the GPS must now reacquire the satellites needed for position determination. This scenario defines two problems. If the aircraft is highly dynamic, there are going to be periods of time where position information is lost. The second problem involves the number of channels in the GPS receiver. A single-channel GPS receiver will take longer to reacquire satellites after a dynamic event than a multichannel receiver. An eight-channel receiver will have an easier time than a four-channel receiver. Total loss, in time, of positioning information is also a function of the equipment being used.

Another limitation of all GPSs is the iteration rate. The update rate of the GPS is 1 sample per second. Because of this limitation, the along-track error (or uncertainty) increases with aircraft speed.

Because of these limitations, we can say that stand-alone GPS receivers are adequate TSPI devices for nondynamic applications. Accuracies expected are 15 m x/y for L1-only receivers and 1 m x/y for L2 receivers. These numbers assume that an external integrity monitor is in place.

The previous discussions have only addressed the position determination of GPS TSPI sources, but there are six other states to consider. These six states are the three for velocity and the three for acceleration. If we had a Doppler GPS, we could directly measure velocity due to the Doppler shift created by our own aircraft movement and the GPS signal (the satellites are moving too). Since most organizations do not have this capability, we will investigate how velocity and acceleration are obtained, and what sort of accuracies we can expect from them.

Velocity can be obtained from the GPS by calculating the rate of change of position with respect to time, or the first-order derivative of the position with respect to time. If the aircraft is flying, the latitude and longitude are always changing. Changing altitude at the same time complicates the matter. If the system provides outputs of position at a high rate, such as an INS of about 60 to 90 Hz, our calculation of velocity would probably be very good. The GPS has an iteration rate of once per second, and a device using a different reference system aids in determining altitude. The velocity that is obtained is really not that accurate, especially in any sort of dynamic environment.

Acceleration is the rate of change of velocity with respect to time, or the second-order derivative of the position with respect to time. Since we have determined that the velocity is really not that good, one can imagine how good the acceleration might be. Not very. This makes perfect sense, as the GPS does not directly measure velocity or acceleration, and measurement of the position has inherent errors. Out of the nine states, only the horizontal position should be used for stand-alone receivers as a TSPI source, and then only after considering the limitations mentioned previously.

2.11 | INERTIAL-AIDED GPS RECEIVERS

As the name implies, inertial-aided GPS receivers obtain navigational aid from either an external or internal inertial reference system (IRS) or inertial navigation system (INS). There are many integrated GPS/INS (IGI), inertially aided GPS (IAG), and GPS-aided INS (GAI) systems on the market today, and many have been adapted for TSPI use. The benefits of these systems are obvious, having read the aforementioned discussions. When GPS information is lost, or degrades to a point lower than the acceptable DOP, the INS takes over as navigator until GPS integrity/accuracy is regained.

Inertial navigation system theory and flight testing are covered in chapter 5 and will not be repeated here. Once again, for those readers not familiar with INSs, I recommend a review of the INS section in chapter 5 prior to continuing on with this discussion. An INS or IRS (combined in this discussion as just INS) is an acceleration and attitude measurement system. That is, the INS directly measures acceleration and attitude (of the aircraft, not the pilot). Note that three of the states in the nine-state vector are acceleration. Similar to the GPS, the INS must calculate the other states.

The INS calculates the three states of velocity by integrating the acceleration over time. Unlike the GPS update rate, the INS will sample accelerations at rates greater than 60 Hz. This high sampling rate, and the fact that accelerations are more or less instantaneous, allows the INS to calculate a very accurate velocity. The problem for the INS is its drift rate. The drift rate is a function of several things—gyro bias, heading misalignment, etc.—and all of these things equate to an inaccurate platform alignment. If the platform is misaligned to true north by 1 mrad, after 100 nm of travel the aircraft would report its position approximately 600 ft from its true position. The error of the INS is time dependent, and that is why most specifications call out a drift error in nautical miles per hour (nm/hr). The good news is that over the short term, the error in the INS position is pretty small and comparable to the GPS.

An inertial-aided GPS may be implemented in one of two ways. The first implementation uses GPS as the primary navigator, and GPS position is used to update the INS at 1-sec intervals. Should GPS information be lost, the INS will take over as navigator until the GPS DOP improves. The second implementation involves the use of a Kalman filter.

In this implementation, all available navigation is used to compute a "system" nine-state vector. The inputs to the Kalman filter are weighted and checked for integrity prior to being used in the algorithm. Should the inputs fail the integrity check, such as a high DOP for the GPS, that input is dropped and is not used in the calculations.

There are different schools of thought as to which system is best for the TSPI user. By far the simplest is the first type, which incorporates GPS positions with INS accelerations and velocities. The second is a bit more complicated, and depending upon the Kalman model implemented, may output degraded GPS positions or INS accelerations as a result of sensor fusion. Check with instrumentation or the range directorate prior to using any of these systems to identify the accuracies and limitations of these systems.

For those organizations that do not have dedicated TSPI assets that incorporate GPS/INS merge, fear not. The implementation can be done on the ground in real time or postflight and can provide the same accuracies. In order to generate this TSPI, you will need access to the aircraft's onboard GPS, INS, and barometric altitude data. You can then write a TSPI program that may say something like

If the DOP of the GPS is 4 or less,
And the INS Q-factor is greater than 5,
Then output GPS latitude and longitude and INS accelerations and velocities with barometric altitude.
If the DOP of the GPS is greater than 4,
And the INS Q-factor is greater than 5,
Then output INS latitude, longitude, accelerations, and velocities with barometric altitude.
If the DOP of the GPS is 4 or less,
And the INS Q-factor is less than 6,
Then output GPS latitude, longitude, accelerations, and velocities with barometric altitude.
If none of the previous conditions exist,
Then output a warning, "TSPI FAIL."

NOTE: The Q-factor is similar to the GPS DOP. The INS will output a status on how healthy the system regards itself. This status is normally a Q-factor. The numerical values run from 0 to 10, with 10 being the best.

Your TSPI program will read the inputs from either the data stream in real time or from a tape playback postflight and output to the user a TSPI nine-state vector. This is graphically illustrated in Figure 2.16.

Nine-State Vector Input Parameters	
Latitude	GPS
Longitude	GPS
Altitude	Barometric
North velocity	INS
East velocity	INS
Down velocity	INS
North acceleration	INS
East acceleration	INS
Down acceleration	INS

FIGURE 2.16 ■
TSPI Nine-State
Vector

■ 2.12 | DIFFERENTIALLY CORRECTED GPS

Differential GPS, which is covered in chapter 5, along with navigation systems and basic theories of operation, will not be covered here. These systems benefit users who require very accurate positioning (latitude and longitude) information. The base station computes a matrix of time errors for all satellites in view. This information may be data-linked to the aircraft for real-time corrections or applied postflight. Most range directorates will differentially correct the information postflight. Corrections can be applied to either stand-alone or aided GPSs. Some systems, such as the Astec Z-12, have been certified by the Federal Aviation Administration (FAA) as truth sources for compliance tests. Accuracies of 1 cm have been demonstrated for nondynamic applications. Users must be aware of the limitations of these systems. The base station can only correct for the satellites that it can see, therefore, the operating range from the base station is a concern. LOS limitations must be known for users uplinking the correction matrix. When the WAAS is fully operational, TSPI users will have access to differentially corrected GPS signals in real time without the need for a base station.

■ 2.13 | SENSOR FUSION

Sensor fusion is a buzzword in and around the military communities. One is led to believe that this is a new development on the cutting edge of technology. In truth, sensor fusion has been around for many years in the range community. Sensor fusion is the process of combining multiple sources of data, in our case truth data, and outputting a result that is more accurate than any one input. The example given in section 2.12 for the INS/GPS program is an example of sensor fusion in TSPI. Historically we have called the output a best estimate trajectory (BET). In the case of a theodolite, it is apparent that the addition of more trackers will provide a more accurate truth. Ground-based tracking radars can also be improved with sensor fusion. There are some ground rules and assumptions that must be made when applying sensor fusion.

For example, if ground-based tracking radar is used, we must first assume that a second radar added to the track will not overlay incorrectly in the same direction at the same time as the first radar. If it did, then there would not be any benefit to adding another radar. Figure 2.17 illustrates the sensor fusion of two tracking radars.

Three radars, assuming that none of the three radars overlay wrong in the same direction at the same time, yields even better results (Figure 2.18).

FIGURE 2.17 ■
Two Tracking
Radars on the Same
Target

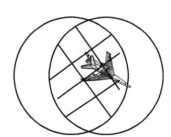

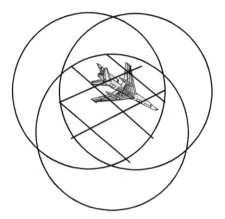

FIGURE 2.18 ∎
Three Tracking
Radars on One
Target

Almost all ranges can provide a BET with multiple TSPI sources. Users should consult with their respective instrumentation and range directorates in order to ascertain the accuracies and capabilities available. The next section addresses other concerns that must be discussed prior to embarking on any avionics test program.

2.14 | RANGES

The general topic of ranges is included here because there are many details that must be addressed prior to prosecuting a test plan. In many cases, a test plan cannot be finalized until TSPI, range assets, and range procedures have been analyzed. As test planning is accomplished, the tester will need to periodically contact range personnel in order to determine if a test can be performed with the assets that are available on the range. In many cases, tests will have to be modified based on the available assets, or worse yet, accomplished at another location. This will impact the fidelity of the test as well as the schedule.

2.15 | RANGE ASSETS

The TSPI requirements as well as other unique range services are identified during the test planning process. Up to this point, we have identified TSPI assets and their expected accuracies. By knowing the requirement for accuracy, we can easily determine if our preferred test area can support our requirements. A secondary task that must be addressed is the ease of scheduling and availability of the required assets. For the expected time period of our test, are there any scheduled down periods for the assets we require? What other programs are expected to compete for those assets? What is our priority in the big-picture scheme of things? All too often, testers expect range assets to be there whenever they need them and are shocked to learn that they are not available when asked for at the last minute.

Other than TSPI, what other assets are we talking about? There are a number of services that we do not normally think of until the day before the test. These services are addressed here in order to assist in the test planning process.

∎ 2.16 I UNIQUE SERVICES

Unique services that are available (or unavailable, as the case may be) range from the obvious to the not so obvious. These services are listed here in no particular order of importance:

- Bomb Drop and Scoring. In any weapons delivery program the test organization will require an area to deliver ordnance and have that ordnance scored. In some cases, it may be desirable to track the store from separation to impact, which will require additional TSPI support.

- Fuel Lab. A funny thing about testers is their incessant desire to add things to an airplane. It could be test equipment, instrumentation, or additional systems. The funny thing about pilots is that they always want to know about weight and the center of gravity (CG) of the airplane. Weight and balance tests are a common requirement.

- Cats and Traps. Most testers will not ever come into contact with catapult and arrestment devices or the requirement to use them. For those that do, this is a specialized test, and equipment is only available at specialized locations.

- Targets and Target Availability. Targets come in all shapes and sizes. They can be land-based targets, airborne targets, or enhanced ocean-going targets. The targets may have to be instrumented or able to carry jamming pods. They will probably require a beacon be installed. They may have to fly high and slow or low and supersonic. They may range from a fraction of a meter RCS to many thousands of meters RCS. They may be manned or unmanned. No range can provide all of the permutations and combinations of targets that testers will require, so it is going to take some logistical work.

- Chaff. It is a deception device that employs plastic (it used to be aluminum foil) cut to the specific wavelengths of threat radars. The unfortunate thing about chaff is that it does not stay airborne. Eventually it falls back to the earth. This creates an environmental hazard, especially for the little creatures that eat it. Chaff can only be dropped in specific areas, which can cause problems for the test community.

- Decoy Deployment. As the name implies, this device is meant to deceive threat radars by either launching or towing a decoy. There are many vendors of these devices and they go by many names—miniature air-launched decoy (MALD), towed air decoy (TAD), etc. These devices are all classified either by their shape or the length of their tether. If these devices are released or cut from the towed aircraft, they must be over a secure area for recovery. I remember one interesting program in New York that wanted the test department to ensure that Sunrise Highway and the Long Island Expressway were closed to traffic when recovering the aircraft. Anyone from that part of New York knows what a nightmare that would create.

- Missile Firings. Missile firings require that the tester ensure safety over the entire footprint of the missile. Missiles do not always go where you tell them, so the tester must map out an area of coverage for any and all contingencies. This can become quite a large area and exceed many range's capabilities.

- Anechoic Chamber. Anechoic facilities are invaluable for antenna pattern testing, electronic warfare integration and testing, and electromagnetic interference and compatibility testing. The size of the facility, signal attenuation capability, and near-field/far-field testing are a few of the requirements that the tester must consider. Scheduling is a major concern, as most facilities are scheduled up to a year in advance.

- Frequency Allocations. Testers do not have carte blanche to radiate signals as they see fit. Radiation into free space will require a license (in the United States anyway) and an available time slot. These are obtained from the Frequency Allocation Council in your region. Your instrumentation group will probably be most familiar with this requirement.

- Data Processing. Turnaround time, formats, and availability of real-time and post-flight processing are major concerns for the flight tester. Any delays incurred in receiving TSPI data are going to have a deleterious effect on the schedule. Remember, a good tester is not going to proceed with the test plan until he knows the results of the current test.

- Control Room. Another area is the availability of a control room with appropriate communication and telemetry capability to permit real-time execution of the mission.

2.17 | INTERFACING WITH THE RANGE

The range is an equal partner in the test planning process. It is imperative that both parties understand what is required of them in order to successfully complete the project. This understanding must be more than an oral agreement made over the telephone. One of the biggest problems that I have seen in the past is merely a lack of communication. There are many examples of tests that had to be redone because of a lack of understanding of what was required for the original test. If the requirements are not specific enough, the data you receive may be data that the provider thought you wanted, but data you cannot use. A few common examples of data screw-ups follow.

Suppose that you required the bearing from target 1 to target 2 as depicted in Figure 2.19.

If you do not tell me what bearing you want, I have to assume a bearing that you want. In Figure 2.19, the relative bearing to target 2 from target 1's cockpit is about 015°. The true bearing, however, is about 105°, and if we had about 15° of east variation, the magnetic bearing would be about 120°. All of these bearings are correct, but they are all different. Obviously, the wrong parameter will significantly impact your results. Let's try another one. In Figure 2.20, what is the aspect angle of the target as seen from Target 1's cockpit?

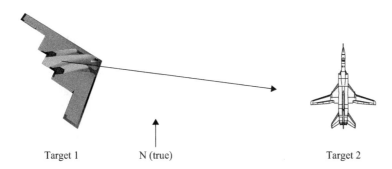

Target 1   N (true)   Target 2

FIGURE 2.19 ■
Bearing to Target 2

FIGURE 2.20 ◾
Target Aspect Angle

TARGET 2

TARGET 1

FIGURE 2.21 ◾
Altitude
Relationships

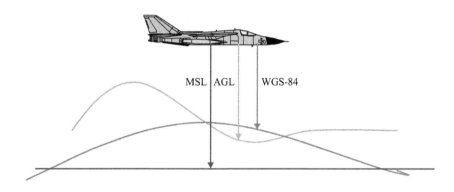

MSL | AGL   WGS-84

If you said that this was a 0° aspect angle, you would be right about 50% of the time. In the United States, for example, the U.S. Air Force designates this as a 0° aspect angle. The U.S. Navy, however, denotes this as a 180° aspect angle. Once again, the wrong selection will impact your results.

What is a WOW switch? It could possibly be weight-on-wheels, but then again it could be weight-off-wheels. Depending on the aircraft manufacturer, WOW can mean different things. Once again, they are both correct and they are both different.

One last example is one of my favorites, and that is altitude. Altitude is one parameter that testers always have a hard time with. Figure 2.21 shows an aircraft in flight with three different altitude references.

The aircraft in Figure 2.21 is at a constant altitude, yet we obtain three different altitude readings depending on the parameter that we choose. The first reference is mean sea level (MSL). The second is above ground level (AGL) and the third is WGS-84, or height above the ellipsoid. All three of these altitude references are available to the pilot and to the tester on the data stream. From the figure, it is apparent that if we expected WGS-84 and were provided MSL, a significant error would be introduced.

There are many examples of these types of "gotcha" errors. The only way to avoid introducing these errors into your data is to specify exactly the parameter you need.

Test planning working groups (TPWGs) dedicated to tester, range, data acquisition, and instrumentation interfaces accomplish this task. A desired outcome of the TPWG should be a range interface memorandum (RIM) that details in writing the requirements for a successful test program. A RIM is by no means a contractual document, but rather it sets guidelines that all parties agree to. I have found this method of documenting requirements to be very useful, especially when original participants move on to different programs and someone new shows up to take his place. At a minimum, the RIM should address the following:

- Test plan general and specific objectives
- Applicable documents, specifically the test plan, interface design specification, and interface control document
- Real-time and postflight requirements, including formats and data turnaround time
- Communication and encryption requirements
- Parameter list, TSPI, and avionics (I have found it advantageous to identify the parameter, give its description, draw a cartoon of what it looks like. See Figures 2.19–2.21, and in the case of calculated parameters, the equation).

2.18 | TIME ALIGNMENT

Another consideration that must be addressed prior to actual testing is your method of time alignment. Since there are multiple sources of data, you must assume that these sources are asynchronous, that is, arriving at different times. Before you attempt to align these times, you must first ensure that all data are time stamped the same. I have told my students for years that it really does not make a difference if you use GPS time, UTC time, wall clock time, or Mickey Mouse time, as long as all data have the same time. The second consideration about time is when the data are actually time stamped. Time should always be inserted as close as possible to the source. If it is not, then a time delay study must be undertaken with the results applied as an offset to the data.

Aircraft clocks, telemetry station clocks, and TSPI clocks insert time into the data streams. It is imperative that all of these clocks are synchronized to the same time. Most systems now synchronize to GPS time. Some systems use a direct insert of the GPS time into the data stream rather than employing a separate clock. If GPS information is lost, so is time, and it only updates at a once per second rate. A common procedure is to utilize a stand-alone clock that is synchronized to the reference time (such as GPS) and then allowed to free run. Most of the clocks used in the instrumentation and test communities are very accurate, with drift rates on the order of 3 msec/hr. Some common brand names of clocks are Arbiter and Kinemetrics. These clocks are connected to time code inserters that time tag the data.

Data that arrives at the collection point (either in a telemetry room or in a postflight playback mode) must be time aligned prior to analysis. Time alignment does not mean arranging data in chronological order. Time alignment ensures that for each time, data from all sources are available. Figure 2.22 illustrates the typical problem of time alignment.

At arrival time 1, there are TSPI data available, but no corresponding avionics data for the same time. Similarly, at arrival time 2, avionics data are present, but no

FIGURE 2.22 ■
Asynchronous
Arrival of Data

Arrival Time	Data Type	
	TSPI	Avionic
1	T1	
2		A2
3	T3	
4		A4
5	T5	
6		A6
7	T7	
8		A8

FIGURE 2.23 ■
Extrapolation
Example

$$\text{LATITUDE}_2 = \text{LATITUDE}_1 + \text{NORTH VELOCITY}_2 * \text{DELTA TIME}$$
$$\text{LONGITUDE}_2 = \text{LONGITUDE}_1 + \text{EAST VELOCITY}_2 * \text{DELTA TIME}$$
$$\text{ALTITUDE}_2 = \text{ALTITUDE}_1 - \text{DOWN VELOCITY}_2 * \text{DELTA TIME}$$

$$\text{NORTH VELOCITY}_2 = \text{NORTH VELOCITY}_1 + \text{NORTH ACCELERATION} * \text{DELTA TIME}$$

$$\text{EAST VELOCITY}_2 = \text{EAST VELOCITY}_1 + \text{EAST ACCELERATION}_1 * \text{DELTA TIME}$$
$$\text{DOWN VELOCITY}_2 = \text{DOWN VELOCITY}_1 + \text{DOWN ACCELERATION}_1 * \text{DELTA TIME}$$

FIGURE 2.24 ■
Interpolation
Example

Given: VALUE V_1 AT TIME T_1 and VALUE V_3 AT TIME T_3
Estimate: VALUE V_2 AT TIME T_2 WHERE:

$$V_1 \leq V_2 \leq V_3 \quad \frac{V_3 - V_1}{T_3 - T_1} = \frac{V_2 - V_1}{T_2 - T_1}$$

$$V_2 = (V_3 - V_1) \times \frac{T_2 - T_1}{T_3 - T_1} + V_1$$

corresponding TSPI data. We must find a way to obtain avionics data at arrival time 1 and TSPI data for arrival time 2. There are two ways of accomplishing this task: interpolation and extrapolation. Interpolation requires two pieces of data (end points) to find the value for any time between them. From Figure 2.22, I can interpolate a value for T2 if I know the values T1 and T3. Extrapolation is predictive and requires a historical point and the present point to predict a value in the future. By knowing the values of T1 and T3, I can predict the value of T4. Which method is best?

The more accurate of the two methods is interpolation because you have data on either side of the desired value. The drawback is that you must wait for more data before you can obtain a value for the present time. This would be unacceptable for real-time telemetry, as the system would have to buffer large amounts of data, thus causing the system to slow down. Real-time telemetry systems use extrapolation in order to keep up with real time, and suffer some degradation in accuracy. Postflight systems use interpolation in order to obtain the best accuracy. Figures 2.23 and 2.24 provide examples of extrapolation and interpolation.

2.19 | SUMMARY

Compliance with applicable specifications or operational requirements requires accurate TSPI. The tester must identify the system accuracies needed to achieve compliance and then choose the correct TSPI source. Part of identifying TSPI sources includes determining the cost to the program. As a general rule, more accuracy will require more money. The truth systems must be at least four times as accurate as the system under test. With today's technology, this task is becoming more difficult, as many systems provide accuracies better than most truth systems can measure. In these cases, the tester, in concert with range and instrumentation personnel, must develop sensor fusion algorithms to assist them.

2.20 | EXERCISES

For each of the following examples, determine the TSPI source required to show compliance.

A. A GPS whose required horizontal accuracy is 15 m in the x/y plane.

B. An AN/ASN-130 inertial navigation system whose stated system accuracy is a 2 nm circular error probable (CEP) positional error, velocity accuracy of 2 ft/sec, and acceleration accuracy of 0.2 ft/sec/sec.

C. An ARN-118 tactical air navigation system (TACAN) which shall provide bearing accuracy to a station within 2° at 300 nm and range information to within 0.2 nm at 300 nm.

D. An ALR-67 radar warning receiver (RWR) must be capable of determining direction of arrival (DOA) to within 2° azimuth and elevation at a range of 10 nm.

E. An APG-XX pulse Doppler (PD) radar with accuracies in the scan mode (measurement accuracy) of 0.5 nm in range, 1° in azimuth and elevation, and 10 knots in range rate. Additional accuracies in the track mode (track accuracy) are 0.2 nm in range, 0.25° in azimuth and elevation, and 4 knots in range rate.

2.21 | ANSWERS TO EXERCISES

The first step in identifying the required TSPI source entails the selection of the accuracy that must be proven. Once this is known, the accuracy can be divided by 4 (since the TSPI system must be four times as accurate as the system under test) to determine the TSPI accuracy. Each of the following explanations will determine system, as well as TSPI, accuracy.

A. The first exercise asked you to prove the horizontal or positional accuracy of the GPS. The desired accuracy is 15 m, which is divided by 4 to obtain the required TSPI accuracy; roughly 13 ft. We then analyze which systems can meet this 13 ft requirement. We know that the ground-based tracking radar's accuracy depends on the range to the target. For a 0.15 mrad radar system, you can obtain a 13 ft accuracy out to about 14 nm. So if you did the entire test within 14 nm of the radar

FIGURE 2.25 ■
GPS Position Data
with Radar
Coverage

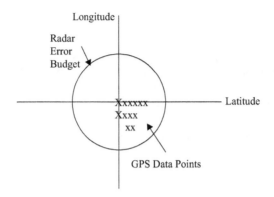

site, you could verify the GPS accuracy. I do not believe that this would be a valid test since the GPS will probably be used as a navigation system and you would want to fly a navigation profile. A navigation profile will surely take you much further than 14 nm distance from the site, which means the accuracy will degrade. The same problem exists with a cinetheodolite system, whose range is very limited. You cannot use a stand-alone GPS receiver unless it is an L2 system. If you do not have an L2 system, you could use an L1 system with differential. What happens if you do not have either? This was surely a problem 15 years ago, when GPS systems were first being installed. This particular problem will arise again as newer, more accurate systems enter the market. You can do this test with older TSPI systems; it just takes a little innovation. Suppose you flew a navigation profile and tracked the aircraft with a 0.15 mrad radar system. If you plotted the GPS position against the radar position and all of our GPS points fall within the error budget of the radar (see Figure 2.25), what can we say about GPS accuracy?

We certainly could not say that the system meets specifications, since our truth data is not accurate enough. We can say that the system is as good as the radar, and, in fact, will probably give us a "warm fuzzy" about our GPS system. A "warm fuzzy" is one of the basic systems tests that are addressed later in the text. The test allows us to see, at very little cost to the program, if the system under test has a good chance of meeting compliance. Since the test was conducted on a navigation profile, the next step would be to test for specification compliance close in to the radar or against a theodolite system.

B. The second exercise asked us about an INS. The first accuracy required was its positional accuracy. The required performance is stated as a 2 nm CEP horizontal. Following our procedure, we divide 12,000 ft by 4, or a required TSPI accuracy of 3000 ft. Any one of the systems discussed can provide this accuracy, and therefore is not a problem. The second and third accuracies of velocity and acceleration are a bit more difficult. An INS directly measures acceleration, and is the only TSPI system that we have discussed which measures acceleration. It is impossible to prove compliance with a positioning TSPI system. Based on our previous discussions, the same can be said for velocity measurement. Unless we had access to a Doppler system, trying to prove velocity accuracy would be futile. We could run the test described in exercise A and make the same conclusions, or we could perform a ground test. If the INS is mounted on a maintenance tilt table, we can insert known accelerations and ensure that the INS reports them correctly. Velocity accuracy is

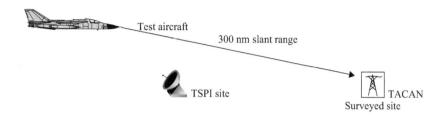

directly proportional to the accuracy of the acceleration measurement. This test is the most accurate way to show specification compliance.

C. The third exercise involves accuracy testing of a radio aid to navigation called a TACAN. The system provides range and bearing from the ground station to a cooperative aircraft. At first glance it appears that our testing will be accomplished at 300 nm away from our TSPI site. This is not true. The ground station must be 300 nm away from our aircraft. In fact, we could be flying directly over our TSPI site while interrogating a station 300 nm away. The test setup is shown in Figure 2.26.

There are two pieces of truth data required for this test. We first need to identify the 3D position of our aircraft (latitude, longitude, and altitude). We need the same information for the TACAN site. We then calculate a range and bearing and compare that to the reported range and bearing. The total error budget includes the uncertainty of our position and the uncertainty of the TACAN position. This total error cannot exceed 1200 ft (0.2 nm) divided by 4, or 300 ft in range. The azimuth accuracy is 2°, which must be converted to ft. At 300 nm, 2° subtends 10 nm, or 60,000 ft; divided by 4 yields a required TSPI accuracy of 15,000 ft. These accuracies are easily accomplished by any of the TSPI systems discussed.

D. The radar warning receiver in the fourth exercise is a system on the aircraft that scans all incoming emissions and attempts to identify their type based on a lookup library. If a match is made, the type of threat is displayed to the aircrew along with its associated bearing. The system is required to display this bearing to within 2° when the threat is at 10 nm. The setup and TSPI requirements are exactly the same as in the previous exercise. The bearing accuracy of 2° at 10 nm equates to 2000 ft. Required TSPI accuracy will be 2000 divided by 4, or 500 ft. This accuracy is easily accomplished by any of the TSPI systems discussed.

E. Our last exercise is a fictitious radar, and we are asked to prove four parameters of range, azimuth, elevation, and range rate. Our TSPI error budget in this exercise involves two aircraft, and the total error budget will be the error for our aircraft plus the error for the target. Figure 2.27 illustrates the setup for this test.

Assume that our aircraft are 100 nm apart and 100 nm from the TSPI site. The specified range accuracy is 0.5 nm and 0.2 nm, respectively, for measurement and track accuracy. Required TSPI accuracy is 750 ft for measurement accuracy and 300 ft for track accuracy. If we used our 0.15 mrad radar, our accuracy at 100 nm would be 90 ft; but since there are two targets, the total error budget is 180 ft. A radar or GPS is perfectly suited to perform this test. Azimuth and elevation call for accuracies of 1° and 0.25° for measurement and track accuracy. These angular measurements translate into 10,000 ft and 2500 ft for the two modes. TSPI accuracy requirements are 2500 ft and 650 ft, respectively. These specifications can

FIGURE 2.27 ■
Radar Test Setup

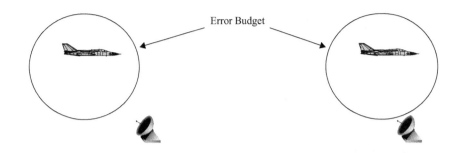

Error Budget

also be proven with our radar or GPS. The final parameter is range rate, specified at 10 knots and 4 knots for measurement and track accuracy. This equates to TSPI accuracies of 2.5 knots and 1 knot for each. Range rate is the velocity of closure between two aircraft. As previously discussed, GPS and radar are positioning systems that do not measure velocity. This radar is a Doppler radar, which does measure velocity directly. We could use INS velocities as truth, but even they are not sufficient. If we assume that the INS tested in exercise B is installed in each aircraft, and we use it as velocity truth, we can expect accuracies on the order of 2 ft/sec, or roughly 1 knot. The total error is the sum of the two aircraft, or 2 knots. With this accuracy we can prove measurement accuracy, but we cannot prove track accuracy. We can only state in a report that the system was tested to the maximum TSPI accuracy, guaranteeing a system accuracy of 8 knots (4 times the TSPI accuracy of 2 knots). This is the type of problem testers are encountering today. It is important to realize that we can only show compliance to the level of our truth sources.

2.22 | SELECTED QUESTIONS FOR CHAPTER 2

1. What does TSPI stand for? What is TSPI?
2. Why do we need TSPI?
3. What is WGS-84?
4. What is the requirement for accuracy when testing avionics?
5. What is an error budget?
6. At what depression angle does a ground-based tracking radar start to become unreliable?
7. Name two types of theodolite systems.
8. What is a GDOP? HDOP? VDOP?
9. Name two types of cinetheodolite systems.
10. What is the accuracy of a 0.2 mil radar system at 100 nm?
11. What is a mil?
12. Can a radar with the accuracy listed in question 10 be used to evaluate a GPS?
13. What is a coordinate conversion? Give one example.
14. What is better, interpolation or extrapolation?

15. What would be considered good accuracy for a ground-based radar?

16. What is a cinetheodolite? How do laser theodolites work?

17. How accurate is a ground-tracking radar (pulse) in velocity? How does it obtain velocity?

18. What is a nine-state vector?

19. What is a DGPS?

20. Describe sensor fusion?

21. Who has access to the GPS L2 frequency?

22. Which is a more accurate measure of velocity, differentiation of GPS position or integration of INS accelerations?

23. Why should you meet with the range directorate before the commencement of testing?

24. Give an example of sensor fusion and explain why it might be a better system.

25. How important is time synchronization?

26. What is an earth model? When is it used?

27. GPS is said to operate in L band. Is that the old or new frequency designation?

28. What is multipath? How can it affect TSPI?

29. Does anyone really believe that the earth is flat?

30. What attributes of target position can a cinetheodolite provide?

31. What attributes of target position can a laser theodolite provide?

32. What are the advantages of using real-time data.

33. What is extrapolation? When is it used?

34. What is interpolation? When is it used?

35. Why do most military ranges require a beacon on the airplane for radar tracking?

MIL-STD-1553 and Digital Data Busses: Data Reduction and Analysis

Chapter Outline

3.0	Overview	61
3.1	Historical Background	62
3.2	1553 System Architecture	62
3.3	A Bit About Bits	65
3.4	1553 Word Types	67
3.5	Data Encoding	67
3.6	Word Formats	68
3.7	Command Words	68
3.8	Anomalous Command Word Conditions	70
3.9	Command Word Summary	71
3.10	Status Words	71
3.11	Data Words	73
3.12	Message Contents	74
3.13	Bus Controller Design	75
3.14	Sample Bus Configurations	76
3.15	The Flight Tester's Task	79
3.16	MIL-STD-1553 Summary	80
3.17	Other Data Busses	81
3.18	Potential Problems for the Analyst	122
3.19	Data Acquisition, Reduction, and Analysis	124
3.20	Selected Questions for Chapter 3	136

3.0 | OVERVIEW

The transfer of data on today's aircraft is accomplished by data bus architecture. Data acquisition is the art of extracting data from these busses and converting them to some meaningful units that can then be analyzed. This text is designed to give every engineer a working knowledge of data bus technology that will allow you to speak intelligently to software design engineers and instrumentation personnel. It will also help you appreciate those personnel who provide data to you. Data acquisition today is a little more complicated than calibrating voltages for strip chart pens.

▮▮▮ 3.1 I HISTORICAL BACKGROUND

Technology made great strides in the area of aircraft avionics during the Vietnam years. Many of these strides were, of course, made by necessity. Unfortunately, many of the cockpits were ill-suited to handle this increased technology. The amount of systems technology added to the cockpits to aid aircrews in performing their missions was directly proportional to human factor problematic issues. Each new system that was added to a cockpit included the sensor, associated cabling, and its own separate controls and display suite. As systems such as radar warning receivers (RWRs), instrument landing systems (ILSs), laser designators, forward-looking infrared units (FLIRs), and television (TV) were added, available real estate declined sharply. Systems were placed in the cockpit where they would fit; many times in less than desirable locations. After all, there is only so much space at the design eye of the aircrew. In addition, all of these systems and their associated accouterments added a great deal of weight. The military set out to try and streamline cockpit and initiated a study looking at these problems from a "systems" approach that was simpler, weighed less, had some sort of standardization, and at the same time was flexible. The military's answer was MIL-STD-1553.

The commercial sector was experiencing many of the same problems as the military; however, their biggest driver was money. A simpler, streamlined system that reduced weight would save money. The commercial study in the United States grew into the ARINC (American Radio Inc.) standards. These standards have been modified and updated and are still in use today. The military still uses MIL-STD-1553, although it is now designated as an interface standard. The military also uses MIL-STD-1760 (Interface Standard for Weapons Systems) and MIL-STD-1773 (Interface Standard for Fiber Optics [vice copper cables]). Commercial industries use a plethora of standards in addition to ARINC. The initial section of this chapter is dedicated to the MIL-STD-1553 protocol; users of 1760 and 1773 will note that these standards are nearly identical to 1553. The United Kingdom version of this standard is Def Stan 00-18 (Part 2) and the NATO version is STANAG 3838 AVS. The latter part of this chapter deals with other data busses and their use in the avionics world.

▮▮▮ 3.2 I 1553 SYSTEM ARCHITECTURE

Avionics and weapons systems test engineers must become familiar with how the military data bus operates. An excellent reference for testers is provided by ILC Data Device Corp., Bohemia, NY. The title of the text is the *MIL-STD-1553 Designer's Guide*, which, at the time of this writing, was in its sixth edition. They have a convenient website at www.ilc-web.com. If you register with them, you can download the guide in Adobe Acrobat. As far as a reference goes, I think that this guide is all you will need. Most technical libraries probably have a copy or two.

The full title of the interface standard is MIL-STD-1553, Aircraft Internal Time Division Command/Response Multiplex Data Bus. To put this in laymen's terms, there is a bus controller (BC) that issues commands on the bus and remote terminals (RTs) that respond on the bus. There is a timing requirement for gaps between messages and response time of the RTs. In some cases where large data flows are needed, a multiplexer is used (called a multiplex) to split the transmission into two or more subchannels. The device accomplishes this by allotting a common channel to several different transmitting devices one at a time (time division). The speed at which the bus can transfer data differs with each type of data bus. The 1553 data bus has a 1 MHz data rate which allows it to transfer 1 megabit/sec. The 1773 data bus can increase the data rate by a factor of 100. 1553 uses serial communication, which means that bits are sent sequentially on a single data channel. The key elements of 1553 architecture are the

- BC
- Bus monitor
- RT
- Stand-alone remote terminal
- Twisted pair shielded cable

The BC provides data flow control for all transmissions on the bus. It has the ability to send as well as receive data. It is the only unit on the bus that initiates a transmission. You may think of the controller as the traffic cop on the bus. It is the controller's responsibility to ensure orderly operations on the bus. The BC uses a command/response method of data transfer. No other device on the bus may talk on the bus unless so ordered by the BC. It also initiates such operations as self-test, synchronization, transmitter shutdown, and overrides for other bus devices. The BC is smart enough to recognize changes in the environment as well as identify failed units. A simplified example of a bus structure is shown in Figure 3.1.

A bus monitor is a passive unit on the bus. The standard defines a bus monitor as a "terminal assigned the task of receiving bus traffic and extracting selected information to be used at a later time." These systems are used for troubleshooting and aircraft instrumentation. They are not active terminals and do not communicate with the bus controller. These devices read all of the traffic on the bus and capture the time and value of selected parameters. These parameters are preprogrammed into the bus monitor.

The first type of monitor is called a bus analyzer. These systems are used extensively in software support facilities and integration labs and can be very helpful to the flight tester. Older models only provided a hexadecimal dump of all messages resident on the bus during a specified time. Newer models provide engineering unit (EU) converted data. Examining a weapons release sequence shows an example of how this device can be used.

A weapons release sequence involves many variables that are time and order dependent. If either the timing is off or events are not in order, the weapon may not release or not arm. A critical test would be to determine proper sequencing of the Stores Management System. A proper sequence is shown in Table 3.1.

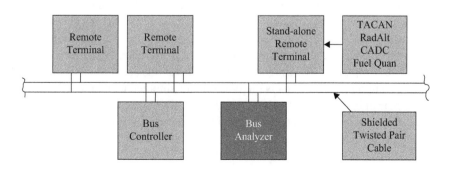

FIGURE 3.1 ■
Simplified Bus
Structure

TABLE 3.1 ▪ Weapons Release Sequence

Time	Event
0	Trigger depress
10 msec	Battery arm
35 msec	Rocket motor arm
50 msec	Ejector arm
150 msec	Roll command sent
250 msec	Frequency set
350 msec	Ejector fire
500 msec	Rocket motor fire

The eight events are completed in 0.5 sec. Normal aircraft instrumentation will probably sample the bus about 25 samples/sec. This sampling rate is too slow to capture the events of interest. The question now becomes, "Where can this test be accomplished?" It should be done in the lab with a bus analyzer. You can trigger the analyzer to capture all data on the bus for 1 sec with the capture initiated at trigger squeeze. The timing and sequencing can then be analyzed.

The second type of bus monitor is common to instrumentation. Most of these systems are called either word grabbers or bus editors. They are normally built by the instrumentation group and are unique to one type of aircraft. The devices are preprogrammed on the ground to look for certain messages (parameters) on the bus and read these data to either a recording device or multiplex them into a telemetry stream. The amount of data these systems can grab for real-time data analysis is limited by the recording pace or the telemetry system.

The RT may be thought of as an avionic component or sensor connected to the bus. An inertial navigation system (INS), global navigation satellite system (GNSS), radar display processor, or electro-optical sensor can be used as RTs. These terminals do not have controller authority unless they are designated as backup bus controllers. There may be a maximum of 32 RTs on a bus (mathematically true, but not true in real life). Interface circuitry inside the sensor is tied directly to the bus, which allows a transfer of data into and out of the sensor.

The stand-alone remote terminal may be thought of as a big analog-to-digital (A/D) converter. Depending on the age of the aircraft under test, there may be several systems on board that are pre-1553 and are still analog. The only way these systems can talk to the rest of the aircraft is to perform an A/D or D/A conversion. Depending on the number of older systems, it may not be feasible to place a converter on each system. This method would take an enormous amount of valuable space and add unwanted weight. Most military designers opt for one converter unit to handle all of the nonstandard (non-1553) systems. These systems are known by a variety of names; the most common are converter interface unit, systems interface unit, digital converter unit, and signal interface unit. Figure 3.1 shows non-1553 systems such as the TACAN, radar altimeter, ILS, fuel quantity and central air data computer (CADC) talking to the bus via this stand-alone remote terminal.

The last component of the bus is the twisted pair shielded cable that carries the data. There are a pair of cables for redundancy purposes. All busses will have an "A" bus and a "B" bus. Shielding is for electromagnetic interference (EMI) reduction. The standard requires the cable shielding provide a minimum of 90% coverage and connectors must have continuous 360° shielding with a minimum of 75% coverage.

3.3 | A BIT ABOUT BITS

We know that data busses transmit data by using bits. Do the systems that we deal with use binary, octal, or hexadecimal? Do we know the difference between the three? If you think back to grammar school math (unless you are a product of the California school system), you may vaguely remember base 2, base 3, etc. If it is too vague, or just not there, continue on. For those who may think that this is too basic, skip to the next section.

Computers use a binary system that is really base 2, as opposed to the base 10 that we are familiar with. If we have two available slots, what numbers can we come up with? Figure 3.2 shows the possibilities for two bits that are available.

The 1553 systems count from right to left, so we shall perform this exercise the same way. Bit 0 is the first bit on the right and equates to 2^0, which equals one. The second bit equates to 2^1, which equals two. The first number we can obtain is zero ones plus zero twos, which equals zero. The next number is one (ones) plus zero twos, which equals one. The third number is equal to zero ones plus one two, which equals two. After the number three in base 2 (11), we can go no farther unless we add another bit, which would equal 2^2, or four. The rationale continues out to the left for as many bits as we have. 1553 uses 16 bits for data, so it could conceivably go as high as 2^{15}.

That was fairly straightforward. What is octal? As the name implies, octal is base 8. Figure 3.3 shows the maximum number you can attain in two bits for a base 8 system. As with binary, you can start at 00, which equates to 0 times 8^0 (0×1) plus 0 times 8^1 (0×8) for a total of zero. The first bit can increment upward until 7. Two bits of 07 would equal 7 (0 eights plus 7 ones). You cannot place an 8 in the first bit because the second bit place is 8^1, or eight. Therefore 8 in base 10 would be written as 10 in base 8. The maximum number that you can identify with two bits in base 8 is 77, which is equivalent to 63 in base 10 ($7 \times 8 + 7 \times 1$). Are there any systems that you are familiar with that use base 8? It may not be readily apparent unless we use four bits. Pilots will see the obvious if we show a four-bit octal system:

7	7	7	7

This is, of course, the Mode 3 Interrogate Friend or Foe (IFF) system. Some of you will know this system by another term: the 4096 code. There are 4096 available squawk codes available in the current Mode 3 IFF system. The code 7777 is octal and is equivalent to 4095 in base 10. The code 0000 is also available, which makes a total of 4096 codes.

The last type of system use would be hexadecimal, or base 16. Since each slot can only contain one character, some other method must be used to identify the units 10 through 15. This is illustrated in Figure 3.4, which shows the maximum number in base 10 that we can obtain by using two bit slots in the hexadecimal system. As with the two previous uses, we start at the top with 00, which equates to 0 in base 10 ($0 \times 16^0 + 0 \times 16^1$). The first bit slot increments upward through 9. Since only one character can occupy a slot, the units 10 through 15 are identified by the alpha characters A through F, where F is the maximum that can be placed in any slot. This makes sense, since the second slot is occupied by 16^1, or 16 in base 10. We can count upward to 15, but then we must move over to the second bit position. The base 10 number 16 would be

2^1	2^0
0	0
0	1
1	0
1	1

FIGURE 3.2 ■ Base 2 Maximum Number for Two Bits

8^1	8^0
0	0
0	1
0	2
0	3
0	4
0	5
0	6
0	7
1	0
1	1
1	2
1	3
1	4
1	5
1	6
1	7
...	
7	7

FIGURE 3.3 ■ Base 8 Maximum Number for Two Bits

16^1	16^0
0	0
0	1
0	2
0	3
0	4
0	5
0	6
0	7
0	8
0	9
0	A
0	B
0	C
0	D
0	E
0	F
1	0
1	1
1	2
1	3
1	4
1	5
1	6
1	7
1	8
1	9
...	
F	F

FIGURE 3.4 ■ Base 16 Maximum Number for Two Bits

written as 10 in hexadecimal, 17 would be 11, and so on. The maximum number that can be written for two bit positions in hexadecimal is FF, or 255 in base 10.

Reading a binary bit string in hexadecimal can be advantageous for flight testers and analysts. The 1553 systems utilize 20-bit words, 16 of which contain usable information. The first three bits are the synchronization pattern and the last bit is a parity bit. Imagine you are an analyst and it is your task to determine if a fault occurs in a system. You have been told that a fault is present if bit 9 registers positive (a 1 is set). Either real-time or postflight processing yields reams of data for this one 16-bit word.

1101010001101111

This 16-bit word would scroll down the page and it would be your task to determine when, if ever, bit 9 went to a 1. This is a difficult task, even for the seasoned analyst, but not impossible. If we convert this binary string to a hexadecimal word, the task can be easily accomplished. In order to read a binary string in hexadecimal, we need to break up the 16-bit word into four equal parts:

1101   0100   0110   1111

We then read each segment in hexadecimal characters. This is done by adding, from right to left, the binary string, and converting the resultant number to hexadecimal. This can be demonstrated by

15	14	13	12	11	10	9	8	7	6	5	4	3	2	1	0
2^3	2^2	2^1	2^0	2^3	2^2	2^1	2^0	2^3	2^2	2^1	2^0	2^3	2^2	2^1	2^0
1	1	0	1	0	1	0	0	0	1	1	0	1	1	1	1

where the top row is the bit number, the second row is the binary multiplier, and the third row is the original 16-bit word. This binary string is then converted to hexadecimal by multiplying row two by row three and then adding across right to left.

15	14	13	12	11	10	9	8	7	6	5	4	3	2	1	0
8	4	0	1	0	4	0	0	0	4	2	0	8	4	2	1
13 = D				4 = 4				6 = 6				15 = F			

read in hexadecimal is D46F.

In this example, we were able to express a 16-bit binary word in a hexadecimal format. The original problem was to identify a failure which would be indicated by bit 9 being set true. By reading in hexadecimal, if all other bit values remain constant, only the 4 will change (to a 6), and the task becomes much easier. As one more example, convert the following binary string to hexadecimal:

0011110011001101

First, break it up into sectors:

15	14	13	12	11	10	9	8	7	6	5	4	3	2	1	0
2^3	2^2	2^1	2^0	2^3	2^2	2^1	2^0	2^3	2^2	2^1	2^0	2^3	2^2	2^1	2^0
0	0	1	1	1	1	0	0	1	1	0	0	1	1	0	1

Multiply row two by row three:

15	14	13	12	11	10	9	8	7	6	5	4	3	2	1	0
0	0	2	1	8	4	0	0	8	4	0	0	8	4	0	1

Add right to left and convert:

15	14	13	12	11	10	9	8	7	6	5	4	3	2	1	0
3 = 3				12 = B				12 = B				13 = D			

Read in hexadecimal: 3BBD.

This concept will become more apparent when we address different types of 1553 words in the coming sections.

3.4 | 1553 WORD TYPES

There are three different types of words on a 1553 bus: command words, status words, and data words. Each of the words contains 16 bits plus a synchronization pattern and a parity bit, for a total of 20 bits. The parity is odd. Command words only come from the bus controller, whereas status words are only sent by the remote terminal. Data words come from either the bus controller or the remote terminal. The bit transmission rate on the 1553 bus is 1 megabit/sec. Fiber optics can increase this rate to 10 to 100 megabits/sec. There are twenty 1 microsecond (μsec) bit times allocated for each word, since each word will contain 20 bits.

3.5 | DATA ENCODING

The 1553 systems use Manchester II biphase encoding as opposed to the more well-known non-return-to-zero (NRZ) encoding such as that used in instrumentation systems. Manchester still uses 1s and 0s, but it is a little more complicated than 1 for high and 0 for low. In Manchester, a 1 is transmitted as a bipolar coded signal high/low (within 1 bit time, the signal transitions from high to low). A 0 is transmitted as a bipolar coded signal low/high (within 1 bit time, the signal transitions from low to high). The transition through 0 occurs at the midpoint of each bit for either 1 or 0. Figure 3.5 illustrates NRZ as compared to Manchester II for a 1 MHz clock.

The synchronization pattern (first three bits of the 20-bit word) uses an invalid Manchester code to aid in identifying the start of a new word. The transition, either low to high or high to low, takes 3 bit times (3 μsec) to complete, with the transition through 0 occurring at 1½ bit times (1.5 μsec). A logic of 1 (high to low) identifies a command or status word, while a logic of 0 identifies the start of a data word. Command and status

FIGURE 3.5 ▪ Data
Encoding: NRZ
Versus Manchester
II Bipolar

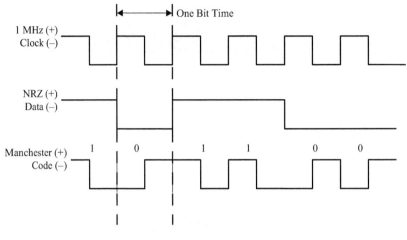

FIGURE 3.6 ▪
Command and
Status Word
Synchronization
Pattern

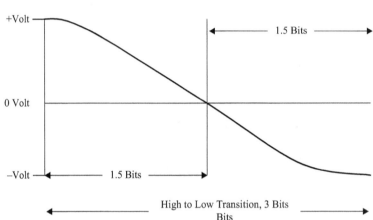

words can use the same synchronization pattern, since command words can only come from the bus controller and status words can only come from the remote terminal. Figure 3.6 shows the synchronization pattern for the command and status words. The data word synchronization pattern is simply reversed.

In fiber-optics systems (MIL-STD-1773), there is a slight difference in the encoding method. Since light cannot have a negative value, the pulses are defined as transitions between 0 (off) and 1 (on) rather than between + and − voltage transitions.

3.6 | WORD FORMATS

The formats of the command, status, and data words are shown in Figure 3.7. A detailed explanation of each of the words follows.

3.7 | COMMAND WORDS

The command word is only sent by the BC. As illustrated in Figure 3.7, the command word is comprised of a 3-bit time synchronization pattern, 5-bit RT address, 1-bit

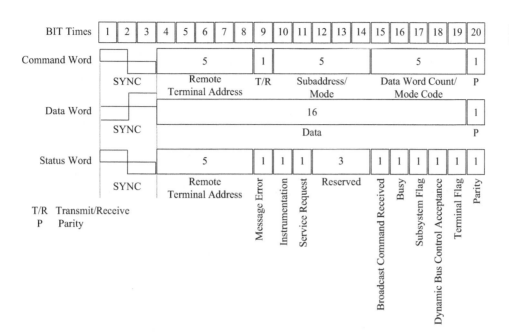

FIGURE 3.7 ■
Word Formats

transmit/receive, 5-bit subaddress/mode field, and 5-bit word count/mode code field followed by a 1-bit parity check, for a total of 20 bits. The 1553 terminals add the synchronization and parity before transmission and remove them during reception. Therefore the nominal word size is 16 bits. By using Figure 3.7, we can construct a representative command word. Suppose that the BC needs nine-state vector information from the INS. The bus controller needs to initiate a command word and send it to the INS for action. Also suppose that the INS happens to be identified as RT 12 on this particular bus. The first 5 bits (after the 3-bit synchronization is removed) of the command word are used to identify the RT address. In our example, this is 12. The address would look like this:

0	1	1	0	0											

The 5 bits are read from right to left, with the first bit representing 2^0 and the fifth bit representing 2^4. Here we have 1 (4) and 1 (8), for a total of 12. The total number of RT addresses that we can have is 32. If all bits are set to 1, we have 31 $(1 + 2 + 4 + 8 + 16)$. We can also have an RT with an address of 0, which allows us to mathematically have 32 remote terminals on any one bus. In practice we are only allowed to have 31 remote terminals on a bus. This anomaly is due to the broadcast mode, which will be explained later. Let's get back to our example. As this command word is sent out on the bus, the INS reads it because it is addressed to it (RT 12). The other remote terminals will ignore the command word.

The next bit in the command word is the transmit/receive bit. Transmit is set to 1, and receive is set to 0. Since the bus controller is requesting the INS send nine-state vector information, this bit will be set to 1. Had the controller been sending information to the INS, then the bit would be set to 0.

0	1	1	0	0	1										

The next 5-bit field is the subaddress/mode and the final 5 bits comprise the word count/mode code field. It is easier to first explain the word count field. These final 5 bits determine how many data words are to be sent or received. If each of the nine-state vector parameters required one data word, then the word count would be nine:

0	1	1	0	0	1						0	1	0	0	1

It should be obvious that the maximum number of data words that can be sent is 31 (all bits set to 1). This limitation on the amount of data that can be exchanged could be deleterious to system operation. Let's examine the information an INS may possess which might be of interest to other systems on the aircraft. The INS either measures or computes latitude, longitude, altitude, velocities (north, east, and down [NED]), accelerations (NED), heading, ground track, groundspeed, pitch, roll, yaw, time-to-go, distance-to-go, cross track and along track errors, and many other parameters. It stands to reason that thirty-one 16-bit data words would be inadequate to transfer all of this information. If this is the case, then how do we expand our capability to transfer this important information? The answer is in the use of the 5-bit subaddress field. This field allows subaddresses from 0 through 31, with each subaddress having 32 associated data words. This now allows 32×32 data words for information exchange, except that there are two reserved fields. Subaddress 0 (00000) and subaddress 31 (11111) are reserved for mode commands, which will be addressed later. In the example, subaddress 5 may contain the nine data words for the state vector:

0	1	1	0	0	1	0	0	1	0	1	0	1	0	0	1

Our completed message shows the command word going to RT 12, transmit subaddress 5, word count nine. Both the BC and the RT are programmed to know where information resides.

3.8 | ANOMALOUS COMMAND WORD CONDITIONS

In section 3.7, I stated that there can be only 31 RTs on a bus even though there are 32 mathematical possibilities. The RT address of 31 (11111) is reserved for broadcast mode. Suppose that there was a parameter on the aircraft that all systems used in their computations. One that comes to mind is altitude. Sensor systems, electronic warfare (EW) systems, weapons systems, and navigation systems all use altitude. Since all systems use this parameter, it would seem to be efficient to broadcast this parameter to all RTs rather than send the parameter individually. This is broadcast mode. When the RT address is set to 31 (11111), all RTs read the message. Unfortunately, the RTs cannot respond stating that they received the message because all of the replies would come back at the same time, making identification of respondents impossible. In order to determine if all RTs did in fact receive the transmission, the bus controller would have to individually ask each RT if they received the transmission. This is a time-consuming process, rendering the broadcast mode unusable. As a result, most aircraft that use MIL-STD-1553 do not employ broadcast mode.

The second anomalous condition is when 0 (00000) or 31 (11111) is found in the subaddress. When this occurs, the command word is said to be a mode command. Earlier in the chapter it was stated that in a 1553 system, RTs may not perform any operation unless commanded to do so by the BC. If the RT needs to perform a self-test, or reset, or synchronize, it must wait until commanded by the BC. These operations are called mode commands. The subaddress of 0 (00000) or 31 (11111) tells the RT that the message is a mode command, and the word count is replaced with the code of the specific operation. There are assigned mode codes, and these codes can be found in the reference material. For example, a mode code of 3 (00011) indicates that the RT is to perform a self-test. In the previous example, if we wanted the INS to initiate a self-test, the command word would look like this:

| 0 | 1 | 1 | 0 | 0 | 1 | 0 | 0 | 0 | 0 | 0 | 0 | 0 | 0 | 1 | 1 |

The reference material will also indicate whether the transmit/receive bit is set to 0 or 1. In the self-test example, the bit is set to 1.

3.9 | COMMAND WORD SUMMARY

The command word is a 20-bit word sent by the BC to an RT. The 3-bit synchronization and 1-bit parity are added before transmission and removed during reception. The remaining 16 bits are comprised of

- A 5-bit field for the RT address; RT 31 (11111) is reserved for broadcast mode. Broadcast is not efficient and therefore not widely used. A total of 31 RTs may occupy a bus.

- A 1-bit field for transmit/receive: 1 = transmit, 0 = receive.

- A 5-bit field for the subaddress. The subaddress allows more data to be exchanged. A 0 (00000) or 31 (11111) in the subaddress field indicates a mode command. The transmit/receive bit in a mode command is per the documentation.

- A 5-bit field for the data word count. If a 0 (00000) or 31 (11111) occupies the subaddress field, the data word count becomes the mode code. Mode codes are assigned per the documentation.

3.10 | STATUS WORDS

Once again, the reader is referred back to Figure 3.7 to aid in the understanding of this section.

After the RT has received a command word from the BC, it builds a status word that informs the bus controller of the RT's status as well as the result of the last transmission. The first 5 bits are the responding RT's address. This allows the bus controller to know from which RT the status word is coming. The next 11 bits are a status field. The good thing about status words is that this 11-bit field will be 0-filled if the message was received without error and the RT's health and welfare are good. Knowing this provides analysts with an excellent tool for monitoring bus communications. This will be described later.

The first bit after the RT address is the message error identification. The receiving RT validates each word and message using the following criteria:

- The word begins with a valid sync
- The bits are a valid Manchester II code
- The word parity is odd
- The message is contiguous

If the word/message does not meet these criteria, the message error is set to 1. The status word, however, is not sent and the bus controller is alerted to a problem due to a timeout. The bus controller can obtain the status word by subsequently issuing a mode command to transmit the status word.

The remaining bits in the status word are called status bits. The instrumentation bit can be used to discriminate the command word from the status word (remember they have the same synchronization). The logic would be 0 in the status word and 1 in the command word. If the instrumentation bit is used this way, then the number of subaddresses is reduced from 30 to 15, and only subaddress 31 (11111) can be used for initiation of a mode command. If the instrumentation bit is not used in this way, then it is always set to 0.

The service request bit is the next status in the field and is used to indicate to the bus controller that the RT requires service. When the bit is set to 1, the bus controller may command a predetermined action or request transmission of a vector word that will identify the type of service requested.

The next three bits are reserved by the U.S. Department of Defense (DOD) and prior approval must be obtained to use them.

The next bit is the broadcast command received status. If the architecture incorporates broadcast mode, this bit will be set to 1 when a broadcast command is received. As described earlier, the status word cannot be sent and is therefore suppressed. If the bus controller wants to know the status of the broadcast command it will have to issue a mode command requesting the status word.

The busy bit indicates to the bus controller that the RT cannot respond to commands due to a busy condition. For example, the terminal may be performing periodic maintenance. Busy conditions are supposed to be the result of a previous command by the bus controller and not due to periodic internal RT conditions. Busy conditions tend to add overhead and disrupt bus communications.

A subsystem flag indicates that a fault exists in the RT's subsystem and that the data requested may not be reliable. This should not be confused with the last status bit, called the terminal flag. The terminal flag indicates a failure within the RT as opposed to its subsystem.

A dynamic bus control acceptance bit indicates to the bus controller that the RT is ready to take control of the bus after receiving a take control mode command. In some applications, this procedure is illegal and the bit will always be set to 0.

It should be apparent that if the message is received without error and the RT status is healthy, the only information in the status word will be the RT address. If we examine the status word, we may see a trend in good versus bad status words. If we assume that the INS in our command word example received the message error free, and the INS is healthy, the status word would look like this:

| 0 | 1 | 1 | 0 | 0 | 0 | 0 | 0 | 0 | 0 | 0 | 0 | 0 | 0 | 0 | 0 |

The message error and status bits are all set to 0. If we read this message in hexadecimal, it looks like:

| 0 | 1 | 1 | 0 | 0 | 0 | 0 | 0 | 0 | 0 | 0 | 0 | 0 | 0 | 0 | 0 |

      6   **0**   **0**   **0**

If the message was received error free, and the RT is healthy, the status word, when read in hexadecimal, will always end in (000) or (800) (when an RT address occupies the 2^0 slot). An analysis program can be written that states: "print on event whenever the status word does not end in 000 or 800." As a result, a record for all failures in communications will then be obtained for the entire test.

3.11 | DATA WORDS

Data words can come from either the bus controller or the RT (Figure 3.7).

The 3-bit synchronization pattern is opposite of the command and status words, and the parity once again is odd. The 16 bits may be packed in many ways and may contain as little as one-half of a parameter or as many as 16 parameters. The tester must be familiar with the latest versions of the interface design specification (IDS) and the interface control document (ICD) in order to determine which parameters are contained in the data words. Let's look at an example of a typical data word (synchronization and parity removed):

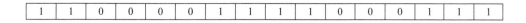

| 1 | 1 | 0 | 0 | 0 | 0 | 1 | 1 | 1 | 1 | 0 | 0 | 0 | 1 | 1 | 1 |

From the ICD, the data word radar altitude (parameter ID R_Alt) is contained in 8 bits: start bit = 0, stop bit = 7; scaling factor = 3 ft. Reading right to left (the first bit received), radar altitude is obtained:

| 1 | 1 | 0 | 0 | 0 | 0 | 1 | 1 | 1 | 1 | 0 | 0 | 0 | 1 | 1 | 1 |

Radar altitude in this example is equal to $(1 + 2 + 4 + 64 + 128) \times 3$ ft, or 597 ft. The parameter of radar altitude only uses the first 8 bits. What is contained in the other 8 bits? It could be another 8-bit parameter, or it could be something else. Imagine looking down on the nose of an aircraft (see Figure 3.8):

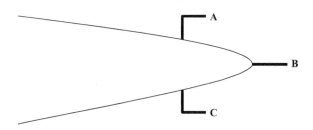

FIGURE 3.8 ▪
Pitot-Static System
Measured Clockwise
From Top (A, B, C)

Many aircraft employ multiple pitot and static sensors on the aircraft. Figure 3.8 shows a typical configuration of three sensors labeled clockwise from the top: A, B, and C. When the aircraft and the free stream are aligned, the three sensors should agree with one another. When the aircraft is yawed, one of the sensors will be slightly different than the other two. In our example, if we apply right rudder, sensor C will be obscured from the free stream and will give a slightly different reading than A and B. Sensor C will be voted out and declared invalid by the air data computer (ADC). The analyst will know this when looking at validity bits on the bus. Going back to our original example with the radar altimeter, the ICD may identify the next three bits as validity bits. The logic is 1 = true and 0 = false.

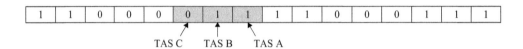

The validity bits in the example show that true airspeed (TAS) A and B are valid and TAS C is invalid. What do the last 5 bits represent? It could be another 5-bit parameter or five more validity bits. The important thing to note is that a data word may contain more than one parameter. Some parameters are "long words." These parameters require two 16-bit data words to comprise the data. Some examples of "long words" are latitude and longitude.

3.12 | MESSAGE CONTENTS

Messages are packets made up of the three word types previously discussed: command words, status words, and data words. There is a response time requirement of between 4 and 12 µsec and a minimum intermessage gap time of 4 µsec. There are three basic types of messages or data transfer:

- BC to RT
- RT to BC
- RT to RT

There are variations in these messages which include mode commands with or without a data word, and broadcast commands with or without a data word. Figure 3.9 shows the sequence of MIL-STD-1553 message formats.

FIGURE 3.9 ■
MIL-STD-1553
Message Formats

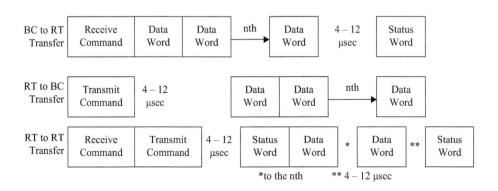

It should be noted that for the three cases, the bus controller initiates the sequence. After any of these three sequences is complete, the bus controller needs to wait a minimum of 4 μsec before initiating the next sequence.

In the first case (BC to RT), the bus controller identifies the receiving RT in the first 5 bits of the command word. The transmit/receive bit will be set to 0 since the RT is expected to receive data. The type and number of data words are identified in the sub-address and word count. The data words follow the command word without any gaps. Figure 3.9 shows the command word followed by data word 1, data word 2, out to the nth data word (identified in the word count). The RT will respond within 4 to 12 μsec with a status word as long as the message meets the criteria described in section 3.10.

For an RT to BC transfer, the bus controller initiates the sequence with a command word to the appropriate RT with the transmit/receive bit set to 1, since the RT will be transmitting data. The type of data and number of data words are once again identified in the subaddress and word count fields. Within 4 to 12 μsec, as long as the criteria for a valid command are met, the RT will send a status word followed by the prescribed number of data words. There are no gaps between the status and data words.

The RT to RT transfer requires the bus controller send two command words, one to the receiving RT and one to the transmitting RT. It must be done in this order to preclude an RT from transmitting data at the same time a receive message is being sent by the bus controller. The transmitting RT responds in the same manner as the RT to BC transfer described in the previous paragraph. The receiving RT has 4 to 12 μsec to respond with a status word after receipt of the last data word (once again assuming that the message was received error free).

Mode commands, with or without a data word, behave like the RT to BC and BC to RT transfers; the subaddress is replaced with a 0 (00000) or 31 (11111) and the word count is replaced by the mode code. The transmit/receive bit is set according to the documentation. Broadcast mode sets 31 (11111) in the RT address and all RTs read the message. The RTs set the broadcast command received status bit in the status word but do not send the status word.

3.13 | BUS CONTROLLER DESIGN

Most test engineers are not going to be tasked with designing system processors, but it is helpful to know a few things about their capabilities and limitations. If you are working with military applications, you may often wonder why the aircraft mission computer has less capability than your laptop. There are a few reasons for this disparity. The first problem, of course, is military standardization. But wait, you say, there are no more military specifications! This is true and it is not true. There is still a need for standards in military applications. Would you allow someone to drop your laptop from 18 inches onto a hardened concrete surface? That type of shock is what military equipment must be able to sustain. It is sometimes called ruggedization of equipment. In the past, it has taken up to 5 years to ruggedize a civilian system. For this reason, a legacy aircraft of the 1990s may have a 486 processor. Some legacy radar systems employ a 286 processor. Things have improved a great deal since about 2005, and military capabilities do not lag their civilian counterparts as much.

A larger problem for military applications is the language that the system uses. Military systems are coded in many languages: machine code, COBAL, Ada, etc.

The language was not standardized until the late 1980s, when the military tried to standardize all programs with Ada. Of course, there were many waivers and this dictum really did not solve the problem. I keep saying problem. It was not a problem in the 1980s because the military accounted for the bulk of all computer software; somewhere on the order of 85%. Today, the military accounts for 10% to 15% of all software. Now can you see the problem? In the 1980s, people were falling all over themselves to work for the military, because that is where the money was. Today, it is not as profitable. Suppose you have an aircraft that is programmed in Ada and you need to modify or add a new function. You must find a company that has expertise in your language that is willing to take on the task. In order for the company to realize a profit from such a specialized task, the cost can be astronomical. It is easier for the company to reprogram the entire processor with a language that is current and for which plenty of expertise is readily available. Of course, when this happens, the test community is in for a rather large regression program.

The last concern for testers is related to the first two. For any digital system, it is desirable to have some redundancy in system operation. In fly-by-wire flight controls, for safety of flight reasons, there is usually triple redundancy for the bus controllers. For the same reasons, Federal Aviation Administration (FAA) Part 25 air carriers have the same redundancy in flight controls as well as the avionics suite. In fighter and bomber military applications, most initial designs call for dual-redundant mission computers (avionics and weapons suite) and triple-redundant flight controls. In addition, most contracts will specify a processor reserve requirement (25–50%) for future growth. Unfortunately these requirements are rarely met. Due to the large number of functions the system is required to perform (controls, displays, navigation, radar, weapons, EW, electro-optical/infrared [EO/IR] sensors, reconnaissance, etc.), the functions wind up being partitioned between the two processors. The only redundancy is in safety-of-flight functions such as controls, displays, and navigation. In the event of a loss of one mission computer in flight, the checklist will more often than not direct the aircrew to "land as soon as practical," since some mission capability has been lost. By the same token, these many functions take up so much processor capacity that the future growth reserve requirement rapidly evaporates. There are some military aircraft in service today that have no growth capability. Legacy aircraft such the F-16, F-18, and F-15 are plagued by this problem; if a new function is to be added to the aircraft, some other installed function must be removed first.

▌3.14 ▌ SAMPLE BUS CONFIGURATIONS

Figure 3.10 shows the simplest bus configuration that a tester may encounter. It describes dual bus architecture: mission bus 1 and mission bus 2. Mission computer 1 is the bus controller for mission bus 1 and the backup bus controller for mission bus 2. Mission computer 2 is the bus controller for mission bus 2 and the backup bus controller for mission bus 1. It should be noted that as discussed earlier, each mission bus contains an A bus and a B bus.

Figure 3.11 contains the basic elements of the simplified structure shown in Figure 3.10 and adds in RTs. There may be 31 RTs on any bus: RT 0 through RT 30. Address 31 is reserved for broadcast mode. The RTs can be connected to more than one bus, and the addresses can be different for each bus. In addition, the mission computers can be connected to each other by an intercomputer bus, which allows direct transfer of data

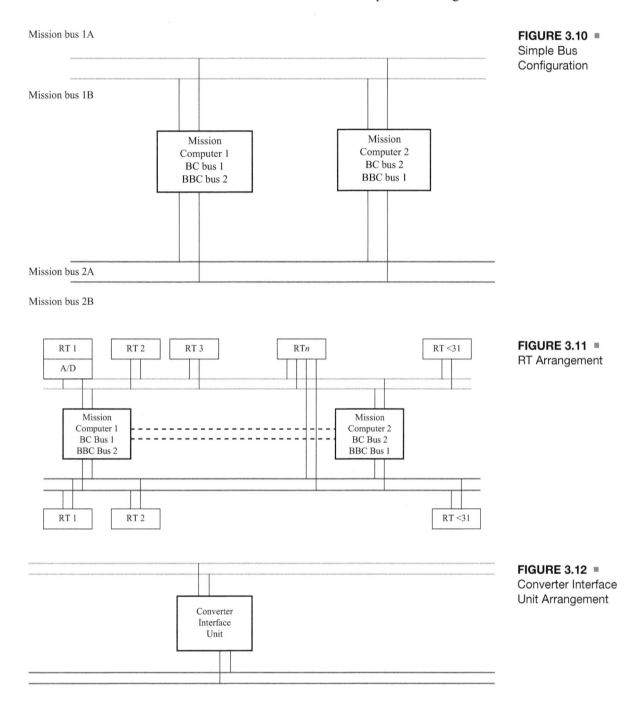

FIGURE 3.10 ▪
Simple Bus
Configuration

Mission bus 1A

Mission bus 1B

Mission
Computer 1
BC bus 1
BBC bus 2

Mission
Computer 2
BC bus 2
BBC bus 1

Mission bus 2A

Mission bus 2B

FIGURE 3.11 ▪
RT Arrangement

RT 1 A/D   RT 2   RT 3   RT*n*   RT <31

Mission
Computer 1
BC Bus 1
BBC Bus 2

Mission
Computer 2
BC Bus 2
BBC Bus 1

RT 1   RT 2   RT <31

FIGURE 3.12 ▪
Converter Interface
Unit Arrangement

Converter
Interface
Unit

between the mission computers. Most of the RTs shown are connected directly to the bus. RT 1 on mission bus 1 is connected to the bus via an analog-to-digital (A/D) converter. Remember that some systems are not 1553 compatible and therefore must go through an A/D converter. In military applications, most systems employ one large converter interface unit which handles all of the A/D conversions required by nonstandard (non-1553) systems. This can be seen in Figure 3.12. All A/D conversions are performed by the converter interface unit. The converter interface unit acts as an RT on both mission busses.

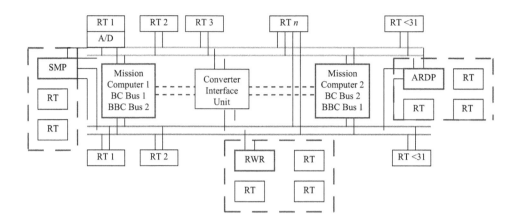

There are other busses on the aircraft and they have to be able to communicate with the mission computers as well as each other. This is accomplished by allowing the bus controllers of other subsystems to also be RTs on the mission busses. This can be seen in Figure 3.13.

Figure 3.13 incorporates our previous architectures, and adds three additional busses: radar, stores management, and EW. The ARDP, located on the right side of the diagram, is the airborne radar data processor, which is the bus controller for the radar bus. The ARDP is also an RT on mission bus 1 and mission bus 2. This allows radar information to be passed along the mission busses for use by other sensors/systems. The ARDP manages its own bus with up to 31 RTs on the bus. By the same token, the SMP (stores management processor), located at the left side of the diagram, is the bus controller for the stores management bus. It is also an RT on mission bus 1 and mission bus 2. The SMP manages its own bus (with up to 31 RTs), and can share information with other sensors/systems via the mission busses. The EW bus is controlled by the radar warning receiver (RWR), in the lower portion of the diagram. Unlike the previous two cases, this bus acts as an RT only on mission bus 2. Information is still shared as long as mission bus 2 remains healthy. Should mission bus 2 go down, there will be no data transfer with the EW bus.

There are a total of six busses depicted in Figure 3.13: mission busses 1 and 2, an intercomputer bus, radar bus, stores management bus, and EW bus. This is a typical arrangement on a fighter-type aircraft. Each bus has the capacity of transferring data at the rate of 1 Mbps, for a total of 6 Mbps.

Even with these data rates, it may become necessary to bypass bus operations in order to increase efficiency. This can readily be seen by investigating data transfer between the INS and radar. The radar must know its spatial relationship with the earth and true north at all times. This is called platform information and is obtained from the aircraft INS. The radar needs this information updated at the highest rate possible. If the INS provides data updates at 90 Hz, then the radar will use a 90 Hz update rate. In our bus architecture, this would require the bus controller to send out a receive and transmit command to the radar and the INS 90 times/sec. Depending on the size of the data words and overhead, this might be the only information on the bus because of the maximum bandwidth of the bus. As a result, it may not be possible to have any other data transfer on the bus. This, of course, is totally unreasonable. The way to fix the problem is to either reduce the INS to the radar transfer rate, or find an alternate means of transferring the data. In this particular application, a direct update (hard wire) from the INS to the radar would likely be implemented, thereby bypassing the bus altogether. This may seem to defeat the purpose of having a data bus at all, but sometimes the operational requirements dictate design.

▮ 3.15 | THE FLIGHT TESTER'S TASK

When evaluating any avionics system that is integrated with a digital data bus, it is the tester's job to verify that data are not being corrupted along the way. The only way this test can be carried out correctly is by first understanding the system design and discussing the bus architecture with the design engineer. Comparing an input to a system to an output of the system does not necessarily yield the answer that we are looking for. Remember, in systems integration, we are not evaluating the box, we are evaluating a correct integration of that box.

In recent system design approaches, an engineer will assess each subsystem from an input and output perspective (i.e., understand what information a subsystem produces and what other systems/subsystems consume that information, as well as what information the subsystem itself consumes). A good example of a parameter that is produced within a system on an aircraft is altitude. A design engineer looks at this parameter, where it is sensed (i.e., produced) and describes its transmission path through the aircraft to other systems that may consume this information. Typically this description is defined in a functional requirements document (FRD). As the name implies, this document describes all of the functional requirements of a given system. There may be an overarching FRD for the overall system (i.e., aircraft), which is then broken into several subsystem specifications that provide details of each subsystem. Typically, a software engineer would then be assigned the task of detailing the derived requirements of the subsystem into a very detailed coding specification that will be used to code the software to meet the requirements.

Again, typically, an interface design specification (IDS) is produced that specifies the data input and output requirements of the subsystem of interest. An ICD is produced that takes the data input and output requirements of a specific subsystem and identifies the types of interfaces required between the other subsystems that consume the data produced by the subsystem of interest.

The job of the tester is to ascertain if the subsystem that has been developed meets the performance of the system/subsystem FRD. To accomplish this, the tester must understand what the FRD says, or, at a minimum, have a very good understanding of the intent of the FRD.

Figure 3.14 shows the parameter altitude, where it is sensed, and where the information flows. This is a simplified version of what a design engineer must do.

We know that altitude is corrected for local pressure in the Mission Computer. In our diagram, altitude that is sensed by the pitot-static system, passed to the ADC, and then to the A/D converter is noncorrected altitude. The mission computer to display processor to display is corrected altitude. In any case, the competent test engineer will want to verify that 1) the pitot-static sensed altitude parameter is accurate (verified by TSPI), 2) the altitude received by the mission computer is not corrupted (verified by data analysis), 3) the conversion completed by the mission computer is accurate and correct (verified by data analysis), 4) the altitude received by the display system is uncorrupted (verified by data analysis), and 5) the altitude displayed to the operator is correct and readable (verified by data analysis and human–machine interface [HMI] assessment). This is just for one parameter and one communication path. Now consider the same for all other data parameters and all other paths.

FIGURE 3.14 ■
Altitude Flow Chart

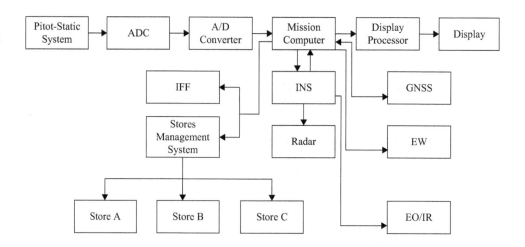

Consider another example, the vertical channel in the INS. In most cases it is not very good, so in many applications INS altitude is aided by barometric altitude. This "corrected altitude" is sent by the mission computer to the INS, and the INS reports back with an INS corrected altitude. INS corrected altitude does not equal corrected altitude or noncorrected altitude; it is something different.

The inherent worst error for GPS is in altitude, so most applications also aid GPS altitude with barometric altitude. In the case of GPS, however, noncorrected altitude is sent to the GPS from the mission computer and the GPS reports back with a GPS corrected altitude. Of course, GPS altitude is not really altitude at all, but a height above ellipsoid. Height above ellipsoid does not equal corrected or noncorrected altitude, or even INS corrected altitude.

Many weapons use an altitude that is more "accurate" than corrected or non-corrected altitude. They may use density altitude, which corrects for temperature and humidity variation from standard conditions. And, yes, you guessed it, it is not the same as any of the other altitudes we have discussed. Other sensors, radar, EO, and EW may also require altitude to aid in their functional performance. Depending upon the installation, the sensors may use any of the aforementioned altitudes. Some sensors, such as radar, compute their own altitude, which is different still from those already mentioned.

So, which altitude is correct? Unfortunately they are all correct. It all depends on where the data are collected (i.e., produced) and where they are used (i.e., consumed). It is incumbent upon testers to have a thorough understanding of the system under test and how data it produces and consumes are being manipulated while passing through the system. Failure to do so will result in an inaccurate assessment of the system under test.

Remember one other thing: even though altitude is altitude (i.e., a data parameter is a data parameter), the parameter must be considered at each stage of the transition from one subsystem (or transport layer) to the next.

▌ 3.16 | MIL-STD-1553 SUMMARY

Data bus architecture has made the sharing and presentation of information in today's cockpit easier than ever. One of the most important features of this technology is that systems have moved from direct interaction with the crew (i.e., direct manipulation of the system) to indirect control of the subsystem through a bus interface. This has allowed the use of multifunction displays (MFDs) that collate information from many sensors and subsystems onto one display while the subsystems themselves are located in various places on the aircraft. Without the use of data bus technology, this would not be possible.

MIL-STD-1553, in particular, offers a standard for the transfer of this information in military applications. Testers, however, must not be lulled into thinking that the testing of such systems is easier than analog, nor should they believe that all tests will be standardized. I cannot emphasize enough the importance of learning the system before testing. Once the system is understood, the tests must evaluate the entire system, not just the box. If these two points are considered with every test plan, the risk of missing a potential problem is minimized.

3.17 | OTHER DATA BUSSES

Up until now, I have only addressed MIL-STD-1553. I have spent a bit of time on this system because it has been the backbone of military avionics and weapons systems for the past 30 years. Modern systems need to move more data between devices and needs have emerged to incorporate audio and video, pushing MIL-STD-1553 beyond the limits of its bandwidth. The search for a bigger, faster bus has yielded some interesting and innovative designs. In the following sections I will examine a number of other digital interface standards testers may come in contact with.

When looking at other systems, it is important to compare them to the advantages of MIL-STD-1553. The primary reason the military uses MIL-STD-1553 is that this system is deterministic. Because of the command/response protocol of MIL-STD-1553, we can always guarantee that the data arrived on time and uncorrupted in real time. There are not that many other interface standards that can make this claim. For this reason alone, we can probably expect to see some form of MIL-STD-1553 in all mission critical systems of the future. Other advantages of MIL-STD-1553 are its local area network (LAN) architecture, which reduces weight and conserves space; its capacity for redundancy (dual-bus redundancy); support for non-1553 devices; a proven history; and component availability. Any replacement for this system must be analyzed with respect to these capabilities.

3.17.1 RS-232C

RS-232C (recommended standard number 232, version C) is one of the oldest standards still in use; it was introduced in 1969 by the telephone industry. It is a standard for short-distance, multiwire, single-ended communication. Bit rates are up to 19 kbps with a 16 m bus and up to 115 kbps over short links. It uses bipolar voltage signaling with simple handshake signals and allows for synchronous and asynchronous operations. The standard specifies pin assignments for a point-to-point topology. It was developed by the Electronic Industries Alliance (EIA) and Bell Laboratories to connect computing equipment, known as data transmission equipment (DTE), to a telephone modem, known as data communications equipment (DCE), so that two computers could communicate over a telephone line. The standard implied that the communication was performed over a short distance on discrete wires with a common return. There was no mention of transmission line parameters such as impedance or attenuation.

Initially designed for communicating through modems, it migrated to communications between computers and peripherals. During the 1980s, RS-232C was the standard of choice for linking computers to devices such as printers, plotters, or other computers. This worked well as long as the data rates were kept to 20 kbps at a 1 m distance or 600 bps at 25 m.

RS-232C became EIA 232E, with the latest version introduced in 1991. Since the standard limited the data rates and distances, newer standards were devised to increase these parameters. EIA 423 allows 100 kbps at distances up to 1200 m and EIA 422 allows data rates of 10 Mbps at distances up to 1200 m.

RS-232 signals are represented by bipolar voltage levels with respect to a common ground between the DTE and the DCE. On the data line, the active condition is represented by a positive voltage between 3 and 15 V and indicates a logic 0. The idle condition is represented by a negative voltage between −3 and −15 V and indicates a logic 1. Figure 3.15 shows the signal levels and character frame defined in RS-232.

Besides the basic transmit and receive data paths, RS-232 specifies handshake lines, primarily for use by modems. The standard defines the function of these handshake lines and the designated pins within a range of connectors. The basic configuration of the RS-232 interface is shown in Figure 3.16. The interconnecting lead between the DTE and DCE is simply wired pin 1 to pin 1, pin 2 to pin 2, etc.

A description of the main signals follows:

- Transmitted data (TXD): the serial-encoded data sent from a computer to a modem to be transmitted over the telephone line.

- Received data (RXD): the serial-encoded data received by a computer from a modem which has, in turn, received it over a telephone line.

- Data set ready (DSR): set true by a modem whenever it is powered on; it can be read by the computer to determine that the modem is online.

- Data terminal ready (DTR): set true by a computer whenever it is powered on; it can be read by the modem to determine that the computer is online.

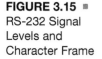

FIGURE 3.15 ■
RS-232 Signal
Levels and
Character Frame

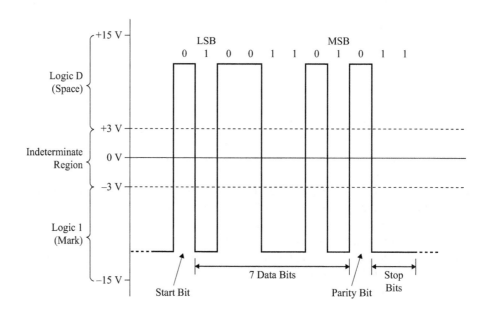

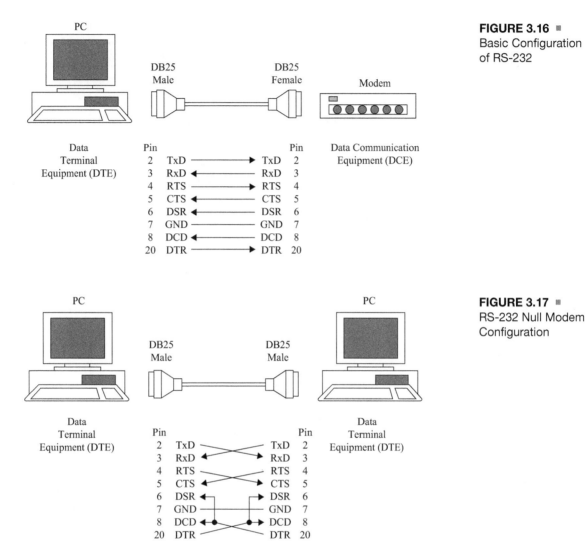

FIGURE 3.16 ■
Basic Configuration
of RS-232

FIGURE 3.17 ■
RS-232 Null Modem
Configuration

- Request to send (RTS): set true by a computer when it wishes to transmit data.

- Clear to send (CTS): set true by a modem to allow the computer to transmit data.

- Data carrier detect (DCD): set true by a modem when it detects the data carrier signal on the telephone line.

There are also clock signals defined for synchronous communications. The DCE extracts a clock signal from the data stream and provides a steady clock signal to the DTE. Computers may be connected to other computers (DTE to DTE) by changing the cable (corresponding pins cannot simply be connected together). Along with the data lines, various handshake lines need to be crossed over. This configuration, known as a null modem, is shown in Figure 3.17.

The depiction in Figure 3.17 assumes that full handshaking capability is implemented in both devices, which may not be true. Cables can be wired so that a device provides its own handshake signals and tells itself to proceed with the transfer regardless of the status of the handshaking signals (i.e., RTS connected to CTS or DSR, and DCD

TABLE 3.2 ■ EIA (RS) 232C Signals

DB-25 Pin No.	DB-9 Pin No.	Symbol	Description (with respect to DTE)	Direction
1		FG	Frame Groud	—
2	3	TXD	Transmitted Data	From DTE
3	2	RXD	Received Data	To DTE
4	7	RTS	Request To Send	From DTE
5	8	CTS	Clear To Send	To DTE
6	6	DSR	Data Set Ready	To DTE
7	5	SG	Signal Ground	—
8	1	DCD	Data Carrier Detect	To DTE
9		—	+ P test pin	—
10		—	− P test pin	—
11		—	unassigned	—
12		SDCD	Secondary Data Carrier Detect	To DTE
13		SCTS	Secondary Clear To Send	To DTE
14		STD	Secondary Transmitted Data	From DTE
15		TC	Transmission Signal Element Timing	To DTE
16		SRD	Secondary Received Data	To DTE
17		RC	Receiver Signal Element Timing	From DTE
18		—	unassigned	—
19		SRTS	Secondary Request To Send	From DTE
20	4	DTR	Data Terminal Ready	From DTE
21		SQ	Signal Quality Detector	To DTE
22	9	RI	Ring Indicator	To DTE
23		—	Data Signal Rate Selector	From DTE
24		—	Transmitter Signal Element Timing	To DTE
25		—	unassigned	—

connected to DTR). In general, DTR and DSR are used to detect if the other device is present and powered up, while RTS and CTS are used to implement hardware flow control. Table 3.2 lists all signals and pin numbers (for 25- and 9-pin connectors) and the direction the signal travels between the equipment.

Other than slow data rates, the RS-232 standard has some other significant problems. These are, in no particular order,

- Only one conductor is used per circuit with a single return for both directions of transmission.

- It can generate cross-talk among component signals.

- The absence or misuse of the handshaking signals can result in buffer overflow or communication lockup.

Newer standards (RS-422, RS-423, and RS-485) attempt to correct some of these problems as well as enhance data rates.

3.17.2 Controller Area Network Bus

The controller area network (CAN) bus uses message transmission based on prioritized bus access with bounded throughput. It operates at 1 Mbps with a 40 m bus and up to 100 kbps with a 500 m bus. The CAN bus uses an NRZ encoding with message lengths of 0 to 8 bytes. It is a serial communications bus developed in the mid-1980s by Robert Bosch

GmbH for the German automobile industry. The needs which drove the requirement are similar to those in aircraft avionics: to provide real-time data communication with reduced cabling size and weight and normalized input/output (I/O) specifications.

There are two versions of CAN: CAN 1.0 and CAN 2.0, with CAN 2.0 divided into two parts known as CAN 2.0A and CAN 2.0B. CAN 2.0 has been formalized in two standards: ISO 11898, for high-speed applications (125 kbps to 1 Mbps), and ISO 11519 Part 2, for low-speed applications. Amendment 1 to both standards covers CAN 2.0B.

Messages transmitted on the CAN bus contain a unique identifier indicating the data content of the message, and the receiving terminals examine the identifier to determine if the data are required by the terminal. The identifier also identifies the priority of the message; the lower the numerical value, the higher the priority. Higher-priority messages are guaranteed bus access, whereas lower-priority messages may have to be transmitted at a later time.

To determine the priority of messages, CAN uses carrier sense, multiple access with collision resolution (CSMA/CR). Collision resolution is by nondestructive bitwise arbitration, in which a dominant logic 0 state overwrites a recessive logic 1 state. A higher-priority terminal having a lower-value identifier transmits a logic 0 before a lower-priority terminal. Transmitting terminals monitor their own transmissions; a competing lower-priority terminal finds one of its logic 1 bits overwritten to a logic 0. The lower-priority terminal terminates its transmission and waits until the bus is free again. There is automatic retransmission of frames that have lost arbitration or have been destroyed by errors in transmission. Nondestructive bitwise allocation allows bus access on the basis of need, which provides benefits not available in other types of architectures such as fixed time schedule applications (1553) or destructive bus allocation (Ethernet). In a CAN bus, outstanding transmission requests are dealt with in order of priority. It is claimed that a CAN will not lock up (Ethernet) because no bandwidth is lost in collisions. Unfortunately, low-priority messages could be permanently locked off the bus, which would cause problems in military applications.

The standard CAN 2.0A message frame is depicted in Figure 3.18 and contains seven different bit fields. The start of frame (SOF) field, consisting of a single dominant bit, indicates the beginning of a message frame. The arbitration field contains the 11-bit message identifier, providing the means for nondestructive bitwise arbitration plus a remote transmission request (RTR) bit. The message identifier allows 2032 unique identifiers to be used for unicast, multicast, or broadcast messages. The RTR bit is used to discriminate between a transmitted data frame and a request for data from a remote node (much like the transmit/receive bit in the 1553 command word). The control field contains 6 bits comprising the format identifier extension (IDE), reserved bit r0, and a 4-bit data length code (DLC). The IDE bit is used to distinguish between the CAN 2.0A format

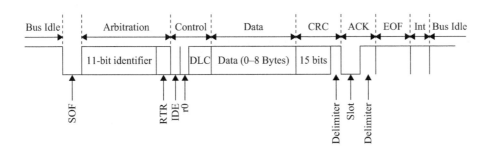

FIGURE 3.18 ■ CAN 2.0A Frame Format

FIGURE 3.19 ▪
CAN 2.0B Frame
Format

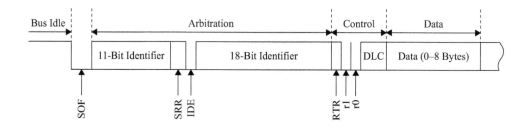

and the CAN 2.0B extended format. The DLC indicates the number of bytes (0–8) in the following data field (much like the data word count in the 1553 command word).

The cyclical redundancy check (CRC) field contains a 15-bit CRC code and a recessive delimiter bit. (**NOTE:** The CRC and what it is used for are described more completely in the MIL-STD-1760 section later in this chapter.) The acknowledge (ACK) field consists of 2 bits: the first is the slot bit, which is transmitted as recessive and is subsequently overwritten by dominant bits transmitted from any node that successfully receives the transmitted message; the second bit is a recessive delimiter bit. The end of frame (EOF) field consists of 7 recessive bits. The intermission (Int) field consists of 3 recessive bits; after the 3-bit intermission field, the bus is again considered free. The bus idle time may be of any arbitrary length, including zero.

The CAN 2.0B format is shown in Figure 3.19. The only difference from CAN 2.0A is found in the arbitration field, which contains two identifier bit fields. The first (base ID) is 11 bits long, for compatibility with version 2.0A. The second field (ID extension) is 18 bits long, to give a total length of 29 bits, allowing more than 5 million unique user identifiers. The distinction between the two formats is made using the identifier extension (IDE) bit. A substitute remote request (SRR) bit is also included in the arbitration field. The SRR bit is always transmitted as a recessive bit to ensure that, in the case of arbitration between a standard data frame and an extended data frame, the standard data frame will always have priority if both messages have the same base (11-bit) identifier. The RTR bit is placed after the 18-bit identifier and followed by 2 reserved bits; all other fields in the 2.0B format are the same as in 2.0A format.

A CAN implements five error detection methods, resulting in a residual (undetected) error probability of 10^{-11}:

- Cyclical redundancy checks (CRCs)
- Frame checks
- Acknowledgment error checks
- Bit monitoring
- Bit stuffing

A CAN discriminates between temporary errors and permanent failures by using error count registers within each terminal; a faulty device will cease to be active on the bus, but communications between the other nodes on the bus can continue.

3.17.3 ARINC 429

The formal name of ARINC 429 is the ARINC 429 Mark 33 Digital Information Transfer System (DITS) specification. This system is simplex (one-way) operating at

100 kbps (high speed) or 12–14.5 kbps (low speed). The message length is 32 bits with a 255 word data block in block transfer mode. ARINC 429 uses RTZ bipolar tristate modulation and can be configured in either a star or bus-drop topology. The standard was developed by the Airlines Electronic Engineering Committee (AEEC) data bus subcommittee, administered by American Radio Inc. (ARINC).

The ARINC 429 specification defines the requirements for a data transmission system based on the use of a single data source and reception of that data by up to 20 sinks or receivers. In the context of MIL-STD-1553, the source can be viewed as the bus controller, whereas the sinks are the RTs. The maximum number of sinks permitted for connection to a source is limited by the specified minimum receiver input impedance.

Either a star (Figure 3.20) or a bus-drop (Figure 3.21) topology may be used. If a full-duplex connection is required, a terminal/line replaceable unit (LRU) may have multiple connectors and separate transmit and receive busses (Figure 3.22). A receiver/sink is not permitted to respond on the bus when it receives a transmission, but as indicated in Figure 3.22, it may respond on a separate bus. LRUs connected to ARINC 429 busses are not assigned addresses; instead, they listen for data words with labels identifying data of interest to the LRU.

ARINC 429 utilizes a twisted pair shielded copper cable for redundancy. Although the bus length is not specified, most systems are designed for less than 175 ft, with some able to handle 20 sinks at distances of up to 300 ft. The signaling is by RTZ bipolar

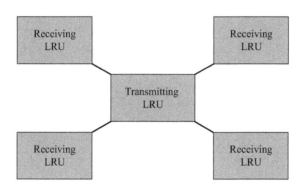

FIGURE 3.20 ▪
ARINC 429 Star Topology

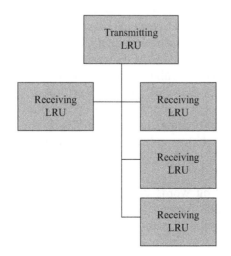

FIGURE 3.21 ▪
ARINC 429 Bus-Drop Topology

FIGURE 3.22 ▪
ARINC 429 Multiple
Bus Topology

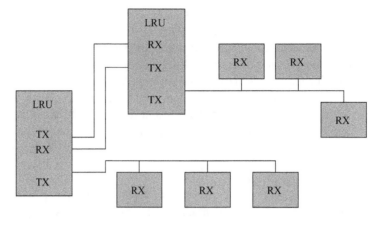

FIGURE 3.23 ▪
Return-to-Zero
Tristate Modulation

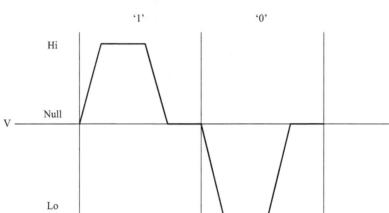

FIGURE 3.24 ▪
ARINC 429 Word
Format

32	31	30	29	28	27	26	25	24	23	22	21	20	19	18	17	16	15	14	13	12	11	10	09	08	07	06	05	04	03	02	01
P	SSM									Data Word												SDI				Label					
		MSB																			LSB	LSB								MSB

tristate modulation and has three states—"Hi," "Null," and "Lo"—which are represented by the differential voltage between the two wires on the cable. A graphical representation of RTZ is shown at Figure 3.23.

A logical 1 is signaled by transmission of a 10 ± 1 V pulse followed by a 0 ± 0.5 V null period. A logical 0 is signaled by the transmission of a -10 ± 1 V pulse followed by a 0 ± 0.5 V null period. Rise and fall times for the pulse are specified in the standard and are a function of bit rate. A drawback associated with RTZ is that unequal numbers of "Hi" and "Lo" bits can result in a transfer of electrical energy into reactive elements of the bus so that the bus may become charged, resulting in end-of-message distortion and ringing. An advantage of tristate modulation is that a clock is not necessary, as the start of new words can be identified by length of null time (words are separated by four bit times of null voltage).

ARINC 429 uses 32-bit words divided into five application domains. The data contained in the words of the message format depend on the application domain and the specific data being transferred. Figure 3.24 represents a typical ARINC 429 word containing five parts.

The source transmits the 32-bit word and each sink is required to inspect the label field, which is encoded in octal to represent the type of information contained within the 32-bit word. The organization of the word is least significant bit (LSB) sent first, except for the label, which is encoded most significant bit (MSB) read first. The label field can have system instructions or data reporting functionality. When performing data block or data file transfers, the label is replaced with the system address label (SAL). Each data item is assigned a label listed in the ARINC 429 specification and each source may have up to 255 labels assigned for its use.

The next two bits are used as a source/destination identifier (SDI). The SDI is optional, and if it is not used, the two bits are added to the data word. The functions of the SDI are any of the following:

- Indicate the intended destination of the data on a multisink bus.

- Identify the source of the data on a multisource bus.

- Add an extension to the label, in which case receiving systems decode the label/SDI combination.

When the SDI is used as a destination indicator, receivers are given binary installation numbers (00,01,10,11), where 00 may be receiver number 4, or it may be used to indicate a broadcast mode (similar to 11111 in MIL-STD-1553).

The data field contains 19 bits and, as previously mentioned, may contain 21 bits if the SDI is not used. The signal/sign matrix (SSM) is also an optional field which is used to indicate sign information: up/down, east/west, north/south, $+/-$, etc. The parity bit is set to make the word parity odd.

Each ARINC 429 word has a predefined rate at which it must be transmitted, as each label has an associated minimum and maximum transmit time defined in the standard. As previously mentioned, the minimum interword gap is four bit times; however, they may be longer.

ARINC 429 allows for block transfer of data between devices. Files are transferred in blocks known as link data units (LDUs) and may contain from 3 to 255 words. The file transfer is handled much the same way as discussed in RS-232. When a source intends to transmit data to a unit, it sends an RTS word containing a destination code and data word count. The receiving unit sends a CTS word containing the same destination code and word count for verification. Upon receipt of the CTS, the source initiates a transfer by sending a start of transmission (SOT) word. The SOT word contains a file sequence number, a general format identifier (GFI), and an LDU sequence number. Up to 253 data words are then sent, followed by an end of transmission (EOT) word that terminates the block. The EOT word includes a CRC and identifies the position of the LDU in the overall file transfer. With successful verification, the receiving unit sends an acknowledgment (ACK) word; the source may then repeat the process. Later versions of the standard allow up to seven LDUs to be transmitted without seeking separate permission from the sink.

3.17.4 ARINC 629

ARINC 629 is a serial, multitransmitter data bus operating at 2 Mbps. It utilizes Manchester II biphase encoding and can transmit up to 31 word strings, each containing one 16-bit label and up to 256 16-bit data words. Access control is via carrier sense multiple

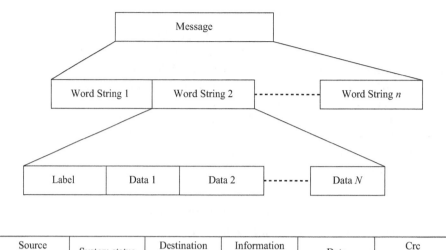

Source Address	System status	Destination Address	Information Identifier	Data	Crc (Optional)

access/collision avoidance (CSMA/CA). Initial research was carried out by Boeing and was implemented on the Boeing 777 in 1989 (ARINC 629). ARINC 629-2 was published in 1991 and ARINC 629-3 in 1994. It was developed because of the need to overcome limitations in ARINC 429.

An ARINC 629 message (Figure 3.25) has variable length up to 31 word strings. A word string is also of variable length and is made up of one label with channel information and data labels for addressing and up to 256 data words.

Figure 3.26 is an example of how a word string may be used. The label is used for the source address. Subsequent words carry system status information, destination address, information identifier, data, and an optional CRC. Each word is 20 bits long: the first three are the synchronization pulse, the next 16 are data, and the last bit is for parity.

In CSMA/CA, bus access control is distributed among all participating terminals, each of which determines its transmission sequence. This is achieved by the use of bus access timers:

- Transmit interval (TI): The TI is the longest timer and operates between 0.5 and 64 msec. It starts every time the terminal starts transmitting and is specified by a 7-bit value in the terminal controller. A message is sent by the terminal from every 1 to 31 TI periods depending on settings held in the terminal's transmit personality. The TI is the same for all terminals.

- Synchronization gap (SG): The SG ensures that the terminals are given access. The SG is chosen to be greater than the maximum terminal gap and the values are 16, 32, 64, or 128 × minimum terminal gap. The SG starts every time bus quiet is sensed; it may be reset before it has elapsed if bus activity is sensed and restarted the next time the terminal starts to transmit. The SG is the same for all terminals.

- Terminal gap (TG): The TG is used to differentiate between terminals and is unique for each terminal. The values of the TG are from 3.6875 to 126.6875 in 1 μsec steps. The TG only starts after the SG has elapsed and the bus is quiet; it may be reset if bus activity is sensed before it has elapsed.

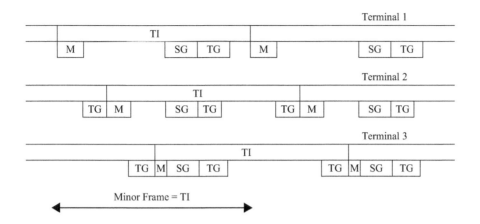

FIGURE 3.27 ◾
Basic Protocol
Periodic Mode

The standard supports two alternative protocols: basic and combined. The two can not operate simultaneously due to fundamental differences in bus access. In both cases the access is coordinated at two levels:

- The basic division of bus time into major and minor cycles may be of a fixed duration and looks a lot like time division multiple access (TDMA) or may be variable depending on the settings of the terminal.

- The access contention by terminals can be accomplished by carrier sensing and observation of the preassigned waiting times, and common bus quiet periods look a lot like carrier sense multiple access (CSMA), but with collision avoidance.

The basic and combined protocols support periodic and sporadic transmissions. Periodic messages are scheduled automatically at the 629 terminal level according to predefined message tables (message contents, destinations, etc.). The way that sporadic transmissions are handled provides the main difference between the two protocols. Figure 3.27 shows the basic protocol periodic mode timing for a simple three-terminal bus.

The order in which the terminals achieve bus access remains the same for all minor cycles and is determined by the initialization sequence and drift between terminal clocks; it does not necessarily equate to the relative duration of the terminals' TG timers. If the sum of all the TGs, transmission times, and SG is less than TI, then the minor cycle time remains fixed and equal to the TI. Periodic access for all terminals can only be guaranteed if the message transmission times (represented by "M" in Figure 3.27) are all equal and constant.

If the sum of all the TGs, message transmission times, and SG is greater than TI, then the system must operate in the aperiodic mode, as shown in Figure 3.28.

For every minor cycle, the terminals transmit in order of their TG durations (shortest first); in Figure 3.28, TG1 < TG3 < TG2. Each minor cycle consists of a series of transmissions separated by the various TG delays, which are followed by an SG delay that provides synchronization. The messages include periodic and sporadic data transmissions that vary in length from minor cycle to minor cycle. The timing stability of the periodic data transmission is dependent on the amount of sporadic data that can cause jitter.

In order to solve the problems associated with aperiodic mode, British Aerospace and Smiths Industries developed a combined protocol to provide a more effective

FIGURE 3.28 ■
Basic Protocol
Aperiodic Mode

FIGURE 3.29 ■
Combined Protocol

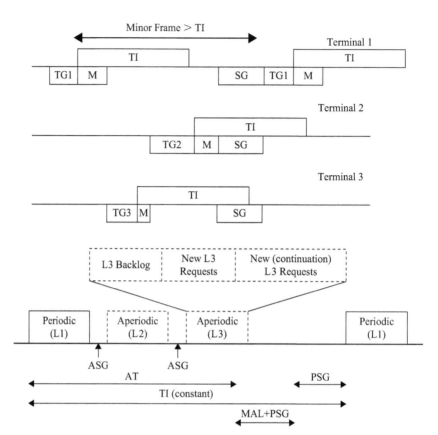

approach to handling periodic and sporadic data. The combined protocol has three levels:

- Level 1: Periodic transmissions
- Level 2: Short but frequent sporadic data
- Level 3: Long but infrequent sporadic data

A new set of timers are also defined for the combined protocol:

- Transmit interval (TI): The TI is the same as in the basic protocol, but is only applicable for the first periodic transmission within each minor cycle.
- Concatenation event (CE): The CE overrides and cancels all subsequent nonelapsed TI timers, effectively forcing all periodic transmissions into a burst separated only by TG delays and the start of each minor frame.
- Terminal gap (TG): Same as in the basic protocol.
- Periodic synchronization gap (PSG): Ensures synchronization at the minor cycle level.
- Aperiodic synchronization gap (ASG): Synchronizes transmissions between transmissions of different levels within minor cycles.
- Aperiodic access time out (AT): Indicates the time to the next periodic (level 1) transmission; any sporadic transmissions that would take longer than this time are prohibited during the current minor frame, thus ensuring fixed-length minor frames.

The combined protocol timing is depicted in Figure 3.29. Level 1 periodic messages are transmitted in a burst (separated by the TG) at the start of the minor frame when the PSG of the previous minor frame elapses. Level 2 aperiodic messages are transmitted (separated by TG delays and in the reverse order of TG duration) after the burst of level 1 periodic data when timer ASG has elapsed, followed by level 3 aperiodic messages.

Each terminal is limited to one level 2 transmission, but may perform multiple level 3 transmissions if time is available within the minor frame. Level 2 sporadic data must be transmitted within the current minor frame or be lost. Level 3 sporadic data may span minor cycles, and backlog level 3 messages take priority. The "MAL" in Figure 3.29 is the maximum aperiodic length, and is equal to 257 words.

Although this standard was once considered for use by Airbus, it is currently used on only one platform (Boeing 777), where it forms part of the fly-by-wire system.

3.17.5 STANAG 3910

STANAG 3910 builds on the basic 1553 bus architecture by adding a 20 Mbps high-speed data bus activated by commands on the low-speed 1553 bus. The standard allows for variable message formats of up to 128 blocks of 32 words, or a total of 4096 16-bit data words. The standard defines three optical (fiber optic) and one electrical high-speed network topology options.

STANAG 3910 was born out of necessity because of the inadequate data transfer capacity of MIL-STD-1553 for the Eurofighter and other future avionics systems where the requirements could not be met by multiple 1553 busses. The designed system was low risk and it relied heavily on existing 1553 designs and protocols.

The first draft standard was created in 1987 by a German standardization committee; the EFAbus (Eurofighter bus) was issued as a project-specific variant of 3910 type A in 1989. An electrical high-speed network (type D) variant was adopted for the French Rafale program in 1990. The standard defines the high-speed fiber-optic and electrical transmission networks, the concept of operation and information flow on the high-speed network, and the optical, electrical, and functional formats to be employed.

The standard defines a 20 Mbps fiber-optic or electrical network controlled by a 1553B or optical equivalent 1 Mbps data bus. Four types of systems are defined:

- Type A
  - Low-speed channel: 1553 data bus
  - High-speed channel: fiber-optic data bus

- Type B
  - Low-speed channel: fiber-optic equivalent of 1553B
  - High-speed channel: physically separate fiber-optic data bus

- Type C
  - Low-speed channel: fiber-optic equivalent of 1553B
  - High-speed channel: wavelength division multiplexed with the low-speed channel onto a common fiber-optic medium

- Type D
  - Low-speed channel: 1553B data bus
  - High-speed data bus: physically separate wire data bus

FIGURE 3.30 ■
3910 Type A
Network

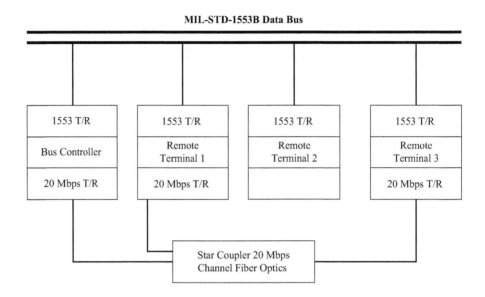

Figure 3.30 shows an example of a type A 3910 network using an optical star coupler. STANAG 3910 defines requirements for the high-speed channel implemented as an optical transmissive star coupler, reflective star coupler, or a linear T-coupled optical bus. EFAbus uses a type A network with optical reflexive star couplers.

All transfers on the high-speed network are initiated via messages on the low-speed bus, with the action required on the high-speed network defined in a 1553B data word and described in 3910 as a high-speed action word. In the Eurofighter, the RT recognizes the high-speed data transfer when a command word with a subaddress of 26 is sent. The high-speed terminal's status is conveyed in 1553B data words, described in 3910 as high-speed status words. Figure 3.31 describes the message formats available in 3910. You will recognize them as being similar to those described in the 1553 section of this chapter. The "**" in the figure refers to the response time as described in 1553, which you will remember is 4 to 12 μsec; the "#" symbol in the figure refers to the intermessage gap, which is 4 μsec. The bus controller receives the status of the high-speed transfer by sending a mode code to the transmit/receive requesting a high-speed status word.

In the Eurofighter implementation of 3910, the bus controller waits for the high-speed message frame to complete before sending the next command word/message on the low-speed 1553 bus.

The formats of the data words (high-speed action word and high-speed status word) are shown in Figure 3.32. The bus controller sends the command word to the transmitting/ receiving RT with a subaddress of 26. The next word (first data word) will be the high-speed action word shown in Figure 3.32. The action word is 20 bits long, with the first 3 bits used as the synchronization pattern. The A bus/B bus (A/B) bit indicates which of the two (redundant busses) networks will be used for the transmission, while the T/R bit indicates whether the RT is to transmit or receive the high-speed transfer. The high-speed message identity/high-speed mode works like MIL-STD-1553 except the subaddress uses 7 bits (128 subaddresses) instead of the 5 bits in 1553. The high-speed word count/mode code is as in MIL-STD-1553, except there are 7 bits for the data word count (128 data words). The parity bit is the same as in 1553 (odd parity). All zeroes (0000000) in the message identity indicates a mode code command.

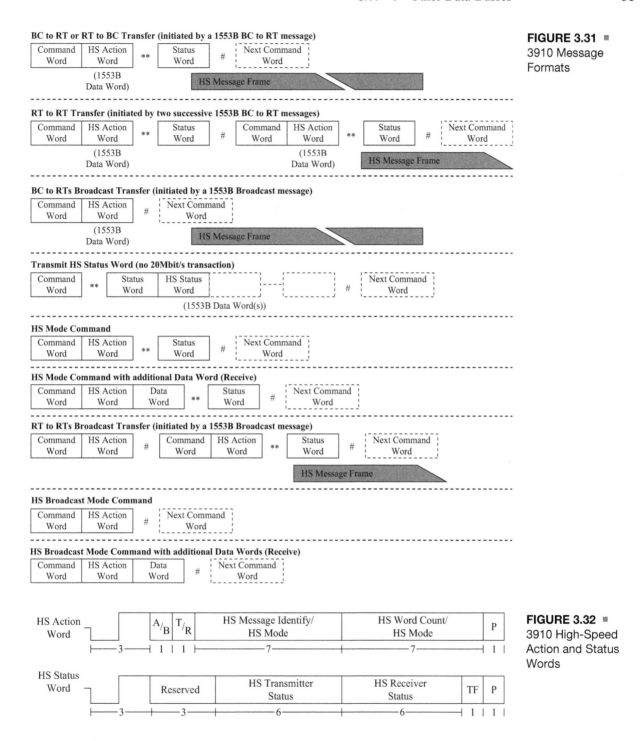

FIGURE 3.31 ■ 3910 Message Formats

FIGURE 3.32 ■ 3910 High-Speed Action and Status Words

The high-speed status word (also shown in Figure 3.32) is comprised of a 3-bit synchronization followed by a 3-bit reserved field. The 6-bit transmitter status field indicates the status of the transmitter section of the high-speed media interface unit (MIU) and the transmission protocol interface unit (TPIU). Only two bits in this field are used: bit 7 indicates that the transmitter is not ready for transmission, and bit 8 indicates

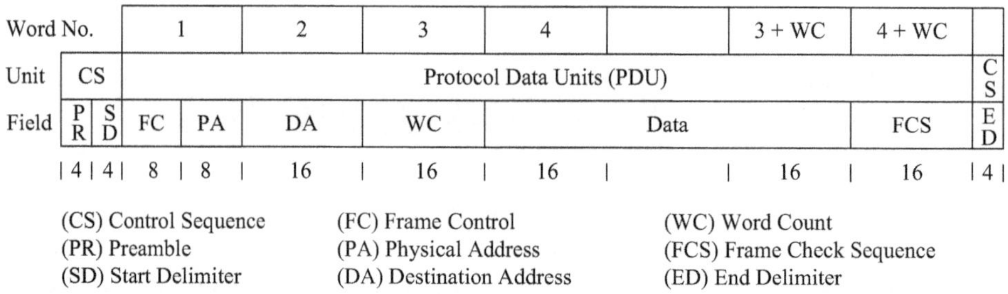

FIGURE 3.33 ■
High-Speed
Message Frame
Format

Word No.	1	2	3	4		3 + WC	4 + WC	
Unit	CS	Protocol Data Units (PDU)						C S
Field	P R | S D	FC	PA	DA	WC	Data	FCS	E D
	|4|4|	8 | 8 |	16 |	16 |	16 |	| 16 |	16	|4|

(CS) Control Sequence   (FC) Frame Control   (WC) Word Count
(PR) Preamble   (PA) Physical Address   (FCS) Frame Check Sequence
(SD) Start Delimiter   (DA) Destination Address   (ED) End Delimiter

that the high-speed transmitter is active (transmitting). The 6-bit high-speed receiver status indicates the status of the receiver section. Only three of these bits are used: bit 16 indicates that the high-speed receiver is not ready for reception, bit 17 indicates that the high-speed receiver is active (receiving), and bit 18 indicates that an error was detected on the most recently received high-speed transmission. Bit 19 is the terminal flag and bit 20 is the parity bit. The high-speed message format is shown at Figure 3.33.

Three control symbol (CS) fields define the boundary of a transmission: the preamble (PR), start delimiter (SD), and the end delimiter (ED). The PR is used by the receiving unit to acquire signal-level synchronization. The SD and ED delimit the start and end of a data transmission sequence and are made up of invalid Manchester code patterns similar to the synchronization pattern. The frame control (FC) field is required to be permanently set to 11000000. The physical address (PA) field is set to the 1553 RT address of the transmitting unit. If the transmitting unit is a bus controller, then the PA is set to 00011111; in all cases the three high-order bits are set to zero (since 1553 only uses a 5-bit RT address). The first bit of the destination address (DA) specifies either logical or physical addressing modes. If physical addressing is used, the first 8 bits are the receiving RT's 1553 address followed by the high-speed subaddress (which is identical to the content of the high-speed message identity field transmitted on the low-speed bus). Logical addressing is used for broadcast mode and is defined by the system designer. The high-speed word count (WC) defines the number of 16-bit data words in the information (INFO) field. The INFO field contains the number of data words defined in the WC field. The frame check sequence (FCS) provides a check for errors in all preceding fields (similar to the CRC).

3.17.6 Universal Serial Bus

The universal serial bus (USB) is a tiered star peripheral interface network operating at 1.5 Mbps, 12 Mbps, or 480 Mbps. The system can transfer messages up to 1024 bytes in length to as many as 127 nodes; it uses non-return-to-zero invert (NRZI) encoding and is capable of isochronous and bulk data transfer. USB is intended to replace the older RS-232 networks for peripherals such as mice, keyboards, scanners, cameras, and printers, and mass storage devices such as CD-ROM, floppy, and DVD drives. There are two versions of the standard: USB 1.1 (September 1998), allowing bit rates of 1.5 and 12 Mbps, and USB 2.0 (April 2000), allowing rates of 480 Mbps.

USB is dedicated to providing communications between a host and a number of slaves; there is no direct communication among the slaves. The standard defines a tiered star, or Christmas tree topology, with a hub at the center of each star. The maximum number of allowable tiers is seven and the maximum number of hubs in a communication path between the host and any device is limited to five. A depiction of the USB topology is shown in Figure 3.34.

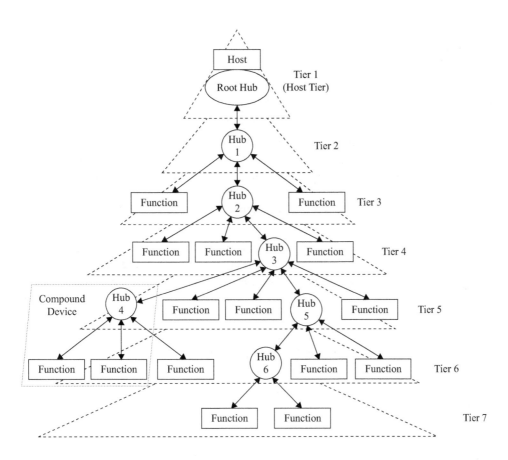

FIGURE 3.34 ■
USB Tiered Star
Topology

There may be up to 127 logical devices (hubs and peripherals) connected via the tiered star topology, with control dedicated to one USB host (normally in the host computer/PC). Plug and play of peripheral devices is managed by automatic reconfiguration of software/firmware within the host. There are three basic units within the topology:

- Host controller—responsible for centralized command and control of the network. It most typically resides within the host computer/PC.

- Hubs—router/repeater devices that provide connectivity to the peripheral devices. Every hub contains a hub controller and a hub repeater; it has one upstream port and one or more downstream ports. Messages arriving at an upstream port are broadcast on all the downstream ports and messages arriving at any downstream port are repeated on the upstream port. A hub with more than one downstream port is called a compound device.

- Peripheral devices—also called functions in the standard, and can be mice, keyboards, scanners, etc.

The data rates within USB are the low-speed peripheral interface (LSPI), at 1.5 Mbps; the full-speed peripheral interface (FSPI), at 12 Mbps; and the high-speed peripheral interface (HSPI), at 480 Mbps.

USB is a polled network in which the host controller initiates all data transfers. USB establishes a 1 msec time base, called a frame, for LSPI and FSPI; it establishes a 125 μsec time base, called a microframe, for HSPI. A frame or microframe may contain several transactions; each transfer type defines the transactions allowed within a frame or microframe for an endpoint. Data carried over the USB network is carried in a USB-defined structure; however, the standard allows device-specific structured data to be carried as well. Formatting and interpretation of the data transported during a bus transaction are functions of the client software. USB defines four transfer types to suit the various real-time requirements of the peripherals:

- Control transfers—provide configuration, command, and status communication flows between client software and its function. They are typically used for command and status information during configuration. They are composed of a setup bus transaction: sending a request from host to device, sending information in the direction indicated by the setup transaction, and waiting for status information from the device.

- Isochronous transfers—a periodic, continuous communication between the host and the device. They are typically used for time critical information that is associated with telephones, speakers, or microphones which preserves the concept of time encapsulated in the data. There are no handshakes or acknowledgments and they are supported by unidirectional flows from the host or device.

- Interrupt transfers—provide low-frequency, bounded-latency communication used by devices such as mice, keyboards, and joysticks.

- Bulk transfers—used with most common USB devices. The transmissions are large nonperiodic packets of data typically used for non-time-critical data that can wait until bandwidth becomes available.

The packet formats for USB are shown in Figure 3.35. All packets begin with a synchronization field, which is not shown in Figure 3.35; this is used by the receivers to align incoming signals with the receiving device's clock. The field is 8 bits long for LSPI and FSPI and 32 bits long for FSPI. Each packet has a packet identifier (PID) specifying the packet type and other information. The start of frame packet contains an 11-bit frame number and a 5-bit CRC covering the preceding fields. The token packet contains the PID identifying a token packet, followed by a device address field identifying the peripheral that is to respond to the token. This is followed by an endpoint field identifying the endpoint within the device that is to be the source or destination of the data to follow. The data packet starts with the PID, followed by a data field that may be from 0 to 1024 bytes long; a 16-bit CRC protects the preceding fields. The handshake packet consists of a PID field identifying it as such, and returns values indicating successful reception of data, flow, control, and stall conditions.

FIGURE 3.35 USB Packet Formats

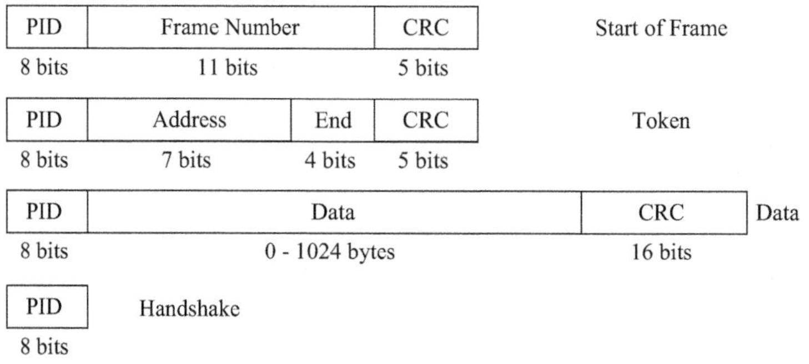

The USB host sends a start of frame packet for every frame (or microframe in the case of HSPI). The available bus bandwidth is shared between devices within the frame or microframe interval. A typical USB data transaction uses a token packet, data packet, and handshake packet. The host controller broadcasts a token packet to a specific device address. The device responds to its unique address and, depending on the contents of the direction field in the PID, will broadcast or wait to receive data. After a data packet has been transmitted or received, a handshake packet is sent by either the host or the device.

3.17.7 Enhanced Bit Rate 1553

Enhanced bit rate (EBR) 1553 is a higher-speed, 10 Mbps version of the 1553 protocol, maintaining the determinism of the basic MIL-STD-1553B system. EBR-1553 is one way to resolve the problem of newer applications requiring higher data rates on legacy systems. It is a preferred method when integrating newer systems, as opposed to rewiring (fiber optics). Another benefit is the ability to reuse software written for 1553 applications, which simplifies the integration task.

The initial topology for EBR-1553 was an active star configuration. In this configuration, a logical hub provides the communications link between the bus controller and the RTs. The active function limits responses from the RTs so the bus controller will only receive data from one RT at a time. A depiction of the EBR-1553 topology is shown in Figure 3.36.

The bus controller, through the logical hub function, is connected to a number of terminals (up to 31; 5-bit address, 11111 reserved for broadcast) with point-to-point physical connections utilizing RS-485 transceivers. These modems work much the same way as the digital subscriber line (DSL) connection works with your home PC. Hub topology is used to improve reliability, and RS-485 transceivers are robust and relatively inexpensive. It can be seen from the figure that RT–RT data transfer rates can now be increased significantly. The bus monitor ensures that the data transfer was completed successfully and monitors for errors in the transmission.

EBR-1553 can extend the terminal address range from 31 to 255 by using a supporting point-to-point CAN bus (see section 3.17.2). The CAN bus assigns RTs or stores an address at power-up. It allows for an 8-bit address that consists of a 3-bit segment address and the 5-bit RT address, effectively allowing eight segments, each containing 32 allowable RTs (32 terminals for seven segments, 31 terminals for the original avionics bus, or 255 total addresses). Figure 3.37 illustrates this expansion (not shown is the bus monitor, which resides on each segment).

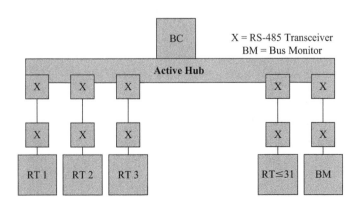

FIGURE 3.36 ■
EBR-1553 Topology

FIGURE 3.37 ▪
EBR-1553 Remote
Terminal Increase
to 255

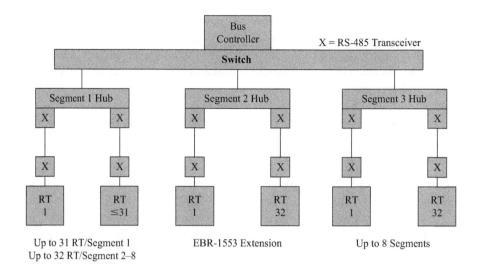

Up to 31 RT/Segment 1
Up to 32 RT/Segment 2–8

EBR-1553 Extension

Up to 8 Segments

One example where EBR-1553 data standards are used is the miniature munitions/ stores interface (MMSI). The MMSI standard uses EBR-1553 as the embedded data bus standard and defines implementation requirements for the electrical interface between the aircraft-carried dispensers (or other carrying devices) and miniature stores.

3.17.8 MIL-STD-1760D

Prior to the implementation of MIL-STD-1760, aircraft and stores were developed independently. This, of course, resulted in unique aircraft/store electrical interconnection requirements which varied from aircraft to aircraft. Unique installations were costly and each addition of a new store required external or internal modifications to the aircraft. The goal of the standard was to develop aircraft that were compatible with a wide variety of stores and stores that were compatible with a wide variety of aircraft.

MIL-STD-1760D defines a standard electrical (and optical) interconnection system for aircraft and stores. The system is based on the use of a standard connector, a standard signal set, and a standard serial digital data interface for control, monitoring, and release of stores. The standard defines implementation requirements for the aircraft/store electrical interconnection system (AEIS) in aircraft and stores. Requirements for mechanical, aerodynamic, logistic, and operational compatibility are not covered. Factors such as size, shape, loads, and clearances are not specified; therefore, full interoperability cannot be assured. Signals for emergency jettison are not included; nor does the standard attempt to establish the electromagnetic interference/electromagnetic compatibility (EMI/EMC) requirements of the aircraft or the store. It does define an interface that is generally capable of meeting the requirements.

The standard provides for a variety of stores interfaces:

- Aircraft station interface (ASI)
- Mission store interface (MSI)
- Carriage store interface (CSI)
- Routing network aircraft system port (RNASP)

These interfaces are shown in Figure 3.38.

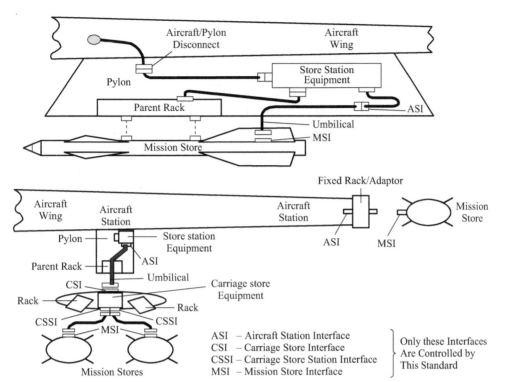

FIGURE 3.38 ▪
1760D Stores
Interfaces

ASI – Aircraft Station Interface
CSI – Carriage Store Interface
CSSI – Carriage Store Station Interface
MSI – Mission Store Interface

} Only these Interfaces Are Controlled by This Standard

Appendix B of MIL-STD-1760D specifies the requirements for message formatting, data encoding, information transfer rules, and other characteristics for utilizing a digital data bus. The data bus interface complies with the requirements of MIL-STD-1553B, Notice 4, with additional requirements. WGS-84 is the specified earth model. SAE-AS15531 may be used in lieu of MIL-STD-1553.

The protocols for MIL-STD-1760D follow the requirements contained in 1553; command, data, and status words are formatted the same, although there are more reserved fields and data dependencies. Messages are formatted the same, however, 1760 allows for mass data transfer. Mass data transfer refers to the transfer of data sets between aircraft and stores (or between stores) where the data set consists of more data words than can be transferred with only a few messages.

All command words are generated by an AEIS bus controller (normally the mission computer or stores management processor). The RT address is the terminal address of the store; broadcast command is authorized. There are 10 subaddress fields reserved in MIL-STD-1760 and they can be found in Appendix B, Table B-1, of the standard; it is reproduced here as Table 3.3. The aircraft may use 00000 or 11111 in the subaddress field to identify a mode code.

Many of the mandatory mode codes are as defined in MIL-STD-1553: transmit status word, transmitter shutdown, override transmitter shutdown, reset remote terminal, transmit vector word, synchronize with data word, and transmit last command. The aircraft shall not transmit dynamic bus control or any of the reserved mode codes. If a store receives a prohibited mode code it shall not alter the state of the store subsystem.

The status word is as defined in MIL-STD-1553; however, status word bits at bit times 10, 12, 13, 14, and 18 shall be set to logic 0. Bit time 10 may not be used to

TABLE 3.3 ■ MIL-STD-1760 Command Words Subaddress Fields

	MIL-STD-1760D APPENDIX B		
	TABLE B-I. Subaddress/mode field application		
	MESSAGE FORMATS 1/ and 2/		
SUBADDRESS FIELD	RECEIVE	TRANSMIT	DESCRIPTION
00000 (00)	B.4.1.1.3	B.4.1.1.3	MODE CODE INDICATOR
00001 (01)	B.4.2.2.6	B.4.2.2.3	STORE DESCRIPTION, AIRCRAFT DESCRIPTION
00010 (02)	USER DEFINED	USER DEFINED	
00011 (03)	USER DEFINED	USER DEFINED	
00100 (04)	USER DEFINED	USER DEFINED	
00101 (05)	USER DEFINED	USER DEFINED	
00110 (06)	USER DEFINED	USER DEFINED	
00111 (07)	B.4.1.5.9	B.4.1.5.9	DATA PEELING
01000 (08)	RESERVED	RESERVED	TEST ONLY 3/
01001 (09)	USER DEFINED	USER DEFINED	
01010 (10)	USER DEFINED	USER DEFINED	
01011 (11)	B.4.2.2.1	B.4.2.2.2	STORE CONTROL/MONITOR
01100 (12)	USER DEFINED	USER DEFINED	
01101 (13)	USER DEFINED	USER DEFINED	
01110 (14)	B.4.1.5.8	B.4.1.5.8	MASS DATA TRANSFER
01111 (15)	USER DEFINED	USER DEFINED	
10000 (16)	USER DEFINED	USER DEFINED	
10001 (17)	USER DEFINED	USER DEFINED	
10010 (18)	USER DEFINED	USER DEFINED	
10011 (19)	B.4.2.2.4	B.4.2.2.5	NUCLEAR WEAPON
10100 (20)	USER DEFINED	USER DEFINED	
10101 (21)	USER DEFINED	USER DEFINED	
10110 (22)	USER DEFINED	USER DEFINED	
10111 (23)	USER DEFINED	USER DEFINED	
11000 (24)	USER DEFINED	USER DEFINED	
11001 (25)	USER DEFINED	USER DEFINED	
11010 (26)	USER DEFINED	USER DEFINED	
11011 (27)	B.4.2.2.4	B.4.2.2.5	NUCLEAR WEAPON
11100 (28)	USER DEFINED	USER DEFINED	
11101 (29)	USER DEFINED	USER DEFINED	
11110 (30)	USER DEFINED	USER DEFINED	DATA WRAPAROUND 4/
11111 (31)	B.4.1.1.3	B.4.1.1.3	MODE CODE INDICATOR

distinguish a command from a status word. The message error and the broadcast command received are the same as 1553; service request bit, busy bit, subsystem, and terminal flag are slightly different. The service request bit is only used for request notification; once a store sets this bit to logic 1, it must ensure that a vector word is immediately available. The vector word format is contained in Appendix B, Table B-II, of the standard and is reproduced here as Table 3.4. The service request bit will be reset to logic 0 when and only when the store has received a transmit vector word mode code for the active service request and the vector word has been transmitted. If the service request bit is still set to logic 1 in a second or subsequent service request, it will be

TABLE 3.4 ▪ MIL-STD-1760 Vector Word Format

MIL-STD-1760D APPENDIX B		
TABLE B-II. Vector word (asynchronous message demand) format		
FIELD NAME	BIT NO.	DESCRIPTION
FORMAT FLAG	-00-	Shall be set to logic 0.
RESERVED 1/	-01-	RESERVED. Shall be set to logic 0.
	-02-	RESERVED. Shall be set to logic 0.
	-03-	RESERVED. Shall be set to logic 0.
	-04-	RESERVED. Shall be set to logic 0.
T/R 1/	-05-	Shall be set to a logic 1 to indicate that the requested message is a transmit command. (Logic 0 indicates a receive command request.)
SUBADDRESS 1/	-06-	MSB = 16
	-07-	
	-08-	— Bits 6 through 10 contain the subaddress of the required message.
	-09-	
	-10-	LSB = 1
WORD COUNT 1/	-11-	MSB = 16
	-12-	
	-13-	— Bits 11 through 15 contain the word count of the required message.
	-14-	
	-15-	LSB = 1

1/ The designated field definitions apply only when bit number 00 is set to logic 0. See TABLE B-III for alternate vector word format.
2/ The vector word shall be set to 0000 hexadecimal, unless the service request notification protocol (B.4.1.5.4) is in progress.

disregarded by the aircraft. Acknowledgment of the receipt of a valid vector word is not a requirement of the aircraft.

The busy bit will only be set to logic 1 when the store cannot move data to or from the store subsystem in compliance with a command, and is only set temporarily in accordance with the standard (50–500 msec). The busy bit is the only indication that a message has been discarded, and no additional notification is implemented. The subsystem flag is set by the store when there is a fault condition within the store or the store-to-terminal interface, which destroys the credibility of the data at the data bus interface. The controller will interpret this as a total loss of store function.

The data words follow the same formatting as with MIL-STD-1553, however, their sequencing follows a more rigid format than previously seen with avionics data. Actual message formatting will be discussed shortly.

There are three primary verifications performed with MIL-STD-1760D implementation:

• Verification of checksum

• Verification of message header

• In the store control message: verification of critical control 1, verification of critical control 2, and critical authority

FIGURE 3.39 ▪
Cyclical
Redundancy Check

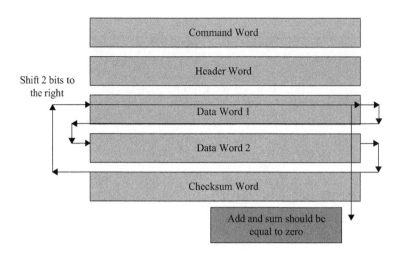

Shift 2 bits to
the right

The use of a checksum is required in the store control, store monitor, store description, transfer control, and transfer monitor messages. The checksum occupies the last word position of both transmit and receive messages. Each data word is rotated right cyclically by a number of bits equal to the number of preceding data words in the message. The resultant rotated data words are summed to each bit with the checksum and the sum shall be equal to zero (bits are not carried). If the checksum fails, the message is discarded; this is also known as the CRC. A graphical representation of the CRC is shown in Figure 3.39.

The first word of each message is the header word for message identification. The format follows the requirements set forth in the standard and is reprinted as Table 3.5. If the header word is missing or incorrectly formatted, the message will be discarded.

In MIL-STD-1760D, the validity of the data words is verified with each store control/store monitor (subaddress 11) transmission. This is accomplished by the use of an invalidity word. Each bit of the invalidity word corresponds to a data word; a logic 1 set in any of the bits indicates that the corresponding data word is invalid. Bits 0 through 15 represent data words 1 through 16. Invalidity word 2 corresponds to data words 17 through 32. An example of invalidity word format is contained in Appendix B, Table XXXI, of the standard, and is reproduced here as Table 3.6.

For a critical control 1 or critical control 2 word not marked invalid (in the invalidity word), both the identifier and address control fields must be correct. For the critical authority word, the coded check field must be correct. As with the header word and CRC discussed earlier, any errors in the critical control or authority will result in the message being discarded; all protocol failures are reported in the store monitor standard message. Examples of the critical word and critical authority are reproduced in Tables 3.7 and 3.8, respectively. These checks are all required to ensure that the correct store is selected and the correct information is received by the store.

In addition to normal transfer of information as defined in MIL-STD-1553B, MIL-STD-1760 allows for specialized transfer and associated messages:

- Store control
- Store monitor
- Store descriptions
- Nuclear weapons control

TABLE 3.5 ■ MIL-STD-1760 Header Word Format

MIL-STD-1760D APPENDIX B	
TABLE B-X. Header word	
HEADER (HEXADECIMAL)	**APPLICATION**
0000 THROUGH 03FF	Used selected
0400	Store control message (B.4.2.2.1)
0401 THROUGH 041F	Reserved for store safety critical control and monitor
0420	Store monitor message (B.4.2.2.2)
0421	Store description message (B.4.2.2.3)
0422	Transfer control message (B.4.2.3.2.1)
0423	Transfer control message (B.4.2.3.2.2)
0424 THROUGH 042D	Reserved for carriage store non-safety critical control and monitor
042E THROUGH 04FF	Reserved for future non-safety critical control and monitor
0500 THROUGH FFFF	User selected

- Nuclear weapons monitor
- Aircraft description
- Mass data transfer

Standard messages follow MIL-STD-1553 format but contain a header word as the first data word; if checksum is used it will occupy the last data word position. The store control message is used for controlling the state of the stores and specifically standardizes the format for safety critical commands. The store control message is a 30 data word receive command with a subaddress of 11 (01011). A description of the store control message is found in Appendix B, Table B-XI, of the standard, and is reproduced here as Table 3.9.

Note the two invalidity words (addressed earlier) that occupy the second and third data words of the message.

The store monitor message is used as a status message to reflect the safety critical condition of the store; the message also includes other nonsafety critical store information. The store monitor message is a 30 word transmit message with subaddress 11 (01011). The format of the store monitor message in contained in Appendix B, Table B-XII, and is reproduced here as Table 3.10.

TABLE 3.6 ■ MIL-STD-1760 Invalidity Word Format

MIL-STD-1760D APPENDIX B		
TABLE B-XXXI. Invalidity word		
FIELD NAME	BIT NUMBER	DESCRIPTION
INVALIDITY	-00-	INVALIDITY OF WORD 1
	-01-	INVALIDITY OF WORD 2
	-02-	INVALIDITY OF WORD 3
	-03-	INVALIDITY OF WORD 4
	-04-	INVALIDITY OF WORD 5
	-05-	INVALIDITY OF WORD 6
	-06-	INVALIDITY OF WORD 7
	-07-	INVALIDITY OF WORD 8
	-08-	INVALIDITY OF WORD 9
	-09-	INVALIDITY OF WORD 10
	-10-	INVALIDITY OF WORD 11
	-11-	INVALIDITY OF WORD 12
	-12-	INVALIDITY OF WORD 13
	-13-	INVALIDITY OF WORD 14
	-14-	INVALIDITY OF WORD 15
	-15-	INVALIDITY OF WORD 16

1/ Invalidity bit set to logic 1 shall indicate that a word is invalid.

2/ For the standard message to/from subaddress 11, the invalidity bits associated with the reserved words and those associated with words that are defined in the system specification or ICD as not used, shall be set to logic 0 (valid).

3/ For the standard message to/from subaddress 11, bits 00 through 15 in the invalidity word shall indicate invalidity of words 1 through 16 in the message and bits 00 through 15 in a second invalidity word shall indicate the invalidity of words 17 through 32 in the message. Bits 14 and 15 in the second word shall only be used during the routing of message to/from a mission store carried on a carriage store and shall be set to logic 0 at all other times.

4/ For user defined messages, utilization and setting of the invalidity bits shall be as defined in the system specification or ICD.

Receive messages with subaddresses 19 and 27 are only used for control of nuclear weapons; transmit messages with subaddresses 19 or 27 are only used to monitor nuclear weapons. The information regarding message content is contained in Aircraft Monitor and Control Project Officers Group System 2 Specification (SYS 2001-01), and is not reprinted here.

The aircraft description message transfers aircraft identity to the store via subaddress 1 (00001). The message includes a header word, invalidity words, country code, aircraft identification words, station number, pylon/bay number, and checksum.

Mass data transfer is initiated by the use of three data message types:

- Transfer control message
- Transfer monitor message
- Transfer data message

The transfer control message is a standard receive message for controlling the transfer of operating modes and the designation of files, records, and block numbers. Each selected file is divided into 1 to 255 consecutive records, with each record further divided into 1 to 255 consecutive blocks. Each block is transferred with one transfer data

TABLE 3.7 ■ MIL-STD-1760 Critical Control 2 Format

MIL-STD-01760D APPENDIX B		
TABLE B-XXXIII. Critical control 2		
FIELD NAME	**BIT NUMBER**	**DESCRIPTION**
STORE CONTROL 1/	-00-	D_{10} = Erase command/authority
	-01-	D_9 = RF jam command/authority
	-02-	D_8 = RF emission activate command/authority
	-03-	D_7 = RESERVED. Shall be set to logic 0.
	-04-	D_6 = RESERVED. Shall be set to logic 0.
	-05-	D_5 = RESERVED. Shall be set to logic 0.
	-06-	D_4 = RESERVED. Shall be set to logic 0.
	-07-	D_3 = RESERVED. Shall be set to logic 0.
IDENTIFIER	-08-	D_2 ⎤
	-09-	D_1 ⎬ — see $\overline{2}$/
	-10-	D_0 ⎦
ADDRESS CONFIRM	-11-	A_4 ⎤
	-12-	A_3
	-13-	A_2 ⎬ — Shall be set to match the logic state of the corresponding interface address discrete lines A4 through A0 as specified in 5.1.6 and 5.2.6
	-14-	A_1
	-15-	A_0 ⎦

1/ Data bits set to a logic 0 shall indicate that the associated function is required to be inactive. Data bits set to a logic 1 shall indicate that the associated function is required to be active. Data bits reset to a logic 0 shall indicate that the associated function is required to be deactivated as applicable.

2/ The IDENTIFIER FIELD shall be set as indicated below.

D_2	D_1	D_0	
0	0	0	RESERVED
0	0	1	Mission Store
0	1	0	Carriage Store
0	1	1	RESERVED
	thru		
1	1	1	RESERVED

3/ Stores shall discard any message found to contain a critical control word that fails one or more of the protocol checks in paragraph B.4.1.4. Stores shall only enable safety critical processes demanded by critical control words which pass the protocol checks detailed in B.4.1.4.

message and contains 30 data words, with the first being the record/block number and the remaining 29 representing file data. The transfer control message is sent as a receive message, subaddress 14 (01110). The format is shown in Table 3.11.

Word 2 of the transfer control message is the instruction word, and it is very similar to a mode code, as it tells the receiving RT what operation it is supposed to initiate. The format of the instruction word is contained in Appendix B, Table XVII, of MIL-STD-1760, and is reproduced here as Table 3.12.

The transfer monitor message is a standard format message for monitoring the status of the mass data transfer operations in the store. The transfer monitor message is initiated

TABLE 3.8 ■ MIL-STD-1760 Critical Authority Format

MIL-STD-1760D APPENDIX B		
TABLE B-XXXIV. Critical authority		
FIELD NAME	BIT NUMBER	DESCRIPTION
CODED CHECK	-00-	$C_{14} = D_{10} + D_9 + D_6 + D_1 + D_0$
	-01-	$C_{13} = D_9 + D_8 + D_5 + D_0$
	-02-	$C_{12} = D_8 + D_7 + D_4$
	-03-	$C_{11} = D_7 + D_6 + D_3$
	-04-	$C_{10} = D_{10} + D_9 + D_5 + D_2 + D_1 + D_0$
	-05-	$C_9 = D_{10} + D_8 + D_6 + D_4$
	-06-	$C_8 = D_{10} + D_7 + D_6 + D_5 + D_3 + D_1 + D_0$
	-07-	$C_7 = D_{10} + D_5 + D_4 + D_2 + D_1$
	-08-	$C_6 = D_{10} + D_6 + D_4 + D_3$
	-09-	$C_5 = D_9 + D_5 + D_3 + D_2$
	-10-	$C_4 = D_{10} + D_9 + D_8 + D_6 + D_4 + D_2 + D_0$
	-11-	$C_3 = D_9 + D_8 + D_7 + D_5 + D_3 + D_1$
	-12-	$C_2 = D_{10} + D_9 + D_8 + D_7 + D_4 + D_2 + D_1$
	-13-	$C_1 = D_{10} + D_8 + D_7 + D_3$
	-14-	$C_0 = D_{10} + D_7 + D_2 + D_1 + D_0$
RESERVED	-15-	RESERVED. Shall be set to logic 0

1/ Coded check bits shall be generated using modulo 2 arithmetic.
2/ D_0 through D_{10} refer to bits D_0 through D_{10} as defined in TABLE B-XXXII and TABLE-B-XXXIII, as applicable.
3/ The coded check bits are based on the BCH 31, 16, 3 polynomial:

$$X^{15} + X^{11} + X^{10} + X^9 + X^8 + X^7 + X^5 + X^3 + X^2 + X + 1$$

as a transmit command, subaddress 14 (01110). The format of the transfer monitor message is found in Appendix B, Table B-XXII, of the standard, and is reproduced here as Table 3.13. Word 3 of the reply from the receiving terminal contains the transfer mode status of the mass data transfer. This indicates to the controller where the RT is in the transfer operation. The format of the transfer mode status word is found in Appendix B, Table B-XXIII, of the standard, and is reproduced here as Table 3.14.

The protocols in MIL-STD-1760 are predicated on MIL-STD-1553B with some variations and enhanced integrity checks. The verification of the data is more stringent in stores applications, as safety is a primary concern. As with MIL-STD-1553, the data transmissions are deterministic, using a command/response method of data transfer.

3.17.9 Fiber Channel

The fiber channel (FC) is a serial, high-throughput, low-latency packet switched or connection-oriented network technology that can operate at 1, 2, or 10 Gbps. It refers to a set of standards developed by the International Committee for Information Technology Standards (INCITS), who are accredited by the American National Standards Institute (ANSI). It was originally developed to handle data storage, providing a means for high-speed data transfer over a serial link between computers and peripherals. Its original use was for high-performance disk/tape interfaces. As development continued, it was

TABLE 3.9 ■ MIL-STD-1760 Store Control Message

MIL-STD-1760D APPENDIX B		
TABLE B-XI. Store control (BC-RT transfer) 1/		
WORD NO.	**DESCRIPTION/COMMENT**	**PARAGRAPH or TABLE**
-CW-	COMMAND WORD (Subaddress 01011 Binary)	B.4.1.1
-01-	HEADER (0400 hexadecimal)	B.4.2.1.1
-02-	Invalidity for words 01-16	TABLE B-XXVI line 2
-03-	Invalidity for word 17-30	TABLE B-XXVI line 2
-04-	Control of critical state of store -	TABLE B-XXVI line 3
-05-	Set 1 with critical authority	TABLE B-XXVI line 5
-06-	Control of critical state of store -	TABLE B-XXVI line 4
-07-	Set 2 with critical authority	TABLE B-XXVI line 5
-08-	Fuzing mode 1	TABLE B-XXVI line 8
-09-	Arm delay from release	TABLE B-XXVI line 12
-10-	Fuze function delay from release	TABLE B-XXVI line 13
-11-	Fuze function delay from impact	TABLE B-XXVI line 14
-12-	Fuze function distance	TABLE B-XXVI line 18
-13-	Fire interval	TABLE B-XXVI line 20
-14-	Number to fire	TABLE B-XXVI line 21
-15-	High drag arm time	TABLE B-XXVI line 16
-16-	Function time from event	TABLE B-XXVI line 17
-17-	Void/layer number	TABLE B-XXVI line 23
-18-	Impact velocity	TABLE B-XXVI line 24
-19-	Fuzing mode 2	TABLE B-XXVI line 9
-20-	Dispersion data	TABLE B-XXVI line 165
-21-	Duration of dispersion	TABLE B-XXVI line 166
-22-	Carriage Store S&RE Unit(s) Select 2/	TABLE B-XXVI line 167
-23-	Separation elements	TABLE B-XXVI line 168 or 169
-24-	Surface delays	TABLE B-XXVI line 170 or 171
-25-		
-26-		
-27-	⎤ — Reserved data words (0000 hexadecimal)	TABLE B-XXVI line 1
-28-		
-29-	⎦	
-30-	Checksum word	B.4.1.5.2
-SW-	STATUS WORD	B.4.1.2

determined that FC could be used as a networking technology as well as a data channel. When used for aircraft, and specifically avionics, the FC standard that is used is the fiber channel–avionics environment anonymous subscriber messaging protocol (FC-AE-ASM).

The FC standards define the topologies, services, and interface details that make up the network. The basic topologies are:

- Point-to-point: used to provide a dedicated bandwidth between two processors, or between a computer and a sensor/display.

- Arbitrated loop: a ring topology that can connect up to 126 nodes on a shared bandwidth network. Each node can arbitrate to obtain control of the loop, establish a

TABLE 3.10 ■ MIL-STD-1760 Store Monitor Message

MIL-STD-1760D APPENDIX B		
TABLE B-XII. Store monitor (RT-BC transfer) 1/		
WORD NO.	**DESCRIPTION/COMMENT**	**PARAGRAPH or TABLE**
-CW-	COMMAND WORD (Subaddres 01011 Binary)	B.4.1.1
-SW-	STATUS WORD	B.4.1.2
-01-	HEADER (0420 hexadecimal)	B.4.2.1.1
-02-	Invalidity for word 01-16	TABLE B-XXVI line 2
-03-	Invalidity for word 17-30	TABLE B-XXVI line 2
-04-	Critical monitor 1	TABLE B-XXVI line 6
-05-	Critical monitor 2	TABLE B-XXVI line 7
-06-	Fuzing/arming mode status 1	TABLE B-XXVI line 10
-07-	Protocol status	TABLE B-XXVI line 25
-08-	Monitor of arm delay from release	TABLE B-XXVI line 12
-09-	Monitor of fuze function delay from release	TABLE B-XXVI line 13
-10-	Monitor of fuze function delay from impact	TABLE B-XXVI line 14
-11-	Monitor of fuze function distance	TABLE B-XXVI line 18
-12-	Monitor of fire interval	TABLE B-XXVI line 20
-13-	Monitor of number to fire	TABLE B-XXVI line 21
-14-	Monitor of high drag arm time	TABLE B-XXVI line 16
-15-	Monitor of function time from event	TABLE B-XXVI line 17
-16-	Monitor of void/layer number	TABLE B-XXVI line 23
-17-	Monitor of impact velocity	TABLE B-XXVI line 24
-18-	Fuzing/arming mode status 2	TABLE B-XXVI line 11
-19-	Monitor of dispersion data	TABLE B-XXVI line 165
-20-	Monitor of dispersion duration	TABLE B-XXVI line 166
-21-	Monitor of carriage store S&RE Unit(s) select	TABLE B-XXVI line 167
-22-	Monitor of separation elements	TABLE B-XXVI line 168 or 169
-23-	Monitor of surface delays	TABLE B-XXVI line 170 or 171
-24-		
-25-		
-26-		
-27-	⎤ — Reserved words (0000 hexadecimal)	TABLE B-XXVI line 1
-28-		
-29-		
-30-	Checksum word	B.4.1.5.2

1/ The message format shown is for RT-BC transfers. The data entities and entity sequence for word numbers 01 through 30 may also be applied to RT-RT transfers provided that the receiving RT is not an AEIS store.

logical connection with another node, and originate traffic. Arbitrated loops do not include any fault tolerance features to handle node or media failure.

- Hubbed loop: a variation of the arbitrated loop in which the connections between adjacent nodes on the loop are made by a central hub. This allows failed or unpowered nodes to be bypassed and a centralized maintenance and management point to be employed.

- Switched fabric: a network capable of supporting multiple, simultaneous full-bandwidth transfers by utilizing a "fabric" of one or more interconnected switches;

TABLE 3.11 ▪ MIL-STD-1760 Transfer Control Message Format

MIL-STD-1760D APPENDIX B		
TABLE B-XVII. Transfer Control (TC) message format		
WORD NO.	DESCRIPTION/COMMENT	PARAGRAPH or TABLE
-CW-	COMMAND WORD (Subaddress 01110 binary)	B.4.1.1
-01-	HEADER (0422 hexadecimal)	B.4.2.1.1
-02-	Instruction	B.4.2.3.2.1.2
-03-	Subaddress select	B.4.2.3.2.1.3
-04-	File number	B.4.2.3.2.1.4
-05-	Record number	B.4.2.3.2.1.5
-06-	Block number	B.4.2.3.2.1.6
-07-	Filerecord checksum	B.4.2.3.2.1.7
-08-	Checksum word	B.4.2.3.2.1.8
-SW-	STATUS WORD	B.4.1.2

TABLE 3.12 ▪ MIL-STD-1760 Transfer Control Message Instruction Word

MIL-STD-1760D APPENDIX B		
TABLE B-XVIII. Instruction word		
FIELD NAME	BIT NUMBER	DESCRIPTION
INSTRUCTION TYPE	-00-	No operation - commands the store to update the TM message with current status of the mass data transfer transactions
	-01-	Select download mode - commands the store to enter (or remain in) the download mass data transfer mode.
	-02-	Select upload mode - commands the store to enter (or remain in) the upload mass data transfer mode.
	-03-	Start new file/record - commands the store to prepare for receiving or transmitting, as applicable. Transfer Data (TD) messages.
	-04-	Erase all files - commands the store to erase data in all store contained memory addresses allocated to mass data transfer storage.
	-05-	Erase designated file - commands the store to erase the designated file.
	-06-	Erase designated record - commands the store to erase the designated record.
	-07-	Select echo mode - commands the store to enter (or remain in) the TD echo mode.
	-08-	Calculate file checksum - commands the store to run the file checksum test.
	-09-	Calculate record checksum - commands the store to run the record checksum test.
	-10-	System start - system start command to the store.
	-11-	Exit transfer mode - commands the store to exit the mass data transfer mode.
	-12-	Select block checksum mode - commands the store to interpret word 30 of the TD message as a message checksum (See B.4.1.5.2.1) if in download mode, or to supply word 30 of the TD message as a message checksum if in the upload mode.
RESERVED	-13-	Shall be set to a Logic 0.
	-14-	Shall be set to a Logic 0.
	-15-	Shall be set to a Logic 0.

TABLE 3.13 ▪ MIL-STD-1760 Transfer Monitor Message Format

MIL-STD-1760D APPENDIX B		
TABLE B-XXII. Transfer Monitor (TM) message format		
WORD NO.	DESCRIPTION/COMMENT	PARAGRAPH
-CW-	COMMAND WORD (Subaddress 01110 binary)	B.4.1.1
-SW-	STATUS WORD	B.4.1.2
-01-	HEADER (0423 hexadecimal)	B.4.2.1.1
-02-	Last received instruction	B.4.2.3.2.2.2
-03-	Transfer mode status	B.4.2.3.2.2.3
-04-	Current selected subaddress	B.4.2.3.2.2.4
-05-	Current file number	B.4.2.3.2.2.5
-06-	Current record number	B.4.2.3.2.2.6
-07-	Current block number	B.4.2.3.2.2.7
-08-	Current file/record checksum	B.4.2.3.2.2.8
-09-	Checksum word position	B.4.2.3.2.2.9

TABLE 3.14 ▪ MIL-STD-1760 Transfer Monitor Message Transfer Status Word

MIL-STD-1760D APPENDIX B		
TABLE B-XXIII. Transfer mode status		
FIELD NAME	BIT NUMBER	DESCRIPTION
MODE STATUS	-00-	In download mode - indicates the store is in download mode.
	-01-	In upload mode - indicates the store is in upload mode.
	-02-	Transfer enabled - indicates the store is ready for transfer of TD messages
	-04-	Erase completed - indicates the store has completed the commanded erase operation.
	-05-	Echo mode selected - indicates echo mode is enabled in the store.
	-06-	Checksum calculation in progress - indicates the store is executing the commanded checksum calculation.
	-07-	Checksum calculation completed - indicates the store has completed the commanded checksum calculation.
	-08-	Checksum failed - indicates the commanded checksum calculation failed.
	-09-	Execution started - indicates the store has initiated execution at the commanded location or the mission store has loaded the location in the TM message that the aircraft is to initiate execution.
	-10-	Exit in progress - indicates the store is exiting the MDT mode.
	-11-	Retransmission request - indicates the store request for retransmission of limited TD data
RESERVED	-12-	Shall be set to a Logic 0.
	-13-	Shall be set to a Logic 0.
	-14-	Shall be set to a Logic 0.
	-15-	Shall be set to a Logic 0.

this provides the highest level of performance. A hybrid topology can also be created by attaching one or more arbitrated loops to the fabric, making it possible to achieve wide connectivity among a group of low-bandwidth nodes.

To support these topologies, four basic port types are defined:

- Node port (N_Port): provides the mechanisms to transport data to and from another port; used in point-to-point and switched topologies.
- Fabric port (F_Port): provides the access point for an N_Port to the switched fabric.
- Loop port (NL_Port): an N_Port with added functionality to provide arbitrary loop functions and protocols.
- Expansion port (E_Port): a port on a switch used to interconnect multiple switches within the fabric.

Figure 3.40 shows the relationship of port types to their respective topologies.

The fabric is responsible for the routing of data between two endpoints, with the connected devices unaware of the routing mechanism or the internal structure of the fabric. The fabric does not passively pass data; the F_Ports perform defined functions and may restrict the communication models and classes of service provided. Before any data transfer occurs, the N_Port must log in to the F_Port and other N_Ports to determine the topology and the relevant capabilities that may affect transfer parameters, as well as to undertake node address assignment. In a loop, all NL_Ports must also initialize before they can log in with each other.

A number of server-like functions are defined within the generic services standards that are used to manage the network and provide functions such as a logic server, fabric controller, and name and time server. Some of the services can reside within an N_Port, although most would reside within a central switch or be distributed across the fabric to reduce the complexity of each node. Each service is assigned a unique, well-known address.

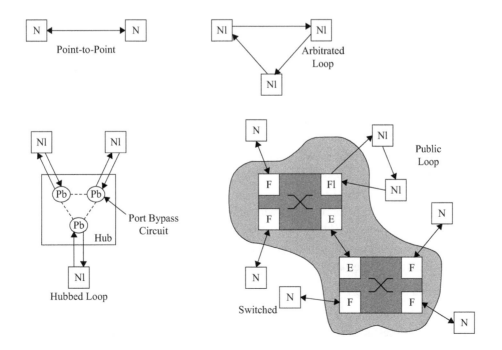

FIGURE 3.40 ▪
Fiber Channel
Port/Topology
Relationships

FIGURE 3.41 ▪
Frame, Sequence,
and Exchange
Structure

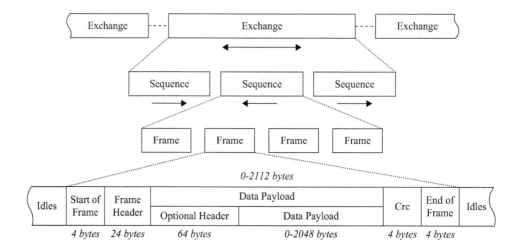

An 8-bit/10-bit encoding scheme is used in FC, which defines 256 data (D) codes, 12 control (K) codes, and 11 reserved codes. Also defined are ordered sets of four words used for low-level link functions such as frame delimiters and port event indications (idle, receiver ready, arbitration request, etc.).

Fiber channel is optimized for large block transfers and includes segmentation and reassembly activities in order to transfer large blocks within the restricted frame data field. These transfers are defined in terms of frames, sequences, and exchanges, with the lowest granularity of messages being frames. Individual frames may contain up to 2112 bytes, and larger data transfers are sent by means of sequences that consist of multiple frames traveling in the same direction. An exchange is usually bidirectional and comprises one or more nonconcurrent sequences. The frame, sequence, and exchange structure is shown in Figure 3.41.

In order to accommodate the requirements of different applications and data types, five classes of connection for data transfers are defined to specify the delivery characteristic for the sequence of frames:

- Class 1: a connection-oriented circuit-switched class of service that establishes a full bandwidth path between two N_Ports through a fabric or across a point-to-point link.

- Class 2: a connectionless packet-oriented service that can be used on any of the basic topologies; it provides confirmation of delivery.

- Class 3: a connectionless packet-oriented service that can be used on any of the basic topologies; it provides unconfirmed delivery service.

- Class 4: a fractional bandwidth connection-oriented service that establishes a virtual connection between pairs of nodes with bandwidth resources allocated to meet specified quality of service (QoS) parameters (delivery confirmation).

- Class 6: a connection-oriented service similar to class 1, but with a multicast facility implemented with the fabric acting as a multicast server.

Fiber channel also provides three forms of multicast/broadcast:

- For an arbitrated loop, the open replicate function provides a broadcast service by allowing the messages to be repeated and stored at all ports on the loop, while the selective replicate function provides a multicast service by enabling a message to be repeated to all nodes on the loop, but only stored by a selected subset of ports.

- For a switched fabric, a class 3 "unreliable" multicast can be used (class 3 does not guarantee delivery).

- A switched fabric can use a class 6 multicast, which uses a multicast server hosted within the fabric to provide confirmation of message delivery.

The performance of an FC network is dependent on network topology and the class of service, as well as basic link rate. The performance may be further degraded by the limitations in bandwidth or latency caused by the fabric switch elements.

Commercially FC is being used to connect servers and storage systems. In the military world, FC has been used in avionics upgrades on the AWACS Extend Sentry, B-1, F/A-18, and Joint Strike Fighter. In the U.S. Navy's F/A-18 tactical area moving map capability (TAMMAC), the connection between the program load device (PLD) and the map generator is FC. For the B-1B mission computer upgrade, FC was used to connect the mission computers and mass storage units, while on the AWACS, FC was used to connect the operator workstations to the main radar distribution system. For the F-35, FC provides the communications between the core processor and sensors; communications, navigation, and identification (CNI), and displays.

3.17.10 Ethernet

Ethernet has evolved since 1980 and is encompassed by the IEEE 802 family of standards. Ethernet can operate at bit rates of 1, 10, 100, or 1000 Mbps or 10 Gbps and uses carrier sense multiple access/collision detection (CSMA/CD) media access (not including full-duplex or switched Ethernet, which is covered in a subsequent section). Several media access control (MAC) standards are defined for a variety of physical media. A logical link control (LLC) standard, a secure data exchange standard, and MAC bridging standards are intended to be used in conjunction with the MAC standards.

All versions of Ethernet use a CSMA/CD media access mechanism with the exception of those using a full-duplex mode of operation. A station connected to a common medium with other stations and needing to transmit data listens until there are no transmissions on the medium (carrier sense) and then starts to transmit in an Ethernet frame; other stations also have the right to transmit (multiple access). While a station transmits, it continuously monitors signals on the media, and if it detects that another station is also transmitting (collision detection), it immediately ceases transmission and waits for a random interval based on a back-off algorithm before it starts transmitting again. On a heavily loaded network, multiple collisions may occur. If repeated collisions occur, then the stations involved begin expanding the set of potential random back-off times in a process known as "truncated binary exponential back-off." In the extreme event of 16 consecutive collisions, the station MAC discards the Ethernet packet it is attempting to send. The higher-level protocols are responsible for ensuring that the data are eventually delivered to the destination station by establishing a reliable data transport service using acknowledgment mechanisms.

The CMSA/CD protocol is half-duplex, so that a station may transmit or receive data, but not simultaneously. The standard defines a full-duplex system without CMSA/CD which allows two stations to transmit and receive data simultaneously over independent transmission paths, thus doubling the aggregate data rate of the link.

FIGURE 3.42 ■
Ethernet Frame:
Fields and Number
of Bytes Per Field

Preamble 7*	SFD 1	DA 2 or 6	SA 2 or 6	L/T 2	Data 0–1500	PAD	FCS 4

*Bytes

The Ethernet frame used to deliver MAC entities is shown in Figure 3.42.

The preamble field consists of 56 bits with alternate "1" and "0" values and allows the receiving station to detect and synchronize on the signal.

The start frame delimiter (SFD) contains a single byte value (10101011) indicating the start of a frame.

The destination address (DA) identifies a unique station to receive the frame, or a broadcast or multicast address (all 1s). The standard allows this field to contain 2 or 6 bytes (in most applications 6 bytes are used).

The source address (SA) identifies the station that originated the frame. As with the DA field, the standard allows 2 or 6 bytes, with most applications using 6 bytes.

NOTE: The IEEE administers the assignment of addresses by assigning 24-bit identifiers, known as organizationally unique identifiers (OUIs), to every organization manufacturing Ethernet interfaces for use as the first half of the station address. A unique 48-bit address is preassigned to each Ethernet interface when it is manufactured, eliminating the need for user administration of addresses. The 48-bit address is variously known as the hardware address, the physical address, or the MAC address.

The length type (L/T) field indicates the number of bytes in the data field or the MAC client protocol type. If the value of the two bytes in this field is less than or equal to 1500, it indicates the number of data bytes. If it is greater than or equal to 1536, it indicates the MAC client protocol type.

The data field carries the data to be transferred between stations, and may contain up to 1500 bytes.

The Ethernet frame, from destination MAC address field to the frame check sequence (inclusive) has a minimum size of 64 bytes. If the size of the frame is less than the minimum size, then a padded field (PAD) is added of sufficient size to make the frame size equal to the 64-byte minimum.

The frame check sequence (FCS) contains a 4-byte CRC value for error checking. The CRC covers all the fields from the destination address to the PAD.

There have been some adjustments to the Ethernet frame based on newer technologies. For virtual local area networks (VLANs) (a defined subset of a LAN), a 4-byte identifier is added between the source address and the L/T fields. This identifier is used to identify which VLAN the frame belongs to; it increases the maximum length of the Ethernet frame to 1522 bytes.

Gigabit Ethernet added an extension field to the end of the frame to ensure that frames are of sufficient duration for collisions to propagate to all stations in the network. The extension field is added and made long enough to ensure that the frame contains at least 512 bytes from the destination address field to the extension field.

Gigabit Ethernet also adds a burst mode of operation that allows a station to transmit a series of frames without relinquishing control of the transmission medium in order to improve performance when transmitting short frames. After successfully transmitting its first frame, a station may transmit additional frames until 65,536 bit times have been transmitted. Interframe gap periods are inserted between the frames and the transmitting station fills these with nondata symbols to maintain an active carrier.

If required, the first frame of a burst will contain an extension field; the remaining fields do not require them.

With the exception of the early standards that used a bus topology, all Ethernet standards use a star topology with a repeater hub at the center. The repeater hub includes a medium attachment unit (MAU), which is essentially a transceiver at the hub end of each segment. The repeater normally does not include MAC or LLC, but retransmits signals received on one segment bit by bit onto the other segments; a repeater may implement a full Ethernet interface for network management purposes. If a collision occurs, the repeater hub propagates the collision by transmitting a jam signal to all segments. The Ethernet bus topology is shown in Figure 3.43.

The repeater hub shown in Figure 3.43 may be replaced with a device known as a switch to implement a switched Ethernet LAN. The switched Ethernet LAN operates without CSMA/CD. The switch examines the destination address in each incoming Ethernet frame and routes it (switches) to the appropriate segment (unlike the repeater, which simply repeats each frame to all segments). A switch may also provide a degree of flow control by temporarily buffering frames. There are three types of switches: cut-through, store-and-forward, and error-free cut-through. There is also a variation on the store-and-forward switch known as a back-pressure switch.

The cut-through switch examines only the destination address in the frame, then consults an address lookup table to determine the port/segment to which it should be forwarded.

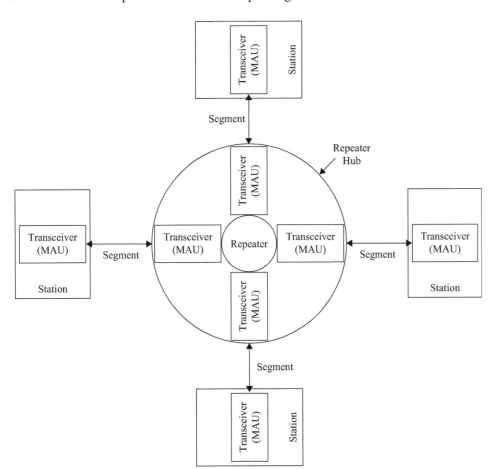

FIGURE 3.43 ▪
Ethernet Star
Topology

It then forwards the frame to the appropriate segment without reading the rest of the frame content. This process results in a minimum delay of packet throughput. Since it does not examine the frame check field, it may forward corrupted data without knowing it. This may result in increased network traffic due to retransmission requests from the destination node.

The store-and-forward switch reads the whole incoming frame into a buffer and checks the frame check field. If the CRC is correct, it consults the address lookup table to determine the port/segment to which the frame should be forwarded. This process is slower than a cut-through switch, but it does not forward corrupted data.

The error-free cut-through switch forwards the frames immediately to the destination port/segment; however, these frames are also read into a buffer and CRC checked. If errors are detected, the ports at which the bad frames arrive are reconfigured to store-and-forward switching. If the errors diminish to preset thresholds, the ports are set back to cut-through switching for higher-performance throughput.

The back-pressure switch is a variation of the store-and-forward switch. This switch overcomes the problem of overflowing buffers in store-and-forward switches by sending the overflow packets back to their source. This effectively slows the workstation transmission rate and thus the arrival of new packets from the source.

3.17.11 IEEE 1394 (FireWire)

IEEE 1394 is a low-cost digital interface intended to integrate entertainment, communication, and computing electronics into consumer multimedia. The standard was conceived in 1986 by Apple Computer Inc. under the trademark name of "FireWire." FireWire was adopted as the IEEE 1394 standard in 1995.

FireWire allows bit rates of 100, 200, and 400 Mbps over cable medium and from 25 to 50 Mbps for backplane operations. The standard allows for asynchronous and isochronous data transfer with message lengths of 512 bytes at 100 Mbps, 1024 bytes at 200 Mbps, and 20,148 bytes at 400 Mbps (asynchronous), or isochronous format data payload of up to 9600 bytes. The standard uses a media access of request/response with physical layer arbitration over a daisy-chained bus and tree topology.

Figure 3.44 shows an example of IEEE 1394 topology made up of point-to-point connections between nodes with multiple ports. Signals received on one port of a node are retransmitted on all other ports so that a transmission from one node is propagated to all others. Node devices have multiple connectors (the standard allows from 1 to 27 per device),

FIGURE 3.44 ▪
IEEE 1394 Topology

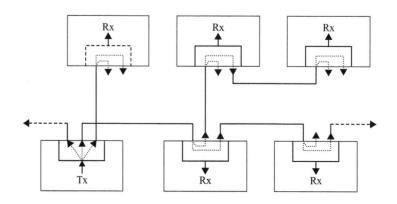

allowing the daisy-chain and tree topologies. Each IEEE 1394 bus segment may have up to 63 devices attached to it. In addition to the cable specification, FireWire includes a backplane specification that can extend the serial bus internally to a device.

IEEE 1394 allows for the use of "plug-and-play" devices; this feature allows devices to be added or removed from the bus with the bus in full operation. Upon altering the bus configuration, topology changes are automatically recognized. This feature eliminates the need for address switches or other user intervention to reconfigure the network. The process of responding to the alteration of a network configuration is known as cable configuration. Cable configuration is accomplished in three phases: bus initialize, tree identify, and self-identity.

Bus initialize occurs when a device is added or removed from the bus or at power-up. A bus reset forces all nodes into a state that clears all topology information and starts the next phase of the process. At this point, the only information known by the node is whether it is a leaf (only one node connected to it), a branch (more than one other node connected to it), or unconnected. Figures 3.45 and 3.46, respectively, show an example of a network after bus initialize and after cable configuration.

The tree identity phase translates the topology into a tree with one node designated as the root and every other connected port in the network designated as either a parent (connected to a parent node [closer to the root]) or a child (connected to a child node [further from the root]). Any unconnected ports are labeled "off" and do not participate further.

The self-identity phase gives each node an opportunity to select a unique physical ID (node number) and identify itself to any management entity attached to the bus. Self-identity uses a deterministic selection process in which the root node passes control of the media to the node attached to its lowest numbered port and waits for that node to send an "identification done" which indicates that all of its children nodes have identified themselves. The root then passes control to its next highest port and repeats the process.

Media access is by means of physical layer arbitration. Referring to Figure 3.46, if node E and node A both require network access, each seeks control of the network by

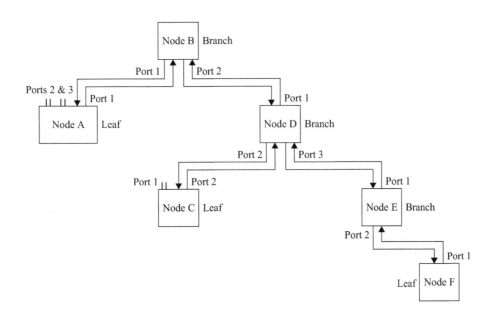

FIGURE 3.45 ■
IEEE 1394 Network
After Bus Initialize

FIGURE 3.46 ■
IEEE 1394 Cable
Configuration
Completion

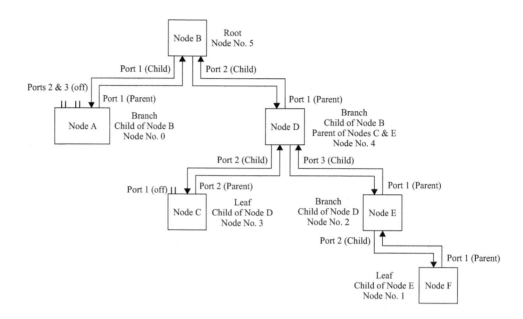

FIGURE 3.47 ■
IEEE 1394
Asynchronous
Subaction

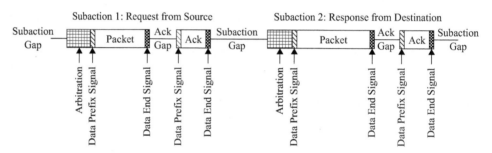

transmitting an arbitration request to its parent (node E to node D and node A to node B). Each parent then forwards its request to its parent and denies access to all of its other children (in this example node D denies access to node C) by transmitting a data prefix signal, including a speed code. The request originating from node A propagates to the root node before the signal originating at node E, since the signal from node E must also propagate through node D. The root node grants access to node A and issues a denial code to node D that will be passed on to node E. The procedure favors the node closest to the root. This unfairness is overcome by the use of a "fair arbitration" protocol that limits the frequency with which any one node may access the network and arranged so that all nodes get an equal opportunity for access.

IEEE 1394 allows for asynchronous and isochronous data transfer, allowing both non-real-time applications such as printers and scanners and real-time applications such as audio and video to operate on the same network. The asynchronous format transfers data and transaction layer information to or from an explicit address. Figure 3.47 shows a single transaction known as an "asynchronous subaction."

To transmit data, the device first gains control of the physical layer, as described in the previous paragraphs. The arbitration process ends with the transmission of a data prefix signal, including a speed code if needed. The packet includes the addresses of both the source and the destination nodes, CRC, and, if appropriate, a data payload.

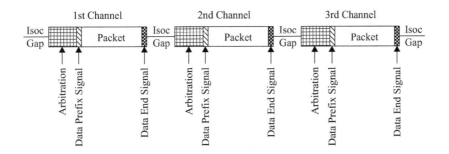

FIGURE 3.48 ▪
IEEE 1394
Isochronous
Subaction

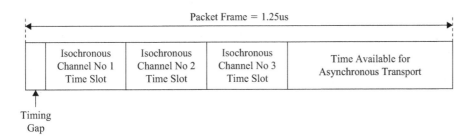

FIGURE 3.49 ▪
IEEE 1394 Packet
Frame

When the receiver accepts the packet, an acknowledgment packet (which may also contain a data payload) is returned to the original sender. The standard defines 13 packet formats, with and without data, for asynchronous transactions. If a negative acknowledgment is returned, error recovery is initiated.

The isochronous format broadcasts data based on channel numbers rather than specific addressing. The sender requests an isochronous channel with a specific bandwidth; isochronous packets are issued on average every 125 μsec. Figure 3.48 depicts an isochronous subaction.

Isochronous channel IDs are transmitted followed by the packet data. The receiver monitors the incoming data's channel ID and accepts only data with a specified ID. User applications are responsible for determining how many isochronous channels are needed and their required bandwidth. Figure 3.49 shows a packet frame with just three channels, although up to 64 isochronous channels may be defined.

The bus is configured to send a start of frame timing indicator in the form of a timing gap. This is followed by the time slots for isochronous channels 1, 2, and 3. The remaining time may be used for any pending asynchronous transmission. Since the slots for each of the isochronous channels have been established, the bus can guarantee their bandwidth, and thus their successful delivery.

IEEE 1394 has achieved wide acceptance in the consumer market, supporting such devices as digital television, video cassette recorders (VCRs), digital video disk (DVD) players, and other multimedia products. In the military avionics world, IEEE 1394 is being used on the Joint Strike Fighter Vehicle Management System.

3.17.12 Avionics Full-Duplex Switched Ethernet

Avionics full-duplex switched Ethernet (AFDX) is essentially a subset of ARINC 644 under development by Rockwell Collins for Airbus. Airbus plans to use AFDX in the new A380 civilian aircraft and in the A400M military transport program. AFDX is a full-duplex, 100 Mbps, switched Ethernet (see section 3.17.10) dual-redundant network with message sequencing numbering. Message frames are sent on both networks, with receiving end systems employing a "first valid wins" policy. This can protect against the complete loss of one network.

3.17.13 Additional Comments on Other Data Busses

This section is not by any means all inclusive of the types of data busses the test engineer may come across. Individual organizations are continually modifying and developing new methods of data transfer in order to best suit their needs. I have tried to cover those systems, or variations thereof, that have the highest probability of being encountered by the evaluator. Before attempting any project, it is imperative that the test engineer perform his or her homework to understand the system as well as data acquisition methods.

TCP/IP (Terminal Control Protocol/Internet Protocol) and video compression is discussed in detail in chapter 12 of the text.

███ 3.18 ███ | POTENTIAL PROBLEMS FOR THE ANALYST

One of the greatest problems the evaluator is going to encounter is the sheer amount of data that will be available to the test community. How do we collect it, and then what do we do with it once we have it? MIL-STD-1553 is a relatively easy system for accessing and retrieving data, and the data rates are somewhat manageable. The newer systems that have just been discussed (FC, Ethernet, etc.) have data rates that are extremely high and the way they are used is completely different than what we have seen with 1553.

Figure 3.50 shows a partial MIL-STD-1553 bus configuration with a bus controller and seven RTs.

In the example, the processing of navigation information is accomplished in the mission computer (bus controller). A Kalman-filtered solution may then be sent to either the stores management system or the radar. The data rates on the bus are limited to 1 Mbps, so not all of the information available from all navigation sensors is being used by the mission computer. In modern system development, all of the information available is being used by a core processor and the answer is sent to the mission computer when requested. Figure 3.51 shows how Figure 3.50 would look like in a modern implementation.

In the revision, each navigational sensor forwards data to a navigation core processor which produces a "system" navigational solution. The data transfer is over a high-speed data bus with rates than can approximate 1 Gbps, although 80 to 100 Mbps is

FIGURE 3.50 ■
Simple MIL-STD-1553 Bus Configuration

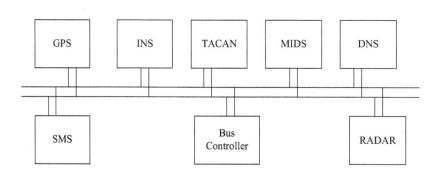

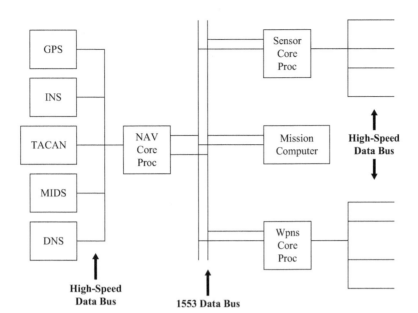

FIGURE 3.51 ■
High-Speed Data
Bus Implementation

probably more realistic. But even at the lower rate, we are still seeing an increase of 100-fold over the basic MIL-STD-1553 system. The first problem for instrumentation and the evaluator is how to capture, store, time tag, and time align these data. The high-speed data busses would also be implemented for a sensor core processor (radar, electro-optics, forward-looking infrared, electronic intelligence, off-board sensors, etc.) and weapons core processor. Each of the core processors is an RT on the 1553 bus; basic command and control is still resident with the mission computer, however, processing of data becomes decentralized. Our original one data bus in 1553 carrying a maximum of 1 Mbps of data is now transformed into four busses, with our three high-speed data busses moving up to 300 Mbps!

The second problem associated with this setup involves the analysis of any particular system. Under the basic MIL-STD-1553 system, an analyst can compare sensor inputs to the system outputs (displayed data) and make an assessment of the performance of the system. In the revised case, all sensor data are forwarded to the core processor, where the decision is made for the best "system" navigation solution, or target solution, or weapons solution, etc. The output of the core processor is a result of artificial intelligence within the processor, and that is what is being reported on the 1553 data bus. The evaluator would have to replicate the core processor in analysis software and provide the sensor inputs to this software. The output of the analysis software could then be compared to the core processor output to the 1553 bus. Sensor errors would be very hard to determine, as they would probably be minimized within the core processor. A test would have to be developed to look at only the sensor of interest (which would entail the shutting down of all other inputs to the core processor); this would be a very laborious task. Of course, a lot of this testing would not have to be done if the output of the core processor always matches the truth data (but how often can that be expected?).

The Joint Strike Fighter uses a mix of four standards: FC for avionics, IEEE 1394b for vehicle systems, 1553 for legacy systems, and EBR-1553 for weapons systems; quite an adventurous task for instrumentation and test and evaluation.

▌ 3.19 ▐ | DATA ACQUISITION, REDUCTION, AND ANALYSIS

In order to successfully execute a test program, data must be collected, analyzed, and submitted to substantiate conclusions. In the chapters 1 and 2 of this text, the importance of valid avionics and TSPI data was stressed. This section will cover the available means of acquiring test data and the tools available to the evaluator for reduction and analysis.

3.19.1 Instrumentation

Data acquisition is accomplished with the use of instrumentation. Instrumentation comes in many shapes and forms and runs the gamut from handheld to data bus monitoring. Handheld instrumentation may be as simple as a stopwatch or writing down GPS coordinates at the time of an event. Data bus monitoring may be in the form of editing the data stream on the aircraft and recording parameters of interest or telemetering an entire data bus to a ground receiver. Telemetry and recording may be analog or digital and may include audio and video. Instrumentation will most certainly be required on your test vehicle, but it may also be required on targets, bombs, or missiles. The development lab and vendor test beds may also have an instrumentation requirement. Current telemetry standards in the United States and used throughout the world are contained in the Range Commanders Council Telemetry Group IRIG Standard 106-04.

3.19.1.1 Pulse Code Modulation

The pulse code modulation (PCM) instrumentation system is the basic instrumentation system on the aircraft and also provides a means of incorporating digital information. A PCM system allows the test community to record parameters of interest and telemeter these parameters to a ground station, where they may be read in real time. PCM allows an analog signal to be converted to a binary field prior to transmission or recording. The basic PCM process is shown in Figure 3.52.

FIGURE 3.52 ▪
PCM Process

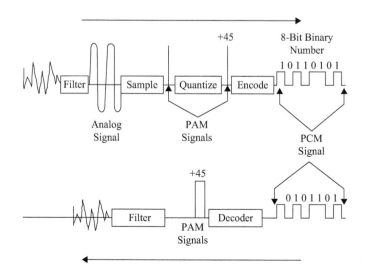

The signal is sensed (voltage, temperature, etc.), amplified, and encoded into a binary field which is sent to a receiving telemetry station. The ground station runs the binary field through a decoder and it is transformed back to its original signal.

PCM is usually associated with analog parameters because the system was designed prior to digitized aircraft. A basic PCM system is still used by the test community to monitor and record analog parameters of interest:

- System health and welfare
- Switch positions/commands
- Instrumentation status
- Interlocks

System health and welfare might include the monitoring of cooling air or voltage to an aircraft system to avoid overheating or undervoltage which may damage the sensor. Instrumentation can be used to monitor correct pressurization of a backup hydraulic system. Switch positions and commands allow the receiving station to monitor mode selections by the aircraft operator. There could be a case where a mode is selected by an operator but the command is not sent on the bus. Instrumentation status may be data time remaining or the position of the data record switch. Interlocks such as weight-on-wheels are normally monitored in most development programs, as they can control or deny many things if not set properly: ground cooling fans, missile fire inhibit, etc. A block diagram of a generic PCM system is shown in Figure 3.53.

The transducer in the block diagram could be a voltage sensor, strain gauge, mass flow sensor, etc.; transducer is used as a generic term. The signal is conditioned and sent to a remote unit that is located in the vicinity of the sensor. The remote unit amplifies the signal and sends it to the PCM master unit, which builds the frame (an instrumentation data set described later) to be sent to either a digital recorder (HDDR in this diagram) or a digital data unit that multiplexes the data and sends it via telemetry to a ground receiver.

As an example, consider the situation depicted in Figure 3.54. An avionics sensor on the aircraft is cooled by fins and forced-air cooling. By engineering analysis, it is

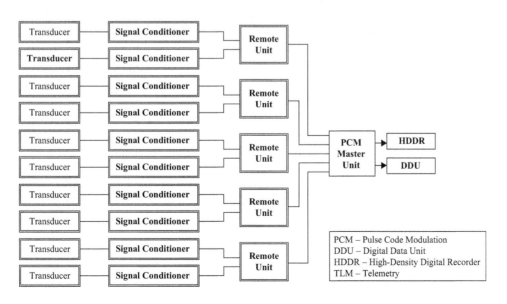

FIGURE 3.53 ■
Generic PCM System

FIGURE 3.54 ■
Instrumentation for
Forced-Air Cooling

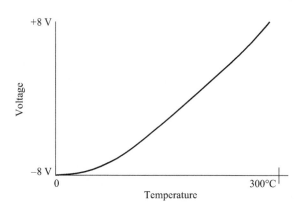

FIGURE 3.55 ■
Calibration Curve for
a Thermocouple

determined that the sensor will fail if the temperature of the sensor exceeds a certain level. Further analysis relates exhaust temperature (cooling air leaving the sensor) to core temperature; therefore, if we monitor the exhaust temperature, we can protect the sensor. A temperature monitoring device (transducer) is placed at the exhaust port of the sensor as part of the instrumentation package on the airplane. The transducer works much like the thermostat on your car; it may consist of a bimetallic element that reacts to temperature. With a current applied to the transducer, the voltage will change based on the change in resistance due to the reaction of the bimetallic element. If we calibrate this device, we can equate a voltage to a temperature.

Instrumentation personnel will try and match as closely as possible the desired range and resolution required by the test engineer to the measurement device. Assume that the thermocouple in this example has a range of 0–300°C for voltage changes of −8 to + 8 V. Instrumentation will provide the test engineer with calibration charts for this device in terms of voltage versus temperature. A typical calibration curve for this device is found in Figure 3.55.

After determining the calibration, it will be necessary to define the resolution of the measurement. Instrumentation uses a term called "counts," which is really the number of bits available in the PCM frame for each parameter. The depiction of the PCM process in Figure 3.52 shows 8 bits available for that example.

Suppose that instrumentation has 10 counts, or bits, available to measure the voltage for our transducer. If every bit was signed (remembering back to section 3.3), it would look like this:

2^9	2^8	2^7	2^6	2^5	2^4	2^3	2^2	2^1	2^0
1	1	1	1	1	1	1	1	1	1

The sum would equal 1023 plus the added measurement of 0, or 1024. This means that we could resolve the voltage down to 16/1024, or approximately 0.015 V. If the curve were linear in Figure 3.55, the resolution of temperature would be 300/1024, or roughly 0.3°C. If this resolution were not good enough for our purposes, we would have to narrow the range of interested temperatures. If we decreased the range to 100°C to 300°C, the new resolution would be 200/1024, or roughly 0.2°C.

Now that the parameter can be quantified, we need a mechanism to deliver the data to the test and evaluation team. In telemetry operations, data are transmitted in a PCM frame. An example of a PCM frame format is shown in Figure 3.56.

In this example, the frame is comprised of 16 subframes with 128 words in each subframe; a word is 10 bits long. The frames are transmitted at a rate of 12.5 frames/sec. The instrumentation system reads from left to right in subframe 1 and then drops down and reads from left to right in subframe 2. The system continues until the entire frame has been read and then transmits the entire frame of data. If we do the math, we will see that 10 bits/word × 128 words × 16 subframes × 12.5 frames/sec equals 256 kbps. This is a low-rate PCM system; most modern systems run at data rates of between 2 and 5 Mbps. It can be readily seen that if the number of available bits per word or the frame rate is increased there will be a corresponding increase in the data rate.

The evaluator also needs to specify the sample rate required of the parameter in question. The sample rate of the parameter is determined by the number of times the parameter appears in the frame. In Figure 3.56, the parameter AAA appears in the first word of all 16 subframes; parameter AAA will be read 16 times in each frame, and with a frame rate of 12.5 frames/sec, will be sampled 200 times each second. Parameter BBB appears in the second word of every other subframe,

FIGURE 3.56 ■
PCM Frame Format

Words 1 Through 128

	1	2	3	4		128
1	AAA	BBB	CCC	GGG		
2	AAA		DDD			
3	AAA	BBB	EEE			
4	AAA		FFF			
5	AAA	BBB	CCC			
6	AAA		DDD			
7	AAA	BBB	EEE			
8	AAA		FFF			
9	AAA	BBB	CCC	GGG		
10	AAA		DDD			
11	AAA	BBB	EEE			
12	AAA		FFF			
13	AAA	BBB	CCC			
14	AAA		DDD			
15	AAA	BBB	EEE			
16	AAA		FFF			

128 Words × 16 Subframes = 1 Frame

Frame rate is 12.5 frames/sec.
Parameter AAA is sent at 200 samples/sec.
Parameter BBB is sent at 100 samples/sec.
Parameter CCC is sent at 50 samples/sec.
Parameter GGG is sent at 25 samples/sec.

which will produce a sample rate of 100 each second (8×12.5). Parameter CCC will be sampled at 50 samples/sec, whereas parameter GGG will be sampled at 25 samples/sec. If a parameter only appeared once in each frame, it would be sampled at the frame rate, or 12.5 samples/sec. This process is known in the instrumentation world as subcommutation.

If a requirement exists for a higher sampling rate than is possible with subcommutation, then instrumentation may elect to use supercommutation. An example of supercommutation is shown in Figure 3.57.

Notice that in Figure 3.57, parameter AAA is read twice in each subframe (as opposed to once) in the subcommutation example. This effectively doubles the sample rate to 400 samples/sec. Parameter BBB is read four times in zach subframe, which yields a sample rate of 800 samples/sec. The parameters must be spaced correctly in the frame. Parameter AAA is repeated twice, so the placement of the second word must be at 128/2, or 64 places, from the original position (word position 65). Parameter BBB is repeated four times, so the placement of the additional words must be at 128/4, or 32 places, from the original position (word positions 34, 66, and 98). Supercommutation is normally used when it is desired to see a certain parameter in real time at a higher rate than normal, usually for troubleshooting purposes. Supercommutation reduces the total number of parameters that you will be able to see in real time.

Test engineers have the responsibility to sit down with the instrumentation group and determine which parameters and what rates are required for each phase of the test program. It can be seen that in the frame provided in the examples, a maximum of 2048 parameters can be sent at a rate of 12.5 samples/sec (assuming all of the parameters use the 10 bits available in each word). Increasing the sample rate on any of the parameters will decrease the total number of parameters that can be seen in real time.

The pilot's voice can also be sent via a PCM frame; in many test organizations this is known as "Hot Mike." Hot Mike will send all audio that is heard by the pilot (UHF, VHF, ICS, etc.). In order to put voice into the PCM frame the audio must first be digitized; the digital information is then sequentially inserted into the PCM frame as it is

FIGURE 3.57 ■
Supercommutation

	1	2	3	4	34	65	66	98	128
1	AAA	BBB	CCC	GGG	BBB	AAA	BBB	BBB	
2	AAA	BBB	DDD		BBB	AAA	BBB	BBB	
3	AAA	BBB	EEE		BBB	AAA	BBB	BBB	
4	AAA	BBB	FFF		BBB	AAA	BBB	BBB	
5	AAA	BBB	CCC		BBB	AAA	BBB	BBB	
6	AAA	BBB	DDD		BBB	AAA	BBB	BBB	
7	AAA	BBB	EEE		BBB	AAA	BBB	BBB	
8	AAA	BBB	FFF		BBB	AAA	BBB	BBB	
9	AAA	BBB	CCC	GGG	BBB	AAA	BBB	BBB	
10	AAA	BBB	DDD		BBB	AAA	BBB	BBB	
11	AAA	BBB	EEE		BBB	AAA	BBB	BBB	
12	AAA	BBB	FFF		BBB	AAA	BBB	BBB	
13	AAA	BBB	CCC		BBB	AAA	BBB	BBB	
14	AAA	BBB	DDD		BBB	AAA	BBB	BBB	
15	AAA	BBB	EEE		BBB	AAA	BBB	BBB	
16	AAA	BBB	FFF		BBB	AAA	BBB	BBB	

Frame rate is 12.5 frames/sec.
Parameter AAA is sent at 400 samples/sec.
Parameter BBB is sent at 800 samples/sec.

received from the A/D converter. It should be noted here that in most development programs in the United States, the pilot's voice must be encrypted if it is sent via a telemetry stream. This is accomplished with an encryption unit, most likely a KY-58, which is fitted between the A/D converter and the instrumentation system. Communications encryption is covered in section 4.2 of this text.

Up to this point we have only discussed analog measurements and how they are sent in the telemetry stream. Digital data from any of the avionics busses on the aircraft can also be sent using the same PCM frame. The PCM process described earlier showed the analog signal converted into a binary format. Digital data busses need no conversion, as they are already in the correct format. All that is needed is a way to extract the desired parameters from the bus. A device which accomplishes this task was identified in section 3.2 and is called a bus monitor.

A bus monitor is a passive unit on the bus. The standard defines a bus monitor as a "terminal assigned the task of receiving bus traffic and extracting selected information to be used at a later time." These systems are used for troubleshooting and aircraft instrumentation. They are not active terminals and do not communicate with the bus controller. These devices read all of the traffic on the bus and capture the time and value of selected parameters. These parameters are preprogrammed into the bus monitor. A depiction of the relationship of the bus monitor to the 1553 data bus is shown in Figure 3.1, repeated on this page for clarity.

Most of these systems are called either word grabbers or bus editors. They are normally built by the instrumentation group and are unique to one type of aircraft. The devices are preprogrammed on the ground to look for certain messages (parameters) on the bus and read these data to either a tape system or multiplex them into a telemetry stream. The amount of data these systems can grab for real-time data transmission is limited by the telemetry system.

The evaluator will also need to determine which digital parameters are required for real-time analysis, but now the ICD will need to be consulted in order to determine the location of those parameters in the data stream. The instrumentation group will program the bus editor on the aircraft to extract the correct parameters. If it is desired to obtain data from multiple busses on the aircraft, then a separate bus editor will be required for each bus.

The biggest limitation on real-time data is the amount of data that can be transmitted in the telemetry stream. Historically avionics telemetry systems have operated in L band and many of the missile telemetry systems have operated in S band. Some video downlink systems operate in C band. L band is restricted, due to bandwidth limitations, in the amount of data that can be sent. Some attempts at video downlink in L band have

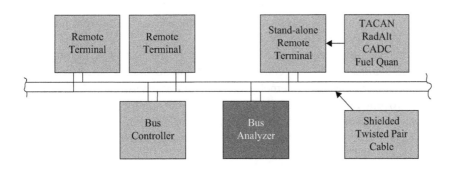

FIGURE 3.1 ■ Simplified Bus Structure (Previously Presented)

resulted in unsatisfactory resolution because of these limitations. The typical telemetry system of today transmits between 2 and 5 Mbps of data on the telemetry stream. With modern aircraft utilizing data busses that can transfer data at up to 8 Gbps, this has become a severe limitation.

3.19.1.2 Advances in Real-Time Telemetry

There are drawbacks in the way data is sent to a receiving station. The greatest problem has been addressed already—the limited amount of data that may be telemetered based on bandwidth limitations. A second problem involves the geometry of the aircraft and, in particular, its telemetry antennas relative to the receiving station. As an aircraft maneuvers, it may block the telemetry signal and the receiving station will experience dropouts. The telemetry signal as well as the bus data are directly recorded on the aircraft and are available for postflight review so the data are not lost. The problem occurs when data are required in order to proceed to the next test; that is, step 2 cannot be attempted unless step 1 is validated as successful. In the past, if a data dropout occurred during step 1, the test would have to be repeated to validate success in real time. This is time consuming and wasteful of your limited budget. A good example would be an envelope expansion test for an aircraft with a new store installed. Safety concerns dictate the need to be absolutely sure that it is safe to proceed to the next step up in airspeed or altitude. If we could retrieve the old data from the recording and resend it to the station, a repeat of the test would not be necessary.

The U.S. Department of Defense has been developing a network architecture that will give the test community new radio-spectrum-enhancing capabilities. The telemetry network system (TmNS) will provide test range computer networks with a wideband wireless capability that can cover hundreds of miles. This is one step in the total upgrade which is known as integrated network-enhanced telemetry (iNET). The goal of iNET is to find a feasible upgrade for the basic architecture of the test ranges' telemetry systems. Under an iNET configuration, time-sensitive data are transmitted immediately using point-to-point telemetry, while less urgent data are collected and sent when bandwidth becomes available. All data are still recorded onboard the aircraft for postflight reduction. This eliminates the problem of lost data during maneuvers, as the data can now be retransmitted. It will also help with near real-time troubleshooting of problems, as additional data (data not normally available to the test community during real-time operations) can be sent while setting up for the next test card. Data received at the test site can then be shared between multiple users, such as other ranges, engineering groups, and contractors, via the iNET network.

It is envisioned that iNET will use L, S, and C band and allocate bandwidth in real time to multiple users. This would be accomplished by an uplink to the aircraft which would be able to change the transmitter and the transceiver in real time, thereby adjusting the spectrum allocation. This would be ideal for multiple test platforms, allocating the spectrum based on need.

The initial architecture of iNET was defined by the Boeing Phantom Works and the proposed standards are near completion and should then be incorporated into the Range Commanders Council Telemetry Group IRIG Standard 106-04. This program is a joint improvement and modernization project managed by the Central Test and Evaluation Investment Program. A limited operational prototype of the system is expected to be in service sometime in 2011. The proposed iNET system interfaces are shown in Figure 3.58.

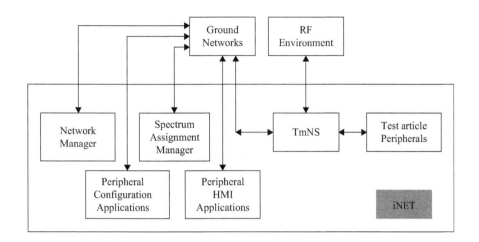

FIGURE 3.58 ■
iNET System
Interfaces

The iNET system is comprised of six contributing elements:

- Test article peripherals
- Peripheral configuration applications
- Peripheral human machine interface (HMI) applications
- Telemetry network system (TmNS)
- Network manager
- Spectrum assignment manager

Test article peripherals are all of the instrumentation devices that reside in the test article: PCM systems, data and video recorders, TCP/IP networks, etc. Peripheral configuration applications allow an instrumentation engineer to remotely configure instrumentation on the test article. Peripheral HMI applications allow test engineers to remotely monitor and control instrumentation on the test article. The TmNS is the communications link between the test article and the ground networks and will transmit data via PCM streams and IP packets. The network management application is an application that allows the monitoring of the health and welfare of the TmNS. The spectrum assignment management application allows a spectrum manager to remotely monitor and control radio frequency (RF) attributes of the TmNS. The system will exchange data much like the tactical targeting network technology (TTNT), which is an improvement on existing data link technology. This system is described in section 5.9.5 of this text. Progress on the iNET system can be followed on their homepage: www.inetprogram.org.

3.19.1.3 Other Instrumentation

The bulk of data used for troubleshooting and analysis does not come from real-time but rather from onboard recording systems. In the early 1980s these systems were multitrack analog recorders and limited in the amount of data that could be captured. Video recorders were VHS and some Beta, which were also limited. Analog data recorders gave way to high-density digital recorders (HDDR) developed to capture MIL-STD-1553 bus data at data rates of 1 Mbps. About the same time, specialized recorders to capture internal radar processing were also developed, and these recorders attained data rates of about 4 Mbps. VHS was replaced by Hi8, which was smaller and produced better quality, and some early work was done on downlinking video along with the telemetry stream. In today's programs, it is not unusual to see data rates in excess of 1 Gbps, which exceed the capabilities of the standard recorder. Data today (avionics as well as video) are written to flash memory, much like the memory sticks that you can carry on your key chain.

The workload for the evaluator has increased significantly with the amount of data that is being collected from the test article. This problem was previously identified and discussed in section 3.18 of this text. The types and sources of data recorded for post-flight analysis are many:

- Avionics data busses (can include 1553, 1760, EBR-1553, Ethernet, etc.)
- Pilot's voice
- Time
- Telemetry stream
- HUD/HMD video
- Display video
- Missile data (avionics and video)
- Stores data (avionics and video)
- TSPI data (onboard and off-board)
- TSPI data differentially corrected
- Satellite data

3.19.2 Real-Time Data Reduction and Analysis

A definition of real-time (or near-real-time) data could be the processing of aircraft telemetry and range data during a flight to provide answers that allow flight profiles to be completed or repeated as necessary to optimize flight test efficiency. The closest you can get to real time is about 0.5 sec, and this is for data that are throughput directly to monitors without any processing or smoothing. These data are used for parameter monitoring or limit checks and are most often used as flight safety checks (altitude, airspeed, temperatures, etc.). Latency for near-real-time parameter comparison used for top-level system performance determination runs about 2 sec.

The importance of real-time data can be seen in five broad categories:

- Multistream merge comparisons
- Test control
- Quick look and specification compliance
- Top-level performance
- Debrief support

Real-time data are merged and time aligned for viewing by the analyst. This operation must be accomplished for the benefits of real time to be realized. This is one of the reasons why data will be near real time, as there is processing required to perform the time alignment; time alignment was discussed in section 2.18 of this text.

Depending on the test being conducted, there may be as little as one telemetry stream and one TSPI stream to be merged, or in the case of complex testing, such as live missile firing, there may be more than five data streams that must be merged and time aligned. Some of the sources of data that may be seen in real-time operations are

- Multiple aircraft avionics telemetry streams
- Multiple missile telemetry streams
- Video downlink
- TSPI ground trackers
- Target video and telemetry

When performing a real-time merge of these data, the control room (an operations center where test control and analysis is performed) will assume that all data are asynchronous; that is, all of the varied data streams arrive at different times. The center will read imbedded time to accomplish the time alignment. Imbedded time is the time stamp on the data which is inserted at the point of collection. For example, a PCM frame will contain a header word that contains time. It is that time that is read, not the time that data arrives at the control room. Figure 3.59 shows this relationship.

The test aircraft in Figure 3.59 inserts time in the PCM header before it is sent to the control room. The control room receives the PCM frame at some time later and marks the time that it was received. Similarly, the TSPI site will determine the test aircraft coordinates and insert the time that the position was sensed. It will send this (via microwave, data link, land line, etc.) to the control room, which will again mark the time it was received. When merging these two streams, the time stamp inserted at the source (test aircraft and TSPI site) will be used, not the time it was received at the control room. This eliminates the need to determine the time lag due to transmission from the two sites.

The responsibility for conducting the test lies with the test conductor. This individual determines flight safety—if testing can proceed, if the systems are functioning as expected—and direct the general flow of testing. For proper control, data need to be provided to the test conductor as expeditiously as possible. For this reason, data provided

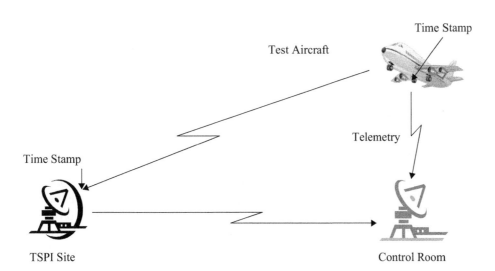

FIGURE 3.59 ■ Time Stamping of Data

to the test conductor are normally nonfiltered or processed. The test conductor receives inputs from the

- Test conductor screen
- Mission display or range control
- Voice communications
- Station analysts

A test conductor screen normally displays critical flight safety parameters such as airspeed, altitude, acceleration, fuel remaining, etc. It may also display parameters identifying test requirements. Sensor status, the selected mode of the system under test, built-in test (BIT) results, and store station selected are possible parameters that may be of concern. In addition, time and instrumentation status should be displayed.

A mission display or range control console provides real-time support by displaying the TSPI-determined position of all of the test participants. In some organizations, this display is used to control the geometry of the test setups. In other organizations, test geometries are controlled by qualified range control officers (much like FAA controllers). This display provides airborne tracks with persistence trails of all test aircraft, the relationship between tracks (range, bearing, and closure), event markers, and the ability to make a hard copy of the display. The display may have an overlay of the restricted area boundaries or even visibility profiles of the TSPI coverage. A replay mode is normally available to assist in flight debriefs and postflight analysis.

Communications also may come from a multitude of sources:

- Hot Mike
- UHF/VHF/HF
- Satellite communications (SATCOM)
- Data link

As described previously, Hot Mike will relay, via telemetry, all voice communications heard by the pilot; this includes the intercommunications system (ICS). This provides situational awareness for the test conductor, allowing him to hear when the pilot is busy with other communications. Voice communications and direction to the aircraft are accomplished over the UHF/VHF/HF frequencies. Over-the-horizon communications can be accomplished with HF/SATCOM/data link when required.

Station analysts can provide the test conductor with detailed information on test results and status based on real-time information. Analysts may have access to information that is not readily available to the aircrew by accessing the aircraft data busses.

Armed with the data described, the test conductor has all the necessary information to vector the aircraft, change test profiles, call an abort, or proceed to a different set of test cards.

Real-time data allow the test team to identify problems early as well as data that need further analysis in postflight, provide for an early look at specification compliance, and allow for an orderly flow to the next objective. Quick look is the term that some test organizations use for this process, as it allows a snapshot of how the test is progressing. Testers have become quite inventive when displaying data in a format that provides status at a glance. Color-coded screens using green for within specification, red for out of tolerance limit, and yellow for a transitional status are extremely helpful for quickly identifying problem areas. By using analog representations of the data, trends can readily be seen. Comparing truth data and system data on the same screen can determine specification compliance. By comparing real-time parameters to expected values, a sense of system performance can be attained.

It should be noted that data viewed in real time is normally at a lower data rate than what is available in the data stream. Whenever indepth analysis is required, data should be taken from the aircraft recording, which has all of the data at the highest data rates.

Data collected in real time provides an excellent tool for flight debrief support. Hard copies of the data screens taken at critical times during the flight, video, and mission display recordings can assist the test team in accurately deconstructing the flight.

3.19.3 Postflight Data Reduction

There are two types of postflight data reduction: batch processing and interactive batch processing. Batch processing involves a request to the data group (personnel tasked with providing data reduction and support) for time slices of data from the aircraft recordings. These may be requests for parameter sets required by the analyst, raw data extraction in hexadecimal, or requests for data files. Interactive batch is a process that allows the user to access aircraft-recorded data which has been engineering unit (EU) converted and resides on a server or workstation. A depiction of these processes is shown at Figure 3.60.

Data obtained is used for troubleshooting of anomalies and determination of specification compliance and system performance. In real time, decisions on whether to continue to the next profile or defer the test to the next flight will be based on the availability of data. During the debrief, data will determine the work schedule: Is new aircraft software required? Can a productive flight be accomplished as planned? Does maintenance need to be scheduled? Postflight processing provides the analyst with all of the data at the highest data rates, and it is these data that will be used to determine specification compliance and true system performance.

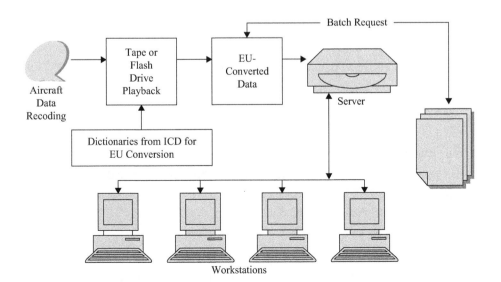

FIGURE 3.60 ■ Batch and Interactive Batch Processing

3.20 | SELECTED QUESTIONS FOR CHAPTER 3

1. How many different types of words can be sent over a MIL-STD-1553 bus? What are they?

2. Name two additional error checks added to the MIL-STD-1553 protocol in MIL-STD-1760.

3. What is a BC? How many sensors can be a BC?

4. Is ARINC 429 a simplex or duplex link?

5. What does FireWire allow a system to do?

6. How does the military routinely handle non-1553-compatible equipment on the 1553 bus?

7. What is an RT? Give five examples of a sensor that might be an RT.

8. What does Manchester code mean? What does NRZ code mean?

9. How come a clock is not used on ARINC 429?

10. What is a message? What types of messages can be sent over the MIL-STD-1553 bus?

11. What does command/response mean to you?

12. What is MIL-STD-1760 used for?

13. What is an interface design specification? What is it used for?

14. How does MIL-STD-1760 differ from MIL-STD-1553?

15. What is asynchronous data? Give one example.

16. What is broadcast mode in MIL-STD-1553?

17. What is a bus analyzer? Why would you need one?

18. Why are header words used with some MIL-STD-1760 messages?

19. What is a return-to-zero tristate modulation?

20. What is isochronous data? Provide two examples.

21. What is meant by deterministic? Which data bus is optimally suited if determinism is required?

22. How is the high-speed data bus activated on the Eurofighter?

23. What is decimal, binary, octal, and hexadecimal?

24. What are weapons A/D converters known as?

25. What is a functional requirements document?

26. Will Ethernet ever achieve a safety assurance level of A?

27. How many RTs can occupy a MIL-STD-1553 bus, mathematically and in reality?

28. What are some problems associated with high-speed data busses that can cause concern for instrumentation and data analysts?

29. What is meant by an action word?

30. Are the status word bits the same in MIL-STD-1760 as in MIL-STD-1553?

31. How many busses can be on an aircraft? What are the limitations?

32. What does contention access mean in the data bus world?

33. What is the data rate on most MIL-STD-1553 busses?

34. Must the ARINC 429 use the source destination identifier as an address?

35. What does a hexadecimal dump of raw data mean to you?

36. What are the three major advantages of employing EBR-1553 on legacy aircraft?

37. What is MIL-STD-1760? Are the protocols different than MIL-STD-1553?

38. What is an intermessage gap? How long is it?

39. What is the instrumentation bit in the MIL-STD-1553 status word used for?

40. What is a mode code in MIL-STD-1553? How is it initiated?

41. If two devices need to share information on an ARINC 429 bus, what needs to be added?

42. What is the most common problem with RS-232?

43. Briefly describe the decentralized processing routines on the next-generation aircraft.

44. What is a cyclical redundancy check? What is it used for?

45. Is there a response time requirement in MIL-STD-553? What is it?

46. What is a bus monitor used for?

47. What is a parity bit? What is it used for?

48. What is a vector word?

49. What is mass data transfer? On which bus would you find this?

50. What is "plug and play"?

51. What is "peer to peer"?

52. What is FC-AE-ASM? Is this a common standard?

53. Is Ethernet deterministic? Why or why not?

54. Can an RT reside on more than one bus?

55. Can a device be an RT on one bus and a bus controller on another?

Communications Flight Test

Chapter Outline

4.0	Overview	139
4.1	Communications Basics	140
4.2	Aircraft Communications Equipment	142
4.3	Test Requirements	144
4.4	The Three Steps in Avionics Testing	144
4.5	Communications Test Plan Matrix	146
4.6	Executing the Matrix	148
4.7	Other Considerations in Communications Tests	149
4.8	Effects of Stores, Landing Gear, and Flaps	149
4.9	Effects of Weather	149
4.10	Logistics	150
4.11	Boredom	151
4.12	Speech Intelligibility	151
4.13	Electromagnetic Interference/Electromagnetic Compatibility	154
4.14	EMI/EMC Testing	158
4.15	EMI/EMC Test Issues	161
4.16	EMI/EMC Elimination	161
4.17	Selected Questions for Chapter 4	164

4.0 | OVERVIEW

The flight testing of communications systems is avionics flight testing in its most basic form, but it does illustrate many of the common factors seen in the more complex systems. The testing of communication systems occurs more often than one would expect. Radios are often upgraded, which requires a flight test program, but these upgrades may be many years apart. What occurs on a more frequent basis is the relocation of antennas or the installation of stores or other obstructions that may affect communications. Some flight testing may be required to ensure that no degradation of communications due to these modifications has occurred. Radios, because they transmit as well as receive radio frequency (RF) energy, are also key players in electromagnetic interference/electromagnetic compatibility (EMI/EMC) testing. This testing will be addressed later in this chapter. Data link systems, which transfer digital information, voice, or data, are tested in a similar way to radios, and are addressed in chapter 5.

▮ 4.1 ▮ COMMUNICATIONS BASICS

It is not the intent of this text to cover the theoretical concepts of any of the avionics systems under test. Rather, a practical usage of the systems will be addressed. There are many texts available to the flight test engineer assigned to a particular system. The engineer should become familiar with any system he is assigned. As mentioned previously, a good flight test engineer knows the system as well as the designer. The text assumes that some basic knowledge is possessed by the tester. Wherever possible, suggested reading which has proven beneficial to the author in the past will be recommended. A good overview of communications theory, for example, can be found in *Modern Electronic Communication*, by Gary M. Miller (Englewood Cliffs, NJ: Prentice Hall).

Our first concern will be the ability to propagate the radio wave; propagation in free space is much different than that seen in wires, cables, or, in the case of radar, waveguides. In wires, the wave does not lose energy as it travels (except for scattering and absorption). In free space propagation, the beam spreads as it travels away from the transmitter; the surface area it occupies increases as the square of the distance traveled. For example, if you were to double the distance from the transmitter, the area occupied by the wave would increase to 2^2, or 4 times the original area. Since energy must be conserved, the energy per unit surface area must also decrease by the square of the distance traveled; this is called the inverse square law, which we will also see when discussing electro-optical (EO) and radar systems. For each doubling of the distance, a 6 dB (power ratio of ¼) loss is experienced. For all of the waves up to millimeter wavelengths, this free space loss is going to be the most detrimental.

The second concern is the interaction of waves with obstructions. When waves become obstructed in free space, new waves will result from the interaction. There are four types of interactions:

- Reflection
- Refraction
- Diffraction
- Scattering

Reflection occurs when a wave encounters a plane object and the wave is reflected back without distortion. Refraction occurs when the wave encounters a change in medium and the direction and speed of the wave is altered. This can be seen when you put your arm into a swimming pool and your arm appears to be broken. Diffraction occurs when the wave encounters an edge; the wave has the ability to turn the corner. Diffraction is dependant on the frequency of the wave; the higher the frequency, the less diffraction. Scattering is the reradiation of waves in multiple directions as the wave impacts an object. Scattering is also frequency dependent, with higher scattering at higher frequencies.

The propagation of waves is also influenced by the atmosphere, and in particular the troposphere and the ionosphere. The troposphere extends from the surface of the earth to about 30 miles in altitude; the ionosphere begins at the edge of the troposphere and

TABLE 4.1 ■ Radio Frequency Designations

Band Designation	Label	Frequency Spread
Extremely Low Frequency	ELF	3–30 Hz
Super Low Frequency	SLF	30–300 Hz
Ultra Low (Voice) Frequency	ULF or VF	300 Hz–3 kHz
Very Low Frequency	VLF	3–30 kHz
Low Frequency	LF	30–300 kHz
Medium Frequency	MF	300 kHz–3 MHz
High Frequency	HF	3–30 MHz
Very High Frequency	VHF	30–300 MHz
Ultra High Frequency	UHF	300 MHz–3 GHz
Super High Frequency	SHF	3–30 GHz
Extremely High Frequency	EHF	30–300 GHz

continues to outer space. The troposphere contains the earth's weather and liquid water, as well as most of the earth's water vapor, gaseous atmosphere, and pollutants. The ionosphere contains oxygen molecules which are ionized during the day by ultraviolet (UV) and x-ray radiation from the sun. At night, the ions recombine to form uncharged oxygen molecules. This ionization converts the ionosphere into an electrically neutral gas of positive and negative charges; this gas is known as a plasma. With low frequencies the plasma behaves much like a mirror and reflects radio waves. As the earth is also a strong reflector of radio waves these frequencies will hop back and forth between the ionosphere and the earth and the wave will propagate very long distances. As the frequency increases to about 50 MHz, the ionospheric effect decreases, and at higher frequencies appears invisible.

The frequencies of interest to us will be the high-frequency (HF) through the extra-high-frequency (EHF) bands (see Table 4.1). The propagation of HF frequencies is accomplished with two waves: a sky wave, which bounces from the ionosphere, and a ground wave, which propagates along the ground. Because of the long wavelengths, HF is not affected by the troposphere and can usually bend around most obstructions. These properties allow HF communications over long distances, including beyond line-of-sight (LOS).

Very-high-frequency (VHF) and ultra-high-frequency (UHF) bands are also not affected by the troposphere, but their frequencies are too high to exploit the ionosphere. Since the wavelengths are relatively small, they are unable to diffract around obstacles, which adversely affects communications. These frequencies are subject to multipath or fading. Multipath was discussed in chapter 2 and occurs when multiple signals (direct transmission from the transmitter and reflected signals from the earth or obstructions on the earth) arrive at the receiver simultaneously. Depending on the phase difference of the signals, they may be constructive (amplifying) or destructive. Thus a listener experiences changes in amplitude and clarity of the received signal.

Microwave and millimetric wave frequencies are affected by the troposphere, require LOS geometries, and are affected above 10 GHz. Above 25 GHz, water vapor and oxygen molecules will severely attenuate a signal due to absorption.

When we use the atmosphere as a transmission medium, we cannot control the noise environment. Electrical noise is defined as any undesired voltages or currents that appear in the receiver and are presented to the listener as static. Static can be annoying, and in severe cases can distort the reception so much that intelligible communications are impossible. Noise can be described in two broad categories: external and internal. External noise includes man-made noise, atmospheric noise, and space noise. Internal noise is noise introduced by the receiver, and can be broken down into thermal noise and shot noise.

Man-made noise is caused by spark-producing systems such as engine ignitions, fluorescent lighting, and electric motor commutators. The noise is radiated from the source and propagated through the atmosphere. If the frequency of the noise is close in frequency to the transmitted signal they will combine. Man-made noise is random and occurs in frequencies up to 500 MHz; man-made noise is dominant in urban environments. In HF and VHF, this noise is usually caused by other signals in the same band. Some early warning radars generate enough noise to be of concern in these frequencies.

Atmospheric noise is caused by naturally occurring disturbances in the earth's atmosphere, such as lightning. Atmospheric noise is evident throughout the entire RF spectrum, however, intensity is inversely proportional to frequency, so lower frequencies will suffer most, with a limited effect on frequencies above 20 MHz.

Space noise is a combination of two sources: the first originates from the sun and is called solar noise; the second is radio noise from outside the solar system, called galactic noise. Solar noise is cyclical, with peaks every 11 years. The next solar peak is expected in 2012. The noise from individual stars from outside the solar system is small, but their cumulative effects are large because there are so many of them. Space noise occurs at frequencies between 8 MHz and 10 GHz and depends a great deal on solar activity.

The directivity or the ability to focus energy is also directly related to frequency. At higher frequencies it is easier to focus a beam of RF energy. For example, a uniformly illuminated circular aperture antenna can generate a beamwidth:

$$\text{Beamwidth} = 1.02 \times \text{wavelength/diameter of the aperture}$$

As the frequency decreases, the wavelength increases, which will cause the beam to become larger, or less directed. The only way to improve our ability to focus energy is to either increase the frequency or increase the size of the antenna. As it turns out, for communications testing we are normally interested in omnidirectional coverage; that is, we should be able to transmit and receive normal communications 360° horizontally about the aircraft.

4.2 | AIRCRAFT COMMUNICATIONS EQUIPMENT

There are many variables to consider when attempting to adequately test an aircraft communications system. The first consideration is the frequencies used by the system. In today's aircraft, it is not unusual for a communications suite to incorporate HF, VHF, UHF, SATCOM, and data link. Each of these frequency bands has its own requirements for range, coverage, modulation, etc. Each of these systems has unique mission requirements which will have to be investigated. A list of the RF bands and their designations are shown in Table 4.1.

When testing RFs it is also common practice to test groups of frequencies within the band. For example, if a VHF radio is installed in an aircraft, you would test low, medium, and high frequencies within the VHF spectrum.

Antennas are also a consideration when testing communications systems. Most aircraft are equipped with an upper and a lower antenna for communications frequencies. This allows communication with a ground station or another aircraft regardless of the orientation of the aircraft. Some radio systems automatically switch antennas based on the received signal strength at a particular antenna. This is true for military tactical air navigation (TACAN) systems. Most communications systems allow selection of the antenna by the aircrew. The selections are upper, lower, and both. In some systems, the antennas are tied and the signal is split when "both" is selected. This fact should be taken into consideration when performing maximum range and intelligibility tests.

The radios may also be capable of operating in different modes or different modulations. It is common for military radios to be able to operate in clear and encrypted modes. The classic case of scrambling requires that the analog voice first be digitized. The analog signal is run through an analog-to-digital (A/D) converter, and the digitized signal is then sent to an encryption unit. The most common encryption units are KY-28s and KY-58s, the latter being the most current. The "program" for the encryption is set with an encryption key, which is classified SECRET. A couple of things that bite testers here should be obvious. All radios involved in the test where encryption is used must have the same, current key and must be functionally checked before the aircraft leaves the ground. Encryption keys are normally good for 24 hours based on Zulu (Greenwich Mean Time [GMT]) time, therefore two keys are normally loaded, one for the current day and one for the next day. The keys should automatically roll to the next day as time passes 2400Z. When an encrypted radio is transmitting, the encrypted signal (called a preamble) is transmitted first and received by another receiver. The receiving system sends the preamble through the encryption unit, where it is decrypted using the encryption key. If the encryption keys for the two radios match, the communications can be achieved. If the keys do not match, receipt of the transmission is impossible. If the keys match, the decrypted signal is run through a digital-to-analog (D/A) converter and is then heard in the headset. The original analog signal is processed quite often before being received by the cooperating system. This process time may result in a degradation of the intelligibility of the signal due to time delay, clipping, etc., and should be noted during testing.

Radios may also operate in a frequency-hopping or antijam mode. The system is known in the United States as Have Quick or Have Quick II. The purpose of the system is to deny access to your conversations by denying an enemy the ability to pinpoint the frequency on which you are communicating. The system utilizes many frequencies within a particular band (UHF, VHF, etc.) and randomly hops among these frequencies at very short intervals. The key to this system is time. All participants must be synchronized to the same time in order to communicate on the proper frequency at any given time. The system establishes a "net" that must be entered and then aligned in time with the other participants. Time can be obtained by either a synchronization pulse from the "net" owner or via an external source such as the GPS. Other than the previously mentioned concerns of intelligibility when using this mode of operation, testers should also pay particular attention to the robustness of the system, and the ease of entering or reentering the "net."

A second radio mode of operation is the ability to modulate the frequency. Most modern radios have the ability to modulate frequencies in either amplitude or frequency (i.e., amplitude modulation [AM] and frequency modulation [FM]). Modulation may be employed based on the operational need. For example, AM provides longer range, while FM provides a clearer signal. Testers need to take note of specification changes based on the modulation that is employed and the inherent characteristics of the modulation used for a given radio.

An additional mode of operation within many communications systems is a direction finding (DF) mode. This mode allows an aircraft to obtain a bearing to any source emanating a selected frequency. The system incorporates a methodology to first locate the frequency and then discern the direction from which it is being transmitted. Some of the more modern systems incorporate a coherent detection feature. This feature allows the system to internally generate the same frequency it is searching for, allowing for detection of the signal in higher noise levels than was previously possible. The aircrew is then given a bearing to or from the selected signal when detected. Depending on the sophistication of the system, it may be able to determine whether the receiver is moving away from or toward the transmitter. The accuracies are not exceptional, but it can help establish situational awareness or even locate a lost wingman. DF is possible with any frequency band available to the aircrew (UHF, VHF, HF, etc.) depending on the DF system being used.

4.3 | TEST REQUIREMENTS

The specification requirements for a communications system are usually defined as clear and effective communication in all directions out to a specified range. A typical specification may read:

> The ARC-XXX will be capable of air-to-air and air-to-ground communications in the UHF mode of not less than 300 nm throughout the omnidirectional range with elevation angles of $+30°$ to $-45°$, clear and encrypted in AM and FM modes.

The military uses a unique letter/number scheme to identify avionics and weapons equipment installed on an aircraft. "ARC" is decomposed as follows: the "A" refers to piloted aircraft, the "R" refers to radio, and the "C" refers to communications. This is based on the Joint Electronics Type Designation System (JETDS). By knowing the nomenclature, it is possible to get a rough idea of what type of system is being discussed. For example, we might be looking at an ASN-130 and an ARN-118. Both of these systems are used on piloted aircraft ("A"). The "S" refers to special or combination and the "R" refers to radio. The "N" in both cases refers to navigation aid. The first is an inertial navigation system (INS), while the second is a TACAN system.

At first glance, it appears that this communications suite will be a relatively simple test. Just fly out to maximum range and perform the required operations. However, this is not a particularly good way to plan this test. In reality, we must plan a logical progression of test points leading to the ultimate goal of maximum range performance. This will require the construction of a test matrix in order to identify all test conditions.

4.4 | THE THREE STEPS IN AVIONICS TESTING

There are three basic steps to take when preparing to evaluate any avionics or weapons system. If these three steps are applied in all test cases, many of the unforeseen pitfalls will be avoided. These unforeseen pitfalls have been called many things; some test programs identify them as "known unknowns." It is the job of the tester to flush out problems, and this can only be accomplished with a complete and logical test plan. The three steps in any test plan are

1. The static test
2. The "warm fuzzy" (or functional evaluation) test
3. Compliance

The static test is just what the name implies, a ground test designed to see if the system works to any degree of confidence while sitting on the ground. It is not a lab test and cannot be replaced by a lab test. There is no guarantee that the way the software/hardware/firmware worked in the lab is going to be the same when integrated with the aircraft. In the case of our communications test, we would exercise all test conditions on the ground with a cooperating aircraft and ground station. You need to be careful as to where the aircraft is located: ensure a clear LOS to the receiver and ensure that the aircraft is far enough away from items that may cause multipath, blockage areas, or electromagnetic interference (EMI) (covered in section 4.13 of this text). If the modes of the radio do not work on the ground, it is a pretty safe bet that they will not work when you are flying 300 miles from home. If some of the functions do not work, fix them and perform another static test. Some testers may argue that if some functions do not work on the ground, then just evaluate the functions that do work in the air and test the nonworking functions at a later date when they are working. Does the term regression test come to mind? What guarantee do we have that fixes to a system have not impacted previously tested functions? The worst-case scenario will have the system completely retested. Thinking back to the first chapter, we must learn to test smarter and more efficiently.

The "warm fuzzy" test is a term used by the author and entails an evaluation that will not cost an inordinate amount of money, will not collect data for statistical analysis, but will give you a nice warm feeling that the system, when evaluated, will produce good results. In technical terms, this test is a true functional evaluation of the system and its capabilities. In our communications test, for example, the specification calls for a maximum range of 300 nm. We may want to test at intermediate ranges of 75 and 150 nm. If volume and clarity are not very good at 150 nm, we probably would not have a "warm fuzzy" about 300 nm. If, on the other hand, our test plan called for us to fly to 300 nm and evaluate the communications system in all modes of operation, what statement could we make about the system if nothing worked at the test conditions? The only statement we could possibly make is that the system does not meet specifications. Why? We have no baseline of past performance and therefore cannot determine the cause of the problem.

Compliance is the easiest of all the tests, and that is why testers tend to gravitate toward these tests when building a test plan. The compliance test places the system at the required specification conditions and gathers enough data to prove to some confidence level that the system is or is not compliant. But here is the problem. Suppose that we put together a test plan which shows compliance only. What happens the first time a system does not work at the test conditions? After the system is reworked, we refly the test conditions. But wait a minute. You have already used up the time and resources that you needed to complete the test. Now you have to refly, which means adding more time and schedule to the evaluation period. The bottom line is that it is easier to remove tests from a test plan than to add tests to a test plan. Build a comprehensive test plan in the beginning and save yourself a lot of aggravation later.

▮ 4.5 ▮ COMMUNICATIONS TEST PLAN MATRIX

Revisiting our specification paragraph for our communications system, we are now ready to devise a test matrix that will cover all of the test conditions necessary to evaluate this system. After the matrix is complete, we can incorporate it into our test plan. All avionics and weapons systems will require the use of a condition matrix.

The matrix in Table 4.2 merely identifies the system specifications. The tester must identify how many test conditions (not test points) are required in order to adequately evaluate this system. Some of the areas are quite easy, while some require additional thinking.

- Direction: There are only two directions, air-to-air and air-to-ground, and each must be tested, so the total test conditions are two.

- Frequency: In our example, there is only one frequency band—UHF—but as mentioned earlier, each band is further subdivided into low, medium, and high. The total test conditions here are three.

- Range: The maximum range is 300 nm, a zero range really equates to the static test. The tester has a few options. One could test at zero and 300 nm, but as addressed earlier, this is not a good option. Suppose a test is completed at every 75 nm. This would equate to five test conditions.

- Azimuth: The system should maintain communications throughout the omnidirectional range, which is a fancy way of saying 360° about the aircraft. We could test at the cardinal headings, but that would probably not be enough. We could test at every degree, but that would be overkill. Each 30° would seem a good compromise, but this will be at the discretion of the testing agency; potential blockage areas from perturbations on the aircraft (propellers, other antennas, or external stores) may call for smaller increments of testing. A test at every 30° would yield 12 test conditions.

- Elevation: The elevation requirement allows the aircraft to bank without losing communications. As with the azimuth parameter, the tester needs to choose a representative number of elevation parameters. Each 15° would seem logical, and this would provide us with six test conditions.

- Encryption: There is either encrypted mode or nonencrypted mode, which provides us with two test conditions.

- Modulation: The operator may select AM or FM, which provides us with two test conditions. **NOTE:** Many terms are used to convey encryption, and confusion may arise when using them. For example, terms used include clear, plain, red, green, black, etc., all of which may mean different things to different communities.

By inserting our test conditions, Table 4.2 now becomes Table 4.3.

TABLE 4.2 ▪ Communications Matrix

ARC-XXX	Direction	Frequency	Range	Azimuth	Elevation	Encryption	Modulation	Total
Specification	AA/AG	UHF	0–300 nm	0–360°	+30° to −45°	Not encrypted/ Encrypted	AM/FM	

TABLE 4.3 ▪ Communications Test Conditions

ARC-XXX	Direction	Frequency	Range	Azimuth	Elevation	Encryption	Modulation	Total
Specification	AA/AG	UHF	0–300 nm	0–360°	+30° to −45°	Not encrypted/ Encrypted	AM/FM	
Conditions	2	3	5	12	6	2	2	8640

The funny thing about these test matrices is that they are multiplicative. That is, in order to calculate the total number of test conditions, you must multiply across the matrix. Table 4.3 shows the total to be 8640 test conditions. It is also important to note that these are *test conditions* not *test points*. You may require six test points, or data points, at each test condition in order to achieve a desired confidence level. In our example, the total of 8640 test conditions would grow to 51,840 test points. To the layman, this would appear to be an unrealistic number of data points to prove a relatively simple system is compliant. However, it is likely not unrealistic, and this is the nemesis of communications system testing. The matrix is a logical test condition buildup and is technically the best way to approach the compliance issue.

The problem that most program or project managers have is equating test points to test flights. If their background is in vehicle flight testing, the problem is more severe. Management may address flight testing with the premise that, historically, test programs accomplish about 10 test points per flight. Using this logic, your communications test will take in excess of 5000 flights. While 10 test points per flight may be valid for envelope expansion or stores separation, it is certainly not valid for avionics testing. Average data rates recorded by instrumentation are approximately 25 samples/sec. If time, space, position information (TSPI) data can be recorded at the same rate, 25 data points are recorded every second at the test condition. This is an important concept that must be understood by all players in the test program. A misunderstanding of what is being collected, and how long it will take, always creates trouble for the avionics tester. This is because metrics are formulated to show progress, and the metrics have no bearing on reality. One of the favorite measures of progress is total test points versus time, or the test point burn-down.

Assume that we are going to show compliance for a navigation system as well as the communications program we just addressed. Assume that for the navigation program we need to accomplish six navigation runs for accuracy. In order to account for the Schuler cycle (the Schuler cycle is discussed in section 5.1 of this text), each navigation run needs to last approximately 84 min. In short, each flight will produce one navigation point, or six flights will be required to complete the navigation portion of the program. Also assume that we can complete our communication matrix in two flights. How will this look on our progress metrics? Figures 4.1 and 4.2 show the two possible metrics, and neither of them provide any worthwhile information. For the example, the navigation flights are flown first, followed by communication flights.

Figures 4.1 and 4.2 show that there was no progress on the test program for the first six flights. In reality, in terms of flights required, each flight accomplished 12.5% of the test program. Unless the program is fully understood, management would expect a burndown of approximately 8000 test points per flight. By the time the second flight is

FIGURE 4.1 ▪ Test
Point Progress

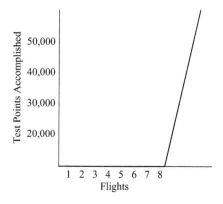

FIGURE 4.2 ▪ Test
Point Burn-Down

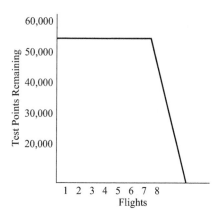

complete, hundreds of hours will have been spent on meaningless discussions of why
you are behind schedule.

4.6 | EXECUTING THE MATRIX

Some thought must also go into executing the test matrix. The first setup is our static
test. Examination of the matrix reveals that 20% of our test conditions can be accomplished here without ever leaving the ground. Having satisfied ourselves that the system
works well enough to fly, we can plan our first flight condition. Our first planned range
is 75 nm, but what do we plan to do once we are there? If we set up an orbit, we can
accomplish most of the required conditions. The one condition that we cannot satisfy is
the elevation change. A constant bank turn will provide a fixed elevation, positive on
one side of the circle and negative on the other side of the circle. We would like to
obtain the full range of elevation change throughout the full range of azimuth change. In
order to accomplish this task we must change the geometry. Depending on aircraft
capabilities, the tester should investigate cloverleaf turns, rolling out on different
headings, or some kind of modified Cuban-8 turns. These geometries will offer ever-
changing elevations with azimuth, thus maximizing the data collection per unit time.

According to our test plan, the next range to be attempted would be 150 nm,
but suppose our communications suite performed flawlessly at the 75 nm condition?

Perhaps we would defer the 150 nm condition and proceed to the 225 nm condition. In this instance, 20% of our required test points would be deferred. That looks good for us, and it is much easier to delete test points from the plan than to add them to a test plan that was not thought out completely. We would progress along this way until the matrix is complete.

4.7 | OTHER CONSIDERATIONS IN COMMUNICATIONS TESTS

There are other considerations that were not specifically addressed in the test matrix. These considerations include

- Effects of stores, landing gear, and flaps
- Effects of weather
- Logistics
- Boredom
- Speech intelligibility
- EMI/EMC

4.8 | EFFECTS OF STORES, LANDING GEAR, AND FLAPS

Because of the location of the antennas on the aircraft, it is entirely possible that certain perturbations such as stores, landing gear, and flaps can have a deleterious effect on radio performance. Aircraft that carry stores externally can be troublesome to the tester due to the many configurations that are possible. In-flight communications tests are normally conducted in the configuration in which the aircraft would normally operate. These tests are rarely conducted in a clean configuration unless there are specific reasons for doing so. Multiple configurations or new configurations (after compliance has been shown) can be accomplished on the ground in some anechoic facilities by performing antenna pattern tests. These tests are described in section 4.10. Communications tests in the landing configuration (gear and flaps down) are normally tested at the nearest range (i.e., in the matrix we discussed at the 75 nm test condition). Testing at the longer ranges is done in the cruise configuration (gear and flaps up).

4.9 | EFFECTS OF WEATHER

As described in section 4.1, weather may have an effect on aircraft communications. Since this is a known phenomenon, most test plans do not identify communications in weather as a separate test condition. What needs to be done is to note the atmospheric conditions for each test. Temperature, pressure, absolute humidity, precipitation, and cloud cover should be addressed for all testing involving RF energy. These conditions also need to be noted for radio aids, radar, EO, and infrared (IR) testing, as atmospherics will affect performance. When analyzing test results, these data can explain differences in day-to-day observations of the same system.

4.10 | LOGISTICS

Some of the logistics areas that need to be addressed for communications testing are crypto keys (similar equipment for cooperative systems) and antenna pattern facilities. Crypto keys will be necessary during military testing when exercising encryption systems such as KY-28, KY-58, and KG-84. The important thing to remember here is that these keys are classified SECRET and will require a communications security (COMSEC) custodian. Military organizations normally do not have too many problems obtaining these keys, but I have seen contractor organizations take as long as 6 months to acquire keys.

Many testers make the mistake of attempting to test new radios with cooperative systems that do not have the same capabilities as the test aircraft. Ground stations, telemetry centers, and cooperating aircraft need to have the same capabilities as the system under test in order to accurately test the communications suite. In addition, the aircrew equipment (helmets, microphones, connectors, etc.) must also be compatible with the system under test.

Antenna pattern testing is a maligned and neglected test in the test community, and is often not scheduled unless there is an actual problem. An antenna pattern is used to describe the relative strength of a radiated field 360° about the antenna at a fixed distance from the receive site. The radiation pattern is a reception pattern as well, since it describes the receiving properties of the antenna. Antenna patterns may be accomplished in an anechoic chamber, with the aircraft mounted on a rotating pedestal, or flight tested against an antenna farm. All three tests will provide the tester with the desired information; which test is used is a matter of availability and cost. The anechoic chamber may be the least available at a lower cost, while flight testing may be available, but it is the most costly. In all of the tests, a transmitted signal of known strength is sent through the aircraft antenna and received by antennas in the test facility. The receivers measure and plot the signal strength as the aircraft changes aspect in space. The radiation pattern is measured in three dimensions; however, since it is difficult to display, most depictions are a slice of the three-dimensional (3D) measurement portrayed in two dimensions. Figure 4.3 is a typical presentation of an antenna pattern. The nose of the aircraft is at the top of the plot, right wing to the right side of the plot, etc.

It can be seen that if Figure 4.3 represented our UHF radio in the previous exercise, we would have some serious problems with reception in the aft quadrant of the aircraft. It would be nice to know that we had these problems before we expended valuable flight test resources. Many programs fail to recognize the need for antenna pattern testing, thinking that it is an additional test that ties up the aircraft. Sometimes they are lucky and no serious problems surface. But other times they are not, and engineering rework must be done and antenna patterns must be flown to validate the rework, all at a serious cost to the schedule. As a rule of thumb, anytime you add, modify, or relocate antennas on an aircraft you need to perform antenna pattern testing. The addition of new stores may require pattern work based on engineering analysis.

There is some good news. As of this date, many anechoic facilities can now perform antenna pattern testing, which will reduce flights and testing offsite.

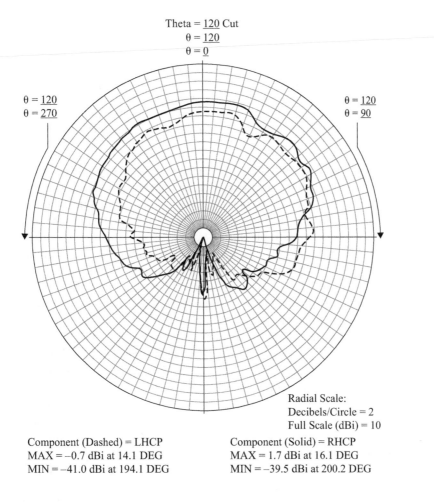

Theta = 120 Cut
θ = 120
θ = 0

θ = 120
θ = 270

θ = 120
θ = 90

Radial Scale:
Decibels/Circle = 2
Full Scale (dBi) = 10

Component (Dashed) = LHCP
MAX = −0.7 dBi at 14.1 DEG
MIN = −41.0 dBi at 194.1 DEG

Component (Solid) = RHCP
MAX = 1.7 dBi at 16.1 DEG
MIN = −39.5 dBi at 200.2 DEG

FIGURE 4.3 ■
Antenna Pattern

4.11 | BOREDOM

I mention boredom because these flights are some of the least liked flights by pilots. If you want to ingratiate yourself to the pilot community, try to add communications testing with other testing so that the entire flight is not dedicated to boring holes in the sky talking on the radio. A bored pilot is not the best candidate to conduct a test. Instead of completing the communications matrix on one flight, try to spread it over three or four flights.

4.12 | SPEECH INTELLIGIBILITY

Have you ever heard pilots or aircrew members speaking over the headsets saying things like "I have you 5 by 5," or "I've got you 3 by 3"? Have you ever wondered what they are talking about? These statements are the listener's perception of the quality of the radios. They use a 5-point scale for loudness and clarity. A 5-by-5 system would be loud and clear, whereas a 3-by-3 system would be weak but readable. I have never heard a 2-by-2 system. But these are only the listener's perceptions and are not quantitative

TABLE 4.4 ■ SINPO Rating Scale

(S)ignal	(I)nterference	(N)oise	(P)ropagation	(O)verall
5 Excellent	5 None	5 None	5 None	5 Excellent
4 Good	4 Slight	4 Slight	4 Slight	4 Good
3 Fair	3 Moderate	3 Moderate	3 Moderate	3 Fair
2 Poor	2 Severe	2 Severe	2 Severe	2 Poor
1 Barely Audible	1 Extreme	1 Extreme	1 Extreme	1 Unusable

enough to judge a system's intelligibility. Testers may come in contact with the SINPO rating, which is used when making reception reports to broadcasters. The code addresses signal strength, interference (from other stations), noise (from atmospheric conditions), propagation disturbance (or fading), and an overall rating. A numerical description of the SINPO code is shown in Table 4.4.

There are a couple of ways of quantifying the intelligibility of a communications system. The first is a rhyme or talker test; the second is the speech transmission index (STI).

Speech intelligibility is measured directly as a percentage of words, rhymes, or sentences that are heard correctly by a number of listeners. A number of talkers will speak words or sentences and a number of listeners will indicate what they hear. The percentage of right answers is used to quantify the intelligibility of the system. A modified rhyme list (Table 4.5) is used for the speech intelligibility test during testing. There are 50 six-word rhyme groups on the rhyme list. The speaker selects one word from each of the 50 groups and the listeners write down the word that they hear. A percentage of right and wrong answers is compiled to determine the speech intelligibility of the system. The Federal Aviation Administration (FAA) recognizes the American Standards Association (ASA) Paper S3.2-1960 as an acceptable rhyme list. A minimum of 75% correct responses is needed for acceptable speech intelligibility.

Another method of determining speech intelligibility is the STI. When determining intelligibility in an auditorium, a specially modulated noise test signal is sent from a loudspeaker at the speaker's location. STI expresses the ability of the channel to carry across the characteristics of a speech signal. STI test signals are based on modulated, speech-shaped noise. A receiver with a microphone gives a direct readout of the STI value at the receiver's location.

The influence that a transmission channel has on speech intelligibility is dependent on

- The speech level
- Frequency response of the channel
- Nonlinear distortions
- Background noise level
- Quality of the sound reproduction equipment
- Echos (reflections with delay >100 ms)
- The reverberation time
- Psychoacoustic effects (masking effects)

TABLE 4.5 ■ Modified Rhyme List

1.	sun	nun	gun	run	bun	fun
2.	kit	kick	kin	kid	kill	king
3.	bust	just	rust	dust	gust	must
4.	pill	pick	pip	pit	pin	pig
5.	ban	back	bat	bad	bass	bath
6.	rent	went	tent	bent	dent	sent
7.	pad	pass	path	pack	pan	pat
8.	bill	fill	till	will	hill	kill
9.	gang	hang	fang	bang	rang	sang
10.	sun	sud	sup	sub	sung	sum
11.	pave	pale	pay	page	pane	pace
12.	safe	save	sake	sale	sane	same
13.	tang	tab	tack	tam	tap	tan
14.	gale	male	tale	pale	sale	bale
15.	test	nest	best	west	rest	vest
16.	pub	pus	puck	pun	puff	pup
17.	pop	shop	hop	cop	top	mop
18.	name	fame	tame	came	game	same
19.	sin	sill	sit	sip	sing	sick
20.	sip	rip	tip	lip	hip	dip
21.	may	gay	pay	day	say	way
22.	sin	win	fin	din	tin	pin
23.	soil	toil	oil	foil	coil	boil
24.	cuff	cuss	cub	cup	cut	cud
25.	wig	rig	fig	pig	big	dig
26.	sap	sag	sad	sass	sack	sat
27.	tick	wick	pick	kick	lick	sick
28.	lot	not	hot	got	pot	tot
29.	park	mark	hark	dark	lark	bark
30.	seen	seed	seek	seem	seethe	seep
31.	dun	dug	dub	duck	dud	dung
32.	beach	beam	beak	bead	beat	bean
33.	did	din	dip	dim	dig	dill
34.	led	shed	red	wed	fed	bed
35.	peas	peal	peach	peat	peak	peace
36.	tease	teak	tear	teal	teach	team
37.	map	mat	match	mad	mass	man
38.	came	cape	cane	case	cave	care
39.	keel	feel	peel	reel	heel	eel
40.	gold	hold	sold	told	fold	cold
41.	paw	jaw	saw	thaw	law	raw
42.	race	ray	rake	rate	rave	raze
43.	bit	sit	hit	wit	fit	kit
44.	fizz	fill	fib	fin	fit	kit
45.	lame	lane	lace	late	lake	lay
46.	bus	buff	bug	buck	but	bun
47.	cook	book	hook	shook	look	took
48.	hen	ten	then	den	men	pen
49.	meat	feat	heat	neat	beat	seat
50.	heal	heap	heath	heave	hear	heat

STI is a numeric representation measure of communication channel characteristics whose value varies from 0 = bad to 1 = excellent.

0	STI	0.3	0.45	0.6	0.75	1.0
BAD		POOR	FAIR	GOOD	EXCELLENT	

In aircraft testing, a digitized signal is broadcast through the aircraft's communication antenna. A receiver on the ground receives and records the transmitted signal. This signal is then sent through a processor where a bit error rate (BER) is determined, and hence the intelligibility of the signal. An application used with some success at NTPS is iSTI, which is an iOS compatible application for measuring the Speech Transmission Index using STIPA test signals. The application complies with IEC-60268-12 4th edition (2011), and features a STIPA (Speech Transmission Index for Public Address Systems) signal generator and a Speech Transmission Index analyzer.

There are two schools of thought aligning themselves with each of the described methods. The STI group will tell you that the STI method is superior to the talking rhyme method because the human element is removed and there is no ambiguity in the assessment. The rhyme school insists that the human element must be considered when assessing communications systems and that STI is better left to evaluate data transfer systems such as automatic dependent surveillance (ADS) or the military data link system. Either method is acceptable in performing speech intelligibility testing on aircraft communications systems.

▌ 4.13 | ELECTROMAGNETIC INTERFERENCE/ ELECTROMAGNETIC COMPATIBILITY

Equipment and systems on modern aircraft need to be able to perform the functions for which they were intended, free from interference from other aircraft systems and equipment. At the same time, their operation should not adversely affect the function or performance of other aircraft systems. This simple concept is the explanation of EMI/ EMC. The interference portion (EMI) is undesirable. The ability to coexist peacefully with other equipment is the desired compatibility (EMC). This section is devoted to the test requirements necessary to ensure EMC.

Prior to addressing the testing required to eliminate EMI and ensure EMC, a small review of electromagnetics must be undertaken. As mentioned in section 4.1, anytime we are tasked with formulating a plan for any system we should first review the system theory and operation. The same is true for EMI/EMC test planning. There are a number of good texts on electromagnetics, electrostatics, flux, etc. There is one text that I find myself referencing constantly which should be a required reference for all flight test engineers. It is called *The Engineer's Manual*, by Ralph G. Hudson (London: John Wiley & Sons).

Remember from your basic electronics courses that current is the electron flow in a circuit. As electrons move in the circuits, wires, and antennas, they produce mutually perpendicular electric and magnetic lines of force, or electromagnetics. These lines of force are known as fields. Electromagnetic (EM) fields radiate outward like wheel spokes and are continuous about the charge. These fields are steady when the current is steady and expand and contract as the current is changed. Alternating current (AC) defines the electron flow as rapidly changing, anywhere from 1 to 1 billion Hz. AC power can be created by alternators or transmitters, or by transforming direct current (DC) to AC. DC is created by batteries, generators, or transformer rectifiers and produces a constant EM field unless switched. High-speed switching (such as computers and data busses) causes the EM field to behave much like the field created by AC.

Electromagnetics are created whenever current is flowing. Electromagnetics can be created internally, by any aircraft-powered equipment (e.g., a sensor, radar, radio, generators, data busses, etc.). Electromagnetics can also be created externally to the aircraft by numerous sources, other aircraft, microwave relays, transmission lines, lightning, and ground-based emitters. When dealing with EMI/EMC, the test community is concerned with two types of designations: emission and susceptibility. Conducted emission (CE) is the noise inducted in a system through connected conductors via connectors and cables. Radiated emission (RE) is the noise added to a system through radiated power. Conducted susceptibility (CS) is improper operation of a system due to conducted noise. Radiated susceptibility (RS) is improper operation of a system due to radiated noise. The task of the test community is to track down sources of electromagnetics and protect the aircraft and its associated systems from them.

Almost all RF transmissions create base frequencies and harmonics. This implies that an RF transmission out of the band of interest can still create interference within the band of interest. An example of this type of interference was found during KC-10 operations. It was determined that HF transmissions caused the digitally controlled refueling boom to move randomly. Aircraft tend to be susceptible to interference due to the aircraft construction and the wealth of good conductors within the aircraft. Another problem unique to aircraft is the close proximity of equipment, especially on smaller fighter-type aircraft where real estate is at a premium. Larger commercial carriers are noticing an increase in EMI as their systems become digitized and the riding public is carrying more and more electronic devices onto aircraft. Cell phones, laser pointers, personal computers with CD-ROMs, and personal video games are a few of the offenders. These systems can interfere with autopilots, nose wheel steering, communications, navigation, and even fire detection systems. Aftermarket installations, such as in-seat personal entertainment systems, must be thoroughly checked for possible interference problems.

The problems can be internally generated by systems on the aircraft. Testers are often amazed that the first time the telemetry is turned on the GPS signal is lost. Telemetry systems for aircraft data normally operate in the L band, which is where GPS resides. Unfortunately there are many other systems that also operate in the L band, so much so that there is very little available bandwidth within this frequency band. Simultaneous operation of these many systems (GPS, TACAN, multifunction information distribution system [MIDS], traffic alert and collision avoidance system [TCAS], air traffic control mode S, etc.) can be a very difficult prospect. Other examples of internally generated interference is the 400 Hz noise (from aircraft generators) on the telemetry signal, static or warble over the headset, or a VHF transmission that causes an instrument landing system (ILS) deviation. Particular attention must be paid to data bus operation in the presence of interference. Pilots can be given erroneous readings on their display suites, or worse yet, noise may enter the digital flight control system (DFCS), causing loss of the aircraft or loss of life.

High-intensity RF fields (HIRFs) can be generated by transmitters either on or off the aircraft. Transmitted energy from microwave stations, radars, electronic warfare jammers, or self-protection pods can be injurious to aircrew and aircraft. For this reason, testers must examine certain tactics and operating doctrines to determine operations that may place the aircraft in the presence of EM fields that are of greater strength than what may normally be seen. An example is a fighter being escorted by an EA-6B employing jamming operations from its ALQ-99 system. A more typical problem may be mutual radar interference between two fighters flying in formation. These fields need to be calculated and added to the test matrix.

Specific examples of EMI in the aviation community are numerous; some with disastrous results. Five crashes of UH-60 Blackhawk helicopters (which killed or injured all on board) shortly after their introduction into service in the late 1980s were found to be due to EMI from very strong radar and radio transmitters affecting the electronic flight control system (in particular, the stabilator system). A U.S. Army UH-60, while flying past a radio broadcast tower in West Germany in 1987, experienced uncommanded stabilator movement, spurious warning light indications, and false cockpit warnings. Investigation and testing showed that the stabilator system was affected by EMI from HIRFs. Apparently when the Blackhawk was initially tested, it was not routinely flown near large RF emitters. The U.S. Navy version of the Blackhawk, the Seahawk, has not experienced similar EMI problems because it is hardened against the severe EM environment found on ships.

An F-16 crashed in the vicinity of a Voice of America radio transmitter because the DFCS was susceptible to the HIRFs transmitted. Since the F-16 cannot fly without the aid of the DFCS, any interruption can be fatal. Many F-16s were modified to prevent this type of EMI, which was caused largely by inadequate military specifications for their avionics systems. The F-16 case history was one of the reasons for the formulation of the FAA's HIRF certification program. A German Tornado also crashed near a Voice of America transmitter in 1984 near Munich.

Weapons systems are not exempt either. During the 1986 U.S. air strike on Libya, several missiles failed to hit their designated targets and one F-111 crashed. The U.S. Air Force concluded that the cause was due to mutual interference within the strike package. The initial development of the F-117 revealed serious problems with the targeting lock-on system. The problem was linked to poor shielding techniques and old hardware designs. In Australia, two warships nearly collided when the radar beams of one disabled the steering of another. The mine hunter HMAS *Huon* (a state-of-the-art coastal mine hunter) went out of control and veered across the bow of the frigate HMAS *Anzac*. The Australian National Audit Office reported the shortcomings in the testing and evaluation of new defense equipment, especially in the areas of EMI and EMC. During a B-52 missile interface unit test, an uncommanded missile launch signal was given. One of the contributing factors was cross-talk in the system wiring; a second factor was not adhering to the EMC control plan requirements. The program only recommended after a year-long redesign and test effort. During the Falklands War, the British ship HMS *Sheffield* sank with heavy casualties after being hit with an Exocet missile. Despite the *Sheffield* having the most sophisticated antimissile defense system available, the system created EMI that affected radio communications with the Harriers assigned to the ship. When the Harriers took off or landed, the defense system was disabled to allow communications, which also allowed an opportunity for the Exocet to target the ship.

The civilian side of the house is not exempt either. A National Oceanic and Atmospheric Administration (NOAA) satellite experienced phantom commands when flying over Europe. Controllers determined that the commands were due to susceptibility to the heavy VHF environment over Europe. A blimp flying near the Voice of America transmitter in Greenville, NC, experienced a sudden double engine failure. An investigation revealed ignition failure due to extreme EMI; the blimp was flying with a new ignition system. Operational frequencies of cellular phones, computers, radios, and electronic games are often EMI sources on air carriers. A DC-10 autopilot was disrupted during final approach by a passenger operating a CD player.

Modern medical equipment has also experienced EMI problems. FAA data show that cell phones appear to be the main culprit, but there are others. An ambulance was transporting a heart attack victim to the hospital. The patient was attached to a defibrillator, unfortunately, every time the technician keyed the radio to request medical advice the defibrillator shut down. The patient died. The cause was traced to the installation of a new fiberglass roof with a long-range radio antenna. Reduced shielding, combined with the strong radiated signal, resulted in EMI affecting the machine. Electric wheelchairs have come under scrutiny due to reports of uncommanded movements and erratic behavior due to EMI.

Car owners have been locked out of their cars by EMI affecting their electronic entry keypads, garage door openers have been activated by overflight of airplanes, and advanced brake systems (ABSs) have failed due to outside transmissions. In all cases, the threat of EMI was not fully understood or envisioned by the evaluator.

The U.S. military uses military standards for the measurement and control of emissions. MIL-STD-461D is the military standard governing the requirements for the control of electromagnetic interference emissions and susceptibility. MIL-STD-462D is the military standard for the measurement of electromagnetic interference characteristics. These two standards have been combined into MIL-STD-461F, which is an interface standard, "Requirements for the Control of Electromagnetic Interference Characteristics of Subsystems and Equipment." This standard describes the required tests and test setups for conducted and radiated emissions.

MIL-STD-464A is the interface standard for the electromagnetic environmental effects requirements for systems. It is concerned primarily with protection from external electrical sources operating from about 10 kHz to about 40 GHz and also includes effects of lightning. MIL-STD-1605 (SHIPS) describes the procedures for conducting a shipboard EMI survey for surface ships.

Another military publication that will be useful is MIL-HDBK-237D, "Electromagnetic Environmental Effects and Spectrum Certification Guidance for the Acquisition Process." This handbook is especially beneficial for test planners, as it identifies all of the electromagnetic environmental effects (E3) considerations throughout the testing process. Each of the military documents also provides the reader with other applicable and relevant documents, including standing NATO agreements (STANAGs).

On the civilian side, HIRF requirements are described in SAE Aerospace Recommended Practice ARP 5583, "Guide to Certification of Aircraft in a High Intensity Radiated Field (HIRF) Environment," and airworthiness requirements are called out in AC 20-158, "The Certification of Aircraft Electrical/Electronic Systems for Operation in the High Intensity Radiated Fields (HIRF) Environment." An additional advisory circular, AC 25.899-1, "Electrical Bonding and Protection Against Static Electricity," pertains to Part 25 compliance with CFR 14, §25.899.

▌ 4.14 ▐ EMI/EMC TESTING

As described in the previous paragraphs, the test procedures for conducting EMI/EMC evaluations are delineated in the appropriate documentation. A generic description of the required testing is provided here for general considerations.

There are four phases of EMI/EMC testing:

- Bonding
- Victim/source
- Field testing (baseline)
- Field testing (HIRF/near field)

The first test that must be accomplished is a bonding test, which is merely a grounding test of the aircraft and equipment. Systems that are not grounded properly will give erroneous results during the tests. Victim/source testing identifies compatibility problems on the aircraft and is a test that most flight test engineers will be involved with at one time or another. It involves the formulation of a matrix (what else?) pitting the system under test against the other aircraft systems.

Assume that we have installed the new UHF radio system that was discussed earlier. We need to set up a matrix that allows a test of this radio's performance in the presence of other system operations. We set up the UHF radio as our victim and identify all sources of possible EMI from other systems on the aircraft. Table 4.6 shows a possible victim/source matrix for the UHF installation.

This matrix is rather simplistic, as it does not identify all sources of potential EMI that may emanate from the aircraft systems. The final matrix will be slightly larger and may include other submodes of operation. Table 4.6 is a good starting point for understanding the basic concept of victim/source. The left column of Table 4.6 identifies our victim, the UHF radio that was installed. As noted previously, we should consider low, medium, and high portions of the bands during our evaluations. The first test will, of

TABLE 4.6 ■ Victim/Source Matrix

Source	Engines	Radar	Radios	Radar Altimeter	Telemetry	IFF	TACAN	EW	TCAS	Data Link
	Off	PS	VHF	On	On	Identify	A/A T/R	RGS	On	TACAN
	On	WX	-Low	Off	Off	Off	A/G T/R	VGS	Off	RelNav
		STT	-Medium					Noise		PPLI
		TF	-High							Off
		TA	HF							
		RWS	-Low							
		TWS	-Medium							
		ACM	-High							
		Off	Off							

Victim
UHF
-Low
-Medium
-High

course, be a static test to obtain the baseline performance of the radio and to ensure that there is no interference from either external power or battery operations.

The method of test would be as follows:

1. Engines off, tune in a low-frequency UHF channel and listen for voice.
2. No other systems in the aircraft shall be turned on.
3. Turn on the radar (if possible with external power only).
4. Select pulse search on the radar.
5. Listen for any change on the headset (static, warble, etc.).
6. Select weather mode (WX).
7. Listen for any change on the headset (static, warble, etc.).
8. Repeat the procedure for the next six modes of the radar.
9. Turn off the radar.
10. Select a low VHF channel and key the mike (transmit).
11. Monitoring UHF, listen for any change on the headset.
12. Select a medium VHF channel and repeat the procedure.
13. Step through the remaining modes of VHF and HF.
14. Turn off the radio (not the UHF radio).
15. Turn on the radar altimeter and note any changes on the headset.
16. Turn off the radar altimeter.
17. Continue the same sequence until exhausting the sources (data link).
18. Still on external power, select a medium UHF channel and listen for voice.
19. Repeat the entire procedure through data link.
20. Still on external power, select a high UHF channel and listen for voice.
21. Repeat the procedure again through data link.
22. Return to the low UHF.
23. Engines on and note any changes in the headset.
24. Repeat the entire matrix of testing for low, medium, and high UHF.

But we are not finished yet. Since combinations of system operations can produce harmonics which are more severe than individual operation, a new matrix of combined operation of sources must be formulated. Unless you are an electrical engineer, you will probably need some help in identifying the worst combinations of systems. The matrix will be relatively small and not nearly as time consuming, but it still must be repeated for low, medium, and high UHF. And we are still not finished.

Since the UHF radio that we installed can generate an RF transmission, there is a possibility that the transmit side of the radio may adversely impact operations of other systems on the aircraft. Remember, compatibility is a two-way street. Compatibility is the ability of a system to perform its intended function without interference from other systems *and* not have an adverse impact on other systems while performing its function. In this case, we must construct another matrix where the UHF radio that was installed is now the source while other aircraft systems become the victims. Table 4.7 shows this new matrix.

TABLE 4.7 ■ Source/Victim Matrix

Victim	GPS	Radar	Radios	Radar Altimeter	Telemetry	IFF	TACAN	Flight Control System	TCAS	Data Link
	On	PS	VHF	On	On	Identify	A/A T/R	L Wing	On	TACAN
	Off	WX	-Low	Off	Off	Off	A/G T/R	R Wing	Off	RelNav
		STT	-Medium					Stab		PPLI
		TF	-High					Rudder		Off
		TA	HF							
		RWS	-Low							
		TWS	-Medium							
		ACM	-High							
		Off	Off							

Source
UHF
-Low
-Medium
-High

For this second series of tests the engines have been removed from the matrix as a possible victim. It has been replaced by GPS, which is a passive receiver that may be affected by radio transmissions. In a similar fashion, the EW jammer has been replaced in the matrix by the flight control system (FCS). As was the case with Table 4.5, this matrix is also not complete, as things such as the display subsystem should also be included. The final test matrix will be somewhat larger, including all systems that may be affected by the radio transmissions. The execution of this matrix is slightly different than what was described for Table 4.6. In this iteration, engines will be running and the source radio as well as the victim systems will be turned off. The individual victim systems will be turned on, verified for proper operation, and then the radio will be keyed. The method of test would be as follows:

1. Start engines.
2. All victim and source systems off.
3. Turn on GPS, acquire satellites, and verify a valid position and groundspeed.
4. Select a low UHF channel and key the radio.
5. Note any deviations in the GPS latitude, longitude, and groundspeed.
6. Stop the radio transmission.
7. Verify GPS.
8. Select a medium UHF channel and key the radio.
9. Note any deviations in the GPS latitude, longitude, and groundspeed.
10. Stop the radio transmission.
11. Verify GPS.
12. Select a high UHF channel and key the radio.
13. Note any deviations in the GPS latitude, longitude, and groundspeed.

14. Stop the radio transmission.

15. All potential victims in the matrix will be tested in the same manner.

16. For the case of the FCS, observers should be posted at the control surfaces.

17. For each radio transmission, the observers are to look for any movement.

It may be necessary to repeat some of the matrix with combinations of the UHF radio and other systems on if the combined harmonics are of concern.

The field testing requirement is performed in accordance with MIL-STD-461E and involves

- Conducted emissions tests
- Conducted susceptibility tests
- Radiated emissions tests
- Radiated susceptibility tests

The applicability of these tests depends on many things, such as the branch of service (Air Force, Navy, or Army) and the intended use of the system and location of use (air, sea, space, ground support, etc.). The strength of the field is also dictated by the interface standard, but tends to a nominal 200 V/m. Operational requirements may dictate that these tests be conducted under more severe field strengths. These procedures then become the near-field tests.

4.15 | EMI/EMC TEST ISSUES

The interface standard also addresses some miscellaneous issues that need to be mentioned here. The field tests must be conducted in an anechoic facility, since clear air will not meet the minimum requirements for noise. I have seen some test organizations perform victim/source testing in a clear-air (outside) environment. I recommend against this procedure. If a potential problem is detected during victim/source testing in a clear-air environment, there is no guarantee that this problem was caused by a controlled emission. External RF sources, not under control of the test team, may be responsible for the anomaly.

Anechoic facilities, like any other test asset, must be scheduled. There are programs that use the facility for months at a time, so entire blocks of a year may be unavailable. Scheduling a chamber block of tests a year or more in advance is not unusual and is something to be considered in the planning phase. Depending on the field requirements, the chamber configuration may have to be changed (larger cones for attenuation, for example). The interface standard provides equations for determining the adequacy of the facility.

4.16 | EMI/EMC ELIMINATION

The simple part of the EMI/EMC puzzle may be finding that you have a problem. It may be more difficult to eliminate the problem. Historically the easiest way to eliminate the problem was by isolation. Equipment can be isolated in two ways: physically and electrically. Physical isolation requires the separation of operating equipment on an aircraft. While this may be a possibility on a large cargo-type aircraft, it is virtually impossible on today's fighter and bomber aircraft. Electrical isolation involves good grounding and electrical shielding. Components can be installed in metal boxes that are well grounded; cables can have grounded shields around the primary conductor.

As more and more technology and equipment is added to the aircraft, more inventive ways have to be devised to eliminate interference. Some of the additional ways to eliminate interference are

- Directional antennas
- Multiple filter assemblies (MFAs)
- Time-sequenced operation or blanking techniques
- Procedural changes

As the name implies, directional antennas allow the antenna to limit its reception to a certain area. This feature reduces noise and can aid in detecting signals that may be at the noise level. The drawback, of course, is that the system is limited in its ability to either direct energy or receive energy from certain areas around the aircraft. In highly dynamic aircraft, this can be a severe handicap. The aircrew may lose the ability to track targets or jam enemy threats based on the aircraft orientation and target aspect.

Another method of reducing interference is to filter out select signals. This method is similar to a Doppler notch on an airborne radar, which notches out certain Doppler returns in order to reduce clutter. An example of this concept is the simultaneous operation of an airborne radar and a radar warning receiver (RWR). Without a filter, the RWR will detect its own radar and display the signal to the aircrew. The display may take the form of video where a "U" (for unknown) may be displayed, or worse, an audio alert may be sounded in the headset. Neither of these indications is desired, as it directs the aircrew away from other tasks to investigate the indication. It also may mask a real threat. We can notch out our onboard radar's signal and eliminate the problem. But what happens when a bona fide threat with very similar characteristics as our own radar paints the aircraft? What happens when our own radar starts to employ frequency agility? Both are examples of where a fix to a problem can cause more problems than the original architecture.

Time-sequenced operation and the employment of blanking techniques are methods of allowing systems to transmit while potential victim systems are rendered unable to receive signals. In the previous example, we could elect to blank the RWR while the radar is transmitting. We would then listen with the RWR between radar pulses. As with the previous example, this method can produce some additional problems.

Figure 4.4 is a simplified bus architecture depicting the mission computer and associated radar, stores, and EW busses. Information between the busses is accomplished by commands from the mission computer. If there is going to be any type of time sequencing or blanking, it must be controlled by the mission computer. As we step through the operations, we will see that the integration that is required is enormous.

1. Radar and RWR operation: Since the radar antenna is not perfect at directing energy, there are some losses through the side and back lobes of the antenna. This energy will be detected by the RWR and a corresponding signal will be displayed to the aircrew. The mission computer can sequence this operation by blanking the RWR when the radar is transmitting and release the blanking signal between radar transmissions.

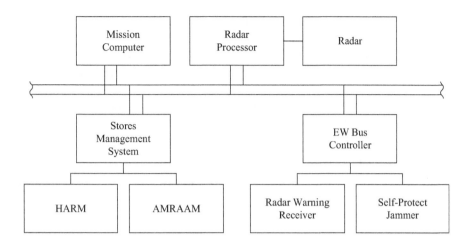

FIGURE 4.4 ■
Simplified Bus
Structure

2. Radar, RWR, and self-protect jammer operations: The self-protect jammer receives incoming RF and compares the signal characteristics to known signals in its threat library (just like the RWR). In the event that a signal is matched to a known threat, the jammer will either automatically attempt to jam the threat or ask the aircrew if they would like to jam the threat. Like the radar, RF from the jammer will be sensed by the RWR, and if in a close enough frequency, by the radar. There are a couple of problems here. First, in order to protect these systems from damage, it is inadvisable to allow high-power RF from a radar or a jammer to enter the receiving antenna of another system. That being known, if the radar is transmitting, the RWR and self-protect system must be off. If the self-protect system is transmitting, the radar and RWR must be off. If the RWR receiver is listening, the radar must be off and the self-protect system must not be transmitting. This sequencing can be accomplished, but certain modes of the radar and jamming effectiveness may suffer.

3. Radar, RWR, self-protect jammer, and high-speed antiradiation missile (HARM) operations: The HARM seeker is a very sensitive receiver that works much like the RWR and jammer receiver. All of the discussion in case 2 holds true here, only now we have added another passive receiver that has the same cautions as the RWR. Since they are both passive, they can listen at the same time and the integration method in case 2 will work here.

4. Radar, RWR, self-protect jammer, HARM, and advanced medium-range air-to-air missile (AMRAAM) operations: The AMRAAM is an improvement over the AIM-7 Sparrow, which was the primary radar-guided missile for many years. During a launch to eject (LTE) cycle, the missile relies on the aircraft radar to update target position. The missile does not expend any power looking for the target during the initial phase of the intercept, which is beneficial for range and battery life. If the missile does not receive updates from the aircraft radar at the specified time interval, it assumes that the radar is gone and the missile goes active. If the missile goes active early, the potential for a target miss is increased. Referring back to case 2, we see that the radar is off whenever the jammer, RWR, or HARM is on. If the radar must remain on to support the launch of the AMRAAM, what can be said about the self-protection of the aircraft? There is none. In a way, the aircrew is lucky that they are carrying an AMRAAM and not an AIM-54C Phoenix, which has a much greater

range and hence a longer period of self-protection loss. Even older missiles such as the AIM-7 can create these same types of problems. The preferred launch method of the AIM-7 is a continuous wave (CW) launch because the AIM-7 needs Doppler shift. CW will provide a continuous shift since the power is continuous. What happens to our self-protection coverage during this launch? It is gone, just like the AMRAAM. An AIM-7 can be launched in a pulse Doppler mode such as track while scan (TWS) or range while search (RWS), but the probability of kill (P_K) is reduced because the Doppler shift in this mode is not as good as in CW.

This leads us to our final method of EMI avoidance.

Procedural changes are a way of eliminating a problem by not placing the operator or the aircraft in a situation where the problem manifests itself. Suppose a problem exists with an aircraft flying at low level. A problem may cause the aircraft to impact the ground. A procedural fix would be to restrict the aircraft from flying at low levels. If the situation is avoided, then the problem goes away. This type of problem avoidance is exhibited in our AIM-7 problem above. We have two options. We can limit AIM-7 launches to pulse Doppler mode only, or we can launch in CW and lose self-protection. There are advantages and disadvantages to both. Pulse Doppler launches will assure us self-protection; however, we may have to launch two AIM-7s at each target to ensure the likelihood of a kill. We will increase the P_K if we launch in CW, but we run the risk of not seeing other potential threats. Engineers involved in this sort of testing are integral to the decision-making process of choosing an option. The decision is a critical one and should not be based on cost or expediency, but rather on total mission requirements and capability.

EMI/EMC testing reaches far beyond testing for interference on radios. Integration of avionics and weapons systems, especially when considering EMI/EMC is a long, involved process and may pose some difficult questions when considering the operational task. We must also consider regression testing after modifications or updates to the aircraft systems. Anytime we add, modify, or relocate equipment or antennas, EMI/EMC tests will be required in order to prove that we have not caused additional problems within the aircraft.

▮ 4.17 ▮ SELECTED QUESTIONS FOR CHAPTER 4

1. Name three range limiting factors of radio waves.
2. What is meant by the term HIRF?
3. Man-made noise dominates at what frequency?
4. What is the relationship between frequency and wavelength?
5. Galactic noise dominates what frequency?
6. What are the three simplified steps in any system test?
7. When building a test matrix, how are test conditions computed?
8. Explain a possible geometry when conducting communications range testing.
9. Briefly describe voice encryption.
10. What is speech intelligibility? What does it determine?

11. Why use amplitude modulation? Frequency modulation?

12. What is the minimum passing grade of a rhyme test?

13. What is EMI? What is EMC?

14. If you wanted to know the requirements for lightning protection, which document would you consult?

15. Identify the unique problems of EMI/EMC in a fighter-type aircraft.

16. What are harmonics and why should we be concerned with them?

17. Is aircraft construction a problem in EMI/EMC?

18. Which type of data bus would be most resistant to EMI? Why?

19. Name two types of EMI avoidance in aircraft? How about two more.

20. What is the problem with employment of MFAs or notch filters?

21. What is the difference between conducted emission and radiated emission?

22. What are some problems with implementing procedural changes to eliminate EMI?

23. Which military specification governs the requirements for EMI/EMC testing? What is the method of test?

24. Name the four steps of EMI/EMC testing.

25. Briefly describe victim/source.

26. Why are antenna patterns important? When should these tests be accomplished?

27. What is near-field testing?

28. What is the difference between electromagnetic interference and electromagnetic compatibility?

29. Blackhawks initially had EMI problems yet Seahawks did not. Why?

30. What are aircraft configuration concerns in communication testing?

31. What are two methods of determining speech intelligibility?

Navigation Systems

Chapter Outline

5.0	Introduction	167
5.1	History	168
5.2	Basic Navigation	168
5.3	Radio Aids to Navigation	175
5.4	Radio Aids to Navigation Testing	184
5.5	Inertial Navigation Systems	191
5.6	Doppler Navigation Systems	207
5.7	Global Navigation Satellite Systems (GNSS)	213
5.8	Identification Friend or Foe	234
5.9	Data Links	238
5.10	Selected Questions for Chapter 5	259

5.0 | INTRODUCTION

Navigation systems span a rather large subject area. This chapter discusses basic radio aids to navigation such as very high frequency (VHF) omnidirectional range (VOR), tactical air navigation (TACAN), nondirectional beacon (NDB), and distance measuring equipment (DME) as well as self-contained navigation systems such as inertial navigation system (INSs), Doppler navigation systems (DNS), and the global positioning system (GPS). Identification friend or foe (IFF), mode S, and data link are also addressed. Integrated navigation systems and future air navigation systems such as wide area augmentation systems (WAAS) and the U.S. military's global air traffic management system (GATM) will be covered in chapter 6. The reference textbook that I use in my courses is *Avionics Navigation Systems,* 2nd ed., by Myron Kayton and Walter R. Fried (New York: John Wiley & Sons, 1997). It does a very good job of explaining the theory and operation of navigation systems. Another text worth reading is *Integrated Navigation and Guidance Systems*, by Daniel J. Biezad (Reston, VA: American Institute of Aeronautics and Astronautics). Flight testers should consult the NATO Research and Technology Organization's *Flight Testing of Radio Navigation Systems* (RTO-AG-300, vol. 18, April 2000).

▮ 5.1 ▮ HISTORY

Kayton and Fried define navigation as "the determination of the position and velocity of a moving vehicle." As we saw in the TSPI chapter, however, the position of an object in space is really a nine-state vector composed of latitude, longitude, altitude, north-east-down velocity, and north-east-down acceleration. The methods of obtaining this information have steadily progressed through the years with ever-increasing accuracy. Dead reckoning, the original navigation method that uses time, speed, and heading, is still a backup mode on all current airborne navigation systems. As we also saw in the TSPI chapter, no single navigation system is capable of directly measuring all nine states. Modern navigation systems may use sensor fusion of multiple navigation systems or employ a navigation computer to calculate the remaining states not directly measured. It is the tester's responsibility to understand not only the accuracy of the system but also the inherent inaccuracies of the measured states. This concept is critical in evaluating modern systems designed for use in congested airspace where the required navigation performance to operate in this airspace is extremely restrictive.

▮ 5.2 ▮ BASIC NAVIGATION

I have found that many times testers are at a loss to effectively plan test missions because they do not speak "pilotese." This is not a slight against testers: they have never been exposed to aircraft flight operations, and most schools do not teach basic earth coordinates anymore. When a pilot asks for latitude, he is not asking for room. Many testers are intimidated by pilots because they cannot effectively communicate with them. The following tutorial is meant to explain some of the basics of navigation. You will not be a navigator after this section, but you will know the difference between a magnetic heading and a true heading. Those readers who are pilots, navigators, or flight literate may want to jump to the next section.

5.2.1 Position

The first concept is that of latitude and longitude. The earth is divided into grids by imaginary lines of latitude and longitude. Latitude lines are horizontal bands around the earth. The equator is 0° latitude, and the North and South Poles are at 90° north and south latitude, respectively. Longitude lines are vertical bands around the earth and start at the Prime Meridian, which runs through Greenwich, England, at 0° longitude. The lines continue east and west and meet at the International Date Line, which is at 180° west and east longitude. Any position on the earth can be identified with a unique latitude and longitude. Figure 5.1 shows these key points.

The distances between any two points may also be determined. Each degree of latitude traverses 60 nm as you move from pole to pole. Degrees are further broken down into minutes. There are 60 minutes in a degree, so each minute will traverse 1 nm (6080 ft). Minutes are then broken down into seconds, and there are 60 seconds in each minute. More commonly, minutes are resolved by tenths, hundredths, and thousandths of minutes. For example, 0.001 min equals 6.08 ft. Each degree of longitude traverses 60 nm, but only at the equator. This is because lines of longitude converge at the poles (Figure 5.1). As we get away from the equator, the distance between each degree of longitude is a function of the cosine of the latitude: 1° of longitude = 60 nm × cosine of the latitude. You can see that at the equator the latitude is 0° and the cosine of zero is one. At the poles, the latitude is 90°, and the cosine of 90 is zero (lines of longitude converge). So if you ever measure distances on a chart, use the tick marks between the lines of latitude, not the lines of longitude (Figure 5.2).

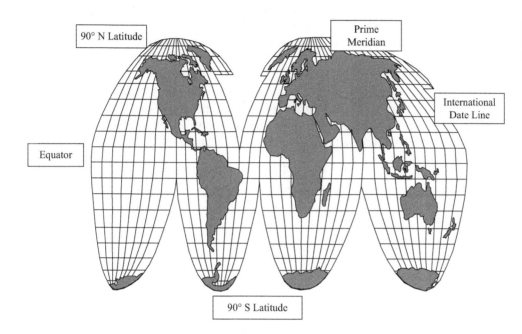

FIGURE 5.1 ■
Spherical
Coordinates

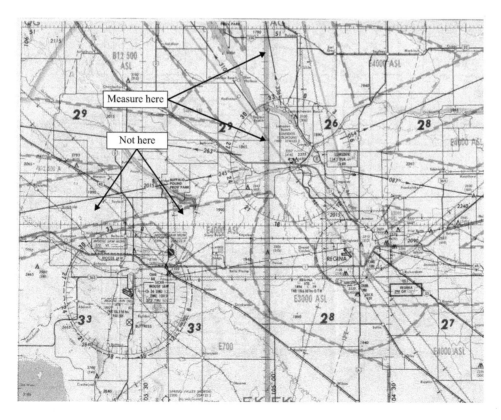

FIGURE 5.2 ■
Latitude/Longitude
Differences

5.2.2 Heading

The next item of interest is the heading between two points. To perform this measurement we will need a chart (map) and a plotter (Figure 5.3 and Figure 5.4). It is important to note here that maps come in all types of resolutions (scales). The amount of resolution, or detail, that we require will determine the required scale. Typical maps for general aviation are called sectional aeronautical charts and have a scale of 1:500,000. The U.S. Geological Survey (USGS) offers 1:100,000 scale topographic maps. The military uses charts ranging from 1:50,000 to 1:1,000,000, depending on mission requirements. Plotters are fairly standard and can provide scales matching the charts in use. Mileage measurements can be read in either statute or nautical miles. There is also a wealth of software programs that will automatically compute headings and distances in preflight planning. Flight management systems deliver real-time measurements and corrections to aircrews using onboard avionics and operator inputs.

All of the charts that we use for flight and flight planning will be referenced to true north. Unfortunately, all of the ground-based radio aids to navigation are referenced to magnetic north. Since the radio aids are in magnetic, pilots fly magnetic. During flight

FIGURE 5.3 ■
Measuring the Heading

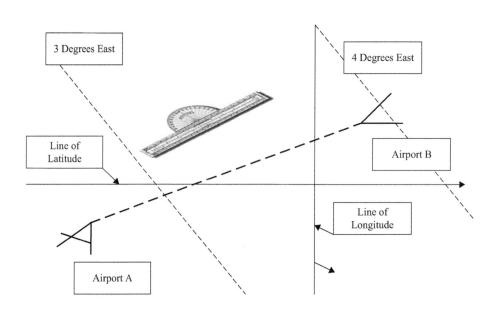

FIGURE 5.4 ■
Reading the Heading Scale

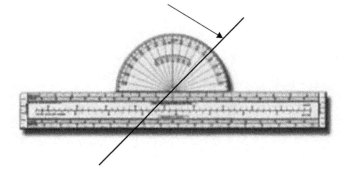

planning, we measure a heading from the chart in true north, but we convert it to magnetic north for the pilots. To accomplish this task, we must understand the concept of magnetic variation and how it affects our measurements.

The plotter is placed with the bottom edge on the line drawn between airport A and airport B (Figure 5.3). The plotter is then positioned until the longitude line intersects the center hole of the plotter. The heading is read on the outside scale of the plotter (Figure 5.4). If a line of latitude is used, as in the case of headings very nearly north or south, the inside scale of the plotter will be used. In our example, the heading read on the plotter scale is 250 or 070°. There are two answers because we can be going from A to B or vice versa (180° out). Since we are traveling ENE from A to B, we know that the correct answer is 070°. This measurement is referenced to true north (since it is read from the chart) and pilots will note the heading as 070° true. We must convert this true heading to a magnetic heading that the pilot will fly.

To convert true heading to magnetic heading, we must apply magnetic variation. Magnetic fields vary around the earth and are constantly changing: we can determine these magnetic fields by consulting a sectional chart, enroute flight information publication, or airport/facility directory (more on these publications later). On the sectional chart depicted in Figure 5.5, lines of magnetic variation are shown via dashed magenta lines and an associated east–west value. The magnetic variation at Salt Lake City is 16° east. In our original example (Figure 5.3), I show a 3° east variation at airport A and a 4° east variation at airport B. The only thing a tester needs to know to correctly apply magnetic variation is *East is least, and West is best.* That is, to convert true heading to magnetic heading we subtract east variation and add west variation.

If our true heading from airport A to airport B was measured as 070° true, the magnetic heading at airport A would be 067° magnetic (070°−3° east variation). As we approach airport B, maintaining the same true heading, our magnetic heading

FIGURE 5.5 ■
Typical Sectional Chart

Controlled Airspace

VOR Salt Lake City

Salt Lake City Airport

VOR Information

VOR Compass Rose

Lines of Magnetic Variation

will change to 066° magnetic. Why? Because the magnetic variation has changed as we progressed along our route. We continue to fly a straight line, yet our magnetic heading changes.

The difference between true and magnetic heading can be quite significant. The true heading from the center runway at Salt Lake City to the Salt Lake City VOR (Figure 5.5) is roughly 360° true. But if we apply the 16° east variation, the magnetic heading becomes 344°. Providing the pilot with the correct heading will determine if he ever gets to the point in space where we would like him to be.

5.2.3 Airspeed

The next variable we need to consider is velocity or, as pilots like to call it, speed. We tend to think of speed much like an automobile. If we travel at 70 mph, in 1 hr we would have traveled 70 statute miles. Aircraft travel is measured in nautical miles per hour, or knots. You would suppose that an aircraft traveling at 300 knots would have traversed 300 nm in 1 hr (1 nm = 6076.12 ft = 1.15078 statute miles). This would be true if the pilot were flying true airspeed (TAS) in a no-wind condition. Unfortunately, pilots usually fly an indicated airspeed (IAS).

The pilot's airspeed indicator is read in knots of IAS. The airspeed is computed by the aircraft's pitot-static and central air data computer system. The system measures impact pressure and compares it to a static pressure: the greater the difference between the two, the larger the IAS. At sea level, IAS and TAS are about the same. As an aircraft climbs, the pressure decreases and so does the impact pressure on the pitot system. The IAS will decrease for a constantly held TAS. As the aircraft configuration changes (angle of attack due to changes in speed, altitude, weight, and flap or gear configurations) the flow across the pitot system will change, thus giving an erroneous reading of pressure. To correct this error, a lookup chart is used to calculate a calibrated airspeed (CAS). If the aircraft is flying at speeds greater than about 200 knots (CAS), the air ahead of the aircraft will become compressed, giving a high reading on impact pressure. An airspeed compressibility lookup chart corrects for this error and produces a knots equivalent airspeed (KEAS). Higher speeds at higher altitudes will create larger differences between CAS and KEAS. For system evaluators, we need only to remember the 2% rule for converting IAS to TAS: For each 1000 ft of mean sea level altitude, TAS will increase by 2% of the IAS. For example, an aircraft flying at 200 KIAS at 20,000 ft will produce a TAS of 280 KTAS (2% of 200 = 4; 4 × 20 = 80; 200 + 80 = 280 KTAS).

All would be fine for us to use true heading and TAS if it were not for that small inconvenience called wind. Systems evaluators are concerned only with our true position over the earth and the speed at which we traverse it. Wind affects our line over the ground and the speed we pass over the ground. These two new terms are called track and groundspeed, and they are the only heading and speed we will ever use. Figure 5.6 shows the effects of wind on our track and groundspeed.

In the first case in Figure 5.6, a wind from the north will drive our aircraft south of the desired point on the nose of the aircraft. You may think of the track as the resultant velocity vector from our own-ship forward velocity and the cross-track component of the wind. The second case demonstrates the effect of wind in the along-track condition. The along-track component of the wind will cause the groundspeed to either increase or decrease, depending on whether it is a tailwind or a headwind.

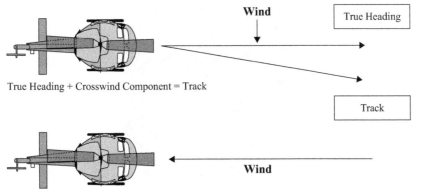

True Heading + Crosswind Component = Track

True Airspeed – Headwind Component = Groundspeed

FIGURE 5.6 ◾
Wind Effects on True
Heading and True
Airspeed

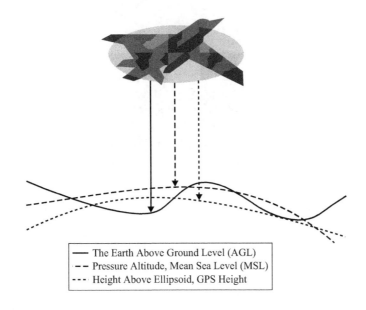

FIGURE 5.7 ◾
Altitude
Measurement

5.2.4 Altitude

The last variable we need to address is altitude. In test and evaluation, we use three different altitudes, or two altitudes and one height, but more on that in a moment. Figure 5.7 shows an aircraft flying over the earth. We can measure its altitude in the three ways that are depicted.

The first measurement of altitude is above ground level (AGL) and is a direct measurement of your altitude in feet. This measurement is accomplished with a radar altimeter that utilizes a frequency modulated (FM) waveform. It can provide readouts to the nearest foot, and its accuracy ranges from 1 ft to 1% of altitude. Figure 5.8 shows a typical aircraft radar altimeter. The second measurement of altitude is the height above mean sea level (MSL). This measurement is obtained with a barometric pressure altimeter (Figure 5.9). Aircraft use the MSL measurement almost exclusively when navigating around the globe. Heights of airports, towers, mountains, and navigation aids are all reported as MSL altitudes. MSL is simply a comparison of the pressure where the aircraft is to the standard-day pressure at sea level. A standard day is defined by the air properties

FIGURE 5.8 ■
Allied Signal Bendix/
King KRA-10A
Radar Altimeter

FIGURE 5.9 ■
Barometric Altimeter

of density, specific volume, pressure, temperature, and viscosity. A standard day would exhibit a temperature of 59°F (15°C) and a pressure of 14.7 psi (101.3 kN/m²). This pressure would also correspond to 29.92 in of mercury (inHg; barometric pressure).

The pilot's barometric altimeter has the capability of setting a pressure for the air mass the aircraft is flying in. Suppose that an aircraft is sitting on the ground at an airport whose field elevation is 0 ft (sea level). If it were a standard day, the pilot would insert 29.92 as a setting for pressure, and the aircraft altimeter would read 0 ft. Since standard days are few and far between, this is rarely the case. Suppose that a high pressure system is sitting over the airfield and the pressure is 30.45 in Hg. If the pilot had 29.92 set in his altimeter, he would be reading a negative 530 ft on his instrument. As he increases the setting to 30.45, the altitude will increase and read 0 ft with the correct setting. You can calculate how far off you will be by merely subtracting the two settings; the hundredths place is worth 10 ft, and the tenths place is worth 100 ft:

$$
\begin{array}{r}
30.45 \\
-\,29.92 \\
\hline
0.53
\end{array}
$$

| | 30 ft |
| | 500 ft |

The knob at the bottom right of the altimeter in Figure 5.9 is where the pilot sets the pressure setting of the day. The window just to the left of the knob displays the setting.

The final altitude is not really an altitude but a height above an ellipsoid. This ellipsoid is the earth model in use, which was discussed in chapter 2.

5.3 | RADIO AIDS TO NAVIGATION

When aircraft fly from point A to point B, they rarely fly a straight line (visual flight rules [VFR] traffic excluded). Aircraft, and their pilots, are directed to highways in the sky called airways. These airways are defined by radio aids to navigation and the radials that emanate from them. When pilots are given directions, they are expected to stay on the line, or as close as inaccuracies in the system will allow. It is fairly important that these radio aids to navigation be checked for accuracy before allowing a pilot to put his life on the line.

Figure 5.10 is a section of the U.S. instrument flight rules (IFR) Enroute High-Altitude Flight Information publication. It defines the allowable highways in the sky for the western United States for aircraft flying under IFR. Using Figure 5.10, let's fly from Los Angeles International Airport to Bakersfield. You can see that we would fly the line identified as J5 to a point called LANDO and then proceed on J5 again to the Shafter VORTAC, which is just to the north of Bakersfield Municipal Airport. This is straightforward enough, but how do pilots get the information to tell them that they are on J5? These lines are not drawn on the ground, so there must be some sort of directional information provided to the pilot's instruments. This directional information is provided by radio aids to navigation. If we take a closer look at Figure 5.10, where J5 starts at Los

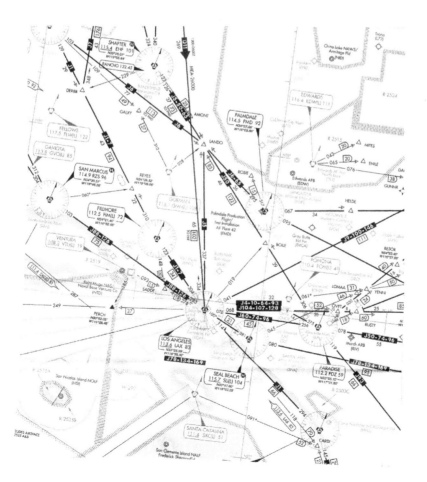

FIGURE 5.10 ■
IFR Enroute High-Altitude Flight Information, Los Angeles Area

Angeles, there is a compass rose around Los Angeles. By following J5, we next see the number 337 superimposed on the airway. The magnetic heading outbound from Los Angeles to LANDO is 337°. But how far is LANDO from Shafter? If we look at the intersection of J5 and Shafter we see another compass rose and a heading of 126°. Further along the route from Shafter there is a little box above J5 with the number 25 in it. This is the distance in nautical miles from Shafter to LANDO. If I departed Shafter on a magnetic heading of 126° and flew for 25 nm, I should be over LANDO, right? This would be true only if there was no wind.

At the center of each compass rose is a six-sided symbol identifying the radio aids to navigation that we will talk about next. Figure 5.11 shows the types of navigation aids

FIGURE 5.11 ■ IFR High-Altitude Chart Legend

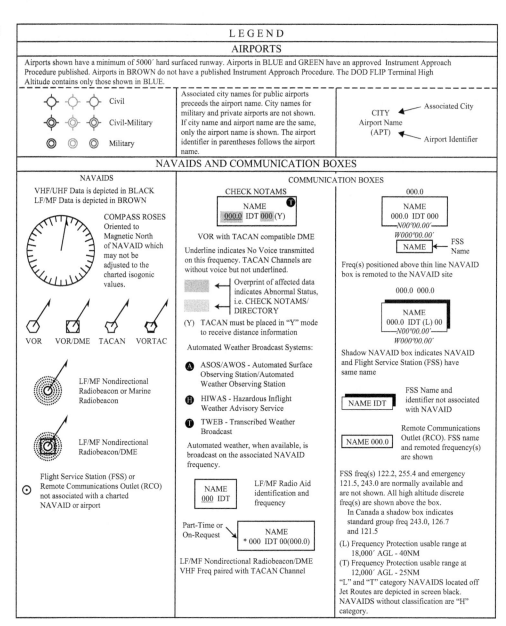

identified by symbols on the high-altitude chart. We will concern ourselves with VHF omnidirectional range (VOR), distance measuring equipment (DME), tactical air navigation (TACAN), and nondirectional beacon (NDB).

5.3.1 VHF Omnidirectional Range

The VOR is a ground-based, short-distance radio aid providing continuous azimuth information in the form of 360 radials emanating from the station. The VOR provides the backbone of the civil airways structure throughout most of the world. Figure 5.12 provides a representation of the concept of radials from a VOR. The triangle represents an aircraft traveling north and east of the VOR. The aircraft is currently on the 045° radial of this VOR station. As the aircraft continues its flight north, it will intercept the 044°, 043°, 042° radials, etc. It will approximate but never intercept the 360° radial unless it turns toward the west.

To determine the radial that the aircraft is on, the aircraft equipment measures and displays the phase difference between two signals that simultaneously originate from the ground station. The reference phase signal is radiated in an omnidirectional circular pattern at a phase that is constant throughout the 360°. The variable phase signal is radiated at 1800 rpm with a phase that varies at a constant rate. As can be seen in Figure 5.12, both signals are in phase at magnetic north, but a phase difference equal to the radial is evident at all other points around the station.

The pilot may have any one of a number of displays for VOR information. Radial information can be displayed on the heads-up display (HUD), multifunction display, or any of the older analog displays. The information is the same, and most digital displays merely replicate the analog displays that they replaced. Figure 5.13 shows two types of displays that are used to display VOR bearing information. The display on the left is a radio magnetic indicator (RMI). It has the ability to display bearings from two radio aids. The compass card will always indicate the aircraft's magnetic heading at the top of the card. The needles will point to the VOR station selected; that is, the pointy end is where the station is. The tail of the needle is the radial that the aircraft is on.

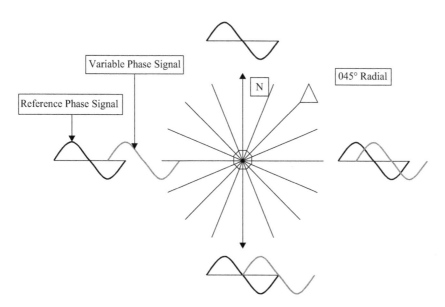

FIGURE 5.12 ■
Radials Emanating from a VOR Station

FIGURE 5.13 ▪

Radio Magnetic
Indicator (Left) and
VOR Display (Right)

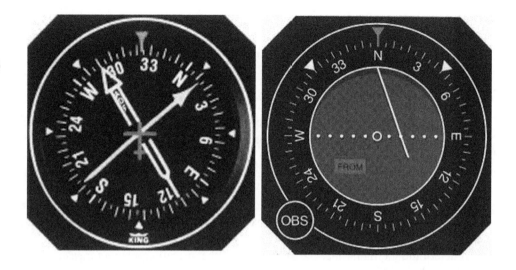

The display on the right is a VOR display commonly found on civil aircraft. The compass card on this display is not slaved to the aircraft compass but is controlled with the omnibearing selector (OBS) knob at the bottom left of the display. The pilot selects the VOR bearing chosen as the reference to fly to or from the station; the compass card rotates to that selection. In the figure, the pilot has rotated the OBS to a desired course of 345°. The vertical white needle on the display is the course deviation indicator (CDI). The needle swings left or right, indicating the direction to turn to return to course. When the needle is left you turn left, and when the needle is right you turn right. When centered, the aircraft is on course. Each dot in the arc under the needle represents a 2° deviation from the desired course. The white arrow in the right center portion of the display is the to–from indicator. In this example, we are going away, or from, the station.

As the name implies, the VOR operates in the VHF radio spectrum. Specifically, the VOR operates within the 108 to 117.95 MHz band. If we refer again to Figure 5.10 and look at the bottom center of the chart, locate the VORTAC of Santa Catalina. Within the box labeled Santa Catalina, there are three other pieces of information. The first is the VHF frequency of 111.4, which is the VOR frequency for Santa Catalina. The pilot would enter this frequency to receive information from Santa Catalina. Once entered, as long as the VOR volume is turned up, the pilot will receive SXC (the three-letter identifier for Santa Catalina) in Morse code. This is how the pilot knows that his system is receiving the correct station. Do all pilots know Morse code? Probably not, but to help them out the high-altitude IFR legend contains Morse code. Sectional charts have the Morse code in the station identifier box on the chart. Within the Santa Catalina box (Figure 5.10), immediately after the SXC identifier is the letter L in parentheses, which identifies the standard service volume (SSV) of the radio aid. The SSV is the volume of space where this station can be utilized. There are three SSVs: terminal (T), low altitude (L), and high altitude (H). In our case, Santa Catalina is a low-altitude service volume. Table 5.1 shows the serviceable areas of the three SSVs. This is an important piece of information, especially in flight testing, because it will determine how far away you will be able to receive the radio aid.

There are two types of VOR systems in use: the older Standard VOR and the newer Doppler VOR (DVOR). In the DVOR system, the carrier is amplitude modulated by the

TABLE 5.1 ▪ VOR, DME, and TACAN Standard Service Volume Classifications

SSV Class	Altitude (ft)	Distance (nm)
(T) Terminal	1000 to 12,000	25
(L) Low Altitude	1000 to 18,000	40
(H) High Altitude	1000 to 14,500	40
	14,500 to 18,000	100
	18,000 to 45,000	130
	45,000 to 60,000	100

reference signal and frequency modulated by the variable signal (the reverse of the standard or conventional VOR [CVOR]). The Doppler principle states that there is a change in frequency of a signal received when the closure between the source and receiver changes. When the closure increases, the frequency increases; the opposite is true when the closure decreases. The rotation of the antenna is reversed in the two systems, and because the variable signal is FM there is less of a chance of destructive multipath.

5.3.2 Distance Measuring Equipment

The VOR will provide the pilot with bearing information to a ground reference but will not provide a range. It is possible to get a positional fix with the intersection of two VOR bearings (called a theta-theta solution in Kayton), but if both VORs are in the same hemisphere with relation to the aircraft ambiguities will exist. Range information will greatly enhance the positional fix. DME is used to establish an aircraft's range from either a ground or airborne beacon. DME is a transponder system requiring that the aircraft and cooperating station possess a transmitter and a receiver. In general operation, the aircraft will generate a pair of interrogation pulses that are received by the cooperating station. After a delay of 50 μsec, the cooperating station sends back a reply. By using pulse-delay ranging, the aircraft computes the slant range to the station.

The DME system works in the UHF band between 962 and 1213 MHz and is capable of replying to 100 airborne interrogations simultaneously. This means that each airborne receiver will have to discriminate its reply from any other aircraft interrogations. The discrimination is accomplished by integrating replies following each interrogation pulse with respect to time. The time between transmission and reception of the receiver's reply will be relatively constant over a small number of sweeps, while all other replies will occur at random time intervals. To assist in discrimination, the airborne equipment has two modes. The first is the search mode, where pulse discrimination is achieved. During this period, the transmitter generates 150 pulse pairs/sec to reduce acquisition time. Once the system is locked onto the range, it enters a tracking mode and generates between 5 and 25 pulse pairs/sec to prevent early saturation of the transponder.

The DME uses 126 channels (frequencies), and to ensure that there is no interaction between the transmitter and receiver, transmit and receive frequencies are separated by 63 MHz. In most civil applications, the pilot need to tune the navigation radios only to the VOR frequency and the equipment will autotune the DME.

In its simplest form, the DME display is simply a digital readout of slant range to the selected station, usually displayed to the nearest 0.01 nm. The displays may be electromechanical or electronic and may be combined with other cockpit displays such as the RMI or heading situation indicator (HSI). In some applications, VOR/DME constant positional fix is used to update the aircraft's navigation system.

5.3.3 Tactical Air Navigation

The TACAN system is an omnidirectional radio aid to navigation developed by the military operating in the UHF band that provides continuous azimuth information in degrees from the station and slant range distance via the DME system. The TACAN system has 126 two-way channels (in reality 126 "X" TACAN and 126 "Y" TACAN) operating in the frequency range of 1025 to 1150 MHz air to ground. Ground to air frequencies are in the ranges of 962 to 1024 MHz and 1151 to 1213 MHz. Our original case of Santa Catalina (Figure 5.10) shows in the information box after the three-letter identifier the number 51. This is the TACAN channel number.

Similar to VOR, the TACAN system measures the time interval between the arrivals of two signals. Two basic signals are produced by the rotation of the inner and outer reflectors of the central antenna. Figure 5.14 shows the antenna array for the TACAN ground station.

A 15 Hz signal is produced once during each rotation of the inner reflector. A 135 Hz signal is produced nine times with each rotation of the outer reflector. When the radio wave lobe of the 15 Hz signal passes through magnetic east, a separate omnidirectional signal is transmitted; this is the main reference signal. When each of the nine lobes of the 135 Hz signal passes through magnetic east, nine additional omnidirectional signals are transmitted and designated as auxiliary reference signals. Figure 5.15 illustrates the main and auxiliary reference signals of the ground station.

To determine the aircraft position in bearing from the station, a phase angle must be electronically measured. This is done between the main reference signal and the 15 Hz signal. This time interval is converted to an angle that isolates one of the 40° segments. The time interval between the reception of the auxiliary reference signal and the maxima of the 135 Hz signal is measured within that segment. The angular difference is converted into degrees magnetic and presented to the pilot. The range is determined from the DME system, which is incorporated into the TACAN system.

Because of the basic construction of the TACAN system (eight auxiliary and one main reference pulse), it is possible to have 40° lock-on errors in azimuth. When the

FIGURE 5.14 ▪ TACAN Ground Station

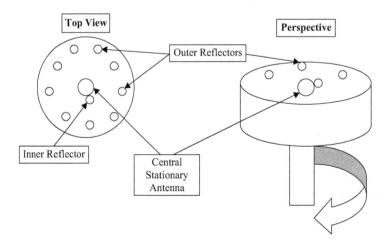

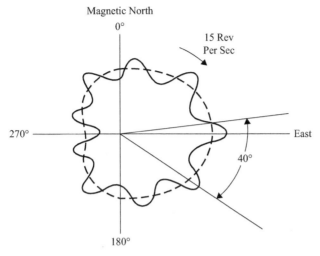

FIGURE 5.15 ▪
TACAN Signal
Pattern

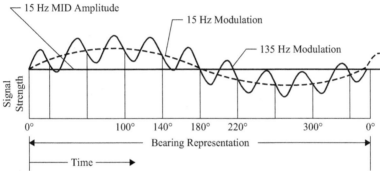

airborne receiver is working correctly, these pulses lock on to the airborne equipment with the main reference at 90° (magnetic east). When the airborne receiver is weak, the main reference pulse may slide over, miss magnetic east, and lock on at one of the auxiliary positions. When this occurs, azimuth indications will be 40°, or some multiple of 40°, in error.

The TACAN within the aircraft also has an air-to-air mode that provides range to a cooperating aircraft. In this mode, the first aircraft will select any TACAN channel (1–126) and direct the cooperating aircraft to select a TACAN channel 63 units away from aircraft 1's selection. For example, if aircraft 1 selected TACAN channel 29, aircraft 2 would select TACAN channel 92. In this scenario, the DME readout will be the distance between the two aircraft. This should obvious if we recall that the DME system separates the transmit and receive frequencies by 63 MHz (section 5.3.3). Table 5.2 provides the transmit and receive frequencies for the 126 channels of X TACAN systems. In newer TACAN systems, the air-to-air mode will provide bearing to a cooperative target in addition to range.

5.3.4 Nondirectional Beacons and Automatic Direction Finders

An NDB installation combines a low-frequency transmitter with an antenna system providing a nondirectional radiation pattern. NDBs are a useful radio aid, as they are relatively low cost and low maintenance. As with VOR, dual receivers can provide a

TABLE 5.2 ■ TACAN Frequency Pairings

Channel	Transmit	REC	A/A REC	Channel	Transmit	REC	A/A REC	Channel	Transmit	REC	A/A REC
1X	1025	962	1088	43X	1067	1004	1130	85X	1109	1172	1046
2X	1026	963	1089	44X	1068	1005	1131	86X	1110	1173	1047
3X	1027	964	1090	45X	1069	1006	1132	87X	1111	1174	1048
4X	1028	965	1091	46X	1070	1007	1133	88X	1112	1175	1049
5X	1029	966	1092	47X	1071	1008	1134	89X	1113	1176	1050
6X	1030	967	1093	48X	1072	1009	1135	90X	1114	1177	1051
7X	1031	968	1094	49X	1073	1010	1136	91X	1115	1178	1052
8X	1032	969	1095	50X	1074	1011	1137	92X	1116	1179	1053
9X	1033	970	1096	51X	1075	1012	1138	93X	1117	1180	1054
10X	1034	971	1097	52X	1076	1013	1139	94X	1118	1181	1055
11X	1035	972	1098	53X	1077	1014	1140	95X	1119	1182	1056
12X	1036	973	1099	54X	1078	1015	1141	96X	1120	1183	1057
13X	1037	974	1100	55X	1079	1016	1142	97X	1121	1184	1058
14X	1038	975	1101	56X	1080	1017	1143	98X	1122	1185	1059
15X	1039	976	1102	57X	1081	1018	1144	99X	1123	1186	1060
16X	1040	977	1103	58X	1082	1019	1145	100X	1124	1187	1061
17X	1041	978	1104	59X	1083	1020	1146	101X	1125	1188	1062
18X	1042	979	1105	60X	1084	1021	1147	102X	1126	1189	1063
19X	1043	980	1106	61X	1085	1022	1148	103X	1127	1190	1064
20X	1044	981	1107	62X	1086	1023	1149	104X	1128	1191	1065
21X	1045	982	1108	63X	1087	1024	1150	105X	1129	1192	1066
22X	1046	983	1109	64X	1088	1151	1025	106X	1130	1193	1067
23X	1047	984	1110	65X	1089	1152	1026	107X	1131	1194	1068
24X	1048	985	1111	66X	1090	1153	1027	108X	1132	1195	1069
25X	1049	986	1112	67X	1091	1154	1028	109X	1133	1196	1070
26X	1050	987	1113	68X	1092	1155	1029	110X	1134	1197	1071
27X	1051	988	1114	69X	1093	1156	1030	111X	1135	1198	1072
28X	1052	989	1115	70X	1094	1157	1031	112X	1136	1199	1073
29X	1053	990	1116	71X	1095	1158	1032	113X	1137	1200	1074
30X	1054	991	1117	72X	1096	1159	1033	114X	1138	1201	1075
31X	1055	992	1118	73X	1097	1160	1034	115X	1139	1202	1076
32X	1056	993	1119	74X	1098	1161	1035	116X	1140	1203	1077
33X	1057	994	1120	75X	1099	1162	1036	117X	1141	1204	1078
34X	1058	995	1121	76X	1100	1163	1037	118X	1142	1205	1079
35X	1059	996	1122	77X	1101	1164	1038	119X	1143	1206	1080
36X	1060	997	1123	78X	1102	1165	1039	120X	1144	1207	1081
37X	1061	998	1124	79X	1103	1166	1040	121X	1145	1208	1082
38X	1062	999	1125	80X	1104	1167	1041	122X	1146	1209	1083
39X	1063	1000	1126	81X	1105	1168	1042	123X	1147	1210	1084
40X	1064	1001	1127	82X	1106	1169	1043	124X	1148	1211	1085
41X	1065	1002	1128	83X	1107	1170	1044	125X	1149	1212	1086
42X	1066	1003	1129	84X	1108	1171	1045	126X	1150	1213	1087

TABLE 5.3 ■ NDB Classifications

Code	Power	Range and Altitude
H	50 W to less than 2000 W	50 nm at all altitudes
HH	2000 W or more	75 nm at all altitudes
MH	Less than 50 W	25 nm at all altitudes

positional fix for the aircraft using the intersection of two NDB bearings. NDB beacons are assigned in the 190 to 535 kHz range and vary in power output. As with the VOR, NDBs are assigned a letter to denote the output power of the beacon and an SSV denoting range and altitude coverage associated with the designations. Table 5.3 shows the NDB power output classification codes.

A 1020 Hz amplitude modulated (AM) audio tone is normally present and transmits Morse code identifications at least once every 30 sec. The more powerful NDBs usually have two-letter identifications, while a single letter identifies those that serve as the marker facility for instrument landing system (ILS) installations. Some NDBs also have voice capability for the transmission of control data and information.

Since the radio signal is nondirectional, a system must be installed in the aircraft to determine bearing to the station; range information is not available. The automatic direction finder (ADF), sometimes called the automatic radio compass (ARC), is an aircraft-installed system provides a bearing to the NDB. The ADF is a low-frequency receiver that operates in the 100 to 1750 kHz range. Since commercial AM radio broadcasts in the 540 to 1650 kHz range, it is possible to use an AM radio station as an NDB. It is also possible to listen to the radio station, which is one very good reason that ADF will probably never be removed from modern aircraft.

The operation of the ADF is dependent on the characteristics of the loop antenna: there is a maximum reception when the loop is parallel to the direction of the wave emanating from the transmitting station. When the loop is rotated, the signal strength gradually decreases until it is at its minimum when the plane of the loop is perpendicular to the direction of the wave; this point is called the null. Further rotation causes the signal strength to increase until it again reaches its maximum when the loop is again parallel to the transmitted wave.

The null property of the loop can be used to find the direction of the transmitting station, provided there is a nondirectional antenna to solve the ambiguity of the loop. The loop alone can be used to locate a transmitting station, but the operator has no way of knowing whether the station is on a specific bearing or its reciprocal; this is known as the 180° ambiguity of the loop. When a signal from a nondirectional antenna is superimposed on the signal from a loop antenna, only one null is received, thus solving the ambiguity. On most military aircraft, the ADF antenna is fixed, so two crossed loop antennas are used and the null position is sensed inductively or capacitively with a goniometer, which is an instrument that measures angles or allows an object to be slaved to a given angle.

Modern aircraft employ synchronous or coherent detection in order to discern the NDB signal from noise. Coherent detection systems internally generate a signal on the same frequency as the tuned station. The receiver then searches through the noise entering the receiver, attempting to match the internally generated signal with the received signal. Once this is accomplished, the system is said to be locked on. Military aircraft also use direction finding in the UHF and VHF frequencies to locate missing wingman, downed pilot's emergency beacons, or friendly forces. The same principles previously covered apply to these systems as well.

Because of the frequency employed and the method in which bearings are obtained, the NDB system has errors not found with VOR and TACAN that can contribute significantly to bearing inaccuracies. Some of the major contributing errors are

- Bank error. Bank error is most predominant at altitude when the aircraft is close to the station. While the aircraft is banking, the bearing pointer points downward toward the station, thus giving an inaccurate reading. The error is greatest on nose and tail bearings when bank is applied.

- Thunderstorm effect. Normally radio waves are distorted by electrical disturbances caused by thunderstorms. There may be erratic fluctuations of the bearing pointer in the direction of the disturbance. It is possible for the bearing pointer to home in on the thunderstorm.

- Night effect. Night effect is caused by the reflection of sky waves from the ionosphere; it is most noticeable at sunrise and sunset, when the height of the ionosphere is changing. The reflected sky waves interfere with the reception of the ground waves and cause the bearing pointer to fluctuate.

- Shoreline effect. Shoreline effect or coastal refraction occurs when radio waves change direction on crossing a shoreline. It is possible to have errors of up to 40° in bearing. The area of maximum error is reached when the bearing from the aircraft to the station is less than 30° to the shoreline.

- Mountain effect. Multipath effects due to mountain reflections will cause fluctuations in the bearing pointer.

- Glinting or quadrantal effect. A quadrantal error is due to the glinting of radio waves off the surfaces of the aircraft. This error is primarily due to the location of the antenna on the aircraft. True effects can be calculated during antenna pattern testing.

5.4 | RADIO AIDS TO NAVIGATION TESTING

5.4.1 Radio Aids to Navigation Ground Testing

The first test in any systems evaluation is the static test. It is a good idea to ensure that the system is functioning properly on the ground prior to expending valuable flight resources. There are many facilities and much test equipment available to the evaluator to make our lives a little easier. We should evaluate the need for testing in the first place. Remember that if we have added, modified, or moved antennas or equipment on the aircraft we will need to perform EMI/EMC testing (previously described in section 4.13). If we have added, modified, or moved antennas, we need to perform antenna pattern testing (addressed in section 4.10).

If we are testing a VOR, it would be advantageous to check the system against a VOR check radial or VOR test facility (VOT). A check radial is a surveyed point on the airfield that informs the pilot that with the correct VOR tuned and identified, the cockpit reading should be as posted. A VOT is a low-power (2 W) VHF omnitest transmitter that permits the ground checking of aircraft VOR equipment without reference to a check radial. The VOT emits an omnidirectional magnetic north radial plus aural identification consisting of a series of dots. It is monitored to a tolerance of ±1°. With the VOR properly tuned to the VOT (108.0 MHz in the United States), the display should indicate a 360° radial indication on the bearing pointer. With 360° set into the track selector window, the track bar should be centered with a "from" indication. The allowable error is 4°.

If we are evaluating a TACAN, a TACAN test set is available. The test set requires entries of channel number, range, and bearing. With the aircraft TACAN turned on, enter the same channel number that has been entered into the test set. Once tuned, the cockpit indications of range and bearing should be the same as those entered into the test set.

As with all avionics systems, controls, and displays, built-in testing (BIT) and self-testing should also be accomplished. Controls and displays testing are covered in chapter 6 and will not be addressed here. A self-test checks the electrical continuity of the system out to the antennas. A BIT, if available, checks the software of the system. Indications, results, and possible logging of errors will be available in the systems' manuals.

5.4.2 Radio Aids to Navigation Flight Testing

Since all of the systems we have discussed provide us with azimuth or range, one would assume that there must be some sort of accuracy requirements associated with these systems, and you would be right. VOR bearing accuracy is on the order of $\pm 4°$, TACAN accuracy is on the order of $\pm 2°$, and NDB bearing accuracy is on the order of $\pm 5°$. DME range accuracy is on the order of 0.2 nm. To evaluate the accuracy of the system, we will need the following information:

- Aircraft's present position (latitude, longitude)
- Altitude (MSL)
- Heading (magnetic)
- VOR/TACAN/NDB position (latitude, longitude)
- VOR/TACAN/NDB altitude (MSL)
- Magnetic variation
- Cockpit reading of bearing and range

Our aircraft's position at the time of the bearing/range reading is provided by TSPI data. The telemetry stream or a recording of the information in the cockpit provides our altitude and magnetic heading. The VOR/TACAN/NDB and magnetic variation are known because the sight does not move. I will discuss where this information can be found in a moment. The cockpit reading is taken from the aircraft's instruments. The geometry should be set up as shown in Figure 5.16. In any avionics test, we try to keep the measurand (the parameter we are evaluating) as constant as possible to reduce measurement errors. For bearing accuracy, we would fly either inbound or outbound on a radial and the bearing will stay constant. For range accuracy, we would fly an arc around the site to keep the DME constant. When we are ready to collect data, we would call a *mark* and note the required measurements. A sample data card for this test is shown in Table 5.4.

The items in the top portion of the data card can be filled out before the flight. The items in the bottom left must be collected during the flight. The items in the bottom right are calculated postflight. The calculation is performed using a coordinate conversion routine, converting delta (Δ) latitude, longitude, and altitude to bearing and slant range. The error that is calculated should be put into absolute value terms. Remember that your initial answer will be a true bearing, which must be converted to a magnetic bearing (since the cockpit reading is in magnetic)—hence the need for magnetic variation.

FIGURE 5.16 ∎
Range and Bearing
Accuracy Geometry

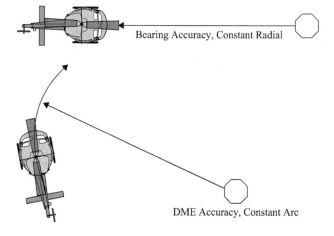

TABLE 5.4 ∎ Radio Aids Range/Bearing Accuracy Data Card

Test #	Station Name	Identifier	Frequency	Station Latitude	Station Longitude	Station Altitude	Magnetic Variation
1							
2							
3							
4							
	Aircraft Latitude	Aircraft Longitude	Aircraft Altitude	Aircraft Heading	Bearing/ Range Read	Bearing/Range Calculated	Error
1							
2							
3							
4							

The magnetic variation that is applied is always the magnetic variation at the radio aid. The test appears straightforward enough; the only missing information is the radio aid data. This information is most easily obtained from an airport facility directory or enroute supplement. A facility directory is shown in Figure 5.17. These publications are distributed by the U.S. Federal Aviation Administration (FAA) or by state authorities in other nations. Before using these publications, ensure that you have a current one. The expiration date is located on the front cover.

A sample of the information contained in the facility directory is shown in Figure 5.18. I have chosen a combination VOR/TACAN station (called a VORTAC) named Hector, which is located in Southern California about 80 mi east of Palmdale or 75 mi northeast of greater Los Angeles. I will explain all of the entries for Hector starting at the upper left and working to the right and then down. The first entry is the station name, and immediately to the right is the latitude and longitude (in WGS-84 coordinates) of the radio aid. This is the information that we would enter on our data card for station latitude and longitude. To the right of the positional information is the statement NOTAM FILE RAL. NOTAM, which means Notices to Airmen. This is a system that informs pilots of changes within the airspace structure and supporting elements. If a runway is closed for construction, if an approach is not authorized, if the GPS signal is unserviceable, or if a radio aid is down for maintenance, these items will be noted in the NOTAM file. It is the pilot's responsibility to check the NOTAM file for all

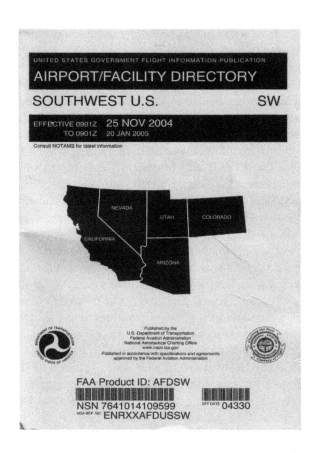

FIGURE 5.17 ■ Airport Facility Directory

HECTOR N34°47.82′ W116°27.78′	NOTAM FILE RAL.		**LOS ANGELES**
(H) **VORTAC** 112.7 HEC	Chan 74	266° 16.5 NM to Barstow–Daggett. 1850/15E.	**H–4H, L–3C**

VOR portion unusable:

340°–055° byd 15 NM blo 10,000′	090°–200° byd 30 NM blo 13,000′
340°–055° byd 28 NM blo 14,000′	200°–231° byd 25 NM blo 9,000′
055°–090° byd 25 NM blo 9,000′	200°–231° byd 32 NM blo 12,000′
090°–200° byd 18 NM blo 11,000′	231°–240° byd 30 NM blo 10,000′
	240°–270° byd 28 NM blo 9,000′

DME portion unusable:

340°–055° byd 15 NM blo 10,000′	090°–200° byd 30 NM blo 13,000′
340°–055° byd 28 NM blo 14,000′	200°–231° byd 20 NM blo 15,000′
055°–090° byd 25 NM blo 9,500′	231°–240° byd 20 NM blo 11,000′
090°–200° byd 18 NM blo 11,000′	240°–270° byd 28 NM blo 9,000′

RCO 122.1R 112.7T (RIVERSIDE FSS)

FIGURE 5.18 ■ Sample Entry from the Facility Directory

---

facilities along his route prior to flight so there will not be any surprises. Of course, testers also need to consult the NOTAMs prior to a test if we plan on using a radio aid during the evaluation. It would not be very bright to brief the pilot that we are going to use Hector VORTAC for our test if Hector is down for scheduled maintenance.

The far right corner tells us that we can find Hector on a Los Angeles sectional chart, Enroute High-Altitude Chart 4H, or Enroute Low-Altitude Chart 3C. The second line starts out with an (H), which means that this radio aid is a high-altitude SSV. The next entry tells us that it is a VORTAC, the VOR frequency is 112.7, three-letter identifier of HEC, and TACAN channel 74. The next item informs us that there is no airport at Hector, and the nearest airport is Barstow-Daggett, which is on the Hector 266° radial for 16.5 nm. The last entry on the line is the station altitude of 1850 ft MSL and a magnetic variation of 15° east.

The next section deals with blockage areas around Hector. Since Hector is situated in a valley bordered by mountains to the north and south, line of sight (LOS) to the station will be a consideration. If you are north of the station you must be at or above 10,000 ft (within 15–28 nm) or at or above 14,000 ft (beyond 28 nm) to have radio LOS. This is another concern for flight testers: Make sure you can see the station during the test. The final entry on the page tells you who to contact for flight following (monitoring). In this case it is the Riverside flight service station and their associated VHF frequencies. This information allows the tester to fill out the data card as well as plan the geometry. Remember the SSV. It will be very difficult to obtain range and bearing accuracies at 80 nm against a terminal SSV radio aid.

In addition to range and bearing accuracies, the radio aid installed on the aircraft must be capable of putting the aircraft in a position to land when the weather is not so good. VOR, TACAN, and NDB systems are authorized for nonprecision approaches. A nonprecision approach allows a pilot to perform the approach while in clouds as long as the ceiling and visibility at the intended airfield is above some weather minimums. In the United States, most nonprecision approaches have weather minimums of 500 ft cloud ceiling and a visibility over the runway of 1 nm. The approach may not be flown if the observations are lower than the minimums. Other countries allow pilots to perform the approach down to the minimums even if the observations are lower than the approach minimums. Figure 5.19 shows a typical nonprecision approach to AF Plant 42, Palmdale, CA. This particular approach is valid for aircraft that have VOR with DME, TACAN, or GPS.

It is not the tester's job to guide the pilot through the approach, but it would be helpful if the tester at least understood what the pilot was doing. The objective of this evaluation is to determine if the radio aid, as installed in the aircraft, can safely place the aircraft in a position to land at the missed approach point. The pilot will make this determination; there are no data to be collected. It is a qualitative evaluation. However, the test needs to be accomplished.

The basic approach plate (Figure 5.19) provides the pilot with information defined in terms of range and bearing from the radio aid (which is not always located at the airfield). The large picture in the center of the page is the bird's-eye view of the approach; the smaller picture at the bottom right is the vertical profile. Below the vertical profile are the weather minima: straight-in approach (400 ft and 1¼ nm) and circling approach (500 ft and 1¼ nm). An airfield diagram is pictured at the bottom left. The approach is as follows: overfly the Palmdale VORTAC at or above 5000 ft; intercept the 070° radial outbound, maintaining altitude at or above 5000 ft; outside of 12 DME, but within 15 DME, execute a right turn to intercept the 085° radial inbound (heading 265°, no wind); once established inbound you are cleared to descend to 4400 ft at 12 DME; at 5.2 DME you are cleared to descend to your minimums and if you cannot safely land out of this approach your missed approach point is at 0.5 DME (1 nm from the airfield). Once again, it is the pilot's call as to whether the system is satisfactory or not.

5.4.3 Other Considerations in Radio Aids Testing

Some other considerations in evaluating radio aids integrations are as follows, in no particular order:

- Aircraft configuration. Aircraft configurations, especially on military fighters, can change on a day-to-day basis, and these configurations can cause different blockage

FIGURE 5.19 ▪
Nonprecision
Approach

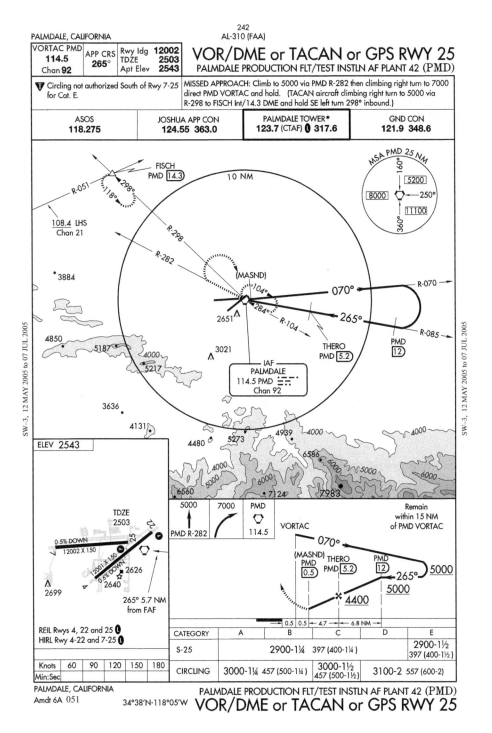

areas for reception. If antenna pattern testing is accomplished, many of these concerns will be alleviated. In flight testing, radio aids to navigation testing as well as communications testing will be accomplished with the most common aircraft configuration.

- Antenna switching. VOR and TACAN systems normally are implemented with an upper and lower antenna that automatically switches due to perceived signal strength at the antenna. This function should be tested by to make certain that there are no dropouts during turning maneuvers.

- Statistical accuracy. I am not going to delve into the black world of statistics, but in everything we do in avionics testing we are concerned with the confidence level of the data. Ideally we would test range and bearing accuracy against three different ground stations. We do this in case one of the stations is out of calibration and gives us erroneous readings. We would only know this if two stations closely agree and the third is out to lunch.

- Atmospheric effects. As previously discussed, NDB/ADF testing will provide varied results based on atmospheric conditions. The tester is responsible for collecting atmospheric data during the test so some sense of the data may be made later.

- Propeller modulation. Care must be taken when evaluating antenna locations on propeller-driven and rotary wing aircraft, as the radio wave will be modulated when passing through these rotating devices.

- Magnetic variation. Magnetic variation is always applied at the ground station; that is, we use the lines of magnetic variation closest to the ground station during our analysis.

- Cone of confusion. Above the ground station there is a cone of confusion within which the equipment receives only DME information. The cone can vary from 60° to 110°, depending on the type of ground installation. The most common is the 90° cone, and bearing information will be lost when the DME reading is equal to the aircraft altitude in DME. For example, an aircraft flying at 18,000 ft will lose bearing information when the DME is equal to 3 nm and will not regain bearing information until the DME exceeds 3 nm.

5.4.4 Civil Certification of Radio Aids

All of the previous tests will be valid toward a civil certification; however, civil authorities require some additional testing as well as variations in methodology. These tests, which may be found in the appropriate advisory circulars (ACs), are noted as follows:

- VOR Testing

  - Test ranges from the station for bearing accuracy are either 160 nm or 80 nm (depends on the aircraft certification altitude).

  - Long-range reception should be accurate during 10° bank turns.

  - VOR nonprecision approaches are initiated with the station 13–15 nm behind the aircraft.

- DME Testing

  - Test ranges, bearing accuracy, and long-range reception are the same as for the VOR.

  - Orbits at 2000 ft, 35 nm from the station should be accomplished; DME should continue to track with no more than one unlock and not to exceed more than one search cycle.

  - A penetration (high-altitude enroute descent) should be accomplished from 35,000 ft down to 5000 ft, 5 to 10 nm short of the facility.

- ADF Testing

    - Pointer reversal should be tested to prove that range equals the altitude flown (cone of confusion validation).

    - Bearing accuracy is shown by flying a minimum of six headings over a known ground point. Accuracy will be shown to be within 5°.

    - Indicator response is a 180° change within 10 sec and accurate to within 3°.

    - Ground reference points must be at least one-half of the service range of the station (Table 5.3).

5.5 | INERTIAL NAVIGATION SYSTEMS

The basic measurement instrument of the inertial navigation system is the accelerometer. There are three accelerometers mounted in the system: one to measure north–south accelerations, one to measure east–west accelerations, and one to measure vertical acceleration. There are two basic types of accelerometers: a moving mass or a pendulum device. The moving mass accelerometer is just a mass on a spring with a ruler attached. The ruler may be an electromagnetic sensor that senses distance. When the vehicle accelerates, the mass moves and the ruler measures the movement. A basic accelerometer is shown at Figure 5.20. The system requires calibrated springs, and these are nearly impossible to make consistent. In a pendulum device, due to inertia, the pendulum will swing off the null position when the aircraft accelerates. A signal pickoff device tells how far the pendulum is off the null position. The signal from this pickoff device is sent to an amplifier and current from the amplifier is sent back into a torquer located in the accelerometer. A torque is generated which restores the pendulum to the null position. The amount of current that is goes into the torquer is a function of the acceleration the device is experiencing.

The acceleration signal from the amplifier is also sent to an integrator, which is a time multiplication device. The integrator starts out with acceleration (which is in feet per second per second) and multiplies it (integrates) by time; the result is a velocity in feet per second. The velocity data are sent to a second integrator, which again multiplies

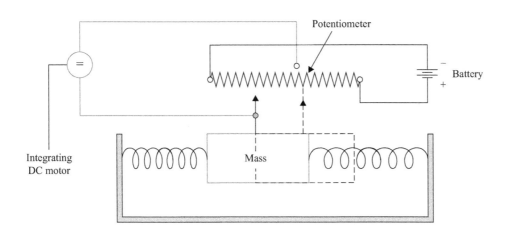

FIGURE 5.20 ■
Basic Moving Mass
Accelerometer

(integrates) by time, with a resultant in feet (distance). Since there are three accelerometers aligned to three axes in the system, there will be three outputs of velocity and three outputs of distance. This is known as the nine-state vector previously addressed in chapter 2: latitude, longitude, altitude, north/east/down (NED) velocity, and NED acceleration. It should be noted that any inaccuracy in the measurement of the acceleration will be magnified as a large error in position.

5.5.1 Gyro-Stabilized Platforms

To accurately measure the acceleration in a given axis, the accelerometer must be constantly aligned to this axis. This is accomplished mechanically or computationally. In the mechanical solution, the accelerometer is mounted on a gimbal assembly, commonly called a platform. The platform, via the use of gyroscopes (gyros), allows the aircraft to go through any attitude change yet maintain the accelerometers level. Figure 5.21 shows a simple platform structure. Figure 5.22 shows how a gyro is used to control the level of the platform.

FIGURE 5.21 ■ INS Platform Structure

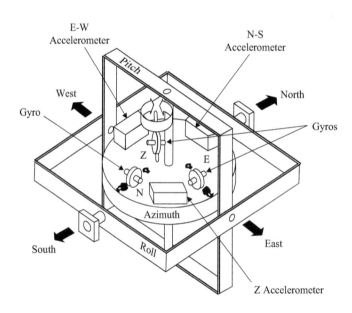

FIGURE 5.22 ■ Gyro Leveling

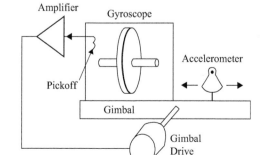

The gyros and the accelerometers are mounted on a common gimbal. When the gimbal is moved off of the level position, the spin axis of the gyro remains fixed; the case of the gyro is moved off level and the amount that the case is tipped will be detected by the signal pickup in the gyro. The signal is amplified and sent to the gimbal drive motor, which restores the gimbal to the level position. Since the accelerometer is always kept level, it does not sense a component of gravity and is able to sense only the horizontal accelerations of the aircraft as it travels across the earth.

This scenario works only for a platform that is fixed in space. Aircraft fly close to the earth, the earth is round, and the earth is rotating. To keep the accelerometers level with respect to the earth so that they sense only horizontal accelerations, these facts must be accounted for. Earth rotation rate compensation is depicted in Figure 5.23. The left side of the figure shows what would happen to the platform if it did not account for the earth's rotation. In this bird's-eye view, the platform maintains its same orientation in space, but from the earth's vantage point would appear to tip over every 24 hours. To compensate for this tipping, the platform is forced to tilt in proportion to the earth's rotational rate. Conversely, with compensation, the bird's-eye view shows the platform tipping over every 24 hours, but with respect to the earth the platform remains level.

The required rate compensation is a function of the latitude of the aircraft because the horizontal component of the earth rotation rate sensed by the gyros is a function of latitude. The maximum earth rotation rate is at the equator and is 15.04°/hr. This value decreases as we move north or south, until it becomes zero at the poles.

Movement of the aircraft around the earth has the same effect on the platform that was caused by the earth's rotation because the earth is round and the aircraft flies an arc as it follows the contour of the earth. The rate of compensation is determined by using the aircraft's velocity. Earth rotation rate and aircraft movement rate compensations are implemented in the system by torquing the gyro. The aircraft movement rate and the earth rotation rate terms are summed and sent to a gyro torquer, which will always keep the platform level with respect to the earth.

A few other compensations are required for correct operation of the INS. The descriptions of these corrections follow; however, the mathematics are left to those industrious engineers. Chapter 7 of Kayton and Fried's text or chapters 3 and 4 of

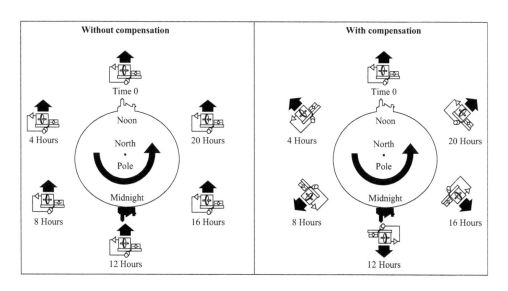

FIGURE 5.23 ■
Earth Rotation Rate
Compensation

Biezad's text can satisfy this particular urge. The first of these additional compensations is a correction for pendulum effect known as Schuler tuning. A pendulum is a suspended mass, free to rotate about at least one axis in a horizontal plane and whose center of gravity is not on the rotational axis. Any pivoted mass that is not perfectly balanced is, by definition, a pendulum. Pendulums align themselves to dynamic vertical when at rest. The pivot axis and the center of gravity are aligned to the gravity vector. Pendulums will tend to break into their natural period of oscillation when they are accelerated. This oscillation is periodic angular motion that has the gravity vector as its midpoint.

A pendulum is said to be at static rest only when its center of gravity and its pivot axis are resting in the same local vertical vector. When a vehicle carrying a pendulum accelerates, the acceleration is introduced to the pendulum via the pivot axis, which moves out of the gravity vector; the center of gravity lags behind. The longitudinal axis of the pendulum now forms some angle other than zero with the local gravity vector, which in turn produces an angular acceleration of the pendulum. During constant velocity, the pendulum seeks to return to the vertical directly under the pivot axis but continually overshoots, manifesting itself as a periodic oscillation. For a given mass, the closer the pivot axis is brought to the center of gravity the lower the period of oscillation and the further the center of rotation.

If the pivot axis and the center of gravity are brought close enough together, the center of turning can be made to coincide with the center of the earth. Once this pendulum is brought to static rest, accelerations of the pivot axis cannot cause the pendulum's longitudinal axis to form any other angle with the gravity vector other than zero. All horizontal velocities will be accompanied by the proper angular velocities to maintain constant alignment of the pendulum to the rotating gravity vector. The pendulum will not oscillate because of horizontal accelerations. To prevent vehicular accelerations from causing an oscillation of the stable element in the INS, the platform is mechanized to have an equivalent length of a pendulum extended to the center of the earth. Any acceleration of the platform is about the earth's center of mass and that of the mechanized pendulum's center of mass. Any errors that introduce an offset in the platform cause the effective mass of the mechanized pendulum to be displaced and introduce an oscillation with a period of 84.4 min. This oscillation causes the platform error to be averaged out over a period of 84.4 min. This is called a Schuler period or cycle. INS platforms that are Schuler tuned are used to bound any errors in the system to acceptable limits so that they do not continue to build.

Two other forces that must be considered are Coriolis and centripetal effects. These are considered external or phantom accelerations. Coriolis forces are apparent because the aircraft is referenced to a rotating earth and appear whenever the aircraft is flying. Due to Coriolis forces, an aircraft moving to the north has an east acceleration, an aircraft moving east has an upward and south acceleration, and an aircraft moving initially upward has a west acceleration. This means that an aircraft flying from the equator to the North Pole will be observed as flying a curved path in space, even though it is flying a straight line. Figure 5.24 shows this phenomenon. For the aircraft to maintain its northerly heading, the accelerometer signals must be corrected according to aircraft velocity and present position.

The oblateness of the earth causes certain centripetal accelerations in certain positions on the earth where the plumb line to the center of the earth does not exactly match the true vertical. The resulting element imbalance causes spurious centripetal accelerations. Compensations for this spurious acceleration must be made before the acceleration is integrated to compute velocity. The oblate earth effect is shown in Figure 5.25.

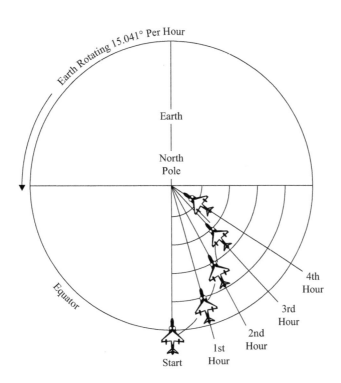

FIGURE 5.24 ■
Coriolis Effect

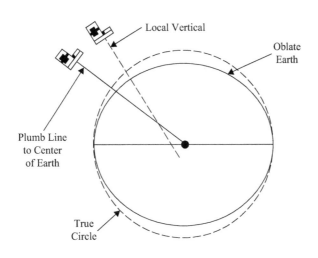

FIGURE 5.25 ■
Oblate Earth Effect

Since we are going to navigate with the INS, it is extremely important that we know our origin and the orientation of true north and the center of the earth. The first step is to determine the gravity vector, which is a relatively simple task. The pilot inserts the present position of the aircraft and enters the first stage of alignment.

In this stage, the gimbal drive motor moves the gimbals until the pendulum of the accelerometer (z axis) is aligned with the gravity vector. At this point there is zero output from the accelerometer, the computer sets the velocity to zero, and attitude information is available. A gyro system used for stabilization (radar, forward-looking infrared [FLIR], backup attitude system) or an attitude, heading, reference system (AHRS) will cease at this point.

FIGURE 5.26 ■
Gyrocompass
Alignment

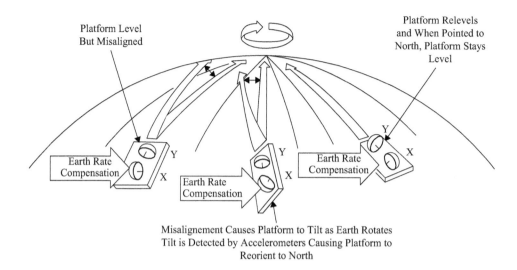

Figure 5.26 depicts how the platform orients itself to true north during the alignment process; this operation is called gyrocompassing. In the far left position, the platform is level and the computer assumes that it is pointing to true north. It only applies earth rotation rate compensation to the east–west (x) gyro. Since the platform is misaligned, the platform will tilt off level as the earth rotates. The tilt is detected by the accelerometers, and a torquing signal is sent to the gyro that controls the platform in the heading axis. The gyro control loop physically reorients the platform toward true north.

Eventually the platform is oriented toward true north (far right), the north–south (y) gyro requires no compensation, and the platform remains level. As the aircraft is flown in the navigation mode, the computer maintains platform alignment using torquing the azimuth gyro via a combination of earth rotation rate and aircraft movement rate.

The gyrocompass system just described has one serious disadvantage: it cannot fly in the polar regions. If the platform were to be flown directly over the poles, the platform would have to rotate 180° at the instant it crossed the pole. This is physically impossible, and in reality, these systems cannot be operated within a few hundred miles of the poles because of the high torquing rates that are required to keep the platform level to north. This problem can be solved with a wander angle inertial system.

The basic fundamentals of a wander angle system are the same as described for the gyrocompass system. To allow the system to fly in the polar regions, the platform takes an arbitrary angle with respect to true north during gyrocompassing. This arbitrary angle is called the wander angle. Figure 5.27 shows the chronology of events for inclusion of the wander angle. In the far left of the figure, the platform is leveled. As with the gyrocompass alignment, the platform is level and all compensation is sent to the east–west (x) axis. Because this assumption is not correct, the platform will tilt off level as the earth rotates, in the same manner as in the gyrocompass example. The tip is detected by the accelerometers as before, but rather than sending all torquing to align to north, the signal is split between the x and y gyros. Eventually the right combination of earth rotation rate compensation to the two gyros is determined for the particular wander angle. The ratio of earth rotation rate compensation is used to compute the initial wander angle.

As this system is flown in the navigate mode, the wander angle will change (hence wander) as a function of the longitude due to the convergence at the poles.

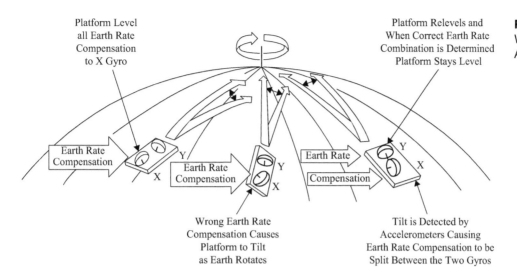

FIGURE 5.27 ▪
Wander Angle
Alignment

The accelerometers are not mounted along true north–south/east–west axes but are offset by the wander angle. This is not a problem for the navigation computer, as it knows the wander angle and can easily compute the true accelerations.

5.5.2 Strap-Down Systems

Lightweight digital computers permit the system to eliminate the gimbals, which create strap-down systems. These sensors are strapped down to the vehicle and computers are used to compute lateral and vertical velocities. A few sensors may be used in strap-down inertial systems, the most common of which in aviation use is the ring laser gyro (RLG).

The laser gyro works on a physical principle discovered by French physicist G. Sagnac in the 1900s. Sagnac found that the difference in time that two beams, each traveling in opposite directions, take to travel around a closed path mounted on a rotating platform is directly proportional to the speed at which the platform is rotating. Although Sagnac and others demonstrated the concept in the laboratory, it was not until the 1960s, with the advent of the laser beam and its unique properties, that the principle could be used in a gyro. The key properties of the laser that make the laser gyro possible are the laser's coherent light beam, its single frequency, its small amount of diffusion, and its ability to be easily focused, split, and deflected.

The RLG is composed of segments of transmission paths configured as either a square or a triangle and connected with mirrors. Figure 5.28 shows a diagram of a triangular RLG. The mirrors in the diagram are located at 5, 6, and 7. One of the mirrors (6) is partially silvered, allowing light through to the detector (9). A laser (8) is launched into the transmission path (4) in both directions, establishing a standing wave resonant with the length of the path. The beams are recombined and sent to the output detector (9). In the absence of rotation, the path lengths will be the same and the output will be the total constructive interference of the two beams. If the apparatus rotates, there will be a difference in the path lengths traveled by the two beams resulting in a net phase difference and destructive interference. The net signal will vary in amplitude depending on the phase shift; the resulting amplitude is a measurement of the phase shift, and consequently, the rotation rate about the body's axis (3).

FIGURE 5.28 ■
Ring Laser Gyro

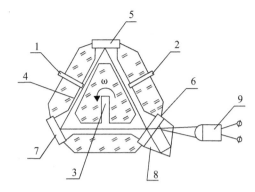

FIGURE 5.29 ■
Ring Laser Gyro INS

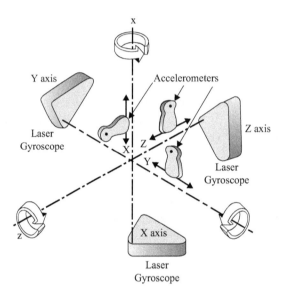

To provide a practical north-seeking gyrocompass, three RLGs are combined with three accelerometers to form a complete navigation, guidance, and control system. Figure 5.29 depicts a common RLG INS system.

One of the significant attributes of the laser gyro is its use of very few moving parts. It is theoretically possible to build laser gyros without any moving components. The spinning mass gyros use gimbals, bearings, and torque motors, whereas the RLG uses a ring of laser light, rigid mirrors, and electronic devices. These differences make the RLG more rugged than the spinning mass gyros, with greater reliability and a longer mean time between failures (MTBF), which translates to lower life-cycle costs. Because the laser gyro uses solid-state components and massless light, it is insensitive to variations in the earth's magnetic and gravitational fields. Shock and vibration have little impact. Unlike a conventional gyro that requires some time to warm up and bring the gyro up to speed, the laser gyro is essentially ready as soon as it is turned on. This is an advantage for military aircraft with alert requirements.

There are some problems with the RLG. When the laser gyro is rotated very slowly, the frequencies of the counterrotating lasers become very close to each other. At this low rotation, the nulls in the standing wave get stuck on the mirrors, locking the frequency of each beam to the same value, and the interference fringes no longer move relative to the detector, which causes the device to no longer track angular position. This phenomenon is known as *lock-in*. This effect is compensated for by adding dithering. With dithering, the entire apparatus is twisted and untwisted about its axis at a rate convenient to the mechanical resonance of the system, thus ensuring that the angular velocity of the system is usually far from the lock-in threshold.

Other gyros may be used in strap-down systems. A device related to the RLG is the fiber-optic gyro (FOG). The FOG works exactly as the RLG, but it utilizes transmission paths with a coiled fiber-optic cable rather than a laser. A vibrating gyro takes Coriolis effects into account; that is, a vibrating element (resonator), when rotated will cause a secondary vibration that is orthogonal to the original vibrating direction. By sensing the secondary vibration, the rate of turn can be detected. For vibration excitation and detection, the piezo-electric effect is often used; therefore vibrating gyros are often called piezo, ceramic, or quartz gyros, even though vibration and detection do not necessarily use the piezo effect. This type of gyro is suitable for mass production and is almost maintenance free. This device has a drawback in that when it is used under external vibration it cannot distinguish between secondary vibration and external vibration. In most applications the system has two mass-balanced tuning forks integrated on a silicon chip arranged handle to handle so that forces cancel. As the forks are twisted about the axis of the handle, the vibration of the tines tends to continue in the same plane of motion. This motion has to be resisted by electrostatic forces from the electrodes under the tines. By measuring the difference in capacitance between two tines of a fork the system can determine the rate of angular motion.

In addition, accelerometer-only systems use four pendular accelerometers to measure all the possible movements and rotations. Usually these are mounted with the weights in the corners of a tetrahedron and therefore are called tetrahedral inertial platforms (TIPs). When the vehicle rolls, the masses at opposite ends will be accelerated in opposite directions. When the vehicle has linear acceleration, the masses are accelerated in the same direction. These devices are inexpensive, but at the current time they suffer from inaccuracy.

5.5.3 Flight Testing of Inertial Navigation Systems

Flight test items of particular interest to the evaluator include the following:

- Time to align
- Align quality
- Alignment types
- Navigation circular error probable (CEP), or drift rate
- Velocity accuracy
- Acceleration accuracy
- Navigation updates
- Navigation hierarchy

5.5.3.1 Time to Align

Time to align is the minimum amount of time that the INS should take to provide an operable navigation system. These times are called out either in the manufacturer's specifications or by the user in the system requirements. It is important to note that the time to

TABLE 5.5 ■ Time to Align Data Card

				Time to Align						
Test No.	Type Alignment	Latitude	Longitude	Start Time	Ready Light	Navigate Selected	Latitude at Navigate	Longitude at Navigate	Groundspeed at Navigate
1									
2									
3									
4									
5									
6									

align is associated with a navigation performance number. In general, the longer the system is left in the alignment mode the greater the performance, or accuracy, of a system. For example, the specification may state that "the AN/ASN-XXX shall provide a drift rate of 1 nm/hr after a normal alignment; normal alignment shall be accomplished in 4 min." The statement ties the accuracy of 1 nm/hr to an alignment time of 4 min. If the system is aligned for, say, 10 min, the accuracy should be even better than 1 nm/hr. Tactical bombers may leave their INS in the align mode for up to 30 min. This is to gain maximum accuracy from the system. I mention this here because if you intend to evaluate the drift error rate of this system and you allow the system to align for 30 min, the data you collect to prove a drift rate will be invalid. I have reviewed test reports of INS evaluations where the evaluator touted a drift rate of 0.5 nm/hr, "far superior to the specification." Upon investigation we find that the aircraft was left in the align mode for a time longer than the 4 min called out in the specification. This constitutes a *no test* as far as specification validation is concerned. A typical data card for this test is shown in Table 5.5. Note the information that is important to this test. The type of alignment is normal, stored heading, carrier, interrupt, or in-flight. These types of alignments will be addressed in more detail shortly. The latitude and longitude are data entered by the pilot or operator. The start time is the time the system is placed into the align mode. The ready light is the time that the operator is apprised of an operable navigator (which must meet the applicable specifications). Navigate selected is the time that the operator places the system into navigate (this time may not exceed the specification time for navigation accuracy flights). The latitude and longitude are again noted after the operator places the system to navigate (it should be the same as entered by the operator). The groundspeed is also noted at the selection of navigate and needs to be equal to zero or the INS will start to drift away rapidly.

The INS will also provide a quality (Q) factor during operation, which is the INS's best guess at how healthy it is. The Q factor is very similar to the figure of merit (FOM) produced in the GPS. The Q factor normally provides a numerical value of 1 through 9, with 9 being the best. In many integrated GPS/INS systems, the Q factor is used to determine whether INS information is good enough to be used in the navigation solution.

5.5.3.2 Alignment Types

There are five general alignment types for the INS. It is important for the tester to note that navigation accuracy is also dependent on the type of alignment used:

- Normal alignment
- Stored heading alignment

- Carrier (CV) alignment

- In-flight alignment

- Interrupted alignment

As the name implies, the normal alignment is the most common INS alignment type and is used most often operationally. The operator moves the INS control from off to standby, thus powering up the system. Depending on the type of installation, the operator may go directly to the align mode or pause in standby until the required entries are made. The operator will be required to enter own-ship latitude and longitude; this allows the INS to perform its transport calculations. If the INS is a spinning mass system, it will require a warm-up period, which is dependent on temperature. The specification will call out how much time is allowed for the warm-up period. During this time the gyros will come up to speed and the operator will be given an indication when the warm-up period has elapsed and the system has entered the first phase of the alignment. An RLG system does not have a warm-up time and will enter the alignment phase immediately. The next indication the operator will receive is completion of the attitude phase. Remember that this is the point where the system has leveled the platform and the z accelerometer is aligned with the gravity vector. We could go to navigate at this point, but we would have platform reference data only (horizon line); the INS will not be able to navigate. The last phase is gyrocompassing, which allows the INS to properly align the north–south/east–west axes. Once accomplished, the operator will receive a ready indication and the system may be placed to navigate.

A stored heading alignment is most often used in alert situations where an aircraft needs to scramble (i.e., take off as soon as possible). A normal alignment is performed as previously described. At the completion of the alignment, the operator requests that the heading be stored in the INS and then shuts down the system. When the operator returns to the aircraft in a rapid reaction scenario, he will power up the INS as normal and the system will advise him that a stored heading is available. The operator has the option of accepting the stored heading or entering into a normal alignment sequence. Since the present position and true heading (and hence true north) are already known by the system, only platform leveling is required. Once the platform is leveled, an operable navigator is available. The advantage is time, since platform lateral referencing does not need to be accomplished. A stored heading alignment usually takes about half the time to perform as a normal alignment. There are some disadvantages to this alignment, however: the accuracy is not as good as a normal alignment and the aircraft cannot be moved in the interim time between the first alignment and the second alignment. The reason for the movement restriction should be obvious; the aircraft applies its earth rotation rate compensation to the gyros based on a known true heading. Moving the platform will cause the gyros to tumble as soon as navigate is selected by the operator.

Aircraft carriers provide some unique problems for INS alignments. As the aircraft sits on the deck of the carrier it is subjected to the forward motion of the ship as well as the pitching and rolling movements of the ocean. A normal alignment would be impossible to perform in such situations. The carrier alignment mode (sometimes called the CV align mode) allows the INS to perform an alignment in these conditions. In pre-GPS days, the ship's INS information was fed to the aircraft either by a hardwire or microwave link. The aircraft INS was said to be aligned when it was in sync with the ship's INS. This method is still used today, but GPS makes it possible for the aircraft to perform an in-flight alignment very quickly, providing similar accuracies.

An in-flight alignment is performed when the INS information is lost or becomes unusable. This can occur when the gyros tumble due to a power failure or precession (an off-axis wobbling of the gyros) or the present position/drift rate becomes unacceptable. In pre-GPS days, an in-flight alignment would only provide the operator with a platform reference (horizon line) and would not provide accurate navigation data. As the aircraft was maintained straight and level, the operator would enter the align mode. The operator would enter present position and heading as well as air data system airspeed. Much like the carrier align, the aircraft would attempt to align the platform using these external inputs. In reality, this alignment could take as long as 20 min to complete before a platform reference was available, and even that was suspect, as it depended on the pilot's ability to fly straight and level during the entire alignment period. Today, GPS can provide the INS with present position, groundspeed, and track at updates once per second. The alignment can take as little as 30 sec and provides accuracies comparable to the normal alignment.

There are two types of interrupted alignments: power interrupt and taxi interrupt. The INS should be able to withstand power interruptions on the ground during the alignment sequence. The INS has a battery that should provide emergency power during transient operations. This alone is a good reason for flight testers to check the battery for proper operation before performing any INS test. This evaluation is easily accomplished by starting the alignment on external power and transferring to internal power midway through the sequence. The transient should be transparent to the INS. The second type of interruption is the taxi interrupt, which is controlled by the position of the handbrake. If the handbrake is released during an alignment, the alignment will be suspended until the handbrake is reset. The INS uses all the information gathered before the suspension to continue the alignment after the handbrake is reset. This works well for a short taxi in the forward direction only. If the aircraft is moved in heading or for any distance, the alignment must be reinitiated. The mode is designed to allow the aircraft to be moved slightly without losing the initial alignment.

Each of these alignment types has a specified accuracy and time limitation. The tester must perform a representative sample of each of these alignments per the data card (in Table 5.5) and then perform a navigation accuracy test following each of these alignment types. The navigation test may be the calculation of a circular error probable (CEP) or the calculation of a drift rate per unit time. The specifications will dictate the accuracy required.

5.5.3.3 Circular Error Probable and Drift Rate

To determine either a CEP or drift rate we need to first construct a navigation route. Most navigation test routes are constructed as right triangles with one leg as a line of longitude and a second leg as a line of latitude. This construction allows us to easily see if the system has a bias in either latitude or longitude. These routes are also constructed to take at least 84.4 min to traverse, or one Schuler cycle. To calculate the accuracy of the system, we will have to know where we really are (TSPI data) and compare that to where the system thinks it is. The calculations for CEP and drift rate will be different, but they will use the same data. Table 5.6 shows a typical data card for navigation accuracy flights. Data card 1 identifies all of the data that must be collected during the flight. Data should be collected at regular intervals along the route with enough frequency to adequately determine accuracies, errors (systemic and bias), and trends of the system. If the data are collected manually, this can be a very busy flight; the minimum sequencing should be about 30 sec. If telemetry or batch processing is used, the timing will depend on the data rate and be limited by the rate of avionics or TSPI, whichever is lower. The heading of the aircraft and the magnetic variation are required to compute

TABLE 5.6 ▪ Navigation Accuracy

Navigation Accuracy (Input Data Card 1 of 2)								
Alignment Type					Time to Align			
Time	Magnetic Heading	Magnetic Variation	True Heading	Airspeed (g)	Latitude Truth	Longitude Truth	Latitude System	Longitude System

Navigation Accuracy (Input Data Card 2 of 2)								
Computed Errors								
Time	Radial	Bearing	N/S	E/W	Cross Track	Along Track	Sx	Sy

cross-track and along-track errors. If using telemetry, track may be provided from the aircraft and these two data columns will not be required. As we fly along the route, the aircraft should vary the airspeed and acceleration to note any anomalies that may present themselves. As was discussed previously, I recommend that the first navigation accuracy flight be flown at a constant airspeed and altitude in order to determine a baseline performance. The latitude and longitude of the truth is the TSPI data time aligned with the avionics (system) data. Remember that the truth data must be four times as accurate as the system under test. Prior to the flight we need to work out our error budget for the TSPI data (e.g., dilution of precision, timing, latency, resolution of the data).

Data card 2 shows the computations to be done on the data. The first two calculated errors are the radial error and the direction of the error (bearing). These data can be used to show the drift rate or calculate a CEP (if we assume that east/west and north/south errors are equal). I will explain the CEP in a moment. The radial error is merely the square root of the sum of the squares of the latitude and longitude errors:

$$\text{Latitude}_{\text{Truth}} - \text{Latitude}_{\text{System}} = \Delta\text{Latitude}$$
$$\text{Longitude}_{\text{Truth}} - \text{Longitude}_{\text{System}} = \Delta\text{Longitude} \qquad (5.1)$$
$$\text{Error}_{\text{Radial}} = \left(\Delta\text{Latitude}^2 + \Delta\text{Longitude}^2\right)^{1/2}$$

FIGURE 5.30 ■
True Error Versus
Schuler Effect

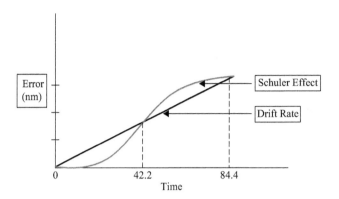

The ΔLatitude and ΔLongitude in equation 5.1 must be converted into feet before we can compute the radial error; a simple conversion if we remember our basic navigation. We can plot this error versus time to determine the drift rate of the system. We can also plot either latitude or longitude error versus time if we are trying to determine a bias in either of the two axes. For Schuler-tuned platforms, it is imperative that we remove the Schuler effect from the radial error that we calculated previously. If we remember that Schuler-tuned platforms are bounded by a period of 84.4 min, then the easiest way to remove the Schuler effect is to fly for one Schuler cycle to determine the true error. Figure 5.30 shows data collected for a Schuler-tuned platform plotting radial error versus time. The Schuler effect (red line) will give an erroneous indication of the error throughout the period except for the times that it crosses through the real system error line (blue line). Notice that these times are at the half-Schuler and full-Schuler period times. A good operator on an aircraft without GPS will only update the aircraft's INS at these times. Updates at other times will actually be detrimental to the accuracy of the INS. The calculated radial errors will produce the red line; testers will add in the blue line using their best engineering judgment. The area above the blue line enclosed by the red line should equal the area below the blue line enclosed by the red line.

If we assume that the east–west and north–south errors are equal, the computation of a CEP for this INS would be relatively easy. Since the CEP in all navigation systems is 50%—that is, half of all data points will be within the stated CEP and the other half will be outside—the midpoint of all the calculated radial errors will be the CEP for the system. If there is an even number of data points in the set, then the CEP will be the average of the two midpoints in the set. If we consider the errors in the two lateral axes to be unique, or bivariant, then we cannot use this approach and must try something a little more cosmic.

One of the approaches used in computing a two-dimensional (2D) CEP of a data set is an approximation that weighs the CEP in favor of the higher standard deviation. This approximation, which may be found in most basic statistics books, is as follows:

If $S_x < S_y$ and $S_x/S_y \geq 0.28$, then the CEP $= 0.562S_x + 0.615S_y$.

If $S_x > S_y$ and $S_y/S_x \geq 0.28$, then the CEP $= 0.615S_x + 0.562S_y$. 　　(5.2)

If neither of the above are true, then the CEP $= 0.5887(S_x + S_y)$.

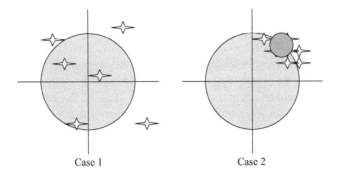

FIGURE 5.31 ▪
CEP About the
Target and the MIP

Case 1   Case 2

In equation 5.2, S_x is the standard deviation of the data set in the x axis, and S_y is the standard deviation of the data set in the y axis. To put this mathematically:

$$S_X = \sqrt{\frac{1}{n-1}\Sigma(x_i - \bar{x})^2}$$
$$S_Y = \sqrt{\frac{1}{n-1}\Sigma(y_i - \bar{y})^2} \tag{5.3}$$

The previous equations will give the standard deviations in x and y, but in relation to a mean impact point (MIP), that is, the center of all the data points. The MIP is simply the mean of the data points in x and the mean of the data points in y. From basic statistics, remember that the mean is defined as

$$\bar{x} = \frac{1}{N}\sum_{i=1}^{N} x_i$$

and

$$\bar{y} = \frac{1}{N}\sum_{i=1}^{N} y_i \tag{5.4}$$

These are the x and y coordinates for the MIP. Should we desire to calculate S_x and S_y for the target as opposed to the MIP (as we would in determining our navigation CEP), the MIP (in equation 5.3) is set to zero. It is always advisable to plot the CEP against the target as well as the MIP because the visualization can help us determine dispersion and whether errors appear to be systemic or a bias. Consider the diagram in Figure 5.31. In each of the two cases, the CEP about the target is the same. However, the CEP about the target in the second case clearly shows that the system is behaving well, but there appears to be a biasing. Case 1 shows the CEP about the target (origin on the x-y axis). Case 2 shows the same CEP about the target but depicts a CEP about the MIP as the smaller circle. The MIP is the center of the smaller circle. This is due to the large dispersion of data in case 1 and the tightness or grouping of data in case 2.

5.5.3.4 Acceleration and Velocity Accuracy

Many flight testers will fall into the trap of flight testing INS velocity and acceleration accuracy. Remember from chapter 2 that (1) the truth must be four times as accurate as the system under test, and (2) the best measure of truth is to use a system that measures the parameter directly. The INS measures acceleration directly and then uses integrators to solve for velocity and position. To prove the accuracy of the INS acceleration, it would be impossible to find a truth source that measures acceleration four times as accurately as the INS. If we had such an accelerometer, it would most likely be in the INS! Acceleration accuracy is a bench test where the INS is placed on a tilt table and known accelerations are input to the INS. The output of the INS is compared to these known inputs to calculate accuracy. We can also look at the output of velocity and compare it to our own calculated velocity with known acceleration inputs. We could compare the INS velocity to a Doppler device (since Doppler measures velocity directly), but these devices in the TSPI world are rare. The bottom line is that you can perform these two tests in the lab. The actual parameters of interest will be positional error and drift, which can, and should, be measured in flight.

5.5.3.5 Navigation Updates

The INS has the capability of being updated (i.e., make its nine-state vector more accurate). The INS may be updated in position or velocity. It is the flight tester's job to ensure that the update functions are implemented and that they work correctly. Velocity updates are normally accomplished with either the Doppler navigation system (DNS) (discussed in the next section) or from the onboard radar. The DNS can accurately calculate north–south and east–west velocities, which are really groundspeeds. This information is fed directly to the INS as a direct substitution for the velocities. These updates will remove any velocity ambiguities it has built up prior to the update. In some aircraft implementations, the INS velocity is constantly updated by the DNS as long as the DNS outputs are valid. The second possibility for updating the INS velocity is via the radar. In the Doppler mode, the radar will identify the closing velocity of terrain on the nose as the ship's groundspeed. By knowing the aircraft's heading (track) the INS can compute the north–south, east–west components of the groundspeed and incorporate them as with the DNS. In many implementations, this update is called the radar's position velocity update (PVU). In some cases, the PVU mode is interleaved (allowed to run intermittently) with another radar mode to allow for continuous INS velocity updates.

Positional (latitude and longitude) updates to the INS include

- Radar/laser
- Radio aids (VOR/DME/TACAN)
- Visual
- GPS
- Off-board (data link)

Some of these updates require computations by the navigation computer, whereas others are a direct substitution of position. A radar or laser update uses the onboard sensors to determine the range and bearing to a known target. Using this information, along with own-ship altitude, the navigation computer can determine (via coordinate conversion) the own-ship position and feed this information to the INS (Figure 5.32). A similar update capability exists with VOR/DME and TACAN. The station's true position is known from a lookup table in the aircraft's mission computers. The radio aid system provides a range and bearing to the ground station, and this information, along with own-aircraft altitude, can be fed to the navigation computer to determine own-ship present position. In some implementations, civil as well as military, a constant INS update is performed as long as a valid signal is received from the radio aid. A visual update is

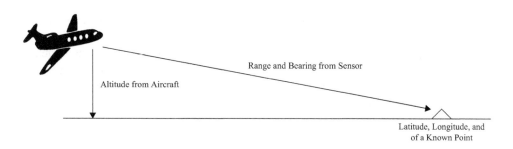

FIGURE 5.32 ■ INS Update via Onboard Sensor

performed by flying over a known point and then inserting the latitude and longitude of the point into the INS when the aircraft is directly over the point. The accuracy of this update is directly proportional to the altitude of the aircraft when over the point. For example, if an aircraft performed a visual update over a point while flying at 500 ft AGL, the pilot could expect to be accurate to within 500 ft. If he were flying at 18,000 ft and performed this update, the accuracy of the update could be expected to be only 3 nm. This implies that visual updates should be performed at the lowest altitude possible.

The easiest of the updates is the direct substitution updates from either the GPS or an off-board sensor through a data link system. Implementation of the GPS update can be either manual or automatic. With a manual update, the pilot forces the GPS position into the INS. With an automatic update, as is the case with an integrated GPS/INS (IGI) or an enhanced GPS/INS (EGI), whenever the GPS position is valid and the FOM is better than a set threshold, the GPS position is sent to the INS at an update rate of once per second. Some systems, such as Link-16, retain a capability to update the aircraft's INS from another platform. A terminal input message (TIM) is received by the aircraft directing a navigation update. The mission computer then forwards the new present position to the navigation computer as an update.

A Note of Caution: Some testers, and indeed operators, may notice that when a navigation update is performed the system does not accept the full update. For example, the pilot may notice that the navigation system places him 1 nm east of his real position. After a valid update, he notices that the navigation system places him ½ nm east of his real position; only half of the update was accepted. The reason is the Kalman filtering techniques and weighting used in modern navigation systems. The system will not accept large changes to any of the states when they exceed allowable predictions. You may find that multiple updates will be required to put the navigation system at the correct position in space.

5.6 | DOPPLER NAVIGATION SYSTEMS

The DNS utilizes the Doppler effect to calculate an aircraft's velocity in the east–west, north–south, and up–down directions. This information, along with a known heading and a starting origin, allows the system to perform dead reckoning navigation. The Doppler effect is covered in detail in chapter 8 of this text. In general, the Doppler effect is the apparent frequency shift whenever relative motion is present between bodies emitting electromagnetic waves. A positive rate of closure (two bodies approaching each other) will compress the wave, shortening the wavelength and thereby increasing the frequency. The opposite is true for opening, or negative rate of closure. The apparent frequency shift is larger at higher operating frequencies than lower frequencies and thus is more able to measure small changes in closure.

If we first take the case of a stationary transmitter radiating toward a moving target, we can describe the relationship of velocity to Doppler shift. If we let f equal the frequency of the transmitter in hertz, V be the velocity of the target in feet per second, and c equal the speed of light in feet per second, the frequency of the reflected energy ($f_{reflected}$) becomes

$$f_{reflected} = f + Vf/c. \tag{5.5}$$

For a moving transmitter, there is an additional change of Vf/c caused by movement of the transmitter during the trip time of the wave. The reflected frequency then becomes

$$f_{reflected} = f + 2Vf/c. \tag{5.6}$$

Since we know the frequency and the speed of light, and we remember the relationship of wavelength and frequency ($\lambda = c/f$), then the Doppler shift can be written as

$$\text{Doppler shift} = 2V/\lambda. \tag{5.7}$$

Consider the case of an aircraft transmitting a narrow beam of radio energy toward the ground. If the beam depression angle is σ (Figure 5.33), then the difference between the transmitted wave and the ground echo is

$$\left[\frac{2Vf}{c}\right] \cos\sigma. \tag{5.8}$$

The choice of the depression angle is a matter of compromise. If the depression angle is too small, then little of the transmitted energy will be reflected back, especially when flying over smooth surfaces such as sand, ice, and water. If the depression angle is too large, its cosine is too small to make an accurate measurement. Typical DNS systems will have depression angles somewhere in the neighborhood of 60° to 70°. When Doppler navigation is combined with onboard radar, the depression angle often cannot be greater than 45°.

Two frequency bands have been allocated to DNSs under international agreement; they are centered about 8.8 GHz and 13.3 GHz. At these frequencies, Doppler shifts of around 10 kHz are experienced and can be measured quite accurately.

Since the aircraft needs to know the velocities in three axes, an ideal system would radiate four beams: two in the forward quadrant and two in the aft quadrant. Left–right and forward–aft velocities can be converted to a groundspeed along a track if the heading is known. Many DNS beam arrangements are used, but most common are Janus Systems (Janus, named for the Roman god who looks ahead as well as behind). Figure 5.34 shows some typical DNS beam arrangements; three of them are Janus-type systems. It is important to note that if the antennas on these systems were fixed, then serious errors would be introduced into the systems during aircraft pitch and roll maneuvers. For this

FIGURE 5.33 ■
DNS Depression
Angle

FIGURE 5.34 ■
Typical DNS Beam
Arrangements

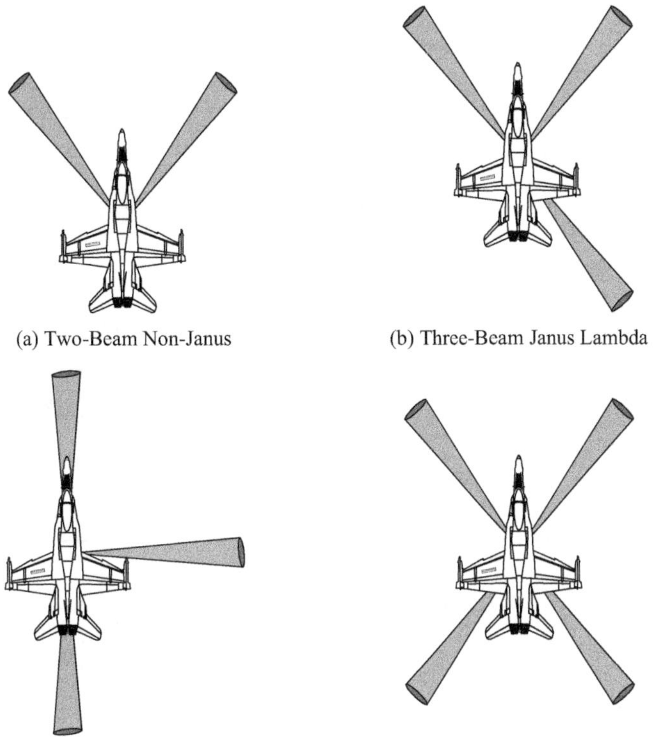

(a) Two-Beam Non-Janus

(b) Three-Beam Janus Lambda

(c) Three-Beam Janus T

(d) Four-Beam Janus X

reason, the antennas must be earth stabilized either mechanically or mathematically. Mechanically stabilized antennas employ a gimbal package, much like the INS, to maintain DNS stabilization during aircraft maneuvers. The gimbals are limited in the amount of tilt and roll they can produce, and there will come a point in time during a maneuver where the system can no longer compensate for the aircraft's attitude. A typical system can be stabilized to 10° in pitch and 45° in roll. Mathematically stabilized systems compute a correction to the measured velocity by knowing the aircraft's pitch and roll attitudes. This system is more restrictive in pitch and roll, as the returned energy decreases and finally disappears on the up-wing antenna during a maneuver.

Figure 5.35 shows the aircraft velocity vectors as a function of x (forward), y (left/right), and z (up/down). The three-beam Janus Lambda system that is depicted directly measures the three velocities of F_p (forward port), F_s (forward starboard), and R_s (rearward starboard). Given this geometry and the three measured Doppler velocities, the three components of aircraft velocity can be calculated. The equations for these calculations are shown in equation 5.9.

$$V_X = \frac{\lambda(\overline{f}_{FP} - \overline{f}_{RS})}{4\sin\beta\cos\phi}$$

$$V_Y = \frac{\lambda(\overline{f}_{FP} - \overline{f}_{FS})}{4\sin\beta\cos\phi} \tag{5.9}$$

$$V_Z = \frac{\lambda(\overline{f}_{RS} - \overline{f}_{FS})}{4\cos\beta}$$

FIGURE 5.35 ■
Aircraft Velocity
Components

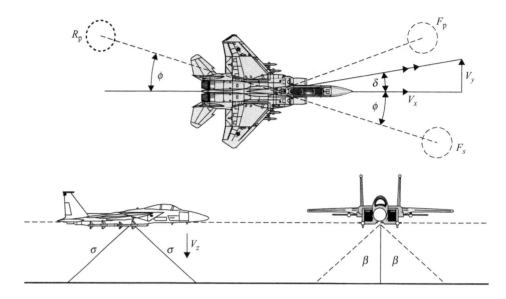

In addition to stabilization in the vertical (earth stabilization), some Doppler navigation antennas are also stabilized with respect to ground track. This can be accomplished by rotating the antennas about the vertical until the cross-track Doppler is nullified. By nullifying the cross-track error, the aircraft drift angle is equal to the antenna offset and the groundspeed is equal to V_x. In systems that are not ground track stabilized, the drift angle and the groundspeed must be computed:

$$\text{Groundspeed} = (V_x^2 + V_y^2)^{1/2}$$
$$\text{Drift angle} = \tan^{-1}(V_y/V_x)$$

(5.10)

The DNS transmissions may be either continuous wave (CW) or pulsed. There are advantages and disadvantages to both types. CW systems employ a transmit and a receive antenna. These systems suffer from leakage (energy from the transmitting antenna seen by the receiving antenna) and, as a consequence, require more power and therefore larger components. The major benefit of CW systems is that they do not experience eclipsing. With single-antenna systems, the antenna must be given a listening time in order to receive echoes from reflective targets. Some echoes will arrive at the antenna at the same time as the antenna is transmitting. Since the antenna cannot process this return, no return for that returned echo (range) can be displayed. These *blind zones* are a function of the pulse repetition frequency (PRF) of the radar and the range to the target. The disadvantages of the CW system can be avoided by frequency modulating the transmitted signal; the ground return will be delayed by a time equal to the distance traveled, whereas the leakage will be undelayed.

The pulsed system avoids the CW leakage problem since there is only one antenna. Signals are generated at a specific length (pulse width) and PRF. As previously mentioned, pulsed systems with a high PRF will experience eclipsing. Some of the eclipsing problems can be solved by varying the PRF, which avoids a constant range being eclipsed, or by making the PRF altitude (and thus range) dependent. Some pulsed systems use coherent detection, which achieves a higher signal-to-noise ratio. The transmitted signal is replicated in the receiver, allowing the receiver to search through the noise to match the returned energy to the internally generated signal.

5.6.1 Doppler Navigation System Errors

As we have seen, an accurate heading and an accurate measurement of x, y, and z velocities are required to maintain an accurate navigator. It should come as no surprise that the errors involved are related to heading and Doppler measurement. If the heading is inaccurate, then the measured drift will be added to that error to obtain the aircraft track, which will also be in error. Groundspeed will, in turn, be applied to a wrong track, which will drive the aircraft position away from true position.

The first cause of error in the measurement of velocity is antenna misalignment. Velocity along the radial axis of the aircraft is proportional to the cosine of the depression angle (equation 5.8). Any error in the depression angle will translate into an error in the velocity measurement. These errors may be caused in boresighting of the system or in the manufacturing process; occasionally variations in waveguide dimensions can affect beam depression angle. If the stabilization of the system is accomplished mathematically, any errors in the attitude measurement of the aircraft will translate into depression angle errors.

The second cause of velocity inaccuracies is due to the type of terrain the aircraft is overflying. Since the radar requires a return from the surface in order to determine velocity, flight over smooth surfaces can create problems, especially for systems employing small depression angles. Flight over sand, snow, lakes, and paved surfaces will tend to reflect the beam away from the transmitter. Flight over the ocean creates unique problems, because unlike the ground the ocean is always moving. The Doppler shift is a combination of aircraft movement as well as ocean movement. Wind across the waves creates additional problems as water droplets are carried by the wind and appear as an additional Doppler shift. Most DNSs provide the operator with a *land* and *sea* mode of operation. The land mode may be considered the normal operation of the DNS. The sea mode employs special processing to account for the ocean's movement and thus velocity errors. The simplest method of correction could be the entry of the system into a memory, or coast, mode. In this mode, the last known drift and groundspeed are used until the operator selects land. This correction is undesirable for long transit times over the ocean. A second correction may be the addition of an average velocity bias applied to the measured Doppler velocity when the sea mode is entered. The bias is a fixed value incorporated into the system and may not be realistic based on changing sea conditions.

A third way to account for errors over the ocean is to use beam shaping or beam switching. Both of these solutions add cost and complexity to a relatively simple system, with the latter being the most effective as well as the most costly. Beam shaping utilizes fan and pencil beams on an alternating basis to seek an average velocity. Beam switching requires a change in depression angles on a scan-to-scan basis. Larger depression angles will give a truer picture of aircraft velocity but are more difficult to measure due to the small cosine of the larger angles. Smaller depression angles provide easier calculations but are more subject to water movement and deflection. A summation of measurements taken at different depression angles can provide a realistic determination of own-ship speed.

5.6.2 Relative Advantages and Disadvantages of Doppler Navigation Systems

The advantages of the DNS system are similar to the INS in that it is entirely self-contained and, as such, can operate anywhere. Similarly, it is not restricted to LOS operations with any ground station. Unlike the INS, there is no alignment required for the DNS, and the DNS measures groundspeed directly as opposed to integrating acceleration.

On the flip side, the DNS requires an independent heading source that must be accurate, as well as an independent attitude reference system. The aircraft will be maneuver restricted due to gimbal limitations of the system. Similar to the INS, the position error of the DNS is unbounded as a result of the time integration of velocity. There will be degraded operation of the system over smooth surfaces like the ocean, and external radiation makes the aircraft vulnerable to enemy detection in wartime.

A DNS-aided INS provides a better navigation system than either system alone, but a GPS/DNS-aided INS provides an even better system. GPS will be addressed in the next section.

5.6.3 Flight Testing of Doppler Navigation Systems

Flight testing of the DNS will be similar in nature to testing of the INS, except the performance of the system will also include flight over different types of terrain. A CEP or drift rate should be calculated for each type of terrain, and in the case of flight over the ocean, different sea states; sea states are covered in section 8.8.5 "Detection of Targets at Sea." As a preview, Table 5.7 provides a description of sea state and the associated wave heights. Velocity accuracy is not normally tested but is evaluated when troubleshooting poor positional accuracies.

The first test of the system will be a mechanical evaluation of the boresight. Performance evaluations should not be attempted prior to this evaluation. Victim/source EMI/EMC and field tests should be conducted prior to flight testing, as described in section 4.13. Prior to departure, the groundspeed should be checked that zero groundspeed is indicated while the aircraft is at rest. Double-check the input of latitude and longitude prior to takeoff. Cross-check the heading input when the aircraft is aligned with the runway. If installed on a helicopter, these checks can be accomplished during the transition to hover.

The navigation route should be designed to fly legs of the route over similar terrain, going from the most benign to the worst case. Legs should also be flown initially at constant speed and altitude to obtain the baseline performance of the system. It does not do any good to fly the first navigation route at varying speeds with large bank angles over varying terrain because you will not know the cause of the inaccuracies. If each leg

TABLE 5.7 ■ World Meteorological Organization Sea States

| Sea State | Wave Height | | Descriptive Term |
	Feet	Meters	
0	0	0	Calm, Glassy
1	0–⅓	0–0.1	Calm, Rippled
2	⅓–1⅔	0.1–0.5	Smooth, Wavelets
3	2–4	0.6–1.2	Slight
4	4–8	1.2–2.4	Moderate
5	8–13	2.4–4.0	Rough
6	13–20	4.0–6.0	Very Rough
7	20–30	6.0–9.0	High
8	30–45	9.0–14	Very High
9	>45	>14	Phenomenal

TABLE 5.8 ■ DNS Test Scenario

Doppler Navigation Flight Sequence				
Flight No.	Type Terrain	Airspeed	Altitude	Maneuvers
1	Diffuse, Good Reflecting	Cruise	Constant	Benign
2	Diffuse, Good Reflecting	Vary, Low to High	Constant	Benign
3	Diffuse, Good Reflecting	Cruise	Vary, Low to High	Benign
4	Diffuse, Good Reflecting	Cruise	Constant	10°, 20°, 30° to Max Bank S Turns
5	Smooth, Sand, Snow, Calm Lakes	Cruise	Constant	Benign
6	Smooth, Sand, Snow, Calm Lakes	Cruise	Constant	10°, 20°, 30° to Max Bank S Turns
7	Ocean Up to Sea State 3	Cruise	Constant	Benign
8	Ocean Up to Sea State 3	Cruise	Constant	10°, 20°, 30° to Max Bank S Turns
9	Ocean, Greater than Sea State 3	Cruise	Constant	Benign
10	Ocean, Greater than Sea State 3	Cruise	Constant	10°, 20°, 30° to Max Bank S Turns
11	Land–Sea Contrast	Cruise	Constant	10°, 20°, 30° to Max Bank S Turns

of the route features a different type of terrain, then a navigation update should be performed at the start of each leg. The easiest way to do this is to use a GPS or fly over a known point on the ground. The data that need to be collected are identical to that needed for the INS (Table 5.6); you will have to add *type terrain* to one of the data fields. A proposed test scenario can be found in Table 5.8.

The general idea is to collect baseline performance data for the system and progress logically to the next level of difficulty. The table is not meant to imply that 11 flights are required for the evaluation. Aircraft with longer flight durations may combine multiple scenarios in one flight. If this is done, remember to update the present position prior to initiating the next scenario. Land–sea contrast is included as a test condition because some systems may incorporate an automatic gain control (AGC) to prevent saturation. If this is the case, strong echoes from land returns may trigger an AGC response in the DNS, which may cause a loss of return from the weaker water echoes. Once again, homework is necessary on the part of the evaluator to understand how the system works prior to embarking on an evaluation.

◼ 5.7 | GLOBAL NAVIGATION SATELLITE SYSTEMS (GNSS)

GNSS is the generic term for all satellite-based navigation systems. Systems currently in use or planned are as follows:

– Global positioning system (GPS) (US)

– Global navigation satellite system (GLONASS) (Russian Federation)

– Galileo (European Union)

– COMPASS BeiDou (North Star) (China)

 – Quasi-zenith satellite system (QZSS) (Japan)
 – Indian regional navigation satellite system (IRNSS) (India)

An introduction to GPS was given in section 2.12 when addressing TSPI systems. As previously mentioned, GPS can provide an excellent source of present position, but operators and evaluators must be aware of some of its limitations. This first section will deal with the basic theory and operation of the GPS.

The GPS baseline constellation is composed of 24 earth-orbiting satellites, with 2 spares, at an altitude of 10,900 nm. There are four satellites in each of six planes that are at 55° inclinations; the orbit time for each satellite is 11 hr, 57 min. For the past several years, the Air Force has been flying 31 operational GPS satellites, plus 3 to 4 decommissioned satellites (*residuals*) that can be reactivated if needed. The optimized coverage is for the mid-latitudes for the simple reason that the United States built it. With this basic knowledge, some things about GPS should become obvious. The first is that you are going to get different constellations on a day-to-day basis even if you fly at the same time each day. Do not make the mistake of thinking that the GPS was great today at 1300 so we will schedule our next flight at the same time next Tuesday to get the same results. The second is that coverage may be lost or not as many satellites may be in view as you get closer to the poles. Parts of Canada, Alaska, and Australia encounter this problem.

The satellites broadcast coded messages to the receivers on the GPS frequencies. Each message contains almanac and ephemeris data, which are the health and position data for all of the satellites. Each message contains five subframes, each of which is 6 sec long, for a total of 30 sec. The almanac is downloaded on the Nav message every 12.5 min. If a GPS receiver does not have the current almanac and ephemeris it must collect it before it can attempt to acquire satellites. A single-channel receiver takes 12 min (for 24 satellites) to initialize. A 12-channel receiver with 12 satellites in view takes about 1 min. It is possible for the GPS to be fooled on its initial position. Most GPS receivers maintain their last known position and continue to maintain time when turned off through battery power. If the GPS is moved without being turned on, when it is reinitialized it will think that it is still in its last known position. It will search the sky for the satellites it thinks should be in view. When it does not find them (because it was moved), it will start to listen for satellites and calculate where it must be with the satellites in view; this may take a bit of time. All GPS units allow the user to enter into a menu to identify present position (or geographic location) to make the GPS search much quicker. Each subframe has a telemetry and handover word that contains a 17-bit time of the week. Table 5.9 provides an illustration of the satellite content.

The GPS operates on two frequencies in the L-band spectrum. The L1 frequency, which contains the coarse acquisition (C/A) code and the precision (P) code is at 1575.42 MHz. The L2 frequency, which contains the P code only, is at 1227.6 MHz. Both frequencies can be acquired by dual-channel frequency receivers (both military and civilian) as long as the P code is not encrypted. Should it become necessary to encrypt the P code, as in time of war, it is renamed the Y code. Access to the Y code can be accomplished only by users with a crypto key, that is, the military.

Three new civil signals will eventually be added to the GPS system via new satellites. L2C (L2 civilian) is designed to meet commercial needs by offering higher accuracy through an ionospheric correction and higher effective power and it will never be encrypted. L5 is designed to meet safety of life civilian requirements and the L1C which is designed for other GNSS systems (Galileo, IRNSS, QZSS) interoperability.

TABLE 5.9 ▪ Satellite Message Content

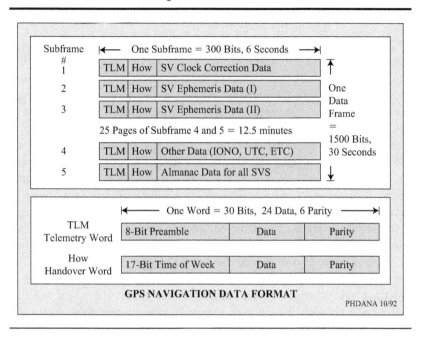

The L2C signal will be available with the Block IIR(M) constellation. The launch of the first Block IIR-M was on 25 September 2005, and the last launch occurred in August 2009. As of December 24, 2013, there were 7 healthy IIR(M) satellites in the GPS constellation, plus one more (SVN-49) designated *unusable*. The Block IIR(M) constellation will comprise 24 satellites with an estimated completion time of 2015. It provides benefits of a second civilian GPS signal (L2C) for improved performance in commercial applications, two new military signals providing enhanced military jam-resistance, and flexible power levels for military signals.

The IIF series expand on the capabilities of the IIR(M) series with the addition of a third civil signal (L5) in a frequency protected for safety-of-life transportation. The F in IIF stands for follow-on. Compared with previous generations, GPS IIF satellites have a longer life expectancy and a higher accuracy requirement. Each spacecraft uses a mix of rubidium and cesium atomic clocks to keep time within 8 billionths of a second per day. The IIF series will improve the accuracy, signal strength, and quality of GPS. The first IIF satellite was launched in May 2010. As of December 24, 2013, there were four healthy IIF satellites in the GPS constellation, with a fifth scheduled for launch in 2014.

The GPS Block III is currently under development by Lockheed Martin. GPS III will provide more powerful signals in addition to enhanced signal reliability, accuracy, and integrity, all of which will support precision, navigation, and timing services. As of April 2013, GPS III Satellite Vehicles (SVs) 03-08 are in the production and deployment phase. Future versions will feature increased capabilities to meet demands of military and civilian users alike. The key improvements are a fourth civilian GPS signal (L1C) for international interoperability and a 15-year design lifespan. Future enhancements include the distress alerting satellite system (DASS) for search and rescue and satellite laser retroreflectors. Updates and status of the GPS system may be found at http://www.gps.gov.

The L2 frequency is normally obtained by first tracking the P code on the L1 frequency using the handoff word. Some military receivers can directly acquire the L2 frequency, but this is outside the norm.

The GPS provides users with a nine-state vector, just as with all other navigation systems. But as we have seen with the other systems, the GPS provides a direct measurement of only one parameter. INS measures acceleration, DNS measures velocity, and GPS measures time. The GPS obtains position by aerotriangulation using spheres of range from three satellites. The range is obtained by using the time difference from when a signal was sent by a satellite to when it was received by the user and multiplying the time difference by the speed of light. The position of the user is defined by the intersection of three spheres around the satellites whose radius is equal to the range just described. The position is three-dimensional (3D) consisting of latitude, longitude, and altitude. Since three parameters are unknown, three equations (satellites) are required for the solution. Time is also unknown, and therefore a fourth satellite is required for the 3D position. In some military applications, barometric altitude is directly substituted into the solution, and only three satellites are required for the 3D position. The biggest problem for the receiver is to calculate the correct time.

If a wrong time is used, there will be an accompanying error in the measured range and therefore position. Pseudoranging is the manipulation of the calculated ranges from each received satellite to minimize errors, primarily the time bias of the receiver. Figure 5.36 provides a graphical representation of this process, and Table 5.10 provides a mathematical example of the pseudoranging process. An important aside is the fact that all measurements are relative to the antenna, not the receiver.

FIGURE 5.36 ■
Pseudoranging

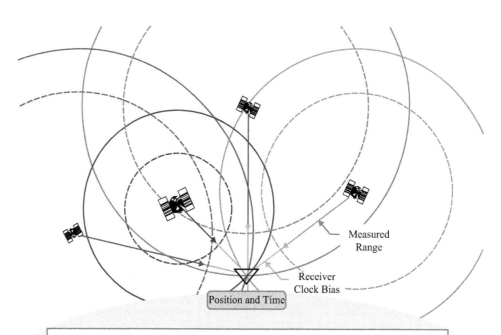

The GPS Navigation Solution
The estimated ranges to each satellite intersect within a small region when the receiver clock bias is correctly estimated and added to each measured relative range.

TABLE 5.10 ■ Pseudoranging Calculation Example

GPS Pseudorange Navigation Example - Peter H. Dana - 4/24/96

Satellite (SV) coordinates in ECEF XYZ from Ephemeris Parameters and SV Time

$SVx_0 := 15524471.175$	$SVy_0 := -16649826.222$	$SVz_0 := 13512272.387$	SV 15
$SVx_1 := -2304058.534$	$SVy_1 := -23287906.465$	$SVz_1 := 11917038.105$	SV 27
$SVx_2 := 16680243.357$	$SVy_2 := -3069625.561$	$SVz_2 := 20378551.047$	SV 31
$SVx_3 := -14799931.395$	$SVy_3 := -21425358.24$	$SVz_3 := 6069947.224$	SV 7

Satellite Pseudoranges in meters (from C/A code epocs in milliseconds)

$P_0 := 89491.971 \qquad P_1 := 133930.500 \qquad P_2 := 283098.754 \qquad P_3 := 205961.742 \qquad$ Range + Receiver Clock Bias

Receiver Position Estimate in ECEF XYZ

$Rx := -730000 \qquad Ry := -5440000 \qquad Rz := 32300000$

For Each of 4 SVs $\qquad i := 0..3$

Ranges from Receiver Position Estimate to SVs (R) and Array of Observed - Predicted Ranges

$$R_i := \sqrt{(SVx_i - Rx)^2 + (SVy_i - Ry)^2 + (SVz_i - Rz)^2} \qquad L_i := \mod[(R_i).299792.458] - P_i$$

Compute Directional Derivatives for XYZ and Time

$$Dx_i := \frac{SVx_i - Rx}{R_i} \qquad Dy_i := \frac{SVy_i - Ry}{R_i} \qquad Dz_i := \frac{SVz_i - Rz}{R_i} \qquad Dt_i := -1$$

Solve for Correction to Receiver Position Estimate

$$A := \begin{bmatrix} Dx_0 & Dy_0 & Dz_0 & Dt_0 \\ Dx_1 & Dy_1 & Dz_1 & Dt_1 \\ Dx_2 & Dy_2 & Dz_2 & Dt_2 \\ Dx_3 & Dy_3 & Dz_3 & Dt_3 \end{bmatrix} \qquad dR := (A^T \cdot A)^{-1} \cdot A^T \cdot L \qquad dR = \begin{bmatrix} -3186.496 \\ -3791.932 \\ 1193.286 \\ 12345.997 \end{bmatrix}$$

Apply Corrections to Receiver XYZ and Compute Receiver Clock Bias Estimate

$Rx := Rx + dR_0$	$Ry := Ry + dR_1$	$Rz := Rz + dR_2$	$Time := dR_3$
$Rx = -733186.496$	$Ry = -5443791.932$	$Rz = 3231193.286$	$Time = 12345.997$

5.7.1 GPS Errors

Since the GPS can measure only time, it stands to reason that the only errors in the GPS are due to errors in the measurement of time. Errors can occur in three areas: the satellites themselves, the propagation of the signal, and the receiver. Each satellite has three atomic clocks to ensure the accuracy of time. U.S. Space Command is responsible for the control segment of the GPS, which includes monitoring the health and welfare of the satellites, positioning spares, correcting ephemeris data, and ensuring the satellites provide accurate time. The master control station is at Schriever Air Force Base, in Colorado, and monitoring stations are situated close to the equator around the world. Even with this control, sometimes errors are missed and are allowed to enter the receiver and hence the navigation solution.

As an example, I offer the following. A significant GPS anomaly occurred on January 1, 2004, beginning at approximately 1833 Z. The anomaly affected precise timing and navigation for users over large portions of Europe, Africa, Asia, Australia, and portions of North America. The anomaly was due to a failed atomic frequency standard (AFS) on satellite 23. The lack of a hard failure indication on the satellite telemetry and satellite visibility limitations caused the satellite to provide hazardously misleading information to GPS users. Figure 5.37 shows the affected area of the anomaly, and Figure 5.38 shows the ranging errors caused in this satellite.

FIGURE 5.37 ■
Affected Coverage
Due to AFS Anomaly

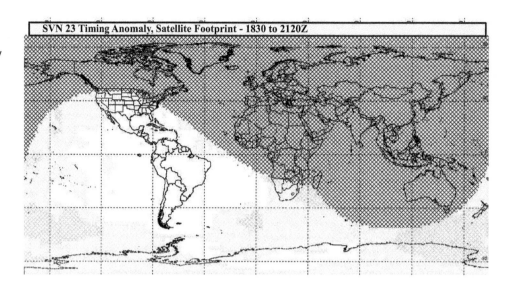

FIGURE 5.38 ■
Ranging Error Due
to AFS Anomaly

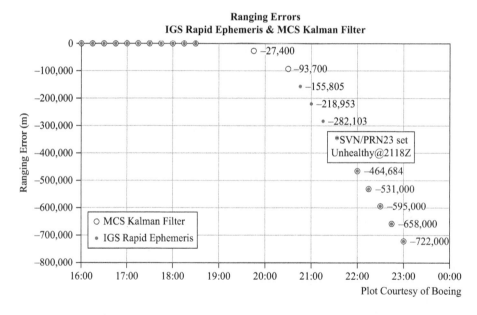

Plot Courtesy of Boeing

Many of satellite errors will be known ahead of time. The control segment may move a satellite within its orbit or take the satellite out of service for a period of time. It is incumbent upon the user to check the notice advisory to NAVSTAR users (NANU) prior to GPS use. The NANU system is much like the notices to airmen (NOTAM) system, which identifies restrictions in the use of navigation aids, airfields, etc. As with the NOTAM system, the NANU system should be checked prior to any flight or test operation scheduled to use GPS. NANU may be accessed in many ways, the most popular being the Internet. One example is the U.S. Coast Guard site at http://www.navcen.uscg.gov. Restrictions on the use of GPS may also be found on the NOTAM system when specified areas may be unavailable to GPS users. This can happen around military installations testing antijam GPS capabilities or radiating energy in the GPS frequencies.

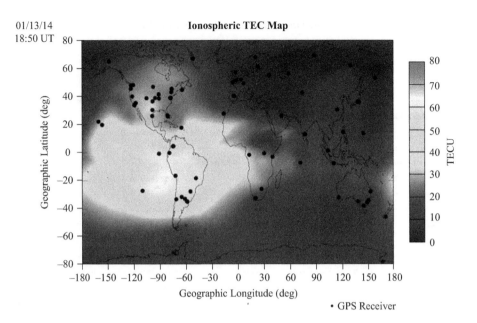

01/13/14
18:50 UT

FIGURE 5.39 ▪
Total Electron
Content Map

Since GPS operates in L band, the earth's atmosphere is almost transparent to the transmission. The largest contributor to time errors in propagation is galactic noise, or solar activity. Programs are available that show possible degradation of the GPS due to solar activity. The Jet Propulsion Laboratory website (http://iono.jpl.nasa.gov/latest_rti_global.html) provides users with global maps of ionospheric total electron content (TEC) in real time. These maps are used to monitor atmospheric weather and predict ionospheric storms that adversely affect GPS accuracy. Figure 5.39 shows an example of an ionospheric TEC map.

By using a little ingenuity, the user sets a threshold for TEC that will identify the maximum degradation of the signal that may be allowable for a particular operation. Other websites and programs are available to the user to assist in determining whether ionospheric scintillation might be a potential problem for their application. Northwest Research Associates produces a product that indicates, by satellite, scintillation levels that exceed certain user-specified thresholds as a function of time. Figure 5.40 provides an example of their product: red indicates levels exceeding user-defined outage levels 60% of the time, yellow more than 30%, green less than 30%, and black indicates that the satellite is below a user-defined elevation angle. These products allow users and evaluators to schedule testing at the optimum times.

Other errors in propagation involve jamming—intentional as well as unintentional. The GPS signal at the antenna is below earth noise; the L1 signal can be measured at −160 dBW. The signal is pulled out of the noise with coherent detection preamp-powered antennas and high-gain receivers, but it does not take much additional noise to obliterate the GPS signal. A 1 W noise jammer centered at L1 can blank the GPS signal for 30 nm. This may be a problem for military applications in hostile areas, but the biggest jamming threat to most users is unintentional jamming; everybody and his brother broadcasts in L band. Systems such as TACAN, JTIDS, MIDS, IFF, and mode S all broadcast in L band. I do not know how many programs lost GPS on the first development flight of a new program because no one considered the adverse effects of telemetry transmissions on the GPS signal (most pulse code modulation [PCM]

FIGURE 5.40 ■
User-Defined
Outage Levels

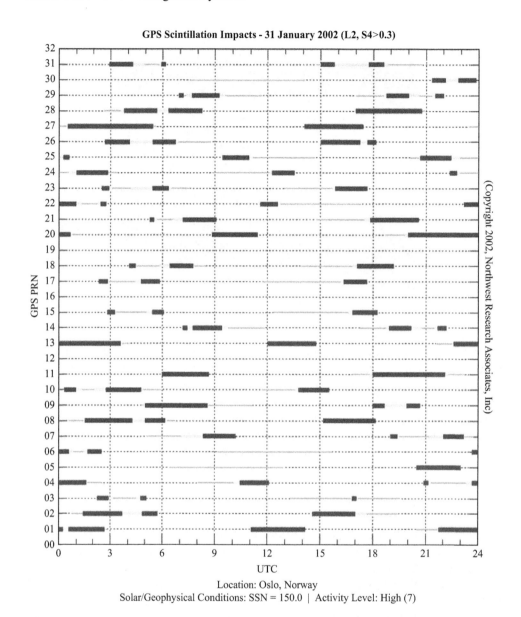

Location: Oslo, Norway
Solar/Geophysical Conditions: SSN = 150.0 | Activity Level: High (7)

telemetry systems operate in L band). Multipath effects in mountainous areas may also be considered as unintentional jamming, as the results are the same.

In October 2011 a series of upgrades were begun to increase the availability of the WAAS LPV service during times of high solar activity; the last peak (11 years) was in February 2013. To ensure safety for users, the WAAS provides information about the ionosphere. This is done via a parameter called the grid ionospheric vertical error (GIVE); the parameter is low for small solar activity and high for greater solar activity. The GIVE parameter ensures WAAS receivers account for delay due to ionospheric activity. The real-time GIVE parameter can be found at http://www.nstb.tc.faa.gov/24Hr_WaasLPV. htm. An example of a daily WAAS GIVE plot may be found at Figure 5.41.

The receiver itself can also have errors in the clock or in the calculation of time. A combination of these errors (satellites, propagation, and receiver) sum to the total

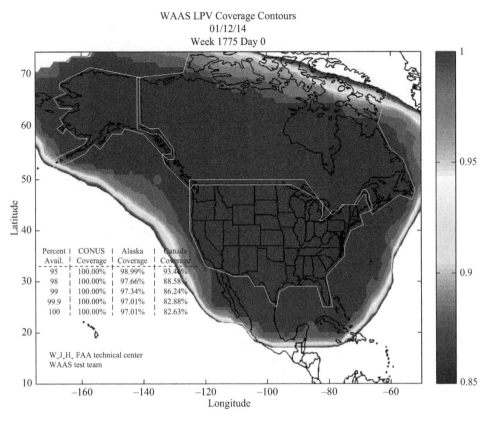

FIGURE 5.41 ■
WAAS LPV
Accounting for GIVE

error in time. This, of course, translates to an error in range from the satellites that carries through to a positional error determined by the receiver. Unless some external monitor is employed, the integrity of the GPS cannot be assured. We will investigate receiver autonomous integrity monitors (RAIMs) and differential systems (which can perform integrity checks) in a later section.

5.7.2 Other Contributors to GPS Inaccuracy

The GPS receiver employs an internal optimum constellation algorithm that selects the four satellites providing the best dilution of precision (DOP). The concept of the geometric dilution of precision (GDOP) was addressed in section 2.5. Briefly, the DOP is a multiplier of accuracy. For example, the GPS provides an accuracy of 15 m in x and y for a DOP of 1. If the DOP increases to 3, the accuracy of the GPS is 3×15 m, or 45 m. It is important to know the DOP for each test of the GPS. The constellation that will provide the best accuracy (GDOP = 1) is three satellites 120° apart at 5° above the horizon and one satellite directly overhead. Since this constellation is not always available, GPS accuracy will degrade as a function of satellite geometry.

The GPS can output a vertical (VDOP), horizontal (HDOP), or positional (PDOP) dilution of precision. The GDOP is the PDOP for a specific user's location. It is important to remember that the GDOP will change based on the user's location, the time of day, and the availability of satellites. Some GPSs output an expected positional error (EPE). The EPE is merely the product of the DOP and the manufacturer's stated accuracy.

FIGURE 5.42 ■
Effect of Satellite
Removal on the
DOP

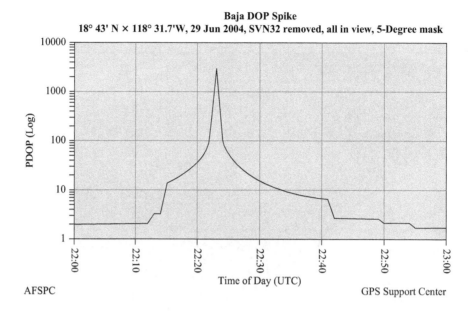

Evaluators should not use EPE as TSPI accuracy or as a measure of GPS performance. The error budget should use 15 m multiplied by the user's GDOP. When satellites are taken offline for a period of time, the user may see a large increase in the GDOP and therefore a large increase in the inaccuracy. Figure 5.42 shows the effect of removing a satellite where only five satellites are in view. This is actual data from the Baja Peninsula on June 29, 2004.

There is also an inaccuracy due to the update rate. The GPS update rate is once per second, and this will contribute to the along-track error; cross-track error will not be affected. Because of the slow update rate, the along-track uncertainty, or error budget, will increase with increasing aircraft speed.

Many of the problems associated with timing accuracy can be eliminated with differential GPS. A ground receiver at a known location compares the computed GPS position to its known location. The error of the GPS location can be eliminated by calculating the clock bias, or Δt, for each of the satellites. As the Δt for each satellite is applied, the GPS position is moved toward the true position. A Δt for all satellites in view is computed to make the GPS position equal to the true position. A matrix of these time corrections can be sent to an aircraft to adjust its GPS position, or it can be corrected in postflight, as in the TSPI case. There are some concerns with this arrangement: relativity effects, LOS considerations, and satellites not in view to the ground receiver.

The error in time will change based on the relative position of the receiver to the satellite. The aircraft may be closer to the satellite than the ground station and may see less of an error. The general consensus is that relativity effects are not a concern as long as the receiver and the ground station are within 300 nm of each other. Since the earth is round, there will come a point where the aircraft is below the horizon and corrections cannot be received from the ground station. The range at which this occurs is a function of the aircraft's altitude and can be calculated using equation 5.11.

$$\text{Range to the radar horizon (nm): } R \sim .86\sqrt{2h}, \qquad (5.11)$$

where h is the altitude of the aircraft.

As the aircraft travels away from the ground station it will begin to receive satellite signals that cannot be seen by the ground station because of the same LOS considerations. The ground station cannot offer corrections for these satellites if the aircraft uses them in its calculations.

5.7.3 GPS Accuracies

Historical data have shown that the accuracies of the GPS are better than specified. Figures 5.43 and 5.44 present actual performance data for the GPS for the 2003 calendar

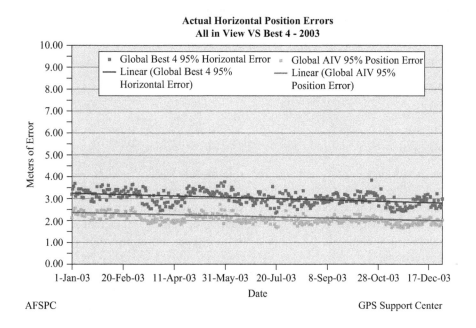

FIGURE 5.43 ■ 2003 Positional Accuracies Best 4 and All Satellites

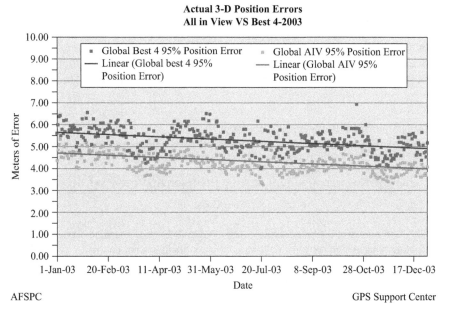

FIGURE 5.44 ■ 2003 Geometric Accuracies Best 4 and All Satellites

year. Although the figures show performance near 1 m, the specified accuracy for L1 (C/A code) is 15 m in the x–y plane (latitude and longitude). Differentially corrected L1 will provide 1 m x–y accuracy. Dual-frequency receivers (L1 and L2) can expect accuracies of 1 m (uncorrected) and 1 cm (differentially corrected), respectively. The accuracies stated are for a nonmaneuvering aircraft. TSPI systems such as the advanced range data system (ARDS) claim accuracies of 1 to 7 m, depending on the maneuver, and velocity accuracies on the order of 1 to 8 knots.

5.7.4 GPS as a Navigation Aid

Since the accuracies appear to be quite good, why is GPS not approved as a sole means of navigation? The answer lies in the requirements for a system to be certified as an approved navigation aid. For any system to be accepted as an onboard navigation system, four criteria must be met:

- Accuracy
- Integrity
- Continuity
- Availability

The performance requirements for sole-means system accuracy are found in Table 5.11. The accuracy required depends on the segment of flight and the type of approach desired. The GPS—on its own without differential corrections—can satisfy most of the basic accuracies (e.g., enroute, terminal, landing). It cannot satisfy the vertical accuracy requirements for more restrictive precision approaches without differential corrections.

Integrity can be defined as the minimum time to alert the operator that the navigation system being used can no longer provide the accuracy required for the task at hand (e.g., enroute, terminal, landing). It can be thought of as similar to the *off* flag on the VOR or TACAN system. Since there is no off flag in the GPS, there is no integrity in the GPS unless a monitoring system is employed.

Three different types of integrity monitors may be employed in the GPS: RAIM, equivalent RAIM, and differential GPS. A RAIM is employed in most civilian and military applications. It must have five satellites in view for integrity monitoring and at least six in view for RAIM with fault detection and exclusion (FDE). The numbers drop

TABLE 5.11 ■ Performance Requirements for Sole-Means Accuracy

	Performance Requirements for Sole Means Accuracy	
	Enroute and Nonprecision Approach	**Precision Approach Cat 1 Landing**
Horizontal Accuracy	100 m	7.6 m
Integrity (Time to Alarm)	8 sec	5.2 sec
Availability	99.999% (5 min/yr of degraded service when continuity cannot be met)	99.9% (9 hr/yr of degraded service when continuity cannot be met)
Continuity	99.999999% (1 in 100 million flights loses service during a 1 hr navigation period)	99.945% (1 in 18,182 flights loses navigation service during a 150 sec precision approach)
Vertical Accuracy	N/A	7.6 m

by one in military systems that directly substitute barometric altitude into the GPS. The basic tenet is that RAIM uses additional satellites (other than the optimum constellation of four satellites) to check the integrity of the navigation solution. By using all of the permutations and combinations of all satellites in view (using a different four in each iteration to calculate a navigation solution), the program is able to identify any satellite that is providing misleading information. The bad satellite will be the common thread in all solutions that differ from the others. Once detected, the bad satellite is excluded from any future calculations. If only five satellites were initially available, then after any exclusion RAIM is not available.

An equivalent RAIM system uses another navigation system to check the output of the GPS. A typical equivalent RAIM system might be a triply redundant INS, which is available on many air carriers. The equivalent RAIM system continuously monitors the output of the GPS solution and issues an alert when the difference between the two systems exceeds a certain threshold.

A differential GPS ground station may also be used as an integrity monitor where unacceptable satellites are identified by the ground station and the information is uplinked to users. The WAAS, European geostationary navigation overlay service (EGNOS), and the multifunction transport satellite (MTSAT) satellite-based augmentation system (MSAS) all satisfy the requirements for integrity monitoring. These systems will be discussed in greater detail later. Current systems can provide a time-to-alert of about 6 sec.

Continuity guarantees a user coverage for a particular phase of flight. The probability of a loss of coverage is more restrictive in the critical phases of flight, such as approach and landing. Due to possible solar activity, multipath effects, or jamming (intentional or unintentional), continuity cannot be assured with the current system. Enhancements have to be made to the current system to allow for replacement of current ILS and microwave landing system (MLS) systems.

Availability simply means that the system is available to the user when the user needs it. Due to the problems already addressed with continuity, availability cannot be assured. Future systems such as the ground-based augmentation system (GBAS) will greatly improve continuity and availability.

5.7.5 Other Satellite Systems

There are five other satellite navigation systems either in service or soon to be in service: GLONASS, Galileo, Beidou/Compass Beidou, Quasi-zenith satellite system, and Indian regional satellite system.

GLONASS is the Russian Federation Ministry of Defense global navigation satellite system initiated in 1982. The GLONASS operational system consists of 21 satellites in three orbital planes with three in-orbit spares. The planes have a separation of 120°, and the satellites within the same orbit plane are separated by 45°. Each satellite operates in circular 19,100 km orbits at an inclination angle of 64.8°. The orbit time is 11 hr, 15 min.

Each GLONASS satellite transmits two types of signals continuously: standard precision (SP) and high precision (HP). The satellites transmit the same code, but on different frequencies within the L band. The SP operates on L1, where L1 is equal to 1602 MHz + n (0.5625 MHz); n is equal to the satellite number (e.g., $n = 1, 2, 3$). Some satellites have the same frequency; however, they are never in the same field of view to the user. The receiver requires four satellites to compute the 3D position and time.

The GLONASS system accounts for leap seconds; however, there is a constant 3 hr bias between GLONASS and the universal time code (UTC) because GLONASS is aligned with Moscow Standard Time.

GLONASS guarantees (99.7% probability) horizontal positioning accuracy of 57–70 m and vertical positioning accuracy of 70 m. In 2001, GLONASS was operating in a degraded mode, with fewer than eight satellites in orbit. With 24 launches in the last 10 years, GLONASS now has 24 operational satellites (full operational status), 3 spares and 1 in testing status. As with GPS, notice advisories to GLONASS users (NAGU) are issued by the Russian Federation Ministry of Defense. Russian geodetic datum PZ-90 is used by the GLONASS system. The status of GLONASS can be followed at http://glonass-iac.ru/en/GLONASS/. Discussions are under way with the United States to modify the GLONASS signal to make it compatible with GPS and Galileo. As a bit of trivia, Vladimir Putin has a GLONASS locater on his dog, Koni (Figures 5.45).

Galileo is the first positioning and navigation system specifically developed for civilian users. It is being developed by the European Space Agency (ESA). Galileo is composed of a constellation of 30 satellites inclined at 56° in three circular orbits of 24.000 km. It is supported by a network of ground monitoring stations. Galileo was been initiated because of concerns regarding GPS integrity, continuity, and availability. The system is touted as providing positioning accuracy of 1 m. Due to constellation geometry, optimum coverage is in the extreme northern and southern hemispheres. The single biggest advantage of Galileo over GPS is the incorporation of an integrity message.

Two Galileo control centers in Europe will control the constellation, time synchronization, integrity signal processing, and data handling. Data transfer to and from the satellites will be performed through a global network of Galileo uplink stations. Each uplink station will have telemetry, communications, and tracking. Galileo sensor

FIGURE 5.45 ▪
Koni (Courtesy
REUTERS/RIA
Novosti/Pool)

stations around the globe will be responsible for monitoring signal quality. Regional components will independently provide integrity for the Galileo services.

Galileo will transmit 10 signals on four L-band carriers: 6 for open service, 2 for commercial service, and 2 for public regulated service. The Galileo message includes ranging codes as well as data. Data messages will be uplinked to the satellites, stored onboard, and broadcast continually. In addition to time, almanac, and ephemeris data, the satellites will also broadcast an accuracy signal giving users a prediction of satellite clock and ephemeris accuracy over time.

Galileo will offer five levels of service: two open and three restricted (Table 5.12).

On January 1, 2007, the Global Navigation Satellite System Supervisory Authority (government) replaced the Galileo Joint Undertaking (private) and the program schedule has slid to the right. Two test satellites, GIOVE A and GIOVE B, were launched and have since been decommissioned; 4 operational satellites (In-Orbit Validation, IOV 1-4) were launched in 2011 and 2012 are being used to validate the Galileo system. Full Operational Capability with 30 satellites ($27 + 3$ spares) is expected in 2019. Initial cost was expected to be Euro 3.2–3.4 billion; logical estimates are double or triple that now (last estimate was Euro 10 billion). In November 2011 the EU commission pledged Euro 7.0 billion to complete and maintain Galileo and EGNOS through 2020.

The Chinese Space Program or Dragon in Space began with the BeiDou Navigation System; BeiDou is loosely translated as North Star. Unlike GPS, GLONASS, and Galileo, the BeiDou-1 system uses geostationary satellites as opposed to orbiting satellites and covers only China. Since the satellites are geostationary, not as many are needed to provide a viable navigation constellation; the BeiDou-1 system contains four satellites and became operational in mid-2007. The compass navigation satellite system (CNSS, or BeiDou-2 or Compass Beidou) will be a true GPS much like the others. The full BeiDou-2 constellation will ultimately include 35 space vehicles: 27 in middle Earth orbit (MEO), 5 geostationary orbits (GSO), and 3 inclined geostationary orbits (IGSO).

The free service will provide 10 m accuracy, and fee for service will provide enhanced service and error messages (8 m in 3D). The program was approved by the Chinese government in 2004; the system was activated on a trial basis on December 27, 2011, and will initially offer high-precision positioning and navigation services to the Asia-Pacific region by late 2014. This will be then expanded into a global coverage by 2020.

The Compass/BeiDou-2 navigation satellite system is currently composed of a total of 15 satellites. The first satellite, Compass-M1, was launched in 2007, followed by Compass-G2 in 2009. Five more satellites were launched in 2010, and four and five satellites were launched in 2011 and 2012, respectively. Compass-IGSO3, launched in April 2011, allowed the system to start providing regional positioning service covering China and neighboring countries, and the network will become fully functional with the launch of the last satellite in 2015.

TABLE 5.12 ■ Galileo Levels of Service

Open Service	Free to air, mass market, simple positioning
Commercial Service	Encrypted, high accuracy, guaranteed service
Safety of Life Service	Open Service + Integrity and Authentication of Signal
Public Regulated Service	Encrypted, Integrity, Continuous Availability
Search and Rescue Service	Near real-time, precise, return link feasible

The Quasi-zenith satellite system (QZSS) system is really an enhancement for the GPS system receivable only in Japan. Three satellites will be launched in a highly elliptical orbit, which will mean that the satellite will be very high on the horizon (almost overhead) and each satellite would be in view for approximately 12 hours. Adding to the GPS constellation, this position will enhance the GDOP. The first satellite, MICHIBIKI, started to provide L1 and L2C signals in June 2011. MICHIBIKI carries out technical and application verification of the satellite as the first phase; then the verification results will be evaluated for moving to the second phase in which the QZ system verification will be performed with three QZ satellites. The system even has a mascot looking down over Japan (Figures 5.46).

The Indian Space Research Organization (IRSO) is in development of its own GNSS called the Indian regional navigation satellite system (IRNSS). The IRNSS system is composed of four geostationary and three orbiting satellites and will cover an area of about 1500 km around India. Coverage will extend from $+/-40°$ Latitude and $40–140°$ E longitude. The system was designed to minimize the maximum DOP using the smallest number of satellites in orbital slots already approved for India. The system will broadcast navigation signals in the L1, L5, and also in S band; correction signals are uplinked in C band.

There will be two kinds of services: special positioning service (SPS) and precision Service (PS). Both services will be carried on L5 (1176.45 MHz) and S band (2492.08 MHz). The navigation signals would be transmitted in the S-band frequency (2–4 GHz) and broadcast through a phased array antenna to keep required coverage and signal strength. The IRNSS is expected to provide navigation accuracies similar to GPS: 10 m over the Indian landmass and 20 m over the Indian Ocean. The system can be augmented with local area augmentation for better accuracy. The first launch of an IRNSS satellite was on July 1, 2013, and the next six will be launched at six-month intervals. By the schedule the system should be operational by 2016. The expected cost of the system is approximately US$230 million.

FIGURE 5.46
QZSS Mascot

5.7.6 Satellite Augmentation Systems

To improve on the basic satellite accuracy, integrity, and availability, a number of augmentation systems have been developed:

- Wide area augmentation system (WAAS) (United States)
- Ground-based augmentation system (GBAS) (Joint Development)
- European geostationary navigation overlay service (EGNOS) (ESA)
- MTSAT satellite-based augmentation system (MSAS) (Japan)
- Indian GPS-aided geo augmented navigation (GAGAN) (India)
- Australian ground-based regional augmentation system (GRAS) (Australia)
- Chinese satellite navigation augmentation system (SNAS) (China)

All of these systems employ the benefits of uplinking differential satellite corrections to users. A simplified depiction of the U.S. WAAS is shown in Figure 5.47.

A series of reference stations located around the United States compute the clock bias for all satellites in view and multiplex this information to master stations which in turn uplink this information to geostationary satellites. The satellites rebroadcast this information on the GPS L1 frequency to all WAAS-compatible receivers. Basic GPS receivers that are not WAAS compatible will need software modifications (if possible) or replacement to take advantage of the WAAS. Phase 1 of WAAS utilized commercial Inmarsat satellites as the geostationary portion of the system. These satellites have since been replaced by three new geostationary satellites: Intelsat Galaxy XV (CRW), Anik F1R (CRE), and Inmarsat I4F3 (AMR). The fourth satellite, GEO-5, to complete full operational capability was contracted in 2012. The current architecture of reference and master stations is shown in Figure 5.48.

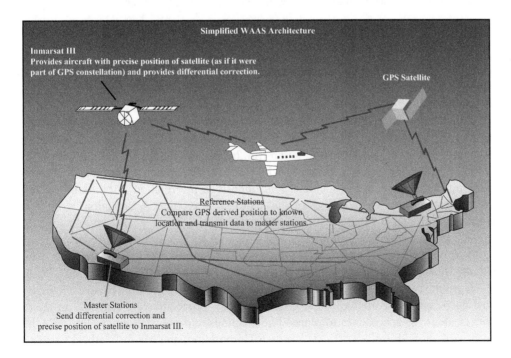

FIGURE 5.47
Simplified WAAS Architecture

FIGURE 5.48 ■
WAAS Reference
and Master Stations

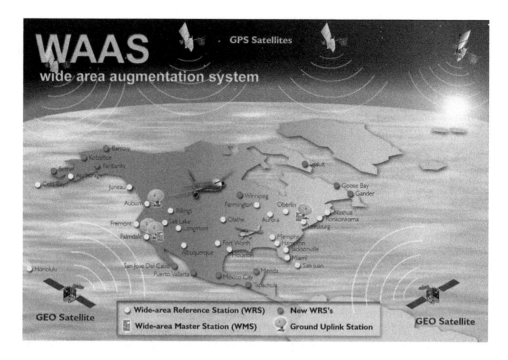

WAAS (Phase 1, initial operating capability [IOC]) was commissioned by the FAA into the NAS in July 2003. WAAS Phase 2 included full localizer performance with vertical guidance capability (LPV) and was achieved in 2008. WAAS Phase 3 is called Full LPV-200 performance (FLP), which calls for WAAS upgrades to enhance coverage and availability; it is scheduled for completion in 2014. WAAS Phase 4, dual frequency operations, began earlier in 2014 and will focus on two major areas:

– Dual Frequency with the addition of L5

– WAAS life-cycle maintenance

This phase is contingent upon the upgrades to the GPS constellation; development will continue through 2018. At the time of this writing, the earliest planned retirement of WAAS is 2028.

With the availability of the WAAS comes a new category of airport approaches called LPV, which allows WAAS-equipped aircraft to fly enhanced approaches down to ILS Cat 1 minima without the need for an ILS. Instead of guiding the aircraft with an RF beam, LPV provides landing guidance with a series of approach waypoints. The course deviation indicator provides localizer and glideslope information in terms of distance away (left/right, above/below) from the approach waypoint. A depiction of an LPV approach is shown in Figure 5.49.

As of January 9, 2014, there are 3,364 WAAS LPV approach procedures serving 1,661 airports; the FAA projects that there will be about 6,000 LPV approaches available by the end of 2018.

Currently, only special procedures are available for helicopters (created for individual users and not available to the general public). Procedures are typically in metropolitan areas for approaches to places like hospitals and police stations. The FAA has been working with Bell (429), Air Methods Corporation, and Mercy Medical Center in

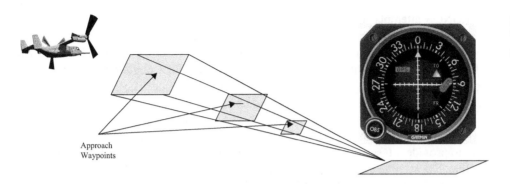

FIGURE 5.49 ▪
WAAS LPV
Approach

Approach
Waypoints

Des Moines, Iowa, and has approved a WAAS-based IFR low-level helicopter structure. The FAA has also been working with Bell, the University of Oklahoma, and CareFlite in Dallas to develop WAAS LPV approaches to medical centers. With WAAS-equipped aircraft, helicopters can fly approved LPV approaches to medical centers and helicopter emergency medical service (HEMS) landing zones to minimums of 300 ft and ¾ nm visibility. Additionally, departures may be performed at minimums of 500 ft and ¾ nm visibility (previous non-WAAS-equipped aircraft were limited to takeoffs with minimums of 700 ft and 2 nm visibility).

NAV Canada is a private sector concern with operations throughout Canada. It provides Air Traffic Control, Flight Information, Weather Briefings, Aeronautical Information Services, Airport Advisory Services and electronic aids to navigation. NAV Canada currently provides LPV service at 36 airports with a total of 57 approaches published. Over the next 12–18 months, NAV Canada has 180 approaches pending design and publication at 92 airports.

Enhancements are also being made to allow users to fly GNSS approaches down to Cat II/III minima. Originally called the Local Area Augmentation System, or LAAS, the system is now designated as the GBAS, which is demonstrated to be 1 m accurate in the horizontal and vertical planes and will support ILS Cat II/III landing requirements. GBAS is in codevelopment with the FAA, Honeywell, and Airservices Australia with assistance and prototypes by DECEA in Brazil, DFS in Germany, and AENA in Spain. The GBAS system will provide services in the airport terminal area to a radius of 20–30 nm by uplinking local differential corrections via data link from the airport (eliminating relativity effects).

Currently, two U.S. locations have obtained operational approval for GBAS use and support revenue airline traffic. These stations are located at Newark Liberty International Airport (EWR) and Houston George Bush Intercontinental Airport (IAH). Internationally, Bremen Airport (BRE) in Germany has also been approved and is being used by airlines for revenue traffic.

Several GBAS stations currently installed outside the United States are expected to receive operational approval shortly. These stations are located in Malaga, Spain, and Sydney, Australia. Additional airports, both foreign and domestic, are considering the installation and approval of GBAS.

EGNOS complements the GPS and GLONASS systems. EGNOS disseminates, on the L1 frequency, integrity signals giving real-time information on the health of GPS and GLONASS in the European theater. EGNOS has been broadcasting a preoperational signal since 2000 and is now operational. EGNOS will be integrated with Galileo when that system becomes fully capable.

EGNOS attempts to correct some of the problems with DGPS: relativity effects, common-view satellites, and receiver algorithms. A total of 34 reference integrity monitor (RIM) stations around Europe monitor GPS and GLONASS and relay their data to four master control centers (MCCs). The MCCs generate a single set of integrity data and wide area differential (WAD) GPS corrections for Europe. The corrections account for clock and ephemeris errors as well as ionospheric effects.

EGNOS operates on the same frequency (L1) and uses the same ranging codes as GPS but uses a different data format. The message cycle follows a 6 sec duty cycle in order to provide a 6 sec time to alert; it also informs users when to exclude a particular satellite. The accuracy is stated as 2 m and is interoperable with all other augmentation systems. EGNOS currently has 34 RIMs and has tested the signal on all three geostationary satellites. EGNOS uses two Inmarsat-3 satellites (eastern Atlantic and Indian Ocean) and the ESA Artemis satellite over Africa.

The first EGNOS LPV approach was accomplished (certified by DFS) (Deutsche Flugsicherung) at Bremen airport on February 12, 2012, by Air Berlin. France, Switzerland, Germany, Italy, and the Channel Islands have approved LPV approach procedures. The publication of these procedures has been possible after the signature of an EGNOS Working Agreement (EWA) between the air navigation service provider and the company ESSP, the EGNOS service provider, which officially declared the start of the EGNOS safety-of-life service intended for Aviation on March 2, 2011. Seven EWAs have been signed in Europe as of 2013.

MSAS is the Japanese MTSAT Satellite-Based Augmentation System. It works on the same principle as the WAAS and EGNOS, utilizing a network of ground monitoring stations in Japan, monitoring and ranging stations outside of Japan, master control stations, and the MTSAT. MSAS has two geostationary satellites (MTSAT-1R and MTSAT-2) and entered operational service on September 27, 2007.

GAGAN (which translates to *sky* in Hindi) is a planned space-based augmentation system developed by the Indian government that uses the geostationary satellite GSAT-4 and 18 total electronic content (TEC) monitoring stations throughout India. Differential corrections are sent to users similar to other satellite-based augmentation systems (SBAS); GAGAN claims 3 m accuracy. The director general of civil aviation (DGCA) of India certified the GAGAN system to RNP 0.1 (Required Navigation Performance 0.1 Nautical Mile) on December 30, 2013. Aircraft equipped with SBAS receivers will be able to use GAGAN signals in Indian airspace for enroute navigation and nonprecision approaches without vertical guidance.

GRAS works a little differently than the other SBASs. The system employs ground-based monitoring stations that calculate differential corrections, which in turn are provided to two master control stations. The master control stations relay the information to a network of VHF transmitters, which broadcast the information to suitably equipped users. The primary reason for this system is the ability to provide LPV approach capability. The system is very similar to the jointly developed GBAS discussed earlier and will use the same equipment developed for GBAS.

The availability of satellite-based augmentation system coverage throughout the world continues to grow. WAAS, EGNOS, and MSAS are operational and allow users required navigation performance (RNP) 0.3 coverage from the central Pacific Ocean eastward through Europe and then again in the Japanese Flight Information Region. Additional SBASs will allow users RNP 0.3 and LPV coverage potentially worldwide. Additional information on these systems, as well as test issues, can be found in chapter 6 of this text. Coverage is graphically depicted in Figure 5.50.

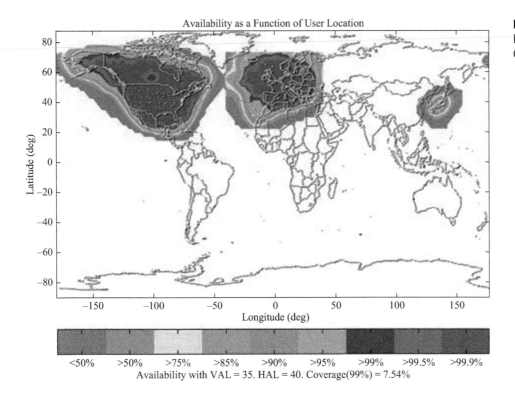

FIGURE 5.50 ■
RNP 0.3 SBAS
Coverage

Availability as a Function of User Location

Availability with VAL = 35. HAL = 40. Coverage(99%) = 7.54%

5.7.7 GPS Recap

The GPS can provide excellent positional (x, y) data to users. Position will be affected by velocity, update rate, and maneuvers. The altitude data of the GPS are normally aided by the barometric altimeter. Integrity, continuity, and availability prevent the GPS from replacing older radio aids to navigation. GPS augmentation systems enhance GPS position determination and make the system more robust. Without a stronger signal or directional antennas, GPS is susceptible to jamming and interference.

5.7.8 GPS Flight Test Objectives

Since the GPS can provide the user with positional accuracy, one would think that this may be a required evaluation. In addition to positional accuracy, the evaluator will be required to test the navigation integration functions (since GPS is rarely a stand-alone system), weapons integration, controls and displays, operational testing, and mission suitability.

It is important to understand that, like every other avionics system, we are not testing the accuracy of the box (since we already know the answer to that question) and are really testing to make sure that we did not degrade the system with our integration/ installation. The type and method of testing will be the same as the methodology explained in section 5.5.3.3. The T&E of a GNSS for use as a positioning system is covered in great detail in section 6.11 of the text.

5.7.8.1 GPS Weapons Integration Testing

Many of the same tests that were accomplished with navigation integration must also be performed for weapons integration. The accuracy and timing of the navigation information sent to the stores management system must be known, as this directly impacts the accuracy of the weapon. Any inaccuracies in the navigation system will add to the error budget of the weapons system. As previously mentioned, the mechanics and integration of the weapon must be known prior to executing a test plan. Some questions that will need to be addressed in the test plan are

- Are GPS positions fed to the weapon?
- Can the weapon acquire and track GPS signals?
- Are there any blockage areas?
- Is there communication with the store after release?
- Can position be updated after release?
- Can the system be retargeted after release?

In-depth weapons testing evaluations are covered in chapter 9 of this text.

5.7.8.2 GPS Operational Testing

The first objective in formulating an operational test plan for the GPS is to determine exactly how the GPS will be employed in a particular aircraft. A transport aircraft may use GPS as a supplemental navigation system to assist in flying from point A to point B and to perform nonprecision approaches. A search and rescue (SAR) helicopter may use the GPS to mark a point and then use GPS steering to return to that point. A fighter may use GPS information to program a *smart weapon* from the stores inventory. Although each of these systems uses the GPS, the accuracy requirements are not the same. The transport aircraft will require the same accuracy as a TACAN system, whereas the SAR helicopter may require accuracy of a few meters. The fighter may hope to attain a weapon CEP on the order of a few feet. Spending an inordinate amount of test time and assets on the transport program to prove GPS accuracy of 1 m just does not make any sense. On the other hand, it is critical that the fighter obtain such accuracy. The test plans will be increasingly more difficult as the intended accuracy for each system increases.

The evaluator needs to set up a simulated task in the test plan that closely resembles how the system will be used in the real world. The plan should start with the most benign conditions and then logically progress to the most stressful evaluation. Some key questions that may be imbedded in the plan are

- Are the controls and displays useful under realistic conditions?
- Is the system usable at all times of day and night, weather or combat conditions?
- Is the basic accuracy of the system good enough to meet operational requirements?
- What alternate displays or displayed information may enhance the probability of success?
- Are update rates satisfactory?
- What problems are associated with loading crypto keys?
- Can the mission be accomplished successfully given the limitations of the system?

▮ 5.8 | IDENTIFICATION FRIEND OR FOE

The IFF system as we know it is part of the FAA Air Traffic Control Radar Beacon System (ATCRBS). The forerunner of this system was developed during World War II

FIGURE 5.51 ■
Typical Surveillance
RADAR

to enable military radar operators to identify aircraft as friendly or enemy. The system's equipment sent a signal to the aircraft transceiver, which in turn replied with a set code depending on how a pilot tuned his selector. Only a few codes were used at that time, and codes were changed daily or more often. When adapted for civilian use, the first transponders had a capability of 64 different codes. There are now 4,096 codes available for monitoring general aviation (four-position octal setting). In addition to responding with a coded reply, the capability exists for encoding own-ship altitude. A mode 3C-equipped aircraft will report altitude and be measured in azimuth and range by the ground facility. By performing a coordinate conversion, the latitude, longitude, and altitude of the aircraft can be determined. A typical airport surveillance radar is shown in Figure 5.51. By looking at the shape of the dish, it should become apparent why altitude must be encoded in the reply.

Four modes are used in the IFF system; three are used by the military and one is used by both civilian and military aircraft. Mode 1 is used by military controllers to determine the type or mission of the aircraft it is interrogating; there are 64 reply codes for mode 1. Mode 2 is also used by military controllers, and, for example, this mode may request the tail number of the interrogated aircraft; there are 4,096 reply codes for this mode. The final military mode is mode 4, which is an encrypted mode and is used to determine friendly aircraft in wartime. Mode 3/A is the standard air traffic control (ATC) mode and provides controllers with the aircraft's unique code. It is used in conjunction with mode 3/C, which provides controllers with the aircraft's uncorrected barometric altitude. The mode 3/A codes are given to the aircraft by ATC and may be changed many times over the aircraft's route of flight. General aviation, flying under visual flight rules, uses a universal code of 1200.

The interrogation signal sent by the ground station consists of two pulses 0.8 μsec in duration, spaced at a precisely defined interval. These pulses are referred to as P1 and P3. Figure 5.52 depicts the spacing of these pulses for each of the IFF modes. In mode 1, the interval between the first and last pulse is 3 μsec, in mode 2 it is 5 μsec, in mode 3/A it is 8 μsec, and in mode 3/C it is 21 μsec. The frequency of the interrogation signal is

FIGURE 5.52 ■ IFF
Interrogation Timing

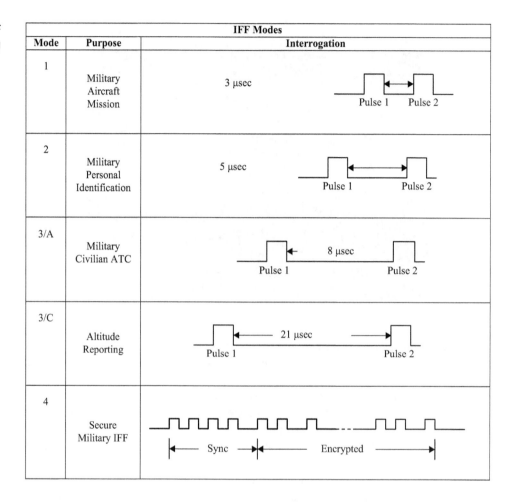

FIGURE 5.52 ■ IFF Interrogation Timing

1030 MHz. The airborne equipment contains circuitry that discriminates between the various timings and automatically sends back the desired reply. The frequency of the replied transmission is 1090 MHz. The transponder replies with a pulse containing 12 bits of data. For civilian systems, the interrogator receives the reply and converts the 12-bit data word to aircraft code and barometric altitude. The resolution of aircraft altitude is limited to 100 ft. Mode 1 replies back with a 6-bit data word and mode 2 replies with a data word of 12 bits.

Since the ground antenna is not perfect, it will emit sidelobes that can trigger a response from an aircraft that is not in the main beam. This can cause *ghosting* on the ATC display, where an aircraft may appear in more than one position. In extreme cases, an effect known as *ring around* occurs, where the transponder replies in such excess that the target is distorted into an arc or circle centered on the radar site. To combat these effects, sidelobe suppression (SLS) is used. SLS employs a third pulse, P2, which is spaced 2 μsec after P1. This pulse is transmitted from an omnidirectional antenna at the ground station rather than from the directional antenna. The power output from the omnidirectional antenna is calibrated so that, when received from an aircraft, the P2 pulse is stronger than either P1 or P3 except when the directional antenna is pointed directly at the aircraft. By comparing the relative strengths of P1 and P2, it can be determined if the antenna was pointed at the aircraft when the reply was received.

Mode 4 is a secure mode of the IFF and is used only by the military. The mode is encrypted because it is essential that enemy forces not be able to identify themselves as friendly. Mode 4 uses a challenge word containing a preamble that tells the transponder it is about to receive a secure message. The challenge is encrypted by a device that uses mathematical algorithms to put it in a secure form. The aircraft transponder receives the message and decodes it using the same mathematical algorithms to determine its validity. A transponder that cannot decipher the challenge cannot respond as a friend, and similarly a transponder will not respond to an enemy challenge.

To prevent unauthorized use of the equipment, a crypto key is used. These keys can be 12 or 24 hr keys based on Zulu time. To eliminate the chance of a random guess by a hostile target corresponding with a correct response, each identification consists of a rapid series of challenges each requiring a different response that must be correct before the target is confirmed as a friend.

5.8.1 Mode S

Mode S, or mode select, is an enhanced data link and reporting system that was designed to replace the current ATCRBS. Mode S is designed to fully interoperate with the ATCRBS; mode S surveillance radars can interrogate IFF transponders, and mode S transponders will reply to IFF interrogations. Mode S was originally developed as a solution to the congestion on 1090 MHz and the minimal number of available transponder codes. The high coverage of radar services available today means some radar sites are interrogating transponders that are within reception range of an adjacent site. This results in a situation called false replies uncorrelated in time (FRUIT), which is the reception of replies at a ground station that do not correspond with an interrogation. With the addition of traffic alert and collision avoidance systems (TCAS) (covered in chapter 6), congestion and reply problems have worsened.

Mode S attempts to reduce these problems by assigning aircraft a permanent mode S address that is derived from the aircraft's tail number. It then provides a mechanism by which an aircraft can be selected, or interrogated, so that no other aircraft reply.

Mode S is really a data link because it is not as limited as the IFF system in the amount of data that it can send. Current transponders are limited to send only 56-bit *standard-length* messages. Mode S has provisions to allow for the transmission and reception of up to 16-segment "extended length" 112-bit messages. As a result, the system can transmit messages faster as well as send messages that could not be sent before. Unfortunately, the capabilities of mode S are not being fully utilized, primarily due to the cost of upgrading all aircraft from the current IFF system. Mode S can be used for enhanced TCAS (TCAS II), which allows for conflict resolution, weather services, and wind shear alerts, differential GPS correction uplinks, traffic information service (TIS), and automatic dependent surveillance broadcast (ADS-B). All of these systems allow for greater situational awareness in the cockpit by providing the operators with large amounts of data relating to the airspace around them. These specialized systems are covered in chapter 6.

5.8.2 IFF Interrogators

Many military aircraft are equipped with airborne IFF interrogators to determine if targets that are detected by onboard or offboard sensors are either hostile or friendly. They are also an excellent device for locating a missing wingman. Military interrogator units have the designation AN/APX, with the most common interrogator being the AN/APX-100, which is employed on many U.S. and foreign military fighters and bombers. Newer systems combine the aircraft's IFF and interrogator units within one LRU. Advances in antenna design (electronically scanned, multiarrayed) and processing (monopulse) have increased the accuracy of the IFF return significantly. While older systems provided the operator with range and azimuth (with varying degrees of accuracy), newer systems such as the AN/APX-109, can provide the operator with range, azimuth, and elevation, with azimuth and elevation accuracies of 2° and range resolution of about 500 ft.

The system searches a predefined scan volume for modes or codes selected by the operator. For example, an operator can select mode 3 and enter a code of interest (say, his wingman). All targets within the scan volume that can reply to a mode 3 interrogation will be displayed to the operator out to the maximum range of the system. Any targets who reply with the correct code will be identified separately. The advantage of such a system that can interrogate mode 4 codes in a hostile environment should be readily apparent. Of course, this system will not display targets that do not have a transponder or whose transponder is in the off or standby position. Targets with bad altitude encoders will reply with an erroneous altitude just as they would to ATC.

5.8.3 Flight Testing of IFF Systems

Since the IFF or mode S transponders are required systems for flight in controlled airspace, these systems must be certified as meeting the minimum operational performance standards (MOPS) for ATCRBS reporting systems. A good place to start is the applicable technical standard order (TSO) that identifies the MOPS for IFF and mode S systems. The TSO for airborne ATC transponder equipment is TSO-C47c, while the TSO for mode S is TSO C-112. The advisory circular for either the IFF or mode S system is the next source of reference to the evaluator. An advisory circular is one means, but not the only means of demonstrating to the FAA or state authority that the installed system meets the requirements of the applicable federal aviation regulations. It may be considered a demonstration test plan, and for civil systems this is more than sufficient. The advisory circular for mode S, for example, is AC 20-131A; civil certifications are covered in depth in chapter 6.

Ground and flight tests will be required for any new installation of an IFF or mode S system. If antenna placement on the new installation is much different than the previous installation, a full series of antenna pattern tests will be required in order to determine possible blockage areas. Victim/source tests will determine the compatibility of the new system with other installed aircraft systems. Other ground tests include pulse spacing and duration of P1, P2, and P3; modes of operation; proper reply and frequency; SLC logic and operation; power; and sensitivity. Flight tests called out in the advisory circular include performance at maximum range and altitude, performance during turns, self-jamming, proper altitude encoding, and performance during approaches.

5.9 | DATA LINKS

The discussions on communications flight testing and mode S are precursors to the evaluation of data links. This section will deal exclusively with military systems: Link 4a,

Link 11/11b, and Link 16. Controller pilot data link communications (CPDLC), SAT-COM and Mode S are civilian versions of data link and are covered in chapter 6.

5.9.1 Tactical Data Link Background

Fighter aircraft, ground forces, sea forces, and antiaircraft weapons all operate within the same battle area. To be effective, all must know their location and their geographic relationship with friendly and hostile forces. It is imperative that accurate information be shared in a timely manner. To meet the command and control needs of fast-moving forces, huge amounts of data must be exchanged between automated systems. This transfer of information is used to enhance the situational awareness of commanders and participants.

The method of obtaining this situational awareness is the tactical digital information link (TADIL). The TADIL is a U.S. Department of Defense (DOD)—approved, standardized communications link suitable for transmission of digital information. The current practice is to characterize a TADIL by its standard message formats and transmission characteristics. TADILs interface two or more command and control or weapons systems via a single or multiple network architecture and multiple communications media for exchange of tactical information.

TADIL-A is a secure, half-duplex, netted digital data link utilizing parallel transmission frame characteristics and standard message formats at either 1364 or 2250 bits/sec. It is normally operated in a roll-call mode under the control of a net control station to exchange digital information among airborne, land-based, and shipboard systems. The NATO equivalent of TADIL-A is called Link 11. In most flight test communities, these data link systems are identified by their NATO designation and not their TADIL number.

TADIL-B is a secure, full-duplex, point-to-point digital data link system utilizing serial frame characteristics and standard message formats at either 2400, 1200, or 600 bits/sec. TADIL-B interconnects tactical air defense and air control units. The NATO equivalent of TADIL-B is called Link 11B. Link 11 (A/B) data communications must be capable of operation in either the HF or UHF bands. Link 11 is used by a number of intelligence-gathering platforms such as RIVET JOINT.

TADIL-C is an unsecure, time-division data link utilizing serial transmission characteristics and standard message formatting at 5000 bits/sec from a controlling unit to a controlled aircraft. Information can be one way (controlling unit to a controlled aircraft) or two way (fighter to fighter). The NATO equivalent of TADIL-C is Link 4. There are two versions of Link 4: Link 4A and Link 4C.

Link 4A plays an important role by providing digital surface-to-air, air-to-surface, and air-to-air tactical communications. Link 4A was designed to replace voice communications for the control of tactical aircraft. Link 4A has been around since the 1950s and has been expanded from its original role to include communication between surface and airborne platforms. The U.S. Navy's automatic carrier landing system (ACLS) operates on the Link 4A system. Link 4A is reliable but is susceptible to jamming.

Link 4C is a fighter-to-fighter data link that was designed to complement Link 4A, however, the two links do not communicate directly with each other. Link 4C offers some electronic protection, as opposed to Link 4A. A typical net includes four fighters as participants. Link 16 was to replace Link 4, however, Link 16 does not support the U.S. Navy's ACLS.

TADIL-J is a secure, high-capacity, jam-resistant, nodeless data link that uses the Joint Tactical Information Distribution System (JTIDS) transmission characteristics and protocols. The NATO equivalent of the TADIL-J is Link 16. An innovation on the JTIDS terminal utilizing advances in technology that allow for lighter systems is the Multifunctional Information Distribution System (MIDS). Link 16 provides improvements over the older Link 4/11 systems in the areas of jam resistance, improved security, increased data rate, larger packets of information, reduced terminal size (MIDS), secure voice, relative navigation, and precise participant location and identification.

5.9.2 Link 16 Operation

Since Link 16 can perform all of the operations of Link 4/11 (with the exception of ACLS) it is reasonable to concentrate on Link 16, as Link 4/11 will merely be subsets of the Link 16 system. For this reason, I have elected to look at the operation of the Link 16 system and then its test and evaluation. I will address the major differences between each of the three systems later.

Link 16 is the DOD's primary tactical data link for command, control, and intelligence, providing critical joint interpretability and situation awareness information. Link 16 uses time demand multiple access (TDMA) architecture and the "J" series message format (TADIL-J). The JTIDS terminal is one of two terminals that provide Link 16 capability to users; the other terminal is the MIDS. JTIDS and MIDS are the communications component of Link 16. JTIDS is an advanced radio system that provides information distribution, position location, and identification capabilities in an integrated form. JTIDS distributes information at high rates, in an encrypted and jam-resistant format.

5.9.2.1 Link 16 J Series Messages

Like other tactical data links, Link 16 conveys its information in specially formatted messages. These message formats are composed of fixed fields similar in nature to the data word structure in MIL-STD-1553. A requesting terminal will ask for specific data and the transmitting terminal will provide the data requested by comparing the request number to a specific action. Remember that in MIL-STD-1553, an RT knows which data to send based on the subaddress in the command word. It works the same with Link 16. The messages exchanged over Link 16 between participating TADIL J units are the J series messages. Each J series message format is identified by a label and a sublabel. Its 5-bit label defines 32 specific formats, while its 3-bit sublabel permits up to eight subcategories for each defined format. Together, they comprise 256 possible message definitions. The TADIL J message catalog is depicted in Table 5.13. In airborne systems, the J series messages are processed by the mission computer.

Since the aircraft mission computer is the bus controller on the 1553 bus, there must be some mechanism to transfer information between the JTIDS terminal and the controller. This is accomplished by the use of terminal input messages (TIMs) and terminal output messages (TOMs). Massive amounts of data are transferred between JTIDS terminals on a timed schedule, and there is no command/response or determinism as defined by the MIL-STD-1553 protocol. The JTIDS terminal will provide data to the mission computer only when directed by a command word; the JTIDS terminal is an RT

TABLE 5.13 ■ TADIL-J Message Catalog

TADIL-J Message Catalog

Network Management
J0.0 Initial Entry
J0.1 Test
J0.2 Network Time Update
J0.3 Time Slot Assignment
J0.4 Radio Relay Control
J0.5 Repromulgation Relay
J0.6 Communication Control
J0.7 Time Slot Reallocation
J1.0 Connectivity Interrogation
J1.1 Connectivity Status
J1.2 Route Establishment
J1.3 Acknowledgment
J1.4 Communicant Status
J1.5 Net Control Initialization
J1.6 Needline Participation Group Assignment

Precise Participant Location and Identification
J2.0 Indirect Interface Unit PPLI
J2.2 Air PPLI
J2.3 Surface PPLI
J2.4 Subsurface PPLI
J2.5 Land Point PPLI
J2.6 Land Track PPLI

Surveillance
J3.0 Reference Point
J3.1 Emergency Point
J3.2 Air Track
J3.3 Surface Track
J3.4 Subsurface Track
J3.5 Land Point or Track
J3.7 Electronic Warfare Product Information

Antisubmarine Warfare
J5.4 Acoustic Bearing and Range

Intelligence
J6.0 Intelligence Information

Information Management
J7.0 Track Management
J7.1 Data Update Request
J7.2 Correlation
J7.3 Pointer
J7.4 Track Identifier
J7.5 IFF/SIF Management

J7.6 Filter Management
J7.7 Association
J8.0 Unit Designator
J8.1 Mission Correlator Change

Weapons Coordination and Management
J9.0   Command
J10.2 Engagement Status
J10.3 Handover
J10.5 Controlling Unit Report
J10.6 Pairing

Control
J12.0 Mission Assignment
J12.1 Vector
J12.2 Precision Aircraft Direction
J12.3 Flight Path
J12.4 Controlling Unit Change
J12.5 Targe/Track Correlation
J12.6 Target Sorting
J12.7 Target Bearing

Platform and System Status
J13.0 Airfield Status Message
J13.2 Air Platform and System Status
J13.3 Surface Platform and System Status
J13.4 Subsurface Platform and System Status
J13.5 Land Platform and System Status

Electronic Warfare
J14.0 Parametric Information
J14.2 Electronic Warfare Control/Coordination

Threat Warning
J15.0 Threat Warning

National Use
J28.0 U.S. National 1 (Army)
J28.1 U.S. National 2 (Navy)
J28.2 U.S. National 3 (Air Force)
J28.3 U.S. National 4 (Marine Corps)
J29   National Use (reserved)
J30   National Use (reserved)

Miscellaneous
J31.0 Over-the-air Rekeying Management
J31.1 Over-the-air Rekeying
J31.7 No Statement

FIGURE 5.53 ■
JTIDS Installation

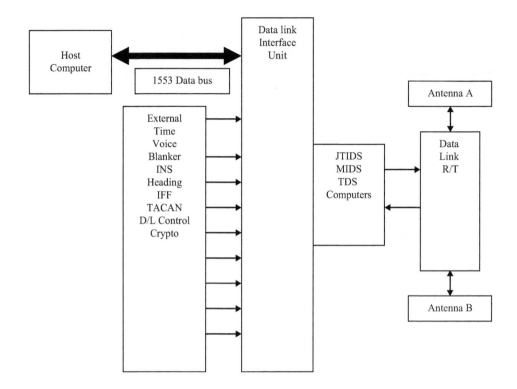

on the avionics bus. A simplified diagram of the JTIDS installation is shown in Figure 5.53.

Terminal input messages are sent from the host computer to the terminal's interface unit (IU) over the 1553 bus. On airborne platforms, the mission computer is the host, whereas on shipboard platforms the command and control processor (C^2P) is the host. The host, as bus controller, issues command words to control the transfer of data. A total of 30 TIMs have been identified for transferring data from the host to the subscriber interface control program (SICP) that executes in the IU. Besides data words, TIMs also contain command and status words; command words from the host and status words from the IU. The host can send TIMs at any time. The host restricts its transfer of TIMs so that no given TIM type is updated more frequently than every 20 msec. In addition, the host restricts its transfer of common carrier messages which are sent in TIMs 2 through 11 at a rate of 10 per 50 msec. Table 5.14 lists the TIMs and their respective functions.

Terminal output messages are used to transfer data from the terminal SICP to the host computer over the 1553 bus. Like TIMs, 30 TOMs have been identified. In addition to the J series messages received from the JTIDS network, TOMs contain terminal status information, performance statistics, and confirmation of control setting changes made by the operator. The SICP has 10 msec to write its TOMs to the buffer, signaling the SICP when it is finished by writing a TIM 29. Table 5.15 lists the TOMs and their respective functions.

The exchange of TIMs and TOMs takes place cyclically during the multiplex cycle. TIMs may be sent to the SICP in the IU any time during this cycle, but access to the TOM buffers is controlled to avoid collisions by the host and SICP. Which system has access is governed by an interrupt called the MUX data transfer complete interrupt (MDTCI).

TABLE 5.14 ■ TIMs and Their Respective Functions

Terminal Input Message No.	Function
TIM 1	Conveys a block of initialization data to the SICP
TIM 2–11	Common carrier messages (data words)
TIM 12	Not used
TIM 13–14	Offers an alternative method for conveying to the SICP for transmission sensor target data and an intent to engage a surveillance track
TIM 15	Conveys the air platform and system status to the SICP for transmission
TIM 16	Used to request information from the SICP
TIM 17	Transfers navigation data and geodetic position to the SICP
TIM 18	Transfers external time reference measurements from the host to the SICP
TIM 19–28	Not used
TIM 29	Controls the process of accessing the MUX port by signaling that the host has completed its request for TOMs for the current cycle
TIM 30	Transfers clock synchronization from the host to the SICP; it is only used when time tags in the navigation data are invalid

TABLE 5.15 ■ TOMs and Their Respective Functions

Terminal Output Message No.	Function
TOM 1	Conveys the system status each MUX cycle
TOM 2–20	Used to transfer messages received from the network to the host
TOM 21–27	Supports the tape recording function; the SICP contains a 700 word buffer that can hold recorded data
TOM 28	Provides the information requested by the host in TIM 16
TOM 29	Provides the status of the net selection request
TOM 30	Provides the terminal's navigation data to the host for use by the host's navigation system

For the 10 msec immediately following this interrupt, the SICP has authorization to update the TOM buffers. The TOM 1 buffer is updated every time; others are updated as required.

Link 16 includes two secure voice channels, called ports. These ports operate at 16 kbps; port 1 is also capable of interfacing with an external voice encoder (vocoder) that can operate at either 2.4 kbps or 4.8 kbps. This allows the system to support encrypted voice. The voice audio is digitized in the IU of the terminal. Control of each voice channel is controlled by the push-to-talk protocol. The digitized voice is not encoded for error correction, yet it is still encrypted, which makes it secure.

When the operator pushes the push-to-talk button, two audio buffers are alternately filled with digitized voice data. As one buffer is transmitted, the other is filled, and vice versa. During the receive portion of the transmission, data are alternately filled in the two buffers and then converted back to audio by the IU. Practical experience indicates that JTIDS secure voice degrades more quickly than encoded JTIDS data; JTIDS voice remains understandable with up to 10% of the transmission in error compared to 50% in error for encoded data.

Link 16 operational frequencies interfere with TACAN and IFF operation. IFF frequencies are notched so as not to interfere with normal IFF operation. With TACAN, however, the physical system is removed from the aircraft and the function is taken over by the JTIDS terminal. The TACAN channel and mode are included in the TIM sent to the SICP and outputs of DME, bearing, and audio are sent via a TOM.

5.9.2.2 JTIDS Architecture

The JTIDS network is a communications architecture known as TDMA. This architecture uses time interlacing to provide multiple and apparently simultaneous communications circuits. Each circuit and each participant on the circuit are assigned specific time periods to transmit and receive.

Every 24 hr day is divided by the JTIDS terminal into 112.5 epochs. An epoch is 12.8 min in duration. An epoch is divided into 98,304 time slots, each with a duration of 7.8125 msec. The time slot is the basic unit of access to the JTIDS network. The time slots are assigned to each participating JTIDS unit (JU) for a particular function; a JU is assigned to transmit or receive during each time slot.

The time slots of an epoch are grouped into three sets: A, B, and C. Each contains 32,768 time slots numbered from 0 to 32,767, and is called the slot index. The time slots belonging to set A are identified as A-0 through A-32767. By convention, the time slots of an epoch are named in an alternating pattern. The time slots of each set are interleaved with those of the other sets in the following sequence:

$$A\text{-}0, \quad B\text{-}0, \quad C\text{-}0$$
$$A\text{-}1, \quad B\text{-}1, \quad C\text{-}1$$
$$A\text{-}2, \quad B\text{-}2, \quad C\text{-}2$$
$$.$$
$$.$$
$$.$$

$$A\text{-}32767, \quad B\text{-}32767, \quad C\text{-}32767$$

This sequence, ending in C-32767, is repeated for each epoch. Figure 5.54 shows the JTIDS TDMA.

The number of slots in each epoch (32,768) is a power of 2 (2^{15}), and remembering back to middle school math, we know that the logarithm (base 2) of 32,768 is 15. This fact is used in JTIDS TDMA to determine the recurrence rate number (RRN). The recurrence rate indicates how many time slots are in the block and how often they occur. The entire set of time slots in A is 32,768, or 2^{15}. It is distributed every third slot and is identified by the notation A-0-15. Half of the slots in A would be 16,384, and would occur at every sixth slot. The recurrence rate for half of the slots would be 14. Similarly, if we halved the number again, every 12th slot would be used, for a recurrence rate of 13.

The relationship between RRN and the interval between slots is shown in Table 5.16. As a consequence of the three-set interleaved structure, three is the minimum spacing between slots in the same set. The time between slots can be calculated by multiplying the number of slots in the interval by 7.8125 msec.

The 12.8 min epoch is too large to be manageable in describing the rapid communication requirements of Link 16, so a smaller, more manageable time interval is required. This basic recurring unit of time is called a frame. There are 64 frames per epoch; each frame is 12 sec in duration and is composed of 1536 time slots. There are

FIGURE 5.54 ▪
JTIDS TDMA

Each day is divided into 112.5 epochs

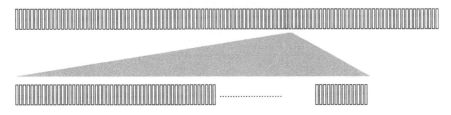

0   Each epoch is divided into 98,304 time slots.

A- B- C- A- B- C- A- B- C- A- B- C- A- B- C-
0   0   0   1   1   1   2   2   2   3   3   3   4   4   4

TABLE 5.16 ▪ Recurrence Rate (Logarithm Base 2 of the Number of Slots)

RRN	No. of Slots Per Epoch	Slot Interval Slots	Slot Interval Time
15	32,768	3	23.4375 msec
14	16,384	6	46.8750 msec
13	8192	12	93.7500 msec
12	4096	24	187.5000 msec
11	2048	48	375.0000 msec
10	1024	96	750.0000 msec
9	512	192	1.50 sec
8	256	384	3.00 sec
7	128	768	6.00 sec
6	64	1536	12.00 sec
5	32	3072	24.00 sec
4	16	6144	48.00 sec
3	8	12,288	1.6 min
2	4	24,576	3.3 min
1	2	49,152	6.4 min
0	1	98,304	12.8 min

TABLE 5.17 ■ Frame Recurrence Rate

| RRN | No. of Slots Per Frame | Slot Interval | | |
		Within the Set	Within Interleaved Sets	Time
15	512	1	3	23.4375 msec
14	256	2	6	46.8750 msec
13	128	4	12	93.7500 msec
12	64	8	24	187.5000 msec
11	32	16	48	375.0000 msec
10	16	32	96	750.0000 msec
9	8	64	192	1.50 sec
8	4	128	384	3.00 sec
7	2	256	768	6.00 sec
6	1	512	1536	12.00 sec

512 slots dedicated to A, B, and C, respectively. The time slots of the frame are numbered and interleaved as the epoch:

$$A\text{-}0,\ B\text{-}0,\ C\text{-}0$$
$$A\text{-}1,\ B\text{-}1,\ C\text{-}1$$
$$A\text{-}3,\ B\text{-}3,\ C\text{-}3$$
$$\cdot$$
$$\cdot$$
$$\cdot$$
$$A\text{-}511,\ B\text{-}511,\ C\text{-}511$$

Since 64 frames comprise an epoch, the number of slots per frame is found by dividing the number of slots per epoch by 64. An RRN of 15 represents every slot in a set; 32,768 slots in an epoch and 512 per frame. Because the three sets are interleaved, the interval between slots in any one set is three and not one. Table 5.17 shows the slot layout per frame and is merely Table 5.16 divided by 64.

Time slots are assigned to each terminal in the network as a block of slots. This block of slots, known as a time slot block (TSB), is defined by three variables: a set (A, B, or C); a starting number, or index (0 to 32,767); and the recurrence rate. Up to 64 TSBs may be assigned. RRNs of 6, 7, and 8, which correspond to reporting intervals of 12, 6, and 3 sec, are most often used. For example, the TSB A-2-11 represents the block of time slots belonging to set A that occur every 16th slot within the set beginning at slot 2. The value of the slot interval (16) for the RRN of 11 can be found in Table 5.15. TSB B-2-11 would represent the block of slots belonging to set B that occur every 16th slot within the slot beginning at slot 2. Although both examples have identical starting slot indices and RRNs, they have no slots in common. These TSBs are said to be mutually exclusive. It is important to remember this fact when making terminal assignments, since only one terminal may transmit at any specific slot. Four rules govern the mutual exclusiveness of TSBs:

- TSBs specifying different sets are mutually exclusive.
- TSBs having the same RRN but different indices are mutually exclusive.

- If one index is even and one index is odd, the TSBs are mutually exclusive regardless of their RRNs.

- TSBs having the same set and index but different RRNs are not mutually exclusive; the one with the smaller RRN is a subset of the one with the larger RRN.

The network structure described so far represents a single net. This is the architecture of the network when the JTIDS terminal is set to either communications mode 1 with the normal interference protection feature (IPF) setting, or to mode 2. Only one frequency is used (969 MHz) and all messages are assigned to time slots in the single net.

Multiple nets can be constructed by stacking several single nets. The time slots of these nets are synchronized so that a time slot of one net coincides with the time slots of every other net. The multinet architecture allows several groups of participants to exchange messages independent of the other groups during the same time slot. There is a 7-bit net number within JTIDS, which allows a network of up to 128 nets to be established. Net number 127 is reserved to indicate a stacked net configuration; the remaining nets are numbered from 0 to 126. In a stacked net configuration, the operator selects which net to use during operations, with each net assigned a unique frequency-hopping pattern. Although it is theoretically possible to construct a stacked network of 127 individual nets, more than 20 can cause degradation in communications.

In order to establish a synchronized network, a single terminal must be designated to provide a net time reference (NTR). The time maintained by this terminal becomes the JTIDS system time. As the reference, this time defines the beginning and end of time slots and ensures time alignment in multiple net architectures. Periodically the NTR transmits a net entry message to assist other terminals in synchronizing with the network.

Synchronization is accomplished in two steps: coarse synchronization is achieved when a net entry message is successfully received, and fine synchronization occurs when the terminal successfully exchanges round-trip timing messages with the NTR. Each terminal maintains a measure of how accurately it knows system time, called its time quality (TQ). It continuously refines its knowledge of system time by periodically transmitting round-trip timing messages and by measuring the time of arrival (TOA) of all received messages. This has become a much simpler task with GPS installations. Unlike other data link systems, JTIDS can continue to operate when the NTR fails or leaves the net.

A platform is assigned either to transmit or receive in each time slot. Several components comprise each time slot:

- Jitter
- Synchronization
- Time refinement
- Message header and data
- Propagation

The slot starts with a delay, or dead time, in which no pulses are transmitted. This is called jitter, and the amount of dead time (jitter) varies with each time slot in a pseudorandom way that is determined by the transmission security (TSEC) crypto variable. Jitter is part of the antijam waveform of JTIDS. Jitter is followed by two sets of predetermined pulse packets called synchronization and time refinement. These patterns are used by the receiver to recognize and synchronize with the signal. The message section is next, which contains header and data information. The end of the slot is marked by a guard period, which allows for the propagation of the signal. The propagation guard time allows for a propagation of 300 nm in the normal range setting or 500 nm in the extended range setting.

5.9.2.3 Link 16 Message Types

Link 16 messages are exchanged during the TDMA time slots of the JTIDS network. Each message consists of a message header and message data. The message header specifies the type of data being sent and identifies the source track number of the transmitting terminal. There are four basic types of messages:

- Fixed format
- Variable format
- Free text
- Round-trip timing

Fixed-format messages are used to exchange the J series messages described in section 5.9.2.1. Variable-format messages provide a general way to exchange any type of user-defined messages. Free-text messages are used for digitized voice, and round-trip timing messages are used for synchronization.

The fixed-format messages consist of one or more words to a maximum of eight. Each word consists of 75 bits: 70 are data, 4 are for parity, and 1 is used as a spare. Three types of words are defined: an initial word, an extension word, and a continuation word. Fixed-format messages consist of an initial word, one or more extension words, and one or more continuation words. If there are an insufficient number of 75-bit words to fill a transmit block, the terminal fills the block with no statement (NS) words. Metering within the host attempts to keep the NS words to a minimum. The word formats of the fixed-format messages are shown in Figure 5.55.

Fixed-format messages are always encoded for parity checking. Bits 4 through 18 of the message header, which contain the source track number, are used in conjunction with the 210 data bits of the 3 data words to calculate a 12-bit parity value. These parity bits are distributed at bit positions 71–74 of each 75-bit word. Bit 70 is reserved as a spare.

Fixed-format messages are also encoded for error detection and correction. The encoding scheme is an algorithm called Reed-Solomon (R-S) encoding. Sixteen error detection and correction bits are added for every 15 bits of data, so that 15 bits of data become 31 bits in the message. R-S encoding can detect and correct up to 8 bits in error.

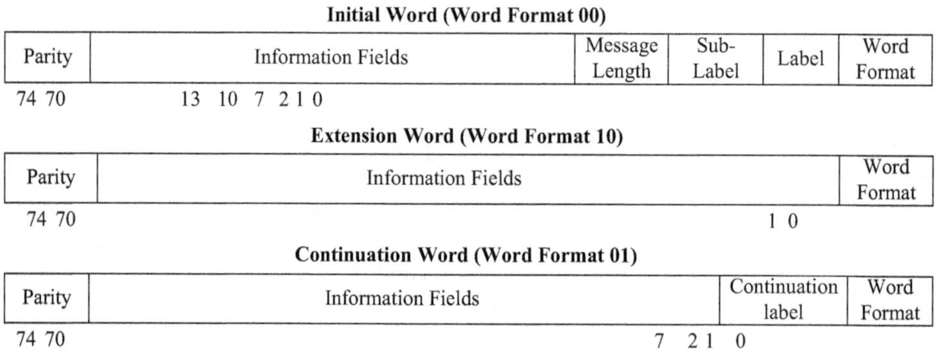

FIGURE 5.55 ■
Word Formats
(Fixed-Format
Messages)

Initial Word (Word Format 00)

Parity	Information Fields	Message Length	Sub-Label	Label	Word Format
74 70	13 10 7 2 1 0				

Extension Word (Word Format 10)

Parity	Information Fields	Word Format
74 70		1 0

Continuation Word (Word Format 01)

Parity	Information Fields	Continuation label	Word Format
74 70		7 2 1 0	

TABLE 5.18 ▪ Assignment of Slots to Match Standard Serial Line Baud Rates

RRN	Slots/Frame	Slots/Sec	Bits/Slot	Bits/Sec
13	128	10-2/3	225 (R-S encoded)	2400
			450 (unencoded)	4800

R-S encoding will transform the 75-bit sequence into a 155-bit sequence. These bits are taken in groups of five to create 31 symbols, transforming each 75-bit Link 16 word into a 31-symbol codeword.

Variable-format messages consist of 75-bit words, similar to the fixed-format messages, and are structured the same as the extension word. Variable-format messages, however, may vary both in content and length, and fields within the message can cross word boundaries. Information within the message itself identifies the fields and their length.

Free-text messages are independent of any message standard; they are unformatted and utilize all 75 bits in the data word and all 225 bits in the three-word block. No parity is associated with free-text messages. They may or may not be R-S encoded for error correction. When R-S encoding is used, the 225 bits of data are mapped onto 465 bits for transmission. When R-S is not used, all 465 bits are available for data, but only 450 are used in order to match standard line rates (Table 5.18). The free-text message format is used for Link 16 voice.

Message words may be taken in groups of 3 words, 6 words, or 12 words to form transmissions. If there are an insufficient number of words to complete a group, the terminal once again fills with NS words. The processing of a group of three words is called standard (STD) format, 6 words is called packed-2 (P2) format, and 12 words is called packed-4 (P4) format.

The message header is used to specify whether the message that follows is fixed format, variable format, or free text. It identifies whether the message is encoded or not and which packing structure has been used. It also identifies the serial number of the secure data unit (SDU) and the track number of the source terminal.

The message header contains 35 bits that are R-S encoded, so the 35 bits become 80 bits. The 80 bits are taken five at a time to form an R-S header codeword containing 16 symbols.

The last of the Link 16 message types is the round-trip timing (RTT) message. The JTIDS time alignment is maintained by the JTIDS NTR. In order to receive and transmit in the net, the terminal must be synchronized with the net. A special set of messages is defined to support synchronization. The RTT message represents the only exception to the rule that a terminal can transmit or receive, but not both, in a single time slot. With the RTT message, a terminal can transmit an interrogation and receive a reply within the same time slot.

The initial exchanges of RTT messages enable the terminal to synchronize with the network. Subsequent exchanges allow the terminal to refine its measurement of system time, or TQ. Each terminal reports its TQ over the network. Each terminal also maintains an internal catalog of terminals within LOS that have the highest value of TQ. These entries assist the terminal in choosing which address to interrogate for the next RTT message.

FIGURE 5.56 ■
RTT Interrogation
and Reply
Messages

RTT Interrogation

Type A

SDU Serial No.	Source Track No.	0	010
34 19	4 3 0		

Type B

SDU Serial No.	TQ	Spare	1	010
34 19 14	4 3 0			

RTT Reply

Interrogator's SDU Serial No.	Time of Arrival (12.5 nsec)
34 19	0

An RTT interrogation may be either addressed (RTT-A) or broadcast (RTT-B). The RTT-A consists of a header message addressed to the unit that reports the highest TQ. It is transmitted in a dedicated time slot, and only the addressed terminal will reply. The RTT-B is not specifically addressed to any terminal but rather to the broadcast. The RTT-B contains the terminal's TQ, and any terminal with a higher TQ can reply.

The RTT interrogation is similar to the message header; it contains 35 bits of R-S encoded data. The interrogation is not followed by any data, and it is transmitted without jitter.

The reply to an RTT interrogation is transmitted 4.275 msec after the beginning of the time slot as measured by the receiving terminal. The reply contains the time that the last symbol of the interrogation was received; this TOA is measured at the antenna and is reported in units of 12.5 nsec. The interrogating terminal uses the reported TOA along with its own measurement of the TOA of the reply to calculate a correction to its system clock. The RTT interrogation and reply formats are shown in Figure 5.56.

5.9.2.4 The JTIDS RF Signal

JTIDS operates in the L-band portion of the UHF spectrum between 960 and 1215 MHz. As mentioned numerous times in this text, this is the frequency where many other navigation and telemetry systems operate. TACAN, DME, GPS, IFF, and telemetry frequencies are but a few of the systems that JTIDS must compete with. Expected ranges for JTIDS are 300 nm air-to-air, 150 nm ship-to-air, and 25 nm ship-to-ship. In order to increase its operational range, JTIDS makes use of relays.

There are three communications modes within the JTIDS terminals: mode 1, mode 2, and mode 4. Mode 1 is the normal JTIDS mode of operation; it uses frequency hopping with full message security and transmission security processing. Mode 2 does not employ frequency hopping, and all pulses are transmitted on a single frequency. Mode 4 also does not employ frequency hopping, and it also eliminates much of the security processing to allow JTIDS to operate as a conventional data link. Mode 1 is the normal operational mode. Modes 2 and 4 represent a reduction in capacity and capability.

There are 51 frequencies assigned to the JTIDS system between 969 and 1206 MHz, 3 MHz apart. Two subbands centered at 1030 MHz and 1090 MHz are excluded because they are used by IFF. In mode 1, each pulse is transmitted on a different frequency in a pseudorandom pattern determined by the crypto variable of the day. Since the frequency is changed on a pulse-to-pulse basis, the nominal frequency-hopping rate is 33,000 frequencies/sec.

In order to prevent interference with other systems, JTIDS employs many protection features. The TACAN system is removed from JTIDS-equipped aircraft and the function is taken over by the JTIDS terminal. Notch filters on the JTIDS terminal transmitter prevent the JTIDS signal from encroaching on the ATCRBS IFF system. To prevent the terminal's RF transmissions from interfering with the normal operations of other systems, the pulse power spectrum and out-of-band transmission characteristics are strictly specified. The pulse power spectrum is unrestricted within 3 MHz of the center frequency, but it must be down 10 dB at 3 MHz, 25 dB at 6 MHz, and 60 dB beyond 15 MHz. The out-of-band emission characteristics, including broadband noise, sideband splatter, harmonics, and other spurious emissions, must be kept below −65 dBm/kHz. In order to adhere to these restrictions, the JTIDS terminals are equipped with an IPF to monitor all terminal transmissions.

The JTIDS IPF monitors the transmitter for out-of-band transmissions, transmissions in the IFF notches, improper frequency-hopping distribution, incorrect pulse lengths, high receive/transmit thermal levels, improper amplifier operation, and time slot duty factor. If the authorized levels are exceeded, the IPF function will automatically disable the terminal's transmissions.

The terminal implements two sets of protection features: peacetime constraints on the terminal operation in mode 1, and the monitoring of transmissions to ensure that they do not interfere with navigation systems. The operator has three IPF settings:

- Normal
- Exercise override
- Combat override

The normal setting is normally required when within 12 nm of land and enforces peacetime constraints on the terminal's transmissions. TDMA is authorized, but the capability and capacity of the system is reduced to prevent interference with civilian and military navigation systems. When the normal IPF setting is in effect, the transmission packing is limited to standard or packed-2 single pulse; high-power output is forbidden. Multinetting and contention access (where simultaneous transmissions may occur) are also not allowed. Time slot usage is minimized to a system level of 40% capacity and a terminal level of 20% capacity. The limitations apply to all time slots, free-text and fixed-format messages, and relays. The permitted time slots are A-0-14, B-1-14, C-0-12, and C-4-11. This restriction on time slot usage, which limits the number of pulses per unit time, is called a time slot duty factor (TSDF) of 40/20. The terminal monitors its performance for compliance with these restrictions. It also measures pulse width, monitors the distribution of pulses to ensure a uniform distribution of pulses across the spectrum, and monitors for out-of-band transmissions. Any violation of the restrictions will cause the terminal to disable all transmissions.

The exercise override setting provides partial interference protection. The peacetime constraints are overridden to allow contention access, multinetting, a time slot duty factor of 100/50, and high power output. Shipboard and AWACS terminals have a high-power amplifier that can boost outgoing transmissions to 1000 W, which is primarily used to overcome jamming. In this setting, the terminal continues to monitor the signal characteristics of the terminal transmissions, including pulse spread, invalid frequencies, and nonuniform frequency distribution.

With the IPF set to combat override, all protection features are overridden. It should only be used when operation is essential to mission success (even with a known IPF failure). It should also be understood by the operator that navigation systems will probably be jammed.

5.9.2.5 Link 16 Participants

This section deals with how the Link 16 network is populated. The assignment of participants in the Link 16 structure is based entirely on operational requirements and the numbers and types of agencies and services involved in executing a mission plan. Many structures are possible and they are designed by agencies specifically organized to make these plans work. The network is broken up into operating groups that have a commonality in mission. The participants are then assigned based on the mission they support. There are also functional groups that lend support to the operation of the network. All of these functional groups are known as network participation groups (NPGs). The transmissions on each NPG consist of messages that support its particular function.

This functional structuring allows JTIDS units (JUs) to participate on only the NPGs necessary for the functions they perform. A maximum of 512 participation groups is possible. NPGs 1 through 29 are associated with the exchange of J series messages; each NPG is associated with a particular set of J series messages. Each NPG and its associated function are described in Table 5.19.

Messages for up to three NPGs can be buffered in the JTIDS terminal by the SICP. Buffering decreases the chance that a message will be lost or dropped when message traffic is heavy. The operator selects which NPGs to buffer.

TABLE 5.19 ■ NPG Assignments and Functions

NPG	Function
1	Initial Entry
2	RTT-A
3	RTT-B
4	Network Management, Redistribution of Network Capacity
5	Location Information, Weapons and Store Status
6	Identification, Location Information, Relative Navigation and Synchronization
7	Air, Surface, and Subsurface Tracks, Land-point SAM Sites, Reference Points, ASW Points, Acoustic Bearings, EW Bearings and Fixes
8	Mission Management, Command Designated Unit to Coordinate Battle Group Weapons and Order Weapons Engagements
9	Air Control, Mission Assignments, Vectors and Target Reports
10	Electronic Warfare
12	Voice Group A
13	Voice Group B
14	Indirect PPLI for Link-11 Participants
18	Weapons Coordination, Similar to NPG 8
19	Fighter to Fighter, Shared Sensor Information
27	Joint Operations PPLI
28	Distributed Network Management
29	Residual Message, Transmission Opportunity for Messages not Assigned to any other NPG

The amount of network capacity given to a particular NPG depends on the communications priorities. These priorities are

- Number and type of participants
- Access requirements to the NPG by the participant
- Expected volume of data
- Update rate of the data

The number of time slots that must be allocated within the NPG to each participant depends on the type of unit and the method of accessing the time slot. Two types of units are participants within Link 16: command and control (C^2) and non–command and control (non-C^2). The two types of units have different missions and different requirements in terms of which NPGs they participate in and how often their data must be updated. The primary C^2 functions are surveillance, electronic warfare, weapons coordination, air control, and net management. The primary non-C^2 functions are target sorting and engagement.

Within each NPG, the time slots assigned to each unit are evenly distributed over time and generally occur every 3 sec, every 6 sec, etc. These sets are assigned using time slot blocks that specify a set, index, and RRN, as previously discussed, and a net number. Time slot blocks are numbered from 1 through 64, so the terminal is limited to a maximum of 64 assignments.

The two major access modes for each time slot are dedicated access and contention access. Dedicated access is the assignment of time slots to a uniquely identified unit for transmission purposes. Only the assigned JU can transmit during that time slot. If there are no data to transmit, then the slot goes empty. The obvious advantage of dedicated access is that it provides each JU on the NPG with a predetermined portion of the network's capacity and a guarantee that there will be no transmission conflicts. One of the disadvantages of dedicated access is that assets are not interchangeable. One aircraft cannot simply replace another in the NPG. If this were a requirement, its terminal would first have to be reinitialized to transmit and receive during time slots matching those of the unit it was replacing.

The assignment of time slots to a group of units as a pool for transmission purposes is known as contention access. In contention access, each unit randomly selects a time slot from the pool during which to transmit. The frequency of the terminal's transmission depends on the access rate assigned to that terminal.

The advantage of contention access is that each terminal is given the same initialization parameters for the corresponding time slot block. This simplifies the network design and reduces the network management burden, since specific assignments for each JU are unnecessary. Since the JUs are now interchangeable, it facilitates the inclusion of new participants and allows units to be easily replaced. The disadvantage of contention access is that there is no guarantee that a transmission will be received.

Because a time slot is not dedicated to one user, simultaneous transmissions are possible with contention access. The likelihood of simultaneous transmissions depends on the number of time slots in the pool, the number of units in a group, and the frequency with which they must transmit. If two units do transmit simultaneously during the same time slot on the same net, whether in dedicated or contention access, it is called time slot reuse. Whenever this occurs, receivers will always hear the unit closest to them.

TABLE 5.20 ■ Network Roles and Associated Functions

Network Role	Function
Network Time Reference	The time established by this unit is, by definition, the network system time. This system time is propagated to all other units by the initial net entry message.
Position Reference	The positional reference must have a positional accuracy of better than 50 ft and should be on a surveyed point. The positional reference is assigned a position quality of 15 and is used as a stable geodetic reference for RelNav.
Initial Entry JU	Assists in the propagation of system time to units that are beyond LOS of the NTR.
Navigation Controller	Used only when a relative grid is desired; establishes a relative grid origin.
Secondary Navigation Controller	Provides stability to the relative grid by working with the navigation controller.
Primary User	All JUs, with the exception of the NTR are primary users. Networks can support upwards of 200 primary users.
Secondary User	A JU operating passively is said to be a secondary user. Units can enter the network passively, or an operator can select passive operations after entering the network actively.
Forwarding JU	A JU designated to forward data between links, such as Link 16 to Link 11.
Network Manager	Unit assigned the responsibility for administering the minute-by-minute operation of the Link 16 network. The network manager monitors force composition, geometry, network configuration, relay, and multilink requirements.

In addition to being C^2 or non-C^2 units, terminals are also identified as having a role in the network. Network roles are functions assigned to a JU, either by initialization or operator entry. Roles are assigned to C^2 units and are based on platform capabilities and expected platform position. With the exception of network manager, all roles are terminal functions that may be changed during operations. The network roles and their associated responsibilities are found in Table 5.20.

5.9.2.6 PPLI and RelNav

Two of the capabilities of the JTIDS terminals which have only been minimally touched upon until now are precise participant location and identification (PPLI) and relative navigation (RelNav). JTIDS terminals employ relative navigation techniques to constantly fix their platform's position. This information is transmitted periodically, along with other identification and status information in the PPLI message.

The PPLI is the friendly unit reporting message and can be used to determine link participants and data forwarding requirements, as well as initiate air control. PPLIs are transmitted by all active JUs on NPG 5/6. PPLIs generated by Link 11 participants are forwarded by the forwarding JU on NPG 14. Location is reported as latitude, longitude, and altitude, as well as the participant's present course and speed. This information is used by the receiving terminal's RelNav function, along with other information such as the quality reports, RTTs, and local navigation inputs, to refine its calculations of its own position.

Each participant in Link 16 is assigned a unique JU number between 00001 and 77777 (octal). In addition, the JTIDS terminals require a source track number of five octal digits. These two numbers must be the same to ensure positive identification. In addition to the JU number, the PPLI message also contains IFF codes and platform types. Periodically the host provides inventory and equipment and ordnance status that is included in the PPLI. The specific numeric inventory of shipboard defense missiles

TABLE 5.21 ■ TQ Assignments

Time Quality	Standard Time Deviation (nsec)	Time Quality	Standard Time Deviation (nsec)
15	<50	7	<800
14	<71	6	<1130
13	<100	5	<1600
12	<141	4	<2260
11	<200	3	<4520
10	<282	2	<9040
9	<400	1	<18,080
8	<565	0	>18,080

TABLE 5.22 ■ Position Quality Assignments

Position Quality	Position Uncertainty (ft)	Position Quality	Position Uncertainty (ft)
15	<50	7	<800
14	<71	6	<1130
13	<100	5	<1600
12	<141	4	<2260
11	<200	3	<4520
10	<282	2	<9040
9	<400	1	<18,080
8	<565	0	>18,080

and aircraft weapons, as well operational, degraded, or nonoperational status of all relevant shipboard and aircraft systems is reported. Status information also includes the air control and JTIDS voice net numbers or other UHF voice frequencies that each platform is using.

Two other pieces of information are provided by the JUs in the PPLI message: time and position quality. Each terminal estimates how well it knows system time. The estimate is based on clock drift, accuracy of the RTT interrogations, and the time since the last RTT reply. This estimate is used to provide the TQ. Values of TQ range from 0 to 15; only the NTR has a TQ of 15. Each terminal provides the TQ in every PPLI message that it transmits. Table 5.21 shows the assignment of the TQ with respect to the time deviation in nanoseconds.

Position quality is also a value of 0 through 15, which indicates the accuracy with which a unit knows its geodetic or relative position. Position is provided by navigation sources, and the value entered should be consistent with the platform's navigation equipment. A geodetic position quality of 15 is assigned to stationary position references and indicates an accuracy of position within 50 ft. When relative grid is used, all positions are measured relative to the navigation controller, and the navigation controller is assigned a position quality of 15. Table 5.22 shows the assignment of position quality with respect to positional errors.

Relative navigation (RelNav) is an automatic function of the JTIDS terminal. It is used to determine the distance between platforms by measuring the arrival times of transmissions and correlating them with reported positions. This information is required by the terminals in the network to remain synchronized. Automatic RelNav is in constant operation in all JTIDS terminals and may be used to improve a terminal's positional accuracy. If two or more terminals have accurate, independent knowledge of their geodetic position, RelNav can provide accurate geodetic information for all participants in the network (aerotriangulation). As a result, the precise geodetic position of every unit can be maintained constantly by every other unit in the network.

5.9.2.7 Determinism

There has been no discussion until now about the determinism of transmissions on Link 16. If we remember back to MIL-STD-1553, we know that one of the biggest advantages of the command/response system is that it is deterministic. We know that the sent data are the received data. What is the mechanism employed in Link 16? In fact, certain messages require acknowledgment to indicate that the message is received. Two levels of acknowledgment, called receipt/compliance (R/C), are performed. The first level of acknowledgment, called machine receipt, is performed automatically by the system. The second level of acknowledgment is a response from the operator and indicates an intention to comply (HAVCO, have complied, or WILCO, will comply) or to not comply (CANCO, cannot comply). In Link 11, the machine portion of the R/C is performed by the tactical data system computers. In Link 16, it is an automatic function performed by the JTIDS terminals. Other responses, such as cannot process (CANTPRO) are possible as well. In Link 11 as well as Link 16, the message is automatically retransmitted if the machine receipt is not received; this process can occur several times. The appropriate operators are informed if a machine receipt has not been sent after several transmissions.

5.9.3 Data Link Comparisons

Protocol. Link 11 uses a polling protocol and a netted architecture. The net operates under a controller that permits access and maintains discipline. Link 4A uses a command/response time division multiplexing (TDM) system. One controller can control multiple aircraft on the same frequency independently. Link 16 uses the principle of TDMA. All JTIDS terminals, or pools of terminals, are assigned specific time slots to transmit or receive data.

   Nodelessness. A node is a unit that is necessary to maintain communications. The node in Link 11 is the net control station (NCS); if the NCS goes down, the entire net goes down. Link 16 is nodeless with the exception of the requirement for a net time reference at startup. Once the net is established, it can run for hours with the loss of the NTR.

   Participants. Each participant, or JTIDS unit, is assigned a unique five-digit octal address from 00000 to 77777. Link 11 is limited to a three-digit octal address, allowing for addresses up to 177.

   Track Numbers and Quality. Link 16 employs a five-character alphanumeric track number, allowing up to 524,284 tracks. Link 11 employs a four-digit octal track number, 0000–7777, allowing 4095 tracks. Link 16 uses track quality, with values ranging from 0 to 15. Each track quality value is defined by a specific positional accuracy, the best of which is 50 ft. Link 11 has track qualities (0–7) that are defined by update rate. A track quality of 7 can have an accuracy worse than 3 nm.

   Track Identification. Link 16 track identification is detailed enough to identify the type of platform, activity, and nationality. Link 11 is limited to three fields. Link 16

PPLI messages can report detailed information on friendly forces: equipment status, ordnance inventory, radar and missile channels, fuel available for transfer, gun capability, ETA/ETD to/from station, and surface missile inventory. Link 16 can provide information on land points and tracks, a category that is not available with Link 11.

Granularity. Granularity is the measure of how precisely a data item can be reported in messages. Link 16 is an order of magnitude better than Link 11.

Graphics and Mapping. Link 16 can report lines, segments, geometric shapes, and maps of any size. Link 11 can report only four basic shapes. Link 16 utilizes a 3D geodetic coordinate system and can report anywhere in the world. Link 11 uses a Cartesian coordinate system and is map limited.

5.9.4 Flight Testing of Data Link Systems

The first thing that will awe flight testers of data link systems is the sheer amount of data that is being transferred among platforms and the mechanisms that programmers have used to maximize the volume and rate of this transfer. As mentioned in every chapter of this text, the first step is to understand how the system is supposed to operate. Testers must understand that there are two operations that are occurring: the transfer of information over the link itself, and the exchange of that information between the host (mission computer) and the data link terminal (JTIDS, MIDS, or tactical data system computers). If it is assumed that the data link is a known entity, then our greatest tasks will be to ensure that a) we have not screwed up the installation, and b) we integrated the terminal successfully into our system architecture. Are the data being presented to the operator correctly, efficiently, and without delay? A typical interface of an airborne data link system was shown in Figure 5.53.

In the following discussion I assume (just as we did in the previous operations section) that we are attempting to integrate a JTIDS/MIDS terminal onto the bus. The data transfer between the mission computer and the SICP is accomplished via TIMS and TOMS (section 5.9.2.1). These messages can be captured on aircraft instrumentation and either telemetered to the ground or captured on aircraft tape. The data rates will be the same as all other MIL-STD-1553/1773 systems. The JTIDS terminal will either receive information from the mission computer (TIMs) or transmit information (TOMs) when directed by a command word. The truth data, however, must be the data that are received by the receive/transmit on the JTIDS side of the diagram. Recall that the SICP can buffer up to three NPGs. What was not mentioned is that there is an optional recorder that is written to by the SICP for all Link 16 data. These data are our truth data. It is the tester's job to ensure that the requested data are in fact the truth data that arrived over the link. Unfortunately, these data are not in the format of the familiar 1553 but in the format described in the previous sections.

In addition to verifying the TIMs and TOMs, the tester will also need to evaluate three other functions associated with the JTIDS/MIDS terminals: PPLI, RelNav, and the determinism of the JTIDS system (HAVCO, CANCO, and CANTPRO). The tester must ensure that the host is providing the correct information to be inserted into the PPLI message: IFF codes, platform type, inventory, ordnance and equipment status, operational or degraded status, JTIDS voice net numbers and other UHF frequencies in use, and time and position quality. In this case, the truth data are resident in the host and can be accessed on the 1553 data bus. RelNav is an automatic function of the JTIDS/MIDS terminal and is in constant operation. RelNav accuracy is obtained by comparing the true position of all participants (TSPI) to the JTIDS-derived position. The determinism of the data link system is accomplished by the machine-generated receipt of messages (much like the status word in the 1553 system). Evaluation of the receipt/compliance function needs to validate that the appropriate operators are informed if a machine receipt has not been sent after several transmissions.

The left side of Figure 5.53 depicts the direct inputs to the data link interface. Each of these systems must be evaluated to ensure that the proper information is being delivered to the interface correctly, in proper format, and in a timely manner. The external clock is the aircraft time, and in most cases will be a GPS time. The voice channels were described in section 5.9.2.1. Evaluation includes support of encrypted voice and push-to-talk protocols. Voice may be encrypted at the host or accomplished in the interface unit, depending on the implemented architecture. The blanker unit allows simultaneous operation of the JTIDS/MIDS and the ATCRBS IFF and TACAN systems. Since these systems all operate on the same frequencies, a series of notch filters and blanking pulses must be employed in order to prevent inadvertent jamming of these systems. The JTIDS system notches out IFF frequencies and prohibits radiation in those frequencies. The TACAN function is assumed by the JTIDS/MIDS terminal, and system transmissions are interleaved to prevent jamming of either system. The tester will be required to ensure that normal operation of the IFF and TACAN is not adversely affected by the JTIDS equipment; this test must be done with the JTIDS equipment on and off. INS and heading information are required by the JTIDS system and this information is directly routed from the systems to the JTIDS interface. This architecture prevents bogging down the 1553 bus with high data rates. A simple test to ensure that the correct information is delivered in the correct format needs to be devised to validate this interface. The JTIDS system uses two crypto keys: one for voice and the other for data. Testing should include the proper loading of these keys as well as the capability to roll over to a new ZULU date. The keys are good for 24 hr, and the time is based on GMT. When flying, there is a probability that you will fly into the next day (2400Z). For this reason, 2 days of keys are loaded into the system and will automatically roll over to the next day when needed. A controls and displays test plan as well as a human factors test plan should also be incorporated.

5.9.5 Advances in C^{4}I

Command, control, and communications intelligence (C^{3}I) has been updated to include computers (C^{4}I). Current systems, including MIDS, are limited by today's standards, especially in the areas of streaming audio and video. The next-generation planned system is the joint tactical radio system (JTRS).

JTRS radios provide additional communications capability: access maps and other visual data, communicate via voice and video, and obtain information directly from battlefield sensors. JTRS provides Internet protocol–based capability and will eventually replace all tactical radios with interoperable LOS and beyond LOS radios. It will allow all forces (ground, sea, and air) to freely exchange tactical information.

There are several versions of JTRS being built for the services. The ViaSat/Data Link Solutions team is responsible for implementing JTRS into the MIDS LVT terminal while maintaining Link 16 and TACAN functions. Boeing is developing the ground mobile radio (GMR), which will support helicopter requirements, U.S. Army and

Marine Corps ground vehicles and U.S. Air Force tactical control. General Dynamics is developing the handheld/manpack version of the JTRS.

The implementation of JTRS will allow the integration of the Tactical Targeting Network Technology (TTNT). TTNT is an advanced, low-latency, high-bandwidth Internet protocol waveform that meets time-sensitive targeting network technology requirements. Because of the TCP/IP architecture plug-and-play protocols are allowed, this eliminates the need for complex net structures (which is one of the biggest problems with Link 16). TTNT has distinct advantages over all current military data link systems:

- Internet-based protocol
- Ad hoc joining to the net (<5 sec)
- Extremely low latency (2 Mbps at 100 nm with a 2 msec delay)
- Low observable compatible
- Antijam capabilities comparable to Link 16
- High data throughput
- Multiple levels of security
- No loss of Link 16 and TACAN functionality

JTRS and TTNT are currently in the development stage; testing will be roughly equivalent to the tests described in the Link 16 section, with the added burden of much more data to capture and analyze.

5.10 | SELECTED QUESTIONS FOR CHAPTER 5

1. What is the difference between magnetic and true heading?
2. What is the Q factor and FOM used for?
3. Name three spinning mass INS error sources.
4. What is total electronic content and what does it affect?
5. What are the four requirements that need to be satisfied for a navigation system to be certified as a short-range navigation system?
6. What does gain mean to you? How can you increase gain?
7. How do you convert true to magnetic?
8. 1° of latitude equals how many miles?
9. What is a space-based augmentation system?
10. 1° of longitude equals how many miles? At 30° N?
11. What is an earth model?
12. What earth model is most commonly associated with GPS? GLONASS?
13. What does NAD-27 mean? Is it used worldwide?
14. What airspeed do we use in a system test?
15. What is meant by the term omnidirectional range?
16. What is an NPG in the MIDS network?

17. What is meant by the term side-lobe cancellation?

18. What measures acceleration in a RLG INS?

19. What is the purpose of the JTIDS/MIDS interface unit?

20. What is the major detriment to accuracy in hyperbolic systems?

21. What problems could you expect when using CW in a DNS system?

22. Which altitude is encoded in mode 3C?

23. What heading do we use in a system test?

24. What is the advantage of the Galileo system over the GPS?

25. What is the difference between MSL AGL and WGS-84?

26. What is a transition altitude?

27. What is a coordinate conversion?

28. Name three types of Link 16 formats.

29. What is a loop antenna? What is the basic principle?

30. A warning flag appears on your TACAN during an approach. What does it mean? Would you expect the same on your GPS system?

31. What is a NOTAM? What is a NAGU?

32. What does a land/sea switch do in a DNS system?

33. How accurate do you think your installed TACAN is?

34. What is frequency hopping? How is it used in Link 16 systems?

35. Are there civil requirements that military aircraft must meet when installing new radio aids equipment?

36. Why do you input a present position prior to aligning the INS?

37. Where would you start if you were to write a test plan for mode S installation?

38. What is ADS-B?

39. NDBs are popular in remote environments. Why?

40. What is meant by the term accuracy in navigation systems? Does it ever change?

41. Who was responsible for cancellation of the OMEGA system?

42. Identify the unique problems of EMI/EMC in a fighter-type aircraft.

43. Can the GPS on its own provide enough accuracy for a Cat II approach?

44. What is the difference between true heading and track?

45. What error does KEAS account for? At what speeds?

46. What is the drawback of Galileo's precision service?

47. How easy is it to replace participants on the Link 16 system?

48. What heading does the INS calculate? How about the GPS?

49. What is an advisory circular? What is it used for?

50. What happens to IAS in a climb with a constantly held TAS?

51. Are GPS weapons really flown with GPS?

52. What is the purpose of employing a data link?

53. Who has access to the GPS L2 frequency? What if the P code becomes the Y code?

54. What is the communications system within Link 16?

55. Briefly explain the differences between Link 16 and other link systems.

56. What does nodeless link mean? Give an example of a nodeless link.

57. What is an NTR? Is one always required?

58. What does a ring laser gyro measure? How is the platform realigned to level?

59. The accuracy of the INS is 1 nm. Which TSPI system would you use?

60. What does granularity mean?

61. What is the difference between FOM and GDOP?

62. If you wanted the latitude, longitude, and altitude of a Link 16 participant, what would you request of the MIDS terminal?

63. What is RelNav?

64. What is an advantage of a directional antenna?

65. How would you convert N 35 09.11 to degrees, minutes, and seconds?

66. What does integrity mean in navigation systems?

67. Which frequency band does Link 16 operate in?

68. Why is TACAN included within Link 16?

69. How easy is it to jam a GPS signal? How can it be prevented?

70. What is the major difference between IFF and mode S?

71. Name two methods of controlling synchronization in Link 16.

72. Briefly describe IFF operation.

73. What is an altitude encoder and why is it used? Which IFF mode is it employed in?

74. What is the optimum GPS constellation?

75. Briefly explain the potential of mode S.

76. How many satellites are required for a 3D position? Are there any exceptions?

77. Magnetic variation for a radio aid is applied where?

78. What is a VORTAC?

79. What information is provided by a VOR? TACAN? NDB?

80. What is an ADF steer?

81. Explain the rho-rho and rho-theta systems.

82. What does high, low, and terminal mean when applied to radio aids?

83. Briefly explain a bearing accuracy test for radio aids.

84. Briefly explain a maximum range test for radio aids.

85. What would you expect for typical INS accuracy? Military or civilian systems?

86. What is the effect of a speed increase on GPS accuracy?

87. What does continuity mean in navigation systems?

88. Briefly explain the operation of an air-to-air TACAN.

89. What is the basic premise of operation of a hyperbolic system?

90. What is the integrity monitor for a radio aid? For GPS?

91. 1° of longitude at the equator equals how many nautical miles? At the poles?

92. Which Doppler system would you prefer, one operating in the L band or one operating in the X band?

93. An aircraft is 2.5° off course at 32 nm. How many miles off course is it?

94. Name the two types of INSs.

95. What is a hyperbolic system? How does it compute a position?

96. What is the relationship between error rate and time-in-align for an INS system?

97. What is a nine-state vector?

98. Can you think of any advantages Link 16 has over Link 4 or Link 11?

99. How would you test acceleration accuracies of an INS?

100. Draw a picture of a BDHI or RMI as the aircraft is heading 020° and is on the 360° radial.

101. An ADF low-frequency receiver operates in the 200 to 1600 kHz range. Why is this important to a pilot?

102. Why are ring laser–type INSs used if they are less accurate than spinning mass gyros?

103. What is a stored heading alignment? Why is it used?

104. Why is the INS selected to NAV at the specification time requirement?

105. How can nets be stacked in a Link 16 system?

106. Are Link 16 messages sent in MIL-STD-1553 format?

107. Why are 1030 and 1090 MHz notched out in Link 16?

108. How is the civilian world protected against MIDS/JTIDS?

109. How does the INS navigate (i.e., compute position)?

110. How are accelerometers employed in an INS?

111. What is a torquer? How is it employed?

112. What does availability mean in navigation systems?

113. Briefly explain how possible alignment errors affect INS performance.

114. What does a ring laser measure? How?

115. Briefly describe a possible navigation route profile and why it is built that way.

116. What is the minimum amount of time required for a true measure of a spinning mass INS performance? Why?

117. Can knowledge of navigation hierarchy become a problem? Why?

118. Describe three types of INS alignments.

119. Describe three types of INS updates.

120. Will the entire update always be accepted by the Kalman-state vector? Why or why not?

121. What is a taxi interrupt in an INS alignment? How does it work?

122. What is the maximum range you would expect between VORs on a jet route if the VOR accuracy was stated as 4°?

123. What is the Doppler effect?

124. What is the primary use of a DNS?

125. What is the importance of depression angle in a DNS?

126. Why is a space-stabilized antenna important in a DNS?

127. What is a PPLI? Give some examples.

128. How can a directional antenna aid GPS operation?

129. Name three advantages of the DNS.

130. Name three disadvantages of the DNS.

131. How does the DNS compute drift?

132. Name the three errors in time in the GPS.

133. What is meant by unintentional jamming?

134. What is continuity in a navigation system?

135. The IFF mode 3 has 4096 codes. Why?

136. Is the DNS bank limited? Why?

137. Give three examples of a global navigation satellite system.

138. What are P1, P2, and P3 in an IFF system?

139. What is a Janus DNS?

140. What does an interrogator do?

141. What is the biggest drawback of the current IFF system?

142. What does the GPS measure?

143. What are the major contributing errors to the GPS?

144. What integrity monitor is employed in the GPS?

145. Why are four satellites required for the GPS?

146. Name four types of tests common to a DNS evaluation.

147. What is a GDOP? What determines the GPS GDOP?

148. Why does the GPS employ two separate frequencies?

149. Are there any advantages of an INS over GPS?

150. What is GPS almanac data? Why is it important?

151. What are the typical errors associated with GPS? CA, P, differential?

152. What does differential GPS correct?

153. Which error of the GPS position is related to aircraft velocity?

154. What is the update rate of the GPS?

155. How long will it take for a brand new GPS to acquire and produce a position?

156. How would you test an installation of a new GPS?

157. You turn on Link 16 and you lose telemetry. Why?

158. Describe the best navigation system you could design having access to all of the systems discussed.

159. Name three accelerations, phantom or otherwise, accounted for in INSs.

160. Name three types of gyros used in INSs.

161. What is a TADIL?

162. What determinism is there in Link 16?

163. How is a DNS platform stabilized?

164. You are flying at 12,000 ft and 160 KIAS. What is the TAS?

165. How does the VOR obtain bearing?

166. How long is a Shuler period?

167. Where would you find the datum used for a map?

168. Name three functions of the interference protection feature on Link 16.

169. How does the TACAN operate in Link 16 systems?

170. What is the difference between track and true heading?

171. How is Link 16 secure?

172. How many VOR stations would you use in a radio aids test?

173. How is DME obtained in a TACAN?

174. Your wingman is missing, you have set TACAN channel 37, what does he need to set?

175. What is pseudoranging in the GPS?

176. What are TIMs and TOMs?

177. What is the most difficult thing about data link testing?

178. What is the basic unit of time in Link 16?

179. What advantages does Galileo have over GPS?

180. Name three SBASs.

181. What is a NANU? Where would you find it?

182. Is an INS a good navigation system to install in a helicopter?

183. Why will ADF/NDB never be removed from aircraft?

184. Other than RAIM, name two other GPS integrity monitors.

185. What is meant by a radial error in an INS test?

186. You notice that the GPS DOP is 3. What is your error budget?

187. You only have a GPS in your aircraft. Can you legally fly IFR?

188. What is coherent detection?

189. What is the difference between a stable platform and a strap-down platform?

190. What is a cone of confusion? When do you encounter it? When do pilots encounter it?

191. Why do we not use a depression angle of 80° or 90° in a DNS?

192. What is a VOT? What does it test?

193. What is an internal optimum constellation algorithm?

194. What effect does flying over snow have on a DNS? Sand?

195. When is the combat setting used on Link 16?

196. Does your aircraft have an atomic clock to measure GPS time? How does it do it?

197. What is the difference between dedicated and contention access?

Part 23/25/27/29 Avionics Civil Certifications

Chapter Outline

6.0	Introduction	267
6.1	FAA Type Certification History	268
6.2	Federal Aviation Regulations in the Code of Federal Regulations	270
6.3	Other Rules and Guidance	272
6.4	FAA Type Certification Process	273
6.5	Avionics Software Considerations in the Technical Standard Order Process	282
6.6	Certification Considerations for Highly Integrated or Complex Systems	284
6.7	Important Notes for Evaluators Concerning Documentation	298
6.8	Differences between EASA and FAA Documentation	300
6.9	Cockpit Controls and Displays Evaluations	301
6.10	Weather RADAR Certification	322
6.11	Airworthiness Approval of Positioning and Navigation Systems	329
6.12	Reduced Vertical Separation Minimums	348
6.13	Proximity Warning Systems	356
6.14	Terrain Awareness and Warning System	382
6.15	Flight Guidance Systems	410
6.16	Landing Systems	459
6.17	Flight Management Systems	504
6.18	Enhanced Vision Systems (EVS)	520
6.19	Summary	527
6.20	Selected Questions for Chapter 6	528

6.0 | INTRODUCTION

The Part 23/25/27/29 in the title of this section refers to the U.S. Federal Aviation Administration (FAA) and European Aviation Safety Administration (EASA) (formerly Joint Aviation Administration) definitions of aircraft types. Part 23 aircraft are defined as normal, utility, acrobatic, and commuter category airplanes; Part 25 defines transport category airplanes. The discussions in this section are applicable to Part 27 (utility helicopters) and Part 29 (transport helicopters) as well as military installations. The systems in this section have, for the most part, been developed for the general and commercial aviation market or mandated for use by the state authorities.

Since the emphasis will be on compliance with the applicable certification requirements, some time will be spent on the civil certification process, its history, and a review of the types of documents that you will need to reference. The discussion will continue with hardware, software, and safety considerations and the requirements for each. Controls and displays and human factors, which are considerations for all test plans, will be discussed in detail and documents to assist in the test planning process will be identified. The systems that will be covered include weather RADAR, global satellite based navigation civil certifications, reduced vertical separation minima (RVSM), terrain awareness warning systems (TAWS), traffic alert and collision avoidance systems (TCAS), flight management systems (FMS), landing systems, autopilots, and integrated navigation systems. Suggested reading and reference material are numerous and will be called out in each of the sections.

6.1 | FAA TYPE CERTIFICATION HISTORY

Each country around the world has created a *state authority* responsible for managing aircraft, pilots, and air traffic management. This authority in the United States is called the FAA. In Europe, it is now known as the EASA for members of the European Union (although each country maintains its own state authority). These organizations have a list of recognized priorities. The first concern is safety and the continued operational safety (COS) of aviation. The second priority is regulatory, where the authorities set safety standards and establish policies and procedures. The final priority is the certification process, which is further broken down into type, production, and airworthiness certifications and will be explained later.

Commercial air transport originated in the United States in 1918 after a demonstration of the capability in 1911. Originally tasked to the U.S. Army, service was transferred to the U.S. Post Office and was transcontinental by 1921. The Airmail Act of 1925 authorized the Post Office to contract with air carriers for mail service. Federal regulations seemed to be a foregone conclusion after the industry suffered 2,000 injuries in two months of service. The Air Commerce Act of 1926 set up the Aeronautics Branch of the Department of Commerce. This branch was responsible for fostering the growth of air commerce, establishing airways and navigation aids, licensing pilots, issuing airworthiness certificates, and investigating accidents. Four original bulletins governing aircraft were issued by the Aeronautics Branch:

- Bulletin 7 (basic handbook, 1926)
- Bulletin 7A (airframes, 1928)
- Bulletin 7G (engines and propellers, 1931)
- Bulletin 7F (aircraft components and accessories, 1933)

The requirements in these initial bulletins were not very severe. If an aircraft could demonstrate a takeoff and landing ground roll of less than or equal to 1,000 feet and a sea level rate of climb of greater than or equal to 250 ft/min it met the requirements for aircraft performance. If it could fly figure-eight patterns around fixed pylons, it demonstrated satisfactory compliance for flying qualities. The first type certificate for an aircraft was granted to the Buhl Airster C-3A on March 29, 1927 (Figure 6.1).

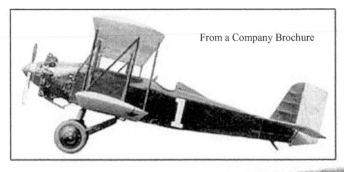

From a Company Brochure

FIGURE 6.1 ▪
Buhl Airster

Its engine developed 200 hp, weighed in at 1686 lb, and took 7.5 min to climb to 2000 feet (266 ft/min, which exceeded the 250 ft/min requirement).

The Black-McKeller Act of 1934 required mail contracts by bid. This caused airlines to turn to passengers for revenue. On May 6, 1935, Senator Cutting from New Mexico was killed in the crash of a DC-2 traveling from Albuquerque, New Mexico, to Kansas City, Missouri. The scheduled TWA flight ran into unforecast weather, ran low on fuel, and crashed while attempting an approach in poor weather conditions. The Accident Board faulted TWA for not carrying reserve fuel, employing unqualified pilots, engaging in instrument flying without effective radios, and trying to land with a low ceiling. TWA countered by faulting the Department of Commerce for poorly maintained navigation beacons and failing to inform pilots of deteriorating weather conditions.

The Civil Aeronautics Act of 1938 was passed as a result of Senator Bronson Cutting's death. This act transferred the civil aviation role from the Commerce Department to a new independent agency called the Civil Aeronautics Authority. The first regulation issued by the authority was Civil Aeronautics Regulation (CAR) 4 "Airworthiness Requirements for Aircraft." This regulation applied to all aircraft, but conflicting requirements for transport and light aircraft caused a separation into two sets of rules in 1948. In 1940, President Franklin D. Roosevelt split the Civil Aeronautics Authority into two agencies: the Civil Aeronautics Board (CAB) and the Civil Aeronautics Administration (CAA).

The CAB was responsible for rulemaking, accident investigation, and the economic regulation of airlines. The CAA was given the responsibility for air traffic control (ATC), airmen and aircraft certification, safety enforcement, and airway development. In 1948, two regulations and two manuals were available for guidance:

- Civil Aeronautics Regulation 3 (CAR 3)—Airworthiness Requirements for Normal, Utility, and Acrobatic Aircraft

- Civil Aeronautics Regulation 4 (CAR 4)—Airworthiness Requirements for Transport Aircraft
- Civil Aeronautics Manuals 3 and 4 (CAM 3, CAM 4)—Flight Test Guidance

In 1956, two regulations were added:

- Civil Aeronautics Regulation 6 (CAR 6) Airworthiness Requirements for Normal Category Helicopters
- Civil Aeronautics Regulation 7 (CAR 7)—Airworthiness Requirements for Transport Category Helicopters

In 1958, the Federal Aviation Act was passed, which transferred the role of the CAA as well as CAB safety rulemaking to the Federal Aviation Agency (FAA). The act added a new role of developing common military and civilian air navigation and air traffic control systems to the new agency. In 1965, the Civil Aeronautics Regulations were recodified into Federal Aviation Regulations.

The recodification of the CARs into the new Federal Aviation Regulations included the following:

- CAR 1 became FAR 21—Certification Procedures
- CAR 3 became FAR 23—Normal, Utility, and Acrobatic Aircraft
- CAR 4b became FAR 25—Transport Aircraft
- CAR 6 became FAR 27—Normal Helicopters
- CAR 7 became FAR 29—Transport Helicopters
- CAR 13 became FAR 33—Engines
- CAR 14 became FAR 35—Propellers

On April 1, 1967, after passage of the Department of Transportation Act the previous year, the Federal Aviation Agency was renamed the Federal Aviation Administration and was placed under the newly formed Department of Transportation (DOT). The Airline Deregulation Act of 1978 phased out the CAB's economic regulation of airlines. The CAB ceased to be an entity in 1984. In 1994, the Federal Aviation Act of 1958 was repealed and the Federal Aviation Regulations were rewritten into the Code of Federal Regulations (CFR). The regulations are meant to "promote safe flight of civil aircraft...by prescribing minimum standards." This is where we are today.

6.2 | FEDERAL AVIATION REGULATIONS IN THE CODE OF FEDERAL REGULATIONS

The FAR are presented in the CFR. The CFR is a codification of general and permanent rules based on the acts of Congress. They are divided into titles that represent broad subject areas, and a few examples are as follows:

- Title 1—General Provisions
- Title 3—President

- Title 7—Agriculture
- Title 14—Aeronautics and Space
- Title 19—Custom Duties

Within Title 14 there are subchapters and FAR parts:

Title 14, Chapter 1
Federal Aviation Administration
Department of Transportation

Subchapter	FAR Part
A. Definitions	1
B. Procedural Rules	11 and 13
C. Aircraft	21 through 49
D. Airmen	60 through 67
E. Airspace	71 through 77
F. Air Traffic and General Operating Rules	91 through 100
G. Air Carriers	121 through 139
H. Schools and Other Certified Agencies	141 through 149
I. Airports	150 through 169
J. Navigation Facilities	170 through 182
K. Administrative Regulations	183 through 191
N. Risk Insurance	198
O. Aircraft Loan Guarantee Program	199

The regulations that govern certification and continued airworthiness can be found in subchapters C, F, G, and K:

Subchapter C—Aircraft

FAR Part 21	Certification Procedures for Products and Parts
FAR Part 23	Airworthiness Standards: Normal, Utility, Aerobatic, and Commuter Category Airplanes
FAR Part 25	Airworthiness Standards: Transport Category Airplanes
FAR Part 27	Airworthiness Standards: Normal Category Rotorcraft
FAR Part 29	Airworthiness Standards: Transport Category Rotorcraft
FAR Part 31	Airworthiness Standards: Manned Free Balloons
FAR Part 33	Airworthiness Standards: Aircraft Engines
FAR Part 34	Fuel Venting and Exhaust Emission Requirements for Turbine Engine Powered Airplanes
FAR Part 35	Airworthiness Standards: Propellers
FAR Part 36	Noise Standards: Aircraft Type and Airworthiness Certification
FAR Part 39	Airworthiness Directives
FAR Part 43	Maintenance, Rebuilding, and Alterations

Subchapter F—Air Traffic and General Operating Rules

FAR Part 91	General Operating and Flight Rules

Subchapter G—Air Carriers, Air Travel Clubs, and Operators for Compensation or Hire: Certification and Operation

FAR Part 121   Certification and Operation: Domestic, Flag, and Supplemental Air Carriers and Commercial Operators of Large Aircraft

FAR Part 125   Certification and Operation: Airplanes Having a Seating Capacity of 20 or More Passengers or a Maximum Payload of 6,000 Pounds or More

FAR Part 127   Certification and Operation: Scheduled Air Carriers With Helicopters

FAR Part 129   Operations: Foreign Air Carriers and Foreign Operators of U.S. Registered Aircraft Engaged in Common Carriage

FAR Part 133   Rotorcraft External Load Operations

FAR Part 135   Air Taxi Operators and Commercial Operators

FAR Part 137   Agricultural Aircraft Operators

FAR Part 139   Certification and Operation: Land Airports Serving Certain Air Carriers

Subchapter K—Administrative Regulations

FAR Part 183   Representatives of the Administrator

How does a proposed rule become a requirement (FAR)? This is accomplished through the rulemaking procedures. The documentation for this process may be found in three guidelines provided by the FAA:

- FAR Part 11—General Rulemaking Procedures
- FAA Order 2100.13—FAA Rulemaking Policies
- Advisory Circular (AC) 11-2—Notice of Proposed Rulemaking Distribution System

The rulemaking process is shown in Figure 6.2.

Any interested person may petition the administrator to issue or amend a rule issued by the FAA. The petition or summary is published in the *Federal Register* with a 60-day comment period. Status reports are provided to the petitioner 10 days after the close of the comment period and every 120 days. If the petition is granted, then a Notice of Proposed Rulemaking is published in the *Federal Register*. If the petition is denied, the petitioner receives a denial of petition signed by the administrator.

Any interested person may also petition the administrator for temporary or permanent exemption from any rule issued by the FAA. A summary is published in the *Federal Register* with a 20-day comment period. Within 120 days a grant or denial of exemption is issued to the petitioner; the notice of this disposition is published in the *Federal Register*.

The Aviation Rulemaking Advisory Committee (ARAC) is composed of members of the aviation community, including air carriers, manufacturers, general aviation representatives, passenger groups, the general public, other government agencies, and other groups or associations. Their purpose is to advise the administrator on rulemaking activities. The ARAC provides a forum to obtain additional input from outside the government on major regulatory issues; all meetings are open to the public.

Rulemaking Process

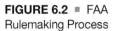

FIGURE 6.2 ■ FAA
Rulemaking Process

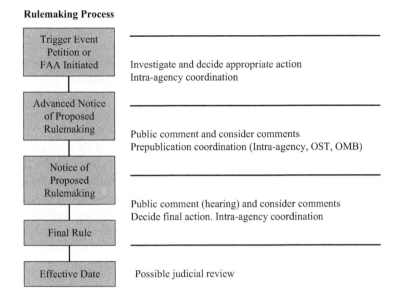

▮ 6.3 | OTHER RULES AND GUIDANCE

- In addition to the FAR contained in the CFR, the evaluator must also be aware of other rules and guidance. The first of these is the Special Federal Aviation Regulation (SFAR), which is issued for a specific trial period and is used to gain knowledge and experience during that period. The SFAR goes through the same rulemaking process as the FAR. The SFAR may eventually be canceled, revised, or incorporated as a FAR. AC 00-44 lists the FARs and SFARs and information on how to obtain them.

Interim airworthiness criteria are issued for a new category for which airworthiness standards do not exist. The tilt rotor is an example where interim criteria had to be issued. The administrator will hold public meetings to solicit comments and work with the manufacturer to establish certification criteria.

Airworthiness directives (ADs) possess the same authority as the FAR; they are issued to correct an unsafe condition in a product. They can be issued only if there is an unsafe condition and the condition is likely to develop in other products of the same design. The AD describes the conditions and limitations under which the product may continue to operate (corrective action). The AD is considered as part of the FAR (by FAR Part 39) and the process is documented in FAA Order 8040.1. An AD can be issued in one of three ways:

- Emergency rule: issued by telegram or priority letter, then published in the *Federal Register*; comments are not solicited.

- Immediate rule with request for comments: published in the *Federal Register* with an effective date 30 days later. The AD may be changed based on the comments received.

- Notice of Proposed Rulemaking, followed by a final rule.

FAA orders and notices are directives for FAA employees and their designees; they do not carry the force of law (e.g., a FAR). The public and certification applicants may find these documents very useful. Orders are permanent directives governing internal operations and provide guidance on a specific subject. Notices are temporary directives that remain in effect for 12 months or less. If they are not replaced, they become indefinite; they are often used to update FAA procedures. The FAA may also issue policy memoranda, which contain statements on policy interpretation. These are also internal to the FAA and provide guidance to a limited audience.

Advisory circulars do not carry the force of law, but provide a means of showing compliance with a FAR requirement. This is an excellent starting point for flight testers, since the AC is basically a ground/flight test plan for specification compliance. The AC is not the only means of showing compliance with a FAR requirement, and an applicant may elect to demonstrate this compliance via the use of another method. The AC is focused on a particular subject and specifically aimed at those regulated by the FAA. The AC, although not required to be used, does encourage standardization.

6.4 | FAA TYPE CERTIFICATION PROCESS

The reference for the certification process is FAA Order 8110.4, "Type Certification Handbook." In a nutshell, an applicant must show compliance with the applicable FARs and the authority will find compliance based on the applicant's submittal. The applicant will show compliance by test, inspection, and analysis. The FAA will find compliance by witnessing tests, inspecting the design, reviewing the analysis, and conducting tests. The applicant is entitled to a type certificate if he shows the type design meets the applicable regulations and any other special considerations and if the authority finds that the type design meets the applicable noise requirements and airworthiness requirements.

Certification is granted via one of three methods:

- Type certificate (TC): for a new aircraft or an aircraft with significant changes.
- Supplemental type certificate (STC): certification of modifications to an aircraft that require substantial engineering and/or flight testing per the FARs; the STC will apply on all similar types of aircraft. (If the applicant held the TC, she would then apply for an amended TC vice an STC.)
- FAA Form 337: certification of a modification that would not affect flight safety for a single aircraft; this is done via an FAA field approval.

The requirement for obtaining certification is found in FAR Part 21, Certification Procedures for Products and Parts. Type certificate requirements are found in subpart B, Provisional Type Certificates; in subpart C, Changes to TCs; in subpart D; and supplemental TCs I, in subpart E. The TC is the basis for all other approvals, and has two major components: production and airworthiness. The TC flight test project will be the most complex and the lengthiest. A TC is required (FAR 21.19) for new aircraft or for aircraft where there are

- Changes in design, configuration, power, speed, or weight that are extensive enough to require a complete investigation of compliance
- Changes in the number of engines or rotors

- Engines or rotors using different principles of propulsion
- Engines or rotors using different principles of operation
- Changes in the number of propeller blades or principle of pitch change operation

In the first line of the previous excerpt, the term *extensive enough* is used, but how big of a change is extensive enough? FAR 21.93 is a good place to start, as this paragraph defines the differences in major and minor changes. According to this paragraph, a major change is any change that has an appreciable effect on weight and balance, structural strength, reliability, operational characteristics, or characteristics affecting airworthiness. A minor change is indicated when none of the previous is true. But what is *appreciable*?

A change is appreciable if the change requires revisions to published data or procedures related to the flight characteristics. This may be a change to the flight manual, weight and balance data, or engine data. A change is also appreciable if tests are required to reestablish compliance with the FARs or to show that hazardous or unreliable conditions have not been introduced as a result of the change. A change may also be appreciable if the change has some effect on characteristics and cannot be classified as minor. If this sounds a bit convoluted, it is, because most of the time a classification of a change will be a judgment call.

In general, anyone can apply for a TC. Normally the TC process is conducted (for the FAA) within the United States, but it may be conducted outside of the United States if no undue burden is involved. The FAA allows five years for transport aircraft applicants and three years for all others (including engines) for the completion of the TC process. There is no expiration date on the TC, and the owner has the right to transfer the TC to another individual. A TC is invaluable since it allows its holder production approval, approval to install certified engines or propellers, or airworthiness approval.

Technically the process for obtaining an STC is the same as obtaining a TC, although the degree of review and involvement is generally less for the STC.

The first step in the TC process is the application submittal. The applicant completes Form 8110-12 and submits it to the Aircraft Certification Office (ACO). Within the United States there are four division chiefs and an international program officer in the FAA certification service organization; they all report to the administration director:

- AIR 100—Aircraft Engineering
- AIR 200—Production and Airworthiness
- AIR 500—Planning and Program Management
- AIR 400—International Airworthiness Program Officer
- Brussels Aircraft Certification—Liaison with Europe, Africa, and Middle East governments

There are four accountable directorates within the certification service organization established in areas of the country where the expertise of particular aircraft resides:

- ANM-100—Transport Airplane, Seattle, Part 25
- ACE-100—Small Airplane, Kansas City, Parts 23 and 31
- ANE-100—Engine and Propeller, Burlington, Parts 33 and 35
- ASW-100—Rotorcraft, Ft. Worth, Parts 27 and 29

They are responsible for policy under a FAR part covering a particular category of aircraft or aeronautical parts and oversee major certification projects wherever they occur involving the category of aircraft they specialize in. Their involvement increases as a certification program becomes significant, and their responsibility is delineated in FAA Order 8100.5, Directorate Procedures. The accountable directorates have final authority on policy for their respective FARs. They also serve as geographical directorates, which are responsible for all the field offices within a geographical area:

- Seattle—Seattle, Long Beach, Denver
- Burlington (Boston)—New York
- Kansas City—Wichita, Atlanta, Chicago, Anchorage
- Fort Worth

The four accountable directorates also handle certification projects for specified countries. For example, the Seattle office is also responsible for projects from

- Brazil
- Denmark
- Finland
- France
- Israel
- the Netherlands
- Norway
- Russia
- Spain
- Sweden

Of course, this can be a bit confusing for the uninitiated. Suppose an applicant in San Diego applies for a helicopter TC. In this case, his field office would be Long Beach (ACO), the accountable directorate would be Ft. Worth (rotorcraft), and the geographical directorate would be Seattle.

Along with Form 8110-12, the applicant must also submit three view drawings along with any preliminary data, engine description and design features, engine operating characteristics, and proposed engine operating limitations. As previously mentioned, the applicant has a specified period to obtain the TC from the date of the application: five years for transport aircraft and three years for all others. If the program slips, the applicant may get an extension, but this will impact the certification basis. Suppose an applicant submits a certification request for a Part 23 airplane on November 1, 2015. By the regulations, she has until November 1, 2018, to complete the process. If the program slips three months, to February 1, 2008, a new application date of February 1, 2016, will be imposed. The applicant is then responsible for meeting any new rules that became effective between November 1, 2015, and February 1, 2016.

The FAA will establish a project team for the new application. The typical makeup will include members of the ACO, which is the local field office, the accountable directorate, the Manufacturing Inspection District Office (MIDO) (which is located at the field office), and the Flight Standards District Office (FSDO) (which is also located at the field office).

The ACO will provide a project manager/lead engineer as well as engineers from different branches (e.g., structures, systems). The ACO will also assign an ACO manager and support staff. The accountable directorate is responsible for assigning a project officer and may be called upon to supply a national resource specialist.

The MIDO conducts conformity inspections throughout the program and reviews the manufacturing process used to build the prototype, which lays the groundwork for production certificate approval. The MIDO is responsible for releasing the aircraft to flight test after verifying conformity with the approved design; airworthiness is certificated by the issuance of an experimental airworthiness certificate. The MIDO will prepare part I of the type inspection report (TIR); flight test prepares part II.

Flight standards will support the project with an Aircraft Evaluation Group (AEG), which is composed of operations and maintenance specialists. This group reviews the flight manual, maintenance instructions, and instructions for continued airworthiness. It is also responsible for evaluating pilot workloads to determine training requirements and for preparing the master equipment list (MEL) for the aircraft.

Within flight test, the pilot and engineer have designated responsibilities. Jointly they are responsible for determining whether performance, flight characteristics, and operations meet FAA requirements. They determine the operational limitations and procedures, review the applicant's flight test results, and help write the type inspection authorization (TIA).

The test pilot has the additional jobs of concurring with the necessary flight tests, conducting the flights, and analyzing the results. He prepares part II of the TIR and briefs the Type Certification Board at the preflight meeting. The engineer coordinates with the applicant, supervises the designees, and reviews data as necessary. She approves the design data, drawings, and test plans and identifies issues for the Type Certification Board. The engineer also coordinates with MIDO and approves airworthiness limitations and reviews instructions for continued airworthiness (FAR 21.50).

6.4.1 Type Certification Board

The function of the Type Certification Board is to provide senior management oversight in the following primary ways:

- Introduction and description of the project
- Discussion of design details and possible problem areas with FAA specialists
- Start the evaluation process
- Establish the basis and need for type certification
- Formulation of special compliance teams to define special conditions
- Establish project schedules
- Resolve problems of major significance and provide for unconventional design solutions

There are three mandatory meetings for the applicant: preliminary, preflight, and final. There may be a familiarization meeting prior to the preliminary meeting for applicants who are not "known" to the authority or do not have a track record of previous applications.

The preliminary meeting is held to collect technical data about the project, determine the proposed certification basis, and establish what information is required to develop a certification program plan. It is at this meeting that any issues will be identified, milestones and schedules will be mapped out, and points of contact for both sides will be identified. The certification program plan, as a minimum, will contain the following information:

- General information
- Certification basis
- Method of compliance
- Compliance checklists (this is not a requirement, but can be of great assistance during the certification effort; this will be addressed later)
- Project schedule
- Delegation

NOTE: The plan is a living document and does get updated during the program.

The general information section contains the name of the applicant, the date of the application, model and project description, and the location of the ACO. The certification basis determines the applicable FARs as well as noise and emissions standards. The certification basis may identify unique situations relevant to the proposed program. These situations could include equivalent levels of safety, issue papers, special conditions, or exemptions. A unique situation could involve programs for which airworthiness regulations do not contain an adequate safety standard due to novel or unusual design features. For example:

- Vertical/short takeoff and landing aircraft—Should airplane or rotorcraft standards be used?
- Two engines driving one propeller
- Canards—What are the icing test requirements?
- Composites

The issue paper process is covered in AC 20-166, *Issue Paper Process*, June 15, 2010. Issue papers provide a structured means for describing and tracking the resolution of significant technical, regulatory and administrative issues that occur during a project. The issue paper process establishes a formal communication for significant issues between the applicant and the Authority. Issue papers form a valuable reference for future TC programs and for development of regulatory changes; by describing significant or precedent-setting technical decisions and rationales employed, they are ideal source documents. The types of issue papers are as follows:

- Method of compliance
- Equivalent level of safety
- Proposed special conditions
- Certification basis
- Determination of compliance
- Environmental considerations
- Export (import) requirements
- New information
- Type validation

- Other types of FAA approvals
- Unsafe features or characteristics
- All other issues

The method of compliance is the most common issue paper, and it defines a particular method of compliance that requires directorate or policy office coordination as a result of peculiarities in the type design or the need to define specific conditions to show compliance. Equivalent level of safety (ELOS) findings are granted when literal compliance with a certification regulation cannot be shown and compensating factors exist which can provide an ELOS (21.21). Compensating factors are normally any design changes, limitations or equipment imposed that will facilitate granting the equivalency. An equivalent level of safety is a determination by the FAA that the applicant's design, while not meeting all FAA requirements, provides for an ELOS. This is usually used for very specific requirements and is a judgment call by the accountable directorate. A request for an equivalent level of safety is discussed in the following example.

The request for an equivalent level of safety ruling was requested by Boeing for its B-777 with respect to auxiliary power unit (APU) monitoring. The issue paper identified FAR 25.1305 as requiring the flight deck instruments to monitor critical APU functions to ensure safe operation within specified limitations as required by FAR 25.1501(b). Continuous APU monitoring with a two-man crew results in too high a workload.

The FAA stipulated that instrumentation must either conform to FAR 25.1305 and FAR 25.1501(b) requirements or accept an equivalent level of safety under FAR 21.21.

The applicant's position stated that automatic APU shutdown provided for fire, rotor speed droop, loss of rotor speed signal, or rotor overspeed. In addition, an *APU limit* caution message was generated for the crew whenever the APU exceeded exhaust gas temperature (EGT), oil pressure, or oil temperature limits. The FAA concluded that this architecture did provide an equivalent level of safety.

The basis for issuing and amending special conditions is found in 21.16. A special condition is issued only if the existing applicable airworthiness standards do not contain adequate or appropriate safety standards for an aircraft, engine or propeller because of novel or unusual design features of the product to be certificated. Novel or unusual applies to features of the product to be certificated when compared to the applicable airworthiness standards. Special conditions are not used to upgrade the applicable airworthiness standards when novel or unusual design features are not involved. Once again, the accountable directorate is responsible for their issuance and they are listed on the type certification data sheet. An example of the mechanism of special conditions is provided below.

Innovative Solutions and Support (IS&S) applied for an STC for the DC-10-30 on July 15, 1997. The STC called for the installation of a digital electronic altimeter. Since altitude is a critical parameter, concern was high due to the possible disruption of the signal by high-intensity radiated fields (HIRF). Interference by HIRF could cause a loss of displayed information or provide hazardously misleading information. The new system was required to meet the standards that were in effect at the time of the original certification (February 1, 1965). Since there were no specific regulations addressing HIRF in 1965, the digital altimeter was considered novel or unusual in terms of type certification basis.

The concern was adverse effects due to external radiation and uncertainty about the adequacy of aircraft shielding. To allay fears, the system's immunity to HIRF had to be established. IS&S was able to prove this immunity via HIRF ground tests. The original TC was amended to include the special conditions provided by the modifications to the airplane by IS&S and that the system met the shielding requirements as demonstrated by testing.

Certification basis issue papers designate the applicable airworthiness and environmental regulations, including special conditions, which must be met for certification per Part 21. The issue paper must provide the definitive justification for selecting the certification basis, including specific amendment levels. An exemption is a temporary or permanent allowable noncompliance with a particular regulation for a specific product.

As previously discussed in section 6.2 of this text, any interested party may petition for a temporary/permanent exemption from any rule issued by the FAA. This is allowed by FAR Part 11 and must follow the rulemaking process. The key for an exemption petition is that it must be in the public interest and not adversely affect safety. The petition may be granted or denied by the accountable directorate or the administrator. An example of a petition for exemption is provided in the following paragraphs.

TRI Phantom, Inc., sought to modify a single-engine Explorer model 552C by increasing the seating from 9 to 13 passengers. The company sought an exemption from FAR 23.3(a), which limits the normal category airplane to nine passengers. In addition, they sought an exemption from FAR Part 135, Appendix A, delineating one engine inoperative requirements for FAR 135 operations in aircraft with 10 or more seats.

The company claimed that the public interest would be served by these exemptions because they could operate the aircraft at a significantly lower cost per passenger, which would equate to lower airfares. Because of the additional seats, fewer flights would have to be flown to serve the same number of passengers, which would translate into lower fuel consumption, conservation of the environment, and smaller workloads for air traffic controllers.

The company argued that safety was not adversely affected since the center of gravity (CG) range and maximum certified gross weight did not change, the propulsion and propeller system were not modified, and no airframe modifications were required. The company also stated that the airplane would not be approved for flight into instrument flight rules (IFR) conditions and this configuration was already in use in foreign countries.

Unfortunately for the company, the FAA stated that the data submitted did not support any of the claims that the applicant made. They (the company) did not show the equivalent level of safety for aircraft certified for more than nine seats in terms of crash worthiness and the number of exits and exit markings. The request for exemption from one engine inoperative requirements failed to offer any reason or argument why the petitioner is unique from the same general class of operators who are also subject to the regulation. In addition, the petitioner failed to provide any logic as to why an equivalent level of safety was provided for aircraft carrying more than nine passengers with one engine versus two. The petition was denied.

Determination of compliance issue papers provide a statement of procedural requirements, including the applicant's responsibilities for showing compliance. They are designed to capture the compliance checklist, which shows the regulatory requirement and the method of compliance proposed by the applicant for each regulation identified in the certification basis. A compliance checklist documents the means of compliance to the applicable airworthiness standards. There is no regulatory requirement for a compliance checklist, so it cannot be required of the applicant. FAR 21.21 requires that the applicant show that his design meets the applicable FAR requirements. Since the applicant is required to show how the plans on showing compliance, a checklist is probably the easiest way of monitoring progress against the plan.

Environmental considerations issue papers are similar in nature but define the applicable environmental regulations required by the applicant to show compliance. Export/import issue papers are written by either the FAA or the CAA, which cites compliance with the airworthiness requirements. A new information issue paper is written by the FAA to examine issues that arise from a better understanding of environmental or other hazards that were not well understood in the past or did not exist previously (e.g., microbursts, cabin ozone hazards). When the FAA is the validating authority, an FAA validation team writes an issue paper for each validation item: a certification item or airworthiness standard of peculiar interest to the validating authority.

For other FAA approvals such as parts manufacturer approvals (PMA) the FAA may use issue papers to document and resolve compliance issues where Directorate or policy office guidance is required. Unsafe features or characteristics issue papers are written for items that could preclude certification. All other issues during the TC project that become controversial or may otherwise require type certification board action to resolve may require an issue paper (e.g., a nonstandard method/means of compliance proposed by the applicant).

FAR 21.33 allows the FAA to make any inspection and test necessary to determine compliance. Conformity inspections were discussed under the MIDO responsibilities in section 6.4. The inspections are accomplished to ensure that the product being certified conforms to design data. The inspections are accomplished by the MIDO or the FAA designee. Designees are engineers, inspectors, and test pilots who do not work directly for the FAA, but are approved to act as their designees. For conformity inspections, the FAA may approve a designated manufacturing inspection representative (DMIR) or a designated avionics representative (DAR). Most common among the designees are the designated engineering representatives (DERs), who specialize in different aeronautical roles. The schedule of events for product inspection is as follows:

- The applicant submits FAA Form 8130-9, Statement of Conformity.
- DER or FAA reviews the data.
- FAA engineering requests conformity inspections via FAA Form 8120-10 or the type inspection authorization.
- ACO sends the request to the appropriate MIDO.
- MIDO conducts inspections (deviations are recorded and dispositioned).
- FAA Form 8130-3, Airworthiness Tag, is issued.
- Completed conformity inspection records are sent back to FAA engineering.
- Results are included in the TIR.

Certification tests are official tests by the FAA or a DER on conformed aircraft to show compliance with certification criteria. The FAA tests are accomplished to verify the applicant's data and are not used for flight test data. Any changes to the aircraft must be documented and conformed, and those tests affected by the modification must be repeated.

The final certification board reviews all data, identifies outstanding issues, and decides on whether to issue a TC. The TIR is issued 90 days after the issuance of the TC. The TIR is an official record of inspections and tests conducted to show compliance with applicable regulations and provide a record of other information pertinent to each TC/STC project.

And you thought that you had it bad with flight test readiness reviews, test review boards, and safety review boards for military products. A good reference for the FAA avionics certification process is an FAA document titled "Description of the FAA Avionics Certification Process," prepared by James H. Williams on April 23, 1997, and available on the FAA website (http://www.faa.gov).

■ 6.5 | AVIONICS SOFTWARE CONSIDERATIONS IN THE TECHNICAL STANDARD ORDER PROCESS

The technical standard order (TSO) system provides test specifications and evaluation considerations that should be employed to evaluate the performance of various types of appliances. Appliances in this case are not toasters and microwaves but avionic systems installed on aircraft. These criteria are called minimum operating performance standards (MOPS). The MOPS may be generated by the government or any of a number of other organizations:

- Society of Automotive Engineers (SAE)
- Radio Technical Commission for Aeronautics (RTCA, and now just called RTCA)
- Manufacturers
- Any other technically oriented body

The MOPS provide reasonable assurance that a TSO-authorized appliance will function reliably and as intended. The MOPS typically consist of

- Test procedures that address operating characteristics
- Test procedures that address environmental conditions
- Test procedures that address software

The TSO system within the U.S. permits almost total delegation of compliance determination to the industry; this is not true under EASA where the European TSO (ETSO) is treated the same as a TC or STC under AMC 21. The exceptions, where FAA approval is required, are found in the approval required for the software validation and verification plan, and sanctioning and monitoring of the manufacturer's quality control procedures.

A finding of compliance is synonymous with approval. What this means in plain English is that when a manufacturer declares that its data and submittal comply with all of the TSO requirements, it renders approval for the submitted data. The FAA responds to the manufacturer with a TSO authorization, not a TSO approval. The manufacturer has already granted itself that approval. Similar principles and procedures are used in the authorization of modified appliances. It is important to note that this is not the case within EASA regulations that consider the TSO as a formal approval and must go through the certification process, similar to the TC, ATC, and STC certificates.

The authorization process for modified appliances depends on the magnitude of the change involved. Two types of modifications apply to hardware: a major change and a minor change. A major hardware change is defined as a change requiring a nontrivial reevaluation to show compliance with the TSO criteria. It may also be a change that, because of new features or functions (or the potential effect on features and functions), requires a nontrivial reevaluation to show compliance with the TSO criteria. A minor change is any hardware change that is not a major change.

A minor change to a TSO appliance is approved in-house by the manufacturer without FAA participation. The required TSO data that have been changed should be submitted to the FAA within a reasonable period of time (usually 30 days). A major change in a TSO appliance requires the manufacturer to reapply for the TSO authorization. The reapplication involves the following:

- Relabeling the article with a new identification description
- Reestablishing compliance with the TSO criteria
- Resubmitting the required TSO documentation
- Redeclaring compliance with the TSO criteria

Major changes may not be implemented until a new TSO authorization is received from the FAA; the FAA has 30 days to respond. The references for the TSO process are FAR 21, subpart O; FAA Order 8150.1A; AC 20-115A; and AC 20-110E.

Changes in software are defined as significant and nonsignificant. The wording is much the same as that found with hardware changes. A significant change is one that requires a nontrivial reevaluation to show compliance with the TSO criteria or a change that, because of new functions or features (or the potential effects on existing functions and features), requires a nontrivial reevaluation to show compliance with the TSO criteria. A nonsignificant change is any software change that is not a significant change. The process for obtaining the new authorization is the same as described for hardware changes.

As previously mentioned, the authority does have the responsibility for approving the manufacturer's software practices as well as monitoring the manufacturer's quality control procedures. The software audit is a means to examine a manufacturer's software practices. The intent is to determine whether the manufacturer is capable of producing software to a stated criticality level. The process addresses software quality as well as a subjective measure to minimize nondetected software errors. The software audit is usually divided into three phases:

- Preaudit activities
- Audit activities
- Postaudit activities

The frequency and intensity of the audits is based on manufacturer's performance, past software record, and FAA experience with the particular manufacturer. The goal of the audit is to ensure that the manufacturer is maintaining its software process as accepted by the FAA.

The preaudit activities involve the formulation and presentation to the authority of required documentation:

- Software management development plan, or software development methodology
- Software quality assurance plan (SQAP)

- Software configuration management plan (SCMP)
- Structural organization of the corporation showing by whom and how the individual software tasks are undertaken, controlled, monitored, and approved for a sample project

The audit activities are geared toward monitoring the manufacturer's response to changes and the effectiveness of its software process. The audit requires that the product be developed through the corporate product development policies, that is, that the plan must be in place and followed. The audit activities include the following:

- A review of problem reports and implementation of changes
- Response resolution of preaudit reviews
- Determining the adequacy and standards for compliance with the validation and verification plan (V&V)
- Evaluating the effectiveness of quality assurance
- Evaluating the effectiveness of the configuration management plan
- Evaluating the effectiveness of any other self-policing functions
- Audit summary and action items

The postaudit activity is a review of all required action items to ensure that the manufacturer has complied with the recommended changes. The postaudit activities include the following:

- Assessment of audit activities
- Resolution of audit action items
- Reporting of changes to software standards, software development methodologies, or policing (control) procedures (SQA, SCM)

6.5.1 Other Considerations in the TSO Process

Three other considerations in the TSO process must be addressed by the manufacturer: interchangeability, the mixing of TSO and nonTSO functions, and component identification. When changes are made to a TSO appliance, the applicant must provide interchangeability methodology. For example, it is the responsibility of the TSO manufacturer to demonstrate functional conformance (one-way or two-way interchangeability) has been substantiated. What this means is that hardware or software changes shall be form, fit, and functional with the previous equipment. Installation factors may need to be taken into account. Repairs and enhancements that do not adversely affect operation and safety should be acceptable for interchangeability.

Software of appliances that provide TSO and nonTSO functions needs to be evaluated as one entity as part of the TSO process, irrespective of whether the software serves TSO or nonTSO functions.

The following items need to be discernible through the manufacturer's component identification scheme:

- Hardware changes (major or minor)
- Software changes (significant or nonsignificant)
- Criticality level (level 1, 2, 3, or multiple)
- Interchangeability (one-way or two-way)

■ 6.6 | CERTIFICATION CONSIDERATIONS FOR HIGHLY INTEGRATED OR COMPLEX SYSTEMS

Today's avionics systems are digital and, by their very nature, are integrated with other aircraft systems. The term *highly integrated* refers to systems that perform or contribute to multilevel functions. *Complex* refers to systems whose safety cannot be shown solely by test and whose logic is difficult to understand without the aid of analytical tools. The reference material for this section can be found in Society of Automotive Engineers (SAE) Aerospace Recommended Practices (ARP) ARP4754, *"Certification Considerations for Highly Integrated or Complex Aircraft Systems,"* and ARP4761, *"Guidelines and Methods for Conducting the Safety Assessment on Civil Airborne Systems and Equipment."* The reference material was developed in the context of Part 25 (civilian transport) but can easily be applied to other categories as well as military applications. The software requirements are referred back to RTCA documents (MOPS) such as RTCA DO-178B and its European counterpart ED-12B. The U.S. military has adopted RTCA DO-178B in lieu of a military standard (DOD-STD-2167A was cancelled). The practices are primarily directed toward systems that integrate multiple functions and have failure modes with the potential to result in unsafe aircraft operating conditions. The goal is safety.

The references are recommended practices and hence do not carry the mandate of law. They are not mutually exclusive and must be taken together when approaching any complex avionics system. The references were developed in response to a request from the FAA to the SAE to define the appropriate nature and scope of system-level information for demonstrating regulatory compliance for highly integrated or complex systems. The documents are designed to provide a common basis for certification and concentrate on meeting the intent of FAR 25.1309 and FAR 25.1301, which is not a particularly easy task.

Specifically, the FARs state:

The airplane equipment and systems must be installed so that:

- Those required for type certification or by operating rules or whose improper functioning would reduce safety, perform as intended under the airplane operating and environmental conditions CFR 14/CS 25.1309 (a)(1)
- Other equipment and systems are not a source of danger in themselves and do not adversely affect the proper functioning of those covered by (a)(1); 25.1309 (a)(2)
- Catastrophic failures are extremely improbable and do not result from a single failure, a hazardous failure condition is extremely remote, and a major failure is remote 25.1309 (b)(1,2,3)
- Information concerning unsafe system operating conditions must be provided to the crew to enable them to take appropriate corrective action. A warning indication must be provided if immediate corrective action is required
- Systems and controls, including indications and annunciations must be designed to minimize crew errors, which could create additional hazards 25.1309 (c)

CFR 14 25.1301 for the associated aircraft types is really a catch-all; each item must be of a kind and design appropriate to its intended function and must be labeled as to its identification, function, or operating limitations, or any combination of these factors. Each item must also be installed according to limitations specified for that equipment. The statement, "Each item must function properly when installed," is shown only in US documentation. There are four major categories to consider when showing compliance with subparagraph (a) to 25.1301. The operating and environmental conditions that installed systems must be evaluated against include the entire operating envelope of the airplane normal and abnormal conditions. These conditions should include external environmental conditions of turbulence, HIRF, lightning and precipitation; the severity of these conditions is dependent on the limitations set by certification standards. In addition to the external environmental conditions previously mentioned, the systems must also be evaluated against the environment within the airplane. These tests include vibration and acceleration loads, variations in fluid pressure and electrical power, fluid or vapor contamination in the normal operating environment as well as accidental leaks or spillage or handling by personnel. RTCA/DO-160G/ED-14G, "Environmental Conditions and Test Procedures for Airborne Equipment," defines the standard series of tests that may be used to support compliance with this subparagraph.

The substantiation of these tests may be shown by test or analysis or reference to comparable service on other aircraft. It must be shown that comparable service experience is valid for the proposed installation. The compliance demonstrations must also show that the normal functioning of such equipment, systems or installations does not interfere with the proper functioning of other systems, equipment, or installations. The other equipment mentioned in subparagraph (a)(2) of 1301 refers to systems that are usually amenities for passengers (e.g., phones, in-flight entertainment). Failure or improper operation of these systems should not affect the safety of the airplane. Tests are necessary to show that their normal or abnormal functioning does not adversely affect proper functioning of aircraft systems and does not adversely influence the safety of the airplane or its occupants (e.g., fire, explosion, high voltage).

Failure conditions need to be identified and their effects assessed to show compliance with subparagraph (b) to 25.1309: catastrophic failures are extremely improbable and do not result from a single failure, a hazardous failure condition is extremely remote, and a major failure is remote. Compliance can be shown by analysis or appropriate simulator, ground or flight testing. When performing the analysis, all possible failure conditions and their causes, modes of failure, and damages from sources external to the system need be considered. These include the possibility of multiple failures and undetected failures or the possibility of requirement, design and implementation errors. The analysis should also note the effect of reasonably anticipated crew errors after the occurrence of a failure or failure condition or the effect of reasonably anticipated errors when performing maintenance actions. From the human factors perspective, the crew alerting cues, corrective action required, and the capability of detecting faults should be examined. Finally, it should be determined what the resulting effects are on the airplane and occupants when considering the stage of flight and operating and environmental conditions.

The maximum allowable probability of the occurrence of each failure condition is determined from the failure condition's effects. But how do we measure probability?

Probability can be quantitative or qualitative; safety assessments that cannot be calculated must be assessed qualitatively. A logical and acceptable relationship exists between the average probability per flight hour and the severity of the failure condition effects. The probability and severity descriptions are taken from FAA and EASA documentation. EASA uses a more detailed description of these terms:

- Failure conditions with no safety effect have no probability requirement
- Minor failure conditions may be probable
- Major failure conditions must be no more frequent than remote
- Hazardous failure conditions must be no more frequent than extremely remote
- Catastrophic failure conditions must be extremely improbable

When using quantitative analyses to determine compliance with the rule the following descriptions have been accepted in terms of average probability per flight hour:

- Probable failure conditions are those having an average probability per flight hour greater than of the order 1×10^{-5}
- Remote failure conditions are those of the order of 1×10^{-5} or less, but greater than of the order 1×10^{-7}
- Extremely remote failure conditions are those of the order 1×10^{-7} or less but greater than of the order 1×10^{-9}
- Extremely improbable failure conditions are those of the order of 1×10^{-9} or less

Where did these numbers come from? Historical evidence indicated that the probability of a serious accident due to operational and airframe related causes was approximately one in one million hours of flight; about 10% of the total were attributed to failure conditions caused by the aircraft systems. It seemed logical that serious accidents caused by aircraft systems should not be allowed a higher probability than this in new aircraft designs. Therefore, it would be reasonable to expect that the probability of a serious accident from all such failure conditions should not exceed one in ten million flight hours or 1×10^{-7}. But how many failure conditions in an aircraft could be catastrophic?

It was assumed, somewhat arbitrarily, that there could be approximately 100 failure conditions in an aircraft that could be catastrophic. If the target allowable probability of 1×10^{-7} was apportioned equally among these 100 conditions, each failure condition could not exceed a probability of 1×10^{-9}. The upper limit for the average probability per flight hour for catastrophic failure conditions would be 1×10^{-9}, which establishes an approximate probability value for the term *extremely improbable*. Failure conditions having less severe effects could be relatively more likely to occur.

When using qualitative analyses to determine compliance with the rule, the following descriptions have been accepted as aids to engineering judgment:

- Probable failure conditions are those anticipated to occur one or more times during the entire operational life of each airplane
- Remote failure conditions are those unlikely to occur to each airplane during its service life, but may occur several times when considering the total operational life of a number of airplanes of the type
- Extremely remote failure conditions are those not anticipated to occur to each airplane during its service life, but may occur a few times when considering the total operational life of a number of airplanes of the type
- Extremely improbable failure conditions are those so unlikely that they are not anticipated to occur during the entire operational life of all airplanes of one type

The associated severity terms must be described as well:

– No safety effect—failure conditions that would have no effect on safety or failure conditions that would not affect the operational capability of the airplane or increase crew workload

– Minor—failure conditions that would not significantly reduce airplane safety and would involve aircrew actions that are well within their capabilities and may include a slight reduction in safety margins or functional capabilities, a slight increase in crew workload such as flight plan changes, or some physical discomfort to passengers or cabin crew

– Major—failure conditions that would reduce the capability of the airplane or the ability of the crew to cope with adverse operating conditions to the extent that there would be a significant reduction in safety margins or functional capabilities, a significant increase in crew workload, or conditions impairing crew efficiency, or discomfort to the flight crew, or physical distress to passengers or cabin crew possibly including injuries

– Hazardous—failure conditions that would reduce the capability of the airplane or the ability of the crew to cope with adverse operating conditions to the extent that there would be:

  – A large reduction in safety margins or functional capabilities

  – Physical distress or excessive workload such that the flight crew cannot be relied upon to perform their tasks accurately or completely

  – Serious or fatal injury to relatively small number of the occupants other than the flight crew

– Catastrophic—Failure conditions that would result in multiple fatalities usually with the loss of the airplane. (The acceptable method of compliance within the EASA documentation notes that a catastrophic failure condition was defined in previous versions of the rule and the advisory material as a failure condition that would prevent continued safe flight and landing.)

The combination of probability and severity, along with the associated verbiage, is depicted in Table 6.1.

6.6.1 The Safety Assessment Process

The Part 25 Airworthiness Standards are based on the objectives and principles or techniques of the fail-safe design concept, which considers the effects of failures and combinations of failures in defining a safe design. The following objectives pertaining to failures apply:

– In any system or subsystem, the failure of any single element, component or connection during any one flight should be assumed regardless of its probability; they should not be catastrophic.

– Subsequent failures during the same flight, whether detected or latent, and combinations thereof, should also be assumed, unless the joint probability with the first failure is shown to be extremely improbable.

TABLE 6.1 ▪ Severity versus Probability

Effect on the Aircraft	No Effect on Operational Capabilities or Safety	Slight Reduction in Functional Capabilities or Safety Margins	Significant Reduction in Functional Capabilities or Safety Margins	Large Reduction in Functional Capabilities or Safety Margins	Normally with Hull Loss
Effect on the Occupants Excluding the Flight Crew	Inconvenience	Physical Discomfort	Physical Distress, Possibly Including Injuries	Serious or Fatal Injury to a Small Number of Passengers or Cabin Crew	Multiple Fatalities
Effect on the Flight Crew	No Effect on Flight Crew	Slight Increase in Workload	Physical Discomfort or a Significant Increase in Workload	Physical Distress or Excessive Workload Impairs Ability to Perform Tasks	Fatalities or Incapacitation
Allowable Qualitative Probability	No Probability Requirement	Probable	Remote	Extremely Remote	Extremely Improbable
Allowable Quantitative Probability; Probability per Flight Hour on the Order of:	No Probability Requirement	$<10^{-3}$ Note 1	$<10^{-5}$	$<10^{-7}$	$<10^{-9}$
Classification of Failure Condition	No Safety Effect	Minor	Major	Hazardous	Catastrophic

Note 1: A numerical probability range is provided here as a reference. The applicant is not required to perform a quantitative analysis, nor substantiate by such an analysis, that this numerical criteria has been met for Minor Failure conditions. Current Transport category aircraft products are regarded as meeting this standard simply by using current commonly accepted Industry practice.

The fail-safe concept uses multiple design principles and techniques to ensure a safe design. The use of one principle or technique is seldom adequate, and normally two or more are required to meet a fail-safe design. The basic fail-safe design will ensure that major failure conditions are remote, that hazardous failure conditions are extremely remote, and that catastrophic failure conditions are extremely improbable.

The following identifies the techniques used in development to ensure a fail-safe design:

– Design integrity and quality (including life limits)—to ensure intended function and prevent failures

– Redundancy or backup systems—to enable continued function after any single (or other defined number of failures such as engines, hydraulics, and flight controls)

– Isolation or segregation of systems, components or elements—so that failure of one does not cause failure of another

– Proven reliability—so that multiple, independent failures are unlikely to occur during the same flight

– Failure warning or indication—to provide detection

– Flight crew procedures—specifying corrective action for use after failure detection

- Checkability—the ability to check a component's condition
- Designed failure effect limits—including the capability to sustain damage and limit the safety impact or effects of the failure
- Designed failure path—to control and direct the effects of a failure in a way that limits the safety impact
- Margins or factors of safety—to allow for any undefined or unforeseeable adverse conditions (known unknowns)
- Error tolerance—considers adverse effects of foreseeable errors during the airplane's design, test, manufacture, operations, and maintenance

Because newer systems are more highly integrated, there is a concern that design and analysis techniques traditionally applied to noncomplex systems may not provide adequate safety coverage. Other assurance techniques such as development assurance, utilizing a combination of process assurance and verification coverage criteria, or structured analysis and assessment techniques applied to the airplane or systems level are applied to more complex systems. Their systematic use increases confidence that errors in requirements or design and integration or interaction effects have been adequately identified and corrected.

Compliance with these two paragraphs can be very subjective unless some specific guidance is offered. The aerospace recommended practices attempt to provide such guidance. The documents provide assistance in the areas of safety assessment, risk analysis, certification planning, requirements determination, and validation.

The safety assessment process includes requirements generation and verification that support development activities. The process provides a methodology to evaluate aircraft functions and the design of systems performing these functions to determine that the appropriate hazards have been addressed; the process can be quantitative as well as quantitative. The guidance states that the process should be managed to provide assurance that all relevant failure modes have been identified, and that significant combinations of failures that may cause those conditions have been considered.

The safety assessment process for integrated systems should account for any additional complexities and interdependencies that arise during the integration effort. The process is required to establish safety objectives for the system and to ensure that the implementation satisfies those objectives. In highly integrated or complex systems, the evaluator will need to use development assurance methods to support system safety assessments. Careful attention must be paid to system architecture: the relationship and interdependency of avionics systems, the flow and sharing of information, latency of data, and software interactions. Limiting the inclusion of systems will obviously simplify the development process; however, there is some risk in eliminating systems that may have a causal effect on safety.

Safety assessments are based on assigned assurance levels that account for the criticality of a potential failure as well as the probability of the failure occurring. More on this later. The safety assessment process provides analytic evidence showing compliance with airworthiness requirements and may include:

- Functional hazard assessment (FHA)
- Preliminary system safety assessment (PSSA)
- Common cause analysis (CCA)
- System safety assessment (SSA)

Typical development cycle

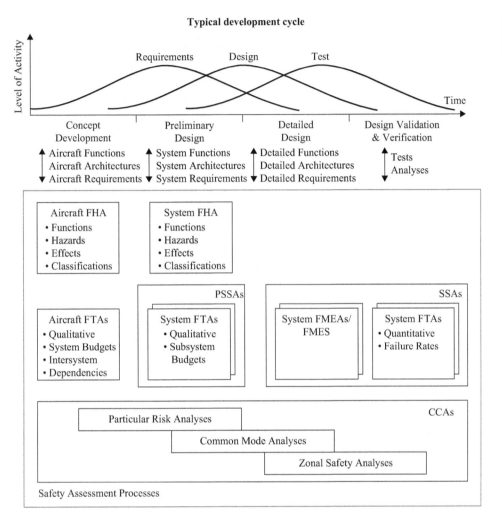

FIGURE 6.3 ■ ARP Safety Assessment Process

The safety assessment process and how safety assessment methods relate to the process are shown in Figure 6.3.

An FHA is conducted at the beginning of the aircraft/system development cycle. It should identify and classify the failure conditions associated with the aircraft functions and combinations of functions. The failure conditions are then assigned a classification; these are in terms of severity and effect. These classifications were shown in Table 6.1. The goal of the FHA is to identify each failure condition and the rationale for its classification. These failure conditions will be used to identify the safety objectives. The output of the FHA is used as a starting point for the PSSA.

Two levels of FHA use the same process: aircraft level and system level. Generation of the FHA at the highest appropriate level is dependent on overall knowledge and experience and will probably require consultation with numerous specialists. Table 6.2 provides an example of the high-level functions and their associated failure conditions that may be considered. The failure conditions identified in Table 6.2 can be further broken down through the use of fault trees. This is commonly called fault tree analysis (FTA) and attempts to identify all causal factors that may result in the failure condition. It is not

TABLE 6.2 ■ High-Level Functions and Associated Failures

Function	Failure Condition
Control flight path	Inability to control flight path
Control touchdown and roll out	Inability to control touchdown and rollout
Control thrust	Inability to control thrust
Cabin control environment	Inability to control cabin environment
Provide spatial orientation	Inability to provide spatial orientation
Fire protection	Loss of fire protection

unusual for the FTA to identify other failure conditions that may result due to initial or subsequent failures.

If we take the first failure condition (inability to control flight path) identified in Table 6.2 and apply fault tree logic, we can come up with various failures that will produce the same condition. Inability to control flight path may be caused by:

- Loss of trim
- Inadvertent trim
- Loss of hydraulics
- Loss of flight control
- Flight control malfunction (hardware or software)
- Loss of the flight control computers

This is obviously not an all-inclusive list but gives the reader an idea of the types of causes that we are looking for. We would then look at what could cause these primary failures. A loss of flight control computers could be caused by a loss of electrical power. In looking to mitigate this failure, it may be decided to install a battery backup for the flight control computers. In the end, the matrix will be quite large and attempts to link individual failures with multiple consequences should be accomplished (common cause analysis, which will be discussed later). It should also be noted that it is desirable to generate an aircraft general hazard list to be used on future projects so that known hazards are not overlooked. As new systems are integrated, the FHA process must be revisited.

The PSSA is an iterative analysis imbedded with the overall aircraft/system development. The PSSA is used to complete the failure conditions list and corresponding safety requirements. It is also used to show how the system will meet the qualitative and quantitative requirements for the hazards that have been identified. The PSSA should identify failures that contribute to conditions identified in the FHA. It then identifies and captures all derived system safety requirements and protection strategies.

Six major categories of protective strategies may be used and identified in the PSSA:

- Partitioning: The practice of isolating software processes; the activation of one process is not dependent on another. This is quite often the case with flight controls, and in the military, with weapons systems. Flight control software will be a criticality level A and it would not be advisable to interleave this software with less critical systems. Many weapons run their software routines independently from the aircraft, thus lessening the probability of noncompatible software.

- Built-in test (BIT): A process that evaluates the functionality of the system software. BIT is one of the most difficult processes to perfect; if the tolerances are too tight, the false alarm rate will increase, and if they are too loose, the probability that an undetected failure will occur increases. Maintenance costs are also a strong driver in BIT development, as false alarms tend to drive maintenance costs higher.

- Dissimilarity: A check and balance provided by a system that operates on dissimilar software. The probability that two independent software routines will arrive at the same wrong answer is remote. With two systems, the operator can only be warned that there is a disparity; with three, the incorrect system can be voted out.

- Monitoring: A system (either hardware or software) that monitors a process for prescribed tolerance levels. Monitoring systems reduce cockpit workload, as the monitoring function is performed by the system, not the operator. The example of the APU monitor discussed previously is an example of monitoring, which as you may remember, was able to provide an equivalent level of safety.

- Redundancy: Two systems performing the same operation. Flight control and mission computers are truly redundant in civil aircraft; FCCs are redundant in military aircraft but the mission computers usually are not.

- Backup: A system (hardware or software) that assumes the operation in the event of a primary system failure. These systems are very common in aircraft hydraulic and electrical systems, and it is not uncommon to have secondary and tertiary backup systems.

The Boeing 777 uses a combination of these processes within its flight control system to meet the required hardware assurance level (of course this can become very expensive) (Figure 6.4).

Possible contributing factors leading to failure conditions may be identified using methods such as fault tree analysis, which was discussed previously. There are other methods as well, such as dependence diagrams and Markov analysis. These methods are not going to be discussed here; however, the reader is referred to SAE ARP4761 for a discussion and examples of these analyses. Hardware failures and possible hardware/software errors as well as faults arising from common causes should be included in the PSSA. Care must be taken to account for potential latent failures and their associated exposure times. Exposure time is defined as the time between when an item was last known to be operating properly and when it was known to be properly operating again.

Common cause analysis is performed to ensure independence between functions. Independence between functions, systems, or items may be required to satisfy safety

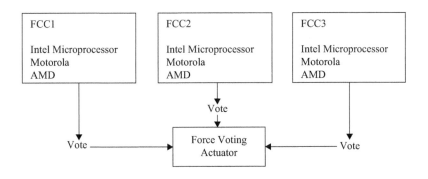

FIGURE 6.4 ▪
Boeing 777 FCS
Protective
Strategies

requirements. It is necessary to ensure that this independence exists, or that the risk associated with dependence is deemed acceptable; CCA provides the tools to verify this independence. CCA identifies failure modes or external events that can lead to a catastrophic or hazardous/severe/major failure condition. Such common causes must be eliminated for catastrophic failure conditions and be within the assigned probability budget for hazardous/severe/major failure conditions.

Common cause analysis is subdivided into three types:

- Zonal safety analysis
- Particular risks analysis
- Common mode analysis

The purpose of zonal safety analysis is to ensure that the equipment installation meets safety requirements. The first step is to verify the basic installation. The installation should be checked against the appropriate design and installation requirements. The next step should be to check for interference between systems. The effects of equipment failures should be considered with respect to their impact on other systems and structures within their physical sphere of influence. The final analysis is in regard to maintenance errors and their effect on the system or aircraft.

Particular risks are defined as those events or influences that are outside the system but may violate independence concerns. Most of these outside influences are common-sense items and can be thought of as the world according to Murphy. Some of the obvious bad things that can happen are:

- Fire
- High-energy devices
- Leaking fluids
- Hail, ice, and snow
- Bird strikes
- Lightning
- High-intensity radiation fields
- Separated parts from the aircraft

A common mode analysis verifies that independent faults identified in the fault-free analysis are independent in the actual installation. The common mode faults that should be analyzed are:

- Hardware or software error
- Hardware failure
- Production or repair flaw
- Installation error
- Requirements error
- Environmental effects
- Cascading faults
- Common external source faults

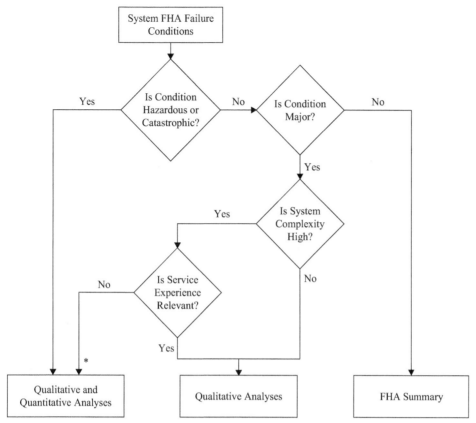

FIGURE 6.5 ■
Guidance in
Determining a
Verification Plan

*Major failure conditions may be satisfactorily analysed with methods that are less rigorous and complete than those for Catastrophic or Hazardous Failure Conditions (e.g. FMEA containing failure rates)

The SSA is a systematic, comprehensive evaluation of the implemented system to show that relevant safety requirements are met; the analysis process is similar to the PSSA. A PSSA is a method to evaluate proposed architectures and derive system/safety requirements. An SSA is verification that the implemented design meets both the quantitative and qualitative safety requirements as defined in the FHA and PSSA. The SSA process flow is generally represented through succeeding levels of verification. RTCA DO-178 procedures should be used to verify that the software implementation meets the required development assurance levels. The hardware development assurance level is verified via procedures that are defined by RTCA DO-254. For each failure condition, it should be determined how the aircraft/system will satisfy the safety objective. The flowchart in Figure 6.5 provides generalized guidance in defining a verification plan for the failure conditions in a given system.

Military designers consider the probability not only of an event occurring but also of being in a flight regime that could have catastrophic consequences if the event occurred. Determination of failures and when they may occur contributes to a probability of unsafe flight (PUF). A PUF is generated for operational flight conditions that are inherently dangerous. For example, if a failure mode in a system is determined in the FHA and the failure is deemed to produce catastrophic results if it occurs when the aircraft is flying at low level, then the percentage of time that the aircraft is flying in those conditions (low level) is taken into account during the safety assessment. The failure mode may not be completely mitigated, but the probability lessens it to an assurance level B or C.

6.6.2 Safety Assessment Certification Plan

SAE ARP4754 defines a process for compliance with FAA/EASA requirements. Although the process is designed for civilian certifications, it is nevertheless a good place to start for military evaluators as well. The certification plan for complex or highly integrated systems includes:

- Defining the product and installation
- Outlining the product development services
- Identifying the proposed means of compliance

Since many of the development assurance activities occur in the development cycle, early action on the certification plan is highly desirable.

The certification plan should address the system and the aircraft environment. The amount of detail in the plan varies with the classification of the associated hazards. At a minimum, the plan should include:

- A functional description of the system to be installed, including hardware and software and an outline of the functional, physical, and informational relationship with other aircraft systems
- A summary of the functional hazard assessment (hazards, failure conditions, and classifications)
- A summary of the preliminary system safety assessment (system safety objectives and classification)
- A description of any novel or design features planned to be used in meeting the safety objectives
- A description of any new technology applications
- A configuration index

The only new subject in the certification plan requirements is the configuration index, which identifies all of the physical elements comprising the system as well as procedures and limitations that are integral to system safety. Any design features or capabilities that enhance safety should also be identified. The index contains the identification of each item, associated software, interconnection of items, required interfaces, and safety-related operational and maintenance practices.

In plain English, the safety assessment certification plan addresses all of the hazards that were identified and the methods to determine that the failures have either been mitigated or reduced to an acceptable level of occurrence probability. The plan should also address the method for declaring that a system is susceptible to a hazard (i.e., what requirement causes the aircraft to be put into a particular situation where hazards may occur?). For example, an airliner may lose all oil pressure in inverted flight, which may have catastrophic consequences, but what requirement exists for an airliner to fly upside down? This is called the validation and verification (V&V) process of system

TABLE 6.3 ■ Development Assurance Levels

Failure Condition	System Development Assurance Level	Software Development Assurance Level	Hardware Development Assurance Level
Catastrophic	A	A	A
Hazardous/Severe	B	B	B
Major	C	C	C
Minor	D	D	D
No Safety Effect	E	E	E

The development assurance level is assigned based on the most severe failure condition associated with the applicable aircraft function. It is determined from the FHA.

requirements. This is done to ensure that the requirements are indeed real. It is a tremendous waste of time and resources to mitigate hazards that will never be encountered. V&V is reserved for the most serious assurance levels. Validation of requirements and specific assumptions is the process of ensuring that specific requirements are sufficiently correct and complete. The verification process ensures that the system implementation satisfies the validated requirements.

Requirements, with their related hazards, provide the basis for the supporting processes. Because hazards have different levels of importance, the allocation of requirements has a significant impact. The requirements associated with a given function define the way the function acts in its environment and, in turn, will determine system development assurance levels (Table 6.3). System development assurance levels are based on the degree of hazard and are generated for software (IAW RTCA DO-178 B/C) and for hardware (IAW RTCA DO-254).

System requirements are generated from many sources:

- Safety requirements
- Functional requirements
- Customer requirements
- Operational requirements
- Performance requirements
- Physical and installation requirements
- Maintainability requirements
- Interface requirements
- Additional certification requirements
- Derived requirements

Validation and verification methods are called out and described in the referenced documents and will not be reproduced here. Tables 6.4 and 6.5 show the V&V methods and their relationship with assurance level. It can be easily seen that the more severe the assurance level, the more intense the V&V process will be. Assurance level A (catastrophic) requires almost all methods be applied and discussed in the certification plan. Levels B and C (hazardous and major) require only slightly less of an examination, whereas levels D and E (minor and no safety effect) are basically negligible for requirements verification.

TABLE 6.4 ■ Validation Methods

Methods	Assurance Level A/B	Assurance Level C	Assurance Level D	Assurance Level E
PSSA	R	R	A	A
Validation Plan	R	R	A	N
Validation Matrix	R	R	A	N
Traceability	R	A	A	N
Analysis/Modeling	R	1 Recommended	A	N
Similarity	A	1 Recommended	A	N
Engineering Judgment	A	1 Recommended	A	N
Implementation Effects	R	A	A	N
Validation Summary	R	R	A	N

R, recommended; A, as negotiated; N, not required.

TABLE 6.5 ■ Verification Methods

Methods	Assurance Level A/B	Assurance Level C	Assurance Level D	Assurance Level E
Verification Matrix	R	R	A	N
Verification Plan	R	R	A	N
Verification Procedures	R	A	A	N
Verification Summary	R	R	A	N
SSA	R	R	N	N
Inspection/Review	R	R	A	N
Test Unintended Function	R	A	A	N
Experience	A	A	A	A

R, recommended; A, as negotiated; N, not required.

The certification plan is good for the as-designed aircraft. Changes to the system, adding new systems, or changing interfaces may require the applicant to revisit the entire safety assessment process.

6.7 | IMPORTANT NOTES FOR EVALUATORS CONCERNING DOCUMENTATION

Problems arise for the uninitiated concerning the requirement flow—that is, where do I go to find the requirements for my test? The rule of law is found in the FARs and the applicant must show compliance with these paragraphs. Without exception, in Avionics Systems certifications, the applicant must show compliance with paragraphs 1301 and 1309: the system must perform its intended function and it must be safe. Compliance with 1309 will require a system safety assessment as just described. To comply with 1301 and demonstrate airworthiness requires knowledge of the document trail (Table 6.6).

Table 6.6 shows the flow of requirements; it starts at the top with the FAR requirements and ends with airworthiness approval by following the AC. An example using the TAWS system is shown in the far right column. In order to determine its

TABLE 6.6 ▪ Requirements Flow

Requirement	Document	Example Terrain Awareness Warning System (TAWS)
Rule of Law	FAR Part Paragraph	CFR 14 25.1301, 1309
Must show Compliance	EASA Certification Codes	CFR 14, Parts 91,121,135 July 2, 2013
Minimum Performance of the System	MOPS	RTCA/DO-161A Minimum Performance Standards-Airborne Ground Proximity Warning Equipment May 27, 1976
How Manufacturers Show Compliance with the MOPS	TSO (FAA) (TSOA is granted) Acceptable Method of Compliance AMC-21 in EASA ETSOA is granted	TSO-C151c Terrain Awareness and Warning System June 27, 2012
Airworthiness	Advisory Circular (FAA) Certification Specification Book 2, EASA, AMC	AC 25-23 Airworthiness Criteria for the Installation Approval of a Terrain Awareness and Warning System (TAWS) for Part 25 Airplanes May 22, 2000

intended function, a MOPS is written; the TSO describes the testing a manufacturer must accomplish to show compliance and receive a TSOA. The AC describes the testing that the installer must perform to receive airworthiness certification. A manufacturer is not required to meet the requirements of a TSO (as it is not mandatory), but failure to do so will result in a failure to obtain a TSOA. Likewise, the installer is not required to follow an AC in his quest to obtain airworthiness certification, as an AC provides "a means, but not the only means to show compliance." In almost all cases, the AC will state that the applicant needs to install a TSO appliance when complying with an AC. One would think that since the top-level requirements are written first and the AC is written last (as the requirements are serial), the dates of publication would follow accordingly. In most cases this is true, as more information is learned about a new system the more detailed the requirements become. This can be frustrating to an evaluator or installer because in many cases the requirements for airworthiness are more stringent than the TSO and functions may not have been evaluated in the TSOA process but are required in the airworthiness process.

In some cases, some documents are revised and others are not, creating even more confusion. If we look at the example of the TAWS system, the reader will note that the least current of the documents is the MOPS and the second least current is the AC. In this case, the AC references the applicant to the TSO for the functionality of the TAWS. In fact, the TSO adds a new installation (Class C), which is not mentioned in the AC. When seeking airworthiness certification for a new TAWS installation, the applicant must follow the guidance in the TSO as well as any unique tests called out in the AC. It is heartily recommended that prior to writing any test plan, the applicant carefully review the entire train of documentation to understand exactly what will be required for demonstration. The requirements will then be written to the certification plan.

■ 6.8 | DIFFERENCES BETWEEN EASA AND FAA DOCUMENTATION

In the United States, the requirements for aircraft are found in the Code of Federal Regulations, Title 14, Aeronautics and Space; for transport category aircraft you would find the regulations for compliance in 14 CFR, Part 25 (paragraphs 1-1587 plus Appendices A–J; commonly referred to as FAR Part 25 requirements). Until the transition to EASA, European member states had to show compliance with JAA regulations; these regulations (JAR) matched the FAR numbering system so transport category requirements would be found in JAR Part 25.

The European Aviation Safety Agency has incorporated the JAR Part 25 requirements into the Certification Specifications for Large Aeroplanes (CS-25). Book 1 of CS-25 contains the JARs, as amended (called the certification codes), and Book 2 contains the Acceptable Means of Compliance (AMC) for the certification codes in Book 1. In general, CS-25 Book 1 certification codes are aligned with FAA and former JAA paragraphs. The AMC is similar to the US Advisory Circulars (AC) and the JAA Advisory Circular Joint (ACJ), which will be addressed later.

Part 21 can be described as the requirements and procedures for the certification of aircraft and related products, parts and appliances, as well as production organization approval; Part M is the applicable continuing airworthiness requirements. In Europe, the Part 21 requirements have been incorporated into commission regulation (EC) No. 1702/2003; the Part M requirements are incorporated into EC No. 2042/2003. The Acceptable Means of Compliance for the Part 21 Implementation Rules is found at Decision No 2003/1/RM, which provides guidance material for airworthiness and environmental certification as well as for the certification of design and product organizations. Part 21 Implementing Rules are contained in two sections: A contains the Requirements for Applicants, and B contains the Procedures for Competent Authorities. Section B of the Part 21 Implementing Rules is also divided into subparts, with the category titles matching section A. Similar to the Implementation Rules, the Acceptable Means of Compliance and Guidance Material is also divided into two sections: A covers the Applicant, and B covers the Procedures for Competent Authorities. This is a novel approach since one document contains the requirements for certification as well as an acceptable method of compliance (no more searching!).

Three basic types of certificates may be issued as a result of a certification project: the type certificate (TC), the supplemental type certificate (STC), and the European technical standard order (ETSO) for parts and appliances. The ETSO is infinitely more rigorous than the TSO. For aircraft that do not meet all of the requirements for a production TC, a restricted TC may be issued if the applicant shows compliance with the applicable TC basis established by the agency to ensure adequate safety with regard to intended use of the aircraft and with applicable environmental protection.

EASA provides more structure in determining major and minor changes. Decision No2003/1/RM provides guidance on what determines a major change. The document provides a decision flow chart as well as specific instances of types of changes to large and small airplanes and a discussion of the rationale for the change classification.

The current cross-referencing of EASA and U.S. documentation is shown in Table 6.7.

TABLE 6.7 ■ EASA and FAA Documents Cross Reference

Former JAR Publication	EASA Publication	Definition US FAR Reference
JAR 21	Implementing Rule Certification EC 1702/2003	Certification Procedures for A/C, Products and Parts FAR Part 21
JAR 145	Implementing Rule Continuing Airworthiness EC 2042/2003 Annex II	Approved Maintenance Organizations FAR Part 145
JAR 147	Implementing Rule Continuing Airworthiness EC 2042/2003 Annex IV	Maintenance Training Organizations FAR Part 147
JAR 66	Implementing Rule Continuing Airworthiness EC 2042/2003 Annex III	Certifying Staff N/A
JAR 22	CS 22	Sailplanes and Powered Sailplanes N/A
JAR 23	CS 23	Normal, Utility, Aerobatic and Commuter Airplanes FAR Part 23
JAR 25	CS 25	Large Airplanes FAR Part 25
JAR 27	CS 27	Small Rotorcraft FAR Part 27
JAR 29	CS 29	Large Rotorcraft FAR Part 29
JAR 34	CS 34	Aircraft Engine Emissions FAR Part 34
JAR 36	CS 36	Aircraft Noise FAR Part 36
JAR 39	Incorporated into Implementing Rule Certification EC 1702/2003	Airworthiness Directives FAR Part 39
JAR E	CS E	Engines FAR Part 33
JAR P	CS P	Propellers Within the A/C Category i.e., Part 23 Subpart E
JAR TSO	CS ETSO	Joint Technical Standard Orders TSO FAR Part 21 Subpart O
JAR VLA	CS VLA	Very Light Aeroplanes No US
JAR VLR	CS VLR	Very Light Helicopters No US
GAI 20	AMC 20	Joint Advisory Material Advisory Circular Joint

■ 6.9 | COCKPIT CONTROLS AND DISPLAYS EVALUATIONS

There is a marked difference between modern cockpits and those of about 20 years ago. Information is now shared among many systems, and information can be retrieved by the operator and presented on flat-panel displays. Displays are no longer directly driven (steam gauges) by independent sensors. As we saw in the previous section, however, the probability of providing hazardous or misleading information that can lead to catastrophic results must be extremely improbable. We must be very careful, then, to evaluate displays not only in how well the information is presented but also with respect to the correctness and timeliness of the displayed information. Similar to displays, controls in modern cockpits have become more automated, a fact that has its benefits as well as disadvantages. It cannot be doubted that a machine is more efficient than a human and in the long run will probably save the operator some money in terms of fuel burn, for example. Many pilots in transport aircraft have become more of a system manager than the guy in charge of the bus.

The pros of modern cockpits are many and carry some major benefits. Safety is enhanced because automation allows more monitoring time on the part of the pilot, produces a less stressful work environment, and is less fatiguing. The crew workload is decreased and human error is reduced. Of course, some would say that human error can be eliminated entirely if the pilot is completely removed from the operation. The bottom line is that it (automation) allows accurate and efficient airline operations.

There is a flip side to all of these good things. Automation may lead to a false sense of security or even a higher sense of insecurity if the automation fails. There is a general temptation to ignore raw data and stick with the directed task (e.g., following the magenta line). Since the pilot becomes peripheral to what is going on, he may feel that he is not involved, which contributes to isolation of the crew. Automation can increase boredom and heads-down time in the cockpit (managing systems), which indirectly decreases safety. In critical situations, the information presented to the pilot may be confusing (multiple modes and automatic mode transitioning). In some architectures, there is a tendency for the computers to take the control authority out of the pilot's hands in critical situations. Clumsy or poorly designed automation can create high workloads at critical times; the workload may be changed but not necessarily reduced. Finally, automation is intended to reduce risks but may encourage aircrews to take greater risks.

6.9.1 Controls and Displays Airworthiness Documentation

FAA airworthiness guidance for controls and displays are found at:

- AC 20-175—Controls for Flight Deck Systems
- AC 23.1311-1C—Electronic Displays in Part 23 Airplanes
- AC 25-11A—Electronic Flight Deck Displays
- AC 25.1322-1—Flightcrew Alerting
- AC 25.1302-1—Installed Systems and Equipment for Use by the Flightcrew
- AC 27.1321—Arrangement and Visibility
- AC 27.1322—Warning, Caution and Advisory Lights
- AC 29.1321—Arrangement and Visibility
- AC 29.1322—Warning, Caution and Advisory Lights

    Additional Guidance when considering the Human Factors impact may be found at:

- AMN-99-2 Reviewing certification plans (Part 25)
- ANM-01-03 Methods of compliance (Part 25)
- PS-ACE100-2001-00 Reviewing certification plans (Part 23)
- GAMA Publication #10 Practices and guidelines for cockpit design (Part 23)

- GAMA Publication #12 Recommended Practices and Guidelines for an Integrated Cockpit/Flightdeck in Part 23 A/C

Human Factors has become a high interest item in the Certification process. ANM-99-2 provides guidance to FAA Certification Teams to conduct effective reviews of the applicant plans with respect to Human Factors (Part 25) and provides assistance to the applicant in developing these plans. ANM-99-2 is to be used by all members of the certification team including Test Pilots and Flight Test Engineers and encourages the development of a Human Factors Certification Plan (AC 21-40). The Authority cannot require the applicant to submit a Human Factors Certification Plan, but like an Advisory Circular, is recommended. AC 25.1302-1 is the newest guidance on the evaluation of controls and displays with respect to human factors considerations. It was published in May of 2013.

Other Guidance is available is available:

- Mil-Std-1472G DoD Standard for Human Engineering
- SAE/AIR 1093, Numeral, Letter and Symbol Dimensions for Aircraft Instrument Displays (mimics much of Mil-Std-1472D)
- SAE/ARP 571C, Flight Deck Controls and Displays for Communication and Navigation Equipment for Transport Aircraft (similar to 1093, but specific to Navigation systems)
- SAE/ARP 1874, Design Objectives for CRT Displays for Part 25 (Transport) Aircraft (specifically lighting, readability, etc.)

6.9.2 Human Factors Concerns

The FAA has recognized that there are problems associated with modern cockpits and automation. The key issue appears to be overreliance on the system. Additional problems identified by the FAA are

- Mode awareness. Aircrews are not always aware of the system mode status.
- Mode behavior. Aircrews are not always aware of mode operations, altitude capture for example.
- Speed and altitude protection. Some modes may not limit altitude or speed.
- Disengagement behavior. Aircrews are not knowledgeable of what happens when automation is disengaged.
- Crew interface. The man–machine interface is sometimes ill conceived.

These problems manifest themselves in aircraft accidents. Since the introduction of glass cockpits, there have been hundreds of accidents and incidents where the man–machine interface has been a contributing factor; fatalities have occurred. The accidents fit into a definite pattern: pilots did not understand what the automation was doing or the aircrew did not receive adequate feedback from the automated system. Some examples are given in the subsequent paragraphs.

The first example is one of an aircrew's temptation to ignore data, or ignorance of what the system was telling them. A "REV ISLN" (reverser isolation-reverser may be unsafe) advisory message illuminated immediately after takeoff to a Boeing 767 aircrew. The message was ignored and the aircrew climbed to FL310 where the thrust reverser deployed and the aircraft broke apart. In another case where the information to the aircrew was confusing there were 47 fatalities. A British Midland Boeing B737-400 experienced a failed engine fan blade and the electronic flight instrument system (EFIS) provided confusing information on which engine failed. The aircrew shut down the wrong engine and the aircraft crashed.

Another case of confusion, but this time mode confusion, was experienced by an Airbus A320 crew making an approach to Strasbourg, France. The aircrew received a change in the approach clearance to a nonprecision approach. The aircrew calculated that a 3.2° flight path would keep them compliant with the step-down fixes. The aircrew input what they thought was 3.2° commanded glideslope; instead, they got a 3200 ft/min descent rate, and the aircraft was lost. A similar event occurred with an Airbus A320 landing in San Diego. The crew credits the ground proximity warning system for the save.

In an instance where disengagement behavior did not act as expected, an Airbus A300 first officer accidentally hit the takeoff and go-around (TOGA) while established on an ILS approach. The aircrew attempted to control the situation by disconnecting the autothrottles and to control the airspeed manually. Since the TOGA commands the aircraft to pitch up during a go-around, the aircrew pushed forward on the yoke. Unfortunately, the autopilot tried to compensate and trimmed the aircraft nose up. The aircraft stalled, and 271 lives were lost.

The automated aircraft sometimes allows the aircrew to enter incorrect information and then uses that information in its computations. An Airbus A320 aircrew entered an incorrect aircraft weight into the FMS. As a result, the approach was flown at 136 knots instead of the correct approach speed of 151 knots. At this speed, the aircraft pitch angle was 8° instead of the standard 3–4°, which substantially reduced tail clearance. The airframe hit the runway surface.

These are but a few of the many instances where the system and the aircrew were on different pages of the script. They are provided as an illustration of what can go wrong if the system and its automation are not completely understood.

6.9.3 Controls Design Principles

With respect to cockpit controls, AC 20-175 provides generalized guidance on flight deck control system design, installation, integration, and approval. Appendix B of the AC provides a template for a control/function matrix, and Appendix C in the same document provides a quick reference to regulations (part, paragraph) for controls related to human factors and applies to all avionics systems being evaluated:

> Each control must operate with the ease, smoothness, and positiveness appropriate to its function. 23.671(a), 25.671(a), 27.671(a), 29.671(a)
>
> Controls must be located {part 23 only: arranged, and identified} to provide for convenience in operation. 23.671(b), 23.777(a), 25.777(a), 27.777(a), 29.777(a)
>
> Controls must be located {part 23 only: arranged, and identified} to prevent the possibility of confusion and subsequent inadvertent operation {part 29 only: from either pilot seat}. 23.671(b), 29.771(b), 23.777(a), 25.777(a), 27.777(a), 29.777(a)
>
> The pilot compartment must allow the pilot to perform his duties without unreasonable concentration or fatigue. 23.771(a), 25.771(a), 27.771(a), 29.771(a)
>
> The aircraft must be controllable with equal safety from either pilot seat. 25.771(c), 27.771(b), 29.771(b)
>
> Vibration and noise characteristics of cockpit equipment may not interfere with safe operation of the aircraft. 25.771(e), 27.771(c), 29.771(c)
>
> Controls must be identified (except where the function is obvious). 23.777(a)
>
> Controls must be located and arranged, with respect to the pilot seat, to provide full and unrestricted movement of each control without interference. 23.777(b), 25.777(c), 27.777(b), 29.777(b)

A control must be of a kind and design appropriate to its intended function. 23.1301 (a), 25.1301(a)(1), 27.1301(a), 29.1301(a)

Controls must function properly when installed. 23.1301(d), 25.1301(a)(4), 27.1301 (d), 29.1301(d)

For alerting lights, installed in the cockpit, the color red must be used for warnings and the color amber for cautions. 23.1322, 27.1322, 29.1322

Use of the colors red, amber, and yellow on the flight deck for functions other than flightcrew alerting must be limited and must not adversely affect flightcrew alerting. 25.1322(e)(1), (f)

Instrument lights must make each instrument and control easily readable and discernible. 23.1381(a)

Instrument lights must make each instrument, switch, and other device, easily readable. 25.1381(a)(1), 27.1381(a), 29.1381(a)

Instrument lights must be installed so that no objectionable reflections are visible to the pilot. 25.1381(a)(2)(ii), 27.1381(b)2, 29.1381(b)2

The minimum flightcrew must be able to access and easily operate controls required for safe operation. 23.1523, 25.1523, 27.1523, 29.1523

A cockpit control must be plainly marked as to its function and method of operation. 23.1555(a), 25.1555(a), 27.1555(a), 29.1555(a)

A typical Part 25 Human Factors compliance matrix is shown in Table 6.8.

AC 20-175 shows a means of compliance with the previously referenced regulatory paragraphs. The guidance in the document applies to the approval of:

- Controls for avionics systems in the flight deck
- Dedicated controls (e.g., switches, knobs)
- Multifunction controls (cursor controls, touchscreens)

The human factors general topics associated with controls are:

- Minimum crew
- Accessibility
- Concentration and fatigue
- Convenience and ease of operation
- Identification and marking
- Confusion and inadvertent operation
- Unrestricted movement
- Intended function

To meet these requirements, the AC addresses a philosophy that should be considered for all designs. It states that a variety of environments should be addressed during the evaluation:

- Appropriate representation of the pilot population
- Lighting conditions
- Use of gloves

TABLE 6.8 ■ Part 25 Human Factors Compliance Matrix

Part 25 Paragraph	General Human Factors Requirements	Method(s) of Compliance AC 20-175 ANM 99-2	Deliverable
25.771(a)	Each pilot compartment and its equipment must allow the minimum flight crew to perform their duties without unreasonable concentration or fatigue	Analysis, Simulator Test, Flight Test	Workload Certification Report
25.771(e)	Vibration and noise characteristics of cockpit equipment may not interfere with safe operation of the airplane	Bench Test	Test Report
25.773(a)(1)	Each pilot compartment must be arranged to give pilots a sufficiently extensive, clear, and undistorted view to enable them to safely perform any maneuvers within the operating limitations of the airplane, including takeoff, approach, and landing	Similarity	Vision Certification Report
25.773(a)(2)	Each pilot compartment must be free of glare and reflections that could interfere with the normal duties of the minimum flight crew	Ground Test	Lighting Certification Report
25.777(a)	Each cockpit control must be located to provide convenient operation and to prevent confusion and inadvertent operation	Simulator Test, Flight Test	Flight Deck Anthropometry Certification Report
25.777(c)	The controls must be located and arranged, with respect to the pilot's seats, so that there is full and unrestricted movement of each control without interference from the cockpit structure or the clothing of the minimum flight crew when any member of the flight crew, from 5′2″ to 6′3″ in height, is seated with the seat belt and shoulder harness fastened	Ground Test	Flight Deck Anthropometry Certification Report
25.1301(a)	Each item of installed equipment must be of a kind and design appropriate to its intended function	System Description, Simulator Demonstration, Flight Test	System Description Document, Demonstration Report, Flight Test Report
25.1309(b)(3)	Systems, controls, and associated monitoring and warning means must be designed to minimize crew errors that could create additional hazards	Hazard Assessment, Simulator Demonstration	Fault Tree Analysis Demonstration Report
25.1321(a)	Each flight, navigation, and power plant instrument for use by any pilot must be plainly visible to him from his station with the minimum practicable deviation from his normal position and line of vision when he is looking forward along the flight path	System Description Analysis, Flight Test	Installation Drawings, Vision Certification Report, Flight Test Report
25.1321(e)	If a visual indicator is provided to indicate malfunction of an instrument, it must be effective under all probable cockpit lighting conditions	Similarity Test, Ground Test	System Description and Statement of Similarity, Flight Test Report
25.1523	The minimum flight crew must be established so that it is sufficient for safe operation, considering **a.** the workload on all individual crew members; **b.** the accessibility and ease of operation of necessary controls by the appropriate crew member; **c.** the kind of operation authorized under § 25.1525. The criteria used in making the determination required by this section are set forth in Appendix D.	Simulator Test, Flight Test	Demonstration Report, Flight Test Report

(Continues)

TABLE 6.8 ■ *(Continued)*

Part 25 Paragraph	General Human Factors Requirements	Method(s) of Compliance AC 20-175 ANM 99-2	Deliverable
25.1543(b)	Each instrument marking must be clearly visible to the appropriate crew member	Analysis, Simulator Test	Vision Certification Report, Demonstration Report
System Specific Human Factors Requirements			
25.1381(a)(2)	The instrument lights must be installed so that no objectionable reflections are visible to the pilot	Ground Test	Flight Test Report
Specific Crew Interface Requirements			
25.773(b)(2)(i)	The first pilot must have a window that is openable... and gives sufficient protection against the elements against impairment of the pilot's vision	Ground Test (to verify no interference with window opening)	Flight Test Report
25.1322	If warning, caution, or advisory lights are installed in the cockpit, they must, unless otherwise approved by the administrator, be	Similarity	System Description Document

a. red, for warning lights (lights indicating a hazard that may require immediate corrective action);
b. amber, for caution lights (lights indicating the possible need for future corrective action);
c. green for safe operation lights; and
d. any other color, including white, for lights not described in paragraphs (a) through (c) of this section, provided the color differs sufficiently from the colors prescribed in paragraphs (a) through (c) of this section to avoid possible confusion

- Turbulence and vibrations
- Interruptions and delays in tasks
- Interference with the motion of a control
- Incapacitation of one pilot
- Use of the nondominant hand
- Excessive ambient noise

With respect to the pilot population, the evaluator should be compared to the anthropometric population. The test should use evaluators from throughout the range of operational users or supplement the study with mock-ups or anthropometric models. Keys to look for are unrestricted motion as well as inadvertent movement. Other contributors could be pilot strength, visual acuity, color perception or pilot cultural characteristics; look at experience with other, similar types of controls.

Controls should be evaluated under all foreseeable lighting conditions that include direct sunlight, indirect sunlight and reflection, sun above the forward horizon and night and/or dark environments. Tactile compensation can be taken into account when evaluating lighting conditions. Gloves will impact tactile feel and sensing and may be a discriminating factor with button spacing.

The test should ensure that the controls are operable during vibration. The evaluator must look at inadvertent actuation as well as the ability to actuate a control in the environment. Vibration and noise should not interfere with the safe operation of the aircraft. These tests should be accomplished inflight as the conditions are difficult, if not impossible, to replicate on the ground. Excessive noise may prohibit the aircrew from receiving aural feedback important to control functions, and can sometimes impact nonaural sensory modes. It may be necessary to incorporate another sensory feedback mechanism in conditions of excessive noise (visual, tactile).

Some control operations involve multiple steps and an interruption or delay may negate a successful result. In environment and use conditions, the evaluator should include interruptions and delays to understand if the controls' behavior results in safety-critical consequences. Aside from the obvious, movement may also be restricted by items brought onto the flight deck by the crew. The evaluator should consider objects that might reasonably be in the flight deck for physical interference, state the objects considered, and show the control is acceptable in the presence of these objects.

Incapacitation of one pilot must be considered in determining the minimum flight crew. Any control required for operation by the pilot in the event of incapacitation of other crewmembers must be viewable, selectable and operable. Controls should be usable by left-handed and right-handed pilots; the evaluator should give special consideration to controls that require speed or precision of force or motion. Gain affects the tradeoff between task speed and error; high gain values tend to favor pilot comfort and rapid inputs, but can contribute to error (overshoot, inadvertent activation). Low gain values tend to favor tasks that require precision, but can be too slow for the task. Special consideration should be given to variable gain controls. In the final analysis, gain and sensitivity should be acceptable for the intended function.

Controls should be designed to provide feedback when actuated. Possible feedback results are:

- Physical state of the control device (force, position)
- State of data construction (text string)
- State of activation or data entry ("enter")
- State of system processing
- State of system acceptance ("error")
- State of system response (page change, zoom, disconnect)

Feedback can be visual, aural, or tactile. If feedback or awareness is required for safety, then the aircrew should be informed of the state of activation or data entry, the state of system processing (downloading . . . .), and the state of system response.

Controls must be identifiable and predictable; they are the action part of the aircraft/ pilot interface. Controls should be easily identifiable to be operated quickly and instinctively by a "pilot under stress" (AC verbiage). Pilots should be able to find controls without looking at them and as such should be distinguishable in some manner, for example, a gear handle in shape of a wheel or flap handles with an airfoil shape. For controls with visible markings that are intended for use in low-light conditions, the markings must be lighted in some way that allows them to be easily read. The lighting of controls needs to consistent with flight crew alerting and need to be dimmable but not so low as to appear inactive. External lighting should not produce bright spots or glare, and automatic adjustment may be used. Night Vision System (NVIS) requirements are covered in Miscellaneous Guidance of AC 27-1 and 29-2.

Controls for data entry must support its intended function; they need to be acceptable for data entry speed, accuracy, error rates, and workload. If data entry involves multiple steps, each step must be discernible and automatic data strings should be discernible from constructed strings (pilot entered). Pilots should be able to easily recover from typical error inputs; this may be envisioned as a "go-back" button. Historically, when pilots found themselves on an unfamiliar display or simply lost, they would power down the display and restart from scratch on a familiar display. Keyboards and cursors should follow standard physical configurations, and cursors should automatically be placed in the first data field. There should be no more than two knobs per knob assembly.

Designers can ask some basic questions prior to implementing any new or improved controls and displays suite:

- Are the controls suitable for the task or the mission?
- Is there adequate control feedback (e.g., moving or nonmoving autothrottles or stick)?
- Are the displays clear, concise, and unambiguous?
- Does the symbology clearly describe the function?
- Is sufficient information provided to allow the operator to know what the aircraft is doing?
- How is automation implemented?
- How much automation is implemented?
- Has complexity been minimized?
- Is there standardization within the cockpit?

Answering these questions will hopefully negate some of the pitfalls that are inherent in today's systems. In attempting to design controls and displays systems some basic tenets must be observed.

6.9.4 Display Design Principles

The basic function of the display suite is to supply information not discernable to an operator's unaided senses. Displays fall into three categories: visual, audio (e.g., warnings), and tactile (e.g., stick shakers). In older display suites, information was simple and standardized. The familiar T layout was developed after World War II and is still in use today. All analog cockpits utilize this arrangement, and almost all digital cockpits replicate the analog displays that they replaced. Figures 6.6 and 6.7 illustrate this concept.

If you look at the figures, you can see the primary flight information displayed from left to right: airspeed, attitude, and altitude. Heading is the fourth parameter and completes the T at the bottom center of the suite. Pilots are taught to develop a scan pattern using the T to rapidly gather information and continuously update their situational awareness. It should also be noted that if pilots are taught this technique and we change the location of these parameters (as in a new digital cockpit) we will actually increase workload and therefore reduce situational awareness. This is one reason why many modern cockpits seem to be just color replicas of the analog suites that are 60 years old.

FIGURE 6.6 ■
Analog Cockpit with
T Arrangement

FIGURE 6.7 ■
Familiar T in a Digital
Design

Although it appears from the previous example that displays have only become prettier, modern display complexity has increased. The previous example only illustrated the pilot's primary flight instruments. Other displays, developed by a multitude of vendors, may have no commonality in the way that information is presented. This information, which does not carry the criticality level that primary flight instruments possess, may be displayed in any fashion the designer desires. This leads to issues affecting the readability of the displays and the aircrew's understanding of what is being presented:

- Continuous versus discrete displays. Some displays will continuously update information while others will only change when a parameter, mode, or condition changes.
- Color displays. The use of colors is dictated by FAR 23/25.1322, but not for all displays and all colors. In some instances, engaged modes and warning annunciators may be the same color. The combinations of colors, such as background and data, may be hard on the eyes or be unreadable in certain lighting conditions.
- Crowded displays. Many times there is just too much information being presented on a single display. This problem is compounded when the display size is too small.
- Symbology. The representation of data by symbols (e.g., next waypoint depicted as a triangle) is not uniform among designers and may change with changes in mode within the same display.
- Information overload. Because the capability exists to depict everything that is going on around the cockpit, some designers have seen fit to display everything that is available. It sometimes becomes difficult to determine which data are important and which are just informational.
- Eye fatigue. As anyone who works with computers already knows, staring at a screen for many hours creates eye strain.
- Memory. Most of the ACs relating to aircraft systems make a point of stating that the operation of a particular system should not rely on pilot's memory, or that operation should be intuitive. With so much information and so many possible modes of operation, this statement becomes meaningless. Take a look at the four F-15 heads-up display (HUD) modes depicted in Figure 6.8 and ask yourself if any memorization by the operator might be required.

Figure 6.8 provides a sample of a fighter implementation and shows how memory plays an important part in understanding information that is provided. There are examples in the civil area as well. Figure 6.9 illustrates the rate of descent versus angle of descent controls discussed in section 6.9.2. In the upper right portion of the display, the selector knob controls either track and flight path angle or heading and vertical speed indicator, depending on the mode indication selected. Not paying attention to the mode select or uncertainty of the inputs could lead to some bad consequences.

6.9.5 Flightcrew Alerting

AC 25-1322-1, Flightcrew Alerting, 12/13/2010 was developed to provide guidance in developing a consistent philosophy for alerting conditions, urgency, and prioritization and presentation. (Part 23 has no guidance within this paragraph, AC 23-8 reserves this paragraph, and Parts 27 and 29 are covered in their respective ACs). The following design concepts have been developed:

- Only nonnormal conditions and operational events that require aircrew awareness should generate an alert.
- All alerts provided to the aircrew must provide the information needed to identify the nonnormal condition and the corrective action required.
- The alerting philosophy should be consistent.

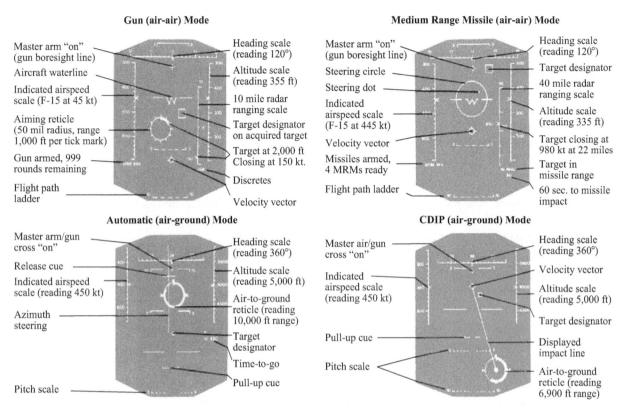

Gun (air-air) Mode

Master arm "on" (gun boresight line)

Aircraft waterline

Indicated airspeed scale (F-15 at 45 kt)

Aiming reticle (50 mil radius, range 1,000 ft per tick mark)

Gun armed, 999 rounds remaining

Flight path ladder

Heading scale (reading 120°)

Altitude scale (reading 355 ft)

10 mile radar ranging scale

Target designator on acquired target

Target at 2,000 ft Closing at 150 kt.

Discretes

Velocity vector

Medium Range Missile (air-air) Mode

Master arm "on" (gun boresight line)

Steering circle

Steering dot

Indicated airspeed scale (F-15 at 445 kt)

Velocity vector

Missiles armed, 4 MRMs ready

Flight path ladder

Heading scale (reading 120°)

Target designator

40 mile radar ranging scale

Altitude scale (reading 335 ft)

Target closing at 980 kt at 22 miles

Target in missile range

60 sec. to missile impact

Automatic (air-ground) Mode

Master arm/gun cross "on"

Release cue

Indicated airspeed scale (reading 450 kt)

Azimuth steering

Pitch scale

Heading scale (reading 360°)

Altitude scale (reading 5,000 ft)

Air-to-ground reticle (reading 10,000 ft range)

Target designator

Time-to-go

Pull-up cue

CDIP (air-ground) Mode

Master air/gun cross "on"

Indicated airspeed scale (reading 450 kt)

Pull-up cue

Pitch scale

Heading scale (reading 360°)

Velocity vector

Altitude scale (reading 5,000 ft)

Target designator

Displayed impact line

Air-to-ground reticle (reading 6,900 ft range)

FIGURE 6.8 ■ F-15 HUD Modes of Operation

FIGURE 6.9 ■ Rate of Descent versus Angle of Descent Controls

The AC further states that cockpit design should include the appropriate combination of alerting elements. These elements may be:

– Master visual alerts

– Visual alert information

– Master aural alerts

– Voice information

– Unique tones (unique sounds)

– Tactile or haptic information (stick/rudder pedal shaker)

The general rules for alerting design state that the designer should ensure the Alert System is synchronized and prioritized. The colors should conform to 25.1322, and if using aural alerts with multiple meanings a corresponding visual, tactile or haptic alert should also be provided. The designer must determine the conditions that will generate the alert and establish how the urgencies and priorities will be generated. Above all, the presentation scheme must be consistent.

Alerts for warnings and cautions must provide cues through at least two senses; advisories are provided through a single sense. Several alert combinations are used to meet this requirement:

– Master visual alert, visual alert and master aural alert

– Master visual alert, visual alert and voice-alert information or unique alert tone

– Voice alert information may be preceded by a master aural alert

– A tactile alert may be combined with a visual or aural alert to meet the two sense requirement

6.9.6 Color Usage

Color usage is dictated by the FARs, as noted previously. The pertinent FAR is 23/25.1322. There are ACs available to assist the applicant in meeting the intent of this paragraph. AC 25-11 and AC 23-1311-1A provide an acceptable means of showing compliance with the FAR. In the subsequent paragraphs I will discuss how designers of one aircraft (MD-11) designed their display color usage and how they are in compliance with the applicable FAR.

FAR 25/23.1322(a) states that warning lights shall be in red and may require immediate action. AC 25-11 and AC 23-1311-1A state that red will be used for warning lights and flight envelope or system limits. The MD-11 procedures state that a red warning light requires immediate crew action. Red is used on the aircraft for autopilot and auto throttle disconnects; high and low airspeed alerts; exceeding pitch, bank, or engine limits; unsafe landing gear; and weather radar precipitation rates of 12 to 50 mm/hr.

According to FAR 25/23.1322(b), caution lights should be in amber and advise the crew that there is a possible need for future crew action. AC 25-11 and AC 23-1311-1A state that cautions, abnormal sources, and weather radar precipitations of 4 to 12 mm/hr shall be in amber. The MD-11 procedures state that the presence of an amber light requires immediate crew awareness and possible subsequent action. Amber is used for level 1 and level 2 alerts, failures and cautions, abnormal sources (such as the captain's static sources on backup), weather radar precipitation rates of 4 to 12 mm/hr, the ILS middle marker, and configuration speed limits.

The color green, according to FAR 25/23.1322(c), is used to indicate safe operation. The AC further stipulates that green should be used for engaged modes, flight director bars, selected data, and weather radar precipitation of 1 to 4 mm/hr. In the MD-11, green is used for data validity, the #2 navigation needle, track diamond, autopilot and autothrottle modes, configuration limits, landing gear safe indication, takeoff and landing checklists complete, and weather radar precipitation of less than 4 mm/hr.

The FAR states that cyan can be used as long as it is not related to red/amber/green light functions and that the color does not cause confusion. The AC calls for cyan to be used for the sky, armed modes, selected data, and selected heading. The MD-11 uses cyan for engine start hydraulic fluid levels, autopilot engage status, the captain's #1 navigation bearing and range display, pitch limit, and systems available for use but not selected.

There is no specified reference in the FAR to magenta; the AC states that magenta should be used for the ILS deviation pointer, flight director bar (also green, see above), crew station #1 selected heading, and active route in the flight plan. It also states that magenta should be used for weather radar precipitation rates greater than 50 mm/hr or turbulence. The MD-11 reserves magenta for electronically generated or derived data such as FMS selected values, flight director commands, radio navigation data, localizer and glideslope deviations, weather radar precipitation greater than 50 mm/hr, and turbulence. Other colors, such as grey, are used for scale shading; white is used for scales and manual entries.

6.9.7 Test and Evaluation of Controls and Displays

There are two types of evaluations associated with controls and displays: functional and evaluation. One of these is done well, while the other is not really done well at all. If you review the previous section regarding colors, you might be led to the erroneous conclusion that the evaluation would be very simple. All we need to do is go down the checklist and make sure that we comply with the FAR/AC. This is only part of the evaluation. I tend to call these evaluations TLAR testing: *that looks about right*. This is the evaluation of the controls and displays suite. The testing is based on human factors and the man–machine interface, but it has nothing to do with evaluating whether or not the displayed information is correct. A functional evaluation evaluates the system's ability to provide accurate, timely, and no misleading information to the aircrew. We do not do functional evaluations very well.

In a functional evaluation, the displayed information and the command/response functions of the controls are measured against the truth data (more on this later). Human factors techniques and validation of the FAR items are components of the evaluation. For both types of tests, video recording of the displays should be accomplished as well as pilot's voice and time code insertion. The pilot's voice will give us insight into how well the suite is designed and will alert the evaluator to potential problems. The time code will allow us to select pertinent time slices for data analysis. When performing the controls and displays evaluations, it is important that more than one operator/pilot be used for the tests. Select a full range of users from low experience levels to senior pilots; if in the military, select users other than all test pilots. Test pilots and senior pilots have many hours of flight time and have developed inherent workarounds to problems. A system that they deem acceptable may be unwieldy for a less experienced pilot.

Much of the testing can be accomplished in the simulator, as long as the simulator emulates the aircraft as closely as possible. Attempt to exercise all failure modes and emergency situations during the evaluations. These scenarios will increase workload and provide a true picture of whether the displayed information is suitable or not.

The first portion of the test should be to evaluate system usability. The displays may be cathode ray tubes (CRTs), liquid crystal diodes (LCDs), flat panel, or a combination of these. Inputs may be made via soft keys, touch screens, keypads, or other controls. The first item to be checked is the feedback mechanism. If an operator selects MENU on a soft key, does the display immediately display the menu, or is there a mechanism that says, "You have selected MENU; is that what you really wanted?" For this type of feedback mechanism, the soft key would have to be depressed again to go to the MENU page. The initial depression may highlight MENU or turn it bold to inform the operator that this mode has been selected. This was a common mechanization on early digital military aircraft. Unfortunately, it took about 20 sec for the pilot to figure out that if he hit the button twice he would get the page right away, thus negating the positive feedback implementation.

Probably the most intelligent design enhancement for modern controls and displays is the "go back" button, which is just like the one you use while surfing the Web. In order to go back to a previous page you simply hit this button and you are there. Imagine that you have just entered a 20-waypoint flight plan into the FMS and just prior to hitting the enter button you notice that some of the entries are incorrect, but your brain does not stop your finger in time. Now what do you do? Without a "go back" button, you might have to go to an edit page, or worse yet, redo the flight plan. With a "go back" button, you merely retrieve the last page and make the corrections. With this type of mechanization, there is no need to have the positive feedback, two-depression system, because you can always recover if you screw something up.

Touch screen displays are much like ATM displays, where you touch the item on the screen rather than using a soft key. These displays may be pressure sensitive or heat sensitive. You would not expect someone to install a heat-sensitive touch screen in a military aircraft because the aircrews wear gloves, but it has been done. Keypad entry or entry via some other control (e.g., lever, knob) should also be evaluated for ease of use and the potential for errors. In many military cockpits, control entry is made via the Mark I, Mod A #2 pencil because the buttons are too small or too close together, so the probability of hitting the wrong button (once again with gloved hands) is high. This is also a good time to evaluate consistency in the controls (is clockwise up and counterclockwise down, etc.?).

As an evaluator, we must also be cognizant of the operational requirements of the operator. In military aircraft, it is possible that the operator may be wearing an outfit that is slightly more constraining than a flight suit. Cold weather gear or antiexposure suits are bad, but the worst is nuclear, biological, chemical (NBC) flight gear. With this equipment, movement, tactile feel, and blood flow are severely restricted. Controls that were acceptable with normal flight gear may no longer be usable. Other operational requirements may be flight with night vision systems. Displays need to be visible in all types of lighting conditions.

The problems with designing displays (wrong information, too much information, etc.) were described in the previous section. During the evaluation you must determine if the designers eliminated these problems or inadvertently contributed to or compounded them. If there is too much information presented on the display, is there a way to

FIGURE 6.10 ▪
Controls and
Displays Suite

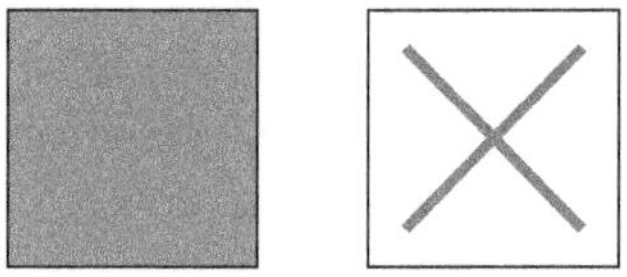

FIGURE 6.11 ▪
Allowable Methods
of Showing
Unreliable Data

declutter the mess? Many displays have submodes (declutter modes) that allow the
operator to remove certain types of information to make the display more readable. Next
the color, font, and character size should be evaluated for compliance with the applicable FAR. The ACs are a good place to formulate these tests.

After we are happy with the displays (TLAR) we can set about determining if the
data are correct and delivered in a timely manner. In order to proceed with this test, we
first need to know how the system is architected. How does the information flow from
where it was sensed to the display? What manipulation is done to the data along the
way? How long does it take from the time the parameter is sensed to when it is presented
to the operator? If these questions are not answered, it does not really make any difference how pretty a display looks because the data cannot be trusted.

We saw in the complex systems section that any safety assessment starts with a
failure hazard analysis (FHA). This FHA is normally initiated with a fault tree analysis.
We must accomplish the same type of test when functionally evaluating controls and
displays. Figure 6.10 depicts a controls and displays suite that consists of five displays,
one of which is a HUD.

Multifunction display (MFD) 3 is the HUD and contains the primary flight instruments for the pilot (airspeed, attitude, altitude, and heading). Since the probability of
failure or providing hazardously misleading information is extremely improbable, the
loss of the MFD 3 hardware due to failure would require that the HUD information be
moved to another MFD. In addition, the pilot must be made aware that the information
on MFD 3 is no longer valid. This can be accomplished in one of two ways: the display
can be blanked or a red "X" can be placed across the display (Figure 6.11).

Assume that the logic in the system states that with a loss of MFD 3, the HUD
information is moved to MFD 2. What happens if MFD 2 fails with a subsequent failure
of MFD 3? Obviously there must be additional logic that states that if these failures
occur in this sequence, then the HUD information will be moved to MFD 4. The logic
that is incorporated is a result of the safety assessment discussed previously. It is our job
as evaluators to ensure that the logic is coded correctly in the software. But this example
is a simple test. What system feeds information to the displays?

In most digital display systems, the displays are controlled by display processors.
We need at least two display processors, because if we only had one and it failed we
would not have any displays. Figure 6.12 shows a typical architecture with the same five
MFDs supported by two display processors. Display processor 1 supports MFD 1, 2, and
3, while display processor 2 supports MFD 3 and 4.

What happens if the aircraft experiences a loss of display processor 1? Since display processor 1 drives MFD 3 (HUD), it stands to reason that display processor 2 must then drive MFD 3 after the loss of display processor 1. What happens if MFD 3 fails with a subsequent loss of display processor 1? Our original logic stated that with the loss of MFD 3 the HUD information would be moved to MFD 2. If display processor 1 then fails, our original logic said that display processor 2 would drive MFD 3, but that display has failed. We must have additional logic that says if display processor 1 fails and MFD 3 fails, then display processor 2 must drive MFD 2. You can see that the testing is starting to get a little more involved.

We are not finished yet, since the mission computers/mission processors drive the display processors. Usually we have at least two mission computers (for the same reason we have two display processors), and either mission computer can drive both display processors. The problem (especially in military applications) is that the mission computers are not truly redundant, and the loss of a mission computer will result in a loss of capability. Figure 6.12 adds the mission computers to the schematic. Mission computer 1 contains the functions of controls and displays, navigation, radar, and weapons. Mission computer 2 contains the functions of controls and displays, navigation, electro-optics, and electronic warfare, so the mission computers are not totally redundant. They are only redundant in the areas of controls and displays and navigation. With the loss of a mission computer, the associated functions will also be lost on the displays. The data are not allowed to freeze, and the correct indications must be given to the pilot. Figure 6.13 shows the loss of mission computer 1 and the associated loss of the radar and weapons information on the displays.

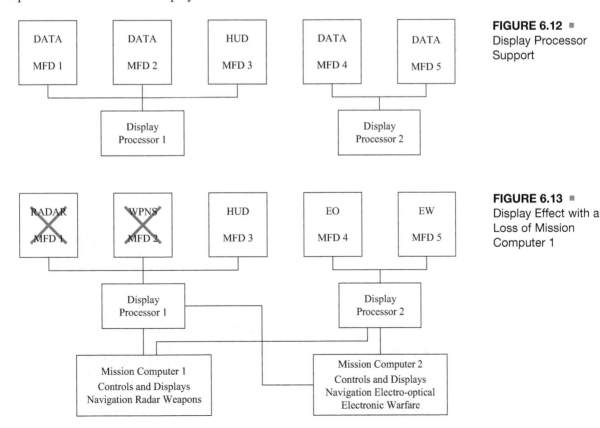

FIGURE 6.12 ▪
Display Processor Support

FIGURE 6.13 ▪
Display Effect with a Loss of Mission Computer 1

The final input to the displayed data is the aircraft sensors (e.g., inertial navigation system [INS], GPS, angle of attack [AOA], air data). These sensors feed into the mission computers; the mission computers determine the validity, perform calculations, and forward the information. This is the "truth data" for the display subsystem. If a sensor is bad, its information will not be allowed on the display and the pilot will receive the same bad data indications. The evaluator will have to devise a test matrix where individual sensor information is failed and validate that the parameter is either removed or X'd out on the display. This will be a rather large matrix and will take some time to accomplish, but it is an extremely important test. We would not like to find out on final approach in marginal weather that the AOA freezes on the display rather than being taken down when the sensor fails.

In addition to validating that the data are correct and not displayed when invalid, the evaluator also needs to determine if there is any latency in the system. The latency is defined as the time between when a parameter is sensed and when it is displayed. The GPS can provide horizontal accuracy of 15 m 90% of the time, but if it takes 2 sec to reach the display, the accuracy could degrade by hundreds of meters, depending on the aircraft speed. This test must be done with all systems to ensure accuracy has not been degraded and that safety has not been compromised. Again, this will be a rather large test matrix and involve quite a bit of time.

A third test involves the ability of the system to accurately display cautions, advisories, and warnings (CAWs) in a timely fashion. In analog aircraft, these advisories are displayed on an annunciator panel in the cockpit with color-coded lights: red for warnings and amber for cautions and advisories. The panel is normally combined with a master caution or master warning light that illuminates when a light appeared on the annunciator panel. This keys the pilot to look at the panel and determine which system is at fault. An annunciator panel is depicted in Figure 6.14. The annunciator panel is analog and analog is passé. With digital systems, faults are now presented on displays. Not all faults will be displayed, and all faults will not appear on all displays. For example, loss

FIGURE 6.14 ■
Caution, Advisory, and Warning Display

of hydraulic pressure is fairly serious, and the fault may appear on all displays, including the HUD. Loss of a radar transmitter is not a warning and may only appear on the radar display; other displays, such as the FMS scratchpad, may inform the pilot to check the radar display. What happens if a failure is supposed to appear on the HUD but a previous failure has caused the HUD information to be moved to a different display? The logic must be present in the system to accommodate this change and properly display the cautions, advisories, and warnings. Another test and another large matrix.

These tests must be run to ensure the integrity of the system and before allowing IFR flight. How long does this testing take? Well, it could take as long as six months, and that is assuming that there are no software changes. What happens when the testing is complete and then you receive a change to the display processor software? It is probably a good idea to run these tests again. Think back a couple of sections to the discussion of software in the TSO process and the concepts of significant and insignificant changes. These tests must be accomplished if there is a significant change, but how confident are you with insignificant changes?

There is another series of tests that also need to be verified and these involve a mechanism called the display override feature. Suppose you are flying and your airspeed is 60 KIAS, AOA is 50°, and your yaw rate is 60°/sec. What can you say is happening to your aircraft? If you said spinning, you are correct. For the pilot of this airplane, the first order of business is to determine in which direction the aircraft is spinning and then apply the appropriate antispin controls. Applying the controls in the wrong fashion may really ruin your day. In some aircraft, help is provided by overlaying the HUD and in some cases all other displays, with spin information. The display will remain until the situation is corrected and then the displays will revert back to their prespin settings. An example of the spin display is shown in Figure 6.15.

The display is telling the pilot that the aircraft spin is to the right and that the stick should be placed in the right forward quadrant. In the civilian world, similar logic is applied on some aircraft for TAWS, which is discussed later in this chapter. If the system detects terrain and issues a pull-up command and the pilot does not have a TAWS display selected, the TAWS display will automatically override one or more of the cockpit displays to provide immediate situational awareness. The logic and implementation of these overrides must also be included in the test plan.

The last portion of the evaluation is the determination that a particular display or suite of displays is suitable to accomplish the mission. We accomplish this test by using tests that are very similar to closed loop handling qualities (CLHQ). The performance and flying qualities testers use a system called the Cooper-Harper chart. A copy of this chart is provided in Figure 6.16. By using a slight modification, we can adapt this same system for determining the utility and usability of a display.

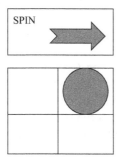

FIGURE 6.15 ■
Spin Display
Override

FIGURE 6.16 ▪
Cooper-Harper
CLHQ Rating Scale

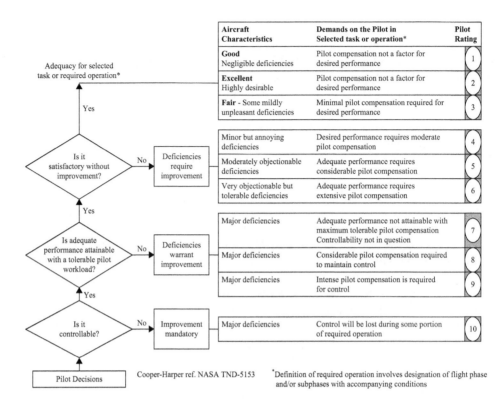

Cooper-Harper ref. NASA TND-5153

*Definition of required operation involves designation of flight phase and/or subphases with accompanying conditions

With CLHQ, the pilot is given a task to perform and is then asked to quantify the task by using to Cooper-Harper scale. Assume that we have changed the control laws in the flight control computers. We then task the pilot with accomplishing an objective that employs these new control laws. We assign the pilot the task of landing the airplane. He attempts to land the aircraft and discovers that stick forces are nonexistent and he has to summon up all of his piloting skills just to get the airplane on the ground without killing himself. The pilot enters the Cooper-Harper scale with the task "land the airplane." The first milestone that he encounters is "Is it controllable?" and the answer is no. We move to the right and pick the most appropriate statement. We have only one choice here and that is a Cooper-Harper rating of 10, major deficiencies, improvement mandatory.

After going back to the lab, the software is reworked and the pilot is tasked again with the new and improved software. Although the control is much more attainable, he finds that he tends to overshoot the nose with corrections to the runway. He enters the chart again and this time passes the first milestone (because it is controllable) but stops at the second milestone, because the workload is still too high, and gives the system a Cooper-Harper rating of a 7. These iterations continue until the task is rated at a Cooper-Harper of 1 to 3.

With a slight modification to the Cooper-Harper CLHQ scale, we can devise a display rating tree that can be used for evaluating the usability of any display. The method is exactly the same as previously described for handling qualities, except in this case we are going to see if the required parameters for accomplishing a task are controllable (as opposed to the aircraft). The display rating tree is depicted in Figure 6.17.

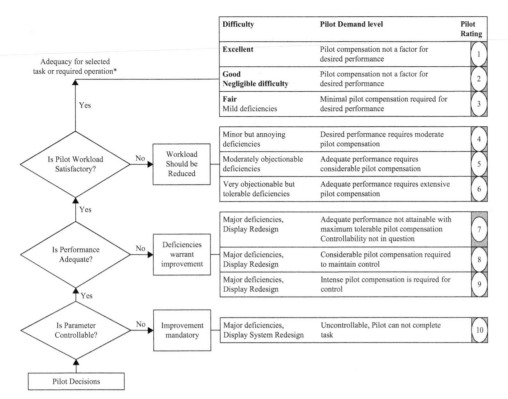

FIGURE 6.17 ■
Display Rating Tree

Assume that we are going to evaluate a HUD and the ability to fly certain maneuvers using only the HUD as reference. We would set up a task and establish certain thresholds for the pilot to meet. The first task may be:

- Fly straight and level 300 KIAS and the thresholds are airspeed ±5 knots, altitude ±25 ft, heading ±1°.

The pilot would then attempt to complete the assigned task using the HUD as the only reference. If the altitude was graduated every 50 ft, this task could not be accomplished, since the assigned threshold was 25 ft. Entering Figure 6.16, we would be stopped at the first milestone, "Is the parameter controllable?" The assigned level would be a 10, much like our original CLHQ example. A software change would be required to the HUD display software to successfully accomplish this objective. The level of complexity would be ever increased until the true usability of the HUD could be established. The same type of test can be devised for any system display, including weapons delivery, radar, and navigation.

6.9.8 Standardization and References for Controls and Displays

Standardization is really a misnomer when it comes to the design of aircraft controls and displays. It would reduce transition time for pilots introduced to new cockpits, and safety levels would be increased if all cockpits were designed the same. However, it would reduce creativity and perhaps stifle the development of better controls and displays. In the civilian world, there are a multitude of documents that are available to assist in the development of these systems. None of these documents carry the force of law and are advisory only. The FARs must be satisfied, and these documents can assist in compliance. The FAR 23/25/27/29 paragraphs that relate to controls and displays are

- 1303—Flight and Navigation Instruments
- 1305—Powerplant Instruments
- 1309—Equipment, Systems, and Installations
- 1311—Electronic Display Instruments
- 1321—Arrangement and Visibility
- 1323—Airspeed Indicating System
- 1335—Flight Director System
- 1337—Powerplant Instruments Installation
- 1541—Markings and Placards

In addition to the ACs, other guidance is available to the applicant:

- ESD-TR-86-278, "Guidelines for Designing User Interface Software" (acceptable standards for digital entry and feedback, acceptable error messages and nomenclatures)
- NASA FTB 3000 (Rev A), "Man Systems Integration Standards" (gives guidance for aural warnings, types of intelligibility, and noise levels)
- SAE AIR 1093, "Numerical, Letter, and Symbol Dimensions for Aircraft Instrument Displays" (mimics much of the work in MIL-STD-1472, see below)
- SAE ARP 571C, "Flight Deck Controls and Displays for Communication and Navigation Equipment for Transport Aircraft" (specific to navigation systems)
- SAE ARP 1874, "Design Objectives for CRT Displays for Part 25 Transport Aircraft" (specifically lighting and readability)
- SAE ARP 4102, "Flight Deck Panels, Controls, and Displays" (how to get the most out of the displays by employing various techniques and standardization)
- SAE ARP 4107, "Electronic Displays" (describes the information that must be available for each particular group of systems, such as primary flight instruments)
- SAE AS 8034, "Minimum Performance Standard for Airborne Multipurpose Electronic Displays" (describes the measurable criteria for, for example, lighting, alignment, width, height)

The U.S. military currently uses one standard:

- MIL-STD-1472D, "Human Engineering Design Criteria for Military Systems, Equipment, and Facilities" (first attempt at military standardization of controls and displays suites)

6.10 | WEATHER RADAR CERTIFICATION

RADAR (radio detection and ranging) is covered in detail in chapter 8 of this text, so the theory of operation will not be discussed here. In this section, we are merely concerned with the evaluations that must occur to show compliance with the applicable regulations. For most avionics installations on the aircraft, the first document the evaluator should investigate is the flight test guide for the applicable aircraft. For transport category aircraft, AC 25-7C, "Flight Test Guide for Certification of Transport Airplanes" (October 16, 2012), is the document to use. For small airplanes we would consult AC 23-8C, "Flight Test Guide for Certification of Part 23 Airplanes" (November 16, 2011). Not all avionics systems are covered by the respective ACs, but they are, nevertheless, the best place to start. In some instances, the AC will refer the reader to another AC or reference document. A new Advisory Circular, AC 20-182, June 17, 2014, Airworthiness Approval for Aircraft Weather Radar Systems consolidates the airworthiness requirements for all aircraft when installing Weather Radar, and forward looking Windshear and Turbulence Detection.

Both ACs cite paragraph 1301 as the fundamental requirement for equipment function and installation; the paragraph should sound familiar, as is has been discussed in previous sections. AC 25-7C summarizes the certification requirements of modern avionics and electrical systems on aircraft by stating that the systems/equipment must

- Perform their intended function (25.1301)
- Be adequately protected for failure conditions (25.1309)
- Be arranged to provide proper pilot visibility and utilization (as appropriate) (25.1321)
- Be protected by circuit breakers to preclude failure propagation and minimize distress to the airplane's electrical system (25.1357)
- Be installed in a manner such that operation of the system will not adversely affect the simultaneous operation of any other system (25.1431)

The AC also makes the case for the test of all installed equipment on aircraft whether or not it is characterized as "optional." Optional equipment would imply that the equipment is not necessary to safely fly the aircraft. The AC makes the point that all equipment installed will obviously be used by the crew; therefore, the statutorily required tests must be applied. Public law requires that the administrator find that no feature or characteristic of the airplane makes it unsafe for the category in which certification is requested. The extent to which that equipment must be tested or evaluated, that the administrator may make the necessary finding with respect to the whole airplane, is a technical determination based on the engineering and operational expertise of the administrator. This logic is more clearly defined in AC 20-168, "Certification Guidance for Installation of NonEssential, NonRequired Aircraft Cabin Systems & Equipment (CS&E)", July 22, 2010.

In light of the preceding paragraph, there are some general guidelines to be followed for all equipment installations (optional or not):

- The criteria and requirements are a function of the applicant's engineering analysis and laboratory (simulator) test program. The combination of analysis, laboratory, and flight evaluation form the whole of the certification requirement.
- The requirement for demonstrating safe operation should normally include induced failures during flight. The types of failures and the conditions under which they occur should be a result of the analysis and laboratory studies.
- The amount of flight testing should be determined through the cooperative efforts of the assigned project personnel. As a rule, flight testing will be required for an initial

certification, while derivative, follow-on items may require less testing. When ground or flight test data are available from similar, previously approved installations and are sufficient to properly evaluate a system's performance, additional testing may not be required.

- Particular attention should be given to those installations where an external piece of gear, such as an antenna, could affect the flight characteristics.

- Installations that can or may change the established limitations, flight characteristics, performance, operating procedures, or any required systems require approval by an FAA ACO flight test branch.

- New installations of equipment should be evaluated as a cooperative effort by the authority and the applicant.

- Throughout the systems/equipment evaluation, the operation of annunciators should be assessed to determine that proper conspicuity and display are provided to the appropriate flight crew member. Any mode of action selected by a manual action, or automatically, should be positively identified. Any submode should be evaluated to determine the need for annunciation.

AC 23-8C is not nearly as verbose as the AC for Part 25 aircraft. The document says that it provides requirements for specific types of equipment (such as weather RADAR). The document only provides the applicant with the additional guidance that particular attention should be given to those installations where an external piece of equipment could affect the flight characteristics. It also states that all installations of this nature should be evaluated by the flight test pilot to verify that the equipment functions properly when installed.

For utility and transport helicopters, AC 27-1B, September 30, 2008, and AC 29-2C, September 30, 2008, address Weather RADAR in the Miscellaneous Guidance (MG) section 1 in chapter 3.

6.10.1 Specific Weather RADAR Tests

AC 25-7C defines 10 specific weather radar tests that need to be completed to demonstrate compliance with FAR 25.1301. Once again, an AC does not carry the force of law; it is a means, but not the only means, of demonstrating compliance. The specific tests are

- Warm-up period
- Display
- Range capabilities
- Beam tilting and structural clearance
- Stability
- Contour display (iso echo)
- Antenna stability, when installed
- Ground mapping
- Mutual interference
- Electromagnetic compatibility (EMC)

Tests described in AC 23-8C, AC 27-1B and 29-2C are inclusive with the exception of warm-up period and display.

Warm-up Period. If a warm-up period is defined by the manufacturer, then all tests will be completed after this time has elapsed. Usually the manufacturer will state in the operator's manual what this time (if any) is. Wording will be something like "a usable display will be available to the operator 2 minutes after selecting standby/operate."

Display. The evaluator should check that the scope trace and sweep display move smoothly and are without gaps or objectionable variations in intensity. As the sweep moves from left to right, there should be no ratcheting across the display. The persistence should be set so that there is always a full display picture rather than returns only under the beam. For color displays, the evaluator should verify that the colors are correct with the proper contrast. This test will be accomplished again when verifying the contour mode. In many systems, BIT, initiated at either start-up or manually by the operator, will run through a color and contrast test.

Range Capabilities. This test verifies the maximum range capabilities of the radar as specified by the manufacturer. The tests should be completed by looking at large radar returns with good contrast from the surrounding area. Excellent choices for this test are land–water contrasts such as coastlines or lakes. As long as the water is not extremely rough, the water will be displayed as a no-show target (no radar energy will be returned from the water). The display will show the water as black, while the surrounding areas of landmass will be bright. The AC states that while maintaining level flight at 90% of maximum approved height for the airplane, set the radar controls so that large radar-identifiable objects such as mountains, lakes, rivers, coastlines, storm fronts, turbulence, etc., are displayed on the radar scope. A little caution about using storm fronts or turbulence: this is supposed to be a test of maximum range, which implies that you know where you are and you know where the object is (truth data). If you do not know the exact location of the weather, then how do you know the radar is displaying the correct range? The paragraph further states that the evaluator should maneuver the airplane and adjust the radar controls so that the tests can be conducted for the range requirements. The radar should be capable of displaying prominent line-of-sight (LOS) targets. What this means is that the tilt will have to be adjusted upward to show targets at the maximum range. On some aircraft installations it is possible to not meet the manufacturer's requirement for maximum range detection. This may be due to the fact that at 90% of the maximum certificated altitude the LOS is less than the stated maximum range. The equation for computing the range to the radar horizon (LOS) in nautical miles is given by Equation 6.1.

$$R = .87\sqrt{2h_R} \qquad (6.1)$$

where R is the LOS and h_R is the aircraft's altitude in feet.

If an airplane is certificated to 12,500 ft and the evaluation is conducted at 12,000 ft, the LOS for the radar will be approximately 135 nm. If the manufacturer's stated maximum range is 200 nm, this test cannot be successfully completed. This does not mean that the radar is not acceptable, as it provided returns at the maximum LOS for this airplane.

Beam Tilting and Structural Clearance. There are two tests within this section: the first is to determine that there is no structural interference between the radar antenna and the radome; the second is to check for bearing accuracy. With the aircraft in level flight, the operator should attempt to find a large radar return with the antenna tilt set at 0°. Once found, maneuver the airplane so that the target is at 0° elevation and 0° azimuth (on the nose). Slowly change the tilt control through an appropriate range and observe that the radar scope presentation does not change erratically, which might indicate structural interference with the antenna. This interference should not occur throughout the entire speed envelope of the airplane. A couple of notes here: If there is any physical interference present, the operator is going to hear it and it will, in all probability, cause a massive failure in the radar. If the structure interferes with the antenna's ability to see the returned energy, then the display will change based on the loss of signal.

In order to check the bearing accuracy, fly under conditions that allow visual identification of a target, such as an island, river, or lake, at a distance within radar range. When flying toward the target, select a course from the reference point to the target and determine the error in displayed bearing to the target at all range settings. Change the heading of the airplane in increments of 10° to ±30° maximum and verify that the error does not exceed ±5° in the displayed bearing to the target. This method is more qualitative than quantitative because it relies on the pilot's initial reference of zero, which could be off by some amount. As discussed in the maximum range test, if we use truth data (aircraft position and target position), a real measure of accuracy can be obtained for the radar.

Stability. The AC directs that while observing a target return on the radar indicator, turn off the stabilizing function and put the airplane through a series of pitch and roll movements, observing the blurring of the display. Turn the stabilizing mechanism on and repeat the pitch and roll movements. Evaluate the effectiveness of the stabilizing function in maintaining a sharp display. This is an evaluation of a system that does not have antenna stabilization, that is, a gyro stabilization package that will maintain the antenna level with respect to the earth regardless of what the aircraft does (to some limit). Gyro or space-stabilized radar tests are covered after the contour tests in this section. The stabilization in this section is more of a point-ahead mode, where the radar will attempt to keep pointing at the position the operator commands and smooth out elevation changes due to aircraft altitude transients in level flight. With the stabilization function turned off, the radar pitch and roll axis is slaved to the aircraft fuselage (or water mark) and the wings. Any change in aircraft space position or attitude will translate to a change in the radar picture, and hence smearing. Figure 6.18 shows the difference between stabilized and nonstabilized radar.

Contour Display (Iso Echo). This is the function of the radar that allows the determination of weather intensity. The AC states that if heavy cloud formations or rainstorms are reported within a reasonable distance from the test base, select the contour display mode. The radar should differentiate between heavy and light precipitation. Of course, you have to know the intensity of the storm to determine if the

FIGURE 6.18 ■
Stabilized versus
Nonstabilized
Platforms

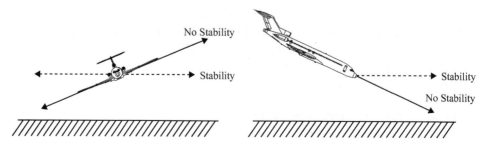

radar displays it correctly. There is that truth data again. Also remember from our discussions of displays that, according to AC 25-11, the color presentations for display of precipitation are

- Green, weather radar precipitation rates of 1 to 4 mm/hr
- Amber, weather radar precipitation rates of 4 to 12 mm/hr
- Red, weather radar precipitation rates of 12 to 50 mm/hr
- Magenta, weather radar precipitation rates of greater than 50 mm/hr or turbulence

Antenna Stability, When Installed. This is the test for gyro-stabilized platforms that was alluded to earlier. Most manufacturers will guarantee stabilization of their platform up to about $\pm 10°$ of pitch and 20° to 30° of bank. When flying within these restrictions, the radar will always provide a clear, undistorted picture of the area of interest selected by the operator. As these limits are exceeded, the gimbal limits of the radar will be exceeded and portions of the display will be lost. For example, if the stabilization limit is specified as 10° of pitch and you start to pitch the airplane up to 10°, picture will not be affected. As you hit 11°, you start to lose information from the bottom of the display (near range) because the stabilization is overcome by the excessive pitch. Similarly, if the aircraft is banked to the right and the bank limit is 20°, the picture will be unaffected up to 20° of bank. At 21° of right bank, you start to lose information on the left side of the display (up wing side) because the stabilization is overcome by excessive bank angle. Stabilization limits are thus determined by noting the pitch and bank angles where display information is lost.

The AC states that while in level flight at 10,000 ft or higher, adjust the tilt 2° to 3° above the point where the ground return was eliminated. Put the airplane through down pitch and then roll movements. No ground return should be present. A note is added that the aircraft should be rolled right and left approximately 15° and pitched nose down approximately 10°.

Ground Mapping. This test is accomplished to see if the weather radar can assist in navigating the aircraft by presenting the operator with a picture of the ground in front of the aircraft. High-resolution radars are able to provide a presentation that looks exactly like a map. Most weather radars suffer from large beam widths and long pulses that can not provide high-resolution ground returns. Large objects should be able to be identified, and therefore aid in the aircrew's situational awareness.

The AC directs that the aircraft fly over areas containing large, easily identifiable landmarks such as rivers, towns, islands, coastlines, etc. The evaluator would then compare the form of these objects on the indicator with their actual shape as visually observed from the cockpit.

Mutual Interference. This test is used to verify that no objectionable interference is present on the radar indicator from any electrical, radio, or navigation equipment, when operating, and that the radar installation does not interfere with the operation of any other equipment. Particular attention should be devoted to radar effect on electronic display systems and on localizer, glideslope, and ADF signals during landing approaches.

Electromagnetic Compatibility. The AC states that with all systems operating in flight, verify by observation that no adverse effects are present in the required flight systems. Both the previous paragraph (mutual interference) and this one (EMC) are really inadequate when considering EMI and EMC testing. We discussed EMI/EMC in sections 4.13 and 4.14 of this text, and I recommend that these procedures always be followed whenever you add, modify, or relocate antennas or equipment.

TABLE 6.9 ■ Weather RADAR Compliance Matrix

No.	Requirement		Subject	Description
1		8(a)	Warm-up Period	All tests should be conducted after the manufacturer's specified warm-up period.
2		8(b)	Display	Check the scope trace and sweep display move smoothly and without gaps or objectionable variations in intensity. Verify appropriate colors and color contrast.
3		8(c)	Range Capabilities	At 90% of maximum approved altitude, objects should be displayed at maximum range specified by the manufacturer; maneuvers should be conducted. The radar should be capable of displaying line-of-sight known prominent targets.
4	AC 25-7C.1301c	8(d-1)	Beam Tilting and Structural Clearance	The structural interference (caused by beam tilting) between the radome and antenna should not occur throughout the operational speed envelope of the airplane.
5		8(d-2)	Bearing Accuracy	Change the heading of the airplane in increments of 10° to ±30°maximum and verify the error does not exceed ±5 degrees maximum in the displayed bearing to the target.
6		8(e)	Stability	While observing a target return on the radar indicator, turn off the stabilizing function and put the airplane through pitch and roll movements. Observe the blurring of the display. Turn the stabilizing mechanism on and repeat the roll and pitch movements. Evaluate the effectiveness of the stabilizing function in maintaining a sharp display.
7		8(f-1)	Contour Display (Iso echo)	If heavy cloud formations or rainstorms are reported within a reasonable distance from the test base, select the contour display mode. The radar should differentiate between heavy and light precipitation.
8	AC 25-7C.1301c	8(f-2)	Contour Display (Iso echo)	Determine the effectiveness of the contour display function by switching from normal to contour display while observing large objects of varying brightness on the indicator. The brightest object should become the darkest when switching from normal to contour mode.
9		8(g)	Antenna Stability When Installed	While in level flight at 10,000 feet or higher, adjust the tilt approximately 2-3 degrees above the point where ground return was eliminated. Put the airplane through pitch (down 10°), then roll (±15°) movements. No ground return should be present.
10		8(h)	Ground Mapping	Fly over areas containing large, easily identifiable landmarks such as rivers, towns, islands, coastlines, etc. Compare the form of these objects on the indicator with their actual shape as visually observed from the cockpit.
11	AC 25-7C.1301c	8(i)	Mutual Interference	Determine that no objectionable interference is present on the radar indicator from any electrical or radio/navigation equipment, when operating, and that the radar installation does not interfere with the operation of any other equipment. Particular attention should be devoted to radar effect on electronic display systems and on localizer, glideslope, and ADF signals during approach to landing.
12		8(j)	EMC	EMC should be evaluated with all systems operating in flight to verify by observation that no adverse effects are present in the required flight systems.

AC 23-8C offers the same type of testing for the installation of weather radar in small airplanes, with some small differences. There are no specific tests called out for warm-up period or display in the AC. The maximum displayed target range for Part 23 installations is 80% of the maximum stated range. Bearing accuracy is $\pm 10°$ in small aircraft, as opposed to the $\pm 5°$ for transport aircraft. Under beam tilting, the Part 23 AC merely states that the antenna should have an elevation travel of $\pm 10°$. There is one addition to AC 23-8B under specific tests and that is that the display must be evaluated under all lighting conditions, including night and direct sunlight. This test is mentioned in AC 25-7C as required for all equipment covered in the section. All other verbiage is the same in Part 23 as it is in Part 25.

An example of a Weather RADAR Compliance Matrix is shown in Table 6.9.

6.11 | AIRWORTHINESS APPROVAL OF POSITIONING AND NAVIGATION SYSTEMS

6.11.1 Background

Conventional aircraft navigation is based on the use of ground-based navigation aids (VOR, DME, NDB, TACAN). This has resulted in a route structure anchored to these navigation aids and dependent on the locations of these aids. The current system is inefficient and does not lend itself to expansion to handle the predicted increase in air traffic. Because of the inability of the current system to expand easily, it became evident that exploitation of emerging technologies was needed. This strategy would, of course, need to maintain the quality and safety of air traffic. One of the concepts that has been pursued is area navigation, or RNAV. RNAV is a method of navigation that allow aircraft to operate on tracks joining any two points within prescribed accuracy tolerances; aircraft would no longer be required to overfly specific ground facilities (radio aids).

In 1990, RNAV was identified by the International Civil Aviation Organization (ICAO) Future European Air Traffic Management System (FEATS) as the future navigation system in the European region. FAA regulations have started to identify minimum navigation performance standards (MNPS) and minimum system performance standards (MSPS) as a condition for certification, as is the case with reduced vertical separation minimums (RVSM). Most importantly, aircraft without RVSM authorization or MNPS will not be allowed in RVSM or RNAV airspace.

RNAV is a method of navigation that permits aircraft operations on any desired flight path within the coverage of station-referenced navigation aids or the limits of self-contained aids. Airborne RNAV equipment automatically determines aircraft position from one or more sensors. Most RNAV equipment can provide lateral displacement from a desired track and provide inputs to the autopilot. B-RNAV is basic area navigation, and satisfies a track-keeping ability of ± 5 nm for 95% of the time. This level of accuracy is comparable to what can be obtained with conventional navigation aids when the navigation aids are less than 100 nm apart. B-RNAV can be equated to required navigation performance 5 nm (RNP 5). This method of track keeping may be expected to be in use until 2018 (phase out period for ground-based navigation aids).

P-RNAV or Precision Area Navigation is being incrementally implemented worldwide; P-RNAV is equivalent to RNP 1 or a Required Navigational Performance of +/−1 nm 95% of the time. All of these implementations are required to implement Free Flight, Free Flight 2000, Random Navigation, Capstone, etc. Air Traffic Control will be replaced with Air Traffic Management (ATM); The U.S. Air Force has determined a need to comply with new FAA directives and has established GATM (Global Air Traffic Management).

6.11.2 AC 20-138D

The backbone of the aforementioned enhancements is borne by the Global Positioning System, GLONASS, Galileo and other GNSS systems. Improvements such as the Wide Area Augmentation System (WAAS), Ground Based Augmentation System (GBAS), EGNOS and SBAS will build upon the basic satellite installations. Aircraft desiring to participate in restricted airspace must comply with the Minimum Performance Standards in the applicable TSOs and the Airworthiness requirements in AC 20-138D, Airworthiness Approval of Positioning and Navigation Systems, March 28, 2014.

AC 20-138 is an effort to consolidate all positioning systems within one document and provides the applicant with a means, but not the only means, of obtaining Airworthiness Approval for the following systems:

- Global Positioning equipment including those using GPS augmentations
- Area Navigation (RNAV) equipment integrating data from multiple navigation sensors
- RNAV equipment intended for Required Navigation Performance (RNP) operations
- Barometric Vertical Navigation (baro-VNAV) equipment

This AC replaces AC 20-138A, which only addressed stand-alone (TSO C-129 (AR)) and WAAS enabled TSO C-145c and TSO C-146c equipment; it also clarifies some details from earlier versions of AC 20-138.

The AC does not address satellite systems planned or in development; all positioning and navigation equipment and RNP Airworthiness guidance will be eventually be included in one AC. AC 20-138 incorporates Airworthiness guidance contained in AC 90-101A (RNP procedures with special aircraft and aircrew authorization required [SAAAR]) and AC 90-105 (RNP Operations and baro-VNAV in the NAS); the AC does not supersede the operational guidance of the RNP 90 series ACs. **Authors note**: SAAAR has since been replaced with the term AR (approval required) to make life simpler. This AC states that deviation from this AC by an alternate means must be found acceptable by the FAA (which is also a new development).

The following documents are cancelled by AC 20-138B:

- AC 20-129 (Airworthiness approval of VNAV systems)
- AC 20-130A (FMS using multiple sensors)
- AC 20-138A (GNSS equipment)
- AC 25-4 (INS Systems)

The AC supersedes the requirements contained in AC 90-45A, Approval of Area Navigation Systems for use in the U.S. NAS; a new section titled Frequently Asked Questions has been added to the version C update.

6.11.3 Positioning Systems TSOA

The applicant may apply for VFR or IFR operations; the normal basis for equipment approval is the TSOA. The AC makes note of the fact that while the TSO process operates on the basis of self-certification by the applicant, it is beneficial to involve the ACO because, as was noted previously:

- Obtaining a TSOA does not ensure that the equipment will satisfy all of the applicable requirements when installed.

- Equipment has become more complex and workload intensive with no standard interface with the pilot; that is, the human factors evaluation of the equipment is subjective.

The AC even goes as far as to recommend that manufacturers elect to obtain an STC for their equipment concurrent with the TSO process. This will ensure that the authority is involved with the process from the beginning. It should be noted that this is also a marked departure from the previous ways of doing business.

The TSOAs will come in many variations with respect to global navigation satellite systems (GNSS). TSO C-129 (AS) and TSO C-196a refer to stand-alone GPS receivers without SBAS and receivers that incorporate many of the processing improvements found in GPS/SBAS systems but without the GPS/SBAS requirements. TSO C-129a has been canceled and TSO C-196a incorporates the improvements of enhanced EMI/EMC protection in the presence of other GNSS systems and a selective availability aware wherein the receiver properly accounts for satellite range error if it is reflected in the user range accuracy index.

The original airworthiness certifications for GPS and GNSS systems came with limitations as the integrity, accuracy, availability, and continuity were not robust enough to offer a primary means of navigation, so other navigation systems were required to be on board. Flight over the ocean and in remote areas was authorized but required the applicant to install redundant systems and install a predeparture RAIM program. SBAS systems automatically meet most of the requirements levied on the earlier certifications, so in a sense, some of the compliance issues are easier for an applicant.

Three equipment classes could be approved under TSO C-129. The classes were based on capabilities (enroute and terminal navigation, nonprecision approach capability), whether or not they included an integrity monitor (RAIM or RAIM equivalent), and level of integration (GPS data to an FMS, or from an FMS to an autopilot). The equipment classes and their capabilities are summarized in Table 6.10.

As opposed to the TSO C-129 approvals, TSO C-145c and 146c equipment do not require other positioning and navigation systems to be on board. RTCA/DO-229D is the source for single-frequency GPS/SBAS MOPS; new MOPS and TSOs will be written for multifrequency/multisystem equipment. TSO C-145c and 146c equipment may be one of four classes within two groupings: functional and operational.

Functional classes are based on the TSOA:

- Class Beta is a TSO C-145c GPS/SBAS sensor and Class Gamma (complete sensor/ navigation system and Class Delta (ILS Look-alike) falls under TSO C-146c

- Class Beta-1 equipment consists of a sensor that determines position (with integrity) and provides this data to an integrated Navigation System (FMS). The equipment also provides integrity in the absence of other navigation sensor inputs such as GPS/WAAS through the use of FDE

TABLE 6.10 ■ GPS TSO C-129(AR) Equipment Classes

Equipment Class	Enroute and Terminal Navigation	Nonprecision Approach Capability	RAIM	Equivalent RAIM	Provides Data to an Integrated Navigation System	Provides Enhanced Guidance to an Autopilot
A1	Yes	Yes	Yes	No	No	No
A2	Yes	No	Yes	No	No	No
B1	Yes	Yes	Yes	No	Yes	No
B2	Yes	No	Yes	No	Yes	No
B3	Yes	Yes	No	Yes	Yes	No
B4	Yes	No	No	Yes	Yes	No
C1	Yes	Yes	Yes	No	Yes	Yes
C2	Yes	No	Yes	No	Yes	Yes
C3	Yes	Yes	No	Yes	Yes	Yes
C4	Yes	No	No	Yes	Yes	Yes

- Class Beta-2 equipment will provide deviations relative to a selected path; in this class of equipment, database capability, display outputs and pilot controls are necessary

- Class Beta-3 equipment includes the above plus VNAV when used with another sensor

- Class 4 Gamma and Delta equipment provides the above and has the capability of performing a precision approach

The operational classes of TSO C-145c/146c equipment may be summarized as follows:

- Class Beta-1 equipment supports oceanic and domestic enroute, terminal, non precision approach and departure operations.

- Class Beta-2 equipment provides a stand-alone navigation system.

- Class Beta-3 equipment performs the stand-alone navigation system when combined with another system.

- Class 4 equipment supports precision approach capability (ILS replacement). and will more than likely satisfy the equipment requirement for Class Beta-1, 2, or 3.

TSO C-161a/TSO C-162a equipment covers systems that use a ground-based augmentation system (GBAS) or systems that uplink GPS/GNSS corrections via a VHF data link. A comparable system would be the Australian ground-based regional augmentation system (GRAS). Some equipment may carry multiple TSOs as they satisfy multiple functions; the AC addresses that multiples are authorized.

TSO-C115c defines an acceptable certification standard for obtaining approval for multisensor navigation systems or FMS that integrate data from multiple navigation systems; TSO-C115b only addressed TSO-C129 GPS systems. If the applicant is integrating LP/LPV capability as in Class Gamma-3 or Class Delta-4, the equipment will require a TSOA IAW TSO-C115c. An upgrade to TSO C-115c is considered a major change.

6.11.4 Positioning Systems Airworthiness

AC 20-138 breaks down the airworthiness approval into three categories:

– Equipment performance
– Installation considerations
– Installed performance—test

It then addresses each functionality with respect to performance:

– Advisory vertical guidance
– GNSS
– RNAV multisensor equipment
– RNP (general)
– RNP approach
– RNP terminal
– Baro-VNAV

6.11.4.1 Equipment Performance

Advisory Vertical Guidance Positioning and navigation equipment may provide vertical path deviation guidance as an aid to pilots to meet barometric altitude requirements. The equipment typically uses GNSS or baro-VNAV but may use any other method to generate the flight path information. Advisory systems do not provide operational credit (i.e., cannot be used) for LNAV/VNAV or LPV approaches. When implemented, a limitation stating that the information is advisory only and that the barometric altimeter must be used to comply with altitude restrictions must be included in the flight manual. The AC states that manufacturers should consider a method to differentiate between vertical guidance and advisory vertical guidance to prevent confusion in the cockpit. There is no TSO guidance or MOPS standards for vertical advisory guidance or how this guidance is generated.

GNSS One of the largest concerns for GNSS equipment is the performance of such systems in the external noise environment. The broadband external interference noise is defined in RTCA/DO-235B. Earlier equipment (TSO-C129, TSO-C145c, TSO-C146c) may experience degradation as more GNSS satellites are launched. Any equipment intended for use beyond 2020 needs to meet the effective noise density requirements contained in RTCA/DO-235B. TSO-C129 and TSO-C196 GPS equipment may be used on RNAV T and Q routes and for RNAV approaches to LNAV minima within the United States (except Alaska) (see AC 90-100a and SFAR 97 for requirements).

T routes (low) are available for use by RNAV equipped aircraft from 1200' above the surface (in some instances higher) up to but not including 18,000' MSL. T routes are depicted on Enroute Low Altitude Charts (Figure 6.19).

Q routes (high) are available for use by RNAV equipped aircraft between 18,000' MSL and FL450 inclusive. Q routes are depicted on Enroute High Altitude Charts (Figure 6.20).

TSO C-129 and 196 sensors are positioning and navigation systems with equipment limitations for the aircraft to have other navigation equipment available (supplemental). These systems have limitations in oceanic and remote areas (FDE), and pre-departure

FIGURE 6.19 ■
T Routes

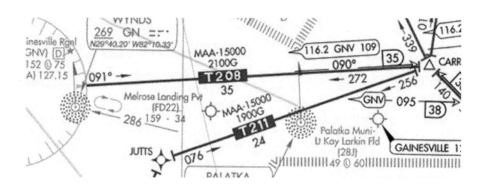

FIGURE 6.20 ■
Q Routes

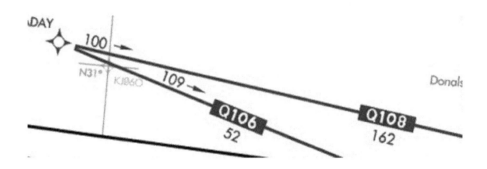

RAIM and FDE checks need to be performed. TSO C-129 equipment may have an FDE program with RAIM; TSO C-196 equipment must have an FDE program. RAIM predictions are available on-line such as the site at http://www.raimprediction.net.

For this equipment, the prediction program must:

- Have the capability of designating satellites that are off-line
- Allow the user to select a route, time and duration
- Identify the maximum RAIM or FDE outage time and any navigation capability outages
- For approaches, the maximum outage can not exceed 5 min
- For RNAV 1 and 2, 5 minutes (using GNSS only)
- For RNP-4, 25 minutes, 34 minutes for RNP-10 and 51 minutes for MNPS operations
- Provide RAIM or FDE availability over an interval of $+/-15$ minutes computed in intervals of 5 minutes or less about the ETA

The basis for GPS antenna certification was contained in TSO C-129a for GPS systems certified under AC 20-138. A newer TSO, TSO C-144, Airborne Global Positioning Antenna, was used with original TSO C-145/146 GNSS SBAS required antennas to be matched to receivers for older approvals. An even newer TSO (TSO-C190) is used for all new SBAS antennas under AC 20-138; manufacturers are also encouraged to use this new TSO when replacements are necessary. The basic functions and performance of the GPS antenna and the sensor must meet the requirements for compliance as outlined in the AC. This guidance is typically for the manufacturer as the equipment does not need to be installed on the aircraft for the evaluation. These tests deal with determining the signal loss along antenna cables, effects of lightning, and antenna minimum radiation pattern gain. A list of acceptable pairings of receiver type and antenna type (by TSO) is contained in Table 2 of AC 20-138.

GNSS/SBAS equipment has two different TSOs depending on whether the equipment is designed as a position sensor only (TSO-C145c, Class Beta) or a sensor/navigation computer combination (TSO-C146c, Gamma Class); TSO-C146c may also offer ILS look-alike approaches (Delta Class). When outside of SBAS coverage, the system may revert to FDE; there are no limits in oceanic and remote areas as long as the operator has a prediction program. Although GNSS/SBAS does not have a requirement for additional navigation systems on board the applicant's operating instructions should encourage the implementation of back-up navigation systems.

GNSS/SBAS equipment that complies with revision a, b, or c of the TSO has the capability of performing LP approaches. LP approaches use the horizontal accuracy and integrity values of the LPV approach but do not provide vertical guidance; the intent is to provide LP approaches only at airports where issues prevent issuing the LPV vertical guidance. LP and LPV approaches are mutually exclusive and will have separate approach plates as opposed to a localizer-only approach on an ILS plate.

For LNAV/VNAV approaches collocated with an LPV approach, the GNSS/SBAS equipment must use the final approach segment data block to define the final approach segment (i.e., uses WGS-84 height above ellipsoid). The vertical flight path for LNAV/VNAV approaches is defined by the threshold location, threshold crossing height and the glidepath angle (values in MSL); DA is always expressed in MSL. This database conversion can be done in the GNSS/SBAS equipment or by the use of lookup tables based on the approach selected. These GNSS/GBAS receivers support Cat 1 precision approaches and some a differentially corrected positioning service (DCPS). All GNSS/GBAS ground stations provide precision approach service; not all stations are expected to provide DCPS. TSO-C190, which references RTCA/DO-301 defines an acceptable antenna (VHF); antennas used for ILS localizer or VOR may be used, but require 360° omnidirectional coverage. For Cat I approaches, the equipment outputs lateral and vertical deviations mimicking an ILS; these parameters may only be output when the aircraft is within the precision approach envelope defined as D_{MAX}. When outside of D_{MAX}, the equipment may not use the broadcast differential corrections. GPS/GBAS approaches will not have LNAV/VNAV minimums published with the GLS procedure; there is no faildown option if guidance is lost during the GBAS final approach.

RNAV Multisensor   In general, the RNAV multisensor system may be thought of as an FMS. Each individual sensor must meet the applicable navigation performance and operational criteria contained in the respective TSO and ACs. For a DME/DME RNAV system to support RNAV1 and RNAV2 operations it must:

– Provide a position update within 30 sec of tuning a station
– Auto-tune multiple DME facilities
– Provide continuous DME/DME position updating
– When switching to a 3rd or second DME pair, there must be no interruption in positioning information

The DME/DME system may only use facilities listed in the Airport Facility Directory; facilities not listed should be excluded. DME facilities to be used should be at a relative angle of between 30° and 150°, expanded to 20°/60° for DME pairs. The limitations for use are:

- Greater than or equal to 3 nm from the facility
- Less than 40° look-up angle from the facility
- If an ARINC FOM is used, the DME usable ranges are provided in Table 3 of
- the AC
- Additional facilities such as a localizer/DME may be used
- Only facilities that meet validity checks may be used

A valid DME facility is defined as one that broadcasts an accurate identifier signal, satisfies the minimum field strength requirement, and is protected from other interfering DME signals. There is no requirement to use another navigation system during normal operation of a DME/DME system. Two DME systems meeting the validity requirements plus one other facility not meeting the criteria must provide a position accuracy of 1.75 nm; FTE is assumed to be 1 nm for RNAV 2 operations and the DME signal in space position error is assumed to be 0.1 nm. The RNAV system must ensure that co-channel DME facilities do not cause erroneous guidance. This may be accomplished by reasonableness checks that are most often accomplished by the FMS; it may also be accomplished by excluding a DME facility when there is a co-channel DME within LOS. The RNAV system must use operational DME facilities; if a system is flagged by NOTAM there must be a method of eliminating that facility from the navigation solution.

Multisensor equipment may also incorporate VOR/DME updating and if used, must meet the performance for the route being flown. Terminal and enroute RNAV implementation in the US does not require using VOR, so any use of the VOR must include reasonableness checks such as cross-checking with a DME/DME system. Positional accuracy for the VOR is 1.75 nm for RNAV 2 and 0.87 nm for RNAV 1. For the special case of using information from a single co-located VOR/DME, the total maximum position fixing error can not exceed the values published in Table 4 of the AC.

An INS/IRU is an acceptable lateral positioning source during a significant portion of flight as long as it meets the requirements described in 14 CFR Part 121, Appendix G, while in the inertial mode; an AHRS does not qualify as an acceptable positioning source. A more commonly implemented use today is an INS/IRU/GNSS integration. As of this time there is a limitation on IGI data output to support enroute through LNAV approach only; higher precision operations would require the integration of an SBAS or GBAS receiver or baro-VNAV installation. Position errors in the GNSS solution should not be able to corrupt the inertial states. All heading and attitude data must comply with the relevant regulations under all foreseeable operating conditions; these conditions should include those addressed in the aircraft FHA. If coasting is employed by the system, the manufacturer must document the coasting performance and limitations.

RNAV equipment may also employ an INS/IRU/DME/DME integration and this integration supports RNAV 1 and RNAV 2 operations. In addition to the DME/DME requirements previously mentioned, the FMS must be able to:

- Provide automatic position updating from the DME/DME
- Accept a position update immediately prior to takeoff
- Exclude VORs greater than 40 nm from the aircraft

The Total System Error (TSE) must be less than or equal to 1 nm throughout the route (95%) and the FTE for the integration on terminal procedures should be limited to 0.5 nm.

RNP General   The AC consolidates the Airworthiness criteria from AC 90-101A (Approval Guidance for RNP Procedures (IAP) with Special Aircraft and Aircrew Authorization Required (SAAAR)) and AC 90-105 (Approval Guidance for RNP Operations and Barometric Vertical Navigation in the U.S. NAS); RNP SAAAR has since been renamed RNP AR (authorization required). SBAS and GBAS equipped FMS systems can provide vertical path guidance for RNP 0.3 operations and RNP AR operations as low as RNP 0.1. Stand-alone GNSS sensors, with FMS, can provide the same operations if these systems are equipped with baro-VNAV. Augmented systems are superior to baro-VNAV systems in that they provide an accurate and repeatable vertical flight path and give a consistent alignment of the glide path angle and visual landing cues. Augmented systems do not need to be temperature corrected, meet RNP 0.1 vertical performance without independent system monitoring, and flight path is not affected by altimeter setting errors.

RNP Approach   GNSS is the primary navigation system to support RNP approach procedures; RNP approaches normally include two lines of minima: LNAV and LNAV/VNAV. Unless the pilot has the visual references in sight to continue the approach, it must be discontinued if:

– The navigation display is flagged invalid

– A loss of integrity function is presented

– The integrity alerting function is flagged before passing the FAF

For systems to be used for RNP Approach there must be a statement of compliance in the AFM, and sensors should be approved IAW their respective TSOs. The accuracy requirements for RNP Approach systems are as follows:

– During operations on the initial and intermediate segments of the approach as well as the missed approach, the TSE and the along track error must be within +/−1 nm (95%)

– During the final segment of the approach, the TSE and along track error must be within +/−0.3 nm (95%)

– The FTE should not exceed 0.5 nm on the initial, intermediate and missed approach segments of an RNP approach procedure; 0.25 nm on the final approach segment

A malfunction of the aircraft navigation equipment that causes the TSE to exceed 2x the RNP value without an annunciation is considered a Major failure condition. The loss of function is considered a Minor failure condition if the operator can revert to an alternate navigation system and safely proceed to a suitable airport. If the lateral error exceeds 2.0 nm/0.6 nm (initial/final) an alert must be given.

An analog deviation display (CDI, EHSI) with TO/FROM indications, failure annunciation, and used as a primary flight instrument should have the following attributes:

– Visible to the pilot in the primary FOV

– Lateral deviation display scaling should agree with alerting and annunciation limits

– Lateral deviation full-scale deflection should match the phase of flight

- The lateral deviation display must be slaved to the RNP path
- As an alternate means, a navigation map may be used if it provides equivalent functionality and is the primary FOV

For systems providing RNP approach, the following capabilities are required:

- Each pilot must be provided the RNP desired path and the aircraft position relative to the path on the primary navigation display
- A navigation database, and the expiry date must be displayed
- A means to retrieve and display data from the database
- A capability to load, by name, the entire RNP approach
- A means to display the distance between waypoints, the distance to a waypoint, along track distances, active navigation sensor type, identification of the active waypoint, groundspeed or time to waypoint and the distance and bearing to the active waypoint

Additional capabilities should include the following:

- A direct-to function
- Automatic leg sequencing with display to pilots
- Capability to execute fly-over and fly-by waypoints
- Fly-over points: Initial fix, track-to-fix, and direct fix
- Display an indication of failure to the pilot
- Indicate the navigation system error alert limit

The AC states that it is recommended that the autopilot or flight director be coupled for RNP approaches; if the lateral TSE can not be demonstrated without these systems then they must be used.

An important question to ask is, "Which minima may I use for the approach?" The answer lies in the installed equipment and which TSOA is assigned. The following TSOA/TSOA combinations allow the operator to use the LNAV minima:

- Stand-alone TSO C-129(AR) (Class A1) and TSO C-146(AR) (Class 1, 2, or 3)
- Multisensor TSO C-129(AR) (Class B1, B3, C1, C3), TSO C-196 sensors, TSO C-145 (AR) (Class 1, 2, or 3) or TSO C-161a sensors using GBAS positioning service

To use the LNAV/VNAV minima, the following TSOA/TSOA combinations are required:

- TSO C-146(AR) Class 2 or 3, TSO C-129(AR) (Class 1) with baro-VNAV
- Multisensor systems using TSO C-145(AR) (Class 2 or 3)
- Multisensor systems using TSO C-129(AR) (Class B1, B3, C1, or C3) with baro-VNAV or TSO C161a sensors with GBAS and baro-VNAV

An example of an RNAV Approach Plate is shown in Figure 6.21. The reader will note the difference in the minima between the GLS approach (lowest minima) to the LNAV approach (highest, not including circling).

RNP Terminal   RNP Terminal deals with RNP 1 operations in the terminal area (obstacle departure procedures, or ODP; standard instrument departures, or SID; and

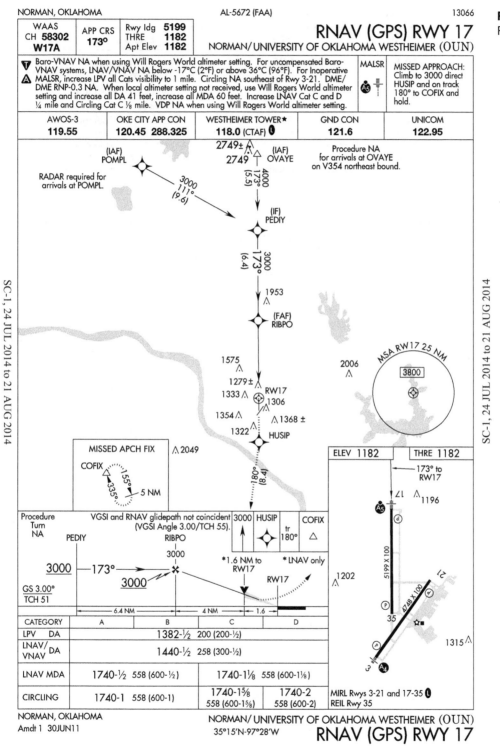

SC-1, 24 JUL 2014 to 21 AUG 2014

SC-1, 24 JUL 2014 to 21 AUG 2014

FIGURE 6.21 ■
RNAV Approach

NORMAN, OKLAHOMA   AL-5672 (FAA)   13066

WAAS CH **58302** **W17A**	APP CRS **173°**	Rwy Idg **5199** THRE **1182** Apt Elev **1182**

RNAV (GPS) RWY 17
NORMAN/ UNIVERSITY OF OKLAHOMA WESTHEIMER (OUN)

▽ Baro-VNAV NA when using Will Rogers World altimeter setting. For uncompensated Baro-VNAV systems, LNAV/VNAV NA below -17°C (2°F) or above 36°C (96°F). For Inoperative
△ MALSR, increase LPV all Cats visibility to 1 mile. Circling NA southeast of Rwy 3-21. DME/ DME RNP-0.3 NA. When local altimeter setting not received, use Will Rogers World altimeter setting and increase all DA 41 feet, increase all MDA 60 feet. Increase LNAV Cat C and D ¼ mile and Circling Cat C ⅛ mile. VDP NA when using Will Rogers World altimeter setting.

MALSR
As

MISSED APPROACH: Climb to 3000 direct HUSIP and on track 180° to COFIX and hold.

AWOS-3 **119.55**	OKE CITY APP CON **120.45 288.325**	WESTHEIMER TOWER★ **118.0** (CTAF) ◐	GND CON **121.6**	UNICOM **122.95**

CATEGORY	A	B	C	D
LPV DA		1382-½ 200 (200-½)		
LNAV/ VNAV DA		1440-½ 258 (300-½)		
LNAV MDA	1740-½ 558 (600-½)		1740-1⅛ 558 (600-1⅛)	
CIRCLING	1740-1 558 (600-1)		1740-1⅝ 558 (600-1⅝)	1740-2 558 (600-2)

MIRL Rwys 3-21 and 17-35 ◐
REIL Rwy 35

NORMAN, OKLAHOMA
Amdt 1 30JUN11

NORMAN/ UNIVERSITY OF OKLAHOMA WESTHEIMER (OUN)
35°15'N-97°28'W
RNAV (GPS) RWY 17

TABLE 6.11 ▪ RNAV FTE Requirements

RNP	FTE	FTE Basis
0.3	0.25	FD and Autopilot
1.0	0.5	FD and Autopilot
	0.8	Manual CDI Operation
2.0	1.0	Manual CDI Operation
4.0	2.0	Manual CDI Operation

Note: Navigation systems providing RNP performance monitoring and alerting (such as GNSS) can satisfy this FTE requirement. These systems must ensure the TSE for each segment is not exceeded; the integration of operational modes and configuration accounts for accuracy and FTE.

standard arrival routes, or STAR). All sensors need to be approved with respect to their applicable TSO; RNP aircraft with P-RNAV approval automatically meet the airworthiness requirements of the AC. The lateral and along-track TSE must be within +/−1 nm (95%); the FTE should not exceed 0.5 nm. A malfunction which is not announced and causes the TSE to exceed twice the RNP value is considered a major failure. The loss of function is considered a Minor failure if the operator can revert to a different navigation system and proceed to a suitable airport. The functional requirements for the displays and the system capabilities are the same as noted for RNP Approach operations. The system eligibility to perform RNP 1 and RNP 2 Terminal operations are the same as noted for RNP approach operations.

A summary of the FTE requirements for RNAV equipment is shown in Table 6.11.

Baro-VNAV   A Baro-VNAV system may be approved for use within the entire NAS; however, unless it is compensated for temperature, this equipment can only be used within the limitations for temperature published on the approach plates (a sample of an Approach Plate limitation is shown in Figure 6.22). Systems with temperature compensation must meet the requirements of RTCA/DO-236B Appendix H.2; limitations must be incorporated into the AFM. Baro-VNAV performance has shown to be inadequate for vertical guidance required for LPV or GLS approaches. Baro-VNAV may be used for vertical guidance during flight operations; the primary barometric altimeter will be used as the primary altitude reference.

The initial certification of a Baro-VNAV system requires an engineering evaluation to verify performance, failure indications and environmental qualifications. For enroute, terminal and approach IFR operations, the Baro-VNAV system must have TSE components in the vertical direction that are less than that shown in Table 6.12.

The altitude associated with the active waypoint should be to the nearest 100' (enroute and terminal) and 10' for the approach phase; if implemented, waypoint horizontal position should be to the nearest 0.1 nm. For VNAV, the ascent or descent gradient angles should be to the nearest 0.1°. Station elevation should be to nearest 1000' for enroute and terminal and to 100' for approach. The capability must exist to enter altitude for up to eight successive waypoints (TO-FROM equipment) or nine waypoints for TO-TO equipment. A means must be available to confirm that input data is correct prior to its use. The system should not give operationally misleading information, and the vertical guidance information should be compatible with the aircraft's flight instrumentation. There should be a continuous display of vertical path deviation with the performance identified in Table 6.13.

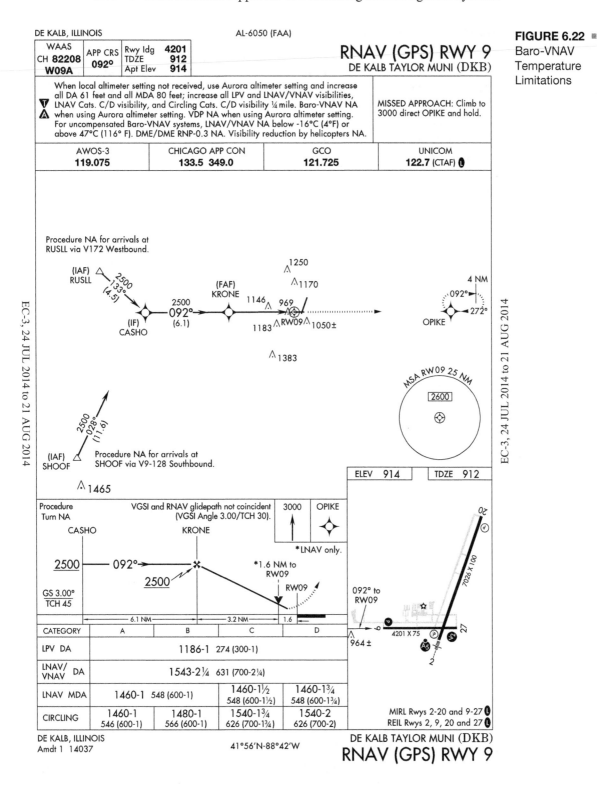

DE KALB, ILLINOIS
AL-6050 (FAA)

FIGURE 6.22
Baro-VNAV
Temperature
Limitations

TABLE 6.12 ▪ Baro-VNAV TSE Requirements

Altitude Region (MSL)	Level Flight Segments and Climb/Descent Intercept of Clearance Altitudes	Flight Along Specified Vertical Profile
At or Below 5000'	150'	160'
5000' to 10000'	200'	210'
10000' to 29000'	200'	210'
Above 29000' to 41000'	200'*	260'

* For aircraft type designs prior to January 1, 1997, the value is 200' in the cruise region and is not to exceed 250' over the full aircraft operating envelope, re RVSM

TABLE 6.13 ▪ Vertical Path Deviation Requirements

	Enroute/Terminal (feet)	Approach (feet)
Minimum Full Scale Deflection	≥500	≥150 *
Readability	≤100	≤30
Minimum Discernable Movement	≤10	≤5

Note: Smaller values of minimum full scale deflection for approach may be acceptable provided the proposed value is found satisfactory by an Engineering evaluation

Alert indications for the system should be located on or near the vertical path indicator and should provide a discernable annunciation for any of the following:

– Inadequate or invalid navigation signals
– Absence of primary power
– Inadequate or invalid displays or sources
– Equipment failures

The Baro-VNAV system should be able to provide navigation guidance within 20 seconds after input of desired vertical track, and navigation guidance should be available within 5 seconds of waypoint data input. The equipment should be demonstrated to function properly throughout the range of environmental conditions (RTCA/DO-160G contains acceptable tests). The crew shall have a means of determining the system status prior to flight, and the equipment shall have a means of annunciating a waypoint alert thus anticipating vertical maneuvering; systems connected to an autopilot or FD should not allow the aircraft to depart an assigned altitude until acknowledged by the crew. If parallel offset track capability is provided, a waypoint alert and vertical maneuver anticipation should be provided prior to arrival at the point where the offset track intersects the angle bi-sector of the parent track. The equipment should operate properly when interfaced with LNAV equipment providing turn anticipation and the software should be validated and verified to at least Level C.

6.11.4.2 General Installation Considerations

As with all avionics systems installations, a safety assessment is required. The failure level classifications for each of the navigation modes are shown in Table 6.14.

The system must comply with paragraph 1301: any probable failure must not degrade or adversely affect the normal operation or performance of other required

TABLE 6.14 ▪ Failure Level Classifications

Hazard \ Mode	Advisory Vertical Guidance	Enroute, Terminal Area Non Precision Approach LNAV or RNP 0.3	Non Precision Approach with Vertical Guidance LNAV/VNAV	LP/LPV Approach	GNSS Precision Approach Cat I
Loss of Navigation	No Effect	Major	Major	Major	Major
Misleading Information	Minor	Major	Major	Hazardous	Hazardous

Note: For RNP values less than 0.3, losing RNP capability constitutes a loss of navigation

equipment or create a hazard. Software is IAW RTCA/DO-178B (AC encourages the applicant to submit a Plan for Software Aspects of Consideration [PSAC]); RTCA/DO-178C has been issued but no current AC or TSO recognizes this revision. Hardware is IAW RTCA/DO-254 (AC encourages the applicant to submit a Plan for Hardware Aspects of Certification [HSAC]). Positioning and navigation manufacturers installation instructions should define compatibility with aircraft systems; manufacturers are encouraged to develop compatibility equipment lists.

Human factors considerations are covered in detail in the AC and the items of interest mimic the compliance matrix presented in section 6.9.3, Table 6.8. As mentioned previously, the applicant can use this compliance matrix with all avionic systems airworthiness certifications.

Antennas are covered in the general installation instructions. Typically, a GNSS antenna is located forward or aft of the wings on top of the fuselage; location should provide the largest, unobstructed view of the satellites. Shadowing by the aircraft structure and/or rotor blades can adversely affect the operation of the GNSS sensor. Location should be optimized to ensure the receiver can take full advantage of the $5°$ mask angle. The antenna should be separated as much as possible from other transmitting antennas and in the case of small aircraft as far away as possible from the windscreen to prevent case to antenna coupling.

The VDB antenna should be located so that the maximum received power from any on-board transmitter does not exceed the desensitization levels of the VDB receiver. For rotary wing aircraft, rotor modulation and multipath considerations may drive the location of the antenna. For multisensor installation, the installer should reduce the likelihood that a single lightning strike would affect all of the sensors (do not place all of the antennas in a straight line). If the aircraft is approved for flight in icing conditions, then the antenna may not be susceptible to ice build-up. Double-shielded cables should be used to prevent interference coupling into the cable.

In regard to the navigation sensors, for positioning and navigation inputs that drive a display, the pilot must be able to select the sensor and be apprised of which sensor is being used as well as what capabilities are being provided. Autopilot navigation modes related to the navigation source should be inhibited if the displayed source is not the same source driving the autopilot. If the displayed navigation source is displayed in the primary FOV, it need not also be displayed on the HUD (if installed). Dual installations should be synchronized whenever possible to reduce crew workload and to prevent confusion as to which system is driving the navigation data.

If the navigation equipment is not synchronized then:

- Either crew member should be able to enter and view data in the offside equipment.
- Workload to manually update both equipment sets should not be too burdensome.
- There should be no possible confusion as to which equipment is driving the displays
- Controls and displays should not lead to confusion or misleading information due to possible inconsistencies.

When integrating navigation sensors with the FGS, a compatibility list of approved FGS/positioning systems should be provided by the manufacturer. A clear indication of which FGS modes are engaged and which modes are armed must be provided to the pilot. The equipment must be compatible with the FGS modes of operation and cannot be limited by a positioning and navigation reduced bank selector during an approach. If buffet protection is not provided by the navigation system, or FGS, limitations must be noted in the AFM. The positioning and navigation should meet the performance requirements of the EFVS if installed.

If a Magnetic/True switch is installed in the aircraft, the navigation system should be driven by the same switch to maintain consistency. If there is an Air Data source switch, a de-selected source may not be used by the navigation equipment. If there is an inertial source switch, a de-selected inertial may not be used by the navigation equipment. If the equipment requires a barometric corrected altitude, the air data system is required to provide that information without additional pilot actions; if the corrected altitude is not available an alert to the pilot must be provided.

The AC addresses the importance of Electromagnetic Compatibility; grounding is essential, and when possible, don't install positioning sensors near a VHF radio. VHF communications and harmonics can cause interference, but can be mitigated by installing filters at the output of the VHF transmitter or isolation (by distance) of the navigation sensors. ELTs can also cause interference and can be mitigated by the use of notch filters on the ELT antenna cable. DME can cause interference, and sometimes a replacement can alleviate the problem. DF equipment may cause interference; relocating the DF antenna on top of the fuselage to the belly of the aircraft may help the problem.

6.11.4.3 Installed Performance-Test

Required Airworthiness tests include ground and flight tests. Since the equipment performance section of the AC is primarily the responsibility of the manufacturer, the applicant's airworthiness certification plan will be focused on the installation considerations and installed performance sections of the AC. If the applicant is installing RNAV multisensor equipment, he will also have to consult AC 25-15, "Approval of Flight Management Systems in Transport Category Airplanes," November 20, 1989. The sample compliance checklist for positioning systems uses multisensor RNAV equipment as an example and may be found in the flight management system section.

Ground Test—GPS, GPS/SBAS, GPS/GBAS   Freedom from interference due to VHF transmissions must be demonstrated; frequencies should be transmitted for 35 seconds while observing its effect on the GNSS system. The AC specifies the transmitting frequencies:

For 25 kHz channels:

121.150 MHz, 121.175 MHz, 121.200 MHz, 131.250 MHz, 131.275 MHz, 131.300 MHz

For 8.33 kHz channels:

121.185 MHz, 121.190 MHz, 130.285 MHz, 131.290 MHz

For installations on rotorcraft, the applicant should ensure that the rotor blades do not interfere with the received signals (will vary with the rotation rate). For equipment that supports LPV or a GLS approach, the MOPS requires an offset compensation for the navigation center to antenna offset (parallax correction). For stand-alone navigation equipment, the ground tests consist of ensuring the displayed parameters (e.g., HSI, CDI) are correct and conducting a failure modes and associated alerts and warnings evaluation.

Flight Test—GPS, GPS/SBAS, GPS/GBAS   Flight test will verify that navigation data is continuous during maneuvering with bank angles up to 30° and an FTE of 1.0 nm for enroute, approach, and transition and 0.25 nm for a nonprecision approach with and without autopilot can be maintained. Additionally, the overall GNSS functionality needs to be evaluated:

– Hold at a designated waypoint

– Intercept and track from a waypoint to a selected course

– Turn anticipation

– Waypoint sequencing

– Selection of an approach

– General presentation of navigation data

– Overall operation of all types of procedures

   If the GNSS is integrated with the FGS, then the following must be evaluated:

– Steering response with the FGS engaged through a series of track and mode changes to include enroute, approach and missed approach

– Proper execution of fly-by turns with varying wind conditions

– Verify that the lateral maneuver anticipation is appropriate for the type aircraft and that an annunciation is given

– Verify a Direct-To execution does not result in an overshoot or hunting

– Evaluate the FGS response to a GNSS fault by pulling the circuit breaker for the GNSS equipment (enroute and approach)

   If the equipment is approved for LPV and Precision Approaches, several approaches using raw data, FD and autopilot as applicable need to flown to ascertain compatibility with the integrated systems. If the autopilot has been modified it needs to be re-tested IAW the applicable autopilot/FD AC. For manual control to the approach flight path, the appropriate display must provide sufficient information to maintain approach path and alignment with the runway. The pilot should verify full-scale deflection while on the approach. The applicant shall evaluate the FGS approach functionality to ensure compatibility with the gain scheduling employed during an LPV approach.

   If the equipment uses barometric altitude input, verify that the equipment properly interprets the barometric reading; if manually input, verify the workload involved. Verify/assess all transfer and switching functions: electrical, alternate navigation sources, air data and inertial. A review of all failure modes and their associated

TABLE 6.15 ■ INS Error with Loss of Update

Time Since Last Update (T) (hr)	INS/IRU 95% Error (nm)
0.0 to 0.5 hr	8*T
0.5 to 1.5 hr	4

annunciations should be verified as operating correctly. A workload analysis when operating GNSS equipment should be conducted during all phases of flight.

If the equipment has GBAS capability, verify in-flight EMC compatibility between VHF communications equipment and the VDB; the test should be accomplished at the highest VDB frequency and transmitting VHF at 100 kHz above the VDB channel. The AC states that the applicant should use 118.0 MHz for VHF equipment that can not transmit below 118.0 MHz; VDB omnidirectional coverage should be demonstrated in-flight.

Flight Test—RNAV Multisensor   Flight tests are accomplished to verify proper operation and accuracy of the multisensor equipment. EMI/EMC tests are required, some of which may be accomplished as a ground test. The applicant must validate multisensor equipment navigation accuracy in each operating mode; the performance of each navigation sensor should be evaluated separately and in combination with other sensors as applicable. Initial certification of VOR/DME or a multiple (scanning) DME sensor requires a demonstration of accuracy by collecting the sensor position at 15 minute intervals and comparing this position to the true aircraft (TSPI) position.

The system should demonstrate its ability to detect, for example, poor signal conditions, inadequate navigation capability, and recovery from in-flight power failure. The auto-tune logic should be reviewed and tested to verify the ground stations are identified and tuned correctly. The growth in position error when reverting to INS/IRU can be expected to be less than 2 nm per 15 minutes; if an applicant desires certification credit for better performance it must be coordinated with the ACO. INS accuracy after a loss of updating is shown in Table 6.15.

Data continuity during normal aircraft maneuvering for the navigation modes must be validated (bank angles up to 30° and pitch angles associated with approaches, departures and missed approaches). The applicant must verify the FTE: 1.0 nm enroute, 0.5 nm approach transition and missed approach and 0.25 nm for a nonprecision approach (with and without FGS). Additionally, verification of all modes of operation, barometric input and switching functions and the operation and annunciation of failure modes must be accomplished.

The interface with the FGS requires an evaluation. The applicant shall evaluate the steering response and display sensitivities with the FGS engaged during a variety of track and mode changes. The test should include several fly-by turns with varying wind conditions to verify the equipment accomplishes the turn as a fly-by waypoint and discourages an overshoot. Lateral maneuver anticipation should be appropriate and the correct annunciation should be provided. Direct-to-steering should not cause an overshoot or hunting. The response to a multisensor fault by pulling a circuit breaker for the equipment should be done for all sensor modes.

For equipment that provides precision approach capability, or for nonstandard displays, conduct several approaches using raw data, FD only and fully coupled to evaluate compatibility. For manual control to the approach path, the appropriate display must provide sufficient information for the approach and runway alignment without reliance on other displays.

Ground Test—Baro-VNAV   The following ground tests should be accomplished with the installation of a Baro-VNAV system:

- Analysis of the manufacturer's procedures for V/V of S/W and associated documentation
- Verification of compliance with RTCA/DO-160F
- Examination of the display capabilities with an emphasis on CAW
- Analysis of failure modes
- Review of reliability data to establish that all probable failures are detected
- Evaluation of the controls, displays, annunciations and system behavior from a Human Factors perspective (see section 6.9.3)
- Evaluation of the pilot's guide
- Review of the installation drawings, wiring diagram and descriptive wire routing
- Analyze data flow diagram to review how data are passed from device to device
- Review the structural analysis report
- Evaluate the integration with the associated lateral navigation equipment

Flight Test—Baro-VNAV   Flight Test of the Baro-VNAV system will include functional checks and a navigation accuracy test. Functional flight tests include the following:

- Evaluation of all Baro-VNAV equipment operating modes
- Examination of the interface function of other equipment connected to the baro-VNAV system
- Review of the failure modes and proper annunciations
- Evaluation of the steering response of the FGS during a series of lateral and vertical track changes
- Evaluation of the displayed baro-VNAV navigation parameters on interfaced Flight Deck instruments (e.g., HSI, CDI)
- Assessment of all switching functions
- Verification of EMC
- Verification of the accessibility of controls and display visibility
- Analysis of aircrew workload when using baro-VNAV equipment

Navigation error is computed by comparing Baro-VNAV position to an accurate TSPI source; tests should be accomplished using a variety of descent rates, angles and lateral navigation source inputs. The accuracy requirement is on a 99.7% probability basis. Flights should verify proper operation of caution indications and lateral navigation interface and normal flight maneuvers should not cause a loss of system sensor inputs. The system dynamic response should be confirmed, and the applicant should note any unusual FTE or errors from using the autopilot and FD with the baro-VNAV system.

Aircraft Flight Manual   The operations manual must address the operation of the equipment and should also cover the operation of related equipment. If there is a limitation associated with the operation of the equipment, a sample airplane/rotorcraft flight manual supplement should be provided. Sample quick reference guides and flight manual supplements can be found in Appendix 5 of the AC.

VFR Installation of GNSS Equipment   GNSS equipment may be installed on a no-hazard basis as a supplement to VFR navigation; such installations need only to verify that the GNSS installation does not introduce a hazard to the aircraft. Loss or misleading VFR navigation information is considered a minor failure; therefore, Software Development Assurance Level D is acceptable. A readable placard stating "GPS limited to VFR use only" must be installed in clear view of the pilot. A change to the flight manual is not required since the placard conveys the limitations. Guidance for VFR use only is contained in Appendix 6 of the AC.

■ 6.12 | REDUCED VERTICAL SEPARATION MINIMUMS

Almost universally, there exists restricted airspace between FL 290 and FL 410 inclusive. Within this airspace, aircraft separation is reduced to 1000 ft vertically; aircraft and operators can only operate in this airspace via RVSM approval. Within the United States, this airspace is known as domestic RVSM (DRVSM) and has been active since January 20, 2005. The guidance material on the authorization of operators and aircraft for RVSM operations is found in AC 91-85, Authorization of Aircraft and Operators for Flight in Reduced Vertical Separation Minimum Airspace, August 21, 2009. The rules for operating in RVSM airspace are found in Part 91, sections 91.180, 91.706 and appendix G (RVSM Operations). It should be noted that aircraft approved per Appendix G are eligible for RVSM operations worldwide. The airspace where RVSM is applied is considered special qualification airspace. In this respect, both the specific operator and the specific type of aircraft the operator intends to use needs to be authorized by the appropriate FAA office before conducting RVSM operations. The reader will note that the operator and aircraft are granted authorization vice an approval; the applicant shows that the aircraft can meet the requirements for operating in this special airspace.

In general, the operator must be able to prove that the aircraft is capable of maintaining altitude to within certain tolerances. These tolerances vary based on where the aircraft operates and whether the aircraft is part of an aircraft group or unique in its configuration. An aircraft group is defined as a group of aircraft that are of nominally the same design and build with respect to all details that could influence the accuracy of height-keeping performance. A nongroup aircraft is an aircraft for which the operator applies for authorization on the characteristics of the unique airframe rather than on a group basis.

For an aircraft to be considered a member of a group for the purpose of RVSM approval, they must meet the following conditions:

- Aircraft should have been manufactured to a nominally identical design and be approved by the same TC, TC amendment, or supplemental TC.

- The static system of each aircraft should be installed in a nominally identical manner and position; the same static source error (SSE) corrections should be incorporated in all aircraft of the group.

- The avionics units installed on each aircraft must meet the minimum RVSM equipment requirements (detailed shortly) and should be manufactured to the manufacturer's same specification and have the same part number (different avionics units may be used if the applicant can demonstrate that the replacement provides equivalent performance).

- The RVSM data package is produced or provided by the airframe manufacturer or design organization.

If an airframe does not meet the four conditions, it must be considered a nongroup aircraft.

In addition to the classification of type of aircraft, the applicant must also state whether the authorization is for the full RVSM envelope or for a basic RVSM envelope. The full RVSM envelope encompasses the entire range of operational Mach numbers, W/δ, and altitude values over which the aircraft can be operated within RVSM airspace. It would be difficult to show all of the gross weight, altitude, and speed conditions that constitute the RVSM envelope on a single chart; a separate chart of altitude versus Mach would be required for each aircraft gross weight. For most jet transports, the required flight envelope can be collapsed to a single chart by the use of the parameter W/δ (weight divided by atmospheric pressure ratio). This is due to the relationship between W/δ and the aerodynamic variables of Mach and lift coefficient. The mathematical explanation of W/δ is shown in Equation (6.2).

$$W/\delta = 1481.4 C_L M^2 S_{REF}, \tag{6.2}$$

where

δ = ambient pressure at flight altitude divided by sea level standard pressure,
W = weight,
C_L = lift coefficient,
M = Mach number, and
S_{REF} = reference wing area.

The basic RVSM envelope encompasses the range of Mach numbers and gross weights within the altitude ranges FL 290 to FL 410 (or maximum available altitude), where an aircraft can reasonably be expected to operate most frequently. The RVSM operational flight envelope for any aircraft may be divided into two zones, as shown in Table 6.16.

TABLE 6.16 ■ Full RVSM Envelope Boundaries

	Lower boundary is defined by	Upper boundary is defined by
Altitude	FL 290	The lower of the following: 1. FL 410 2. Airplane maximum certified altitude 3. Altitude limited by cruise, thrust, buffet, or other aircraft limitations
Mach or Speed	The lower of the following: 1. Maximum endurance (holding) speed 2. Maneuver speed	The lower of the following: 1. M_{MO}/V_{MO} 2. Speed limited by cruise, thrust, buffet, or other aircraft limitations
Gross Weight	The lowest gross weight compatible with operation in RVSM airspace	The highest gross weight compatible with operation in RVSM airspace

The boundaries for the basic RVSM envelope are the same as those for the full RVSM envelope except in regard to the upper Mach boundary, which may be limited to a range of airspeeds over which the aircraft group can reasonably be expected to operate most frequently.

6.12.1 RVSM Minimum Equipment Requirements

The minimum equipment requirements for RVSM operations are

- Two independent altitude measurement systems, each of which must have the following elements:
  - Cross-coupled static source/system provided with ice protection if located in areas subject to icing
  - Equipment for measuring static pressure sensed by the static source, converting it to pressure altitude and displaying the pressure altitude to the aircrew
  - Equipment for providing a digitally coded signal corresponding to the displayed pressure altitude for automatic altitude reporting
  - Static source error correction, if needed
  - Reference signals for automatic control and alerting at selected altitude
- One altitude reporting transponder. If only one is fitted, it must have the capability of operating from either of the altitude measurement systems.
- An altitude alert system.
- An automatic altitude control system.

In addition to the minimum required equipment, the aircraft must satisfy RNP 5.

6.12.2 RVSM System Performance

The requirements for group aircraft are as follows:

- At the point in the basic RVSM envelope where the mean altimetry system error (ASE) reaches its largest absolute value, the absolute value should not exceed 80 ft (25 m).
- At the point in the basic RVSM envelope where the mean ASE plus 3 standard deviations reaches its largest absolute value, the absolute value should not exceed 200 ft (60 m).
- At the point in the full RVSM envelope where the mean ASE reaches its largest value, the absolute value should not exceed 120 ft (37 m).
- At the point in the full RVSM envelope where the mean ASE plus 3 standard deviations reaches its largest absolute value, the absolute value should not exceed 245 ft (75 m).

The requirements for nongroup aircraft are as follows:

- For all conditions in the basic RVSM envelope, | residual SSE + worst case avionics | $\leq$160 ft (50 m).
- For all conditions in the full RVSM envelope, | residual SSE + worst case avionics | $\leq$200 ft (60 m).

The term ASE is defined as the difference between the pressure altitude displayed to the flight crew when referenced to ISA standard ground pressure setting and the free stream pressure altitude. In order to determine the mean and three sigma values of the ASE it is necessary to take into account the different ways in which variations in ASE can arise. The factors that affect ASE are

- Unit-to-unit variability of avionics
- Effect of environmental operating conditions on avionics
- Airframe-to-airframe variability of SSE
- Effect of flight operating condition on the SSE

The assessment of the ASE may be measured or predicted, but must address all factors. The effect of flight operating condition as a variable can be eliminated by evaluating ASE at the most adverse flight condition in an RVSM flight envelope. It is possible for an aircraft group to obtain RVSM authorization when the ASE or three sigma ASE does not meet the requirement in certain flight regimes of RVSM airspace. In this case, the restriction must be identified in the data package and annotated in the appropriate aircraft flight manuals; a visual or aural warning system noting this restriction is not required to be installed on the aircraft.

For nongroup aircraft, there is no mean or three sigma value of the ASE because there can be no group data to identify airframe-to-airframe variability. A single ASE value is established for nongroup aircraft, which is the simple sum of the altimetry system errors. In order to control the overall population distribution, the limit is set less than that for group aircraft.

The residual SSE is the amount by which SSE remains uncorrected or overcorrected after the application of static source error correction (SSEC). Worst case avionics is the combination of tolerance values, specified by the manufacturer for the altimetry fit into the aircraft, which gives the largest combined absolute value for residual SSE plus avionics errors.

There are additional tolerances that must be demonstrated to obtain RVSM authorization. An automatic altitude control system is required and should be capable of controlling altitude within ±65 ft (20 m) about the acquired altitude when operated in straight and level flight under nonturbulent, nongust conditions. This requirement is relaxed to ±130 ft (40 m) for aircraft types with certification or major changes in type design on or before April 9, 1997, which are equipped with an automatic altitude control system or FMS that allow variations of up to 130 ft. Where an altitude select/acquire function is provided, the altitude select/acquire control pane must be configured such that an error of no more than ±25 ft (8 m) exists between the display selected by the flight crew and the corresponding output to the control system.

The altitude deviation warning system should signal an alert when the altitude displayed to the flight crew deviates from the selected altitude by more than a nominal value. For aircraft for which application for type certification or major change in type design is on or before April 9, 1997, the nominal value shall not be more than ±300 ft (90 m). For aircraft with an application for type certification or major change in type design after this date, the nominal value shall not be more than ±200 ft (60 m). The overall equipment tolerance in implementing these nominal threshold values shall not exceed ±50 ft (15 m).

During the RVSM approval process it must be verified analytically that the projected rate of occurrence of undetected altimetry system failure does not exceed

1×10^{-5} per flight hour. All failures and failure combinations whose occurrences would not be evident from cross-cockpit checks and which would lead to altitude measurement/display errors outside the specified limits need to be assessed against this budget.

6.12.3 RVSM Airworthiness Authorization

The RVSM airworthiness authorization occurs in two steps: the manufacturer or design organization develops the data package that is sent to the ACO; once approved, the operator applies the procedures defined in the package to obtain approval from the CMO or FSDO to conduct RVSM operations.

As previously discussed, the initial work is to define the group and type envelope under which the applicant is seeking approval. The data package will specify the group and envelope and contain the data needed to show compliance. It will also contain the compliance procedures to be used to ensure that all aircraft submitted for airworthiness approval meet RVSM requirements and the engineering data to be used to ensure continued in-service RVSM approval integrity.

Since the ASE will vary with flight condition, the data package should provide coverage of the RVSM envelope sufficient to define the largest errors in the basic and full RVSM envelopes. Where precision flight calibrations are used to quantify or verify altimetry system performance, they may be accomplished by any of the following methods:

- Precision tracking radar in conjunction with pressure calibration of the atmosphere at test altitude
- Trailing cone
- Pacer aircraft (the pacer aircraft must be calibrated to a known standard; it is not acceptable to calibrate a pacer aircraft with another pacer aircraft)
- Any other method acceptable to the FAA or approving authority

NOTE: If you recall the TSPI section, you will realize that the first method has an inherent error in the radar. The error budget will grow as the distance from the tracking radar increases. Care should be taken when using this method, as the error budget of the radar may, in fact, exceed the maximum allowable error for the ASE.

When authorization is sought for group aircraft, the data package must be sufficient to show that RVSM requirements have been met. Because of the statistical nature of the requirements, data packages will vary from group to group. Other considerations for aircraft group approval include:

- The mean and airframe-to-airframe variability of ASE should be established based on precision flight test calibration of a number of aircraft. Where analytical methods are available, it may be possible to enhance the flight test database and to track subsequent change in the mean and variability based on geometric inspections and bench tests or any other method acceptable to the approving authority.
- An assessment of the aircraft-to-aircraft variability of each error source should be made. For some error sources (especially small ones) it may be acceptable to use specification values to represent the three sigma values; for other error sources (especially larger ones), a more comprehensive assessment may be required.
- In many cases, one or more of the ASE sources may be aerodynamic in nature (such as variations in the surface contour in the vicinity of the static pressure source).

If evaluation of these errors is based on geometric measurements, substantiation should be provided that the methodology used is adequate to ensure compliance.

- In calculating the ASE, all error sources should be summed and an algebraic solution of the mean should be used to show compliance.

When authorization is sought for nongroup aircraft, the following guidelines should be followed:

- The data package should specify how the ASE budget has been allocated between residual SSE and worst case avionics. The applicant and approval authority agree on what data is needed to satisfy the approval requirements.

- Precision flight test calibration of the aircraft to establish its ASE or SSE over the RVSM envelope should be required. Flight calibration should be performed at points in the flight envelope(s) as agreed by the certifying authority.

- Calibration of the avionics used in the flight test should be accomplished as required to establish the residual SSE; the number of test points should be agreed by the certifying authority.

- Specifications for the installed altimetry avionics equipment indicating the largest allowable errors will be included in the data package.

- If subsequent to aircraft approval for RVSM operations avionics units that are from a different manufacturer are fitted, it should be demonstrated that the standard of avionic equipment provide equivalent altimetry system performance.

In all cases, the data package must contain a definition of the procedures, inspections/tests, and limits that will be used to ensure that all aircraft approved against the data package "conform to type," guaranteeing that future builds continue to meet the budget allowances stated in the data package.

For continued airworthiness, the following items should be reviewed and updated as appropriate to include the effects of RVSM implementation:

- The structural repair manual, with special attention to the areas around the static source, angle of attack sensors, and doors if their rigging can affect airflow around these sensors.

- Master minimum equipment list (MMEL).

- For nongroup aircraft where airworthiness approval has been based on flight test, the continuing integrity and accuracy of the altimetry system shall be demonstrated by periodic ground and flight tests of the aircraft at intervals agreed on with the approving authority. (Exception to this requirement may be given if it can be adequately demonstrated that the relationship between any subsequent airframe/system degradation and its effects on altimetry system accuracy are understood and adequately compensated/corrected for.)

- In-flight defect reporting procedures should be defined to facilitate identification of altimetry system error sources.

- For group aircraft where approval is based on geometric inspection, there may be a need for periodic reinspection, and that interval needs to be defined.

- Any variation/modification from the initial installation that affects RVSM approval should require clearance by the airframe manufacturer or design organization and be cleared by the FAA to show that RVSM compliance has not been compromised.

Each operator requesting RVSM operational approval should submit a maintenance and inspection program that includes any maintenance requirements necessary for continuous airworthiness. At a minimum, the following items need to be reviewed as appropriate for RVSM maintenance approval:

- Maintenance manuals
- Structural repair manual
- Standard practices manuals
- Illustrated parts catalog
- Maintenance schedules
- MMEL/MEL

The purpose of the maintenance and inspection program is to ensure the integrity of the approved RVSM altimetry system. The maintenance plan addresses the need for inspections and tests anytime the system is repaired or modified. The AC calls out the particulars that must be addressed in the plan.

Those aircraft positively identified as exhibiting height-keeping performance errors that require investigation should not be operated where RVSM is applied until the following actions have been taken:

- The failure or malfunction is confirmed and isolated by maintenance.
- Corrective action is carried out as required to comply with the ASE requirements and verified to ensure RVSM approval integrity.

6.12.4 RVSM Operational Approval

As previously mentioned, operational approval is granted to each individual operator as well as the aircraft. This entails operator knowledge of operational programs, practices, and procedures when operating in RVSM airspace. There is an RVSM homepage (http://www.faa.gov/ats/ato/rvsm1.htm) that provides current guidance and information on the aircraft and operator approval process as well as the RVSM monitoring program. Before an operational approval is granted, the FAA should be satisfied that operational programs, flight crew training and knowledge, and operations manuals are adequate. RVSM approval is an approval for RVSM operations worldwide.

Operations training programs and operating practices and procedures should be standardized for the following:

- Flight planning
- Preflight procedures at the aircraft for each flight
- Procedures prior to RVSM airspace entry
- In-flight procedures
- Flight crew training procedures
- Wake turbulence procedures
- TCAS/ACAS operations in RVSM airspace
- Oceanic contingency procedures
- Postflight

These procedures are discussed in detail in Appendices 4 and 5 of AC 91-85 and are also available on the RVSM homepage. The training of operators must be reviewed by the approval authority. 14 CFR Part 121, 125, and 135 operators should submit training syllabi and other appropriate training items related to RVSM operations to show that the practices and procedures are included in initial and recurrent training programs. 14 CFR Part 91 operators should show the FAA that pilot knowledge of RVSM operating practices and procedures is adequate to warrant approval for RVSM operations. Acceptable proof is described in the AC.

Operations manuals and checklists must also be submitted as part of the operational approval process. They are checked to ensure that the standardized practices and procedures previously identified are included. An operating history should also be included in the application, including any events or incidents related to poor height-keeping performance that may indicate weaknesses in training, procedures, maintenance, or the aircraft group intended to be used. The applicant should also provide a plan for participation in the RVSM monitoring program. In some cases, the FAA may request a validation flight as the final step in the approval process. The FAA may accompany the operator on a flight through airspace where RVSM is applied to verify that operations and maintenance procedures and practices are applied effectively.

Authorization is issued to operators by one of two means: for 14 CFR Part 121, 125, and 135 operators, approval to operate in RVSM airspace is granted by a change in the flight manual, operations specification paragraph Part B (enroute) and Part D (aircraft maintenance); 14 CFR Part 91 operators are granted approval to operate in RVSM airspace via a letter of authorization (LOA), which is valid for 24 months.

The operator should provide a plan for participation in the RVSM monitoring program. This program should normally entail at least a portion of the operator's aircraft in an independent height-monitoring program. This independent system may be either a Height Monitoring Unit (HMU), aircraft geometric height measurement element (AGHME) or a GPS monitoring unit (GMU). HMUs are located at Linz in Austria, Nattenheim in Germany, and Geneva in Switzerland. The HMU radius of operation is 30 nm for Linz, 45 nm for Nattenheim, and 45 nm for Geneva. The HMU is an automated height measuring device and is scheduled by coordination through the flight plan and London Control.

AGHMEs are ground-based radar tracking units that compare reported IFF altitude to the radar-based derived altitude. There are six sites located in North America: Atlantic City, New Jersey, Wichita, Kansas, Cleveland, Ohio, Phoenix, Arizona, Ottawa, Ontario, and Lethbridge, Alberta. An AGHME application needs to be forwarded to the appropriate constellation site (available on the AGHME website); no coordination with ATC is necessary.

The GMU is a portable GPS system that is carried on board the monitored aircraft. The unit records GPS position, which is compared postflight to IFF received altitude enroute. In the United States, the GMU program is overseen by ARINC. The aircraft must have an operable mode S transponder to take advantage of the AGHME.

6.12.5 Error Reporting and Loss of RVSM Authorization

Since the incidence of altimetry errors that can be tolerated in the RVSM environment is very small, it is incumbent upon operators to immediately take corrective action to rectify the conditions that caused the error. The operator must report the event to the

FAA within 72 hr of the occurrence with initial analysis of causal factors and measures to prevent further events; the requirement for follow-up reports is at the discretion of the FAA. Errors that should be reported are total vertical error (TVE) equal to or greater than ± 300 ft (90 m), ASE equal to or greater than ± 245 ft (75 m), and assigned altitude deviation (AAD) equal to or greater than ± 300 ft (90 m).

Height-keeping errors fall into two categories: equipment malfunction and operational errors. An operator who consistently commits errors of either variety may be required to forfeit approval for RVSM operations. If a problem is identified that is related to one specific aircraft type, then RVSM approval may be removed for that one specific aircraft type.

The FAA may consider removing RVSM operational approval if the operator's response to a height-keeping error is not effective or timely. The operator's past performance may be a factor in deciding to remove operational approval. Authorization may be removed until the root causes of the errors are shown to be eliminated and RVSM programs and procedures are shown to be effective.

▌ 6.13 ▐ | PROXIMITY WARNING SYSTEMS

This section will cover the evaluation and approval of aircraft proximity warning systems, either with the earth or other aircraft. We will look at the traffic alert and collision avoidance system (TCAS), traffic advisory system (TAS), automatic dependent surveillance broadcast (ADS-B), ground proximity warning system (GPWS), and terrain awareness warning system (TAWS).

6.13.1 Traffic Alert and Collision Avoidance System

The TCAS is designed to alert aircrew to the potential for conflicts with other aircraft within the area. The system uses the existing ATCRBS system and the capabilities of mode S transponders to coordinate with other TCAS-equipped aircraft. The TCAS provides two types of advisories to the aircrew:

- Traffic advisory (TA): informs the crew that there is traffic in the area.
- Resolution advisory (RA): advises the aircrew that corrective action is necessary to avoid a collision with an intruder aircraft.

Efforts to establish a collision avoidance system started in 1955 within the air transport industry. The basic TCAS concept was implemented by the FAA in 1981, and TCAS was evaluated aboard Piedmont and United Airlines aircraft through the mid-1980s. The operational evaluation lasted until 1988. In 1991, the FAA mandated use of TCASII (Version 6.0) in all aircraft with 30 or more seats as a result of a DC-9/private aircraft midair over CA; the deadline was extended to 1993. An Operational Evaluation of 6.0 was conducted and suggested improvements were incorporated into Version 6.04a (1993). The principal changes included the reduction of nuisance alerts and altitude crossing logic.

The results of the 6.04a evaluation indicated that the actual vertical displacement of an RA response was much greater than 300 feet, which was having an adverse effect on controllers and the ATC system. The required changes resulted in the release of Version 7.0

in 1999. Version 7.0 incorporated numerous changes and enhancements to the collision avoidance algorithms, aural annunciations, RA displays and pilot training programs. The main purpose of these changes was to reduce the number of RAs issued and to minimize the altitude displacement when responding to an RA.

TCAS Version 7.0 performance has been analyzed since 2000 and has resulted in Version 7.1; the MOPS was released in 2008. This release attempts to improve the RA logic that was causal in the near mid-air over Japan in 2001 and the fatal mid-air on the German-Swiss border in 2002. In these cases, the pilots maneuvered opposite to the displayed RA; studies showed that pilots occasionally maneuvered in the opposite direction from the Adjust Vertical Speed, Adjust RA. To mitigate this problem all Adjust Vertical Speed RAs have been replaced with Level Off, Level Off. Versions 6.04a and 7.0 are expected to remain in operation for the foreseeable future.

All US aircraft with more than 30 seats must have TCASII, and all installations of TCASII in the United States after 1995 must be with Version 7.0. Since 1995, all European civil fixed-wing aircraft of 5700 kg or maximum seating of 19 or more must have TCASII Version 7.0. Other countries including Argentina, Australia, Chile, Egypt, India, and Japan have also mandated TCASII within their respective airspace. US military, under GATM, have also installed TCASII on their transport aircraft. TCASII Version 7.0 or higher must be installed if used in RVSM airspace.

TCAS equipment works the same way as the ground-based IFF system described in section 5.8.2 of this text. A TCAS I system sends out an omnidirectional interrogation pulse and waits for replies from other aircraft within the range of the signal (on the order of 50 nm). We know from previous discussions that the range to an aircraft can be determined from the time it takes to receive a reply. The bearing of the return is measured much like the ADF systems mentioned previously. By successive interrogations, the system can determine range rate (by computing the rate of change of range with respect to time). Similarly, the system knows the altitude of the replying aircraft by examining the altitude-encoded reply. By successive interrogations, the aircraft can also compute the vertical velocity (by computing the rate of change of altitude with respect to time).

The TCAS II system has the added benefit of communicating with other aircraft via the mode S system. The communication between two TCAS II receivers allows maneuvering commands to be issued to the aircrew in the case of a conflict ("you climb and I will descend"). The relationship between the TCAS II processors and the mode S interrogators for two cooperating TCAS II aircraft is shown in Figure 6.23.

The warnings are given to the aircrew based on a minimum reaction time, which is called the tau (τ) concept in some documentation. Simply put, there is a minimum

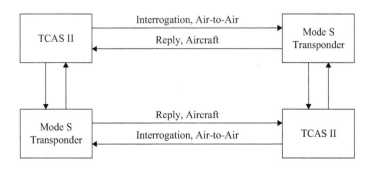

FIGURE 6.23 ▪
Interrogation/Reply between TCAS II Systems

reaction time that an aircrew would need to avoid a potential collision with another aircraft. The time that it would take for an intruder on a collision course to hit my aircraft (assuming we are at the same altitude) is a function of the distance to the aircraft and the total velocity of closure between our aircraft. The time that this event would occur is given by Equation (6.3):

$$T_{\text{collision}} = R/V_{\text{C}}, \tag{6.3}$$

where

$T_{\text{collision}}$ = time to a collision with an intruder aircraft,
R = range to the intruder aircraft, and
V_{C} = total velocity of closure between the two aircraft.

For example, suppose an aircraft is on my nose on a reciprocal heading at 30 nm traveling at 300 KTAS and I am traveling at 300 KTAS as well (no wind). The distance is 30 nm and the V_{C} is 600 knots. Substituting into Equation (6.3):

$T_{\text{collision}} = 30/600 = 0.05$ hr $= 180$ sec.

If we then wanted to receive an advisory based on reaction time, we could calculate a distance to issue an alert based on velocity of closure. For example, if we assume for our altitude, with an intruder in the forward quadrant we need a 48 sec reaction time, how far away must the intruder be to issue the alert? Once again, using Equation (6.3) and substituting the reaction time for the collision time:

$$48 \text{ sec} = .013 \text{ hr} = R/600$$
$$0.013(600) = 8 \text{ nm} = R.$$

The alert would be issued at 8 nm from the intruder aircraft. This is the way TCAS computes when advisories should be issued. A depiction of the TCAS traffic and resolution advisory zones is shown in Figure 6.24.

The right side of the figure shows the altitude clearances that will trigger the traffic and resolution advisories. The traffic advisories are given if the intruder aircraft is within 850 to 1200 ft of own-ship altitude (depending on altitude and version of the software). The resolution advisories are given when the intruder aircraft is within 600 to 800 ft of own-ship altitude (once again, depending on altitude and version of the software).

FIGURE 6.24 ■
TCAS II Traffic and Resolution Advisory Zones

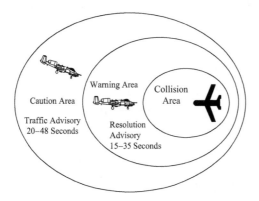
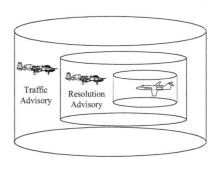

TABLE 6.17 ▪ Revised TA and RA Alert Thresholds

- The following are the existing and revised alert thresholds that apply to operations at high altitude. The only changes in the thresholds are reductions in the altitude threshold for the issuance of TAs and RAs. The values that change between FL 300 and FL 420 are highlighted here.

TCAS Advisory		Range Threshold (sec)		Altitude Threshold (ft)		Fixed Range Threshold Used With Slow Closure Rates (nmi)	
		V6.04a	V7	V6.04a	V7	V6.04a	V7
Traffic Advisory	FL 200–FL 300	48	48	850	850	1.3	1.3
	FL 300–FL 420	48	48	1200	*850*	1.3	1.3
Corrective RA	FL 200–FL 300	35	35	600	600	1.1	1.1
	FL 300–FL 420	35	35	700	*600*	1.1	1.1
Preventive RA	FL 200–FL 300	35	35	700	700	1.1	1.1
	FL 300–FL 420	35	35	800	*700*	1.1	1.1

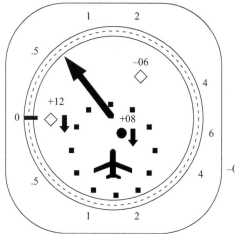

● = Traffic Advisory - Solid (Amber) Circle
- Intruder aircraft entering caution area.
(20–48 seconds from entering collision area)

■ = Resolution Advisory - Solid (Red) Square
- Intruder aircraft entering warning area
(15–35 seconds from entering collision area)

◆ = Proximate Traffic - Solid Diamond (Blue or White)
- Traffic within 6 nautical miles and +/–1,200 feet

◇ = Other Traffic - Hollow Diamond (Blue or White)
- Shown to enhance situational awareness

–02 ↓ = Data Tag. Two digit number with plus or minus sign represents the intruders altitude difference above or below the TCAS aircraft in hundreds of feet. The arrow appears if the intruder is climbing or descending at a rate greater than 500 fpm.

FIGURE 6.25 ▪
Stand-Alone TCAS Display

The vertical τ is defined as the relative altitude divided by the rate of change of altitude; the system calculates if the intruder will pass through the altitude thresholds. The high-altitude advisory thresholds were revised in the software due to the implementation of RVSM. These changes are shown in Table 6.17.

Figure 6.25 illustrates the stand-alone TCAS display. Four types of traffic can be displayed for TCAS, and they are illustrated to the right side of the display. The rounded dial bordering the TCAS information is a VVI display; the aircraft in the picture is currently climbing at a rate of approximately 750 ft/min. If a resolution advisory is issued requiring a climb or descent, a bar will appear on the outside of the scale indicating a commanded climb or descent rate.

In many aircraft installations, the TCAS information is superimposed on existing cockpit displays. Figure 6.26 shows a TCAS implementation on an MD-11 navigation display. There is a 5 nm range ring around the aircraft in the picture and there are three

FIGURE 6.26 ▪
Navigation Display
with TCAS

other aircraft within the display field of view. Number 1 is other traffic, number 2 is proximate traffic, and a traffic advisory has been issued for number 3, who is overtaking us from the rear.

Figure 6.27 is from the MD-11 as well; however, we now have the navigation display set to the TCAS mode. The addition of the starred ring (6) around the aircraft is the aircraft's nonintrusion zone. If any intruder's closest point of approach will take the aircraft inside the starred ring and its altitude is within the threshold, the TCAS will issue traffic and resolution advisories at the threshold times. The traffic information displayed is other traffic (1), proximate traffic (2), traffic advisory (3), and two resolution advisories (4 and 5). Figure 6.28 shows TCAS information superimposed on a 25 nm weather radar display.

The aural alerts that are available within the TCAS for software versions 6.04a, 7.0, and 7.1 are shown in Table 6.18.

The TCAS resolution advisories are treated just as an emergency procedure; that is, upon receiving a resolution advisory the pilot initiates the maneuver and then advises ATC of what he has done as opposed to the other way around.

To reduce false alarms and to prevent aircraft from entering unsafe regimes, a number of inhibits may be applied to the TCAS depending on its installation. A summary of the inhibits that have been programmed into the TCAS computer is shown in Table 6.19.

Additional inhibits are addressed in Table I of AC 20-131A, "Airworthiness Approval of Traffic Alert and Collision Avoidance Systems (TCAS II) and Mode S Transponders," March 29, 1993. These inhibits are based on individual aircraft performance and are applied when a maneuver will cause an unsafe reduction in airspeed. Evaluators will have to consult this table to see if additional inhibits need to be implemented in their installation.

FIGURE 6.27 ◾
Navigation Display
in the TCAS Mode

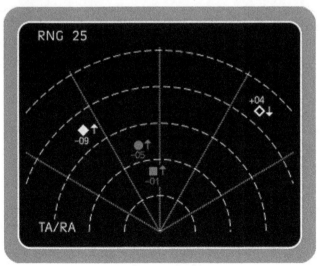

FIGURE 6.28 ◾
TCAS Information
Superimposed on
the Weather Radar

6.13.2 Certification of the TCAS II System

The five documents that dictate the performance and airworthiness of the TCAS in regards to certification are

- RTCA DO-185A, "Minimum Operational Performance Standards for Traffic Alert and Collision Avoidance System II (TCAS II) Airborne Equipment," December 16, 1997

TABLE 6.18 ▪ TCAS Advisories

TCAS Advisory	Version 7.1 Annunciation	Version 7.0 Annunciation	Version 6.04a Annunciation
Traffic Advisory	Traffic, Traffic		
Climb RA	Climb, Climb		Climb, Climb, Climb
Descend RA	Descend, Descend		Descend, Descend, Descend
Altitude Crossing Climb RA	Climb, Crossing Climb, Climb, Crossing Climb		
Altitude Crossing Descend RA	Descend, Crossing Descend, Descend, Crossing Descend		
Reduce Climb RA	Level Off, Level Off	Adjust Vertical Speed, Adjust	Reduce Climb, Reduce Climb
Reduce Descent RA	Level Off, Level Off	Adjust Vertical Speed, Adjust	Reduce Descent, Reduce Descent
RA Reversal to Climb RA	Climb, Climb, NOW, Climb, Climb, NOW		
RA Reversal to Descend RA	Descend, Descend, NOW, Descend, Descend, NOW		
Increase Climb RA	Increase Climb, Increase Climb		
Increase Descent RA	Increase Descent, Increase Descent		
Maintain Rate RA	Maintain Vertical Speed, Maintain		Monitor Vertical Speed
Altitude Crossing, Maintain Rate RA Climb and Descent	Maintain Vertical Speed, Crossing Maintain		Monitor Veritcal Speed
Weakening of RA	Level Off, Level Off	Adjust Vertical Speed, Adjust	Monitor Vertical Speed
Preventative RA	Monitor Vertical Speed		Monitor Vertical Speed, Monitor Vertical Speed
RA Removal	Clear of Conflict		

TABLE 6.19 ▪ TCAS System Inhibits

INHIBIT	PARAMETERS
Increase Descent RA	Inhibited below 1450 ft AGL
Descend RA	Inhibited below 1200 ft AGL while climbing and inhibited below 1000 ft AGL while descending
RA and TA Voice Messages	Inhibited below 400 ft AGL while descending and below 600 ft AGL while climbing (TCAS automatically reverts to TA only)
Close Range Surveillance	May not function at separations of less than approximately 900 ft
Self-Test	Can be inhibited when airborne
Advisory Priority	Can revert to TA only when higher priority advisories (GPWS, wind shear, etc., occur)
Climb RA	1500 ft/min can be inhibited based on aircraft performance capability
Increase Climb RA	2500 ft/min can be inhibited based on aircraft performance capability
Altitude Limit for Climb RA	Can be inhibited based on aircraft performance capability

- TSO-C119c, "Traffic Alert and Collision Avoidance System (TCAS) Airborne Equipment, TCAS II with Optional Hybrid Surveillance," April 14, 2009

- AC 20-131A, "Airworthiness Approval of Traffic Alert and Collision Avoidance Systems (TCAS II) and Mode S Transponders," March 29, 1993

- AC 20-151A, "Airworthiness Approval of Traffic Alert and Collision Avoidance Systems (TCAS II) Version 7.0 and 7.1 and Associated Mode S Transponders," September 25, 2009

- AC 120-55C, "Air Carrier Operational Approval and Use of TCAS II," February 23, 2011

NOTE: AC 20-131A provides guidance for TCAS II installations with version 6.04. AC 20-151A provides guidance for TCAS II installations with versions 7.0 and 7.1. There are three major differences between the two: aural alerts, inhibits, and revised TA and RA thresholds. The basic ground and flight tests are the same, although AC 20-151A does not address flight tests required when changes are made to the surveillance function of the TCAS II system. The revised TA and RA alerts have been previously discussed and are shown in Table 6.18. The inhibits listed in Table 6.19 change slightly with version 7.0 and 7.1. If your installation is for these versions, you must consult AC 20-151A for the changes. If your aircraft is equipped with TCAS and flies in RVSM airspace it must be with version 7.0 or higher. As AC 20-131A and AC 20-151A deal with airworthiness approval, AC 120-55C deals with operational approval for TCAS use in Part 121 operations. Operational approval pertains to training and maintenance programs, manuals, operational procedures, minimum equipment lists (MELs), and other areas necessary for safe and effective TCAS use and the qualification of aircrews through the approved training programs. The following discussion covers the airworthiness approval for version 6.04 systems and does not address the operational approval.

This is a good place to note that TCAS II will only provide resolution advisories in the vertical axis. The International Civil Aviation Organization (ICAO) had originally planned that aircraft meet the requirements of air collision and avoidance system III (ACAS III), which calls for advisories in both the horizontal and vertical axes. At the time of this writing, there was no work being performed on the development of a TCAS III system, and it is unknown if this requirement will ever be met.

The TCAS installation requires the applicant to apply through the type certification or supplemental type certification process. For airworthiness, the applicant may follow the requirements of AC 20-131A or AC 20-151A, although these documents are only one means of obtaining certification; a certification plan is required.

As TCAS II is an integrated system, certain types of equipment installations are required. TCAS II must have mode S installed to handle the communication requirements necessary for coordinating resolution advisories. In addition, the pilot must be able to control mode S and the TCAS II independently or simultaneously:

- Operation of mode S only

- Operation of mode S and TCAS II in the RA/TA mode simultaneously

- Operation of the ATCRBS transponder only if installed

- Operation of mode S and TCAS II in the TA mode simultaneously

- Operation of TCAS II in the standby mode

- A means to select the ATCRBS code
- A means to "IDENT"
- A means to initiate TCAS II self-test
- A means to indicate that TCAS II is in self-test
- A means to suppress altitude reporting

There are also optional controls that may be provided:

- Selection of weather radar only
- Selection of weather radar and traffic display simultaneously
- Selection of actual, flight level, or relative altitude of traffic

The active mode S transponder shall have a top and bottom omnidirectional antenna. The TCAS II shall have a top directional antenna and either an omnidirectional or directional bottom antenna. The pressure altitude information must come from the most accurate source and correspond to that being transmitted by mode S. Information should be provided to indicate an invalid pressure altitude.

Radar altitude information is provided to the TCAS II to trigger the inhibits listed in Table 6.19. Discrete inputs from aircraft configuration sensors (e.g., gear, flaps) are also sent to the TCAS II to trigger inhibits. Discrete information shall be provided to the mode S transponder pertaining to the aircraft's unique identification code and maximum airspeed capability. Heading information may be provided for use as a surveillance reference, but must be flagged when the heading is invalid. An indication shall also be provided to indicate that resolution advisories are not possible due to failure of the TCAS II equipment or any of its sensors or displays.

6.13.3 TCAS Test and Evaluation

The testing and evaluation of TCAS II is dependent on whether it is a first-time installation or a follow-on (same TCAS II has been installed previously on other types of aircraft). The test for the latter is a subset of the first-time installation (this discussion will address the first-time installation). It should also be noted that in many cases, a new TCAS installation may require a first-time installation of a mode S system; the AC addresses this fact and calls out the requirements for mode S testing.

6.13.3.1 TCAS Ground Test, Version 6.04a

The first test called out in the AC is a bearing accuracy test. The AC addresses two methods: rotating the aircraft to a fixed transponder site or testing on an antenna range capable of radiating the aircraft from 360°. I have found that a good way of performing this evaluation is to use a cooperating aircraft either taxiing around the airfield or in the overhead pattern; the GPS position of the cooperating target, when compared to the test aircraft position, will yield a true bearing. This can be compared to what is displayed on the TCAS II display. Basic accuracy requirements are 15°, however, larger errors are allowed in the area ±45° from the tail (blockage areas). This test will also allow the evaluator a quick look at the display performance (smoothness of the aircraft track, possible TA/RA, inhibits, etc.).

Simulated failures of the aircraft sensors integrated with TCAS II should be evaluated to see if the resulting system failure states agree with the predicted results.

The EMI test called out in the AC advises the tester to determine that there is no objectionable conducted or radiated interference to previously installed equipment. It also advises to pay particular attention to interference to TCAS equipment from the weather radar, particularly if the radar is operating in C band. This is hardly a description of a good EMI/EMC evaluation, and I recommend that the evaluator revisit chapter 3 of this text to conduct a complete investigation. TCAS operates in L band, and as I have said before, everyone and his brother operates in L band; particular attention should be given to, for example, GPS, TACAN, JTIDS, and harmonics from VHF transmissions.

The general arrangement and operation of the controls and displays, placards, warning lights, self-test, and aural tones should be evaluated for functionality and visibility (lighting conditions) and against human factors principles. Interfaces that should be evaluated are pressure and radar altitude, mode S identification codes, and maximum airspeed. The evaluator needs to verify that wind shear, TAWS, and TCAS II voice alerts are compatible. The wind shear alert must be clearly understood with simultaneous wind shear and TCAS II alerts; if aural warnings are prioritized, the alert priorities should be wind shear, TAWS, then TCAS II.

6.13.3.2 TCAS Basic Flight Tests, Version 6.04a

As with the ground tests, the evaluator is directed to observe the aircraft systems to determine that there is no objectionable mutual interference with the installation of TCAS II. The aural alerts should have sufficient loudness and clarity during low and high cockpit noise levels (e.g., idle descent versus high power at V_{MO}) with and without headsets. If the TCAS II test function is used to simulate voice announcements, ensure that the audio level is not changed by use of the test function.

The traffic information should remain valid and usable during aircraft maneuvers ($\pm15°$ pitch and up to 30° of bank). For accuracy, this test can be performed with the planned encounter flight tests explained in the next section. The effective surveillance range of the system must also be evaluated; this can be done against targets of opportunity. The easiest way to accomplish this test is to fly to a point that equates to the maximum surveillance range away from an airport or airway where known traffic is evident and verify the traffic is displayed on the TCAS II display.

Unless previously demonstrated during ground tests, configuration discretes associated with the TCAS II logic, including inhibits, must be validated as functioning correctly. The TCAS must also be evaluated for noninterference during coupled autopilot and flight director approaches down to aircraft minimums. All selectable modes of the TCAS II should be selected to ensure that they perform their intended function.

6.13.3.3 Planned Encounter Flight Tests, Version 6.04a

These tests are done with a cooperative aircraft with compatible equipment (e.g., similarly equipped TCAS II for TCAS–TCAS coordination, mode C only replies) to demonstrate adequate surveillance and to verify smooth, predictable performance. Before any of these tests are attempted, safety rules of engagement and "knock it off" criteria must be briefed and understood by all participants. A simple method is to declare that all aircraft will have a 500 ft safety bubble around the aircraft. The test aircraft will own all even altitudes and the intruder aircraft will own all odd altitudes. Neither aircraft will leave their owned altitude without a visual contact on the other aircraft. On the transition to the working area, the test aircraft and the intruder aircraft will fly in close formation at the same altitude to verify altitude readouts; a nominal difference of 100 ft may be acceptable, any error greater than 100 ft will cause an abort of the test. The aircraft will also verify the mode S reporting altitudes are the same by coordinating with ATC and confirming the readout on the ground. Errors greater than 100 ft as reported by ATC will also cause an abort of the mission.

Knock it off is simply an expression that means that all testing must cease and all aircraft must immediately return to their assigned altitudes. Some examples of a knock it off call includes the following:

- Loss of situational awareness by either aircraft
- Violation of an aircraft's 500 ft safety bubble
- Any aircraft enters the weather
- Another aircraft enters the area
- Test objective is accomplished
- Further testing is pointless
- Any emergency in either aircraft

These are a few of the common criterion; it will be up to the individual test organizations to determine an inclusive list. The following flight geometries are called out in the AC:

- Intruder overtaking TCAS aircraft from the aft quadrant
- Head-on, low and high closure speeds
- Head-on, above climb limit, TCAS II to TCAS II
- Head-on, TCAS II against mode C with TCAS II aircraft above the intruder and above the climb limit (force TCAS aircraft to descend)
- Head-on, at 300 ft above calm water (multipath protection)
- Converging flight
- Crossing, intruder above TCAS II aircraft and descending, or vice versa
- Evaluate similar setups in TA-only mode
- A mix of intruder transponder modes (A, C, S) should be evaluated, but primary emphasis should be on TCAS II to TCAS II coordination and mode C replies
- Flights should be set up with TCAS II both above and below the intruder

TCAS II performance in flight should also be demonstrated in the presence of electrical transients. The TCAS system should not see any adverse affects, generation of false targets, or false alerts. Normal TCAS II functions and displays should be restored within 3 sec.

6.13.3.4 Surveillance Flight Tests, Version 6.04a

The MOPS dictates the surveillance performance of TCAS II, and the bench tests that are performed are done in controlled conditions. A surveillance flight test is done to completely validate a TCAS II surveillance design. A flight test should be performed for each new TCAS surveillance design and whenever major modifications are made to the surveillance function of a previously certified system. The intent of the flight is to evaluate the system under environmentally stressful conditions (e.g., multipath, high-density traffic, multiple ground interrogations). The AC recommends an area of the United States where this test should be accomplished, as well as the time of day and day of the week. It is unusual to see such specific test information. However, the chosen area has all the elements for producing stress on any TCAS II, a key being a traffic density of 0.1 transponder-equipped aircraft per square mile.

The test should be accomplished in the Los Angeles basin area in the vicinity of the Long Beach and Santa Ana airports, on weekends between 10:00 A.M. and 3:00 P.M., when the ground visibility is greater than 5 nm with a ceiling of at least 10,000 ft to ensure the highest traffic densities. The flight paths should include a representative mixture of the following conditions:

- Overland flights at an altitude of 3000 to 6000 ft

- Over water flights between altitudes of 3000 and 6000 ft for a duration that is at least 20% of the required flight duration

- Approach and departure flights to the Long Beach and Santa Ana airports

The AC allows the applicant to choose another area to perform this evaluation, but it is up to the applicant to demonstrate that the area is similar in traffic density to Los Angeles. In addition, the area must also have a minimum of three FAA or military secondary surveillance radars located within 30 nm of the TCAS II aircraft to provide an interference environment similar to Los Angeles.

The objective of this test is to collect TCAS II reporting data and compare it to the actual traffic in the area. The applicant is responsible for collecting the data (at least 10,000 sec of tracking data, or about 1 hour of recording in the Los Angeles basin) and forwarding these data to the FAA for an independent analysis. The AC contains the type of information required and the analyses that must be performed to verify the integrity of the TCAS II surveillance function. This test is only required for a new TCAS II or an existing system with a major change to the surveillance function. The test is normally accomplished by the vendor on a test aircraft and is only mentioned here for completeness. Descriptions of TCAS II displays and the required verbiage for the aircraft flight manual are included as attachments in the AC.

6.13.3.5 TCAS Airworthiness Versions 7.0 and 7.1

As noted, the guidance for obtaining Airworthiness for the installation of a version 7.0 or 7.1 TCAS may be found in AC 20-151A. The AC adds requirements for equipment and deletes some of the tests required of earlier installations. The document states that a pilot control of TCAS equipment shall be provided and shall allow the selection of TCAS in TA/RA mode with Mode S simultaneously, TCAS in TA mode with Mode S simultaneously, or TCAS in standby mode.

Features that should be provided are the ability to select an assigned 4096 code, the ability to initiate TCAS self-test and a means to suppress altitude reporting. Features that may be provided include the selection of Weather RADAR only or Weather and TCAS, the display of traffic within selected altitude bands, the selection of altitude or relative altitude of traffic, and the selection of TCAS information on MFDs.

With respect to antennas, the AC states that the Mode S transponder shall have a top and bottom omnidirectional antenna and the TCAS shall have a top directional antenna and a bottom omnidirectional or directional antenna. The antennas should be mounted as close to the centerline as possible and should have at least 20 dB isolation from other L band frequency antennas. The processor must have a TSOA IAW TSO-C119b or 119c and may include hybrid surveillance as an optional capability.

Traffic and resolution advisory displays should conform to the standard symbology provided in RTCA/DO-185A or B, section 2.2.6. Optional Caution/Warning lights may be installed that are separate from the display; two types of annunciators have been used in acceptable installations are a discrete amber annunciator for a TA (optional) located in the pilot's primary FOV and inhibited below 400' AGL, a discrete red warning annunciator for an RA, in the pilot's primary FOV, inhibited below 900' AGL. A VSI with a lighted red arc or alphanumeric message is also acceptable.

Each TCAS aural alert should be annunciated by a dedicated message over the cockpit speaker system at a volume adequate for clear understanding at high cockpit noise levels. The AC mentions that the evaluation should include the case where the flight crew member is wearing a headset covering the outboard ear when appropriate. The appropriate alerts are listed in Table 6.18.

Certification is via the TC or STC process; any change in system equipment requires either an initial or follow-on approval. A Certification Plan is required, and a ground test showing TCAS to TCAS communication and TCAS to Mode S interoperability is required.

In some cases, TCAS may command maneuvers that may significantly reduce stall margins or result in stall warnings. Conditions where this may occur include bank angles greater than 15 degrees, weight/altitude/temperature considerations, initial speeds, one engine inoperative, etc. The AC provides a table of maneuvering conditions that can be encountered with TCAS; the applicant should verify safe operation of the aircraft during these maneuvers with an associated RA. If unsafe conditions exist, the applicant should consider adding a discrete inhibit during these conditions (AC 20-151A, Table 1). These are in addition to the required inhibits listed in Table 6.19.

The ground tests and planned encounter tests described for 6.04a are the same for Versions 7.0 and 7.1. A typical TCAS version 7.1 compliance matrix is shown in Table 6.20.

6.13.4 Traffic Advisory System

This section covers any active traffic advisory equipment that may be installed in aircraft to provide situational awareness. These systems are very similar to TCAS I, which provide an alert (TA) on potential conflicting aircraft and display of other traffic (OT) within the surveillance area. It does not have the capability of providing resolution advisories, and there is no aircraft-to-aircraft communication via mode S.

The system must meet the MOPS set forth in RTCA DO-197A, "Minimum Operational Performance Standards for an Active Traffic Alert and Collision Avoidance System (Active TCAS I), section 2," September 12, 1994. Manufacturers must meet the minimum performance standards called out in TSO-C147, "Traffic Advisory System (TAS) Airborne Equipment," April 6, 1998. AC 20-131A and 20-151A do not address these types of systems since they are used for situational awareness, but the TSO states that

It is the responsibility of those installing this article either on or within a specific type or class of aircraft to determine that the aircraft installation conditions are within the TSO standards.

TABLE 6.20 ▪ TCAS version 7.1 Compliance Matrix

No.	Requirement		Subject	Description
1	AC 20-151A	3-3a	Bearing accuracy	Demonstrate the bearing estimation accuracy of the TCAS II system as installed in the aircraft. A maximum error of ±15 degrees in azimuth is acceptable; however, larger errors are acceptable in the area of the tail (e.g., within ±45 degrees of the tail) when that area is not visible from the cockpit.
2	AC 20-151A	3-3b	Sensor failures	Evaluate simulated failures of the aircraft sensors integrated with TCAS II to determine that the resulting system failure state agrees with the predicted results. These tests should be part of the ground test plan.
3	AC 20-151A	3-3c	EMI	Survey the flight deck EMI to determine that the TCAS II equipment is not a source of objectionable conducted or radiated interference to previously installed systems or equipment, and that operation of the TCAS II equipment is not adversely affected by conducted or radiated interference from previously installed systems and equipment. Pay attention for possible interference with TCAS II equipment from weather radar, particularly if operating in the C-band.
4	AC 25-151A	3-3d	Human Factors	Determine that they are designed and located to prevent inadvertent actuation. Evaluate TCAS displays and annunciations to determine that they support flight crew awareness of TCAS status changes that could result from TCAS mode selections, intentional pilot actuation of other installed systems, or inadvertent pilot actions with TCAS or other installed systems. Evaluate TCAS displays to ensure all information is, at a minimum, legible, unambiguous, and attention-getting (as applicable). In particular, where transponder functions are integrated with other system controls, ensure that unintended transponder mode switching, especially switching to STANDBY or OFF, is not possible. Pay close attention to line select keys, touch screens or cursor controlled trackballs as these can be susceptible to unintended mode selection resulting from their location in the flight deck (e.g., proximity to a foot rest or adjacent to a temporary stowage area)
5	AC 20-151A	3-3e	self-test	Evaluate the TCAS II self-test features and failure mode displays and annunciators
6	AC 20-151A	3-3f	interface	Verify that the pressure altitude source and radio altimeter are properly interfaced with the TCAS II equipment
7	AC 20-151A	3-3g	alerts compatibility	Wind shear TAWS warning and TCAS voice alerts are compatible
8	AC 20-151A	3-3h	performance	Observing any available area traffic.
9	AC 20-151A	3-3i	visibility	Evaluate the TCAS II system installation for satisfactory identification, accessibility, and visibility during both day and night conditions
10	AC 20-151A	3-3j	logic with configuration change	Determine that any configuration of discretes associated with the TCAS II logic, including inhibits of climb RAs, operate properly. (Changes in logic or function with aircraft configuration, altitude, or speed.)

(Continues)

TABLE 6.20 ■ (Continued)

No.	Requirement		Subject	Description
11	AC 20-151A	3-3k	maximum airspeed	Verify that the ICAO 24-bit aircraft address and maximum airspeed are correct. Additionally, verify that other features, which may be optional, such as extended squitter, aircraft identification reporting, hybrid surveillance or other data link uses also function correctly. Verify that the transponder and data sources meet the requirements of the failure condition classifications associated with the features. For example, an un-annunciated failure of the transponder extended squitter resulting in erroneous information being transmitted is at least a major failure condition.
12	AC 20-151A	3-3l	altitude Averter	If connected, verify that the altitude alert is providing correct data to TCAS and that the TCAS II version 7.0 or 7.1 logic, as applicable, correctly weakens or strengthens the displayed RA using the altitude alert input.
13	AC 20-151A	3-3m	air/ground	Verify that the air/ground inputs are connected properly.
14	AC 20-151A	3-4a	Interference	During all phases of flight, determine if there is any mutual interference with any other aircraft system.
15	AC 20-151A	3-4b	aural message	Evaluate TCAS II aural messages for acceptable volume and intelligibility during both low and high cockpit noise levels (idle descent at low speed and high power at maximum operating limit speed V) with headset covering outboard ear only (when appropriate) and without headsets. In the case of turbo-prop aircraft where the aircrew utilizes headsets via the aircraft audio distribution panel, the aural messages should hold the same acceptable volume and intelligibility during both low and high cockpit noise levels. If the TCAS II TEST is used to simulate voice announcements, ensure that the audio level is not changed by use of the TEST function.
16	AC 20-151A	3-4c	traffic information	Demonstrate that traffic information remains valid and usable when the aircraft is pitched ±15 degrees and rolled approximately 30 degrees during normal maneuvers by observing area traffic in the traffic advisory display
17	AC 20-151A	3-4d	surveillance range	Evaluate the effective surveillance range of the traffic display, including target azimuth reasonableness and track stability. Use of targets of opportunity or a nontransport category (low speed) aircraft as a target for these tests is permissible
18	AC 20-151A	3-4e	logic/inhibit	Determine that any configuration discretes (changes in logic or function with configuration, altitude, or speed) associated with the TCAS II logic, including inhibits of climb RAs, operate properly unless previously demonstrated during ground tests
19	AC 20-151A	3-4f		Perform the additional flight tests in RTCA/DO-185A or RTCA/DO-185B, paragraph 3.4.4 as applicable, unless previously accomplished under TSO-C119b or TSO-C119c, respectively

(Continues)

TABLE 6.20 ▪ *(Continued)*

No.	Requirement		Subject	Description
20	AC 20-151A	3-4g	interference	Evaluate TCAS II for noninterference during coupled auto-pilot and flight director approaches to the lowest minimums approved for the aircraft
21	AC 20-151A	3-4h	altimetry check	Before any cooperative flight tests at any altitude involving the TCAS II-equipped aircraft and another aircraft, fly both aircraft in close formation to ensure matched altimetry readouts. These checks should be flown at the speeds and altitudes to be used for the tests
22	AC 20-151A	3-4i	modes	Evaluate all selectable modes of the TCAS II to determine that they perform their intended function and that the operating mode is clearly and uniquely annunciated
23	AC 20-151A	3-4j	regression	Reevaluate any previously installed aircraft systems that required changes as a result of the TCAS II installation. (e.g., electronic flight instrument system (EFIS), flight director (FD), PFD, navigation displays (ND), IVSI, interface etc.)
24	AC 20-151A	3-4k	Hybird surveillance	If hybrid surveillance functionality is included, perform the flight tests in RTCA/DO-300, MOPS for TCAS II Hybrid surveillance, dated December 13, 2006, paragraph 3, unless previously accomplished under TSO-C119c
25	AC 20-151A	3-5a	intruder	Intruder overtaking TCAS II aircraft (from the aft quadrants)
26	AC 20-151A	3-5b	intruder	Head-on
27	AC 20-151A	3-5c	Converging	Converging

1. Crossing (intruder above TCAS II, descending or vice versa.)

2. Evaluate the TA-only mode during planned encounters.

3. Evaluate a mix of intruder transponder modes (A, C, S and S with extended squitter) but primary emphasis should be on TCAS II-to-TCAS II coordination, and on Mode C replies from the intruder aircraft.

4. Evaluate a mix of encounters with TCAS II both above and below the intruder.

5. If a flight test is necessary to ensure compatibility with other designs, verify correct air-to-air coordination between the test TCAS II and another manufacturer's previously approved equipment (refer to paragraph 2-15).

6. Evaluate the effect of electrical transients (bus transfer) during encounters. The TCAS II should not experience adverse effects. No false TAs or RAs should be generated as a result of electrical transients. Normal TCAS II functions and displays should be restored within approximately three seconds.

No.	Requirement		Subject	Description
28	AC 20-151A	3-6b	Antenna	Verify the installed antenna(s) are compatible with the Mode S transponder and provide an adequate response to ground radar interrogations during normal aircraft maneuvers.

(Continues)

TABLE 6.20 ■ (Continued)

No.	Requirement		Subject	Description
29	AC 20-151A	3-6c	Mode S	Demonstrate that the Mode S transponder functions properly as installed and does not interfere with other aircraft electronic equipment
30	AC 20-151A	3-6d	Mode S	If the Mode S transponder uses a top mounted antenna in addition to a bottom mounted antenna installed at, or near, the same location used by a previously approved ATCRBS transponder antenna. Conduct a comprehensive ground test and evaluation in accordance with Appendix B and perform a functional flight test
31	AC 20-151A	3-6e	Mode S	If a Mode S transponder is installed in an aircraft that does not have a previously approved ATCRBS transponder installation, or that uses a bottom mounted antenna location that differs significantly from that used by a previously approved ATCRBS transponder antenna, conduct the following ground and flight tests: (1) Conduct ground tests and evaluations per Appendix B. (2) Climb and Distance Coverage (3) Long Range Reception (4) High Angle Reception (5) High Altitude Cruise (6) Surveillance Approach (7) Holding and Orbiting Patterns (8) Altitude Reporting

This implies that there must be some type of verification (ground/flight tests) to ensure that the system performs its intended function.

There are two classes of TAS equipment defined in the TSO:

- Class A equipment incorporates a horizontal situation display that indicates the presence and relative location of intruder aircraft, and an aural alert informing the crew of a traffic advisory.

- Class B equipment incorporates an aural alert and a visual annunciation informing the crew of a traffic alert.

The TAS is only required to have one directional antenna, which may be mounted on either the top or bottom of the aircraft. There are only two advisories that are given to the pilot: traffic advisories and other traffic. The system may be designed with the same color coding as previously discussed with TCAS, but may also use shapes only, so the system may be implemented with monochrome displays.

6.13.4.1 TAS Class A Equipment

As a minimum, the TAS display, which must be readable under all lighting conditions, must provide the following information to the pilot:

- A differentiation between other traffic and traffic advisories
- Bearing
- Relative altitude (above or below [±] and a numerical value)
- Vertical trend of intruder aircraft (up/down arrow when the intruder's vertical speed is ≥500 ft/min)

- Range (selected range will be depicted)
- An own-aircraft symbol, with aircraft heading at the 12:00 position
- A range ring of 2 nm around the aircraft when a range of less than 10 nm is selected
- The traffic advisory will be a filled rectangle and other traffic will be an open rectangle
- Overlapping traffic symbols should be displayed with the appropriate information overlapped. The highest-priority traffic should appear on the top of other traffic symbols
- The display should be capable of depicting a minimum of three intruder aircraft simultaneously. As a minimum, the display shall be capable of displaying aircraft that are within 5 nm of own-ship position; the display may provide multiple crew-selected ranges.
- If the intruder range of a TA target is greater than the selected range display, no less than one-quarter of the TA symbol shall be placed at the edge of the display at the proper bearing.
- The traffic symbol size shall be no less than 0.2 in. high and the height of the relative altitude characters shall be no less than 0.15 in.
- No bearing advisories shall be presented for an intruder generating a TA when the intruder's bearing cannot be derived. The advisory shall be in tabular form and centered on the display below the own-aircraft symbol; the display will have provisions for displaying at least two no-bearing TAs.

As with TCAS II, the TA aural alert is "Traffic, Traffic," spoken once. The TA will be inhibited below 400 ± 100 ft AGL when TAS is installed on an aircraft equipped with a radar altimeter. For aircraft without a radar altimeter, the TA will be inhibited when the landing gear is extended. For aircraft with fixed gear and no radar altimeter, the TA is never inhibited.

The TA is issued based on τ, which was discussed in section 6.12.1.

6.13.4.2 TAS Class B Equipment

Since a display is not required equipment for the class B TAS, there are some unique requirements not found in the class A equipment. A visual "Traffic" annunciation should be provided for the duration of the TA. All TAS aural alerts should be inhibited below 400 ± 100 ft for aircraft fitted with a radar altimeter. As with class A, for aircraft without a radar altimeter, the aural annunciations shall be inhibited when the landing gear is extended. When installed on fixed-gear aircraft without a radar altimeter, the aural annunciations are never inhibited. Aural annunciations may be inhibited on the ground when the aircraft is fitted with a weight-on-wheels system.

The aural alert messages shall be announced in threat priority sequence, greatest threat first. The initial aural traffic advisories shall be spontaneous and unsolicited. The unsolicited annunciation should be as follows: "Traffic, x o'clock," spoken once (where x is the clock position of the intruder). If surveillance bearing information is not available on the intruder, "Traffic, No Bearing" shall be annunciated. The current relative bearing to the intruder shall be annunciated as a traffic update upon aircrew command. Additional information such as relative altitude, range of intruder, and vertical trend may also be annunciated.

The acceptability of the annunciations must be reviewed during flight testing. The following factors, at a minimum, must be evaluated for acceptability:

- Quantity of unsolicited annunciations
- Duration of annunciations
- Annunciation clarity
- Volume of the annunciations

The evaluation should occur under normal cockpit workload conditions during departure, cruise, approach, and landing phases of flight and should include evaluation of suitability in a normal ATC voice communication environment. There shall be a means to request a traffic advisory update, mute a current aural advisory, and cancel or restore aural advisories (turning the system off is an acceptable means of cancellation, where the default condition of the equipment in the on position is aural advisories active).

The criteria for issuing the traffic alert is the same as mentioned for class A TAS and TCAS II. The active TAS equipment shall provide logic to inhibit traffic alerts of altitude reporting intruders that are on the ground. This logic shall be used when the TAS-equipped aircraft is below 1700 ft AGL with a hysteresis of +50 ft. Of course, if the aircraft does not have a radar altimeter, this requirement goes away. If the aural memory storage capacity is exceeded, the system will delete the least threatening intruders first.

6.13.5 Automatic Dependent Surveillance Broadcast (ADS-B)

To improve the safety, efficiency, and capacity of the Nation Airspace System (NAS), the FAA is transforming the current ground-based radar air traffic control system to a satellite-based system called ADS-B. As the cornerstone of the next generation air transportation system (NextGen), ADS-B supports these improvements and enables the NAS to accommodate growth expected from future demand.

An ADS-B-equipped aircraft determines its own position and periodically broadcasts this position and other relevant information to potential ground stations and other aircraft with ADS-B-In equipment. Position data is usually derived from GNSS or from an aircraft's inertial reference system. ADS-B can be used over several different data link technologies, including mode-S extended squitter (1090 ES) operating at 1090 MHz, universal access transceiver (978 MHz UAT), and VHF data link (VDL Mode 4); the FAA has chosen 1090 ES and UAT. With ADS-B, an A/C periodically broadcasts its own state vector and other information without knowing what other vehicles or entities might be receiving it; there is no acknowledgment or reply. ADS-B is *automatic* in the sense that no pilot or controller action is required for the information to be issued. It is *dependent surveillance* in the sense that the surveillance-type information obtained depends on the suitable navigation and broadcast capability in the A/C.

ADS-B consists of three components:

- A transmitting system that includes message generation in the A/C
- A signal protocol: 1090 ES, UAT, VHF data link
- A receiving system that includes message reception and report assembly functions

There are two commonly recognized types of ADS for aircraft applications:

- ADS-addressed (ADS-A), also known as ADS-contract (ADS-C)
- ADS-broadcast (ADS-B)

ADS-B differs from ADS-A in that ADS-A is based on a negotiated one-to-one peer relationship between an aircraft providing ADS information and a ground facility requiring receipt of ADS messages. ADS-A reports are employed in the Future Air Navigation System (FANS) using the Aircraft Communication Addressing and Reporting System (ACARS) as the communication protocol. During flight over areas without radar coverage (e.g., oceanic and polar), reports are periodically sent by an aircraft to the controlling air traffic region.

The ADS-B link can be used to provide other broadcast services, such as TIS-B and FIS-B. Traffic Information Services-Broadcast (TIS-B) supplements ADS-B air-to-air services to provide complete situational awareness in the cockpit of all traffic known to the ATC system. The ground TIS-B station transmits surveillance target information on the ADS-B data link for unequipped targets or targets transmitting only on another ADS-B link. TIS-B uplinks are derived from the best available ground surveillance sources: Primary and secondary surveillance RADAR, ADS-B reports, and airport surface traffic reports.

Flight Information Services-Broadcast (FIS-B) provides weather text, weather graphics, NOTAMs, ATIS, and similar information over the data link. FIS-B is inherently different from ADS-B in that it requires sources of data external to the aircraft or broadcasting unit, and has different performance requirements such as broadcast rate. In the US, FIS-B services will be provided over the UAT link in areas that have a ground surveillance infrastructure.

The ADS-B data link supports a number of airborne and ground applications; each application has its own operational concepts, algorithms, procedures, standards, and user training. A Cockpit Display of Traffic Information (CDTI) is a generic display that provides the flight crew with surveillance information about other aircraft, including their position. A CDTI function might also display current weather conditions, terrain, airspace structure, obstructions, detailed airport maps, and other information relevant to the particular phase of flight.

ADS-B has the enhancing feature of providing airborne collision avoidance. ADS-B is seen as a valuable technology to enhance ACAS/TCAS operation, which has been previously addressed. Incorporation of ADS-B can mitigate the deficiencies of TCAS and provide benefits such as the following:

- Decreasing the number of active interrogations required by TCAS, thus increasing effective range in high density airspace
- Reducing unnecessary alarm rate by incorporating the ADS-B state vector, aircraft intent, and other information
- Use of the ACAS display as a CDTI, providing positive identification of traffic
- Extending collision avoidance below 1000 feet above ground level, and detecting runway incursions

Eventually, the ACAS function may be provided based solely on ADS-B. Other applications that benefit from ADS-B are as follows:

- Air Traffic Management
- Improved search and rescue

 – Enhanced flight following
 – Lighting control and operation
 – Airport ground vehicle and aircraft rescue and firefighting vehicle operational needs
 – Altitude height keeping performance measurements
 – General aviation operations control

There are two signals for ADS-B: ADS-B out and ADS-B in. ADS-B out is the ability to transmit a properly formatted ADS-B message from the aircraft to ground stations and other ADS-B equipped aircraft. ADS-B in is the ability of the aircraft to receive information transmitted from ADS-B ground stations and other aircraft. On January 1, 2020, when operating in the airspace designated in 14 CFR §91.225, you must be equipped with ADS-B out avionics that meet the performance requirements of 14 CFR §91.227. ADS-B in is not mandated by the new rule; if an operator chooses to voluntarily equip their aircraft, ADS-B In will also require the installation of a compatible display. The new rule applies to much of the designated airspace classes.

Class A airspace extends from FL180 to FL600. Class B airspace is defined around key airport traffic areas and has the shape of an inverted wedding cake, Class B airspace normally begins at the surface with an upper limit of 10,000 ft. Class C space is structured in much the same way as Class B airspace, but on a smaller scale. Class C airspace is defined around airports of moderate importance that have an operational control tower and is in effect only during the hours of tower operation at the primary airport; the vertical boundary is usually 4,000 ft. Class D airspace is generally cylindrical in form and normally extends from the surface to 2,500 ft. AGL. The outer radius of the airspace is generally 4 nautical miles; Class D airspace reverts to Class E or G during hours when the tower is closed. Controlled airspace that is not classified as A, B, C or D is designated as Class E. In most areas of the United States, Class E airspace extends from 1,200 ft. AGL up to, but not including, 18,000 ft. Class F is not used in the United States and Class G airspace includes all airspace below flight level 600 not otherwise classified as controlled. Class G airspace is typically the airspace very near the ground (1200 ft or less). A diagram of these classes is shown in Figure 6.29.

FIGURE 6.29 ■
Airspace Classes

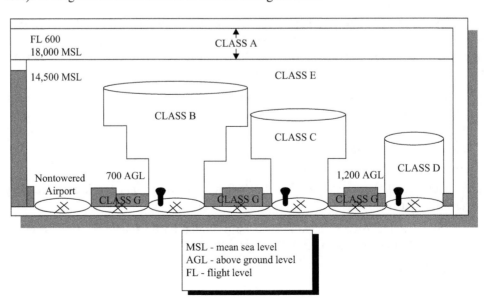

The rule covers operation in all Class A, B, and C airspace. It also covers operation in Class E airspace within the 48 contiguous states and the District of Columbia at and above 10,000 ft MSL, excluding the airspace at and below 2,500 ft above the surface, and Class E airspace at and above 3,000 ft MSL over the Gulf of Mexico from the coastline of the United States out to 12 nautical miles. It is also mandated around those airports identified in 14 CFR part 91, Appendix D.

6.13.5.1 ADS-B Certification

The rule specifies ADS-B out equipment be certified to either TSO-C154C (universal access transceiver) or TSO-C166b (1090ES). Equipment certified to TSO-C166b will be required to operate in Class A airspace and equipment certified to either TSO can be used while operating in the designated airspace outside Class A.

The applicable documentation for ADS-B airworthiness includes:

– RTCA/DO-260B, "Minimum Operational Performance Standards for 1090 MHz Automatic Dependent Surveillance-Broadcast (ADS-B)," December 13, 2011

– RTCA/DO-282B, "Minimum Operational Performance Standards for Universal Access Transceiver (UAT) Automatic Dependent Surveillance-Broadcast (ADS-B)," December 13, 2011

– TSO C-154c and TSO C-166b, previously identified, December 2, 2009

– TSO C-195b, "Avionics Supporting Automatic Dependent Surveillance-Broadcast (ADS-B) Aircraft Surveillance Applications," Draft June 2014

– Advisory Circular AC 20-165, "Airworthiness Approval of Automatic Dependent Surveillance-Broadcast ADS-B Out," May 21, 2010

– Advisory Circular AC 20-172A, "Airworthiness Approval for ADS-B In Systems and Applications," March 23, 2012

AC 20-165 covers the airworthiness approval for ADS-B out systems. The AC covers two link options: 1090 extended squitter (1090 ES) and universal access transceiver (UAT) (978 MHz); 14 CFR 91.225 requires 1090 ES in Class A airspace, and FIS-B is available only on the UAT link. The body of the AC provides specific information on installation and test and evaluation guidance. The appendices of the AC provide information on the message elements, ADS-B position sources and a latency analysis.

AC 20-172A covers the airworthiness approval for ADS-B In Systems and Applications. This AC provides guidance material for the reception of ADS-B, traffic information services-broadcast (TIS-B) and automatic dependent surveillance-rebroadcast (ADS-R) messages. This AC does not cover flight information services-broadcast (FIS-B) messages; this application will be covered in a future AC. The AC states that it is not mandatory, but if it is used, it must be used in its entirety. All installations of ADS-B in should also provide ADS-B out.

6.13.5.2 ADS-B Out Airworthiness

ADS-B equipment broadcasts a system design assurance (SDA) parameter within its message; this parameter indicates the probability of a system malfunction that would provide false or misleading information of the aircraft's position. Aircraft that are not complex (for all intents and purposes, not applicable for the majority of newer aircraft) would have a preset SDA; aircraft that are complex or highly integrated would require a

System Safety Assessment (SSA). The applicant must calculate the position and velocity latency of the installed system: total latency and uncompensated latency. This is because this information will be used by ATC and other aircraft to ensure safe separation.

Total latency is defined as the difference between the time the position is measured (sensed) to the time the position is transmitted from the aircraft; the total latency must be less than or equal to 2.0 seconds. Uncompensated latency is the difference between the time of applicability for the transmitted position and the actual time the position is transmitted from the ADS-B system (senescence); the uncompensated latency must be less than or equal to 0.6 seconds.

AC 20-165 describes the ADS-B out message and the correct location of the message parameters. The parameters encoded in the ADS-B out message are as follows:

- Latitude and Longitude
- Horizontal velocity; N/S E/W velocity while airborne and a combination of ground speed and ground track or heading while on the surface
- System integrity level (SIL); described previously
- System integrity level supplement (SILSUPP); identifies the SIL as being calculated on a per hour or per sample basis
- Navigation integrity category (NIC); integrity of the position source
- Call sign/flight ID
- IDENT/emergency status
- Mode 3/A
- Navigation accuracy category for position (NACP); HDOP
- Navigation accuracy category for velocity (NACV); $1 = < 10$ m/s, $2 = < 3$ m/s, 3 or $4 =$ GNSS velocity accuracy
- Geometric altitude; WGS-84 height above ellipsoid
- Geometric vertical accuracy; VDOP
- Ground track angle (or heading)
- Barometric altitude
- Barometric altitude integrity code (NICBARO)
- TCAS status source

With respect to pilot interface and human factors, the system must display the operational status to the crew (Mode 3A code, IDENT, emergency codes). If unable to transmit, or in the event of a loss of function, the ADS-B must provide an appropriate annunciation to the crew. Regulations require that aircraft equipped with ADS-B Out operate with the equipment on at all times; if equipped with an on/off control it must be located to prevent inadvertent actuation. An anonymous function is available but may be used only if not requesting ATC services (Mode 3/A 1200). A single point of entry should be used for Mode 3/A, IDENT and emergency codes, and the design should minimize errors by the crew. The design should also enable the crew to detect and correct errors. Transmission of a false position is considered a major failure and should not occur at a rate of greater than 1×10^{-5} per hour; the safety analysis should consider the potential of all pilot errors.

During flight test, EMI/EMC is to be evaluated during all phases of flight. Other systems that may have been modified due to ADS-B installation should be verified to operate properly. A flight profile to show the ADS-B equipment performs properly with the FAA ground system needs to be accomplished; the AC provides the test area and flight profile to be flown for this evaluation.

6.13.5.3 ADS-B In Airworthiness

There are three sources currently available for ADS-B in:

- ADS-B direct (1090 MHz) direct from other aircraft
- ADS-R (rebroadcast over 1090 MHz from ATC; original report sent to ATC via UAT)
- TIS-B (1090 and UAT format sent from ATC on non–ADS-B targets, that is transponder replies)

Flight information services-broadcast (FIS-B) provides weather text, weather graphics, NOTAMs, ATIS, and similar information. FIS-B is currently not available, but will be a future fourth source for ADS-B In. The AC provides guidance for the display of traffic information while on the airport surface and while airborne. TSO C-195a defines the MOPS that provide the basis for installation; the TSO defines 3 classes of equipment:

- Class A—cockpit display of traffic information (CDTI) (surface only)
- Class B—CDTI (no restriction)
- Class C—airborne surveillance and separation assurance processing (ASSAP)

Class A equipment displays ADS-B traffic while ownship is on the surface and moving slower than 80 kts; it must deactivate the CDTI when in the air or at speeds greater than 80 kts. Class B equipment supports the display of ADS-B traffic without restriction. Class C equipment processes ADS-B messages to generate traffic for a CDTI. A summary of ADS-B in equipment classes and their allowable functions is shown in Table 6.21.

The ADS-B in criticality for a loss of function or in the case of providing hazardous or misleading information is shown in Table 6.22.

The airborne application displays traffic from a bird's-eye view relative to ownship. Each aircraft symbol displayed conveys position, direction and altitude information; optional information, such as identity may also be displayed. The enhanced visual

TABLE 6.21 ■ ADS-B In Equipment Classes

Avionics / Applications	CDTI Surface Only A	CDTI B	ASSAP C
Enhanced Visual Acquisition	Not Permitted	Class B1	Class C1
Surface (Runways Only)	Class A2	Class B2	Class C2
Surface (Runways and Taxiways	Class A3	Class B3	Class C3
Visual Separation on Approach	Not Permitted	Class B4	Class C4
Airborne	Not Permitted	Class B5	Class C5
In-Trail Procedures	Not Permitted	Class B6	Class C6

TABLE 6.22 ▪ ADS-B In Equipment Criticality

Avionics Applications	Loss of Function	Hazardous Misleading Information
Enhanced Visual Acquisition	Minor	Major
Surface (Runways Only)	Minor	Major (>80 kts)
		Minor (< 80 kts)
Surface (Runways and Taxiways	Minor	Major (>80 kts)
		Minor (< 80 kts)
Visual Separation on Approach	Minor	Major
Airborne	Minor	Major
In-Trail Procedures	Minor	Major

approach application allows the pilot to select an aircraft to follow on the approach; additional information about the selected aircraft, including range and groundspeed, is displayed to enhance the pilot's situational awareness. The surface application with runways and taxiways displays ADS-B traffic superimposed on a map of the airport surface; different symbols indicate aircraft on the ground and airborne traffic. The surface application for runways does not include taxiway traffic. The In-Trail Procedure (ITP) application enables aircraft that desire FL changes in procedural airspace to achieve these changes more efficiently. The ITP permits a climb through, or descent through, or maneuver between, properly equipped aircraft using a new distance based longitudinal separation minimum. The ITP requires specific application unique processing and display parameters; ITP requires operational approval (aircraft and crew, AR).

The CDTI may be presented on a dedicated display or integrated with the aircraft's EFIS, MFD or electronic flight bag. CDTI equipment should be compliant with the Class A or Class B equipment requirements of TSO-C195. The display should be mounted in the forward FOV for best situational awareness; side mounted displays may be used but may be limited in more advanced applications. The crew must have an unobstructed view of the display. ITP applications must have a CDTI and it is recommended to have a vertical profile displayed during the maneuver; the required traffic symbology is contained in Appendix 2 of the AC (AC 20-172A). The CDTI may use a dedicated control panel or be integrated with existing aircraft systems. A means must be available to adjust the display range between the minimum and maximum values, adjust the altitude bands between the minimum and maximum values, and adjust the brightness of the display.

Optional controls may also be provided:

– Relative and actual altitude

– Selection of at least one traffic element

– Selection of alternate display criteria

– Declutter and return with a single action

– A means to indicate if auto-declutter is on

– Pan/zoom

– A means to designate traffic for an application (such as enhanced visual approach)

The ASSAP system accepts inputs from ADS-B reports, ADS-R reports, and TCAS tracks; it correlates sources, creates tracks, and performs application-specific processing.

The tracks and alerts are then sent to the CDTI; ASSAP equipment must be compliant with TSO-C195a Class C requirements.

Electrical bonding and load analysis are required. The total latency to receive, process and display traffic must be less than 3 seconds (time from receipt of ADS-B message). The total latency of ownship position at the display must be less than 3.5 seconds; ownship time of applicability must be within 1 second of the time of the display. A latency analysis must be performed (appendix 1 of the AC provides methodology). The same position source used for ADS-B out should be used to provide position information to the ASSAP equipment. Equipment should be installed IAW manufacturers' instructions; parallax corrections to the GNSS antennas should be known. The applicant needs to verify the equipment's environmental qualifications (RTCA/DO-160G) are compatible with the aircraft. Any limitations must be noted in the AFM. As with all avionics systems in complex aircraft, a safety assessment is required: unannunciated failures or hazardous misleading information is improbable for Class B, remote for Class C and Probable for Class A.

Ground tests and flight tests are required to show airworthiness; the ground tests entail the following:

- Verification of proper receipt and processing of messages
- Verification of proper integration with TCAS
- Verification of proper integration with the position sensor
- Verification of accuracy of CDTI displayed information
- Simulated messages from test equipment and targets of opportunity should be used for the ground evaluation

During ground tests, the following traffic information should be validated as received correctly:

- Relative horizontal position
- Ground speed/differential ground speed
- Directionality
- Pressure altitude and relative pressure altitude
- Vertical trend
- Air/ground status
- Flight ID
- TIS-B/ADS-R service status
- TCAS inputs

Ground tests will additionally verify:

- EMI/EMC
- Correct detection and annunciation of failures
- General arrangement of controls and displays
- Human factors emphasizing FAA key areas of concern
- Display symbology IAW appendix 2 of the AC
- ADS-B self-test

The reader will note that the required ground tests are nearly identical to those found in the TCAS airworthiness testing.

Flight tests need to be conducted with cooperative ADS-B aircraft and conducted within the TIS-B and ADS-R coverage areas. The flight tests should verify:

- Aircraft flight ID
- Ability to select a desired target aircraft
- Ability to display ground speed
- Bearing/distance to aircraft
- Altitude and relative altitude
- Direction
- Appropriate display of targets during maneuvers

During maneuvers, the displayed targets should not exhibit jitter or ratcheting and should not blur or shimmer. Filtering (of track data) should not induce positioning errors and false or redundant tracks should not occur regularly. If installed, surface target information on the CDTI should match the real world. If ITP is installed, the functionality should be evaluated in accordance with the maneuver criteria described in Table 2 of AC 20-172A.

6.14 | TERRAIN AWARENESS AND WARNING SYSTEM

The terrain awareness and warning system (TAWS) really encompasses a family of terrain avoidance/awareness equipment. The basic system is known as a ground proximity warning system (GPWS), which depends on own-ship parameters and a radar altimeter. The enhanced ground proximity warning system (EGPWS) adds a digital terrain database to the basic system so the aircraft can look ahead, much like an onboard sensor such as radar or forward-looking infrared (FLIR). Helicopter TAWS (HTAWS) is a specialized version, and only applies to Part 27 and 29 rotorcraft. The primary reason for the installation of any TAWS is to prevent controlled flight into terrain (CFIT), which is an accident or incident in which the airplane, under the flight crew's control, is inadvertently flown into terrain, obstacles, or water without either sufficient or timely flight crew awareness to prevent the event. Since March 2005, the FAA has required the installation of TAWS for all turbine-powered airplanes with six passenger seats or more operating under Parts 91, 121, and 135. Larger aircraft have required TAWS equipment since 1992. Although not required, the FAA also provides guidance for general aviation aircraft with fewer than six passenger seats desiring TAWS installation.

The documentation that is relevant to the TAWS in relation to aircraft certification includes

- RTCA/DO-161A, "Minimum Performance Standards—Airborne Ground Proximity Warning Equipment," May 27, 1976
- TSO-C151c, June 27, 2012 "Terrain Awareness and Warning System," June 27, 2012
- AC 23-18, "Installation of Terrain Awareness and Warning System (TAWS) Approved for Part 23 Airplanes," June 14, 2000

- AC 25-23, "Airworthiness Criteria for the Installation Approval of a Terrain Awareness and Warning System (TAWS) for Part 25 Airplanes," May 22, 2000

- RTCA/DO-309, Minimum Operational Performance Standards (MOPS) for Helicopter Terrain Awareness and Warning System (HTAWS), March 13, 2008 contains the MOPS for HTAWS

- TSO-C194, Helicopter Terrain Awareness and Warning System (HTAWS), December 17, 2008, defines the standards manufacturers must meet for obtaining a TSOA for their product

- AC 27-1B and AC 29-2C, Miscellaneous Guidance (MG 18) contains a means of showing Airworthiness compliance

  The basic GPWS provides the following functions:

- Excessive rates of descent

- Excessive closure to terrain

- Negative climb rate or altitude loss after takeoff

- Flight into terrain when not in landing configuration

- Excessive downward deviation from an ILS glideslope

- Descent of the airplane to 500 ft above the terrain or nearest runway elevation (voice call out, "Five Hundred")

  The enhanced portion of TAWS (EGPWS) adds the following functions:

- Forward-looking terrain avoidance (FLTA), which includes reduced required terrain clearance (RTC) and imminent terrain impact. The FLTA function looks ahead of the airplane along and below the airplane's lateral and vertical flight path and provides suitable alerts if a potential CFIT threat exists.

- Premature descent alert (PDA). The PDA function of the TAWS uses the airplane's current position and flight path information as determined from a suitable navigation source and airport database to determine if the airplane is hazardously below the normal (typically 3°) approach path for the nearest runway as defined by the alerting algorithm.

The TAWS must provide an appropriate aural and visual signal for FLTA and PDA caution and warning alerts.

   There are three classes of TAWS equipment. Class A equipment contains all of the basic GPWS functions, FLTA, and PDA. Class B equipment must provide alerts for excessive rate of descent, negative climb rate, or altitude loss during takeoff, a 500 ft voice callout when the airplane descends to 500 ft above the nearest runway elevation, FLTA, and PDA. Class C equipment is designed for general aviation aircraft that are not required to have TAWS installed. Class C TAWS equipment must meet all the requirements for class B TAWS with the small exceptions noted in TSO-C-151b. A summary of Class A and Class B requirements is shown in Table 6.23.

   Class A equipment must provide terrain information to be presented on a display system. The following information must be provided on the display:

- The terrain must be depicted relative to the airplane's position such that the pilot may estimate the relative bearing to the terrain of interest.

TABLE 6.23 ▪ Summary of TAWS Requirements

Class A and Class B TAWS Requirements								
TAWS Class	Operating Rule	Passenger Seats (minimum)	FLTA	PDA	GPWS	FMS/RNAV or GPS	Display	Terrain Database
A	121	—*	Yes	Yes	1–6	FMS or GPS	Yes	Yes
A	135	>9	Yes	Yes	1–6	GPS	Yes	Yes
B	135	6–9	Yes	Yes	1, 3, 6	GPS	No	Yes
B	91	≥6	Yes	Yes	1, 3, 6	GPS	No	Yes

*There is no seat threshold for 14 CFR Part 121. All 14 CFR Part 121 airplanes affected by the TAWS rules must install TAWS regardless of the number of seats.
The GPWS equipage requirements in 14 CFR 121.360 and 123.153 expired and are superseded by the TAWS requirements in 14 CFR 121.354 and 135.154. TSO C-151c Class A equipment must meet the TSO C-92c requirements, but a separate TSO authorization for TSO-C92c is not required.

TABLE 6.24 ▪ TAWS RTC by Phase of Flight

Phase of Flight	TERPS (ROC)	TAWS RTC Level Flight	TAWS RTC Descending
Enroute	1000 ft	700 ft	500 ft
Terminal Intermediate Segment	500 ft	350 ft	300 ft
Approach	250 ft	150 ft	100 ft
Departure	48 ft/nm	100 ft	100 ft

- The terrain must be depicted relative to the airplane's position such that the pilot may estimate the distance to the terrain of interest.
- The terrain must be oriented to either the heading or the track of the airplane; a north-up orientation may be added as a selectable format.
- Variations in terrain elevation depicted relative to the airplane's elevation (above and below) must be visually distinct. Terrain that is more than 2000 ft below the airplane's elevation need not be depicted.

Operators required to install Class B equipment are not required to include a terrain display; however, the equipment must be capable of driving a terrain display function in the event that the installer wants to include a terrain display.

Class A and B equipment must provide suitable alerts when the airplane is currently above the terrain in the airplane's projected flight path but the projected amount of terrain clearance is considered unsafe for the particular phase of flight. The required obstacle (terrain) clearance (ROC) as specified in the U.S. Standard for Terminal Instrument Procedures (TERPS) and the Aeronautical Information Manual (AIM) have been used to define the minimum requirements for obstacle/terrain clearance (RTC) appropriate to the FLTA function. The TAWS RTC as a function of the phase of flight is shown in Table 6.24.

Class A and B equipment must provide suitable alerts when the airplane is currently below the elevation of a terrain cell along the airplane's lateral projected flight path and based on the vertical projected flight path the equipment predicts that the clearance will be less than that depicted in Table 6.20. The equipment must provide the alerts for reduced RTC and imminent terrain impact in straight and level and turning flight.

Class A and B equipment must also provide an alert when it determines that the airplane is significantly below the normal approach flight path to a runway. Alerting criteria may be based on height above runway elevation and distance to the runway; it may also be based on height above the terrain and distance to the runway. The documentation allows the manufacturer to determine the means for mechanizing premature descent. The PDA should be available for all types of instrument approaches: straight-in, circling, and approaches not aligned within 30° of the runway heading. The TAWS equipment should not generate PDAs for normal operations around the airport: for example, traffic patterns, circling minimums, VFR enroute to the airport.

A standard set of visual and aural alerts has been defined for the TAWS, and these are repeated in Table 6.25.

In addition to providing the aural and visual alerts listed in Table 6.21, the TAWS must perform the following functions:

- The required aural and visual alerts must initiate from the TAWS simultaneously, except when suppression of aural alerts are necessary to protect pilots from nuisance aural alerting.

- Each aural alert must identify the reason for the alert, such as "Too Low Terrain" and "Glideslope" or other acceptable annunciation.

- The system must remove the visual and aural alert once the situation has been resolved.

- The system must be capable of accepting and processing airplane performance-related data or airplane dynamic data and providing the capability to update aural and visual alerts at least once per second.

- The visual display of alerting information must be immediately and continuously displayed until the situation is no longer valid.

Class A equipment must have an interactive capability with other external alerting systems so an alerting priority can be automatically executed for the purpose of not causing confusion on the flight deck from different alerting systems. Class B equipment does not require prioritization with external systems; if prioritization with those functions is provided, the prioritization scheme must be the same as for Class A equipment. An approved prioritization scheme is depicted in Table 6.26.

Class B equipment must establish an internal priority system for each of the functions. The priority scheme must ensure that more critical alerts override the presentation of any alert of lesser priority. Class B equipment need only consider the TAWS functions required for Class B equipment. Table 6.27 provides a TAWS internal priority scheme.

Since the TAWS is based on a digital terrain database, two things have to be correct for the system to work properly. The first is the onboard navigation system; the navigator places the aircraft on the map. The second is the map itself. Both of these systems must be accurate to issue timely alerts and reduce the number of false alarms. Horizontal position for TAWS must come from a GNSS source meeting TSO C-129(AR) or any revision of TSO C-145, TSO C-146 or TSO C-196. The TSO (C-151c) notes that Part 121 operators may operate on a nonGNSS source. Vertical position may come from a barometric or GNSS source; Class A requires a RADAR altimeter.

Class A and B equipment that use a GPS internal to the TAWS for horizontal position information and are capable of detecting a positional error that exceeds the

TABLE 6.25 ■ Standard Set of TAWS Visual and Aural Alerts

Standard Set of Visual and Aural Alerts		
Alert Condition	**Caution**	**Warning**
FLTA Functions Reduced Required Terrain Clearance and Imminent Impact with Terrain Class A & Class B	**Visual Alert** Amber text message that is obvious, concise, and must be consistent with the aural message. **Aural Alerts** Minimum selectable voice alerts: "Caution, Terrain; Caution, Terrain" and "Terrain Ahead; Terrain Ahead"	**Visual Alert** Red text message that is obvious, concise, and must be consistent with the aural message. **Aural Alerts** Minimum selectable voice alerts: "Terrain, Terrain; Pull-Up, Pull-Up" and "Terrain Ahead, Pull-Up; Terrain Ahead, Pull-Up"
Premature Descent Alert (PDA) Class A & Class B	**Visual Alert** Amber text message that is obvious, concise, and must be consistent with the aural message. **Aural Alert** "Too Low Terrain"	**Visual Alert** None Required **Aural Alert** None Required
Ground Proximity Envelope 1, 2, or 3 Excessive Descent Rate Mode 1 Class A & Class B	**Visual Alert** Amber text message that is obvious, concise, and must be consistent with the aural message. **Aural Alert** "Sink Rate"	**Visual Alert** Red text message that is obvious, concise, and must be consistent with the Aural message. **Aural Alert** "Pull-Up"
Ground Proximity Excessive Closure Rate (Flaps not in Landing Configuration) Mode 2A Class A	**Visual Alert** Amber text message that is obvious, concise, and must be consistent with the aural message. **Aural Alert** "Terrain, Terrain"	**Visual Alert** Red text message that is obvious, concise, and must be consistent with the aural message. **Aural Alert** "Pull-Up"
Ground Proximity Excessive Closure Rate (Landing Configuration) Mode 2B Class A	**Visual Alert** Amber text message that is obvious, concise, and must be consistent with the aural message. **Aural Alert** "Terrain, Terrain"	**Visual Alert** Red text message that is obvious, concise, and must be consistent with the aural message for gear up. **Aural Alert** "Pull-Up"—for gear up None Required—for gear down
Ground Proximity Altitude Loss after Takeoff Mode 3 Class A & Class B	**Visual Alert** Amber text message that is obvious, concise, and must be consistent with the aural message. **Aural Alerts** "Don't Sink" and "Too Low Terrain"	**Visual Alert** None Required **Aural Alert** None Required

(Continues)

TABLE 6.25 ■ (Continued)

Alert Condition	Caution	Warning
	Standard Set of Visual and Aural Alerts	
Ground Proximity Envelope 1 (Gear and/or flaps other than landing configuration) Mode 4 Class A	**Visual Alert** Amber text message that is obvious, concise, and must be consistent with the aural message. **Aural Alerts** "Too Low Terrain" and "Too Low Gear"	**Visual Alert** None Required **Aural Alert** None Required
Ground Proximity Envelope 2 Insufficient Terrain Clearance (Gear and/or flaps other than landing configuration) Mode 4 Class A	**Visual Alert** Amber text message that is obvious, concise, and must be consistent with the aural message. **Aural Alerts** "Too Low Terrain" and "Too Low Flaps"	**Visual Alert** None Required **Aural Alert** None Required
Ground Proximity Envelope 3 Insufficient Terrain Clearance (Gear and/or flaps other than landing configuration) Mode 4 Class A	**Visual Alert** Amber text message that is obvious, concise, and must be consistent with the aural message. **Aural Alert** "Too Low Terrain"	**Visual Alert** None Required **Aural Alert** None Required
Ground Proximity Excessive Glideslope or Glidepath Deviation Mode 5 Class A	**Visual Alert** Amber text message that is obvious, concise, and must be consistent with the aural message. **Aural Alert** "Glideslope" or "Glidepath"	**Visual Alert** None Required **Aural Alert** None Required
Ground Proximity Altitude Callout (See Note 1) Class A & Class B (See Note 3)	**Visual Alert** None Required **Aural Alert** "Five Hundred"	**Visual Alert** None Required **Aural Alert** None Required

Note 1: The call out for ground proximity altitude is considered advisory.
Note 2: Visual alerts may be put on the terrain situational awareness display, if doing so fits with the overall Human Factors alerting scheme for the flight deck. This does not eliminate the visual alert color requirements, even in the case of a monochromatic display. Typically in such a scenario, adjacent, colored annunciator lamps meet the alerting color requirements.
Note 3: Additional callouts can be made by the system, but the system is required to make the 500 ft voice callout.

appropriate alarm limit for the existing phase of flight are considered acceptable. When the alarm limit is activated, the GPS computed position is considered unsuitable for the TAWS function and an indication must be given to the aircrew indicating that TAWS is no longer available due to navigation error.

TABLE 6.26 ▪ Alert Prioritization Scheme

ALERT PRIORITIZATION SCHEME			
Priority	Description	Alert Level[b]	Comments
1	Reactive Windshear Warning	W	
2	Sink Rate Pull-Up Warning	W	continuous
3	Excessive Closure Pull-Up Warning	W	continuous
4	RTC Terrain Warning	W	
5	V_1 Callout	I	
6	Engine Fail Callout	W	
7	FLTA Pull-Up warning	W	continuous
8	PWS Warning	W	
9	RTC Terrain Caution	C	continuous
10	Minimums	I	
11	FLTA Caution	C	7 s period
12	Too Low Terrain	C	
13	PDA ("Too Low Terrain") Caution	C	
14	Altitude Callouts	I	
15	Too Low Gear	C	
16	Too Low Flaps	C	
17	Sink Rate	C	
18	Don't Sink	C	
19	Glideslope	C	3 s period
20	PWS Caution	C	
21	Approaching Minimums	I	
22	Bank Angle	C	
23	Reactive Windshear Caution	C	
Mode 6[a]	TCAS RA ("*Climb*", "*Descend*", etc.)	W	continuous
Mode 6[a]	TCAS TA ("*Traffic, Traffic*")	C	continuous

[a]These alerts can occur simultaneously with TAWS voice callout alerts.
[b]W = Warning, C = Caution, A = Advisory, I = Informational.

TABLE 6.27 ▪ TAWS Internal Alert Prioritization Scheme

TAWS Internal Alert Priority Scheme	
Priority	Description
1	Sink Rate Pull-up Warning
2	Terrain Awareness Pull-up Warning
3	Terrain Awareness Caution
4	PDA ("Too Low Terrain") Caution
5	Altitude Callout (500 ft)
6	Sink Rate
7	Don't Sink (GPWS Mode 3)

As a minimum, terrain and airport information must be provided for the expected areas of operation, airports, and routes to be flown. The terrain and airport information must be of an accuracy and resolution suitable for the system to perform its intended function. Terrain data should be gridded at 30 arc seconds with 100 ft resolution within 30 nm of all airports with runway lengths of 3500 ft or greater, or whenever necessary (this is true in mountainous areas). It should be gridded at 15 arc seconds (or even 6 arc seconds) with 100 ft resolution within 6 nm of the closest runway. Terrain may be gridded in larger segments over oceanic and remote areas.

NOTE: There are 60 arc seconds in an arc minute, and an arc minute is 1/60 of 1°. One degree of latitude equals 60 nm. One arc second is $1/3600 \times 60 = 1/60$ nm = 100 ft; 30 arc seconds = 3000 ft; 15 arc seconds = 1500 ft; and 6 arc seconds = 600 ft (in latitude); for longitude, $1° = \cos$ latitude (60) nm.

Class A and B equipment must include a failure monitor function that provides reliable indications of equipment condition during operation. It must monitor the equipment itself, input power, input signals, and aural and visual outputs. A means must be provided to inform the flight crew whenever the system has failed or can no longer provide the intended function. The equipment must also have a self-test function to verify system operation and integrity. Failure of the system self-test must be annunciated.

Class A equipment must have the capability, via a control switch to the flight crew, to inhibit only the FLTA function, the PDA function, and the terrain display. This is required in the event of a navigation failure or other failures that would adversely affect the system. The basic TAWS (GPWS) required functions should remain active when the inhibit function is used. Automatic inhibit is acceptable, but must be annunciated to the crew.

TAWS search volume and alerting logic is based on the aircraft's phase of flight. The TSO defines four phases of flight for TAWS: enroute, terminal, approach, and departure. The phases are defined by distances from and height above the nearest airport.

- Enroute. The enroute phase exists anytime the aircraft is more than 15 nm from the nearest airport or whenever the conditions for terminal, approach, or departure are not met.

- Terminal. The terminal phase exists when the airplane is 15 nm or less from the nearest runway while the range to the nearest runway threshold is decreasing and the airplane is at or below a straight line drawn between the two points specified in Table 6.28 relative to the nearest runway.

- Approach. The approach phase exists when the distance to the nearest runway threshold is equal to or less than 5 nm and the height above the nearest runway threshold is equal to or less than 1900 ft and the distance to the threshold is decreasing.

- Departure. The departure phase needs to be determined by some reliable parameter indicating that the aircraft is on the ground. This could be a weight-on-wheels sensor or logic, which states that if airspeed is less than 35 knots and altitude is ±75 ft, then the airplane must be on the ground. Similarly, logic may state that if airspeed is greater than 50 knots and altitude is greater than 100 ft above field elevation, then the airplane must be flying. At any rate, once the airplane has reached 1500 ft above the departure runway, the departure phase is terminated.

TABLE 6.28 ▪ Height Above Versus Distance to the Runway

Distance to Runway	Height Above Runway
15 nm	3500 ft
5 nm	1900 ft

6.14.1 TAWS Evaluations

Each of the pertinent ACs (AC 25-23, AC 23-18) is similar in the proposed method of airworthiness certification. The number of seats and the operating rules of the applicant will determine the class of TAWS equipment required (see Table 6.23). This section addresses the evaluations necessary for Class A and Class B TAWS installation. TAWS Class C installation will be addressed in the next section.

As noted in earlier sections of this chapter, the proposed certification plan will be multifaceted. Since this is an integrated and highly complex system, a systems safety assessment will be required. The probability of an unannounced failure, hazardously misleading information (HMI), or false alerts at the box level is 10^{-5}. In order to accomplish this, the TAWS needs to meet the following criteria:

- The probability of a failure that would lead to the loss of all TAWS functions shall be $\leq 10^{-3}$ per flight hour.
- The probability of a false caution or warning alert due to undetected or latent failures shall be $\leq 10^{-4}$ per flight hour.
- The probability of an unannounced failure of the system to provide the required alerting functions due to undetected or latent failures shall be $\leq 10^{-4}$ per flight hour.
- The probability of the system providing HMI to the TAWS display due to undetected or latent failures shall be $\leq 10^{-4}$ per flight hour.
- Failure of the installed TAWS shall not degrade the integrity of any essential or critical system installed in the airplane with which the TAWS interfaces.

In addition to the safety assessment, the applicant will be expected to submit a human factors plan, testing plan, conformity and continued airworthiness plans, and compliance documentation. Simulators are mentioned in the ACs as viable tools as an aid to certification. Some of the characteristics of a TAWS installation and flight deck integration that may be evaluated via simulation are

- Displays
- Alert prioritization
- Mode transitions
- Pop-up displays
- Autoranging
- Self-test
- Operational workload issues
- Accessibility and usability of the TAWS controls
- System failure modes

Two subjects that have not been mentioned previously have been brought up in this list. Autoranging is not required, but may be installed. A TAWS alert at a range greater than the selected display range will automatically provide the aircrew with a new display range so the threat can be seen. If implemented, it must be obvious to the crew that the range scale has changed; switching back to a manually selected range should require minimal effort. A pop-up feature is similar to display override, which was previously discussed. When a TAWS display is not being presented to the crew, a TAWS alert pop-up feature automatically displays TAWS-related information.

6.14.2 TAWS Ground Test

The ACs state that a ground evaluation should be conducted for each TAWS installation. Once again, simulators may be used to enhance/reduce the amount of on-aircraft ground testing. In addition, computer-generated test equipment accepted by the TAWS manufacturer may be used in place of flight testing for basic TAWS Class B equipment and minimize the extent of flight testing required for Class A TAWS equipment. Some of the items that should be considered for the ground test are as follows:

- Acceptable location of the TAWS controls, displays, and annunciators
- Exercise self-test functions
- Evaluation of identified failure modes
- Evaluation of all discretes and TAWS interfaces
- Electromagnetic interference (EMI) and EMC
- Electrical transient testing
- Visibility and characteristics of any provided display under all lighting conditions

6.14.3 TAWS Flight Test

AC 25-23 and AC 23-18 provide identical flight test requirements for the installation of TAWS Class A and B equipment. The level of flight testing is determined by the type of aircraft, the level of integration and system architecture, and whether or not the system has been certified in another aircraft. The requirements for flight testing are determined for each installation. Any first-time installation or new sensor inputs require flight testing. The ACs provide flight test guidelines by examining six examples of certification and which modes of the TAWS must be evaluated in flight. Table 6.29 relates the type of installation to the required flight test.

- Example 1 is a first-time installation of the vendor's equipment under the TC/STC process; thorough ground and flight testing must be conducted.
- Example 2 is a follow-on installation of a previously approved TAWS where a required sensor input has not been previously approved—a new barometric altitude input (or equivalent) that was not previously approved.
- Example 3 is a follow-on installation of a previously approved TAWS system where the terrain display has not been previously approved.
- Example 4 is a follow-on installation of a previously approved TAWS system where the sensor that provides the horizontal position input has not been previously approved (e.g., perhaps upgrading the GPS to a GPS with WAAS capabilities).

TABLE 6.29 ■ Flight Test Matrix for TAWS Examples

TAWS Functions	Example 1	Example 2	Example 3	Example 4	Example 5	Example 6
FLTA	X	X	X	X		X
PDA	X	X			X	X
Basic GPWS	X				X	
Terrain Display (Class A)	X		X	X		X
Horizontal Position Source	X			X		X

Example 5 does not apply to Class B equipment.
Class B equipment are not required to provide a display.

- Example 5 is a follow-on installation of a previously approved TAWS where the sensor that provides radar altitude has not been previously approved.

- Example 6 involves the installation of a vendor's TAWS into an airplane previously fitted with basic GPWS from the same vendor and the algorithms and sensors for the basic GPWS functions of TAWS remain the same.

The ACs are quite nebulous in discussing the flight test considerations for the TAWS airworthiness evaluations. For the FLTA function, the ACs only state that

- All alerts (cautions and warnings) are given at an appropriate point in the test run.

- All pop-up, autorange, and other display features are working correctly.

- The display depicts the terrain accurately.

The ACs advise that test runs be made at 500 ft above the terrain or obstacle of interest and at least 15 nm from the nearest airport (enroute phase of flight), and if this is not possible, lower the fly-over altitude to 300 ft (terminal/intermediate phase of flight). The clearance numbers (500 and 300 ft) are only good for an aircraft that is descending. The clearances numbers for level flight in the same phases of flight are 700 and 350 ft respectively (see Table 6.24). For an accurate determination of whether your TAWS system is installed correctly or not you would have to revisit TSO-C-151c. An accurate evaluation of the FLTA function requires test points for level and descending flight during the four phases of flight previously discussed: enroute, terminal, approach, and departure. The warnings that are provided to the aircrew will ensure that the RTC altitudes in Table 6.24 can be met. (The TSO assumes a 1 sec pilot response time and 0.25 g constant g pull-up at various descent rates.) Table 6.30 shows the test criteria for TAWS cautions and warnings for FLTA during a descent in the enroute phase of flight. Figure 6.30 depicts the 100 ft/min case.

It should be fairly obvious that this can only work correctly if the altitude of the obstruction, its horizontal position, and the aircraft's position and altitude are known. We must also compare apples to apples. If the database is in WGS-84, then our TAWS must measure altitude (height) in WGS-84. If we try and compare barometric altitude with a WGS-84 database we will run into some severe problems, especially at higher altitudes or in extremely cold climates.

During enroute level flight (±500 ft/min), a terrain alert should be posted when the airplane is within 700 ft of the terrain and is predicted to be equal to or less than 700 ft within the prescribed alerting time or distance.

TABLE 6.30 ■ TAWS Enroute Descent Alerting Criteria

Enroute Descent Alerting Criteria					
Descent Rate (ft/min)	Alt lost with 3 sec Pilot Delay	Alt Required to L/O with 0.25g	Total Alt Lost Due to Recovery Maneuver	Maximum TAWS Caution Alert Height Above Terrain: "Caution Terrain," "Terrain Ahead"	Minimum TAWS Warning Alert Height Above Terrain: "Terrain, Terrain, Pull-up, Pull-up," "Terrain Ahead, Pull-up, Terrain Ahead, Pull-up"
1000	50	17	67	1200 ft	567 ft
2000	100	69	169	1400 ft	669 ft
3000	200	278	478	1800 ft	978 ft

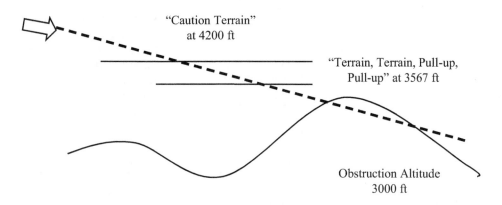

Aircraft Altitude 5000 ft
1000 ft/min Descent

"Caution Terrain" at 4200 ft

"Terrain, Terrain, Pull-up, Pull-up" at 3567 ft

Obstruction Altitude 3000 ft

FIGURE 6.30 ■
Enroute Descent
Alert at 1000 ft/min

TABLE 6.31 ■ TAWS Terminal Descent Alerting Criteria

Terminal Descent Alerting Criteria					
Descent Rate (ft/min)	Alt Lost with 1 sec Pilot Delay	Alt Required to L/O with 0.25g	Total Alt Lost Due to recovery Maneuver	Maximum TAWS Caution Alert Height Above Terrain: "Caution Terrain," "Terrain Ahead"	Minimum TAWS Warning Alert Height Above Terrain: "Terrain, Terrain, Pull-up, Pull-up," "Terrain Ahead, Pull-up, Terrain Ahead, Pull-up"
1000	17	17	34	700 ft	334 ft
2000	33	69	102	900 ft	402 ft
3000	50	156	206	1100 ft	506 ft

For the terminal phase of flight (descent), a terrain alert must be provided in time as to ensure that the airplane can level off with a minimum of 300 ft altitude clearance over the terrain/obstacle. The pilot reaction time, pull-up maneuver, and descent rates are the same as previously described for the enroute test. Table 6.31 shows the terminal descent alerting criteria for TAWS.

The test is run the same way as depicted in Figure 6.30 with the aircraft within 15 nm of an airport. For level flight during the terminal phase, a terrain alert should be posted when the airplane is less than 350 ft above the terrain and is predicted to be within 350 ft or less for the prescribed alerting time or distance.

During the approach phase (distance to nearest runway threshold is ≤5 nm, height above threshold is ≤1900 ft, and distance to runway is decreasing), a terrain alert must be provided in time to ensure that the airplane can level off with a minimum of 100 ft altitude clearance over the terrain/obstacle. The assumed pilot response is the same as for enroute and terminal conditions; the vertical velocity cases are somewhat lower. Table 6.32 provides the conditions for the TAWS alerts for descent in the approach phase.

During level flight operations at the minimum descent altitude (MDA), a terrain alert should be posted when the airplane is within 150 ft (it is allowable to issue an alert as much as 200 ft) of the terrain and is predicted to be ≤150 ft within the prescribed alerting time or distance.

The final function of FLTA is imminent terrain impact. Class A and B TAWS equipment must alert the aircrew when the aircraft is currently below a terrain cell along

TABLE 6.32 ▪ TAWS Approach Descent Alerting Criteria

Approach Descent Alerting Criteria					
Descent Rate (ft/min)	Alt Lost with 1 sec Pilot Delay	Alt required to L/O with 0.25g	Total Alt Lost Due to Recovery Maneuver	Maximum TAWS Caution Alert Height Above Terrain: "Caution Terrain," "Terrain Ahead"	Minimum TAWS Warning Alert Height Above Terrain: "Terrain, Terrain, Pull-up, Pull-up," "Terrain Ahead, Pull-up, Terrain Ahead, Pull-up"
500	8	4	12	350 ft	112 ft
750	12	10	22	400 ft	122 ft
1000	17	18	35	450 ft	135 ft
1500	25	39	64	550 ft	164 ft

TABLE 6.33 ▪ FLTA Imminent Terrain Impact Alerts

Imminent Terrain Impact Alerting Criteria				
Groundspeed (knots)	Height of Terrain Cell (MSL)	Distance from Runway (NM)	Test Altitude Run-in (MSL)	Alert Criteria
200–300	10,000	30	9000	Must Alert
400–500	10,000	30	8000	Must Alert
150–250	2000	10	1500	Must Alert
100–140	600	5	500	Must Alert
100–160	600	4	200	Must Alert
160	600	5	500	Must Alert

the aircraft's lateral flight path and based on the vertical projected flight path the equipment predicts that the RTC (Table 6.24) cannot be met. The assumed pilot response is as previously described. Table 6.33 contains the TSO recommended test conditions for FLTA imminent terrain impact.

So, as a recap of the flight tests required for FLTA:

- Enroute descent alert
- Enroute level alert
- Terminal descent alert
- Terminal level alert
- Approach descent alert
- Approach level alert
- Imminent terrain impact

The second mode of TAWS is the Premature Descent Alert (PDA). The purpose of the evaluation is to verify that the pilot will be alerted to a low altitude condition at an altitude that is defined by the specific design PDA alert surface. In other words, the TSO does not specify unique "pass/fail" criteria for this mode. When applying for a TSO, the applicant defines these criteria along with the proposed recovery procedures. When performing the flight test of this function, the evaluator must first determine the PDA alert surface used by the vendor and perform the recovery procedures used in the vendor's assumptions.

TABLE 6.34 ▪ PDA Test Conditions

Premature Descent Alert Criteria				
Groundspeed (knots)	Vertical Speed (ft/min)	Distance from Runway Threshold (nm)	PDA Alert Height (MSL)	Recovery Altitude (MSL)
80	750	15	Determined	Determined
100	1500	15		
120	750	15		
140	1500	15		
160	750	15	By	By
200	1500	15		
250	2000	15		
80	750	12		
100	1500	12		
120	750	12	Vendor	Vendor
140	1500	12		
160	750	12		
80	750	4		
100	1500	4		
120	750	4		
140	1500	4		
80	750	2		
100	1500	2		
120	750	2		
140	1500	2		

Flight testing of the PDA function can be conducted in any airport area within 10 nm of the nearest runway. The airplane should be configured for landing at approximately 1500 ft AGL along the final approach segment of the runway at approximately 10 nm from the runway. At the 10 nm point, a nominal 3° glide path descent should be initiated and maintained until the PDA occurs.

AC 25-23 and AC 23-18 both note that the runway that is selected should be relatively flat along the extended centerline, as terrain/obstacles will trigger the FLTA function. The PDA caution will sound "Too Low Terrain," as opposed to the "Caution Terrain" or "Terrain Ahead" for the FLTA function. The AC also notes that if it is not possible to find a runway that is free of terrain/obstacles, then the evaluator may increase the barometric altitude by 1 inHg to allow radar altimeter inputs to trigger the PDA. Since the recovery altitude is based on groundspeed and vertical velocity, the TSO recommends that this evaluation be conducted with multiple combinations of these two parameters. Table 6.34 defines these conditions.

In addition to evaluating the FLTA and PDA functions, the TAWS should be demonstrated as free from nuisance alerts. In general, the aircraft should be able to conduct normal flight operations without nuisance TAWS alerts. The TSO specifies that that for each phase of flight

- It must be possible to descend at 4000 ft/min in the enroute phase and level-off 1000 ft above the terrain using a normal level-off procedure without a TAWS caution or warning.

- It must be possible to descend at 2000 ft/min in the terminal phase and level-off 500 ft above the terrain using normal level-off procedures without a TAWS caution or warning.

FIGURE 6.31 ▪
Mode 1, Excessive
Rate of Descent

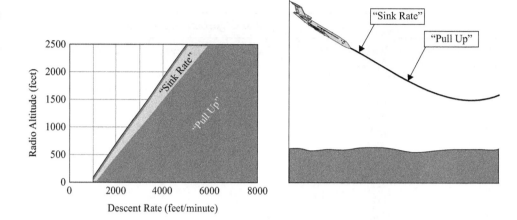

- It must be possible to descend at 1000 ft/min in the approach phase and level-off at the MDA using normal level-off procedures without a TAWS caution or warning.

The TSO calls out specific instrument approaches that should be flown with lateral and vertical errors to ensure adequate protection by the TAWS. In addition, the TAWS should be evaluated against known accident cases to ensure integrity. These tests are performed in simulation by the vendor and will not be addressed here.

TAWS class A equipment must contain the basic GPWS functions: excessive rate of descent, excessive closure to terrain, negative climb rate or loss of altitude after takeoff, flight into terrain when not in landing configuration, excessive downward deviation from an ILS glideslope, and voice callouts "500 feet." Class B TAWS equipment is only required to have three of the basic GPWS functions: excessive rate of descent, negative climb rate or altitude loss after takeoff, and voice callouts.

The minimum operational performance standards for the basic GPWS is contained in RTCA DO-161A and the governing TSO is TSO-C92c. Airworthiness guidance is the same as for TAWS (AC 25-23 and AC 23-18). There are some slight differences between the Class A and Class B functions of the basic GPWS and they will be addressed shortly.

Excessive rate of descent (mode 1) should be tested over level terrain while verifying the system correlates to the RTCA DO-161A envelopes. The mode 1 envelope for the caution "Sink Rate" and the warning "Pull Up" are shown in Figure 6.31.

For Class A TAWS equipment, the aircraft should initiate the test above the AGL altitude selected for the test and "fly" into the envelope noting the altitude (AGL) and vertical velocity at the callouts. For example, establish the aircraft at 2500 ft AGL and initiate a descent rate of 3000 ft/min. The TAWS should signal "Sink Rate" at 1250 ft AGL and "Pull Up" at 100 ft AGL. This test verifies the proper installation of barometric altitude (and the corresponding calculation of barometric altitude rate) and the radar altitude.

For Class B TAWS equipment, the TSO presents a slightly different chart for mode 1 and is shown at Figure 6.32. The y-axis of the chart in the TSO is labeled "Height Above Terrain" as opposed to radar altitude in Figure 6.31. This is because (in Class B installations) the height is determined by subtracting terrain elevation (from the terrain database) from the current QNH barometric altitude (or equivalent). Remember apples to apples! In both cases, only one test run is required to verify the installation.

The test of excessive closure rate to terrain (mode 2) verifies the installation/ integration of the radar altimeter. The envelopes for the caution "Terrain, Terrain" and

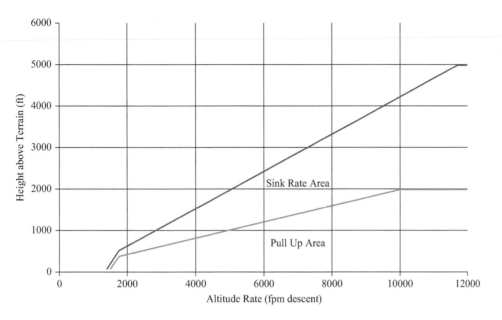

FIGURE 6.32 ■
Class B Excessive
Descent Rate

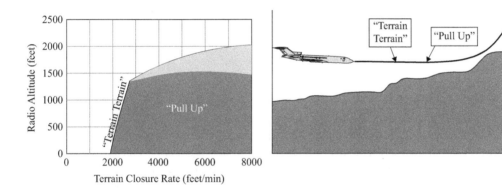

FIGURE 6.33 ■
Mode 2A Excessive
Terrain Closure,
Flaps Up

the warning "Pull Up" vary when the flaps are in a landing configuration or not. This difference is called out as mode 2A (flaps not in a landing configuration) and mode 2B (flaps in a landing configuration). These modes are for class A equipment only. The evaluation should be flown over an area of known rising terrain where the slope and airspeed combination will trigger the aural alerts. Mode 2A and code 2B envelopes are depicted in Figures 6.33 and 6.34, respectively.

Mode 3, negative climb rate or altitude loss after takeoff, is the same for TAWS class A and class B equipment, however, like mode 2, class B equipment determines a height above the runway using barometric altitude (or equivalent) and runway elevation in lieu of radar altitude. This test is conducted immediately after takeoff before climbing above 700 ft AGL. If a descent is initiated following a takeoff or go-around, the TAWS stores the altitude value at which the descent began and compares successive altitude data to the stored value. Activation of the "Don't Sink" or "Too Low Terrain" is induced when a minimum terrain clearance as a function of altitude loss is exceeded. This warning area is depicted in Figure 6.35. The flight test would entail performing a

FIGURE 6.34 ■
Mode 2B Excessive
Terrain Closure,
Flaps for Landing

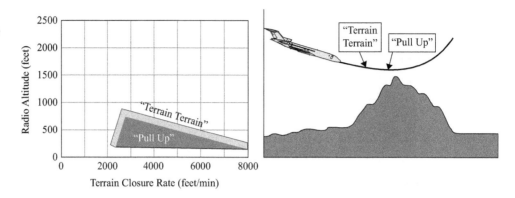

FIGURE 6.35 ■
Mode 3 Altitude
Loss After Takeoff

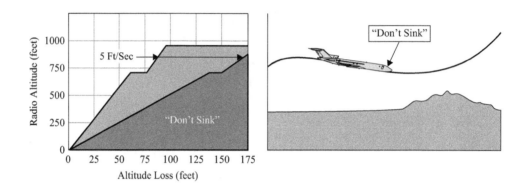

normal takeoff and, at 500 ft AGL, initiating a descent rate of 5 ft/sec and noting the altitude lost and current altitude when the alert is given.

Mode 4, flight into terrain when not in landing configuration, is only required for class A equipment and provides alerts and warnings for insufficient terrain clearance based on phase of flight and airspeed. This mode is broken down into three submodes: mode 4A, mode 4B, and mode 4C. Mode 4A is active during the enroute and approach phase of flight, with gear not in landing configuration, and has two warning areas initiating a "Too Low Terrain" or "Too Low Gear." Figure 6.36 shows these warning regions.

Mode 4B is also active in the enroute and approach phase of flight with gear in a landing position, but flaps not in a landing configuration. Similar to mode 4A, there are two warning regions described by the TSO that will issue the alerts "Too Low Terrain" or "Too Low Flaps." Figure 6.37 shows the warning areas for mode 4B.

Mode 4C is active during the takeoff phase of flight with either gear or flaps not in landing configuration. It is designed to prevent CFIT on takeoff climb into terrain that produces an insufficient closure rate for a mode 2 warning. Mode 4C is based on a minimum terrain clearance that increases with radar altitude during takeoff (Figure 6.38). Modes 4A and 4B should be tested while conducting a visual approach to a runway, planning to enter the warning areas as depicted. Mode 4C should be conducted from a runway with rising terrain on the extended departure centerline.

Mode 5 provides alerts when the airplane descends below the glideslope on an ILS approach. The alerts are issued in two distinct tones depending on the aircraft's position on the glideslope. Figure 6.39 depicts the alert areas for mode 5. When the airplane is at

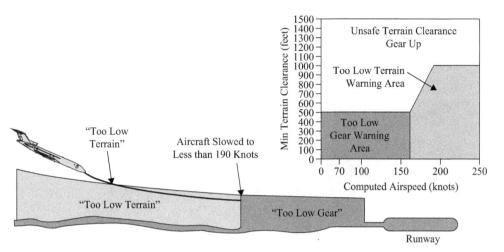

FIGURE 6.36 ■
Mode 4A Unsafe
Terrain Clearance,
Gear Up

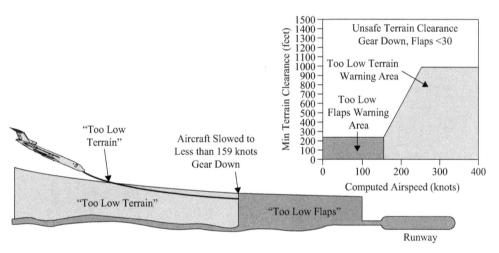

FIGURE 6.37 ■
Mode 4B Unsafe
Terrain Clearance,
Gear Down, Flaps
Not In Landing
Configuration

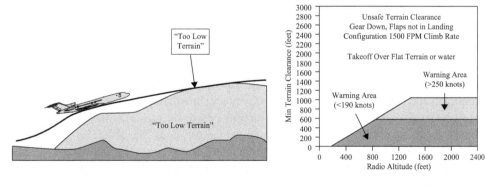

FIGURE 6.38 ■
Mode 4C Unsafe
Terrain Clearance,
Minimum Terrain
Clearance

least 1.3 dots (but not more than 2 dots) below the glideslope, a soft "Glideslope" is issued. If the aircraft continues the descent and deviates more than 2 dots below the glideslope, a hard "Glideslope" is issued. The evaluation is performed on an ILS approach first setting the aircraft 1.5 dots below the glideslope and then increasing the deviation to just over 2 dots.

FIGURE 6.39 ■
Excessive
Downward Deviation
from an ILS
Glideslope

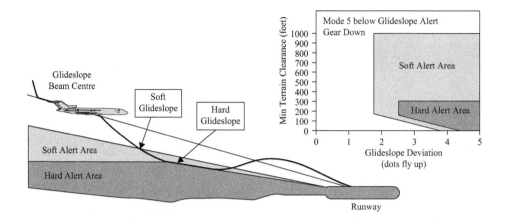

FIGURE 6.40 ■
Mode 6 Advisory
Callouts

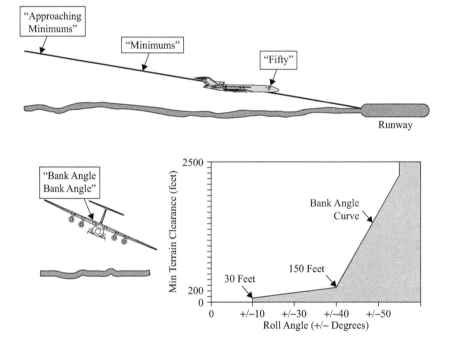

Mode 6 is the advisory callout. Only one callout is mandatory for class A and class B equipment and that is the "500" that will be seen and heard during all of the tests that have been previously discussed whenever the aircraft passes through 500 ft. Unlike class A, which uses a radar altimeter, class B equipment determines the 500 ft height by subtracting runway elevation (from the airport database) from the current barometric altitude (or equivalent). When the height above terrain first reaches 500 ft, a single voice alert "500" or equivalent must be provided.

Many aircraft have elected to incorporate other advisory callouts into the TAWS. They may be, but are not limited to, "Approaching Minimums," "Minimums," "50" (ft), or "Bank Angle." Although not required, if they are incorporated they must be proven as correctly performing their intended function. Figure 6.40 presents examples of the mode 6 advisory callouts.

TABLE 6.35 ■ TAWS Required Terrain Clearance

Phase of Flight	Small Aircraft Required Obstacle Clearance	TAWS Required Terrain Clearance, Level Flight	TAWS Required Terrain Clearance, Descending
Cruise	500 ft	250 ft	200 ft
Takeoff	48 ft/nm	100 ft	100 ft
Landing	250 ft	150 ft	100 ft

The TAWS may also be used to issue wind shear warnings. This is accomplished by looking at increasing and decreasing headwinds, and increasing and decreasing updrafts and downdrafts. In some aircraft, this is called TAWS mode 7, wind shear warnings. An evaluation of this mode would be performed in a simulator or by using vendor-approved computerized test equipment. Mode 7 is not required by the TSO, but if installed, it must be able to perform its intended function.

6.14.4 TAWS Class C Evaluations

There are some modifications to the TSO regarding TAWS class C equipment. Class C equipment is intended for small general aviation airplanes that are not required to install class B equipment.

The first change to class B requirements involves a change in the description of phase of flight. Class C defines phases of flight as

- Takeoff: positive rate of climb, inside the airport traffic control area, distance to the nearest runway threshold is increasing, and the airplane is below 1000 ft.

- Cruise: anytime the airplane is outside of the airport traffic control area.

- Landing: the airplane is inside the airport traffic control area, the distance to the nearest runway threshold is decreasing, and the airplane is below 1000 ft.

NOTE: The vendor is the determining factor of the airport control area, which is usually out to 10 nm from the nearest airport.

For the TAWS to work properly, a means must be provided to compute an actual aircraft MSL altitude that is immune to temperature errors and manual correction errors that would otherwise prevent the TAWS from performing its intended function. If the TAWS includes a terrain display output. This reference altitude value used for TAWS alerts should also be output for display. Since the altitude value is necessarily based on GPS-derived MSL altitude, which is required for horizontal position in all class B and C TAWS, the displayed value must be labeled MSL/G or MSL-G, or another obvious acronym that relates to the pilot that altitude is GPS-derived MSL altitude.

The RTC also differs from the TAWS class A and B equipment. Table 6.24 is revised for class C and is shown in Table 6.35.

The aural and visual alerts are also slightly modified from Table 6.25. Table 6.36 shows the TAWS class C standard set of visual and aural alerts. The test conditions specified in the TSO are appropriate to small aircraft.

A terrain alert during cruise descent must be issued in time as to ensure that the airplane can level off with a minimum of 200 ft clearance over the obstacle when descending toward terrain within the operational flight envelope of the airplane. The test

TABLE 6.36 ■ TAWS Class C Visual and Aural Alerts

STANDARD SET OF VISUAL AND AURAL ALERTS		
Alert Condition	Caution	Warning
Terrain Awareness Reduced Required Terrain Clearance	**Visual Alert** Amber text message that is obvious, concise, and must be consistent with the Aural message.	**Visual Alert** Red text message that is obvious, concise and must be consistent with the Aural message.
	Aural Alert Minimum Selectable Voice Alert: "Caution, Terrain; Caution, Terrain"	**Aural Alert** Minimum Selectable Voice Alert: "Terrain; Terrain"
Terrain Awareness Imminent Impact with Terrain	**Visual Alert** Amber text message that is obvious, concise, and must be consistent with the Aural message.	**Visual Alert** Red text message that is obvious, concise and must be consistent with the Aural message.
	Aural Alert Minimum Selectable Voice Alert: "Caution, Terrain; Caution, Terrain"	**Aural Alert** Minimum Selectable Voice Alert: "Terrain; Terrain"
Terrain Awareness Premature Descent Alert (PDA)	**Visual Alert** Amber text message that is obvious, concise and must be consistent with the Aural message.	**Visual Alert** None Required
	Aural Alert "Too Low; Too Low"	**Aural Alert** None Required
Ground Proximity Excessive Descent Rate	**Visual Alert** Amber text message that is obvious, concise, and must be consistent with the Aural message.	**Visual Alert** Red text message that is obvious, concise and must be consistent with the Aural message.
	Aural Alert "Sink Rate"	**Aural Alert** "Pull-Up"
Ground Proximity Altitude Loss after Take-off	**Visual Alert** Amber text message that is obvious, concise, and must be consistent with the Aural message.	**Visual Alert** None Required.
	Aural Alert "Don't Sink"	**Aural Alert** None Required.
Ground Proximity Voice Call Out (See Note 1)	**Visual Alert** None Required	**Visual Alert** None Required.
	Aural Alert "Five Hundred" or selected altitude	**Aural Alert** None Required

Note 1: The call out for ground proximity altitude is considered advisory.
Note 2: Visual alerts may be put on the terrain situational awareness display, if doing so fits with the overall Human Factors alerting scheme for the flight deck. This does not eliminate the visual alert color requirements, even in the case of a monochromatic display. Typically in such a scenario, adjacent, colored annunciator lamps meet the alerting color requirements.

conditions assume a descent along a flight path that has terrain that is 500 ft below the expected level-off altitude. If the pilot initiates the level-off at the proper altitude, no TAWS alert should be expected. If he delays the level-off, a TAWS alert is required to recover to level flight in a safe manner. The assumed pilot reaction time is 3 sec with a 1 g incremental pull-up at descent rates of 500, 1000, and 200 ft/min. Table 6.37 depicts the cruise descent class C alerting criteria.

TABLE 6.37 ■ Cruise Descent Class C Alerting Criteria

Cruise Descent Alerting Criteria					
Descent Rate (ft/min)	Alt Lost with 3 sec Pilot Delay	Alt Required to L/O with 1g Pullup	Total Alt Lost Due to Recovery Maneuver	Maximum TAWS Caution Alert Height Above Terrain: "Caution Terrain, Caution, Terrain"	Minimum TAWS Warning Alert Height Above Terrain: "Terrain, Terrain"
500	25	1	26	550 ft	226 ft
1000	50	4	54	600 ft	254 ft
2000	100	17	117	800 ft	317 ft

TABLE 6.38 ■ Approach Descent Class C Alerting Criteria

Approach Descent Alerting Criteria					
Descent Rate (ft/min)	Alt Lost with 1 sec Pilot Delay	Alt Required to L/O with 1g Pullup	Total Alt Lost Due to Recovery Maneuver	Maximum TAWS Caution Alert Height Above Terrain: "Caution Terrain, Caution, Terrain"	Minimum TAWS Warning Alert Height Above Terrain: "Terrain, Terrain"
500	8	1	9	300 ft	109 ft
750	12	2	14	325 ft	114 ft
1000	17	4	21	350 ft	121 ft

During cruise level flight conditions (vertical speed is ±200 ft/min), a terrain alert should be posted when the airplane is within 250 ft of the terrain and is predicted to be ≤200 ft above the terrain.

A terrain alert during approach descent must be given in time to ensure that the airplane can level off with a minimum of 100 ft clearance above the terrain/obstacle. Table 6.38 provides the alerting conditions for TAWS class C approach descent. The pilot conditions are the same as for cruise descent with the exception of descent rates of 500, 750, and 1000 ft/min.

The approach level flight requirements for alerting are the same as for class B equipment. During level flight operations at the MDA, a terrain alert should be posted when the airplane is within 150 ft of the terrain.

The criteria for FLTA imminent terrain impact is the same as for class B equipment (Table 6.33) except only speeds of 100 to 250 knots should be used and pilot recovery is an incremental 1 g. The PDA conditions are the same as for class B TAWS equipment (Table 6.34).

It must be possible to descend at 2000 ft/min and level off 500 ft above the terrain using a normal level-off procedure without a caution or warning alert. Further, it must be possible to descend at 1000 ft/min in the final approach segment and level off at 250 ft using normal level-off procedures without a caution or warning alert.

The basic GPWS functions of excessive descent rate, negative climb rate or altitude loss after takeoff, and altitude callouts are the same as written for class B equipment. A compliance matrix for a TAWS Class A installation is shown in Table 6.39.

TABLE 6.39 ■ Sample TAWS Class A Compliance Matrix

No.	Requirements AC Paragraph		Subject	Description
1	AC 25-23	13	Display Presentation	**a.** Terrain Display.
				b. Terrain Display Presentation.
				c. Pop-Up Mode-Switching Functionality.
				d. Auto-Range Switching Mode.
2	AC 25-23	14	Alerts	**a.** In addition to being complaint with the requirements of § 25.1322, the TAWS alerts should be clear, concise, and unambiguous.
				b. The visual and aural alerts are consistent with the alerting philosophy of the airplane flight deck in which the TAWS equipment is installed
3	AC 25-23	16	System Inhibit	**a.** A means for one of the flight crew to inhibit the following TAWS functions must be provided:
				• FLTA
				• PDA
				• Basic GPWS—Flight into terrain when not in the landing configuration
				• Basic GPWS—Excessive downward deviation from the ILS glideslope
				b. Care should be taken to ensure that instinctive, inadvertent, or habitual reflexive action by the flight crew does not inhibit these functions.
4	AC 25-23	18	System evaluation with simulators	Simulators may be used as a tool to evaluate specific installations of TAWS. The level of simulation fidelity required will depend on the type of credit being sought. Some of the characteristics of a TAWS installation and flight deck integration that may be evaluated via simulation are:
				• displays,
				• alert prioritization,
				• mode transitions,
				• pop-up displays,
				• auto-ranging,
				• self-test,
				• operational workload issues,
				• accessibility and usability of the TAWS controls,
				• and systems failure modes.

#				
5	AC 25-23	19 a	Ground Test Considerations	A ground test should be conducted for each TAWS installation. Some items to consider for ground test should include: • an acceptable location of TAWS controls, displays, and annunciators; • exercise of self-test functions; • evaluation of identified failure modes; • evaluation of all discrete and TAWS interfaces; • Electro-magnetic interference (EMI)/electro-magnetic compatibility (EMC) testing; and electrical transient testing.
6	AC 25-23	19b	Ground Test Considerations	Considerations can also be made for evaluating display characteristics if it can be shown that all of the performance aspects of the display that are available during flight can be evaluated on the ground.
7	AC 25-23	20b(1)	FLTA	Flight testing to verify the proper operation of the FLTA function can be conducted in an area where the terrain or obstacle elevation for the test runs is known to be within approximately 300 feet.
8	AC 25-23	20b(2)	FLTA	Test runs are recommended to be level flight at approximately 500 feet above the terrain/obstacle of interest.
9	AC 25-23	20c(1)	PDA	Flight testing to verify the proper operation of the PDA function can be conducted in any airport area within 10 NM of the nearest runway.
10	AC 25-23	20c(2)	PDA	The airplane should be configured for landing at approximately 1500 feet AGL along the final approach segment of the runway at approximately 10 NM from the runway.
11	AC 25-23	20c(3)	PDA	At the 10 NM point, a normal three degree flight path angle descent can be initiated and maintained until the PDA alert occurs.
12	AC 25-23	20(d)(1)	GPWS	Excessive Rates of Descent.
13	AC 25-23	20(d)(2)	GPWS	Excessive Closure Rate to Terrain. This test must be conducted in an area of known rising terrain.
14	AC 25-23	20(d)(3)	GPWS	Negative Climb Rate or Altitude Loss after Takeoff. This test is conducted immediately after takeoff before climbing above 700 AGL or above runway elevation.
15	AC 25-23	20(d)(4)	GPWS	Flight into Terrain when Not in Landing Configuration. This test should be conducted while on a visual approach to a runway.

(Continues)

TABLE 6.39 ■ (Continued)

No.	Requirements AC Paragraph	Subject	Description
16	AC 25-23	GPWS	Excessive Downward Deviation from an ILS Glideslope. This test should be conducted during an ILS approach.
17	AC 25-23	GPWS	Voice Callout "Five Hundred Feet." This test is conducted during an approach to a runway.
18	AC 25-23	Terrain Display	The terrain must be depicted relative to the airplane's position such that the pilot may estimate the relative bearing to the terrain of interest.
19	AC 25-23	Terrain Display	Flight testing to verify the proper operation of the terrain Display should be conducted while verifying all the other required TAWS functions.
20	AC 25-23	Terrain Display	Emphasis should be placed on verifying compliance with the provisions specified in paragraph 13.a. (Terrain Display) of this AC during normal airplane maneuvering during all phases of flight. Pop-up and auto-ranging features should be evaluated, if applicable.
21	AC 25-23	Terrain Display	The FAA recommends that the applicant perform sustained turns to evaluate: – symbol stability, – flicker, – jitter, – display update rate, – color cohesiveness, – readability, – the use of color to depict relative elevation data, – caution and warning alert area depictions, – map masking, and – overall suitability of the display.
22	AC 25-23	Added Features Flight Test Considerations	Flight testing may be required to verify the proper operation of added features such as: – windshear detection, – bank angle, – altitude call outs "Approaching Minimums," or – Other features not required by TSO-C151a.

| 23 | AC 25-23 | 20g | Pressure Altitude Variations in Cold Weather | The TAWS may be designed to account for the effects of cold weather on barometric altitude, while determining vertical position. Flight testing may be required, unless a suitable verification procedure can be conducted. This will depend on the design of the cold weather compensation. |

| 24 | TSO-C151c | Appendix 3 1.3 | Enroute Level Flight Requirement | |

GS (kts)	Height of Terrain Cell (MSL)	Test Altitude	Alert
200	5000	6000	No
250	5000	5800	No
300	5000	5800	No
200	5000	5700(0/-100)	Must
250	5000	5700(0/-100)	Must
300	5000	5700(0/-100)	Must
400	5000	5700(0/-100)	Must
500	5000	5700(0/-100)	Must

| 25 | TSO-C151c | Appendix 3 1.2 | Enroute Descent Requirement | TSO-C151c, Appendix 3 1.2, TABLE A, ENROUTE DESCENT ALERTING CRITERIA |

| 26 | TSO-C151c | Appendix 3 1.4 | Terminal Area (Intermediate Segment) Descent Requirement | Refer to TSO-C151c, Appendix 3 1.4, TABLE C Terminal Descent Area Alerting Criteria. |

| 27 | TSO-C151c | Appendix 3 1.5 | Terminal Area (Intermediate Segment) Level Flight Requirement | Refer to TSO-C151c, Appendix 3 1.5, TABLE D Terminal Area Level Flight Alerting Criteria. |

| 28 | TSO-C151c | Appendix 3 1.6 | Final Approach Descent Requirement | Refer to TSO-C151c, Appendix 3 1.6 TABLE E Final Approach Descent Alerting Criteria. |

(Continues)

TABLE 6.39 ■ *(Continued)*

No.	Requirements AC Paragraph		Subject	Description
29	TSO-C151c	Appendix 3 1.7	Final Approach Level Flight Requirement	Refer to TSO-C151c, Appendix 3 1.7 TABLE F Final Approach Level Flight Alerting Criteria.
30	TSO-C151c	Appendix 3 2.0	FORWARD LOOKING TERRAIN AVOIDANCE IMMINENT TERRAIN IMPACT TEST CONDITIONS	Refer to TSO-C151c, Appendix 3 2.0 TABLE G Imminent Terrain Impact Alerting Criteria.
31	TSO-C151c	Appendix 3 4.0	NUISANCE ALERT TEST	The test conditions mentioned in the TSO-C151c Appendix 3 4.0 must be conducted to evaluate TAWS performance during all phases of flight.
32	TSO-C151c	Appendix 3 3.1	PREMATURE DESCENT ALERT	Refer to TABLE H Premature Descent Alerting Criteria

6.14.5 HTAWS

HTAWS equipment, which is specific to Parts 27 and 29 helicopters, is divided into two classes: A and B. HTAWS provides two TAWS functions and five GPWS functions:

- FLTA, reduced required terrain clearance and imminent terrain or obstacle impact
- Excessive rates of descent (optional) (1)
- Excessive closure rate to terrain (2)
- Excessive altitude loss after takeoff (3)
- Flight into terrain when not in landing configuration (4)
- Excessive downward deviation from ILS glide slope (5)

Although HTAWS is not required by operational rules for helicopters, installation is consistent with TSO-C151c. Class A equipment must have a display, and may be either a WX RADAR, EFIS or other compatible MFD; orientation of the display is Track Up. Similar to TAWS, a GNSS provides the position source. Class A may provide automatic autorotation detection when excessive descent rate warning functions are included. The implementation should make provisions for a low altitude mode to allow VFR operations at less than enroute altitudes with minimal nuisance alerts.

Class B equipment provides only two principal alerts: imminent terrain or obstacle impact, and excessive altitude loss after takeoff. Class B does not require a display; if it does it must meet the requirements of Class A installations. Class B is required to interface with an approved GPS for horizontal position, does not require an interface to a RADAR altimeter, and should include a low altitude mode as described for Class A equipment.

Certification is accomplished through the TC, ATC, or STC process. The significant difference between TAWS and HTAWS is the lack of specific TSO requirements that will necessitate greater engineering and flight test efforts in the Airworthiness certification. In this light, much of the required testing is consistent with TSO-C151b, AC 23-18 and AC 25-23, adjusted for helicopter operations. Human factors, displays, alerts, auto pop-ups and priorities are as with the fixed wing requirements.

The Miscellaneous Guidance Section 18 (MG-18) of AC 27-1B and AC 29-2C provides an alternate table for Required Obstacle Clearance and Required Terrain Clearance for Helicopters (Table 6.40).

MG-18 offers flight test considerations and requirements for HTAWS evaluations. Excessive rate of descent (1) is tested over level terrain or water to validate correct interface of the Barometric and RADAR altimeters; only one test run is required to verify proper installation. Excessive closure rate to terrain (2) should be conducted in an area of known rising terrain; it is recommended that one level test run at 200' above the terrain peak elevation be conducted. For Class A this will validate proper installation of the RADAR Altimeter. Excessive altitude loss after takeoff (3) is conducted after a normal takeoff profile has been established but before reaching pattern altitude. This test verifies proper installation of barometric altitude, barometric altitude rate and RADAR altitude. Flight into terrain when not in landing configuration (4) is conducted on a visual approach over level terrain or water. For Class A equipment, this test verifies proper integration of airspeed, radar altitude, and gear sensor inputs. Excessive downward deviation from an ILS glide slope is conducted during an ILS approach. This test verifies proper integration of the GS to HTAWS; the test must be repeated for dual ILS installations.

TABLE 6.40 ▪ Helicopter RTC

Phase of Flight	TERPS (ROC)	TAWS (RTC) Level Flight	TAWS (RTC) Descending
Enroute	1000 ft	150 ft	100 ft
Terminal Intermediate	500 ft	150 ft	100 ft
Approach	250 ft	150 ft	100 ft
Departure See note	48 ft/nm	100 ft	100 ft

During the departure phase of flight, the FLTA function of Class A and B equipment should alert if the aircraft is projected to be within 100 ft vertically of terrain. Class A and B equipment should not alert if the aircraft is projected to be more than 150 ft above the terrain.

MG-18 provides the applicant with a list of applicable paragraphs from TSO-C151b (TAWS) and RTCA/DO-161A (GPWS) that will satisfy the airworthiness regulations.

6.15 | FLIGHT GUIDANCE SYSTEMS

The basic autopilot has been around in one form or another since 1910 when Lawrence Sperry developed a two-axis (roll and pitch) autopilot based on work done on torpedoes and ships. His son provided an airborne demonstration of the autopilot to the French in 1914 and marketed the autopilot for commercial use in 1932. Today's autopilots comprise many modes and now include flight director and autothrust systems, as well as interfaces with stability augmentation systems (SAS). Typical modes of operation in each of the three axes may include the following:

Pitch Channel	Roll Channel	Yaw Channel
Attitude Hold	Attitude Hold	SAS
– Turbulence	– Turbulence	Turn Coordination
– Wind Shear	– Wind Shear	Engine Out
VNAV (FMS)	LNAV (FMS)	
Control Wheel Steering	Control Wheel Steering	
Altitude Hold/Capture	Heading Hold/Select	
Vertical Speed Hold	Bank Limiting	
Glideslope	VOR/GPS/INS Nav	
Flight Path Angle	Localizer	
Airspeed/Mach Hold	Auto Land	
– Speed/Pitch	Holding	
Go-Around	Back-Course	
Auto Land		
Take-Off		
Flight Level Change		
Airspeed Protection		
Auto Pitch Trim		
SAS		

The purpose of incorporating these modes is many: workload and fatigue for the pilot is reduced; an autopilot is more precise than a human, which in turn makes it more efficient (i.e., saves money); the autopilot can multitask; and it has built-in redundancy and enhances safety.

As previously discussed, the autopilot is a highly integrated, complex system. It can interface with the radar altimeter, CADC, flight control computers, FMS, INS, flight director, autothrottles, elevator feed and flap limit sensors, flight control position sensors, and SAS. These interfaces, along with the controls and displays and human factors issues, can be a challenging task for even the most experienced evaluator.

Because of the dramatic changes in technology and system design, there has been a marked increase in automation, complexity, and level of integration. The FAA has recognized this fact and has recently revamped the FAR paragraph covering the FGSs for Part 25 aircraft. Autopilot has been renamed flight guidance systems and includes the functions of autopilots, flight directors, and automatic thrust control, as well as any interactions with stability augmentation and trim functions.

The applicable documentation for the FGSs includes the following:

- The minimum performance standards, contained in SAE Aeronautical Standard AS-402A, "Automatic Pilots," February 1, 1959
- The TSO for autopilots, contained in TSO-C9c, "Automatic Pilots," September 15, 1960
- The AC for Part 23 aircraft, contained in AC 23-17C, "Systems and Equipment Guide for Certification of Part 23 Airplanes and Airships;" paragraph 23.1329, "Automatic Pilot System;" and paragraph 23.1335, "Flight Director Systems," November 17, 2011
- The AC for Part 25 aircraft, AC 25.1329-1B Change 1, "Approval of Flight Guidance Systems," October 16, 2012
  - The AC for Part 27 aircraft, AC 27-1B, "Certification of Normal Category Rotorcraft (Changes 1–3 incorporated)," paragraph 1329, September 30, 2008
  - The AC for Part 29 aircraft, AC 29-2C, "Certification of Transport Category Rotorcraft" (Changes 1–3 incorporated), paragraph 1329, September 30, 2008

In addition, other requirements are called out in the following:

- AC 23-8C, "Flight Test Guide for Certification of Part 23 Airplanes," November 16, 2011
- AC 25-7C, "Flight Test Guide for Certification of Transport Category Aircraft," October 16, 2012

The flight director requirements are incorporated into the autopilot section of AC 25-7C; the flight director section is AC 23-8C is marked "reserved," and the reader is referred to AC 23-17C for autopilots.

AC 23-17C is an attempt to consolidate compliance material within one document; it contains policy guidance through 2011. As with all ACs, it does not contain the force of law but shows a means of compliance for TC and STC applicants. AC 25.1329-1B provides the reader with a cross-reference of the CFR 14 Part 25.1329 subparagraphs to a compliance paragraph in the AC. For example, Part 25.1329(a) calls out the requirement for quick disengagement for the autopilot and autothrust systems. The matrix in the AC points the reader to chapter 3, paragraphs 27 and 29 of the AC for the means of compliance, which is very handy.

6.15.1 Components of the Flight Guidance System

For the purposes of AC 25.1329-1B, the term FGS includes all the equipment necessary to accomplish the FGS function. This includes the sensors, computers, power supplies, servo motors/actuators, and associated wiring as well as any indications and controllers necessary for the pilot to manage and supervise the system.

The individual elements of the system are as follows:

- Flight guidance and control (autopilot or flight director displayed heads up or heads down)
- The autothrottle/autothrust systems (autothrust is a generic term and includes power control systems for propeller-driven airplanes)
- Interactions with stability augmentation and trim systems
- Alerting status, mode annunciations, and situation information associated with flight guidance and control functions
- The functions necessary to provide guidance and control in conjunction with an approach and landing system

  - Instrument landing system
  - Microwave landing system
  - Global navigation satellite system
  - GNS landing system

- The FGS also includes those functions necessary to provide guidance and control in conjunction with an FMS; it includes the interface between the FMS and the FGS necessary for the execution of flight path and speed commands.
- The design philosophy should support the intended operational use regarding FGS behavior, modes of operation, pilot interface with controls, indications, alerts, and mode functionality.

6.15.2 Part 23 Airworthiness Certification of the Autopilot

An acceptable means of compliance is found in AC 23-17, paragraph 23.1329. AC 23-8B defers autopilot testing to AC 23-17C. The AC separates the testing into five major areas:

- Cockpit controls
- Malfunction evaluations
- Recovery of flight control
- Performance flights
- Single-engine approach

The AC also provides some ground rules and considerations for all autopilot testing:

- A single malfunction may not result in a hard-over signal in more than one axis. When the result of this single malfunction is shown to be nonhazardous (from the safety assessment) (defined as no hard-over signals, slow-over is acceptable if it is demonstrated that they are easily controllable without exceptional skill) then multiple axes being affected is acceptable if

- The malfunction evaluations are acceptable even with the maximum drive signal due to the limited rate of change authority of the powered controlling element and flight control surfaces.
- The monitor/limiting device is independent of the autopilot element.
- The signal is less than the hard-over signal due to the monitor/limiting device.
- An acceptable fault analysis shows the functional hazard of a combined monitor failure and automatic pilot malfunction is not catastrophic.
- The functional hazard of a failure of a lock-out device/system to inhibit autopilot engagement until the pre-engagement check is successfully completed is hazardous or less.
- Pre-engagement check of the monitor system is mandatory with either a manual or automatic activation.
- Automatic pilot authority is not greater than necessary to control the airplane.
- Alterations of increased engine horsepower or major changes in exterior cowlings and surfaces should consider the effect/compatibility with the autopilot. An increase in engine horsepower beyond 10% may adversely affect the autopilot system malfunctions, performance, controllability, and longitudinal stability characteristics. Flight testing may be required to verify the original approval of the autopilot system is still valid.
- Malfunction testing should be accomplished at aft CG (most critical).
- Performance and controllability evaluations should be accomplished at the most forward CG.
- The force required to overcome the autopilot as called out in 14 CFR Part 23.143 are maximums.
  - The maximum temporary force to overpower the autopilot has not been allowed to exceed 30 pounds in roll, 50 pounds in pitch, and 150 pounds in yaw. These forces are applicable only to initially overpowering the autopilot system.
  - The maximum prolonged force to overpower the autopilot should not exceed 5 pounds in roll, 10 pounds in pitch, and 20 pounds in yaw.
- A reasonable period of time has been established for pilot recognition from the time a failure is induced in the autopilot system and the beginning of pilot corrective action following hands-off or unrestrained operation:
  - A 3 sec delay following pilot recognition of an autopilot system malfunction through a deviation of the airplane from the intended flight path, abnormal control movements, or by a reliable failure warning system in the climb, cruise, and descent flight regimes.
  - A 1 sec delay is assumed in approach and maneuvering flight regimes.

6.15.2.1 Cockpit Controls

In order to show compliance, the evaluation of the cockpit controls should include the following:

- The location of the autopilot system controls should be readily accessible to the pilot, or both pilots, if a minimum of two pilots are required.

- Annunciators must conform to the proper colors (Part 23.1322).
- Controls must be usable under bright sunlight to night lighting conditions (Part 23.1381).
- Either a quick disconnect or interrupt switch for the autopilot system are located on the side of the control wheel opposite the throttle(s) and are red in color. A disconnect switch stops all movement of the autopilot system; an interrupt switch momentarily interrupts all movement of the autopilot system.

NOTE: A quick release control must also be included on each aircraft control stick (for those Part 23 aircraft that do not use a control wheel) that can be operated from either pilot seat.

- Any automatic disconnects of the autopilot system must be aurally annunciated. If warning lights are used to supplement the aural warning, they should meet the requirements of Part 23.1322. Use of a visual warning as the sole means of annunciating automatic disconnects is not acceptable.
- Motion and effect of autopilot cockpit controls should conform to the requirements of Parts 23.1329 and 23.779.

6.15.2.2 Malfunction Evaluations

There are some basic ground rules that need to be abided by when performing malfunction evaluations. Malfunction evaluations should be conducted with the airplane loaded at the most critical weight or the most critical weight/CG combination. Maximum untrimmed fuel imbalance should be considered during the evaluation. If autothrottles are installed, they should be operating and autopilot servo torque should be set to the upper tolerance limit. The simulated malfunctions should be induced at various airspeeds and altitudes throughout the airplanes operating envelope. The envelope should include the maximum operating altitude for turbocharged or high-altitude airplanes or be within 10% of the service ceiling for normally aspirated engines. Vertical gyro failures should not be considered.

The simulated failures and resulting corrective actions are not acceptable if they result in any of the following:

- Loads that exceed the substantiated structural design limit loads
- Acceleration outside of 0 to 2 g (the positive g limit may be increased up to the positive design limit maneuvering load factor if it has previously been determined analytically that neither the simulated failure nor resulting corrective action would result in loads beyond the design limit loads of the airplane)
- Speeds in excess of V_{NE} (never exceed speed) or for airplanes with an established V_{MO}/M_{MO} (maximum operating limit speed), a speed midway between V_{MO}/M_{MO} and the lesser of V_D/M_D (design diving speed), or a speed demonstrated under Part 23.253 (high-speed characteristics)
- Deviations from the flight path including bank angle in excess of 60° or a pitch attitude in excess of ±30° deviation from the attitude at which the malfunction was introduced
- A hazardous dynamic condition

NOTE: Since I have introduced some abbreviations for different speeds, I have included all the abbreviations you may come across during civil certifications in the following table for completeness (Table 6.41).

TABLE 6.41 ▪ Civil Certification Speed Definitions

Abbr	Meaning	Abbr	Meaning
V_A	Design Maneuvering Speed	V_{MO}/M_{MO}	Maximum Operating Limit Speed
V_B	Design Speed for Maximum Gust Intensity	V_{MU}	Minimum Unstick Speed
V_C	Design Cruise Speed	V_{NE}	Never Exceed Speed
V_D	Design Diving Speed	V_{NEI}	Helicopter Instrument Flight Never Exceed Speed
V_{DF}/M_{DF}	Demonstrated Flight Diving Speed	V_{NO}	Maximum Structural Cruising Speed
V_F	Design Flap Speed	V_O	Maximum Operating Maneuver Speed
V_{FC}/M_{FC}	Maximum Speed for Stability Characteristics	V_R	Rotation Speed
V_{FE}	Maximum Flap Extended Speed	V_S	Stalling Speed or the Minimum Steady Flight Speed at which the Airplane is controllable
V_H	Maximum Speed I Level Flight with Max Continuous Power	V_{SO}	Stalling Speed or Minimum Steady Flight Speed in Landing Configuration
V_{LE}	Maximum Landing Gear Extended Speed	V_{SI}	Stalling Speed or the Minimum Steady Flight Speed Obtained in a Specific Configuration
V_{LO}	Maximum Landing Gear Operating Speed	V_{SSE}	Safe, Intentional, One Engine Inoperative Speed
V_{LOF}	Lift-Off Speed	V_{TOSS}	Takeoff Safety Speed for category A Rotorcraft
V_{MC}	Minimum Control Speed with Critical Engine Inoperative	V_X	Speed for Best Angle of Climb
V_{MCA}	Airborne Minimum Control Speed with Critical Engine Inoperative	V_Y	Speed for Best Rate of Climb
V_{MCG}	Minimum Control Speed During TO Ground Run when Critical Engine is Suddenly Made Inoperative	V_{YI}	Helicopter Instrument Flight Climb Speed
V_{MCL}	Minimum Control Speed During Landing Approach when the Critical Engine is Suddenly Made Inoperative	V_1	Takeoff Decision Speed
V_{MCL-2}	For A/C with 3 Engines, the Minimum Control Speed During Landing Approach with One Critical Engine Inoperative and a Second is Suddenly Made Inoperative	V_2	Takeoff Safety Speed
V_{MIMI}	Helicopter Instrument Flight Minimum Speed	V_{2min}	Minimum Takeoff Safety Speed

6.15.2.3 Normal Flight Malfunctions

The airplane's performance should be evaluated when the effect caused by the most critical single failure condition that can be expected to occur to the system and can be detected by the pilot is induced into the autopilot system. Hidden or latent failures, in combination with detectable failures should be considered when determining the most critical failure condition. Normal flight includes climb, cruise, and descent flight regimes with the airplane properly trimmed in all axes. Airplane configurations (combinations of gear and flaps), speeds, and altitudes should be evaluated for unsafe conditions. Reaction time for normal flight malfunctions is 3 sec. The more critical of the simulated malfunctions are

- A simulated malfunction about any axis equivalent to the cumulative effect of any failure or combination of hidden failures, including manual or electric automatic trim, if installed

- The combined signals about all affected axes, if multiple axis failures can result from the malfunction of any single component

6.15.2.4 Maneuvering and Approach Malfunction

Maneuvering flight tests should include turns with the malfunction induced at the maximum bank angle for normal operations up to the autopilot authority limits. Airplane configurations (combination of gear and flaps), airspeeds, and altitudes should be evaluated to determine if unsafe conditions exist. The simulated malfunctions described under section 6.13.2.3 are applicable here as well. The resultant accelerations, loads, and speeds should be within the limits described for normal flight malfunctions. The reaction time for maneuvering and approach malfunctions is 1 sec. Malfunctions induced during coupled approaches should not place the airplane in a hazardous attitude that would prevent the pilot from conducting a missed approach or safe landing. Recovery from malfunctions should be shown either by overpowering or by manual use of a quick disconnect device after the 1 sec delay. Altitude losses resulting from simulated malfunctions are to be measured accurately and presented in the limitations section of the AFM or approved manual material.

Accurate measurement of altitude loss because of an autopilot malfunction during an instrument landing approach is essential. This altitude loss during a critical phase of flight provides the basis for establishing the minimum approach altitude during autopilot-coupled approaches. The loss should be determined by measuring from the altitude at which the malfunction is induced to the lowest altitude observed during the recovery maneuver, unless instrumentation is available to measure the vertical deviation from the intended glide path to the lowest point in the recovery maneuver.

Figure 6.41 graphically illustrates an acceptable means of compliance for determining altitude loss during approach. In performing this test, the airplane should be established on the ILS glideslope and localizer with the configurations and approach speed(s) specified by the applicant for the approach. Simulated automatic flight control system malfunctions should be induced at critical points along the ILS, taking into consideration all design variations and their limits in sensitivity and authority; the malfunctions should be induced in each axis. A 3° glideslope should be used for these tests to determine the malfunction effects to be expected in service.

For use during a coupled ILS approach, the automatic control system should not fail in such a way that it causes the airplane wheels to descend below a limit line lying below

FIGURE 6.41 ■
Acceptable Method for Determining Altitude Loss on Approach

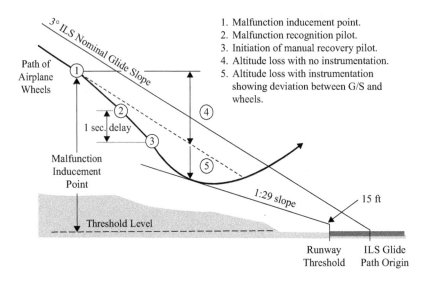

1. Malfunction inducement point.
2. Malfunction recognition pilot.
3. Initiation of manual recovery pilot.
4. Altitude loss with no instrumentation.
5. Altitude loss with instrumentation showing deviation between G/S and wheels.

the glideslope, sloping upward at 29:1 from a point 15 ft above the runway threshold. With the airplane established on the glideslope in approach configuration, at the approach speed, the most critical malfunction is induced at a test altitude referenced to the runway threshold. Measure the altitude loss between the test altitude and the lowest point of the manual recovery unless instrumentation is available to measure the vertical deviation from the intended glide path to the lowest point in the recovery maneuver. The altitude loss and the known distance to the threshold from the lowest recovery altitude are compared to the limit line. The lowest test altitude from which malfunction and manual recovery can be completed without the airplane wheels descending below the limit line is considered the minimum height for use of the automatic flight control system. Normal and emergency operating procedures must be included in the AFM for the automatic flight control system to include minimum altitude for use.

6.15.2.5 Alternate Means of Compliance

An alternate means of compliance is available to those systems that employ limiters or monitors when the function of these devices is necessary to prevent the airplane from exceeding the malfunction limits defined in section 6.13.2.2. There are three alternate means of compliance, and the major difference is the criticality level as the result of a failure: (1) addresses failure effects less than major; (2) addresses failure effects less than hazardous; and (3) addresses failure effects less than catastrophic.

Alternate Means 1   With the monitor/limiter inhibited, autopilot malfunction flight testing may not cause any of the following:

- Roll to exceed 80°
- Pitch to exceed +45°, −35°
- Accelerations outside of the 0 g to 2.5 g envelope
- Airspeed exceeding V_{NE} for an airplane having an established V_{MO}/M_{MO}, a speed not greater than a speed midway between V_{MO}/M_{MO}, and the lesser of V_D/M_D or the speed demonstrated under CFR 23.253.

In addition, a fault analysis should show the failure effect of a monitor failure combined with an autopilot malfunction is less than major. A pre-engagement check of the monitor is mandatory. No credit is allowed for a pilot-activated pre-engagement check unless there is a lockout device or system.

Alternate Means 2   The limits in alternate means 2 are the same as for means 1 with a slight exception in acceleration. With the monitor/limiter inhibited, autopilot malfunction flight testing may not cause accelerations outside of the −2 g to 2.5 g envelope. An acceptable fault analysis showing the failure effect of a combined monitor failure and an autopilot malfunction must be less than hazardous. In addition, the failure effect of the failure of a lockout device to inhibit autopilot engagement must be less than major. Pre-engagement check of the monitor is mandatory with either manual or automatic means and the autopilot engagement must be inhibited until the pre-engagement check is successful completed.

Alternate Means 3   For this method, flight tests with monitors inhibited are not required. An acceptable fault analysis must show that the failure effect of a combined monitor failure and autopilot is less than catastrophic. The failure of a lockout device or system to inhibit autopilot engagement must be less than hazardous. A pre-engagement check of the monitor is mandatory with either automatic or manual means. Autopilot engagement must be inhibited until the pre-engagement check is successfully completed and autopilot authority cannot be greater than that needed to satisfactorily control the airplane.

6.15.2.6 Recovery of Flight Control

This section states that the ability to recover flight control from the engaged autopilot system either by manual use of a quick disconnect or by physically overpowering the system must be evaluated.

6.15.2.7 Performance Flights

Performance evaluation tests should be conducted with the airplane loaded to its most adverse CG and weight conditions. Autopilot performance with the servo torque values at the lowest production torque tolerance limit should be used to demonstrate safe controllability and stability. Flight tests are necessary to ensure the autopilot system performs its intended function, including all modes presented for approval (Part 23.1301).

6.15.2.8 Single-Engine Approach

For multiengine airplanes, an engine failure during a normal ILS approach should not cause a lateral deviation of the airplane from the flight path at a rate greater than 3°/sec or produce hazardous attitudes. The rate should be measured and averaged over a 5 sec period. If approval is sought for ILS approaches initiated with one engine inoperative, the autopilot should be capable of conducting the approach.

6.15.3 Airworthiness Certification of Part 25 Flight Guidance Systems

An acceptable means of compliance is found in AC 25.1329-1B, "Approval of Flight Guidance Systems," October 16, 2012, and further guidance in AC 25-7C, "Flight Test Guide for Certification of Transport Category Airplanes," October 16, 2012.

AC 25.1329-1B did an excellent job of cross-referencing the subparagraphs of 25.1329 to the related means of compliance in the AC; however, in the latest iteration, Appendix 2 "Flight Test Procedures" was removed and incorporated into AC 25-7C! I will follow the same scenario addressing 25.1329 subparagraphs (a) through (n).

6.15.3.1 Quick Disengagements

Subparagraph (a) of paragraph 1329 states that quick disengagement controls for the autopilot and the autothrust must be available to each pilot. A means of compliance is contained in paragraphs 27 and 29 of chapter 3 in the AC. The purpose of the quick disengagement control is to ensure the capability for each pilot to manually disengage the autopilot quickly with a minimum of pilot hand/limb movement. The quick disengagement control must be located on each control wheel or equivalent. It should be within easy reach of one or more fingers/thumb of the pilot's hand when the hand is in a position for normal use on the control wheel or equivalent. The criteria for the quick disconnect control are that it

- Must be operable with one hand on the control wheel or equivalent and the other hand on the thrust levers

- Should be accessible and operable from a normal hands-on position without requiring a shift in hand position or grip on the control wheel or equivalent
- Should be easily locatable by the pilot without having to locate the control visually
- Should be designed so that any action to operate the quick disengagement control should not cause an unintended input to the control wheel or equivalent
- Should be designed to minimize inadvertent operation and interference from other nearby control wheel (or equivalent) switches/devices such as radio control or trim
- A note in the AC adds that when establishing the location of the quick disconnect control, consideration should be given to its accessibility with large displacements or forces on the control wheel and the possible need to operate the quick disconnect with the other hand

If the safety assessment dictates a need for an alternate means of disengagement, the following needs to be addressed:

- Independence from the primary quick disconnect control
- Whether alternative means are accessible to the pilot
- Reliability and the possibility of latent failures of those alternative means

The following means of providing an alternate means of disengagement have been found to be acceptable:

- Selecting the engagement control to the off position
- Disengaging the bar on the mode select panel
- Activating the trim switch on the yoke
- The AC notes that the use of circuit breakers as an alternate means of disengagement is not acceptable

Autothrust quick disconnect controls must be provided for each pilot on the respective thrust control (thrust lever or equivalent). A single-action, quick disconnect switch must be incorporated on the thrust control so that switch activation can be executed when the pilot's other hand is on the control wheel (or equivalent). The disengagement control should be positioned so that inadvertent disengagement of the autothrust function is unlikely. Positioning the control on the outboard side has been shown to be acceptable for multiengine aircraft. Thrust lever knob, end-mounted disengagement controls available on both sides to facilitate use by either pilot have shown to be preferable to those positioned to be accessible by the pilot's palm.

6.15.3.2 Failure of the FGS to Disengage

The effects of the failure of the system to disengage the autopilot or autothrust system when manually commanded by the pilot are to be assessed (in accordance with Part 25.1309). The AC points the applicant to paragraphs 27, 29, and 30 in chapter 3 of the AC as well as chapter 8. Chapter 8 is the safety assessment chapter of the AC. The safety assessment process has already been covered in section 6.6 of the text and will not be repeated here, but the AC does specifically mention failure to disengage the FGS in paragraph 89 of chapter 8. It states that the implication of failures that preclude the quick disengagement from functioning should be assessed. The assessment should consider the effects of failure to disengage the autopilot or autothrust functions during the approach using the quick disconnect controls; the feasibility of using an alternative means if disengagement should also be assessed. If the assessment indicates that the aircraft can be landed manually with the autopilot/autothrust engaged it should be demonstrated in flight testing.

6.15.3.3 Transients Caused by Engagement

Subparagraph (c) of 1329 states that engagement or switching of the flight guidance system, a mode or a sensor may not cause a transient response of the airplane's flight path any greater than a minor transient. A minor transient is defined as one that would not significantly reduce safety margins and would involve aircrew actions that are well within their capabilities. A minor transient may involve a slight increase in aircrew workload or some physical discomfort to passengers or cabin crew.

With an automatic engagement of the autothrust system, it should be clear to the aircrew that the engagement has occurred. The automatic engagement may not cause a transient larger than a minor transient. The transition to the engaged state should be smooth and should not cause large unexpected changes in pitch attitudes or pitching moments. The reason for automatic engagement (such as "wake up" mode to protect for unsafe speeds) should be clear and obvious to the aircrew.

A note in the AC states that the design should consider the possibility that the pilot may attempt to engage the autothrust function outside of the normal flight envelope or at excessive or too low engine thrust. It is not expected that the autothrust feature should compensate for situations outside the normal engine operation range unless that is part of the intended function of the autothrust system.

For the autopilot, engagements should be free of perceptible transients. Under dynamic conditions, including maneuvering flight, minor transients are acceptable; under normal operating conditions, significant transients may not occur.

6.15.3.4 Transients Caused by Disengagements

Similar to the engagement scenario, paragraph 25.1329(d) states that under normal conditions the disengagement of any automatic control function of a flight guidance system may not cause a transient response of the airplane's flight path any greater than a minor transient. The AC contains the same words when referencing manual or automatic disengagement of the autopilot. It also states that the disengagement should be free of out-of-trim forces that are not consistent with the maneuvers being conducted by the airplane at the time of the disengagement. If multiple autopilots are engaged, any disengagement of an individual autopilot may not adversely affect the operation of the remaining engaged autopilots.

For autothrust systems, the disengagement should not preclude, inhibit, or interfere with timely thrust changes for go-around, landing, or other maneuvers requiring manual thrust changes.

6.15.3.5 Disengagements under Rare and Nonnormal Conditions

The FAR allows for greater than minor transients for disengagements under rare and nonnormal conditions. For these conditions, a transient no greater than a significant transient may occur. A significant transient may lead to a significant reduction in safety margins, an increase in flight crew workload, discomfort of the flight crew or physical distress to the passengers or cabin crew, including nonfatal injuries. Significant transients do not require, to remain within or recover to the normal flight envelope, any of the following:

- Exceptional piloting skill, alertness, or strength
- Forces applied by the pilot greater than those specified in 25.143(c) (controllability and maneuverability)
- Accelerations or attitudes in the airplane that might result in further hazard to secured or nonsecured occupants

The flight crew should be made aware (via a suitable alerting or other indication) of conditions or situations (e.g., continued out-of-trim) that could result in a significant transient at disengagement. The number and type of alerts required should be determined by the unique situations that are being detected and by the crew procedures required to address those situations. Alerts should be clear and unambiguous and should be consistent and compatible with other flight deck alerts. Care should be taken with thresholds of these alerts so that they do not become a nuisance to the aircrew. The three conditions to be considered for alerts are sustained lateral control command, sustained pitch command, and bank and pitch angles.

If the autopilot is holding a sustained lateral control command, it could be indicative of an unusual operating condition for which the autopilot is compensating. Examples of such unusual operating conditions are asymmetric lift or drag due to asymmetric icing, fuel imbalance, or asymmetric thrust. In the worst case, the autopilot may be operating at or near its full authority in one direction. If the autopilot were to disengage while holding this lateral trim, the aircraft could undergo a rolling moment that could take the pilot by surprise. A timely alert should be considered to permit the crew to manually disengage the autopilot and take control prior to any automatic disengagement that might result from the condition.

If the autopilot is holding a sustained pitch command, it could be indicative of an unusual operating condition such as inoperative automatic horizontal trim. If the pilot were to disengage while holding this pitch command, the aircraft could undergo an abrupt change in pitch that could possibly surprise the pilot. A timely alert should be considered to permit the crew to manually disengage the autopilot and take control prior to any automatic disengagement that might result from the condition.

Most autopilots are designed with operational limits in both the pitch and roll axes such that those predetermined limits will not be purposely exceeded. If the airplane exceeds those limits, it could be indicative of a situation where the pilot is required to intervene. A timely alert should bring this situation to the attention of the aircrew and permit the crew to manually disengage the autopilot and take control prior to any automatic disengagement.

It is preferable that the autopilot remain engaged during out-of-trim conditions. If there is an automatic disengagement feature due to excessive out-of-trim, an alert should be generated and precede any automatic disengagement with sufficient margin to permit timely crew recognition and manual disengagement.

These alerts are not meant for all automatic disengagements, but only for those which if not addressed would lead to a significant transient (or greater) for which the pilot would be unprepared. There are also cases where the FGS will provide intentional out-of-trim conditions (parallel rudder operations in engine failure compensation modes, pitch trim operation during approach/landing, or certain types of speed/Mach trim systems). Alerts are not required for these modes. Other indications, such as mode and status annunciations, should be provided.

6.15.3.6 Controls, Indications, and Alerts

The man–machine (or human–machine if you prefer) interface with the FGS is essential in ensuring safe, effective, and consistent FGS operation. The manner in which information is presented to the aircrew is vital to aircrew awareness as to which automation is being utilized. The features and controls must be optimized to minimize flight crew errors and confusion. Indications and alerts should be presented in a manner compatible with the procedures and assigned tasks of the flight crew and provide the necessary information to perform those tasks. Indications must be grouped and presented in a logical fashion and should be visible from each pilot's station under all lighting conditions. Colors, fonts, font size, location orientation, and movement should all contribute to the effectiveness of the system.

The function and direction of motion of each control must be readily apparent or plainly indicated on or adjacent to each control to prevent inappropriate use or operation. In general, the design of the FGS should consider the following:

- Differentiation of knob shape and position
- Design to support correct selection of target values
- Commonality across aircraft types
- Positioning of controls and related indications
- Inadvertent operation

Engagement of the FGS functions must be suitably announced to each pilot; the operator should be provided with appropriate descriptions of the FGS modes and their behaviors. Mode annunciators must indicate the state of the system and be presented in a manner compatible with flight crew procedures and tasks and be consistent with the mode annunciation design for the specific aircraft type. The mode selector switch position or status is not acceptable as the sole means of mode annunciation. Modes and mode changes should be depicted in a manner that achieves flight crew attention and awareness.

The FGS mode annunciations must effectively and unambiguously indicate the active and armed modes of operation. The annunciation should convey, in explicit and simple terms, what the FGS is doing (active modes) and what it will be doing (armed modes) as well as target information such as selected speed, heading, and altitude. Mode annunciations should be located in the forward field of view, such as on the primary flight display. Engaged modes should be annunciated at different locations than armed modes to assist in mode recognition.

Color, font type, font size, location, highlighting, and symbol flashing have historical precedent as good discriminators when implemented appropriately. The fonts and font sizes should be chosen so that the annunciation of FGS mode and status information is readable and understandable without eye strain when viewed by the pilot seated at the design eye position; the use of graphical or symbolic indications is acceptable. Implementation of discriminators should follow accepted guidelines (see AC 25-11). Color should be used in a consistent manner and ensure overall compatibility with the use of color on the flight deck. FGS color should be similar and consistent with other aircraft systems such as the FMS. The use of monochrome displays is not precluded, provided that the aspects of flight crew attention and awareness are satisfied. Engaged modes should be annunciated with different colors than armed modes to assist in mode recognition.

Mode changes that are operationally relevant, especially mode reversions and sustained speed protection, should be clearly and positively annunciated to ensure flight crew awareness. Altitude capture is an example of an operationally relevant mode because pilot actions during that brief time the mode is operational may have different effects on the airplane. The FGS submodes that are not operationally relevant, such as transition from ILS capture to ILS track, need not be annunciated. The transition from an armed mode to an engaged mode should provide an additional attention-getting feature, such as boxing and flashing on an electronic display for a brief period (about 10 sec) to assist in flight crew awareness. Aural notification of mode changes should be limited to special considerations.

In-service experience has shown the mode annunciation alone may be insufficient (either unclear or not compelling) to communicate mode changes to the aircrew. The safety consequences of the aircrew not recognizing the change should be considered and, if necessary, an alert should be issued.

Mode information provided to the pilot should be sufficiently detailed so that the consequences of the interaction between the FGS and the flight crew can be determined unambiguously. The FGS should provide timely and positive indication when the FGS deviates from the pilot's direct commands (target altitude, speed commands) or from the pilot's preprogrammed set of commands (waypoint crossing). The interface should provide a clear indication when there is a difference or conflict with pilot-initiated commands. The failure of a mode to engage or arm when selected by the pilot should be apparent.

6.15.3.7 Protection from Hazardous Loads and Deviations

Subparagraph (g) of 25.1329 states that under any condition of flight appropriate to its use the FGS may not produce hazardous loads on the airplane, nor create hazardous deviations in flight path. This applies to both fault-free operation and in the event of a malfunction, and assumes that the pilot begins corrective action within a reasonable period of time. The AC calls out three conditions to be evaluated in FGS operation:

- Normal conditions
- Rare normal conditions
- Nonnormal conditions

For each of the conditions (which are generic in nature and do not represent all conditions that an aircraft may experience over its lifetime), an expected level of performance is described. For normal conditions, the expected levels of performance are

- Any performance characteristics that are operationally significant or operationally limiting must be identified with an appropriate statement or limitation in the AFM.
- The FGS should perform its intended function during routine airplane configuration or power changes, including operation of secondary flight controls and landing gear.
- Evaluation of the FGS performance for compliance should be based on the minimum level of performance needed for the intended functions.
- There are certain operations that dictate a prescribed level of performance. When the FGS is intended for operations that require specific levels of performance, the use of the FGS should be shown to meet those specific levels of performance (low visibility operations, Cat II and III operations, RVSM operations, RNP, etc.).

- The FGS performance should be at least equivalent to that expected of a pilot performing a similar task. When integrated with navigation sensors or the FMS, the FGS should satisfy the FTE tolerances expected for use of those systems in performing their intended functions.
- The autopilot should provide smooth and accurate control without divergent or perceptible sustained nuisance oscillation.
- The flight director in each available display presentation should provide smooth and accurate guidance and be appropriately damped to achieve satisfactory control task performance without pilot compensation or excessive workload.
- The autothrust function should provide smooth and accurate control of thrust without significant or sustained oscillatory power changes or excessive overshoot of the required power setting.
- The automatic pitch trim function should operate at a rate sufficient to mitigate excessive control surface deflections or limitations of control authority without adverse interactions with automatic control of the aircraft. Automatic roll and yaw trim functions, if installed, have similar requirements.

Table 6.42 provides examples of normal conditions for FGS operations.

TABLE 6.42 ▪ Examples of Normal Conditions

Examples of Normal Conditions	
Terms	**Descriptions**
No Failure Conditions	All airplane systems that are associated with airplane performance are fully operational. Failure of those systems could impair the FGS's ability to perform its functions
Light to Moderate Winds	Constant wind in a specific direction that may cause a slight deviation in intended flight path or a small difference between airspeed and groundspeed
Light to Moderate Wind Gradients	Variations in wind velocity that may cause slight erratic or unpredictable changes in flight path
Light to Moderate Gusts	Nonrepetitive momentary changes in wind velocity that can cause changes in altitude or attitude but the aircraft remains in positive control at all times
Light Turbulence	Turbulence that momentarily causes slight, erratic changes in altitude or attitude
Moderate Turbulence	Similar to light turbulence, but of greater intensity. Changes in altitude or attitude occur but the aircraft remains in positive control at all times
Light Chop	Turbulence that causes slight, rapid, and somewhat rhythmic bumpiness without appreciable changes in altitude or attitude
Moderate Chop	Similar to light chop, but of greater intensity. It causes rapid bumps or jolts without appreciable changes in altitude or attitude
Icing	All icing conditions covered by 14 CFR Part 14, Appendix C, except asymmetric icing, which is a rare normal condition

The AC notes that representative levels of the environmental effects should be established consistent with the airplane's intended operation.

For rare normal conditions (Table 6.43), the expected levels of performance are

- The FGS should be designed to provide guidance or control for the intended function of the active modes in a safe and predictable manner within the flight envelope and for momentary excursions outside of the normal flight envelope.

- Operations may result in automatic or pilot-initiated autopilot disengagement close to the limit of the autopilot authority.

- It is not necessary that the FGS always be disengaged when rare normal conditions that may degrade its performance or capability are encountered; the FGS may significantly help the flight crew during such conditions. The design should address the potential for the FGS to mask a condition from the flight crew or otherwise delay appropriate flight crew action.

- Operating an airplane in icing conditions can have significant implications on the aerodynamic characteristics of the airplane. During autopilot operation, the flight crew may not be aware of the gradual onset of icing conditions. A means should be provided to the aircrew when icing conditions begin to have an effect on FGS performance.

- The implication of icing on speed protection should also be assessed. If the threshold of the stall warning system is adjusted due to icing conditions, appropriate adjustments should also be made to low speed protection.

TABLE 6.43 ■ Examples of Rare Normal Conditions

Examples of Rare Normal Conditions	
Terms	**Descriptions**
Significant Winds	Constant wind in a specific direction that may cause a large change in intended flight path or groundspeed or cause a large difference between airspeed and groundspeed
Significant Wind Gradients	Variation in wind velocity that may cause large changes in the intended flight path
Wind Shear/Microburst	A wind gradient of such magnitude that it may cause damage to the aircraft
Large Gusts	Nonrepetitive momentary changes in wind velocity that can cause large changes in altitude or attitude; aircraft may be momentarily out of control
Severe Turbulence	Turbulence that causes large, abrupt changes in altitude or attitude. It usually causes large variations in indicated airspeed; aircraft may be momentarily out of control
Asymmetric Icing	Icing conditions that result in ice accumulations that cause the FGS, if engaged, to counter the aerodynamic effect of the icing conditions with a sustained pitch, roll, or yaw command that approaches the maximum authority

The AC notes that airplanes intended to meet paragraph 121.358 for wind shear warning and guidance need FD wind shear guidance. The FGS may also provide suitable autopilot control during wind shear. It refers the reader to AC 25-12 and AC 120-41.

TABLE 6.44 ▪ Examples of Nonnormal Conditions

Examples of Nonnormal Conditions	
Terms	**Descriptions**
Significant Fuel Imbalance	Large variation in the amount of fuel between the two wing tanks (and/or center and tail tanks if so equipped) that causes the FGS, if engaged, to counter the aerodynamic effect with a pitch, roll, or yaw command that is approaching maximum system authority
Nonstandard Ferry Flight Configurations	Possible aerodynamic drag (symmetrical and unsymmetrical) caused by nonstandard ferry flight conditions such as locked high-lift devices, landing gear locked in the deployed position, or an extra engine carried underneath one wing in an inoperative position
Inoperative Engine(s)	Loss of one or more engines that causes the FGS, if engaged, to counter the aerodynamic effect of the difference in thrust with a pitch, roll, or yaw command that is approaching maximum system authority
Loss of One or More Hydraulic Systems	Loss of one or more hydraulic systems down to the minimum number of operational systems with which the FGS is certified to operate
Inoperative Ice Detection/Protection System	Loss of the ice detection/protection system on an airplane so equipped, where the FGS is certified for operation in icing conditions with that failure present

For nonnormal conditions (Table 6.44), the expected levels of performance are as follows:

- The FGS should be designed to provide guidance or control for the intended function of the active modes in a safe and predictable manner, both within the normal flight envelope and for momentary excursions outside the normal flight envelope. If a determination is made that there are nonnormal conditions in which the FGS cannot be operated safely, appropriate limitations should be placed in the AFM.

- Operations in nonnormal conditions may result in automatic or pilot-initiated autopilot disengagement close to the limit of the autopilot authority.

- It is not necessary that the FGS always be disengaged when nonnormal conditions that may degrade its performance or capability are encountered; the FGS may significantly help the flight crew during such conditions. The design should address the potential for the FGS to mask a condition from the flight crew or otherwise delay appropriate flight crew action.

6.15.3.8 Speed Protection

Paragraph 25.1329(h) states that when an FGS is in use, a means must be provided to avoid excursions beyond an acceptable margin from the speed range of the normal flight envelope. If the airplane experiences an excursion outside of this range, a means must be provided to prevent the FGS from providing guidance or control to an unsafe speed. The intent of the paragraph is that the FGS should provide a speed protection function for all operating modes so that the airspeed can be safely maintained within an acceptable margin of speed range of the normal flight envelope. The FGS may use any of the following ways, or a combination of ways, to provide acceptable speed protection:

- The FGS may detect the speed protection condition, alert the flight crew, and provide speed protection control or guidance.
- The FGS may detect the speed protection condition, alert the flight crew, and then disengage the FGS.
- The FGS may detect the speed protection condition, alert the flight crew, and remain in the active mode without providing speed protection, control, or guidance.
- Other systems such as the primary flight control system or the FMS (when in a VNAV mode) may be used to provide equivalent speed protection.
- If compliance is shown by the use of alerts alone, then the alerts should be shown to be appropriate and timely, ensure aircrew awareness, and enable the pilot to keep the airplane within an acceptable margin of the speed range of the normal flight envelope.

The design of the speed protection function should consider how and when the speed protection is provided for combinations of autopilot, flight director, and autothrust operation. Care should be taken to set appropriate values for transitioning into and out of speed protection such that the flight crew does not consider the transitions a nuisance. The speed protection should integrate pitch and thrust control, and consideration should be given to automatically activating the autothrust function when speed protection is invoked. If an autothrust is either not provided or unavailable, speed protection should be provide by pitch control alone.

The role and interaction of autothrust with elements of the FMS, primary flight control system, and the propulsion system should also be accounted for in the design of speed protection. For example, consideration should be given to the effects of an engine inoperative condition on the performance of the speed protection function.

Speed protection covers both the low-speed and high-speed flight regimes. When the FGS is engaged in any mode (approach mode will be addressed shortly) for which the available thrust is insufficient to maintain a safe operating speed, the low-speed protection function should be invoked to avoid unsafe speed excursions. V_{MO}/M_{MO} mark the upper speed/Mach limit of the normal flight envelope, however, high-speed protection may also come into play based on aircraft configuration (gear or flaps extended).

Activation of speed protection should take into account such factors as the phase of flight, turbulence and gusty wind conditions, and compatibility with speed schedules. The low-speed protection should activate at a suitable margin to stall warning that will not result in an unacceptable level of nuisance alerts. The applicant needs to consider the operational speeds for all-engine and engine-inoperative cases during the following phases of flight:

- Takeoff
- Departure, climb, cruise, descent, and terminal area operations (during these phases, airplanes are normally operated at or above the minimum maneuvering speed for the given flap configuration)
- During high-altitude operations it may be desirable to incorporate low-speed protection at the appropriate engine-out drift-down speed schedule, if the FGS or other sensors can determine that thrust deficiency is due to engine failure

- Approach

- Transition from approach to go-around and go-around climb

- The AC notes that a low-speed alert and a transition to the speed protection mode at approximately $1.13V_{SR}$ (stall reference speed) for the landing flap configuration has been found to be acceptable

Low-speed protection during approach operations and recovery from wind shear are two special cases of this function. Speed protection should not interfere with the approach and landing phase of flight. It is assumed that with autothrust operating normally, the combination of thrust and pitch control during the approach will be sufficient to maintain speed and the desired vertical flight path. In the cases where it is insufficient, an alert should be provided in time for the flight crew to take appropriate corrective action.

For approach operations with a defined vertical flight path (ILS, MLS, GLS, LNAV, and VNAV), if the thrust is insufficient to maintain the desired approach path and approach speed, the intent of low-speed protection may be satisfied by the following:

- The FGS may maintain the defined vertical path as the airplane decelerates below the desired approach speed until the airspeed reaches the low-speed protection value. The FGS would then maintain the low-speed protection value as the airplane departs the defined vertical flight path. The FGS mode reversion and low-speed alert should be activated to ensure pilot awareness. (The AC notes that the pilot is expected to take corrective action and add thrust and return the airplane to the desired vertical path or go-around, as necessary.)

- The FGS may maintain the defined vertical path as the airplane decelerates below the desired approach speed to the low-speed protection value. The FGS will then provide a low-speed alert while remaining in the existing FGS approach mode. (The AC notes again that the pilot is expected to take corrective action to add thrust to cause the airplane to accelerate back to the desired approach speed while maintaining the defined vertical path or go-around, as necessary.)

- The FGS may maintain the defined vertical path as the airplane decelerates below the desired approach speed until the airspeed reaches the low-speed protection value. The FGS will then provide a low-speed alert and disengage. (The pilot is expected to take corrective action when alerted to the low-speed condition and disengagement of the autopilot, and add thrust and manually return the airplane to the desired vertical path or go-around.)

If the speed protection is invoked during approach such that the vertical path is not protected, the subsequent behavior of the FGS after speed protection must be considered. Activating low-speed protection during the approach, resuming the approach mode, and reacquiring the defined vertical path may be an acceptable response if the activation is sufficiently brief and does not cause large speed or flight path deviations.

The interaction between low-speed protection and wind shear recovery guidance is the second special case of low-speed protection. If the wind shear recovery guidance meets the criteria of AC 25-12 and AC 120-41, then it also meets the requirements of subparagraph (h) of 1329. The autopilot should be disengaged when the wind shear recovery guidance activates unless autopilot operation has been shown to be safe in these conditions and provides effective automatic wind shear recovery as noted in the mentioned ACs.

Some of the factors to consider for high-speed protection are

- Operations at or near V_{MO}/M_{MO} in routine atmospheric conditions are safe; small, brief excursions above V_{MO}/M_{MO} by themselves are not unsafe.

- The FGS design should strive to strike a balance between providing adequate speed protection margin and avoiding nuisance activation of high-speed protection.

- Climbing to control airspeed is not desirable, as departing an assigned altitude can be disruptive to ATC and potentially hazardous (RVSM airspace). As long as the speed does not exceed a certain margin beyond V_{MO}/M_{MO} (about 6 knots), it is better that the FGS remain in the altitude hold mode.

- The autothrust function, if operating normally, should effect high-speed protection by limiting its speed reference to the normal speed envelope (i.e., at or below V_{MO}/M_{MO}).

- The basic airplane high-speed alert should be sufficient for the pilot to recognize the overspeed condition and take corrective action to reduce thrust. If the airspeed exceeds a margin beyond V_{MO}/M_{MO} (about 6 knots), the FGS may transition from altitude hold to overspeed protection mode and depart (climb) the selected altitude.

- When the elevator channel of the FGS is not controlling airspeed, the autothrust function, if engaged, should reduce thrust as needed to prevent sustained airspeed incursions beyond V_{MO}/M_{MO} down to the minimum appropriate value.

- When thrust is already the minimum appropriate value or the autothrust function is not operating, the FGS should begin using pitch control as needed for high-speed protection.

- If conditions are encountered that result in airspeed excursions above V_{MO}/M_{MO}, it is preferable for the FGS to smoothly and positively guide or control the airplane back to within the speed range of the normal flight envelope.

6.15.3.9 FGS Considerations for the HUD

If the HUD is designed as a supplemental system, it will not replace the requirement for a standard heads-down display (HDD). If the HUD is used as the primary flight instrument, it needs to be shown that the HUD is satisfactory for controlling the airplane and for monitoring the performance of the FGS. The following characteristics apply:

- It should be shown that if the HUD guidance cues are followed regardless of the appearance of external visual references, they do not cause the pilot to take unsafe actions.

- It should be shown that there is no interference between the indications of primary flight information and the flight guidance cues. In takeoff, approach, and landing FGS modes, the flight guidance symbology should have occlusion priority (not obscured by primary flight information).

- The HUD guidance symbology should not excessively interfere with the pilot's forward view, ability to maneuver the airplane, ability to acquire opposing traffic, or ability to see the runway environment.

- The mechanization of guidance on the HUD should be no different with the HDD.

- It should be shown that the guidance remains visible and that there is a positive indication that it is no longer conformal to the outside view (high crosswind and yawing conditions).
- The location and presentation of the HUD information cannot distract the pilot or obscure the outside view.
- The HUD display should present flight guidance information in a clear and unambiguous manner; clutter should be minimized. Some flight guidance data elements are essential or critical and should not be removed by any declutter function.

The HUD FGS symbology should be compatible and consistent with the symbology on other FGS displays, such as the HDD; they need not be identical, but should not be prone to misinterpretation. The heads-up and heads-down primary flight display (PFD) formats and data sources need to be compatible to ensure that the FGS-related information presented on both displays has the same intended meaning:

- Symbols should be in the same format.
- Information (symbols) should appear in the same general location relative to other information.
- Alphanumeric readouts should have the same resolution, units, and labeling.
- Analog scales and dials should have the same range and dynamic operation
- The FGS modes and status state transitions should be displayed on the HUD using consistent methods (except for the use of color).
- Information sources should be consistent between the HUD and HDD used by the same pilot.
- When FGS command information (like a flight director command) is displayed on the HUD in addition to the HDD, the HUD depiction and guidance cue deviation scaling needs to be consistent with the HDD.
- The same information concerning current HUD system mode, reference data, status state transitions, and alert information that is displayed to the flying pilot on the HUD should also be displayed to the nonflying pilot on the HDD.

Although HUDs are typically not intended to be classified as integrated caution and warning systems, they may display cautions, warnings, and advisories as part of their FGS function. In this regard, HUDs should provide the equivalent alerting functionality as the heads-down PFDs. Warnings that require continued flight crew attention on the PFD also should be presented on the HUD (e.g., TCAS, TAWS, wind shear). If master alerting indications are not provided within the peripheral view of the pilot while using the HUD, the HUD should provide annunciations that inform the pilot of a caution or warning condition. Some additional considerations for HUD alerts are as follows:

- For monochrome HUDs, appropriate use of attention-getting properties such as flashing, outline boxes, brightness, size, or location are necessary to compensate for the lack of color normally associated with cautions and warnings.
- For multicolor HUDs, the use of red, amber, and yellow for symbols not related to caution and warning functions should be avoided.

- Single HUD installations rely on the fact that the pilot not flying will monitor the heads-down instruments and alerting systems for failures of systems, modes, and functions not associated with the PFDs.

- If master alerting indications are not provided within the peripheral field of view of each pilot for a dual HUD installation, then each HUD should provide annunciations that direct the pilot's attention to heads-down alerting displays. The types of information that should trigger the HUD master alerting display are any cautions or warnings not duplicated on the HUD, as well as any engine or system cautions and warnings. The intention is to avoid redirecting the attention of the flying pilot to other displays when an immediate maneuver is required (e.g., TCAS, wind shear).

- If TAWS, wind shear detection, or guidance or TCAS is installed, then the guidance, warnings, and annunciations required to be a part of these systems and normally required to be in the pilot's primary field of view should be displayed on the HUD.

Upsets due to wake turbulence or other environmental conditions may result in near instantaneous excursions in pitch and bank angles and a subsequent unusual attitude. If the HUD is designed to provide guidance for recovery from upsets or unusual attitudes, recovery steering guidance commands should be distinct from and not confused with orientation symbology such as horizon pointers. If the HUD is designed to provide only orientation during upsets or unusual attitudes, those orientation cues should be designed to prevent them from being mistaken for flight control input commands.

6.15.3.10 Characteristics of Specific Modes

There have been many types of implementations of specific FGS modes on aircraft in the past. The AC provides guidance and interpretive material that clarifies the operational intent of the modes and criteria shown to be acceptable in current operations. These specific modes are repeated here.

Lateral Modes

- In the heading or track mode, the FGS should maintain the aircraft heading or track. When the airplane is in a bank when the mode is engaged, the FGS should roll the airplane to a wings level position and maintain the heading or track when the wings are level (typically less than 5° of bank angle).

- In the heading or track select mode, the FGS should acquire and maintain the selected value; when initially engaged, the aircraft should turn in the shortest direction to the selected heading or track. Once in the mode, any changes will result in changes of heading or track; the FGS will turn the airplane in the direction selected by the pilot (i.e., clockwise selection will turn the aircraft to the right, regardless of shortest distance). The target heading or track value should be presented to the crew.

- In the LNAV mode, the FGS should acquire and maintain the lateral flight path commanded by the FMS (or equivalent).

- If the airplane is not on the desired lateral flight path or within the capture criteria the FGS, the LNAV mode should enter an armed state; it will transition to the engaged state at a point where the lateral flight path can be smoothly acquired and tracked.

- For FGS incorporating LNAV during TOGA, the design should specify maneuvering capability immediately after TO and any limits that may exist. After TOGA, maneuvering should be based on airplane performance, preventing excessive roll (wingtip impact with runway), and satisfying operational requirements where terrain or thrust limitations exist.

Vertical Modes

- To avoid unconstrained climbs or descents for altitude transitions, the altitude select controller should be set to a new target altitude before the vertical mode can be selected. If it is allowed, the applicant should consider the consequences of unconstrained climb or descent (CFIT or ATC altitude restrictions). Consideration should also be given to appropriate annunciation of deviation from a selected altitude and pilot action to reset the selected altitude.
- In the vertical speed mode, the FGS should smoothly acquire and maintain a selected vertical speed. Considerations should be given to selecting a value outside of the performance capability of the airplane as well as the use of the vertical speed mode without autothrust (these situations could lead to potential high- or low-speed protection conditions).
- In the flight path angle mode, the FGS should smoothly acquire and maintain the selected flight path angle. The same considerations previously mentioned for the vertical speed mode apply here as well. Acceptable means of compliance have included a reversion to an envelope protection mode or a timely annunciation of the situation.
- In the airspeed/Mach hold mode, the FGS should maintain the airspeed or Mach at the time of the engagement.
- In the airspeed/Mach select mode, the FGS should acquire and maintain a selected airspeed or Mach; it may be either preselected or synchronized to the airspeed or Mach at the time of the engagement.
- In the flight level change mode, the FGS should change altitude in a coordinated way with thrust control on the airplane. The autopilot/FD will typically maintain speed control through the elevator; the autothrust function, if engaged, should control the thrust to the appropriate value for climb or descent.

Vertical Modes/Altitude Capture

- The altitude capture mode should command the FGS to transition from a vertical mode to smoothly capture and maintain the selected target altitude with consideration of the rates of climb and descent experienced in service.
- The altitude capture mode should be automatically armed to ensure capture of the selected altitude. Annunciation of the armed status is not required if the mode is armed at all times; if the FGS is in the altitude capture mode, it should be annunciated.
- The altitude capture mode should engage from any vertical mode if the computed flight path will intercept the selected altitude and the altitude capture criteria are satisfied (except as specified during an approach; glide path for approach mode active).
- Changes in the climb/descent command references (vertical speed command), with the exception of those made by the flight crew using the altitude select controller, should not prevent capture of the target altitude.
- The altitude capture mode should smoothly capture the selected altitude, using an acceptable acceleration and pitch attitude with consideration for occupant comfort.
- The acceleration may result in an overshoot. To minimize the overshoot, the normal acceleration limit may be increased consistent with occupant safety.
- Pilot selection of other vertical modes at the time of altitude capture should not prevent or adversely affect the level-off at the target altitude. One means of

compliance is to inhibit other vertical modes during altitude capture unless the target altitude is changed; if glide path criteria are satisfied during altitude capture, then the FGS should transition to glide path capture.

- The FGS must be designed to minimize flight crew confusion concerning the FGS operation when the target altitude is changed during altitude capture. It should be suitably annunciated and appropriate for the phase of flight.

- Adjusting the datum pressure at any time during altitude capture should not result in the loss of the capture mode; the transition to the pressure altitude should be accomplished smoothly.

- If the autothrust function is active during the altitude capture, the autopilot and autothrust functions should be designed so that the FGS maintains the reference airspeed during the level-off maneuver.

Vertical Modes/Altitude Hold

- The altitude hold mode may be entered either by aircrew selection or by transition from another vertical mode.

- When initiated in level flight, the altitude hold mode should provide guidance or control to maintain altitude at the time the mode is selected.

- When initiated by pilot action when climbing or descending, the FGS should immediately initiate a pitch change to arrest the climb or descent and maintain the altitude when level flight is reached. The intensity of the leveling maneuver should be consistent with occupant comfort and safety.

- When initiated by an automatic transition from altitude capture, the altitude hold should provide guidance or control to the selected altitude.

- Any airplane response to a change in pressure datum should be smooth.

Vertical Modes/VNAV

- In the VNAV mode, the FGS should acquire and maintain the vertical flight path commanded by the flight management function (i.e., FMS). If the airplane is not on the desired FMS path when VNAV is selected, the FGS should enter an armed mode or smoothly provide guidance to acquire the FMS path. The FGS should transition from arm to engaged at a point where the FGS can smoothly acquire the desired FMS path.

- The deviation from the VNAV flight path should be displayed in the primary field of view, such as the PFD, navigation, or other acceptable display.

- When VNAV is selected for climb or descent, the autothrust function (if installed and engaged) should maintain the appropriate thrust setting and maintain target speed when leveling.

- The FGS should preclude a VNAV climb unless the mode select panel altitude window is set to an altitude above the current altitude.

- The FGS should preclude a VNAV descent unless the mode select panel altitude window is set to an altitude below the current altitude.

- The FGS should not allow the VNAV climb or descent to pass through a mode select panel altitude except when on final approach to a runway.

Multiaxis Modes

- In the takeoff mode, the vertical element of the FGS should provide vertical guidance to acquire and maintain a safe climb-out speed after initial rotation for takeoff.

- In the takeoff mode, the lateral element of the FGS, if implemented should maintain runway heading/track to wings level after liftoff; a separate lateral mode annunciation should be provided.

- If rotation guidance is provided, the use of this guidance should not result in a tail strike. It should be consistent with takeoff performance methods necessary to meet takeoff performance up to 35 ft AGL. If no rotation guidance is provided, pitch command bars may be displayed during takeoff roll, but not considered as providing guidance.

- The autothrust function should increase and maintain engine thrust to selected thrust limits.

- The FGS design should consider all engine and engine-inoperative requirements consistent with system performance characteristics after takeoff.

- Takeoff system operation should be smooth and continuous through transition from the runway portion of the takeoff to the airborne portion and the reconfiguration for enroute climb. The pilot should be able to use the same display for the operation; changes in guidance modes and display formats should be automatic.

- The vertical axis guidance of the takeoff system during normal operation should result in the appropriate pitch attitude and climb speed for the airplane.

- Normal-rate rotation of the airplane to the commanded pitch attitude at $V_R - 10$ knots for all engines operative and $V_R - 5$ knots for engine out should not result in a tail strike.

- The system should provide pitch commands that lead the airplane to smoothly acquire a pitch attitude that results in capture and tracking of the all-engine takeoff climb speed: V_2 (takeoff safety speed) $+ x$. x is the all-engine speed additive from the AFM (normally 10 knots or higher). If pitch-limited conditions are encountered, a higher climb airspeed may be used to achieve the required takeoff path without exceeding the pitch limit.

- For engine-out operation, the system should provide commands that lead the airplane to smoothly acquire a pitch attitude that results in capture and tracking of the following reference speeds:

  - V_2, for engine failure at or below V_2; this speed should be attained by the time the airplane has reached 35 ft AGL.

  - Airspeed at engine failure for failures between V_2 and $V_2 + x$.

  - $V_2 + x$ for failures at or above $V_2 + x$. (The airspeed at engine failure may be used if it has been shown that the minimum takeoff climb gradient can be achieved.)

- If implemented, the lateral element of the takeoff mode should maintain runway heading/track or wings level after liftoff and a separate lateral mode annunciation should be provided.

- Characteristics of the go-around mode and resulting flight path should be consistent with a manually flown go-around.

- The vertical element of the FGS go-around mode should initially rotate the airplane or provide guidance to rotate the airplane to arrest the rate of descent.

- The FGS should acquire and maintain a safe speed during climb-out and aircraft configuration changes (typically V_2, but a different speed may be found safe for wind shear).

- The autothrust function, if installed, should increase thrust and either maintain thrust to specific thrust limits or maintain thrust for an adequate, safe climb; the autothrust should not exceed thrust limits. The autothrust should not reduce thrust for wind below the minimum value required for an adequate, safe climb, nor reduce the thrust lever position below a point that would cause a warning to activate.

- The initial go-around maneuver may require a significant change in pitch attitude; it is acceptable to reduce thrust or lower pitch attitude for the comfort of the occupants after a safe climb gradient has been established. It should be possible for the pilot to reselect the full thrust value if needed.

- The go-around mode should engage when selected by the pilot, even if the mode select panel altitude is at or below the go-around initiation point. The airplane should climb until another vertical mode is selected or an altitude is set above the current aircraft altitude.

- The FGS design should consider an engine failure resulting in a go-around and an engine failure occurring during the go-around.

- In the approach mode, the FGS should capture and track a final approach lateral and vertical path, if applicable, from a navigation (FMS) or landing system (e.g., ILS, MLS, GLS).

- The FGS should annunciate all operationally relevant approach modes; modes that are armed and waiting for capture criteria to be satisfied should be indicated in addition to the active precapture mode. A positive indication of the capture of a previously armed mode should be provided.

- If the FGS contains submodes that arm without further action (flare) and are significant, they should be annunciated.

- Glideslope capture may occur prior to localizer capture. In this case, it is the flight crew's responsibility to ensure proper terrain clearance in a descent without lateral guidance.

Autothrust Modes

- In the thrust mode, the FGS should command the autothrust function to achieve a selected target thrust value.

- In the speed mode, the FGS should command the autothrust to acquire and maintain the selected speed target value (assuming the selected speed is within the speed range of the normal envelope). The autothrust may fly a higher airspeed than the selected target speed during an approach in winds or turbulence.

- In the retard mode, the FGS should work in the same manner for automatic and manual landings when the autothrust function is engaged.

Selected Target Altitudes in Approach Conditions

- The FGS vertical modes should allow the pilot to set the target altitude to a missed approach value prior to capturing the final approach segment. This should be possible for capturing from both above and below the final approach segment.

- It should be possible to define a descent path to the final approach fix and another path from the final approach fix to the runway with the target altitude set for the missed approach altitude. Appropriate targets and descent points should be identified by the FMS.

Control Wheel Steering

- In the control wheel steering (CWS) mode, the FGS allows the flight crew to maneuver the airplane manually through the autopilot control path. During CWS, the pilot, rather than the FGS, is in control of the airplane.
- It should be possible for the pilot to maneuver the airplane using normal flight controls with the CWS engaged and to achieve the maximum available control surface deflection without using forces so high that controllability requirements are not met.
- The maximum bank and pitch attitudes that can be achieved without physically overpowering the autopilot should be limited to those necessary for the normal operation of the airplane.
- The AC notes that normal operational limits are typically 35° in roll and +20° to −10° in pitch.
- It should be possible to perform all normal maneuvers smoothly and accurately without nuisance oscillation; it should also be possible to counter all normal changes in trim due to a change in configuration or power.
- The stall and recovery characteristics of the airplane should remain acceptable; it is assumed that recovery from a stall is with CWS in use unless automatic disengagement of the autopilot is provided.
- Consideration should be given to adjustments to trim that may be made by the autopilot in the course of maneuvers that can be reasonably expected.
- If CWS is used for takeoffs and landings, it must be shown that sufficient control (amplitude and rate) is available, reasonable mishandling is not hazardous, runaway rates and control forces are such that the pilot can overpower the autopilot without a significant deviation in flight path, and that any lag in the CWS response is acceptable.
- The autopilot, when engaged should not automatically revert to the CWS mode by applying a control input to the column or wheel unless the autopilot is in a capture mode (altitude or localizer capture).
- CWS is different from touch control steering, which is a temporary disengagement of the FGS via a button or paddle switch, allowing the pilot to hand-fly the aircraft. When the switch is released, the FGS is reengaged. A supervisory override works the same way, only the disengagement is caused by a force on the flight deck controls; as the force is released, the FGS reengages.

FGS and Fly-by-Wire Systems

- If speed protection is implemented in both the FGS and fly-by-wire flight control system they need to be compatible; the FGS speed protection (normal flight envelope) should operate to or within the limits of the flight control system (limit flight envelope).
- Information should be provided to the crew about the impact on the FGS following degradation of the fly-by-wire system.

6.15.3.11 Compliance Demonstration by Simulation and Flight Test

As a reminder, the criteria that govern successful compliance are shown in three paragraphs of CFR 14: 25.1301, 25.1309, and 25.1329. Compliance with paragraphs 1301 and 1309 includes specifics in AC 25-7C and AC 25.1309 (remember, you will be required to perform a safety assessment and design a human factors test plan for these systems). The FGS criteria are found in AC 25.1329-1B and have been addressed in this section.

The certification plan encompasses the evaluation of all modes of the FGS in all anticipated flight conditions and for all phases of flight. The human factors aspects must address the integration, man–machine interface, controls and displays, and visual and aural alerts and indications. The safety assessment is as discussed earlier in this chapter; AC 25.1329-1B devotes an entire chapter (chapter 8) to safety assessment considerations for the FGS.

Other ACs that may have to be consulted for completion of the certification plan include AC 120-29A, "Criteria for Approval of Category I and II Weather Minima for Approach," and AC 120-28D, "Criteria for Approval of Category III Weather Minima for Takeoff, Landing, and Rollout."

A simple way of looking at the plan is to take the characteristics of the specific modes (identified in AC 25.1329-1B, chapter 6) and exercise them against the conditions described in section 6.15.3.7 of this text. During the evaluations, human factors, controls and displays, and alerts and warnings will be embedded in your results. Additional testing, such as HUD considerations (if installed), precision landings (ILS, MLS, and GLS), and determination of the minimum use height (MUH) for the autopilot will also be included in the plan. Not all of the testing will be accomplished in flight; representative simulators can be used to demonstrate compliance for controls and displays and the man–machine interface portions of the certification plan. The level and fidelity of the simulator used should be commensurate with the certification credit being sought, and its use should be agreed upon with the certification authority.

AC 25.1329-1B does an excellent job of providing the required tests to show compliance. Chapter 9 describes the compliance demonstration using flight test and simulation. The actual flight test procedures have been moved to AC 25-7C, and flight cards can be developed directly from the procedures given. It is not the author's intent to replicate those procedures here, as they are already provided for you. At a minimum, the following test objectives need to be incorporated into the plan:

- Under normal and other than normal conditions (rare normal and nonnormal) the FGS should be evaluated to determine the acceptability of
  - Stability and tracking of automatic control elements
  - Controllability and tracking of guidance elements
  - Acquisition of flight paths for capture modes
  - Consistency of integration of modes
  - If the FGS provides wind shear escape guidance, perform demonstration requirements consistent with AC 25-12, "Airworthiness Criteria for the Approval of Airborne Windshear Warning Systems in Transport Category Aircraft"
- Specific performance conditions that require evaluation by flight test or simulation include
  - FGS performance and safety in icing conditions

- Low speed protection at high altitude with simulated engine failure, climb to altitude capture with simulated engine failure, vertical speed with insufficient climb power, and approach with speed abuse
- High speed protection at high-altitude level flight with autothrust function, high-altitude level flight without autothrust function, and high-altitude descending flight with autothrust function
- Evaluation of the go-around mode to assess the rotation characteristics of the airplane and the performance of the airplane in acquiring and maintaining a safe flight path. The go-around mode needs to consider the following factors:

  - Airplane weight and CG
  - Various landing configurations
  - Manual thrust or autothrust
  - Consequences of thrust rates with selection of go-around
  - Engine failure at initiation of go-around
  - Engine failure after go-around power is reached
  - Initiation altitude (e.g., ground effect, flare)

- In addition, during go-around, the following characteristics need to be evaluated:

  - Pitch response of the airplane during the initial transition
  - Speed performance during airplane reconfiguration and climb-out
  - Integrated autopilot and autothrust operation
  - Transition to missed approach altitude
  - Lateral performance during an engine failure
  - Where height loss during a go-around maneuver is significant or is required to support specific operational approval, demonstrated values for various initiated heights should be include in the AFM

- The flight test or simulator program should demonstrate that the FD or HUD guidance elements provide smooth, accurate, and damped guidance in all applicable modes to achieve satisfactory control task performance without pilot compensation or excessive workload. The FD guidance should be evaluated under the following conditions:

  - Stability augmentation off
  - Alternate fly-by-wire control modes (direct law), if any
  - Engine inoperative

- The FGS considerations for the HUD, delineated in section 6.15.3.9, need to be incorporated into the test plan.
- A flight evaluation needs to be conducted to demonstrate compliance with the FGS override and disengagement functions (sections 6.15.3.1 through 6.13.3.6).
- The safety assessment process will identify any failure condition that would require pilot evaluation to assess the severity of the effect and the validity of any assumptions used for pilot recognition and mitigation. The classification of a failure condition can vary according to flight condition and may need to be confirmed by simulator or

flight testing. Assessment of failure conditions contains four elements to be used in the evaluation:

- Failure condition insertion
- Pilot recognition of the effects of the failure condition
- Pilot reaction time (time between pilot recognition of the failure condition and the initiation of the recovery)
- Pilot recovery

- Failures should include

- Failures of the autopilot, including multiaxis and autotrim failures
- Failures of the FD if the conditions are relevant to the safety assessment
- The most critical flight condition (CG, weight, flap setting, speed, power, altitude, and thrust)
- Failures on takeoff. The worst case will be defined by the safety assessment with regards to the net effect on the flight path of the airplane at takeoff and the aircraft's attitude and speed during climb-out. Failures that cause the airplane to pitch up or pitch down or bank should be evaluated. If the FGS provides on-runway guidance for takeoff, then the effect of any failures on that guidance should be evaluated.
- Failures in climb, cruise, descent, and holding should be evaluated for height loss.
- Maneuvering failures as introduced at the bank angle used for normal operation. The bank angle should not exceed 60° at the recovery.
- The MUH must be determined for the type of approach performed (ILS, MLS, GLS, RNAV, nonprecision, etc.)

- Pilot recognition cues have been identified as

- Hardover (recognition time of about 1 sec)
- Slowover (usually seen in a deviation; may take some time)
- Oscillatory

- Pilot reaction time is dependent on many variables, but the following times are shown to be acceptable:

- Climb, cruise, descent and holding—recovery action should not be initiated until 3 sec after the recognition point
- Maneuvering flight—recovery action should not be initiated until 1 sec after the recognition point
- Approach—recovery action should not be initiated until 1 sec after the recognition point
- For the final phase of landing (below 80 ft), the pilot can be assumed to react with no delay.
- For phases of flight where the pilot is exercising manual control using CWS, if implemented, it can be assumed that the pilot will recover at the recognition point.

- An incremental normal acceleration on the order of 0.5 g is considered the maximum pilot recovery technique. During the pilot recovery maneuver, the pilot may over-power the autopilot or disengage it.

- The MUH for the autopilot must be determined from flight testing and therefore needs to be incorporated as a test objective.

6.15.3.12 Determination of the Autopilot MUH

There are two methods used for determination of the MUH for the autopilot. The first method is used for approaches with a vertical path reference; the second is used for those approaches that do not incorporate a vertical path reference.

Figure 6.42 provides an illustration of the *deviation profile method*, which is used for assessment of the MUH for approaches with a vertical path reference. The first step is to identify the worst-case malfunction. The approach should be flown in representative conditions (e.g., coupled ILS) with the malfunction initiated at a safe height. The pilot should not initiate recovery from the malfunction until 1 sec after the recognition point. The delay is intended to simulate the variability in response to effectively a hands-off condition. It is expected that the pilot will follow through on the controls until the recovery is initiated.

A deviation profile can be drawn from the results of the flight test. The deviation profile is slid down the glide path until it is tangential to the 1:29 line or the runway. The failure condition contribution to the MUH approach may be determined from the geometry of the aircraft wheel height determined by the deviation profile relative to the 1:29 line intersecting a point 15 ft above the runway threshold. The method of determination may be solved graphically or analytically.

The MUH approach is based on the recovery point for the following reasons:

- It is assumed that, in service, the pilot will be hands-off until the autopilot is disengaged at the MUH in normal operation.

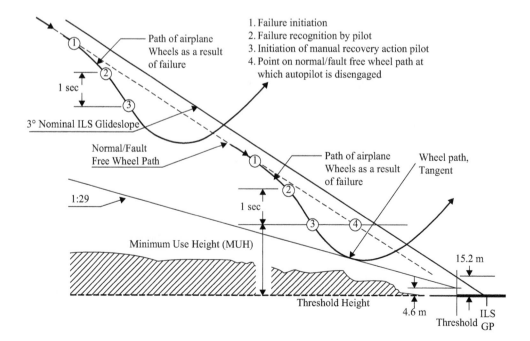

FIGURE 6.42 ■ Deviation Profile Method for Determining the MUH

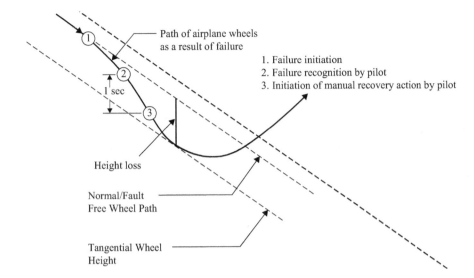

FIGURE 6.43 ▪
Height Loss Method
for Determining the
MUH

- The test technique assumes a worst case based on the pilot being hands-off from the point of malfunction initiation to the point of recovery.

- A failure occurring later in the approach than the point of initiation of the worst-case malfunction is assumed to be recovered earlier and, consequently, less severe.

Figure 6.43 depicts the height loss method for determining the MUH of the autopilot for an approach without a vertical path reference. A descent path of 3° with a nominal approach speed should be used unless the autopilot is to be approved for significantly steeper descents. The vertical height loss is determined by the deviation of the aircraft wheel height relative to the nominal wheel flight path.

A sample Part 25 FGS compliance matrix is shown in Table 6.45.

6.15.4 Part 27 and Part 29 FGS

The advisory circulars for helicopters have not incorporated paragraph 1335 (Flight Directors) into paragraph 1329; the guidance for airworthiness in AC 27-1B and 29-2C are the same. Stability augmentation systems (SAS) are covered in Appendix B of the respective ACs. In accordance with paragraph 1301, the autopilot should be evaluated to show that it meets its intended function. If the autopilot uses the same servos and servo amplifiers as the SAS, the autopilot cannot produce a more severe hardover than the SAS. If this is the case, then no other testing than that of the SAS in Appendix B is required.

The evaluation requirements are similar to those addressed for Parts 23 and 25. If monitors are used to meet the hardover malfunction requirements, then the Certification Plan is a collaborative effort with the authority. There must be some method to indicate the alignment of the actuating devices to the pilot. Depending on complexity, the autopilot may be interlocked without the completion of a pre-flight test. Controls should be labelled and easily accessed by the pilot, and the annunciator colors must conform to AC 27/29.1322. Controls and displays should be readable in all lighting conditions, and movement of controls should be correct in design and sense. Any disconnect should be annunciated to the crew.

TABLE 6.45 ■ Part 25 FGS Compliance Matrix

No.	Requirement AC Paragraph	Subject	Description
1	27.a(1)	Single Switch Action	Each pilot should be able to select the autopilot function of the FGS with a single switch action. The single switch action should engage both the pitch and roll axes.
2	27.a(2)	Engagement Status of Autopilot	The autopilot system should provide positive indication to the flight crew that the system has been engaged. [See § 25.1329(i)]
3	27.a(3)	Multiple Autopilots	If a single autopilot within a multiple autopilot installation can be individually selected by the flight crew, the engagement annunciation should reflect the flight crew selection. It should not be possible for multiple autopilots to be engaged in different modes. For modes that use multiple autopilots, the additional autopilots may engage automatically at selection of the mode or after arming the mode. A means should be provided to determine that adequate autopilot capability exists to support the intended operation.
4	AC 25.1329-1B 27.a(4)	Acceptable Transients	With normal operating conditions, significant transients may not occur. [See § 25.1329(d)]
5	27.a(5)	Flight Director Not Engaged	Without a FD engaged, the initial lateral and vertical modes should be consistent with minimal disturbance from the flight path.
6	27.a(6)	FD Engaged	With a FD engaged, the autopilot should engage in a mode consistent with (i.e., the same as, or if that is not possible, then compatible with) the active FD mode of operation. Consideration should be given to the mode in which the autopilot will engage when large commands are present on either or both FDs.
7	27.b(1)(a)	Normal Conditions	Under normal conditions, automatic or manual disengagement of the autopilot should be free of perceptible transients or out-of-trim forces that are not consistent with the maneuvers being conducted by the airplane at the time of disengagement. Disengagement in normal conditions may not result in a transient any greater than a minor transient. If multiple autopilots are engaged, any disengagement of an individual autopilot may not result in a transient any greater than a minor transient. [See § 25.1329(d)]
8	27.b(1)(b)	Other Than Normal Conditions	Under other than normal conditions (i.e., nonnormal or rare normal conditions), disengagement of the autopilot may not result in a transient any greater than a significant transient. [See § 25.1329(e)] The flight crew should be able to respond to a significant transient without using any of the following: 1. Exceptional piloting skill, alertness, or strength; 2. Forces greater than those given in § 25.143(c), or; 3. Accelerations or attitudes in the airplane that might result in a hazard to secured or non-secured occupants.

#	Reference	Topic	Description
9	27.b(1)(c)	Potential for Significant Transient	The flight crew should be made aware (via a suitable alerting or other indication) of conditions or situations (e.g., continued out-of-trim) that could result in a significant transient at disengagement.
10	27.b(2)(a) AC 25.1329-1B	Alert Type	Since it is necessary for a pilot to immediately assume manual control following disengagement (manual or automatic) of the autopilot, a warning (both visual and aural) must be given. [See § 25.1329(j)]
11	27.b(2)(b)	Aural Warning Specifications	A timely aural warning must be provided and must be distinct from all other cockpit warnings. [See § 25.1329(j)] The aural warning should continue until silenced by one of the following means: 1. Activation of the autopilot quick disengagement control; 2. Reengagement of the autopilot; 3. Another acceptable means.
12	27.b(2)(c)	Multiple Autopilot System	1. Disengagement of an autopilot channel within a multiple-channel autopilot system, downgraded system capability, or a reduction in the level of system redundancy that requires immediate flight crew awareness and possible timely action should cause a caution level alert to be issued to the flight crew. 2. Disengagement of an autopilot channel within a multiple-channel autopilot system that requires only flight crew awareness should cause a suitable advisory level alert to be issued to the flight crew.
13	27.b(3)(b)	Quick Disengagement Control Location	The quick disengagement control must be located on each control wheel or equivalent. [See § 25.1329(a)] It should be within easy reach of one or more fingers/thumb of the pilot's hand when the hand is in a position for normal use on the control wheel or equivalent.
14	27.b(3)(c) AC 25.1329-1B	Criteria	1. Must be operable with one hand on the control wheel or equivalent and the other hand on the thrust levers. [See § 25.1329(a)] 2. Should be accessible and operable from a normal hands-on position without requiring a shift in hand position or grip on the control wheel or equivalent. 3. Should be easily locatable by the pilot without having to first locate the control visually. 4. Should be designed so that any action to operate the quick disengagement control should not cause an unintended input to the control wheel (or equivalent). 5. Should be designed to minimize inadvertent operation and interference from other nearby control wheel (or equivalent) switches/devices, such as radio control or trim.

(Continues)

TABLE 6.45 ■ *(Continued)*

No.	Requirement AC Paragraph	Subject	Description
15	27.b(4)(b)	Alternative Means of Autopilot Disengagement	The following means of providing an alternative disengagement have been found to be acceptable: 1. Selecting the engagement control to the "off" position. 2. Disengaging the bar on mode select panel (MSP). 3. Activating the trim switch on the yoke.
16	27.b(5)	Flight Crew Pitch Trim Input	If the autopilot is engaged and the pilot applies manual pitch trim input, and the autopilot is designed to disengage because of that flight crew action, the autopilot must disengage with no more than a minor transient. [See § 25.1329(c)]
17	28.a(1)(a)	FD Selection	A means may be provided for each pilot to select (i.e., turn on) and de-select (i.e., turn off) the FD for display on his or her primary flight display (PFD).
18	28.a(1)(b)	FD Engagement	An FD is considered "engaged" if it is selected and displaying guidance cues.
19	28.a(1)(c)	Engagement Status of FD	The selection status of the FD and the source of FD guidance should be clear and unambiguous. Failure of a selected FD should be clearly annunciated.
20	28.a(1)(e)	Autopilot Engaged	An FD should engage into the current modes and targets of an already engaged autopilot or FD, if any. With no autopilot engaged, the basic modes at engagement of the FD functions should be established consistent with typical flight operations.
21	28.a(2)(a)	Display	The FD command guidance cue(s) will typically be displayed under the following conditions: 1. When the FD is selected and valid command guidance is available, or; 2. When the FD is automatically providing guidance
22	28.a(2)(b)	Engagement Indication	The display of guidance cue(s) (e.g., FD bars) is sufficient indication that the FD is engaged.
23	28.a(2)(b) AC 25.1329-1B	Invalid Guidance	The FD guidance cue(s) should be removed when guidance is determined to be invalid.
24	28.b	FD Disengagement	There may be a means for each pilot to readily de-select his or her on-side FD function. Flight crew awareness of disengagement and de-selection is important. Removal of guidance cue(s) alone is not sufficient indication of de-selection because the guidance cue(s) may be removed from view for a number of reasons, including invalid guidance or autopilot engagement. Therefore, the FD function should provide clear and unambiguous indication to the flightcrew that the function has been deselected.
25	29.a(1)	Autothrust Engagement Indication	The autothrust function should provide the flight crew positive indication that the system has been engaged.
26	29.a(2)	Accessibility of Controls	The autothrust engagement controls should be accessible to each pilot.

No.	Document	Paragraph	Topic	Description
27		29.a(3)	Inadvertent Engagement/Disengagement	The autothrust function should normally be designed to prevent inadvertent engagement and inadvertent application of thrust for both on–ground and in–air operations. For example, separate arm and engage functions may be provided.
28		29.a(4)	Automatic Engagement	Intended automatic engagement, such as a "wake up" mode to protect for unsafe speeds, may be acceptable. If such automatic engagement occurs, it should be clear to the flight crew that automatic ngagement has occurred. The automatic engagement may not cause a transient larger than a minor transient. [See § 25.1329(c)]
29		29.b(1)	Autothrust Disengagement Indication	Positive indication of disengagement of the autothrust function must result in a caution level alert to the flightcrew. [See § 25.1329(k)] The autothrust "engage" status annunciations should be deleted.
30	AC 25.1329-1B	29.b(1)(a)	Visual Indication	1. Following automatic disengagement: Visual indication of disengagement should persist until canceled by flightcrew action. 2. Following manual disengagement: If an aural alert is provided, visual indication of disengagement should persist for some minimum period. If an aural alert is not provided, visual indication of disengagement should persist until canceled by flight crew action.
31		29.b(1)(b)	Aural alert	If provided, the aural disengagement alert should be of sufficient duration and volume to assure that the flight crew has been alerted that disengagement has occurred. However, an extended cycle of an aural alert is not acceptable following disengagement, if such an alert can significantly interfere with flight crew coordination or radio communication.
32		29.b(2)	Inadvertent Disengagement	The autothrust normally should be designed to preclude inadvertent disengagement during activation of autothrust modes of operation.
33		29.b(3)	Consequence of Disengagement	Under normal conditions, autothrust disengagement may not cause a transient any greater than a minor transient. [See § 25.1329(d)]
34		29.b(4)	Autothrust Quick Disengagement	Autothrust quick disengagement controls must be provided for each pilot on the respective thrust control (thrust lever or equivalent). A single–action, quick disengagement switch must be incorporated on the thrust control, so that switch activation can be executed when the pilot's other hand is on the control wheel (or equivalent). [See § 25.1329(a)]
35	AC 25.1329-1B	30.b(1)(a)	Autopilot Override Override Force	The autopilot should disengage when the flight crew applies a significant override force to the controls. The applicant should interpret "significant" as a force that is consistent with an intention to overpower the autopilot by either or both pilots. The autopilot should not disengage by minor application of force to the controls, such as a pilot gently bumping the control column while entering or exiting a pilot seat during cruise.

(Continues)

TABLE 6.45 ▪ (Continued)

No.	Requirement AC Paragraph	Subject	Description
36	30.b(1)(b)	Transients Resulting from Override	In normal operating conditions, a transient larger than a minor transient may not result from autopilot disengagement when the flight crew applies an override force to the controls. [See § 25.1329(d)]
37	30.b(1)(c)	Sustained Override Force Below level Required for Automatic Disconnect	Sustained application of force below the disengagement threshold may not result in a potential hazard. [See § 25.1329(l)]
38	30.b(2)(a)	Potential Hazard	If the FGS is not designed to disengage in response to any override force, then the response to an override may not result in a potential hazard. Sustained application of an override force may not result in a potential hazard, such as when the flight crew abruptly releases the force on the controls. [See § 25.1329(l)]
39	30.b(2)(b)	Transients Resulting from Override	In normal operating conditions, a transient larger than a minor transient may not result from manual autopilot disengagement after the flight crew has applied an override force to the controls. [See § 25.1329(d)]
40	30.c(1)	Autothrust Override Force	It should be possible for the pilot to readily override the autothrust function and set thrust by moving the thrust levers (or equivalent) with one hand.
41	30.c(2)	Response to Override	The autothrust response to a flightcrew override may not create a potential hazard. [See § 25.1329(m)]
42	30.c(3)	Engagement Status with Override	Autothrust functions may be designed to safely remain engaged during pilot override. Alternatively, autothrust functions may disengage as a result of pilot override, provided that the design prevents unintentional autothrust disengagement and adequately alerts the flight crew to ensure pilot awareness.
43	31.a	FGS Mode Engagement Philosophy	A description of the philosophy used for the mode at engagement of the autopilot, FD, and autothrust functions should be provided in light crew training material.
44	31.a AC 25.1329-1B	Engagement Mode Compatibility	It should not be possible to select incompatible FGS command or guidance functions, such as commanding speed through elevator and autothrust at the same time.
45	42.a	HMI	The human-machine interface with the FGS is crucial to ensuring safe, effective, and consistent FGS operation. The manner in which FGS information is depicted to the flightcrew is vital to flightcrew awareness and, therefore, to safe operation of the FGS.
46	42.b	Control, Indication and Alert	These features must be designed to minimize flight crew errors and confusion. [See § 25.1329(i)]
47	42.c	Information Provided by FGS	It is recommended that the applicant evaluate the adequacy and effectiveness of the information provided by the FGS interface, that is the controls, indications, alerts, and displays, to ensure flight crew awareness of FGS behavior and operation.

48		FGS Controls	The FGS controls must be designed and located to prevent crew errors, confusion, and inadvertent operation. [See § 25.1329(i)] They should be designed to provide convenience of operation to each crewmember. The function and direction of motion of each control must be readily apparent or plainly indicated on or adjacent to each control, if needed to prevent inappropriate use or confusion. [See § 25.1329(f)] Sections 25.777(b) and 25.779(a) provide requirements regarding direction of motion of flight deck controls. These requirements apply for FGS command reference controls that select target values, such as heading select or vertical speed. Section 25.781 also provides requirements for the shapes of the knobs.
49		Differentiation of Knob Shape and Position	Errors by the flight crew have included confusing speed and heading knobs on the MSP.
50	AC 25.1329-1B	Design to support correct of target value	Use of a single control, such as concentric controls, for selecting multiple command reference targets has resulted in erroneous target value selection.
51		Positioning of controls and related indications	Individual FGS controls, flight mode annunciator (FMA), and related PFD information should be positioned so that, as much as reasonably practical, items of related function have similarly related positions. Misinterpretation and confusion have occurred due to the inconsistent arrangement of FGS controls with the annunciations on the FMA.
52		Inadvertent operation	The FGS controls must be located to discourage or avoid inadvertent operation, such as inadvertent engagement or disengagement. [See § 25.777(a)]
53		Annunciation of engagement of FGS	Engagement of the FGS functions must be suitably annunciated to each pilot. [See § 25.1329(i)]
54		Description of FGS modes	The operator should be provided with appropriate descriptions of the FGS modes and their behaviors.
55		FGS mode annunciations	Mode annunciation must indicate the state of the system. [See § 25.1329(i)]
56		Active and armed modes	The FGS mode annunciations must effectively and unambiguously indicate the active and armed modes of operation. [See § 25.1329(i)]
57		Location	Mode annunciations should be located in the forward field of view.
58		Discriminator	Color, font type, font size, location, highlighting, and symbol flashing have historical precedent as good discriminators, when implemented appropriately.
59	AC 25.1329-1B	Color	Color should be used in a consistent manner and assure compatibility with the overall use of color on the flight deck.
60		Operationally relevant mode changes	Mode changes that are operationally relevant—especially mode reversions and sustained speed protection—should be clearly and positively annunciated to ensure flightcrew awareness.

(Continues)

TABLE 6.45 ■ *(Continued)*

No.	Requirement	AC Paragraph	Subject	Description
61		44.d(2)	Attention-getting features	The transition from an armed mode to an engaged mode should provide an additional attention–getting feature, such as boxing and flashing on an electronic display (see AC 25-11) for a suitable, but brief, period
62		44.d(3)	Use of alert	The safety consequences of the flight crew not recognizing mode changes should be considered. If necessary, an appropriate alert should be used.
63		44.e	Failure to engage or arm	The failure of a mode to engage or arm when selected by the pilot should be apparent.
64		44.f	FGS mode display and indications	Mode information provided to the flight crew should be sufficiently detailed, so that the consequences of the interaction between the FGS and the flight crew can be determined unambiguously.
65		45.b(1)	Speed protection alerts	To assure crew awareness, an alert should be provided when a sustained speed protection condition is detected.
66		45.b(2)(a)	Low speed protection	Low speed protection alerts should include both an aural and a visual component.
67		45.b(2)(b)	High speed protection	High-speed protection alerts need include only a visual alert component because of existing high-speed aural alert requirements.
68	AC 25.1329-1B	45.b(3)	Consistency	Alerts for speed protection should be consistent with the protection provided and with the other alerts in the flight deck.
69		45.b(4)	Nuisance alert	Care should be taken to set appropriate values for indicating speed protection that would not be considered a nuisance for the flight crew.
70		45.c	Loss of autopilot approach mode	The loss of the Approach mode requires immediate flight crew awareness. This may be accomplished through autopilot disengagement
71		45.d(2)	Flight crew alerts	To help ensure flight crew awareness and timely action, appropriate alert(s)—normally a caution or warning—should be provided to the flight crew for conditions that could require exceptional piloting skill or alertness for manual control following autopilot disengagement (e.g., significantly out of trim conditions). The number and type of alerts required should be determined by the unique situations that are being detected and by the crew procedures required to address those situations. Any alert should be clear and unambiguous and should be consistent and compatible with other flight deck alerts. Care should be taken to set appropriate thresholds for these alerts so that they are not considered a nuisance for the flight crew.
72		45.d(3)(a)	Sustained lateral control command	If the autopilot is holding a sustained lateral control command, a timely alert should be considered to permit the crew to manually disengage the autopilot and take control prior to any automatic disengagement that might result from the condition.

#	Ref	Topic	Description
73	45.d(3)(b)	Sustained pitch command	If the autopilot is holding a sustained pitch control command, a timely alert should be considered to permit the crew to manually disengage the autopilot and take control prior to any automatic disengagement that might result from the condition.
74	45.d(3)(c)	Bank and pitch angle	A timely alert should bring this condition to the attention of the flight crew and permit the crew to manually disengage the autopilot and take control prior to any automatic disengagement that might result.
75	45.d(4)	Automatic disengagement	If there is an automatic disengagement feature due to excessive out-of-trim, an alert should be generated and should precede any automatic disengagement with sufficient margin to permit timely flight crew recognition and manual disengagement.
76	52.a	Intended function	The FGS should provide guidance or control, as appropriate, for the intended function of the active mode(s) in a safe and predictable manner within the airplane's normal flight envelope.
77	52.b	Effect of system tolerances	Where system tolerances have a significant effect on autopilot authority limits, consideration should be given to the effect on autopilot performance.
78	53.c(2)	Performance: configuration changes	The FGS should perform its intended function during routine airplane configuration or power changes, including the operation of secondary flight controls and landing gear.
79	53.c(6)	Autopilot	The autopilot should provide smooth and accurate control without divergent or perceptible sustained nuisance oscillation.
80	53.c(7)	Flight Director	The FD, in each available display presentation (e.g., single cue, cross–pointer, flight path director) should provide smooth and accurate guidance and be appropriately damped, to achieve satisfactory control task performance without pilot compensation or excessive workload.
81	53.c(8)	Autothrust	The autothrust function should provide smooth and accurate control of thrust without significant or sustained oscillatory power changes or excessive overshoot of the required power setting.
82	53.c(9)	Automatic trim	Automatic roll and yaw trim functions, if installed, should operate without introducing adverse interactions with automatic control of the aircraft.
83	57.a(2)	Speed protection	1. The FGS may detect the speed protection condition, alert the flight crew, and provide speed protection control or guidance. 2. The FGS may detect the speed protection condition, alert the flight crew, and then disengage the FGS. 3. The FGS may detect the speed protection condition, alert the flight crew, and remain engaged in the active mode without providing speed protection control or guidance.

AC 25.1329-1B

(Continues)

TABLE 6.45 ■ (Continued)

No.	Requirement	AC Paragraph	Subject	Description
84		57.b(3)(a)	Low speed protection during approach	Speed protection should not interfere with the approach and landing phases of flight.
85	AC 25.1329-1B	57.b(3)(b)	Autothrust operation	It is assumed that with autothrust operating normally, the combination of thrust control and pitch control during the approach will be sufficient to maintain speed and desired vertical flight path. In cases where it is not sufficient, an alert should be provided in time for the flight crew to take appropriate corrective action.
86		57.b(3)(c)	Defined vertical path	For approach operations with a defined vertical path if the thrust is insufficient to maintain both the desired flight path and the desired approach speed: 1. The FGS mode reversion and low speed alert should be activated to ensure pilot awareness. 2. The FGS will then provide a low speed alert while remaining in the existing FGS approach mode. 3. The FGS will then provide a low speed alert and disengage.
87		57.c(2)(a)	High speed protection: Speed excursion	Operations at or near V MO/M MO in routine atmospheric conditions (e.g., light turbulence) are safe. Small, brief excursions above V MO/M MO by themselves are not unsafe. The FGS design should strive to strike a balance between providing adequate speed protection margin and avoiding nuisance activation of high-speed protection.
88	AC 25.1329-1B	57.c(2)(b)	High speed protection while in altitude hold mode	Climbing to control airspeed is not desirable, because departing an assigned altitude can be disruptive to air traffic control (ATC) and potentially hazardous (e.g., in RVSM airspace). As long as the speed does not exceed a certain margin beyond V MO/M MO (e.g., six knots), it is better that the FGS remain in Altitude Hold mode. The autothrust function, if operating normally, should effect high-speed protection by limiting its speed reference to the normal speed envelope (i.e., at or below V MO/M MO). The basic airplane high-speed alert should be sufficient for the pilot to recognize the overspeed condition and take corrective action to reduce thrust. However, if the airspeed exceeds a margin beyond V MO/M MO (e.g., six knots), the FGS may transition from Altitude Hold to the Overspeed Protection mode and depart (i.e. climb above) the selected altitude.
89		57.c(2)(c)	High speed protection during climbs and descents	When the elevator channel of the FGS is not controlling airspeed, the autothrust function, if engaged, should reduce thrust, as needed to prevent sustained airspeed excursions beyond V MO/M MO (e.g., six knots) down to the minimum appropriate value.

When thrust is already the minimum appropriate value or the autothrust function is not operating, the FGS should begin using pitch control, as needed, for high-speed protection.

If conditions are encountered that result in airspeed excursions above V MO/M MO, it is preferable for the FGS to smoothly and positively guide or control the airplane back to within the speed range of the normal flight envelope.

No.	Ref	Topic	Description
90	63.a	Heading or track hold	In the Heading or Track Hold mode, the FGS should maintain the airplane heading or track.
91	63.b	Heading or track select	In the Heading or Track Select mode, the FGS should expeditiously acquire and maintain a "selected" heading or track value consistent with occupant comfort.
92	63.c(1)	Lateral flight path	In the LNAV mode, the FGS should acquire and maintain the lateral flight path commanded by a flight management function.
93	63.c(2)	Automatic mode transitions	If the airplane is not on the desired lateral path or within the designed path capture criteria when LNAV is selected, the FGS LNAV mode should enter an armed state. The FGS should transition from the armed state to an engaged state at a point where the lateral flight path can be smoothly acquired and tracked.
94	63.c(3)	Takeoff or go-around (TOGA)	For an FGS incorporating the LNAV mode during the TOGA phase, the design should specify maneuvering capability immediately after takeoff and any limits that may exist. After TOGA, maneuvering should be based upon aircraft performance with the objective to prevent excessive roll attitudes where wingtip impact with the runway becomes probable, yet satisfy operational requirements where terrain and/or thrust limitations exist.
AC 25.1329-1B			
95	64.a	Target altitude selection	To avoid unconstrained climbs or descents for any altitude transitions when using applicable vertical modes, the altitude select controller should be set to a new target altitude before the vertical mode can be selected.
96	64.b	Vertical	In the Vertical Speed mode, the FGS should smoothly acquire and maintain a selected vertical speed.
97	64.c	Flight path angle	In the Flight Path Angle mode, the FGS should smoothly acquire and maintain the selected flight path angle.
98	64.d	Indicated airspeed IAS/Mach hold	In the Airspeed/Mach Hold mode, the FGS should maintain the airspeed or Mach at the time of engagement.
99	64.e	IAS/Mach select	In the Airspeed/Mach Select mode, the FGS should acquire and maintain a selected airspeed or Mach.
100	64.f	Flight level change	In the FL Change mode, the FGS should change altitude in a coordinated way with thrust control on the airplane.

(Continues)

TABLE 6.45 ■ *(Continued)*

No.	Requirement AC Paragraph		Subject	Description
101		64.g(1)	Altitude capture: mode transition	The Altitude Capture mode should command the FGS to transition from a vertical mode to smoothly capture and maintain the selected target altitude with consideration of the rates of climb and descent experienced in service.
102	AC 25.1329-1B	64.g(2)(a)	ALT capture: automatic arming of mode	The Altitude Capture mode should be automatically armed to ensure capture of the selected altitude. Annunciation of the armed status is not required if the Altitude Capture mode is armed at all times. If the FGS is in the Altitude Capture mode, it should be annunciated.
103		64.g(2)(b)	Engagement from any vertical mode	The Altitude Capture mode should engage from any vertical mode if the computed flight path will intercept the selected altitude and the altitude capture criteria are satisfied, except as specified during an approach.
104		64.g(2)(c)	Changing climb/descent command references	Changes in the climb/descent command references with the exception of those made by the flight crew using the altitude select controller, should not prevent capture of the target altitude.
105		64.g(2)(d)	Capturing selected altitude	The Altitude Capture mode should smoothly capture the selected altitude, using an acceptable acceleration limit and pitch attitude with consideration for occupant comfort.
106		64.g(2)(e)	Minimizing acceleration overshoot	The acceleration limit may, under certain conditions, result in an overshoot.
107		64.g(2)(f)	Selecting other vertical mode	Pilot selection of other vertical modes at the time of altitude capture should not prevent or adversely affect the level off at the target altitude.
108		64.g(2)(g)	Changing target altitude	The FGS must be designed to minimize flight crew confusion concerning the FGS operation when the target altitude is changed during altitude capture.
109		64.g(2)(h)	Barometric pressure adjustment	Adjusting the datum pressure at any time during altitude capture should not result in loss of the capture mode. The transition to the pressure altitude should be accomplished smoothly.
110		64.g(2)(i)	Maintaining reference airspeed	If the autothrust function is active during altitude capture, the autopilot and autothrust functions should be designed such that the FGS maintains the reference airspeed during the level-off maneuver.
111	AC 25.1329-1B	64.h(1)(a)	ALT hold: pilot selection entering mode	**1.** Level flight. When initiated by pilot action in level flight, the Altitude Hold mode should provide guidance or control to maintain altitude at the time the mode is selected. **2.** Climbing or descending. When initiated by pilot action when the airplane is either climbing or descending, the FGS should immediately initiate a pitch change to arrest the climb or descent and maintain the altitude when level flight (e.g., less than 200 ft per minute) is reached. The intensity of the leveling maneuver should be consistent with occupant comfort and safety.

	AC 25.1329-1B			
112		64.h(1)(b)	Automatic transition	When initiated by an automatic transition from Altitude Capture, the Altitude Hold mode should provide guidance or control to the selected altitude.
113		64.h(2)	Mode transition annunciation	Automatic transition into the Altitude Hold mode from another vertical mode should be clearly annunciated for flight crew awareness.
114		64.h(3)	Barometric pressure adjustment	Any airplane response due to an adjustment of the datum pressure should be smooth.
115		64.i(1)(a)	FMS acquire vertical path	In the VNAV mode, the FGS should acquire and maintain the vertical flight path commanded by a flight management function (i.e., FMS or equivalent).
116		64.i(1)(b)	Deviation from vertical path	If the aircraft is flying a vertical path (e.g., VNAV path), then the deviation from that path should be displayed in the primary field of view, such as the PFD, navigation display (ND), or other acceptable display.
117		64.i(2)(a)	Climb/Descent: autothrust function	When VNAV is selected for climb or descent, the autothrust function (if installed and engaged) should maintain the appropriate thrust setting.
118		64.i(2)(b)	Preclude VNAV climb	The FGS should preclude a VNAV climb, unless the mode select panel (MSP) altitude window is set to an altitude above the current altitude.
119		64.i(2)(c)	Preclude VNAV descent	The FGS should preclude a VNAV descent, unless the MSP altitude window is set to an altitude below the current altitude, except when on a final approach to a runway.
121		64.a(1)	Takeoff mode: vertical guidance	In the Takeoff mode, the vertical element of the FGS should provide vertical guidance to acquire and maintain a safe climb out speed after initial rotation for takeoff.
122	AC 25.1329-1B	64.a(2)	Lateral guidance	In the Takeoff mode, the lateral element of the FGS, if implemented, should maintain runway heading/track or wings level after liftoff. A separate lateral mode annunciation should be provided.
123		64.a(3)	Rotation guidance	If rotation guidance is provided, the use of the guidance should not result in a tail strike. It should be consistent with takeoff methods necessary to meet takeoff performance requirements up to 35 ft above ground level (AGL). If no rotation guidance is provided, the pitch command bars may be displayed during takeoff roll but should not be considered as providing rotation guidance, unless it is part of the intended function.
124		64.a(4)	Autothrust	The autothrust function should increase and maintain engine thrust to the selected thrust limits (e.g., full takeoff thrust, de-rate).
125		64.a(5)(a)	Transitions between flight phases	Takeoff system operation should be continuous and smooth through transition from the runway portion of the takeoff to the airborne portion and reconfiguration for enroute climb. The pilot should be able to continue the use of the same primary display(s) for the airborne portion as for the runway portion. Changes in guidance modes and display formats should be automatic.

(Continues)

TABLE 6.45 ■ *(Continued)*

No.	Requirement AC Paragraph	Subject	Description
126	64.a(5)(b)	Pitch attitude and climb speed/normal operation	1. Normal rate rotation of the airplane to the commanded pitch attitude at V R (takeoff rotation speed)—10 knots for all engines operative and V R—5 knots for engine out should not result in a tail-strike. 2. The system should provide commands that lead the airplane to smoothly acquire a pitch attitude that results in capture and tracking of the all-engine takeoff climb speed, V 2 (takeoff safety speed) + X. X is the all-engine speed additive from the AFM (normally 10 knots or higher).
127	AC 25.1329-1B 64.a(5)(c)	Engine-out operation	For engine-out operation, the system should provide commands that lead the airplane to smoothly acquire a pitch attitude that results in capture and tracking of the following reference speeds: 1. V 2, for engine failure at or below V 2. This speed should be attained by the time the airplane has reached 35 feet altitude. 2. Airspeed at engine failure for failures between V 2 and V 2 + X. 3. V 2 + X, for failures at or above V 2 + X. Alternatively, the airspeed at engine failure may be used, provided it has been shown that the minimum takeoff climb gradient can still be achieved at that speed.
128	64.a(5)(d)	Lateral commands during takeoff mode	If implemented, the lateral element of the Takeoff mode should maintain runway heading/track or wings level after liftoff and a separate lateral mode annunciation should be provided.
129	64.b(1)	Go-around mode: vertical elements	The vertical element of the FGS Go-Around mode should initially rotate the airplane or provide guidance to rotate the airplane to arrest the rate of descent.
130	64.b(2)	Speed	The FGS should acquire and maintain a safe speed during climb out and airplane configuration changes. Typically, a safe speed for go-around climb is V 2, but a different speed may be found safe for windshear recoveries (See AC 25-12).
131	64.b(3)(a)	Autothrust during climb	The autothrust function, if installed, should increase thrust and either maintain thrust to specific thrust limits or maintain thrust for an adequate, safe climb.
132	64.b(3)(b)	Pitch attitude during go-around	The initial go-around maneuver may require a significant change in pitch attitude.
133	64.b(4)	Engagement	The Go-Around mode should engage when go-around is selected by the pilot, even if the MSP selected altitude is at or below the go-around initiation point.
134	AC 25.1329-1B 64.b(5)	All engine and engine out capability	The FGS design of the Go-Around mode should address all engine and engine-out operation. The design should consider an engine failure resulting in a go-around and the engine failure occurring during an all engine go-around.
135	64.c(1)	Final approach path	In the Approach mode, the FGS should capture and track a final approach lateral and vertical path, if applicable, from a navigation (NAV) or landing system, for example, ILS, MLS, GLS, RNP, area navigation (RNAV).
136	64.c(2)	Mode annunciations	The FGS should annunciate all operationally relevant approach modes.
137	64.c(3)	Submodes	The FGS may have submodes that become active without additional crew selection.

#	Reference	Paragraph	Topic	Description
138		64.c(4)	Mode engagement sequence	Glideslope capture mode engagement may occur prior to localizer capture.
139		66.a	Autothrust mode	In the Thrust mode, the FGS should command the autothrust function to achieve a selected target thrust value.
140		66.b	Speed mode	In the Speed mode, the FGS should command the autothrust function to acquire and maintain the selected target speed value, assuming that the selected speed is within the speed range of the normal flight envelope.
141		66.c	Restart mode	If a Retard mode is implemented in the FGS, it should work in the same manner for both automatic and manual landings when the autothrust function is engaged.
142	AC 25.1329-1B	67.a	Approach operations	The FGS vertical modes should allow the pilot to set the target altitude to a missed approach value prior to capturing the final approach segment. This should be possible for capturing from both above and below the final approach segment.
143		67.b	VNAV path operations	It should be possible to define a descent path to the final approach fix and another path from the final approach fix to the runway with the target altitude set for the missed approach altitude. Appropriate targets and descent points should be identified by the FMS.
144	AC 25-7C.1329	b(2)(b-1)	High altitude cruise low speed protection evaluation	Low speed protection is intended to prevent loss of speed leading to an unsafe condition. (aa) At high altitude at normal cruise speed, engage the FGS into an Altitude Hold mode and a heading or lateral navigation (LNAV) mode. (bb) Engage the autothrust into a speed mode. (cc) Manually reduce one engine to idle power or thrust. (dd) As the airspeed decreases, observe the FGS behavior in maintaining altitude and heading/course. (ee) When the low speed protection feature becomes active, note the airspeed and the associated aural and visual alerts including possible mode change annunciations for acceptable operation.
145	AC 25-7C.1329	b(2)(b-2)	Altitude capture evaluation at low altitude	Low speed protection is intended to prevent loss of speed leading to an unsafe condition. (aa) At a reasonably low altitude (e.g., approximately 3000 ft above MSL where terrain permits) and at 250 knots, engage the FGS into Altitude Hold and a heading or LNAV mode. (bb) Engage the autothrust into a speed mode. (cc) Set the altitude pre-selector to 5000 ft above the current altitude. (dd) Make a flight level change to the selected altitude feet with a 250 knots climb at maximum climb power or thrust.

(Continues)

TABLE 6.45 ■ *(Continued)*

No.	Requirement AC Paragraph	Subject	Description
			(ee) When the FGS first enters the Altitude Capture mode, reduce thrust/power on one engine to idle.
			(ff) As the airspeed decreases, observe the airplane trajectory and behavior.
			(gg) When the low speed protection condition becomes active, note the airspeed and the associated aural and visual alerts including possible mode change annunciations for acceptable operation.
146	AC 25-7C.1329	b(2)(b-3)	High vertical speed evaluation	Low speed protection is intended to prevent loss of speed leading to an unsafe condition.
			(aa) Engage the FGS in the Vertical Speed mode with a very high rate of climb.
			(bb) Set the thrust/power to a value that will cause the airplane to decelerate at approximately 1 knot per second.
			(cc) As the airspeed decreases, observe the airplane trajectory and behavior.
			(dd) When the low speed protection condition becomes active, note the airspeed and the associated aural and visual alerts including possible mode change annunciations for acceptable operation.
147	AC 25-7C.1329	b(2)(b-4)	Approach evaluation	Low speed protection is intended to prevent loss of speed leading to an unsafe condition.
			(aa) Conduct an instrument approach with vertical path reference.
			(bb) Couple the FGS to the localizer and glideslope (or LNAV/VNAV, etc.).
			(cc) Cross the final approach fix/outer marker at a reasonably high speed at idle thrust/power until low speed protection activates.
			(dd) As the airspeed decreases, observe the airplane trajectory and behavior.
			(ee) When the low speed protection becomes active, note the airspeed and the associated aural and visual alerts including possible mode change annunciation for acceptable operation.
			(ff) Note the pilot response to the alert and the recovery actions taken to recover to the desired vertical path and the re-capture to that path and the acceleration back to the desired approach speed.
148	AC 25-7C.1329	b(4)(b)	FGS Climb, Cruise, Descent, and Holding Modes	Examination of the following modes are considered appropriate for inclusion in this section: 1. Altitude Hold/Select 2. Area Navigation 3. Backcourse

4. Heading Hold/Select
5. IAS Hold/Select
6. Lateral Navigation
7. Level Change
8. Localizer (only)
9. Mach Hold/Select
10. NonPrecision Approach
11. Pitch Attitude Hold
12. Roll Attitude Hold
13. Turbulence
14. Vertical Navigation
15. Vertical Speed Hold/Select
16. VOR
17. VOR Navigation

149	b(4)(c)	Special characteristics	Operation of the system should not result in performance for which the pilot would be cited during a check ride (i.e., exceeding a speed target of 250 knots by more than 5 knots, if appropriate, during operations below 10,000 feet altitude, or overshooting a target altitude by more than 100 feet during capture of the pre-selected altitude). Resetting the datum pressure or the selected altitude at any time during altitude capture should not result in hazardous maneuvers.
150	b(6)(b)	Airworthiness approval to CAT1 minimums	1. Conduct a series of approaches (usually 4 or more) on Type I rated ILS beams to a radio altitude of 160 ft. (20% below the CAT I decision height of 200 ft).

2. Conduct the approaches with and without automatic throttles, with and without yaw damper, alternate flap setting and so forth, for all configurations and combinations of equipment for which the applicant seeks approval.

3. At least three Type I beams should be included in the evaluation, one of which should exhibit very noisy localizer and glideslope characteristics.

4. Failure modes/conditions described in paragraph 181b(7)(f) of this AC, appropriate to ILS approach modes, should be conducted.

5. The definition of a successful approach is one that positions the airplane at the decision height (DH) such that the airplane can be safely landed without exceptional piloting skill or strength.

AC 25-7C.1329

(Continues)

TABLE 6.45 ■ (Continued)

No.	Requirement AC Paragraph		Subject	Description
151		b(6)(d)	ILS with one engine inoperative	If approval is sought for ILS approaches initiated with one engine inoperative, and with the airplane trimmed at the point of glide path intercept, the automatic flight control system should be capable of conducting the approach without further manual trimming.
152		b(7)(a)	Control wheel steering	It should be possible for the pilot to overpower the automatic pilot system, and achieve the maximum available control surface deflection, without using forces that exceed the pilot control force limits specified in § 25.143(d).
153		b(7)(b)	CWS Maximum bank and pitch attitude	The maximum bank and pitch attitudes that could be achieved without overpowering the automatic pilot system should be limited to those necessary for the normal operation of the airplane. Typically, these attitudes are ±35° in roll and +20° to −10° in pitch.
154		b(7)(d)	Stall	The stall and stall recovery characteristics of the airplane should remain acceptable with control wheel steering in use.
155	AC 25-7C.1329	b(9)(a)	Failure and malfunction	Investigations should include the effects of any failure conditions identified for validation by a system safety assessment conducted to show compliance with § 25.1309(d). After the recognition point, action by the test pilot should be delayed to simulate the time it would take for a line pilot to take control after recognizing the need for action. Satisfactory airplane response to autopilot hardovers should be shown throughout the entire certificated airspeed/altitude flight envelope. If an autothrottle is installed, the malfunctions should be examined with and without the autothrottle operating.
156		b(9)(c)	Failure and malfunction: different flight phase	Corrective action should not be initiated until three-seconds after the pilot has become aware, either through the behavior of the airplane or a reliable failure warning system, that a malfunction has occurred. The altitude loss for the cruise condition is the difference between the observed altitude at the time the malfunction is introduced, and the lowest altitude observed in the recovery maneuver.
157	AC 25-7C.1329	b(9)(d)	Maneuvering flight	Maneuvering flight tests should include turns with the malfunctions introduced when maximum bank angles for normal operation of the system have been established, and in the critical airplane configuration and stages of flight likely to be encountered when using the automatic pilot. A one second delay time following pilot recognition of the malfunction, through the behavior of the airplane or a reliable failure warning system, should be used for maneuvering flight malfunction testing. The altitude loss, for maneuvering flight testing, is the difference between the observed altitude at the time the malfunction is introduced, and the lowest altitude observed in the recovery maneuver.

For malfunction evaluations, the ACs state that the more critical of the following should be inserted into the autopilot:

– A signal in any axis that is equivalent to any single failure (including autotrim)
– The combined signals about all affected axes if multiple axes failures can occur from the malfunction of a single component

The reaction times are given in appendix B of the ACs:

– Hover, takeoff and landing—normal reaction time
– Maneuvering and approach—1 sec
– Climb, cruise, descent—3 sec

Simulated failure and corrective action should not exceed the aircraft structural loads. Reasonable flight path deviation is explained in appendix B (normally 0–2 g, no unacceptable workload or strength), and the MUH (same as previously discussed) must be computed and added to the RWFM.

A sample Part 29 autopilot/SAS compliance matrix is shown in Table 6.46.

6.16 | LANDING SYSTEMS

This section addresses the XLS-type landing systems: ILS, MLS, and GNSS Landing System (GLS). Each of these systems are different from the nonprecision approaches that were discussed in chapter 5 in that they provide vertical as well as horizontal guidance for landing.

6.16.1 Instrument Landing System

The ILS is a precision approach system that allows safe landing of the aircraft from 200 ft ceiling and 2400 ft runway visibility range (RVR) down to autoland and rollout with taxi guidance. Ceiling is the bottom of the cloud deck at the runway. If you were standing on the runway and looked up at a cloud deck, the vertical distance between you and the clouds would be the ceiling. The RVR is the distance that you can see down the runway from the approach end down to the departure end. If it is foggy, you will not be able to see as far as on a clear day. Pilots use the ceiling and RVR to see if they have conditions that allow them to make the approach. The ceiling and RVR are the minimum atmospheric conditions under which you are allowed to perform the approach. A nonprecision approach in the United States has, on average, minimums of 500 ft and 0.5 nm (ceiling and RVR, respectively). If the airport called the weather conditions as 400 ft broken (ceiling) and 0.5 nm, you cannot legally attempt a nonprecision approach, as the ceiling is below the minimum. You would have to select an approach that provides more guidance/accuracy and allows the approach to be performed with lower minima than the nonprecision approach. The ILS provides this accuracy.

There are five categories of ILS:

• Category I (Cat I) allows manual approaches down to a 200 ft decision height (DH; continue the landing or go-around), weather minima are 0.5 miles visibility or 2400 ft

TABLE 6.46 ■ Part 29 AP/SAS Compliance Matrix

AC 29-2C	Subject	Sub Topic	Description
§ 29.1329 B1 (i)	AP	Hardover test	In demonstrating malfunctions of the autopilot system, generally servo actuator hardovers are the most critical malfunction. If this is the case and the autopilot system utilizes the same servos and servo amplifiers as the stability augmentation system (SAS) and the autopilot function cannot produce a more severe hardover than the SAS, then no additional consideration is required for this malfunction. An evaluation using the guidance in paragraph AC Appendix B would be sufficient.
§ 29.1329 B 1 (i)	AP	Altitude Hold	The automatic pilot system should be evaluated to demonstrate that it can perform its intended function of flying the rotorcraft
§ 29.1329 B 1 (i)	AP	BC	The automatic pilot system should be evaluated to demonstrate that it can perform its intended function of flying the rotorcraft
§ 29.1329 B 1 (i)	AP	Heading Hold	The automatic pilot system should be evaluated to demonstrate that it can perform its intended function of flying the rotorcraft
§ 29.1329 B 1 (i)	AP	VRT	The automatic pilot system should be evaluated to demonstrate that it can perform its intended function of flying the rotorcraft
§ 29.1329 B 1 (i)	AP	NAV	The automatic pilot system should be evaluated to demonstrate that it can perform its intended function of flying the rotorcraft
§ 29.1329 B 1 (iii)	AP	Servo alignment monitoring/indication	The rule specifies that unless there is automatic synchronization, there should be some method to indicate the alignment of the actuating device to the pilot. The intent of this requirement is to provide a means such that the pilot does not inadvertently engage the system into a hardover condition. One method of achieving this has been the use of servo force meters. These meters monitor the current into the servo motor and indicate to the pilot if a signal is being sent to the servo prior to system engagement.
§ 29.1329 B 2 (i)	AP controls	AP control location	Location of the automatic pilot system controls are such that their operation is properly labeled and is readily accessible to the pilot(s).
§ 29.1329 B 2 (ii)	AP	Annunciation	Annunciator colors conform to the colors specified in § 29.1322 (Reference paragraph AC 29.1322).
§ 29.1329 B 2 (iii)	AP Displays	Labeling	A determination is made that the controls, control labels, and placards are readable and discernible under all expected cockpit lighting conditions.
§ 29.1329 B 2 (iv)	AP	Trim direction	Motion and effect of the autopilot cockpit controls should conform with the requirements of § 29.779.
§ 29.1329 B 2 (v)	AP annunciations	Annunciation	Any disconnect of the autopilot should be annunciated.

§ 29.1329 C 1	AP	AP/SAS dropout	To preclude hazardous conditions that may result from any failure or malfunctioning of the autopilot the following failures should be evaluated. This evaluation should also account for any hazards that also might be caused by inadvertent pilot action. The guidance in paragraph AC 29 Appendix B should be used to determine the appropriate reaction times of the human pilot to an autopilot malfunction. Climb, cruise, and descent flight regimes. The more critical of the following should be induced into the automatic pilot system. (i) A signal about any axis equivalent to the cumulative effect of any single failure, including autotrim (if installed). (ii) The combined signals about all affected axes, if multiple axes failures can result from the malfunction of any single component.
§ 29.1329 C 2	AP	Loads	The simulated failure and the subsequent corrective action should not create loads in excess of structural limits or result in dangerous dynamic conditions or deviations from the flight path. Additional guidance regarding the method of determining pilot recognition times and reasonable flight path deviation due to these simulated failures is contained in paragraph AC 29 appendix B b(6). Resultant flight loads outside the envelope of zero to 2 g will be acceptable provided adequate analysis and flight test measurements are conducted to establish that no resultant aircraft load is beyond limit loads for the structure, including a critical assessment and consideration of the effects of structural loading parameter variations (e.g., center of gravity, load distribution, control system variations, maneuvering gradients).
§ 29.1329 C 5	AP	Recovery	To aid in recovery of the rotorcraft, after a malfunction occurs, one pilot should be able to physically overpower the autopilot and then disengage it with ease, and it should remain disengaged until further pilot action to reengage. The pilot should be able to return the rotorcraft to its normal flight attitude under full manual control without exceeding the loads or speed limits defined in paragraph c(2) and without engaging in any dangerous maneuvers during recovery. The maximum servo authority used for these tests should not exceed those values shown to be within the structural limits for which the rotorcraft was designed.
§ 29.1329 C 5	AP controls	AP Disconnect	The control to disconnect the autopilot should be easily available to the pilot who is now resisting the malfunctioning force of the autopilot. It is recommended that the disconnect button be placed on the cyclic control. It should be red and conspicuously marked "Autopilot Disconnect."

(Continues)

TABLE 6.46 ■ (Continued)

AC 29-2C	Subject	Sub Topic	Description
§ 29.1329 C 6	AP interlocks	AP Interlocks	The autopilot system should have appropriate interlocks to its engagement to ensure it does not operate improperly as a result of information furnished by an external device or system. An example of this is the navigation receivers and the compass system. If for a particular mode of operation the autopilot uses signals from these systems, the autopilot should be interlocked from operating in those modes if invalid information is being received from that system.
§ 29.1329 D 1	AP	AP ILS Malfunctions	Throughout an approach, no signal or combination of signals simulating the cumulative effect of any single failure or malfunction in the automatic pilot system, except vertical gyro mechanical failures, should provide hazardous deviations from flight path or any degree of loss of control.
§ 29.1329 D 2	AP	AP ILS malfunctions	The aircraft should be flown down the instrument landing system (ILS) in the configuration and at the approach speed specified by the applicant for approach. Simulated autopilot malfunctions should be induced at critical points along the ILS, taking into consideration all possible variations in autopilot sensitivity and authority. The malfunctions should be induced in each axis. While the pilot may know the purpose of the flight, the pilot should not be informed when a malfunction is about to be or has been applied except through aircraft action, control movement, or other acceptable warning devices.
§ 29.1329 D 3	AP	Engine failure on approach	An engine failure during an automatic ILS approach should not cause a lateral deviation of the aircraft from the flight path at a rate greater than 3° per second or produce hazardous attitudes.
§ 29.1329 D 4	AP	Engine failure on approach	If approval is sought for ILS approaches initiated with one engine inoperative, the automatic pilot should be capable of conducting the approach.
§ 29.1329 D 5	AP	Engine failure on approach	Deviations from the ILS flight profile should be evaluated as follows: **(i)** The rotorcraft should be instrumented so the following information is recorded: **(A)** The path of the rotorcraft with respect to the normal glide path **(B)** The point along the glide path when the simulated malfunction is induced **(C)** The point where the pilot indicates recognition of the malfunction **(D)** The point along the path of the rotorcraft where recovery action is initiated.
§ 29.1329 G	AP	AP interlocks	There should be a means of sequencing actions or interlocking engagement with sensor inputs to prevent autopilot initiated maneuvers that could result in hazardous operations due to: **1.** Engagement of the autopilot; **2.** Malfunctions of autopilot input or feedback signals that could result in unbounded output commands.

Appendix B (6) (i)	SAS	Fly Through	If a SAS installation stabilizes the rotorcraft by allowing the pilot to "Fly through" and perceive a stable, well-behaved vehicle, it qualifies as a SAS, and if reliable, receives credit under Sections III through IV of Appendix B for use in complying will all-handling qualities requirements.
Appendix B (6) (ii)	SAS	Single Failures	Reasonable single failures of the SAS must be evaluated and the resultant handling qualities must be evaluated to assure that in this degraded configuration, (1) handling qualities have not been degraded below "VFR" levels defined in FAR Part 29, Subpart B,
Appendix B (6) (ii)	SAS	Single Failure	Reasonable single failures of the SAS must be evaluated and the resultant handling qualities must be evaluated to assure that in this degraded configuration, the rotorcraft is free from any tendency to diverge rapidly from stabilized flight conditions
Appendix B (6) (ii)	SAS	Single Failure	Reasonable single failures of the SAS must be evaluated and the resultant handling qualities must be evaluated to assure that in this degraded configuration, the rotorcraft can be flown IFR throughout its endurance capability without undue difficulty by the minimum flight crew.
Appendix B (6) (iii)	SAS	Reliability	If no restriction on the SAS is applied, and if credit is to be given for system reliability and the applicant exempted from consideration of malfunction, hardover and oscillatory functions (AP excluded), a thorough system evaluation is needed. Malfunctions should be tested in all fases of flight including takeoff, climb, cruising, landing, maneuvering and hovering. Appropriate time delay should be applied. Appendix B (6) (v-vi) for consideration for test and limits.
Appendix B (6) (iii)	SAS	Reaction Time	A good method to accurately determine pilot recognition and reaction time is to establish typical climb, cruise, descent, and approach conditions and instruct a subject pilot to react as soon as he recognizes individual hardover conditions in pitch, roll.
Appendix B (6) (v)	SAS Controls	SAS Disconnect	All cockpit emergency controls including emergency quick disconnect should be "red"
Appendix B (6) (vii)	SAS	Multiple failures	Following a single failure, subsequent failures and probable combinations of failures must be considered. Probability analysis of combinations of failures must be conducted by systems experts to determine the need for testing of advanced failure scenarios.

(Continues)

TABLE 6.46 ■ (Continued)

AC 29-2C	Subject	Sub Topic	Description
Appendix B (6) (viii)	SAS	Beep trim	Other areas for investigation include beep trim and auto trim failures. The delay times of paragraph b(6)(iii) are appropriate for all such failures. System malfunctions may also include component failures that result in oscillatory outputs of the actuator(s). These should be sustainable at least as long as the specified hardover delays, should be manageable thereafter with hands on the controls, and should allow disconnect of the malfunctioning system.
§ 29.6721	SAS	Failure identification	This rule requires that the pilot be made aware of stability augmentation, automatic or power-operated system failures that could lead to an unsafe condition. Examples of clearly distinguishable warnings include, but are not limited to, an obvious aircraft attitude change following the failure or an audio warning tone. A visual indication itself may not be adequate since detection of a visual warning would normally require special pilot attention. The use of devices such as stick pushers or shakers is not acceptable as a warning means since the automatic flight control systems (AFCS) may provide a hands-off capability or normal helicopter vibrations could mask a control shaker.
§ 29.6722	SAS	Direction of corrective action	The corrective flight control input following a system failure should be in the logical direction. For example, a malfunction resulting in a nosedown pitch of the aircraft should require a corrective cyclic control input in the aft direction.
§ 29.6722	SAS	Location of SAS release controls	The system deactivating means does not have to be located on the primary flight control grips; however, it should be easily accessible to the pilot.
§ 29.6722	SAS	Inadvertent release of SAS	Consideration should be given to the consequences of inadvertent de-selection of the automatic stabilization system, especially if the deactivation control is mounted on a primary control grip.

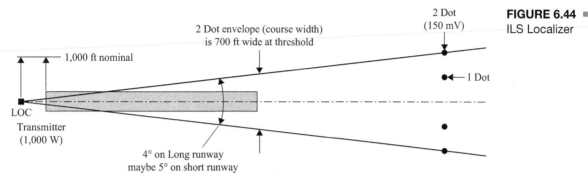

FIGURE 6.44 ■
ILS Localizer

Localizer

Function: Provides horizontal guidance.
Antenna: Optimum (a) 1,000 ft from end of runway and on centerline. Horizontal polarization.
Building: Transmitter building (B) is offset 2,000 ft minimum for the center of the antenna array
              and within 90° to 120° for the approach end.
Frequency: 108.1 to 111.9 add. tenths only.
Modulation: Navigation modulation depth on course 20% for 90 Hz and for 150 Hz. Code identification.
              1020 Hz at 5%.

RVR, which can be decreased to 1800 ft RVR with touchdown zone and centerline lighting at the runway.

- Cat 2, coupled simplex autopilot, 100 ft DH, 1200 ft RVR
- Cat 3A, autoland with rollout control, 100 ft DH and 700 ft RVR
- Cat 3B, autoland with taxi visibility, 50 ft DH and 150 ft RVR
- Cat 3C, autoland with taxi guidance, 0 ft DH and 0 ft RVR with redundant systems

The ILS provides the aircrew with course (localizer) and altitude deviations. The localizer and glideslope information is determined by measuring received signal strength from two beams operating at different modulations. The aircrew is apprised of the deviations by deflections on a course deviation indicator (CDI). It is critical that there are no obstructions to beam reception (blockage/multipath) or the indications will be erroneous. Figure 6.44 describes the localizer installation and geometry, and Figure 6.45 provides the cockpit indications.

The glideslope portion of the ILS works in much the same way as the localizer. The installation and the geometry of the ILS glideslope are shown in Figure 6.46, and the cockpit indications are found at Figure 6.47. It should be noted that the deviations that are calculated by the aircraft equipment are angular; the further away from the transmitter you go the larger the distance subtended by the angle.

The ILS system also provides range to the runway threshold. This is accomplished by the use of marker beacons. There are three possible beacons on the ILS approach:

- Outer marker: 400 Hz signal at 4 to 7 miles from the touchdown point; illuminates a blue light on the cockpit display.
- Middle marker (which is not on all systems): 1300 Hz signal at 200 ft on the glideslope (approximately 3500 ft from touchdown); illuminates an amber light on the cockpit display.
- Inner marker: 3000 Hz signal at 100 ft on the glideslope; illuminates a white light on the cockpit display.

Figure 6.48 details the installation and geometry of the ILS marker system.

FIGURE 6.45 ■
Localizer CDI
Indications

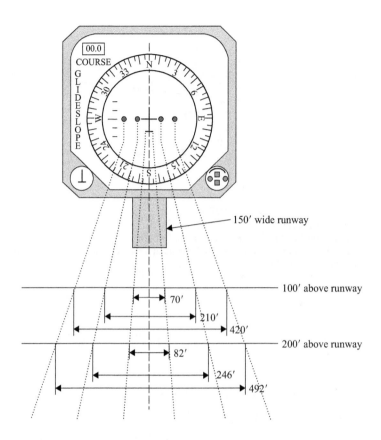

150′ wide runway

100′ above runway

70′

210′

420′

200′ above runway

82′

246′

492′

FIGURE 6.46 ■
Glideslope
Installation and
Geometry

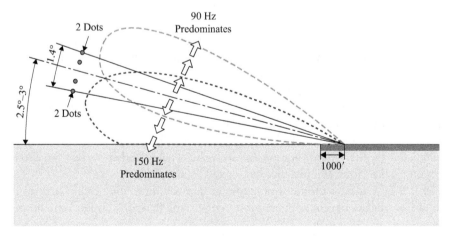

90 Hz
Predominates

2 Dots

1.4°

2.5°–3°

2 Dots

150 Hz
Predominates

1000′

Glideslope

Function:	Provides vertical guidance.
Antenna:	Sited (D) to provide 50 ft(+10/−3 ft) runway threshold crossing height. Horizontal polarization.
Building:	Transmitter building (E) is located 250 to 600 ft from centerline of the runway.
Frequency:	329.3 to 335.0 MHz.
Modulation:	Navigation modulation on path 40% (each) for 90 Hz and for 150 Hz.
Path:	Established at an angle between 2° and 3° (3° optimum).
Path Width:	Path width (F) approximately 1.4° (full scale limits).

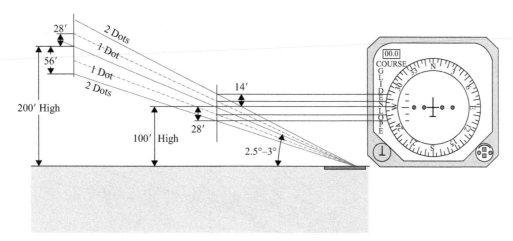

FIGURE 6.47 ▪
Cockpit Glideslope
Indications

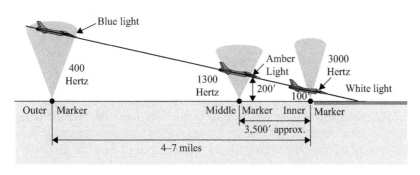

FIGURE 6.48 ▪ ILS
Marker Beacons

Middle Marker

Function:	Indicates vicinity of cat I decision height point.
Location:	At decision height point (G) ± 500 ft longitudinal ± 300 ft lateral.
Frequency:	75 MHz.
Modulation:	1300 Hz.
Keying:	Alternate dot and dash.

Inner Marker

Function:	Indicates decision height for cat II approach (normally 100 ft above TZE). Marks progress reference point for cat III approach.
Location:	Between middle marker and end of the runway.
Frequency:	75 MHz.
Modulation:	3000 Hz.
Keying:	6 dots/sec.

Outer Marker

Function:	Provides a fix and altitude of glideslope at that fix.
Location:	At or past (H) the glideslope intercept point.
Frequency:	75 MHz.
Modulation:	400 Hz at 95%.
Keying:	2 dashes/sec.

6.16.2 Microwave Landing System

There are definite drawbacks to the current ILS. ILSs are good for only one runway and have only one ideal flight path. Parallel runways require two separate installations, and even then run into the potential of interference. In order to remove the potential of multipath, large amounts of money may have to be spent on construction and earth moving. One system that was examined for possible use in the United States and is used

FIGURE 6.49 ■
MLS Geometry

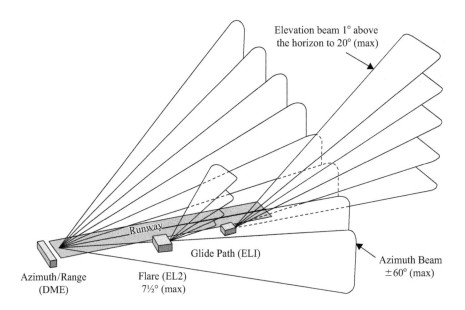

operationally in other parts of the world is the time reference scanning beam (TRSB) MLS, or just MLS.

The MLS uses three scanning beams to provide azimuth and elevation, and a transponder to provide range. The aircraft computer determines deviation from course and glideslope by relating the timing of the received pulses from horizontally and vertically scanned beams. The MLS setup is shown in Figure 6.49.

In the figure, EL1 is the glideslope reference and EL2 is a second lower-power beam used for the flare prior to touchdown. The MLS uses the same CDI as the ILS to provide the aircrew with lateral and vertical deviations. Since the beams are not fixed and can span a relatively large volume, an infinite number of approaches can be designed for the single ground station. The accuracies for the MLS are stated as 100 ft in range, 0.1° in elevation, and 0.2° in azimuth. A depiction of the MLS coverage volume is shown in Figure 6.50.

6.16.3 GNS Landing System

The last of the XLSs is the GNSS landing system (GLS). The GLS is composed of a satellite-based navigation constellation (e.g., GPS, Galileo, GLONASS), a ground-based augmentation system such as GBAS and a means of sending information (differential corrections and approach waypoints) to the aircraft (e.g., a data link). Originally called the local area augmentation system (LAAS), it is now designated as the ground-based augmentation system (GBAS). GBAS is demonstrated to be 1 m accurate in the horizontal and vertical planes and will support ILS Cat II/III landing requirements. GBAS is in co-development with the FAA, Honeywell and Airservices Australia with assistance and prototypes by DECEA in Brazil, DFS in Germany and AENA in Spain. The GBAS system will provide services in the airport terminal area to a radius of 20–30 nm. The components of this system are shown in Figure 6.51.

GBAS advantages include high accuracy, availability and integrity, and will support departure procedures, guided missed approaches and terminal area operations with

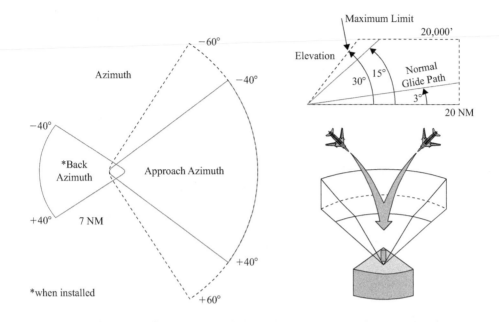

FIGURE 6.50 ▪
MLS Coverage Volume

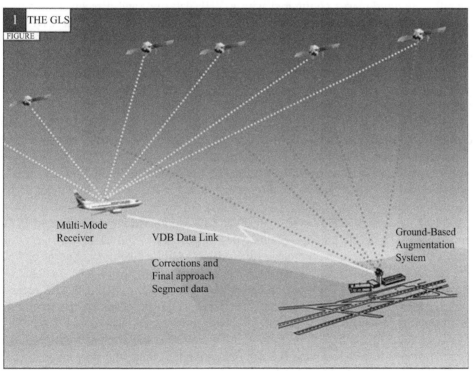

Boeing commercial airplanes GNS landing system

FIGURE 6.51 ▪
GLS Components

flexible curved approaches. GBAS increases the efficiency of arrival and departure procedures, and eliminates capacity constraint due to ILS bottlenecks. The system supports offset landing thresholds and provides a navigation solution that supports the most demanding RNP requirements.

The original US GBAS development site was in Memphis, Tennessee, which was canceled, and two sites are now in development: one in Newark, New Jersey, and the other in Houston, Texas. The initial testing at Newark encountered EMI from passing truckers on the NJ Turnpike, which required siting and software modifications be performed to mitigate the problems. Six GBAS landing systems approaches were approved for Houston's George Bush Intercontinental Airport on May 31, 2012. Internationally, Bremen Airport in Germany has also been approved and is being used by airlines for revenue traffic. Several GBAS stations are currently installed outside the United States and are expected to receive operational approval shortly. These stations are located in Malaga, Spain, Agana, Guam, Frankfurt, Germany, Rio de Janeiro, Brazil, and Sydney, Australia. Additional airports, both foreign and domestic, are considering the installation and approval of GBAS.

The original plan for the GPS was to replace all of the ground-based navigation systems, including the ILS. Because of the problems with vertical accuracy, integrity, continuity, and availability, it was not feasible. The first attempt at using GPS for a precision landing system was the special use category I (SCATI) GPS landing system, which was developed by a consortium of manufacturers and air carriers. The configuration used a differential GPS receiver at the airfield that uplinked differential GPS corrections and approach glideslope waypoints to a cooperative and suitably equipped aircraft. The aircraft was configured with a VHF data link and a receiver capable of processing the differential corrections to aid the navigation solution. The CDI in the airplane was driven by a position in space rather than the angular measurements used in ILS and MLS. The aircraft processor determines its position in space (latitude, longitude, and altitude) and compares it to where it is supposed to be on the localizer and glide path (approach glideslope waypoints). The deviation then drives the pitch, bank, and steering bars on the CDI.

With the advent of the WAAS, EGNOS, MTSAS, and other SBAS systems, differential corrections can now be delivered to the user from geostationary relay satellites. Because of this accuracy, a new family of approaches has been approved for use called localizer performance with vertical guidance (LPV). WAAS is now approved to provide guidance down to 200 ft above the airport's surface for these vertically guided instrument approaches.

As of the end of 2013 more than 3000 LPV approaches were approved in the United States The basic premise of this type of approach is similar to the SCATI just described; only for the LPV approaches there is no additional equipment at the airfield. The differential corrections for the GPS are received from the WAAS system. The waypoints for flying the approach are contained in the aircraft's GPS/WAAS receiver and have been preapproved (TERPS) by the authority. TERPS guarantee the accuracy of the waypoints for the approach as well as clearance from obstructions.

When the pilot chooses to perform an LPV approach, the airplane treats the information just as if it were receiving ILS information; pitch, bank, and steering on the CDI are displayed as explained previously.

6.16.4 Other Category I Landing Systems

There are two other "landing systems" identified in the AC: RNP and area navigation systems. RNP is the statement of the navigation performance necessary for operations within a designed airspace. RNP is specified in terms of accuracy, integrity, and availability of navigation signals and equipment for a particular airspace, route, procedure, or operation. Area navigation systems employ an FMS with VNAV and multiple navigation sensors to output a position. The accuracy of these systems is covered by other AC 20-138C, which is addressed in section 6.11 of the text.

6.16.5 Airworthiness Certification of Landing Systems

The guidance for the ILS (Cat I/II), MLS, and GLS is found in AC 120-29A, "Criteria for Approval of Category I and Category II Weather Minima for Approach," August 12, 2002. The criteria for Cat III ILS is found in AC 120-28D, "Criteria for Approval of Category III Weather Minima for Takeoff, Landing, and Rollout," July 13, 1999. These ACs contain operational approval to operate at the minimums and airworthiness approval for the systems and equipment installed in the airplane. The following discussion concentrates on the latter of the two. Appendix 2 of AC 120-29A contains the airworthiness criteria for category I systems and appendix 3 covers the airworthiness criteria for category II systems. Appendix 2 of AC 120-28D covers the airworthiness criteria for category III systems.

AC 27-1B only addresses guidance for localizer and glide slope systems; it does not refer to AC 120-29. AC 29-2C addresses guidance for ILS Cat II in chapter 3, Miscellaneous Guidance; it does reference AC 120-29, especially performance criteria in appendix 1.

6.16.5.1 Airworthiness of Category I Systems

In general, the applicant will submit a certification plan that describes how any non-aircraft elements of the approach system relate to the aircraft system from a performance, integrity, and availability perspective. Standard landing aids (ILS, MLS) can be addressed by reference to ICAO standards and recommended practices. In addition, the plan should address

- The system concepts and operational philosophy to allow the regulatory authority to determine whether criteria and elements other than those contained in the AC are necessary.

- Approach system performance should be established considering the environmental and deterministic effects that may reasonably be experienced for the type of operation for which certification and operational approval are being sought.

- Where reliance is placed on the pilot to detect a failure of engagement of a mode when it is selected, an appropriate indication or warning must be provided.

- The effect of failures of the navigation facilities must be considered, taking into account ICAO and other pertinent state criteria.

- The effect of the aircraft navigation reference point on the airplane flight path and wheel-to-threshold crossing height shall be assessed.

Approach System Accuracy Requirements   The following are general system accuracy requirements that pertain to all of the category I landing systems:

- The performance of the system should be validated in flight testing using at least three different representative facilities for a minimum of nine approaches, with a representative range of environmental and system variables that have an effect on overall performance.

- Performance assessments should take into account the following variables applied according to their expected distribution in service:

  - Configuration (flaps)
  - CG
  - Landing weight
  - Conditions of wind, turbulence, and wind shear
  - Characteristics of ground- and space-based systems and aids
  - Any other parameter that could affect system performance (e.g., airport altitude, approach path slope, variations in approach speed)

- The criteria for acceptable approach performance are based on acquiring and tracking the required flight path to the appropriate minimum altitude for the procedure. The acquisition should be accomplished in a manner compatible with instrument procedure requirements and flight crew requirements for the type of approach being conducted.

- An approach guidance system shall not generate command information (FD, HUD) that results in flight path control that is oscillatory or requires unusual effort by the pilot to satisfy the performance requirements.

- An approach control system shall not generate flight path control (autopilot) with sustained oscillation.

- The approach system must not cause sustained nuisance oscillations or undue attitude changes or control activity as a result of configuration or power changes or any other disturbances to be expected in normal operation.

Specific performance for the ILS includes

- The performance standards for signal alignment and quality are contained in ICAO annex 10 (an equivalent state standard may be used).

- Lateral tracking performance from 1000 ft height above touchdown (HAT) to 200 ft HAT should be stable without large deviations (within ±50 mamp deviation) from the indicated path.

- Vertical tracking performance from 700 ft HAT to 200 ft HAT should be stable without large deviations (within ±75 mamp deviation) from the indicated path.

Specific performance for the MLS includes the following:

- The performance standards for signal alignment and quality are contained in ICAO annex 10 (an equivalent state standard may be used).

- Lateral tracking performance from 1000 ft HAT to 200 ft HAT should be stable without large deviations (within ±50 mamp deviation) from the indicated path.

- Vertical tracking performance from 700 ft HAT to 200 ft HAT should be stable without large deviations (within ±75 mamp deviation) from the indicated path.

Specific performance for the GLS is the same as for the ILS/MLS.

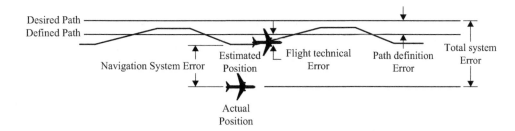

FIGURE 6.52 ▪
Navigation Lateral
Error Components
of the TSE

The accuracy criteria for RNP are designed to enable a seamless transition from an enroute RNP to an approach RNP. RNP operations are based on the accuracy of the airplane flight path in absolute terms with respect to the defined flight path over the ground. The total system error (TSE) is characterized by the combined performance of the airplane systems and any navigation aids. The certification plan should identify any navigation aids on which the RNP performance will be established and how the airplane performance interacts with the navigation aids to meet the TSE performance requirements. The certification plan should identify the assumed relationship between airplane performance and any navigation aid performance. The relationship of all errors summed into the TSE is shown in Figure 6.52.

The components of Figure 6.52 are

- The desired path: the path that the pilot, or pilot and ATC, is expected to fly.

- In order for an aircraft to follow the desired path it is necessary that the navigation system (airborne or on the ground) generate a defined path.

- The difference between the desired path and the defined path is called the path definition error.

- The difference between the estimated aircraft position (from the aircraft navigation system) and the desired flight path is the path steering error. This error includes display errors and the flight technical error (FTE, the accuracy with which the aircraft is controlled either by the pilot or the autopilot).

- The error in the estimation of the aircraft's position s called the navigation system error.

- The sum of the path deviation error, navigation system error and the path steering error (FTE plus any display error) is the TSE, which is the difference between the desired flight path and the actual flight path.

Particular levels of RNP can be satisfied using various navigation aids such as ILS/MLS, or by using combinations of navigation sensors (i.e., GPS/INS), or in concert with an FMS. When a computerized path (series of waypoints) is used as the basis for an approach operation, the desired flight path must typically be defined by a series of 3D earth-based coordinates for the applicable waypoints or path definition points. Approach or missed approach operations can be approved by demonstration of the capability to meet the required navigation performance for a specific approach procedure, for a set of particular procedural types, or for a set of RNP levels.

Associated with an RNP level is a containment limit that is specified as "two times the level of RNP" ($2 \times$ RNP). The system performance integrity provided by this RNP containment limit is intended to support its application as a basic element for either aircraft separation or obstacle or terrain clearance assessment.

TABLE 6.47 ▪ Required RNP Levels for Approaches

RNP Level	Applicability/Operation (Approach Segment)	Normal Performance (95%)	Containment Limit
RNP 1	Initial/Intermediate Approach	±1 nm	±2 nm
RNP 0.5	Initial/intermediate/final approach: supports limited Cat I minima	±0.5 nm	±1 nm
RNP 0.3	Initial/intermediate/final approach: supports limited Cat I minima	±0.3 nm	±0.6 nm
RNP 0.3/125 ft*	Initial/intermediate/final approach with specified barometric vertical guidance: supports limited Cat I minima	±0.3 nm ±125 ft	±0.6 nm ±250 ft
RNP 0.03/45 ft	Final approach with specified vertical guidance: supports Cat I minima	±0.03 nm ±45 ft	±0.06 nm ±90 ft
RNP 0.01/15 ft	Final approach with specified vertical guidance: supports Cat I/II minima	±0.01 nm ±15 ft	±0.02 nm ±30 ft
RNP 0.003/15 ft	Final approach with specified vertical guidance: supports Cat I/II/III minima	±0.003 nm ±15 ft	±0.006 nm ±30 ft

*A second value for the RNP indicates a required altitude accuracy as well as the lateral accuracy.

The expression RNP level is used to describe a specific value or level of navigation performance. Table 6.47 provides RNP levels that can support initial, intermediate, final, and missed approach segments. These values are recommended in the AC but are not yet approved as international standards.

Specific performance requirements for area navigation systems are specified in AC 25-15, "Approval of Flight Management Systems in Transport Category Airplanes," and AC 20-138C, "Airworthiness Approval of Positioning and Navigation Systems," May 8, 2012.

Approach System Integrity Requirements   The applicant shall provide the certification authority with an overall operational safety assessment plan for the use of systems other than ILS or MLS for "path in space" guidance. This plan shall identify the assumptions and considerations for the nonaircraft elements of the system and how they relate to the airplane system certification plan. The onboard components of the landing system need to comply with 14 CFR Part 25.1309 (hence the safety assessment plan). The following are published in the AC as guidance criteria with respect to 1309:

● The aircraft system response to the loss of ILS guidance signals (localizer and glideslope) shall be established.

● The aircraft system response to the loss of MLS guidance signals (azimuth and elevation) shall be established.

● The aircraft system response to the loss of GLS guidance signals shall be established.

● The aircraft system response to the loss of navigation services used to conduct RNP operation shall be established. In particular,

  – The aircraft system response to any switchover to alternate navigation services shall be established.

- It shall be demonstrated that the airplane will maintain the required flight path within the containment limits when unannunciated failures not shown to be extremely remote are experienced.

• The integrity requirements for area navigation systems are as specified in AC 20-138C.

Approach System Availability Requirements Below 500 ft on approach, the demonstrated probability of a successful landing should be at least 95% (no more than 5% of the approaches result in a go-around due to a combination of failures in the landing system and the incidence of unsatisfactory performance). A dual or single area navigation approach system installation should meet the availability requirements consistent with the operational objective of 14 CFR Part 121, section 121.349 (as applicable to standard operations specifications [ops specs]).

Go-Around Requirements A go-around may be required following a failure in the approach system as required by the flight crew or air traffic service at any time prior to touchdown.

• It should be possible to initiate a missed approach at any point during the approach until touchdown on the runway. It should be safe to initiate a missed approach that results in a momentary touchdown on the runway.

• A go-around should not require unusual pilot skill, alertness, or strength.

• The proportion of approaches terminating in a go-around below 500 ft due to a combination of failures in the landing system and the incidence of unsatisfactory system performance may not be greater than 5%.

• Information should be available to the operator to ensure that a safe go-around flight path can be determined.

Flight Deck Information, Annunciation, and Alerting Requirements The controls, indicators, and warnings must be designed to minimize crew errors that could create a hazard. Mode and system malfunction information must be presented in a manner compatible with the procedures and assigned tasks of the aircrew. The indications must be grouped in a logical and consistent manner and be visible under all expected normal lighting conditions.

For manual control of approach flight path, the appropriate flight displays, whether heads-down or heads-up, must provide sufficient information without excessive reference to other cockpit displays to enable a suitably trained pilot to do the following:

• Maintain the approach path

• Make the alignment with the runway, and if applicable safely flare and roll out

• Go around

Sufficient information should be provided to the air crew to monitor the progress and safety of the approach operation. The required flight performance monitoring capability should include at least

• Unambiguous identification of the intended path for the approach and, if applicable, safety flare and roll out ILS/MLS approach identifier/frequency, and selected navigation source

• Indication of the position of the aircraft with respect to the intended path (raw data localizer and glideslope, or equivalent)

A positive, continuous, and unambiguous indication should be provided for the modes actually in operation as well as those armed for engagement. Where the engagement of a mode is automatic (localizer, glideslope acquisition), clear indication must be given when the mode has been armed by either action of a member of the flight crew or automatically by the system.

Alerting requirements are intended to address the need for warning, caution, and advisory information for the flight crew. CFR 14 section 25.1309 requires that information must be provided to alert the crew to unsafe system operating conditions and to enable the crew to take appropriate corrective action. A warning indication must be provided if immediate corrective action is required. The design should account for crew alerting cues, corrective action required, and the capability of detecting faults. A caution is required whenever immediate crew awareness is required and timely subsequent crew action may be required. A means shall be provided to advise the flight crew of failed airplane system elements that affect the decision to continue or discontinue the approach. For RNP systems, the guidance or control system shall indicate to the flight crew when the actual navigation performance (ANP) exceeds the RNP.

Appropriate system status and failure annunciations suited to the guidance system and navigation sensors used and any related aircraft systems (autopilot, flight director) should be provided to the operator prior to departure to determine system capability to perform the approach. While enroute, the failure of each airplane component affecting the approach capability should be indicated via an advisory (unless a warning is warranted) without flight crew action. A means should be provided to advise the aircrew of failed airplane system elements that affect the decision to continue to the destination or divert to an alternate. System status indications should be identified by names that are different than operational authorization categories (e.g., do not use a status of "Cat I").

Multiple Landing Systems and Multimode Receivers   Many aircraft have the capability of conducting approach and landing operations using multiple landing systems; for some, this capability is established by using a multimode receiver (MMR). Where practicable, the flight deck approach procedure should be the same irrespective of the navigation system selected. For these systems:

- A means should be provided to confirm that the intended approach aid has been correctly selected.
- During the approach, an indication of a failure in each nonselected airplane system element must be provided to the aircrew as an indication of system status.
- The loss of acceptable deviation data shall be indicated on the display; it is acceptable to have a single failure indication for each axis common to all navigation sources.
- The navigation source selected for the approach shall be positively indicated in the primary field of view at each pilot station.
- The data designating the approach (e.g., ILS frequency, MLS channel, GLS approach identifier) shall be unambiguously indicated in a position readily accessible and visible to each pilot.
- A common set of mode "arm" and "active" indications (e.g., LOC, GS) is preferred for ILS, MLS, and GLS operations.
- A means should be provided for the crew to determine a failure of a nonselected navigation receiver function in addition to the selected navigation receiver function.

- The capability of each element of a multimode approach and landing system shall be available to the aircrew to support dispatch of the airplane.

- An enroute failure of a multimode approach system element shall be annunciated to the aircrew as an advisory.

- A failure during an approach of a multimode approach system element may be annunciated to the aircrew as a warning, caution, or advisory.

If installed, the MMR has some unique certification requirements. These systems are normally "ILS look-alike" and are replacement units for aircraft that have had a single ILS or MLS receiver. When these systems are installed, the following general certification guidance is used:

- An impact assessment should identify any new functionality (differences between current and new systems, functionality being added).

- Credit can be taken for the existing approval.

- TSO/MOPS compliance should be demonstrated.

- An impact on airplane system safety assessments should be accomplished.

- EMI/EMC testing is required.

- Electrical loading needs to be accomplished.

- Flight data requirements need to be assessed.

- AFM changes are to be incorporated.

- The certification plan needs to address the functional integration aspects of the receiver with respect to other systems, controls, warnings, and displays.

Airborne System Requirements In order to safely conduct approach and landing operations, there are aircraft systems that need to operate correctly and in concert with the approach and landing system.

- All aircraft systems must comply with the basic performance, integrity, and availability requirements stated in the previous section.

- Autopilot criteria is specified in AC 25.1329-1C and detailed in section 6.15 of this text.

- The following criteria are applicable to HUD systems:

  - Workload associated with the use of the HUD should be considered in showing compliance with section 25.1523 (minimum crew)

  - The HUD must not significantly obscure the pilot's view through the cockpit window.

  - The HUD must provide sufficient guidance information, without excessive reference to other cockpit displays, to enable a suitably trained pilot to maintain the approach path and perform a go-around.

  - The pilot should be able to align with the runway without the HUD adversely affecting the task.

  - If command information is provided for the flare and landing, it must not be misleading and should be consistent with the characteristics of normal manual maneuvers.

- If only one HUD is installed, it should be installed in the pilot-in-command station.
- The HUD guidance must not require exceptional piloting skill to achieve the required performance.
- The HUD system performance and alerting should be consistent with the intended operational use for duties and procedures of the flying pilot and the pilot not flying.
- If the autopilot is used to control the flight path of the airplane to intercept and establish the approach path, the point during the approach at which the transition from automatic to manual flight takes place shall be identified and used for the performance demonstration.
- Any transition from autopilot to HUD guidance must not require exceptional piloting skill, alertness, strength, or excessive workload.

- Hybrid systems, such as enhanced vision systems (EVSs) require a proof of concept evaluation to establish suitable criteria.
- If a HUD is used to monitor an autopilot system, it should be shown to be compatible with the autopilot system and permit a pilot to detect unsuitable autopilot performance.
- Satellite-based systems should be shown to provide equivalent or better capability than ground-based radio aids for comparable operations or meet provisions applicable to RNP.
- Satellite-based systems should not exhibit adverse characteristics during acquisition or loss of satellites.
- Area navigation systems should comply with criteria contained in the applicable ACs previously noted.
- If an RNAV system's operational software program can be modified, a "version" identification must be provided and available for display to the pilot or maintenance personnel.
- The RNAV system must have a database that is suited for the specific aircraft and navigation system and must be assessed as having current data.
- The RNAV system pilot input/output functions, keys, and displays should have standard functions available and operate with industry standard convention and practice.
- Single RNAV systems must be accessible and usable by either pilot located at a pilot or copilot crew station.
- Dual RNAV systems must have a convenient and expedient way of cross-loading and updating data.
- The RNAV system performance must be consistent with the operational levels sought or be consistent with an identifiable performance standard, such as for various levels of RNP.
- If credit is sought for operating on complex and closely spaced multiple waypoint paths, an interface with a suitable "track up" or "heading up" navigation map display is necessary.
- The RNAV system must provide a means to monitor lateral and vertical deviation and a means to ensure suitable operation and updating; if RNP is included, the system

must be able to identify the RNP level to be used and the actual navigation position or expected positional error.

- If autothrottle capability is installed, the applicant should identify any necessary modes, conditions, procedures, or constraints that apply to its use.

- Use of autothrottle should not cause unacceptable performance of any of the autopilot modes intended for use, and vice versa.

- If a data link is used to provide data to the airplane, then the integrity of the data link should be commensurate with the integrity required for the approach.

- The role of the data link in the approach system must be addressed as part of the aircraft system certification process until such time as an acceptable national or international standard for the ground system is established.

AC 27-1B addresses localizer and glide slope systems; for the localizer:

- At minimum 10 nm from the transmitter, the signal should be received for 360° of rotorcraft heading at bank angles up to $+/-10°$, at all normal pitch angles at an altitude of approximately 2000 ft

- Localizer should be checked for rotor modulation in the approach by varying the RPM throughout its normal range

- Intercept angles of 50° at ranges of at least 10 nm should be demonstrated

- At least three acceptable front and back course flights should be conducted to 200 ft or less above the threshold

For the glideslope:

- The glide slope signal should remain valid at all aircraft headings while at least 10 nm from the transmitter at $+/-30°$ of the localizer course

- Verify correct tracking and command

- At least three approaches to 200 ft above the threshold should be accomplished

- Glide slope should be checked for rotor modulation and EMI/EMC; the AC notes that some interference from the VHF is acceptable

The AC also specifies correct operation of the marker beacons. A sample fixed-wing compliance matrix for an ILS Cat 1 installation is shown in Table 6.48, and a compliance matrix for a Part 29 installation is shown in Table 6.49.

6.16.5.2 Airworthiness of Category II Systems

In general, the airworthiness requirements for category II landing systems are quite similar to the airworthiness requirements detailed in the previous section for category I landing systems. In this section, I will address only the differences between the two. The general requirements are identical for category I and category II. Area navigation landing systems do not meet the criteria for category II systems, so they are not discussed in this section.

Approach System Accuracy Requirements Performance shall be demonstrated by flight testing, or analysis validated by flight testing, using at least three different representative facilities for a minimum of 20 total approaches, with a representative range of environmental and system variables that have an effect on overall performance.

TABLE 6.48 ■ ILS Cat I Compliance Matrix

No.	Requirements AC	Paragraph	Subject	Description
1	AC 120-29A	5.1.3.1	Lateral	Lateral Tracking Performance from 1000' HAT to 200' HAT should be stable without large deviations from indicated path
2	AC 120-29A	5.1.3.2	Vertical	Vertical Tracking Performance from 700' HAT to 200' HAT should be stable without large deviations from indicated path
3	AC 120-29A	5.1.3.4	Typical Wind and Wind Gradient Disturbance Environment	**a.** Should be capable of coping with at least, • Headwind—25 kts • Tailwind—10 kts • Crosswind—15 kts **b.** Wind Gradients/Shear—at least 4 kts per 100 ft. from 500 ft. HAT to the surface; **c.** Recommended Capability—Ability to cope with 8 kts per 100 ft for 500 ft, moderate turbulence, knife edge shears of at least 15 kts over 100 ft, 20 kts lateral directional vector shears of 90° over 100 ft, and ability to cope with a 20 kt logarithmic shears between 200 ft and the surface.
4	AC 120-29A	5.5	Flight Director System	• Should be compatible with autopilot • Situational information displays of navigation displacement must be provided to both pilots • Displays must be appropriately scaled and readily understandable in the modes or configurations applicable, to ensure unacceptable deviations and failures can be detected
5	AC 120-29A	5.9.1	Instruments, Systems and Displays	• The following must be provided to both pilots ○ Attitude Indicator/EADI/Primary flight displays ○ HSI/EHSI/ND • T format instrument panel layout, conventional airspeed and altitude scale • Location and placement of instruments must be appropriate for both pilots, with appropriately scaled and readily understandable presentations and mode of display • Redundant lateral and vertical path displacement information from the final approach course and specified glide path must be provided ○ For DA(H) below 250 ft, lateral and vertical displacement must be provided on PFD/EADI ○ DA(H) should be provided to both pilots—use RH/RA or Baro for indication of altitude/height

6	AC 120-29A	5.10	Annunciations	• Appropriate system status and failure annunciations • Automatic audio callouts as described in 5.11 • Suitable rain removal method Mode annunciation labels should not be identified by landing minima classification (CAT I/CAI II)

7	AC 120-29A	5.11	Auto Aural Alerts	1. Should not interfere necessary crew communication or coordination procedures
				  • At 500 ft
				  • At flare, 50 ft/30 ft/10 ft as appropriate
				2. Low alt radio alt callouts should address the situation of higher than normal sink rate during flare, or an extended flare

8	AC 120-29A	5.20	GPWS/TAWS	Airborne equipment should have appropriate interface or compatibility with GPWS and TAWS
9	AC 120-29A	5.21	FDR Interface	Airborne equipment should have appropriate interface or compatibility with FDR/Cockpit Voice Recorder

10	AC 120-29A	App 2/6.2	Approach System Accuracy Requirements	1. Performance shall be demonstrated by flight test, or analysis validated by flight test, using
				  • Three different representative facilities
				  • A minimum of nine total approaches
				  • With a representative range of environmental and system variables
				  • Configuration of the airplane
				  • Center of gravity
				  • Landing weight
				  • Conditions of wing, turbulence, and wind shear
				  • Characteristics of ground and space based systems and aids
				  • Parameters such as airport altitude, approach path slope, approach speed, which may affect system performance
				2. An approach guidance system shall not generate oscillatory flight path control or requires unusual effort by the pilot
				3. The approach system must cause no sustained nuisance oscillation or undue attitude changes or control activity as a result of configuration or power changes or any other disturbance

(Continues)

TABLE 6.48 ■ *(Continued)*

No.	Requirements AC Paragraph	Subject	Description
11	AC 120-29A	ILS Approach	• Lateral tracking performance—from 1000' HAT to 200' HAT should be stable without large deviations • Vertical tracking performance—From 700' HAT to 200' HAT should be stable without large deviations
12	AC 120-29A	Approach System Integrity	• The aircraft system response to loss of ILS guidance signals (Localizer and Glide Slope) shall be established • Able to maintain the required flight path within the containment limits when un-annunciated failures not shown to be extremely remote are experienced
13	AC 120-29A	Approach System Availability	• Below 500 ft on approach, the demonstrated probability of a successful landing should be at least 95%—no more than 5% of the approaches result in a go-around, due to the combination of failures in the landing system and the incidence of unsatisfactory performance
14	AC 120-29A	Go-Around	• Must be able to initiate a missed approach at any point during the approach until touchdown on the runway, or that results in a momentary touchdown on the runway • A go-around should not require unusual pilot skill, alertness and strength • No more than 5% of the approaches result in a go-around below 500 ft, due to the combination of failures in the landing system and the incidence of unsatisfactory performance • Information (speed/altitude/attitude) should be available to the operator to assure that a safe go-around flight path can be determined

15	AC 120-29A	Flight Deck Information	• For manual control of approach flight path, the appropriate flight displays must provide sufficient information, without excessive reference to other cockpit displays, to enable a suitably trained pilot to
			  ○ Maintain the approach path
			  ○ Make the alignment with runway, safely flare and roll out
			  ○ Go-round

- Sufficient information should be provided to the pilots to monitor the progress and safety of the approach
  - Unambiguous identification of the intended flight path (approach identifier/frequency, and navigation source)
- Indication of the position of the aircraft with respect to the intended path (raw data localizer and glide path, or equivalent)

#	Ref	Section	System	Content
16	AC 120-29A	App 2/6.6.2	Annunciation	• Positive, continuous and unambiguous indication should be provided for the modes actually in operation, as well as those armed for engagement. ○ Clear indication must be given when the mode has been armed either by flight crew or automatically
17	AC 120-29A	App 2/6.6.3	Alerting	• Warning—A warning indication must be provided if immediate corrective action is required • Caution—a means shall be provided to advise the flight crew of failed airplane system elements that affect the decision to continue or discontinue the approach, the guidance system shall indicate to the flight crew when the Actual Navigation Performance exceeds the RNP
18	AC 120-29A	App 2/8.2	Auto Pilot	• Criteria applicable to A/P systems is as specified by section 25.1329
19	AC25-7C	1301,c,(2)	Localizer System	a. Antenna pattern b. Localizer Intercept c. Localizer Tracking
20	AC25-7C	1301,c,(3)	Glideslope System	a. Antenna pattern b. Glideslope Intercept c. Glideslope Tracking
21	AC25-7C	1301,c,(4)	Marker Beacon System	a. In low sensitivity, the marker beacon annunciator light should be illuminated for a distance of 2,000 to 3,000 ft when flying at an altitude of 1,000 ft on the localizer centerline in all flap and gear configurations. Glideslope Intercept b. If a high/low sensitivity feature is installed and selected, the marker beacon annunciator light and audio will remain on longer than when in low sensitivity. c. The audio signal should be of adequate strength and sufficiently free from interference to provide positive identification. d. As an alternative procedure, cross the outer marker at normal ILS approach altitudes and determine adequate marker aural and visual indication. e. Illumination should be adequate in bright sunlight and at night.
22	AC25-7C	1301,c,(4)	Marker Beacon System	f. EMC

TABLE 6.49 ■ Part 29 ILS Cat I Compliance Matrix

AC 29-2C	Subject	Sub Topic	Test Item Description
AC 29-2C MG1 4 (ii)	ILS	LOC Signal reception	The signal input to the receiver presented by the antenna system should be of sufficient strength to keep the malfunction indicator out of view when the rotorcraft is in the approach configuration and at least 10 NM from the station. This signal should be received for 360 degrees of rotorcraft heading at all bank angles up to 10 degrees left or right at all normal pitch altitudes, and at an altitude of approximately 2,000 ft.
AC 29-2C MG1 4 (iii)	ILS	LOC Course Guidance	The deviation indicator should properly direct the aircraft back to course when the rotorcraft is right or left of course.
AC 29-2C MG1 4 (iv)	ILS	LOC Signal identification	The Station identification signal should be of adequate strength and sufficiently free from interference to positive station identification, and voice signals should be intelligible with all electric equipment operating and pulse equipment transmitting.
AC 29-2C MG1 4 (v)	ILS	LOC Rotor modulation	Localizer performance should be checked for rotor modulation in approach while varying rotor RPM throughout its normal range.
AC 29-2C MG1 4 A	ILS	LOC Intercept	Localizer Intercept. In the approach configuration and a distance of at least 10 NM from the localizer facility, fly toward the localizer front course, inbound, at an angle of at least 50 degrees. Perform this maneuver from both left and right of the localizer beam. No flags should appear during the time the deviation indicator moves from full deflection to on course. If the total antenna pattern has not been shown by ground checks or by VOR flight evaluation to be adequate, additional intercepts should be made.
AC 29-2C MG 1 4 B	ILS	LOC Tracking	While flying the localizer inbound and not more than 5 miles before reaching the outer marker, change the heading of the rotorcraft to obtain full needle deflection. Then fly the rotorcraft to establish localizer on course operation. The localizer deviation indicators should direct the rotorcraft to the localizer on course. Perform this maneuver with both a left and a right needle deflection. Continue tracking the localizer until over the transmitter. At least three acceptable front course and back course flights should be conducted to 200 ft or less above threshold.

Reference	System	Item	Description
AC 29-2C MG1 5 (i)	ILS	Glide slope Signal Reception	The signal input to the receiver should be of sufficient strength to keep the warning flags out of view at all distances to 10 NM from the facility. This performance should be demonstrated at all aircraft headings from 30 degrees left to 30 degrees right of the localizer course.
AC 29-2C MG1 5 (ii)	ILS	Glide slope Intercept	While flying the localizer course inbound in level flight, intercept the glide slope below path at least 10 NM from the station. Observe the glide slope deviation indicator for proper crossover as the aircraft flies through the glide path. There should be no flags from the time the needle leaves the full scale fly-up position until it reaches the full scale fly-down position.
AC 29-2C MG1 5 (iii)	ILS	Glide slope Tracking	While tracking the glide slope, maneuver the aircraft through normal pitch and roll attitudes. The glide slope deviation indicator should show proper operation with no flags. At least three acceptable approaches to 200 ft or less above threshold should be conducted.
AC 29-2C MG1 5 (iv)	ILS	Glide Slope EMC	With all rotorcraft electrical equipment operating and all pulse equipment transmitting, determine that there is no interference with the glide slope operation (some interference from the VHF may be acceptable), and that the glide slope system does not interfere with other equipment.
AC 29-2C MG1 5 (i)	ILS	Glide Slope Guidance	The deviation indicator should properly direct the aircraft back to path when the aircraft is above or below path.
AC 29-2C MG1 5 (i)	ILS	Glide Slope EMI	Interference with the navigation operation should not occur with all rotorcraft equipment operating and all pulse equipment transmitting. There should be no interference with other equipment as a result of glide slope operation.
AC 29-2C MG1 5 (v)	ILS	Glide Slope Rotor Modulation	Glide slope performance should be checked for rotor modulation in approach while varying rotor RPM throughout its normal range.
AC 29-2C MG1 6 (i)	ILS	Marker Beacon	The marker beacon annunciator light should be illuminated for a period of time representing 2000 to 3000 ft distance when flying at an altitude of 1,000 ft as it passes over a marker beacon.

Specifically, the airborne system should be demonstrated in at least the following conditions, taking into account manual/coupled autopilot and autothrottle configurations:

- Wind conditions
  - 20 knots—headwind component
  - 10 knots—crosswind component
  - 10 knots—tailwind component

  Specific accuracy performance standards for the ILS are as follows:

- Lateral tracking performance from 300 ft HAT to 100 ft HAT should be stable without large deviations (within ±25 mamp deviation) from the indicated course for 95% of the time per approach.
- Vertical tracking performance from 300 ft HAT to 100 ft HAT should be stable without large deviations (within ±35 mamp deviation) from the indicated path or ±12 ft, whichever is greater, for 95% of the time per approach.

  Specific accuracy performance standards for the MLS are as follows:

- Lateral tracking performance from 300 ft HAT to 100 ft HAT should be stable without large deviations (within ±25 mamp deviation) from the indicated course for 95% of the time per approach.
- Vertical tracking performance from 300 ft HAT to 100 ft HAT should be stable without large deviations (within ±35 mamp deviation) from the indicated path or ±12 ft, whichever is greater, for 95% of the time per approach.

  Specific accuracy performance standards for the GLS are the same as above for the ILS and MLS.

  Specific accuracy performance standards for the RNP are the same for category I and category II systems.

Approach System Integrity Requirements   The requirements for category II systems are the same as for category I with the following addition:

- For ILS and MLS, the aircraft system response during a switchover from an active localizer or glideslope (or active elevation or azimuth) to a backup transmitter will be established.
- For GLS, the aircraft system response during any switchover to alternate differential augmentation, pseudolites, and data services, as applicable, shall be established.

Approach System Availability Requirements   The availability requirements are the same for ILS, MLS, and GLS for category I and category II systems.

Go-Around Requirements   The go-around requirements are the same for ILS, MLS, and GLS for category I and category II systems.

Flight Deck Information, Annunciation, and Alerting Requirements   The display requirements are the same for category I and category II, with the following addition for category II systems:

- Although excessive deviation alerting is not required, the authority will approve systems with this function as long as it meets appropriate criteria.

- If a method is provided to detect excessive deviation of the airplane laterally and vertically during approach to touchdown, and laterally after touchdown, then it should not require excessive workload or undue attention.

- The provision does not require a specific deviation method or annunciation, but may be addressed by parameters displayed on the ADI, EADI, HUD, or PFD.

- When a dedicated deviation alerting is provided, its use must not cause excessive nuisance alerts.

The following criteria from CS-AWO 236 are an acceptable means of compliance for category II systems:

- Excess deviation alerts should operate when the deviation from the ILS or MLS glide path or localizer centerline exceeds a value from which a safe landing can be made from offset positions equivalent to the excess deviation alert without exceptional piloting skill and with the visual references available in these conditions.

- Excess deviation alerts should be set to operate with a delay of not more than 1 sec from the time that the deviation thresholds are exceeded.

- Excess deviation alerts should be active from at least 300 ft HAT to the DH, but the glide path alert should not be active below 100 ft HAT.

Multiple Landing Systems and Multimode Receivers   The category I and category II requirements are the same for multiple landing systems and MMRs.

Airborne System Requirements   There are soma additions/modifications to the airborne system requirements for category II systems.

- The autopilot must not have normal features or performance, or performance in typical adverse environmental conditions that would cause undue crew concern and lead to disconnect (inappropriate response to ILS beam disturbance or turbulence, unnecessary abrupt flare or go-around attitude changes, unusual or inappropriate pitch or bank attitudes or side slip responses).

- The autopilot must maintain the approach path or, if applicable, align with the runway, flare the airplane within the prescribed limits, or promptly go-around with minimum practical loss of altitude.

- Autopilot performance must be compatible with either manual speed control or, if applicable, autothrottle speed control.

- Autopilot mode definition and logic should be consistent with the appropriate industry practice for mode identification and use (labeling, mode arming, engagement).

- Definition of new autopilot modes or features not otherwise in common use should be consistent with their intended function and consider the potential for setting appropriate or adverse precedent.

- The autopilot alerting system should be consistent with the intended operational use for duties and procedures for both the flying pilot and the pilot not flying.

- If the autopilot is used to control the flight path of the airplane to intercept and establish the approach path, the pilot should be able to transition from automatic to manual flight at any time without undue effort, attention, or control forces and with a minimum of flight path disturbance.

- If a HUD is installed, any transition from autopilot to HUD guidance, or vice versa, must not require exceptional piloting skill, alertness, strength, or excessive workload.

- An FD system or alternate form of guidance, if used, must be compatible with the autopilot.

- A fault must cause an autopilot advisory, caution, or warning as necessary. If a warning is necessary, the pilot must be able to detect the warning with a normal level of attention and alertness expected during an approach or go-around.

- With heads-down displays, an FD system or alternate form of guidance must be designed so that the probability of display of incorrect guidance commands is remote.

- Whenever practical, a fault must cause guidance information to be immediately removed from view on the heads-down displays.

AC 29-2C notes that aircraft and systems, certification and continuation training of flight crews and a continuing maintenance program for the aircraft and Cat II required systems are required for Airworthiness. The additional equipment necessary for Cat II approval is the flight control guidance system; either a flight director system or an automatic approach coupler. A flight director system needs only to present computed steering data for the localizer and raw glide slope data on the same instrument. A single axis steering autopilot could be used if it is coupled to the ILS localizer.

Contemporary rotorcraft flight director and autopilot systems use at least two axes command guidance or coupling and some provide guidance in three axes: localizer, glide slope and airspeed. A marker beacon system or a RADAR altimeter is required for operations with decision heights of 150 ft or less. The AC states which data must be recorded for every approach and provides a suggested data card (AC 29 MG 3-1 and 3-2). In order to certify the FGS, a demonstration of 50 ILS approaches with a 90% success rate must be accomplished. If the FGS has not been previously certified in any other rotorcraft, then a certification program on the FGS must be completed before the 50 ILS approaches.

The installed equipment must meet the performance requirements specified in AC 20-129 appendix 1; the accuracy requirements for tracking equipment are also in accordance with the Appendix. Coupler systems that require manual trimming by the pilot to center the AFCS actuators should be evaluated in turbulent conditions; these systems may not meet the trim requirements at the 100' DH or provide sufficient tracking accuracy without an unusually high workload. Hardover malfunctions need to be evaluated in all axes and the MUH needs to be determined and incorporated into RWFM.

6.16.5.3 Airworthiness of Category III Systems

AC 120-28D provides an acceptable means, but not the only means, for obtaining and maintaining approval of operations in category III landing weather minima and low-visibility takeoff, including the installation and approval of associated airborne systems. As with category I and category II systems, the AC addresses the operational approval as well as the airworthiness approval. This section will address only the airworthiness approval of airborne systems used during a takeoff in low-visibility weather conditions and systems used to land and rollout in low-visibility conditions. The airworthiness criteria are found in appendices 2 and 3 of the AC.

6.16.5.4 Airworthiness of Systems for Takeoff in Low Visibility

These systems are required when visibility conditions alone may be inadequate for safe takeoff operation. The airworthiness criteria provide the requirements to track and maintain the runway centerline during a takeoff from brake release on the runway to liftoff and climb to 35 ft AGL, and from brake release through deceleration to a stop for a rejected takeoff. The use of this takeoff system must not require exceptional skill, workload, or pilot compensation. The takeoff system shall be shown to be satisfactory with and without the use of any outside visual references, except that outside visual references will not be considered when assessing lateral tracking performance. The airworthiness evaluation also determines whether the combination of takeoff guidance and outside visual references would unacceptably degrade task performance or require exceptional workload and pilot compensation during normal and nonnormal operations with system and airplane failure conditions. For the purpose of the airworthiness demonstration, the operational concept for coping with the loss of takeoff guidance is based on the availability of some other method for the flight crew to safely continue or reject the takeoff.

The intended takeoff path is along the axis of the runway centerline. This path must be established as a reference for takeoff in restricted-visibility conditions. A means must be provided to track the reference path for the length of the runway to accommodate both a normal takeoff and a rejected takeoff. The required lateral path may be established by a navigation aid (ILS, MLS) or by other methods if shown to be feasible by a proof of concept (the AC calls this a PoC). Methods requiring a PoC include

- The use of ground surveyed waypoints, either stored in an onboard database or provided by data link to the airplane with a path definition by the airborne system
- The use of inertial information following initial alignment
- Sensing of the runway surface, lighting, or markings with a vision enhancement system
- Deviation displays with reference to the navigation source (ILS/MLS receiver)
- Onboard navigation system computations with corresponding displays of position and reference path
- Vision enhancement system

In addition to indications of the aircraft position, the takeoff system should also compute and display command information (flight director) as lateral guidance. Parameters that should be provided are airplane position, deviation from the reference path, and deviation rate. Takeoff systems that provide only situational information in lieu of command information might be found acceptable but do require a PoC.

General Requirements for All Takeoff Systems   There are three basic types of takeoff systems: ILS, MLS, and GLS. GLS can be used in concert with a differential system, ground-based augmentation system, or data link. The airplane elements of the takeoff system must be shown to meet the performance, integrity, and reliability requirements identified for the types of operation for which approval is sought. The relationship and interaction of the aircraft elements with nonaircraft elements must be established and understood.

When international standards do not exist for the performance and integrity aspects of any nonaircraft elements of the takeoff system, the applicant must address these considerations as part of the airworthiness process. A means must be provided to inform the operator of the limitations and assumptions necessary to ensure safe operation. It will be the responsibility of the operator and the certification authority to ensure that appropriate criteria and standards are applied.

General requirements that need to be applied to all takeoff systems are as follows:

- Systems should ensure that a takeoff or a rejected takeoff can be safely completed on the designated runway, runway with clearway, or runway with stopway, as applicable.
- The system performance must be satisfactory, even in nonvisual conditions, for normal operations, aircraft failure cases, and recovery from displacements from nonnormal events.
- The system should be easy to follow and not increase workload significantly compared to the basic airplane.
- The display should be easy to interpret in all situations. Cockpit integration issues should be evaluated to ensure consistent operations and pilot responses in all situations.
- The continued takeoff or rejected takeoff operation should consider the effects of all reasonable events that would lead a flight crew to make a continued takeoff or a rejected takeoff decision.
- The airplane must not deviate significantly from the runway centerline during takeoff while the takeoff system is being used within its established limitations.
- The performance of the system must account for the differences, if any, between the runway centerline and the intended lateral path; compliance may be demonstrated by flight testing, or by a combination of flight testing and simulation.
- Flight testing must cover those factors affecting the behavior of the airplane (e.g., wind conditions, ILS characteristics, weight, CG).
- In the event that the airplane is displaced from the runway centerline at any point during the takeoff, the system must provide sufficient lateral guidance to enable the flying pilot to control the airplane smoothly back to the intended path in a controlled and predictable manner without significant overshoot.
- The performance envelope for evaluating takeoff systems includes

  - Takeoff with all engines operating
  - Engine failure at V_{ef} (continued takeoff)
  - Engine failure just prior to V_1 (rejected takeoff)
  - Engine failure at a critical speed prior to V_{mcg} (rejected takeoff)
  - Wind and runway conditions consistent with basic aircraft takeoff performance demonstrations

- The aircraft response to a permanent loss of localizer (azimuth) signal shall be established, and the loss of the localizer (azimuth) signal must be appropriately annunciated to the crew.
- The aircraft system response during a switchover from an active localizer (azimuth) transmitter to a backup transmitter shall be established.
- The takeoff system should provide required tracking performance with satisfactory workload and pilot compensation under foreseeable normal conditions.

Takeoff System Integrity   The onboard components of the low-visibility system and associated components considered separately and in relation to other systems should be designed to meet the requirements of Part 25.1309. The elements not on the airplane should not reduce the overall safety of the operation to unacceptable levels. The following criteria are provided for the application of Part 25.1309 to takeoff systems:

- The system design should not possess characteristics, in normal operation and when failed, that would degrade takeoff safety or lead to a hazardous situation.

- Any single failure of the airplane that could disturb the takeoff path must not cause loss of guidance information or give incorrect guidance information.

- Failures that would result in the airplane violating the lateral confines of the runway while on the ground should be detected by the takeoff system and promptly annunciated to the pilot.

- If pitch or speed guidance is provided, failures that would result in rotation at an unsafe speed, pitch, or rate angle should be detected by the takeoff system and promptly annunciated to the pilot.

- There may be failures that result in erroneous guidance yet are not annunciated. For these failures, other information and visual cues allow the pilot to detect the failure and take corrective action. These failures must be identified and the ability of the pilot to detect them and mitigate their effects must be verified by analysis, flight testing, or both.

- Whenever the takeoff system does not provide valid guidance appropriate for the takeoff operation it must be clearly annunciated to the crew and the guidance removed. The removal of the guidance alone is not an adequate annunciation.

- The probability of the takeoff system providing misleading information is improbable when the crew is alerted to the condition by a suitable fault annunciation or by information from other sources within the primary field of view.

- The probability of the takeoff system providing misleading information is extremely improbable if the takeoff system does not have the capability of detecting the failure or if no information is provided to the pilot to immediately detect the malfunction and take corrective action.

- In the event of a probable failure (e.g., engine failure, electrical source failure), if the pilot follows the takeoff display and disregards external visual reference, the airplane performance must meet the requirements shown in Figure 6.53.

- Loss of any single source of electrical power or transient condition on any single source of electrical power should not cause the loss of guidance to the flying pilot or loss of information that is required to monitor the takeoff to the pilot not flying.

- Takeoff systems that use navigation aids other than ILS/MLS require an overall safety assessment of the integration of aircraft and nonaircraft elements to ensure the takeoff system is acceptable.

Takeoff System Availability   When the takeoff operation is predicated on the use of a takeoff system, the probability of a system loss should be remote (10^{-5} per flight hour).

Flight Deck Information, Annunciation, and Alerting Requirements   The controls, indicators, and alerts must be designed to minimize crew errors that could cause a hazard. Mode and system malfunction indications must be presented in a manner

FIGURE 6.53 ■
Takeoff Command
Guidance
Performance
Envelope

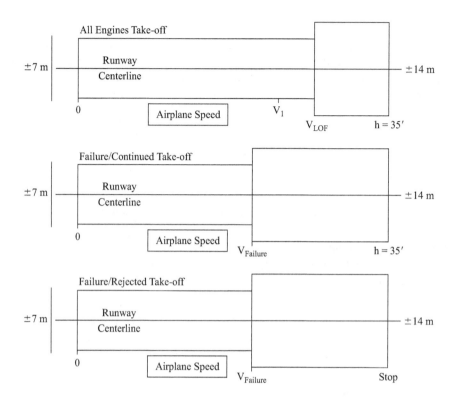

compatible with the procedures and assigned tasks of the flight crew. The indications must be grouped in a logical and consistent manner and be visible under all expected normal lighting conditions.

The system shall be demonstrated to have no display or failure characteristics that lead to degradation of the crew's ability to adequately monitor takeoff performance, conduct the entire takeoff, and make the appropriate transition to enroute climb speed and configuration for all normal, abnormal, and emergency situations.

Prior to takeoff initiation, and during takeoff, positive, continuous, and unambiguous indications of the following information about the takeoff system must be provided and made readily evident to both pilots:

- System status
- Modes of engagement and operation
- Guidance source

The takeoff system must alert the aircrew whenever the system suffers a failure or any condition that prevents the system from meeting the takeoff system performance requirements of Figure 6.53.

Warnings shall be provided for conditions that require immediate pilot awareness and action. Warnings are required for the following conditions:

- Loss of takeoff guidance
- Invalid takeoff guidance
- Failures of the guidance system that require immediate pilot awareness and compensation

During takeoff, whenever the takeoff system does not provide valid guidance appropriate for the takeoff operation, it must be clearly annunciated to the crew and the guidance must be removed. The removal of the guidance alone is not an adequate annunciation. The annunciation must be located to ensure rapid recognition and must not distract the pilot making the takeoff or significantly degrade the forward view.

Cautions shall be provided for conditions that require immediate pilot awareness and possible subsequent pilot action. The cautions need not generate a master caution light, which would be contrary to the takeoff alert inhibit philosophy; they should not cause flight crew distraction during takeoff roll.

Advisories shall be provided for conditions that require pilot awareness in a timely manner; they should not be generated after the takeoff has commenced. The status of the takeoff guidance system shall be provided to the crew.

Performance Evaluations   For new systems and any significant changes to an existing system, the performance of the airplane and its systems must typically be demonstrated by flight testing, which must include a sufficient number of normal and nonnormal operations conducted in conditions that are reasonably representative of actual expected conditions and must cover the range of parameters affecting the behavior of the airplane.

The performance evaluations must verify that the takeoff system meets the center-line tracking requirements shown in Figure 6.53.

The system performance must be demonstrated in nonvisual conditions for

- Normal operations
- Engine failure cases
- Recovery from displacements from nonnormal events

The performance shall be demonstrated to have a satisfactory level of workload and pilot compensation. The demonstration of system performance should comprise at least the following:

- 20 normal, all-engine takeoffs
- 10 completed takeoffs with simulated engine failure at or after the appropriate V_{ef} for the minimum V_1 for the airplane; all critical cases will be considered
- 10 rejected takeoffs, some with simulated engine failure just prior to V_1, and at least one run with simulated engine failure at a critical speed less than V_{mcg}

Other areas that need to be addressed by the certification plan include

- Engine failures should be assessed with respect to workload and pilot compensation throughout the entire takeoff phase.
- Demonstrated winds during normal, all-engine takeoffs should be at least the head-winds for which credit is sought and at least 150% of the crosswinds and tailwinds for which credit is sought but not less than 15 knots of headwind or crosswind.
- The takeoff system shall not exhibit any guidance or control characteristics that would cause the flight crew to react in an inappropriate manner.
- The system shall be demonstrated to have no display or failure characteristics that lead to degradation of the crew's ability to adequately monitor takeoff performance.
- The system must be demonstrated as meeting the annunciation and failure annunciations noted in the previous paragraphs.

- For takeoff systems that use an ILS localizer signal, the airplane response to a loss of localizer or switchover to a backup transmitter shall be demonstrated and properly annunciate to the crew.
- For takeoff systems that use an MLS, the airplane response to a loss of azimuth or switchover to a backup transmitter shall be demonstrated and properly annunciate to the crew.
- Failure cases should typically be spontaneous and unpredictable on the subject's or evaluation pilot's part.

Safety Assessment   A safety assessment is required as part of the certification plan.

Airborne System Requirements   To safely perform takeoff operations in limited-visibility conditions, there may be other aircraft systems that need to operate correctly and in concert with the takeoff system. Although most systems use either ILS or MLS, there are some systems that may have potential uses on future systems. These types of systems will require a proof of concept.

Peripheral Vision Guidance System   Peripheral vision systems have not been shown to be suitable as the primary means of takeoff guidance. Such systems may be used as a supplemental means of takeoff guidance only if a suitable minimum visual segment is available. A PoC will be required if the applicant is seeking approval as a primary means of takeoff guidance with this system.

HUD Takeoff System   The following criteria are applicable to heads-up display takeoff systems:

- The workload must be satisfactory.
- The HUD installation and display presentation must not significantly obscure the pilot's outside view.
- Appropriate transition from lateral takeoff guidance (35 ft AGL) through transition to an enroute climb for a takeoff and from brake release through deceleration to a stop for an aborted takeoff should be ensured.
- For the entire takeoff, for all normal and nonnormal situations (except for the loss of the HUD), it must not be necessary for the flying pilot to make any immediate change of primary display reference for continued safe flight.
- The HUD takeoff system must provide acceptable guidance and flight information to enable the pilot flying to complete or abort the takeoff as required.
- The HUD shall provide information suitable to perform the intended operation; the current mode of the HUD and FGS shall be clearly annunciated on the HUD unless they can be acceptably displayed elsewhere.
- A proposed system that displays situational information in lieu of command information (lateral deviation as a cue) requires a PoC.
- If the system is designed as a single HUD configuration, then the HUD shall be installed at the pilot-in-command's station.
- Associated cockpit information must be provided to the pilot not flying to monitor the flying pilot's performance and perform other assigned duties.

Satellite-Based Systems   The performance, integrity, and availability of any ground station elements, any data links to the airplane, any satellite elements, and any database considerations when combined with the airplane system should be at least equal to the performance, integrity, and availability of an ILS-based system. The role of the satellite-based elements should be addressed as part of the airplane system certification process until such time as an acceptable national or international standard is established for the satellite-based system.

Onboard Database   Unless there is a means to upload the path definition via data link, the required lateral ground path should be stored in an onboard database for recall and incorporation into the guidance/control system when required to conduct the takeoff. In addition,

- A mechanism should be established to ensure the continued integrity of the takeoff path designators.

- Corruption of the data is considered hazardous and therefore the probability of its occurrence is extremely remote.

- The flight crew should not be able to modify (either intentionally or inadvertently) the database.

- The integrity of the onboard database should be addressed as part of the certification process.

Data Link   Data may be sent to the airplane via data link so that the takeoff flight path can be defined with the required accuracy. The required takeoff path may be stored in a ground station database that is uplinked to an airplane; the airplane guidance and control system may then incorporate this information to conduct the takeoff. If using such a system, the following must be determined:

- The integrity of the data link must be commensurate with the integrity required for the operation.

- Satellite systems used during takeoff must support the required performance, integrity, and availability (including the assessment of satellite failures and the effect of satellite geometry [GDOP]).

- The capability of the takeoff system failure detection and warning system must preclude an undetected failure or combination of failures that are not extremely remote from producing a hazardous condition.

- The safety assessment should include failure mode detection coverage and adequacy of monitors and associated alarm times.

- The effect of aircraft maneuvers on the reception of signals and reacquisition of the signal after the loss must be assessed.

Enhanced Vision Systems   AC 120-28D does not really address the approval of enhanced vision systems. This topic is covered in section 6.18 of this text.

6.16.5.5 Airworthiness of Systems for Landing and Rollout in Low Visibility

The low-visibility landing system is intended to guide the airplane down the final approach segment to a touchdown in the prescribed touchdown zone with an appropriate sink rate and attitude without exceeding the prescribed load limits of the airplane.

The rollout system is intended to guide the airplane to converge on and track the runway centerline from the point of touchdown to a safe taxi speed. The general considerations for this system are the same as noted for the takeoff system described in section 6.14.5.4.

Types of Landing and Rollout Operations   There are four types of category III landing and rollout operations typically considered:

- Fail operational landing with fail operational rollout
- Fail operational landing with fail passive rollout
- Fail passive landing with fail passive rollout
- Fail passive landing without rollout system capability

NOTE: A fail operational system is a system capable of completing the specified phases of an operation following the failure of any single system component after passing a point designated by the applicable safety analysis. A fail passive system is a system that, in the event of a failure, causes no significant deviation of the aircraft flight path or attitude.

Each of these types may be considered with landing with an engine failure prior to the initiation of the approach or landing and rollout with an engine failure after the initiation of the approach but prior to the decision or alert height. The latter is considered as a basic requirement for any system demonstration intended for category III.

The types of sensors used for category III landing systems are the same as those identified for the takeoff system, and the airworthiness considerations will be the same, although adding vertical deviation requirements. MLS and ILS are supported by international standards. GNSS can provide lateral deviation, however, it must be augmented to meet the vertical accuracy requirements for category III systems. This augmentation may include LAAS, or GNSS and WAAS augmented with equivalent DME distance or marker beacon determination, all of which will require a proof of concept.

Landing and Rollout System Performance   The performance, integrity, and availability are applied to both the landing portion of the system as well as the rollout portion of the system. There are some common requirements for both portions and they are addressed first in all three sections. The following define the common requirements for landing and rollout system performance:

- If the landing system is designed to perform an alignment function prior to touchdown, to correct for crosswind effects, it should operate in a manner consistent with a pilot's technique typically using the wing low-side slip procedure.
- Nonavailability of the alignment mode or failure of the alignment mode must be easily detectable and annunciated to the pilot.
- Although derotation is not a requirement, if it is provided, the FGS should provide derotation, consistent with pilot technique.
- Systems that provide derotation must avoid any objectionable oscillatory motion or nose wheel touchdowns, pitch up, or other adverse behavior as a result of spoiler deployment or thrust reverser operation.
- Automatic control during the landing and rollout should not result in any airplane maneuvers that would cause the flight crew to intervene unnecessarily.

- Guidance provided during landing and rollout should be consistent with a pilot's manual technique.

Landing System Performance All types of low-visibility landing systems, including FGS guidance, guidance for manual control, and hybrid shall be demonstrated to achieve the performance accuracy and associated probability described below:

- Longitudinal touchdown earlier than a point on the runway 200 ft from the threshold to a probability of 1×10^{-6}
- Longitudinal touchdown beyond 2700 ft from the threshold to a probability of 1×10^{-6}
- Lateral touchdown with the outboard landing gear more than 70 ft from the runway centerline to a probability of 1×10^{-6}
- Structural limit load to a probability of 1×10^{-6}. An acceptable means of establishing that the structural limit load is not exceeded is to show separately and independently that

  - The limit load that results from a sink rate at touchdown of not greater than 10 ft/sec, or the limit rate of descent, whichever is greater, does not exceed the structural limit load
  - The lateral side load does not exceed the limit value determined for the lateral drift landing condition

- Bank angle resulting in a hazard to the airplane (wing, high-lift device, engine nacelle touches the ground) to a probability of 1×10^{-7}
- Airspeed control must be to within ±5 knots of the approach speed, except for momentary gusts, up to a point where the throttles are retarded to idle for landing.

Rollout System Performance The rollout system, if included, should control the airplane or provide guidance via the FGS for manual control from a point of landing to a safe taxi speed. The loss of rudder effectiveness as the airplane is slowed could be a factor in the level of approval that is granted to a system. The applicant should describe the system concept for rollout control so that the absence of low speed control, such as nose wheel steering, can be assessed. The rollout system should be demonstrated to

- Not cause the outboard tires to deviate from the runway centerline by more than 70 ft starting from the point at which touchdown occurs and continuing to a point where safe taxi speed is reached to a probability of 1×10^{-6}
- Capture the intended path or converge on the intended path in a smooth, timely, and predictable manner.
- Minor overshoots are acceptable; sustained or divergent oscillations or unnecessary aggressive responses are unsatisfactory.
- Promptly correct any lateral movement away from the runway centerline in a positive manner.
- Following touchdown, if not already on a converging path, cause the airplane to initially turn and track a path to intercept the runway centerline at a point far enough in front of the airplane that it is obvious to the aircrew that the rollout system is working properly.
- The rollout system should intercept the centerline sufficiently before the stop end of the runway and before the point at which taxi speed is reached.

Variables Affecting Performance The performance assessment shall take into account at least the following variables, with the variables being applied based on their expected distribution:

- Configuration (flaps/slats)
- CG
- Landing gross weight
- Conditions of headwinds, crosswinds, turbulence, and wind shear
- Characteristics of the applicable navigation systems and navigation aids with regard to variations in flight path definitions (e.g., ILS, MLS, GLS)
- Approach airspeed and variations in approach airspeed
- Airport conditions (elevation, runway slope, runway condition)
- Individual pilot performance (for systems with manual control)
- Any other parameter that may affect performance

Landing and Rollout System Integrity A safety assessment is required as part of the certification plan. The plan should identify the assumptions and considerations for the nonaircraft elements of the system and how these assumptions and considerations relate to the airplane system certification plan. The effect of the navigation reference point on the airplane on flight path and wheel to threshold crossing height shall be assessed.

Landing System Integrity The applicant should show that landing system design meets the intent of section 25.1309. The following criteria are provided for the application of 1309 to landing systems:

- A single malfunction or combination of malfunctions of the landing system that could prevent a safe landing or go-around must be extremely improbable, unless it can be detected or annunciated as a warning to allow pilot intervention, and then it must be extremely remote.
- Failure to detect and annunciate malfunctions that could prevent a safe landing or go-around must be extremely improbable.
- The exposure time for assessing failure probabilities for fail passive systems is the average time required to descend from 100 ft HAT or higher to touchdown, and for fail operational landing systems it is the average time to descend from 200 ft HAT or higher to touchdown.
- For a fail passive landing system, a single malfunction or any combination of malfunctions must not cause a significant deviation of the flight path or attitude following system disengagement.
- A fail operational landing system, following a single malfunction, must not lose the capability to perform vertical and lateral path tracking, alignment with the runway heading, or flare and touchdown within the safe landing requirements.

Safe Landing Requirements For the purpose of analysis, a safe landing may be assumed if the following requirements are achieved:

- Longitudinal touchdown no earlier than a point on the runway 200 ft from the threshold
- Longitudinal touchdown no further than 300 ft from the threshold (not beyond the end of the touchdown zone lighting)

- Lateral touchdown with the outboard landing gear within 70 ft of the runway centerline (assumes a 150 ft wide runway; lateral touchdown performance limit may be appropriately increased if operation is limited to wider runways)

- Structural limit load (as described previously)

Rollout System Integrity   As with the landing system integrity, the rollout system, if provided, is expected to meet the intent of section 25.1309; a safety assessment is required as part of the certification plan. The following criteria are provided for the application of 1309 to rollout systems:

- A fail operational rollout system must meet the safe rollout performance requirements (no deviation greater than 70 ft from the centerline) after any single malfunction or any other combination of malfunctions not shown to be extremely remote.

- For any rollout system below 200 ft HAT, unannounced malfunctions that would prevent a safe rollout must be shown to be extremely improbable.

- For a fail passive rollout system, the loss of a fail passive automatic rollout function after touchdown will cause the FGS to disconnect; the loss of a fail passive rollout system after touchdown will be shown to be improbable.

- Whenever a fail passive landing rollout system does not provide valid guidance, an annunciation should be provided to both pilots and the guidance information removed. Removal of the guidance alone is not an adequate annunciation unless other systems within the primary field of view provide additional information.

- For any rollout system, malfunctions that only affect low-speed directional control should not cause the airplane wheels to exceed the lateral confines of the runway from touchdown to safe taxi speed for a probability of 1×10^{-7}.

Landing System Availability   Below 500 ft on approach, the probability of a successful landing should be at least 96% for approach demonstrations conducted in the airplane. Compliance with this requirement typically should be established during flight test with approximately 100 approaches. For an airplane equipped with a fail passive landing system, the need to initiate a go-around below 100 ft HAT due to an airplane failure condition should be infrequent (1 per 1000 approaches). For a fail operational landing system below 200 ft HAT, the probability of total loss of the landing system must be extremely remote. For any annunciation that is provided, that annunciation must enable a pilot to intervene in a timely manner to avoid a catastrophic result. Total loss of the system without annunciations shall be extremely improbable.

Rollout System Availability   For a full passive rollout system, from 200 ft HAT through landing and rollout to a safe taxi speed, the probability of a successful rollout should be at least 95%. For a fail operational rollout system, during the period in which the airplane descends below 200 ft HAT to a safe taxi speed, the probability of degradation from fail operational to fail passive shall be fewer than 1 per 1000 approaches and the probability of total loss of rollout capability should be extremely remote. After touchdown, complete loss of fail operational automatic rollout function or any other unsafe malfunction or condition shall cause the FGS to disconnect. The loss of a fail operational rollout system after touchdown shall be extremely remote.

Automatic Braking System   If automatic braking is used, the following criteria apply:

- The automatic braking system should provide antiskid protection and have manual reversion capability.
- It should provide smooth and continuous deceleration from touchdown through full stop.
- Disconnect of the automatic braking system must not create additional workload or distraction.
- It should not interfere with the rollout control system.
- Manual override must be possible without excessive brake pedal forces.
- It should not be susceptible to inadvertent disconnect.
- A positive indication of system disengagement or failure should be provided.
- Braking distances with the automatic braking system on wet and dry runways need to be demonstrated and presented in the AFM performance section.

Flight Deck Information, Annunciation, and Alerting Requirements   The same general rules for controls, indicators, and warnings for the takeoff system apply to the landing and rollout system as well. There are some additional flight parameters for monitoring capability:

- Unambiguous identification of the intended path for the approach, landing, and rollout
- Indication of the position of the aircraft with respect to the intended path (situation information localizer and glide path)

After beginning the final approach, the loss of a fail passive or fail operational system shall be annunciated. Whenever a fail passive system, for manual control, does not provide valid guidance, a warning must be given to both pilots and the guidance removed.

After initiation of the final approach, a fail passive system (landing, rollout, or guidance) shall alert the flight crew to any malfunction or condition that would adversely affect the ability of the system to safely operate or continue the approach and landing.

After initiation of the final approach (between 1000 ft HAT and the alert height), a fail operational system shall alert the flight crew of any failure that would adversely affect the ability of the system to safely operate or continue the approach and landing or any malfunction that degrades the system to a fail passive status. Below the alert height, these malfunction alerts are inhibited.

Although excessive deviation alerting is not required, the authority will approve a system with this function as long as it meets appropriate criteria. If a method is provided to detect excessive deviation of the airplane laterally and vertically during approach to touchdown, and laterally after touchdown, then it should not require excessive workload or undue attention. The provision does not require a specific deviation method or annunciation, but may be addressed by parameters displayed on the ADI, EADI, HUD, or PFD. When a dedicated deviation alert is provided, its use must not cause excessive nuisance alerts.

A means must also be provided to inform the flight crew when the airplane has reached the operation alert height or decision height, as applicable.

Performance Evaluations For new systems and any significant changes to an existing system, the performance of the airplane and its systems must typically be demonstrated by flight testing. Flight testing must include a sufficient number of normal and non-normal operations conducted in conditions that are reasonably representative of actual expected conditions and must cover the range of parameters affecting the behavior of the airplane.

The reference speed used as the basis for certification must be identified; the applicant should demonstrate acceptable performance within a speed range of −5 to +10 knots with respect to the reference speed. The transition from landing to rollout must be smooth, without characteristics that would lead the flight crew to react in an inappropriate manner.

Landing systems for manual control with guidance must meet the same requirements for touchdown footprints, sink rates, and attitudes as automatic systems. The landing and rollout system must be shown to be acceptable with and without outside cues.

For the initial certification of a landing and rollout system for manual control with guidance in a new type of airplane or new type of HUD installation, at least 1000 simulated landings and at least 100 actual aircraft landings are typically necessary. If demonstrating engine-out performance, and the procedures are the same as the all-engine case, then compliance may be demonstrated with 10 to 15 landings. If the procedures are different, then the certification authority will determine the number of engine-out approaches case by case.

For aircraft intending to use an automatic landing system at high altitudes, the airport elevation at which satisfactory performance may be attained must be demonstrated. The AC provides detailed test procedures in appendix 3 to perform this evaluation.

Simulation may be used in part for the airworthiness certification process of category III landing systems. These systems are typically called "pilot-in-the-loop" and require high-fidelity, engineering-quality simulation. AC 120-40B, "Airplane Simulator Qualification," provides a means to qualify simulators for pilots. Meeting these requirements provides a known basis for acceptance of simulator capability, and is desirable, but may not necessarily be sufficient, to meet the requirements of an engineering simulation to demonstrate landing system performance. AC 120-28D imposes, again in appendix 3, the required simulation capabilities to demonstrate airworthiness for category III landing systems.

A flight test performance demonstration should be conducted to confirm the results of the simulation. The principal performance parameters are as follows:

- Vertical and lateral flight path with respect to the intended path
- Lateral deviation from centerline during rollout
- Altitude and height above the terrain during approach
- Air data vertical speed
- Radar altitude sink rate
- Airspeed and groundspeed
- Longitudinal and lateral runway touchdown point

Data acquired during demonstration flights should be used to validate the simulation. Unless otherwise agreed by the certification authority, the objective of the flight test program should be to demonstrate performance of the system to 100% of the steady-state wind limit values (25 knot headwind, 15 knot crosswind, and 10 knot tailwind). The simulation can be considered validated if at least four landings are accomplished during flight testing at no less than 80% of the intended steady-state wind value and a best effort has been made to achieve the full steady-state wind component values. It must be shown that the landing system is sufficiently robust near the desired AFM wind-demonstrated values.

Safety Assessment   A safety assessment is required as part of the certification plan.

Airborne Systems   The performance, integrity, and availability requirements of the on-aircraft systems must meet the requirements applicable to the type of operation intended. Specific requirements for airborne systems used for category III landing and rollout operations are contained in the following paragraphs.

Automatic Flight Control Systems

- When established on the final approach path below 1000 ft HAT, it must not be possible to change the flight path of the airplane with the FGS engaged except by initiating an automatic go-around.
- It must be possible to disengage the automatic landing system at any time without significant out-of-trim forces.
- It must be possible to disengage the automatic landing system by applying a suitable force to the control column, wheel, or stick.
- Following a failure or inadvertent disconnect of the autopilot or loss of the automatic landing mode, when it is necessary for the pilot to assume manual control, a visual alert and an aural warning must be given.

Autothrottle System

- An automatic landing system must include autothrottles unless airplane speed can be controlled manually without excessive workload and the touchdown limits can be achieved in normal and nonnormal conditions.
- Approach speed can be selected manually or automatically. In either case, each pilot must be able to determine that the aircraft is flying an appropriate speed.
- Thrust and throttles should be modulated at a rate consistent with pilot expectations.
- An indication of the pertinent automatic throttle system engagement must be provided.
- An appropriate alert or warning of autothrottle failure must be provided.
- It must be possible for each pilot to override the automatic throttles.
- Automatic throttle disengagement switches must be mounted on or adjacent to the throttle levers, where they can be operated without moving the hand from the throttles.
- Following a failure, failure disconnect or inadvertent disconnect of the automatic throttle, or uncommanded loss of a selected automatic throttle mode, a suitably clear and compelling advisory or indication should be provided.

HUD Guidance   The requirements for HUD performance previously mentioned in sections 6.16.5.1 and 6.16.5.4 apply here as well.

Hybrid HUD/Autoland Systems   Hybrid systems (fail passive autoland system used in combination with a monitored HUD FGS) may be acceptable for category III if each element of the system alone is shown to meet its respective suitability for category III and, if taken together, the components provide the equivalent performance and safety to a nonhybrid system. Hybrid systems with automatic landing capability should be based on the concept of use of the automatic landing system as the primary means of control, with the manual FGS serving as a backup or reversionary mode.

Hybrid systems must be demonstrated to the certification authority via a proof of concept evaluation during which specific airworthiness and operation criteria will be developed that meet the requirements of nonhybrid systems. Combining an automatic landing system that meets the fail passive criteria with a HUD that also meets the criteria does not necessarily ensure that an acceptable fail operational system will result. These systems may be combined to establish a fail operational system for low-visibility operations provided certain considerations are addressed:

• Each element of the system alone meets the requirements for a fail passive system.

• The automatic landing system will be the primary means of control, with the manual system a backup or reversionary mode.

• Manual rollout guidance must be provided for hybrid systems that do not have automatic rollout capability.

• The transition between automatic and manual modes should not require extraordinary skill, training, or proficiency.

• The transition between automatic and manual modes must be safe and reliable.

• The capability of the pilot to use a hybrid system to safely accomplish the landing and rollout following a failure of one of the hybrid system elements below alert height must be demonstrated (even if operational procedures call for the pilot to initiate a go-around).

• Appropriate annunciations must be provided to the flight crew to ensure a safe operation.

• The combined elements of the system must be demonstrated to meet fail operational criteria.

• The overall system must be shown to meet accuracy, availability, and integrity criteria suitable for fail operational systems.

• Demonstrations are necessary for each element of the hybrid system for low-altitude go-around in the altitude range of 50 ft HAT and touchdown.

• Hybrid system demonstrations must be conducted in the following conditions:

  – Without external visual reference

  – With visual reference

  – With the presence of external visual reference that disagrees with instrument reference

Other Airborne Systems   The requirements for other proof of concept systems (GLS, data link, and enhanced vision systems) were addressed in earlier sections, and the aforementioned requirements are the same for category III operations.

▋ 6.17 | FLIGHT MANAGEMENT SYSTEMS

A FMS is just that. It takes inputs from many sensors within the aircraft and assists the pilot with the management of the aircraft flight plan, departure, enroute flight, transition, approach, fuel, and time. The system improves crew performance, reduces the crew workload, and assists air carriers with increased efficiency. The system relies heavily on automation, which is a good thing for efficiency and a decrease in workload, but it tends to take the pilot out of the loop, which is not a good thing.

A FMS is a generic term that encompasses a wide variety of capabilities and subsystems:

- Flight planning
- VNAV and LNAV
- Fuel management
- Cautions, advisories, and warnings
- Moving map
- Weather radar
- Communications
- Outputs to the electronic flight instrument system (EFIS)

A typical FMS interface is shown in Figure 6.54a, where the FMC is the central processor or flight management computer for the FMS.

The FMS can provide guidance for the specific departure and transition, climb with fuel and thrust management, and navigation for the route. It can provide guidance for the best endurance and fuel economy and performs continuous evaluation and prediction of fuel consumption along the route. During the descent the FMS will automatically comply with speed and altitude restrictions and provide the transition to the automatic landing system. The pilot is advised of the proper landing speed and the approach minimums. The FMS can also provide the proper guidance for missed approach/go-around, the missed approach holding point, and the proper direction of the turn in holding. A typical FMS sequence is shown in Figure 6.54b.

The typical FMS will use two or three multifunction display units (processors) that provide the operator with the ability to program the flight. The departure, transition, route, arrival, approach, runway, and alternates are all entered by the crew prior to flight. Additional information such as fuel on board, number of passengers and their associated weight (with baggage), and runway condition are also input by the crew to determine takeoff roll, refusal and rotation speeds, fuel consumption, and best climb and cruise speeds.

The navigation sensors on board, such as GNSS, INS, VOR/DME, TACAN, and ADF are fed to each of the processors to compute the best estimate of position. The setup allows complete redundancy between the processors. The processors output display information as well as guidance to the autopilot/autothrust systems. The FMS also provides cautions, advisories, and warnings to the aircrew as required. Informational messages may also be provided on some systems. The FMS is capable of providing LNAV, and some systems also are capable of providing VNAV. A typical FMS architectural design is shown in Figure 6.55.

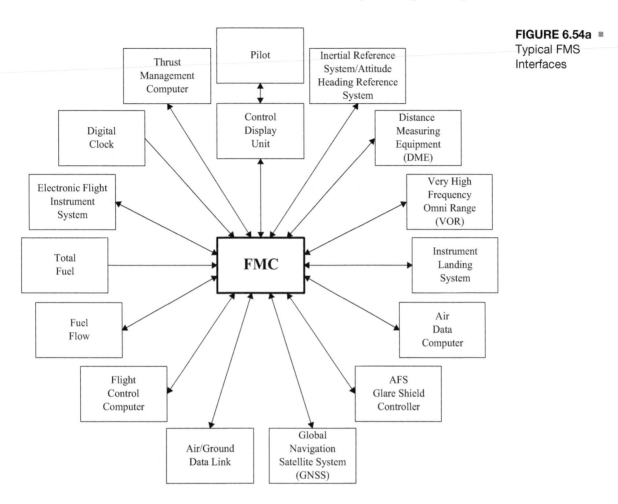

FIGURE 6.54a ■
Typical FMS
Interfaces

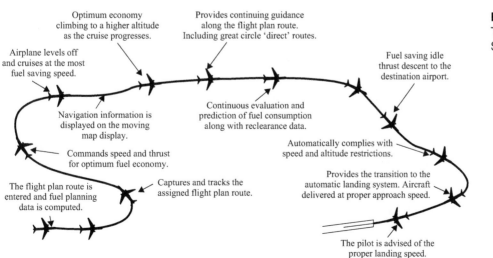

FIGURE 6.54b ■
Typical FMS
Sequencing

FIGURE 6.55 ■
Typical FMS
Architecture

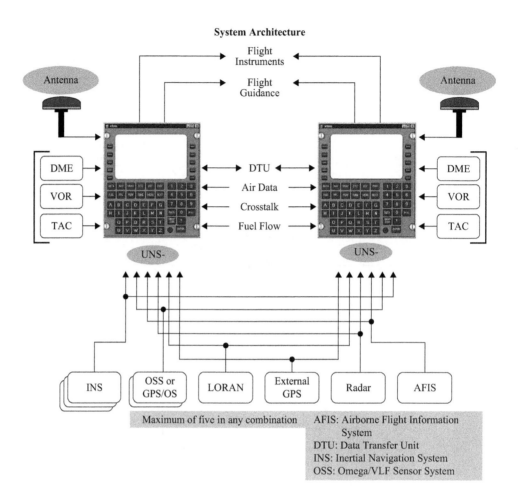

6.17.1 Airworthiness Certification of the FMS

AC 25-7C refers you to AC 20-129, "Airworthiness Approval of Vertical Navigation Systems", September 12, 1988 (canceled by AC 20-138C), and AC 20-130A, "Airworthiness Approval of Navigation or FMS Systems Integrating Multiple Navigation Sensors," June 14, 1995 (canceled by AC 20-138C). The basic FMS AC, AC 25-15, "Approval of FMS in Transport Category Aircraft," November 20, 1989, remains in effect and will be used in conjunction with AC 20-138C (described in detail in section 6.11 of the text. AC 23-8C, paragraph 1307, Miscellaneous Equipment is marked *reserved*.

AC 27-1B and 29-2C do not specifically cover FMS; information regarding test is contained in the Controls and Displays and IFR sections. AC 20-138C, "Airworthiness Approval of Positioning and Navigation Systems," May 8, 2012, cancels AC 20-129 and AC 20-130A and incorporates this information. The MOPS is contained in RTCA/DO-283A, and the applicable TSO is TSO-C115C, January 9, 2012.

The certification program for extensively integrated FMS should be through the TC or STC process; a comprehensive certification plan should be developed by the applicant. System reliability and integrity should be evaluated IAW AC 25.1309-1E. As with any integrated system, the FMS installation must meet the system reliability and

TABLE 6.50 ■ NAS Lateral Navigation Requirements

	Enroute (Random Routes)	Enroute (J/V Routes)	Terminal	Nonprecision Approach
Equipment Error, Along and Cross Track (2σ, 95% Probability)	+3.8 nm	+2.8 nm	+1.7 nm	+0.3 nm (+0.5 nm for navigation data that comes from a single collocated VOR/DME station)

integrity requirements of 14 CFR 25.1309. Software development must be done in accordance with RTCA DO-178B/C, and the environmental conditions and test procedures for airborne equipment (RTCA DO-160G) must be satisfied.

6.17.2 AC 25-15 Airworthiness Requirements

The FMS should be integrated with other onboard systems and functionally mechanized with the basic airplane such that all lateral, vertical, and autothrottle modes perform their intended function and permit the crew to safely, readily, and efficiently operate the airplane within its approved flight envelope.

The navigation accuracy requirements vary with the airspace the aircraft is operating in. For aircraft intending to operate in U.S. airspace, the accuracy requirements in Table 6.50 must be satisfied.

Other lateral navigation requirements apply:

- Systems that use multiple navigation sensor inputs should meet the requirements of AC 20-138C; the FMS is in reality an RNAV system, and therefore must meet the requirements for RNAV systems previously defined in section 6.11.

- FMS equipment may provide the capability to conduct various nonprecision approaches utilizing data from a variety of sensors.

- Other sensor inputs may be used to enhance the navigation solution, however, the specified reference facility for the approach being conducted must be used in the navigation solution; the applicant should demonstrate that the system accuracy is not degraded below that established by the use of the specified reference facility.

- FMS equipment may be used to conduct nonprecision approaches where the specified reference facility that defines the approach is not used as long as the applicant shows that the accuracy and failure modes are compatible with the approach.

- To provide acceptable agreement between the FMS track line and the ILS/MLS course centerline, the FMS should use ILS/MLS sensor data to compute FMS position.

- FMS equipment should provide the capability to proceed to a selected waypoint and hold on a specified inbound course to a waypoint with a repeated crossing of the selected waypoint.

- Proper magnetic variation must be applied within the FMS to accurately maintain a course from a designated navigation aid.

The vertical navigation function relates to a vertical profile defined by two waypoints. The FMS should meet the vertical performance criteria defined in section 6.11 under RNAV equipment.

A navigation database is used in the FMS and should be updated every 56 days. The aircrew should be provided with the expiration date of the database and it should contain all of the information required for successful operation of the system. The pilot must not be able to alter the contents of the database. Flight plans built by the crew should not be erased with electrical transients or a loss of electrical power.

The following controls and displays functions should be available in the FMS:

- Automatic frequency tuning may be employed in the FMS, but must not override a manual selection by the pilot.
- The display of time or distance to go should relate to a navigation aid or waypoint located on the programmed airplane course unless adequate annunciation is provide to the contrary.
- A means should be provided to indicate on demand the sensors currently being used in the navigation solution as well as the status of all sensors or stations being tracked.
- It is recommended that a manual update function of the FMS be provided.

The FMS should annunciate the status of the selected navigation mode and alert the aircrew in the event of failures that could affect the navigation of the airplane; a lack of adequate navigation in the approach mode should be annunciated by means of a flag on the primary navigation display. Other required annunciations are as follows:

- Adequate annunciation should be provided for system failures that result in a loss or reduction of function.
- When the aircraft is being controlled in pitch, roll, thrust, or airspeed by FMS functions, the annunciation of these modes or submodes shall be presented in the primary field of view.
- It is desirable to have the FMS provide a means to annunciate a significant position deviation from the computer programmed flight path of the airplane; a hazardous deviation is marked by a full-scale deflection.
- A means should be provided to annunciate an approaching waypoint crossing.
- The system should annunciate when the accuracy for which the navigation mode was approved can no longer be assured.
- The probability of providing hazardously misleading information in LNAV, VNAV, long-range navigation, position information, or in the navigation database must be shown to be improbable.

The FMS may provide the function of computing and displaying achieved airplane takeoff performance information to the crew in real time. It is desirable that the accuracy of the individual performance parameter displays fall within the following ranges:

- Takeoff distance: ±100 ft or $\pm2\%$ (whichever is less)
- Airspeed: ±4 knots or $\pm2\%$ (whichever is greater)
- Longitudinal acceleration: ±0.2 ft/sec^2 or $\pm2\%$ (whichever is greater)
- Distance to rotation speed: $\pm5\%$

6.17.3 FMS Flight Tests

The following tests should be included in the applicant's flight test plan to verify that the FMS performs its intended function and that there are no adverse affects to other airplane sensors and systems:

- Evaluation of all normal operating modes such as speed command, attitude/thrust in combination with lateral and vertical navigation modes for takeoff, climb, cruise, descent, and approach.
- Evaluation should include a sufficient number of simulated engine-out conditions. At least one actual engine shutdown should be accomplished.
- FMS-controlled buffet boundary margin should be checked.
- Autothrottle function should be tested for approach and takeoff modes.
- Fuel state computations should be compared with flight planning data.
- Data obtained from flight demonstration should be referenced to flight deck instrumentation, not corrected for position error.
- Evaluation requirements for RNAV systems listed in AC 20-138C.

    When performing these evaluations, it is important that the following correct operation of function be noted:

- Smooth course guidance information transition between different sources
- Verify nonuse of a navigation aid for nonprecision approaches
- Evaluate small intercept angles transitioning from FMS navigation to ILS final approach course
- Holding patterns hold over the same point/remain within protected airspace
- Appropriate alerts for VNAV
- If pilot can select speed, cannot exceed V_{MO}/M_{MO} or go below $1.2V_{SI}$
- Current database date displayable
- Flight plans not lost or changed due to power transients or loss
- Automatic frequency tuning should not override pilot-selected frequency
- Should be able to determine sensors/stations being used for navigation
- Should be possible to manually update FMS position
- Proper FMS annunciation
    - Loss of function
    - Mode
    - Excessive deviations
    - Waypoint alerts
    - Visual cue or continuously displayed map mode
- FMS engine-out detection and behavior
- Proper reference speed with an engine failure during takeoff
- Observance of 250 knots below 10,000 ft as appropriate for aircraft weight
- Input units should be clearly discernible

- Proper computation of maximum altitudes considering buffet margin
- FMS should not reduce speed (i.e., reduce power) before safe levels with anti-ice systems on
- Should be able to adjust V_{app} but not V_{ref}
- Use of FMS/ATS (automatic thrust system) to set "clamp speeds"; ATS should be able to advance throttles to required limit before hitting 80 knots
- ATS should be free of "hunting" when in either speed or thrust control modes
- FMS should automatically advance ATS to go-around mode if go-around selected

6.17.4 Multisensor Flight Tests

Flight tests are conducted to verify proper operation and accuracy of the multisensor equipment as installed in the aircraft. The following VFR flight tests recap some the evaluations previously called out in section 6.11:

- Evaluation of the installed multisensor equipment to verify that it is functioning safely and operates in accordance with the manufacturer's specifications
- If coupled to an autopilot, evaluation of the steering response while the autopilot/flight director is coupled to the system during a variety of different track and mode changes
- Evaluation of all available display sensitivities
- Verify the installation does not adversely affect other aircraft equipment
- Accessibility of the controls
- Visibility of the controls in all types of lighting
- Validate multisensor equipment navigation accuracy in each mode (e.g., oceanic, enroute, terminal) for which approval is requested

For certification under IFR, multisensor flight testing will require all of the evaluations listed for VFR plus the following:

- Required flight evaluations for any new sensor should be conducted according to the AC for installation of that sensor
- Review of the various failure modes and associated annunciations
- Evaluation of displayed multisensor equipment parameters on interfaced cockpit instruments such as HSI, CDI, and moving map
- Assessment of all switching and transfer functions, including electrical bus switching
- Evaluation of satisfactory EMC
- Analysis of crew workload
- Verify continuity of navigation data during normal maneuvering, including holding patterns and up to at 30° of bank for left and right 360° turns
- Verify that the FTE can be maintained
- For equipment including an approach mode, conduct a sufficient number of approaches using the navigation database to verify the proper operation of annunciations, waypoint sequencing, and display sensitivities

A sample compliance matrix for Part 25 FMS installation is shown in Table 6.51.

TABLE 6.51 ■ A Sample Compliance Matrix for Part 25 FMS Installation

No.	Requirements AC	Paragraph	Subject	Description
1	AC 25-15	5d	System integrity	Determine the detectability of a failure condition
				Determine the required subsequent pilot actions
				Make a judgment if satisfactory intervention can be expected by a properly trained crew
				Failure of the FMS should not degrade the integrity of other safety related systems installed in the airplane
2	AC 25-15	5e(1)(i)	Long range navigation	Sensors should be evaluated to demonstrate accuracy suitable for the airspace for which the approval is sought
3	AC 25-15	5e(1)(ii)	U.S. NAS navigation	Systems that are intended for enroute, terminal and approach operations in the NAS should meet the following accuracy criteria.
4	AC 25-15	5e(1)(ii)A	Sensor qualification	
5	AC 25-15	5e(1)(ii)B(1)	Non-precision Approach with use of Navaid	Display of course guidance information should be smooth during transition among the various sensor inputs. For NDB approaches, data from the NDB that defines the approach procedures should be displayed in the pilot's primary field of view.
6	AC 25-15	5e(1)(ii)B(2)	Nonuse of Navaid	The system accuracy and failure modes are compatible with the performance requirements and terminal instrument procedures criteria
7	AC 25-15	5e(1)(ii)C	Precision approach	ILS raw data must be displayed in the ADI/EADI or primary flight display (PFD). Transition from FMS to ILS by the AP/FD should be smooth and without significant overshoot
8	AC 25-15	5e(1)(ii)D	Holding patterns	The equipment should provide the capability to proceed to a selected waypoint and hold on a specified inbound course to a waypoint with a repeated crossing of the selected waypoint. If holding patterns are pilot selectable or included in the database, they should be defined.
9	AC 25-15	5e(1)(ii)E	North Reference Effects	If the FMS has course selection in degrees ("TO-FROM" operation), such as charted airways, should have technical or operational provision to ensure that the proper north reference is used to define the desired course.
10	AC 25-15	5e(2)	Vertical navigation	The error of the airborne VNAV equipment should be less than the special value.
11	AC 25-15	5e(2)(i)	Cruise Altitude Departure	Adequate annunciation of an impending automatic departure from a cruise altitude should be provided.
12	AC 25-15	5e(2)(ii)	Speed Selection	If the system provides for pilot selection of speed, it should not be possible to select a speed above V_{mo}/M_{mo} or a speed below the speed 1.2Vs1 for the existing airplane configuration.
13	AC 25-15	5e(3)	Navigation database	FMS should include information such as flight plan magnetic variation and victor and Jet airways and so on.
				Displaying to the flight crew the dates for which the data is current and a means for updating the database.

(Continues)

TABLE 6.51 ■ (Continued)

No.	Requirements AC	Paragraph	Subject	Description
14	AC 25-15	5e(3)(i)	Flight Plan	The FMS should provide a means to allow the flight crew to easily define additional waypoints and insert them into the flight plan. The accuracy of the data defining-waypoints should be the location to the nearest 0.1 minute of latitude and longitude or the equivalent in terms of bearing and distance from a defined location or Nav aid. The flight plan should not be changed or erased by electrical transients or the loss of electrical power.
15	AC 25-15	5e(3)(ii)	Data Base Modification	The flight crew should not be able to modify information in the database.
16	AC 25-15	5e(4)(i)	Automatic frequency tuning	Automatic frequency tuning feature should not override the manually selected frequency.
17	AC 25-15	5e(4)(i)	Remote tuning	If remote tuning of communication transceivers through FMS is available, it should provide a means to manually tune the radios to the emergency frequencies (121.5/243 MHz).
18	AC 25-15	5e(4)(ii)	Time and Distance Display	The display of time or distance to go should relate to a Navaid or waypoint located on the programmed airplane course unless adequate annunciation is provided.
19	AC 25-15	5e(4)(iii)	System Status Display	A means should be provided to indicate the specific sensors and/or stations currently used in the navigation calculations as well as the status of all sensors and/or stations being tracked.
20	AC 25-15	5e(4)(iv)	Position Update	It is recommended that the ability to manually update the FMS position be provided.
21	AC 25-15	5e(5)	Status Annunciation	The FMS should provide the means to annunciate the status of the selected navigation mode and to alert the flight crew to system failures.
22	AC 25-15	5e(5)(i)	Loss or reduction of function	Adequate annunciation should be provided for system failures that result in a loss or reduction of functions
23	AC 25-15	5e(5)(ii)	Mode Annunciation	When the aircraft is being controlled in pitch, roll, thrust or airspeed by FMS functions, the annunciation of these modes or submodes of FMS operation shall be presented in a clear and unambiguous manner in the flight crew's primary field of view.
24	AC 25-15	5e(5)(iii)	Excessive Deviation Annunciation	It is desirable to have the FMS provide a means to annunciate a significant position deviation from the computer programmed flight path of the airplane.
25	AC 25-15	5e(5)(iv)	Waypoint Alert	A means should be provided to annunciate an approaching waypoint crossing.
26	AC 25-15	5e(5)(v)	Sensor and Station Quality	The system should annunciate when the quality of the navigation signals or sources, including signal to noise ratio and station geometry, has degraded to the level where the accuracy for which the navigation mode was approved can no longer be assured.
27	AC 25-15	5e(6)	Failure modes	An equipment installation failure modes and effects analysis should be provided.
28	AC 25-15	5e(7)	Navigation integrity	The computation and display of hazardously misleading navigation information should be shown to be improbable.
29	AC 25-15	5f	Performance management	The performance management logic should provide protection against exceeding 250 knots IAS below 10,000 ft MSL, however pilot override of this feature should be provided.

30	AC 25-15	5f(1)	Ground procedures	The computer should contain reasonableness logic to prevent the airplane from being operated beyond its certificated envelope limits.
31	AC 25-15	5f(2)	Takeoff mode	The takeoff mode generally displays some or all of the takeoff speeds and appropriate engine thrust settings based upon the airport ambient temperature, airplane weight and thrust reduction.
32	AC 25-15	5f(2)i	Takeoff auto throttle	The auto throttle servo configuration may impact the performance of the system when operating in some modes. Such a system would be usable only as an aid for setting takeoff power.
33	AC 25-15	5f(2)ii	Engine failure	The method of detecting an engine failure should be reviewed to establish the validity of the scheme. Low EGT sensing may not account for a slowly failing engine or simulated failure during training. Automatic bleed-shut-off sensing may not account for MEL item, for example. This type of review will aid in establishing the AFM limitations and procedures, and possible MEL revisions.
34	AC 25-15	5f(2)iii	Takeoff pitch attitude	Airspeed must be used as the primary reference and this should be clearly stated in the AFM.
35	AC 25-15	5f(2)iv	Engine-out flap retraction	The FMS should capture the target speed for each flap/slat configuration.
36	AC 25-15	5f(2)v	Takeoff speed command modes	The system should provide commands that lead to the airplane smoothly acquiring a pitch attitude that results in a takeoff reference speed of $V_2 + 10$ knots by the time the airplane has reached approximately 35 ft altitude. The FMS may provide automatic takeoff thrust control system (ATTCS) functions that will allow for reductions in takeoff thrust up to 10% of the takeoff thrust.
37	AC 25-15	5f(2) (v)(A)	Takeoff	The system should provide commands that lead to the airplane smoothly acquiring a pitch attitude that results in a takeoff reference speed of $V_2 + 10$ knots by the time the airplane has reached approximately 35 ft altitude.
38	AC 25-15	5f(2) (v)(B)	Takeoff noise thrust cutback	The system should provide commands that result in an indicated airspeed of $V_2 + X$ at all takeoff flap settings prior to thrust cutback.
39	AC 25-15	5f(2) (v)(C)	Reduced thrust takeoff	The thrust should be at least 75% of the takeoff thrust or derated takeoff thrust if such is the performance basis, for the existing ambient conditions.
40	AC 25-15	5f(2) (v)(D)	Automatic takeoff thrust system	The FMS may provide automatic takeoff thrust control system (ATTCS) functions that will allow for reductions in takeoff thrust up to 10% of the takeoff thrust.
41	AC 25-15	5f(2) (v)(E)	De-rated takeoff thrust	A derated takeoff thrust is a takeoff thrust less than the maximum takeoff thrust and for which a set of independent takeoff limitations and performance data may exist in the FMS computer memory. When these data are used for a derated takeoff, the thrust setting parameter, which establishes thrust for takeoff, is considered a takeoff operating limit.
42	AC 25-15	5f(2) (v)(F)	Transition	The system should provide commands during periods of transition that are commensurate with what would be expected from manual operation during the same condition.

(Continues)

TABLE 6.51 ■ *(Continued)*

No.	Requirements AC Paragraph		Subject	Description
43	AC 25-15	5f(3)	Climb Mode	The performance management function may provide a menu for manual or automatic selection of the climb speed schedule.
44	AC 25-15	5f(4)(i)	One-Engine-Inoperative Drift down	The performance management function may control the drift down with one engine inoperative.
45	AC 25-15	5f(5)	Descent Mode	The computer should provide logic that would prevent the power being automatically reduced below safe levels or the minimum required when descending with ice protection systems operating per AFM instructions.
46	AC 25-15	5f(6)	Landing Approach	The system should provide commands that result in an indicated airspeed of V_{ref} + wind speed for all landing flap settings. This speed should be maintained even if the critical engine fails during stabilized approach.
47	AC 25-15	5f(7)	Landing Approach Go-Around	1. The airplane performance following addition of go-around thrust, during the landing approach go-around maneuver, with or without the simultaneous loss of an engine, should be such that the indicated airspeed is not reduced below what existed upon initiation of the maneuver.
				2. The automatic and manual go-around procedures and speeds should be the same.
48	AC 25-15	5f(8)	Performance Information Display	1. The performance management primary flight modes selected by the flight crew should be displayed in the pilot's primary field of view.
				2. Airplane performance information that was processed or computed by the FMS should be readily retrievable for display by the control display unit during any phase of the flight.
49	AC 25-15	5g(1)	ATS Take-off Mode	1. The ATS design should provide an auto throttle clamp (inhibit) speed during takeoff.
				2. Adequate annunciation should be provided for the system failure condition where the throttles either fails to clamp at the takeoff inhibit speed or the throttles become unclamped during the takeoff mode below 400 ft AGL.
				3. Verify there is no throttle stagger that may result in unacceptable thrust differential after throttle clamp.
50	AC 25-15	5g(2)	ATS Climb Mode	The ATS logic (operate in either speed or thrust control modes) should provide flight envelope speed protection based upon airplane configuration.
51	AC 25-15	5g(3)	ATS Cruise Mode	The system should be free of excessive throttle oscillations/hunting and it should be compatible with the autopilot pitch axis relative to speed on pitch/speed on throttles. Engine over boost, and flight envelope speed protection should be provided.
52	AC 25-15	5g(4)	ATS Descent Mode	The ATS logic (operate in either speed or thrust control modes) should provide flight envelope speed protection based upon airplane configuration.

#	Source	Ref	Topic	Description
53	AC 25-15	5g(5)/5g(6)	ATS Landing Approach Mode & Go-Around mode	1. The ATS (both at speed or angle of attack control modes) should be designed to control speed to within +5 knots of the target speed but not less than the higher of V_{ref} or $1.3V_s$ of the landing configuration. 2. In the FMS go-around mode, the ATS should operate in the thrust control mode. The system should automatically transition from the speed mode to the thrust control mode upon selection of the FMS go-around mode.
54	AC 25-15	5h	Takeoff performance monitor	The FMS may provide the function of computing and displaying achieved airplane performance information to the flight crew in real time.
55	AC 25-15	5h(1)	System accuracy	System tolerances will affect error in takeoff distance, airspeed and longitudinal acceleration.
56	AC 25-15	5h(2)	Runway considerations	If the time to a specific takeoff speed is computed and displayed, that speed should be sufficiently below decision speed V1 to allow the flight crew adequate time to decide if the difference between actual and computed performance is acceptable. The FAA feels that approximately 5 sec is sufficient for this purpose.
57	AC 25-15	5h(3)	System threshold	The system should alert the flight crew when the achieved performance is less than the reference performance by 5%.
58	AC 25-15	5h(4)	Takeoff thrust	The implementation of the takeoff monitoring function should consider the manner in which the flight crew will set thrust at the beginning of the takeoff roll. The thrust used for the AFM should be used for this purpose.
59	AC 25-15	5h(5)	Airmass analysis	The applicant should perform air mass analysis to show the effects of a typical runway gust condition such as 10 knots/gust to 20 knots to show the suitability of the concepts used.
60	AC 25-15	5h(6)	Annunciation and display	A real time digital or analog display of achieved and reference performance should be provided in the flight crew's primary field of view.
61	AC 25-15	5h(7)	Function integrity	Annunciated system failure conditions are considered major in accordance with AC 25.1309 since the display of the computed data no matter how it is presented will influence the flight crew's decision to continue the takeoff roll or initiate a rejected takeoff.
62	AC 25-15	5i	Fuel State	The accuracy and resolution of the displayed fuel state computations should be equal to or better than that provided by the flight deck instrumentation.
63	AC 25-15	5j	Flight deck checklist	The FMS may provide additional features such as takeoff, landing and emergency flight deck checklists.
64	AC 25-15	5k	Flight management system data link	Airborne data links can be used by the FMS to provide a medium for the automated acquisition of flight plan navigation and airplane performance related data.
65	AC 25-15	5l	Software identification	If the FMS or its sensors are designed to display a software program identifier to the flight crew, the program identifier should be displayed by the FMS control display unit and the AFM should indicate the approved identifier.

(Continues)

TABLE 6.51 ▪ *(Continued)*

No.	Requirements AC	Paragraph	Subject	Description
66	AC 25-15	5m	Equipment installation	An equipment installation failure modes and effects analysis should be provided, the extent of which is dependent upon the degree of integration of the FMS with new or existing airplane systems and sensors.
67	AC 25-15	5n	Test and evaluation	
68	AC 25-15	5n(1)	Environmental Tests	The major components comprising the FMS should be qualified to the appropriate sections of RTCA document DO-160B, "Environmental Conditions and Test Procedures for Airborne Equipment," or equivalent.
69	AC 25-15	5n(2)	Ground Tests	The applicant should provide a ground test plan that includes the tests necessary to verify that the FMS as installed in the airplane performs its intended function and that there are no adverse effects to existing airplane systems and sensors.
70	AC 25-15	5n(3)i	Reevaluation	Flight test evaluations should be made to determine that prior approvals of existing airplane systems have not been compromised.
71	AC 25-15	5n(3)iii	Navigation systems employing single or multisensory inputs.	If approval is sought for certain sub-(reversionary) operating modes utilizing single or a reduced 24 combination of sensor inputs, the evaluation must include an appropriate number of data points for each configuration for which the approval is sought.
72	AC 25-15	5n(3)iv	Reference to the flight deck	The data obtained from the flight demonstration should be referenced to the flight deck instrumentation for airplane configuration and airspeed that the flight crew normally uses to determine airplane performance when operating the airplane manually.
73	AC 25-15	5n(3)vi	Take off engine out	The evaluation should include a sufficient number of simulated engines out conditions for the takeoff mode to demonstrate that the AFM engine out speed is smoothly captured and maintained.
74	AC 25-15	5n(3)vii	Engine out Electrical load	At least one actual engine shutdown should be accomplished to evaluate bus or generator load switching effects on the FMS.
75	AC 25-15	5n(3)viii	Buffet Boundary Protection	An FMS controlled buffet boundary protection margin should be demonstrated for all applicable flight modes by commanding a relatively large speed change at a point close to the boundary
76	AC 25-15	5n(3)xii	Fuel computation	An evaluation of the fuel state computations should be made for reasonableness when compared with flight planning data.
77	AC 25-15	5o	Airplane flight manual supplement	The AFMS should provide the appropriate system limitations and a comprehensive description of all normal and submodes of system operation including what actions are expected by the Flight crew for each case.
78	AC 20-138C	6-3.d1	GNSS	RNAV multisensor equipment must provide a smooth guidance transition between baro-VNAV and GPS/SBAS or GPS/GBAS vertical guidance and a clear mode annunciation to the pilot
79	AC 20-138C	6-4.a(1)	DME/DME	1. Provide position update within 30 sec of tuning DME navigation facilities 2. Auto-tune multiple DME facilities 3. Provide continuous DME/DME position updating

#	Document	Ref	Topic	Description
80	AC 20-138C	6-5	VOR/DME	Multisensor incorporate VOR/DME meet the performance for the route to be flown
81	AC 20-138C	15-2	Interface to primary navigation display	Navigation data must be displayed on a lateral deviation display. The display of navigation parameters data should meet the requirements
82	AC 20-138C	15-2.a		The navigation system in use must be clearly and unambiguously indicated for installations where multisensor outputs can drive a display that is shared in common with other navigation equipment
83	AC 20-138C	15-2.b		When in terminal and approach modes, the bearing and distance to the active waypoint must be displayed with a minimum resolution of 1° and 0.1 NM up to a range of 99.9 NM.
84	AC 20-138C	15-2.c		Navigation data, including a TO/FROM indication and a failure indicator, must be displayed on a lateral deviation display such as a CDI or HIS.
85	AC 20-138C	15-2.1a	Nonnumeric deviation displays	Nonnumeric cross-track deviation must be continuously displayed in all navigation modes
86	AC 20-138C	15-2.1b(1)(2)		Nonnumeric deviation displays with a TO/FROM indication and a failure annunciation should have the following attributes: 1. Distance to the active waypoint must be displayed either in the pilot primary field of view or on a readily accessible display page unless there are alternate means to indicate waypoint passage. 2. Nonnumeric vertical deviation must be continuously displayed when in an approach mode containing vertical guidance
87	AC 20-138C	15-2.1c		The lateral deviation scaling must be consistent with any alerting and annunciation limits, if implemented.
88	AC 20-138C	15-2.1d		The lateral deviation scaling must also have a full-scale deflection suitable for the current phase of flight and must be based on the required total system accuracy
89	AC 20-138C	15-2.1e		The full-scale deflection value must be known or must be available for display to the pilot commensurate with enroute or terminal values
90	AC 20-138C	15-2.1f		The lateral deviation display must be automatically slaved to the RNAV computed path. The course selector to the deviation display should be automatically slewed to the RNAV computed path or the pilot must adjust the Omni bearing selector (OBS) or HSI selected course to the computed desired track.
91		15-2.2	Map displays	Map displays should give equivalent functionality to a nonnumeric lateral deviation display described above, and be readily visible to the pilot with appropriate map scales (scaling may be set manually by the pilot.)

(Continues)

TABLE 6.51 ■ (Continued)

No.	Requirements AC Paragraph		Subject	Description
92	AC 20-138C	15-3a	Interface to Remote Annunciator	Multisensor equipment may drive remote annunciators in the primary field of view.
93	AC 20-138C	15-3b		Visual annunciations must be consistent with the criticality of the annunciation and must be readable under all normal cockpit illumination conditions. Visual annunciations must not be so bright or startling as to reduce pilot dark adaptation.
94	AC 20-138C	15-3c		Waypoint sequencing, start of a turn, turn anticipation, TO/FROM indication, approach mode annunciation, and automatic mode switching should be located within the pilot's primary field of view or on a readily accessible display page.
95	AC 20-138C	15-3d		The multisensor equipment should provide annunciations for loss of integrity monitoring, loss of navigation, TO/FROM indication, and approach mode annunciation within the primary field of view.
96	AC 20-138C	15-4	Unique Software Considerations	
97	AC 20-138C	21-1	Ground test	
98	AC 20-138C	21-2.a	EMC test	Evaluation to determine satisfactory EMC between the multisensor equipment installation and other onboard equipment
99	AC 20-138C	21-2.b	Multisensor accuracy	Validate multisensor equipment navigation accuracy in each operating mode.
100	AC 20-138C	21-2.c	VOR/DME DME/DME certification	Initial certification for systems including a VOR/DME or multiple (scanning) DME sensor that has not been previously certified must be based upon a demonstration of system accuracy by recording the VOR/DME or DME/DME sensor position and comparing it to the actual position during evaluation flights.
101	AC 20-138C	21-2.d	Evaluation of INS/IRU	
102	AC 20-138C	21-2.e	Data continuity during the normal maneuvers	Verify navigation data continuity during normal aircraft maneuvering for the navigation modes to be validated (e.g., bank angles of up to 30° and pitch angles associated with approaches, missed approaches and departures).
103	AC 20-138C	21-2.f	Normal operation of FMS	Verify the overall operation of the multisensor equipment
104	AC 20-138C	21-2.g	FTE	Verify that FTE does not exceed 1.0 NM for enroute, 0.5 NM for approach transition and missed approach operating modes, and 0.25 NM for nonprecision approach, both with and without autopilot or FD use, as applicable.
105	AC 20-138C	21-2.h	Barometric input	If the equipment uses barometric input, verify that the equipment properly interprets the barometer reading.
106	AC 20-138C	21-2.i	Switching and transfer functions	Verify/assess all switching and transfer functions, including electrical bus switching, pertaining to the multisensor installation.

107	AC 20-138C	21-2.j	Failure modes and associated annunciations	Various failure modes and associated annunciations, such as loss of electrical power and loss of signal reception. Verify that a warning associated with loss of navigation is accompanied by a visible indication within the pilot's primary field of view.
108	AC 20-138C	21-2.1a	Steering response FD/AP coupled with FMS	Evaluate steering response while the FD or autopilot is coupled to the multisensor equipment during a variety of different track and mode changes.
109	AC 20-138C	21-2.1b	Fly by turns	Several fly-by turns should be accomplished with varying wind conditions for FD and autopilot. Verify the equipment accomplishes the turn as a fly-by waypoint and discourages overshoot.
110	AC 20-138C	21-2.1c	Annunciation of impending waypoint crossing	Verify that the lateral maneuver anticipation supplied by the multisensor equipment is appropriate for the aircraft type. Verify that if the multisensor equipment is coupled to an autopilot or on approach, an appropriate annunciation of impending waypoint crossing is provided.
111	AC 20-138C	21-2.1d	Direct-To function	Verify that execution of the Direct-to function with a resultant aircraft heading change does not overshoot and cause S turns that exceed the lateral path tracking FTE requirements.
112	AC 20-138C	21-2.1e	Autopilot response to a multisensor	Evaluate the autopilot response to a multisensor equipment fault by pulling the circuit breaker for the equipment. This test should be done in each of the sensor modes, if applicable.
113	AC 20-138C	21-2.2	GNSS precision approach	For manual control to the approach flight path, the appropriate flight displays must provide sufficient information without excessive reference to other cockpit displays, to enable a suitably trained pilot to maintain the approach path, make alignment with the runway or go-around.
114	ANM 99-2		Human factors	

▋ 6.18 ▋ | ENHANCED VISION SYSTEMS (EVS)

The generic term for systems that utilize sensors to enhance the vision of the operator is enhanced vision systems. All EVS Systems share a commonality in providing Situational Awareness to the operator. Some of the benefits of EVS systems:

- Aid in CFIT avoidance in conjunction with TAWS
- Runway incursion detection
- In-flight and on-ground safety enhancements
- Helicopter heavy-lift operations
- Helicopter EMS services
- Helicopter oil rig servicing
- Helicopter desert landing dust blow up

Enhanced Flight Vision System (EFVS) is a subset of the EVS, however, EASA uses the term EVS as an equivalent to the FAA EFVS (which is a little confusing). EFVS systems use active or passive sensors to see along the flight path and may be used as an aid in nonprecision approaches and meet the requirements of 14 CFR § 91.175. They are used to provide enhanced flight visibility while flying a straight-in standard approach procedure. The EFVS provides enhanced flight visibility at the Decision Height or Minimum Descent Altitude in lieu of flight visibility by using a HUD to display sensor imagery of the approach lights or other visual references for the runway environment at a distance no less than the visibility prescribed in the instrument approach procedure being used. Simply stated, the system provides enhanced visibility to the pilot.

An EVS provides a display of the forward external scene through the use of imaging sensors such as FLIR, MMW RADAR or low light image intensification. An EVS is different than an Enhanced Flight Vision System in that it provides Situational Awareness, but does not allow CFR 14 § 91.175 (l) Operational credit. EFVS systems are integrated with an FGS and provide additional symbology and steering cues to the pilot via the HUD.

All EVS systems, including EFVS comprise the following elements:

- Sensor system
- Sensor display processor
- EVS display
- Pilot interface

EVS sensors may be passive or active. Passive sensors utilize the IR spectrum; within the IR band there are four areas of interest for the designers of these systems:

- 0.72–1.5 microns for image intensifying devices (NIR, night vision systems, night vision goggles)
- 1–2 microns for detecting light bulb filaments (SWIR, not including LED)
- 3–5 microns for cultural returns at greater range in high moisture (MWIR)
- 8–12 microns for maximum cultural definition (LWIR)

Systems designed as an aid to the pilot can use a combination of the NIR, SWIR, MWIR or LWIR frequency bands. EVS may also use an active sensor; millimetric wave radar

(MMWR) may be used in lieu of or in addition to an Infra-red (IR) imaging device, MMW RADARs are less susceptible to attenuation (scattering and absorption) than IR systems and therefore have superior performance in weather penetration.

As opposed to the use of sensors, the synthetic vision system (SVS) uses a computer-generated image of the external scene from the perspective of the cockpit. The image is derived from aircraft attitude, precise positional accuracy and a database of terrain, obstacles and relevant cultural features. SVS provides situational awareness but does not provide operational credit (lower minimum approaches). The SVS may be presented on a primary flight display from the perspective of the flight deck or on a secondary display as a moving map; the AC calls these views egocentric and exocentric. The elements of the SVS:

- Display
- System interface
- Terrain and obstacle database
- Position source
- Altitude source
- Attitude source
- Heading and track source

The terrain database is normally referenced to a NIMA (national imagery and mapping agency) Digital Terrain Elevation Data (DTED) model. There are three DTED levels of varying resolutions:

- DTED Level 0 uses 30 arc second spacing (nominally 1 km)
- DTED Level 1 uses 3 arc second spacing (approx 100 m)
- DTED Level 2 uses 1 arc second spacing (approx 30 m)

When evaluating an SVS, some characteristics must be addressed. Display factors such as conformality, refresh rate, FOR and FOV (field-of-regard and field-of-view) and display placement such as primary FOV versus secondary FOV, Heads-up versus Heads-down are critical. The method of terrain presentation such as depiction, grid, color and shading, resolution and range indications must be evaluated. System modes of operation and functionality such as alerts (terrain, obstacles, traffic and airspace), declutter modes, mode reversion and image control are called out as high interest items. The integration of guidance symbology (if incorporated) such as guidance cues, pathway markings, navigation information, crosswinds and unusual attitude recovery must also be evaluated.

A combined vision system (CVS) involves a combination of synthetic and enhanced systems. An example of CVS may be a database driven images combined with real-time sensor images superimposed and correlated on the same display. The initial portion of an approach may be provided by SVS with sensors providing the imagery as the aircraft is nearer to the runway. CVS will provide situational awareness but not necessarily operational credit. The AC usually considers CVS as a combination of EVS and SVS; if the applicant combines SVS and EFVS he is expected to meet the applicable performance criteria for SVS and EFVS.

6.18.1 EFVS

The FAA defines enhanced flight visibility as the average forward horizontal distance, from the cockpit of an aircraft in-flight, at which prominent topographical objects may be clearly distinguished identified in day or night by a pilot using an EFVS. This enhanced visibility allows the pilot to continue the approach from the DH or MDA down to 100 ft above the touchdown zone. At this point and below, certain things would have to be visible to the pilot without EFVS to proceed to a landing on the intended runway.

The use with an EFVS with a HUD improves the level of safety by improving situational awareness; it provides visual cues to maintain a stabilized approach and minimizes missed approach situations. Additionally, an EFVS allows the pilot to observe obstructions on runways, such as aircraft or vehicles earlier in the approach and during ground operations in reduced visibility conditions. Command guidance, path deviation indications, flight path vector and flight path reference cue must be displayed on the HUD. In order to perform its intended function, the flight path reference cue needs to be set by the pilot to the desired value for the approach. The pilot needs to see the cue in the context of pitch scale to verify that it's correctly set, and it needs to be shaped and located so as to allow the pilot to monitor the airplane's vertical path (glide path anchored to TD).

The EFVS imagery and external scene topography must be presented so that they are aligned with and scaled to the external view. When using an EFVS, the approach light system (if installed), or the runway threshold (lights or markings), and the runway touchdown zone (lights or markings) must be distinctly visible and identifiable to the pilot before descending below the DH or MDA for the pilot to continue the approach. The FAA requires that several components be provided by an EFVS to provide an adequate level of safety. The EFVS sensor imagery must be presented on a HUD that is centrally located in the pilot's primary field-of-view and in the pilot's line of vision along the flight path. The imagery must be real time, independent of the navigation solution derived from the aircraft avionics, and clearly displayed so that it does not adversely obscure the pilot FOV through the cockpit window. Aircraft flight symbology that must be provided on the HUD includes the following:

- Airspeed
- Vertical speed
- Attitude
- Heading
- Altitude
- Command guidance
- CDI
- Reference cue

Cues that are referenced to the imagery must be aligned with and scaled to the external view.

The all weather window system was the first system to be certified for lower landing minimums by the FAA. Gulfstream is the A/C integrator using a Kollsman IR sensor and

Honeywell HUD. The sensor is a staring array, cryogenically cooled InSb sensor optimized in the 1–5 micron region. It has been certified for Gulfstream IV/V as well as the G300-G550 series models and by FedEX for their entire fleet of MD-10/11 and A300/310 A/C.

6.18.2 EVS Airworthiness

AC 20-167, "Airworthiness Approval of Enhanced Vision System, Synthetic Vision System, Combined Vision System and Enhanced Flight Vision System Equipment," June 22, 2010, provides a means, but not the only means, of showing compliance with airworthiness regulations. EFVS approaches are approved by a change to the operating rules:

- Part 91 para 91.175 (General Flight Rules)
- Part 121 para 121.651 (Domestic Flag Carriers)
- Part 125 para 125.381 (More Than 6000# and 20 Passengers)
- Part 135 para 135.225 (Commuter and On-Demand Carriers)

The AC covers equipment installed in Part 23/25/27/29 aircraft. As with all of the newer advisory circulars, AC 20-167 provides the applicant with a compliance matrix, flight test considerations, Safety assessment and human factors requirements. The AC is arranged as follows:

- Intended function
- General requirements
- Specific requirements
- Installation considerations
- HIRF
- Performance demonstration
- Environmental qualification
- Compliance matrix
- Systems safety
- Flight test considerations (includes human factors)
- Sample AFM supplement

Intended Function

For EVS, SVS or CVS, the airworthiness package should contain the intended functions of the system: what features are displayed and the criticality of pilot decision making using the display. Any intended additional functions (i.e., terrain alerting) should be defined according to AC 25-11A, AC 20-175 and other category unique ACs in accordance with 14 CFR § 2X.1301. For EFVS, the functions must be clearly defined: its use to visually acquire the references to operate below the MDA/MDH or DA/DH as described in CFR 14 § 91.175 and the criticality of the pilot decision making using the display.

General Requirements

EVS, SVS or CVS functionality can be superimposed on the EFIS System (i.e., the PFD) as installed in the flight deck; the information may also be superimposed on a HUD or equivalent display system. EVS, SVS or CVS may also be selected on an MFD or navigation display as one of the stand-alone formats on the flight deck. If the applicant would like to display this functionality on side-mounted displays (electronic flight bag, or EFB) they should follow the guidance in AC 120-76A. Utilization of the EVS display should not be used during ground operations if sensor proximity to the taxiway surface causes a distraction.

EVS/SVS/CVS systems are required to:

- Have a means to control the brightness
- Not degrade the presentation of essential flight information
- Meet the requirements of the original approval
- Not adversely affect any other installed system
- Perform its intended function in all conditions for which approval is being desired
- Display the mode of operation to the flight crew
- Not have any undesirable characteristics

If the EVS/SVS/CVS is implemented on the primary flight display (PFD) the following requirements must also be met:

- The image, or loss of image may not adversely affect the PFD functionality
- The imagery should be aligned with the aircraft's flight path and not be confusing to the aircrew
- The FOR may be variable but cannot distract the aircrew
- The display must not impede a visible zero pitch reference line (the AC recommends the incorporation of a velocity vector)
- The imagery may not provide the pilot with misleading information

The EVS/SVS/CVS may be implemented on a HUD, or equivalent. If this implementation is chosen then the system must not conflict with the pilot's compartment view (for Part 25 airplanes, the FAA will issue special conditions to achieve the intended level of safety in § 25.773). It must also not impede the pilot performing any task during all phases of flight. The imagery must be conformal to the real world, and the design of the HUD should be IAW Aerospace Recommended Practices. The EVS/SVS/CVS may be implemented on a secondary display. If so, the orientation of the imagery must be clear to the pilot and if PFD information is displayed, it should meet PFD integrity and availability requirements. The imagery, or loss thereof, must not adversely affect other approved display functionality.

Specific Requirements

EVS installations must meet the following requirements:

- The imagery must be de-selectable; if implemented on a HUD the switch must be on the yoke or thrust control
- Status of the EVS must be obvious to the crew

- FOR should be sufficient for the intended functions
- SAE design standards should be incorporated
- The display should not have any undesirable characteristics

The AC states that displays for Part 25 should apply the characteristics listed in AC 25-11A, Part 23 should use AC 23-1311-1B; other aircraft (Part 27/29) should also use these ACs as recommended guidance. On an HDD, the FOR should be suitable for the pilot to transition from heads-down to out the window or HUD. The image must be refreshed at 15 Hz or better, and the latency cannot exceed 100 msec. On a HUD, a control for brightness/contrast must be available. The presentation should not be distracting or impair vision, not mask hazards, and should not degrade task performance or safety.

For SVS implementations, the design and installation safety levels should be appropriate for the system's intended function; ACs 25-11A and 23.1311-1 should be used as guidance. Failure or de-selection of the imagery should obvious to the crew, and the FOR should be appropriate. All dominant topographical features should be present and identifiable by the crew.

For terrain presentations, a potential obstacle should be obvious and not conflict with TAWS or HTAWS. Topographical features must not intersect approach guidance, and terrain that generates an alert should be displayed above the artificial horizon zero-pitch line. SVS based primary displays should be clear and unambiguous when unusual attitude recovery is required. The FOR should be appropriate for use, and water should be displayed and distinguishable from the sky. An update rate of 15 Hz or better is required and the pilot's view must not be depicted below the earth's surface.

The scene range should be the natural horizon, except in systems used for departures and approaches where it should be the lesser of the following: the natural horizon, 40 nm or 10 min at maximum cruise. The scene range must not be misleading (i.e., no mountains in the distance); obstacles taller than 200 ft must be displayed conformably. The horizontal position source for the SVS should at least meet the criteria for TAWS or HTAWS. Any aircraft incorporating an egocentric perspective must also provide a TAWS, HTAWS, or terrain warning system.

Additionally, for SVS implementation:

- The attitude, heading and track sources cannot conflict with information on the navigation display
- Resolution and accuracy of the database must comply with that contained in TSO-C151c or TSO-C194
- Runway elevations must be accurate at the approach end, not just the center

CVS requires a real-time imaging sensor and display that provides demonstrated vision performance for its intended functions; CVS must meet the respective requirements of EVS and SVS. EVS and SVS must be conformal to each other and the data fusion requires the alignment to be within 5 mrad laterally and vertically at the boresight of the display. Image discrepancies between SVS and EVS due to failures must be obvious to the crew.

EFVS General and Specific Requirements

In addition to the EVS requirements, EFVS design:

- Should mitigate system failures more frequent than extremely improbable, which produce unsafe conditions

- Requires a HUD or equivalent display
- FOR should be sufficient for the system to operate in all anticipated flight, configuration and environmental conditions
- Should display the required flight instrument data
- Include brightness/contrast control, which is effective with dynamically changing background conditions
- Controls must be visible and accessible to the pilot from any normal seated position

The controls must be adequately illuminated in all lighting conditions, and not create objectionable glare. Unless the applicant can show that fixed illumination is satisfactory under all lighting conditions for which approval is sought, there must be a readily accessible control to immediately deactivate or reactivate the EFVS image on the HUD. This control must be on demand without the pilot removing his hands from the yoke or thrust control.

The AC devotes some time to the minimum detection range required of the sensor. The detection range will be a function of the sensor: detector element(s), Detector Angular Subtense and transmissivity of the atmosphere on a particular day. The discussion in the AC regarding performance is rather simplistic although it does mention utilizing the LOWTRAN and MODTRAN models.

The EFVS image must be compatible with the FOV and head motion box of a HUD designed against SAE ARP 5288 (Transport Category HUD Systems). The EFVS display criteria must meet the airworthiness certification requirements in 14 CFR parts 23/25/27/29; these requirements are identified by part/paragraph number in appendix I of AC 20-167. The guidelines for the HUD displays apply to EFVS: MIL-STD-1787C, AC 25-11A, SAE AS 8055, SAE ARP 5288, and SAE ARP 5287.

The EFVS image and installation:

- Must be able to perform its intended function
- Must allow for the accurate identification of visual references
- Must not degrade safety of flight
- Must not have unacceptable display characteristics
- Must have a control of brightness
- Must have an accessible control for removing the imagery
- Must not degrade the presentation of the flight information
- Must not be misleading or cause confusion
- Must conform to the external scene
- Must not distort the compartment view
- Must not cause fatigue
- Must not significantly alter the color of the external view
- Must allow the pilot to recognize a nonconformal view

Additionally, the EFVS should have a latency not to exceed 100 msec (latency may exceed 100 msec if it demonstrated not to be misleading). The minimum fixed FOR must be 20° horizontal and 15° vertical; in applications where the FOR is centered about the velocity vector, the minimum vertical FOR must be 5° (+/−2.5°). A variable FOR is allowed using a slewable sensor if centered about the velocity vector with a minimum FOV of 5° (+/−2.5°).

Installation Requirements

For EVS/SVS/CVS, in addition to those already noted:

- BIT is required
- A complete fault hazard analysis and systems safety analysis (SSA) should be conducted to identify failure modes and classify the hazard levels
- Any failure or malfunction that could cause misleading imagery should be immediately annunciated and the imagery removed
- EVS/SVS/CVS operation may not adversely affect other aircraft systems and vice versa
- The system must meet the environmental conditions specified in RTACA DO-160G to include HIRF and EMI testing

All requirements previously mentioned for EVS/SVS/CVS are applicable to EFVS as well. Any mode of EFVS operation must be annunciated on the flight deck and visible to the crew; the modes of EFVS operation must be made available to the flight data recorder as required. During installation, it is required to validate all alleviating flight actions that are considered in the safety analysis for incorporation in the AFM limitations section or for inclusion in type-specific training. The applicant must demonstrate a Design Assurance Level for hardware and software to be no less than that required for non-EFVS precision and nonprecision approaches with decision altitudes of 200' or above. A sample EFVS systems safety requirements guidance is provided in appendix 2 of AC 20-167; an example of an FHA is given in Table 4.

Performance Evaluation

The performance evaluation will require bench testing, flight testing, data collection, and reduction to show the proposed performance criteria can be met. Minimum performance standards require an evaluation of the system used during anticipated operational scenarios (i.e., taxi, takeoff, approach, missed approach, failure and crosswind conditions and approaches into specific airports appropriate for the system's intended function). For EFVS, the applicant must demonstrate performance at the lateral and vertical limits for the type of approach credit being sought. Appendix 4 of the AC provides sample EFVS Flight Test considerations.

No specific test procedures are cited; system performance tests as they relate to operational capability are the most important tests. The applicant should specify the individual verification methods to be used in the Certification Plan; this should be confirmed with the appropriate certification office as an acceptable course before evaluation begins. Human Factors testing needs to be accomplished; a sample evaluation matrix is found in section 4-15 of appendix 4 in the AC. The requirements for design assurance levels for hardware (RTCA/DO-254) and software (RTCA/DO-178B) should be followed; appendix 5 of the AC contains a sample AFM supplement.

6.19 | SUMMARY

After completing this section, it should be easy to understand why the certification authority grants 3 to 5 years to complete the certification process. There are some items to remember when dealing with civil certifications. The first is that, when dealing with avionics systems, the critical subparagraphs are 1301 and 1309: Can the system perform its intended function, and can it do so without adversely impacting other aircraft systems? In attempting to answer these questions, the evaluator will first be led to the design of the system. The key to avionics design is software, and the overriding guidance on software is RTCA DO-178B/C, "Software Considerations in Airborne Systems and Equipment Certification." More recently, and not really called out in the previous discussions, is guidance on avionics hardware design. The overriding documentation in this area is RTCA DO-254, "Design Assurance Guidance for Airborne Electronic Hardware." These two documents, along with RTCA DO-160G, "Environmental Conditions and Test Procedures for Airborne Aircraft," are going to be referenced in every avionics system civil certification.

And how do we comply with the intent of these documents? That is where the best practices of the SAE documents come in. Software, by its very nature, is highly complex and integrated, and therefore we turn to SAE ARP-4754, "Certification Considerations for Highly Integrated or Complex Aircraft Systems." This document states the need for a safety assessment, and that assessment is described in SAE ARP-4761, "Guidelines and Methods for Conducting the Safety Assessment Process on Civil Airborne Systems and Equipment."

When dealing with individual avionics systems, the chain of command is the MPS or MOPS for the equipment (usually an RTCA document), followed by the TSO (for the equipment manufacturer), and lastly the AC (a means, but not the only means, of showing airworthiness compliance). If done right, all three of these documents should basically say the same thing, but due to the published date and changes in the technology, each document may addend the previous one.

These documents, together with the human factors guidance offered by the authority, serve as the basis of your certification and flight test plan. As with any flight test program, it is incumbent upon the evaluator to do her homework prior to designing a plan.

Only deal with military applications and think that this will never apply to you? Think again. The U.S. Department of Defense is continually dropping MIL-STDs and adopting the civilian versions (like RTCA DO-178B). In addition, military aircraft routinely fly in civilian airspace and must abide by the same rules (read certification) as civilian aircraft do.

▋ 6.20 ▋ | SELECTED QUESTIONS FOR CHAPTER 6

1. What system, when installed, can achieve integrity within the GPS?
2. What role does mode S play in the TCAS II system?
3. Can TCAS II provide horizontal resolutions?
4. Briefly describe the RAIM predictive algorithm.
5. What is the major deficiency of the basic ground proximity warning system?
6. What is the major deficiency of the enhanced ground proximity warning system?
7. Name the four factors that must be considered when using a system as a navigation reference.

8. What is TAS? What can it be used for?

9. Why are numerous pilots required for controls and displays evaluations?

10. What is RNAV? What does RNAV-5 mean?

11. What is the difference between a major hardware change and a minor hardware change according to FAA/JAA regulations?

12. What are the two types of software changes according to FAA/EASA regulations? Who makes this determination?

13. In dealing with civil certifications, what does MOPS mean?

14. Which agency controls the TSO process?

15. What are the advantages of installing an FMS in an aircraft?

16. Why is EMI/EMC an important consideration in TCAS installation?

17. Is the military exempt from RVSM requirements?

18. What is the purpose of RVSM?

19. Would you employ a TAWS in a military aircraft? Why or why not?

20. What is flight technical error?

21. What is the minimum horizontal accuracy required for IFR GPS installations?

22. What is the minimum vertical accuracy required for IFR GPS installations?

23. How many major equipment classes of GPS are called out in the specifications (TSO C-129A)?

24. What would be a possible alternative to the TCAS II system?

25. What problems are associated with faulty altitude encoders in regard to TCAS advisories?

26. What is ASE in the RVSM approval process?

27. What are the three acceptable methods of determining SSEs in the RVSM approval process?

28. Why is human factors a consideration when installing a GPS/FMS/EFIS in an aircraft?

29. What is the standard color scheme for warnings and alerts in the cockpit?

30. How does TCAS compute a closest point of approach?

31. Why is waviness of the aircraft skin a consideration when applying for RVSM certification?

32. Briefly describe a means of determining altitude loss during an approach when experiencing an autopilot malfunction.

33. What are the military standards for altitude and heading hold during autopilot operations?

34. What are the five categories of ILS?

35. How does the aircraft determine glideslope on an ILS approach?

36. List five test objectives for a possible ILS installation.

37. What is the difference between ATC and ATM?

38. What is the specified accuracy of the FMS during enroute and terminal operations?

39. List four possible test objectives for a possible FMS installation.

40. Are there any significant drawbacks (disadvantages) to cockpit automation?

41. What must be checked immediately upon FMS initialization?

42. What are performance charts used for in FMS installations?

43. Are operators allowed to access and change the FMS database?

44. What can the basic TCAS provide to the aircrew?

45. What are the two types of TCAS advisories?

46. Are there any concerns about how aural tones are implemented in an avionics system?

47. At what frequency does the TCAS operate?

48. Can TCAS be saturated?

49. What information is displayed to the aircrew by TCAS?

50. TCAS II, version 7, reduces the TA from 1200 to 850 ft. Why?

51. What does the VSI or VVI provide the aircrew in the TCAS mode?

52. List four potential problems with the current TCAS.

53. What is the potential problem with the TA aural tone in the cockpit?

54. Briefly describe how the GPWS operates.

55. What are the seven alert modes of the GPWS system?

56. What are the four additional modes of the EGPWS system?

57. What is CFIT? Can it be eliminated by TAWS?

58. Can the TAWS alerts be overridden?

59. What is a display override?

60. What is an aerospace recommended practice?

61. In civil certifications, what is a highly integrated system? Give an example.

62. What altitude bands are incorporated in RVSM?

63. What are some of the altitude errors that need to be considered under ASE?

64. What is an AC (FAA or EASA regulations)?

65. What is SSE? What three variables contribute to SSE?

66. What is the difference between the basic envelope and full envelope under RVSM?

67. What is the difference between group and nongroup aircraft under RVSM?

68. What errors contribute to the total height keeping accuracy as defined by RVSM?

69. Is RVSM approval, once granted, always in force? What would rescind an RVSM approval?

70. Why is the aircraft structural repair manual a part of the RVSM submittal?

71. Are avionics controls and displays standardized? Why?

72. Which lighting tests must be conducted when evaluating controls and displays?

73. What is meant by "positive feedback"? Why is it important?

74. In a controls and displays functional test, what is being evaluated?

75. What are the two acceptable means of informing the operator that the data are no longer acceptable?

76. What are some of the drawbacks in allowing engineers to develop controls and displays?

77. What three things need to be recorded during controls and displays evaluations?

78. What is the relationship between criticality and probability of failure?

79. What are digital-only displays not used in cockpits?

80. Describe one method for quantifying display usability.

81. What does an RNAV 0.5 approach mean to you?

82. Each major class of GPS equipment (A, B, C) is further subdivided. What are these subdivisions?

83. What is the basic premise in paragraph 1301 and 1309 of the FARS/EASA (Part 23/25)?

84. What is meant by maneuver anticipation for GPS evaluations?

85. When shall GPS units certificated for VFR/IFR flight provide a navigation warning flag?

86. When shall the GPS provide an integrity alarm (certificated equipment)?

87. Which documents are the overriding authorities for software development? Environmental conditions?

88. What is a RAIM-equivalent system?

89. What is a predeparture RAIM program?

90. What is FDE in a RAIM system?

91. What is a pseudorange step error in a RAIM system?

92. What is the primary reason for installing an autopilot in an aircraft?

93. What is an autoland approach?

94. Why are disconnects (auto or manual) required in autopilot systems? Give three reasons for these safety devices.

95. Name three ways that an autopilot may be disconnected

96. In evaluating autopilot malfunctions, a 1 sec delay is typically used under what conditions?

97. A 3 sec time delay for autopilot failure modes is used in which flight conditions?

98. ILS and MLS provide the aircrew with what type of information?

99. What is the significance of the deviation dots for the autopilot localizer?

100. What is the significance of the deviation dots for the autopilot glideslope?

101. You are attempting to certify a transport category helicopter. What would be your basis for certification?

102. What is a development assurance level?

103. What impact would a new software delivery to the display processors have on your controls and displays subsystem?

104. You are tasked with evaluating a new FMS installation. What instrumentation would you require?

105. FAA regulations have eliminated the requirements for verifying the accuracy of a GPS. Is this a sound judgment?

106. The FMS you are evaluating displays present position to the nearest 0.1 min. The navigation function must provide 0.124 nm horizontal accuracy. How do you perform this test?

107. The system that you are evaluating digitally displays EGT to the nearest 1°. You have collected five data points (display versus truth) with values of 0.1, 0.2, 0, 0.1, and 0. What is the accuracy of the display?

108. The system that you are evaluating digitally displays bearing to the nearest degree. You have collected six data points with values of $-2°$, $0°$, $1°$, $2°$, and $-1°$. What is the accuracy of the system?

109. What is the relationship between assurance levels and probability of failure?

110. The result of a potential problem would cause a loss of aircraft and loss of life. What is the assurance level and what probability can we live with?

111. What does RNP 0.3 mean?

112. Do antennas have to be matched with GPS/GNSS equipment?

113. Are there any cases where the MOPS have been amended by the TSO?

114. What is an airworthiness directive?

115. Who can ask for an exemption to FAA requirements?

116. What is a notice of proposed rulemaking? Why is it done?

117. What is the difference between a TC and an STC?

118. What is an equivalent level of safety?

119. You are attempting to certify a normal, utility aircraft. What is the basis for certification?

120. Where would you go to find information on a human factors certification plan?

121. Does *optional* equipment need to be evaluated? If yes, to what degree?

122. What are the basic requirements to allow oceanic and remote operations with the GPS?

123. Can it be assumed that all TSO equipment meets the MOPS?

124. Where would you find a flight test guide for transport category aircraft?

125. Where would you find a flight test guide for small airplanes?

126. What is LNAV? VNAV?

127. Where would you find requirements for workload assessment for Part 25 aircraft?

128. What is workload assessment used for in civil certifications?

129. Are all FMS systems built the same? Are they all intuitive?

130. Why does TCAS II change the TA threshold for higher altitudes?

131. Are GPS and SATCOM compatible?

132. Can a system depend on pilot memory for correct operation?

133. Where would I find EMI/EMC test considerations for avionics systems installations?

134. What is the difference between a fly-by and flyover waypoint?

135. Give two examples of a flyover waypoint.

136. What is a predictive RAIM program? When is it exercised?

137. What is barometric-aided RAIM? What is it good for?

138. Are civil certifications an easy process?

Electro-optical and Infrared Systems

Chapter Outline

7.0	Introduction	535
7.1	Infrared History	536
7.2	IR Radiation Fundamentals	547
7.3	IR Sources	552
7.4	The Thermal Process	554
7.5	Atmospheric Propagation of Radiation	556
7.6	Target Signatures	562
7.7	EO Components and Performance Requirements	567
7.8	Passive EO Devices	581
7.9	Laser Systems	588
7.10	Passive EO Flight Test Evaluations	594
7.11	Active EO Systems	614
7.12	Selected Questions for Chapter 7	618

7.0 | INTRODUCTION

This chapter deals with the flight testing of electro-optical (EO) systems (e.g., day television [TV], image intensification [I^2] systems, etc.) and infrared (IR) systems (e.g., forward-looking infrared [FLIR], IR line scanners, etc.). There have been vast improvements in both of these types of systems as a result of miniaturization, production techniques, and processing technology since their introduction to the military world in the 1960s. As accuracies in detection and identification improve, the method of testing these systems becomes more exact. As with radar testing, the evaluator must be cognizant of the target environment and how changes in this environment alter the results of testing. There are many texts that explain the basics of EO systems. I would recommend reading *Electro-Optical Imaging System Performance, 5th Edition*, by Gerald C. Holst (Bellingham, WA: International Society for Optical Engineering, 2008). This text does a very good job of explaining FLIR models and the environmental factors that affect system performance. Other texts that may prove helpful to the evaluator are *Introduction to Sensors for Ranging and Imaging*, by Graham M. Booker (Raleigh, NC: SciTech Publishing, 2009), and *Introduction to Infrared and Electro-Optical Systems*, by Ronald G. Driggers, Paul Cox, and Timothy Edwards (Norwood, MA: Artech House, 1999).

The Advisory Group for Aerospace Research and Development (AGARD), an advisory group for the North Atlantic Treaty Organization (NATO), published the *Avionics Flight Test Guide* (March 1996), which may be of use for military testing. AGARD has since been renamed the Research and Technology Organization (RTO). John Minor, formerly of the U.S. Air Force's test pilot school published and presented an excellent paper at the Society of Flight Test Engineers 33rd Annual International Symposium, August 19–22, 2002, entitled "Flight Test and Evaluation of Electro-Optical Sensor Systems," which is a good companion to the AGARD text.

7.1 | INFRARED HISTORY

It is hard to believe that IR studies began as early as 1800, when Sir William Hershel used a prism and thermometer to show an increase in temperature below the red light in the visible spectrum. This increase in temperature, and thus energy, below the visible red light was dubbed infrared, or below red (or sometimes beyond red, depending on how you are viewing an electromagnetic spectrum chart). It was also demonstrated that there was a relationship between temperature and wavelength, and scientists spent the next hundred years or so trying to define this relationship. By 1830 it was recognized that IR radiation emitted by all bodies whose temperature was above absolute zero is thermal radiation. Gustav Kirchhoff initiated the studies of radiation in 1859 by showing the proportionality of absorption and emission. Steady progress was made through 1900 by the scientists whose names are near and dear to engineers everywhere: Stefan, Boltzmann, Wien, Rayleigh, and Planck.

7.1.1 Applications

R. D. Hudson listed more than 100 separate applications for thermal imaging systems in 1969. He identified four major categories: military, industrial, medical, and scientific. These major categories were further subdivided into six subcategories: search, track and ranging, radiometry, spectroradiometry, thermal imaging, reflected flux, and cooperative sources. In the test world, two broad categories can be identified—military and civilian—although the boundaries between the two are becoming blurred. Since the mission requirements of military and civilian systems differ, one can infer that their performance will also differ. The design of the system will depend on the application. Design parameters must take into account atmospheric transmittance, available optics, and detector spectral response. For example, the design of a system to detect a missile plume is very different from one that is designed to detect people or vehicles involved in illegal poaching.

Due to atmospheric spectral transmittance, electronic imaging system design is partitioned into seven spectral regions:

- Ultraviolet (UV) region (optical sensors for missile warning systems)
- Visible spectrum (new-generation night vision goggle [NVG] systems)
- Near-IR imaging region (most NVG systems, laser)
- Short-wavelength IR imaging band (laser systems, FLIR, IR line scanner, missile seekers)

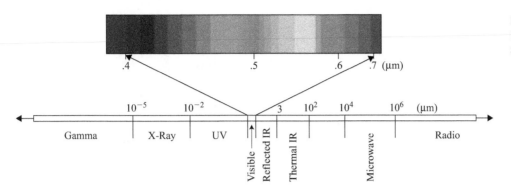

FIGURE 7.1 ▪
Electromagnetic
Spectrum

- Mid-wavelength IR imaging band (civilian thermal imagers, FLIR, newer focal plane arrays)
- Long-wavelength IR imaging band (FLIR)
- Very-long-wavelength IR imaging band

Terminologies among the UV, visible, and IR communities are further segregated and will be addressed later. Figure 7.1 shows the area of interest for electronic imaging systems within the electromagnetic spectrum.

The EO part of the spectrum is broken out in relation to wavelength, expressed in micrometers (μm) or microns:

- Infrared   0.72 μm to 1000 μm
  - Very far IR   15 μm to 1000 μm (sometimes called extreme IR)
  - Far IR   6 μm to 15 μm
  - Middle IR   1.5 μm to 6 μm (reflected IR $<$ 3 μm)
  - Near IR   0.72 μm to 1.5 μm
- Visible   0.39 μm to 0.72 μm
- Ultraviolet   0.01 μm to 0.39 μm

Figure 7.2 shows a more detailed relationship between wavelength and the EO spectrum.

Military and civilian EO avionics systems attempt to exploit the best and worst attributes in each of the spectral wavelengths. Near-IR systems were developed to assist with normal night vision. Common names for such systems are night vision devices (NVDs) and night vision goggles (NVGs). There are a variety of these types of systems. They may be scopes (as applied to rifles) or monocular visual devices (used in driver or dismounted soldier applications) or binocular systems (mostly used in aviation). These systems greatly enhance night operations, but they do not turn night into day. There are limitations to these systems, and when used in aviation, these limitations must be well understood. Failure to know and heed these limitations can have disastrous results.

Night vision goggles are devices that make an object more visible at low light levels. Often called image intensifiers, these devices make use of available ambient light, either visible or near-IR energy that is reflected off the object, and amplifies the light. The performance of these systems is directly related to the amount of available ambient light. The measure of light is the lux, or in some cases foot-candles, which is a measure of illumination (1 lux = 0.0929 ft-candles). For example, direct sunlight

FIGURE 7.2 ■
EO Wavelength
Relationship

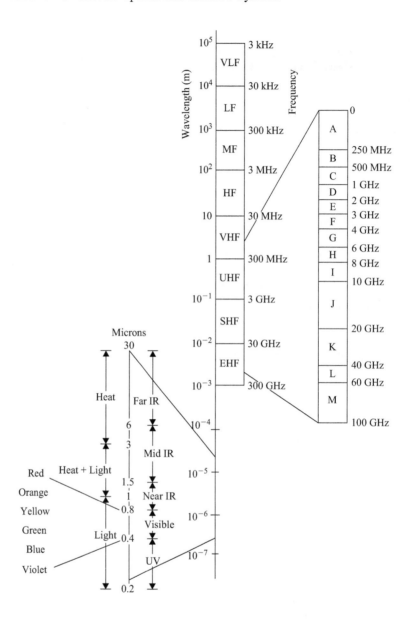

illumination is approximately 100,000 lux. The illumination of the full moon on a clear night is approximately 0.1 lux. Common practice when discussing NVGs is to use millilux (mlux), which is equal to 0.001 lux. Table 7.1 shows common illumination conditions.

There have been many advances in NVG technology since the first military applications in the 1960s; each major advancement is marked by a new generation of systems. Current technology is defined as generation IV, which uses on-chip processing and is sometimes called "filmless" technology, which will be discussed later.

Generation I (Gen I) systems were introduced as aids for military riflemen in the 1960s. The common name for such systems was the star scope or starlight scope (Figure 7.3). They were rather large (more than 17 inches in length), very heavy, and required considerable power to operate. The amplification was based on electron

TABLE 7.1 ▪ Illumination Levels

Human Vision	Lux	Foot-candles (approximate)	Condition
Photopic (Good Acuity, Color)	10^5	10^4	Sun or Snow
	10^4	10^3	Full Daylight
	10^3	10^2	Overcast Daylight
	10^2	10^1	Very Poor Daylight
	10^1	10^0	Twilight
	10^0	10^{-1}	Deep Twilight
Scotopic (Poor Acuity, No Color)	10^{-1}	10^{-2}	Full Moon
	10^{-2}	10^{-3}	Quarter Moon
	10^{-3}	10^{-4}	Starlight
	10^{-4}	10^{-5}	Overcast Starlight

FIGURE 7.3 ▪ Generation I Starlight Scope

acceleration, and the light amplification was on the order of 1000 times ambient lighting. While the amplification would be considered exceptionally good, even by today's standards, the excessive size, weight, and power requirements rendered Gen I NVDs unacceptable for aviation use.

Generation II systems were introduced in the late 1960s and used electron multiplication via a microchannel plate (MCP). The gain was increased to approximately 20,000 times ambient lighting and improvements reduced the size, weight, and power requirements. The U.S. Army AN/PVS-5 NVG system employed a full face mask and the user could not wear eyeglasses (Figure 7.4). This system was modified (AN/AVS-5A) for use in aircraft.

Generation III systems improved on MCP performance by using an aluminum oxide coating. Spectral response was increased by switching to a gallium arsenide (GaAs) photocathode. These systems have a gain of approximately 30,000 to 50,000 times ambient lighting and are used extensively in ground and airborne applications (Figure 7.5). Common applications of the Gen III systems are CATS EYES (U.K.) and Aviator Night Vision Imaging Systems (ANVIS) (U.S.). The latest U.S. iteration is the AN/AVS-9 or ANVIS 9.

The difference in performance of the three generations is illustrated in the Figure 7.6 comparison.

Thermal imaging systems (exploiting middle and far IR emissions) developed in much the same way as the near-IR and visible light systems. These thermal imagers (TIs)

FIGURE 7.4 ▪
Generation II
AN/PVS-5

FIGURE 7.5 ▪
Generation III
Aviation System

may also be categorized by "generations," with current technology identified as third generation. Each generation is categorized by the type of scan, electronics and processing, and wavelength of interest. Most first-generation TIs use a bidirectional scan with a linear array comprising 60, 120, or 180 vertical elements and employ a 2:1 interlace, much like a TV tube. They use analog electronics and filtering and are optimized for a single waveband. Examples of first-generation TIs are the common module and EOMUX systems.

The first TI systems took advantage of the thermal emissions in the far-IR region; the detectors were optimized for the 8 to 12 μm range. The common module FLIR (Figure 7.7) is an example of a first-generation TI which allowed a user to directly view the intensity sensed by individual detector elements. Because of multiple optical paths, there were significant signal losses, resulting in poor picture quality.

FIGURE 7.6 ▪
Generation I, II, and
III Comparison

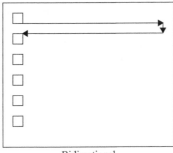

Analog Electronics
Amplified and Filtered

Single Wave Band

Bidirectional
Linear Array

FIGURE 7.7 ▪
Common Module
FLIR

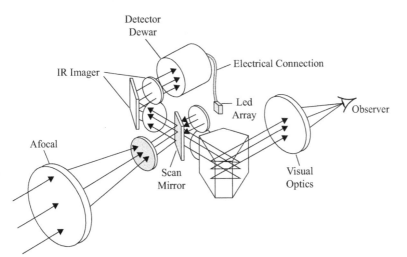

FIGURE 7.8 ■
EO MUX FLIR

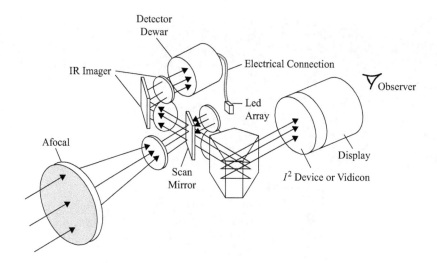

The initial advance in this first-generation TI device was the ability to put the optical signal into a TV format. The system was still analog and the picture was obtained by passing the optical signal through either a vidicon (optical image scanner) or employing image intensifier technology. The basic system was still the same, and losses from the multiple optical paths were still present. The quality of the display was better, and allowed for multiple observers, but the additional equipment increased the size and weight of the system. Figure 7.8 shows the improved common module system, now identified as an EO MUX FLIR.

The next logical step in TI advancement was the addition of the electrical signal processor, or the digitization of the optical signal. In this system, the detector directly feeds an electrical signal processor. The FLIR image is then put into a standard TV format. This architecture provides a greatly improved signal due to a reduction in signal losses. By eliminating unnecessary equipment, this system is relatively small and much lighter. These improvements are known as an EMUX FLIR. Figure 7.9 illustrates these improvements.

The EMUX FLIR is classified as second generation and is different from first-generation TIs in that it employs a unidirectional scan usually with a 4×480 detector array (4 horizontal detectors by 480 vertical detectors). Since every pixel in the vertical on the display is matched to a vertical detector in the array, the optics need only be scanned in the horizontal. Since there are four detectors in the horizontal, time delay integration (TDI) is used, which improves the signal-to-noise ratio (SNR) and reduces blurring. TDI is a method of scanning in which a frame transfer device produces a continuous video image of a moving object by means of a stack of linear arrays aligned with the object to be imaged in such a way that as the image moves from one line to the next, the captured image moves along with it. It is also used in radar as well as charge-coupled devices (CCDs) (to integrate more light from the scene).

Most of the iterations of FLIR technology for first-generation systems incorporated flat mirrors for scanning. These limitations were overcome with second-generation technology. Scanning systems that use flat mirrors have relatively slow maximum scanning rates. At the higher scanning rates, the mirrors tend to bend because of inertia, causing image distortions. By using a rotating, multifaceted mirror, scan rates can be increased without suffering a loss of resolution. Since the FLIR image is presented in a

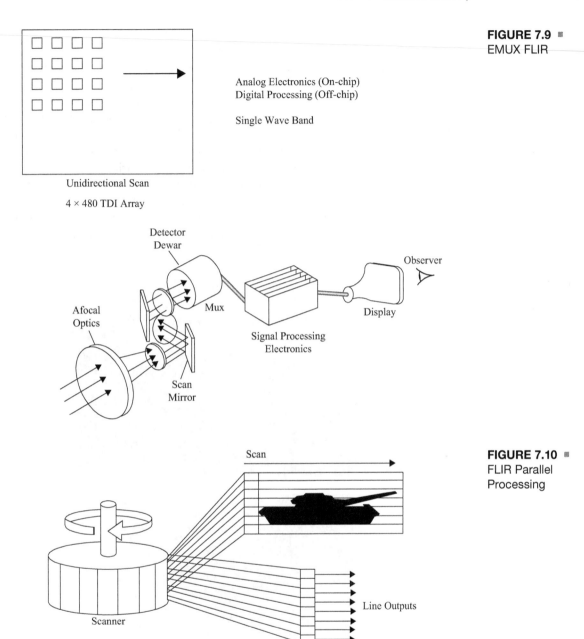

FIGURE 7.9 ■
EMUX FLIR

Analog Electronics (On-chip)
Digital Processing (Off-chip)

Single Wave Band

Unidirectional Scan

4 × 480 TDI Array

Detector
Dewar

Observer

Afocal
Optics

Mux

Display

Signal Processing
Electronics

Scan
Mirror

Scan

FIGURE 7.10 ■
FLIR Parallel
Processing

Line Outputs

Scanner

standard TV format, a picture resolution of 480 × 640 lines is needed (for U.S. systems). A series of 60, 120, or 180 detector elements have to be scanned throughout this dimension by adjusting the vertical tilt of the scanning mirror or by employing a secondary nodding mirror to account for all of the horizontal lines. By increasing the number of detectors to 480 (vertical), a single horizontal scan will populate the entire frame (Figure 7.10). An array of 480 × 640 detectors would not have to scan, but rather stare, in the direction of interest.

Up until the late 1990s, U.S. fighter forces relied on the low-altitude navigation and targeting infrared for night (LANTIRN) pod for low-altitude navigation and weapons delivery. As the Cold War ended and improved "smart" weapons entered the inventory, the need developed for a TI sensor that could be used for these weapons in high-altitude strikes. The LANTIRN, as a first-generation IR device, was ineffective for this role. The U.S. Navy launched an advanced targeting FLIR (AT-FLIR) program for its F-18s, while the U.S. Air Force launched its own program, known as the advanced targeting pod (ATP). The outcome of these development programs was the third-generation FLIR and CCD camera targeting system. Pods that are in use include the Sniper, Litening, and Terminator systems. Figure 7.11 shows the Litening and Sniper targeting pods.

The third-generation TI devices use a focal plane staring array with indium antimonide (InSb), with a peak response in the 3 to 5 μm region, or mercury cadmium telluride (HgCdTe), also called merc-cad or MCT, which has a peak response near 12 μm. The detectors are arranged in a mosaic with common arrays of 240×320 and 480×640, but are available up to 2048×2048. An array of 480×640 would have a

FIGURE 7.11 ■
Sniper (top) and
Litening (bottom)
Targeting Pods

detector for each pixel on a standard display, whereas in a smaller array, such as a 240×320, the scene would have to be shifted on successive looks to fill in the picture (much like a raster scan on a radar). This scanning process is called a dither or micro-scan. Because the image must be scanned across the array, there will be an overall reduction in the scene dwell time that results in a loss of sensitivity and increases the possibility of smearing.

Another improvement in the third-generation TIs is the incorporation of digital processing on the same chip as the detector. The savings on weight and required real estate realized by this technology are evident. Digital visible imagers, such as cameras and video recorders, may use a CCD or a complementary metal oxide semiconductor (CMOS) (more on these later) as the detector element. These detectors are fabricated from silicon but are only sensitive in the visible and near-IR spectra. The InSb or MCT detectors need some sort of analog-to-digital (A/D) converter to transmit the voltage, resistance, or charge of each of the detectors to the measurement circuitry. This is accomplished by the incorporation of a readout integrated circuit (ROIC), which mechanically and electrically interfaces the individual detector outputs in the focal plane array (FPA) with the external digital electronics. The ROIC transforms the relatively small electrical output of the detector into a relatively large, measurable output voltage.

The ROIC is typically fabricated in silicon using the same processes as CMOS and then is either hybridized (hybrid mosaic) or bonded to the detector array; the resultant assembly is called the FPA. The FPA may also be constructed as one combined (monolithic) array, as shown in Figure 7.12.

Detector Array

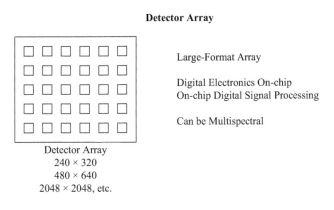

Large-Format Array

Digital Electronics On-chip
On-chip Digital Signal Processing

Can be Multispectral

Detector Array
240×320
480×640
2048×2048, etc.

FIGURE 7.12 ■
Focal Plane Array

Focal Plane Array

Monolithic

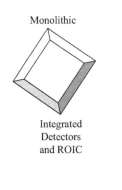

Integrated
Detectors
and ROIC

Hybrid

ROIC Chip

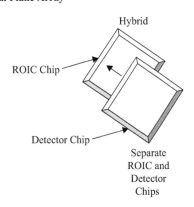

Detector Chip

Separate
ROIC and
Detector
Chips

The FPA may also have the capability of being multispectral, which is sometimes called two-color. This capability offers the benefits of both the mid-IR and long-IR spectra. The FPA does have some disadvantages; probably the greatest being the tendency of the electrical responses of the individual detector elements to be nonuniform. When the array is at rest (not receiving any thermal energy), each of the elements has a different baseline or zero-signal level (i.e., there are offsets to each of the elements). This also occurs when the element is radiated and the delta in the signal is also different. If uncorrected, it would make the array unpractical. Fortunately these deltas can be measured in controlled conditions and then corrections can be made in the software or even directly in the ROIC. The process is much like an instrumentation calibration done for strain gauges on a test aircraft. The arrays are more difficult to fabricate than the scanning arrays, which increases the cost, and they still need to be cooled down to eliminate thermal noise.

In contrast to passive TV systems, which use reflected energy (sunlight, moonlight, or starlight), passive IR depends on emitted energy or emittance; ambient light does not directly affect the detection of this energy. IR detection sensors have become an integral element of modern avionic systems because of their ability to provide information that complements and improves the effectiveness of the overall system. The IR system also has some advantages over other airborne sensors. First, the sensor is independent of artificial or natural illumination. Second, the sensor is virtually impervious to most camouflage techniques and the classification of targets can be made on the basis of their respective radiation characteristics.

Each IR system is designed to perform a specific task, and its components are chosen to optimize system performance for a particular wavelength region, maximum detection, high resolution, etc., depending on the type of source. Regardless of the task, all IR systems include the following components:

- Optics to collect IR emissions
- Detectors to convert radiant energy into electrical signals
- Electronics and processors to amplify the signals
- Displays to permit the operator to see the information

7.1.2 IR Terminology

Infrared emissions have many properties that should be understood prior to beginning any evaluation. The first thing that should be addressed is radiometric terminology so as to alleviate any confusion over discussion on performance or characteristics. Table 7.2 is a summary of the basic radiometric terminology, including quantities, definitions, units, and symbols.

As can be readily seen from the terms in Table 7.2, the words commonly end in *ance* or *ivity*. Those terms that end in *ance* represent a term that describes the property of a specific sample. Those terms that end in *ivity* represent a term that describes the property of the generic material. If we want to describe a process, then the terms would end in *ion*, and we would have emission, absorption, and reflection.

TABLE 7.2 ▪ Radiometric Terminology

Symbol	Term	Meaning	Unit
Q	Radiant energy	Energy	joule (J)
Φ	Radiant flux	Radiant energy per unit time, also radiant power	watt (W)
M	Radiant emittance/exitance	Power emitted from a surface	Wm^{-2}
I	Radiant intensity	Power per unit solid angle	Wsr^{-1}
L	Radiance	Power per unit solid angle per unit projected source area	$Wm^{-2}\,sr^{-1}$
E,I	Irradiance	Power incident on a surface	Wm^{-2}
ε	Emissivity	Ratio of radiant emittance of source to that of a blackbody at the same temperature	None
α	Absorptance	Ratio of absorbed radiant flux to incident radiant flux	None
ρ	Reflectance	Ratio of reflected radiant flux to incident radiant flux	None
τ	Transmittance	Ratio of transmitted radiant flux to incident radiant flux	None
E_λ	Spectral irradiance	Irradiance per unit wavelength interval at a particular wavelength	$Wm^{-2}\,\mu m^{-1}$
L_λ	Spectral radiance	Radiance per unit wavelength	$Wm^{-2}\,sr^{-1}\,\mu m^{-1}$

NOTE 1: Terms qualified by "spectral" and symbols subscripted with λ indicate a quantity specified per unit wavelength interval.
NOTE 2: The processes of absorption, reflection (including scattering), and transmission account for all incident radiation in any particular situation; the ratio of total incident flux is equal to one; i.e., $\alpha + \rho + \tau = 1$ (total incident radiation).

7.2 | IR RADIATION FUNDAMENTALS

As mentioned throughout the text, it is not the author's intention to derive mathematical equations. Equations are used where they can help the reader to understand a particular concept, or if the text would suffer if the equations were not included. Temperature is the basic concept, however, in EO, several temperatures must be defined. An object at a particular temperature does not necessarily radiate that temperature.

Emissivity is generally discussed in terms of how much IR energy an object radiates as compared to that energy radiated by a blackbody. A blackbody is defined as an object that absorbs all IR energy incident upon it. If part of the energy incident to the object is reflected rather than absorbed, the object becomes a graybody (the term graybody does not refer to the color perceived by the eye). Kirchhoff's law shows that a good absorber is also a good emitter and that the emissivity (ε) of a surface can be determined as follows:

$$\varepsilon = \frac{\text{Total radiant emittance of a graybody}}{\text{Total radiant emittance of a blackbody}}, \tag{7.1}$$

with both bodies being at the same temperature.

The emissivity factor of a blackbody is equal to one since all IR energy incident upon the object is absorbed and reradiated. If part of the energy is reflected, the object becomes a graybody and the emissivity factor becomes less than one. The more energy that is reflected, the grayer the body becomes and the lower the emissivity factor becomes. The emissivity factor is, therefore, an indication of the grayness of the body. Different types of surfaces have different emissivity factors. Typically a dull, dark surface will absorb and reradiate most of the energy incident upon it and will have a high emissivity factor, while a bright shiny surface will reflect much of the energy and will have a low emissivity factor. Suppose that two identical aircraft are parked on the

FIGURE 7.13 ▪
Radiator Response
Curves

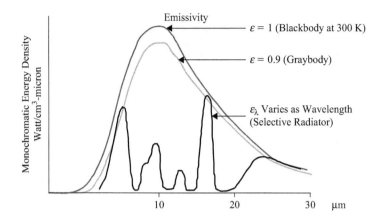

TABLE 7.3 ▪ Emissivity of Common Materials

Material	ε
Highly Polished Silver	0.02
Highly Polished Aluminum	0.08
Polished Copper	0.15
Aluminum Paint	0.55
Polished Brass	0.60
Oxidized Steel	0.70
Bronze Paint	0.80
Gypsum	0.90
Rough Red Brick	0.93
Green or Gray Paint	0.95
Water	0.96

taxiway and are allowed to be solar heated throughout the day. Also suppose that one aircraft is painted blue and the other is burnished aluminum. If we were to view these two aircraft through an IR imaging device late in the day we would see quite different pictures. The aircraft that is painted blue will show up as a "hot" object, whereas the shiny aircraft will show up as a "cool" object. This is because the painted aircraft is a good emitter and a poor reflector. The polished surface is a good reflector, but a poor emitter. Figure 7.13 shows the radiator response curves for a blackbody, a graybody, and a selective radiator, and Table 7.3 lists the ε for some common materials.

The Stefan–Boltzmann law states that every object in the universe is constantly receiving and emitting thermal radiation from every other object. The amount of radiation emitted is a function of its temperature, absorption, or emittance and its surfaces. The Stefan–Boltzmann law is expressed by equation 7.2:

$$M = \varepsilon \sigma T^4, \tag{7.2}$$

where

M = rate of emission per unit area (W/cm^2),

ε = emissivity of the radiating surface,

σ = Stefan–Boltzmann constant = $5.670\,373 \times 10^{-12}$ W cm^{-2}S^{-1}K^{-4}, and

T = absolute temperature (K).

FIGURE 7.14 ■
Wien's
Displacement Law

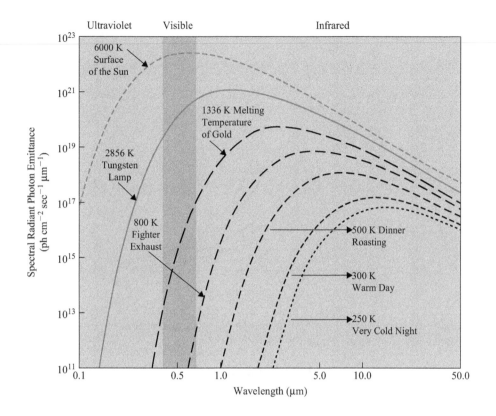

So if an object has an emissivity of 0.6 and its temperature is 300 K, what is the rate of emission? *Answer: 0.027 W/cm².*

Wien's displacement law states that maximum radiation occurs at specific frequencies that are temperature dependent. The plot of spectral distribution of energy indicates that the wavelength of peak radiation decreases as temperature increases. Peak wavelength and temperature are related in Wien's displacement law. This law states that the wavelength (λ) at which M_λ is multiplied by the absolute temperature (in K) of the blackbody is equal to a constant.

$$\lambda_m T = 2898 \ \mu mK, \tag{7.3}$$

where

T = absolute temperature (K), and

λ_m = wavelength of maximum energy (μm),

or as an approximation:

$$\lambda_m = 3000 \ (\mu mK)/T(K). \tag{7.4}$$

Figure 7.14 illustrates Wien's displacement law.

Planck's law defines the spectral radiation from a blackbody. This law states that the spectral exitance is a function of wavelength and absolute temperature only.

FIGURE 7.15 ■
Planck's Law

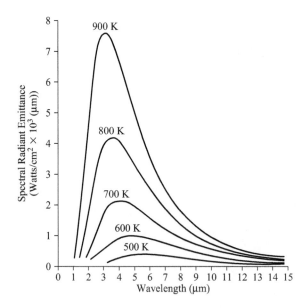

The relationship between the radiation intensity, spectral distribution, and temperature of a blackbody is given by equation 7.5:

$$M_\lambda = \frac{C_1}{\lambda^5} \left[\frac{1}{e^{(C_2/\lambda T)} - 1} \right],$$ (7.5)

where

M_λ = radiation emitted by the blackbody per unit surface area per unit wavelength (in W/cm^2/unit wavelength),

T = absolute temperature of the blackbody (K),

λ = wavelength of emitted radiation, and

e = base of natural logarithms = 2.718.

C_1 and C_2 are constants; their values are dependent on the unit of wavelength used. If λ is in meters, then

$C_1 = 3.7418 \times 10^8$ W-μm^4/m^2 and

$C_2 = 1.4388$ μmK.

The effect of a change in temperature may be graphically illustrated by plotting radiated energy versus wavelength at specific blackbody temperatures (Figure 7.15).

Infrared radiation has the same transmission properties as all other electromagnetic radiation. It will radiate outward from the source in all directions. The power available at a receiver in a nonatmospheric case (i.e., transmission is not affected by the atmospheric medium) will vary inversely as the square of the distance to the source. This is known as the inverse square law. This law states that the intensity of radiation emitted from a point source varies as the inverse square of the distance between the source and a receiver. The equation relating power and distance from the source is as follows:

$$M_{\text{receiver}} = \sigma T^4 / 4\pi d^2.$$ (7.6)

Thus it can be seen that doubling the distance (d) will cause the power available at the receiver (M) to decrease by a factor of 2^2, or 4. Tripling the distance will decrease power

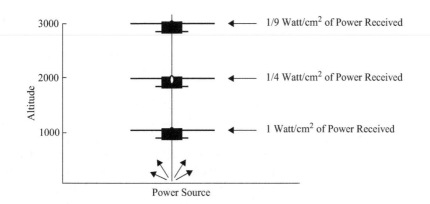

FIGURE 7.16 ▪
Power Available at
the Receiver

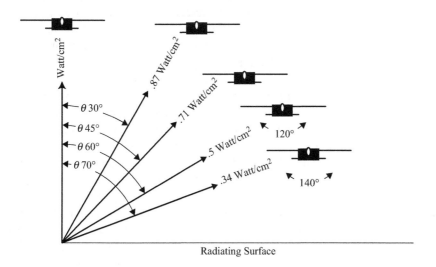

FIGURE 7.17 ▪
Effect of Viewing
Angle on Power
Received

by a factor of 3^2, or 9. This is graphically illustrated in Figure 7.16, and once again assumes that the transmission is not affected by the atmosphere. Conversely, the total radiated energy is a function of the fourth power of temperature and is represented by the area under the exponential curve for specific temperature. The value of the peak radiation is found to be a function of the fifth power of temperature. Therefore doubling the temperature of a blackbody will increase total radiation 16-fold and peak radiation 32-fold.

Another factor influencing the amount of power available at the receiver is the angle at which the radiating source is viewed. The amount of power is a function of the cosine of the angle from which the surface is viewed (θ). This is known as Lambert's law of cosines and is graphically illustrated in Figure 7.17. Lambert's law of cosines is expressed in equation 7.7:

$$I = \frac{MA}{2\pi d^2} \cos \theta, \qquad (7.7)$$

where

  $I =$ radiant intensity received at the detector,

  $M =$ radiant emittance, and

  $A =$ area of the source.

7.3 | IR SOURCES

It has been noted that all bodies with temperatures above absolute zero radiate energy, and the brief physics tutorial in the previous section cannot predict the type of radiation that the body will radiate or reflect. IR sources can be either controlled or natural. Controlled sources may be a target of interest or a calibration device. The temperature, size, aspect, etc., of the object is known. Natural sources are those that contribute to the background. They can be terrestrial, atmospheric, or celestial. Terrestrial sources can be either man-made or those that occur naturally. Some examples of man-made natural sources are road material, paints, oil slicks, and construction material. Naturally occurring emitters and reflectors are water, sand, rock, metals, and foliage.

Determining the true reflectance or emittance of natural sources can be very difficult. This is because natural sources are rarely uniform: planar, homogeneous, or without losses. In many cases, we try to determine radiation by looking at surface and bulk reflectance. Surface reflectance is that radiation reflected from the surface without penetration. Bulk reflectance is the part of the radiation refracted at the surface and transmitted into the body and scattered randomly, with a portion returning to the surface. At certain incident radiation frequencies, the surface reflectance becomes maximum and bulk reflectance approaches zero. These regions are called reststrahlen bands and are due to minerals in rocks, sands, and soils. Reststrahlen bands populate the 8 to 14 μm region.

Another determinant for natural radiation is the effect of wet and dry stacking. This type of reflectance applies to a commonly occurring condition in nature in which a rough diffuse reflecting material is covered by a thin sheet of smooth dielectric such as water. In this case, the surface reflectance is based on the specular reflectance of the dielectric. The bulk reflectance is less than that for the dry material alone. This is commonly seen against wet soil, even though the film is not planar. Figure 7.18 is from Holst (*Electro-Optical Imaging System Performance*) and shows the differences in reflectance of wet and dry soils taken from four different areas of the United States.

Similar to wet and dry stacking is layering. This is formed by foliage and canopy cover. Many vegetative canopies tend to exhibit a layered structure, with particular botanical features occupying the top, middle, and lower layers. Soil typically forms the lower boundary, and the top tends to provide the dominant influence on bidirectional reflectance. The reflectance of a canopy when viewed remotely will be the sum of directly illuminated components, shadowed components, and components indirectly illuminated by scattering.

Whenever the IR device looks above the horizon, the sky provides the background radiation. The prime IR sources in the sky are celestial and atmospheric sources. The radiation characteristics of celestial sources depend primarily on the source's temperature and the characteristics undergoing modification by the earth's atmosphere. In addition, the received radiation characteristically changes depending on the altitude of an observer.

The sun being an approximate blackbody radiator at a temperature of 6000 K has its radiant energy peak at 0.5 μm. Half of its radiant power occurs in the IR wavelengths, as shown in Figure 7.19, and the distribution of radiant power falls off as the wavelength increases.

The next most important celestial source of IR is the moon. The bulk of the energy received from the moon is solar radiation, modified by reflection from the lunar surface,

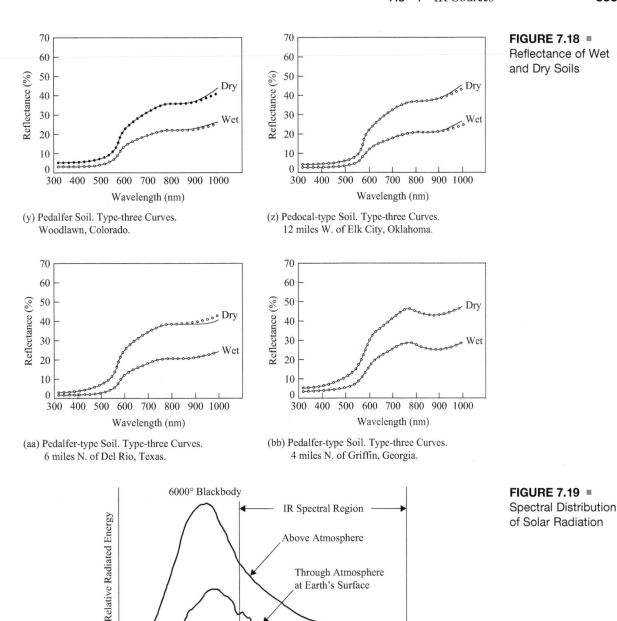

FIGURE 7.18 ■
Reflectance of Wet and Dry Soils

(y) Pedalfer Soil. Type-three Curves. Woodlawn, Colorado.

(z) Pedocal-type Soil. Type-three Curves. 12 miles W. of Elk City, Oklahoma.

(aa) Pedalfer-type Soil. Type-three Curves. 6 miles N. of Del Rio, Texas.

(bb) Pedalfer-type Soil. Type-three Curves. 4 miles N. of Griffin, Georgia.

FIGURE 7.19 ■
Spectral Distribution of Solar Radiation

slight absorption by any lunar atmosphere, and the earth's atmosphere. The moon is also a natural radiating source. During the lunar day its surface is heated to as high as 373 K, and during the lunar night the surface temperature falls to 120 K. Other celestial sources are very weak point sources of IR radiation, when compared to the sun and moon, and can be disregarded except in special astronomical applications.

Because the sun is such an intense source of radiation, the background of the sky has to be considered separately for day or night. The primary differences between night sky

FIGURE 7.20 ■
Spectral Energy
Distribution of
Background
Radiation From the
Sky

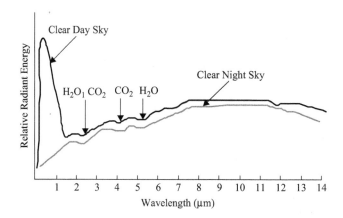

background radiation normal to the earth's surface are shown in Figure 7.20, which is a plot of the spectral distribution of energy for clear night and clear day skies. At night, the short-wavelength background radiation, caused by the scattering of sunlight, air molecules, dust, and other particles, disappears. In fact, at night there is a tendency for the earth's surface and the atmosphere to blend with a loss of the horizon, since both are at the same ambient temperature and both have approximately the same high emissivity factor. Radiation from the clear night sky approximates that of a blackbody at 273 K, with the peak intensity occurring at a wavelength of about 10.5 μm, and as Figure 7.20 illustrates, the overall energy level is slightly lower than that of a clear daytime sky.

One of the main constituents of the lower regions of the atmosphere is water vapor, and a special effect on background radiation occurs when the water vapor condenses into clouds. Clouds produce considerable variation in sky background, with the greatest effect occurring at wavelengths shorter than 3 μm. This is caused by solar radiation reflected from the cloud surfaces. Early IR homing missiles showed a greater affinity for cumulus clouds than the target aircraft. Discrimination from this background effect requires the use of not only spectral filtering, which eliminates the shorter wavelengths, but also spatial filtering, which distinguishes the smaller area of the target from the larger area of the cloud edge.

Terrain produces a higher background-energy distribution than that of the clear sky. This is caused by reflection of sunlight in the short-wavelength region and by natural thermal emission at the longer wavelengths. The absence of sharp thermal discontinuities and breaks in the radiation pattern between an object and its surroundings complicates the problem of the detection of terrestrial targets. While generally good pictures are obtained, occasionally objects with different emissivities and temperatures produce the same radiant emittance, and thus there is no contrast in the picture. This is known as thermal crossover and results in a loss of picture information.

7.4 | THE THERMAL PROCESS

In steady state, the heat flux (heat per unit area) that is incoming must equal the outgoing heat flux. From the previous section, remember that an object receives radiation from multiple sources: sunlight (direct and reflected), diffuse skylight, and deep space. These sources constitute the incoming heat flux. The outgoing heat flux

TABLE 7.4 ▪ Thermal Properties of Common Materials

Material	$k/c_h D$ (m²/sec)	k (W/m°C)	Diurnal Depth, d (m)
Stone concrete	4.8×10^{-7}	0.92	0.115
Granite	12.7×10^{-7}	1.9	0.187
Pine wood	0.7×10^{-7}	0.1	0.043
Lime stone	8.1×10^{-7}	0.7	0.149
Ice	11.2×10^{-7}	2.2	0.176
Damp soil	5.0×10^{-7}	2.6	0.116
Dry soil	3.1×10^{-7}	0.35	0.093
Building brick	4.4×10^{-7}	0.63	0.11
Cast iron	121.0×10^{-7}	57.0	1.73
Aluminum	860.0×10^{-7}	203.0	4.48

has four components: surface exitance, convective heat, conductive heat, and eva-poration. The surface exitance is always positive, while the other three depend on relative temperatures. All three vary with the surface temperature, which is the variable of interest.

The thermal emission of a source depends on the absolute temperature of the material and its spectral emissivity. The temperature of the emission need not be the surface or contact temperature; it might be the core temperature of an object. Thermal emission can be written as

$$T_R = \varepsilon^{\frac{1}{4}} T, \tag{7.8}$$

where

T_R = thermal emission,

ε = emissivity, and

T = temperature of emission.

For very thick material, the diurnal depth of heat penetration might be surprisingly small. With these small diurnal depths, the contact or surface temperature can be used as the temperature of emission. Table 7.4 shows the diurnal depth of some common materials.

Thermal inertia is a result of the delayed transfer of heat energy between the surface and the bulk interior. The maximum surface temperature occurs at some time after maximum income or sun loading. The maximum diurnal temperature will peak some 3 hr after solar noon. In the case of large bodies of water, the shortwave daylight is absorbed deep within the bulk of the body. In this case, there will only be a slight diurnal surface temperature change. This is because the surface temperature is relatively close to the bulk temperature. This is not true with a swimming pool, where surface temperature can be quite a bit higher than the bulk temperature during daylight heating. Soil moisture below the surface increases both heat capacity and conductive heat exchange, and will therefore exhibit greater thermal inertia than dry soil.

There are also special cases that can alter the thermal emission of a material. Buried heating pipes will obviously add additional heat to the surroundings. Volcanic regions are also causes for changes in surface temperature, largely due to convective heating. In sub-Arctic regions in winter (when the surface is covered by snow), the soil temperature below the snow is normally higher than the temperature of the snow itself. Differences in surface temperature will be observed based on differences in subsurface convective heating. The temperature of snow lying over a frozen large body of water will be slightly higher than the snow lying over soil due to the higher temperature of the subsurface water. Compressed snow from vehicles will have a higher temperature than noncompacted snow due to the increase in the heat transfer coefficient.

▌ 7.5 | ATMOSPHERIC PROPAGATION OF RADIATION

Electromagnetic radiation, when propagated through the atmosphere from the source to the receiver, is subjected to three modifying phenomena. These modifiers include

- A reduction in radiation intensity
- Nonscene path radiance scattered into the field of view (FOV), reducing the target contrast
- Forward small-angle scattering caused by aerosols and turbulence, reducing image fidelity

All three of these modifiers will be examined as to their exact nature and cause, and how they will affect the viewed image.

As previously noted, EO system performance is a function of the radiation temperature (T_R) of an object. Also recall that the radiation temperature of an object is related to either the core or surface temperature of the object in view and its emissivity (see equation 7.8).

The temperature or temperature differential measured at the aperture of the viewing system depends on the slant range and the atmospheric transmittance. Simply put, the radiation is going to be attenuated as it travels through the atmosphere. In an equation form, it looks like

$$\Delta T = \tau_{atm}{}^{R} \Delta T_R. \tag{7.9}$$

It is useful to identify this attenuation in terms of a coefficient of reduction of radiation. For IR systems, the most common term is called the coefficient of extinction. Extinction is the total reduction of radiation along the line of sight. The two contributors to this reduction are absorption and scattering. Absorption may be defined as energy being removed from the photon field, and scattering implies an altering of the radiation propagation. By using the Beer–Lambert law for transmittance (τ):

$$\tau_{atm}(\lambda) = e^{-\gamma(\lambda)R}, \tag{7.10}$$

where

$$R = \text{path length/slant range},$$
$$\tau_{atm}(\lambda) = \text{transmittance } f(\lambda), \text{ and}$$
$$\gamma(\lambda) = \text{spectral extinction coefficient } f(\lambda).$$

Because absorption and scattering are independent:

$$\gamma(\lambda) = \sigma(\lambda) + \kappa(\lambda), \tag{7.11}$$

where

$\sigma(\lambda)$ = scattering coefficient $f(\lambda)$, and

$\kappa(\lambda)$ = absorptive coefficient $f(\lambda)$.

For IR systems, an unusual coefficient has been adopted:

$$\tau_{atm}(\lambda) = e^{-\gamma(\lambda)}$$

or

$$[\tau_{atm}(\lambda)]^R = e^{-\gamma(\lambda)R}, \tag{7.12}$$

where typical values are

$\gamma(\lambda) = 0.001 - 0.4/km$

$\tau_{atm}(\lambda) = 0.65 - 0.95/km$

In transparent media, $\gamma = 0$, $\tau_{atm} = 1$.

The factors affecting absorption and scattering are many. They can be aerosols, pollutants, fog, rain, snow, water vapor, etc. High absolute humidity, for example, reduces transmittance by causing particulate growth or particle clusters. Water vapor concentration may range from a few tenths of a gram per cubic meter in desert areas to 40 or 50 g/m^3 in the tropics or near the sea surface. Unfortunately for the tester, extinction is not linearly related to vapor content. Some normalizations and assumptions must be made with any IR detection test. Since the IR spectrum contains wavelengths 5 to 20 times the visible spectrum, IR detectors are not susceptible to many of the particulates that may affect EO systems or normal vision. As particle sizes grow, the IR and EO systems will be affected equally.

As with any system, noise will affect the capability of a system to differentiate targets of interest. Atmospheric self-emission is independent of the source and is seen even if the source is not present. This noise is called path or background radiance. The magnitude of the background radiation varies with the direction of the observation, air density, location, time of day, and meteorological conditions. Path radiance reduces the SNR and, for background limited systems, introduces noise. Background limited system is the name given to an idealized detector. It is one that is limited only by the noise in the incoming photon stream. It is also called a background limited IR photodetector (BLIP).

Refractive index fluctuations in the medium of transmission create optical turbulence, which in turn causes image distortions. These fluctuations are caused by density gradients, temperature and humidity gradients, and pressure differences. Atmospheric turbulence affects image quality by deviating the path of scene photons away from the receiver.

The concentration of gases, particulates, and water concentration vary from location to location and from day to day. It is important to understand how these concentrations affect transmittance.

The atmosphere is composed of many gases and aerosols. In order of concentration, they appear as follows: nitrogen, oxygen, ozone, water vapor, carbon dioxide, carbon

FIGURE 7.21 ▪
Transmittance
Reductions Due to
Absorption

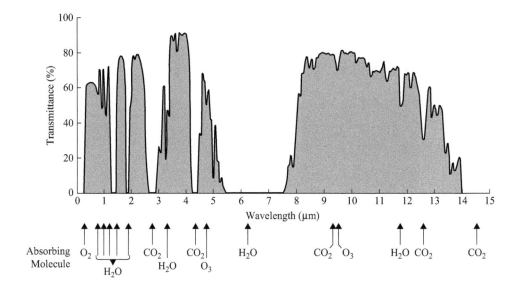

monoxide, nitrous oxide, and other trace gases. Aerosols that are suspended particulates include dust, dirt, carbon, minute organisms, sea salt, water droplets, smoke, and pollutants. Absorption is wavelength and absorbing molecule dependent; that is, absorption is at a maximum at the natural frequency of a particular molecule. Figure 7.21 shows transmittance reductions versus wavelength and absorbing molecule.

The dominant absorber in the 3 to 5 μm region is carbon dioxide. The absorption band is obvious after a few meters of path length, and is zero for any reasonable path. Water vapor determines the upper and lower limits for both the mid-wave infrared (MWIR) and long-wave infrared (LWIR) regions. In the 8 to 12 μm region, water vapor is the dominant absorber. Water vapor affects transmittance more dramatically in the LWIR region; as water vapor increases, transmittance decreases. Because of this fact, it may be inferred that an MWIR system may be better for maritime or tropical use. Noise would have to be considered prior to making such a decision. Knowing this information is nice, but how can we apply this knowledge when evaluating IR system performance?

There have been many normalizations developed over the years relating transmissivity to atmospheric effects, and Holst reviews these normalizations in chapter 15 of *Electro-Optical Imaging System Performance*. The first model relates transmittance as a function of meteorological range (R_{vis}). For a wavelength of 0.555 μm:

$$\tau_{atm} = e^{-\sigma R} = e^{3.912/R_{vis}(R)} \tag{7.13}$$

and

$$R_{vis} \approx (1.3 \pm 0.3) R_{vis-obs}, \tag{7.14}$$

where

σ = scattering cross section,

R_{vis} = meteorological range, and

$R_{vis-obs}$ = observer visibility.

Observer visibility is obtained from the International Visibility Code (Table 7.5).

TABLE 7.5 ■ International Visibility Code

DESIGNATION	VISIBILITY
Dense fog	0–50 m
Thick fog	50–200 m
Moderate fog	200–500 m
Light fog	500 m–1 km
Thin fog	1–2 km
Haze	2–4 km
Light haze	4–10 km
Clear	10–20 km
Very clear	20–50 km
Exceptionally clear	>50 km

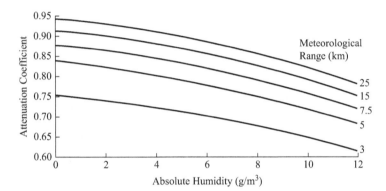

FIGURE 7.22 ■ Transmissivity as a Function of AH and Meteorological Range

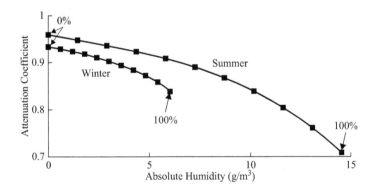

FIGURE 7.23 ■ Transmissivity as a Function of Humidity

The transmittance or attenuation coefficient may also be obtained as a function of absolute humidity (AH) and meteorological range (R_{vis}). For a low visibility coupled with a high AH, the transmittance will suffer; similarly, a low AH with a high visibility will yield a very good transmittance. This relationship is depicted in Figure 7.22. Since the AH is derived from the relative humidity and the absolute temperature, it would make sense that transmissivity will suffer more in summer than in winter. This relationship is shown in Figure 7.23, and calculation of the AH is shown in Figure 7.24.

There are models that can provide the extinction (transmittance) for given meteorological conditions. The most common models in use today are low, moderate, and high

FIGURE 7.24 ■
Conversion of
Relative Humidity to
Absolute Humidity

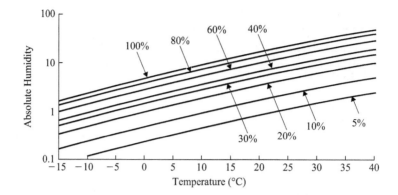

FIGURE 7.25 ■
Extinction
Coefficients for
Different Aerosols

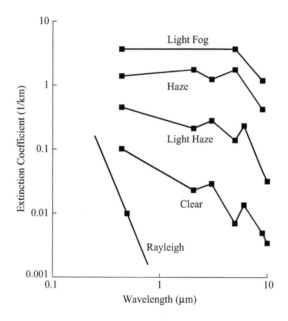

transmittance (LOWTRAN, MODTRAN, HITRAN) models. The LOWTRAN model has been replaced by the MODTRAN model, but many books and test procedures still address the LOWTRAN model. The extinction coefficient values in the models are per kilometer. For wavelengths of 7 to 10 μm, typical values are

- Clear air, 0.1 to 0.005/km
- Light haze, 0.6 to 0.05/km
- Haze, 2 to 0.6/km
- Light fog, 6 to 1.5/km

What this says is that for every kilometer of travel, it is possible to lose a percentage of the radiated energy. In the worst case for clear air, 99.9% of the radiated energy will be seen at 1 km, whereas the same energy can be attenuated by 6% (94% seen) in light fog. This attenuation factor will be applied for every kilometer of path travel. This relationship is shown in Figure 7.25.

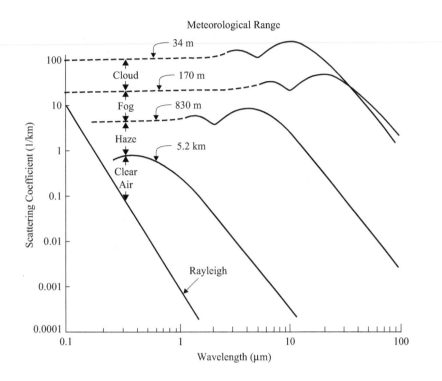

FIGURE 7.26 ▪
Scattering
Coefficients for
Spherical Particles

Scattering and absorption by particles depends on the radius of the particle, its shape, the wavelength of the incident radiation, the angle between the radiation and the viewing direction, and the complex indices of refraction. When the particles are small (diameter of the particle is much less than the wavelength), so-called Rayleigh scattering occurs, with the scattering proportional to λ^{-4}. For most naturally occurring low-density aerosols and artificial aerosols, significant scattering occurs in the visible wavelengths and minimum scattering in the IR band. As the particles grow in size, the region of maximum scattering occurs in the IR bands; of course, very large particles will affect all regions equally. The relative scattering coefficients for different-size spherical particles is shown in Figure 7.26. As with extinction, the coefficient is per kilometer and percentage lost.

In the LOWTRAN model, the scattering coefficient has been related to rain rate in millimeters per hour and is defined as

$$\sigma_{\text{rain}} \approx 0.365(\text{rain rate})^{0.63}. \tag{7.15}$$

If it is assumed that the raindrops act as scatterers and the attenuation coefficient is assumed to be independent of wavelength, then the transmittance is defined as

$$\tau_{\text{avg}} = e^{-\sigma}_{\text{rain}}. \tag{7.16}$$

The rain rate in millimeters per hour associated with the definition of rain intensity is found in Table 7.6.

The average attenuation values (τ) from data collected over 5 to 10 years yield the values shown in Table 7.7.

TABLE 7.6 ▪ Representative Rain Rates

Rain Intensity	Rain Rate (mm/hr)
Mist	0.025
Drizzle	0.25
Light rain	1.0
Moderate rain	4.0
Heavy rain	16
Thundershower	40
Cloudburst	100

TABLE 7.7 ▪ Average Attenuation Values

Weather Quality	Average Attenuation	Approximate Percentage of Time with Better Weather
Poor	0.70/km	80%
Fair	0.80/km	65%
Average	0.85/km	50%
Good	0.90/km	25%
Excellent	0.95/km	2%

For IR systems, knowing the environmental conditions of our area of operations and how this will influence the performance of the system under test is a very important consideration. MWIR systems will be affected by scattering caused by natural aerosols more than LWIR systems, but LWIR systems are more adversely affected by absorption due to water vapor. The key extinction molecule for MWIR systems is carbon dioxide. Before acquiring a system for operational use, the area of operation must first be investigated in order to determine which detector (MWIR/LWIR) is best suited to the environment.

7.6 | TARGET SIGNATURES

A target is an object that is to be detected, located, recognized, or identified, whereas the background is any distributed radiation that offsets the target. Target signatures use the following features that distinguish them from the background:

- Spatial (size, length, width, height, or orientation)
- Spectral
- Intensity features

For IR system performance predictions, the target signature is represented with an area-weighted differential temperature (Δt). For optical systems such as cameras or closed-circuit TV (CCTV), the performance predictions are based on a differential contrast (ΔC), which is a function of lighting intensity. The weakness in using an area-weighted system is that the sum of hot-cold or black-white of a target can mathematically equal zero. This approach works well for passively heated targets and backgrounds, but starts to fall apart with active targets that have a distribution of

FIGURE 7.27 ▪
FLIR Image of
Military Transport
Truck

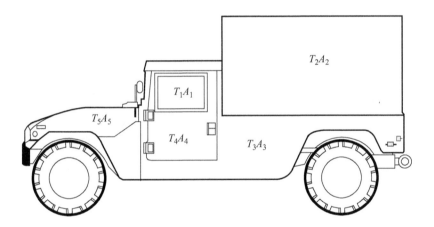

FIGURE 7.28 ▪
Calculation of T_{avg}

temperatures due to internal combustion and friction that render the results of the Δt approach less accurate. For IR systems, this distribution leads to target detection, recognition, and identification. This section will focus on thermal signatures; however, path radiance and methods to analyze data are generic to all spectral bands.

Because of target signature complexity, an area-weighted target temperature will be used:

$$T_{\text{avg}} = \frac{\sum_{i=1}^{N} A_i T_i}{\sum_{i=1}^{N} A_i}, \tag{7.17}$$

where the target contains N subareas A_i, each having a temperature of T_i.

If the background has an average temperature of T_{B}, then $\Delta T = T_{\text{avg}} - T_{\text{B}}$. This is illustrated by averaging the temperatures in the FLIR image of Figure 7.27 by using the areas of Figure 7.28.

Natural backgrounds are heated passively through the absorption of solar energy with daily heating beginning at sunrise. After midday, solar heating declines and backgrounds begin to cool, and after sunset, the backgrounds approach air temperature. Low thermal inertia objects tend to track the solar radiation; this is called a diurnal cycle, which is shown for various natural elements in Figure 7.29. All target signatures are a function of

- Absorption coefficient at the solar wavelength
- Emissivity
- Thermal inertia

The diurnal cycle can also apply to the fixed object's orientation, as shown in Figure 7.30.

FIGURE 7.29 ■
Diurnal Cycles of
Natural Elements

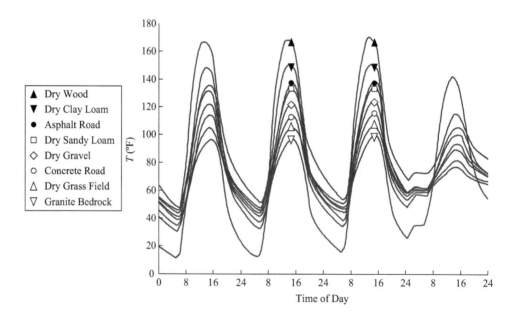

FIGURE 7.30 ■
Diurnal Cycle as a
Function of
Orientation

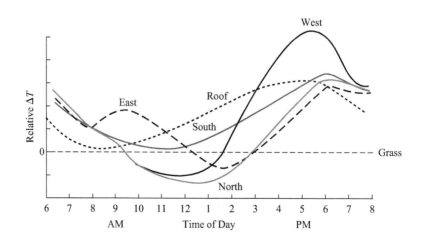

Normally the sky is very cold with an effective temperature of 20 K (deep space); if the atmospheric transmission is high the sky will appear cold. As the transmittance decreases, the path radiance increases and the sky appears warmer. For example, an aircraft flying at high altitude may be comparing targets of interest against a 20 K background, but when near the horizon the same target may be against an effective 293 K background.

The environment can also modify the Δt that is seen by an IR sensor. Passing clouds can modify target signatures; a heavy overcast may obliterate any Δt. When raining or snowing, the solar loading is often zero. The high thermal conductivity of water aids in heat dissipation and can wash out the scene. Water and mud reduce frictional cues through cooling and insulation, respectively. Wind aids in heat transfer so that under moderate wind conditions target temperatures will be lower and the surrounding areas will be hotter. This phenomenon is illustrated in Figure 7.31 in which the wind is blowing from left to right.

Optical turbulence is the fluctuation of the index of refraction resulting from atmospheric turbulence. It is responsible for a variety of effects, including

- Temporal intensity fluctuations (scintillation)
- Beam wander and broadening in lasers
- Image dancing
- Image blurring

The turbulence-induced fluctuations in the refractive index produce a phase distortion of the wavefront. The distorted wavefront continues to propagate and is itself further distorted.

There are two central issues that will influence the overall performance of optical systems utilized in an aerodynamic environment: aeromechanical and aero-optical effects. Aeromechanical effects arise from the interactions of the external flow fields with the airborne platform and combine with vibrations to cause jitter. Jitter results in image blurring and general optical misalignment and spurious laser beam motion on targets. Since jitter will be a result in almost all installations, it will be one of the major drivers for airborne testing. Aero-optical effects arise from index of refraction changes induced by moving through a flow field. Aero-optical effects will cause a loss of contrast and resolution, beam spread and wander for outgoing wavefronts (lasers), and reduced far-field peak intensity.

FIGURE 7.31 ■ IR Scene Modified by Wind

When high-energy lasers (HELs) are used, the wave propagates through the atmosphere and small amounts of energy are absorbed by matter in the air. This absorbed energy heats the air, forming a distributed thermal lens along the atmospheric path. The lens has the following effects on the laser:

- Beam spreading
- Beam bending
- Distortion

This thermal blooming limits the maximum power that can be efficiently transmitted through the atmosphere.

There are two major classes of targets that are of interest to optical sensors. They are surface targets (including subsurface) and airborne targets. The surface group contains all land vehicles and water vessels. The specific radiation characteristics of these types of targets are seldom, if ever, reported in open literature. Some things to consider are factors that will affect the radiation characteristics:

- Weathering and general deterioration (such as surface paint)
- Dust, dirt, and mud accumulation
- Different parts of the vehicle are at different temperatures, with the exhaust areas being the hottest
- Different materials are used in the manufacture of the vehicle that have different emissivities
- Use of heat suppression devices may be used to reduce the IR signature

Most of the comments made about surface vehicles are equally applicable to aircraft, except that aircraft tend to emit large quantities of hot gasses. A comparison of the IR signature from an aircraft's IR exhaust plume to the entire aircraft signature is shown in Figure 7.32. It should be noted that the exhaust plume shows peak responses at about 4.2 and 4.5 μm. These are the spectral bands called the blue spike plume (4.1 to 4.3 μm) and the red spike plume (4.3 to 4.6 μm). Because of atmospheric attenuation of the combustion gases, they become a negligible contributor to the total aircraft IR signature at ranges of a few kilometers from the aircraft. The aircraft also experiences skin friction during flight, which heats up the skin and produces a response in the 8 to 15 μm spectrum.

In order to judge the true performance of any IR system, it is paramount to know the attributes of the target as well as the current and historical atmospheric conditions. Knowledge of the true attributes of a target constitutes a calibrated target, whereas knowledge of the target within a set of atmospheric conditions constitutes a calibrated signature. Unfortunately this is almost always very difficult to accomplish in flight test environments.

For IR data to be called calibrated, the following two requirements need to be met:

- The radiance or the apparent temperature of any given surface area on the target in the real world must be retrievable from the collected data.
- The geometry of the measurement setup (or the relationship between the pixel size in the image and the actual target area) must be known.

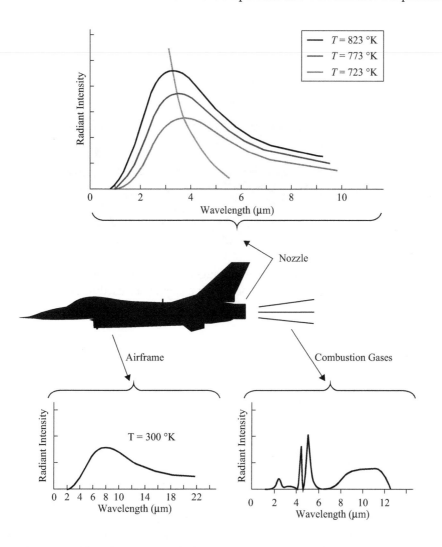

FIGURE 7.32 ■
IR Signatures
of Aircraft

It is important that the entire measurement chain is in the loop; for example, video errors must be known and external reference blackbodies must be used. There are many organizations who perform these measurements. The Georgia Tech Research Institute (GTRI) performs measurements of both ground-based and airborne targets. The target is placed on a turntable at a known bearing and range from the sensors. A blackbody is placed at an equal bearing and range and comparisons of the radiance of the target and blackbody are analyzed. A picture of the target setup is shown in Figure 7.33.

7.7 | EO COMPONENTS AND PERFORMANCE REQUIREMENTS

The major components of EO sensors are the scanning mechanisms, detectors (for the IR systems), optics, and displays. We will briefly examine these components and highlight some of the key principles.

FIGURE 7.33
IR Turntable
Measurement
Facility

Any scanning system is a complex arrangement of optical, mechanical, electrical, and electronic systems. Scanners and scanning systems are used in one or more of the following applications:

- Warning systems
- Tracking
- Pointing or designating
- Communications
- Imaging

7.7.1 Scanning Techniques

Image-forming IR systems fall into two categories: those in which an image is formed directly and those in which an optical scanning principle is employed. Direct imaging systems normally exploit a comparatively long exposure time, especially if the object of interest is stationary with respect to the detector. However, they all suffer from one or more disadvantages, such as poor angular resolution due to thermal spreading or temperature gradients within the detector, insufficient dynamic range, low contrast, limited spectrum coverage, and poor temperature resolution. The best method, at the present time, for providing high-quality imagery is to use arrays of cooled quantum detectors that are fabricated into FPAs, as described in section 7.1.1 (Figure 7.12).

The simplest method for obtaining a thermal image using scanning techniques is to use a single, small IR detector to scan all the points in a scene and use the output to modulate a display. The use of a single detector means scanning in two dimensions. If a linear array of detectors is used, scanning can be confined to one dimension; both these techniques are used in airborne systems. A two-dimensional (2D) array or mosaic of detectors can be used without scanning, which comes back to the idea of a direct image-forming device. These types of mosaics are called staring FPAs. A typical two-axis scanning system is shown in Figure 7.34.

Infrared radiation from the scanned scene enters the system's optics through the IR window. It passes through the IR optics and is focused onto the detector array. This IR energy is absorbed by the detectors and transformed into a very small electrical signal. This very small signal is then used as the input signal for the preamplifiers. In the

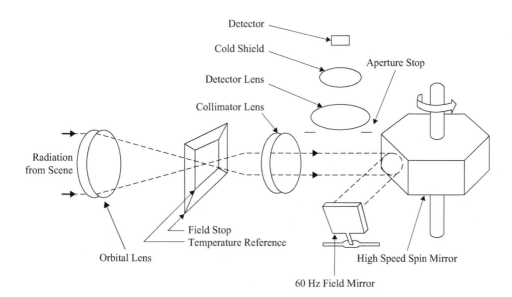

FIGURE 7.34 ■
Two-Axis Scanner
(Loral Polygon
Scanner)

earlier-generation systems there is one preamplifier channel for each detector. The signals generated by the detectors are unusable as a composite video signal because of their low level. For this reason, several stages of amplification are required to produce a usable signal. In the case of the FPA, the amplification is done on the chip via a readout integrated circuit (ROIC) which can be integrated with the detectors or as a separate ROIC chip. This miniaturization is one reason why newer generations are superior (smaller and lighter) to their predecessors.

As the scan mirror moves across the scene, IR energy is reflected onto the detectors. At certain points in the scan cycle the mirror will be in a position where IR energy passing through the optics will not be reflected onto the detectors. This part of the scan is referred to as the inactive part of the scan cycle, and that part of the cycle where video is reflected on the detectors is called the active part.

On many of the IR systems, the operator can select either white hot or black hot by utilizing a polarity switch located on the control panel. This switch activates an additional inverter stage on each of the postamplifiers and reverses the polarity of the video signal. Thus hot IR targets can be made to appear as either black or white targets on the display. The IR level (brightness), video gate, video polarity, and gain control circuits are all located in the auxiliary electronics. These signals are mixed with the video signal in the postamplifiers to produce a composite video output signal. This composite signal is used to control the brightness and contrast of the light-emitting diodes (LEDs) during the active part of the scan cycle. The LED output is used to reproduce the image of the scanned scene.

7.7.2 IR Detectors

The responsive element of the detector is a radiation transducer. Transducers can be separated into two groups: photon detectors and thermal detectors.

Photon detectors rely on four processes to accomplish IR detection:

- Photoemissive effect
- Photoconductive effect

- Photovoltaic effect
- Semiconductors

Thermal detectors rely on one of five basic processes to accomplish IR detection:

- Thermocouple
- Bolometer
- Calorimeter
- Fluid expansion thermometer
- Evaporograph

Photon detectors operate according to the photoelectric effect. They are the most sensitive and have been the most widely used optical detectors. All electromagnetic radiation occurs in discrete quanta called photons. Each photon contains an amount of energy (E) determined solely by the frequency of the radiation according to equation 7.18:

$$E_g = h\nu = hc/\lambda \text{(joules)}, \tag{7.18}$$

where

h = Planck's constant = 6.6256×10^{-34} (W/sec^2),

c = velocity of propagation = 3×10^8 (m/sec),

ν = frequency of radiation (Hz), and

λ = wavelength of radiation (m).

The incident radiant flux (P) is related to the photon rate (φ_γ) by equation 7.19:

$$\begin{aligned} P &= \varphi_\gamma E_\gamma \text{ (watts)} \\ P &= hc\varphi_\gamma/\lambda. \end{aligned} \tag{7.19}$$

The electrons surrounding the nucleus of an atom exist in energy bands as illustrated in Figure 7.35. Electrons in the lower energy state, E_0 through E_n, are bound to the atom in sharply defined energy levels and take no part in electrical conduction. Electrons in the nondiscrete energy band above E_n, but below that required to escape the material ($E_n + \varphi$), are not bound to the atom and are therefore free to take part in electrical conduction. Electrons that have energies above the escape level take part in emission

FIGURE 7.35 ▪
Energy Levels of
Atomic Electrons

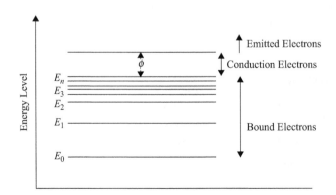

(escape from the surface of the material). The work function (φ) is the minimum incremental energy required to cause a bound electron to be emitted from the material.

An incident photon of radiation can interact directly with an atom to raise the energy level of a bound electron to the conduction band or even to the point of emission. Both of these events are termed the photoelectric effect. When the freed electron remains within the material, the process is called photoconduction. When the freed electron is emitted from the surface of the material, the process is called photoemission. In either case, the photon of radiation is annihilated.

As previously indicated, the energy contained in a photon of radiation is proportional to the frequency of the radiation. In order for photoconduction to occur, the incident photon must possess enough energy (the radiation must be of high enough frequency) to raise the energy level of a bound electron into the conduction band. In order for photoemission to occur, the incident photon energy must be even higher, as indicated in Figure 7.36. Thus photon-type radiation detectors exhibit a cutoff wavelength beyond which they do not respond. There is a corresponding cutoff frequency below which they do not respond.

The existence of a cutoff frequency is an unfortunate characteristic of photon detectors; many targets of military significance are at temperatures so low (IR radiation frequency so low) that photon detectors do not respond well to their radiations. Figure 7.37 shows the below and beyond cutoff frequencies for common detector elements.

D^*, shown as the y-axis in Figure 7.37, is a normalized detectivity that is particularly convenient in comparing the performance of detectors. Since the conditions of measurement can affect the value of D^*, these conditions are normally quoted in parentheses with any statement of D^* as follows:

$$(77 \text{ K}, 3.7 \text{ μm}, 1000 \text{ Hz}).$$

This would mean that the value of D^* quoted was obtained at an operating temperature of 77 K, at a wavelength of 3.7 μm, and using a frequency of 1000 Hz.

The interaction between an absorbed photon and the absorbing atom (the photoelectric effect) is essentially instantaneous. Photon detectors are therefore characterized by a relatively fast time constant (on the order of microseconds). This characteristic generally makes photon detectors more suitable for rapid-scan sensors than the much slower thermal detectors.

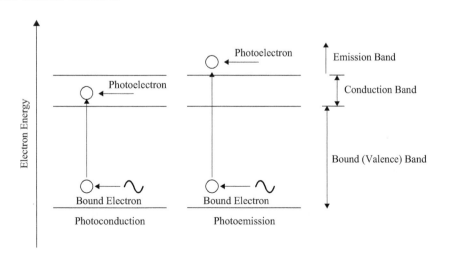

FIGURE 7.36 ■
Photoelectric Effect

FIGURE 7.37 ▣
Cutoff Frequencies
for Various
Detectors

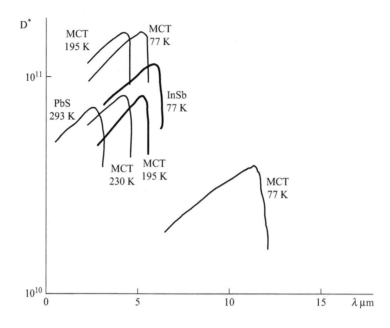

FIGURE 7.38 ▣
Photoemissive
Radiation Detector

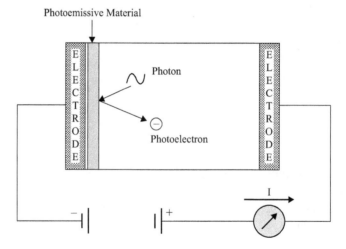

The SNR at the output of a photon-type radiation detector is generally larger than that for a thermal-type detector. Thus the detectivity (a direct measure of SNR) is, in general, higher.

Most photon-type detectors perform well only at extremely low (cryogenic) temperatures. Since detectivity generally decreases rapidly with increasing temperature, most sensors operating in the IR band that utilize photon-type detectors must incorporate relatively elaborate cryogenic cooling mechanisms.

A photoemissive radiation detector consists of an evacuated vessel in which two electrodes are located. One of the electrodes, the cathode, is coated with a photoemissive material. The external circuit consists of a voltage supply (the battery) and an electrical current measuring device (the ammeter). Photons of sufficient energy striking the surface of the coated electrode produce photoemissive photoelectrons that flow to the positive electrode (anode), thus causing an electrical current in the external circuit, indicated by the ammeter. This type of detector is shown in Figure 7.38.

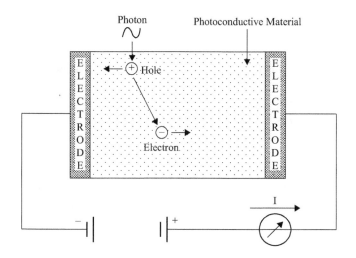

FIGURE 7.39 ■
Photoconductive
Radiation Detector

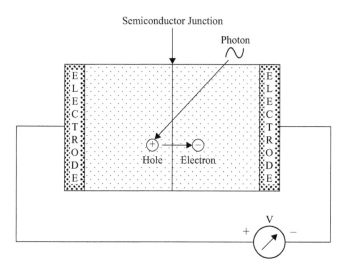

FIGURE 7.40 ■
Photovoltaic
Radiation Detector

A photoconductive radiation detector consists of a piece of semiconductor material with two electrodes attached, as shown in Figure 7.39. The external circuit consists of a voltage source (the battery) and a current measuring device (the ammeter). Photons entering the semiconductor material produce photoconductive photoelectrons within the semiconductor material (and corresponding "holes" in the bound states of the atoms). Under the influence of the externally applied electric field, the photoelectrons move toward the positive electrode and the "holes" move toward the negative electrode, thus causing a measurable electrical current to flow in the external circuit. Since the energy required to induce photoconduction is less than that required to induce photoemission, photoconductive detectors can respond to longer-wavelength radiation.

A photovoltaic radiation detector consists of a semiconductor diode as shown in Figure 7.40. The external circuit consists of an electrical voltage measuring device (the voltmeter). Photons entering the semiconductor material produce photoelectrons and corresponding "holes." As a result of the difference in energy levels in P-type and N-type semiconductor materials, the photoelectrons tend to migrate into the N-type material and the holes tend to migrate into the P-type material, thus producing an electric potential (voltage) across the P-N junction. This voltage is measured by the external voltmeter. The photon energy required for photovoltaic detection is essentially the same as that for photoconduction. Therefore the cutoff wavelengths are approximately the same.

Photovoltaic detectors are the most commonly used processing in the third-generation FPA systems. Photoconductive detectors are most commonly used in earlier-generation IR systems and most civilian applications since they are capable of operating out into the far-IR region. Two kinds of semiconductors are available: intrinsic semiconductors and extrinsic semiconductors.

Intrinsic semiconductors are obtained by refining the material to a high degree of purity. They are insulators at absolute zero since the valence band is full and the conduction band is empty. As the temperature is increased, an increasing number of electrons obtain sufficient energy to cross the energy gap between the valence and conduction bands. If the semiconductor is irradiated with photons of the correct energy, electrons can again be excited into the conduction band, and this is the photoconductive process. The thermally excited electrons are a prime source of detector noise, and this effect can be reduced by cooling the detector to very low temperatures. The characteristics of common intrinsic detectors are given in Table 7.8.

Progress toward operation at longer wavelengths has been based mainly on the exploitation of the small amount of energy required to ionize impurity atoms in the pure host material. The deliberate introduction of impurity atoms into the host material is known as doping, and quite small concentrations of impurity substantially increase the conductivity of the material, which results from irradiation. This process is called extrinsic detection; the lower the ionization energy, the lower the detector's operating temperature if thermal ionization (noise) is to be kept acceptably low. The characteristics of common extrinsic detectors are given in Table 7.9.

A thermocouple consists of one hot and one cold junction. The hot junction (thermoelectric junction) generates a voltage in response to incident radiation. The cold junction is always in thermal equilibrium with the environment. The voltage (V) generated across the two junctions is related to their temperature differences by equation 7.20:

$$V = C_t(T_h - T_c), \tag{7.20}$$

where

T_h = the temperature of the hot junction (K),

T_c = the temperature of the cold junction (K), and

C_t = the thermoelectric power coefficient (V/K).

TABLE 7.8 ■ Characteristics of Common Intrinsic Detectors

Detector Material (Operating Mode)	Approximate Cutoff Wavelength (μm)	Operating Temperature (K)	Typical Time Constant (sec)	D* at 1000 Hz (cmHz$^{0.5}$/W)
Lead Sulfide (PbS) (pc)	4	77	3×10^{-3}	2×10^{11}
Indium Arsenide (InAs) (pv)	4	77	5×10^{-7}	4×10^{11}
Lead Selenide (PbSc) (pc)	6.5	77	4×10^{-5}	3×10^{10}
Indium Antimonide (InSb) (pc)	5.5	77	6×10^{-6}	8×10^{10}

TABLE 7.9 ▪ Characteristics of Common Extrinsic Detectors

Detector Material (Operating Mode)	Approximate Cutoff Wavelength (μm)	Operating Temperature (K)	Typical Time Constant (sec)	D* at 1000 Hz (cmHz$^{0.5}$/W)
Mercury-Doped Germanium (Ge:Hg) (pc)	14	27	2×10^{-7}	2×10^{10}
Mercury Cadmium Tellurium (Hg:Cd:Te) (pv)	13	77	$<1 \times 10^{-8}$	5×10^{9}
Cadmium-Doped Germanium (Ge:Cd) (pc)	20	4.2	1×10^{-7}	2×10^{10}
Antimony-Doped Silicon (Si:Sb) (pc)	23	4.2	1×10^{-7}	1×10^{10}
Copper-Doped Germanium (Ge: Cu) (pc)	27	4.2	5×10^{-7}	3×10^{10}
Zinc-Doped Germanium (Ge:Zn) (pc)	40	4.2	2×10^{-8}	1.5×10^{10}

The bolometer is the thermal detector most frequently employed in airborne systems. A bolometer senses the change in resistance of a sensitive element caused by the temperature change. The sensitive element may be a metal, a semiconductor, or a superconductor. Metallic and semiconductor bolometers are the most widely used. The resistance change in a metallic bolometer is linearly dependent on changes in temperature. The resistance change in a semiconductor bolometer is an exponential rather than a linear function of temperature. For small temperature variations, however, the resistance change can be assumed to be linearly dependent upon temperature changes.

The calorimeter consists of a radiation-absorbing mass, the temperature increase of which is measured, directly, as an indication of total radiant energy absorbed.

The fluid-expansion thermometer consists of a radiation-absorbing liquid or gaseous mass, the temperature rise of which is determined indirectly by measuring the change in volume of the fluid.

The evaporograph consists of a radiation-absorbing screen, the temperature rise of which is indicated indirectly by the rate of evaporation of a thin film of volatile substance. By focusing an image of the sensor FOV upon the screen, a thermal image is obtained.

7.7.3 Detector Cooling

For the reasons previously noted (i.e., avoidance of internal noise and the ability to sense small temperature differences), many detectors must be cooled, although some of the newer bolometers can operate at room temperature. There are four types of cooling processes used:

- Liquid cooling: nitrogen (77 K), hydrogen (20 K), and helium (4.2 K)
- Joule-Thomson cooler: gas transferred into liquid
- Cryogenic/Stirling cooler: like a refrigerator, contraction and rapid expansion of helium or freon
- Thermoelectric cooler: works on the opposite principle of a thermal detector

7.7.4 Optics

The function of the optics within any of the EO systems is to collect and focus energy onto a detector. In some cases, such as IR, the optics may also function as a filter allowing only frequencies of interest to be passed on to the detector. The glass window may be coated or fabricated of materials that only allow certain IR frequencies to be passed through, or they may be designed to reflect or absorb all IR frequencies and only allow visible light to pass. Some of the common materials used in these windows may be germanium sapphire, zinc sulfide, silicon, zinc selenide, or calcium fluoride. In all cases these windows are specially manufactured to very high tolerances and add cost to the system.

When designing an EO system, probably the two biggest concerns when considering the optics are the size and the performance. The amount of space required for the optics and its associated weight will be directly related to the cost. Performance requirements will drive the resolution and sensitivity of the optics.

Figure 7.41 depicts the thin lens approximation used for analysis in most optical subsystems. The clear aperture is not necessarily the diameter of the optical element; D is the effective aperture diameter. The effective aperture may best be visualized with a camera. Although the lens may have a fixed diameter, the operator can select how much light is allowed in by reducing or expanding the pupil diameter by adjusting the camera f-stop. The focal length (f) is the distance from the aperture to the focal point of the collected energy. A magnifying glass can collect the sun's energy and focus it to a single point on the ground where it is strong enough to start a fire. Moving the glass closer or farther away from the sun will move the focal point away from the ground surface. The f-stop, or focal ratio, expresses the diameter of the effective aperture in terms of the focal length of the lens. If the focal length is 16 times the effective aperture, then the f-stop would be $f/16$; the greater the f-stop, the less light (energy) per unit area will reach the image plane of the system. Images will be fainter as the focal ratio is increased.

This information will come into play when discussing the EO system FOV, which is directly related to the focal length of the optics used. In IR systems for a given detector size, as the focal length decreases, the FOV increases and the resolution decreases.

7.7.5 Field of View

As noted previously, the optical design will dictate the area seen by the system; this area is known as the system FOV. A FLIR system will most often have a wide FOV, narrow FOV (and sometimes a medium FOV), and a zoom. The wide FOV is typically on the order of 25° horizontal and 20° vertical to match the display resolution, which for U.S.

FIGURE 7.41 ■
Focal Ratio

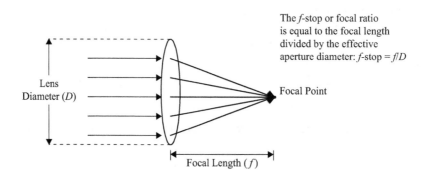

The f-stop or focal ratio is equal to the focal length divided by the effective aperture diameter: $f\text{-stop} = f/D$

Lens Diameter (D)

Focal Point

Focal Length (f)

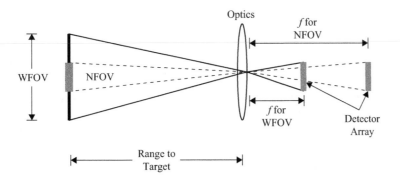

FIGURE 7.42 ■
Field of View

systems is 480×640 pixels (or in a ratio of 4 vertical:3 horizontal; V/H). The narrow FOV will usually give a 4:1 resolution increase, for a reduction in the FOV to about 6.25° horizontal and 5° vertical. The zoom function may be optical (a real zoom) or digital (brings the picture closer with no increase in resolution) and is normally on the order of a 2:1 increase in resolution. TV systems normally use a pan and zoom function that allows the operator to run the system from zero magnification out to the limits of the optics. In slewable systems, the field of regard (FOR) is the area with unobstructed viewing as the system is slewed throughout its azimuth and elevation limits. In fixed systems, the FOR equals the FOV.

An instantaneous FOV (IFOV), or the detector angular subtense (DAS), as noted by Holst (*Electro-Optical Imaging System Performance*), is used to describe the FOV of a singular detector element. The IFOV can be calculated by dividing the detector dimension by the effective focal length. The relationship of FOV to the focal length (f) can be seen in Figure 7.42. In the figure, the optics are the complete set of optics for the system and f is the effective focal length of the complete system. By knowing the IFOV for a staring array, the FOV can be obtained for the horizontal and vertical by multiplying the IFOV by the number of elements (either horizontal or vertical) in the array.

7.7.6 Displays

As with all technology, the way information is displayed to an operator has changed significantly over the years. Like the home TV, displays have gone from large boxy units with relatively small viewing areas to large flat screens that are extremely thin and can be hung on a wall like a painting. Older displays are of a cathode ray tube (CRT) design, which is a vacuum tube containing an electron gun and a fluorescent screen, while newer displays use liquid crystal display (LCD) technology. Both will be described here for completeness.

7.7.6.1 CRT Displays

The construction and principles of operation of a typical TV display are the same as those associated with any CRT. A raster scan is used for imaging detail, whereby the spot formed by the electron gun is moved over the whole area of the screen in a regular series of parallel horizontal lines. The image is built up by varying the brightness of the spot in synchronization with the raster. The process is repeated at a sufficient refresh rate and usually involves an interlace method so as to produce an apparently continuous and dynamic image.

The TV raster used for EO systems is produced by deflecting an electron gun in the aircraft TV picture tube in a series of horizontal scans and vertical deflections (Figure 7.43). The aircraft electron gun and the weapon electron gun are aligned or synchronized by the master oscillator sync signals. U.S. Air Force EO systems use a horizontal scanning frequency of 15,750 Hz and a vertical deflection period of 60 Hz. The product of these parameters is a TV picture or field consisting of 262½ lines of video. An entire field is generated every 1/60th of a second.

Apparent resolution of the displayed TV presentation is improved by using a technique called a 2:1 interlace. The lines of each successive field are interlaced by scanning first the odd lines of the display and then filling in the even lines when the following field is generated (Figure 7.44).

Two interlaced fields comprise one frame consisting of 525 lines of video whose repetition rate is 60/2, or 30 times/sec. The frame constitutes a complete TV picture element, but its relatively low repetition rate does not generate flicker because of the two vertical scans required to produce it. The result is an improved apparent video resolution. The human eye then uses the property of image retention to tie successive frames together and create apparent motion, in the same manner as a motion picture.

The detail of a TV picture in the vertical direction is determined by the number of scanning lines. Resolution in the horizontal is similarly determined by the speed with which the TV system is able to respond to abrupt changes (i.e., by the frequency

FIGURE 7.43 ■
TV Raster

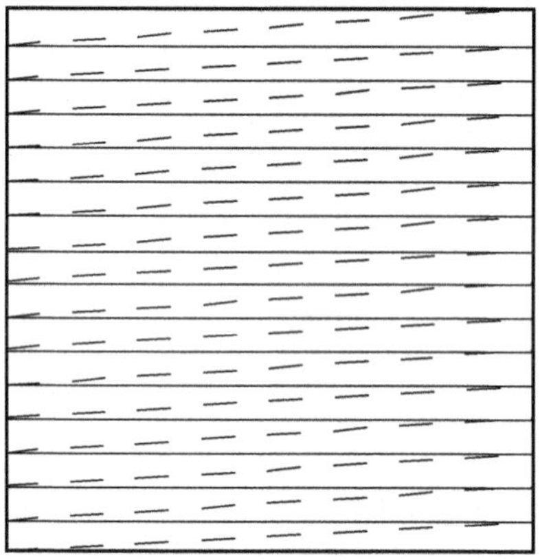

FIGURE 7.44 ■
TV Raster with 2:1
Interlace

band that the TV system transmits). Using the U.S. domestic TV system as an example, and assuming that in scanning a horizontal line the transition from black to white is accomplished in a distance along the line equal to the spacing between adjacent lines, then

- For active lines with a width-to-height ratio of 4:3, the distance is

  $\times 1.33 = 0.00155$ of the active length of one line.

- The spot moves from left to right across the scope in 53.5 μsec; hence, the time interval for resolution is

  $\times 53.5 = 0.08$ μsec.

- In situations where the overshoot is less than 5%, the relationship between rise time and bandwidth (B) is given by

  Rise time $= 0.35/B$.

- Hence a video system that will permit the response to rise from a minimum to a maximum in this length of time without overshoot must have a bandwidth of not less than

  $B \geq 4.375$ MHz.

  A larger bandwidth than this will give greater horizontal resolution, while a reduced bandwidth degrades the horizontal resolution.

All photon sensors are characterized by some form of limiting resolution, that is, the smallest detail that can be seen will be limited by some sensor parameter. An optically perfect lens will create a divergent beam; thus, once a lens size has been chosen for a system, the maximum angular resolution is then determined and system resolution can never be better than the diffraction limitation of that lens. (This aspect is more applicable to IR rather than visual systems.) In the case of film, the limit is due to the photosensitive particles, which are of a finite size. With TV, as already discussed, vertical resolution is set by the number of horizontal lines employed, horizontal resolution is determined by the video bandwidth, and the size of the electron beam affects both horizontal and vertical resolution.

When the TV camera itself has a resolution limit, greater scene detail can be viewed at a given range only by increasing the lens focal length (f). However, such an increase in f without a corresponding increase in lens diameter may result in a decrease in lens resolution and a loss of image light level resulting in a less than hoped for scene resolution. In low light level conditions, this method of increased focal length will probably result in a loss of scene resolution. Therefore, when designing a system, the required scene resolution must be determined first, and then the particular elements are sized to meet the required performance.

7.7.6.2 Liquid Crystal Displays

There is an excellent tutorial available online describing the components, uses, problems, and resolutions of LCD technology. It was put together by the 3M Company and can be viewed at http://solutions.3m.com/wps/portal/3M/en_US/Vikuiti1/BrandProducts/secondary/optics101/.

Each individual pixel in a black-and-white LCD and each subpixel in a color display is created from a source of light and a light valve, which can turn the light on, off, or to some intermediate level. In an LCD, millions of light valves form the display panel, while a backlight and various display enhancement films create the illumination.

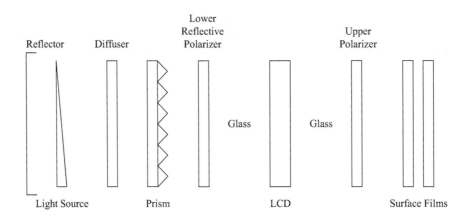

FIGURE 7.45 ▪
LCD Components

The LCD panel is at the center of the display, which is where the liquid crystal itself is located. Transparent electrodes patterned on each pane of glass encompass the liquid crystal. The orientation of the liquid crystal can be changed in subtle ways by applying a voltage to the electrodes in order to change the level of illumination displayed in each subpixel. A simplified drawing of the construction of an LCD is shown in Figure 7.45.

The panel is sandwiched on one side by various front surface films that enhance the display properties and on the other side by a source of light known as a backlight. The LCD panel is the essential component in a display that controls the amount of light reaching the viewer. Light passes through a bottom polarizer that orients the light to a single state of polarization by absorbing more than 50% of the incoming unpolarized light. The polarized light then passes through the liquid crystal layer. The orientation of the liquid crystal can be controlled by applying a voltage to the transparent electrodes on the glass encompassing it. The liquid crystal's degree of orientation controls to what degree it will rotate the incoming polarized light with a typical LCD. When there is no voltage applied to the electrodes, the liquid crystal is naturally oriented to rotate the light 90°, allowing it to pass freely through the top polarizer. However, if a voltage is applied, the liquid crystal aligns to the electric field and does not rotate the light, allowing the top polarizer to block it completely. By applying an intermediate voltage, the liquid crystal can be partially oriented to partially rotate the incoming light, creating shades of gray.

Adding a color filter to the LCD panel creates color displays. In a color LCD, each red, green, and blue subpixel is individually controlled, allowing varying amounts of red, green, and blue light through to the viewer for each pixel.

Active matrix addressing (each pixel is individually controlled) allows for very-high-resolution LCDs and is used for nearly every laptop, most cell phones, LCD monitors, and TVs. Each subpixel of the display is individually controlled by an isolated thin-film transistor (TFT) which is patterned onto the glass and allows the electrical signal for each subpixel to be locally stored without influencing adjacent elements. For a standard notebook computer screen that has a resolution of 1024×768 pixels, there are $3 \times 1024 \times 768$ subpixels (more than 2 million), and each one of them is individually controlled by a TFT.

The surface films are used to enhance or restrict viewing of the display. On an automatic teller machine (ATM), for example, the screen cannot be read unless the viewer is directly in front of it; this is done for obvious security reasons. In aircraft, we may have quite the opposite problem. The display may not be mounted directly in front of the operator and therefore must be able to be read off-boresight; a film is available that diffuses the light to make this viewing possible. Another film may be used to enhance viewing capability in direct sunlight or to reduce glare.

7.8 | PASSIVE EO DEVICES

7.8.1 IR Line Scanners

An analog IR line scanner (IRLS) is an imaging device that forms images by successive scans of a rotating mirror. The scans are transverse to the line of flight of the vehicle carrying the IRLS. The second scan needed for a 2D image is provided by forward motion of the vehicle along its flight path. An IRLS in a high-speed reconnaissance aircraft is one of the most efficient means of gathering pictorial information.

The velocity-to-height (V:H) ratio is one of the dominant parameters in IRLS systems because it determines the required scan rate. The units are in radians per second, and the IRLS may scan with a single detector or multiple detectors in the along track axis to keep rotation rates at practical values. The IRLS typically uses either single-facet or multifacet axe head scanners.

Simple trigonometry determines ground coverage once the maximum FOV has been determined from the maximum V:H ratio and scan rate. Figure 7.46 shows the relationship between ground coverage and V:H/scan rate.

It can be seen that for use on low-level flights, the system requires a very large FOV in order to obtain the desired coverage. The low flight altitudes and the need to view in both directions drive the need for a 180° FOV.

The IRLS must also be stabilized against aircraft roll motions, because high roll rates and large roll excursions of reconnaissance aircraft will interfere with image interpretation and can cause a loss of image. The roll angles, rates, and accelerations can all be very high. Pitch excursions will be encountered in those aircraft that fly at very low levels, especially in a terrain-following mode. Since the terrain is very rarely flat, the IRLS requires a signal from the aircraft so the film or tape recording rate can be

$$FOV = 2\theta_{max}$$
$$\theta_{max} = FOV/2 = \tan^{-1}(X/H)$$
$$R_{MAX} = H/\cos\theta_{max}$$
$$X_{MAX} = H\tan\theta_{max}$$

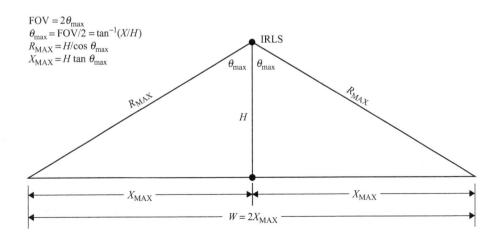

FIGURE 7.46 ■ IRLS Ground Coverage

adjusted to the proper V:H ratio. The signal also permits channel selection in multi-channel designs (i.e., variable number of along-track detector elements selected as a function of airspeed). Only some detectors in the along-track axis may be needed at a low V:H ratio, but more will be required at a high V:H ratio.

The digital IRLS scans along the wing line, much like the analog IRLS, but now uses on-chip processing as described in section 7.1.1. The digital IRLS is one of three classes of reconnaissance sensors currently in use. The IRLS is called a whisk broom sensor because of the sweeping action along the wing line. It has the same requirements of line-of-sight (LOS) stabilization as noted for the analog IRLS, but has a higher resolution over the same patch of terrain. This is because the detector array may be comprised of thousands of elements stacked in one axis (vertical array swept horizontally). This provides a unique problem of collecting so much data that it cannot be easily data linked in real time. Data compression or sending only a subset of the data are options that are used. The F-18 IRLS and sensors aboard the land remote sensing satellite (LANDSAT) and advanced very high resolution radar (AVHRR) satellites are examples of the whisk broom reconnaissance sensor.

There are two other types of reconnaissance sensors, which are not classic IRLS. These are called push broom and framing sensors. The push broom sensor uses the same type of array as the whisk broom sensor, but orients it along the wing line and pushes the array along the flight path (horizontal array pushed vertically). With this setup there is no scanning mechanism required. Push broom sensors are generally lighter and less expensive than whisk broom sensors. The Advanced Tactical Airborne Reconnaissance System (ATARS) is an example of a push broom reconnaissance system. The framing system uses a very large focal plane array (1000×1000 up to 5000×5000) to take snapshots of the area of interest (hence the picture frame). In some applications these frames can be patched together to produce a picture of a larger area; the Global Hawk uses a framing sensor. Each of these sensors produces large amounts of data due to the resolution obtained, and the transmission of large data packets is a problem. LOS stabilization is not as large a factor in framing systems as it is with whisk broom or push broom sensors.

7.8.2 Forward-Looking Infrared

A FLIR imaging system is the IR analog of a visible TV camera. The purpose of the FLIR is to provide a scene image by the detection and appropriate processing of the natural radiation emitted by all material bodies. A system that operates in the 8 to 14 µm region is usually referred to as an LWIR FLIR. A system that operates in the 3 to 5 µm region is usually referred to as an MWIR FLIR. A modern FLIR forms a real image of the IR scene by detecting the variation in the imaged radiation and creating a visible representation of this variation. Characteristics such as FOV (degrees), resolution element size (milliradians), and spatial frequency (cycles/milliradian) are expressed in angular units. Noise is essentially constant and is usually expressed in terms of temperature and is dominated by the large background. The noise equivalent temperature difference (NETD) is the internal noise figure converted to a minimum resolvable temperature difference (MRΔT).

The FLIR systems are divided into two categories: staring systems and scanning systems. Staring systems fill up the desired focal plane area with detectors, thus scanning is not required.

7.8.3 Television Systems

7.8.3.1 Analog TV Systems

Television has been applied to a wide variety of uses in avionics systems ranging from aerial reconnaissance cameras to missile homing devices. The airborne TV versions differ in size and ruggedness from their commercial counterparts, but their principles of operation are identical.

Television is accomplished by scanning different parts of a scene with a camera tube which converts the incident light into photoelectric energy with a voltage proportional to the light intensity of the scene under examination. The varying voltage resulting from this scanning is amplified and can either be passed directly to a display or it can be modulated on a radio frequency carrier wave that is radiated to another area.

The display, or TV receiver, synthesizes the original fluorescent scene of a CRT by causing a cathode ray spot to trace over successive portions of the reproduced scene in accordance with the scanning of the original scene. At the same time, the CRT varies the brightness of the spot in accordance with the envelope amplitude of the received signal. Because the scanning process is carried out with sufficient rapidity, an illusion of continuous, nonflickering motion is achieved.

As stated previously, a TV camera receives light energy from a scene and converts it into photoelectric energy. The TV camera is more sensitive than the human eye since it has higher quantum efficiency and a larger light-sensitive area. There are two processes by which the conversion of light energy into electrical energy is obtained: photoemission and photoconduction. These processes are the same as previously described for IR detectors.

There are two basic types of TV camera tubes: the vidicon and the orthicon. The vidicon TV camera tube, shown in Figure 7.47, is an electron beam scanning imaging sensor that scans the overall FOV by deflecting an internal electron beam.

In the vidicon, the incoming radiation is focused onto an internal screen composed of photoconductive material. An image is thus formed on the screen consisting of

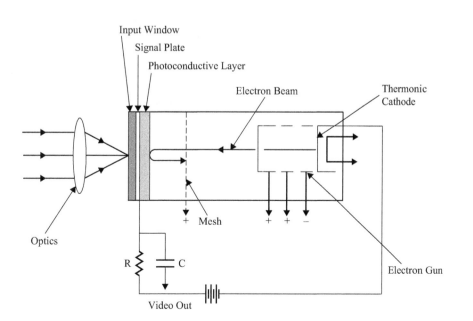

FIGURE 7.47 ■
Vidicon TV Tube

areas of greater or lesser conductivity. A narrow beam of electrons is deflected in such a way as to scan the back of the photoconductive screen, as shown in the figure. When the electron beam impinges upon an irradiated area of the screen, the electrons are conducted through the screen to a transparent, conductive signal plate and into an external circuit. The resulting current in the external circuit constitutes the output signal from the sensor.

One limitation of the basic vidicon that must be emphasized is its tendency to smear moving images, an example of the effect known as lag. Lag occurs in all TV tubes to some extent and becomes worse at low light levels; it can represent the real limit to tube performance in airborne applications. The causes of lag vary considerably from tube to tube. In the vidicon, the main lag effect is target element capacitance.

As a rule of thumb, if the human eye can discriminate a specific target or contrast feature, the EO weapon vidicon will also be able to define and display that particular feature. However, notable exceptions occur when we deal with variations of the same basic color groups. Since EO vidicons deal only in black, white, and shades of gray, certain camouflage schemes when viewed against a similar background will be discernible by color contrasts to the human eye. The same color contrasts, however, may be converted to the same shade of gray and be indistinguishable on a TV display. Low ambient light conditions present special problems for an EO system. The vidicon system normally lacks the low light level TV system's filter wheels and iris-type devices used to operate over a wide range of lighting conditions. The photoconductive material used by the vidicon needs a minimum level of light intensity to function. Under low ambient light conditions, normally around sunrise and sunset, the SNR between the light and dark contrast areas becomes too close for the video processor to resolve. The result is insufficient video contrast for video identification and target tracking.

Extremely bright light conditions (the sun in particular) also cause special problems for the vidicon system. A direct view of the sun, focused by the lens system onto the signal plate, will actually burn the photosensitive material. This damage shows up as a "scar" on the video and will severely degrade weapon performance if the scar covers any part of the tracking gate.

Another type of electron beam scanned TV camera tube is the image orthicon, shown in Figure 7.48. The orthicon is the basic photoemissive tube. In an image orthicon, the incoming radiation is focused on a photocathode, thus producing an electron-cloud "image" of the FOV. These photoelectrons are accelerated toward a target screen, as shown in the figure. The accelerated electrons strike the target screen, producing secondary emissions and leaving areas of the screen deficient in electrons. The screen is

FIGURE 7.48 ■
Orthicon TV Tube

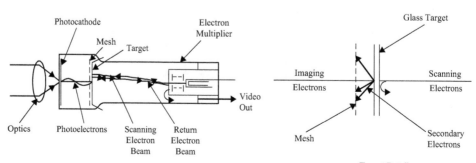

then scanned from behind by a narrow beam of electrons. Where the scanning beam impinges upon an electron-deficient area of the target screen, the beam loses electrons to the screen. The electrons remaining in the beam (not absorbed by the target) constitute the output signal and are returned to an electron multiplier at the base of the camera tube. The electron multiplier, a device that amplifies current by cascading secondary emissions, greatly increases the output current. The image orthicon is therefore much more sensitive than a vidicon. However, it has limitations as a low light tube because of its poor noise performance at low light levels.

7.8.3.2 Digital TV Systems

Television systems have followed a pattern of development similar to the IR devices described earlier. Modern TV and camera systems use one of two types of digital sensor technology, with a third in early development. The sensor is a silicon semiconductor that is composed of an array of photosensitive diodes, called photosites, that capture incoming photons and convert this energy to an electrical signal. The electrical signal is then converted to digital data as pixels. The pixels are stored in their correct order as a file that can be viewed or retained in memory. The types of sensors include CCD, complementary metal oxide semiconductor (CMOS) (sometimes referred to as active pixel sensors), and single-carrier modulation photo detector (SMPD).

Charge-coupled devices were developed by Bell Labs in the 1960s and are silicon-based integrated circuits consisting of a dense matrix of photodiodes that operate by converting light energy in the form of photons into an electronic charge. This is essentially the photovoltaic radiation detector described previously in section 7.7.2. CCD sensors derive their name from how the charge is read after an image is captured.

Utilizing a special manufacturing process, the sensor is able to transport the built-up charge across itself without compromising image quality. The first row of the array is read into an output register, which in turn is fed into an amplifier and an A/D converter. After the first row has been read, it is dumped from the readout register and the next row of the array is read into the register. The charges of each row are "coupled" so as each row moves down and out, the successive rows follow in turn. The digital data are then stored as a file that can be viewed and manipulated.

Complementary metal-oxide semiconductor sensors were first discussed at length in a paper by Dr. Eric Fossum, a scientist at NASA's Jet Propulsion Laboratory (JPL), in 1992. The JPL developed much of the technology that would be implemented into the CMOS sensor. CMOS sensors are created using the same manufacturing technology that is used to make microprocessors. These chips contain transistors at each pixel position that amplify and move the charge using conventional wires, which is a more flexible technology, as every pixel can be read individually. Because the transistors occupy space on the array, some of the incoming light hits the transistors and not the photosites, which leads to picture noise. CMOS sensors function at a very low gain, which may also contribute to noise.

Each of these sensors has its advantages and disadvantages. CCD sensors are expensive to produce because of the special manufacturing processes that must be employed. CMOS chips can be fabricated easily, so they tend to be extremely inexpensive compared to CCD sensors. CCDs use more power than their CMOS counterparts; an increase of up to 100%. This limits the use of CCDs where power consumption is an important factor. CMOSs are susceptible to noise, whereas CCDs can produce high-quality, low-noise images. CCDs suffer from vertical smearing on bright light sources; CMOSs do not. Since CMOS sensors have amplifiers built in, the uniformity (same response for a given lighting condition) is low. CCDs have better sensitivity in low light situations because part of the CMOS chip is taken up by transistors, which are not sensitive to light. The entire surface of the CCD is used for light capture. All camera functions can be placed on a single chip in the case of CMOS sensors, which is not the case for CCDs.

A newer technology that is in development is the SMPD. The SMPD is a variation of the CCD sensor that reacts to light levels of less than 1 lux and can capture low light video at a rate of 30 frames/sec, which is faster than either CCDs or CMOSs. Companies who make these chips claim that SMPD is the first, full-color, high-sensitivity image chip for taking pictures in the dark without the use of a flash. The sensor is 2000 times more sensitive than a CMOS or CCD sensor, is half the size, and is projected to be more cost effective.

7.8.4 IR Search and Track

An IR search and track (IRST) system is a specific type of FLIR that is used to passively and autonomously detect, track, classify, and prioritize targets of interest. Because IR systems can only provide azimuth and elevation attributes of targets, signal and data processing algorithms must be used to calculate the other attributes of the target that are required to build a track file (range, range rate). The advantage over active systems such as radar is that they collect all information passively, and hence do not highlight themselves.

The technologies used in FLIR and IRST are similar in the actual detection of targets of interest, but since longer-range detections are desirable for identification and weapons employment, some additional processing is used in IRST. Detection of small, point-size targets is not possible in a cluttered environment unless the SNR of the target is increased; the same data must be postprocessed to decrease false alarms.

Infrared search and track systems calculate the range (and from the range, range rate) by using kinematics ranging calculations or by incorporating a laser range finder (LRF). Ranging calculations require some intelligence of the target and rather large processors. The ranging calculations are preferred, as they can calculate range at distances beyond the capabilities of the LRF and do not highlight the search aircraft. Another concern is that an LRF will add considerable size and weight to the IRST system.

The simple range formula for a point detection system such as an IRST is given by

$$R = \sqrt{\frac{\Delta W \times \tau(R)}{NEI \times K}},$$
(7.21)

where

R = range to target,

$\tau(R)$ = atmospheric transmittance over the range,

ΔW = target radiant intensity contrast,

NEI = noise equivalent irradiance of the sensor, and

K = required SNR for a given detection/false alarm probability.

The range and type of target can be calculated by the IRST by estimating the atmospheric transmittance and background spectral radiance from available meteorological data. In each consecutive scan or frame, the contrast irradiance in the direction of a target is measured as well as the azimuth and elevation. A target type is selected from the library of the IRST computer based on the measured irradiance and the target range is calculated based on the selected target type, estimated atmospherics, and measured contrast irradiance. When the target is not radially inbound, the range is calculated based on the target type selected, the angular movement of the target (aerotriangulation), and the range estimate based on the last scan; the estimates are constantly combined to form a best estimate of range.

The F-14D AN/AAS-42 IRST was one of the first IRST systems to be fielded in the United States; a comparable system was fielded by the Soviets on the Su-27. The AN/AAS-42 operated in the 8 to 12 μm spectrum using a MgCdTe detector array, aligned vertically, and scanned horizontally. The system provided track file data on all targets while simultaneously providing IR imagery to the cockpit displays. The U.S. Air Force's airborne laser (ABL) program is currently equipped with a version of the AN/AAS-42.

The F-18 IRST set being developed by Lockheed Martin uses slewable, liquid-cooled, focal plane array MgCdTe detectors that are mounted in the forward section of the aircraft's centerline fuel tank. The location gives the system a desirable $70° \times 70°$ FOR. The system can track multiple targets in a track-while-scan (TWS) mode or provide a single target track (STT); modes are integrated with the cockpit hands-on stick and throttles (HOTAS).

There are two IRST systems that have been developed in Europe: passive infrared airborne track equipment (PIRATE) and the infrared optical tracking and identification system (IR-OTIS). The PIRATE was developed by a European consortium led by Thales for the Typhoon. IR-OTIS was developed by Saab Bofors Dynamics for the Saab JAS39 Gripen. Both systems are Gen III sensors utilizing focal planar arrays of HgCdTe detector elements. The PIRATE claims long-range (up to 100 km) detection, tracking, and prioritization of multiple airborne targets with high resolution and scene imagery. IR-OTIS has three operating modes: IRST, in which the system uses several different search FOVs; FLIR, in which the system LOS is directed by the aircraft; and track, in which the tracker controls the LOS. It is also possible to automatically switch from IRST to track.

There are three other systems that are in concurrent development in the United States: the electro-optical targeting system (EOTS), the electro-optical distributed aperture system (EO-DAS), and the distributed aperture system infrared search and track system (DAS-IRST). The EO-DAS is a common module in EOTS and DAS-IRST. EOTS is being developed for the F-35 Joint Strike Fighter (JSF), whereas the DAS-IRST is being developed by the U.S. Navy for fleet defense. Northrop-Grumman and Lockheed are developing the EOTS and EO-DAS and the Naval Research Laboratory's Optical Sciences Division is coordinating the efforts on DAS-IRST.

The EOTS is based on the Sniper pod and will be carried internally on the F-35 under the nose. The EOTS comprises a Gen III FLIR, laser, and digital CCTV. The system will provide long-range target detection and identification, automatic tracking, IRST and laser designation, range finding, and spot tracking. The DAS-IRST is comprised of a 512×2560 HgCdTe focal plane array and two FOVs: $3.6° \times 48°$ and $10° \times 48°$. The system also employs a combination of filters to provide spectral band coverage in the 3.4 to 4.8, 3.8 to 4.1, and 4.6 to 4.8 μm ranges.

7.9 | LASER SYSTEMS

The name laser is an acronym derived from the phrase "Light Amplification by Stimulated Emission of Radiation." It functions on the same basic principle as its forerunner, the maser, and as such is often known as an optical maser. In both devices, atoms that have had their energy increased are stimulated, causing them to release this excess energy. Stimulation is the key to the principle of laser operation and is the basis for the unique characteristics that distinguish the laser from other optical radiation.

The process of stimulated emission of radiation was first described, theoretically, by Albert Einstein in 1917. It is a process whereby a light beam of a specific color or wavelength interacts with an atom in such a way that the atom emits additional light that moves in the same direction and has the same wavelength as the original beam. If a beam is fed into a suitable laser amplifier, the amplifier preserves the original direction and wavelength of the input beam and the intensity of the light increases. An oscillation can be produced by placing mirrors at the ends of the amplifier, forcing the light to bounce back and forth between the mirrors. The intensity of the light in such an oscillator can build up to many orders of magnitude greater than the intensity of the original light signal, producing a highly directional laser beam of an extremely pure color.

The laser is based on the fact that electromagnetic radiation can interact directly with the atomic structure of a material in such a way as to result in the absorption or emission of electromagnetic radiation by the material. The structure of an atom is represented by the Bohr model shown in Figure 7.49.

The protons and neutrons are assumed to be located in a nucleus at the center of the atom. The electrons are assumed to orbit about the nucleus in "shells," as shown in the figure. In actuality, the "shells" are energy levels, as shown in the energy diagram in Figure 7.49. Each "level" represents a fixed, discrete energy and can be occupied by no more than a specific number of electrons. An electron can be raised from one energy level to a higher energy level by absorbing energy from a photon of electromagnetic radiation (in a laser, this process is called pumping). Conversely, it can fall to a lower energy state by emitting energy in the form of a photon of electromagnetic radiation (this process is called radiative decay). Other non-radiative processes exist by which the energy level of the electrons can be changed. In radiative absorption or decay, the

FIGURE 7.49 ■
Bohr Model

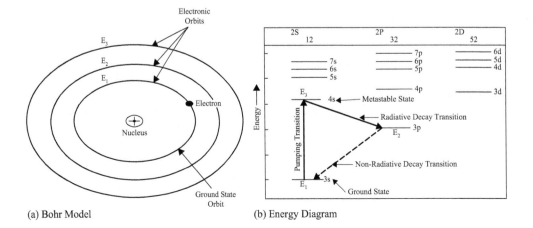

(a) Bohr Model   (b) Energy Diagram

frequency of the radiation involved is determined entirely by the difference in energy of the two levels between which the transition occurs:

$$\Delta E = h\nu \text{ (joules)(energy or frequency)}, \tag{7.22}$$

where

ν = frequency of radiation absorbed or emitted (Hz),

h = Planck's constant = 6.63×10^{-34} (W/sec^2), and

ΔE = difference in energy between transition levels (joules).

Thus, in Figure 7.49, the "pumping transition" shown requires an energy input equal to $(E_3 - E_2)$.

In a laser, the electrons in the lasing material are raised to higher energy states (pumped) by one of several methods. Some of the electrons in the higher energy levels then fall back to lower energy levels by emitting a photon of electromagnetic radiation (radiative decay). In order for the radiant flux contributed by newly emitted photons to add to that of the existing photons (constructive interference), the emitted photons must be coherent and in phase with the existing photons. Such synchronous emission is achieved in the laser by passing the previously emitted radiation through the lasing material. The electromagnetic fields created by the passing photons encourage the radiative decay of electrons in a manner that produces photons in phase with the passing photons. In most implementations, the laser photons are passed repeatedly through the lasing material, inducing ever-increasing rates of radiative decay (increase of flux) on each pass.

There are three important electromagnetic processes that can take place in a particle:

• Absorption

• Spontaneous emission

• Stimulated emission

Absorption occurs if electromagnetic energy is incident on a particle in the ground state. Part of this energy can be extracted from the radiation field to cause the ground-state electron to make a transition to one of the higher energy states.

During spontaneous emission, the transition the electron makes is from a higher energy level to a lower one, and the difference in energy between the initial and final states of the particle is converted into electromagnetic energy in the form of a single photon. This process forms the basis of operation for the ordinary fluorescent light bulb. Since there is no constraint on the phase or direction of the photons resulting from spontaneous emission, the radiation is noncoherent.

The process of stimulated or induced emission is similar to absorption in that an electron undergoes a level-changing transition under the influence of an external radiation field. However, in this case the induced transition is downward, from a higher to a lower energy level, so that there is a net loss in energy of the particle accompanied by an increase in the energy of the radiation field. The most significant point here is that, unlike spontaneous emission, the radiation resulting from stimulated emission propagates in the same direction and has the same frequency as the stimulating field. Imagine starting with a single photon that interacts with a particle to stimulate the emission of a second, similar photon, and these two interact to stimulate two additional photons, and

FIGURE 7.50 ▪
Absorption,
Spontaneous, and
Stimulated Emission

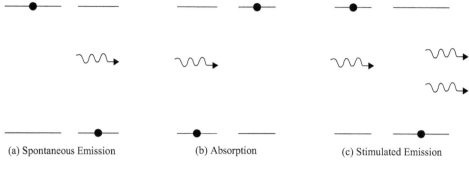

(a) Spontaneous Emission   (b) Absorption   (c) Stimulated Emission

FIGURE 7.51 ▪
Three-Level Laser
Scheme

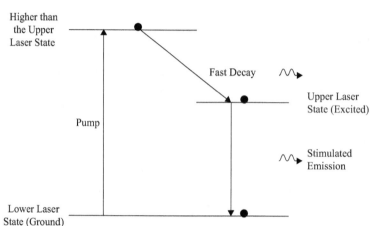

these four interact to stimulate four more. Eventually a large radiation field will build up and every photon will have the same characteristics as the one that started the process. Figure 7.50 shows this absorption, spontaneous, and stimulated emission.

What has been described so far is a two-level process; the particles have only two available energy levels: the grounded state and the excited state (sometimes called the lower laser and upper laser states). The difference between the two states is referred to in terms of photon energy, or $h\nu$. Since the probabilities for stimulated absorption and emission processes are equal, it is detrimental to have any particles in the ground state. For this reason, two-level lasers are not practical; it is not generally possible to pump more than half of the molecules into the excited state. There are two methods for avoiding this problem: three- and four-level laser schemes.

The three-level scheme first excites the particles to an excited state that is higher in energy than the upper laser state. The particles then quickly decay down to the upper laser state. It is important for the pumped state to have a short lifetime for spontaneous emission compared to the upper laser state. The upper laser state should have as long a lifetime (for spontaneous emission) as possible so that the particles live long enough to be stimulated and contribute to the gain of the laser. Figure 7.51 depicts the three-level laser scheme.

The four-level laser scheme goes a step further by also depopulating the lower laser level by a fast decay process. This greatly decreases the loss of laser photons by stimulated absorption processes since the particles in the lower laser level have a short lifetime for spontaneous emission. Figure 7.52 depicts the four-level laser scheme.

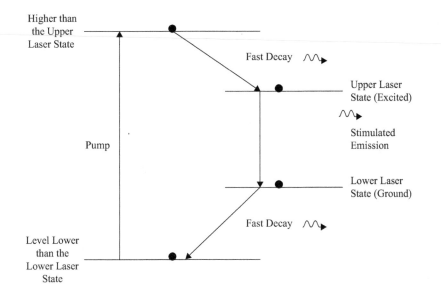

FIGURE 7.52 ▪
Four-Level Laser
Scheme

The main attributes of lasers, as compared to other sensors, are

- Collimation
- Monochromaticity
- Short pulse capability
- High spectral radiance

Laser beams can be collimated to approximately the diffraction limit; that is, the rays are nearly parallel to one another and diverge only slightly as they travel. In fact, a plane-wave laser source radiates a beam that is almost constant in width for a distance (D) given by the following relationship:

$$D = d^2/4\lambda, \tag{7.23}$$

where

d = the diameter of the source, and

λ = the wavelength of the radiation.

Another important feature of laser light is that it is monochromatic, or a single color; all the light waves in the beam have the same wavelength. Almost all conventional light sources emit light that is a mixture of wavelengths, or colors, and as a result, the light appears white. A few conventional light sources, such as mercury-vapor lamps and neon lights, emit nearly monochromatic light, but their light is neither coherent nor inherently collimated. A laser assembled with sufficient care to operate at all, operates at a Q in excess of 10^4. Q is a measure of the power distribution or efficiency, and is calculated as

$$Q = f/BW, \tag{7.24}$$

where

f = central radiation frequency (Hz), and

BW = bandwidth of radiation (Hz).

Because the triggering mechanism for energy release in a laser is the passage of photons rather than transport of mass particles, stored energy can be released rapidly to provide short-duration, high-energy pulses as compared to microwave or other visible or IR pulsed power sources.

Perhaps the most important limitation of ordinary light sources is their inherent low brightness. No matter how high their temperature, they cannot emit more energy than a perfect radiator. The theoretical output of a perfect radiator, a blackbody, is given by the blackbody radiation curve first derived by Planck. The visible surface of the sun, for example, behaves much like a blackbody with a temperature of about $6000°C$. The sun's total radiation, at all wavelengths, is 7 kW/cm^2 of its surface, and no matter how we collect and concentrate sunlight it is impossible to achieve any greater radiation. If desired, the laser's power can be focused to produce intense heating. For instance, a lens with a focal length of 1 cm will focus the beam to a spot only 10^{-2} cm in diameter, corresponding to an area on the order of 10^{-4} cm^2. In this spot the laser beam will deliver power at a density of 100 million W/cm^2. Brief though the flash is, its power is thousands of times greater than that which could be obtained by focusing sunlight, and is enough to melt or vaporize a spot on the surface of even the most refractory material. This ability to deposit huge amounts of energy must be carefully considered anytime the laser is fired. Very expensive test equipment has been destroyed in the past because of operators failing to appreciate this fact. This high spectral radiance is a result of the spatial coherence of the laser light.

It should be noted that the output power of the laser is quite high and that many of the operating wavelengths are in the visible spectrum, which creates a problem for possible eye damage. The cornea is the transparent window at the front of the eyeball and transmits light with wavelengths between 0.4 and 1.4 μm. Laser light in this band can penetrate the retina of the eye and be a potential source of damage. Light entering the eye is brought to a point on the retina producing an optical gain of between 2×10^5 and 5×10^5; if a gain of 5×10^5 is assumed, a beam of energy 1 mJ/cm^2 at the cornea becomes 500 J/cm^2 at the retina. The light passes through the nerve layers of the retina and reaches the light-sensitive layers, which rapidly heat. Causing damage to these layers produces a blind spot. The extent of the damage depends on the energy reaching the light-sensitive layers and the area irradiated. Short-pulsed lasers (less than 10 msec) may give rise to a different damage mechanism owing to the large amount of energy being delivered in a very short time. A rapid rise in temperature occurs such that the liquid components of the cells are converted to gas, with a resultant swelling of the cells. In most cases these changes are so rapid that they are explosive and the cells rupture. The pressure transients produced result in an annular blast zone around the thermal damage; the immediate damage is severely compounded if a blood vessel is ruptured and blood flows into the vitreous.

NATO STANAG 3606 gives maximum permitted exposure levels of irradiance that are agreed to provide an acceptably low probability of optical damage for neodymium: yttrium aluminum garnet (Nd:YAG) lasers. These are 50 mJ/m^2 for single-pulse lasers and 16 mJ/m^2 for a train of pulses at 10 Hz. The eye can be protected from laser light by the use of absorbing plastic filters, dynamic eye protectors, or eye patches. Absorbing plastic filters incorporated as a visor are the most practical current method of aircrew eye protection. Organic dyes within the plastic are used to selectively absorb specific wavelengths. However, there will always be a trade-off between optical density and visual transmittance; as one increases, the other decreases. Low visual transmittance, and hence color attenuation, leads to eye fatigue and difficulty with reading cockpit instruments. The ideal dynamic eye protector is nearly transparent except when activated by a hazardous light source, at which time it becomes nearly opaque. Such helmet visors have been demonstrated for use against nuclear flash, but have reaction times on the order of 150 msec, which is too slow for use against laser pulses on the order of nanoseconds.

One other method of making lasers eye safe is the process of Raman shifting, where the laser output is sent through a different medium, such as helium. The wavelength is then shifted out of the 0.4 to 1.4 μm region.

The major areas in which lasers have found military application are

- Communications
- Range finding
- Target marking
- Optical radar
- Direct fire simulators
- Thermonuclear fission
- High-energy laser (HEL)

Speech communication equipment requires a bandwidth of approximately 4 kHz/channel to avoid mutual interference. In the overcrowded radio frequency bands there is insufficient bandwidth available to avoid this interference. Since lasers operate at optical frequencies, there is sufficient bandwidth available to accommodate all the information links in the world without overlap, and recent developments have made it possible to achieve signal modulation at these high frequencies. Unfortunately laser light is attenuated and scattered by the atmosphere in the same way as ordinary light; clouds, haze, and fog limit the long-range possibilities. The high monochromaticity of the laser makes it unwise to use wide-band transmission models such as LOWTRAN to estimate the atmospheric propagation of a laser beam. A GaAs laser is being developed by the U.S. Navy for secure, short-range, ship-to-ship communications. Fiberoptic communications, as described previously in chapter 3, use lasers extensively.

If a pulse of light is timed from initiation to reception on reflection from a target, then, since the velocity of light is known, the range can be determined with an accuracy set by the pulse width. In pulsed lasers, the pulse width is as low as a few tenths of a nanosecond, and ranges can be determined to within 10 m, as shown in Table 7.10.

Laser target markers (LTMs) have been developed to pinpoint specific targets in a confusing battlefield target array. Such LTMs, which usually double as a range finder, can be used to reflect laser energy into an airborne laser sensor or a laser-guided weapon. The laser signal normally has a coding superimposed such that the airborne sensor can reject other laser reflected signals.

TABLE 7.10 ■ Typical Laser Range Finder

Minimum range	100 m	Maximum range	20 km
Range accuracy	±5 m	Angular divergence	1 mile
PRF	6 to 10 ppm	Pulse width	10 nsec
Weight	20 kg	Energy output	200 mJ

The higher frequency and shorter wavelength of the laser compared with conventional radar permits its use for the determination of position and velocity in circumstances where radar clutter is normally a limitation. In theory, it is possible to detect high-speed, low-flying aircraft by optical radar; however, in practice it is difficult to direct the laser beam on to such a target and then track it.

The narrow laser beam has been used with GaAs IR lasers to simulate gunfire for training. The weapon is provided with a laser whose LOS is collimated with the weapon and is activated whenever blank ammunition is fired. Targets are fitted with IR detectors. Each laser is coded for identification and the signal is decoded by the target to determine the probability of a kill. Normally, after a hit, the target's laser transmitter is automatically deactivated.

The precision focusing of high-intensity radiation afforded by the laser has been exploited to achieve the high temperatures in deuterium discharges necessary to initiate nuclear fission by thermal means.

The U.S. Department of Defense (DOD) defines an HEL as any laser device having an average level greater than 20 kW or a single pulse of energy greater than 30 kJ.

7.10 | PASSIVE EO FLIGHT TEST EVALUATIONS

The primary purpose of any EO system is to discriminate targets. For these systems, it is the ability to break out a target from the background, either by contrast or a difference in temperature. Three basic descriptions are used for EO systems:

- Detection—an object is present (a blob is seen and has a reasonable chance of being an object of interest)

- Recognition—the class the object belongs to (the blob can be distinguished as a tank, truck, destroyer, fighter, man, etc.)

- Identification—the object is discerned with sufficient clarity to determine type (the blob can be specifically identified as T-52 tank, an F-18, an AEGIS class friendly ship, etc.)

This is best illustrated by Figures 7.53 and 7.54, which are taken from chapter 21 of Holst's text (*Electro-Optical Imaging System Performance*). The first figure is that of a Soviet Bear bomber. View A is a silhouette of this bomber in its clearest detail. View B is what would be seen with the least angular resolution. Each one of the pixilated blocks in each picture is called a resolution cell, or detector angular subtense (DAS); this concept was previously discussed in section 7.7.5 of the text. The DAS is defined as the detector size divided by the focal length; it can be thought of as the angular cone over which the detector senses radiation. Since this is an angular measurement, the detector will see more of the scene as the range increases. Since the object of interest will become smaller with range, it should be obvious that the resolution will also decrease with increased range. As the object is covered by more resolution cells, the image becomes clearer and more distinguishable until Figure F, when it is covered by 64 contiguous DASs across the wing.

The same process can be seen with the Sherman class destroyer in Figure 7.54. Figure A is what the destroyer would look like with total clarity. At some longer range, fewer of the resolution cells will cover the object and less clarity will be obtained (Figure B). Eventually, as the destroyer moves closer to the sensor, more and more

Soviet Bear Bomber

(A) Silhouette
(B) 4 contiguous DASs across wingspan
(C) 8 contiguous DASs across wingspan
(D) 16 contiguous DASs across wingspan
(E) 32 contiguous DASs across wingspan
(F) 64 contiguous DASs across wingspan

For a staring array, the DAS and pixel are the same. Eight levels of gray are used.

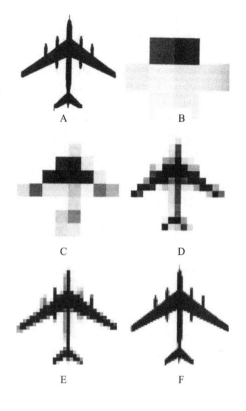

FIGURE 7.53 ▪ Discrimination Versus DAS Coverage

Sherman Class Destroyer

(A) Silhouette
(B) 8 contiguous DASs across length
(C) 16 contiguous DASs across length
(D) 32 contiguous DASs across length
(E) 64 contiguous DASs across length

For a staring array, the DAS and pixel are the same. Eight levels of gray are used.

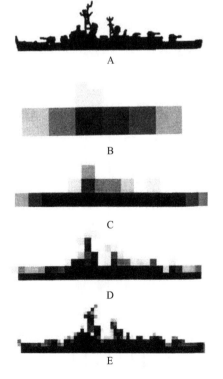

FIGURE 7.54 ▪ Discrimination of a Sherman Class Destroyer

resolution cells fall upon the object until the type can be identified. In both cases we would be able to claim a "detection" with Figure 7.53B and 7.54B, a "recognition" with Figures 7.53C and 7.54C, and an "identification" with Figures 7.53E and 7.54E.

The detection, recognition, and identification process just described is true only for the static case. In reality, there are many external clues that will aid in our discrimination levels. In all cases, there is a human observer involved, and the human image interpretation process is very difficult to model.

One of the key parameters that determine the true resolution of an EO system is called the modulation transfer function (MTF). Literally, the MTF is a measure of how well the system reproduces, or displays, the actual scene. Mathematically, the MTF is defined as the displayed image modulation divided by the input scene modulation. The system MTF is really a summation of the transfer functions of each of the components. In a FLIR system, for example, there would be an optical MTF, a detector MTF, and a display MTF; taken together they represent the FLIR MTF.

An easy way to interpret the MTF is to imagine an optical system looking at an equivalent bar chart. The chart is comprised of alternating black and white bars evenly spaced at 100% contrast. A lens cannot fully transfer this contrast (to the CCD, CMOS, or IR detector) because of the diffraction limit. As the bar spacing is decreased, the frequency is increased and it becomes more difficult for the lens to efficiently transfer the image contrast. This scenario is shown in Figure 7.55, which shows that for a large spacing (low frequency), the contrast is replicated very well with minor loss in contrast. As the bar spacing is decreased (frequency increased), the contrast suffers. The MTF is the measurement of a system's ability to transfer contrast at a particular resolution level from the object to the image, or a way to incorporate resolution and contrast into a single specification. An MTF graph will normally be supplied with the documentation of

FIGURE 7.55 ■
Percentage of Contrast Transferred to Image

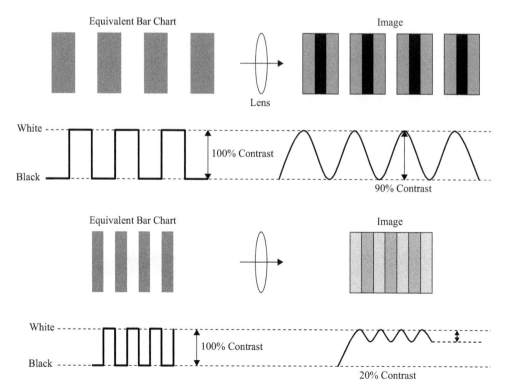

the system from the vendor and plots spatial frequency (cycles/mrad) versus MTF (percent contrast). The MTF is used to approximate the best focus of an IR system and is usually set for the spatial frequency associated with corresponding 0.5% contrast (50%).

Sometimes the discrimination levels are fairly easy: an aircraft with a sky background or a moving boat on the water. Sometimes they are fairly difficult: a cool tank in a high-clutter background, natural objects viewed during the diurnal crossover, etc. At any rate, a process for equating range predictions versus system angular resolution must be devised.

This process is similar to taking an eye test. If we are told that our vision is 20/40, this really means that what we can discriminate at 20 paces from the chart is what a "normal" person can discriminate at 40 paces; or to put it a different way, your eye's angular resolution is not as good as a normal person's. Conversely, if your vision is measured at 20/15, you can discriminate at 20 paces what the "normal" person can discriminate at 15 paces; your angular resolution is better than the "normal" person's. Poor vision can be aided in most cases by improving the optics of your eyes (wearing glasses or contact lenses). The same is true for EO systems.

A very smart man by the name of Mr. John Johnson is credited with determining the relationship between discrimination and DAS. His approach is called the equivalent bar pattern and was first developed for NVGs. He used scale models of eight military vehicles and one soldier against a bland background. Multiple observers were then asked to view the targets through NVGs and ascertain the distances at which they were able to detect, recognize, and identify the targets. Air Force tribar charts were viewed with similar contrast and the maximum resolvable bar pattern frequency was determined. A typical bar chart is shown at Figure 7.56. For optical systems (TV, NVG), the length of any bar is five times the width (L:W is 5:1); for IR systems (IRST, FLIR), the length of any bar is seven times the width (L:W is 7:1). One cycle is defined as one black bar and one white bar, or twice the width of the bar. For optical systems, the discrimination is based on contrast, whereas with IR the discrimination is based on temperature

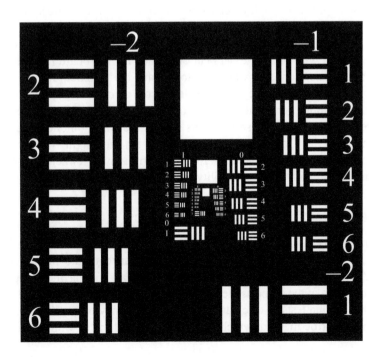

FIGURE 7.56 ■
Equivalent Bar Chart (1951 U.S. Air Force Resolution Target)

FIGURE 7.57 ■
Johnson's
One-Dimension
Equivalent Bar
Pattern

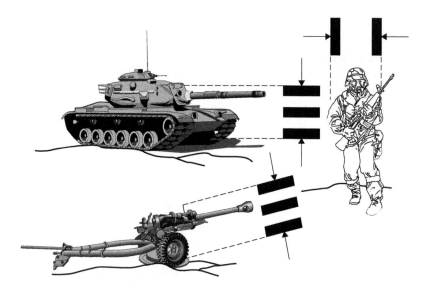

TABLE 7.11 ■ Johnson Criteria

Discrimination Level	Meaning	Cycles Across Minimum Dimension
Detection	An object is present	1.0
Recognition	The class to which the object belongs	4.0
Identification	The object is discerned with sufficient clarity to specify the type	8.0

difference (ΔT). For optical systems, the board is painted black and white, for 100% contrast. There are other boards that are distributed shades of gray which enable the evaluator to plot spatial frequency versus minimum resolvable contrast (100% contrast boards will only provide spatial frequency plots versus lighting levels). For IR systems, the board can either be temperature controlled, as with hot water, or it can be made with black-painted metal with the white bars cut out, then only the board needs to be heated.

Through analysis, Johnson was able to equate the number of cycles, from an equivalent bar pattern (see Figure 7.57), needed across an object in order to discriminate as detection, recognition, or identification. Johnson used the smallest dimension of the object as the critical dimension, so all of his studies are based on one dimension.

Johnson's criteria (Table 7.11) were revised by the Night Vision Laboratory, yielding the current industry standard for a one (minimum dimension)-dimensional target discrimination with 50% probability level (Table 7.12).

7.10.1 Determination of Spatial Frequency

Spatial frequency is used to predict detection, recognition, and identification ranges. Conversely, if we know the size of the target, we can compute the required spatial frequency. As previously mentioned, since the DAS is an angular measurement, it will subtend a larger area of the scene as the distance increases. There are two methods that can be used during ground tests to calculate the spatial frequency of an EO system. The first method is the scientific method, which entails using a collimator with calibrated targets.

TABLE 7.12 ▪ One-Dimensional Target Discrimination (50% Probability)

Discrimination Level	Meaning	Cycles Across Minimum Dimension
Detection	An object is present (object versus noise)	1.0 ± 0.025
Orientation	The object is approximately symmetrical or unsymmetrical and its orientation can be discerned (side view versus front view)	1.4 ± 0.35
Recognition	The class to which the object belongs (e.g., tank, truck, fighter, destroyer, man)	4.0 ± 0.8
Identification	The object is discerned with sufficient clarity to specify type (e.g., T-52 tank, friendly jeep)	6.4 ± 1.5

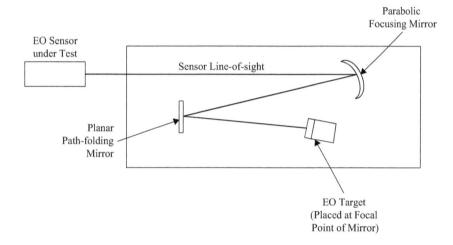

FIGURE 7.58 ▪ Optically Collimated EO Test Target

This method is used by larger testing organizations who can readily access specific lab equipment. The second method can be accomplished if better equipment is not available and can achieve results very close to the scientific method.

7.10.1.1 Scientific Method of Determining Spatial Frequency

Spatial frequency can be determined scientifically by using a combination of the bar chart and a collimator. This procedure is described in detail in the *Avionics Guide to Flight Test* (section 4.2.9), from the Advisory Group for Aerospace Research and Development (AGARD). AGARD has since been renamed the Research and Technology Organization and is under the auspices of NATO. The procedure is also described in detail in John Minor's article, "Flight Test and Evaluation of Electro-Optical Sensor Systems." It will also be a requirement to test the spatial frequency while airborne, as this will enable the evaluator to determine if there are any aeromechanical or aero-optical effects (see section 7.6 of this text).

The EO collimator consists of a spherical collimating mirror and a planar path-folding mirror mounted on an optical bench with an EO target. The target can consist of a radiation source, such as a laser or other illuminator, a reflective target to be illuminated by an external source, or a thermal bar target for IR sensor testing, so the same setup can be used for any EO system. The arrangement of the collimated target components is as shown in Figure 7.58.

The EO target is positioned at the focal point of the collimating mirror. The radiation from a single point on the target is thus reflected from the collimating mirror with all rays parallel to the line of sight from the EO sensor under test to the point on the target. The image of the target therefore appears to the sensor as if it is at an infinite range. The purpose of the collimated source is to provide calibrated targets for long-range sensors without the difficulties associated with actually providing targets at large ranges.

The thermal bar target consists of a temperature-controlled surface (IR source) with a slotted template mounted in such a way as to block part of the radiation from the source. The temperature of the template is held constant at ambient temperature by mounting on a heat sink. A bar of the target is formed where a slot in the template allows the sensor under test to view the temperature-controlled source through the template. The spaces of the target are formed where the spaces between the slots in the template block the radiation from the temperature-controlled source. The spatial frequency of the collimated bar target image is adjusted by changing bar target templates with different bar and space widths (smaller will yield a higher frequency and larger will yield a lower frequency). The typical temperature-controlled source is electrically heated and thermoelectrically cooled to produce a bar-space temperature differential of 20°C to +20°C. The temperature differential must be configurable to allow sufficient testing of the system under test at its design limits (e.g., 0.2°C to about 0.05°C). A representation of an IR sensor under test is shown in Figure 7.59.

FIGURE 7.59 ■
Collimated Thermal
Bar Target

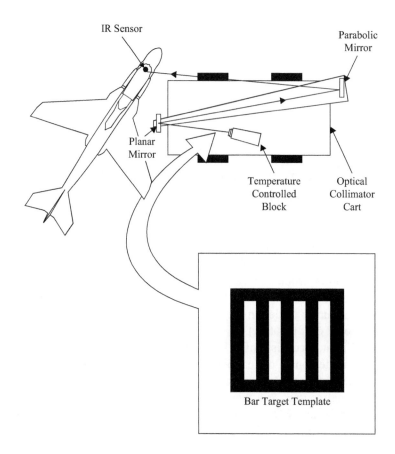

7.10.1.2 A Secondary Method of Determining Spatial Frequency

By using the equivalent bar chart (or spatial frequency board), we are in effect calculating the angular resolution. On the ground, the board is placed perpendicular and in front of the sensor at a given distance (Figure 7.60). The actual board will have many different-size bar patterns, but all in the correct ratios, similar to the chart depicted in Figure 7.56.

In the optical case, the lighting at the board (with a light meter) is measured and the bars are viewed at the sensor display in wide FOV or minimum zoom. The board is moved either toward or away from the sensor until the bars become indiscernible. The next step is to refine the exact point that is the maximum distance at which the bars are seen clearly. Measurements that need to be made are the distance to the board, light intensity at the board, width of one full cycle (one white plus one black bar), and FOV selected. The exercise is repeated for different FOVs and different lighting levels. The test is repeated using contrast boards to determine the minimum resolvable contrast. Everything is the same for the IR sensor, except temperature measurements are substituted for lighting levels. The temperature differences (ΔT) are obtained by heating the metal board with an appropriate device (heat gun, heating elements, etc.) and measuring with a thermocouple (board temperature minus ambient temperature). A data card for the test is shown in Table 7.13.

The angular resolution and the spatial frequency are calculated from the recorded data. If we call the width of one cycle d and the distance to the board R, then the spatial frequency (f_S) is given by

$$f_S = R/d \times 0.001 \text{ cycles/mrad.} \tag{7.25}$$

The angular resolution (θ_r) is equal to $1/f_{S(cutoff)}$. It should be noted that the angular resolution provided by the manufacturer is normally for one detector element or DAS. The spatial frequency that we calculate and plot is based on 1 cycle (1 black/1 white or 2 DASs). If the angular resolution is provided by the manufacturer, then the $f_{S(cutoff)}$ will be equal to $1/2\theta_r$.

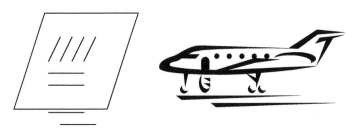

FIGURE 7.60 ■ Spatial Frequency Ground Test

TABLE 7.13 ■ Spatial Frequency Ground Test Card

Sensor Type/Installation:					Date:		
Lighting	**ΔT**	**Time**	**FOV**	**Cycle Width**	**Distance to Board**	**Angular Resolution (Calculated)**	**Spatial Frequency (Calculated)**

FIGURE 7.61 ▪
Spatial Frequency
Results

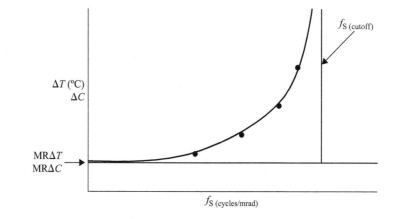

FIGURE 7.62 ▪
Airborne Spatial
Frequency Test

The spatial frequency can then be plotted as a function of ΔT for IR systems, or lighting level (i.e., contrast) for optical systems. A series of plots are then developed for all FOVs available. If a system has a wide, narrow, and zoom FOV, then three spatial frequency plots will be developed. These plots become the installed spatial frequency and are compared to the manufacturer's stated spatial frequency to calculate the degradation due to installation. An example of the spatial frequency obtained from the ground test is shown in Figure 7.61. The figure is bounded horizontally by the MRΔT; at lower spatial frequencies the system is sensitivity limited by its ability to resolve temperature differences. The figure is bounded vertically by the cutoff spatial frequency, which is the resolution limit of the system.

To be done correctly, the airborne test should be done against a calibrated EO ground target. A representation of the setup is shown in Figure 7.62.

The ΔT of the IR board is known. The aircraft flies toward the board and the operator calls "mark" when the bars are discriminated. This test is repeated for different ΔT, FOV, polarity settings, and automatic versus manual level and gain. The airborne spatial frequency test for EO systems can be flown the same way, but the EO target board may be painted on the ground, on a wall, or on mobile boards of known contrast. Truth data are required for this test, as I have to know where I am in space when the "mark" is called. I also need to know the surveyed position of the target board and my altitude, as well as the atmospherics for the test so that data can be corrected for transmissivity (see section 7.5). For TV and camera systems, the lighting conditions must also be known; a photometer may be used to obtain this information. A sample test matrix is found in Table 7.14. The airborne spatial frequency that is calculated is then plotted against the manufacturer and installed spatial frequencies to obtain the aeromechanical and aero-optical effects.

The end result should look something like Figure 7.63.

A Note about Spatial Frequency: It is possible to obtain a horizontal spatial frequency and a vertical spatial frequency. This happens when either the vertical or

TABLE 7.14 ■ Flight Test Matrix for Spatial Frequency

System under Test	ΔT	FOV	Level and Gain	Polarity
	Approximately six 0°C–15°C	Wide medium narrow zoom	Manual and automatic	White hot and black hot

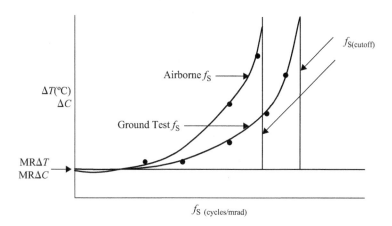

FIGURE 7.63 ■ Differences Between Installed and In-Flight Spatial Frequency

horizontal bars merge while the other set can still be discriminated. If we remember the setup with standard U.S. TV systems, there are 480×640 pixels that make up the screen. This would indicate that the vertical bars on the display should be present even as the horizontal bars start to merge. In order to use a combined spatial frequency, take the square root of the product of the horizontal and vertical spatial frequencies.

A second note about Spatial Frequency: If you are attempting to calculate the spatial frequency of an EO system which incorporates an automatic iris, the automatic iris must first be disabled before changing lighting levels. If it is not, the evaluator will find that the spatial frequency will remain constant for all lighting levels.

7.10.2 Range Predictions

By knowing the spatial frequency, the atmospherics, and some of the attributes of the target, we can make pretty good estimations on detection, recognition, and identification ranges. By performing the following steps, you too can make these predictions.

Step 1. Determine the spatial frequency (described in section section 7.10.1). Remember that the ratio of width to height in the bar pattern is 5:1 for EO (minimum resolvable Δ contrast) and 7:1 for IR (minimum resolvable ΔT).

Step 2. Determine the critical dimension. As previously discussed, and shown in Figure 7.57, and Tables 7.11 and 7.12, the one-dimensional (1D) target discrimination is based on the smallest dimension of the target. A 2D discrimination is based on the square root of the area. If we let h_c be the critical dimension, then

$$h_c = \text{smallest dimension (1D)}$$
$$h_c = \sqrt{LW} \text{(2D)}$$

(7.26)

It should be noted that as the aspect of the target changes, the apparent length and width (as seen by the viewer) changes. This will affect the range at which the target is seen. This is graphically represented in Figure 7.64.

FIGURE 7.64 ■
Aspect Change
versus Recognition
and Identification
Ranges

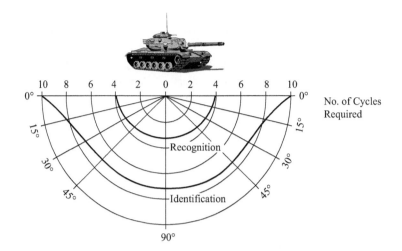

Step 3. Determine the critical distance. This is really the process of overlaying the required bar patterns over the target with the bars aligned to the longest dimension. The parameter is designated as d_c and is defined as

$$d_c = h_c/N. \tag{7.27}$$

The critical dimension is known from step 2, and N is the required number of cycles across the target for each phase of discrimination. From Table 7.12, we see that the number of required cycles for detection, recognition, and identification for the 1D target discrimination is 1, 4, and 8, respectively. Should we be interested in a 2D discrimination level, we would need another table. For military applications, two sets of tables are provided which call out "extended" discrimination levels. These tables try to define the gray areas between detection, recognition, and identification and are included as Tables 7.15 and 7.16 for completeness.

Step 4. Calculate the predicted ranges for specific targets. The basic range prediction formula is given by

$$R_{\text{Pred}} = d_c f_S \times 1000. \tag{7.28}$$

This formula is rarely used since it is only good for a 50% probability of detection in a moderate clutter environment. (These are the conditions for Tables 7.15 and 7.16.) What would happen if you wanted a 75% probability of detection, or if the target was in a high-clutter environment? For these cases, the range prediction formula becomes

$$R_{\text{Pred}} = \frac{h_c}{NxTTPF} f_S \times 1000. \tag{7.29}$$

The only variable not known in the second form of the equation is the target transfer probability function (TTPF). Think of the TTPF as an adjustment of the number of cycles required due to changes in the probability of detection requirement or clutter environment. The TTPF due to changes in the probability of detection is shown in Table 7.17. A combined TTPF for clutter region and probability is shown in Table 7.18.

TABLE 7.15 ■ Extended 1D Discrimination Levels

Extended 1D Discrimination Levels		
Task	Description	Recommended No. of Cycles
Detection	A blob has been discerned which may warrant investigation	0.5
	The blob has a reasonable chance of being the object sought	1.0
	The blob has a high probability of being the object sought due to location, motion, etc.	1.5
Orientation	The object is roughly symmetric or orientation can be discerned	2.0
Clutter Rejection	The object is a potential target and not clutter	2.25
Classification	The broad class of object types to which the object belongs can be determined	2.5
Type Recognition	Object discerned with sufficient clarity that the class can be determined	3.0
Classical Recognition	Object discerned with sufficient clarity that its specific class can be determined	4.0
Identification Friend or Foe	The country of manufacture can be determined	6.0
Identification	Object discerned with sufficient clarity to determine specific type within the class	8.0
Target Selection	Real targets can be distinguished from replica decoys	10.0
Operational Success	Ownership of the object can be determined and probable hostile/nonhostile intent established	12.0

TABLE 7.16 ■ Extended 2D Discrimination Levels

Extended 2D Discrimination Levels		
Task	Description	Recommended No. of Cycles
Detection	A blob has been discerned which may warrant investigation	0.38
	The blob has a reasonable chance of being the object sought	0.75
	The blob has a high probability of being the object sought due to location, motion, etc.	1.13
Orientation	The object is roughly symmetric or orientation can be discerned	1.5
Clutter Rejection	The object is a potential target and not clutter	1.69
Classification	The broad class of object types to which the object belongs can be determined	1.88
Type Recognition	Object discerned with sufficient clarity that the class can be determined	2.25
Classical Recognition	Object discerned with sufficient clarity that its specific class can be determined	3.0
Identification Friend or Foe	The country of manufacture can be determined	4.5
Identification	Object discerned with sufficient clarity to determine specific type within the class	6.0
Target Selection	Real targets can be distinguished from replica decoys	7.5
Operational Success	Ownership of the object can be determined and probable hostile/nonhostile intent established	9.0

TABLE 7.17 ■ TTPF for Probability of Detection

Probability of Discrimination	Multiplier
1.00	3.0
0.95	2.0
0.80	1.5
0.50	1.0
0.30	0.75
0.10	0.50
0.02	0.25
0	0

TABLE 7.18 ■ TTPF for Clutter and Probability of Detection

Probability of Detection	Low Clutter SCR > 10	Moderate Clutter 1 < SCR < 10	High Clutter SCR < 1
1.0	1.7	2.8	**
0.95	1.0	1.9	**
0.90	0.90	1.7	7.0*
0.80	0.75	1.3	5.0
0.50	0.50	1.0	2.5
0.30	0.30	0.75	2.0
0.10	0.15	0.35	1.4
0.02	0.05	0.1	1.0
0	0	0	0

*Estimated.
**No data available.

The SCR in Table 7.18 is the signal-to-clutter ratio; there are three defined SCRs used in the evaluation of EO systems. The SCR is defined as

$$SCR = (\text{max target value} - \text{background mean})/\sigma_{\text{clutter}}, \tag{7.30}$$

where

$$\sigma_{\text{clutter}} = \sqrt{\left(\frac{1}{N}\sum_{i=1}^{N}\sigma_i^2\right)},$$

σ_i = the pixel value root mean square (rms) in a square cell that has side dimensions twice the target minimum dimension, and

N = the number of adjoining cells.

This multiplier can now be inserted into the previous equation and the predicted ranges can be computed.

Step 5. Adjustments to the predicted ranges based on atmospherics. If you remember from the previous discussions, the transmissivity, or extinction coefficient, changes with respect to the current atmospherics. The ranges will have to be adjusted for the loss of energy due to scattering and absorption. If the transmissivity is 80%, or a loss of 20%, then all of the predicted ranges will be adjusted (reduced) by 20%. The new range will be calculated by multiplying the calculated range by 0.80. These adjustments can be found in section 7.5 of this text, or for a better explanation, see chapter 15 of Holst's text, *Electro-Optical Imaging System Performance*.

Step 6. Operational considerations when performing range predictions. There are some other considerations that must be dealt with when calculating detection, recognition, and identification ranges. In normal operations, search would be performed in a wide FOV; therefore, the spatial frequency used in calculating the detection range will be the one calculated in wide FOV. Once something is detected, the operator will normally select a medium FOV or narrow FOV for recognition; the spatial frequency used here is the one for the appropriate FOV. Since the spatial frequency will increase for the same ΔT, it is not uncommon to calculate a recognition range that is higher than the detection range (see the range prediction formula in step 4). When this occurs, use the recognition range equal to the detection range due to the change in FOV, and hence the spatial frequency. The same considerations apply to identification range if another FOV is available.

In looking at the information required to perform range predictions, we can readily see that if we do not have a calibrated target, then we have to make some assumptions. We will have to know (or assume) the target size and the associated ΔT or contrast of the target with its background. Some evaluators make the mistake of using a very high ΔT or a very high brightness level in their predictions, especially when looking at targets that generate their own radiation (combustion engines, etc.). Remember that detectors have a cutoff frequency, and there will be a temperature at which the detector will not respond very well with increased temperature (see section section 7.7.2). These detectors will be resolution or gain limited. The area-weighted ΔT approach described in section 7.6 works well for the range prediction problem. The answer obtained may be slightly conservative, but it is better than being overly optimistic.

The same is true for target size. Theoretically, as target size increases, the apparent detection range has no bounds, but one has to remember how the spatial frequency is calculated. As the target size (d) grows, the spatial frequency approaches zero; with inhibitory response, the detection range actually decreases as target size grows. When performing range calculations, it is best to concentrate on a portion of large targets rather than the entire complex (smokestacks of a factory, the locomotive of a train, etc.).

The last consideration is how the operator will use the system. The inexperienced operator may elect to have the system provide the displayed picture. This is an automatic level and gain select feature of EO systems. The experienced operator may elect to manually limit what is displayed to him and control the level and gain. The level and gain of an IR system can be somewhat equated to the brightness and contrast of a TV system. The level is associated with the minimum resolvable temperature. The operator may elect to see all targets (temperatures) that the system is capable of seeing (maximum level). If a system can detect targets above 1°C and can resolve targets to within 0.1°C, maximum level will show all targets in the FOV whose temperature is 1°C or higher. During the daytime, probably everything in the FOV is above 1°C, and the display will be washed out. As the operator levels down, the minimum resolvable temperature (the temperature of objects displayed) is increased and the noise is removed from the display. In order to get definition (contrast), the operator uses the gain control. Maximum gain will show contrast based on the maximum gain of the system; in this case 0.1°C. This will once again wash out any definition, since most objects will be within 0.1°C of each other. As the operator gains down, the contrast is based on higher ΔT and the picture has more definition. When spatial frequency is being determined, the more experienced operator, using manual settings, will invariably yield higher spatial frequency values and better results.

7.10.3 Field of View and Field of Regard

Field-of-view and FOR testing are important for two reasons. The first is that we want to ensure that wide FOV covers an area large enough to aid in detections, but not so large as to degrade resolution. Narrow FOV should be small enough to provide enhanced resolution, but not so small to cause objects to leave the FOV with small slew inputs. The second is to ensure that the FOR is suitable for the mission in regards to view and tracking requirements.

The easiest way to calculate the FOV is to position the aircraft with the sensor at boresight pointing toward a wall. For each FOV, have an assistant mark on the wall the edges of the FOV as called out by the operator. This test is shown pictorially in Figures 7.65 and 7.66.

If the sensor is slewable, the FOR is calculated by rotating the sensor through 360° (or the slew limits) in azimuth or elevation, noting any obstructions. The FOR should then be plotted, noting these obstructions.

7.10.4 Boresight

Boresight testing is accomplished to ensure that all sensors (including the pilot) are using the same plane of reference and are free from parallax errors. Slight errors in boresight may prevent an operator from seeing a target in a narrow FOV display because the target is outside of the azimuth or elevation limits of the display. Corrections to the boresight can be accomplished either mechanically or in software, and ground tests will determine if the corrections were implemented successfully.

The alignment of the EO sensor LOS reference (boresight) with the aircraft armament datum line is normally determined by employing a special test fixture incorporating an IR

FIGURE 7.65 ■
Horizontal FOV
Determination

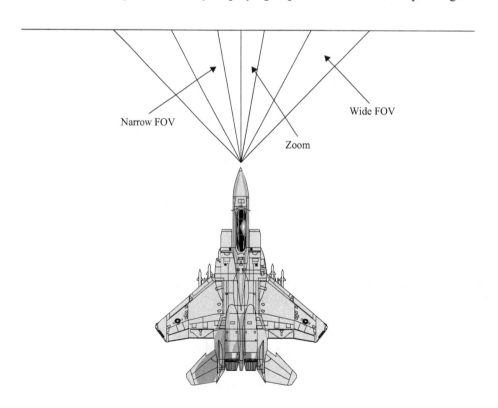

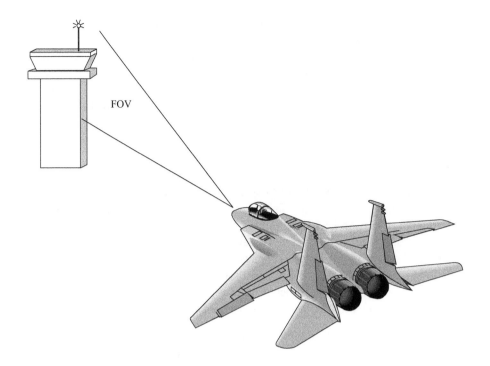

FIGURE 7.66 ■
Vertical FOV
Determination

target and collimator (or autocollimator) that bolts to the airframe and provides an on-boresight target. The following procedure is employed:

- Mount the test fixture on aircraft.
- Set sensor LOS reference to zero azimuth and zero elevation.
- Note the indicated position of the test target in relation to the sensor LOS reference (boresight).

The relative alignment of the boresight of an EO sensor with respect to the boresights of other sensors and systems can be determined by simultaneously acquiring a known target on the boresight of the respective sensors and comparing their indicated positions. Both on-the-ground and in-flight procedures are employed.

For operational considerations, and if a fixture is not available, use the same wall that was used for the horizontal FOV test and note the relative positions of the pilot design eye point and the boresight (0° azimuth, 0° elevation), then calculate the angular difference from the pilot point of reference. It may be found that the difference is significant enough (especially in narrow FOV or zoom) that the system is unusable (i.e., the target is out of the sensor's FOV). The same procedure is carried out for the horizontal and vertical FOV (shown in Figure 7.66).

7.10.5 Sensor Slew Rate

Slew rates are tested to determine if the operator can adequately control the movement of the sensor. Slew rates are critical in manual track systems; overly sensitive controls will cause operator-induced oscillation, whereas highly damped controls may prevent tracking entirely. Excessive slew rates in the narrow FOV will cause the target to leave the FOV and the operator may be forced to return to the wide FOV to reacquire the target.

The maximum sensor LOS slew rates are most readily determined by commanding sensor head slew from one limit to the opposite limit and timing the transit with a stopwatch. The test should be accomplished in both directions (left/right and up/down) and should be repeated for repeatability and statistical significance. The evaluator should note if the slew rate is pressure sensitive, that is, more pressure on the switch commands a higher slew rate. The test should also be repeated in all FOVs, since the slew rate decreases with decreasing FOV.

7.10.6 Line-of-Sight Drift Rate

Drift is a significant problem when EO systems are used to either slave other sensors, such as a laser designator or range finder, or when they are used in weapons employment. Pointing errors in an LRF will result in erroneous ranges, and bearing miscues to a weapon may result in an errant bomb. The evaluator needs to determine the drift rate of the system and determine its impact on mission requirements.

The long-term LOS instability (i.e., drift rate) of an EO sensor is best determined on the ground, using the following procedure, illustrated in Figure 7.67. This test can be accomplished in conjunction with the horizontal FOV test covered in section 7.10.3.

- Zero the drift with the sensor control as per manufacturer instructions.
- Mark the sensor LOS reference (cross-hairs) on the wall.
- After various time intervals, note the positions of the points in the target plane at which the sensor reference is then directed.
- Calculate the sensor drift rate according to the formula:

$$\theta = \frac{\tan^{-1}[D(t_1)/R]}{t_1 - t_0},\tag{7.31}$$

where

$D(t_1) =$ linear drift distance at range R at time t,

$R =$ range from sensor to target plane,

$t_1 =$ time of measurement, and

$t_0 =$ time of initial alignment.

The same procedure is employed to determine drift in elevation.

FIGURE 7.67 ▪
Sensor LOS Drift
Rate

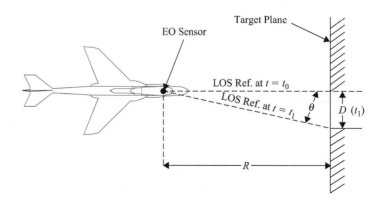

7.10.7 Bearing Accuracy

Bearing accuracy is tested in conjunction with the FOV and drift rate tests described in sections 7.10.3 and 7.10.6. Place the sensor boresight (cross-hairs) at increments of 10° (as indicated by the display) and have an assistant mark the wall where the sensor is pointing. Compare the calculated bearing to the displayed bearing. In-flight determination of bearing accuracy can be performed according to the following procedure:

- Fly the prescribed flight path, at constant altitude, toward a well-defined target of known position.

- Acquire the target on the sensor and vary the bearing to target and record the indicated bearing as a function of time.

- Note the aircraft position using appropriate TSPI data and calculate test aircraft to target bearing.

- Compare the target bearing indicated by the sensor to truth data by time correlating the data.

7.10.8 LOS Jitter

The short-term LOS instability (jitter) of an EO sensor must be determined in flight because in-flight vibration is a major contributor to LOS jitter. The usual method of measurement of LOS jitter is indirect and utilizes the measurable loss of sensor angular resolution due to LOS jitter (delayed detection ranges). A direct method of determining the LOS jitter of an EO imaging sensor is to record the displayed image of a well-defined source (using a camera or video recorder) and measure the actual jitter in the displayed image.

7.10.9 In-flight Detection, Recognition, and Identification Ranges

The maximum ranges at which an EO sensor provides detection, recognition, and identification of a target must be determined in flight and can utilize an EO sensor bar target or predetermined operational-type targets. The EO bar target has the advantage of known, controlled characteristics. Mission-related targets have the advantage of realism. Although the radiant intensity of some mission-related targets can be measured by independent instrumentation, quantitative sensor evaluation generally requires the use of a specially designed target such as the bar target. The procedure for determining the maximum detection and recognition ranges is as follows:

- Fly the prescribed flight path, at constant altitude, toward a well-defined target at a known location (turn-in ranges will have been established by the range predictions made previously).

- Determine the aircraft's position (range to target) by range (ground) instrumentation or by an onboard ranging system (laser range finder/radar) and record as a function of time.

- Align the sensor LOS to place the EO target of known anticipated position in the wide FOV.

- Acquire the target (detection) on the sensor (noting the time of detection).

- Switch to the medium FOV or narrow FOV and continue to observe the EO target on the sensor until the target features, or bars for a bar target, become distinguishable.

- Note the time of feature definition as the time of recognition and note the same parameters as for detection.
- Continue the run until identification of the target is obtained and note the same parameters as for detection.
- Correlate aircraft-to-target range times with detection and recognition times to obtain maximum ranges for detection, recognition, and identification.

NOTE: The maximum detection, recognition, and identification ranges are a function of the radiant and spatial characteristics of the target, the ambient background illumination, and the atmospheric conditions at the time of the test. To be meaningful, the maximum ranges must be associated with specific target characteristics and test conditions.

7.10.10 Target Tracking

There are three fundamentally different methods of "position tracking" employed with EO sensors. True tracking systems use the video signal produced by the target and track some feature of the target image as seen by the tracker (the centroid of the image, edge [contrast] of the image, etc.). Inertial position tracking uses information from an inertial navigation system and tracks a fixed location on the surface of the earth and not necessarily a specific feature. Target designator tracking tracks illumination from a target designator (i.e., a laser designator). In any case, the tracking error can be determined by the following:

- Fly a prescribed flight path toward a prominent EO target of known location.
- Determine the sensor/aircraft position and velocity with suitable range or onboard instrumentation.
- Acquire the EO target on the sensor.
- Initiate target track at the earliest practicable instant.
- Record the corresponding range as maximum track acquisition range.
- Alter the flight path and attitude of the aircraft in order to subject the EO sensor tracking LOS to specified LOS rates.
- Measure and record the EO sensor video signal and tracking error signal as a function of time (the EO sensor display can also be recorded using a cockpit camera).
- Correlate the derived tracking errors with LOS slew rates by time correlating the data obtained in the previous steps.

It is usually a good idea to first test the tracking accuracy on the ground, working from very low LOS rates (such as an assistant walking across the FOV) to higher LOS rates (such as an aircraft in an overhead pattern).

7.10.11 Other Lab Tests

There are other tests that are accomplished in the lab, normally by the engineering group; they are included here for completeness. It is important to note, however, that these tests assess the inherent (i.e., design) characteristics of the system under test. In most cases, these characteristics are advertised and well known. The assessment of these characteristics really only need be addressed when a system is in development and is being prepared for production, or when a system is consistently failing in performance and a decision has been made to determine if the advertised performance is, well, as advertised.

7.10.11.1 Minimum Resolvable Temperature Differential (Thermal Resolution)

The minimum target temperature differential resolvable by an imaging thermal detector is best determined by the combined angular resolution/thermal resolution test. A simpler procedure for determining MRΔT also employs a bar target, as described below:

- Set the spatial frequency of the bar target to a value well below one-half of the sensor cutoff frequency.
- Adjust the bar target temperature differential to a value well above the expected MRΔT of the sensor under test.
- Focus the sensor on the bar target and optimize the sensor control settings.
- Gradually reduce the bar target temperature differential, allowing sufficient time for stabilization between changes, until the variations in radiant intensity due to the individual bars are no longer discernible on the sensor display.
- Gradually increase the bar target temperature differential until the variations in radiant intensity due to the individual bars are just discernible.
- Record the bar target temperature differential attained as the MRΔT for the sensor.

7.10.11.2 Noise Equivalent Temperature Differential

The noise equivalent input temperature differential (NEΔT) for a thermal detector can be determined by the following procedure:

- Set the spatial frequency of a thermal bar target to a value well below one-half of the sensor cutoff frequency.
- Adjust the bar target temperature differential to a value well above MRΔT for the sensor under test.
- Focus the sensor on the target and optimize the sensor control settings.
- Adjust the bar target temperature differential to a value that produces a target signal just large enough to be measured accurately in the presence of the noise.
- Record the target temperature differential and resulting sensor video output voltage response attained in the previous step.
- Block the target input to the sensor and measure the rms noise on the sensor video output voltage.
- Calculate the sensor NEΔT according to

$$NE\Delta T = N\Delta T/S, \tag{7.32}$$

where

N = rms noise voltage on sensor video output signal (volts),

S = peak-to-peak variation in sensor video output voltage due to bar target (volts), and

ΔT = bar target temperature differential producing the output signal S (K).

The response of an EO sensor to time-related variations in target radiant intensity can be determined as a steady-state (sinusoidal) time-frequency response or a transient (impulse or step-function) response. The time-frequency response can be determined by subjecting the sensor to a sinusoidal intensity-modulated source and measuring the amplitude and phase variations in the video output signal from the sensor. The transient response can be determined by subjecting the sensor to a pulsed-intensity modulated source and recording the sensor video output signal response as a function of time.

7.11 | ACTIVE EO SYSTEMS

7.11.1 Lasers

NOTE: The following test procedures are attributed to Dr. George Masters as printed in section 4.4 of *Electro-Optical Systems Test and Evaluation*, revised September 2001, from the U.S. Navy Test Pilot School systems course syllabus. These procedures are in use in many test organizations around the world.

The performance testing methods for passive EO sensors are also applicable to the performance testing of laser ranger/designators, with the comments, modifications, and exceptions noted below. The following test procedures are applicable without exception to the testing of laser designators:

- LOS slew limits
- LOS slew rates

7.11.2 Boresight Accuracy

The boresight alignment of a laser designator is best determined by ground testing using a collimated target. The test procedure is as follows:

- Acquire and focus the EO sensor on a point-source target aligned with the aircraft armament datum line (ADL).
- Place a paper shield over the target and mark the ADL reference position on the paper (materials other than paper can be used).
- Fire the laser, burning a small spot on the paper shield.
- Measure the displacement of the laser burn spot from the ADL reference mark.
- Convert linear laser boresight error (D_e) to angular laser boresight error (δ_e) according to the equation

$$\delta_e = \tan^{-1}[D_e/FL_c](\text{radians}), \tag{7.33}$$

where

    D_e = linear laser boresight error, and

    FL_c = focal length of the collimator.

The pointing accuracy of a laser designator can be determined by using the above test procedure to check laser alignment with points off-boresight (displaced from the aircraft ADL).

7.11.3 LOS Drift Rate

The long-term LOS instability (drift rate) of a laser designator is best determined using the collimated target test setup employed in the boresight test. The procedure is as follows:

- Conduct the first four steps of the boresight test, obtaining a laser burn spot on the paper target.
- After a suitable designated time, fire the laser again, obtaining a second burn spot on the target.
- Measure the relative displacement between the two laser burn spots.
- Calculate the laser designator LOS drift rate, θ_d, according to the equation:

$$\theta_d = [1/t_d]\tan^{-1}[D_d/FL_c](\text{rad/sec}) \tag{7.34}$$

where

D_d = relative linear displacement between two burn spots,

FL_c = focal length of the collimator, and

t_d = time interval between two laser firings.

7.11.4 LOS Jitter

The short-term LOS instability (jitter) of a laser designator must be determined by flight testing because in-flight vibration is a major contributor to LOS jitter. The test procedure consists of the following steps:

- Fly the prescribed flight path, at constant altitude and airspeed, on a heading designed to pass directly over the EO target and normal to the face of the target.
- Acquire the EO target with on an onboard sensor.
- Designate the target (fire the designator laser); observe the (designated) target using the ground-based TV camera and record the resulting video as a function of time.
- Determine the airborne designator-to-target range using suitable onboard or range instrumentation.
- Determine the laser designator LOS angular jitter by time correlating the jitter-induced movement of the laser spot on the target with the designator-to-target range data.

7.11.5 Laser Ranging Accuracy

The ranging accuracy of a laser designator is best determined by flight test employing an EO target. The procedure is as follows:

- Fly the prescribed flight path, at constant altitude and airspeed, on a heading designed to pass directly over the EO target and normal to the face of the target.
- Acquire the EO target with an onboard sensor.
- Designate the target and laser range on the target.

- Record the resulting data as a function of time.
- Independently determine and record the aircraft-to-target range using onboard or range instrumentation (the highly accurate ranging provided by laser ranging makes it extremely difficult to obtain "truth" data of sufficient accuracy).
- Time correlate the laser range data with the "truth" data to determine ranging error.

The static ranging accuracy of a laser ranger can be determined by ground tests using optical retroreflectors positioned at known ranges from the test aircraft position.

7.11.6 Laser Beam Divergence

The divergence of a laser beam is best determined by ground tests employing the collimating apparatus. The laser is fired into the collimator, as shown in Figure 7.68.

Templates with various circular apertures are placed at the focal point of the collimator mirror. The size of the circular aperture is varied until the aperture just begins to occlude the beam, as indicated by a reduction in the radiant flux received by the radiometer. A reduction to 90% of the no-aperture value is taken as the reference point. The angular beam width (divergence) of the laser beam is then equal to the diameter of the (90% flux) aperture divided by the focal length of the collimator. The geometry of the test is illustrated in Figure 7.69.

If all of the rays in the laser beam were parallel ($a = 0$), they will all be focused to a point at the focal point of the mirror. If the most divergent rays (edge of the beam) make

FIGURE 7.68 ■
Laser Beam
Divergence Test
Setup

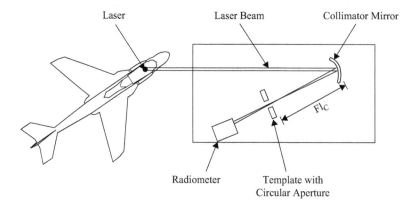

FIGURE 7.69 ■
Laser Beam
Divergence Test
Geometry

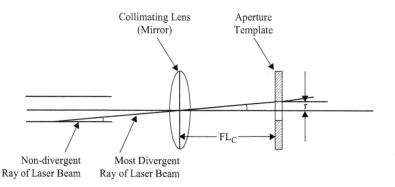

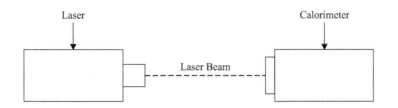

FIGURE 7.70 ■
Laser Output Power
Measurement

an angle α with the beam axis, they will just graze the edge of the hole in an aperture plate of radius r, as shown. Thus the half-angle of the beam is equal to

$$\alpha = r/FL_c \text{ (radians)}, \tag{7.35}$$

where

FL_c = focal length of the collimator (m), and

r = radius of the aperture that just grazes the beam (m).

7.11.7 Laser Output Power

The average output power of a laser ranger/designator is best determined by ground testing employing the test setup depicted in Figure 7.70.

The test procedure is as follows:

- Align the laser output beam with the center of the calorimeter.
- Fire the laser.
- Record the calorimeter output reading (laser average output power).

The pulse peak power and pulse energy of a pulsed laser can be calculated from the average power by means of the following equations:

$$P_{\text{peak}} = P_{\text{avg}}/DC, \tag{7.36}$$

$$DC = \tau_p/PRI, \tag{7.37}$$

$$E_{\text{pulse}} = P_{\text{peak}}\tau_p, \tag{7.38}$$

where

DC = duty cycle,

E_{pulse} = energy in a single pulse (W-sec),

P_{avg} = laser pulse average power (W),

P_{peak} = laser pulse peak power (W),

PRI = laser pulse repetition interval (sec), and

τ_p = laser pulse width (sec).

7.11.8 Laser Pulse Amplitude

The pulse amplitude and pulse amplitude stability of a pulsed laser ranger finder/designator are best determined by ground tests employing the test setup depicted in Figure 7.71.

FIGURE 7.71 ▪
Laser Pulse
Parameter Test
Setup

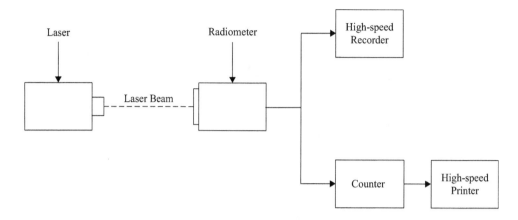

The laser is fired into the radiometer and the output of the radiometer is recorded on the high-speed recorder, as shown in the figure. Pulse amplitude stability can be determined by comparing the output recording over a specified time interval. Laser misfire rate (missing pulse rate) can also be determined by examination of the recorded radiometer output.

7.11.9 Laser Pulse Width

The pulse width of a pulsed laser ranger finder/designator is best determined by ground tests employing the test setup depicted in Figure 7.71. The laser is fired into the radiometer and the output of the radiometer is recorded, as shown. Because of the small pulse widths and the extremely fast rise and fall times associated with a laser pulse, an accurate determination of pulse width requires the use of special, fast-response detectors and recorders. In addition, corrections are generally made for the response times associated with the detector and recorder.

7.11.10 Laser Pulse Repetition Interval

The pulse repetition interval (or frequency) of a laser range finder/designator is best determined by ground tests employing the test setup depicted in Figure 7.71. The laser is fired into the radiometer and the output pulses are counted and timed by the counter, as shown. The output of the counter (pulse count and pulse rate) is recorded by a high-speed printer. Analysis of the high-speed printout yields pulse repetition frequency (interval) and PRF stability.

7.12 | SELECTED QUESTIONS FOR CHAPTER 7

1. What is the relationship between frequency and wavelength?
2. What attributes of a target can an optical system provide?
3. What frequency bands do most military applications of FLIR use?
4. What characteristics would one find in the MWIR band?
5. What characteristics would one expect to see in the LWIR band?

6. What band would you expect to see NVGs operate in?

7. What is an IRLS?

8. How many detectors would you expect to see on a staring array system?

9. Briefly explain how a scanning array operates.

10. Why are FLIR arrays grouped in factors of 480×640?

11. Why do air-to-ground missiles come in versions of EO/IR and laser?

12. If you wanted to launch an IR missile from a distance of 3 nm, and the error of the system was 1 mile, how small could the target be?

13. You want to track missile plume and you have a choice of detectors: indium antimonide (InSb) or mercury cadmium telluride (HgCdTe). Which would you choose?

14. What are the critical parameters to know when employing an IRLS?

15. How is scan rate related to velocity, height, and FOV in an IRLS?

16. In which frequency band would you expect to find a staring array?

17. How is background radiation compensated for in a FLIR system?

18. How are detectors cooled?

19. What is an IRST system? What advantages are there over a FLIR system?

20. How does an IRST system determine range?

21. What is transmittance?

22. What is a blackbody?

23. What is the relationship between a blackbody and a white body?

24. For an MWIR system, there is a dead zone of zero transmittance at 4.2 μm. Why?

25. What is the relationship between absorbed, scattered, and transmitted power?

26. What are the two types of reflectance?

27. What is surface reflectance?

28. What is bulk reflectance?

29. Name three sources of radiation.

30. Name three methods of heat loss from a body.

31. When identifying the temperature of an object we can use either surface or core temperature. On which types of material would we use one or the other?

32. What is meant by the sun's zenith angle?

33. How is the spectral irradiance of the sun affected by zenith angle?

34. Why is compressed snow of a higher temperature than noncompressed snow?

35. What is spectral extinction?

36. Name the two major components of extinction.

37. Give four examples of atmospheric phenomena that contribute to extinction.

38. How does humidity contribute to transmittance reduction?

39. Explain the relationship of scattering and particulate size.

40. How does atmospheric turbulence affect image quality?

41. What is refraction? How does it affect image quality?

42. How do pressure gradients affect image quality?

43. What is the biggest atmospheric contributor to extinction in the LWIR region?

44. Can we determine an attenuation coefficient by knowing the international visibility code? How?

45. Is attenuation as a factor of humidity linear? Do seasons affect this relationship? How?

46. With rain, is the size of the raindrop the only consideration in radiation scattering? If not, what else affects scattering?

47. Which would be the best sensor choice for a mid-latitude, winter operating environment? A tropical, over-water environment?

48. All other variables normalized, a mer cad and InSb detector operate about the same with a background temperature of about 300 K. Which performs better with a higher background temperature? Lower temperature?

49. Explain a diurnal cycle. How many diurnal cycles are there in a day? At what times?

50. What type (medium) is most common in military lasers?

51. What frequencies can lasers operate in?

52. What are the eye safe considerations in laser operations?

53. How can we operate on the eye with an IR laser without damaging the optic nerve?

54. What is the difference between optical zoom and digital zoom?

55. Why is stabilization important in laser systems?

56. What is meant by thermal bloom?

57. Would it be easy to determine closure rate with a laser? Why or why not?

58. What is an edge tracker system?

59. You intend to attack a factory complex at night using laser-guided bombs (LGBs). If you are the designator, which direction would you attack from?

60. Your actual detection ranges are less than predicted. You have accounted for clutter, transmittance, and actual ΔT. What is the probable cause?

61. What does the inverse square law mean to you?

62. What does Lambert's law of cosines mean to you?

63. Name two problems associated with flat mirrors in FLIR systems.

64. What problems are associated with single-detector systems?

65. Why are multiple detectors used in some serial scanning arrays?

66. What is a Nyquist frequency?

67. Briefly explain what is meant by diurnal depth. What do we use this information for?

68. What temperature in Celsius is 300 K? What temperature is this most commonly associated with?

69. What is the relationship, if any, between temperature and peak spectral response?

70. At what wavelength would we find a blue spike plume? Red spike?

71. It is a clear day, meteorological range 25 nm, and absolute humidity 9 g/m³. What is the attenuation coefficient?

72. What is an aerosol in relation to optical systems?

73. You are interested in a target that has a spectral response of 4.3 μm. Which detector element would you choose?

74. If you use the FLIR 92 detection model with a 2D critical dimension. Do you need to consider MRΔT due to aspect ratio?

75. A CO_2, pulsed, 20 nsec laser is to be tested by your organization. Is there an eye hazard?

76. Calculate the critical dimension in 1D for a man when his dimensions are 2 ft × 6 ft. In 2D?

77. What happens to spatial frequency with increasing ΔT?

78. What is a lux? What is a foot-candle?

79. What is the basic concept of NVG/NVD?

80. What is emissivity?

81. Total radiation is a function of temperature by what relationship?

82. I double the target temperature. What can I say about the total radiation?

83. How is peak radiation related to temperature?

84. I double the target temperature. What can I say about the peak radiation?

85. Target temperature is 300 K. What is the peak spectral radiance? At what wavelength?

86. What could be a problem with an IR MAWS system?

87. Briefly describe a FOV test for an optical system.

88. Briefly describe a slew rate test for an optical sensor.

89. Why is the display itself an important part of FLIR/EO testing?

90. How is core temperature of an object related to radiation temperature?

91. Briefly explain the effect of wind on potential IR targets.

92. What is the relationship between extinction coefficient and transmittance?

93. Would you expect longer-range detections with an MWIR or an LWIR system? Why?

94. What problem would you expect to see (aero_optical) with an EO detection system as an aircraft increases speed?

95. What two things must be known for IR data to be called calibrated?

96. What are the significant differences between DT and OT testing of an EO system?

97. What is the normal sequence of events when detecting, recognizing, and identifying targets?

98. What are the differences in spatial frequency boards for optical and IR tests?

99. What is the TTPF for low clutter 90% probability?

100. Using FLIR 92, 2D, how many cycles are required to ensure operational success? Using 1D Johnson criteria?

101. How does an LGB discriminate between own-ship and spurious directed energy?

102. Explain the importance of boresighting. What is parallax correction?

103. What is the effect of recent rain on IR targets?

104. You have a laser ranger, FLIR, and pulse Doppler radar on board your aircraft. How would you optimize a target track?

105. When looking at a sky background, what is the effective background temperature?

106. What happens to background temperature as the target approaches the horizon?

107. Name three environmental modifiers that will affect ΔT between the target and background.

108. Explain how Arnold Schwarzenegger defeated the predator in the movie *Predator*.

109. Explain jitter. How important is jitter in flight testing?

110. Explain centroid track in video tracking systems. What is the primary purpose of an AVTS?

111. What is ATR? How does it work? What are the human interface requirements with ATR?

112. List the three higher-order discrimination levels.

113. What is detection?

114. What is recognition?

115. What is identification?

116. How clear are the lines between the three?

117. Which FOV is selected for each of the three?

118. Johnson criteria are an equivalent bar pattern approach for angular resolution. How does the FLIR determine resolution? How does an EO (optics) determine resolution?

119. What is the level in a FLIR system?

120. What is the gain in a FLIR system?

121. Which parameter (level or gain) relates to sensitivity? Which relates to resolution?

122. If a FLIR system operates in a detection range of $-10°C$ to $100°C$ and the MRT is $1°C$, what is the sensitivity limit? What is the resolution limit?

123. Briefly explain a spatial resolution test.

124. In order to accurately chart spatial frequency, what must the tester do?

125. Why is it important for more than one operator to perform spatial frequency tests?

126. What is the difference between minimum dimension and critical dimension? What are each used for?

127. Can the aspect angle alter lab tests? Field tests?

128. What clutter environment does Johnson criteria assume?

129. How can we adjust the basic Johnson criteria for an increase in clutter?

130. What is FLIR 92?

131. How are range predictions for a system made? How accurate are they?

132. Can we use the same tests for a TV system as we do for the FLIR system?

133. How would you set up a test of a postlaunch guided EO munition to ensure that the munition received the desired commands?

134. What would be a good countermeasure against an EO-guided munition?

135. What would be a good countermeasure against an IR-guided munition?

136. Explain FOV and FOR.

137. What considerations regarding FOV must be observed when employing flares?

138. Is EMI and EMC a consideration in EO and FLIR testing?

139. Identify three human factors considerations when evaluating an EO/FLIR system.

140. What workload considerations should be addressed when considering an EO/FLIR system?

141. Why do you think, or do you, that it would be very difficult to obtain statistical numbers for detection, recognition, and identification?

Radio Detection and Ranging – Radar

Chapter Outline

8.0   Introduction 625
8.1   Understanding Radar 626
8.2   Performance Considerations 629
8.3   Radar Utility 630
8.4   Radar Detections 636
8.5   Maximum Radar Detection 641
8.6   Radar Sample Applications 645
8.7   Pulse Delay Ranging Radar Modes of Operation and Testing 652
8.8   Doppler and Pulse Doppler Modes of Operation and Testing 668
8.9   Air-to-Ground Radar. 694
8.10 Millimetric Wave Radar. 710
8.11 Miscellaneous Modes of Radar 712
8.12 Some Final Considerations in Radar Testing 712
8.13 Selected Questions for Chapter 8. 714

8.0 | INTRODUCTION

Radar evaluation is perhaps one of the most exhaustive, complex, and challenging types of testing that the flight test engineer will encounter. As with all other avionics and weapons systems, it is imperative that the evaluator possess a basic knowledge of radar and an in-depth knowledge of the system under test. For the tester who is new to radar, I recommend that you start with *Stimson's Introduction to Airborne Radar, 3rd Edition* (Raleigh, NC: SciTech Publishing, 2014). This is a wonderful book, an easy read, and a must reference for the radar engineer. A companion text is Merrill I. Skolnik's *Introduction to Radar Systems, 3rd edition* (New York: McGraw-Hill, 2002). This book is more of the textbook that you have seen in your college or university, but it gives an excellent treatment of clutter and detections at sea. Another reference is one given in previous chapters: the NATO Aerospace Group for Avionics Research and Development (AGARD), now called the NATO Research and Technology Organization. They have two excellent guides: RTO-AG-300, volume 16, *Introduction to Airborne Early Warning Radar Flight Test*, and basic radar flight testing information is contained in

AGARD's *Introduction to Avionics Flight Test*. Attendance in a radar and radar flight test course is also highly recommended.

The purpose of this section is not to repeat what is contained in the aforementioned references. Radar theory is covered only to make the test procedures understandable. The section will review radar fundamentals, identify radar modes of operation, examine the methods of test for these modes, and address any special test considerations.

8.1 | UNDERSTANDING RADAR

Radar is an acronym for radio detection and ranging, and as the name implies, radio energy is used to determine the attributes of a target. The information that we hope to obtain from the radar is range, range rate, azimuth, and elevation. Most objects reflect radio waves. The intensity of the reflections is dependent on some key parameters. The first is the polarization of the radio signal. Polarization is a term used to express the orientation of the wave's field (in particular, the electric field). If the electric field is vertical, it is said to be vertically polarized; if the electric field is horizontal, it is said to be horizontally polarized. A circular polarization can be developed by simultaneously transmitting horizontally and vertically polarized waves 90° out of phase. A receiving antenna can extract the maximum amount of energy from a received signal if the signal's polarity and the antenna's polarity are the same. If the polarizations are different, the energy is reduced in proportion to the cosine squared ($\cos^2$) of the angle between them (i.e., a power loss factor is said to be equal to $\cos^2\theta$). Table 8.1 shows the reduction in extracted energy due to polarization differences.

A linearly polarized antenna will only pick up the in-phase component of a circularly polarized wave; as a result, the antenna will have a polarization mismatch loss of 50% (-3 dB), no matter what angle the antenna is rotated to.

The "dB" noted in the second column is a loss expressed in decibels. The unit originated as a measure of attenuation in telephone cable and is named after Alexander Graham Bell. It is a ratio of power input at one end of a cable to that received at the other end; a decibel was the attenuation of 1 mile of telephone cable. The decibel is a logarithmic unit and has the advantage of reducing the numbers needed to describe a gain or loss of radio energy. It is extremely useful in radar discussions as well as for data links, and will be addressed again in the UAS chapter. For our discussions, a gain is described as the output (final) power/input (initial) power (P_2/P_1), and a loss is described as the input power/output power (P_1/P_2). Converting power gain ratios to decibels and decibels to power ratios is given by

Converting power ratios to decibels: Power ratio (in dB) $= 10\log_{10}P_2/P_1$ (8.1)

TABLE 8.1 ■ Signal Losses Due to Differences in Polarization

Polarization Angular Difference	Signal Loss (dB)	Signal Loss (%)
0	0	0
30	−1.3	25
45	−3.0	50
60	−6.0	75
90	Infinite	100

TABLE 8.2 ▪ Basic Power Ratios

Basic Power Ratios	
dB	Power Ratio
0	1
1	1.26
2	1.6
3	2
4	2.5
5	3.2
6	4
7	5
8	6.3
9	8

and

$$\text{Converting from decibels to power ratios: } P_2/P_1 = 10^{\text{dB}/10}. \tag{8.2}$$

Stimson provides the reader with an easy way to perform these conversions by remembering some simple basic power ratios. These basic power ratios are provided for reference in Table 8.2.

If we want to express a power ratio in decibels (equation 8.1), we can do so by following a couple of easy steps. Suppose that we had a gain of 63,000 and we want to express this in terms of decibels to make the numbers more manageable. The first step is to express the gain in scientific notation. Thus

$$63,000 \text{ becomes } 6.3 \times 10^4.$$

In this expression, 6.3 is the basic power ratio and 10 is raised to the fourth power. The number four will become the first term of our decibel expression and 6.3 is converted to decibels (Table 8.2) and applied as the second term, equating our gain of 63,000 to 48 dB:

$$63,000 = 6.3 \times 10^4 = 48 \text{ dB}.$$

In order to convert a d term to a power ratio, we merely work in reverse and convert the second term of the decibel expression to a basic power ratio:

$$34 \text{ dB} = 2.5 \times 10^3 = 2500.$$

Negative decibels are used to determine power ratios less than 1 (loss). A ratio expressed in decibels can be inverted by putting a negative sign in front of it. For example, beamwidth is commonly measured at the -3 dB point (the distance away from the antenna where power is measured to be 50%).

$$-3 \text{ dB} = \frac{1}{3} \text{ dB} = \frac{1}{2} \times 10^0 = \frac{1}{2} \text{ (power ratio)}.$$

A loss of 1,000,000/1 can be expressed in decibel terms by

$$1/1,000,000 = 1/1 \times 10^6 = 1/60 \text{ dB} = -60 \text{ dB}.$$

A more in-depth explanation with further examples can be found in Stimson's chapter 6.

The second parameter of concern in our ability to see reflected energy is the frequency that we employ. The frequency of the signal is represented by (f) and has the relationship

$$f = c/\lambda, \tag{8.3}$$

where

f = frequency (Hz),

c = speed of light (3×10^8 m/sec), and

λ = wavelength (m).

The choice of frequency will influence many of the performance parameters of the radar as well as impact the design. These considerations will be addressed shortly. Most of the airborne systems that we employ are called metric radars because of the frequencies that are employed; some applications utilize the millimetric (millimeter wavelength) spectrum. Table 8.3 provides the frequency bands radars that are discussed in this chapter. It is interesting to note that there is an "old" designation and a "new" designation. The old designations were given during early radar development, and in the hopes of confusing nosy-buddies, the lettering scheme followed no rhyme or reason. The new designations further define the frequency spectrum and follow an ascending order. You will find that most radar people still use the old designations, while folks involved with electronic warfare use the new designations.

The evaluator may find it useful to be able to identify the center frequency for common airborne radar. The center frequencies can be found in Table 8.4.

TABLE 8.3 ■ Radar Frequency Designations

IEEE US (Old Radar Designation)	Origin	Frequency Range	Wavelength	NATO, US ECM (New Radar Designation)
W	W follows V in the alphabet	75–111 GHz	400 mm–270 mm	M (60–100 GHz)
V	Very Short	40–75 GHz	700 mm–400 mm	L (40–60 GHz)
K_A	Kurtz (above)	26–40 GHz	1.1 cm–0.7 cm	K (20–40 GHz)
K	Kurtz (which is German for short)	18–26 GHz	1.6 cm–1.1 cm	J (10–20 GHz)
K_U	Kurtz (under)	12.4–18 GHz	2.5 cm–1.6 cm	
X	Used in World War II for fire control as an "X" for crosshairs	8–12.4 GHz	3.7 cm–2.5 cm	I (8–10 GHz)
C	Compromise between S and X	4–8 GHz	7.5 cm–3.7 cm	H (6–8 GHz)
				G (4–6 GHz)
S	Short Wave	2–4 GHz	15 cm–7.5 cm	F (3–4 GHz)
				E (2–3 GHz)
L	Long Wave	1–2 GHz	30 cm–15 cm	D (1–2 GHz)
UHF		0.3–1 GHz	<1 m–30 cm	C (0.5–1 GHz)

TABLE 8.4 ▪ Radar Center Frequencies

Radar Center Frequencies		
Frequency Band (Old)	**Center Frequency (GHz)**	**Center Wavelength (cm)**
Ka (above)	33	0.8
Ku (under)	16	2
X	10	3
C	6	5
S	3	10

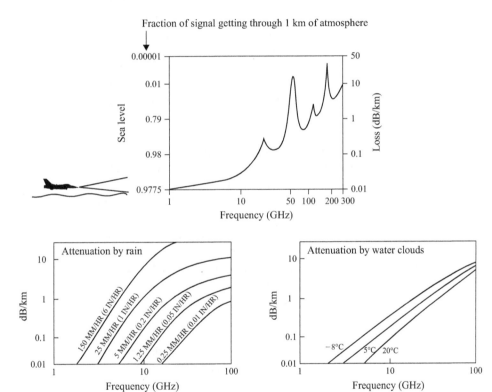

FIGURE 8.1 ▪ Atmospheric Absorption Versus Frequency

8.2 | PERFORMANCE CONSIDERATIONS

As mentioned previously, the choice of frequency will have a direct effect on the performance and design of the radar. The first thing to be considered is the attenuation of the signal as it propagates through space. Radio waves are attenuated by two properties: absorption and scattering. The absorption is due mainly to oxygen and water vapor, and the scattering is due to condensed water vapor (such as raindrops) and other particulates in the atmosphere: smoke, soot, pollution, oil vapor, etc. As the frequency is increased, the transmittance through the atmosphere is decreased (more energy is absorbed and reflected). Figure 8.1 shows the dramatic effect of absorption with increasing frequency.

Because of this relationship between absorption and frequency, it should be evident that increasing amounts of power would have to be generated for higher frequency signals in order to obtain greater ranges. Note that at 50 GHz (millimetric radar), fully 99% of the signal is attenuated by the atmosphere.

Electric noise from sources outside the radar is high in the high frequency (HF) band (3–30 MHz) and decreases with increasing frequency, reaching a minimum somewhere between 0.3 and 10 GHz, depending on the level of galactic noise. Beyond 10 GHz, man-made noise dominates, becoming stronger at K band (20–40 GHz) and higher frequencies.

Lower frequencies (longer wavelengths) are directly proportional to the size of the equipment required, that is, lower frequencies require larger equipment (e.g., wave-guides) to generate and transmit the signal. At lower frequencies, the equipment is large and heavy, whereas higher-frequency radars can be placed in smaller packages and weigh substantially less. Figure 8.2 shows three types of radar operating in three different bands: millimetric, X band, and S band. The millimetric radar is found on helicopters or actively guided ordnance, while the X-band radar is found on a typical fighter and the S band is used for early warning. The size differences in these three types are obvious.

The width of the radar beam is directly proportional to the ratio of the wavelength to the diameter of the antenna. This phenomenon will play a large part in low probability of intercept (LPI) and beam shaping. Narrow beamwidths provide higher power densities and excellent angular resolution, but the antenna normally has to fit in the nose of an aircraft, thus limiting size (diameter). In order to make the beam narrower, you must increase the frequency (shorter wavelength) or increase the size of the antenna. Figure 8.3 shows the PAVE Phased Array Warning System (PAVE PAWS) which operates in the UHF Band. Originally designed to track incoming ballistic missiles during the Cold War, its purpose now is to track all of the junk in space down to the size of a beer can. It uses a low frequency to obtain range, but in order to obtain angular accuracies it must employ an extremely large antenna.

Doppler shifts are proportional to the rate of closure as well as frequency. The higher the frequency, the larger the Doppler shift for a given closure. Doppler sensitivity to small differences in closing rate can be increased by selecting higher frequencies. The performance considerations are summarized in Table 8.5.

■ 8.3 | RADAR UTILITY

When using radar for target detection, ground mapping, weather avoidance, or any of the other modes of the radar, the operator is in search of four basic variables:

- Range
- Range rate
- Azimuth
- Elevation

The simplest method of determining range to a target is to measure the time it takes for energy to be returned from the target. Since the radio wave travels at the speed of light, range can easily be determined by multiplying the time it takes to return by

FIGURE 8.2 ▪
Radar Size as a
Function of
Frequency

Millimetric Wave Radar, 0.8 cm

X-band Radar, 3 cm

S-band Early Warning Radar, 10 cm

the speed of light and then dividing the result by two, since the path of the wave is a round-trip.

$$R = \frac{c}{2}\Delta t. \qquad (8.4)$$

The time is measured in microseconds, where 1 μsec equates to a wave travel (to the target) of 150 m, or roughly 500 ft. The accuracy of the range measurement is dependent on the shape of the pulse (Figure 8.4). The radar can measure the range to the

FIGURE 8.3 ▪
PAVE Phased Array
Warning System
(PAVE PAWS)

TABLE 8.5 ▪ Radar Performance Considerations

f	λ	Attenuation	Component Size	Power Required for Range	Beam Shaping	Doppler Measurements
Low	Long	Low	Large	Low	Poor	Poor
High	Short	High	Small	High	Good	Good

FIGURE 8.4 ▪
Accuracy
Dependent on Pulse
Shape

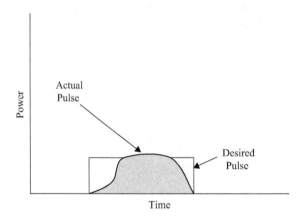

target by using the leading edge or the trailing edge of the pulse. In an ideal world, the pulse would be perfectly square and the leading and trailing edge would be easily observed. Since there is going to be some ramp-up and decay time in the transmitter, the pulse will not be exactly square, which results in some uncertainty and therefore some inaccuracy in the measurement.

Pulse delay ranging works well as long as there is only one pulse to be measured. If there are two pulses in space because the target is very far away, the radar will not know which measurement to use. Figure 8.5 shows an airborne radar detecting a target that is slightly further than the distance traveled by one interpulse period. Simply put, the echo from the target is not received from the first pulse before a second pulse is sent by the radar. In Figure 8.5, the true distance to the target is 60 nm, but the interpulse period will

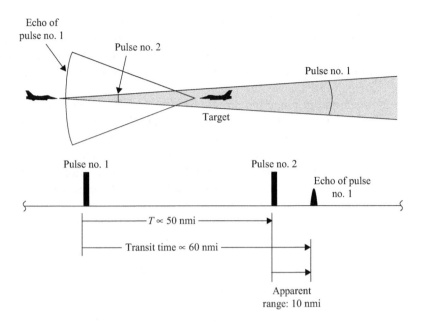

FIGURE 8.5 ▪
Range Ambiguity

only allow detections up to 50 nm before another pulse is sent. The radar will compute the transit time based on pulse no. 2 and provide an incorrect range; in this case 10 nm.

This situation can be determined beforehand by computing a maximum unambiguous range. For a given pulse repetition frequency (PRF; number of pulses per second represented as Hz), the longest range that may be observed without being ambiguous (no more than one pulse in space) is called the maximum unambiguous range (represented as R_u). Since the interpulse period (called the pulse repetition interval [PRI]) is equal to one divided by the PRF, we can substitute our PRI in place of Δt in equation 8.4 and solve for R_u:

$$R_u = cT/2, \tag{8.5}$$

where T equals the interpulse period (PRI).

We can also put equation 8.5 in terms of PRF by a simple substitution for T:

$$R_u = c/2PRF. \tag{8.6}$$

A rule of thumb is R_u in nautical miles is equal to 80 divided by the PRF (in kHz); R_u in kilometers is equal to 150 divided by the PRF (in kHz).

As you can see, ambiguous range returns can severely limit the range at which we can detect targets if the ambiguities cannot be solved. Fortunately there are ways to resolve some of these problems. The most common method of resolving a range ambiguity is called PRF switching or sometimes called PRF jittering. This technique works by taking into account how much a target's apparent range changes when the PRF is changed. Knowing this, and the amount the PRF has changed, it is possible to determine the number of whole times, n, that R_u is contained in the target's true range.

Determining n is best illustrated by a hypothetical example provided in Stimson's chapter 12 (p. 156). We will assume that for reasons other than ranging, a PRF of 8 kHz has been selected. Consequently the maximum unambiguous range, R_u, is $80/8 = 10$ nm. However, the radar must detect targets out to ranges of at least 48 miles—nearly five times R_u—and undoubtedly it will detect some targets at ranges beyond that as well. The apparent ranges of all targets will, of course, lie between 0 and 10 nm. To span this 10 mile interval, a bank of 40 range bins (range gates) has been provided. Each bin position represents a range interval of ¼ mile. The range gate width is maintained constant while switching PRFs.

A target is detected in range bin no. 24, 64, 104, etc. The target's apparent range is $24 \times ¼ = 6$ miles. On the basis of this information alone, we know only that the target could be at any one of the following ranges: 6, 16, 26, 36 nm, and so on. To determine which of these is the true range, we switch to a second PRF. To keep the explanation simple, we will assume that this PRF is just enough lower than the first to make R_u ¼ mile longer ($R_u = 10.25$ nm) than it was before. (A practical system would not be mechanized with PRFs so closely spaced.)

What happens to the target's apparent range (closest range) when the PRF is switched will depend on the target's true range. If the true range is 6 miles, the switch will not affect the apparent range. The target will remain in bin no. 24, but will also appear in bins 65 (16.25 nm), 106 (26.5 nm), etc. But if the true range is greater than R_u, for every whole time R_u is contained in the target's true range, the apparent range will decrease by ¼ mile.

We can find the true range by 1) counting the number of bin positions the target moves, 2) multiplying this number by baseline R_u (10 nm), and 3) adding the result to the apparent range (shortest range return).

Suppose the target moves from bin no. 24 (apparent range 6 miles) to bin no. 21, a jump of three bin positions; the target's true range is $(3 \times 10) + 6 = 36$ miles.

$$R_{\text{true}} = nR_u + R_{\text{apparent}}. \tag{8.7}$$

There will come a point with very high PRFs where range ambiguities will not be able to be resolved. Another method of calculating range will have to be used; this method will be discussed later in the pulse Doppler radar section.

Range rate can be calculated from the rate of change of range with respect to time from our pulse delay ranging calculations. Range rate can be directly measured by sensing shifts in Doppler frequencies. As illustrated in Figure 8.6, a wave radiated from a point source is compressed in the direction of motion and spread out in the opposite direction. If the radar and the target are both in motion, the radio waves are compressed (or stretched) at three points: transmission, reflection, and reception. As seen previously in section 5.6:

$$\text{Doppler shift} = 2(f/\lambda).$$

The Doppler shift is defined as positive when the target is approaching the radar (compression) and negative, or opening, when the target is receding from the radar.

FIGURE 8.6 ▪
Doppler
Compression

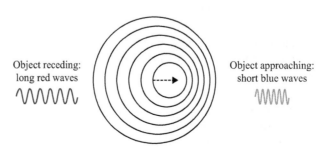

Object receding:
long red waves

Object approaching:
short blue waves

In X-band radars, one knot of range rate is measured approximately as 35 Hz of Doppler shift.

In most airborne applications, the azimuth and elevation are measured as the pointing angle of the radar with respect to a fixed reference. The radar may have either an earth or aircraft stabilization. With earth-stabilized systems, the radar maintains level with the center of the earth and true north. It accomplishes this with information from the aircraft navigation system (INS, AHRS, etc.), or in some cases, a dedicated radar stabilization package. The radar scans with reference to the earth regardless of aircraft attitude. In aircraft stabilization, the radar scans with reference to the fuselage reference line (FRL), or boresight, and the aircraft wing line. Aircraft stabilization is normally used in a dogfight or heads-up display (HUD) mode so aircraft maneuvers will not cause a loss of stabilization or loss of target track. In Figure 8.7, the operator has selected a 0° tilt (level scan) on the radar. With earth stabilization selected, the operator will not see the target because the radar will scan level with respect to the earth. With aircraft stabilization selected, the target is now in the scan volume of the radar.

When azimuth and elevation accuracy are required, a narrow beam must be employed. Angular measurement is derived from the antenna pointing angle when the echo is received and is a direct function of the beamwidth. The beamwidth can be measured in two ways: the first is the null-to-null beamwidth, and the second is the measurement of the beam at the −3 dB point.

Radar energy cannot be directed entirely through the main lobe of the radar. There are sidelobes and back lobes to the radar beam, and they rob as much as 25% of the available radar energy from the intended direction. Figure 8.8 is a three-dimensional (3D)

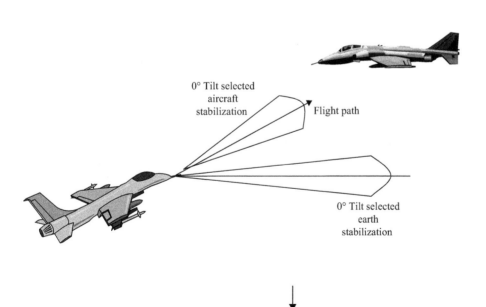

FIGURE 8.7 ■
Earth Versus Aircraft
Stabilization

0° Tilt selected
aircraft
stabilization

Flight path

0° Tilt selected
earth
stabilization

FIGURE 8.8 ■
Antenna Beamwidth
Measurements

First null
location

−3 dB
measurement

representation of a radar beam. The edges of the main lobe (beamwidth) are most easily defined as the nulls on either side. This is called the null-to-null measurement of the beamwidth. From the standpoint of the radar, it is more realistic to define the beam at a point in space where the power has dropped to some arbitrarily selected fraction of the center of the beam. Beamwidth is commonly measured between points where power has dropped to one-half of the maximum (-3 dB).

For either a linear array or a rectangular aperture over which the illumination is uniformly distributed, the null-to-null beamwidth (in radians) is twice the ratio of the wavelength to the length of the array:

$$\theta_{nn} = 2(\lambda/L), \tag{8.8}$$

where

λ = wavelength of the radiated energy, and

L = length of the aperture (same units as λ).

The -3 dB beamwidth is a little less than half of the null-to-null beamwidth:

$$\theta_{-3\,dB} = 0.88(\lambda/L). \tag{8.9}$$

For a uniformly illuminated circular aperture of diameter d, the -3 dB beamwidth is

$$\theta_{-3\,dB} = 1.02(\lambda/d). \tag{8.10}$$

If the antenna has tapered illumination, the beamwidth becomes

$$\theta_{-3\,dB} = 1.25(\lambda/d). \tag{8.11}$$

The rule of thumb for X-band radars is that the -3 dB beamwidth for tapered illumination is 85° divided by the diameter in inches and 70° divided by the diameter in inches for untapered illumination.

8.4 | RADAR DETECTIONS

There are many factors that will determine a radar's ability to detect targets. Optimization of these parameters will affect sizing and performance. Some of the parameters for consideration are

- Power
- Antenna size
- Reflecting characteristics of the target
- Length of dwell time on the target
- Wavelength/frequency of the transmission
- Intensity of background noise and clutter
- Processing

Knowledge of the parameters that affect target detection will allow the determination of a probability of detection. Examining the energy received from a target can help

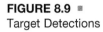

FIGURE 8.9 ■
Target Detections

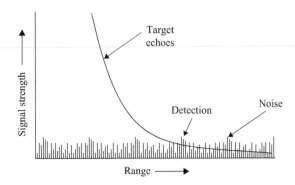

us to understand a detection range. A detection range is defined as the point where the returned echo signal is stronger than competing noise. Figure 8.9 shows the relationship between range to a target and the received signal. The detection occurs when the target echo return exceeds that of the noise.

When the radar antenna is directed at the target, a portion of the radiated energy will be incident upon the target. A portion of that energy (since some is reflected away and some is absorbed) is returned to the antenna, and only a portion of the returned energy is captured by the antenna. This process can be expressed mathematically in the general range equation:

$$\text{Received signal energy} = \frac{P_{\text{avg}} G \sigma A_{\text{e}} t_{\text{int}}}{(4\pi)^2 R^4},\tag{8.12}$$

where

P_{avg} = average transmitted power,

G = antenna gain,

σ = radar cross section of target,

A_{e} = effective antenna area,

T_{int} = integration time, and

R = range.

If we substitute in the minimum detectable signal energy, S_{min} (as depicted in Figure 8.9), we can solve for the maximum detectable range. This solution is found in equation 8.13:

$$R_{\text{max}} = \sqrt[4]{\frac{P_{\text{avg}} G \sigma A_{\text{e}} t_{\text{int}}}{(4\pi)^2 S_{\text{min}}}}.\tag{8.13}$$

Equation 8.13 is only good for an antenna trained on a target, that is, nonscanning. If we assume that the gain is the same throughout the entire scan volume of the radar, we merely have to replace integration time (t_{int}) with time on target (t_{ot}) to determine the maximum range detection with a single scan of the radar, which is slightly more useful.

If we equate S_{min} to a signal-to-noise ratio (SNR) of 1, S_{min} will be equal to the noise, which is simply $K T_{\text{s}}$ (K is Boltzmann's constant and T_{s} is the equivalent system noise temperature). By also knowing that gain is equal to $4\pi (A_{\text{e}}/\lambda^2)$, the terms in

FIGURE 8.10 ◾
Peak Versus
Average Power

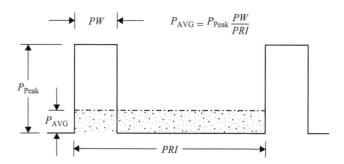

equation 8.13 can be rearranged to make more sense of what is contributing to target detection:

$$R_0 = \sqrt[4]{\frac{P_{\text{avg}} \sigma A_e^2 t_{\text{ot}}}{(4\pi) k T_s \lambda^2}}, \tag{8.14}$$

where R_0 is the range at which the integrated SNR is 1 and λ is the wavelength.

We will next look at the individual terms to see how they affect range detections with radar. The first term in the equation is the average power. Average power is defined as the peak power multiplied by the ratio of the pulse width (PW) over the PRI. The PW is the length of time the transmitter is on. An infinitely long pulse would approximate continuous wave energy, where peak power equals average power. The ratio of PW/PRI is called the duty factor, or sometimes the duty cycle. Figure 8.10 shows these relationships.

We can see from equation 8.14 that increasing the average power will increase the detection range, if only by the fourth root of the factor. There are three ways of increasing the average power; each of the methods carries with it some problems. The first method is to increase the peak power. Referring back to Figure 8.10, it is easy to see that if the peak power grows, the average power will also grow. Unfortunately this would require larger equipment, which is not really an option for most airborne applications.

A second method for increasing the average power is to increase the PRF. If more pulses are sent in the same period of time, the transmitter must fire more often and the average power will increase. Increasing the PRF is an option, but the ambiguities in range must be accounted for. As previously mentioned, there will be a PRF that is so high that all range ambiguities cannot be eliminated.

A third way to increase the average power is to increase the length of the pulse. A longer PW causes the transmitter to remain on for a longer period of time for each pulse. If the PRF is not decreased, the average power will increase. This method affects a capability of the radar not previously addressed, that is, the ability of the radar to resolve two targets separated in range, called, interestingly enough, range resolution. Figure 8.11 shows an airborne radar detecting two targets flying in trail (i.e., separated only in range). In order for the radar to see both targets, the trailing edge of the transmitted pulse must have passed target A before the leading edge of the echo from target B reaches target A. In Figure 8.11, L represents the PW. If the PW in this example is 1 μsec, the separation between targets A and B must be at least 500 ft in order for the radar to resolve both targets (1 μsec is equal to 300 m or 1000 ft distance traveled

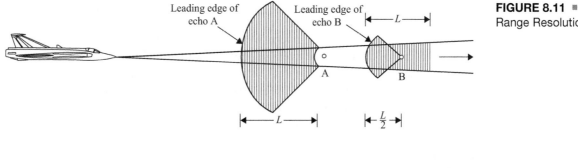

FIGURE 8.11 ▪
Range Resolution

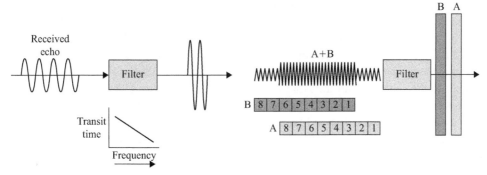

FIGURE 8.12 ▪
Pulse Compression
(Chirp)

through space). A rule of thumb therefore is range resolution is equal to 500 ft/μsec. If the PW of the radar is 5 μsec, the targets must be separated in range by at least 2500 ft in order to be resolved as two targets.

At very long ranges, resolution of targets within the radar's resolution cell is not as important as the actual detection of the targets themselves. As the range decreases, it becomes more important to determine the actual number of targets the radar is seeing. In order to accomplish this, the PW will have to be shortened. But if we shorten the PW, we shorten the range of detection. In many radars, the PW is determined by the range scale selection. If a long range is selected by the operator, the radar will employ a long PW; the opposite is true for a short range selection. There will be some cases, however, where the PW will be too long and targets will not be identified due to a merged return.

It is possible to code successive increments of the transmitted pulse with phase or frequency modulation. Then, when receiving the echo, the modulation is decoded and successive increments are progressively delayed. The radar now superimposes one increment on top of the other (and so on) and in this way a range resolution can be achieved as if a pulse was transmitted with the width of an increment. This method is called pulse compression; one example is linear frequency modulation, also known as "chirp."

With chirp, the frequency of each transmitted pulse is increased at a constant rate throughout its length. Every echo has the same linear increase in frequency and is passed through a filter that introduces a time lag that decreases linearly with frequency at exactly the same rate as the frequency of the echoes increases. In this way, the trailing portion of the echo takes less time to pass through than the leading portion. Successive portions thus tend to bunch up. When the echo emerges from the filter, its amplitude is much greater and its width is much less than when it entered. The pulse has been compressed. Figure 8.12 shows what happens when the echoes from two closely spaced

CHAPTER 8 I Radio Detection and Ranging – Radar

targets pass through the filter. The incoming echoes are merged indistinguishably; in the output, however, they appear separately.

Noise shows up in the bottom term of equation 8.14, and it tells us that we can have the same effect of doubling the average power if we halve the mean level of the background noise (KT_s). The time on target term (t_{ot}) is in the numerator, and if the scan rate was halved, it would have the same effect as doubling the average power or halving the noise. It would appear that if the antenna size is doubled, the detection range will also be doubled. This would be true except that, if the antenna size is doubled, the beamwidth is halved, which implies that the scan rate will also have to be halved in order to maintain the same t_{ot}.

The radar cross section (RCS, σ) appears in the numerator, and corresponding increases act just like the average power. RCS is not, however, just dependent on the size of the target. There are three attributes to the RCS of an object: geometric cross section, directivity, and reflectivity. Geometric cross section is a fairly easy concept: it is the area as seen by the radar. The area will change based on the aspect of the target with respect to the radar. An aircraft that is coming directly at the radar will have a smaller geometric cross section than the same target at a beam aspect (90° to the radar). The directivity of the RCS is the ability, or inability, of the target to redirect energy back to the radar. The reflectivity of the target is the term for the fraction of the intercepted power that is radiated back to the radar. The RCS of any target is dependent on the transmitted frequency and changes based on the angle of intercept.

An example of how geometric shape interacts with the directivity is shown in Figure 8.13. All incident energy on the cube is going to be returned to the radar since all points on the cube face are orthogonal to the radar energy. There is only one point on the sphere that is orthogonal to the transmitted wave, so only a very small portion of the incident energy is reflected back to the radar. The figure tells us that we should not have shapes like F-4 intakes mounted on airplanes.

Aircraft designers reduce their radar cross section by eliminating strong echo sources such as corner reflectors, aircraft antennas, and engine inlets. The designs include rounded edges with no corners and variable dimensions on aircraft access panels and doors. The two aircraft shown in Figure 8.14 are stealth aircraft; the B-2 employs a geometric sphere at the nose of the aircraft and appears devoid of any straight edges.

FIGURE 8.13 ■
Radar Cross Section

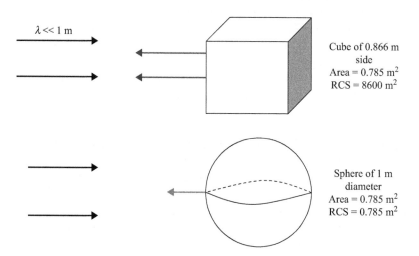

$\lambda \ll 1$ m

Cube of 0.866 m side
Area = 0.785 m^2
RCS = 8600 m^2

Sphere of 1 m diameter
Area = 0.785 m^2
RCS = 0.785 m^2

FIGURE 8.14 ■
RCS Applied to
Aircraft Design

The recently retired F-117, on the other hand, appears to be composed of all flat plates joined with rounded corners. The F-117 took advantage of directivity rather than geometric shape; radar energy incident on the F-117 is directed away from the transmitting antenna by the angular construction of the plates.

Reflectivity is used by both aircraft, as they are coated with radar absorption material (RAM). RAM is applied in a multilayer, wafer structure and can reduce the transmitted signal by as much as 60 dB.

Because aircraft are made up of many complex scatterers, the reflections from airborne targets will not be uniform. The reflections will appear to make the return scintillate, with peaks and valleys in the return varying with aspect angle. Some aspects may lead to early or late detection. Since this phenomenon is dependent on the wavelength, an airborne radar can help solve this problem by periodically switching frequencies.

8.5 | MAXIMUM RADAR DETECTION

In airborne applications, we strive to obtain detections at the maximum range possible. In this regard, we must determine the probability of obtaining that detection. The most commonly used probability is the blip-scan ratio (BSR); it is the probability of detecting a given target at a given range any time the antenna beam scans across the target. It is sometimes referred to as single-scan or single-look probability, and obviously the higher the probability desired, the shorter the detect range. The BSR is used in the evaluation of surveillance-type radars, and may be associated with fighter- and attack-type radars.

8.5.1 Blip-Scan Ratio

NATO document RTO-AG-300, volume 16, *Introduction to Airborne Early Warning Radar Flight Test*, provides an excellent reference for BSR and BSR flight testing. In simple terms, a sample radar scans the search volume, sending out pulses at a rate equal to its PRF. Each pulse is an opportunity for the radar to receive an echo from a target. Since the beamwidth, scan rate, and PRF of the radar are known, the number of opportunities on a given target is also known. Detection must imply that not only does the radar receive energy from a target, but the received signal must be recognized as a target. When the radar processor determines that the energy received meets the requirements to be declared a detection, the operator is presented with a "blip," or return, at the target range, azimuth, and elevation. In a real radar display (not synthetic video), the operator is responsible for determining if returns are true targets.

If we define a blip as a positive indication of a target, and the scan as the passage of the beam through the search volume containing the target, the BSR is the quotient of the number of integrated detections divided by the number of detection opportunities. The BSR value must be associated with a range to be of value, so it is typical to define bins of radial ranges where this ratio is calculated. For surveillance systems, the typical value of interest for the BSR is 0.5. This may be equated to a probability of detection of 50%. The expression is written as R_{50}, and represents the range for which the probability of detection is 50%.

The simplest way to perform the calculation is to divide the search volume into discrete range bins of equal length; the bin is identified by its start and end range. For example, if a surveillance radar had a 300 nm range capability, we can describe 30 range bins of 10 nm each. If a target remains in the 150 to 160 nm range bin long enough for the radar to sweep its position 20 times and a blip is present 10 times, the BSR for this bin would be 10/20, or 0.5. The R_{50} for this radar and target would be 155 nm.

It is possible to see a BSR higher for some longer ranges than some intermediate ranges. This may be due to changes in PW, PRF, etc., based on the range scale selection. It is best to graphically represent the results of a series of blip-scan detection tests run against the same target and the same test conditions. Figure 8.15 shows a typical plot of BSR for a surveillance radar.

As can be seen in Figure 8.15, there are significant decreases in the detection capability between range bins 180 and 200 and again at 100 to 120. Although an R_{50} is achieved at 225 nm, it is not sustainable until 100 nm.

8.5.2 Prediction of Maximum Range Detection

Prior to embarking on any flight testing, it is mandatory that the evaluator gather as much information as possible regarding the capabilities of the system. You may find that predictions indicate that there is no possible way that the radar will meet the requirements (we have never seen that before). Stimson (chapter 11) provides a method of predicting radar range performance for any desired probability of detection.

Video type: Synthetic

FIGURE 8.15 ■
Blip-Scan Ratio
Results

	Test A/C			Target A/C	
Date:	1 Jan 97	Model:	E-9C	Model:	Lear 36
Flight:	486	Buno:	179000	Buno:	150000
Run:	#1	Altitude:	24,000 ft MSL	Altitude:	28,000 ft MSL
Terrain:	Near-land	GND Speed:	205 KTS	GND Speed:	320 KTS
Profile:	44				

Range bins	30	40	50	60	70	80	90	100	110	120	130	140	150	160	170	180	190	200	210	220	230	240	250
Blips	6	13	17	18	18	26	24	18	7	6	18	24	29	12	11	13	4	8	13	13	9	5	3
Scans	8	16	20	18	20	28	29	26	20	29	26	48	48	20	20	21	21	21	21	22	24	23	30
B/S Ratio	.75	.81	.85	1	.90	.92	.82	.80	.35	.30	.69	.60	.72	.60	.55	.61	.19	.28	.61	.59	.37	.21	.10

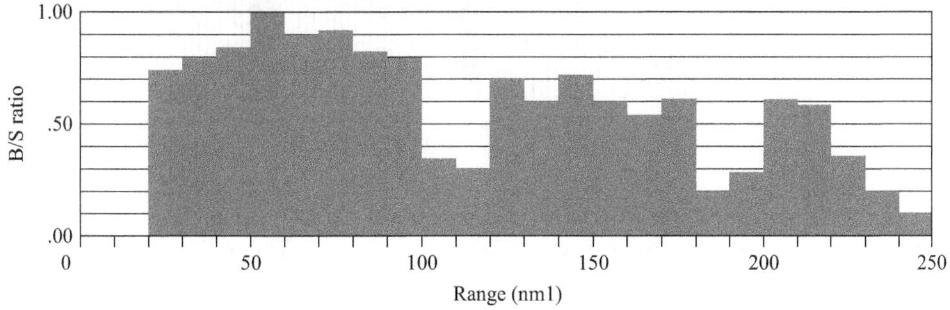

In this analysis, for a given probability of detection, a corresponding detection range can be computed:

- Choose an acceptable false alarm rate.
- Calculate the false alarm probability for the individual threshold detectors.
- Find the corresponding threshold setting.
- Determine the mean value of the SNR for which the signal plus the noise will have the specified probability of crossing the threshold.
- Compute the range at which this SNR will be obtained.

The false alarm rate is the average rate at which false alarms appear on the radar display. The false alarm time is merely the reciprocal of the false alarm rate. The false alarm rate must be chosen based on mission requirements. A very low false alarm rate can be obtained by raising the threshold; however, detection ranges will suffer. For fighter-type aircraft, a false alarm time of about 1 min is considered acceptable.

Each time the antenna beam sweeps over the target, a stream of pulses is received. In the radar's signal processor, the energy contained in the stream is added up. This process is called predetection integration. As the target range decreases, the strength of the integrated signal increases. Because of the random character of the noise, both in amplitude and in phase, the mean strength of the noise remains about the same. At the end of each integration period, the output is applied to a detector, and when this output exceeds a certain threshold (false alarm [FA] threshold), a synthetic target blip is presented on the display.

The completely random noise occasionally exceeds the threshold and the detector will falsely indicate that a target has been detected (Figure 8.16). This is called a false alarm. The higher the detection threshold relative to the mean level of the noise energy, the lower the false alarm probability, and vice versa.

FIGURE 8.16 ■
Signal-to-Noise
Threshold

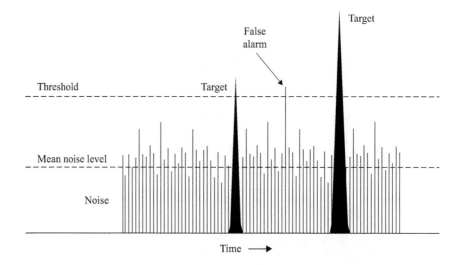

The setting of the threshold is crucial. If it is too high, detectable targets may go undetected. If it is too low, too many false alarms will occur. The optimum setting is just high enough above the mean level of the noise to keep the false alarm probability from exceeding an acceptable level. The mean level of the noise, as well as the system gain, may vary over a wide range. Consequently the output of the radar's signal processor must be continuously monitored to maintain the optimum threshold setting. If the false alarm rate is too high, the threshold should be raised; if it is too low, the threshold should be lowered. For this reason, the automatic detectors are called constant false alarm rate (CFAR) detectors.

The mean time between false alarms is related to the false alarm probability of the radar's threshold detectors. In general, a bank of Doppler filters is provided for every resolvable increment of range (range gate) and a threshold detector for every filter. The probability of false alarm is the integration time divided by the product of the number of range gates, the number of filters per bank, and the desired false alarm time.

$$P_{\text{fa}} = \frac{t_{\text{int}}}{t_{\text{fa}} x N_{\text{RG}} x N_{\text{DF}}}, \qquad (8.15)$$

where

P_{fa} = probability of false alarm,

T_{int} = integration time,

t_{fa} = false alarm time,

N_{RG} = number of range gates, and

N_{DF} = number of Doppler filters per bank.

As seen in equation 8.15, the probability of false alarm can be decreased by increasing the number of range gates or Doppler filters. There is a limit to increasing these terms beyond a certain amount.

The required SNR versus detection probability for a wide range of false alarm probabilities has been calculated with standard mathematical models. Where specific RCS data are not available, or a rigorous calculation is not required, curves can be used

TABLE 8.6 ■ Swerling Cases

| | Fluctuations | | |
Case	Scan to Scan	Pulse to Pulse	Scatterers
I	X		
II		X	Many independent
III	X		
IV		X	One main

to identify the appropriate SNR. Swerling curves are commonly used to match detection probability with the required SNR. These curves can be found in almost any radar text. Stimson provides these curves in chapter 11, whereas Skolnik addresses them in chapter 2. There are four cases to which the Swerling models apply (Table 8.6): cases I and II assume a target is made up of many independent scattering elements, as in a large (in comparison to wavelength) complex target, such as an airplane; cases III and IV assume a target is made up of one large scattering element. Cases I and III assume that the RCS fluctuates from scan to scan, while cases II and IV assume that the target fluctuates from pulse to pulse.

Once the required SNR is determined to obtain the desired detection probability, the range can be easily calculated by returning to equation 8.14 and inserting the required SNR into the denominator. The range of the required probability of detection is rewritten as equation 8.16:

$$R_{\text{Pd}} = \sqrt[4]{\frac{P_{\text{avg}}\sigma A_{\text{e}}^2 t_{\text{ot}}}{(4\pi)^2 (S/N)_{\text{REQ}} kT_{\text{s}}\lambda^2}}. \tag{8.16}$$

To account for the effects of closing rate (which is almost always the case in airborne applications), the detection range is expressed in terms of a cumulative probability of detection. The cumulative probability of detection (P_{c}) is defined as the probability that a given closing target will have been detected at least once by the time it reaches a certain range. The cumulative probability of detection is related to the single-scan probability of detection by the relationship

$$P_{\text{c}} = 1 - (1 - P_{\text{d}})^n, \tag{8.17}$$

where n is the number of scans.

We will see a variation in the cumulative probability of detection when we address R_{90} flight testing a little later. It is very important that evaluators run through the previous equations with the manufacturer's data to see if the system can realistically meet the specification requirements.

8.6 | RADAR SAMPLE APPLICATIONS

Armed with a basic knowledge of radar and its properties, we should be able to "build" a radar based on known mission requirements. The following examples are generic systems and are meant to give the reader a feel for the physics behind the build.

The first type of system is one designed for tactical air-to-air (Table 8.7). The mission of this radar is to detect targets at long range and employ missiles at beyond visual range (BVR). The old, now retired F-14A/D was tasked with fleet defense and possessed such a radar. The F-15C Eagle possesses a similar type of radar.

The choice of frequency provides the ability to shape the beam and take advantage of the Doppler content because of the relatively low attenuation by the atmosphere. Better resolution and beam shaping can be accomplished with a higher frequency, such as K_a or K_u, but the attenuation would be unsatisfactory. The PW is large in order to increase the average power, which will increase the detection range, but decrease range resolution (if a version of pulse compression is not used).

At long ranges, range resolution is not as important as the detection itself. At shorter ranges, the PW is shortened in order to enhance range resolution. The PRF is high in order to decrease Doppler ambiguities as well as enhance detection by increasing the average power. Because the PRF is high, this radar will not be able to calculate range from simple pulse delay ranging and will have to employ another method (called frequency modulation [FM] ranging), which will be addressed later. The beamwidth is large in order to scan large volumes in space. A raster scan allows the radar to scan in elevation as well as azimuth, thereby increasing the scan volume.

Most air-to-air radars have multiple scan selections. Figure 8.17 shows three raster scan selections. The first is an 8-bar raster scan; the antenna slews one elevation beamwidth from left to right through the azimuth sector. As it reaches the maximum positive azimuth, the antenna steps up one elevation beamwidth and slews to the left.

TABLE 8.7 ▪ Tactical Air-to-Air Radar Parameters

Radar Parameter	Value	Reason
Frequency	X band (9–10 GHz)	Good Doppler resolution Beam shaping Attenuation not too high
Pulse Width	Large (>2 μsec)	Range
PRF	High (>300 kHz)	Range
Beamwidth	Large (>4°)	Search volume
Scan	Raster	Scan volume
Display Type	B scope range vs. azimuth or range rate vs. azimuth	Range or range rate information
Stabilization	Earth or aircraft	Search or dogfight modes

FIGURE 8.17 ▪
Radar Raster Scan

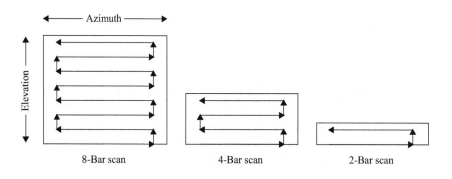

8-Bar scan   4-Bar scan   2-Bar scan

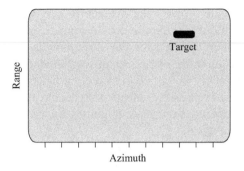

FIGURE 8.18 ▪
B Scope
Presentation

TABLE 8.8 ▪ Tactical Air-to-Ground Radar Parameters

Radar Parameter	Value	Reason
Frequency	K_u band (16–35 GHz)	Beam shaping Equipment size not a factor
Pulse Width	Small (<0.4 μsec)	Range resolution
PRF	Low (<1 kHz)	Little or no range ambiguity
Beamwidth	Narrow ($<1.5°$)	Azimuth resolution
Scan	Azimuth only (although some terrain-following radar will use a raster scan)	Update rates
Display Type	Sector PPI	Range and azimuth information in a true representation
Stabilization	Earth	Search and track modes Weaponry will require information based on an earth coordinate system

The process continues until the selected bar pattern is reached. If the elevation beam-width of the radar is $2°$, an 8-bar raster scan will search $16°$ in elevation. Detection range will decrease, since the t_{ot} will decrease (assuming the scan rate remained the same).

A B scope is a particular method of displaying target information to an operator. The B scope provides either range versus azimuth or range rate versus azimuth to the operator. It is not a true depiction of the volume as seen through the window because the bottom of the display is distorted. Figure 8.18 shows a B scope.

The stabilization may be either earth stabilized (normal search and track modes) or aircraft stabilized for close-in, highly dynamic tracking.

Our second system is of a type that is designed for tactical air-to-ground. These types of systems are designed to accurately identify ground targets and resolve multiple targets for the purpose of delivering some sort of weaponry. One would expect to suffer maximum range for maximum resolution (Table 8.8).

The frequency employed here is fairly high and will suffer much greater attenuation than the X-band system seen in the previous paragraph. The range at which these systems are used, however, is generally very short, so attenuation is not generally a problem; there can be severe problems when attempting to identify targets obscured by rain or snow. This is a penalty that is endured in order to obtain very narrow radar beams for enhanced azimuth accuracy and resolution.

The PW is small, which lowers the average power, which in turn decreases the detection range. As mentioned in the previous paragraph, since operational ranges are small, this is not a problem. The narrow PW does provide enhanced range resolution, which is an important requirement of tactical air-to-ground radars. The 0.4 μsec PW will provide a range resolution of two targets separated in range of 200 ft. The PRF is kept low, which also adversely impacts the detection range, but at the same time it allows for a longer maximum unambiguous range. Ambiguities are easily solved, as the maximum range of detection is kept small.

Most air-to-ground radars will only sweep in azimuth, and the scan rate is kept fairly high at short ranges. The system has the capability of using a fan or pencil beam, depending on the accuracy that is required. Some systems that use this type of radar for terrain following (TF) may employ a pencil beam in a raster scan in order to obtain accurate height of terrain measurements at greater distances from the aircraft.

Many of today's multimode radars operate in X band but use special processing to obtain the resolutions needed for accurate weapons delivery. These processing modes are synthetic aperture radar, Doppler beam sharpening, and inverse synthetic aperture radar, and are covered in section 8.9.4 of this text.

The display is called a plan position indicator or present position indicator (PPI) (depending on which text you are reading). The display is an accurate representation of the world as seen by the pilot through the windscreen. Unlike the B scope, the picture is not distorted at the bottom of the display. The display presents azimuth versus range, where range is marked by range rings emanating from the bottom of the display. The airplane is at the bottom of the display and aircraft heading or track is at the top of the display. Figure 8.19 shows a PPI presentation.

A ground mapping radar is commonly employed as an aid to navigation. Unlike the tactical air-to-ground system, range is important to this system. The parameters for the ground mapping radar are shown in Table 8.9.

The frequency for this system is chosen to reduce attenuation and enhance the available range. It is usually combined with a weather detection system. As noted in the tactical air-to-ground system, weather can cause severe problems with a K_u-band radar. A high-frequency radar can easily detect weather; the problem is seeing through the weather to find out what is on the other side. An X-band system has the capability of detecting weather, but since the wavelength is larger than the object (raindrop), only a portion of the radiated energy is reflected back to the transmitter. Some of the energy

FIGURE 8.19 ■ PPI Scope Presentation

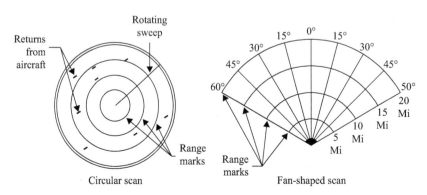

Plan position indicator

TABLE 8.9 ▪ Ground Mapping Radar Parameters

Radar Parameter	Value	Reason
Frequency	X band (10 GHz)	Frequency not susceptible to large attenuations Size can fit in most tactical aircraft
Pulse Width	Small (<1 μsec)	Range resolution
PRF	Low (<300 Hz)	Little or no range ambiguity
Beamwidth	Narrow (~3°) (enhanced azimuth accuracy and resolution with Doppler beam sharpening [DBS])	Azimuth resolution
Scan	Azimuth only (although some terrain-following radar will use a raster scan)	Update rates
Display Type	Sector PPI	Range and azimuth information in a true representation
Stabilization	Earth	Search and track modes Navigation based on earth coordinates

TABLE 8.10 ▪ Airborne Early Warning Radar Parameters

Radar Parameter	Value	Reason
Frequency	As low as 450 MHz (UHF)	Frequency not susceptible to attenuations; very long range
Pulse Width	Very large (>13 μsec)	Increased average power for greater range
PRF	Low (<200 Hz)	Range ambiguities easily resolved; long maximum unambiguous range
Beamwidth	Very wide (>7°); function of frequency	Large scan volume
Scan	Azimuth only, 360° scan	Large beamwidth negates elevation scan
Display Type	PPI or sector PPI	Range and azimuth information in a true representation
Stabilization	Earth	Targeting information is provided on WGS-84 earth models

passes through the weather and is reflected back by objects (such as mountains) that lie beyond the weather system.

The PW remains small for range resolution, and the PRF remains low to resolve range ambiguities. The beamwidth is fairly large, but this is a result of the frequency that we have chosen. There is a special processing method called Doppler beam sharpening (DBS), which effectively narrows the beamwidth an order of magnitude, thus producing excellent azimuth accuracy and resolution. DBS will be discussed later. The scan and the display are as with the tactical air-to-ground radar.

An airborne early warning radar (sometimes called an airborne early warning system [AEWS]) is designed to detect targets at the maximum range possible and direct airborne interceptors to engage these targets. The AEWS does not concern itself with fine resolution in range and azimuth; there may be cases where two or three targets are identified by the AEWS as a single target. Table 8.10 presents the AEWS radar parameters.

The frequency is extremely low in order to get maximum range for the available power. Since the frequency is so low, a very large antenna is required to shape the beam to a usable beamwidth. The PW is very large in order to increase the average power, and therefore increase the detection range. Since the PW is large, range resolution suffers. With a 13 μsec PW, targets would have to be separated by more than 6500 ft in range in order to be classified as two targets by the AEWS. The PRF is kept low in order to resolve range ambiguities; it also provides for a greater maximum unambiguous range. The scan is normally 360°, but can be sectored by the operator if desired.

The presentation can be the same as shown in Figure 8.19, or it can display the entire 360° volume, as shown in Figure 8.20.

The aircraft is at the center of the display, and in this figure north is up, so the display correlates to a map in the normal position. The operator can also display track up or heading up as the top of the display.

Because the elevation beam is so wide, there is no need to scan in elevation; a 7° beam will subtend 210,000 ft at 300 nm. There will be quite a bit of ambiguity in the altitude (elevation), on the same order as seen with the azimuth.

The basic weather radar as installed in civil aircraft looks much like the ground mapping radar previously identified. The only significant difference between the two is that the weather radar employs a longer PW to add to detection range capability (Table 8.11). The range resolution will suffer, but this is not a huge loss when searching for weather cells.

FIGURE 8.20 ■
360° PPI Display

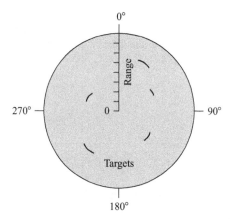

TABLE 8.11 ■ Weather Radar Parameters

Radar Parameter	Value	Reason
Frequency	X band (10 GHz)	Frequency not susceptible to large attenuations
Pulse Width	Large (>5 μsec)	Range
PRF	Low (<200 Hz)	Little or no range ambiguity
Beamwidth	Narrow (~3°)	Azimuth resolution
Scan	Azimuth only	Update rates
Display Type	Sector PPI	Range and azimuth information in a true representation
Stabilization	Earth	Navigation based on earth coordinates

8.6.1 Other Display Types

There are other presentations of radar data that the evaluator may come across in testing. The first of these display types is called an A scope. This presentation provides raw radar video in terms of range versus amplitude of the radar return. It is up to the operator to determine the real target echoes from the noise. This type of display is not seen much anymore. Many of the older surface-to-air missile (SAM) tracking radars utilize this display. Figure 8.21 shows the A scope presentation.

Another display type is called a C scope, and it is used to relate radar information to the HUD. The C scope displays azimuth versus elevation without range. This 2D presentation is exactly what the pilot sees when looking through the HUD. In many tactical aircraft, the pilot can select an autoacquisition mode (sometimes called autoguns, HUD mode, air combat maneuvering (ACM) mode, boresight, and others) and the radar will enter an aircraft stabilization mode and refine its search pattern in azimuth and elevation to match the HUD field of view (FOV). A representation of the C scope is shown in Figure 8.22.

The final display is called a patch map, and is the result of synthetic aperture radar (SAR), inverse SAR (ISAR), or DBS processing; the output of the display in many cases looks like a photograph. The radar dwells over an operator designated area of interest and collects data to put together a collage; the smaller the resolution cell of the processing, the higher the resolution in the display. SAR, ISAR, and DBS will be covered later. In older systems, the data were written to disk and the presentations were not available until postflight. Modern aircraft have the capability of processing and displaying the area of interest in near real time—on the order of a couple of seconds. Pilots can use these presentations for precision bombing, target designation, and even airborne radar approaches to a runway. A patch map display is shown in Figure 8.23.

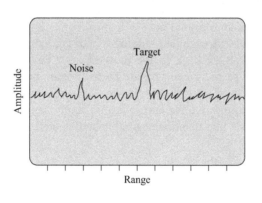

FIGURE 8.21 ■
A Scope Display

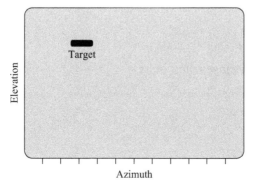

FIGURE 8.22 ■
C Scope Display

FIGURE 8.23 ▪
Patch Map Display

8.7 | PULSE DELAY RANGING RADAR MODES OF OPERATION AND TESTING

The simplest type of pulse radar is the weather radar installed in many of the civil aircraft in operation today. The required evaluation of this type of radar was covered in section 6.9 and will not be repeated here.

The pulse radar gives us excellent ranging to a target by measuring the time it takes a transmitted pulse to return back to the receiver after being reflected from a target (equation 8.4). The pulse radar can provide us with azimuth and elevation to a target referenced to either earth or aircraft coordinates. It should come as no surprise then that much of our testing will involve the accuracies of range, azimuth, and elevation.

There are other tests that must be accomplished as well when evaluating the pulse radar:

- Built-in test (BIT)
- Stabilization
- Observations
- Detections
- Maximum detection range (R_{90}/R_{50})
- Minimum detection range (short pulse)
- Measurement accuracy
- Track initiation range
- Track accuracy
- Resolutions
- Transition time

8.7.1 Pulse Radar Built-In Test

The built-in test (BIT) function of the radar comes in many types: power-up BIT (PBIT), initiated BIT (IBIT), operational readiness test (ORT), periodic BIT (PBIT), and maintenance BIT (MBIT). All of these tests are designed to alert the operator to the health and welfare of the system. The key to all built-in testing is that BIT results should be understandable, retrievable, and timely. The BIT will check all radar interfaces (navigation, stabilization, own-ship parameters, etc.), transmitter operation, video injection, antenna drive, and mode selections. It will run software routines and compare the results to the expected results. If there are any failures that will degrade mission capability, then the operator is supposed to be apprised of such situations, usually through the pilot fault list (PFL). Other failures may be written to the maintenance fault list (MFL) in memory for maintenance review at a later date. The preferred method of test to evaluate the radar's ability to recognize and capture faults is through deliberate fault injection. This test is more easily accomplished in the lab, as access to the interfaces is available and failures are easily introduced. It is more difficult, if not impossible, to run these tests on the aircraft because of limited access and the inability to inject software failures into flight-worthy software. On-aircraft testing may be limited to shutting down systems, such as the global positioning system (GPS), that interface with the radar and observing that those failures are logged.

One of the primary concerns is the number of false alarms, or conversely, the number of faults not detected. This particular test is part of the operational test and evaluation (OT&E) reliability and maintainability evaluation. When maintenance detects a failure in the MFL, the failure should point to a line replaceable unit (LRU) as the guilty party with some degree of certainty (on the order of 99.999%). What this means is that when a failure is logged in the MFL, that failure is cross-referenced to possible LRU problems. The cross-reference will identify three possible LRUs that may be at fault and the probability that one of them is the cause is very high. Maintenance will pull these three LRUs and rerun the BIT to see if the failure is cleared. If the fault has cleared, it now becomes a task of determining which of the three LRUs was at fault. This can involve local-level, depot-level, or vendor-level testing. What happens on some occasions is that none of the LRUs appear to be at fault. The cards are returned for inventory (RFI), and at some later date they are back in another airplane. And then what do you think happens? Of course, the same failure returns. This is one reason why configuration control is such an important aspect of flight testing. The LRUs may also be good, and in this case, the radar BIT has provided us with a false alarm. The other side of the coin is where the radar fails either in total or in some mode and the BIT does not report a failure; this is a fault not detected, and is also not good.

The last portion of the BIT is the time it takes to execute the BIT and display the results. There will be a specification for each available BIT and the amount of time it should take to execute. A power-up BIT of around 3 min is probably the norm, whereas initiated BIT (during flight) is an abbreviated BIT and may run only 30 sec. The time in BIT should be noted every time it is executed during the flight test program.

8.7.2 Pulse Radar Stabilization

The stabilization tests are similar to those explained in section 6.9.1; however, with a military application, the radar receives platform stabilization reference from a navigation system or separate gyro package dedicated to the radar platform. The primary purpose of this test is to ensure that the radar can maintain stabilization throughout the operational envelope of the aircraft up to the physical limitations of the radar. The radar can scan physically (i.e., hydraulic or electric drive motors) or electrically, and the evaluation demonstrates that the radar points to the position commanded by the operator. In cases of physically scanned antennas, the antenna drive unit will detect a failure if the commanded position does not match the actual position and the failure is announced on the PFL.

The initial tests should be conducted in pulse search with the smallest raster available (1-bar or 2-bar raster) and earth stabilization selected. The stabilization limits of the radar can be obtained in the same manner as with the weather radar. Incremental changes of pitch will establish the elevation limits, and incremental changes in bank will establish the roll limits. The limits are established when the returns start to disappear from the top (pitch down) and bottom (pitch up) and when the returns start to fade from the up-wing side of the display during bank maneuvers.

Probably of more importance in military applications is the effect of "g" on the radar stabilization. This test determines if the antenna can slew to the correct position under aircraft maneuvers. This test may require additional instrumentation if antenna position information cannot be obtained from the bus. The operator should select the smallest raster available, earth stabilization, and 0° tilt. From a shallow dive, start a pull-up with ever-increasing g; the aircraft needs to be recovered before the pitch stabilization limits of the radar are reached. This test needs to be repeated to maximum g for all raster scans available. The radar should be able to maintain the correct scan pattern in azimuth and elevation throughout all of these maneuvers. In addition, there should be no interference noted between the antenna and the radome. You can see that this test is of some importance to the operational capability of the radar system. If the scan collapses under certain maneuvers, target illumination will not be maintained, and hence all information about that target will be lost.

8.7.3 Pulse Radar Observations and Detections

An observation is defined as the radar's ability to see targets in the pulse search mode. These returns (echoes from objects within the radar scan volume) are not displayed to the aircrew, but rather are forwarded to the airborne radar signal processor (ARSP) for processing. The only way to evaluate this function is to analyze internal radar processing data. This can be a difficult undertaking as these data are usually proprietary and not normally available to the evaluator. The test team must work with the radar vendor to retrieve this information. If an observation passes the screening criteria of the radar, the return is declared a detection and sent to the airborne radar display processor (ARDP) for display to the operator.

This evaluation is a ground test and is normally accomplished with the vendor during radar development. Most radar developers will utilize a roof house antenna where the test item is mounted atop a tower and is slid out on rails so the radar can view targets of opportunity. Range ambiguity resolution, pulse compression, detection thresholds, and false alarm rates are tested and adjusted to obtain maximum operational usefulness of the system. The truth data are usually FAA data tapes (IFF tracks) of targets in the area. Using these tapes, the radar developer can sort out the true airborne targets from the false alarms.

Additional logic is employed before radar observations are declared detections. The logic may state that an observation must be seen in three of five consecutive frames before it is declared a detection (a frame is one complete scan of the radar; an 8-bar scan requires the entire 8 bars be complete to produce a frame). The logic may state that observations must be seen on two of three consecutive bars in three consecutive frames. Each radar is different and the radar evaluator needs to be aware of, and understand, the logic in order to make a proper evaluation. These tests will be ongoing throughout the entire development cycle of the radar.

The key to this evaluation is to determine if the correct information is reaching the operator. In some cases, detections are claimed by the ARDP but not displayed to the aircrew. In other cases, the logic is flawed and observations that do not meet the criteria are displayed anyway. It is extremely important to complete these tests prior to attempting installed system ground and flight testing. The evaluator will find that radar development is extremely incremental in nature (i.e., software driven). For this reason, as mentioned in chapter 1, plan on regression testing. As a corollary, strive to find and fix large problems early in the program, as this will decrease the need for performing large amounts of repeat testing.

One last point about detections: For a radar developer, a detection occurs when the radar says it occurs. For the user, a detection occurs when the operator sees it on the display. Is this ever a problem? You bet it is, and sometimes it takes a court to determine who is right. This problem occurs routinely for the aircraft integrator, since the displays and the system are not necessarily built by the same company. When developing a test plan and exit criteria (what data need to be collected and what must the results be in order to satisfy the requirement), ensure that there is agreement among all parties about what will be classified as a detection.

8.7.4 Pulse Radar Maximum Range Detections

The maximum detection range is expressed in terms of probabilities, and the most common method of determining the probability of detection is the BSR, as explained in section 8.5.1. This method is called the single-scan or single-look probability and is used for surveillance radars. In fighter aircraft, it is more common to determine the R_{90}, or the 90% cumulative probability range of detection (the range at which the cumulative probability equals 90%). The method of determining the R_{90} is best illustrated by an example.

Suppose you were given a specification that stated that the "APG-XXX radar shall have the capability of detecting a 5 m^2 target with a velocity of closure of 1075 knots flying at an altitude of 35,000 ft by the test aircraft flying at 15,000 ft at a range of 60 nm." It is further stated that 60 nm shall be the 90% probable range of detection. The first thing to be agreed upon is what a detection consists of. For the radar engineer, a detection might be when a detection occurs in the radar instrumentation data; an aircrew member may declare a detection when he observes a target on the radar or tactical situation display. Once this discussion is over, we can set about developing the geometry for the test. The intercept should be set at a 180° aspect (head on) at a range equal to 120% of the specification value (in our case, 120% of 60, or 72 nm). This is done to prevent late detections (caused by early turn-in or observer latency) from skewing the statistics. When the detection is observed, either in radar data or by the observer, the truth data of the test aircraft as well as the target are noted and a true range is calculated. This range becomes one data point to be used in calculating the mean, or $\overline{X}$. The R_{90} is defined in equation 8.18:

$$R_{90} = \overline{X} - ts,$$
(8.18)

where

R_{90} = range at which cumulative probability equals 90%,

$\overline{X}$ = mean value of the test data,

s = standard deviation, and

t = standardized Student's t variable.

A simple review of statistics tells us that the sample mean, $\overline{X}$, is defined as

$$\overline{X} = \frac{1}{N} \sum_{i=1}^{N} x_{i},$$

(8.19)

where

$\overline{X}$ = mean value of the test data,

X_{i} = value of the specific measured variable, and

N = number of test points.

The sample standard deviation, s, is defined in equation 8.20:

$$s = \left[\left(\frac{1}{N-1}\right) \left(\sum_{i=1}^{N} x_{i}^{2}\right) - \left(\frac{N}{N-1}\right) \overline{X}^{2} \right]^{\frac{1}{2}}.$$

(8.20)

The variable t is the standardized Student's t variable, which is used to account for sample sizes that are very small when compared to the population. We normally use the Student's t distribution when our sample size is less than 30. Since this evaluation takes a bit of time to accomplish for one data point, it is a safe assumption that we will not be doing more than 30 setups. The Student's t distribution can be obtained from any statistics book for a 90% probability. The value of t would be 3.078 for 2 data points, 1.383 for 10 data points, and 1.328 for 20 data points.

We can now begin the test process and start to collect data toward specification compliance. Figure 8.24 describes the geometry and data required for the maximum detection range evaluation.

This test is relatively easy as long as the variables of RCS, closure, and altitude match the specification. If they do not, then we must normalize the conditions and adjust the setup. Suppose that the RCS of the target provided is 10 m^2 instead of the 5 m^2 called out in the specification. What are the ramifications of this change? We remember from the radar range equation that an increase in RCS will increase the maximum range of

FIGURE 8.24 ■
Maximum Detection
Range

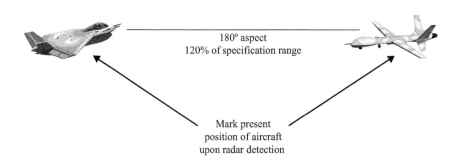

180° aspect
120% of specification range

Mark present
position of aircraft
upon radar detection

detection, but by how much? The normalized range for a target RCS other than 5 m^2 is given empirically by equation 8.21:

$$x_{is} = \left(\frac{\sigma_T}{\sigma_{T'}}\right)^{\frac{1}{4}} \times x_i, \qquad (8.21)$$

where

x_{is} = range normalized for target size,

x_i = actual range demonstrated in flight,

σ_T = specification-defined RCS (5 m^2), and

$\sigma_{T'}$ = actual test target RCS.

If we turn inbound at 72 nm and note a detection at 70 nm, we would normalize this range using equation 8.21:

$$x_{is} = \left(\frac{1}{2}\right)^{\frac{1}{4}} \times 70 = 58.8 \text{ nm.}$$

The answer of 58.8 nm does not meet the requirements of the specification (60 nm) and cannot be included in the data set. We need to reverse engineer the equation in order to determine what our new turn-in range must be for this different RCS target. By setting x_{is} to 72 nm (our original 120% specification value) and solving for x_i (our new turn-in range) we find:

$$72 = \left(\frac{1}{2}\right)^{\frac{1}{4}} \times x_i; \quad x_i = (72)/(0.84) = 85.7 \text{ nm} (\sim86 \text{ nm}).$$

In order for our geometry to meet the requirements of the specification, our new turn-in range for a 10 m^2 target is now 14 nm greater than the original turn-in range. This reverse engineering can be accomplished for any RCS size target and a series of graphs plotting RCS versus turn-in range can be generated. This is a good tool for any test conductor on a radar development program.

As with RCS, changes in the closure rate will also affect our detection range. The example specification calls for a closure of 1075 knots. If our test aircraft flies at a comfortable 400 knots, that means that our target would have to fly at 675 knots in order to meet the closure specification. This is not going to happen; and even if you could get the target up to that speed, it would probably be out of gas after the first pass. We have to normalize our test conditions for velocity of closure differences in a fashion similar to RCS normalization.

The radar range equation shows the term time on target in the numerator. With more time on target, the detection range will increase and with less time it will decrease. If the scan rate of the radar remains the same, a slower closure rate will mean more scans across the target for the same range traversed. Assume that the radar completes one frame every second. A closure rate of 1000 frames/sec will yield one radar hit on the target every 1000 ft. If the closure is reduced to 500 frames/sec, the radar will have two hits on the target every 1000 ft, thus increasing the time on target, and hence the detection range.

The x_{is} will be scaled if the range rate differs from the nominal value using equation 8.22:

$$x_{isR} = \frac{x_{is}}{S_R}, \qquad (8.22)$$

where

x_{isR} = range normalized for target size and range rate,

S_R = scaling factor that is a function of range rate (NOTE: S_R is provided by the radar vendor), and

x_{is} = range normalized for target size (calculated in equation 8.21).

The data must also be scaled for atmospheric attenuation due to altitude changes. Our specification calls for detections at 35,000 ft. What would happen if there are clouds at 35,000 ft and the test has to be accomplished at 20,000 ft? Because the density increases at lower altitudes, the attenuation will increase, causing shorter detection ranges. This relationship for normalization purposes is shown in equation 8.23:

$$x_{iA} = x_i - \text{inv } \log\left[\text{Log}(x_i) - S_A \times \frac{x_i}{20} \right],$$ (8.23)

where

x_{iA} = incremental range for atmospheric attenuation, and

S_A = scaling factor that is a function of altitude (NOTE: S_A is provided by the radar vendor).

The completely normalized range (x_i') that will be used for statistical calculations is obtained by

$$x_i' = x_{isR} + x_{iA},$$ (8.24)

where x_i' is the range normalized for RCS, closure rate, and altitudes.

Table 8.12 provides a possible data card for pulse search R_{90} detections.

The number of data points to be collected to satisfy the R_{90} requirements is really determined by equation 8.18. The key is to have a small standard deviation and a small t. Since the standard deviation will not become evident until the collection of data, the only variable the tester can control in the planning process is t. The t distribution does not change much after 20 samples for 90% ($t = 1.328$), so planning to accomplish more than 20 samples would be counterproductive. The test plan would call out 20 passes to successfully accomplish the requirement.

8.7.4.1 Other Considerations in R_{90} Testing

The previously described testing positioned the target in a clear or clutter-free region, since the target was located at 35,000 ft. There are other considerations that will require

TABLE 8.12 ▪ Pulse Search R_{90} Detections Data Card

Radar Pulse Search R_{90}									
					Detection				
Test Run No.	Turn-in Range	Target RCS	Target Altitude	Target Groundspeed	Range (per radar)	Target Latitude	Target Longitude	Test Aircraft Latitude	Test Aircraft Longitude

evaluation by the test team. The first of these considerations is look-down, or clutter testing. The specifications will be less stringent, as the target is now competing with a strong ground return. The method of test is identical to the clear region R_{90} test, and the number of passes required, normalizations, etc., are the same.

There may also be a requirement to test look-down geometry over terrain with differing backscatter coefficients. The most common evaluation is over land and over water, as the coefficients differ greatly.

These conditions are added to the test matrix and the total number of conditions required is accomplished by multiplying across the matrix. If we need to test clear and clutter, over land and over water, the number of test conditions (and hence the number of test points) required would be multiplied by four.

8.7.5 Pulse Radar Minimum Detection Range

This test is really a validation of the physics of radar. The minimum detection range is the shortest range to the target at which the radar can provide usable information to the operator. This test is sometimes called short pulse detection, since the radar uses the shortest PW available in order to get as close as possible to the target. In many radars this is an automated feature tied to range scale selection by the operator.

As previously discussed, the resolution range is equal to the PW (in microseconds) times 500 ft. The minimum range is computed in exactly the same fashion. If the range to the target is too close, the trailing edge of the pulse will not arrive at the target before the leading edge of the pulse is already back at the receiver. This range is equal to one-half of the trip time of the pulse, or 500 ft/μsec. If the PW of the radar is 1 μsec, the minimum detection range of the radar is 500 ft.

The test is set up as shown in Figure 8.25, and a data card is provided in Table 8.13. The test aircraft sets up behind the target aircraft at the same speed and at a range that is 1000 ft greater than the predicted minimum detection range. The test aircraft marks when there is a detection in pulse search. On the test conductor's call, the test aircraft increases speed by 10 knots and calls when the target is lost. On that mark, the true

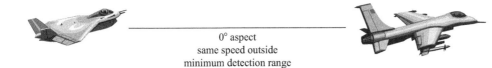

0° aspect
same speed outside
minimum detection range

FIGURE 8.25 ■
Pulse Search
Minimum Detection
Range Geometry

TABLE 8.13 ■ Pulse Search Minimum Detection Data Card

Test Run No.	Initial Detection				Loss of Detection			
	Target Latitude	Target Longitude	Test Aircraft Latitude	Test Aircraft Longitude	Target Latitude	Target Longitude	Test Aircraft Latitude	Test Aircraft Longitude

positions of the test and target aircraft are recorded. The test aircraft then reduces speed by 20 knots and calls when a detection is seen; the true positions of both aircraft are recorded again. This procedure is repeated as necessary. It is imperative that the test be initiated in a trail position, as opposed to head on.

8.7.6 Pulse Radar Measurement Accuracy

The pulse delay ranging radar provides the user with the range, azimuth, and elevation of the detections. It would make sense therefore to assume that there should be some specification defining the accuracy of these variables. And there is. There will be an accuracy specified for each of these parameters. The accuracies must be shown throughout the operating limits of the radar. For example, if the radar can scan ±60° in azimuth, then range, azimuth, and elevation accuracies must be shown throughout the azimuth scan volume. The same holds true for elevation and range volume. Another matrix of test variables will be made to determine the test conditions required.

The ranges should be broken out into range bins, much like what was accomplished during the R_{50} blip-scan tests. The accuracies are stated for the midpoint of the bin. For example, if the range bin is 21 to 30 nm, the accuracies would be stated for 25 nm, the midpoint of the bin. A sample test matrix for measurement accuracy is shown in Table 8.14. In this example, the maximum detection range is 60 nm, and the radar scans ±60° in azimuth and +45°/−30° in elevation.

As shown in Table 8.14, a total of 2400 test conditions are required for this test if we agree that testing every 5° is the way to go. If we assume that we would need about six test points for every condition, we are looking at 14,400 test points. Here is another case where the program manager is going to be quite unhappy upon hearing the bad news, and you are about to receive a serious tongue lashing; that is, unless you can describe how this test is not going to take 100 flights to complete.

The key here is to remember the data rates, and how many data points can be collected with every pass. This matrix can be accomplished in relatively quick fashion (about four flights) if it is set up accurately and sufficient radar and truth data are available. Figure 8.26 shows the geometry of how this test should be run.

TABLE 8.14 ■ Pulse Search Measurement Accuracy Test Matrix

Pulse Search Measurement Accuracy				
	Range Bin (0–60 nm)	Azimuth (+60°/−60°)	Elevation (+45°/−30°)	
Accuracy Required	500 ft	1°	1°	
Test Variables	10 nm bins (6 bins)	Every 5° (25)	Every 5° (16)	Total test conditions: $6 \times 25 \times 16 = 2400$
	0–10 nm			
	11–20 nm			
	21–30 nm			
	31–40 nm			
	41–50 nm			
	51–60 nm			

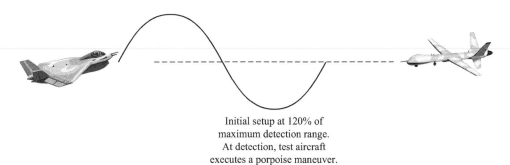

FIGURE 8.26 ■
Measurement
Accuracy Geometry

Initial setup at 120% of
maximum detection range.
At detection, test aircraft
executes a porpoise maneuver.

The initial setup is performed exactly as the maximum detection range test. After the detection is accomplished, the test aircraft executes a porpoise maneuver that will continue through crossover; the operator must maintain the detection. You can see that during this maneuver, the range and elevation are continuously changing, filling in the matrix; the azimuth is constant. If the data rate for the radar data is 25 samples/sec and there is matching TSPI data at the same rate, 25 data points will be collected every second. These passes are accomplished a number of times until the matrix is filled in.

In order to collect the azimuth data, the same setup is used; however, this time the test aircraft executes a saw-tooth maneuver in the horizontal to vary the azimuth with range; elevation is held constant. It is important that the test aircraft perform the maneuvers and not the target. When the target is maneuvering, the test aircraft will lose control, and valuable time and effort will be wasted.

Measurement accuracy is normally assessed using ensemble statistics. Ensemble statistics allow the weighting of data collected at different times by the total amount of data collected at that time. For example, if we collect five data points today and tomorrow we collect 500 data points for the same evaluation, the mean and standard deviation for tomorrow's data will hold more weight than today's. For measurement accuracy, the absolute value of the ensemble mean plus twice the ensemble standard deviation are compared to the specification:

$$\text{Measurement accuracy} = (|\overline{X_E}|) + 2s_E. \tag{8.25}$$

The ensemble mean and ensemble standard deviation are further explained in equations 8.26 and 8.27:

$$\overline{X_E} = \frac{N_1\overline{X_1} + N_2\overline{X_2} + \cdots N_K\overline{X_K}}{N_1 + N_2 + \cdots N_K}, \tag{8.26}$$

where

$\overline{X_E}$ = ensemble mean of the total sample space,

$\overline{X_K}$ = sample mean of the Kth test data sample, and

N_K = number of test points in the Kth sample mean; and

$$s_E = \left[\frac{1}{N_T - 1}\sum_{j=1}^{k}\left(\sum x_j^2\right) - \left(\frac{N_T}{N_T - 1}\right)\overline{X_E}^2\right]^{\frac{1}{2}}, \tag{8.27}$$

where

s_E = ensemble standard deviation of the total sample space,

N_T = total number of test points equal to $N_1 + N_2 + \ldots N_K$,

$\overline{X}_E$ = ensemble mean of the total sample space,

k = total number of test points in the sample space, and

$\sum x_j^2$ = sum of the square of the specified measured variable and is equal to

$$\sum x_j^2 = (N_j - 1)s_j + N_j\overline{X}_j^2,$$

where

N_j = number of test points in the jth sample,

s_j = standard deviation of the jth sample, and

$\overline{X}_j$ = mean of the jth sample.

8.7.7 Pulse Radar Track Initiation Range

A pulsed delay ranging radar can acquire and track only one target; this track is known as a pulse single-target track (PSTT). There have been many methods of tracking targets with airborne radars, and now is perhaps the best time to review some of those modes. A target is said to be tracked by a radar when the radar can predict where the target will be in the next time frame and smoothly slew the antenna so that the target is always within the tracking beam.

The earliest type of tracking was called lobing. The radar operator designates a target of interest via acquisition symbols and through some means (trigger detent, button depression, etc.) informs the radar that he wants the system to track this target. The radar switches to a pencil beam and lobes the target (places the beam to the left and right of the target) (see Figure 8.27).

This type of tracking works alright as long as the target only maneuvers in the horizontal plane. If the target maneuvers in the vertical plane, the radar is unable to detect the movement and tracking is lost. In order to combat this problem, a conical track was developed. The same pencil beam is utilized, but now the antenna is nutated in

FIGURE 8.27 ■
PSTT Lobing

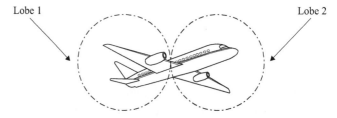

Radar lobes left then right on the target
and drives the antenna to equalize the
returns in both lobes. When the returns
are always equal the target is being
tracked.

order to develop a conical lobing about the target. This type of scan can cover the horizontal and vertical movements of the target (Figure 8.28).

Versions of these two types of tracks are found with surface-to-air tracking systems. These systems use two antennas: one transmitter and one receiver. The transmitter radiates the target while the receiving antenna rotates or lobes from side to side. The results are the same as just described; however, the target does not know it is being tracked because all tracking is passive. These systems are called conical scan receive only (COSRO) and lobe-on receive only (LORO).

Modern radars employ a method of tracking called monopulse, which is employed with planar array–type antennas. In theory, it works the same as the conical scan; however, the antenna is not rotated. The echo from a target is received by the antenna and partitioned into four quadrants. The return is analyzed for the strength of the return in each of the four quadrants and is steered to obtain an equal return in all four quadrants (Figures 8.29 and 8.30).

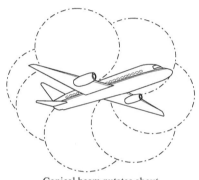

Conical beam nutates about
the target seeking an equal
return in each lobe position.

FIGURE 8.28 ■
PSTT Conical Track

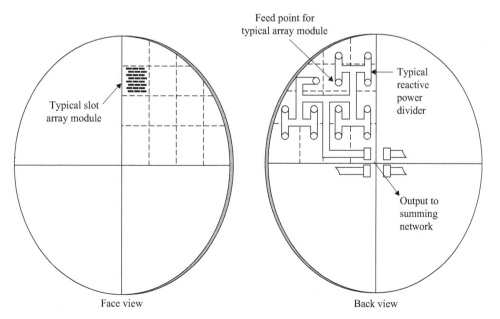

Face view

Back view

FIGURE 8.29 ■
Planar Array
Mechanism

FIGURE 8.30 ▪
Monopulse Tracking

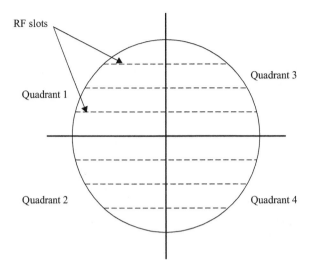

The array in Figure 8.30 contains many radio frequency (RF) slots, each of which transmits and receives radar energy. The slots may work independently or all together, depending on the task at hand. The array in the figure is divided into four quadrants labeled 1 through 4. The radar processor sums the quadrants in hemispheres and compares the return to the opposite hemisphere. If the sum of one and two is higher than the sum of three and four, then the antenna must be moved to the left (from our perspective). If the sum of one and three is higher than the sum of two and four, then the antenna must be moved upward. These comparisons are continuously made in order to smoothly drive the pointing angle of the radar to the centroid of the target.

The first evaluation regarding PSTT is the track initiation range. The track initiation range is that range where a track file can be initiated and maintained all the way through target crossover. This statement is made because there is a probability that the track may drop out after some time and then be reacquired at a shorter distance. The shorter distance becomes the track initiation range.

The track initiation range tests are conducted the same way that we initially tested R_{90}. There are two exceptions: one, as previously noted, is that after the track is initiated, it must be held through crossover; the second exception is that the range need not be normalized for range rate. This is because of the high rate of track initiation opportunities (the antenna is trained on the target). The range must still be normalized for RCS and density differences. The specification may also require the variables of clutter and terrain type be included in the evaluation.

8.7.8 Pulse Radar Track Accuracy

This test duplicates the evaluation of measurement accuracy detailed in section 8.7.5. This test evaluates the radar's ability to determine the range, azimuth, and elevation of a target that is being tracked (as opposed to accuracies in the search mode). Since the target will be tracked in PSTT, all radar energy is devoted to this one target and the accuracies will be very good. TSPI data which was used for the measurement accuracy tests may not be good enough for this evaluation, and is something that must be considered. The test matrix in Table 8.14 can be used again for this test with the appropriate changes to the specifications.

8.7.9 Pulse Radar Resolutions

Since a pulse radar can provide attributes of range, azimuth, and elevation, it should also be able to resolve two targets separated by any of these variables. This evaluation is a verification of the radar's resolution cell. If two targets lie entirely within the radar's resolution cell, the radar will only detect one target. Two targets that lie outside any of the resolution parameters of range, azimuth, and elevation will show up as two targets. Figure 8.31 depicts the radar resolution cell.

Range resolution is dependent upon the PW and was explained previously in section 8.4 and depicted in Figure 8.11. Remember that many radars will change the PW based on range scale selection, while others may employ pulse compression (chirp; Figure 8.12), which will enhance the resolution. In order to perform this test, we must first be aware of the PW and if it is changed with range selection. Remember also the rule of thumb on range resolution: 1 μsec is equal to 500 ft. A typical radar set of PW parameters is shown in Table 8.15.

In this particular radar, the range resolution is 250 ft on a 5 nm range display and 10,000 ft on a 200 nm display. If all range scale selections have to be evaluated, this radar will require six test conditions (one for each range scale selection).

The test geometry is relatively simple, as depicted in Figure 8.32.

All aircraft should be at the same speed and on the same heading. The targets should be at a 0° aspect angle and separated by 1 PW of distance plus 3000 ft. The range resolution is not range dependent, so the distance between the test aircraft and the targets

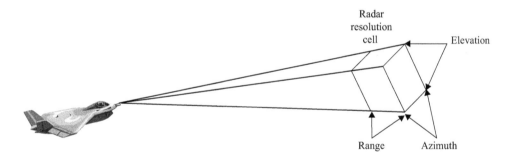

Radar resolution cell

Elevation

Range   Azimuth

FIGURE 8.31 ◾ Radar Resolution Cell

TABLE 8.15 ◾ PW Variations with Range Scale Selection

	Range Scale (nm)					
	5	10	25	50	100	200
PW (μsec)	0.5	1	2	5	10	20

Test aircraft 300 knots 0° aspect

Target aircraft 300 knots in trail, separated by 1 PW of resolution plus 3000 ft

FIGURE 8.32 ◾ Range Resolution Geometry

is not critical. The distance will only become a factor with the ability to see the targets on the display. Prior to performing the flight test, do some homework on the number of available pixels on each display setting. In some cases, the display resolution is not adequate to meet the specification, especially in the largest range scale settings.

On the test conductor's mark, and after the test aircraft has two positive detections, the trailing target will increase his airspeed by about 10 knots and stabilize. The test aircraft will mark when the two detections merge to a single return. The trailing target will then decelerate 20 knots and the test aircraft will mark when two detections reappear. This test should be repeated multiple times as data requirements dictate. The distance between the targets can be adjusted based on the range scale selected by the test aircraft. A sample data card is shown in Table 8.16.

Azimuth resolution is a function of the beamwidth and is dependent on the range to the targets. The beamwidth was described previously in section 8.3, and the determination of the beamwidth is given in equations 8.8 through 8.11. Figure 8.33 shows the effect of beamwidth on azimuth accuracy and resolution. A large beamwidth will mask two targets (two targets will appear as one), while a narrow beamwidth will break out the two targets at the same range. It is important to note that even with large beamwidths, radars can resolve two targets in azimuth as the targets get closer to the radar.

TABLE 8.16 ■ Range Resolution Sample Data Card

		2 Targets Displayed				1 Target Displayed			
Run No.	Time	Target 1 Latitude	Target 1 Longitude	Target 2 Latitude	Target 2 Longitude	Target 1 Latitude	Target 1 Longitude	Target 2 Latitude	Target 2 Longitude

FIGURE 8.33 ■ Antenna Beamwidth Versus Resolution

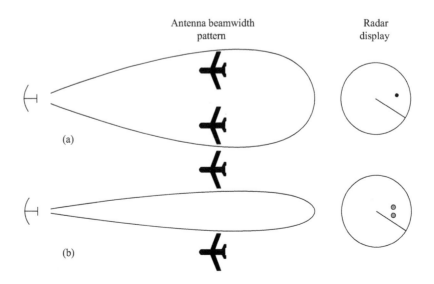

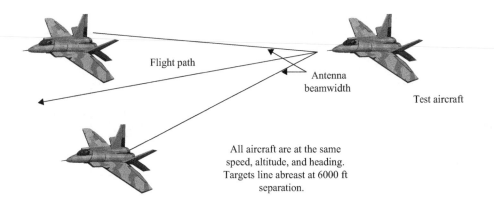

FIGURE 8.34 ▪
Azimuth Resolution
Geometry

Flight path

Antenna
beamwidth

Test aircraft

All aircraft are at the same
speed, altitude, and heading.
Targets line abreast at 6000 ft
separation.

Similarly, the azimuth accuracy is a function of the beamwidth. A narrower beam will be able to more accurately depict the target's location. With a wide beam, the target can be anywhere within the beam as it sweeps across the target. The depiction on the display will be an apparent smearing of the target, making it appear larger than it really is. The width of the target will appear to be equal to the arc of the beamwidth, and this error will increase as the range to the target increases.

The azimuth resolution tests are conducted in a similar fashion to the previously discussed range resolution tests. Figure 8.34 presents the geometry of the azimuth resolution evaluations.

The target aircraft should be set up at 6000 ft, line abreast, in front of the test aircraft. All aircraft should be on the same heading, altitude, and airspeed. The test aircraft should be exactly between the two targets; the flight path will intersect the beamwidth. The distance that the test aircraft should be in trail depends on the beamwidth. We can determine this distance by using the formula

$$\text{Range (nm)} \times \text{Inclusive Angle } (^\circ) = \text{Separation (ft)} \times 100. \qquad (8.28)$$

The formula is best illustrated by using an example. Suppose that the beamwidth of the antenna is 2° and the targets are separated by 6000 ft. At what range does the separation equal the beamwidth? Using equation 8.28,

$$R \times 2 = 6000/100,$$
$$R = 60/2,$$
$$R = 30 \text{ nm}.$$

For this set of parameters, we would set up our test aircraft 35 nm in trail of our two targets. At this range, the operator will see only one target on the display. At the call of the test conductor, the test aircraft increases his speed and drives toward the targets. The operator will call mark when the return splits into two separate detections. At the call of the test conductor, the test aircraft then decreases speed to less than the target's speed and calls mark when the two detections merge to a single return. This procedure can be run as often as the data requirements dictate. For this test, the position of all three aircraft must be recorded at the mark. Table 8.17 provides an example of an azimuth resolution data card.

The final resolution test is the ability of the radar to resolve two targets separated in altitude. This is a difficult test to perform in the air for multiple reasons. The first is that the geometry is difficult to attain, as the targets must be one atop the other. The second reason is

TABLE 8.17 ■ Azimuth Resolution Sample Data Card

Azimuth Resolution Data Card							
		Position at Mark Call					
Run No.	Time	Target 1 Latitude	Target 1 Longitude	Target 2 Latitude	Target 2 Longitude	Test Aircraft Latitude	Test Aircraft Longitude	No. of Displayed Targets

the display's inability to depict this information in a usable manner. If the test is accomplished using a one-bar scan, two targets will never be seen unless the operator manually adjusts the tilt. If a multiple raster scan is used, the operator will see both targets, but not at the same time, since each target (if separated by an elevation beamwidth or more) will reside on a separate bar of the scan. This test is best accomplished in the lab using aircraft/radar representative hardware and software and utilizing a radar target generator. Trying to accomplish these tests in flight testing generally turns out to be a time waster.

8.8 | DOPPLER AND PULSE DOPPLER MODES OF OPERATION AND TESTING

8.8.1 Doppler Characteristics

When an RF signal is pulse modulated it will appear in the frequency spectrum as peaks on both sides of the original RF spaced by the PRF. Now it is possible for a target to return a Doppler shift greater than the PRF. If that happens, we do not know if this is a shift in the original RF or a shift in one of the harmonics, that is, the Doppler shift is ambiguous. To resolve Doppler ambiguities, we must have some way of telling what whole multiple of the PRF, if any, separates the observed frequency of the target echoes from the carrier frequency. If not too great, this multiple (n) may easily be determined. There are two common ways: range differentiation, and PRF switching.

Generally the simplest way to determine the value of n is to make an approximate initial measurement of the range rate by the differentiation method. From this rate we compute the approximate value of the true Doppler frequency. Subtracting the observed frequency and dividing by the PRF yields the factor n.

Suppose, for example, that the PRF is 20 kHz and the observed Doppler frequency is 10 kHz. The true Doppler frequency then could be 10 kHz plus any whole multiple of 20 kHz, up to 70 kHz. The true Doppler could have one of these values: $-10, 10, 30, 50,$ and 70 kHz. Figure 8.35 illustrates this example.

The approximate value of the true Doppler frequency computed from the initial range rate measurement was computed as 50 kHz. The difference between this frequency and the observed Doppler frequency is $50 - 10 = 40$ kHz. Dividing the difference by the PRF, we get

$$n = 40 \div 20 = 2.$$

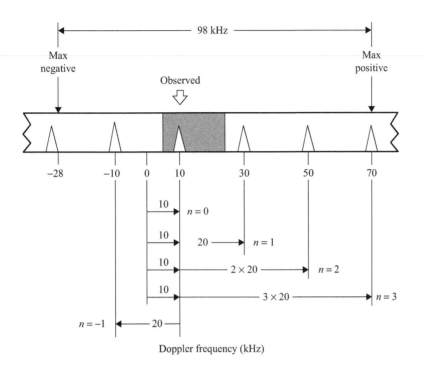

FIGURE 8.35 ■
Computation of the
Multiple (*n*)

The echo's carrier is separated from the observed Doppler frequency by two times the PRF.

Although we assumed that the initial range rate measurement was fairly precise, it need not be particularly accurate. As long as any error in the Doppler frequency computed from the initial rate measurement is less than half the PRF, we can still tell in which PRF interval the carrier lies. The initially computed "true" Doppler frequency, for example, might have been only 42 kHz, almost halfway between the two nearest possible exact values (30 and 50 kHz).

Nevertheless, this rough, initially computed value (42 kHz) is still accurate enough to enable us to find the correct value of *n*. The difference between the initially computed value of the Doppler frequency and the observed value is $42 - 10 = 32$ kHz. Dividing the difference by the PRF, we get $32 \div 20 = 1.6$. Rounding off to the nearest whole number, we still come up with $n = 2$. After having determined the value of *n* once, we can, by tracking the target continuously, determine the true Doppler frequency, and thus compute range with considerable precision based solely on the observed frequency.

The value of *n* can also be determined with a PRF switching technique similar to that used to resolve range ambiguities. In essence, this technique involves alternately switching the PRF between two relatively closely spaced values and noting the change, if any, in the target's observed frequency.

Switching the PRF will have no effect on the target echo's carrier frequency, f_c. The target carrier frequency equals the carrier frequency of the transmitted pulses plus the target's Doppler frequency, and is completely independent of the PRF. However, this is not the case with the sideband frequencies above and below f_c.

Because these frequencies are separated from f_c by multiples of the PRF, when the PRF is changed, the sideband frequencies correspondingly change (Figure 8.36). Which direction a particular sideband frequency moves (up or down) depends on

FIGURE 8.36 ▪
Relationship of
Changing PRF
with Sideband
Frequencies

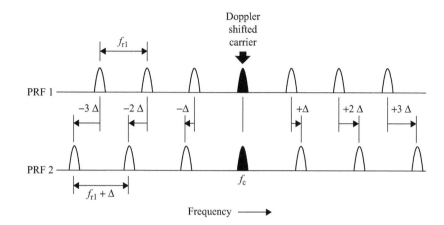

FIGURE 8.37 ▪
Change in Observed
Frequency with PRF
Switching

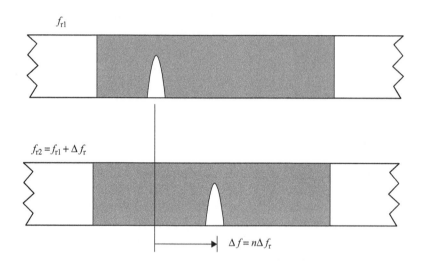

two things: 1) whether the sideband frequency is above or below f_c, and 2) whether the PRF has been increased or decreased. An upper sideband moves up if the PRF is increased and down if it is decreased. A lower sideband, on the other hand, moves down if the PRF is increased and up if it is decreased.

If the PRF is changed by 1 kHz, the first set of sidebands on either side of f_c will move 1 kHz, the second will move 2 kHz, and the third, 3 kHz. If the PRF is changed by 2 kHz, each set of sidebands will move twice as far.

By noting the change, if any, in the target's observed Doppler frequency, we can immediately tell where f_c is relative to the observed frequency. If the observed frequency does not change, we know that it is f_c. If it does change, we can tell from the direction of the change whether f_c is above or below the observed frequency. And we can tell from the amount of the change by what multiple of the PRF f_c is removed from the observed frequency (Figure 8.37).

The factor n by which the PRF must be multiplied to obtain the difference between the echo's carrier frequency f_c and the observed frequency is

$$n = \Delta f_{obs}/\Delta f_r \qquad (8.29)$$

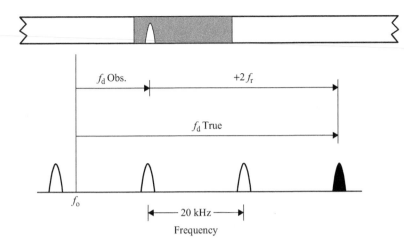

FIGURE 8.38 ■
Computation of the
True Doppler
Frequency

where

Δf_{obs} = change in target's observed frequency when PRF is switched, and

Δf_r = amount PRF is changed.

If, for example, an increase in PRF (Δf_r) of 2 kHz caused a target's observed Doppler frequency to increase by 4 kHz, the value of n would be $4 \div 2 = 2$. In order to avoid the possibility of "ghosts" when returns are simultaneously received from more than one target, the PRF must generally be switched from one to another of three values, instead of two, just as when resolving range ambiguities. Switching the PRF has the disadvantage of reducing the maximum detection range due to shorter integration times.

Having determined the value of n by either of the methods just described, we can compute the target's true Doppler frequency, f_d, simply by multiplying the PRF by n and adding the product to the observed frequency (Figure 8.38).

8.8.2 Doppler Clutter Regions

Ground return falls into three categories: main lobe return, sidelobe return, and altitude return, which is sidelobe return received from directly beneath the radar. Main lobe return is the signal used in many ground applications, but it is clutter for radars that must detect airborne targets.

The principal means of discerning target echoes from ground clutter is Doppler resolution. The way in which the clutter is distributed over the frequency band is its Doppler spectrum.

The range rate of a patch of ground is entirely due to the radar's own velocity. The projection of the radar's velocity on the LOS to the patch can be substituted for ΔV. For a ground patch directly ahead, this projection equals the radar's full velocity, V_R. For a ground patch directly to the side or directly below, the projection is zero. In between, it equals V_R times the cosine of the angle (L) between V_R and the LOS to the patch.

The Doppler frequency of the return from a patch of ground is

$$f_d = 2V_R \cos L / \lambda \quad (\lambda = c/f_c). \tag{8.30}$$

The radar has a look-down angle that is also a factor in the calculation; this is disregarded in the formula above. Since returns can be received from ground patches in

FIGURE 8.39 ■
MLC With Radar
Look Angle

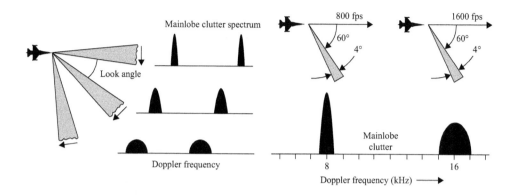

any direction, the ground return generally covers a broad range of frequencies. It can also be seen that for a return at an angle of 0° (on the nose of the aircraft), the Doppler shift will be equal to groundspeed, and for an angle of 90° (under the aircraft or abeam), the Doppler shift will be equal to zero. For this reason, main lobe clutter (MLC) is commonly addressed as groundspeed and sidelobe clutter (SLC) is commonly addressed as zero closure. We will see that in reality clutter is evident for the full 360° around the aircraft.

Main lobe clutter is produced whenever the main lobe intercepts the ground. Because the ground area intercepted by the main lobe can be extensive and the gain of the main lobe is high, MLC return is generally quite strong. Each type of surface has a typical backscattering coefficient, which is a measure of how much clutter is returned. Another factor is the grazing angle. Smooth surfaces at small grazing angles produce little clutter, but at large grazing angles they produce a strong return (sea, lakebeds, flat terrain). This will be covered later in the air-to-ground section.

The spectrum of the MLC varies continually. As the antenna look angle increases, the center frequency of the spectrum decreases, and the width increases from nearly a line to a broad hump. As the speed of the radar increases, both the frequency and the width increase; if the aircraft speed is doubled, the center frequency and the width of the spectrum will also double. Figure 8.39 shows an MLC and its relation to radar look angle.

The SLC is the radar return received through the antenna's sidelobes, and it is always undesirable. Excluding the altitude return, SLC is not nearly as concentrated as MLC, but it covers a much wider band of frequencies. Its strength is governed by the gain of the particular sidelobe within which the ground patch lies, by the backscattering coefficient, and by the grazing angle. Man-made structures can produce strong returns.

Sidelobes extend in all directions, even to the rear. Therefore, regardless of look angle, there are always sidelobes pointing ahead, behind, and at virtually every angle in between. As a result, the frequency band covered by the SLC extends from a positive shift corresponding to the radar's velocity to an equal negative shift. The area illuminated by the SLC is extremely large, as shown in Figure 8.40, and much of it comes from relatively short ranges.

Beneath the aircraft there is usually a region of considerable return where the ground is so close to a single range that the sidelobe return from it appears as a spike on a plot of amplitude versus range. This return is called the altitude return, since generally its range equals the radar's absolute altitude. The altitude return is not only much stronger than the surrounding SLC, but may be as strong as the MLC. The area from which it comes may be very large and is often at extremely close range. This point is illustrated in Figure 8.41. A radar is at an altitude (*h*) of 6000 ft over flat terrain. As you

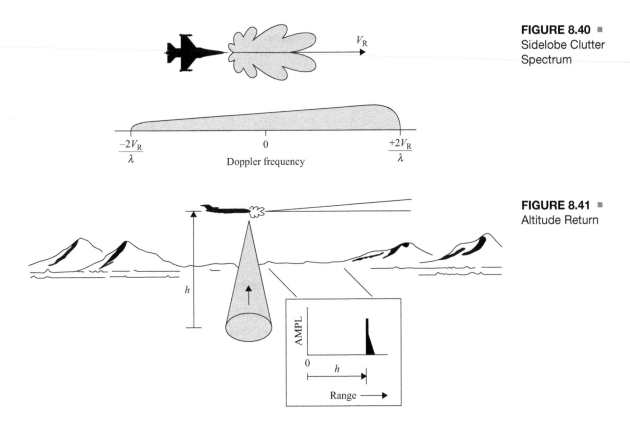

FIGURE 8.40 ■
Sidelobe Clutter
Spectrum

FIGURE 8.41 ■
Altitude Return

can see, the slant range to the ground at an angle of incidence of θ equals $h/\cos \theta$. Because the cosine of a small angle is only slightly less than one, the slant range to the ground is only 500 ft greater than the altitude (vertical range), even when θ is as much as 22°. If the slant range at an angle of incidence of 22° is rotated about the vertical, it traces a circle on the ground having a diameter of roughly 5000 ft. A circle this size contains nearly 20 million ft^2. Thus the radar receives all of the return from a 20 million ft^2 area, at a range of just 1 nm, in the round-trip transit time for a range increment of 500 ft. That is only 1 μsec. At near vertical incidence, the backscattering coefficient tends to be very large.

The altitude return also peaks in a plot of amplitude versus frequency, but not sharply. The projection of the radar velocity, V_R, on the slant range to the ground equals $V_R \sin \theta$. Unlike the cosine, the sine of an angle changes most rapidly as the angle goes through zero.

While the Doppler frequency of the clutter is zero when θ is zero, it increases to nearly 40% of its maximum value $(2V_R/\lambda)$ at an angle θ of only 22°. The return from a circle of ground that produces a sharp spike in a plot of amplitude versus range therefore produces only a broad hump in a plot of amplitude versus Doppler frequency. Normally the Doppler frequency of the altitude return is centered at zero. However, if the altitude of the radar is changing, such as when the aircraft is climbing, diving, or flying over sloping terrain, the return varies about the zero Doppler. For a dive, the Doppler frequency will be positive and for a climb it will be negative. Even though the frequency is fairly low, it can be considerable. In a 30° dive, for instance, the altitude is changing at a rate equal to half the radar velocity. Figure 8.42 depicts the math behind this explanation.

FIGURE 8.42 ■
Zero Doppler Return
as a Function of Dive
Angle

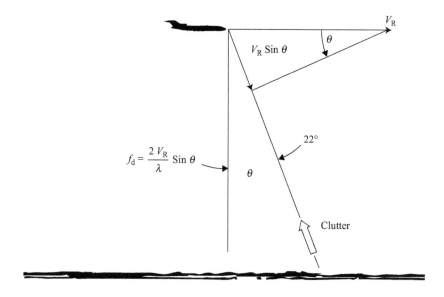

Despite its strength, the altitude return is usually less difficult to deal with than the other ground returns. Not only does it come from a single range, but its range is predictable and its Doppler frequency is generally close to zero. Unfortunately this is also the Doppler frequency of a target pursued at constant range (e.g., a tail chase with zero closing rate). Some radars apply a notch for the zero Doppler (±100 knots for rising and falling terrain), but only at a range equal to the aircraft's altitude. This way, targets on the nose with zero closure will be detected as long as they are inside or outside of the range equal to the aircraft's altitude.

Having addressed the characteristics of MLC, SLC, and altitude return individually, the next step is to view the composite clutter spectrum and its relationship to the frequencies of the echoes from representative airborne targets. It is assumed that the PRF is high enough to avoid Doppler ambiguities.

Stimson does an excellent job of explaining the composite Doppler clutter spectrum in chapter 22 of his book, and some of his pertinent figures on the subject are included here for completeness. The relationship between target and clutter frequencies for a nose-aspect approach is illustrated in Figure 8.43. The target's Doppler frequency is greater than that of any of the ground return. This target is said to be in the "clear region."

Figure 8.44 depicts a tail chase with a low closure rate. Because the closing rate is less than the radar's velocity, the target's Doppler frequency falls within the band of frequencies occupied by the SLC. Just where it falls depends on the closing rate.

Figure 8.45 depicts the target in the beam (perpendicular to the LOS of the radar), where the closure is equal to the groundspeed (MLC). As a result, the target is hidden in the MLC.

In Figure 8.46, the target's closing rate is zero. In this situation, the target echoes have the same Doppler frequency as the altitude return.

Figure 8.47 shows two opening targets. Target A has an opening rate that is greater than the radar's groundspeed, so this target appears in the clear beyond the negative-frequency end of the SLC spectrum. On the other hand, target B has an opening rate less than the radar's groundspeed, so this target appears within the negative frequency portion of the SLC spectrum.

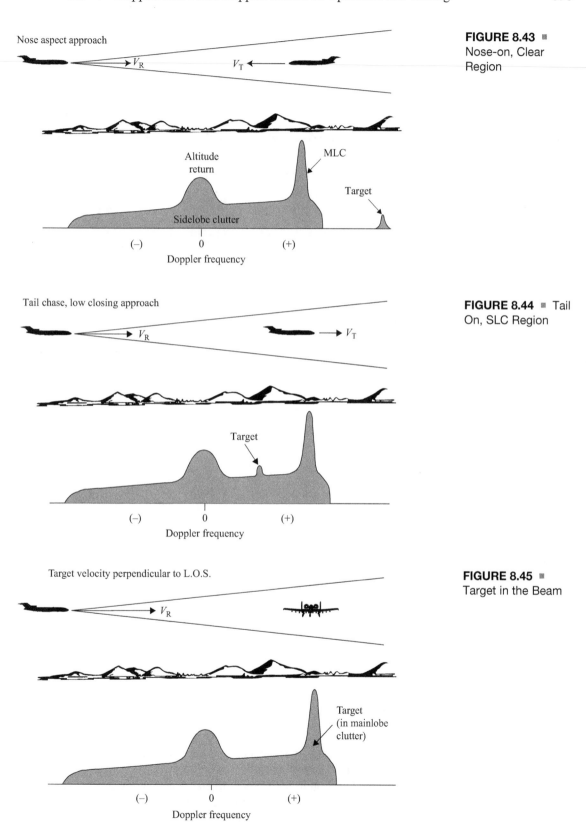

FIGURE 8.43 ■ Nose-on, Clear Region

Nose aspect approach

V_R V_T

Altitude return

MLC

Target

Sidelobe clutter

(−)   0   (+)
Doppler frequency

FIGURE 8.44 ■ Tail On, SLC Region

Tail chase, low closing approach

V_R V_T

Target

(−)   0   (+)
Doppler frequency

FIGURE 8.45 ■ Target in the Beam

Target velocity perpendicular to L.O.S.

V_R

Target (in mainlobe clutter)

(−)   0   (+)
Doppler frequency

FIGURE 8.46 ■
Target in the Altitude
Line

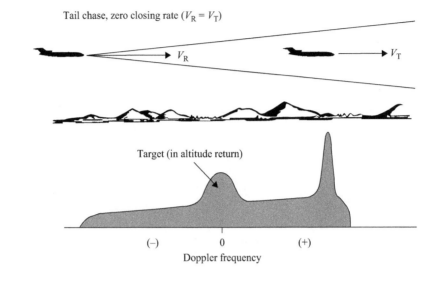

FIGURE 8.46 ■
Target in the Altitude
Line

FIGURE 8.47 ■
Effects of Opening
Target Velocity

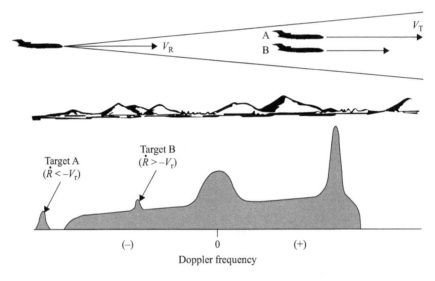

FIGURE 8.47 ■
Effects of Opening
Target Velocity

With these situations as a guide, the relationship of the Doppler frequencies of target return and ground return for virtually any situation can easily be pictured (Figure 8.48). Remember that at lower PRFs, Doppler ambiguity can occur which may cause a target and a ground patch having quite different range rates to appear to have the same Doppler frequency.

8.8.3 Doppler Ambiguities

The disturbing effect of clutter is even more pronounced when range or Doppler ambiguities exist. For this we take a look at an aircraft flying at low altitude over terrain from which a considerable amount of clutter is received. Assume that we have the situation depicted in Figure 8.49. Two targets, A and B, are in the antenna's main lobe. Target A is being overtaken from the rear and has a low closing rate. Target B is approaching the radar head on and has a high closing rate. Target A is at a range from

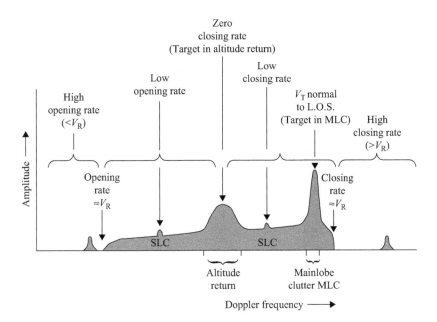

FIGURE 8.48 ■
The Doppler Clutter Spectrum

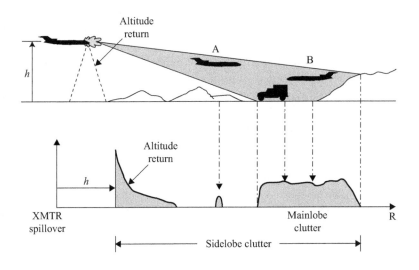

FIGURE 8.49 ■
Ambiguities

which only sidelobe return is being received; target B is at a range from which both main lobe and sidelobe returns are being received. Within the ground patch illuminated by the main lobe is a truck, heading toward the radar.

If we look at the range diagram of Figure 8.49, we see that SLC extends outward from a range equal to the radar's altitude. It decreases rapidly in amplitude when the range increases. The echoes from target A stand out clearly above the SLC. The echoes from target B and the truck are completely obscured by the much stronger MLC. Even though we know where to look for these echoes, we cannot distinguish them from the clutter. Toward the left end you will notice two strong spikes. The one at zero range is due to what is called transmitter spillover—energy from the transmitter that leaks into the receiver during transmission, despite all efforts to block it. The second spike is the altitude return. The Doppler profile for this example is shown in Figure 8.50.

FIGURE 8.50
Doppler Profile for
the Example

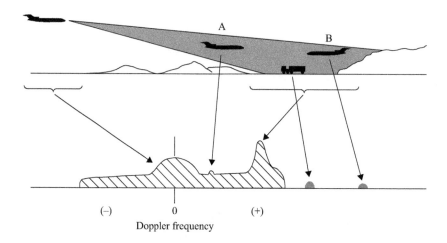

In Figure 8.50, the aircraft is shown with the Doppler profile beneath it. Target A is being overtaken, so its Doppler frequency falls below that of the MLC, in the band of frequencies blanketed by SLC. Because a good deal of this clutter comes from shorter ranges than the target's range, the target echoes barely protrude above the clutter. If the target were smaller or at a much greater range, its echoes might not even be discernible. Since target B and the truck are approaching the radar, they have higher Doppler frequencies than any of the clutter.

As was seen earlier, when transmissions are pulsed, each element of the radar return has sideband frequencies separated from the Doppler-shifted carrier frequency by multiples of the PRF. The altitude return that has zero Doppler frequency also appears to have Doppler frequencies of ±PRF, 2 PRF, 3 PRF, etc. The same is true for every return in the Doppler profile. The entire profile is repeated at intervals equal to multiples of the PRF. The true profile is called the central line return, and the repetitions are called PRF lines.

If the PRF is sufficiently high, the repetitions of the Doppler spectrum will in no way affect the ability of the radar to discriminate between target echoes and ground clutter. However, if PRF is low, repetitions begin to overlap (Figure 8.51). This can result in a target's echoes and ground clutter passing through the same Doppler filter(s), even though the true Doppler frequencies of the target and the clutter may be quite different.

Examples of this phenomenon can be seen with target B and the truck. If the PRF is reduced, the repetitions of the profile increasingly overlap; more and more SLC occupies the space between successive MLC lines and the MLC lines move closer together. Since the width of these lines is independent of the PRF, reducing the PRF causes the MLC to occupy an increasingly larger percentage of the receiver pass-band and causes the altitude return and other close-in SLC to bunch up in the space between.

As the percentage of the pass-band occupied by MLC increases, it becomes increasingly difficult to reject even the MLC on the basis of its Doppler frequency without at the same time rejecting a large percentage of the target echoes. The lower the PRF, the more severe the effects of Doppler ambiguities on ground clutter.

The formula used to express the maximum unambiguous Doppler is

$$\text{Maximum unambiguous Doppler} = \text{PRF} - 2V_R/\lambda. \tag{8.31}$$

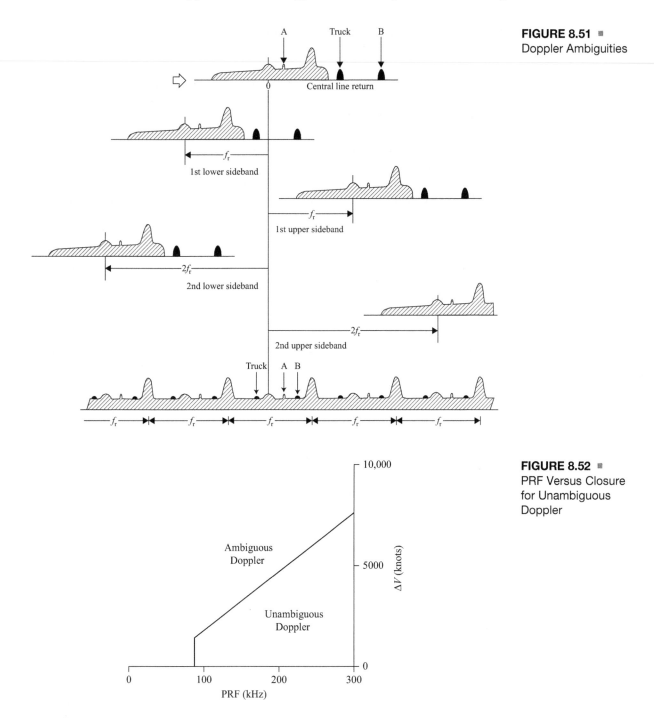

FIGURE 8.51 ■
Doppler Ambiguities

FIGURE 8.52 ■
PRF Versus Closure
for Unambiguous
Doppler

In Figure 8.52, the maximum closing rate for which the Doppler frequency will be ambiguous in a clutter environment is plotted versus PRF. A wavelength of 3 cm and a nominal radar velocity of 1000 knots are assumed. The plot decreases linearly from a closing rate of about 8000 knots at a PRF of 300 kHz to a closing rate of 1000 knots (the radar groundspeed) at a PRF of 70 kHz. (It is terminated at this point since, at lower PRF, the maximum positive and negative SLC frequencies overlap.) The area beneath the curve encompasses every combination of PRF and closing rate for which the observed Doppler frequencies are unambiguous. Conversely, the area above the curve encompasses every combination for which the Doppler frequencies are ambiguous.

For example, at a PRF of 250 kHz, and a closing rate of 5000 knots, the Doppler frequency is unambiguous, whereas at a PRF of 150 kHz and the same closing rate, the Doppler frequency is ambiguous. If the radar-carrying aircraft's groundspeed is greater than 1000 knots, the area beneath the curve will be correspondingly reduced, and vice versa. Since Doppler frequency is inversely related to wavelength, the shorter the wavelength, the more limited the region of unambiguous Doppler frequencies.

To illustrate the profound effect of wavelength on Doppler ambiguities, Figure 8.53 plots the maximum closing rate at which the Doppler frequency will be unambiguous for a 1 cm wavelength. Not only is the area under the curve comparatively small, but even at a PRF of 300 kHz, the maximum closing rate at which the PRF is unambiguous is less than 2000 knots.

The regions of unambiguous range and unambiguous Doppler frequency (for $\lambda = 3$ cm and $V_R = 1000$ knots) are shown together in Figure 8.54. Drawn to the scale of this diagram, the region of unambiguous range (first range zone) is quite narrow. To the right of it is the comparatively broad region of unambiguous Doppler frequencies. In between is a region of considerable extent within which both range and Doppler frequency are ambiguous. Clearly the choice of PRF is a compromise. If the PRF is increased beyond a relatively small value, the observed ranges will be ambiguous, and unless the PRF is raised to a much higher value, the observed Doppler frequencies will be ambiguous.

Because of the tremendous impact the choice of PRF has on performance, it has become customary to classify radars in terms of the PRF they employ. The following is a widely used, consistent set of definitions:

- A low PRF is one for which the maximum range the radar is designed to handle lies in the first range zone; range is unambiguous.

- A medium PRF is one for which both range and Doppler frequencies are ambiguous.

- A high PRF is one for which the observed Doppler frequencies of all targets are unambiguous.

FIGURE 8.53 ■
PRF Versus Closure
for a 1 cm
Wavelength

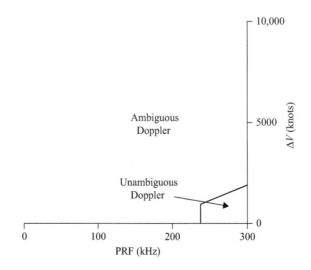

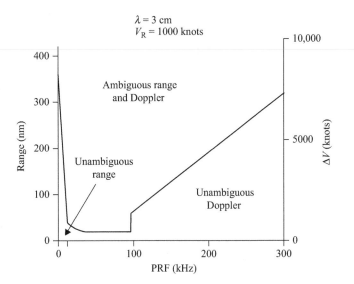

FIGURE 8.54 ■
Unambiguous
Range and Doppler
for PRF Versus
Closure

8.8.4 FM Ranging

From the previous discussions it has been shown that if we want to increase detection ranges, one of the best ways is to increase the PRF. As the PRF is increased, there will come a time where the range ambiguities cannot be resolved and pulse delay ranging will be impossible. Even at short ranges there will be a distance where pulse delay ranging will be rendered inoperable due to the pulse length. This is due to the fact that, at short ranges, the leading edge of the pulse is returned to the transmitter before the trailing edge of the pulse is incident on the target. This is best illustrated in the case of the radar altimeter.

8.8.4.1 Radar Altimeters

The earth is a very good reflector of radio energy, and since aircraft are flying relatively close (a few miles), altitude can be measured very accurately. A low-powered system can give a constant reading of altitude above ground level by measuring the range to the ground. A version of TF can be accomplished by using the radar altimeter and following a preprogrammed, timed course across known terrain.

Radar altimeters are continuous wave (CW) and employ FM ranging to obtain range (as opposed to pulse radars). These altimeters operate in either the UHF or the C band. In the C band, frequencies of 4.2 to 4.4 GHz are reserved for operation. We can use frequencies this low because the directivity of the beam is not critical and the power required is not high; therefore, the components needed are quite small.

There are disadvantages as well to employing these types of frequencies in altimeter operation. These frequencies are good for cloud penetration, but this same advantage makes flight over ice and snow unfavorable. Under some conditions there may be no echo from the top of the ice, and the ice may be deep enough that there is no reflection from the bottom. If there is a reflection from the bottom, the distance is measured by the altimeter as if the propagation was through the air (ice, 0.535; snow, 0.8). If flying over ice, for example, your altitude would read roughly double your true altitude. The range to the ground can be measured indirectly with CW radars. The most common method in use in altimeters is linear frequency modulation, or FM ranging. FM ranging

FIGURE 8.55 ■
Simple FM Ranging

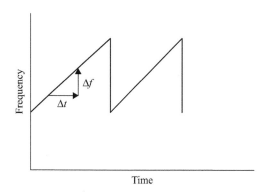

converts the time lag between transmission and reception to a frequency difference. By measuring the frequency difference, the time lag (range) is determined. The mechanics for implementing FM ranging are illustrated in Figure 8.55. The transmitter frequency is increased at a constant rate so that each successive pulse has a slightly higher frequency than the previous one. The modulation is continued for a period of time in order to reach the most distant target and the instantaneous differences between transmitted and received frequencies are measured. The transmitter is then returned to the starting frequency.

For the static case—target not moving (as in a same-speed tail chase, or in the case of the radar altimeter, flat terrain)—the determination of ranges is simple. Since we know the rate of change of frequency of the transmitter, we can compare the frequency received at any period of time and equate that change to a time, and hence a range. The rate of change of frequency is given by k, where

$$k = \Delta f / \Delta t. \tag{8.32}$$

Once again referencing Stimson, I again use one of his examples (chapter 14). Suppose the measured frequency difference is 10,000 Hz and the transmitter frequency has been increasing at a rate of 10 Hz/μsec. The transit time can be calculated using equation 8.32 and solving for Δt:

$$\Delta t = 10{,}000 \text{ Hz}/10\text{Hz}/\mu\text{sec} = 1000 \ \mu\text{sec}.$$

Since 12.4 μsec of round-trip transit time corresponds to 1 nm of range, the target's range is equal to $1000/12.4 = 81$ nm. Since there are truly no static cases, the effects of Doppler must be removed. A constant frequency can be inserted at the end of the rise cycle and the target echo will report the Doppler shift. A more common way of eliminating Doppler effects is to use a two-slope cycle where the two slopes are identical, yet rising and falling. The net Doppler effect in the rising slope is cancelled out in the falling slope. Examples of these two methods are shown in Figures 8.56 and 8.57.

The accuracy of FM ranging depends on the slope (rate of change of frequency [k]) and the accuracy with which the frequency difference is measured. The greater the slope, the greater the frequency difference for a given period of time; the greater this difference and the greater the accuracy with which the frequency can be measured, the greater the accuracy (see equation 8.32). The ability to accurately measure the frequency is dependent on the length of time in which the measurement is made. We have seen this parameter in the radar range equation (t_{int}), and it is limited by the amount of time it

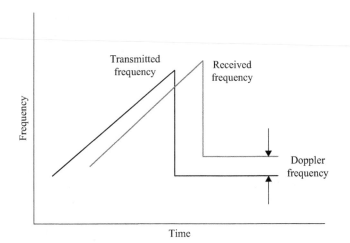

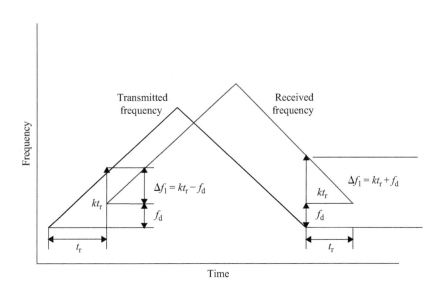

FIGURE 8.57 ◾
Two-Slope
Modulation with
Positive and
Negative Doppler

takes for the antenna to scan across the target. For radar altimeters, k is very high, and very accurate measurements of height above the ground can be made. For air-to-air pulse Doppler search modes, k is made fairly shallow to reduce clutter. Because k is shallow, range accuracy suffers. Pulse delay ranging accuracy is on the order of feet, but FM ranging accuracy is on the order of nautical miles.

8.8.4.2 Flight Testing of Radar Altimeters

Radar altimeters are used to measure the aircraft's height above the ground. They are normally accurate up to about 5000 ft. There are some radar altimeters, such as those used for TSPI, that are calibrated and may be used to 10,000 ft. The accuracy of radar altimeters is on the order of

- 0–100 ft (1 ft)
- 100–5000 ft (1% of altitude).

Since the radar altimeter directly measures time, and range is computed, it will be difficult to obtain a TSPI source that is four times as accurate as the range measurement. Normally cinetheodolite or laser theodolite systems will be required for the demonstration testing. The "warm fuzzy" testing can be accomplished by performing a cursory check with the barometric altimeter with flight over known terrain. A detailed check can be performed by accomplishing a tower fly-by. A detailed description of FAA requirements can be found in appendix 1 to 14 CFR FAR 91.

Other checks that will be required are bank angle effect and flights over a variety of terrain types. Since the radar altimeter is not gyro stabilized (such as the Doppler navigation system), there will come a point during a bank where the altitude information is no longer reliable. Because the beam is wide and the distance is relatively small, bank angles of up to 60° will still produce accurate altitude information on most platforms. The true bank angle limitation must be determined during testing.

Flight over snow and ice (as previously mentioned) may yield inaccurate altitude information. Flight over smooth surfaces such as lakes, sand, or snow may not produce a return due to scattering, and flight over the ocean may provide multiple Doppler returns that corrupt the altitude solution. Other sources of error can be caused by leakage from the transmitting antenna to the receiving antenna, also corrupting the altitude solution.

8.8.5 Special Cases: Detection of Targets at Sea

Because of the unique characteristics of Doppler, detections of targets against certain backgrounds can be difficult. The case of detection of targets at sea is a prime example. For a thorough discussion on this topic, I recommend Merrill Skolnik's *Introduction to Radar Systems*, sections 2.7 and 13.3. As with previous sections, I will refer to Skolnik in the interest of completeness and not to repeat the entire text.

As with land or airborne targets, the RCS of the target must first be known. The RCS of the target is a function of its geometric shape, reflectivity, and directivity. The ocean offers a unique background for the radar, and many factors, not the least of which are RCS and grazing angle, will affect what the radar can see. Normally a 50th percentile is used when describing the RCS of a ship, with the peak at the broadside being omitted. For low grazing angles, the following formula is used to describe the ship's RCS:

$$\sigma = 52f^{1/2}D^{3/2}, \tag{8.33}$$

where

$\sigma = $ RCS (m^2),

$f = $ radar frequency (MHz), and

$D = $ ship's full-load displacement (ktons).

Equation 8.33 is only good for targets at the grazing incidence or very close to it. For other than grazing angles, the ship's 50th percentile RCS can be approximated as equivalent to the ship's displacement in tons. RCS may change by an order of magnitude by adjusting the incidence angle. Of course, ships are even going stealth, as illustrated in Figure 8.58.

Since grazing angle has been introduced, there are a few other considerations for the radar. At grazing angles approaching zero, the return received from the target is nearly cancelled by a return from the same target reflected off the water. This cancellation

FIGURE 8.58 ■ The Stealth Sea Shadow

is due to a 180° phase shift occurring when the return is reflected. As the grazing angle increases, a difference develops between the direct and reflected returns, and the cancellation decreases. The shorter the wavelength, the more rapidly the cancellation decreases.

The echo from the surface of the sea is dependent on a number of variables as well:

- Wave height
- Wind speed
- Length of time and distance (fetch) over which the wind has been blowing
- Direction of the waves relative to the radar beam
- Whether the sea is building or decreasing
- Presence of swell as well as sea waves
- Presence of contaminants
- Radar frequency
- Polarization
- Grazing angle
- Radar PW
- Sea ice

Given all of the variables that feed into clutter, about the only thing we can say with certainty is that we are uncertain about the clutter. Since the state of the background is never completely known, an uncertainty exists for the background. Data have been gathered to measure sea clutter; however, due to uncertainties, the data contain inaccuracies on the order of ± 3 dB. Sea state is commonly used to describe sea clutter, but it is not a complete measure in itself. Table 8.18 contains sea state information as defined by the World Meteorological Organization.

Some test organizations use the sea state to define the clutter region when performing evaluations over the ocean, but as noted previously, sea state does not address all of the issues that affect detection. The following paragraphs highlight the variables that need to be considered when performing detection at sea.

TABLE 8.18 ■ World Meteorological Organization Sea States

| Sea State | Wave Height | | Descriptive Term |
	Feet	Meters	
0	0	0	Calm, Glassy
1	0–⅓	0–0.1	Calm, Rippled
2	⅓–1⅔	0.1–0.5	Smooth, Wavelets
3	2–4	0.6–1.2	Slight
4	4–8	1.2–2.4	Moderate
5	8–13	2.4–4.0	Rough
6	13–20	4.0–6.0	Very Rough
7	20–30	6.0–9.0	High
8	30–45	9.0–14	Very High
9	>45	>14	Phenomenal

There are three scattering regions defined by the grazing angle:

- Quasispecular: at angles near vertical which are a result of facet-like surfaces.
- Interference region: at angles that are very low and the scattered wave produces destructive interference.
- Diffuse region: at all other angles, where the chief scatterers are components of the sea that are of a dimension comparable to radar wavelength (Bragg's scatter, similar to rough surface).

It appears that sea clutter is essentially independent of frequency for vertical polarization except at the low grazing angles. With horizontal polarization, the clutter echo decreases with decreasing frequency and is also most pronounced at lower grazing angles. The effect of polarization on sea clutter also depends on wind.

Except for grazing angles near normal, sea clutter will increase with wind speed. Over the range of 10 to 20 knots, the effect is approximately 0.5 dB/knot; at wind speeds greater than 20 knots, the effect is approximately 0.1 dB/knot. Backscatter is generally higher when looking into the wind as opposed to downwind or crosswind. Wind affects higher frequencies and horizontal polarizations.

It has been found that when the resolution is relatively small, the sea echo is not spatially uniform, causing spiking. It has also been noted that small ships stand out much better in a 2° beamwidth than in a 0.85° beamwidth. This would imply that higher-resolution radars may have a higher probability of false alarms.

There is little backscatter energy from smooth sea ice, and it is much smaller than for the sea; rough ice behaves much like land clutter. Similar to radar altimeters, lower frequencies will penetrate sea ice more deeply and give erroneous indications.

Oil slicks have a smoothing effect on breaking waves and damp out the capillary and gravity waves normally developed by wind. This phenomenon allows high-resolution radars to detect oil spills, sometimes as small as 400 liters. Vertical polarization appears to give more contrast than horizontal polarization.

By knowing the effects of the variables on detections over water, the radar designer can optimize performance. Optimization includes

- Moving target indicator (MTI)
- Frequency

- Polarization
- Pulse width
- Beamwidth
- Scan rate
- Observation time
- Frequency agility

As described in the airborne case, employing a pulse Doppler radar will aid detections, as sea clutter is generally stationary. Even without MTI, detecting ships should be a relatively minor problem due to the large target-to-noise ratios. Aircraft without MTI will experience problems while flying over land–sea contrast, as land clutter may enter through the sidelobes.

Although lower frequencies enjoy less effect from sea clutter, higher frequencies have other advantages that are more important: beam shaping and direction, size, reduction of multipath effects, and resolution. Horizontal polarization results in less sea clutter than vertical polarization.

With all other factors constant, the shorter the PW, the shorter the maximum range. Decreasing the PW also decreases the amount of clutter the target must compete with. If peak power is limited, the target may be lost in clutter. The shorter the resolution cell, the higher probability of false alarms, and a short pulse may only see part of a large vessel.

Some final considerations are beamwidth, scan rate, observation time, and frequency agility. A smaller beamwidth is not better against sea clutter. Since the sea is relatively frozen with a time required for the clutter to decorrelate of 10 msec, high scan rates can take advantage of this fact. A longer observation time is required to pull small targets out of clutter. The benefit of frequency agility is decreased since the sea spikes are correlated over a relatively long period and appear as target echoes.

8.8.6 Velocity Search

Some fighter aircraft employ a mode called velocity search. This mode detects and presents targets as a function of azimuth versus range rate. This search mode employs a high PRF waveform and processing to provide increased detection ranges. Unfortunately this high PRF processing results in a loss of range information for targets. Normally this mode only detects targets that have a positive velocity vector along the own-ship radar LOS.

Unlike other radar search modes, tail aspect (tail chase) targets will not be detected. The displayed velocity is the target velocity; own-ship velocity does not influence the displayed target velocity. Velocity search rejects ground clutter (groundspeed, GS notch) and is thereby suited for look-up or look-down operations. Target range is not available until a spotlight search or track has been established. A moving target rejection (MTR) feature is normally included.

Velocity search should blank all sidelobe returns. Velocity search processing will not normally filter out jet engine modulation (JEM) (engine sidebands) and therefore may display multiple returns on targets. The minimum detectable velocity varies in proportion to antenna LOS, and may be higher than the selected MTR velocity. Velocity search is not as discriminatory as other search modes and often allows spurious signals to pass through to processing. These spurious signals can be the result of mutual interference from own-ship or wingman avionics equipment.

FIGURE 8.59 ■
Velocity Search Test
of MTR Feature

Test
aircraft

Target

Since the primary use of velocity search is for maximum range detection of closing targets, it can be assumed that this would be our primary flight test. Maximum range detection tests are planned in the same manner as R_{90} or R_{50} detection tests. In velocity search testing, however, target airspeed is varied throughout the entire range of detectable target speeds (1200 to 2400 knots maximum). Azimuth accuracy is the only other variable provided in this mode and can be piggy-backed onto minimum detectable velocity tests.

Other testing that needs to be accomplished in this mode includes

- Moving target rejection
- Elimination of sidelobe returns
- JEM effects
- Minimum detectable target velocity

The MTR feature can be tested in a head-on geometry as a continuation of the maximum range detection tests. This test has to be done head-on, as tail chases are not displayed in this mode. After the maximum range detection, adjust the test bed heading in increments of 10° and note the closure when the target is dropped from the display (Figure 8.59). As we turn away from the target, the LOS produces smaller and smaller closure solutions. The target should disappear when the LOS closure is equal to or less than the selected MTR. This test is also run with the MTR deselected in order to determine the minimum detectable target velocity.

During the maximum range detection and bearing accuracy tests, the ability of the radar to eliminate sidelobe returns will also be evaluated. If the testing is in a controlled area, all aircraft positions will be known and hence spurious detections can be correlated against antenna position. Since we know that JEM may not be filtered out, this test becomes an aircrew determination of the impact of JEM false alarms on the display, and whether or not these returns adversely impact the pilot's situational awareness. Remember from previous discussions that JEM produces varying range rates about the same azimuth and range of the true target. On a range rate versus azimuth display, this can be very confusing (see Figure 8.60).

8.8.7 Pulse Doppler Search and Track

In order to obtain range, range rate, azimuth, and elevation, many of today's airborne radars employ a pulse Doppler mode. Depending on the mission, these radars may use either a high or medium PRF. In order to take advantage of high and medium PRF performance, the ideal solution may be to interleave the two based on mission requirements.

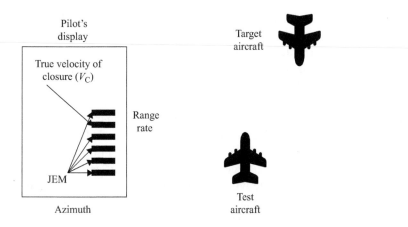

FIGURE 8.60 ◾
JEM Lines on a
Velocity Search
Display

The two most common implementations of pulse Doppler on today's airborne radars are range-while-search (RWS) and track-while-scan (TWS). Both of these modes are incorporated together on many radars. The good news for testers is that the processing is similar enough that data points for one mode will often satisfy the requirements for the other. In most cases, RWS detections are fed to TWS for track processing.

As the name implies, RWS displays Doppler targets to the aircrew in a B scope format of range versus azimuth. The aircrew may elect to "hook" or "bracket" a target to enter the single target track (STT) mode. As with the velocity search mode, RWS allows the aircrew to select an MTR feature, which eliminates some moving targets by employing a Doppler notch. In addition, the MTR may be selectable for a high or low notch. Typical notches are on the order of 90 to 120 knots (high) and 50 to 75 knots (low).

Range-while-search is a search mode only, and the aircrew has to select STT or TWS in order to obtain detailed information about the target. RWS provides the aircrew with range, range rate, azimuth, and elevation. The mode may also allow an altitude line blanker function, which eliminates zero Doppler at a range equal to altitude (medium PRF only). In medium PRF applications, RWS operation at low altitude will cause a clutter automatic gain control (CAGC) to kick in. This prevents saturation of the radar and results in shorter detection ranges. There are some variations on the basic RWS function.

- Up-look search does not disable MLC in medium PRF operations. This allows for longer-range detections since clutter processing is eliminated. Of course, false alarms are increased (weather, antenna elevation too low, etc.).

- Air combat maneuvering (ACM) and pilot assist lock-on (PAL) modes are designed to initiate an STT on targets within a selected scan volume. The scan volume can be fixed, slewable, or slaved to boresight.

Track-while-scan is designed to automatically track multiple airborne targets. Track information is obtained from repetitive search target detections as the antenna sweeps the scan volume. Detection is achieved through RWS processing. Since the time on target, or dwell, is significantly less than STT, TWS can be expected to be less accurate. The aircrew does have the option of transitioning from TWS to STT by "hooking" or "bracketing" the target.

Depending on the radar implementation, TWS can track from 5 to 25 targets simultaneously. These tracks are automatically prioritized or can be overridden by the aircrew. One of the tracked targets is designated as the primary, or "target of interest," by the operator. This is the only target that can be designated as an STT or requested for more information. An expand mode may be available for TWS where the range and azimuth about the primary target can be expanded (about 4 or 5 to 1).

There are some other pulse Doppler modes and functions that may be incorporated into the radar under test. Raid assessment (RA), expanded data request (EDR), or raid count request (RCR) allows the operator to request information about the STT concerning the true number of targets within the tracked cell. RA may be entered from TWS; however, a dwell time will be required and other tracked targets may extrapolate. RA can be extremely helpful during resolution testing, especially on range scales that do not allow sufficient pixels to display multiple targets.

Some radars have the capability of interleaving modes, thus providing the optimum amount of information to the aircrew. The situation awareness mode (SAM) on the F-16, for example, allows an interleave of TWS and a pseudo-STT. The main difference from TWS is that the radar actually leaves the scan pattern to dwell for a period of time on the designated target. STT has a higher-quality track than SAM; however, SAM has a higher-quality track than TWS. In flight testing, the evaluator will be tasked to prove the accuracy of each of these modes.

8.8.7.1 RWS, TWS, and STT Flight Testing

Maximum detection range testing is the same for pulse Doppler search as it is for pulse search. R_{90} should not be attempted in TWS, as a single bar scan is not normally selectable in TWS; R_{50} may be computed. If true maximum range detection is desired (as in the case of the F-15/F-14), the proper way to test would be detections and measurement accuracy in RWS and track initiation range and track accuracy in TWS, STT, and SAM (if available). Remember that the accuracies in TWS will not be as good as in SAM, and SAM accuracies will not be as good as STT.

The parameter of range rate must be added to our measurement accuracy matrix in pulse Doppler. Range accuracy specifications for high PRF operations will not be as restrictive as medium PRF due to FM ranging. Range rate accuracy specifications for medium PRF operations will not be as restrictive as high PRF due to range rate ambiguities. Track accuracy and track initiation range must be tested in TWS, as well as STT and SAM if available; STT accuracies will be superior. Table 8.19 replicates Table 8.14 from section 8.7.5 with the addition of a range rate column, which is our newest measured parameter in pulse Doppler search. Also note that the number of test conditions is greatly increased (72,000 versus 2400). This is not a huge problem, for two reasons. The first is that as the saw-tooth or porpoise maneuver is being flown, the range rate is constantly changing, so there is no additional geometry to be flown. Data points can be collected at rates equal to the bus rate. The second is that there is no way that you are going to be able to flight test range rates of 2700 knots closing or 1200 knots opening. These test points are excellent candidates for the lab, and that is exactly where they are done. In flight testing, we only test the heart of the envelope (where most operational geometries are expected), the others are evaluated in the lab with a target generator.

Resolution testing in the pulse Doppler modes requires an additional resolution test for targets differing in range rate. In testing resolutions for pulse Doppler modes, the tester will find that the geometries are very difficult to set up and most of the geometries will resolve in range rate. The proof may be found in internal radar data if the resolution

TABLE 8.19 ▪ RWS Measurement Accuracy

Pulse Doppler Search (RWS) Measurement Accuracy					
	Range Bin (0–60 nm)	Azimuth (+60°/−60°)	Elevation (+45°/−30°)	Range Rate (−1200/2700 knots)	Total Test Conditions
Accuracy Required	1 nm = 1% of range	1°	1°	10 knots	$6 \times 25 \times 16 \times 30$ = 72,000
Test Variables	10 nm bins (6 bins)	Every 5° (25)	Every 5° (16)	Every 100 knots (30)	
	0–10				
	11–20				
	21–30				
	31–40				
	41–50				
	51–60				

FIGURE 8.61 ▪ Range Rate Resolution Setup

of the display is worse than the resolution of the radar. You may also want to try raid assessment to assist in the testing.

Figure 8.61 describes the setup for range rate resolution testing. Targets 1 and 2 should be in close formation at the same speed (assume 300 knots). The test aircraft is in trail with the formation on the same heading and speed. The range between the test aircraft and the formation is not important, as range rate resolution is not dependent on the range. The range should be consistent with the ability of the test aircraft to maintain a visual to control the test. At the test aircraft mark, aircraft 1 should increase speed by 2 knots and stabilize. The procedure is repeated in 2 knot increments until two targets are resolved by the test aircraft. Notice that if we attempt to run the range resolution test as described in section 8.7.8, the targets will resolve in range rate before they are resolved in range. A pulse Doppler radar's best attribute is range rate, since it is measured directly. Almost any setup for resolution testing (range, azimuth, elevation, or range rate) will result in the radar resolving targets in range rate long before any of the other variables. For this reason, I recommend that range rate resolution tests for pulse Doppler radars be accomplished in flight testing, with all others accomplished in the lab.

Other pulse Doppler tests that may be required include

- Up-look search
- Air combat maneuvering (ACM) modes
- Interleaving
- RA/raid cluster resolution
- Noncooperative target recognition (NCTR)
- Multiple track accuracy (TWS)

Up-look search will provide the operator with longer detection ranges than RWS because the main lobe clutter processing is not disabled. The testing is accomplished in the same way as previously described in pulse search R_{90} testing (section 8.7.4). This test will be for head-on, look-up (target altitude greater than the test aircraft altitude) targets only. The evaluator will need to ensure that the target altitude is high enough to avoid the radar painting the ground at these long ranges.

Air combat maneuvering mode testing is the pilots' favorite type of radar testing because they are actually able to fly the airplane rather than droning along in straight and level flight. By selecting an ACM mode, the system will switch to aircraft stabilization and the pilot will be presented with a C scope display. This display was mentioned in section 8.6.1 and is shown in Figure 8.22. The radar will only search within the selected area and it will establish an STT on the first detection. For example, if the pilot selects a HUD mode, the radar will search the same volume depicted by the HUD. If the HUD presents the pilot with a FOV of 25° horizontal by 20° vertical centered about the design eye point, the radar will search the same area in a raster scan. In a boresight mode, the radar will run the range gate out to some distance, perhaps 5 nm, along the boresight line. Any target along the boresight line and within the specified range will be acquired. The test is initiated with the test aircraft 3000 to 9000 ft in trail with the target aircraft. On the test aircraft's call, the target aircraft begins the prebriefed set of maneuvers. The test aircraft cycles through the ACM modes of interest, breaking and reacquiring the STT. Of particular interest in the evaluation is the system's ability to acquire and maintain lock throughout the maneuver and within the scan volume, harmonization between the HUD and the radar, minimum and maximum ranges, ease of reacquisition, and ease of cursor movement (as in the case of a slewable ACM scan). The primary analysis tools for these tests are pilot comments and the HUD video.

Interleaving is the ability of the radar to operate and display two modes of the radar to the operator. The modes do not have to operate at the same rate (i.e., alternate scans). In most interleave operations, the secondary mode is operated at a slower rate than the primary mode. For example, suppose the pilot is interested in airborne search as well as weather detection. The primary mode may be RWS or TWS interleaved with the secondary mode of weather detection. The system may be mechanized to switch to weather every 20th frame. The pilot would receive updated information on TWS for 19 frames and on the 20th frame would receive weather information. The display may physically shift to a weather presentation for the update, or simply overlay the TWS display with weather information. The interleave may be transparent to the user, such as a position velocity update (PVU) interleaved with RWS. If you remember from chapter 5, some radars have the ability to measure the aircraft's groundspeed and update the navigation system with this information. In this scenario, the operator does not see any display change other than an advisory that the PVU is in operation. The purpose of this evaluation is to check proper operation and update rate and to ensure that the primary mode is not adversely affected (i.e., TWS targets extrapolating) when the secondary mode is interleaved. One of the more common interleaves is TF with terrain avoidance (TA), which will be discussed later.

The next three modes may be evaluated in flight; however, most organizations tend to evaluate them in the lab with only a cursory flight test. Raid assessment may come in very handy when performing resolution testing. Remember from section 8.7.8 that systems may be display limited; although the radar sees two targets, the display is not capable of displaying more than one due to resolution limits. In these cases, the operator can bracket the single target, request raid assessment, and the radar will reply with a "2" next to the target, indicating two targets within the cell.

Noncooperative target recognition (NCTR) is not resident in all multimode radars. The radar processor examines the JEM returns from a target and attempts to classify the target (either by engine or airframe) and display this information to the operator. The distribution of the JEM returns (variances in range rate) will change as the radar look angle on the target changes. The radar must have sufficient intelligence (prior knowledge) about the target in order to make this classification. Evaluations include flight against targets while varying aspect and radar look angle to the target, and measuring the success rate of the radar in its ability to classify the target.

Multiple track accuracy is a required test of the TWS mode. The tests for track accuracy described earlier in this section for TWS, STT, and SAM need to be repeated for multiple TWS tracks. If the system you are evaluating can track up to 32 targets simultaneously, do not attempt to perform a flight test with 32 controlled targets—it is not going to happen. Track accuracy for the maximum number of TWS tracked targets is a lab evaluation. You may be required to track two or three targets simultaneously and compare the accuracy to the previously determined single TWS track accuracy. The amount of testing and number of tracks required is a discussion item to be resolved between you and your customer.

8.8.8 Beacon Mode

In some mission scenarios, it is critical to obtain information on friendly locations. Cooperative targets can be airborne refueling tankers, air and ground controllers, friendly ground forces, or ground targets. Most multimode airborne radars employ a beacon mode that is similar in design to the FAA radar beacon reporting service (interrogate friend or foe [IFF]). A discussion of IFF and airborne interrogators can be found in section 5.8 of this text. In beacon mode, the radar sends an interrogation pulse and then goes into a listening mode for a time equal to the maximum expected range of the cooperative reply. Even in an interleave mode, sufficient listening time must be allowed or you risk the chance of eclipsing the reply. The reply is provided to the operator in a B scope format providing a range and bearing to the target. In order to prevent erroneous replies from targets other than the desired target, the interrogate and reply frequencies are matched and have a unique delay setting.

8.8.8.1 Beacon Mode Flight Testing

The primary evaluation of this mode consists of functionality (to include proper blanking pulses to the radar) and range and azimuth accuracy. Detection ranges for this mode are superior to the search modes, as the reply is initiated at the target and is effectively a point source. Maximum detection ranges are run as described in the pulse search R_{90} section (section 8.7.4) of this text. Where the radar look angle to the target is critical in maintaining a constant RCS in pulse search and pulse Doppler search R_{90} testing, such is not the case here, as we are looking at a transponder versus a radar reflection. The turn-in is still accomplished at 120% of the specification value; that is, if the specification calls for a maximum range detection of 100 nm in the beacon mode, the setup will be initiated at 120 nm.

The azimuth accuracy test is performed the same way as previously described in section 8.7.5 (pulse radar measurement accuracy), noting that the test is controlled from the test aircraft and all maneuvering is done by the test aircraft.

8.9 | AIR-TO-GROUND RADAR

When looking at an air-to-ground ranging radar, the same range equation described in section 8.4 (equation 8.13) may be used.

$$R_{\max} = \sqrt[4]{\frac{P_{\text{avg}} G \sigma A_e t_{\text{int}}}{(4\pi)^2 S_{\min}}}.$$

(8.13, previously referenced)

The RCS value in the numerator is replaced with a backscattering coefficient (σ_0) which is the RCS of a small increment of ground area (ΔA). If you remember our discussion on RCS in section 8.4, it was stated that the RCS is a function of geometric shape, reflectivity, and directivity. For the air-to-ground case, the geometric shape is the projection of the incremental area (ΔA) onto a plane perpendicular to the LOS of the radar. This projection upon the plane depends on the incidence angle, which is represented by θ, and is illustrated in Figure 8.62.

The projected area is reduced in proportion to the cosine of the incidence angle. As θ increases, the projected area perpendicular to the beam decreases and less energy is returned to the receiver. This was discussed earlier in the special case of the RCS of ships (section 8.8.5). In that section we used grazing angle, which is equal to $90 - \theta$ (i.e., a grazing angle of zero would be looking broadside at a boat). In many cases, σ_0 is normalized by dividing it by the cosine of θ. This normalization is given a new symbol, γ (or in some texts η), and is called the normalized backscattering coefficient. Over most angles of incidence, γ is more or less constant, and that is why it is used.

Stimson provides a table of typical backscattering coefficients, and it is included here for completeness as Table 8.20.

We have seen how the Doppler return changes when looking at an off-boresight position. A quick review of section 8.8.2 provides us with the Doppler frequency of the return from a patch of ground:

$$f_d = 2V_R \cos L / \lambda \ (\lambda = c/f_c).$$

(8.30, previously referenced)

The radar has a look-down angle, which is also a factor in the calculation (this is disregarded in the formula above). Since returns may be received from ground patches in any direction, the ground return generally covers a broad band of frequencies. It can also be seen that for a return at an angle of 0° (on the nose of the aircraft) the Doppler

FIGURE 8.62 ■
RCS of a Ground
Patch of Incremental
Area ΔA

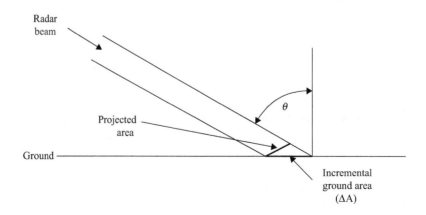

TABLE 8.20 ■ Typical Backscattering Coefficients

Terrain	Coefficients (dB)	
	σ_0	$\gamma\ (\eta)$
Smooth Water	−53	−45.4
Desert	−20	−12.4
Wooded Area	−15	−7.4
Cities	−7	0.6
Values are based on a 10° grazing angle and 10 GHz frequency		

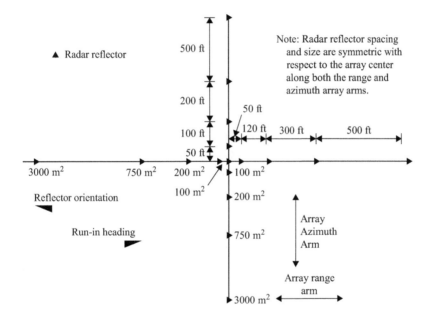

FIGURE 8.63 ■
Radar Ground Array

shift will be equal to groundspeed, and for an angle of 90° (under the aircraft or abeam) the Doppler shift will be equal to zero.

8.9.1 Air-to-Ground Radar Test

Of concern to the tester is maximum range detection, range, and azimuth resolution. Since most air-to-ground radars scan only in azimuth, no elevation accuracy is required. Maximum range is determined the same way as for the weather radar, trying to identify targets at maximum range. Once again, targets such as land–water contrast make excellent maximum range targets. If there is a specification of range versus RCS, the target dimensions and backscattering coefficient must be known. Range and azimuth accuracy are tested on a radar reflector ground array. A typical array is shown in Figure 8.63; these arrays can be found at most major test facilities.

As run-ins are accomplished on the array (from left to right in Figure 8.63), the operator calls out when and if any of the individual reflectors become visible. Remember that range resolution testing is not a function of range, but rather a function

FIGURE 8.64 ∎
Air-to-Ground
Range Resolution
Test

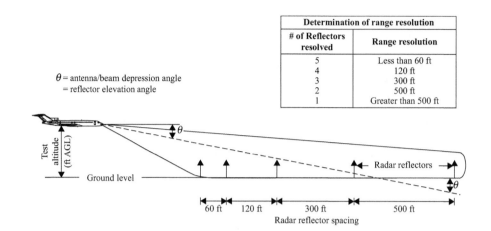

Determination of range resolution	
# of Reflectors resolved	Range resolution
5	Less than 60 ft
4	120 ft
3	300 ft
2	500 ft
1	Greater than 500 ft

θ = antenna/beam depression angle
= reflector elevation angle

Test altitude (ft AGL)

θ

Ground level

Radar reflectors

θ

60 ft   120 ft   300 ft   500 ft

Radar reflector spacing

of PW. We will know before the test is accomplished how many of the reflectors should be visible as we turn inbound toward the array. A range resolution test is shown in Figure 8.64. This test may have to be accomplished a number of times if the PW changes with range scale selection. This concept was already covered in section 8.7.8.

Azimuth resolution can be accomplished on the same run-in, noting when the horizontally spaced arrays become visible. Remember that azimuth resolution is a function of range, as the beam subtends less distance as range decreases, thus making more of the reflectors visible. The range at which the reflectors become visible can be calculated before flight if the beamwidth is known by applying equation 8.28, which was discussed in section 8.7.9:

$$\text{Range (nm)} \times \text{Inclusive Angle (}^{\circ}\text{)} = \text{Separation (ft)} \times 100.$$

(8.28, previously referenced)

8.9.2 Ground Moving Target Indicator and Track

The ground moving target indicator (GMTI) mode is designed to detect moving targets on land or sea. Moving vehicles, including cars, tanks, trucks, ships, and aircraft while taxiing or in flight, can be detected at very low speeds. A background map is normally available for navigation and detection of stationary targets. A moving target rejection feature is incorporated to eliminate false targets due to stationary ground returns. The MTR velocity changes as the target becomes further from the own-ship velocity vector.

Most land-based targets can be considered point targets. Due to their size, the possibility exists that the radar will jump from target to target (same range cell). In order to assist tracking only targets of interest, an expand mode (perhaps a 4:1 increase in range resolution) is incorporated. The expand mode may cause a loss of track against large targets such as ocean-going tankers.

Ground moving target track (GMTT) is designed to automatically maintain an accurate track of a moving vehicle on land or sea for weapons delivery. GMTT is entered from the GMTI mode and will track the target as long as the return of the target exceeds that of clutter. Operation and track in the GMTT mode is similar to the TWS/STT air-to-air modes and provides the aircrew similar information. Detection and track is assisted by positioning the target along the own-ship velocity vector.

Some airborne radars also incorporate a mode that will track a designated fixed point on the ground. Fixed target track (FTT) is designed to automatically maintain an accurate track of a stationary discrete target for fix-taking and weapons delivery. The radar searches the area of a cursor designation for returns that exceed the background clutter. The radar may be mechanized to change its PW if the target is too large or small for the existing PW set.

Another mode associated with GMTI and GMTT is a position velocity update (PVU), which was addressed briefly in chapter 5. The radar looks at the MLC along the ground track of the aircraft and reports the velocity of closure as groundspeed to the navigation system. The navigation system uses this information to update the nine-state vector parameters of velocity.

8.9.2.1 GMTI and GMTT Flight Testing

Testing is normally accomplished against enhanced targets at known locations and speeds. It is not uncommon to augment cars, trucks, and barges with radar reflectors in order to separate them from other ground moving targets that may affect the test. Some organizations have constructed test vehicles from dune buggies, augmenting the basic vehicle with radar reflectors and outfitting them with GPS and radios. However, operational testing would not be predisposed to such setups, as real-world targets will not be augmented.

The testing that needs to be accomplished includes

- Maximum range detection
- Velocity cutout
- Track accuracy
- FTT acquisition and track
- FTT track accuracy
- Navigation update accuracy

Since it is difficult to find an area where the target of interest is the only target in the area, consideration has to be given to enhancing the target of interest. When performing maximum range detection, it is best to start with the easiest task for the radar before graduating to the higher levels of difficulty. The first test should be on an augmented target located along the radial axis of the radar, maximum level of closure (target speed), and at reciprocal headings. The setup should be initiated at 120% of the maximum expected range. Subsequent test runs should be run at decreased target speed to just above the velocity cutout.

The velocity cutout test is performed as described in section 8.8.6 for the velocity search moving target rejection feature. The measurement accuracy (GMTI) and track accuracy (GMTT) tests are conducted as previously described for pulse search and pulse Doppler search measurement accuracy (section 8.7.5). A word of caution for this test: Do not turn away so far from the original heading as to cause the target radial velocity to drop below the velocity cutout. If this occurs, this will cause the detections (GMTI) to be lost or the tracks (GMTT) to extrapolate.

Fixed target track, acquisition, and accuracy are tested in a similar fashion, only this time with known ground targets. Once again the testing should be structured with benign testing first (large RCS targets free from competing clutter) and difficult testing last. Large, clutter-free targets could include radar reflectors, mountain peaks, and large vessels at anchor. The most difficult targets may include a particular building in an industrial area. In addition to evaluating the accuracy of the track, ease of acquisition, and the ability to maintain track on the original target without a transfer of lock needs to be evaluated. If implemented, the ability to adjust the PW and its impact on the ability of the radar to track large and small targets also needs to be tested.

Positional velocity update may be transparent to the user, so most of the work will be accomplished postflight during the analysis phase. The obvious errors (incorrect velocities being sent to the navigation system) will be fairly easy to detect. Timing errors such as data latency and senescence or errors within the Kalman filter will be more difficult to detect and identify. With the addition of groundspeed (which is directly measured by the radar), the accuracy of the navigation system will improve, and your testing needs to verify this. Compare the baseline navigation accuracy to the accuracy obtained with PVU enhancement.

8.9.3 Terrain Following and Terrain Avoidance

Terrain following radar provides accurate information on the terrain ahead of the aircraft for input to the TF processor. TF accomplishes this task by measuring the height of the terrain within the scan volume of the radar. TA provides situational awareness of the terrain in front of the aircraft within the scan volume of the radar. TF provides an accurate radar-measured height of objects in front of the aircraft, whereas TA provides layers of terrain clearance displayed similar to a weather radar.

The height of the terrain is determined by a submode of the radar called height above terrain (HAT). In order to accomplish this task, the radar must have accurate information about its own height above the terrain. Because of the accuracies required,

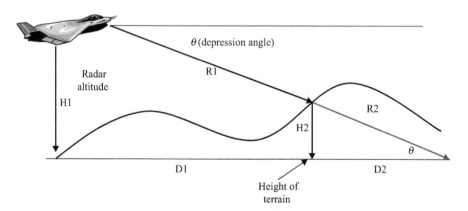

FIGURE 8.65 ▪ HAT Determination

Discussion of Figure 8.65: The radar in Figure 8.65 sees a return from the terrain for the given depression angle (θ) and calculates a range equal to R1. It knows its own altitude from the radar altimeter. If the terrain is flat, for the given depression angle and altitude the distance measured for the return should be R1 + R2; R2 can therefore be determined. By using similar triangles, the height of the terrain (H2) can also be determined by their ratios:

$$(R1 + R2)/R2 = H1/H2.$$

Similarly, the distance (D1) to the terrain may also be computed either by the tangent function or Pythagoras' theorem ($a^2 + b^2 + c^2$).

this measurement is done with a radar altimeter. The calculations are rather straightforward and are depicted in Figure 8.65.

The antenna will scan a narrow azimuth in front of the aircraft and calculate the distance and height of each return. Some systems will raster scan in elevation as well as azimuth. In this case, the narrowest beam and shortest PW is employed to obtain maximum accuracy and resolution. Each of the calculations is fed to a terrain following processor (TFP) which uses this information and own-ship parameters (groundspeed, altitude, track. etc.) to plot a path for the aircraft while maintaining a set altitude (commonly called a set clearance plane) over the terrain. These commands are sent to the autopilot, which will fly the path prescribed by the TFP. The crew has the option of selecting a "hard" ride or a "soft" ride in the TF mode. In the "hard" ride, the TFP attempts to maintain the exact set clearance plane, even though the g excursions are high. In a "soft" ride, the TFP eases the aircraft over the terrain, smoothing the path to make the ride more comfortable by limiting the aircraft g.

Because the TF mode employs a relatively high frequency (usually K_a or K_u for beam shaping), short PW, LPI, and high processing workload, the maximum range in this mode is restricted. The calculations of height and distance may only extend out to 6 or 8 miles and in a narrow azimuth in front of the aircraft. In order to provide situational awareness to the crew, TA is interleaved with TF. In many cases, TA uses a digital terrain database very similar to a GPS moving map, although the military database resolution is significantly better than what is available to the civilian world. The display that is provided to the aircrew is much like an A scope (see section 8.6.1), where the amplitude is replaced by height of the terrain versus range. The first 6 to 8 nm of range is a true presentation of the terrain as determined by the radar. The additional information is filled in by the TA mode. A representative TF display augmented by TA is shown in Figure 8.66.

The figure depicts the aircraft altitude (y-axis) versus the range to upcoming terrain (x-axis). The white lines indicate the height of the terrain and the red line indicates the

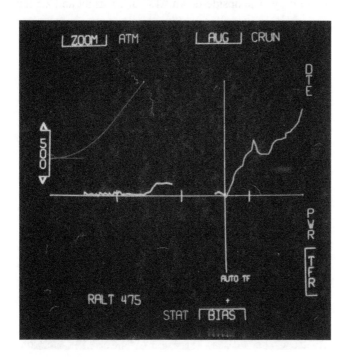

FIGURE 8.66 ■ Representative TF Display

projected flight path to maintain the 500 ft set clearance plane (caret on the left side of the display). The aircraft's radar altitude is 475 ft (bottom left of the display) and the right side of the display is augmented with digital terrain elevations (DTEs). The information to the left of the vertical white line is actual radar information and the information to the right of the vertical line is from the digital terrain database. The scale reference marks are at 5 nm increments. The gap between 9 and 12 nm is due to a valley that the radar cannot see due to the obstruction at 9 nm.

Terrain following must also be able to determine the type of object detected by the radar. Precipitation, towers, cables, and terrain must be accurately identified, their height determined, and presented to the aircrew. TF is not normally a long-range detection system, and the area of interest is normally small. Based on the mission, the TF mode must also possess an LPI. LPI means that the radar should not radiate energy in such a way that it highlights itself. If an aircraft is flying in the TF mode, there must be some mechanism to prevent the radar from radiating energy at a positive elevation setting that will not intercept terrain. Due to the varying signal return of different types of terrain, the TF mode must be able to perform in many background environments.

Terrain avoidance may be provided by the radar sensor or the digital terrain database. Figure 8.67 depicts a typical TA display that is using the digital terrain database. Note that it is very similar to a weather display, but in a B scope format. There are four colors on the cockpit display: black for terrain significantly below own-ship altitude, green for below own-ship altitude, yellow for terrain at the aircraft altitude, and red for terrain higher than own-ship altitude. Figure 8.67 shows black for well below and increasing intensity for higher terrain. These displays are not to be used for TF, but are used for pilot situational awareness.

When the radar sensor is used, the display information is presented in a PPI format giving the pilot a true picture of the area in front of the aircraft. The display shows terrain that is either above (red) or below (blue) the selected altitude. Figure 8.68 shows the TA mode using the radar sensor only. The upside-down "Ts" at 15° right and 2.5 nm are towers.

FIGURE 8.67 ■
Terrain Avoidance
With Digital Terrain
Database

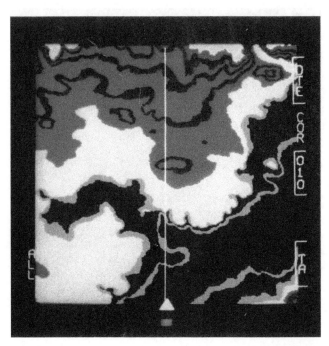

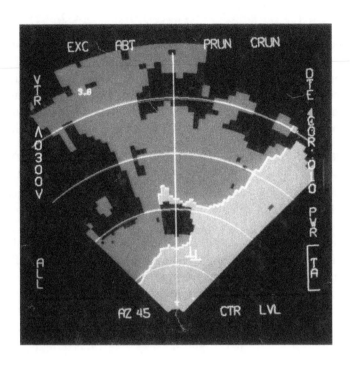

FIGURE 8.68 ▪
Terrain Avoidance
with Radar Sensor

There are some inherent problems with the TF and TA modes. The first is the inability of the radar to scan large volumes in azimuth. The azimuth scan is intentionally kept small so that the radar can stay ahead of the aircraft. You can see that if the radar scans a full 120° as it does in most other modes, as well as raster scan in elevation, the scan rate will have to be increased significantly. All of the data need to be processed in the TFP and commands need to be issued to the autopilot. Since this is flight critical information, and the probability of it being in error must be exceptionally small (1×10^{-8} to 1×10^{-9}), a deterministic (i.e., 1553) data bus must be used. We know from previous discussions that the 1553 data bus is limited in the amount of data that can be sent, and this will slow down the data transfer rate. The amount of data, and hence the scan rate, must be scaled to the bus data rate. By keeping the scan narrow, we can live within these restrictions. What would happen if the aircraft, while in TF mode needed to take evasive action (break right or left)? If the TA display is not up in the cockpit, the pilot would be turning into a blind and perhaps hazardous area. Since the radar beam is fixed ahead of the aircraft (scanning left and right of boresight), it is not going to be of any help in this situation. This problem becomes more hazardous with increasing speed of the aircraft. Pulling straight up will disengage the TF system, but this maneuver may not be the most expedient one to avoid the threat. This can be a very real problem and may have to be investigated during the flight test evaluations.

A second problem is the basic limitations of the radar due to physics. We discussed earlier the concept of backscattering coefficients and their relation to grazing angle. At the altitudes we are flying, the grazing angle is very shallow; in some scans, close to broadside. What would happen in flight over snow, sand, or calm water? In these conditions it will be very difficult for the radar to obtain an accurate measure of the height of the terrain.

Another problem is the determination of man-made objects and precipitation. Since we are using a high-resolution radar in the K band, precipitation may look much like terrain. This presents an obvious problem while flying TF, as the aircraft will attempt to maintain the clearance plane over the precipitation and eventually disengage. The determination of the height of towers and the classification of cables that support them are other unique problems for the radar. The tapering of the tower may cause the radar to provide an inaccurate measurement of the height, and the polarity of the return from cables may cause them to be invisible to the radar. These problems can be mitigated with special processing or polarity changes, but the solutions must be evaluated in flight testing.

8.9.3.1 Terrain Following and Terrain Avoidance Flight Testing

Safety is the primary concern during TF/TA flight test evaluations. For all of the testing, the following sequences shall be observed: buildup in risk, buildup in speed, build-down in terrain, and build-down in altitude. In layman's terms, we are going to start with the least risk and progress to the higher risk test conditions. The initial testing will probably be accomplished at 10,000 ft so that if the system does not behave as planned there is plenty of room for recovery. The initial testing is accomplished over reflective terrain, probably rolling hills, at the best maneuvering speed. Subsequent tests are performed at the same altitude and speed with varying terrain; the tests are repeated at varying speeds. These tests must be approached cautiously, noting and correcting any anomalous conditions. A typical test sequence is shown in Table 8.21. The table assumes an airspeed

TABLE 8.21 ■ Terrain Following Step-Down Logic

Terrain Following Flight Test Conditions			
Condition Number	Aircraft Altitude	Aircraft Speed	Terrain Type
1	10,000 ft	300 knots	Gently rolling hills
2	10,000 ft	300 knots	Slightly rugged
3	10,000 ft	300 knots	Mountainous
4	10,000 ft	300 knots	Land/water contrast
5	10,000 ft	300 knots	Ocean > sea state 4
6	10,000 ft	300 knots	Ocean < sea state 4
7	10,000 ft	300 knots	Sand
8	10,000 ft	300 knots	Snow
9–16	10,000 ft	350 knots	Repeat terrain types in conditions 1–8
17–24	10,000 ft	400 knots	Repeat terrain types in conditions 1–8
25–32	10,000 ft	450 knots	Repeat terrain types in conditions 1–8
33–40	5,000 ft	300 knots	Repeat terrain types in conditions 1–8
41–48	5,000 ft	350 knots	Repeat terrain types in conditions 1–8
49–56	5,000 ft	400 knots	Repeat terrain types in conditions 1–8
57–64	5,000 ft	450 knots	Repeat terrain types in conditions 1–8
65–72	1,000 ft	300 knots	Repeat terrain types in conditions 1–8
73–80	1,000 ft	350 knots	Repeat terrain types in conditions 1–8
81–88	1,000 ft	400 knots	Repeat terrain types in conditions 1–8
89–96	1,000 ft	450 knots	Repeat terrain types in conditions 1–8
97–104	500 ft	300 knots	Repeat terrain types in conditions 1–8
105–112	500 ft	350 knots	Repeat terrain types in conditions 1–8
113–120	500 ft	400 knots	Repeat terrain types in conditions 1–8
121–128	500 ft	450 knots	Repeat terrain types in conditions 1–8

Repeat as necessary for any lower set clearance planes.
Tests should not be conducted in precipitation or in areas of towers or cables.

range of 300 to 450 knots and a set clearance plane down to 500 ft. The table will have to be expanded accordingly for increased speeds or lower set clearance planes.

Before progressing to each successive level of difficulty a measurement accuracy test must be performed. Of interest in this test is the ability of the system to determine accurate height of the terrain as well as bearing accuracy and range to the terrain. These tests must be conducted in straight and level as well as turning flight.

In order to accomplish this test successfully, an accurate model of the terrain being overflown must be known. For every position of latitude and longitude there must be a corresponding altitude that has been accurately measured. Most ranges that provide this type of truth data can guarantee the measurements to about 4 m. In addition to knowing the position of the terrain, we must also know the accurate position of our aircraft. Once again, most ranges that provide this data will use differential GPS, which is corrected postflight. As the antenna scans in front of the airplane, the range, height of the return, and bearing to that return are calculated and sent to the TFP. For each calculation a true range, height, and bearing angle to the terrain is calculated from the TSPI data and the two answers are compared to obtain the true accuracy of the calculations.

The following additional tests must also be accomplished in the TF and TA modes:

- TF letdowns over varying terrain
- TF/TA tower, rain, and cable classification
- Transverse flight profiles
- Flight over snow, sand, water, and ice
- Horizon detect algorithms
- Interleaving
- Pull-up cues to aircrew
- Primary and degraded navigation performance

There will be occasions when it will be necessary to let-down in the TF mode to the desired set clearance plane. This would be the case for high-low-high profiles where the aircraft ingresses at high altitude until a desired let-down point where the aircraft will descend to the low level altitude, complete the mission, and then egress at high altitude. If the initial descent is done in instrument flight rules (IFR) conditions, it is critical that the TF mode detect the height of the terrain from a high altitude and allow a smooth transition down to the desired set clearance plane. These let-downs will be evaluated against the same types of terrain described in Table 8.21.

Tower classification is accomplished in both TF and TA modes. The problem with towers (other than that previously mentioned regarding the taper) is that they are not all constructed the same. Towers can be made of concrete, wood, aluminum, steel, or any combination thereof. They can be solid, or those that look like erector sets. They come in all sizes; some are as short as 50 ft and some may be as tall as 1500 ft. Towers must be detected, classified, and displayed to the crew in sufficient time to allow for any evasive maneuvering. In almost all cases the aircraft will be commanded to fly over, rather than around, the tower due to the possible nondetection of support cables around the tower. The specifications normally call for correct detection and height determination of small, medium, and large towers constructed of a variety of materials. Table 8.22 provides a possible breakdown of test conditions for tower detection and classification.

TABLE 8.22 ■ TF/TA Tower Detection and Classification Sequence

TF/TA Tower Detection and Classification			
Test Condition	Tower Height	Construction	Geometry
1	Small (50–150 ft)	Wood	Solid
2	Small (50–150 ft)	Metal	Solid
3	Small (50–150 ft)	Concrete	Solid
4	Small (50–150 ft)	Wood	Open
5	Small (50–150 ft)	Metal	Open
6	Small (50–150 ft)	Concrete	Open
Test conditions 1–6 are repeated against medium (150–400 ft) and large (400–1500 ft) towers.			

Rain classification is to be displayed in the same manner as in weather radar. The TF mode should be able to detect and classify rain as

- Light: 1 to 4 mm/hr
- Moderate: 4 to 12 mm/hr
- Heavy: 12 to 50 mm/hr
- Cloudburst: >50 mm/hr

In all fairness, the TF mode will not be able to differentiate terrain from a cloudburst, but it is not unusual to see it quoted in the specifications. As previously noted, it is extremely difficult for a radar operating in the K band to operate effectively in weather because it cannot penetrate the weather to see what terrain lies beyond. In order to claim the return as weather rather than terrain requires some very special processing, which will not be introduced here. Flight in rain and the ability to detect rain is a requirement and will be performed in flight testing; unfortunately we cannot schedule rain. If you are testing the radar in a desert locale, you might be waiting an awful long time to fly in weather. It may be a requirement to deploy off-site to find the appropriate weather conditions, so take this into consideration when formulating the test plan.

Like rain, cable detection may be a fight with physics. There may be a means of switching to circular polarization to aid in the detection of cables as well as aid in weather performance.

Transverse flight is quite an interesting test, and is sure to give the pilots an interesting ride. It is a situation where the terrain in front of the airplane is sloped upward along the wing line, almost like flying in a "V" notch. Figure 8.69 shows this situation.

In Figure 8.69 the aircraft is flying toward you with a set clearance plane safely established, but if the pilot looks out of his right window the terrain seems very, very close. For a transverse slope, the TFP calculates a safe horizontal separation as well as maintains the selected set clearance plane. The test is accomplished against terrain with varying slopes, starting with shallow or gently sloping terrain and increasing the slope with successive test points. As with weather, the scheduling of this type of test may necessitate the aircraft to fly off-site.

Flight over known no-show areas (dry sand, snow, ice, and smooth water) is accomplished to ensure that the system properly initiates corrective action and advises the crew when accurate measurement of the terrain height can no longer be guaranteed. Similarly, tests must be accomplished against terrain where the radar cannot detect the horizon. This case is illustrated in Figure 8.70.

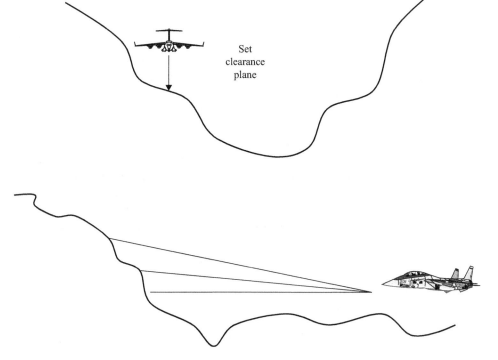

FIGURE 8.69 ▪ Terrain Following Transverse Slope

FIGURE 8.70 ▪ No Horizon Detected

The aircraft in Figure 8.70 still detects terrain in the uppermost vertical raster of the radar. The aircraft will be commanded to pitch up. While in the pitch, the radar will continue to scan and attempt to detect a horizon (i.e., no return from the radar beam), but because of the height of the terrain, the radar will still not detect a horizon. At this point, the system will issue an audio warning to the pilot, usually a "No Horizon Detect," while simultaneously issuing a pull-up cue, then the TF mode will disengage. The timing of the scenario described is based on own-ship parameters such as altitude, airspeed, g available, and range to the terrain. The flight test of this function cannot be accomplished without validating the algorithms in the lab. As with all of the previous TF/TA testing, this function is tested with a build-up in risk and with the most benign conditions tested first.

Tests of the interleave function are as described previously; interleaving a second mode cannot adversely affect the performance of the primary mode. Common interleave situations are TF with TA and TA with weather.

The final evaluation of TF and TA involves degraded navigation and its impact on the ability of the aircraft to continue safely in the mode. Of primary interest is the loss of the primary radar stabilization input. Suppose that the primary navigation source for the airplane is an integrated GPS/INS. The stabilization information, which is key to establishing a depression angle for the radar, comes from the INS. What happens if the INS fails? We know that GPS on its own cannot provide this information, so maybe the aircraft is outfitted with a second INS. But what if the second INS fails? Maybe a secondary backup is provided by an attitude heading reference system (AHRS). What are the implications to the TF system if one of these backups is employed? In order to answer these questions the range, height, and bearing accuracy tests previously described must be reflown with each change of navigation sensor.

8.9.4 Synthetic Aperture Radar, Inverse SAR, and Doppler Beam Sharpening

Synthetic aperture radar (SAR) is a Doppler mode of the radar that uses sophisticated processing to produce a high-resolution image of the ground patch of interest. It can only be used by a moving radar beam over a relatively immobile target (stationary or as close as possible to stationary). The process involves looking at an object from multiple angles with a small antenna, and when combined can replicate a picture as if taken by a very large antenna. When combined with very small increments of range (small PWs), the picture can be of quite high resolution.

In a typical airborne application, a single antenna is attached to the side of the aircraft. The beamwidth will be as small as allowable based on the size of the antenna and its operating frequency; the PW will be as small as the desired resolution. The radar emits a series of pulses as the aircraft travels forward and the amplitude and phase of the returns are recorded and combined within the processor. The result is as if they were made simultaneously from a large antenna; this process creates the "synthetic aperture" much larger than the transmitting antenna or the aircraft carrying it. Suppose we want a 5 ft resolution cell over a target at a 25 nm range from the aircraft. We can obtain a range resolution of 5 ft by narrowing the PW to 0.01 µsec. If we calculate the beamwidth we need for a 5 ft azimuth resolution at 25 nm and then use the X-band beamwidth rule of thumb:

$$R \times BW = \text{resolution } (\times 100 \text{ ft})$$
$$25 \times x = 0.05$$
$$x = 0.05/25$$
$$x = 0.002°$$

$$\theta_{3dB} = 85°/d, \text{ where } d \text{ is the diameter in inches}$$
$$0.002° = 85°/d$$
$$d = 85/0.002$$
$$d = 42,500 \text{ inches} = 3542 \text{ ft}$$

In order to obtain this resolution we would need an antenna ½ mile long. The 3000+ ft of the synthetic aperture is accomplished by the aircraft's forward motion. In reality, the beamwidth of the synthetic array is about one-half of the equivalent antenna. So in our example, the aircraft would have to examine the target for about 1500 ft of travel in order to produce a 5 ft resolution cell. Stimson provides the following achievable resolution relationships in his text (chapter 30):

$$\text{Resolution}_{\text{range}} = 500(\tau) \text{ ft } (\tau \text{ is the PW}). \tag{8.34}$$

For a real array:

$$\text{Resolution}_{\text{azimuth}} = \lambda/LR \ (L \text{ is the length of the array and } R \text{ is range}). \tag{8.35}$$

For a synthetic array:

$$\text{Resolution}_{\text{azimuth}} = \lambda/2LR. \tag{8.36}$$

NASA and the military have long employed SAR techniques for mapping and imagery obtained from space-based vehicles. Figure 8.71 is a NASA photo identifying a container ship.

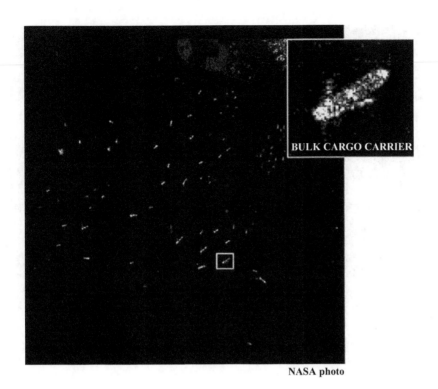

FIGURE 8.71 ■
SAR Imagery of a
Container Ship

BULK CARGO CARRIER

NASA photo

FIGURE 8.72 ■
Tank Convoy

Sandia labs

Figure 8.72 is a photo of a column of tanks and Figure 8.73 is a SAR image of the same tanks taken from an aircraft. These photos were taken by Sandia Labs.

One other type of SAR processing, not normally found on aircraft, is 3D SAR or interferometric SAR, which was developed to add a third dimension to SAR maps.

FIGURE 8.73 ▪
SAR Image from an
Aircraft

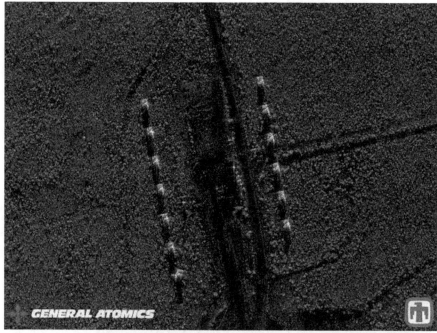

Sandia labs

FIGURE 8.74 ▪
Interferometric SAR

NASA has made extensive use of this technique for space-based imaging of the earth and planets. By having two antennas side by side, these systems can make stereo pairs of SAR images of the earth, providing maps with elevation information or actual 3D presentations requiring special glasses for viewing. Figure 8.74 is an example of an interferometric SAR.

Inverse SAR (ISAR) techniques were developed for imaging of ships and space-based objects. For ships, it would be next to impossible for a slow moving ship to

FIGURE 8.75 ▪
USS *Crocket*

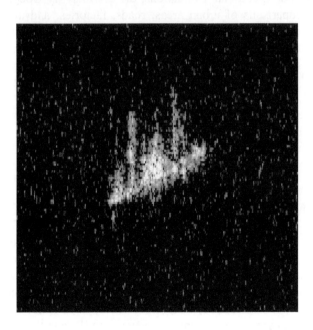

FIGURE 8.76 ▪
ISAR Image

employ the SAR techniques previously covered. In ISAR, the radar records the echoes from moving targets such as ships, spacecraft, and rotating planets. The recordings are made over many viewing angles, forming a microwave hologram of the moving or rotating target. Like SAR, but in reverse. Figures 8.75 and 8.76 are from the U.S. Navy's Radar Analysis Branch showing the USS *Crocket* as an ISAR image.

A commonly used technique for SAR systems is called Doppler beam sharpening (DBS). Because the real aperture of the radar antenna is so small, the beam is relatively wide and spreads energy over a wide area in a direction orthogonal to the direction of the platform. DBS takes advantage of the motion of the platform in that targets ahead of the platform return an up-shifted Doppler signal and targets behind the platform return a down-shifted Doppler signal. The amount of shift varies with the angle forward or backward from the orthonormal direction (target orientation). By knowing the speed of the platform, the target signal return is placed in a specific angle bin that changes over time. Signals are integrated over time and the radar beam is synthetically reduced to a much smaller beamwidth. Based on the ability of the radar to see small Doppler shifts, the system can effectively have hundreds of very small beams acting concurrently, which dramatically increases azimuth resolution. This is a common mode on many fighter and bomber aircraft and is most effective when the target of interest is placed at the 11 o'clock or 1 o'clock position.

8.9.4.1 Flight Testing of SAR Modes

The most common way to evaluate these modes is against a ground radar reflector array, as depicted in Figure 8.63. These modes are designed to provide the user with very small resolution cells. With run-ins against the reflector array, the user should be able to identify the number of reflectors with spacing equal to the resolution selected. Most systems allow the operator to select a cell from as large as 250 ft down to a cell as small as 4 to 5 ft. For operational evaluations, the user may try Google Earth as an aid to truth data. Comparisons of urban areas, roads, factories, and airfields can all be accomplished by comparing the radar presentation to the satellite view provided by this software. It is not useful for finding vehicles, aircraft, or ships, since Google Earth is not a real-time system. If there is a need to evaluate the system against these targets, then the evaluator will have to arrange to have them placed at known locations, as described in section 8.9.2.1.

Synthetic aperture radar testing is a validation of the mechanization of the mode within the radar. Figure 8.77 depicts what is going on inside the radar in order to obtain the azimuth resolution required.

▮ 8.10 | MILLIMETRIC WAVE RADAR

Millimetric wave (MMW) radar operates in the 30 to 300 GHz frequency bands; it is a cross between the metric radar and IR. MMW radars have advantages over IR systems in that they are not as susceptible to aerosols and particulates due to their wavelength. They also have advantages over metric radars in that they can be made much smaller and are able to produce very narrow beams, thus aiding in resolution. They have disadvantages as well. MMW radars are active systems (as opposed to IR) and thus can be detected by other sensors. Because of the frequency employed, there is very high atmospheric attenuation. Figure 8.1 shows the relationship of frequency and attenuation. At an operating frequency of about 50 GHz, only 1% of the signal will pass through the atmosphere. This indicates that MMW radars are probably not good for long range unless a huge amount of power is added.

Generally MMW radars have less of a problem with multipath effects because the beam is more directive, and ground reflections are more diffuse because of the shorter wavelength. Rain backscatter is appreciable in the millimetric region, which limits radar performance. Although precipitation is the principle source of backscattering, it will occur with chaff, birds, and sometimes insects. The reflectivity of various grasses, crops, and trees produces variability in the terrain backscatter and change with the season (tall crops, harvested area, leaves or no leaves on the trees, etc.). Reflectivity changes with moisture content; the reflectivity of snow depends on the free water content, with dry

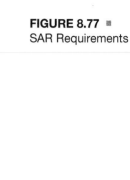

FIGURE 8.77 ■
SAR Requirements

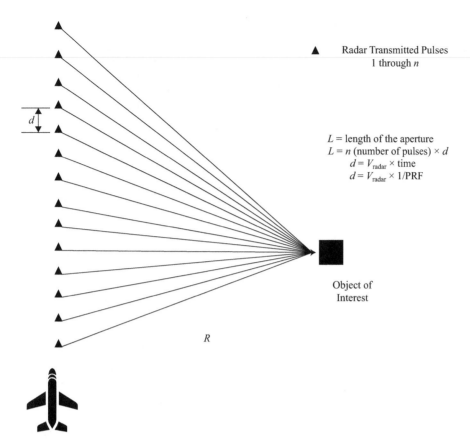

▲ Radar Transmitted Pulses
1 through n

L = length of the aperture
L = n (number of pulses) $\times d$
$d = V_{\text{radar}} \times$ time
$d = V_{\text{radar}} \times 1/\text{PRF}$

Object of
Interest

R

Range (R)	Required L	Resolution
25 nm	750 ft	
50 nm	1500 ft	5 ft
100 nm	3000 ft	

Remember that the L required is roughly one-half of that required for a real beam.

snow being more reflective than wet snow. Sea clutter behavior is much like that encountered with metric radars.

Some typical applications of MMW radar are

- Seeker and missile guidance
- Airborne radar (ground surveillance and fire control)
- Ground-based radar (surveillance, fire control, track of air and space-based targets)
- Instrumentation radar (TSPI)
- Automotive warning and braking systems

In many applications, the MMW radar is part of a sensor-fused weapons system. IR applications use MMW radar as a second color to discriminate between targets and decoys (directed IR countermeasures flares). MMW radar is also used in civilian enhanced flight vision systems in concert with IR to detect airfields and the runway environment as an aid to landing in marginal weather (see section 6.16).

FIGURE 8.78 ▪
Radome Test Fixture

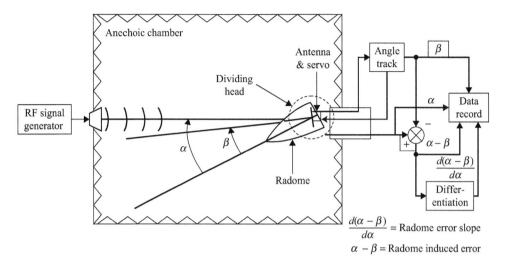

$$\frac{d(\alpha - \beta)}{d\alpha} = \text{Radome error slope}$$

$$\alpha - \beta = \text{Radome induced error}$$

8.11 | MISCELLANEOUS MODES OF RADAR

All radars have built-in electronic protection (formally known as electronic counter-measures). These functions include automatic gain control (AGC), sidelobe blanking or cancellation, sensitivity time constant (STC), and pulse length measurement, or they employ a guard channel. There are more than 70 functions that may be part of the airborne radar's electronic protection package. Most of these functions are discussed in chapter 9.

In addition, all of the modes of the radar should be evaluated with loss or degradation of the navigational inputs. This has been emphasized throughout this section and is repeated here as a reminder.

One other test that may or may not be the responsibility of flight testing is the effects of the radome. Since the manufacturing process is not perfect, each radome will be slightly different than the next. Imperfections can cause a loss of transmissivity or diffraction that will cause angular errors in the radar measurements. These errors can be calculated in a lab and then applied in software. In some extreme cases, the radar and the radome are a matched set and must be replaced as a unit. This is an important aspect of configuration control that needs to be understood by the evaluator. Figure 8.78 shows the lab setup and Figure 8.79 shows the results of the test.

8.12 | SOME FINAL CONSIDERATIONS IN RADAR TESTING

In these last few paragraphs I will address some of the issues that evaluators will come across if they are ever involved in the flight testing of a radar system. The first lesson is that radar testing is probably going to be dreadfully boring to the flight crew (with the possible exception of the ACM modes). There will be a lot of straight and level flying and boring holes in the sky, but for the analyst, these tests are some of the most data intensive and time consuming. Also understand that, unlike other tests, pilot comments are minimally useful in radar testing. When planning radar flight tests, do not overbook

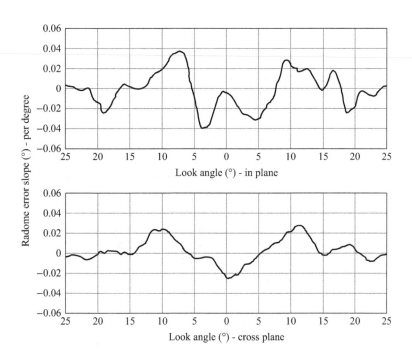

FIGURE 8.79 ▪
Radome Errors

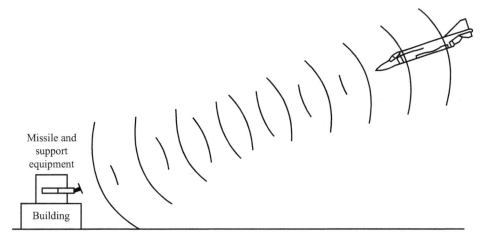

FIGURE 8.80 ▪
Radar Roof House
Facility

your flights, for the two previous reasons mentioned. Allow enough time between missions to properly analyze the data and ensure that the system is functioning as expected.

Without exception, all radar vendors will utilize either a flying test bed or roof house (or both) to develop and validate radar software. It is imperative that the customer/aircraft integrator be involved in this testing to understand exactly how the radar works. Remember that the systems flight tester has to be as knowledgeable about the system as the design engineer. An example of a roof house facility is shown in Figure 8.80.

When building a test plan, pay particular attention to the types of targets and facilities that will be required. Do the targets require a particular RCS? Is there a requirement for very small, high-speed targets? Do you need high altitude, bomber-size targets? Can you augment the targets you possess to mimic the requirements (radar reflectors, Luneberg lens). Is there a requirement for jamming targets or chaff-dispensing targets?

Do you need ground or oceangoing targets? Do I need to instrument the targets or obtain surveyed points? Is there a requirement for a radar ground array? Who can supply these targets and how much lead time is required? These are the types of questions that should be asked for every test event that is planned.

Once the types of targets are determined we must then determine where we can accomplish these tests. Is the range available for the time frame anticipated? Is range space adequate for the setup ranges we have determined? Is adequate TSPI available? What is the data turnaround time? How many flights can be supported, and at what rate? Does the range require radar beacons or laser reflectors? Is the control room adequate (types of data, number and types of displays, communication, etc.)? What is the method for range control and geometry setups? What is the cost? These are questions that program management will also be asking you to answer.

TSPI may be a problem, especially when attempting to prove functions such as range rate accuracy. Remember that the velocity of closure is measured directly by the radar. There is no way that the range will be able to provide velocity data that is four times as accurate as the radar, which will force you into testing to the limits of the truth data.

Finally, radar testing is not going to be done overnight, no matter who dictates it. Plan on regression testing, as the software will be incremental; especially so with radar. Attempt to solve the big problems first. It is no fun to be nearing the end of a program and have a massive software change thrown at you and you realize that the last 200 flights have to be repeated.

8.13 | SELECTED QUESTIONS FOR CHAPTER 8

1. What does radar mean?
2. What is the relationship between component size and frequency?
3. In the radar range equation, what does σ stand for?
4. What are the components of σ?
5. Why is gain important to an antenna?
6. Convert the power ratio of 1000 to decibels.
7. Convert 45 dB to a power ratio.
8. What is the relationship between size, radar frequency, and wavelength?
9. What is the relationship between wavelength and the ability to detect targets?
10. Does RCS change with wavelength?
11. What attributes of a target can a pulse radar determine?
12. How does a pulse radar determine range?
13. What is the relationship between range accuracy and PRF?
14. What is PRF?
15. What is the relationship between pulse length and range accuracy?
16. For a radar with a pulse length of 12 μsec, what would be the expected range resolution?
17. What would happen if you closed on a target for a gunshot with this radar?

18. How can this problem be avoided?

19. Can a radar altimeter be designed by using a pulse radar?

20. What are the characteristics of the radar altimeter? How does it work?

21. What does Doppler measure? How?

22. What attributes of a target can a pulse Doppler provide?

23. Which target attribute of a pulse Doppler radar is the most accurate? The least?

24. What is a Doppler notch?

25. What are the advantages of a notch? The disadvantages?

26. What is clutter?

27. Can a Doppler radar pull a target out of clutter?

28. In what areas can a pulse Doppler radar resolve targets?

29. Which is the most accurate? (Reference question number 28.)

30. Given a 3 dB beamwidth of a radar as 5°, how far apart would two targets have to be at 60 nm in order to be resolved?

31. What is JEM?

32. What does JEM provide the aircrew?

33. What sort of frequency would you expect to see in terrain following radars? Why?

34. What are three variables that should be taken into account when testing the detection range of a radar?

35. Which detection variable has the most effect?

36. Which detection variable has the least effect?

37. In which direction are the ranges affected by each of the detection variables?

38. What is beacon mode?

39. On an F-15 radar, what would you expect to see for PRF, pulse, beamwidth, etc.?

40. What is the difference between peak power and average power?

41. How can we boost the average power?

42. Why would we want to boost the average power?

43. What are the disadvantages of employing low frequencies in a radar altimeter?

44. What is the relationship between atmospheric absorption and frequency?

45. How would you calculate the RCS of a ground target?

46. What is LPI? Why is it important in choosing a radar operating frequency?

47. How is target ranging accomplished in a high PRF radar?

48. What is PRF stagger? What is it used for?

49. What is chirp? What is it used for?

50. Name two types of pulse compression.

51. What is a radar PRI?

52. Explain the difference between earth- and aircraft-stabilized radar.

53. A radar dish looks like a big banana, what is it used for?

54. What is SLC on a pulse radar?

55. What is SLC on a PD radar?

56. What is MLC on a pulse radar?

57. What is MLC on a PD radar?

58. What are the advantages of applying a Doppler notch at the earth's Doppler return?

59. What are the disadvantages of applying a Doppler notch at the earth's Doppler return?

60. What can be done to eliminate the problems associated with the zero notch?

61. What is velocity search mode?

62. What can be said about resolution testing done at long ranges?

63. What is a velocity reject feature?

64. What problems are associated with the MLC notch?

65. Why are low and high velocity reject switches employed?

66. Briefly describe a monopulse tracker.

67. Briefly describe a conical scan tracker.

68. How can a radar employing very low frequency establish fine resolution?

69. What is the relationship between strength of target return and range?

70. As V_c increases, what happens to the probability of detection?

71. In a circular antenna, what are the two factors that affect gain?

72. Briefly explain the amount of energy returned to the radar versus what was sent.

73. How can RCS fluctuations be accounted for in predicting range?

74. What is the aspect rule of thumb when conducting maximum detection testing?

75. What is a false alarm rate?

76. In which frequency would you find WX radars?

77. Would you use a K band radar as a WX radar?

78. Which frequency would the California Highway Patrol (CHiP) use to deter speeders?

79. Name two applications for MMW radar.

80. Name three advantages of MMW radar.

81. Name three disadvantages of MMW radar.

82. What is a PPI scope?

83. What is a B scope?

84. What is the determining factor in range accuracy of a pulse radar?

85. How is the beamwidth of a radar measured?

86. What is meant by the −3 dB point?

87. What is a resolution cell?

88. What should be the turn-in range for maximum detection tests?

89. Name three parameters that must be normalized in maximum detection tests.

90. Why is multipath harmful to a radar?

91. What is an ensemble mean?

92. If our sample size is small, what must be done to the standard deviation?

93. What parameters must be included in our test matrix when performing measurement accuracy?

94. Are azimuth and elevation resolution range dependent?

95. What TSPI data should be used when performing range rate accuracy tests?

96. What did you learn about radar testing from this module?

97. Are there any radar tests that are inappropriate for flight testing?

98. What is a backscattering coefficient?

99. What is the purpose of a radar reflector array?

100. What problems are associated with flying over snow and ice with a radar altimeter?

101. Using statistics, what is meant by knowing when to stop collecting data?

102. Name two problems with radar altimeters for obtaining accurate altitude.

103. How is Doppler shift accounted for in FM ranging?

104. What is the relationship between Doppler shift and frequency?

105. If the Predator was equipped with MMW radar instead of IR, would he have been able to kill Arnold Schwarzenegger?

106. What is an advantage of a DNS?

107. At best, what can we say about target detections at sea?

108. Name three factors that affect detection of targets at sea.

109. For other than grazing angles, what would be the RCS of a ship at sea?

110. Name four factors that affect sea clutter.

111. In sea detect, which would perform better, high- or low-resolution radars?

112. What is the major problem associated with velocity search mode?

113. What are transition tests?

114. What is the advantage of employing CW in a Doppler radar?

115. What is the disadvantage of employing CW in a Doppler radar?

116. If we wanted to install an MMW radar for long range detect, what problems would we encounter?

117. Which parameter must be known when conducting TF radar tests?

118. What is TA? How is it used?

119. Why is RWS used in maximum range detection versus TWS?

120. What is the good news for testers about RWS/TWS testing?

121. What is an STT?

122. Which has better accuracy, TWS or STT tracks?

123. What is an advantage of up-look search?

124. What should be done to targets when conducting GMTI/GMTT tests?

Electronic Warfare

Chapter Outline

9.0	Introduction	719
9.1	Electronic Warfare Overview	720
9.2	The Threat	724
9.3	Air Defense Systems	728
9.4	Electronic Attack	729
9.5	Noise Jamming	734
9.6	Deception Jamming	738
9.7	Chaff Employment	742
9.8	Flare Employment	743
9.9	Electronic Protection Measures	745
9.10	Electronic Warfare Systems Test and Evaluation	748
9.11	Finally	760
9.12	Selected Questions for Chapter 9	760

9.0 | INTRODUCTION

For the most part, technical or operational discussions of electronic warfare (EW) and electronic countermeasures (ECM) are classified and are restricted to a "need to know" basis. This chapter will not discuss classified material, but will try to guide the evaluator through a basic knowledge of EW systems and provide a generic series of tests that can be tailored to a specific system. When performing EW testing, the flight tester may want to reference *Radar Electronic Warfare*, by August Golden, Jr. (Reston, VA: American Institute of Aeronautics and Astronautics, 1987), for a good overview of radar and EW methods. The Advisory Group for Aerospace Research and Development (AGARD)/ Research and Technology Organization (RTO) has published *Electronic Warfare Test and Evaluation* (RTO-AG-300, volume 17). This document is more of a tool for managing an EW program and is similar to the "Electronic Warfare Test and Evaluation Process: Direction and Methodology for EW Testing," AFMAN 99-112, March 27, 1995. The U.S. Air Force (USAF) also uses a supplementary text called *Electronic Warfare Fundamentals* which is handled by Detachment 8 at Nellis Air Force Base; the most recent publication is November 2000. This document has been in publication for many years in various shapes and sizes and I have included some of its artwork in this chapter.

This chapter will cover the functional areas of EW and provide some examples for each category. ECM will be addressed, looking at passive and active ECM techniques. As promised in the previous chapter, we will cover electronic protection measures (EPM), mostly as they pertain to the onboard radar. Finally, we will cover EW systems test and evaluation.

9.1 | ELECTRONIC WARFARE OVERVIEW

Electronic warfare has been around as long as the telegraph. In those days, in order to prevent the enemy from using the electromagnetic (EM) spectrum, you cut the wires. The concept of EW has evolved much like the old MAD comic book "Spy vs. Spy." Whenever a new technology is developed, somebody else is developing a new technology to defeat it. There is a really good book on the history of EW. It is by Alfred Price and is called *Instruments of Darkness: The History of Electronic Warfare* (London: Macdonald and Jane's Publishers Ltd, 1978). Mr. Price is also the author of the three volume *The History of US Electronic Warfare* (Alexandria, VA: Association of Old Crows, 2000).

The first real use of EW was in World War I when the British used direction finders to locate German fleet movements. With increased use of radar came increased use of countermeasures. World War II introduced jamming, chaff (called rain), early warning systems (e.g., Chain Home self-defense), night fighters, and deception. In addition, in the Korean War, Allied losses were estimated to have been three times higher had they not used EW.

A boom in technology during the Vietnam War years accelerated the use of ECM and electronic counter-countermeasures (ECCM). USAF losses during the Linebacker II campaign were 15 B-52 bombers. Without ECM, losses against surface-to-air missiles (SAMs) and antiaircraft artillery (AAA) were estimated to have been 75 to 100 bombers. In the same conflict, the U.S. Navy had 85 losses due to missiles. Had they not employed ECM techniques, their losses were estimated at 425, with an additional 200 losses to AAA. Against man-portable air defense (MPAD) heat-seeking SA-7 missiles, aircraft employing ECM techniques could expect to survive against 80 launches before being hit. The same aircraft without ECM could only expect to survive seven launches before being hit.

During the Yom Kippur War in 1973, the Israelis used chaff and jammers against 51 surface-to-air missiles and there were no hits. Egyptian aircraft did not use EW and lost nine aircraft to Israeli missile shots.

The Yugoslavia–Kosovo action of 1999 by NATO forces was against a Soviet inspired integrated air defense system. This system consisted of widely dispersed fixed and mobile EW radar systems with multiple SA-3 and SA-6 missiles. The Federal Republic of Yugoslavia (FRV) utilized emission control (EMCON) engagement tactics (acquisition radars were not used) of SA-7, SA-14, SA-16, and SA-18 infrared (IR) missiles as well as barrage AAA. They used "old" but very effective deception tactics utilizing camouflage, constant relocation of SAM and AAA assets, decoys of SAM systems and aircraft in the open, and radio frequency (RF) radiators mimicking SAMs.

NATO employed saturation tactics, force protection and force integration, stand-off jamming (SOJ), and self-protection jamming (SPJ) by all NATO aircraft. Suppression of enemy air defenses (SEAD) was carried out by the EA-6B prowler, F-16CJ, and

Tornado with high-speed antiradiation missiles (HARMs) and air-launched antiradiation missiles (ALARMs). Kosovo saw the operational debut of the B-2 bomber with the joint direct attack munition (JDAM). Communications jamming (chatter and spoofing), communication antijam, and communications intelligence (COMINT) operations comprised the air order of battle and engagement orders.

One F-117 "Stealth" and one F-16C were downed by 1961-era SA-3s; unfortunately most SAM systems were still operational by the end of the war.

9.1.1 Electronic Warfare Terminology

Electronic warfare is military action involving the exploitation of the EM spectrum to attack the enemy or render their operations ineffective. There are three major divisions of EW:

- Electronic attack (EA)
- Electronic protection (EP)
- Electronic warfare support (ES)

Electronic attack uses EM or directed energy to attack an enemy's capability. It is comparable to the older term of electronic combat. The intent is to neutralize, degrade, or destroy the enemy's combat capability. It uses a combination of nondestructive actions (intrusion, jamming, and interference) and the destructive capabilities of antiradiation missiles and directed energy weapons. The following terms are associated with EA:

- Antiradiation missile (ARM): a missile that homes passively on a radiation source.
- Directed energy (DE): use of DE weapons (soft kill [scramble the electronic brains]/ hard kill [destruction]), devices, and countermeasures to cause direct damage or to prevent use of the EM spectrum (e.g., airborne laser [ABL] in the Boeing 747).
- Jamming: radiation, reradiation, or reflection of EM energy to degrade or neutralize an enemy's capabilities.
- Deception: manipulation of the enemy's EM spectrum for the purposes of deception.

  Some representative systems included in EA include

- EA-6B (ALQ-99 SOJ)
- HARM, ALARM (Figure 9.1)
- ALQ-131 jamming pod
- AN/ALQ-165 airborne self-protection jammer (ASPJ) (Figure 9.2)

Electronic protection is that division of EW involving actions taken to protect personnel, facilities, and equipment from the effects of friendly or enemy employment of EW, including ECCM and infrared counter-countermeasures (IRCCM). EP focuses on the protection of friendly forces against enemy employment of EW and any undesirable effects from friendly employment of the same. EP includes protection from destructive and nondestructive threats.

The following terms are associated with EP:

- Emission control (EMCON): selective and controlled use of emitters to optimize command and control and to prevent detection by enemy sensors.

FIGURE 9.1 ■
High-Speed
Antiradiation Missile

FIGURE 9.2 ■
Self-Protection
Jammer

- Operations security (OPSEC): EM hardening; action taken to protect personnel, facilities, and equipment by filtering, shielding, grounding, or attenuating friendly and hostile emissions.
- Frequency agility, diversity, and deconfliction (frequencies).
- Other procedures and techniques such as EW expendables (chaff, flares, decoys), war reserve modes (radar), and OPSEC.

Some representative systems in EP include

- ALE-44 chaff/flare dispenser (Figure 9.3)
- ALE-50 towed decoy
- MJU-8 (flare) (Figure 9.4)
- F-117 low observability/stealth
- ALE-55 fiber-optic towed decoy (FOTD)

Electronic warfare support is that division of EW involving actions to intercept, identify, and locate sources of EM energy for the purpose of immediate threat recognition (electronic support measures [ESM]). ES provides the information to guide threat avoidance, targeting, and homing. ES data can be used to produce signal intelligence (SIGINT), communications intelligence (COMINT), and electronic intelligence (ELINT).

The following terms are associated with ES:

- Combat direction finding: actions taken to obtain bearings to emitters.
- Combat threat warning: actions taken to survey and analyze the EM spectrum in support of the operational commander's information needs.

FIGURE 9.3 ■
ALE 44 Chaff/Flare
Dispenser

FIGURE 9.4 ■
MJU-8 Flare

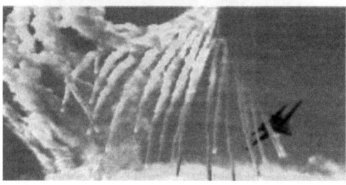

Some representative ES systems include

- AAR-47 missile warning system (MWS)
- ALR-56C radar warning receiver (RWR)
- ALR-67(v)3 RWR (Figure 9.5)
- ALR-69A digital RWR
- Joint surveillance target attack radar system (JSTARS) (Figure 9.6)
- On-board EW simulator (OBEWS)

FIGURE 9.5 ■
ALR-67 Radar
Warning Receiver

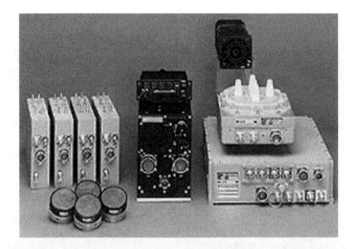

FIGURE 9.6 ■ Joint
Surveillance Target
Attack Radar
System

9.2 | THE THREAT

Radar systems have the inherent capability to determine accurate range, range rate, azimuth, and elevation for airborne targets. Military commanders have taken advantage of
these capabilities by employing radar systems to provide air defense for high-value targets.
The primary missions of radar systems employed for air defense are attack warning and
threat engagement. The EW community groups threat radars in three classifications:

- Indirect threats
- Direct threats
- Environmental radars

  Early warning radars (EWRs) or long-range radars (LRRs) are indirect threats and
are used to alert air defenses against incoming aircraft. These radars are characterized by

- Long pulse widths (high average power)
- Low pulse repetition frequency (PRF) (long listening time)

FIGURE 9.7 ▪
Height Finder Radar

- VHF, E band (30 to 3000 MHz)
- Normally a slow circular scan; 2° to 3° in azimuth beamwidth and up to 10° in elevation beamwidth

Acquisition radars are a variation of the EWR and LRR systems that are used to support AAA and SAM batteries by providing range, azimuth, and elevation information. Compared to the early warning systems they have

- Shorter pulse width (lower power, better range resolution)
- Higher PRF (accuracy, better velocity resolution)
- Higher frequency (I band for beam shaping), faster circular scan rate

Another indirect threat radar is the ground controlled intercept (GCI). GCI is any site that can accurately fix a target and direct an airborne interceptor to the area. EWR in concert with height finders can accomplish this task via VHF/UHF communications.

Height finder radars (HFRs) are indirect threats used in conjunction with EWR to determine the altitude of an incoming target. In order to accomplish this it employs a narrow beam in azimuth and an elongated (~4°) beam in elevation. In order to shape the beam without the antenna being too large, the frequencies are high. A height finder radar is shown in Figure 9.7. Height finder radar properties include

- Medium pulse widths
- Moderate PRF (compatibility with EWR)

- E band

- Rapid, large vertical scan (~30°)

V-beam radar are a type of EWR that employ two beams in a circular search pattern. The first beam is vertical and the second beam is at some angle away from vertical. As the beam sweeps, a time delta (Δt) is obtained when the target intercepts the second beam. By knowing this Δt, the distance up the second beam, or altitude, can be determined. The V-beam radar is shown in Figure 9.8.

Direct threat radar systems are used for acquisition and tracking, constantly feeding updated information to weapons systems computers. Conical scan (Figure 9.9) radars utilize a conical beam rotated in a circular fashion. The beam overlaps itself in the center of the scan, thereby reducing the effective beamwidth. The target is tracked in this overlay. General characteristics include

- Short pulse widths

- High PRF

- High frequency

- F–I band (beam shaping)

Airborne intercept (AI) radars are direct threats that utilize techniques similar to ground-based radars for tracking targets. The exception is that these radars must

FIGURE 9.8 ■
V-Beam Radar

FIGURE 9.9 ■
Conical Scan Radar

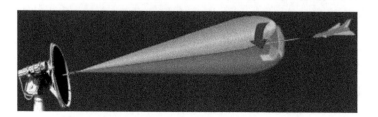

employ higher frequencies (smaller wavelengths) in order to keep the components small. Frequencies are normally in the H through K bands. Types of AI radars include

- Monopulse
- Track-while-scan (TWS)
- Doppler, continuous wave (CW)/frequency modulation CW (FM-CW)/pulsed

Monopulse radars accomplish tracking by comparing the ratio of two or more signals received by two or more feeds which transmit together but receive independently. By comparison of the energy returned in each of the beams, azimuth and elevation corrections can be made. A monopulse radar has advantages over conical scan in that it gathers data more quickly and has the ability to track a target even through pulse-to-pulse fading. Monopulse is used by SAM batteries. A complete description of monopulse tracking can be found in section 8.7.6 of this text.

Track-while-scan is not a true tracking radar in the normal sense in that it is not centered on the target. TWS as it relates to ground-based tracking radars employs a different mechanization from that of airborne radars (section 8.8.7). It transmits two beams on different frequencies sectored so that they overlap in the same region of space. The horizontal beam provides azimuth and range, while the vertical beam provides elevation and range. A separate beam is transmitted to guide an associated missile to the target. The SA-2 Fan Song (NATO designates threat radars by a two-word code; the first letter of the first word indicates its frequency band) which is a typical TWS radar.

Doppler radar exploits the frequency shifts from changes in relative velocities. CW Doppler cannot provide range, but is an excellent tool for measuring closure. Since it is CW, it is difficult to obtain the high RF power required to detect targets at long range because of the closeness of the transmitting and receiving antennas to each other. This is a problem in all CW systems. FM-CW Doppler radar is similar to CW radar except that the frequency is modulated. This modulation allows the system to determine range as well as relative velocity. The problems relating to power as previously mentioned for CW still apply to the modulated system. Pulsed Doppler radar is similar to CW systems with the exception that it employs a pulse modulation. By modulating the pulses, a higher peak power is generated, providing greater range detection. Since all information is detected on one pulse, the system is difficult to detect, is less susceptible to unfriendly interdiction, and components can be made smaller.

Although not a specific type of radar, all systems previously mentioned can gain azimuth and elevation information on an aircraft that is radiating RF energy. This is accomplished in a receive-only or "sniff" mode (angle-on-jam [AOJ], home-on-jam [HOJ]). This mode of operation is a serious threat because it is impossible to detect. Jamming may be ineffective since the radar type is also concealed.

Nonthreat radars include navigation, mapping, and space surveillance systems. Over-the-horizon radars are primarily used to detect missile launches at ranges beyond line of sight (LOS). These radars use an RF scatter system that can detect disturbances in the ionosphere caused by missile penetration. These types of radars are rare and normally one does not come in contact with them.

Airborne navigation radars operate in the HF band (I–K band) and provide a map of the terrain below the aircraft. Scan patterns can be conical or sector and are used in conjunction with inertial navigation systems (INS), global navigation satellite systems (GNSS), or Doppler to provide extremely accurate presentations.

Side-looking radars (SLARs) are Doppler ground mapping radars that take advantage of a zero Doppler shift to create an extremely small "artificial aperture." Each successive beam position is stored, and when put together they form a moving map of the area of interest.

Space surveillance radars are low frequency (UHF) radars with extremely long pulse widths and peak power in excess of 3 MW. These radars can track 1 m^2 targets to ranges in excess of 2000 nm. Targets are analyzed by their roll and tumble rates. The Space Debris and Tracking System (SPADATS) in northern Florida is an excellent example of this type of radar.

Airport surveillance radars (ASRs) are located at most major airports. Their primary role is to track arriving and departing flights. They are precise, short pulse (near detection), high PRF systems with detection ranges of 50 to 90 nm. Frequencies of operation are normally in the E and F bands.

▬▬ 9.3 | AIR DEFENSE SYSTEMS

The obvious problem affecting today's aircrews is how to penetrate the enemy's defenses without being detected, or detected late, and at the same time be able to utilize the EM spectrum without interference. We will briefly examine a typical air defense system. The system that is modeled most often is one designed by Russia, known as the Soviet Block Integrated Air Defense System (IADS). The IADS network is shown in Figure 9.10.

The IADS is a network of interdependent entities that can be divided into four categories:

- Passive detection
- Active detection
- Weapons systems
- Command and control

FIGURE 9.10 ■
Integrated Air
Defense System

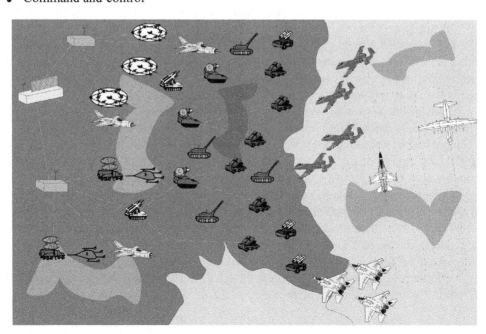

The most important thing to remember is that no one piece of the defensive system is isolated. Each entity is tied to the others by communication links, voice, or data link. This transmission of information among units is described as "cross-tell." The system's efficiency is wholly dependent on effective, clear communications throughout the network. Initial detection of incoming forces can be accomplished with passive detection. Passive detection is that element of the system which is tasked to intercept RF energy emanating from an attacking force. The key to determining an enemy force's position is aero-triangulation (angle only ranging [AOR]), similar to automatic direction finding (ADF), GPS, or IR ranging. The range of this equipment is variable; however, it is normally the longest range detection system.

Active detection is that category which employs active radar emissions in order to determine the range, azimuth, and elevation (altitude) of an incoming force. The EWR and LRR have the longest range and are employed first to determine the range and azimuth of the threat. This information is handed off to the HFR to determine altitude. GCI controllers can then direct airborne interceptors to an optimum location for intercept. Location information is not precise enough from EWR and HFR to direct the intercept. Because of the lack of precision information from the EWR and HFR, acquisition and target tracking radars for terminal defense weapons are employed. Acquisition radars provide precise azimuth and range information to the terminal defense radars. This handoff negates long search periods for the terminal defense trackers.

Weapons systems attempt to destroy an incoming force by the use of

- Interceptor aircraft
- SAM batteries
- AAA batteries

The command and control system is the final component of the air defense system. Clear, concise, and accurate communications allow the network to be extremely efficient. Areas are normally broken into sectors, with each sector commander reporting to an area commander. Without reliable communications, the network will collapse.

An additional component of these networks may be forward search radars mounted on ships or other platforms. Two detection platforms used by the United States and NATO are the airborne warning and control system (AWACS) and JSTARS. AWACS provides battle commanders with information on airborne traffic, while JSTARS provides commanders with ground movement information. Similar information is obtained by the U.S. Navy via the E-2C.

Radar systems are the cornerstone of a modern IADS. Radar and IR threat systems operate at frequencies that span most of the EM spectrum. In order to effectively employ air power on the modern battlefield, the systems that support the IADS must be negated. A basic knowledge of how radar and IR systems operate, their capabilities, limitations, and available countermeasures is the key to defeating these systems.

9.4 | ELECTRONIC ATTACK

Electronic countermeasures is defined as the employment of electronic devices and techniques for the purpose of destroying or degrading the effectiveness of an enemy's aids to EW. This definition encompasses all actions taken to deny the enemy full use of the EM spectrum, including jamming, deception, deployment of expendables, suppression, and destruction. ECM is now called electronic attack (EA).

There are two general ECM techniques: passive and active. Passive techniques may be employed against all systems:

- Radar
- IR
- Electro-optical (EO)

Active techniques against radar include

- Noise jamming
- Deception jamming (repeater and transponder)

What types of countermeasures to employ, if any, and when to employ them is contingent on many factors. Some of the factors that need to be considered are

- Signal-to-noise ratio (SNR)
- Antenna beamwidth
- Frequency agility
- Receiver threshold
- Sensitivity of the receiver
- Bandwidth
- Signal processing
- Gain
- Atmospheric conditions
- Day/night
- Identification friend or foe (IFF)

We will see that countermeasures are most effective when they are used at the appropriate time. This may require some previous intelligence about the threat.

9.4.1 Passive Techniques

As the name implies, EA passive techniques require no radiation or reradiation from your aircraft to destroy or limit the enemy's ability to use the EM spectrum against you. Some of the techniques that fall into this category are

- Avoidance
- Saturation
- Destruction
- Intimidation
- Emission control and reduction (EMCON)
- Radar cross section reduction
- Chaff, decoys, and flares
- Color and camouflage

- Minimize/reduce size

- Speed and maneuverability

Avoidance separates your aircraft from the threat. Terrain masking is a technique of avoiding enemy radars by taking advantage of no-show areas due to LOS limitations. Diversionary raids meant to occupy the threat operator's time are also a method of avoidance. Other measures include changes in heading, speed, and altitude and reversing track.

Saturation is a technique of employing a large strike force that will saturate the enemy's defenses to the point where all targets cannot possibly be tracked. Destruction calls for elimination of the defense network support radars. Obviously, if the target loses radar support, the site cannot track you. This technique is also called suppression of enemy air defenses (SEAD) and was most notably accomplished during Desert Storm by Tornados and EA-6Bs.

Intimidation causes the defense network to fear for its own safety if detection radars are turned on. Imminent destruction by HARM or similar missiles can cause the defensive network to remain silent. This technique was used in the Iraq no-fly zone during the time between the two Gulf Wars. Emissions reduction is a technique of reducing RF emissions in order to deny the enemy passive detection capability. This technique may also involve using a narrower radar beam during own-ship navigation. This technique can also be described as low probability of intercept (LPI). Radar cross section reduction will reduce the range at which a threat radar can detect and track your aircraft (low observable, stealth).

Expendables may also be used in your arsenal of passive techniques. These expendables include chaff, decoys, and flares. Chaff is used to either hide your aircraft from the threat radar or create false targets and saturate the enemy's radar. A chaff corridor can be dispersed ahead of the attack force which will shield the incoming strike package from being detected by the enemy's radar. The actual mechanics of chaff employment are covered in a later section.

Corner reflectors are small hardware devices that have a very large radar cross section (RCS). They appear as very large targets to a low PRF radar. In most applications they are used in overland scenarios to appear as large ground-based targets. They may also be used at sea to mimic small ships or gunboats.

Towed decoys are used as a self-protection device for strike aircraft. The RCS of the decoy as well as the length of the tether deceive enemy radar into believing that the towed decoy is in reality the strike aircraft. Unlike expendable decoys, the towed decoy remains with the strike aircraft and is only released while over a friendly airfield. The towed decoy can consist of a radar reflector, RF decoy, or even towed flares to aid in defense of IR threats.

Decoys may also be expended/launched from an aircraft. Most of these decoys use a reflector to enhance their size and fool the threat radar. These decoys can be as large as a cruise missile or very small, as is the case with the miniature air-launched decoy (MALD). A MALD is shown in Figure 9.11.

There are two passive IR techniques that are in use: false target creation and suppression of radiation. False target creation is done with flares; the flare is used to shroud the aircraft's IR signature. As with chaff, the mechanics of flare employment is covered in a later section. Suppression of IR radiation can be accomplished by shielding and cooling of jet engines. An example would be helicopter exhaust directed up into the cool rotor downwash or by running cool bypass air in fan engines. Ground/sea installations

FIGURE 9.11 ▪
Miniature Air-
Launched Decoy

FIGURE 9.12 ▪
CF-18 With Canopy
Painted on the
Bottom of the
Aircraft

use screening agents such as smoke composed of plastic particles. Cooling fluid is used in U.S. Navy installations around hot stacks of ships.

Color and camouflage can delay recognition or, if conditions are right, deny recognition by enemy observers and EO systems. Color schemes that have worked well are "ghost" paint jobs on air-superiority aircraft, and black, mottled green and brown, or tan and brown on strike aircraft. Canopies have been painted on the underside of aircraft to confuse enemy air interceptors. This has been done on all Canadian CF-18s (Figure 9.12). If the color is subdued or has a matted finish, reflections will be reduced. Shiny objects or surfaces highlight an aircraft when struck by the sun, such as the rearview mirrors in the rear cockpit of the F-4. At night, both external and internal lighting should be dimmed to decrease detection ranges. When initiating an attack, it is always best to come from the sun or moon and utilize available shadows.

Size, speed, and maneuverability play an important role in avoiding detection. As was shown previously, the smaller the target, the harder it is to detect. The B-1B was designed to ingress to targets at very low altitude and very high speed. Speed in combination with its defensive electronics countermeasures suite provides the aircraft with a degree of safety from threat detection.

In summary, we can say that passive ECMs work and are relatively cheap, with the exception of stealth, and effective. They require a conscious effort to be effective, and work best when used intelligently and against the threats for which they are designed. For example, uncoordinated flare deployment tends to highlight you rather than protect you.

9.4.2 Active Techniques

Active techniques in EA include noise and deception jamming. Radar jamming is the intentional radiation or reradiation of RF signals to interfere with the operation of the victim radar. Radar jamming creates confusion and denies critical information to negate the effectiveness of enemy radar systems. Noise jamming is a series of pulses, each of which has an amplitude unrelated to the amplitude and phase of any other pulse. In deception jamming, the system intercepts and returns target-like pulses. One or more characteristics of the radiated signals are different from the signals scattered by the target. With a repeater or transponder, the system receives the radar pulse of energy that strikes the aircraft, amplifies it, modulates it, delays it or augments it, and retransmits it back to the radar.

To be effective, radar jamming must have detailed information about the threat radar. The electronic support portion of EC is responsible for gathering this detailed data. This information is then programmed into onboard EC systems. There are currently three employment options for noise and deception jamming techniques:

- Stand-off jamming (SOJ)
- Escort jamming
- Self-protection jamming

To counter EW radar, GCI, and acquisition radars, noise and deception jamming techniques are employed by stand-off jammers. SOJ aircraft deny the enemy information on the size of the attack package, ingress and egress routes, and critical acquisition information. SOJ aircraft can employ noise techniques to display strobing on the victim radar or generate false targets to create confusion.

Escort jamming requires the jamming aircraft become an integral part of the strike package. EA-6B Prowlers are currently being utilized by all services in the United States; however, they will soon be replaced by the EA-18G Growler. Using noise (ALQ-99) jamming, the escort jammer attempts to deny range and azimuth information by injecting high-power signals in the main radar beam and sidelobes.

Self-protection jamming systems use deception jamming techniques for a variety of reasons. The systems are small—noise requires power, and less power requires less weight—and can be called upon to jam multiple threats.

9.5 | NOISE JAMMING

Noise jamming is produced by modulating an RF carrier wave with noise, or random amplitude changes, and transmitting that wave at the victim radar's frequency. It relies on high power levels to saturate the radar receiver and deny range and occasionally azimuth and elevation information. The effectiveness of a noise jammer is dependent on the jamming-to-signal (J:S) ratio, power density, quality of the noise signal, and polarization of the transmitted signal. In order to achieve a high J:S ratio, noise jammers usually generate high-power signals. The signals are introduced into the main beam to deny range information or into the sidelobes to deny angle information.

When detecting a target with a radar, a portion of the energy radiated through your antenna is incident upon the target. A portion of that energy is reflected back to your antenna and a portion of that energy is intercepted by your antenna. If the SNR is greater than the prescribed value (something slightly higher than one to reduce false alarms) a target is detected and displayed to the aircrew. Now imagine that same target radiating noise energy in your direction. If his noise jammer put out the same amount of power as your radar (and assume similar gains), the received jammer power would be much higher than the signal of the reflected return (because the jamming signal does not have to make the round-trip). In order for you to see the target, the reflected radar return must be higher than the jamming signal. This is the J:S ratio, provided mathematically below (for the jammer as the target).

$$P_{RT} = \frac{P_R G_R^2 \sigma_T \lambda^2}{(4\pi)^3 R^4}$$

or

$$\frac{J}{S} = \frac{P_J G_J B_R 4\pi R^2}{P_R G_R B_J \sigma_T} \tag{9.1}$$

where

P_{RT} = received signal power
P = power
G = antenna gain
B = bandwidth
R = range
σ = radar cross section (at the wavelength of the radar)
λ = wavelength
J/S = jammer-to-signal ratio

Subscripts:

R = receiver
T = target
J = jammer

Equation (9.1) can be simplified to

$$\frac{J}{S} = \frac{P_J G_J}{P_T G_T} \times \frac{4\pi R^2}{\sigma} \tag{9.2}$$

where

J/S = jamming-to-signal ratio
P_J = jamming power transmitted
G_J = jamming antenna gain
P_T = peak power transmitted by the radar
G_T = radar antenna gain
R = range from the jammer to the radar
σ = aircraft RCS

From this equation it is readily apparent that if a noise jammer is to be effective we should employ high power and gain. It also shows us that the jammer will not be effective at short ranges unless the power is increased proportionately. These relationships are shown in Figures 9.13 and 9.14.

A radar noise jamming system is designed to generate a disturbance in a radar receiver to delay or deny target acquisition. Since thermal noise is always present in the receiver, noise jamming attempts to mask the presence of targets by substantially adding to the noise level.

If the noise jamming signal is centered on the frequency and bandwidth of the victim radar, the jamming signal has a high power density. This can only be accomplished with some intelligence. If the generated jamming signal has to cover a wide bandwidth or frequency range, the power density at any one frequency is reduced. Radar systems that use frequency agility or employ a wide bandwidth can reduce or negate the effectiveness of the noise jammer.

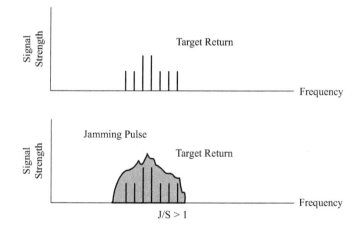

FIGURE 9.13 ■
J:S Ratio Greater Than One

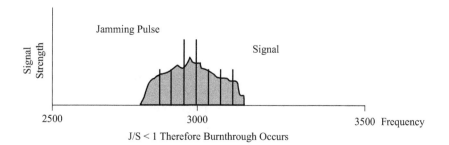

FIGURE 9.14 ■
J:S Ratio Less Than One

FIGURE 9.15 ◼
Swept Spot
Jamming

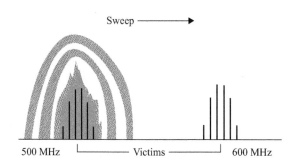

The quality of the jamming signal has an impact on effectiveness. The jamming signal must exactly match the victim radar's thermal noise profile in order to be effective As previously mentioned in chapter 8, polarization mismatches will have a profound impact on effective radiated power (ERP). A 100% loss will occur if the jamming radar is orthogonal to the victim radar.

There are many variations on the basic noise jammer. Some of them include:

- Spot or broad-band noise
- Barrage noise (low average power, continuous)
- Swept spot noise (high peak power, intermittent)
- Stepped pulse frequency modulation
- Polarization modulation (effective against monopulse radars)
- Multiple noise/target sources (for centroid tracking)
- Cover pulse
- Sidelobe jamming

Barrage jamming is a type of noise jamming used to cover a large bandwidth or range of frequencies. The power density will be lower than a spot noise jammer whose bandwidth is narrow. This type of jamming is effective against frequency agility or multiple-beam radars; barrage jamming was used extensively in World War II.

The obvious advantage of spot jammers (centering the jamming on one or a series of closely spaced frequencies) is the high power density that can be accomplished over a narrow bandwidth. But what happens when the victim radar shuts down or changes frequencies? The jammer then becomes a beacon highlighting itself to the rest of the world. An operator or a computer must constantly monitor and tune the jamming signal to the target radar's frequency. The complexity of this task increases when attempting to jam frequency agile radars, which can change frequencies with every pulse.

When high power is required over a large bandwidth, one solution is to take spot jamming and sweep it across a wide frequency range. A rapid sweep can approximate barrage jamming at a higher power density by using a high sweep rate. Victim radars can experience ringing with fast swept spot jamming. Figure 9.15 shows the sequence of swept spot jamming.

Cover pulse jamming is a modification of swept spot jamming. A repeater jammer acts as a transponder. It receives several pulses and determines the PRF. Using an oscillator which predicts the next arrival time, it responds with an amplified return, thus giving erroneous range information. The cover pulse is shown in Figure 9.16.

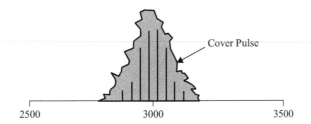

FIGURE 9.16 ■
Noise Cover Pulse

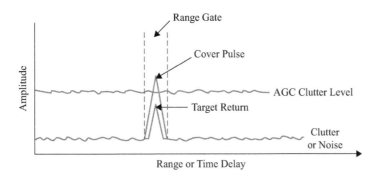

FIGURE 9.17 ■
Range Gate Pull-Off

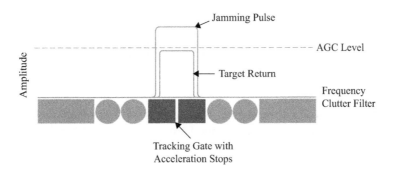

FIGURE 9.18 ■
Velocity Gate
Pull-Off

The range gate pull-off (RGPO) utilizes the cover pulse. The deception jammer transmits a cover pulse that is much stronger than the target return. The cover pulse raises the automatic gain inside the range gate and the range gate tracking loop initiates tracking on the cover pulse. The jammer then increases the time delay, driving the range gate off of the true target. This sequence is shown in Figure 9.17.

Against CW and pulse Doppler radars, a type of cover pulse jamming is used to initiate velocity gate pull-off (VGPO). The cover pulse is a strong signal with the same Doppler shift as the aircraft return. As with the RGPO, this false return steals the velocity gate and drives it off of the true target. This sequence can be seen in Figure 9.18.

Modulated jamming alters the noise jamming signal at a frequency that is related to the scan rate of the radar. The frequency of the sine wave is slightly higher than the scan rate. This results in constantly varying phase differences between the jammer and the radar. A strong signal to the radar will occur wherever the phases reinforce each other, causing the radar to track those false returns; the scan rate must be known. Modulated jamming against a conical scan radar is depicted in Figure 9.19.

FIGURE 9.19 ▪
Conical Scan
Modulated Jamming

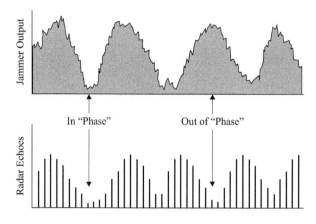

FIGURE 9.20 ▪
TWS Modulated
Noise Jamming

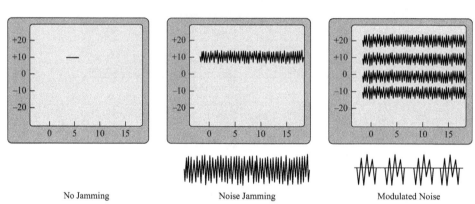

No Jamming   Noise Jamming   Modulated Noise

Against a TWS radar, a rectangular waveform is used to modulate the noise signal. The PRF of the modulation is set to some harmonic of the TWS rate. In the example shown in Figure 9.20, a modulating signal frequency is four times the scan rate of the radar, resulting in four jamming strobes. If the jamming is slightly out of phase with the scan rate, the jamming strobes will roll across the screen.

9.6 | DECEPTION JAMMING

Deception jamming systems are designed to inject false information into a victim radar. This false information can be range, range rate, azimuth, or elevation. To be effective, a deception jammer receives the victim radar signal, modifies the signal, and retransmits this altered signal back to the victim. Unlike the noise jammer, deception jammers need precise intelligence on the victim radar signal characteristics.

The following are two basic methods of deception jamming:

- False target generation (which causes confusion)
- Track breaker (which are deception techniques used to induce errors in range, angle, elevation, and velocity)

Deception jamming is accomplished by making modifications to the pulses. A change in the time of arrival will result in a deceptive range. A change in the amplitude of the pulse will result in angle deception. A change in the frequency of the pulse will result in erroneous velocity/range rate. A change in polarization or sending multiple pulses from multiple sources will result in angular errors.

False target generators are effective against search and track and GCI radars. There are two objectives with the employment of false target generators: create an inability to ascertain the real target from the false target, and make the victim believe that perhaps a mass attack is occurring in a particular sector. This is done by generating constant range targets over a sector using a straight-through repeater or by generating many targets over an azimuth by using a transponder.

Track breakers can be used to break the lock of the victim radar in range, angle, or range rate. Range deception is accomplished by covering the skin paint with a jamming pulse; the automatic gain control decreases the sensitivity of the radar. The jammer then slowly increases or decreases the time delay to walk off in range. This is the previously described RGPO.

Angle deception works against sequential lobing radars. The jammer makes use of several antenna beams in conical scan tracking radars. The jammer amplifies weaker pulses more than stronger pulses which confuses the tracker nulling into believing a false position.

9.6.1 Special Types of Deception Jammers

Range deception jammers are effective against high PRF (HPRF) and low PRF (LPRF) radars. Against HPRF radars, the following techniques can be employed:

- Multiple frequency repeater
- Narrowband repeater noise
- Pseudorandom noise
- Pulse repeater
- Repeater swept amplitude modulation

    Against LPRF radars, the following techniques can be employed:

- Range gate stealer (range gate pull-off)
- Velocity gate pull-off
- Chirp gate stealer

Range deception jamming against HPRF radars that employ frequency modulation ranging (FMR) methods is accomplished by transmitting a number of evenly spaced frequency components. The spacing of these components determines the range error, which is usually kept small in order to generate a number of false targets around the true target.

The multiple frequency repeater (MFR) generates false ranges in the radar by transmitting a fixed set of frequency components, usually centered about the true target value. The false ranges are not only designed to confuse the operator, but the short ranges may also raise automatic thresholds, thus denying further observations.

The narrowband repeater noise jammer (NBRN) is designed to create a noiselike spectrum about the target return, denying the radar the capability to detect in Doppler. Since the noise is very narrow, the jamming detection circuits may also be unable to detect the target, thus denying all tracking capability.

The pseudorandom noise jammer (PRN) is designed to create a noise-like spectrum about the target return. The difference between an NBRN and PRN is the method of generation and the resulting spectrum. The NBRN spectrum is created by a number of closely spaced frequency components whose pattern varies with time. The PRN spectrum is generated rapidly enough that it does not appear to change from one radar look to another.

The pulse repeater (PR) generates a set of closely spaced frequency components in the same way as the MFR. The difference is that the minimum MFR spacing is set to the range error that the jammer will create. The PR spacing can be varied over a much wider range; the MFR can be thought of as a specific type of PR.

The repeater swept amplitude modulation (RSAM) jammer is designed to be an angle deception device, but the on/off switching creates closely spaced frequency components. When RSAM is swept across lobing frequencies that match the false range values, false observations are created. While the MFR frequencies are fixed, the RSAM frequencies are constantly changing.

The chirp gate stealer (CGS) is designed to break the track against an LPRF single-target track mode that employs pulse compression. The jammer transmits a pulse that has been shifted in frequency. When this pulse is compressed in the radar, the shift in frequency corresponds to a shift in range. As the frequency shift increases, the range increases, and the result is the same as for the range gate stealer previously discussed.

The random Doppler (RD) jammer is designed to create many false Doppler targets. The jammer transmits a randomly varying frequency component around the true target. Each target will thus appear at about the same range, but each will have a different range rate. This is designed to confuse the signal processing to an extent where true range rate cannot be determined (Figure 9.21).

Angle deception jamming is designed to prevent the radar from tracking the target in single target track mode. The jamming can disrupt the tracking by either corrupting the signal processing loop or by taking advantage of the hardware limitations of the antenna. In the first case, the jammer must be designed for the specific tracking method used by the radar.

The RSAM jammer is designed to break the angle tracking of a radar that uses iterative lobing. It is a pulse repeater which varies the on/off frequency across the lobing frequencies of the radar in an attempt to induce false drive signals in the angle tracker. The amplitude modulation (Figure 9.22) is varied from the lowest expected value to the highest expected value every jammer cycle.

Cross-polarization (X-POLE) (Figure 9.23) jamming is designed to break the track against single target track modes. The jammer transmits a signal that is about 90° out of polarity with the transmitted signal. This results in the radar attempting to track in an antenna cross-polarized lobe, which does not have the same angular characteristics as

FIGURE 9.21 ■
Doppler Noise
Jammer

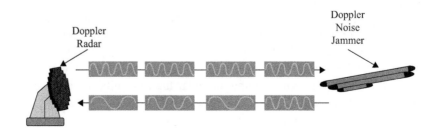

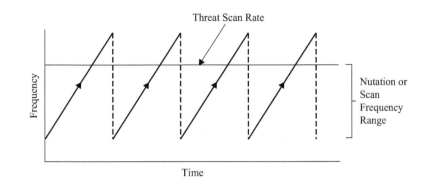

FIGURE 9.22 ◾
Angle Deception

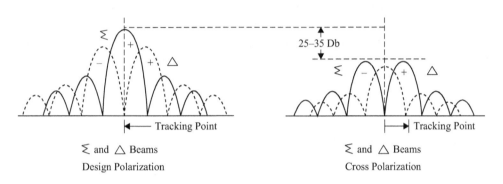

FIGURE 9.23 ◾
Cross-Polarization
Jamming

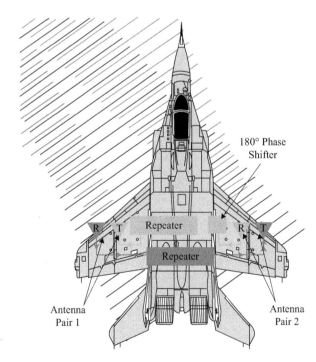

FIGURE 9.24 ◾
Cross-Eye Jammer

the normal main lobe. The cross-polarized waveform is usually superimposed on another type of jammer.

Cross-eye (X-EYE) (Figure 9.24) jammers are designed to break the angle track against radars that utilize a phase-dependent monopulse tracking method. The jammer uses two isolated receive/transmit antenna pairs that retransmit the radar signal 180° out of phase with respect to each other. This results in a phase-distorted wave front arriving at the radar antenna, which causes an angular drive off.

9.7 | CHAFF EMPLOYMENT

Chaff was first used by British bombers in World War II. It worked so well that it was used for the duration of the war. Today, it is still the most widely used EA expendable. The most important chaff characteristics are RCS, frequency coverage, bloom rate, Doppler content, polarization, and persistence. The RCS of a chaff bundle is dependent on the frequency of the victim radar and the dispensing aircraft's aspect angle. Chaff RCS is greatest when the chaff bundle and the dispensing aircraft are abeam of the threat radar and smallest when they are nose- or tail-on. The optimum size of the chaff is one-half the wavelength of the victim radar's RF. Since a single size is restricted in effectiveness to a narrow range of frequencies, different lengths are normally packaged together. A typical chaff cartridge (RR-180) is shown in Figure 9.25.

Within the cartridge there are a number of different length dipoles of varying number. The length and number of dipoles is wholly dependent on the expected threat. A typical grouping for the RR-180 chaff dispenser is shown in Table 9.1. Notice that the frequency coverage for this grouping covers the S through K_u bands or the acquisition through fire control radars. In order for the chaff to be effective, there will have to be some intelligence of the known threat so the cartridges can be constructed properly.

FIGURE 9.25 ▪
RR-180 Chaff
Cartridge

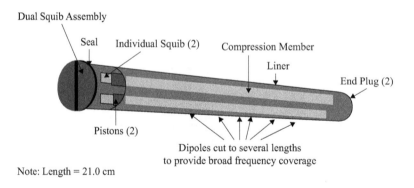

Note: Length = 21.0 cm

TABLE 9.1 ▪ RR-180 Frequency Distribution

Cut Number	Center Fre. (GHz)	Cut Length	No. of Dipoles Per Cut	Total No. of Dipoles
1	2.84	2.00″ (5.10 cm)	360,000	360,000
2,6	9.15	.62″ (1.57 cm)	360,000	720,000
3,7	8.10	.70″ (1.78 cm)	360,000	720,000
4	5.15	1.10″ (2.80 cm)	360,000	360,000
5	14.92	.38″ (.96 cm)	360,000	360,000
8	6.44	.88″ (2.24 cm)	360,000	360,000
				2,880,000

The ability of the chaff to effectively defeat a target tracking radar is directly related to the chaff dispense and blooming rate, which determines the chaff RCS. The chaff RCS needs to be larger than the aircraft's RCS and it needs to bloom within the radar field of view (FOV) or resolution cell. Chaff release with a nonmaneuvering dispensing aircraft in low turbulence will produce a small chaff cloud, whereas release in turbulence or during maneuvers will produce a large chaff cloud. The Doppler content of the chaff will reside at the main lobe clutter (MLC) for a beam engagement and move toward the target velocity as the aspect moves toward nose-on. It will never equal the target velocity, as the chaff speed is reduced at bloom. Polarity plays an important role based on the polarity of the victim radar.

Chaff may be employed against most tracking radars. Chaff employed against a TWS radar is designed to put multiple targets, with an RCS greater than the aircraft, in the resolution cell of the horizontal and vertical beams. Since the tracking loop tracks the strongest return, TWS will switch. Against a conical scan radar, chaff puts multiple, large RCS targets within the separate scans of the radar. These returns generate error signals in the tracking loop and drive the radar off. Like the conical scan, chaff against a monopulse radar is designed to put multiple targets in at least two of the tracking beams, generating azimuth and elevation errors, thus driving the track signal off of the target.

9.8 | FLARE EMPLOYMENT

Self-protection flares were developed to counter threat systems operating in the IR spectrum. Chaff and flare dispensers such as the ALE-40, ALE-45, and the ALE-47 are designed to allow the pilot to dispense flare cartridges when engaged by an IR threat. The most important characteristics of the flare's ability to decoy a threat are IR frequency matching, flare rise time, and flare burn time.

The MJU-7 flare, for example, is used in both the ALE-40 and ALE-47. The flare grain is composed of magnesium and tetrafluoroethylene, which burns at 2000 to 2200 K. As the flare burns, it emits IR energy at different wavelengths from the luminous zone. It also produces smoke, which can highlight an aircraft. The wavelength coverage and associated energy of the MJU-7 is shown in Figure 9.26. Its coverage is good in the 1 to 2.5 μm range, where most older threats reside. The 2.5 μm range is

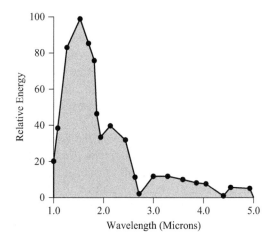

FIGURE 9.26 ▪
MJU-7 Frequency Coverage

FIGURE 9.27 ▪
MJU-7 Burn Times

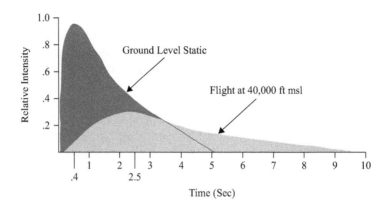

found aft of the aircraft and is comprised of hot exhaust gases. Older IR missiles did not have cooled seeker heads and therefore were only good for aft engagements. The response in the 3 to 5 µm region, where newer threats reside, is reduced; newer threats require that multiple flares be dispensed in order to defeat them.

A flare must reach peak intensity shortly after ejection or it will not be effective. The MJU-7 shows a rise time of 0.4 sec in the static case. In order to counter a short missile engagement or missiles with a narrow FOV, the rise must be as short as possible. As altitude increases, the rise time also increases (Figure 9.27). At 40,000 ft the MJU-7 reaches peak intensity at 2.5 sec. The flare takes longer to attain peak intensity at altitude; however, it burns longer (5 sec at sea level/10 sec at 40,000 ft). A longer burn time increases the probability of success with a single flare, but once again it must be in the missile's FOV.

The flare must ignite within the missile's FOV or the decoy will not work. Newer missiles have incorporated countermeasures that make it more difficult to decoy the threat. Since the missile is passive, most aircraft need to detect a launch visually in order to employ flares. Missile attack warning systems, or Missile approach warning systems (MAWS), that can detect missiles via IR, ultraviolet (UV), or millimetric wave (MMW) radar have been deployed in some aircraft. However, many suffer from a high false alarm rate.

9.8.1 Infrared Protective Measures

There are two important characteristics of IR missiles that influence the effectiveness of self-protection flares. The first is the ability of the missile to discriminate between the aircraft IR signature and background clutter. The second is the flare rejection capability built into the missile seeker and missile guidance section. The detector influences the ability of the missile to discriminate the target IR from background IR. Older missiles with uncooled seekers had a peak sensitivity of about 2 µm and were therefore limited to stern shots. By cooling the detectors with an inert gas such as argon, missiles can track longer wavelength radiation.

Flare rejection capability, or IRCCM, allows the newer IR missiles to track the aircraft while rejecting multiple flares. Flare rejection is based on two computer functions: "switch" and "response." Both functions must operate successfully to reject flares.

An IR missile using a rise time switch monitors the IR energy level of a target. A sharp rise in received energy within a specified period of time indicates a flare. The flare response is turned off when energy returns to its original levels. This switch can be decoyed with multiple flares using slow rise times, but flares currently in use have rapid rise times.

Infrared missiles using a two-color switch sample IR returns in two different wavelength bands. A sudden increase in the lower bandwidth without a comparable rise in the second bandwidth triggers a flare response. Advanced missiles may employ two separate detectors to monitor the bands; target tracking may be accomplished by either detector.

A kinematic switch takes advantage of the fact that flares separate very quickly from the dispensing aircraft due to drag. In a beam engagement, the seeker transfers to the flare and sees a dramatic change in LOS and triggers a flare response. In a head or tail engagement, the LOS is not as dramatic and will probably not trigger a response. Current flares can defeat this switch in a beam engagement by dispensing multiple flares at short intervals.

The seeker's response to a switch is to reject the flare or limit its effect on target track. As long as the flare remains in the seeker FOV, the missile will operate in a degraded mode. The following four responses are possible to a switch:

- Simple memory
- Seeker push-ahead
- Seeker push-pull
- Sector attenuation

A simple memory response is merely a coast mode. The seeker assumes that the flare will exit to the rear and maintains its motion relative to the target. The missile will continue to reject track data until the switch times out. If time out occurs before the flare leaves the FOV, the missile will track the flare.

The seeker push-ahead response causes the seeker gimbals to drive the seeker forward in the direction of target movement. This causes the flare to leave the FOV more rapidly. The faster the bias, the faster the flare will depart the FOV.

The seeker push-pull causes the gimbals to push the seeker away from the higher IR source. When the lower energy from the target is detected, it pulls the seeker to that source. The sector attenuation response is accomplished by placing an attenuation filter across part of the seeker.

9.9 | ELECTRONIC PROTECTION MEASURES

The continuing battle for unrestricted use of the EM spectrum has resulted in more than 150 radar EP techniques. The most common techniques are

- Radar receiver protection
- Jamming avoidance
- Jamming signal exploitation
- Overpowering the jamming signal
- Angle discrimination
- Doppler discrimination
- Time discrimination

Radar receiver protection is meant to limit the amount of energy entering the receiver. Sensitivity time control (STC) is used to counter close-in chaff or clutter. Receiver gain is normal for long range and reduced for close-in targets. Automatic gain control (AGC) is used to counter chaff, clutter, and most types of transmitted jamming. The AGC senses the signal output of the receiver and creates a back-bias to maintain a constant output level (fast AGC, instantaneous AGC). Automatic noise leveling (ANL) counters noise jamming and modulated or unmodulated constant wave jamming by sampling receiver noise at the end of each PRF and sets the gain accordingly for the next pulse interval.

Jammer avoidance can be accomplished in several ways. Noise jammers can be rendered ineffective by the use of frequency agility or frequency diversity which will cause a noise jammer to sweep across a greater bandwidth, reducing its power density. Jammers can be countered by changing the polarity or switching to circular polarization. As discussed earlier in chapter 8, the radar has the ability to track via the leading or trailing edge of the pulse. By varying this method, manipulated pulses can be detected and discarded. The radar may also use coherent detection. The radar internally generates the signal that is sent and compares the returns to best match the internally generated attributes.

The jammer itself can be exploited by a number of techniques. Home-on-jam (HOJ) and angle-on-jam (AOJ) are two methods employed against airborne interceptor noise jammers. A noise jammer will display a strobe on the radar display at the azimuth of the jammer; range will be unknown. In the simplest case, the strobe is held at a constant azimuth which results in your radar rolling out in trail with the AI. The rollout distance is dependent on the azimuth where the strobe is held. This concept is illustrated in Figure 9.28.

The initial contact on the target (blue airplane, black strobe on radar display) is at 50° right and you turn to the right to place the target at 45° right (red strobe on radar display). You maintain the strobe at 45° right and you will eventually roll out in trail and offset 45°. Radar and some missiles have AOJ and HOJ capability and follow this same scenario.

The radar may also employ aural recognition to recognize threat signals. Most radars will incorporate a "sniff" mode, which is simply an idle mode where the radar

FIGURE 9.28 ■
HOJ Operational
Scenario

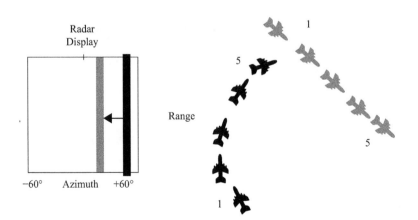

ceases to radiate and only listens for signals. A jammer that continues to radiate while the radar is in "sniff" mode will highlight itself. As a last resort, the radar operator may resort to a manual gain selection on the radar, controlling the amount of energy seen by the receiver.

As seen earlier in the discussion of noise jammers, the target will eventually burn through as the target return (SNR) exceeds that of the jammer (J:S ratio). Employment of a narrowband long pulse may increase the burn-through range.

The radar may also discriminate the attributes of waveform in pulse, angles, or Doppler. A fast time constant (FTC) used in radars is a circuit with a short time constant used to emphasize signals of short duration to produce discrimination against extended clutter, long-pulse jamming, or noise. Based on time, the radar receiver will only allow pulses equal to its own pulse width to pass and be displayed as targets. A pulse width discriminator (PWD) is similar to an FTC in that it is used to discriminate against received pulses that do not have the same duration as the radar transmitted pulse. A PWD is used in eliminating the effects of pulse-type interference when the interference pulses are not the same length as the real radar pulse. Since the circuit generates a blanking gate that shuts off the receiver when a pulse of improper length is sensed, a loss of valid target data can result. It offers good discrimination against long-pulse jamming and jamming signals employing low-frequency noise modulation. It offers little protection against short pulses and high-frequency noise modulation. A clutter elimination (CE) circuit discriminates against any target echo that exceeds three times the transmitted pulse width. It is normally employed on the lower beams of a high-frequency radar; this allows targets above a preset signal strength to be presented, while the clutter (land) is eliminated.

Sidelobe blanking (SLB), sidelobe cancellation (SLC), and sidelobe suppression (SLS) are three EPMs used to block angular interference. SLB is a device that employs an auxiliary wide angle antenna and receiver to sense whether a received pulse originates in the sidelobe region of the main antenna and to gate it from the output signal if it does. The technique uses an omnidirectional antenna and compares relative signal strength between the omnidirectional antenna and the radar antenna. The omni-channel (plus receiver) has slightly more gain than the sidelobes of the radar, but less gain than the main lobe of the radar. Any signal that is greater in the omni-channel must have been received by a sidelobe and is blanked. SLC is a device that employs one or more auxiliary antennas and receivers to allow linear subtraction of interfering signals from the desired output if they are sensed to originate in the sidelobe of the radar. This technique employs the same configuration as the SLB, except that gain matching and cancellation takes place; extraneous signals entering the sidelobes are cancelled, while the targets remain. SLS is the suppression of that portion of the beam from a radar antenna other than the main lobe.

Angle discrimination may also be accomplished by varying the antenna position or scan rate. Antenna manual positioning allows the operator to slew the antenna to the desired angle of interest. Antenna jog allows the antenna to deviate from the normal sector scan, much like was seen in the discussion of situational awareness mode in chapter 8. Antenna slow scan varies the scan rate of the radar, causing the jammer to radiate at the wrong time.

Doppler discrimination can be accomplished by varying the PRF, either jitter or stagger. It can also be accomplished by pulse-to-pulse or beam-to-beam correlation.

▮ 9.10 I ELECTRONIC WARFARE SYSTEMS TEST AND EVALUATION

The USAF has put together an acceptable process for EW test and evaluation. It is outlined in AFMAN 99-112, "Electronic Warfare Test and Evaluation Process: Direction and Methodology for EW Testing," March 27, 1995. This process is similar to all avionics systems testing, however, it stresses the need to predict, test, and compare as opposed to fly-fix-fly. This is critical since many of the required tests are difficult to fly and can be extremely costly. The process mandated by the USAF requires the military as well as contractors to comply. The initial planning must be coordinated with the EW Single-Face-to-Customer Office located at Eglin Air Force Base in Florida.

Rigorous ground testing is required in any EW test program. The use of modeling and simulation is a requirement in USAF testing. Simulations are used to predict ground and flight test results; ground tests can be used to validate simulation results and provide a high confidence level for flight test results. By validating ground test results at key points within an operating envelope, flight testing will not have to be done throughout the complete envelope.

The U.S. Department of Defense (DOD) has also emphasized commonality of test resources to reduce cost and correlate system results, including

- Modeling and simulation
- Measurement facilities
- System integration labs
- Hardware in the loop
- Simulations
- Installed system test facilities
- Open air ranges

Public law requires that testing and evaluation be performed at each of these facilities for first-level EW systems. The names and locations of these facilities can be found in the previously referenced AFMAN 99-112.

When considering the evaluation of an EW system, it is important for the evaluator to develop a systems approach to testing. All of the systems within the EW suite will be interconnected and any changes in one will very likely impact others. Some of the systems that should be considered in the EW evaluation are

- Fire control radar
- Radar warning receiver
- Missile approach warning system
- IR/radar jammer
- Chaff/flare/decoy dispenser
- Towed decoys
- Weapons (antiradiation missile, precision guided munitions)

The basic architecture of the EW system should include

- Integration of all available EW systems on a single dedicated avionics bus
- Minimize electromagnetic interference (EMI) and minimize electromagnetic compatibility (EMC)
- Reduce pilot workload
- Minimize reaction time
- Automate the deployment of decoys
- Automate the selection of the jamming program

Almost all fighter, bomber, and attack helicopters with a self-protection system employ a dedicated 1553 data bus for the EW system. Newer-generation aircraft use the Ethernet or fiber-channel for bulk data transfer and the 1553 for command and control. I recommend that the reader revisit section 4.16 of this text for an example of the problems that may be encountered when trying to integrate an EW suite and the concerns that arise with EMI and EMC. Human factors issues are important, especially when dealing with a single-pilot cockpit. You can imagine the workload of integrating the ALQ-99 from a four-man cockpit to a one-pilot cockpit. The only way that this switch could be feasible is if the defensive/attack system is automated to the maximum extent possible, like in the Joint Strike Fighter. The same problems that have been discussed earlier are relevant here as well. What is the possibility of the system triggering a false alarm? What is the possibility of a threat not being detected? What is the possibility of a threat being detected and the wrong type of jamming being employed? What indications are given to the aircrew for situational awareness? Remember that automation is a good thing except when the pilot is out of the loop, does not understand what the automation is doing, or is not apprised when something is wrong. We certainly do not want to salvo chaff and flares over a threat area when no threat is looking at us, for it soon will be.

When evaluating each of the EW systems/functions, we must look not only for the level of protection that the system offers, but also for any degradation of capabilities caused by the installation of such systems. If we add EP to the radar, we would like to know (a) Does it work? and (b) How does it adversely affect the basic performance of the radar? In the previous chapter, radar capabilities and performance requirements were covered. These included

- Detection range (look-up/look-down)
- Target angle/angular accuracy (azimuth/elevation)
- Target range/range accuracy
- Target Doppler (target detection, tracking, and identification)
- Range, angle, and Doppler resolution
- False alarm rate
- Self-test and fault reporting (built-in test [BIT])
- Controls and displays
- Response to crew commands

For each of these functions we determined the basic performance and then could predict with some certainty how the radar would perform in operational scenarios. With the

FIGURE 9.29 ▪
ULQ-21
Countermeasures
Set

FIGURE 9.29 ▪
ULQ-21
Countermeasures
Set

FIGURE 9.30 ▪
ALQ-167
Countermeasures
Set (Inboard)

addition of EP, these tests must be revisited, much like regression testing which was described earlier.

The targets for these tests will be equipped with an ECM threat simulation pod. Most of these pods are built upon one of two basic systems: the AN/DLQ-3C or the AN/ULQ-21(V), commonly known as DLQ-3 and ULQ-21. These systems are capable of generating noise, deception (repeater), transponder, and combination ECM techniques. The ULQ-21 system is shown in Figure 9.29.

These two basic systems are normally built into a pod that is capable of being carried on an aircraft. These systems are cockpit controlled and contain recorders and a time code to retrieve data postflight. The U.S. Navy uses an AN/ALQ-167(V) which houses both the DLQ-3 and ULQ-21. The ALQ-167 mounted on a U.S. Navy aircraft is shown in Figure 9.30. All of the radar maximum range and accuracy testing described in chapter 8 will be repeated for radar EP where all targets carry some form of counter-measures set. The tests measure the effectiveness of the radar EP system as well as the

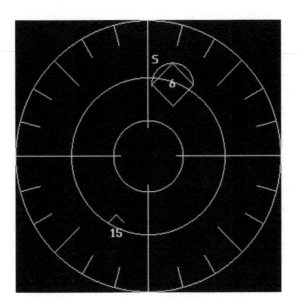

FIGURE 9.31 ■
Typical RWR Display

degradation of the radar performance in the presence of jamming. Range and accuracy tests will be performed against each type of jamming being countered by the radar EP.

A radar warning receiver (RWR) is designed to alert the aircrew of an impending radar threat (enemy fighter, missile) in a timely, clear, and unambiguous manner. Most RWR systems will provide the aircrew with

- Threat azimuth
- Threat range
- Threat identification
- Threat prioritization
- Missile launch warning

In some of the documentation (especially the older documentation), RWR systems are identified as radar homing and warning (RHAW) systems; aircrew call them RHAW gear. The RWR is a passive detection system that provides radar warning and display. The system monitors the EM spectrum over the limits of its receivers and identifies emissions by matching received characteristics to known threat signature characteristics stored in its internal memory/library:

- Radio frequency
- PRF/PRI
- PW
- Scan rate

Some systems can estimate threat range by direction finding (DF) and triangulation, or by means of lethal signal strength curves. Threats are automatically prioritized (depending on flight altitude and on an aircrew-selected priority table). These threats are displayed to the aircrew via numbers and symbols on a GCI-type display (own aircraft is at the center of the display. Figure 9.31 shows a typical display implementation of RWR information.

The following paragraph describes a generic approach to an RWR display as presented in Figure 9.31. The symbology and layout of the screen is typical, but not representative of any specific system.

The "S" in the figure is a search radar and is about 5° right of the nose just beyond the second range ring. The "6" is an SA-6 15° right of the nose on the second range ring and is the highest priority threat based on the diamond and hemisphere graphics. The "15" is an F-15 at 150° left at the second range ring. The range rings can be either true target range or threat lethality based on the system's capabilities. If it is true target range, each range ring will be equivalent to some range. If the system displays threat lethality, the inner ring is the most lethal, whereas the outer ring is the least lethal. If this is the case for Figure 9.31, and we continue on our same heading, the "6" will move toward the innermost (most lethal) ring. As it crosses the inner ring (meaning you are surely within launch parameters) it starts flashing and a loud launch audio is heard from the integrated communications system (ICS) as the SA-6 launches the missile. The "S" will also move toward the center of the display, assuming the same heading is maintained. Assuming that the F-15 is your wingman, it will not move because your wingman should be maintaining a tactical trail formation.

When evaluating the RWR, there are a number of objectives to be accomplished. The system must be able to detect the threat at some minimum range. It should be able to determine the mode or function of the threat (search, acquisition, target tracking, fire control). It should provide a direction of arrival (DOA) of the threat (bearing accuracy) and range to the threat or threat lethality. It should be able to resolve ambiguities between threats possessing similar attributes. False alarms should not exceed some desired number and the information should be presented in an unambiguous, concise, and timely fashion. Audio warnings should be clear and easily understood.

When conducting any EW testing it is a good thing to first check on the capabilities of the EW ground test facility. The parameters of the emitters that you are flying against must match the parameters in the RWR library. You will have no control over these parameters, as they are encoded by either the vendor or the military command, but they must match. If not, the display will only provide you with a "U" (unknown) and perhaps the frequency band that the threat resides in.

In addition to matching library parameters, the evaluator will use this information to determine the geometries of the setup as well as the maximum range of detection expected. A typical set of parameters provided by the ground facility is shown in Table 9.2.

The first column of the table contains the name of the emitter. If the name of the emitter has two words with the first word starting with a "B" then it emulates a known NATO threat. The second word's first letter indicates which band the threat resides in. The second column also gives the threat's operating band, but it is given with the old radar designation series. The first column uses the standard NATO, or new, designation series. For example, Brand Iron is in the "I" band when discussing standard NATO frequency designations or "X" band when using the old radar frequency designations. The third column provides the emitter's center frequency, and the fourth column indicates the pulse width. The table is rounded out with the threat antenna's polarity, beamwidth, and transmitter's peak and average power.

One additional piece of information the evaluator will have to obtain from the EW ground facility is any obstructions or blockage areas that affect the LOS of the threat

TABLE 9.2 ▪ Typical EW Test Range Parameters

EW Test Range Emitter Parameters						
Name	Band	Center Frequency (MHz)	Pulse Width (μsec)	Polarity	Beamwidth (°)	Extreme Power Peak/Avg.
G2	VHF	216	5.0	H	30	2 kW/5 W
G4	UHF	450	6.0	H	28	5 kW/15 W
G6	L	1600	6.0	V	28	5 kW/15 W
Blind Eyes	S	2840	0.82	Rt Circ	4	200 kW/280 W
Bright Fire	S	3170	1.5	H	1.2	1 MW/500 W
Bad Find	S	3315	1.47	H	1.1	1 MW/1.5 kW
Best Guess	C	5570	0.30	V	1.1	180 kW/53 W
Big George	C	5600	2.0	V	1.5	250 kW
Blue Grass	C	5690	1.0	V	1.6	250 kW
Bleak Hope	C	6525	0.49	H	1.6	175 kW/55 W
Black Ink	X	9350	0.25	V	1.1	250 kW/51 W
Brand Iron	X	9425	0.25	V	0.9	200 kW/32 W
Burst Joint	K_u	16,530	0.80	V	1	24 kW/34 W
Big Job	K_u	17,530	2.46	V	.4	125 kW/77 W
CW7	S	3300	—	V	3.6	100 W
CW8	C	6500	—	V	2.2	100 W
CW9	X	9225	—	V	1.3	100 W

emitters. We do not want to set up our aircraft in blockage areas during the tests, for obvious reasons.

The first test of the RWR system could be DOA testing. The aircraft is equipped with four receiving antennas in order to provide 360° omnidirectional coverage about the aircraft. The antenna also provides coverage in the vertical. The specifications will probably call for two accuracies: one for straight and level flight, and the other for a maneuvering aircraft; the accuracies will be much less for a maneuvering aircraft. The geometry of the setup is performed as described for communications testing in chapter 4 of this text. Remember to perform antenna pattern testing before you spend a lot of time flight testing. For the bearing accuracy you will need your present position, the threat's position, and a record of the indicated bearing in the cockpit. If the system provides range versus lethality, you will also need to record the displayed range to the threat. Postflight, perform a coordinate conversion of latitude, longitude, altitude to range, azimuth, and elevation and compare the data to the recorded bearing and range. If the specifications call for the system to classify the threat at maximum range, then the evaluator will have to do a little homework to calculate the range by using the radar range equation. You will need to know the threats average power and the gain of your receiving antennas.

If your system displays lethality versus range, then you will need to have additional information from the EW ground facility. The time the threat changes modes (acquisition, target tracking, and fire control) must also be recorded and compared to the movement of the threat on the display and any change in symbology.

The system will also have a specification for the amount of time it takes to correctly identify a threat and display it to the crew; this is called the system response time. Once again, this parameter will be classified. This test is performed by flying a route that

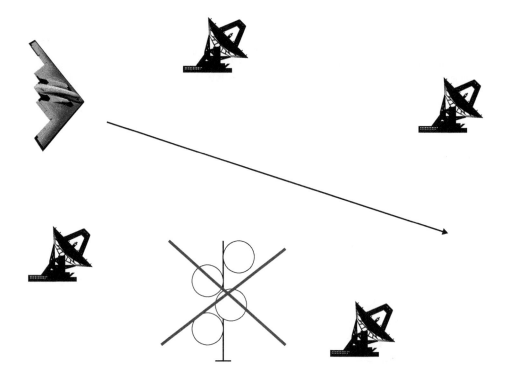

places your aircraft in the radar beam of the threat and noting how long it takes the system to recognize, classify, and display this threat. You will have to know when you entered the beam and the time from beam entry to the time it was correctly displayed (i.e., not as an unknown). This is why you need the beamwidth attribute of the threat. Another option is to have the ground station track the aircraft by some passive means and then engage the radar system. This may alleviate having to know the radar beamwidth.

This system response also needs to be tested against multiple threats in multiple frequency bands and in multiple quadrants. Figure 9.32 shows the proper way to conduct this test. A route should be flown that places the aircraft among multiple threats in different quadrants. The purpose of the test is to stress the system's capabilities to perform multiple classifications using all of the receiving antennas on the aircraft. Many test facilities use simulators to emulate the threats, and one pedestal can contain as many as four antennas emulating four different threats. Although this is good from an economic standpoint, all threats will reside in one quadrant, which will not properly test the capability of the RWR.

While on the subject of simulated threats it should be noted that the real threats aircrews encounter are probably not going to be brand new, out-of-the-box systems. The pulse from the actual threat may differ from the manufacturer's stated specifications (due to age, calibration, etc.). If we test against a simulated threat where the pulse looks exactly like the manufacturer's specifications (called a clean emitter) we may get great results; but aircrews encountering a real threat may only see an "unknown" (called a dirty emitter). For this reason, we need to intentionally degrade the simulators to replicate real-world threats. Most operational test and evaluation (OT&E) organizations require testing against real threats as opposed to simulations.

Another problem that has been seen in RWR testing is the failure of the system to detect threats above the altitude of the aircraft; not only AI threats, but ground-based threats as well. Imagine the scenario of flying down a valley with a SAM or AAA on the overlooking hills. It must be a consideration for your testing if the aircraft will be flown in these conditions (e.g., attack helicopter, etc.).

Additional testing includes threat display and prioritization, which can be accomplished automatically by the system or selected by the operator. The operator has the ability to display all threats (up to the limits of the system) or just a particular class of threats (AI, ship-borne, ground-based, unknowns). The operator may also classify which threats should be the highest priority. This type of testing is ideally suited for the lab or the anechoic facility, as it require a huge amount of resources to accomplish in flight testing. Threats can be injected (in the lab) or radiated (in the anechoic facility) as the operator steps through the available selections and notes the changes in the displayed information.

In most electronic systems that have any critical operational function, there will be three types of BIT: power-up BIT (PBIT), operator initiated BIT (IBIT), and continuous BIT (CBIT). PBIT is completed at system power-up and completes both a continuity check of the subsystems and primary system as well as a software check to determine if the software has been corrupted. Typically the display will indicate a "pass" or "fail" and potentially some error codes. In addition, anytime the software is checked, there will be a "checksum" number or alphanumeric sequence displayed to the operator that the operator should compare to the correct value. CBIT is a BIT process that runs on a noninterference basis (i.e., low priority) continuously while the system is operating. Anytime the system is performing operational processing, the CBIT processes get dropped while the operational processing is occurring. IBIT is an operator initiated BIT that is purposely selected by the operator. This will typically be used if, during the CBIT process, some error indication is provided to the operator. IBIT is typically intrusive, meaning it takes priority over any other process when initiated. However, there will likely be some sequence the operator can use to stop the IBIT process in the event that the system is required for operational use. All three types of BIT should be verified during testing, including the injection of known fault conditions to determine if the BIT processes function properly.

BIT and self-test are normally run at power-up or by operator selection. The BIT portion tests the software, whereas the self-test portion checks the electrical continuity through to the antennas. Software faults that adversely affect mission capability are reported to the pilot on the display; all others are written to the maintenance fault list (MFL). The self-test tests the capability to detect threats at each antenna for each of the operational bands (normally four). A typical band structure for an RWR is shown in Table 9.3.

TABLE 9.3 ■ Typical RWR Frequency Bands

RWR Frequency Band	EW Frequency Band
Band 0	C and D Bands
Band 1	E and F Bands
Band 2	G and H Bands
Band 3	I and J Bands

FIGURE 9.33 ▪
RWR Self-Test
Band 1

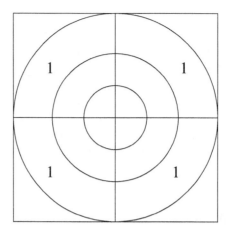

As the self-test runs, the operator can monitor the progress on the display. Figure 9.33 shows a successful self-test of band 1 in all four quadrants. The system will cycle through each of the four bands and advise the pilot of the success or failure in each of the four quadrants. For example, if band 1 failed in the right forward quadrant, the "1" would be missing in the top right section of the display.

The last portion of the controls and displays testing is the aural warnings and cues. Each system is mechanized in a slightly different way, but there are some common features that should be present. There should be what is called a "new guy" tone that sounds whenever a new threat is displayed. It cues the operator to note that a new radar has been detected and is displayed. Typically there will also be some momentary flashing enhancement of the displayed symbol that provides situational awareness to the aircrew. A missile launch audio is just that: when the system determines a launch has occurred it produces a distinct and annoying audio tone that will not go away. The corresponding display shows the suspect threat, typically with a flashing symbol enhancement of the basic threat symbol (e.g., flashing diamond or circle). The aircrew can adjust the volume or can adjust the system to only issue audio warnings on the highest priority threats. As with the displays, exercises can be run in the lab or anechoic facility to validate the logic of the audio warnings.

Missile approach warning systems (MAWS) are used to alert the aircrew, in a timely and persuasive manner, of an approaching missile. These systems provide the aircrew with a bearing and range rate (some systems also provide range) and typically are installed with a countermeasure dispenser system (CMDS) to initiate an automatic selection of appropriate countermeasures, often chaff and flares. The single biggest problem with these systems is the false alarm rate.

The systems can operate in the ultraviolet (UV), IR, or radar spectrums. Two typical MAWS are shown in Figure 9.34. The photo in the upper right of the figure is an AN/AAR-54 missile warning system made by Northrop-Grumman which operates in the UV spectrum. The photo at the bottom left of the figure is a Soviet Azovsky MAK-UT MAWS which operates in the IR spectrum; this installation is on a Bear Bomber.

The IR and UV system variants look for emissions from the missile plume to identify the threat and estimate range rate by rate of change of intensity. Radar variants employ Doppler detection and can provide range, range rate, azimuth, and elevation. The IR versions are susceptible to false alarms created by flares, explosions, truck backfires, own-ship missile launches, and unguided missile launches. This can be a

FIGURE 9.34 ▪
Two Types of
MAWS Systems

problem, especially at low altitude or in an airport environment. You can imagine an aircraft on approach, say into Los Angeles International Airport, when a false alarm is received when the aircraft is over the 405 freeway and a salvo of flares is commanded to be released. UV systems are a little more discriminating, but are susceptible to other sources, such as arc-welders. Radar systems need to detect the target first, and the target is competing with the ground. A delayed detection time severely reduces the crew's available reaction time.

When evaluating a MAWS, the test engineer needs to pay particular attention to

- System search volume
- Maximum/minimum detection range
- Bearing accuracy
- Velocity of closure measurement
- False alarm rate

The search volume evaluation should be accomplished in an anechoic facility and is merely an extension of antenna pattern testing. The evaluator needs to ensure that the entire area of interest below the aircraft (detection footprint) is seen by the antennas without any detection holes.

The maximum and minimum detection ranges can be calculated on the ground as long as the attributes of the sensors are known (FOV, detector sensitivity, target RCS, target velocity, Δt, etc.). The actual tests for the IR and UV sensors are conducted the same as previously described in chapter 7 of this text. Testing of radar detectors will be next to impossible to evaluate in the air, as the test requires missile shots directed toward the aircraft.

Bearing accuracy for IR and UV systems is a simple task, as it only requires a point source (at a known location) on the ground; the tests are conducted as described in chapter 7. The same comments for radar detection ranges apply here as well.

Velocity of closure measurements are a tough test for any of these systems. How do you measure the accuracy of V_c without launching a missile at the test aircraft? Many test organizations have used some clever test setups with varying degrees of success. One organization built a ramp and installed flashbulbs at incremental points to simulate a missile flight toward the aircraft. Another organization installed flashbulbs along the side of a mountain and had the aircraft fly a parallel course with the string of bulbs. At any rate, this will be a very difficult exercise.

The false alarm rate will probably not meet the desired specifications, as there are too many competing objects that could possibly be a missile signature. The evaluator may find that special procedures/restrictions must be applied when using a MAWS. For example, the aircraft may be required to turn the autodispense feature (chaff and flares) off when flying an approach due to the probability of false alarms.

Self-protection jammers work in a similar fashion to the RWR, only in this case, when the jammer classifies a threat it automatically selects the best jamming mode to counter the threat. The aircrew may elect to manually override the system, which will prevent the jammer from activating until compliance is received from the operator. As with the RWR, the library is critical in classifying the threat; if the library is wrong, the threat may be misrepresented or declared an unknown. The jammer may also be synchronized with the chaff and flares dispensing system and initiated by the jammer if selected as the appropriate jamming technique. The requirements for testing are similar to those described for the RWR:

- Self-test and fault reporting (BIT)
- Cockpit controls and feedback displays
- Autonomy system
- Area coverage
- Interference blanking
- Synchronization with dispensers
- Aircraft performance implications
- Store certification (carriage limits)
- Miscellaneous (power, cooling requirements)

The self-test checks the continuity through the forward and aft transmit and receive antennas. The cockpit indications will be very similar to those described for the RWR. In the automatic mode, the display will indicate to the pilot when jamming has been selected against a detected threat. In the manual mode, the system will inform the operator that a threat has been detected and a recommended jamming package has been selected, but not initiated. The pilot may opt to allow the selected jamming package, or in some systems he may be allowed to select a different jamming scheme.

The system does not indicate its effectiveness, only that it is operating (type of jamming, frequency band, and observed threat). The jammer will incorporate a look-through to ensure that jamming ceases if the threat goes silent.

Area coverage can be accomplished during antenna pattern testing in the anechoic facility. Remember, we are not testing the coverage of the unit (that is already known from the manufacturer), we are testing the coverage as the unit as installed.

Interference blanking is listed as a test because this system is an active system, as opposed to the RWR, which is passive. Once again I refer the reader to the test described in section 4.16 of this text, which fully describes the scope of EMI/EMC testing of EW systems. In short, the jammer will not be able to radiate continuously because it will adversely affect other installed systems (e.g., RWR, radar, HARM, and radar-guided missiles). For this reason, blanking pulses must be sent to the jammer whenever other onboard systems need access to the spectrum. Similar pulses are sent to the radar to allow the jammer access, and so on. CW operation of the radar is prohibited with jammer operations, for obvious reasons. These blanking pulses, inhibits, and, in some cases, notch filters are controlled by either the mission computer or EW bus controller. They are aircraft unique and must be evaluated as part of any EW program.

When required, and if implemented, the jammer will send a dispense signal to the chaff and flare dispensing unit. This is also an aircraft-unique implementation and must be included in testing. The signal can be monitored on the bus for proper operation or, in the case of integrations that are hardwired (jammer to dispensing unit), by observing the firing of squibs (Figure 9.25). For a ground test, we can remove the chaff and flares from the dispensing unit and only leave the firing squibs (which is like a fuse that initiates the charge). As a fire signal is sent to the dispensing unit, the appropriate squib will fire. If the squib does not fire, then the signal was not sent. Similarly, the number of squibs fired will determine if the correct program was issued.

The last three items relate to the testing required whenever a store is placed on the aircraft. In every contract you deal with, the first rule of the contract will state that "the addition of store XXX will not adversely impact the performance of the aircraft." In reality, this is nonsense, yet every contract will make this statement. Before the systems group gets the airplane and system, the performance and flying qualities and loads and flutter groups will have to clear the airplane to fly with this store attached. We will cover some of these requirements in chapter 10.

Expendables (chaff, flares, and decoys) are used to delay or negate a weapons launch. They can accomplish this by target screening, such as a chaff corridor, or confusing the operator with multiple targets or delaying target acquisition. In a tracking situation they can cause a break lock, increase miss distance after launch, or, as a last line of defense, shift the intercept point or affect weapons fusing.

Some of the considerations in expendables testing are

- Total available quantity of each
- System response time
- IR/radar frequency coverage
- Available salvo/interval options
- Ejection parameters
- Dispenser location
- Ejection trajectory
- Velocity
- Blooming/IR rise time
- Aircraft performance implications/maneuver limits
- Separation/dispense testing
- Store certification (carriage limits, jettison limits)

The quantity available is a fixed parameter over which we have no control. The purpose of this test is to ensure that the pilot is properly apprised of the current conditions of the system (quantity remaining, program set, automatic/manual, slaved to jammer, low quantity, and empty advisories). All of these tests can be accomplished on the ground by changing the physical conditions of the expendables and ensuring the information is properly relayed to the operator.

System response time is important in as much as we are testing the latency or aging through the system. Remember that the chaff bloom or flare peak intensity should occur within the resolution cell of the threat system. Any latency could adversely impact this objective. This test is also a ground test and can be piggy-backed with the jammer integration test described previously by monitoring and timing the squib fire. The pilot should also have the option of selecting a program or salvo and automatic or manual modes. The program is based on the optimum dispense rate (of chaff or flares) to defeat a particular threat. For example, it may be determined that in order to defeat an SA-XX missile in its tracking mode, chaff should be dispensed two bundles at a time for a total of eight bundles spaced at 0.2 sec. The pilot would then input the bundles per drop, interval, and total bundles into the programmer. This applies to the flare program as well. In addition, automatic and manual operation should be checked. This test can also be accomplished on the ground and piggy-backed with the previous test.

All of the other tests mentioned are either engineering design problems or performance and flying qualities tests for the addition of a new store. We may be called upon to evaluate the success or failure of an installation by flying an operational scenario against EW trackers.

9.11 | FINALLY

I cannot emphasize enough the importance of EMI/EMC and anechoic chamber testing for these systems. The basic tests of these systems are relatively simple, yet the evaluation of their effects on other aircraft systems and the possible degradation of capabilities is not a simple test and is critical for successful operation of the aircraft as an integrated system. Perform a good series of antenna patterns before embarking on any flight test program; failure to do so will result in poor results and a waste of time and money. Whenever possible, attempt to run as many tests as possible in an anechoic facility; it is a lot cheaper than flight testing, and real-time troubleshooting can be accomplished. It may also be a necessity because of security considerations. We do not want to let them know what we know. Be prepared to spend some time in the chamber; experience has shown that these tests can run for as long as 3 to 6 months.

9.12 | SELECTED QUESTIONS FOR CHAPTER 9

1. What are the three major divisions of electronic warfare?
2. Name three direct threat radars.
3. Name three indirect threat radars.
4. Give three representative examples of electronic attack.

5. Give three representative examples of electronic protection.

6. Give three representative examples of electronic support.

7. What is an IADS? Where can they be located?

8. What is a radar warning receiver? What is it used for?

9. Name four parameters that must be known about threat emitters prior to using them in flight testing.

10. How does antenna polarization affect the outcome of EW testing?

11. How does a V-beam radar obtain height?

12. What is target burn-through? When does it occur?

13. How is a range gate pull-off accomplished?

14. How is a velocity gate pull-off accomplished?

15. Name three types of deception jamming.

16. Name three types of passive deception.

17. What are the major benefits of employing passive deception?

18. Is intelligence about the threat a requirement for deception jamming? Why or why not?

19. What is DOA? How is it tested?

20. What EW testing would you schedule in the anechoic chamber if you were to perform an RWR integration?

21. What is a threat library? Where does it reside?

22. Explain the difference between a "clean" and a "dirty" threat.

23. Explain the difficulty in evaluating a MAWS.

24. What is interference blanking used for?

25. What are some of the concerns in the geometric setup of receiver response testing?

26. If you required an airborne jammer, what type of system might you employ on the target?

27. What type of display information can the RWR provide to the crew? How about a self-protection jammer?

28. Would you normally have a selection for noise jamming on a self-protection jammer? Why or why not?

29. Does the onboard radar possess inherent countermeasures?

30. Why would you paint a canopy on the bottom of your aircraft?

31. What is a MALD? How does it work?

32. How is home-on-jam accomplished?

33. What is a false target generator? How does it work?

34. What is a track breaker? How does it work?

35. What is the effect on a victim radar if you employ cross-polarization jamming?

36. What are the problems encountered when using flares as a countermeasure?

37. How does chaff affect a victim radar?

38. What is sidelobe blanking? How about sidelobe cancellation?

39. What does AGC do for the receiver?

40. What are some problems associated with using simulated threats to evaluate your EW system?

41. Are there any concerns with EMI/EMC for EW installations?

42. Is it possible to lose mission capability with the installation of an EW suite? If so, why?

Air-to-Air/Air-to-Ground Weapons Integration

Chapter Outline

10.0   Introduction   763
10.1   Weapons Overview   764
10.2   Stores Management System   767
10.3   Air-to-Air Missiles.   785
10.4   Air-to-Ground Weapons.   803
10.5   MIL-HDBK-1763 Test Requirements   812
10.6   AGARD Flight Test Techniques Series, Volume 10 Requirements.   820
10.7   Weapons Delivery Considerations for Helicopters   824
10.8   Selected Questions for Chapter 10.   828

10.0 | INTRODUCTION

In this chapter we will discuss the system aspects of weapons integration. The chapter is not concerned with loads, flutter, captive carry, or station clearance issues, although they will be touched upon during the discussions. As usual, some references are available to the evaluator, and these will be quoted where necessary. The first reference is the MIL-STD-1760D, "Interface Standard for Aircraft-Store Electrical Interconnection System," August 1, 2003. This document covers the digital data bus requirements for aircraft and stores, and was covered at some length in chapter 3 of this text. The second is MIL-HDBK-1763, "Aircraft/Stores Compatibility: Systems Engineering Data Requirements and Test Procedures," June 15, 1988. A detailed examination of the requirements in 1763 is covered in section 10.5. A third reference is from the NATO Aerospace Group for Avionics Research and Development (AGARD), now called the NATO Research and Technology Organization. AGARD Flight Test Techniques Series, Volume 10, "Weapon Delivery Analysis and Ballistic Flight Testing," July 1992, is perhaps one of the better documents available to evaluators for the test and evaluation of weapons systems. The highlights of this AGARD document are explained in section 10.6. A general document used for all aircraft is Mil-A-8860B, "General Specifications for Airplane Strength and Rigidity," May 20, 1987.

For rotary wing operations, Air Standard 20/21, "Airborne Stores Ground Fit and Compatibility Criteria," Change 1, April 21, 1991, establishes the requirements for carriage compatibility; electrical power interfaces are covered in Mil-Std-704F, "Aircraft Electrical

Power Characteristics," March 12, 2004. Four documents cover the airworthiness requirements: Army Regulation 70-62, "Airworthiness Qualification of Aircraft Systems, May 21, 2007; ADS-20 HDBK, "Armament and Fire Control System Survey for Army Aircraft," December 19, 2005; ADS-45 HDBK, "Data and Test Procedures for Airworthiness Release for US Army Helicopter Armament Testing," December 19, 2005; and ADS-62-SP, "Data and Test Requirements for Airworthiness Release for Helicopter Sensor Data and Testing Requirements in Development Stage," June 29, 2001. Rotary wing weapons testing is covered in section 10.7.

10.1 | WEAPONS OVERVIEW

Some general characteristics apply to all air-to-air and air-to-ground weapons. The broad categories are as follows:

- Seeker head/guidance section
- Target detection section/fuse
- Control section
- Warhead section
- Propulsion section
- Battery/gas grain generator
- Umbilical cable/arming and option wires

The seeker head in guided weapons is composed of a sensor whose purpose is to acquire and track a target. The sensor may be a passive radar, as in the case of a semiactive homing missile (AIM-7 Sparrow), and/or active radar using terminal guidance (advanced medium range air-to-air missile [AMRAAM]). The seeker may be infrared (IR) or ultraviolet (UV), as in the AIM-9 Sidewinder or stand-off land attack missile (SLAM), or it may use optics (television), such as the glide bomb GBU-15. The systems will normally contain a gimbaled gyro package for space stabilization and may contain an inertial navigation system (INS) or global navigation satellite system (GNSS) for precision navigation.

Associated with the sensors are the antennas, lenses, processors, and detector elements required for operation. There will be associated physical limiting factors such as the field of view (FOV) and field of regard (FOR) (FOV plus gimbal limits), as well as operational limitations such as weather, day or night operations, smoke, or any electronic protection measures employed. The guidance section houses the navigation computer, autopilot guidance, and homing techniques.

The following common fusing (arming and detonation) mechanisms may be employed:

- Time
- Proximity
- Impact or delay
- Hydrostatic (pressure)

Time is the simplest and is self-explanatory; a timing sequence is initiated at release, and the weapon is armed after the prescribed amount of time. A time of fall or

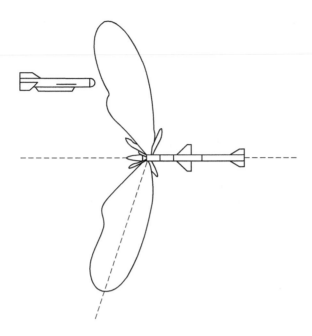

FIGURE 10.1 ▪
Doppler Proximity
Fusing

travel less than the prescribed time results in a dud. Proximity fusing is normally a Doppler function; when the weapon sees a step in the velocity of closure up from zero or near zero, the weapon detonates. Figure 10.1 depicts this concept.

The missile in Figure 10.1 radiates a beam of energy slightly forward of orthogonal to the direction of flight. At the point shown, there is no Doppler return, but as soon as the target enters the beam, the missile sees a step input of positive Doppler and detonates. The missile can also use an active sensor, such as radar or laser, and detonate on range. Another form of proximity fuse is one that calculates the proximity to the earth; this is accomplished with either a barometric or radar altimeter. Most weapons incorporate an impact fuse (when high g forces are sensed, the weapon detonates). Another version of the impact fuse incorporates a delay in detonation to allow the weapon to penetrate the surface.

The control section of the weapon allows it to "fly" by utilizing control surfaces or vectored thrust. The control surfaces can be wings, fins, or canards, and the movement is powered either electrically or hydraulically. The weapon is self-contained, so there is limited power available to the servos unless a ram air turbine (RAT) is outfitted to the system. Electric power is supplied by batteries, whereas hydraulic power is usually supplied by pressurized reservoirs. There are limitations on the maximum rate and torque of the servos, which limits the performance (or g available) of the weapon.

The warhead incorporates a series called an explosive train. This train consists of a primer, delay, detonator, booster, and bursting charge. The weapon causes its damage by fragmentation, blast, or incendiary. All weapons have safety mechanisms in place that prevent inadvertent release on the ground and in the air or harmful detonation too close to the delivery vehicle.

There are five basic classes of weapons:

- Free fall/unguided/single trajectory: these consist of general purpose bombs (GPs)

- Free fall/unguided/multiple trajectory: these consist of cluster bombs (CBUs)

- Free fall/guided: these consist of laser-guided bombs (LGBs) and inertial-aided munitions (IAMs)
- Propelled/unguided: these consist of rockets
- Propelled/guided: these consist of air intercept missiles (AIMs) and air-to-ground missiles (AGMs)

The designations of munitions tell you what they do if you know the code. The codes are shown for missiles and munitions in Tables 10.1–10.4.

TABLE 10.1 ■ Missile Designations

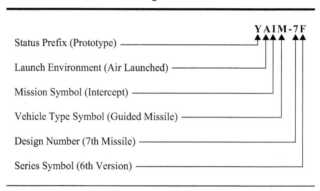

TABLE 10.2 ■ Missile Designation Codes

Status	Environment	Mission	Type
J-Special Test/Temp	A-Air	D-Decoy	L-Launch Vehicle
N-Special Test/Perm	B-Multiple	E-Special Electronics	M-Guided Missile
X-Experimental	C-Coffin	G-Surface Attack	N-Probe
Y-Prototype	F-Individual	I-Intercept Aerial	R-Rocket
Z-Planning	G-Runway	Q-Drone	
	H-Silo Stored	T-Training	
	L-Silo Launched	U-Underwater Attack	
	M-Mobile	W-Weather	
	P-Soft Pad		
	R-Ship		
	S-Underwater		

TABLE 10.3 ■ Munitions Designations

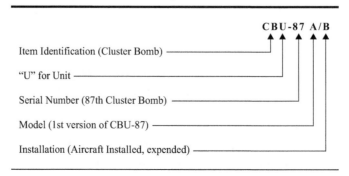

TABLE 10.4 ■ Munitions Designation Codes

Identification	Identification	Identification	Identification	Installation Designator
AD-Certain Adapting Items	CN-Misc Containers	LM-Ground Based Launcher	PW-Internal Dispenser	A-Aircraft Installed/Fixed
AB-Explosive Items	DS-Tgt Directing Device	LU-Illuminating Units	RD-Dummy Rocket	B-Aircraft Installed/Expendable
BB-Simulated Bombs	FM-Fuses	MA-Misc Items	RL-Rockets	E-Ground Item Moveable
BD-Bombs and Mines	FS-Fuse Safety	MD-Misc Simulated	SA-Sights	
BL-Bomb Racks	FZ-Fuse Related Item	MH-Munitions Handling	SU-Store Suspension and Release	
BR-Retarding Device	GA-Aircraft Gun	MJ-Munitions CM	TM-Tanks	
BS-Stabilizing Device	GB-Guided Bomb	MT-Mount	TT-Test items	
CB-Cluster Bomb	GP-Pod Gun	PA-External Device	WD-Warheads	
CC-Actuator Cartridges	LA- Aircraft Launcher	PD-Leaflet Dispenser	WT-Training Warheads	
CD-Clustered Munitions	MK-USN Bomb Designation	PG-Ammunition	M-US Army Bomb Designation	

▮ 10.2 ▮ STORES MANAGEMENT SYSTEM

As with other avionics systems, a conversion from analog to digital was necessary in order to save weight and space and allow the sharing of information across a distributed bus. The stores management system (SMS) is controlled by the stores management processor (SMP), which is normally a remote terminal (RT) on mission or avionics busses 1 and 2 and is the bus controller of the weapons or armament bus (Figure 10.2).

Most military aircraft utilize a 1553 bus for avionics and a 1760 bus for weapons. The 1760 bus utilizes the same protocol as the 1553 bus, with two additions: header words and cyclical redundancy checks (CRCs). In more state-of-the-art applications, such as miniature munitions/stores interface, EBR-1553 may be used to increase speed as well as the total number of RTs. The reader is again referred to chapter 3 of this text for a review of 1553 and 1760 data busses.

The general requirements for the SMS are to

- Provide the control logic necessary to prepare, release, fire, and jettison all stores
- Provide the interface with the mission computer via the 1553 bus and be capable of monitoring, interpreting, and providing store status
- Provide the interface with aircrew controls and displays and issue alerts
- Provide test capabilities for built-in tests (BITs), installed weapons tests, and ground tests for all stations

The major functions of systems management include system initialization, stores inventory, missile preparation, power distribution and control, and avionics and

FIGURE 10.2 ■
SMS 1553
Relationship

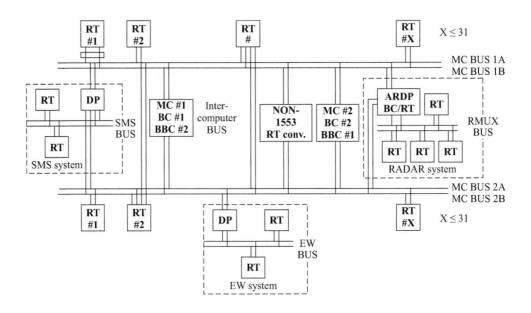

armament bus communications. The stores jettison function includes emergency jettison, air combat maneuvering (ACM) jettison (which is a prebriefed selective jettison feature), selective jettison, and auxiliary jettison. Systems tests include SMS BIT, weapons test, ground test, and system reporting. In many cases an in-flight training mode is available that inhibits weapons release but provides normal cockpit indications.

The typical SMS will comprise the following as a minimum:

- Stores management processor
- Missile power relay unit (MPRU)
- Decoders (type 1, 2, etc.)
- Fuel tank jettison unit (FTJU)
- Gun control unit (GCU)

A typical SMS architecture is shown in Figure 10.3. The SMP is a programmable digital computer that provides control of the SMS. It is the bus controller of the ARMUX (armament bus) and an RT on the AVMUX (avionics bus). It is the electrical interface between the SMS avionics systems and the weapons/stores interface of the SMS. Normally the SMP will contain an emergency jettison generator circuit.

The MPRU functions as an RT on the ARMUX. It provides the distribution of alternating current (AC) and direct current (DC) power to all store stations under control of the SMP. The MPRU provides AC and DC power to the GCU and provides AC and DC release power to the decoders and FTJU. It also provides AC and DC power to onboard missiles.

Decoders function as an RT on the ARMUX and perform all of the analog-to-digital (A/D) and digital-to-analog (D/A) conversion functions to allow the SMP to talk to non-1553 stores. A type 1 decoder can interface with the AIM-9M, AIM-7F/M, or AIM-120, or any combination of two of these missiles. A type 1 decoder provides interlock return

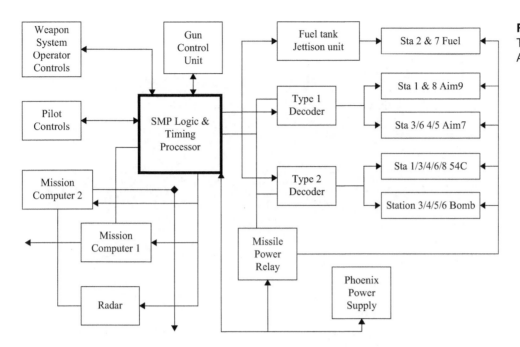

FIGURE 10.3 ▪
Typical SMS
Architecture

and release consent circuits for the AIM-120. A type 2 decoder can service only one weapon at a time and interfaces with the BRU-32A or triple ejector rack (TER) bomb racks. A type 2 decoder provides the release or separation signals to bomb racks as well as circuitry for mechanical arming (e.g., nose/tail) of MK80 series bomb types.

The FTJU

- Provides firing voltages to squibs to jettison a specific fuel tank

- Provides relay logic for selective jettison

- Provides a separate relay logic in the event of an emergency jettison (emergency jettison commands are received directly from the SMP emergency jettison circuit)

The GCU provides the interface between the SMP and gun system. The GCU incorporates a burst limit switch to select maximum rounds fired during trigger squeeze and provides orderly firing termination at detection of trigger release, burst limit, or last round.

10.2.1 SMS Processes

There are two basic processes within the SMS: cold start and warm start. The cold start process will power up, initialize, command BIT, and perform inventory for all stations. Warm start will normally perform a safety check only; inventory will not be performed. The SMS is required to maintain status and inventory after a warm start.

At start-up, the SMP commands the MPRU to initialize the decoders (i.e., supply power). The SMP also polls the FTJU and GCU; the MPRU responds back with BIT results. Each individual decoder replies back with BIT results and polls stations to determine if a store is present, and if the store is compatible. If a missile or smart weapon is present, the store is prepped after the inventory process.

FIGURE 10.4 ■
SMS Inventory

A complete inventory is performed at cold start. The SMP checks the MPRU for the
SMS weapons replaceable assembly (WRA) complement. All decoders are powered up,
BIT checked, and then checked for stores present. If a decoder finds a store present, but
the store is not compatible, power is removed from the decoder and the store is rendered
inoperable. The SMP reports inventory to the mission computer for display to the
operator. A display of a weapons inventory page is shown in the upper left (multifunction
display 1) of Figure 10.4. It should come as no surprise that the total of all permutations
and combinations of stores that can be loaded is a very large number. Just look at any of
the pictures that accompany any aircraft (Figure 10.5). The testing to validate inventory
and compatibility issues on the aircraft should never be attempted on the airplane, as the
amount of time required to upload and download stores would be excessive.

The SMP is smart enough to identify invalid/unauthorized weapons configurations.
Concerns include weight, drag, and hazardous stations (unsafe firing or release of certain
stores from multiple aircraft stations). Depending on the implementation, the SMP may
display a red X across a station, issue an aural alert, or display a master caution/warning.
Power is removed from the station until the condition is resolved and another cold start
is performed. The same results will occur when an actual aircraft to preloaded configuration conflict is present.

Once a valid inventory is established, missiles and smart weapons can be prepped. A
store declared invalid during inventory cannot be prepped. Weapons prep is an interface
check between the weapon and other required avionics systems on the aircraft. Missiles
are tuned, BIT checked, or run through a missile on aircraft test (MOAT). Weapons
requiring navigation information are checked using synthetic inputs from the aircraft.
Prebriefed target coordinates with associated weapon IDs are verified as being accepted
correctly. Stores can be selectively prepped by the operator; a store will not be declared
ready until prepped.

FIGURE 10.5 ▪
Combinations of
Weapons

Tuning is accomplished on semiactive homing missiles and tests the ability of the missile to "tune" to the onboard radar's frequency. On U.S. and Allied platforms, missile tuning applies to the AIM-7 Sparrow. Each missile on the aircraft must be tuned separately to be declared prepped and ready for use.

A test request is received from the mission computer to execute a missile tune. The mission computer commands radar to continuous wave (CW) or pulse Doppler (PD) mode. If the missile was previously tested, the SMP cycles off power to the station for 6 sec and then on. The SMP sends a CW/PD select command to the missile and then transmits an injected video (simulated Doppler) signal to the missile. The radar applies tuning to the missile and sweeps until rear automatic frequency control (AFC) and speed gate lock are achieved. The missile transmits "Missile Ready" (tuned) to SMP and the SMP transmits "Missile Tested" and "Missile Ready" to the mission computer for status display. Missile tuning is not really comprehensive, as it tests only the ability of the missile to lock onto the speed gate. In some applications, this test can only be accomplished in CW mode. A complete test takes approximately 3 sec.

The MOAT is performed on missiles that receive postlaunch communication from the mother ship. When MOAT is selected on a missile, the SMP receives a test request from the mission computer; the SMP provides the missile test control. The SMP then commands the radar to select semiactive channels and transmit missile messages. The SMP receives the missile responses and processes the test data. It provides test results to the mission computer and saves the test results for postlaunch examination.

There are specific tests that MOAT commands:

- Prelaunch tests: verify initialization and frequency lock to the selected semiactive channels
- Postlaunch tests: verify the rear message link and the ability to acquire a simulated target

- Autopilot tests: verify the system interface and autopilot functions dependent on weapons system for stimuli
- Built-in self-test (BIST): initiates the BIST and collects test results
- Nonselected channel tests: tests phase-lock and radio frequency (RF) message reception capability of the remaining semiactive channels

A complete MOAT takes about 15 sec; the results are used to determine weapon system and missile capability and fault isolation of failures. Results are normally displayed in hexadecimal format to the pilot and recorded in the maintenance fault list.

The AMRAAM BIT also has commanded logic from the SMS. When AIM-120 BIT is selected, the SMP receives a BIT request from the mission computer. It then determines if duty cycle timeout has been satisfied. If yes, the SMP initiates AMRAAM BIT by applying power to the missile and outputting launch initiate discretes to the radar and mission computers. The SMP transmits system time and targeting and uncertainty data blocks to the missile. The SMP interprets the missile status message to determine when BIT is complete. Power is removed from the missile when a BIT complete message is received from the missile, 5.5 sec has elapsed from the start of the BIT, or the BIT has been terminated.

10.2.2 SMS Weapons Interfaces

The following examples will be covered in this section:

- AIM-9
- AIM-7
- AIM-120
- Gun
- Bombs

AIM-9 Sidewinder

A typical AIM-9 Sidewinder interface to the SMS is shown in Figure 10.6. A normal launch of an AIM-9 consists of the pilot first selecting a cooled AIM-9 (Figure 10.7). The newer AIM-9 series missiles have cooled seeker heads for better temperature differential (ΔT) discrimination. As was discussed in chapter 7 of this text, the detector element must be cooled down before a usable system is ready. Missiles on the aircraft are cooled by a gas or thermoelectric cooler and take a little time before a missile is ready. The cool-down of the detector is normally initiated during weapons prep. A hot trigger or "Ready" light will appear when "Master Arm" is on. Trigger activation starts the launch-to-eject (LTE) cycle in the SMP. The SMP transmits a fire command to the LAU-7 launcher and an engine derich command to the engine midcompression bypass (this is so the engine does not compressor stall with an ingestion of the missile exhaust gases). The launcher puts the AIM-9 in self-track and activates the gas grain generator. The missile guidance section transmits a motor fire signal to the rocket motor through the LAU-7 and the missile is rail launched from the station.

AIM-7 Sparrow

The AIM-7 Sparrow is a semiactive homing missile that homes in on reflected radar energy from a target with a particular Doppler shift. This scenario is shown in

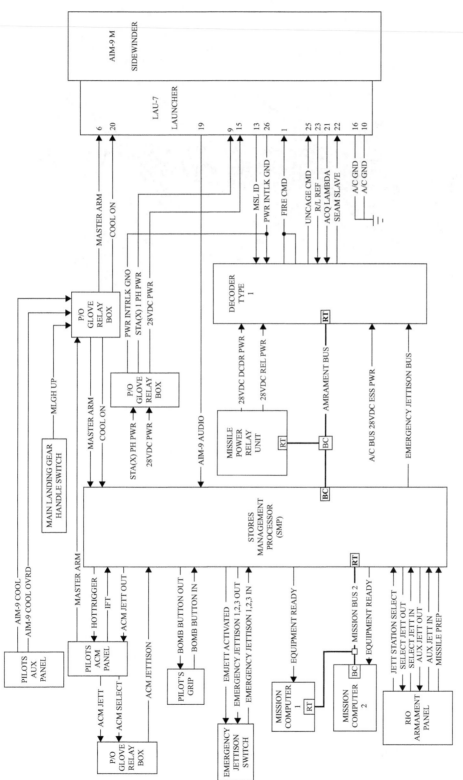

FIGURE 10.6 ■ Typical AIM-9 Sidewinder Interface

FIGURE 10.7 ▪
AIM-9 Sidewinder

FIGURE 10.8 ▪
AIM-7 Semiactive
Homing

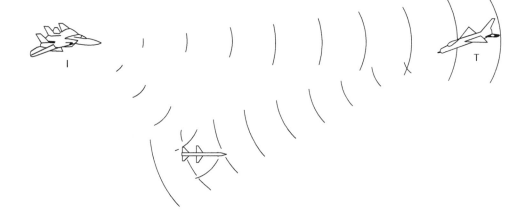

Figure 10.8. Based on the geometry of the intercept, the velocity of closure between the radar and the target will change. The radar tells the missile to look for a Doppler shift either up or down (remember that the missile knows the frequency of the radar since it was tuned during weapons prep). This Doppler shift is called a speed gate.

The aspect of the target, and hence the range rate, is sent from the radar to the missile. The missile then searches within its FOV for a return equal to the carrier frequency plus the Doppler shift; it then homes in on that signal. The speed gate logic within the missile is represented in Table 10.5. These are not the real numbers of any missile, but they serve as an example. If we tell the missile to look for a head-on speed gate, it will look for a velocity of closure (Doppler) equal to 900 to 1200 knots; a beam speed gate would be 200 to 500 knots. If the initial geometry at missile firing was a head-on and the target then turns to the beam, what happens? The missile initially guides

TABLE 10.5 ■ Semiactive Homing Speed Gates

Aspect (Speed Gate)	Target Velocity of Closure
Head On	1200 - - - - - 900
Quarter Head	900 - - - - - 500
Beam	500 - - - - - 200
Quarter Tail	200 - - - - - 0
Tail	0 - - - - - (−300)

on the high closure return, but then it disappears. It searches through the speed gates from high to low until it sees a return. The target return will appear in the beam speed gate, but so does the main lobe clutter. There is therefore a chance that the missile will guide to the ground, especially in a look-down geometry.

In a normal AIM-7 launch the pilot selects a "Ready" AIM-7 (tuned); a hot trigger/launch button will appear when the master arm is on and the mission computer transmits "Launch Enable." When the pilot pulls the trigger/launch button the activation results in "Launch Initiate" being sent to the mission computer and radar, and the start of LTE in the SMP. The radar transmits in CW or PD mode. The mission computer sends injected video and "PD/CW Select" to the missile via the SMP. The SMP transmits "Missile Battery and Hydraulic Activate" to the missile and the mission computer sends antenna pointing commands, steering commands, target data, and own-ship speed to the missile via the SMP.

If the missile "tunes," passes BIT, and detects battery up, it transmits "Commit to Launch" to the SMP. The SMP removes power from the missile and transmits an engine derich command to the engine midcompression bypass. The SMP then sends an eject command to the LAU-92 launcher; the SMP sends a "Motor Fire" command to the missile after the ejector foot has extended.

The AIM-7 and its components are shown in Figure 10.9. A typical AIM-7 Sparrow interface is shown in Figure 10.10.

AIM-120 AMRAAM

In a typical AMRAAM launch the pilot selects a "Ready" AIM-120. A hot trigger/launch button appears when master arm is on and the mission computer transmits "Launch Enable." The trigger/launch button activation results in "Launch Initiate" being sent to the mission computer and radar, and the start of LTE in the SMP. The radar transmits in single target track (STT) or track-while-scan (TWS) mode. The SMP applies station power and transmits "Release Consent" to the missile, which starts the batteries. The SMP transmits system time to the missile and receives a wakeup time from the missile, which in turn is transmitted to the mission computer. The mission computer transmits targeting and uncertainty data to the missile via the SMP. The SMP receives "Commit to Launch" from the missile if the missile is operable. The SMP removes station power and the "Release Consent" from the missile and transmits an engine derich command to the engine midcompression bypass. The SMP then transmits an eject command to the LAU-92 launcher.

The AIM-120 is shown in Figure 10.11, and the interface is depicted in Figure 10.12.

FIGURE 10.9 ∎
AIM-7 Sparrow

AIM-7 Components

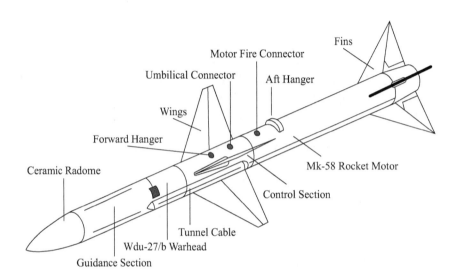

Guns

Gun firing is initiated by an SMP detection of a trigger discrete with the master arm on and gun selected. A concurrent receipt of gun discrete at the GCU enables the hydraulic drive command enable relay. While the hydraulic drive command and the low rate drive are enabled, the gun clearing command is disabled. The SMP transmits a trigger select discrete to the GCU and an engine derich command to the midcompression bypass via the MPRU. The hydraulic drive command is generated by the GCU if

- Gun clearing command is false
- Last round is false
- Gun arm enable is true
- Trigger select is true

If all the interlocks are correct, the purge command and interlock drive are activated and the air control is set to zero (full air flow). The GCU receives interlock compliance and applies power to the gun. The GCU detects and uses the feed rate signal to generate rounds fired and rounds remaining to the SMP. Rounds/pulse count is used to accumulate the burst count, which will terminate the firing cycle. The latest rounds remaining count received by the SMP is reported to the mission computer in the current inventory message.

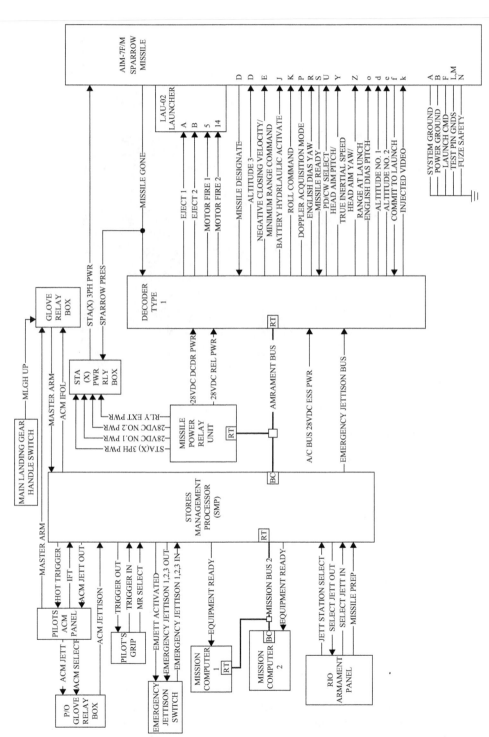

FIGURE 10.10 ■ Typical AIM-7 Sparrow Interface

FIGURE 10.11 ■
AIM-120 AMRAAM

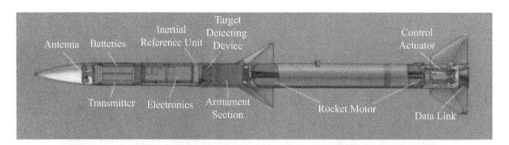

Gun firing is terminated under the following conditions:

- Removal of trigger discrete
- Burst limit exceeded
- Last round detected
- Removal of master arm
- Loss of interlock compliance
- Loss of gun command enable
- Receipt of weapons clear or other weapon selected
- Loss of trigger interlock

The Vulcan cannon is shown in Figure 10.13, and a typical gun interface is shown in Figure 10.14.

Bombs

The delivery of bombs is initiated when the pilot selects the bomb type. The weapon program is accessed by the SMP; the program includes quantity, interval, arming, and fusing. Placing the master arm on provides power to the bomb release unit (BRU) or TER. The hot trigger/bomb button appears when master arm is on and the mission computer transmits "Release Enable" to the SMP. The SMP transmits an arming command to the type 2 decoder and, when the bomb button is activated, allows release power to be transmitted to the decoder. The SMP transmits release pulse and eject

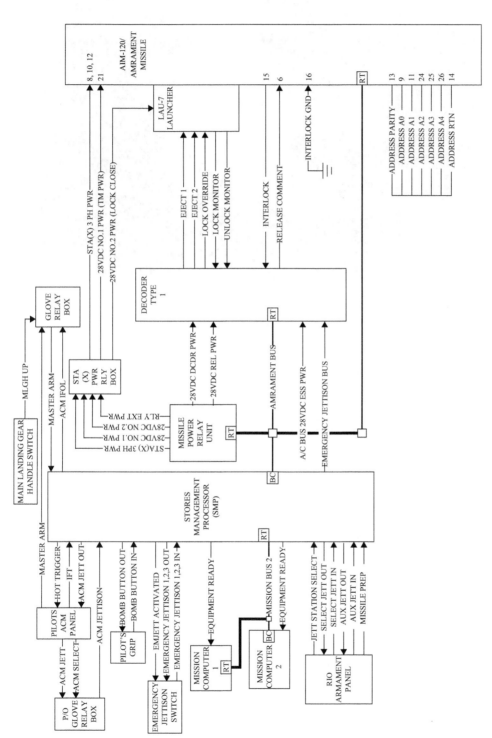

FIGURE 10.12 ■ Typical AIM-120 AMRAAM Interface

FIGURE 10.13 ▪
Vulcan Cannon

Characteristics

Gun	Externally powered, six barrel, 20 mm Gatling
Weight (Total)	841 lb (381 kg)
Feed System	270 lb (122 kg)
Gun	252 lb (114 kg)
Ammunition	319 lb (145 kg)
Firing Rate	4000/6000 shots per minute
Ammunition	
Capacity	570 rounds
Type	20 mm M50 series, PGU-28/B Saphei
Feed System	Rotary linkless, double-ended
Drive System	Hydraulic
Barrel Life	40,000 rounds
Reliability	15,000 mean rounds between failure
System Life	250,000 rounds
Gun Life	150,000 rounds minimum

commands to the BRU/TER, energized nose/tail solenoids capture arming wires, and fuse voltage is applied to the bomb as it is ejected from the aircraft.

Representative bombs are shown in Figure 10.15, and a typical bomb interface is shown in Figure 10.16.

10.2.3 Other SMS Interfaces

There are three other interfaces that are common to most SMSs:

- Sidewinder expanded acquisition mode (SEAM)
- SMS interlocks
- SMS jettison

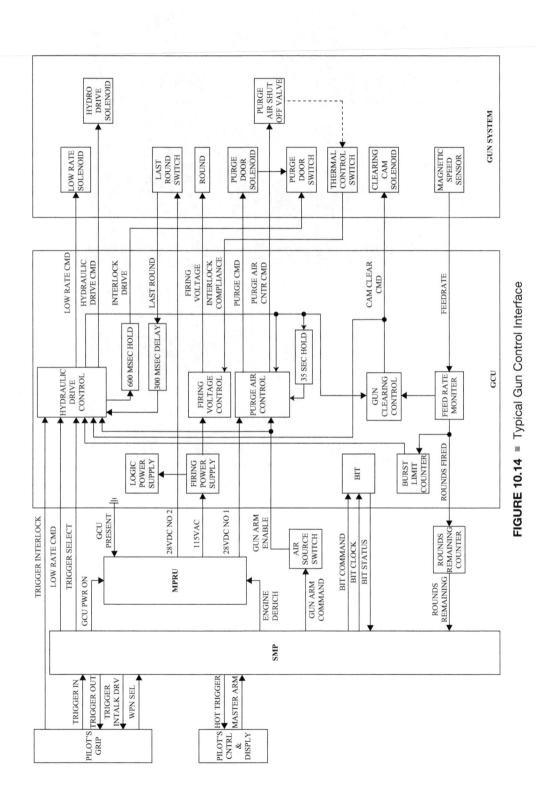

FIGURE 10.14 ■ Typical Gun Control Interface

FIGURE 10.15 ▪
MK-82 (top) and
BLU-107 Durandal
Runway Buster
(bottom)

The SEAM has two modes of operation: cage and SEAM. SEAM mode is automatically entered with an AIM-9 selection; the operator can toggle between SEAM and cage. Detection of a valid false/true transition causes the SMP to switch from SEAM to cage mode. Each subsequent detection causes the SMP to switch back and forth between SEAM and cage.

The Sidewinder message is received from the mission computer and is examined for valid track true indications. If the track is read as true, full SEAM mode is engaged. The SMP responds by applying a manual uncage command to the missile. The SMP then allows the missile seeker head to be slaved to external commands (system target or system track). The SMP mode process entails the extraction of seeker head position

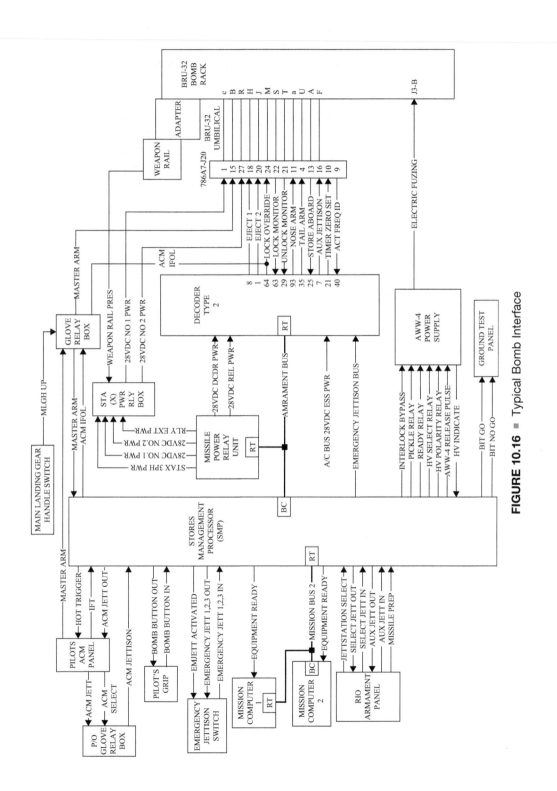

FIGURE 10.16 ■ Typical Bomb Interface

parameters from the missile and transmission of these coordinates to the mission computer. The SMP also receives pointing commands from the mission computer and compares commanded seeker head position to actual seeker head position (x, y to polar coordinates). The comparison of commanded versus actual generates error signals that are used to correct to the desired pointing angle.

In the SEAM lock-on process, the SMP detects the missile audio tone as exceeding the specified threshold and transmits an audio true discrete to the mission computer. The mission computer determines if the target position and the seeker head position coincide within tolerance. If it does, the mission computer transmits "Coincidence True" to the SMP. The SMP removes SEAM enable from the missile and allows the missile to self-track while the SMP continues to extract seeker head position parameters.

If the audio falls below the specified threshold for more than 1 sec, the system reverts to SEAM slaving. A SEAM break lock condition will also exist if the cage/SEAM switch is depressed, commanding the missile to be caged to boresight, or if a "Break Lock" command is received from the mission computer. The mission computer transmits a missile masking command when the seeker head is pointing to the nose of its own aircraft. The SMP will step to the next missile station, providing the station is on the opposite side of the aircraft and the current missile is not in self-track. If the conditions are not met, the mission computer transmits a break lock command to the SMP.

The SMS is responsible for monitoring the entire weapons system for clearance to release a store (safety, arming and fusing considerations, aircraft performance, etc.). Some key interlocks are

- Weight-on-wheels (WOW) switch. On most aircraft, power cannot be supplied to stores stations when the aircraft is on the ground (however, some past accidents have shown that this is not always the case).
- Gear handle position. The gear handle must be in the up position to fire/release belly-mounted missiles/bombs.
- Boarding steps. Some aircraft have self-contained boarding ladders or steps that are retracted into the aircraft once the crew is aboard. If these steps extend in flight, or the discrete is not properly set, power will normally be removed from forward-firing ordnance.
- Type weapon select. Depending upon the aircraft mechanization, some systems will not allow the firing or release of a weapon that the SMS believes is incompatible for the situation. For example, if AIM-120s and AIM-9s are on the aircraft, the SMS will not allow the pilot to fire an AIM-9 at a target on the nose at 20 nm (incompatible range).
- Station selected. The station selected must match the type of weapon desired; a mismatch will usually cause a red X to be drawn over the tactical information display.
- Master arm. Think of the master arm as the master power switch; weapons cannot be released without placing the master arm to the "on" position.
- Consent switch. This is usually associated with weapons that require two-man control, such as nuclear devices.

10.2.4 SMS Test and Evaluation

The following evaluations are required of the SMS:

- Weapons inventory
- Weapons prep
- SMS sequencing
- SMS interlocks
- MIL-STD-1760D messages

Most of the SMS validation tests are going to be accomplished in the lab. As seen in Figure 10.5, and previously described in the text, the combinations and permutations of possible stores configurations may seem endless. On-aircraft inventory should only be accomplished with the most common load-out, which will be determined by the user. All other weapons combinations should be demonstrated by lab testing. Remember that even invalid loads must be evaluated in order to determine if the SMS can detect improper and incompatible configurations; proper feedback must be available to the aircrew.

Weapons prep is accomplished in the lab and on the aircraft. Depending on the store(s) being carried, the SMS must demonstrate the ability to successfully prep the weapon for delivery. Weapons prep must provide the aircrew with positive indications of whether the store is available for firing/release. If a weapon fails to prep, the aircrew must be properly advised that the store is unavailable.

Weapons sequencing for firing/releasing stores must be demonstrated to operate properly. The initial testing is done in the lab by recording all the weapons bus data during the LTE or release sequence. A bus analyzer is required for this testing. Initiate the record cycle at the initiation of the sequence (e.g., trigger squeeze, bomb button depression) and record all data on the weapons bus for the expected period of the sequence (usually not more than 1 or 2 sec). The data should then be reviewed to ensure that the sequence is performed in the correct order and that timing requirements are met. Sequencing or timing that is incorrect will result in either a hung store, release of an unarmed store, or an unguided munition.

Interlocks must be evaluated to show that release is inhibited if one or more of the required interlocks are not set. The easiest way to conduct this test is to repeat the sequencing tests in the previous paragraph with interlocks set and not set and verify the outcome.

Although not mentioned specifically in this chapter, MIL-STD-1760D messages, formatting, and timing must be demonstrated successfully. The 1760 bus is described in detail in chapter 3 of this text. Safety is an overriding concern when evaluating the message content and format of the armament bus.

10.3 | AIR-TO-AIR MISSILES

The basic configuration of an air-to-air missile is shown in Figure 10.17.

The airframe type will determine the initial response and the amount of torque required to execute a maneuver. Typically missiles with wing control (controls at or near the center of gravity [CG] of the missile) will provide a rapid initial response to control

FIGURE 10.17 ■
Basic Air-to-Air
Missile
Configuration

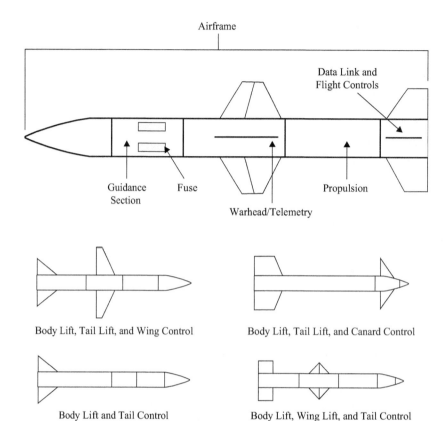

FIGURE 10.18 ■
Airframe Types

inputs, but large servo torque is required. Missiles with tail control or canard control will provide a slow initial response, but require relatively low levels of torque. Since less torque is required, the electromechanical servos can be kept small, requiring less space with less of a weight penalty. Four typical missile airframe types are shown in Figure 10.18.

The coordinate systems used by missiles varies; it can be an earth-stabilized coordinate system, such as an INS, it can be referenced to the aircraft coordinate system, or it can use a polar coordinate system. The thing to remember here is that if the missile is using a coordinate system other than what the aircraft is using, a coordinate conversion has to be made prior to sending target location, pointing angle, etc., to the missile. Pitch, roll, and yaw conventions will also have to be determined and converted, if necessary, to ensure a successful release. The standard convention system for missile attitude is shown in Figure 10.19.

The autopilot used in the missile may be a basic lateral autopilot that uses attitude and rate gyros as well as accelerometers, or it may be a lateral autopilot that uses a pseudoattitude reference. The latter eliminates the attitude gyros, which saves on weight and space.

The missile can use any of the tracking methods previously described in chapters 7 and 8 of this text and will not be repeated here. A radar-guided missile can use continuous semiactive homing, sampled data midcourse with updates from the launch vehicle, flight by INS with updates, or active terminal guidance. Some typical radar-guided missiles and their guidance modes are shown in Table 10.6. The Skyflash, a

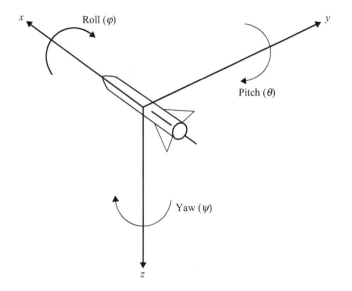

FIGURE 10.19 ■
Missile Coordinate
System

TABLE 10.6 ■ Air-to-Air Guided Missile Modes

Missile Type	Continuous Semiactive (CW and PD)	Sampled Data Midcourse w/Updates	Inertial Midcourse w/Updates	Terminal Active Guidance
Skyflash Aspide	X			
Sparrow AIM-7	X			
Phoenix AIM-54A	X	X		X
Phoenix AIM-54C	X	X	X	X
AMRAAM AIM-120			X	X

European semiactive homing missile, and the AIM-54 A/C, a retired, long-range missile requiring postflight missile messages, are included in the table only to show other types of guidance.

The way the missile navigates to the target also varies from missile to missile. In the simplest case, the missile can use pure pursuit, which constantly keeps the target on the nose. This is the least optimum route, as it takes the long way around. Proportional navigation is the shortest route to the target but requires a prediction of where the target will be at missile impact. Some missiles use a combination of the two: initially using proportional navigation, then transitioning to pure pursuit. These two navigational modes are shown in Figure 10.20.

The propulsion section determines the maximum velocity and time of flight of the missile. A boost provides the highest velocity in the shortest period, while a sustained propulsion design provides a constant velocity over the entire time of flight. The relationship between thrust, velocity, and time for various propulsion sections is shown in Figure 10.21.

The warhead section is selected based on the type of damage that is desired. A continuous rod warhead is designed to cut structural members of the target; it has a narrow fragment field with relatively low fragment velocity. The warhead is designed with side-by-side interconnected rods (like an accordion or chain-link fence) that expand

FIGURE 10.20 ▪
Missile Navigation

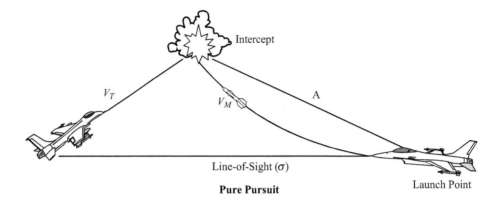

Pure Pursuit

Proportional Navigation

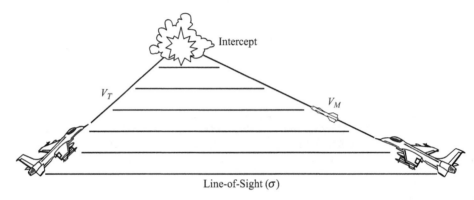

FIGURE 10.21 ▪
Propulsion
Section Profiles

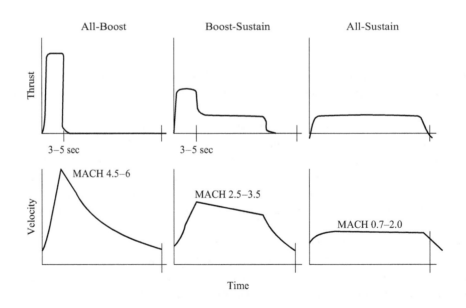

as they leave the missile. Annular blast fragmentation (ABF) is designed to penetrate internal target components with high-velocity fragments (BBs, discrete rods, and fleshettes). The weapon can also use a zirconium disk as a secondary incendiary damage mechanism. A selectively aimable warhead (SAW) uses several detonators fired in a

TABLE 10.7 ■ Warhead Characteristics

Warhead Type	Fragment Type	Fragment Velocity	Kill Mode	Miscellaneous
Continuous Rod	Continuous Rod	4000–5000 fps	Cuts Surface Structure Members	• Padding required to ensure continuous rod – reduces explosive content • Narrow field • Low velocity fragments; week against high velocity target
Fragmentation	Small Fragments, 30–2000 Grain	6000–8000 fps	Penetration of Internal Components; Pilot, Fuel, Engine	• Wide spread of fragments, 60°–90° • High velocity fragments good against velocity target
Annular Blast Fragmentation (ABF)	Discrete Rods, Light-Weight	6000–8000 fps	Attack Structural Members	• Good against high velocity targets
Selectively Aimable Warhead (SAW) in Design Development	Discrete Rods	Very High Velocity in Aimed Direction >8000 fps	Attack Structural Members with Penetration Capatbility	• Requires target direction/aspect in fuzing

sequence to direct the force of the warhead explosion in the direction of the target. The sequence delay is a function of target closure to maximize damage. This warhead requires intelligence of the target's location. A summary of warhead characteristics is found in Table 10.7.

10.3.1 Air-to-Air Missile Test and Evaluation

A good starting point for the test and evaluation of air-to-air missiles can be found in MIL-HDBK-1763, "Aircraft/Stores Compatibility: Systems Engineering Data Requirements and Test Procedures," June 15, 1988. Do not be fooled by the title of the document, as it provides a lot more guidance than just stores compatibility. It applies to every activity of the system life cycle involving aircraft and stores, including development, manufacturing, deployment, operations, support, training, and disposal. The document is for guidance only and cannot be cited as a requirement. The handbook defines all the engineering data package requirements and provides standard test methodology, test instrumentation, and data from each test to establish the extent of aircraft/stores compatibility. There are four separate compatibility situations that must be considered when certification of a store on an aircraft is required:

- Adding inventory/stores (or stores configurations/flight envelopes not previously authorized) to the authorized stores list of inventory aircraft
- Adding inventory/stores (or stores configurations) to the authorized stores list of modified or new aircraft

- Adding new or modified stores (or stores configurations) to the authorized stores list of modified or new aircraft
- Adding new or modified stores (or stores configurations) to the authorized stores list of inventory aircraft

The handbook provides a good flow diagram for the certification process as well as a list of applicable documents. Section 3 of the handbook is a glossary, and sections 4 and 5 contain general and detailed test requirements. Appendix A contains detailed ground test procedures, and appendix B details flight test procedures. Appendix C describes the compatibility engineering data package (CEDP), which is a standardized list of reference data essential for determining the extent of aircraft/stores compatibility and is the source from which service technical publications are derived.

The tests described in the handbook comprise all of the tests necessary to ensure compatibility. As this text is concerned with systems integration, many of the tests described in the handbook are beyond the scope of this text (loads and flutter testing, for example) and are only mentioned here for completeness. It is best to consult with the experts if you are responsible for this type of testing. Some of the tests, particularly the lab and ground evaluations, will be accomplished by the engineering department and will have been accomplished long before you receive the missile for integration testing. The following discussion refers to missiles; bombs, and their related 1763 requirements, are addressed in a subsequent section.

The general objectives include

- Analytical study
- Laboratory tests
- Analytical simulation
- Hardware integrated simulation
- Captive carry tests
- Live missile launches

Analytical studies provide a review and understanding of the technical documents. They also determine the strong and weak points of the proposed system. An analytical study provides an assessment of the potential performance and defines the specifics for testing.

Laboratory tests provide a controlled open-loop laboratory environment for collection of specification data, threshold data, and basic subsystem performance.

Analytical simulation is a mathematical representation of the missile system/subsystems in closed-loop operation. A part of the analytical model is trajectory simulation, which is a nonlinear 5 or 6 degree of freedom (DOF) aerodynamic and kinematic model. It provides a detailed representation of the guidance/autopilot functions and a simplified representation of seeker and signal processing as well as a lethality simulation (probability of kill [P_K]).

Hardware integrated simulation is a closed-loop analytical/hardware representation of the missile system used to evaluate missile performance as affected by guidance, seeker, and signal processing.

The systems evaluator will be intimately involved in captive carry and live launch testing. Captive carry involves tests where the missile is carried aloft into representative environments (e.g., live targets, ECM, IRCM, decoys, chaff, flares, clutter, atmosphere, weather). Live missile launches are performed in combat representative environments:

- Clutter backgrounds (desert, sea, rural, cultural, look-up/look-down)
- Target maneuvering
- Multiple targets (RCS, aspect, speed/altitude)
- Weather
- Electronic attack (ECM, IRCM)

10.3.2 Air-to-Air Missile Lab and Ground Tests

These tests are described in detail in appendix A of MIL-HDBK-1763. Many, if not most, of these tests are accomplished long before the missile is accepted for aircraft integration. The tests are normally accomplished by the vendor or, in the case of software tests, by the aircraft integrator in the systems integration lab. Some of these tests include

- Seeker tests
- Fuse tests
- Warhead tests
- RCS tests
- Radome tests
- Trajectory simulations
- Launch acceptability region determination

Seeker tests come in many forms. They can involve laboratory tests with RF stimulation in an anechoic chamber or signals hardwired to the seeker. Field tests can be accomplished with ground-mounted target representations or ground-based seeker acquisition and tracking of airborne targets in a setup similar to radar roof house testing. An example of a laboratory test is shown in Figure 10.22. An RF signal generator is used

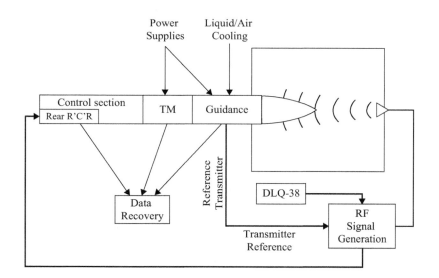

FIGURE 10.22 ■
Seeker Test Setup

FIGURE 10.23 ▪
Warhead
Characterization and
Lethality Test Setup

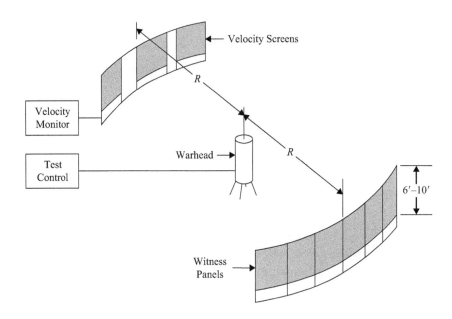

to generate targets to send toward the missile seeker head. The DLQ-38 in Figure 10.22 is a jammer that is used to test the missile's electronic protection capabilities. The ability of the seeker to acquire and track the generated target is monitored either in real time or postflight from the missile telemetry signal (a telemetry package is installed in the section where the warhead has been removed). Additional cooling may be required, as the missile will be powered up for much longer periods of time than it would experience operationally.

Fuse tests verify fuse design thresholds and assess fuse sensitivity. These tests determine fuse detection capability as a function of target type, size, and aspect. They also determine vulnerability to electronic attack (premature fusing or failure to arm).

Warhead tests are accomplished with two goals: characterization and lethality. Characterization tests determine the fragment pattern and velocity; lethality tests determine the damage mechanisms and P_K. A typical warhead characterization and lethality test setup is shown in Figure 10.23.

Radar cross-section measurements of missiles are accomplished the same way as for aircraft, which was previously described in chapter 8 of this text. The same can be said for radome refraction tests; it is important that the refractive errors be known and accounted for. It is even more critical with missiles, as the sensor FOV is much smaller than an aircraft radar.

Analytical trajectory simulations are used to determine launch envelopes (at required P_K); these envelopes go by many names, depending on who built the system: dynamic launch zone (DLZ), missile launch envelope (MLE), launch acceptability region (LAR), or no-escape envelope. These simulations are also used to investigate aerodynamic performance and look at the separation trajectory during employment/ launch and jettison.

Launch envelopes are provided to the crew on the heads-up display (HUD) or tactical situation display when a valid track file is selected and all SMS interlocks are satisfied. The envelopes are built based on own-ship and target attributes, missile

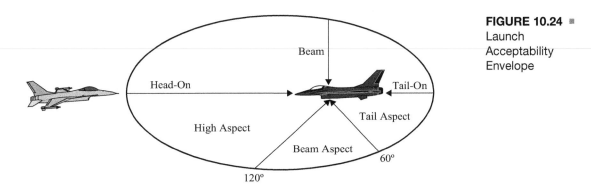

FIGURE 10.24
Launch
Acceptability
Envelope

performance parameters, and the desired P_K. The most common parameters used in the formulation of the envelope are

- Probability of kill (P_K)
- Missile internal electrical power supply, which is limited by either a thermal battery or gas grain generator
- Missile speed profile, which is based on average missile Mach for the missile time of flight, target range, and minimum missile airspeed or Mach
- Own-ship airspeed, altitude, and heading
- Target airspeed, altitude, heading, and aspect
- Missile end-game performance (g available)
- Minimum missile closure for fuse functioning
- Missile seeker head gimbal limits
- Actuator servo limitations

The SMS computes the launch envelope for the weapon selected and then displays it to the crew. The envelope itself looks much like a traffic collision avoidance system (TCAS) envelope. The envelope is dynamic and varies based on changes in the input parameters (Figure 10.24).

The analytical simulation addresses the separation region around the airplane for the first 1 to 3 sec after release/firing. The flow field around the airplane will change based on airspeed, altitude, angle of attack, weight, acceleration, and presence of stores on the airplane. The missile, when fired, may not behave as expected, based on changes in the aforementioned parameters. The loads, flutter, and separations group (engineering, flight testing, or both) has the responsibility to ensure safe separation is possible for the operating envelope of the airplane. The initial data modeled in the analytical simulation will come from wind tunnel results or analysis by similarity. The model will be refined based on flight test results. The end result of the analysis is an operational employment envelope (firing/launch as well as jettison) available for the missile (Figure 10.25). More will be said about these types of tests in the air-to-ground section of this chapter.

10.3.3 Air-to-Air Missile Launcher Qualification Tests

The delivery system of the missile will be either an ejection launcher or a rail launcher. If the missile to be integrated requires modifications to an existing launcher

FIGURE 10.25 ▪
Separation Region

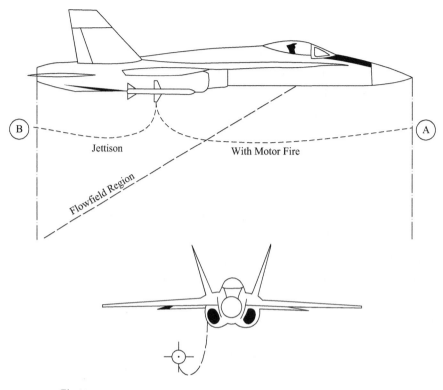

FIGURE 10.26 ▪
Ejection Launcher

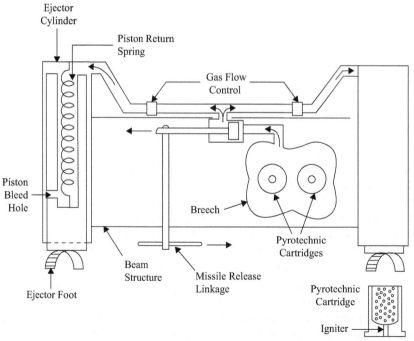

or the development of an adapter or new launcher entirely, a full series of tests similar to those of RTCA/DO-160G will be required; all other implementations will require a subset of the full testing. The two types of launchers are shown in Figures 10.26 and 10.27.

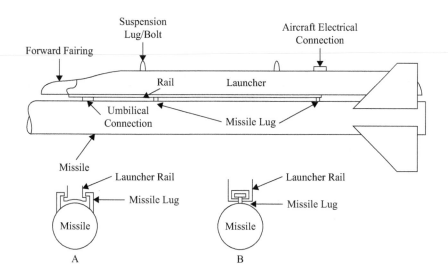

FIGURE 10.27 ■
Rail Launcher

There are five types of testing associated with launchers. The first set of testing comprises the qualification tests that are the responsibility of the vendor. The vendor is responsible for meeting the performance requirements set in the environmental standards:

- Temperature (high and low, shock and storage)
- High-g load
- Lock-shut firing (ejection launcher, cartridge ignition without missile release)
- Structural static loads (worst-case captive carry and landing loads)
- Dynamic loads (vibration, shock, and acoustic)
- Fatigue (long-term static loads program)
- Rain/moisture intrusion
- Humidity
- Salt spray
- Icing
- Fungus
- Sand and dust

The second set of testing is on-aircraft installation tests, which check compatibility, electrical connections, clearance, etc. The third test is called a "pit" test or sometimes a "catcher's mitt" test, where the launcher physically ejects the missile from the aircraft while the aircraft is on the ground. This is part of the safe separation tests and allows the program to progress to the fourth set of testing, airborne launch. The "pit" test is not accomplished with rail launchers, for obvious reasons. The final tests are the lab integration tests, which are an ongoing series of tests throughout the program development. The full series of ejection launcher test procedures are summarized in Table 10.8. The rail launcher testing matrix is similar to the ejection launcher matrix, with the exception of the "pit" tests, which are not required.

TABLE 10.8 ▪ Ejection Launcher Test Requirements

Parameters	Ejection Launcher Test Parameters				
	Tests				
	Qualification	Aircraft Install	Pit	Air Launch	Laboratory
1. Linear velocity imparted to the missile	×		×	×	×
2. Angular rate imparted to the missile	×		×	×	×
3. Missile trajectory	×		×	×	×
4. Ejector firing	×		×	×	×
5. Umbilical separation	×		×	×	×
6. Missile acceleration	×		×	×	×
7. Sequence of events	×		×	×	×
8. Foot pressure applied to the missile	×		×	×	×
9. Cartridge ignition	×		×		×
10. Gas residue effects	×		×	×	×
11. Maintenance analysis	×				×
12. Loading	×	×	×	×	
13. Unloading	×	×		×	
14. Access space	×	×	×	×	
15. Electrical connections		×	×	×	
16. Visibility of indicators	×	×	×	×	
17. Physical compatibility	×	×	×	×	
18. Handling		×	×	×	
19. Safety	×	×	×	×	
20. Electrical circuity performance	×	×	×	×	×

The length of time required to clear the aircraft for integration testing is a function of how many stations and combinations of launchers/missiles are desired. For example, the AMRAAM can be fired from either a rail or ejector launcher. There may be a requirement to carry the missile on two wing stations (rail launcher) and four belly stations (ejector launcher), for a total of six stations. Historical data show that just to clear the safe separation and jettison envelopes would require approximately 60 assets (10 separations per station). These tests do not include the "live fire" program, which is the responsibility of the systems flight test group.

Do not be lulled into believing that you can combine separations and systems testing to save time and money. The objectives for each of these tests are unique and are mutually exclusive of each other and therefore are not easily piggybacked with each other. The data requirements, geometries, targets, and special assets such as jammers are not the same for separations and systems testing. All too often, decisions are made early on to combine testing to save time and money. The end result is always the same: out of budget and out of time without a qualified system.

10.3.4 Air-to-Air Missile Reliability and Maintainability

Reliability and maintainability are normally an operational test and evaluation (OT&E) requirement, and the systems world rarely gets involved in this type of testing. Reliability is the probability that the missile will provide satisfactory performance for a specified period of time when used under certain conditions. The maximum time of

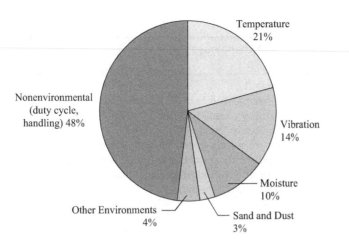

FIGURE 10.28 ■
Distribution of
Missile Service
Failures

consideration is the mean time between failure (MTBF) and the minimum time of consideration is the flight time of a typical operational mission. One would expect that the reliability would be very high for the short time of an operational mission and then steadily decrease out to the MTBF time. Typical numbers used in the United States are about 97% for the operational flight and 35% at the MTBF.

Service failures (failure of the missile to provide satisfactory performance) can be narrowed down to a few primary causes: nonenvironmental, temperature, vibration, moisture, sand, and dust. The pie chart in Figure 10.28 shows the typical spread of service-related failures. It should be noted that the primary cause of these failures is nonenvironmental, which is a combination of shipping, handling, and component failure.

Similar to the qualification testing, the reliability and maintenance group performs a series of reliability ground tests. These tests are designed to detect failures that may be resident in the missile. A breakdown of these tests and the method of evaluation is presented in Table 10.9.

A final test commonly used to determine reliability is the all-up-round test (AURT), which is designed to provide as complete a test as possible of the assembled, ready-to-use missile without activation of the rocket motor, fuses, and warhead. This test is performed at contractor facilities, depots, weapons stations, and test sites. The ideal goal is to detect 100% of all bad missiles and not reject any good missiles. In reality, quality figures for the AURT are

- Percentage of malfunctioning (bad) missiles detected is approximately 95%
- Percentage of functioning (good) missiles rejected is approximately 5%

A typical AURT stand is shown in Figure 10.29.

10.3.5 Air-to-Air Missile Systems Test

After the missile has been successfully cleared for carriage on our aircraft we can begin true systems integration testing. As has been said numerous times in this text, we are not testing the missile (unless, of course, it is a new design). We are testing the integration of the missile into the aircraft. Missile prep, inventory, tune, MOAT, BIT, and message sequencing are all tested as previously described under the SMS portion of this chapter.

TABLE 10.9 ■ Reliability Ground Tests

Environment	Failure Source Percent	Test Techniques
Acceleration, Captive Carry	<2	Centrifuge
Humidity	5	Combined Temperature, Altitude, and Humidity Chamber
Salt Fog	<2	Salt Fog Chamber
Shock, Captive Carry	<5	Shakers or Flight Test
Shock, Ignition	<5	Drop Test or Electro-Dynamic Shaker
Temperature, Altitude	40	Combined Temperature, Altitude, and Humidity Chamber
Temperature, Shock	<5	Soak at One Temperature and Place in a Chamber at Lower Temperature
Vibration, Captive Carry	40	Shaker and Acoustic
Vibration, Transportation	<2	Shakers
Altitude	<5	Altitude Chamber

FIGURE 10.29 ■
AURT Test Stand

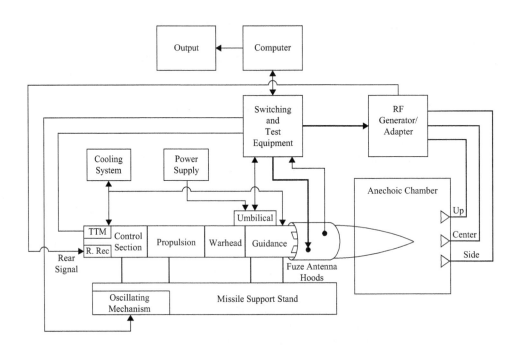

Further testing includes "captive" flights, where the missile stays with the airplane, and "live" fire, where the missile is operationally directed at a target.

Captive carry tests are used to ensure that the integration between the missile and aircraft systems has been accomplished successfully. For these tests, an inert, telemetry-equipped missile is used. In most test organizations, these inert test vehicles (ITV) are known as silverbirds and goldenbirds. They have empty warhead and propulsion sections and are modified to carry electronics, cooling, hydraulics, telemetry, and recording equipment that are required for captive test flight operations. These ITVs make the launch aircraft believe that the missile has been launched after a pilot launch command. Because of the telemetry on board the missile, analysts in real time or postflight can compare what was sent to the missile (aircraft telemetry) to what was received by the missile (missile telemetry).

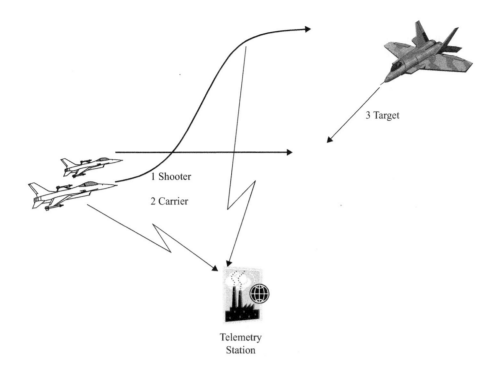

FIGURE 10.30 ■
Three-Plane Captive
Carry Geometry

The most common geometries used are called three-plane captive carry tests. In this scenario, one aircraft acts as the "shooter," one aircraft (carrying an ITV) acts as the missile, and one aircraft acts as the target (Figure 10.30).

At the initial setup, planes 1 and 2 detect plane 3 (target) and establish a track file (e.g., TWS, STT). Once the track file is valid, planes 1 and 2 select a missile for launch (this is the ITV on each airplane). Plane 1 will count down . . . 3, 2, 1, fire. Planes 1 and 2 both initiate an LTE and plane 2 turns off its radar and initiates a climb ahead and above of plane 1. Plane 1 continues to track the target and update the ITV on plane 2. Postlaunch missile messages issued by plane 1 are received by the ITV on plane 2 and the data are telemetered to the telemetry (TM) station. Messages that are sent by plane 1 as well as target information are sent to the TM station via the aircraft telemetry. All the information that is necessary to evaluate the performance of the missile integration is available in the TM station. Although this discussion is centered around air-to-air guided missiles, this geometry works for all missiles—AMRAAM, Walleye, SLAM, Harpoon—that have some type of postlaunch control or update.

The way the information is used is depicted in Figure 10.31.

Working from the top left box in Figure 10.31, the radar detects the target and telemeters target information (range, range rate, and azimuth). The time, space, position information (TSPI; see chapter 2) system (right top of Figure 10.31) knows where the target is and where your aircraft is so it can calculate a true range, range rate, and azimuth to the target. These values can be compared in real time or postflight and the accuracy of the radar can be determined. With a weapon selected, the SMS provides steering commands for the optimum engagement envelope as well as provides the aircrew with a LAR. On the ground, we can see the steering commands and the LAR envelope either by down-linked video of the displays or from bus traffic (left side of second and third boxes in Figure 10.31). Since we have our own-ship parameters

FIGURE 10.31 ▪
Data Analysis of a
Missile Launch

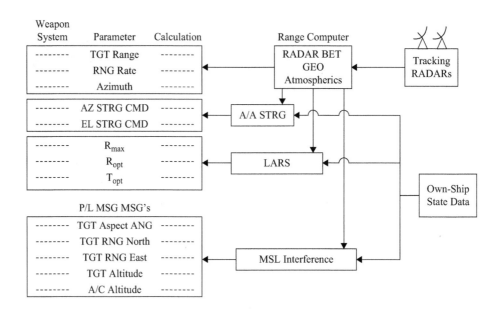

available on the ground (from the aircraft telemetry) and the own-ship parameters of the target (from TSPI), we can run them through the steering command and LAR algorithms and output the true steering and launch envelope. They, in turn, can be compared to the aircraft-generated data and steering and launch envelope accuracy can be determined. The final box in the figure is the postlaunch missile messages and commands. We know what was sent to the missile because this information comes to us in the aircraft telemetry. We also know what was received by the missile since that information comes to us via the missile telemetry. We can compare the two to identify any communications problems.

The test just described would be initially done under the most benign conditions: for example, target not maneuvering or jamming, firing at optimum range, in the heart of the firing envelope. Subsequent tests will measure the true missile–aircraft performance. Some modifications to the basic test include

- Continuous missile support (illumination, data link) during fire control computer master mode transitions (AA-NAV-AG)
- Continuous missile support during radar mode transitions (STT-TWS-SAM)
- Continuous missile support during master arm switch manipulations
- Continuous missile support during simultaneous multitarget engagement
- Frequency deconfliction (missile/radar)
- Data link interference
- Continuous missile support during periods of target fading, or after break lock
- Cluster target engagement
- Uncertainties (radar, navigation)
- Jamming targets

When the test team is sure that the missile–aircraft interface is correct (by validation of the captive program), it can progress to the "live fire" portion of the program.

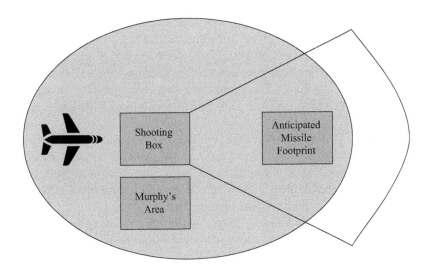

FIGURE 10.32 ■
Missile Safety
Footprint

The primary concern of any live fire program is the safety of flight. The test plan that you write for this event will be the most scrutinized plan that you will ever produce, and do not be surprised if it takes up to 6 months to get approval; so think ahead. The plan should be designed just as the captive program was; start with the most benign of launches and progress to the most difficult ones. Live fire will probably be accomplished by all of the testing agencies involved from contractor through OT&E.

The first consideration is the footprint of the missile. In most operations, the launch will be accomplished in a designated shooting box on a specified heading and the missile's maximum range can be plotted (Figure 10.32). But what happens if the missile loses a fin during the launch or if Murphy decides to ride along with you? The missile safety footprint is built on the notion that the missile can maneuver 360° around the desired heading. Based on the delivery conditions, the maximum range of the missile is plotted for every possible track the missile may take. The combination of desired missile track and possible missile track becomes the safety footprint for the launch.

It should be obvious that this footprint can become quite large when dealing with long-range missiles. While the along-track range for older AIM-9 missile variants may only be a few miles, an AIM-120 may include 50 to 60 mi. Very-long-range missiles may require a footprint on the order of hundreds of miles. This was certainly the case with the now retired AIM-54C Phoenix missile. Knowing the footprint tells us two things: where we can fire and how much area needs to be cleared. Missiles can only be fired in areas that are controlled and that can be cleared of intruders on the day of the launch. Edwards Air Force Base, in California, is a large flight test center for the U.S. Air Force, yet it does not control enough area to test even the smallest missiles. Flight test programs at Edwards must deploy to the large water ranges in order to perform live fire exercises.

And how do we clear the ranges of traffic on the day of the exercise? This is easier said than done in many cases. The safety office determines what percentage of traffic must be cleared prior to launch, and it may be 100%. This will require NOTAMS, notices to mariners, and maybe boats and aircraft to search the area and chase away unwanted guests. This operation may commence many hours before the actual launch time, and a clearance to launch will depend on their success.

Targets for the evaluation must match the characteristics of the target called out in the test plan. Targets will be drones that are either ground or air launched. Drones can be enhanced with lenses or flare pots, depending on what target it is trying to mimic. Most organizations try not to "kill" the drone but rather measure miss distance to the target without detonating a warhead. The drones are then recovered for reuse. Kills are determined by miss distance telemetry in the missile or via TSPI sources (normally cinetheodolites).

As with the ITV, the missiles used for live fire, especially for contractor and development test and evaluation (DT&E) testing, will have the warhead removed and a telemetry package installed in the missile. OT&E, concerned more with operational realism, will use full-up rounds without telemetry. The drone may also have some telemetry either for position or own-ship parameters.

You might be thinking to yourself, "This sounds like a lot of money!" It is. Live fire exercises require a lot of time and money in preparation, range clearance, and assets. Many organizations run a full dress rehearsal a day or two prior to the actual launch to ensure all aspects of the test have been addressed. This is sound thinking, but it effectively doubles the cost. Since these operations do cost a lot of money, you want to make sure that the probability of an abort is low. This is normally accomplished by redundancy: backup drone, backup shooter, and backup missile. Redundancy does indeed reduce the probability of an aborted mission due to a failure, but it introduces more assets, players, and once again more money. The operation is going to be some combination of all of these things, and close cooperation with program management is imperative.

Aside from a wayward missile, the biggest safety concern during live fire is communications. There are many players involved in the exercise: area clearance, range safety, test conductor, TM station, TSPI site, drone control, ground control intercept (GCI), shooter, backup shooter, safety chase, and photo chase. All of these players have communications and are an integral part to the success of the mission. If radio discipline is not exercised, the likelihood of a mishap is significantly increased. Suppose a "clear to launch" call is given to the drone delivery aircraft and is misinterpreted by the shooter as a "clear to launch" his missile. Shooting down the drone delivery vehicle would certainly ruin someone's day. The entire exercise must be scripted, with distinct calls made for each event along a timeline. That way everyone can follow along and know exactly the status of that timeline.

The amount of data acquired during these evaluations is enormous, but more importantly, it will come from many different asynchronous sources. If you remember back to chapter 2, asynchronous implies that the data arrive at different times and therefore must be time aligned and merged. The data station that is accustomed to handling three or four streams of telemetry data is now faced with time aligning and merging five or six streams, and it must be done in real time. It must be done in real time because an abort must be called if an analyst detects a problem with the missile, the aircraft, or a sensor.

Because of the complexity and cost of live fire exercises, they are usually fairly small in number. They are not small in duration and cost for the reasons previously mentioned. Table 10.10 provides some historical data on previous missile live fire programs. New missile programs typically average about 90 live fires over the life of the program, whereas modifications entail a slightly lower number.

Although there will be many other things going on, the time to start working on the live fire plan is about 1 year before the proposed start date. The individuals assigned to

TABLE 10.10 ■ System Missile Launches

System		Contractor Demonstration Test (CDT)	U.S. Navy Test and Evaluation/Joint Test and Evaluation (JTE)	Operational Test and Evaluation (OPEVAL)	Total
AIM-9L Sidewinder		15[1]	20	30[1]	65
AIM-7F Sparrow		6[2]	36[2]	1–25[3] 11–25	92
AIM-54A Phoenix		63	29	30	122
AGM-84 Harpoon		26	18	24	68
Skyflash Aspide		12	11	5	28
F-15/AIM-7F		7	15	34	56
F-14/AIM-7F Active Fuse			15	6	21
AIM-7M Sparrow	Prototype 5[4]	9 (13)[6]	29 (38)[6]	35 (45)[6]	96
AIM-120 AMRAAM	Valid 20[5]	10 (13)[6]	42 (55)[6]	22 (29)[6]	97

[1] Approximate values.
[2] Twenty tests originally planned for a combined CDT/NTE.
[3] Negative results of initial OPEVAL resulted in a second OPEVAL.
[4] Two tests per contractor.
[5] A total of 16 assets per contractor originally planned but reduced to 10 assets per contractor during validation.
[6] Test (planned) assets based on estimated achievement ratio of approximately 1.3.

coordinate this effort should have no other responsibilities during this time frame, should be able to work with program management, and should have a lot of patience.

10.4 | AIR-TO-GROUND WEAPONS

For air-to-ground weapons evaluation, especially ballistics testing and accuracy, I recommend AGARD-AG-300, volume 10, "Weapons Delivery Analysis and Ballistic Flight Testing," July 1992. This AGARD document provides an historical look at computing ballistics errors and measuring accuracies of unguided weapons. It provides an excellent baseline for the evaluator in how to go about measuring errors within the system and how to account for them in the final delivery. Although the document is based on the method of test by the 3246th Test Wing at Eglin Air Force Base, in Florida, appendix A contains a list of questions regarding how ballistics analysis and testing are performed in other nations; responses are provided by Canada, Germany, and France.

Appendix B of MIL-HDBK-1763, "Aircraft/Stores Compatibility: Systems Engineering Data Requirements and Test Procedures," June 15, 1998, provides flight test procedures for the following:

- Ballistic tests (test 290)
- Weapon free-stream ballistics test (test 291)
- Operational flight program (OFP) ballistics evaluation test (test 292)
- Separation effects derivation test (test 293)
- Operational flight program (OFP) ballistics verification test (test 294)

FIGURE 10.33 ■
Vertical Plane Bomb
Trajectory

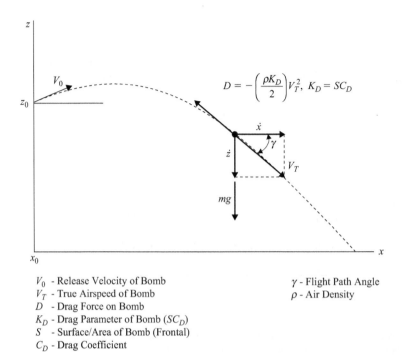

$$D = -\left(\frac{\rho K_D}{2}\right)V_T^2, \quad K_D = SC_D$$

V_0 - Release Velocity of Bomb
V_T - True Airspeed of Bomb
D - Drag Force on Bomb
K_D - Drag Parameter of Bomb (SC_D)
S - Surface/Area of Bomb (Frontal)
C_D - Drag Coefficient

γ - Flight Path Angle
ρ - Air Density

The purpose of this section is not to repeat the excellent work already done in the referenced documents, but to look at and try to understand the basic theory of air-to-ground weapons delivery, know the inputs required by the weapons system computer to solve the delivery problem, and be familiar with the process used in air-to-ground testing. In the end, it is desired that the reader understand the concept of error budget analysis.

The basic problems in air-to-ground weapons delivery are threefold. The aircraft delivery system must determine the position of the target, predict the trajectory of the weapon (missile, bomb, bullet, or rocket) and provide guidance, and release the weapon in such a way that its trajectory coincides with the target position. In order to make the problem as simple as possible we can assume a few things. We can assume that there are no normal forces (i.e., no angle of attack or sideslip) and that there are negligible gyroscopic forces (unlike a bullet). We can also assume, due to the low ejection velocity, that there is no motion with respect to the air mass in the vertical plane and nonaerodynamic forces are only due to gravity. With these assumptions we can plot the trajectory of a bomb, for example, in the vertical and horizontal planes (Figure 10.33 and Figure 10.34).

The drag parameter of the bomb (K_D) is calculated separately for each store and can be different with changes in the fin assemblies. These are known as low-drag and high-drag configurations. A representative diagram of the drag parameter versus Mach for a low-drag and high-drag configuration is shown in Figure 10.35.

The bomb equations of motion come directly from Newton. We know that mass times the acceleration equals the sum of the applied forces, therefore mass times acceleration in the x plane equals the sum of forces in the x plane, and mass times acceleration in the z plane equals the sum of forces in the z plane. The mathematical derivation of the resulting equations of motion is not important to this discussion. The equations (since they represent acceleration) are integrated once to get velocity. The velocity of the bomb is added to the velocity of the air mass; this resultant velocity (with respect to the earth) is integrated to compute trajectory (position). This process can

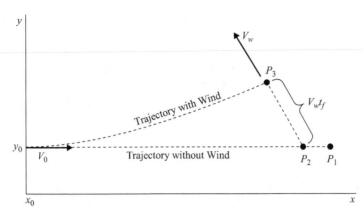

FIGURE 10.34 ▪
Horizontal Plane
Bomb Trajectory

P_1 = Impact Point in Vacuum
P_2 = Impact Point without Wind
P_3 = Impact Point with Wind
t_f = Time of Flight of Bomb
V_0 = Release Velocity of Bomb
V_w = Wind Velocity
x = Range Displacement
y = Cross-Range Displacement

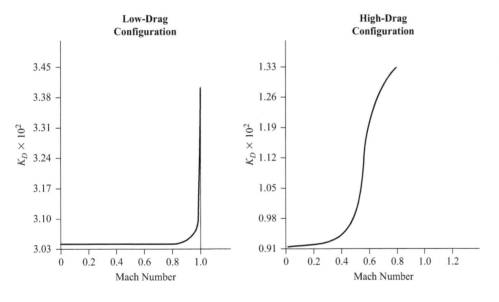

FIGURE 10.35 ▪
Bomb Drag
Parameter (K_D) for
the Mk 81 Bomb

be calculated in real time in the weapons computer and repeated until the computed bomb position coincides with the target position at the time of impact.

In air-to-ground gunnery, the equations of motion are nearly the same as for bombs, with the following exceptions:

- Large release velocity gives much a shorter time of flight and a much different drag profile

- Reduced importance of aircraft velocity

- Gyroscopic effects are not negligible but are unpredictable and are included in nominal dispersion

At any rate, the differential equations are integrated to determine the trajectory exactly as with bombs.

For rockets, the equations of motion differ from bombs and bullets because of a time-varying force (thrust) along the longitudinal axis, and the mass of the rocket is also changing with the depletion of propellant. Significant normal (nonlongitudinal) forces exist, but like the gyroscopic forces on a bullet they are included in the dispersion. Trajectory is calculated the same as with bombs and bullets.

10.4.1 Air-to-Ground Release Geometry

We know the forces that act upon a store as it is released from an aircraft, but how do they affect what the pilot will visualize? Figure 10.36 shows the geometry for one given set of conditions. That is, that an aircraft at a given range to the target at a known velocity and one set of wind conditions, at a given altitude, known groundspeed, track, aim-off-distance, etc., which releases a store will have an impact point as shown in the figure. Any changes to the parameters without compensation elsewhere will lead to the bomb impacting at a different point. In the predigital days (e.g., F-4, A-7), pilots used fixed sights to deliver weapons. The delivery conditions were prebriefed, and the pilot made every attempt to meet the delivery conditions as closely as possible to get an accurate bomb impact. Pilots learned to compensate based on the conditions at hand. If the pilot was too fast on the delivery he would release early because he knew that being fast would result in a long bomb. He would adjust the sight depression for head- or tailwinds and offset the airplane into the wind for crosswinds. Today a computer does the compensation and tells the pilot where the bomb will hit, but input errors to the calculations will still produce a bad bomb. The evaluator needs to understand the mechanics of the trajectory to troubleshoot bad bomb circular error probable (CEP), even with an automated system.

Manual weapons delivery is simply the procedures and techniques that allow the pilot to compute a weapons release point without the aid of a computer. In manual weapons delivery, the pilot uses a limited set of cues to first determine how to initially aim his plane at the target and then determine the actual release point. To do this, the

FIGURE 10.36 ▪
Release Geometry

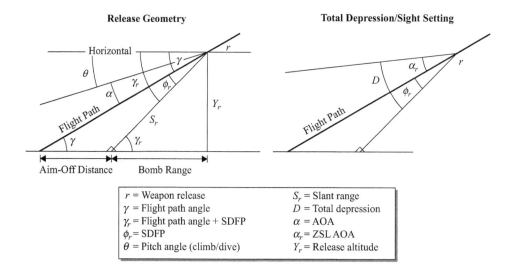

r = Weapon release	S_r = Slant range
γ = Flight path angle	D = Total depression
γ_r = Flight path angle + SDFP	α = AOA
ϕ_r = SDFP	α_r = ZSL AOA
θ = Pitch angle (climb/dive)	Y_r = Release altitude

pilot uses two sets of cues: the gun sight aiming reticle (pipper) and the aircraft instruments (airspeed, dive angle, attitude, and altitude). To understand manual bombing, the pilot has to understand the basics of weapons ballistics.

Every weapon that is dropped from a plane is affected by gravity; when the weapon is released, it moves forward and down. During weapons delivery, a pilot selects an aim-off point at some distance beyond the target; if the pilot were to maintain a constant dive angle, the aircraft would impact the ground at this point. If a weapon is dropped or fired from this plane, it will never impact at the aim-off point, but at some distance short of this point. A gun sight depression angle (mil setting) is used to advise the pilot of where the real impact point will be (as long as all other flight parameters are correct).

In looking at other parameters of the delivery equation, we can see the effects of not releasing in the correct flight conditions:

- Altitude (using sight picture): higher than planned will result in a shorter bomb, lower than planned will result in a longer bomb (i.e., higher than planned gives early sight picture on target and results in a short bomb; lower than planned gives late sight picture on target and results in a long bomb)

- Altitude (manual bombing, i.e., no sight picture and relying solely on proper release parameters/conditions): higher than planned results in a long bomb, lower than planned results in a short bomb

- Airspeed: higher than planned gives a longer bomb, slower than planned gives a shorter bomb

- Roll: right wing down throws the bomb to the right, left wing down throws the bomb to the left

- Yaw: the airplane track needs to cross the target

- Pitch: a shallow dive angle gives a shorter bomb, a steep dive angle gives a longer bomb (see the explanation for altitude; a shallow dive angle gives an early sight picture)

- g: a negative g gives a longer bomb, a positive g gives a shorter bomb

- Range: a shorter slant range gives a longer bomb, a longer slant range give a shorter bomb

- Winds: wind tends to affect lighter and higher drag weapons more adversely than heavier and lower drag weapons; a tailwind gives you a longer bomb, a head wind gives a shorter bomb

- Head add/tail subtract: for a headwind you add to the depression angle, for a tailwind you subtract from the mil setting; a nominal 1 mil/knot is commonly used

- Crosswind: you need to offset the aim-off point into the wind; a nominal 10 ft/knot offset is commonly used; you can also just "fly your butt over the target" (crab)

In modern weapons delivery systems, the computations for release are made in the weapons release computer or in the stores management processor. The computer uses own-ship inputs such as airspeed, altitude, and attitude; an accurate measure of range to the target (radar, laser range finder [LRF], and GNSS); and ballistics information to compute where the weapon will impact. Other information the computer will need is the fudge factor determined during separation effects testing, which will be discussed later.

The continuous computing impact point (CCIP) is an automatic bombing mode wherein the computer continuously solves the bomb's free-fall equations and displays in the HUD the predicted impact point. Thus the pilot simply "flies" the pipper to the target and depresses the bomb release (pickle) button. To make the calculations, the computer needs to know the aircraft's speed, dive angle, and height above target, and the bomb's drag coefficients.

Once the CCIP is selected, a bomb fall line (BFL) is drawn between the flight path marker and the pipper. The pilot's job is to fly the CCIP pipper across the target. When the pipper is coincident with the target, the pilot depresses and holds the pickle button for release. Normally, at pickle button depression the bomb is released immediately. However, if the instantaneous bomb range is shorter than can be displayed in the HUD (bomb impact point below the nose), the CCIP pipper is actually a pseudo-CCIP pipper and the weapon system will automatically enter a delayed continuous computed impact point (DCCIP) mode and a delayed bomb release cue (horizontal bar) will appear on the BFL. Assuming the pilot continues to keep the pickle button depressed (providing release consent), when the horizontal bar reaches the CCIP pipper, the flight path marker flashes and the bomb is released. Between pickle and release the pilot must follow the horizontal steering guidance by keeping the flight path marker on the postdesignate BFL.

There are other factors that can affect the outcome of any weapons delivery:

- Bomb physical properties
- Atmospheric conditions
- Bomb separation effects

The bomb physical properties are the differences in mass and weight, and therefore "ballistics," in bomb-to-bomb comparisons. Atmospheric conditions include unknown winds, shear, and pressure gradients. The store is usually considered to be free of separation effects (aircraft influence) 1 to 3 sec after release when it reaches steady-state conditions. What happens during the first 3 sec identifies the true trajectory of the store. AGARD-AG-300, volume 10, presents a concise summary of data grouped by each of the four error sources (the three just noted as well as errors resulting from aircraft release conditions). The table is repeated here (Table 10.11) for completeness.

As can be seen from the table, the two largest contributors to errors are release conditions and separation effects. Release conditions for the most part are now handled in software, and there is better quality control in the fabrication of weapons. Separation effects need to be determined by the joint test team, so they too may be accounted for in software.

Separation effects need to be accounted for in manual as well as automated weapons deliveries. There are two ways of determining separation effects of weapons: statistical

TABLE 10.11 ▪ Predicted Error Sources for Mk 82 Low-Drag Deliveries (TAS/Altitude/Dive Angle)

Error Source/Test Conditions	450 knots/ 5000 ft/Level	360 knots/ 5000 ft/Level	450 knots/ 8000 ft/45°	360 knots/ 8000 ft/45°
Aircraft Release Conditions	57%	40%	53%	54%
Bomb Physical Properties	10%	30%	7%	5%
Atmospheric Conditions	2%	8%	1%	2%
Separation Effects	31%	22%	39%	39%

and analytical. If an infinite number of bombs were dropped while trying to hold all other release conditions the same, the CEP would identify the separation effect on the weapon. This method is sometimes called the "offset correction method." This method is pooh-poohed by purists, as it could take quite a bit of time and many weapons deliveries.

The other method is determined analytically by first observing the separation effects in wind tunnel testing where the initial corrections are made. Follow-on flight separations using high-speed stores separation cameras and TSPI (normally cinetheodolites) tracking from release to impact will determine true trajectories, and adjustments to the release algorithm or manual depression settings can be made. This method is detailed in MIL-HDBK-1763 test 293; the test describes the requirements, assets, data collection, and analysis necessary to obtain separation effects.

It is important to note that these tests would have to be carried out with all weapons loads and aircraft configurations, as the separation effects will change. Once again, the reader is referred to AGARD-AG-300, volume 10, which is covered in section 10.6.

10.4.2 Air-to-Ground Sources of Errors

The first source of error is weapon boresighting, or where the weapon is looking. Correct weapon boresighting is critical for guns, rockets, and missiles and less critical for free-fall (nonpropelled) bomb deliveries. But boresight errors may affect weapon separation. Abnormal weapon separation might cause a flight safety hazard or affect the bomb range. Weapon boresighting is verified on the ground (carriage installation) and modeled for flight. The weapon boresight may change during differing flight conditions. Wings may twist in certain flight regimes, which may affect the boresight of wingtip-mounted munitions; boresight changes in flight are accounted for in the OFP software.

The second source of error is the system boresighting. By revisiting Figure 10.38, we can see that a failure of any of the systems onboard the aircraft to provide accurate range, altitude, depression angle, attitude, and groundspeed will produce an error in the impact point. System boresighting is performed in order to optimize weapon delivery accuracy by

- Minimizing pilot aiming errors
- Minimizing sensor LOS errors
- Minimizing sensor ranging errors
- Minimizing postdesignate aim-point drift

Pilot aiming errors can be caused by a number of things. The first is the design eye point (DEP) in the cockpit. The DEP has been addressed several times already in this text. One can imagine that a pilot who is sitting too high in the cockpit will have a later sight picture and the bomb will go long; the reverse is true for a pilot who is sitting lower than the DEP. Any parallax errors in the HUD will also cause aiming errors; the true world may not be what the pilot sees through the HUD. Parallax errors may also show up in the HUD video recording system. Since the HUD video is a prime tool in data analysis and error budgeting, it is imperative that the evaluator determine any differences between the recorded HUD picture and the picture that the pilot sees. This is accomplished by comparing the HUD video to the verbal pilot feedback during flight testing.

FIGURE 10.37 ■
F-15 References to
Boresight

Reference lines:
A - Gun cross line, zero sightline
B - Water line (WL), Fuselage reference line
     (FRL), Radar boresight line
C - Aim-9 missile boresight line
D - Aircraft flight path
E - Depressed pipper sightline

Angles:
1 - Angle of attack
2 - Gun cross elevation angle
     (2° or 35 Mils above WL)
3 - Pipper depression angle from zero sightline
4 - Pipper depression angle from flight path
5 - Aim-9 weapon depression angle
     (0.5° or 9 Mils)

FIGURE 10.38 ■
F-16 References to
Boresight

Reference lines:
A - Gun cross line, zero sightline
B - Water line (WL), Fuselage reference line
     (FRL), Radar boresight line
C - Aim-9 missile boresight line
D - Aircraft flight path
E - Depressed pipper sightline

Angles:
1 - Angle of attack
2 - Gun cross elevation angle
     (2° or 35 Mils above WL)
3 - Pipper depression angle from zero sightline
4 - Pipper depression angle from flight path
5 - Aim-9 weapon depression angle
     (0.5° or 9 Mils)

Sensor errors due to incorrect boresighting (or incorrect parallax correction within the OFP) will also cause impact errors. The onboard radar may be used to calculate the range to the target and the height of the target. Range is accomplished by pulse delay operations, and determination of height is obtained by use of the radar depression angle. If the radar is not boresighted to the "system" (everyone is referenced to the same point in space), errors in range and height of the target will occur. This is demonstrated in Figures 10.37 and 10.38, which depict the references to boresight in the F-15 and F-16, respectively. You will notice that in both cases the fuselage reference line (FRL), DEP, and radar bore line are all aligned. In the case of the F-15, the AIM-9 bore line is offset from the FRL, and the same is true for the centerline and wing stations on the F-16. These are parallax errors, and are corrected for in software to allow the stores, sensors, and pilot to look at the same point in space.

Like the radar, a laser is used to calculate range to the target and altitude. The pointing angle is critical in determining these measurements. Similarly, the laser "spot" must be shown correctly on the HUD for pilot awareness. An incorrectly displayed laser spot will be corrected by the pilot, resulting in an impact error. Other electro-optical (EO) systems such as TV and FLIR provide the weapons solution with azimuth and elevation (pointing angle) of the target position. An error in this input will slave the ranging device (laser or radar) to the wrong position, again resulting in range and altitude errors. Even small angular errors can contribute to large errors in the weapon impact point if released at long ranges (angular error subtends more range at greater distances).

A targeting pod may house the EO systems (laser, FLIR, TV), and incorrect boresighting of the pod or incorporation of the parallax corrections will contribute to a bad bomb. Not thought of normally as a weapon's sensor, the INS provides key inputs to the weapons release algorithms. Groundspeed (cross-track and along-track), attitude (pitch, roll, and yaw), g (acceleration in three axes), and rates are all provided by the INS. You know from the discussions in chapter 5 of this text that the INS is only as effective as its alignment to each of the three axes. Although this has become much easier in recent years with GPS/INS integrated systems, where the INS can align from the GPS, there are basic mechanical problems that can contribute to gross errors.

There have been documented instances where the INS was not installed properly and the case did not seat properly in its compartment. Imagine a box that is not square with the corners but rather "rides up" against the compartment wall. The INS, because of its mechanical alignment, will induce an error equal to the angle of the case. Also, remember that the vector velocities and accelerations are also formulated from these angles.

A second item worth noting regarding the INS (and, in truth, the aircraft navigation system) is the possible mismatch between cursor placement and aircraft position. This is a case where the pilot places the cursor over a target such as a HUD, FLIR, or radar, and the target position coordinates (obtained from the navigation system) are fed to the weapons release computers. If the sensor presentation is not perfectly aligned with the navigation system, an error in position will be sent to the processors, resulting in a poor delivery. An easy test is to designate known target locations with the sensor systems and read on the data bus the coordinates that are being sent to the weapons release computers.

The boresighting process must be initiated as part of the flight test preparation period. The evaluator needs to account for all possible errors and endeavor to minimize their impact on the ballistic accuracy tests. In many cases, the errors are known and are accounted for in the weapons release algorithms in the OFP. The evaluator needs to know which "fudge factors" are modeled into the software; yet another instance where the evaluator must be as smart as the engineer that designed the system. Only after you have a firm grasp of potential problems should the test team embark on weapons delivery flight testing.

The final contributor to errors is the pilot. In manual weapons delivery profiles, the pilot needs to meet the delivery parameters exactly or the weapon will not impact at the desired location. In a computerized delivery, the software will compensate for not being on the exact parameters, but the pilot still needs to stabilize on parameters in order to allow the processor to perform its job. The weapons release computers use information provided by the sensors; if the delivery is dynamic rather than completely stabilized, the fire control solutions will still be in error due to system lag.

10.4.3 Minimum Release Altitude

When evaluating the live fire of missiles we considered the safety footprint of the missile. With air-to-ground weapons we concern ourselves with a minimum release altitude (MRA). The MRA must be determined before the release of live weapons. The MRA is defined as the greatest of the following:

- Altitude lost during pullout plus some safety buffer
- Minimum altitude required for safe escape
- Vertical drop of the weapon required for fuse arming

The altitude lost during pullout is a function of

- Gross weight
- True airspeed
- Dive angle
- Pullout load factor (g)

The safety office must take into consideration

- Delivery stick length
- Pilot response
- Delay (1 to 2 sec)
- AOA limiters
- Aircraft G limits

The safety buffer that is applied is generally given as a function of release altitude, release speed, and dive angle.

The aircraft flight path has to remain outside the weapon fragmentation envelope during weapon delivery. Weapon fragment travel charts (1/1000), which depict vertical and horizontal fragment travel range and the frag full up/frag full down time (time it takes for the maximum vertical travel of the bomb fragments), are used to compute the envelope. Fragment travel is a function of the type of weapon and warhead, the target density altitude, and the impact angle. Safe escape should take all possible weapons failure modes into consideration.

The weapon has to be armed before fuse functioning criteria are reached (impact, proximity, etc.). A minimum fuse arming time is required to allow safe separation of the weapon from the aircraft during release (bird hits, bomb-to-bomb collisions, etc.). Low-drag munitions generally have high minimum arming delays (4 to 6 sec minimum). The vertical drop required for fuse arming is a function of the weapon/fuse combination, release altitude, release speed, and dive angle.

10.5 | MIL-HDBK-1763 TEST REQUIREMENTS

An integral part of validating the aircraft–stores interface is the assurance that the weapon, when employed, will hit the target. MIL-HDBK-1763 addresses this validation as ballistics tests. The handbook is for guidance only; it cannot be cited as a requirement. If it is cited as a requirement, the contractor does not have to comply. The handbook applies to all Agencies within the U.S. DoD and Industry with determining the extent of compatibility between military aircraft and aircraft stores.

Four separate compatibility situations must be considered when certification of a store on an airplane is required:

- Adding inventory/stores (and/or configurations/flight envelopes not previously authorized) to an inventory A/C
- Adding inventory/stores (and/or configurations) to the authorized stores list of modified or new aircraft
- Adding new or modified stores (configurations) to the authorized stores list of modified or new aircraft
- Adding new or modified stores (configurations) to the authorized stores list of inventory aircraft

The certification process is not an easy one; the average duration for the aircraft/store certification process lasts just over 3 years. As with all of the military and civilian guidance there is a long list of reference material provided to the reader: for example, Mil-B-81006B, which is the demonstration of dispersion requirements for free-fall bombs, is noted (this spec will be covered later and is mandatory). Another disclaimer in the document states that in the event of a conflict between the text of the document and the references cited the text takes precedence. Nothing in the document, however, can supersede applicable laws and regulations unless a waiver has been obtained. The document provides a glossary of terms and notes that there are some differences between USAF and USA terminology, such as store certification versus airworthiness qualification.

The handbook is divided into two sections: general requirements, and detailed requirements, which are further broken down into ground and flight requirements. The procedures for each of the ground and flight tests are defined in appendices A and B of the handbook; appendix C describes the engineering data package, which must be submitted for stores certification consideration. Many of the general and specific requirements are covered by engineering, loads, flutter and P&FQ groups within the test organization before the aircraft and store arrive at the systems group for evaluation; this can be different from organization to organization. Tests that are nonspecific to systems will only be treated as an overview in this section.

10.5.1 MIL-HDBK-1763 General Test Requirements

Physical compatibility review is the first consideration and looks at the store to see if it can be interfaced properly:

- Mechanical compatibility (e.g., suspension equipment, connectors, ejectors, lug spacing)
- Electrical compatibility (e.g., power, pin assignments, wire size, routing, antenna requirements)
- Optical compatibility (e.g., wavelength, line run, fiber optic size)

An aid in this evaluation is the Aircraft Stores Interface Manual (ASIM), which is owned by Naval Air Systems Command in PAX River, Maryland.

Structural analysis is an engineering function and a detailed analysis is required if the store is not similar enough in nature to other previously qualified stores. This analysis is covered in Mil-HDBK-244A, "Guide to Aircraft/Stores Compatibility."

An analysis of the structural integrity of the store itself as well as the loading effects on the aircraft and the suspension equipment must be undertaken. Myriad effects may have to be analyzed such as strength envelopes (maneuvering, takeoff, and landing), change of CG due to sloshing liquid, fatigue, ground vibration, and dynamic hang-fire loads.

Aerodynamic analysis is also in the purview of engineering, and their greatest concern, other than loads defined previously, is the change of performance and handling qualities of the carrying aircraft. In every contract that's adding a store to the aircraft, it will state in the ground rules and assumptions that "the addition of the store will have no impact on the performance of the aircraft." This, of course, is rarely the case. Changes in the performance and handling qualities could be influenced by weight and CG changes, changes in lift or increases in drag, asymmetric loads or changes with gust conditions, and the location of the store with relation to the engine inlet or loss of trim function. The United States requires the reader to use Mil-A-8860B (AS), "General Specification for Airplane Strength and Rigidity."

Aeroelasticity is really the study of interactions on the airplane between aerodynamic, inertial and elastic forces. With the addition of a new store, it would be important to see how that store reacts with other stores and flow fields on the airplane. Aeroelastic consideration looks at multiple aircraft configurations (with the new store) and their impact on the aeroelastic stability of the entire system, including freedom from flutter. Multiple load-outs would be evaluated during TO and landings, maneuvering and cruise, throughout the operating envelope of the aircraft. Since this would entail multiple aircraft configurations, comparing the new configurations to similar configurations previously cleared provides a low cost method for a preliminary analysis.

One of the most important pre-flight analyses is that of determining the separation characteristics of a store, because it affects aircraft safety and weapon accuracy. Proper use of predictive methods for store separation enhances the safety of store delivery and jettison during flight testing. Aircraft-store certification flight test programs can be reduced in scope and cost by reducing full-scale based on analysis and positive correlation of predictions with the flight test data (test-evaluate-analyze). Environmental analysis of the store, suspension equipment, and any other affected aircraft components should be qualified to the limits in their respective specifications. These limits may be substantiated by the procedures described in Mil-Std-810G or RTCA/DO-160G.

There is a need to analyze the effects of gas and other by-products produced by thrust-augmented stores. The analysis is concerned with two major effects of these stores:

– The ingestion of the plume (expended gas and suspended foreign material) by the aircraft engine(s)
– The impingement of the plume on the aircraft surfaces or other stores

Engine ingestion analysis is intended to address the gas plume and any foreign material suspended within the plume. In a perfect world, the engine would not ingest any of these gases or by-products. The analysis looks at the maximum amount of gases that may be ingested at varying flight conditions (e.g., altitude, attitude, A/S) and seeks to find which of these cases will cause detrimental effects on the aircraft such as flameouts, overspeed, and non–self-recoverable surge stalls. After the store has been satisfied with respect to the engine ingestion problems, an analysis must be made of the impingement effects of the gas plume on the aircraft or on other stores. Factors that should be considered are:

– Residue build-up and its chemical composition

– Corrosion

– Blast overpressure

– Shock wave

– Thermal effects

– Erosion due to the velocity of the exhaust by-products

Similar to exhaust gas impingement an analysis must also be performed on the effects of gun gas. Gunfire generates large volumes of gun gas, which is generated at a rapid rate and is a direct function of the propellant per round, firing rate, and burst length; gun gas also becomes flammable and explosive when mixed with the outside atmosphere. The effects considered during exhaust gas impingement are also considered for gun gas impingement.

An analysis of the ballistic behavior of a store in the presence of separation effects must be accomplished if accuracy requirements are to be achieved and is assessed for two store flight regimes. The first is the free-stream ballistics, which are independent of the releasing aircraft and occur after initial store separation disturbances have decayed. The second is for the entire store flight trajectory and is characteristic of the weapon, the releasing aircraft, specific aircraft configuration, and release conditions. Ballistics validation is accomplished in one of two ways. Undisturbed trajectories can be analyzed to calculate a ballistic drag coefficient and a ballistic dispersion (CEP); the CEP must be less than the allowable maximum compatible with an estimate of system accuracy desired for the worst anticipated delivery system. If ballistics and CEP of the store is not known it will be necessary to conduct and analyze drops to establish aircraft independent ballistics; if such drops are required, Mil-B-81006 should be employed to minimize the number of drops required.

The assessment of ballistic accuracy for the store/aircraft release system requires a much larger envelope of release conditions and configurations than is required to establish aircraft independent ballistics. Maximum ballistic CEP values for each weapon/aircraft combination should be assigned based on P_K required, system accuracy, and the ratio of ballistic error to system error. Free-flight drop tests will determine what limitations must be placed on aircraft release envelopes and store loading configurations. If employed, an analysis of missile, rocket, and gun accuracy should be conducted.

The effects of missile system errors (background clutter, bore-sighting, installation, own-ship, and target parameters) should be included in the analysis. For rockets, bore-sight effect and system errors should be included in the analysis. The error budget and prediction should address subsystems that interface with the missile system that may affect hardware and software performance.

1763 addresses the variables affecting gun performance as system errors and target errors. System errors are ammo dispersion; gun-pointing error; fire control prediction error; aircraft position, orientation, speed, and acceleration errors; and any other errors like boresight, target track, and range. Target errors include target type, maneuvers, and range.

Safe escape analyses are required on all munitions systems that could potentially be hazardous to the aircraft due to munitions fragmentation. For the USAF, the analysis is conducted IAW methodology described in the Advanced Safe Escape Program (ASEP), for the USN, IAW the Probability of Aircraft Tactical Hazard (PATH) manuals, and for the USA, IAW with AIR STD 20/21 for the worst case within the predicted launch envelope.

EMI, EMC, EMP, and lightning analyses are conducted IAW Mil-Std-464A.

10.5.2 MIL-HDBK-1763 Detailed Test Requirements

Stores certification tests are conducted for two reasons. The first is to validate the results of the analysis; obviously, if the test results compare favorably to the predictions, the test program can be significantly shortened. The second reason is to provide data where no qualitative prediction can be made; data must be examined to determine the acceptability of the aircraft-store configuration that is being evaluated. Ground and flight tests for store certification are contained in appendices A and B of 1763. When determining the extent of testing required, it is important to consult with the using agency to determine the mission requirements:

– Airspeed and altitude
– Acceleration for carriage, employment, and jettison
– Range
– Delivery modes
– Delivery accuracy
– Loading configurations
– Mission profile

Ground and flight tests are usually required for certification. Much of the ground testing must be completed before flight tests begin for reasons of safety. In most programs, captive carry, loads, flutter, and jettison tests should be completed before systems integration tests can begin. Modeling flight configurations in wind tunnel tests is extremely useful as it avoids risks; it also can provide data on the aerodynamic effects on stores which is important in determining store separation characteristics and freedom from flutter. With this data, the quantity of flight testing is greatly reduced by defining areas of the operational envelope critical to certification.

For the purposes of this discussion, we will cover only tests that we may be involved with:

Ground Tests

– 101 and 102, Fit and Function
– 110, Static Ejection
– 160, Gun, Rocket, Missile Firing

Flight Tests

– 290, Ballistics

As emphasized previously throughout the text, all modern aircraft will undergo software modifications with the addition of new stores/weapons; modifications may affect the mission computer, stores management system, or any of the aircraft sensors.

The regression testing required as a result of S/W upgrades is normally covered under Ground Test 102 (Function). When developing the certification plan for any aircraft-store configuration, dedicated mission effectiveness or human factors should be evaluated. Some human factors to consider are as follows:

- Munitions preparation
- Loading, aircraft checkout, and preflight
- Cockpit weapon switchology
- Takeoff and ingress
- Target acquisition and tracking
- Target engagement
- Safe escape maneuvers
- Target egress
- Landing
- Integrated combat turn capability
- Downloading

The purpose of the fit test is to investigate the compatibility of the mechanical and electrical interface between the store and the aircraft. The test will ensure:

- Adequacy of the loading procedures
- Sufficient clearance exists between the store and its surroundings
- Satisfactory servicing, carriage, and deployment
- Proper store alignment
- Required electrical, electromechanical, and optical connections can be made

Data requirements, Instrumentation and Test Procedures are IAW AIR STD 20/21, Airborne Stores Ground Fit and Compatibility Criteria. The second half of the Fit and Function Test is the Function Test which validates the electrical functions for all of the ground and flight modes of operation including:

- Jettison
- SMS failure reporting
- Weapon inhibits
- Boresight procedures
- Armament control

The final ground test of interest is the gun/rocket/missile firing test. The test is written from the point of view that both the store and the aircraft are in development; other cases would require less testing. As with all weapons testing where there may be a series of unknowns, the test plan should be a build-up in risk, starting with the most benign conditions first. Once system integrity has been established flight testing can be accomplished to qualify the weapon system throughout the flight envelope. Ground gun firing tests are conducted to determine, verify, evaluate, and authorize the safety, compatibility, and performance requirements of the gun system/aircraft prior to flight tests. The gunfire induced environments and their effects:

- Muzzle blast and overpressures
- Effectiveness of muzzle and blast deflectors
- Recoil, counter–recoil and loads
- Vibration and acoustic frequencies and amplitudes
- Gun gas impingement
- Proper gun operation in various attitudes and g loadings
- Accuracy
- Interaction with other aircraft systems
- Bullet clearance from rotor blades and other surfaces IAW AIR STD 20/21

The test is performed at an approved facility where the projectile will be contained within the test enclosure. Test instrumentation, conditions, and procedures are delineated in the appendix. The test is considered successful if all data is successfully collected, and a demonstration of the proper operation of the gun system (IAW the design intent relating to compatibility and safety). The data should indicate the gun system is compliant with the applicable specs and has no adverse impact on other aircraft systems.

Similar to gun firing tests, rocket/missile firing tests are accomplished to verify safe operating characteristics and establish compatibility with the aircraft. The ground tests will:

- Provide data on the launchers and associated intervalometers
- Determine plume blast, pressure profile, thrust/impulse, exhaust chemical content, and temperature distribution
- Determine structural effects on the aircraft
- Determine the effects of a malfunction
- Measure trajectory and attitude changes
- Measure clearances between trajectory and rotor blades and aircraft surfaces
- Assess the effect of firing in ripple mode on aircraft engines

The test is performed at an approved facility where the projectile will be contained within the test enclosure. The test instrumentation, conditions, and procedures are delineated in the appendix. The test is considered successful if:

- All data is successfully collected
- Proper operation of the system IAW the design intent relating to compatibility and safety
- No damage to the launching aircraft that would affect airworthiness
- Data indicate that a launch will not result in an inadvertent release; unacceptable damage to the aircraft structure, launcher, or other store; or adverse effect on other systems or equipment

An integral part of validating the aircraft/stores interface is the assurance that the weapon when employed will hit the target; 1763 addresses this validation as ballistics tests.

There are four tests called out in the handbook to basically verify accuracy. The first test is called the weapons free-stream ballistics test (listed as test 291 in MIL-HDBK-1763). This test is run when the baseline aircraft independent ballistics have not been established for the weapon. The second test called out in the handbook is the operational flight program (OFP) ballistics evaluation test (test 292). The objective of this test is to evaluate the initial accuracy of the OFP basic algorithms. Evaluation of the ballistics includes a CEP test and a range bias test. If both CEP and range bias evaluations meet the acceptance criteria, then the ballistic accuracy of the OFP for the particular aircraft/weapon configuration will have been verified. If not, the next two tests (293 and 294) must be accomplished.

The third test is called the separation effects derivation test (test 293), and as the name implies, it is meant to derive the separation effects coefficients for the preferred aircraft–store configurations.

The fourth test is identical to test 292 and is designed to verify the ballistics accuracy of the OFP. It is called the OFP ballistics accuracy test (test 294).

If the sequence of these tests sounds a little ambiguous, it is. The decision tree in Figure 10.39 will attempt to clear up any uncertainty that you may have.

Separation effects need to be accounted for in manual as well as automated weapons deliveries. The store is usually considered to be free of separation effects

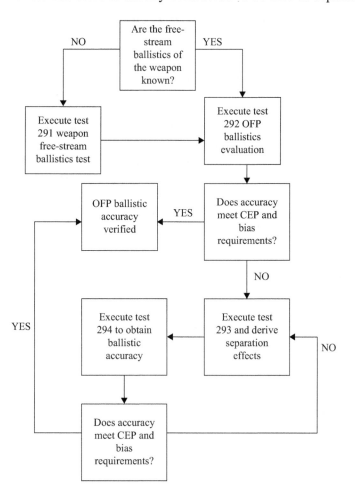

FIGURE 10.39 ■ Ballistics Accuracy Decision Tree

(aircraft influence) 1–3 sec after release when it reaches steady state conditions. What happens during the first 3 sec identifies the true trajectory of the store. AGARD presents a concise summary of data grouped by each of four error sources (release conditions, physical properties, atmospheric conditions, and separation effects) (refer to Table 10.11).

MIL-HDBK-1763 does an excellent job of identifying the instrumentation and data requirements for all of these tests but does not really go into detail as to why these tests are required or what the outcome should be. This information is found in AGARD-AG-300, volume 10, section 4.0, which will be addressed next.

10.6 | AGARD FLIGHT TEST TECHNIQUES SERIES, VOLUME 10 REQUIREMENTS

According to AGARD-AG-300, volume 10, there are three basic objectives of ballistic test programs:

- To obtain flight test data necessary to establish store free-stream flight characteristics
- To obtain the flight test data necessary to establish separation effects
- To obtain the flight test data necessary to establish the weapons delivery accuracy of the aircraft's OFP

It is necessary to perform flight testing to obtain the data necessary to establish or verify the store's drag, event times, and other factors that affect the store's flight characteristics. As previously mentioned, the free-stream drag characteristics are generally independent of aircraft and mode of delivery 1 to 3 sec after release, as the store is no longer influenced by the aircraft's flow field. In general practice, the store is first tested in a wind tunnel and then verified during flight testing.

Separation effects testing (what happens to the store during the first 1 to 3 sec after release prior to entering free-stream) involves releasing stores from an aircraft, one at a time, under very controlled conditions. Truth data are obtained for both the releasing aircraft and the store and are analyzed to quantify changes in the store trajectory due to the aircraft flow field. The effects are incorporated into the weapons delivery algorithms or applied to mil settings in the case of manual deliveries. In the F-16, separation effects are modeled by adjusting the velocity vectors in the along-track and vertical directions before they are fed to the air-to-ground integration routine in the OFP. The velocity adjustments (deltas) are derived from test drops and are curve fitted to a function of Mach number and normal acceleration. The mission computer uses these functions to compensate for separation effects for given store configurations and delivery conditions. Because the look-up tables for these corrections will grow as a function of multiple configurations and delivery profiles, it is important to prioritize the most important set of conditions.

AGARD notes that not all configurations and delivery modes need to be tested, as the accuracy may be good enough without extensive flight testing to determine what minor deviations may exist. The answer will be obvious after the completion of OFP accuracy testing.

Operational flight program accuracy testing provides data for the analysis of the entire weapons system. The pilot will attempt to deliver the store within a specified CEP on simulated targets. The entire weapons system comprises all of the inputs to the weapons release calculations: radar, forward-looking infrared (FLIR), laser, HUD, radar altitude, etc. The pilot will attempt to deliver the store in a controlled, noncompensating manner, attempting to repeat the deliveries the same way each pass rather than attempting to score a "shack" (bull's-eye). Truth data and aircraft instrumentation, including HUD video, are required for a complete analysis. Analysis of OFP accuracy testing will enable the evaluator to determine an error budget that will identify the sources of errors.

If the test results satisfy the user's accuracy criteria, testing is considered complete; if the user requires a higher accuracy or OFP accuracy-verification, additional testing is required. This process is usually accomplished in three phases. Phase I uses a pre-production OFP with the validated free-stream characteristics of the store; a sufficient number of stores is dropped for confidence and statistical significance. If the user is satisfied, Phase I is complete and a production OFP is released. If the user is not satisfied with the results and it has been determined that there are no problems with the non-ballistic portion of the system, Phase II is initiated. In this phase, more data is collected to refine the modeling coefficients in the OFP and a new pre-production OFP (patch) is generated. Stores are then dropped to obtain a new CEP and bias data (Phase III). If the user is satisfied, the OFP is productionized; if not, the process is repeated from Phase II, usually at increasing expense and schedule. AGARD makes the point of using good judgment when attempting to improve accuracy and cites an example of an F-16 CBU-58 test. The lesson learned was that accuracy and resource expenditure trade-offs need to be considered before testing begins. This consideration would constrain natural tendencies of users to keep demanding more accuracy regardless of the expenditure involved.

When computing the accuracy of the system, CEP and range bias are considered. CEP is the radius of a circle centered on the target, or mean impact point (MIP), which contains 50% of all bombs dropped at a given set of delivery conditions for a specific loadout. Range bias is the distance that bombs hit long or short of the target. Separation effects will affect accuracies, but so will errors in the avionics systems, which provide inputs to the weapon release computer. AGARD considers the fire control RADAR, CADC, INS, and rate sensor unit as the primary inputs to weapons release or fire control computer (WRC, FCC); it mentions a targeting pod (LANTIRN), but only as a future enhancement. In fact, DGPS, targeting pods, LRF, EO/FLIR, and off-board tracks are used today and can be the sources of error. Sensors will provide: range to target, height above target, depression angle, altitude, airspeed/groundspeed, winds, heading, acceleration, attitude, and rates.

Range is the easiest to understand; if the sensed range (from radar, laser, DGPS) is longer than the real range, the bomb will fall short; if the range sensed is closer than actual the opposite is true. Radar and laser may also be used to calculate the height above the target, which requires an accurate measurement of the depression angle (see HAT mode in the RADAR section). Errors in the stabilization (INS, AHRS) or errors in the RADAR mount will cause errors in the depression angle which in turn create errors in height, which in turn will create errors in release range. Cross-checking sensors or using Best Estimate processing can alleviate some of these concerns.

Errors in airspeed due to inaccurate CADC, INS, or GPS sensors will affect the range bias. If the sensed speed is lower than actual, the delivery will be at a higher speed causing a long bomb for example. The RSU is responsible for computing the normal acceleration of the aircraft; a bad RSU will have two effects on the delivery. A delivery under a higher g than expected will cause a short bomb and will also affect the separations algorithms. Failure to account for winds (or incorrectly calculate them) will result in cross-track errors due to cross winds and short or long bombs due to head winds and tail winds respectively. Depending on the implementation, the WRC may assume that the calculated winds are constant for the time of flight of the weapon or may use some algorithm to linearly decrease the winds to zero at the target altitude; it could use an average wind for the time of flight. Errors in azimuth and elevation from other on-board sensors will cause errors in target position (left/right), which will prevent the aircraft from flying over the target.

Errors in display presentations, symbology, and cursors are not sensors, but can contribute to errors in the delivery. The steering and aiming symbols are projected on the HUD; if the HUD plate is not aligned properly the symbols will be improperly located and aiming errors will occur. For example, if the HUD is angled too low, the projected pipper or target box will appear lower on the HUD causing the pilot to release the weapon closer to the target for a long bomb. If the symbology has been corrected for canopy distortion the wrong corrections will be used since now the information is being projected on a different area of the canopy. Two other errors are the design eye and parallax errors. In a fixed sight bombing system, pilot seating height will have the same impact as previously discussed with an improperly aligned HUD plate. If the HUD has been properly aligned and the pilot can view the HUD, then he is at the design eye viewing point. Parallax errors come into account with HUD recording; symbology that has been corrected for the pilot's viewpoint may not necessarily be corrected for the HUD camera due to different distortion errors.

As with avionics system tests, a test matrix must be developed to prove some level of performance, or, in the weapons case, to determine separations effects or OFP accuracy. Factors to be considered in the development of this matrix are type of weapon, weapon functioning envelope, and number of weapons required to satisfy test objectives.

There are two basic types of free-fall weapons: intact and functioning. An intact weapon is one that hits the ground in the same configuration as it was released. A functioning weapon will undergo changes in configuration (dispenser) after it leaves the aircraft. A MK 82 LDGP is an example of an intact weapon; a MK-20 Rockeye dispenser is an example of a functioning weapon because at some time after release the dispenser separates releasing multiple sub-munitions. A GBU-15 is an intact weapon but since wings deploy after launch they will affect the store trajectory. Functioning weapons add more complexity because sub-munitions form a pattern that must also be modeled. This pattern will change as release conditions and time of dispenser functioning change. Because of these two variables there could conceivably be an endless number of conditions to evaluate.

The flight conditions at which the weapon will properly work as designed is called the weapons functioning envelope and is usually determined during the DT&E evaluation period and must be considered when designing a test matrix. The document again provides two horror stories where the test team wasn't aware of the functioning envelope and tested in areas where it wasn't possible to arm the weapon (speed and altitude). It is also possible that a weapon design will exceed the capabilities of the carrying aircraft; both the store functioning envelopes and aircraft operating envelopes must be properly considered.

Flight tests require the pilot to make a series of weapons deliveries against simulated targets on a scorable range. Required TSPI and the data that must be collected can be found in MIL-HDBK-1763, appendix B. Deliveries will be accomplished in all delivery modes, such as visual, radar, laser, FLIR, TV, targeting pod) and a CEP will be computed for each of the modes. A good question to ask is how many deliveries must be made in each mode to achieve a desired confidence interval? Well, lucky for us. AGARD addresses this problem in section 6.1.5 "Number of Weapons Required for OFP Accuracy Testing." As should be expected, CEPs with small dispersion (small standard deviations) require fewer deliveries for the same desired confidence levels.

The number of weapons required for store free-stream testing depends on the type of weapon and functioning envelope and on the type of testing to be performed. For intact weapons, a minimum of 36 stores is required to fully characterize ballistic performance:

- Level flight, loft, dive
- Lowest operational speed, medium, and highest speed
- Minimum four bombs per test condition for confidence
- 3 profiles × 3 speeds × 4 bombs = 36

For functioning weapons, the numbers will be increased to ensure that the store fuze functions as designed in the time and altitude (proximity) modes. Added to the matrix will be three timer values and three altitudes for dispenser functioning:

$$3 \text{ profiles} \times 3 \text{ speeds} \times 4 \text{ bombs} \times 3 \text{ timer value} \times 3 \text{ altitudes} = 216$$

The required numbers of deliveries can be determined statistically, or by analyst experience; the numbers provided above are from the latter. If you wanted to determine the free-stream drag coefficient to the 85% accuracy level with a 95% confidence level, you would need 19 deliveries vice the quoted 4 bombs per test condition.

Stores separation effects are highly dependent on the aircraft loadout. Because of the limitations noted earlier, usually only one or two stores configurations are tested (most critical defined by the user). As in the case of free-stream testing, data are required at a minimum of three airspeeds and at load factors that cover the "g" range sufficiently to permit modeling between the data points. The example given in AGARD is an F-15E loaded with 12 MK-82s loaded on the fuselage conformal rack stations; 6 bombs are loaded symmetrically on each side in two rows of three. The weapons may be delivered in four modes: level (1 g), dive (< .5 g), toss (2.5 g), and loft (4.0 g). Data are required for each mode at three airspeeds and a minimum of four releases per station are required; because the load is symmetrical, two data points are obtained for each station when all bombs are released. Therefore:

$$4 \text{ profiles} \times 3 \text{ speeds} \times 12/2 \text{ stations} \times 4 \text{ bombs per station} = 288 \text{ stores}$$

A second example given in the document is an F-16 with 2 × 2 Durandals; the loadout is symmetrical. The weapon is delivered either in level flight or 10° dive, so using the same logic:

$$2 \text{ profiles} \times 3 \text{ airspeeds} \times 4/2 \text{ stations} \times 4 \text{ bombs per station} = 48 \text{ stores}$$

AGARD notes a couple of other considerations when trying to model separation effects. Although three airspeeds are recommended, the evaluator needs to be aware that there is a risk in modeling the intermediate airspeeds and this risk would be high if the evaluator does not have a historical database for reference. The need to obtain adequate data to model g is equally important; separation effects can be substantial for some conditions and loadouts. Separation effects are different for each station and should therefore be modeled that way.

Once separation effects have been modeled, it is necessary to perform testing to validate the OFP. The testing provides an end-to-end system assessment of the overall accuracy of the weapons delivery system; a minimum of 12 stores is required for each release mode. In the F-15E example noted previously:

$$4 \text{ release modes} \times 1 \text{ combat airspeed} \times 12 \text{ bombs} = 48 \text{ bombs}$$

A CEP and range bias is performed and the results are compared to the accuracy criteria. If the criteria are met, then the testing is terminated.

Eglin AFB, which authored this volume of AGARD and also volume 5, *Stores Separation Analysis*, presumes that a range bias does not exist for probabilities greater than 90% using a one-tailed cumulative binomial test (based on engineer experience). Table III in AGARD indicates that if 5 of 15 stores impact long of the target and the balance short, then the probability of a range bias does not exist (short bias). If only 4 of the 15 impact long, then a probability of a bias is assumed to exist. The table provides users with probabilities for other combinations of bombs that hit long or short. Figure 15 of the same document shows the relationship between number of stores, confidence level, and % of acceptable error.

The balance of the document spends considerable time on TSPI, avionics systems checks, and data analysis including CEP and accuracy analysis, which have all been addressed over the course of the text. Some examples of the output of the analysis such as ballistics tables and safe escape charts are provided at Tables VIII through X. References and a bibliography are provided, and the Appendices contain questions and answers from Canada, France, and Germany on how they perform ballistics analysis and testing.

10.7 | WEAPONS DELIVERY CONSIDERATIONS FOR HELICOPTERS

The overriding document for helicopter armament testing is ADS-45-HDBK, "Data and Test Procedures for Airworthiness Release for Helicopter Armament Testing," December 19, 2005. This document mimics much of the testing described in MIL-HDBK-1763 and will not be repeated here. Air Standard 20/21, "Airborne Stores Ground Fit and Compatibility Criteria, Change 1," April 21, 1991, is referenced in ADS-45 and establishes the requirements for carriage compatibility.

The certification process is not a short one and can last for years. The average duration for the helicopter/store certification process is roughly three years. This handbook contains guidance for the performance of a survey on armament and fire control subsystems integrated on the aircraft. A survey is the act of collecting information to determine the current state of the design with respect to established performance requirements. The survey will consist primarily of ground and flight tests; unless otherwise specified in the contract, the survey will be used to find needed improvements or problems areas that need resolution. Surveys are normally conducted on new or major modifications to reduce program risk.

The handbook stresses that test articles (items considered for test) may not be that obvious. *Armament* or *weapons* are easily identified; however, the fire control test articles are less obvious. The fire control system will consist of any hardware and software that is necessary to perform:

– Target acquisition/designation
– Target estimation
– Aircraft state sensing
– Environment sensing
– Sensor input processing
– Ballistic solution processing
– Stores management
– Aiming
– Launching/firing/dispensing
– Ballistic solution processing and the stores management
– Postlaunch controlling of the munitions

The general requirements and engineering analysis called out in the handbook are nearly identical to those previously described in section 10.5.1. The document, however, does address software considerations. Weapons inhibits, limits, and interrupts (WILIs) should be validated to show that weapons are prohibited from interfering with each other and are prohibited from firing when firing constraints are exceeded (e.g., activation of a LASER on the ground or lazing the cabin). The aircraft flight test phase should be conducted after S/W integration tests, hazard analyses, safety-of-flight (SOF) analysis, and tests. Aircraft ground tests are conducted to substantiate that the armament and fire control subsystem are safe for flight.

The rationale for conducting stores certification tests is described as:

– Validate the results of the analysis; obviously, if the test results compare favorably to the predictions, the test program can be significantly shortened
– The second reason is to provide data where no qualitative prediction can be made

Data must be examined to determine the acceptability of the aircraft store configuration that is being evaluated.

The contractor will prepare an armament and fire control system survey (AFCSS) test plan that describes a systematic ground and flight test program designed to proceed in an orderly manner from installation of the system through determination of the armament and fire control capabilities.

ADS-62-SP, "Data and Test Requirements for Airworthiness Release for Helicopter Sensor Data and Testing Requirements in Development Stage," June 29, 2001, will need to be consulted prior to development flight test. This standard establishes design and documentation requirements that shall be completed prior to issue of a flight release or airworthiness release (AWR) to conduct helicopter flight testing pertaining to the pilotage and target acquisition/designator system sensors. It includes statements and analyses that should be performed to guarantee a safety standard; the requirement may be tailored for each test.

10.7.1 Helicopter Specific Weapons Types

Guns

Armor and armament, especially the latter, must be evaluated in contexts other than survivability. For attack helicopters, the ability to survive a hostile environment, although essential for the mission completion, is not in itself a measure of the worth of vehicles' armament. In other words, the possible trade-offs among performance, armor, and armament always affect survivability but may involve directly such other primary variables as performance.

Several types of guns are available for helicopter use. They include air-cooled, gas-operated 7.62 mm weapons, air-cooled automatic 20 mm and 30 mm weapons, and air-cooled electric weapons for firing 40 mm grenade ammunition. For some applications, the gun is installed on recoil adapters. This installation minimizes the effect of gun recoil forces upon the helicopter structure. In selecting and designing recoil structure adapters, careful consideration should be given to gun muzzle energy; gun weight including attaching feeder and drive motor; and the response dynamics of the helicopter. Feed mechanisms for these weapons vary, and each presents a special set of design problems. In some instances, the ammunition feed must start from rest and must reach the peak rate of fire within the time required for the firing of one round; this imposes high acceleration forces upon the feed train and the storage containers. In other applications, a feed mechanism is used that extracts cartridges from a recycling conveyor belt. The integration feeder is adapted for this purpose by replacing the feed cover with special link guides. The high acceleration forces imposed upon the belt and cyclical deviation of the gun from a nominal rate of fire require careful design of the ammunition feed train to ensure equivalent belt tension on both sides of the gun feeder.

There are minimum requirements imposed by the nature of the weapon. The location must provide accessibility, unimpeded projectile flight paths, debris ejection paths, and the ability to jettison externally mounted gun pods. The vehicle must be able to withstand gun muzzle blast effects. Beyond these minimum requirements, the degree of optimization must be related to the overall effectiveness criteria used in evaluating the aircraft. The trade-offs will involve structural and geometrical limitations arising from the desire to optimize cost and flight performance versus the optimum location to maximize weapon systems performance.

Gun location must be such as to avoid intersection of the extremes of the projectile flight path envelope with the helicopter structure, including the main rotor and externally carried stores. The projectile flight path envelope is described by a circular dispersion of the firing projectile, with the circle center being coincident with the gun barrel centerline. Factors to consider in determining the dispersion envelope include: gun and ammunition dispersion, aerodynamic forces acting upon the projectile, and deflections of the gun mount and helicopter structure.

Guns shall be located as far as is possible from helicopter structure to minimize the effects of muzzle blast. The aircraft skin near the muzzle and adjacent aircraft structure must be strong enough to prevent gun blast damage. The reinforcement requirement of the aircraft skin and structure is determined by the distance between gun and skin, the thickness of the skin, and the density of the frames and stringers. The gun muzzle shall never be located near enough to canopies, radar antennas, or door frames to cause or create a hazardous condition. For some guns, a muzzle brake can be incorporated to reshape the blast pressure field. This device distorts the blast field so that the peak pressures and impulses are rotated and displaced from their normal positions relative to the gun barrel thereby reducing recoil forces. Consideration also should be given to the relationship of different weapons systems to each other, such as machine guns versus rockets or missiles; both from stationary firing position (of machine guns) and during the trajectory of rockets, for example.

Ejected ammunition cases and links shall not impinge upon helicopter structure, control surfaces, rotors, or externally carried equipment. The trajectories of the ejected debris can be determined from gun ejection velocity. Debris ejected velocities can be increased by the use of accelerator mechanisms; the selection of this design technique must consider the available space, and the attitude, kinematics, and shape of the ejected debris. Deflector plates may be used and can be strategically placed to redirect the case ejection path.

Rockets

The current rocket type qualified for use on U.S. military rotary wing aircraft is the 2.75 in. (70 mm) folding fin aircraft rocket (FFAR). The FFAR was developed at the Naval Ordnance Test Station, China Lake in the late 1940s and is available in a variety of warhead/fuze combinations to suit specific helicopter mission requirements. It is carried in and fired from the helicopter by means of tubular launchers.

The current 2.75 in FFAR launchers for helicopters consist of a fixed, forward-firing, rearward venting, open-breech tube cluster. The individual tubes may be reusable or replaceable, or the cluster may be expendable. The launchers normally are installed with the launcher axis (boreline) parallel to the line of flight under specific flight conditions. The primary function of the rocket launcher is to release the rocket safely from the helicopter without disturbing the rocket from its intended flight path. The launcher design should provide for accurate alignment of the launcher boreline with the helicopter aiming reference under all tactical deployment conditions.

The rocket system shall include methods to allow both automatic and manual firing of rockets in a predetermined quantity, sequence, and timing interval. The requirements for rocket selection and firing sequence will be defined by the basic helicopter system specification. At a minimum, the number of rockets to be fired (single, in pairs, or ripple) at a preset time interval shall be selectable. The timing mechanism shall be designed for reliability and accuracy under all firing conditions.

Guided Missiles

Guided missile launchers and guidance control equipment of the number and type described by the governing helicopter specification shall be installed. The missile currently being used for helicopter applications is the AGM-114 Hellfire which was developed primarily for anti-armor applications originally for use with the AH-64 Apache, and now used on a number of platforms. Helicopter missile launchers generally will be installed offset from the helicopter centerline on armament pylons or stub wings to protect the tail control surfaces and rotor system from possible immersion in the exhaust wake of the missile.

Good design practices include location of the launcher on the helicopter to prevent:

- Engine compressor stall or flameout as a result of exhaust gases entering the engine intake ducts

- Exhaust gas impingement upon or ignition debris collision with the airframe and all rotor systems

- Harmful corrosion effects as a result of deposits of missile exhaust residue within the engine or upon other components that are not readily accessible for cleaning
- Impairment of pilot's or gunner's vision by flash during firing
- Excessive acoustic noise in the crew compartment during firing
- Pitting or coating of the canopy by exhaust gas and debris
- Aerodynamic interference between launchers and control surfaces, sensors, and adjacent stores

The design and location of the launcher installation should be such as to minimize corrosive effects resulting from the exhaust particles inherent to solid propellant missiles.

Proper consideration of preventive or corrective methods, including cleansing of affected parts, can significantly reduce the possibility of structural corrosion or surface damage caused by motor exhaust. Ground safety cutout switches in order to prevent accidental firing during ground operations are required. The firing circuit may be interrupted by landing gear extension, by landing skid compression under helicopter weight, or by insertion of an interlock pin at each launcher by the ground crew. A manual-operated override switch shall be provided to permit ground checkout of the firing circuit.

10.8 | SELECTED QUESTIONS FOR CHAPTER 10

1. What is the most important objective to fully understand when conducting weapons testing?

2. You are about to conduct a weapons accuracy test. What factors will affect your CEP?

3. What truth parameters must be known when conducting accuracy tests?

4. What is a parallax correction?

5. What impacts will parallax corrections have on the outcome of an accuracy test?

6. Why are boresighting tests performed?

7. Why are sensors boresighted? What are they boresighted/referenced to?

8. How does the pilot/operator performance contribute to the error analysis of weapons accuracy?

9. List the three measurements that must be taken in order to calculate an error budget.

10. List five potential pilot errors during a weapons accuracy test.

11. List five potential errors in the aiming system (HUD, displays).

12. List five potential errors in the sensor systems.

13. Is video an important tool in error analysis of a weapons delivery system? Why or why not?

14. List three reasons for computing a minimum release altitude prior to actual delivery.

15. What are the benefits of wing control air-to-air missiles? What are the drawbacks?

16. Why is a three-plane captive carry test used for missiles with postlaunch control?

17. What is a roof house used for?

18. What is the importance of conducting a radome test?

19. Why are reliability and maintainability important factors in considering a new weapons system?

20. Why are live fire tests so expensive?

21. How does Doppler fusing work?

22. What is sensor fusion? Briefly describe a typical fused weapons system.

23. Define free-stream ballistics and separation effects.

24. If certification by analogy (similarity) is NOT possible, why is the important next step in the certification process to conduct a physical compatibility review?

25. What is the primary purpose of static ejection ground tests?

26. List the three primary reasons why we conduct wind tunnel tests as part of the stores certification process.

27. What are three limiting factors when using wind tunnels for stores certification testing?

28. Which documents would you consult prior to any weapons integration test?

29. What are the general requirements for the SMS?

30. What function does a decoder perform?

31. What is an inventory function? How is it accomplished?

32. A weapon fails prep. What indications are provided to the pilot by the SMS?

33. What is a MOAT? What functions does it perform?

34. What is a missile tune? What functions does it perform?

35. Name four common weapons interlocks.

36. Explain the primary purpose of flight testing in the stores certification process.

37. List the typical instrumentation required for (a) release tests, and (b) free-stream ballistics tests.

38. (a) When certifying a well-understood store (e.g., Mk 82) on a new aircraft, is it necessary to collect free-stream ballistics data? Why? (b) When certifying a well-understood store (e.g., Mk 82) on a new aircraft, is it necessary to collect separation effects data? Why?

39. Explain the weapons accuracy error budget and why it is important to know it before beginning a stores certification program.

40. What is the purpose of a speed gate in a semiactive homing missile?

41. Name two types of navigation used by air-to-air missiles.

42. Why would the SMS command an engine derich as part of an LTE?

43. What is a LAR? How is it constructed?

44. Name five instances where gun firing would be terminated.

45. What is a SEAM mode? What is it used for?

46. What is the data bus architecture for the SMS and associated weapons systems?

47. What is meant by a postlaunch missile message?

48. What is a pit or catcher's mitt test used for?

49. What is the largest contributor to missile service failures?

50. What is a safety footprint? How is it constructed?

A Typical Avionics Integration Flight Test Program

Chapter Outline

11.0 Introduction 831
11.1 Vehicle Test Requirements 831
11.2 Avionics Test Requirements 833
11.3 Test Planning 835
11.4 Responsibilities of the Test Team 843
11.5 Analysis and Reporting 848
11.6 Selected Questions for Chapter 11 849

11.0 | INTRODUCTION

As the name in the title implies, this chapter is a recap of the text put into the context of a real integration program. It attempts to take the reader from "womb to tomb" on the test program, attempting to touch on all of the variables the evaluator will be forced to confront. For this exercise, I have elected to integrate a high-speed antiradiation missile (HARM) into the F-14D (this should not hurt anyone's feelings since the F-14D is now retired from service). The reader can assume that she is sitting in her cube and her boss has just dropped this requirement on her desk. "I need you to estimate this job for me. I'm briefing the Director in an hour and I need a SWAG (scientific wild-ass guess) on the program. You know, what's required, length of the program, assets required, etc. Don't worry, it isn't a firm estimate and the company won't hold you to it." Our fearless tester should be fearful because everyone knows that this SWAG will be a firm proposal by the end of the day.

11.1 | VEHICLE TEST REQUIREMENTS

Prior to the systems integration phase, and in some cases parallel to the phase, the aircraft must be cleared for carriage and release. In other words, the systems integration group cannot fly the HARM on the F-14D until the loads, flutter, and performance groups say it is OK. A description of their requirements can be found in MIL-HDBK-1763 (revisit chapter 10). The major tests are

- Carriage envelope clearance
- Clearance for captive carry

FIGURE 11.1 ▪
Allowable Flight
Envelope for HARM
Carriage

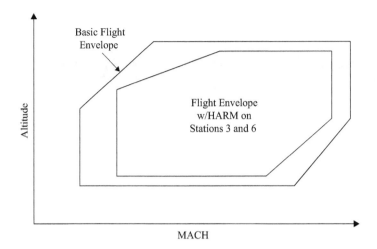

- Measure loads for separation analysis
- Demonstrate freedom from flutter
- Survey the vibration environment for the missile
- Determine aircraft performance

It is interesting to note that all programs that involve the integration of a store on an airplane will have a statement in the ground rules of the contract that states "the addition of this store will not adversely affect the basic performance of the aircraft." This statement is, of course, ridiculous, as there will always be some degradation of performance or capability (range, endurance, g available, speed, etc.). The loads and flutter groups basically perform an envelope expansion flight test program each time a store is integrated on an aircraft. The result is a modified aircraft performance diagram. Figure 11.1 shows a generic aircraft performance envelope for aircraft altitude versus Mach. With the addition of the HARM on stations 3 and 6 (assume belly stations), the aircraft becomes more restricted in terms of absolute minimum and maximum speeds and minimum and maximum speeds at altitude. These flight restrictions will become part of the basic flight manual (Dash-1 and Dash-34 for the U.S. Air Force and naval air training and operating procedures standardization (NATOPS) for the U.S. Navy). There is a possibility of asymmetric store loading restrictions that occasionally impose further impacts on the flight envelope and subsequent flight testing that need to be investigated.

Once the aircraft is cleared to carry the missile, it must then be determined that the missile can be released/fired safely. In order to accomplish this, further testing is required (again, see MIL-HDBK-1763). The major tests here include

- Store separation tests
- Clear missile firing envelope
- Clear launcher/missile jettison envelope
- Measure skin temperature due to missile exhaust plume impingement
- Evaluate effect of missile exhaust ingestion on engine

Separation tests require special instrumentation and video equipment. Both the Aerospace Group for Avionics Research and Development (AGARD) and the previously mentioned MIL-HDBK-1763 do an excellent job of delineating the types and quantities of required data. These tests will validate previous wind tunnel results, and may require nose and wingtip cameras (high speed, 3 to 5 sec), missile telemetry, a chase camera, and a safety chase. A store trajectory analysis program that analyzes flow field anomalies, tumble, and roll rates is normally used during postflight data analysis; there are vendors (in addition to military units) who can supply the cameras and the analysis programs.

In some instances, additional tests may be required. U.S. Navy integrations will require carrier suitability tests, which evaluate the installation under extreme operational conditions. These extreme conditions could include a demonstration of the structural adequacy of the HARM installations during high-sink landings, arrestments, and catapults.

The vehicle test and evaluation community will develop the test plan and flight requirements for the aforementioned tests. Using historical data, the number of flights required would look like the following:

Typical estimated flight requirements:

• HARM envelope clearance (2 stations)	12 flights
• Aircraft performance	6 flights
	6 flights
• Launches	6 flights
• Jettison (HARM/launcher)	30 flights

11.2 | AVIONICS TEST REQUIREMENTS

Avionics flight test requirements involve the integration and performance of the HARM implementation. The test team will

- Define the program
- Initiate the test planning process
- Determine what is needed (assets, instrumentation, data)
- Execute the plan
- Provide analysis and reporting

In all test programs, there are some basic assumptions that are made. Some of the typical assumptions are that the weapons system performance baseline is the basic aircraft performance (the integration of the HARM will neither improve nor degrade the basic aircraft; see Figure 11.1). Another assumption is that the performance demonstration requirements are defined as the basic HARM capabilities (i.e., the HARM must function properly). For this exercise I am also going to add the assumption that full-scale integration will be conducted as a participatory program and contractual compliance is the responsibility of the contractor. The government will provide as government furnished equipment (GFE) all target/support/chase aircraft and facilities. Also assumed is that HARM vehicle/system firings will be conducted at a customer designated location.

It is extremely important to state these assumptions because it tells the customer that the cost and length of the program is contingent upon these assumptions holding true. You can see how much the cost would increase if the contractor was on the hook to pay for support aircraft, facilities, and fuel.

The first task of the avionics test team is to identify the general test objectives. Are we to evaluate, demonstrate, or show actual performance? An evaluation may mean the functionality or feasibility of using a system or a particular mode of a system. Demonstration or qualification requires the collection of data to statistically show compliance with a requirement. Performance is a demonstration of system usability under operational conditions. Misidentifying the basic objective can be catastrophic in terms of time and money by underestimating the real task.

For the HARM integration, some potential evaluation tasks include prelaunch missile interface functions, air-to-ground attack steering, HARM launch envelope, usability of HARM target of opportunity (TOO) video, human factors crew/system interface, and controls and displays.

A potential demonstration is a proof that there are no adverse effects on mission capability with the incorporation of this store. Other demonstrable items include launch modes of the HARM (TOO, prebrief, and anticompromise), and the effectiveness of built-in tests (BITs).

Some tasks are not as obvious as others. One additional task is the development of NATOPS/Dash 1/34 inputs. Whenever a new or modified system is installed on the aircraft someone needs to write descriptions of normal and abnormal procedures for the flight and tactics manuals. Most contractors and military organizations have an integrated logistics service that is ultimately responsible for publishing the changes; but who is responsible for writing them in the first place? Unfortunately, in most organizations this task falls to flight test personnel, since they are writing the reports anyway. It is one of those jobs that everyone knows must be done but is never budgeted for.

System performance will normally be shown against representative targets during realistic operational scenarios. Data are collected, but are not normally analyzed to show compliance. HARM multiple target discrimination, tracking during maneuvers, and target prioritization are examples of performance objectives for this exercise.

After the generalized objectives for the program are defined, the responsible test agency and locations for the tests can be established. This is another important aspect of the test planning process that is also quite often overlooked. If a responsible organization is not identified during all phases of testing, destructive in-fighting among agencies/organizations will occur. You do not want engineering supervising the development/operational test (DT/OT) portions of the test, and in a similar vein, you do not want flight test supervising the development effort.

Table 11.1 provides an example of how the different phases of the test program can be broken down. It provides the phase, a brief description of what is to be accomplished during that phase, where the work is to be accomplished, and who has ultimate responsibility/authority over the phase. Some of the lines are very clear: engineering is responsible for software development and flight test is responsible for demonstration. Some of the lines are not as clear. Identification of hardware and software problems in the functional phase is accomplished on the airplane as well as in the lab (part of the fly-fix-fly process has been described in chapter 1)—a sharing of responsibility. It is important to establish up front that when the system is delivered to test and evaluation (T&E) for the start of testing, then T&E becomes the responsible organization.

After understanding the generalized objectives and identifying the responsible organization, the next task for the avionics group is homework. Remember, it is the tester's job to be as knowledgeable about the system under test as the design engineer. This task will require many hours in the lab, speaking with engineering, and researching

TABLE 11.1 ▪ Responsible Test Organizations

Phases	Purpose	Location	Responsible Organization
Development	Develop software and subsystem interfaces	Lab	Engineering
Integration	Develop total system software and hardware functions	Lab	Engineering
Functional	Identify hardware and software design and integration problems	Aircraft, Lab	Test and Evaluation
Evaluation	Optimize system software functions (as required)	Aircraft, Lab	Test and Evaluation
Demonstration	Demonstrate specification compliance	Aircraft, Lab	Test and Evaluation
Performance	Ability to perform the mission	Aircraft	Operational Test and Evaluation

the documentation. During this phase the tester will determine the test methodology and the data required to satisfy objectives. The first document the tester will want to examine is the program performance specification (PPS), which gives the reader the "big picture." This document explains what capabilities the HARM has, how they are implemented, and what modes are incorporated. It provides the tester with an overview of HARM system operation and the enhancements provided to the operator.

The functional requirements document (FRD) provides an in-depth discussion of the HARM, its modes of operation, and its capabilities. The interface design specification (IDS) defines how information is exchanged between the HARM and other systems on the aircraft. The interface control document (ICD) defines the location of the parameters on the data bus. It is also the time to review controls and displays, radar warning receiver, and stores management system (SMS) documentation.

With a firm understanding of system operation and the interoperability requirements of other aircraft systems, the tester can proceed with test plan preparation and data acquisition requirements. Instrumentation needs to be consulted at this point in order to determine if special equipment or processing routines need to be developed for the program. Remember that special equipment will be an added cost, but more importantly its development will take time.

11.3 | TEST PLANNING

Rather than attempting to write a test plan from scratch, many organizations utilize a system of test information sheets (TIS) or test summary sheets (TSS). These sheets are a script for each required test and contain the following information: test title, objective, profile, prerequisites, test site, test method, data requirements, data reduction, and analysis and acceptance (exit) criteria. Two examples of the test summary sheet format are shown in Figures 11.2 and 11.3.

Test summary sheets (TSSs) are a useful tool for many reasons. The first is that any test can be duplicated by an interested party at a later date just by following the scripted format.

FIGURE 11.2 ■
Test Summary
Sheet, Example 1

Test Summary Sheet H1

Test Title: HARM Control Panel (HCP) Built-in Test (BIT).

Objective: Verify proper indications during HCP BIT.

Profile: Static.

Prerequisites: None.

Test Site: Ground (Internal Power).

Test Method: Prior to powering up the HCP, the Missile Select (MSL SEL) switch will be placed in the "OFF" position. Electronic Countermeasures Officer 1 (ECMO 1) will select CDI repeat mode (S1) and toggle the HCP PWR switch from "OFF" to "ON" to "BIT." Pilot and ECMO 1 will monitor the HCP, CDI, and ADI.

Data Requirements: Kneeboard

Data Reduction and Analysis: BIT shall result in a "GO" lamp on the HCP and no "FAIL" messages on the CDI. The ADI vertical needle shall be driven from its left side (stowed) position to three viewable test positions: full left, center, and full right. The ADI horizontal needle shall be driven from its top side (stowed) position to three viewable test positions: full up, center, and full down. The needle shall dwell for 1.5 sec at each position.

FIGURE 11.3 ■
Test Summary
Sheet, Example 2

Test Summary Sheet H13

Test Title: HARM Range Known Mode – Off Boresight (100° Rel Brg).

Objective: Demonstrate the ability to release a HARM off boresight in Range Known mode.

Profile: Level flight, 350 KTAS, 10,000 ft.

Prerequisites: H11.

Test Site: Within TM and tracking range of the facility.

Test Method: ECMO 2/3 will create a platform track and then hook and assign that track as a HARM target. ECMO 1 will select Missile Station (1, 2, 4, or 5) and designate the HARM threat data to that selected station. Pilot will fly toward the target (±20°) via the HIS steering cues. At a predetermined range (see CONFIDENTIAL appendix C), pilot will turn the aircraft to position the target at approximately 100° relative bearing and initiate the launch sequence. Pilot will loft the aircraft as commanded by the ADI horizontal steering needle.

Data Requirements: Range to target, target bearing, and time to flight at release.

Data Reduction and Analysis: Postflight comparison of avionic, missile, and TSPI data.

Acceptance Criteria: Upon completion of the HARM selection and assignment process, a missile ready indication ("MRDY") shall be displayed on the CDI (S1), DDI (Zone 4), and HCP. Upon completion of the HARM release sequence, normal launch indications shall be displayed on the CDI ("DRDY") and DDI ("LNCH"). At release, TOF displays on the CDI and DDI (Zone 5) shall transition to TTI and begin countdown. TOF at release data shall be approximately equal to predicted performance values (see CONFIDENTIAL appendix C).

There is no guessing involved as would be the case if you had to interpret the test from reading a test plan or final report. If written correctly, the flight cards are already done and can be found in the test method paragraph. TSSs provide exit criteria which are so important in flight testing. The exit criteria answer the question, "When can I say the test is completed?" TSSs also note what data are required and what if any manipulation must be done to the data.

The first example of a TSS in Figure 11.2 is a stand-alone ground test. We know this since there are no prerequisites and the test is performed in a static condition on the ground. The data requirements for this test state "kneeboard," which simply means that the evaluator is noting the indications and checking off the box on the flight cards. For a pilot, the flight cards are normally strapped to his knee in flight, hence "kneeboard."

The second example of the TSS in Figure 11.3 is a little more involved; it is to be accomplished in flight and does have a prerequisites. Note that in the test method section there is a sentence, "a predetermined range (see CONFIDENTIAL, appendix C)." This means that the range is a classified value, and if the number is placed here, the TSS and hence the entire test plan would become classified. Whenever possible, strive to keep the test plan unclassified. Classified documents must be safeguarded (stored in a safe when not in use), numbered (accounted for), cannot be copied, and can only cause problems for anyone who has ownership of them. There is no fun in receiving two weeks unpaid vacation for a security violation. Most people who are on a distribution for a classified document will lock it in the safe immediately when received, never to be seen again. Test plans are going to be used on a daily basis and need to be accessible. Reference classified values to the classified documents from whence they came and keep the test plan (and the test report) unclassified.

From the test plan, the evaluator can now determine which test facilities are required to complete the plan. Table 11.2 provides an example of the required test facilities and their use for this example.

Anechoic chambers have been mentioned throughout the text and will be used in this example for many of the basic evaluations. HARM/airframe glinting is a phenomenon much like multipath effects. Since the HARM has an active seeker, it is possible that the missile may see the target (a radiating threat antenna) as glinted off the airplane.

TABLE 11.2 ■ Test Facilities and Facilities Requirements

Facility	Used For
Anechoic Chamber	HARM/Airframe Glinting
	System Integration
	Intrasystem Electromagnetic Interference and
	Compatibility (EMI/EMC) (Victim Source)
Electronic Warfare Test Range	In-flight Evaluation and Demonstration
	of HARM Integration
Warning Area (Military Restricted Area)	In-flight Evaluation and Demonstration
	Captive Carry Testing
Electronic Warfare Integrated	Integrated Testing with the Anechoic Chamber
System Test Laboratory (EWISTL)	
Catapult/Arresting Gear	System Carrier Suitability
Naval Air Warfare Center, China Lake	HARM Missile Firings

FIGURE 11.4 ∎
Glinted Target

This will give the HARM the wrong polar coordinates to the target, and the glinted target will be your own airplane.

Figure 11.4 shows how aircraft glinting can affect the HARM. All tactics manuals warn operators to never launch against a glinted target... for obvious reasons.

For all of the reasons mentioned in chapter 4 of this text, antenna patterns (of the installed missile), victim/source, and field testing are accomplished in the anechoic chamber. As a benefit, many full-size chambers also incorporate a data link hookup to a simulator in an adjacent building where the pilot can see firsthand how the aircraft and systems react during testing in the anechoic chamber. The U.S. Navy (Patuxent River NAS, Maryland) calls their system the Electronic Warfare Integrated Systems Test Laboratory (EWISTL), whereas the U.S. Air Force (Edwards AFB, California) calls their system the Integrated Facility for Avionic Systems Testing (IFAST).

The electronic warfare test range allows the HARM to engage against real and simulated threats. These ranges were described in chapter 9 of this text. The warning area is a military restricted airspace covered by TSPI assets where captive carry and simulated launches are evaluated. Live missile firings need to be accomplished at a range that has sufficient area to cover the missile safety footprint described in chapter 10. For this exercise we have chosen the Naval Air Weapons Center, China Lake, which has sufficient coverage.

As this is a U.S. Navy program, I have to consider the effects on the HARM during catapults and arrestments. In the United States, this can be accomplished at Patuxent River or Lakehurst, New Jersey.

11.3.1 Data Requirements

The initial review of data requirements should not be specific down to the parameter level, but rather should be a review of the types of instrumentation that are required. We cannot define specific parameters until the test plan is actually written. There are two general types of data that are required for any avionics or weapons integration test program: avionics data and truth data.

Avionics data are of two types: analog and digital. Analog data are what we used to call pulse code modulated (PCM) data, and consist of things like switch positions, voltages, and the health and welfare of the weapon. Digital data are bus data, and in our case will be comprised of 1553 and 1760 formats. Digital data can be sent as selected parameters, or in some special cases, the entire bus can be telemetered to the control station. Care needs to be taken that classified data are not put on the data stream, as well as information that is not classified which could be used to determine classified information.

Truth data are defined as TSPI and allow us to know where everyone really are, as opposed to where the system thinks they are. TSPI is covered in chapter 2 of this text. Almost all ranges in the United States have come to rely almost exclusively on the global positioning system (GPS), and in particular, differentially corrected GPS (DGPS) for TSPI data. But if we need to follow the HARM during its launch we will also need some sort of cinetheodolites.

During the initial review, we also need to check what is to be provided by each of the parties (government/contractor). The two organizations will contribute government furnished equipment (GFE) and contractor furnished equipment (CFE). The identification of this equipment is found in a preliminary list as an attachment to the statement of work (SOW). Like ground rules and assumptions, this list needs to be checked to avoid missteps in estimating the scope of the evaluation. For this exercise, I have identified the GFE as

- Avionics development lab
- Instrumented HARM with telemetry and cooling
- Aircraft support

The avionics development lab is the aircraft software systems support facility and is responsible for the development, coding, and integration of aircraft software. The instrumented HARM is a goldenbird, described in chapter 10, which will be used for the captive carry program. Cooling is required as a separate environmental control system (ECS) package so the missile can be run continuously without overheating and destroying the sensitive HARM electronics. Aircraft support is the routine maintenance, preflight, postflight, and fueling of the test aircraft.

The preliminary instrumentation list should contain the items necessary for collecting the data needed. Some of the items that should be considered include

- Power and control system
- Pulse code modulation (PCM) system
- Telemetry/tape system
- Time code
- Video recorder (2 hr) and downlink
- Onboard positioning
- C-/X-band transponder
- Missile simulate unit
- Inadvertent jettison monitoring system

Instrumentation, data acquisition, and PCM systems were covered in detail in section 3.19.1 of this text, which describes the collaboration that is necessary between the test engineer and the instrumentation group. Pilot's voice, or Hot Mike, can also be sent to the telemetry station via a PCM frame. In most development test programs, the data sent to the ground, including communications, are encrypted. We know from chapter 4 of this text that in order to encrypt communications we must first digitize them. Since the voice is now digitized, it is an easy matter to insert it into a PCM frame.

You might be saying right about now that this is really interesting, but what does it have to do with test planning. The PCM frame is designed by the instrumentation group, but the determination of which parameters need to be accessed is the responsibility of the flight test engineer. The test engineer must sit down with instrumentation and decide which parameters are to be collected and at what sample rates. The PCM frame will then be filled in accordingly. If the test engineer does not know how the PCM system works he will be unable to perform this task.

Just as important as the telemetry system is the tape recorder system. The important thing here is to try and size the recorder capability based on data rates and test time. For example, suppose you wanted to record six busses of 1553 data (1 Mb/sec/bus) and the average test time in flight is 3 hr. If you choose a recorder that has a 2 hr record capability at a max rate of 6 Mb/sec, you are going to run out of record time before the test is complete. This means the pilot will have to turn the recorder off when it is not in use and back on when it is required. Unfortunately, Murphy says that whenever a problem occurs in the aircraft, the position of the record switch will be "off."

The time code, the PCM, and the recorder format all must be in accordance with the Range Commanders Council Telemetry Group, IRIG (Interrange Instrumentation Group) Standard 106-07, "Telemetry Standards," September 2007. The standard allows interoperability among test facilities and allows the easy exchange of data. The standard is common across all branches of the service in the United States and is used at many other test facilities around the world. Remember, IRIG does not provide the correct time; IRIG provides the standard for formatting the time inserted into data.

Video recording can be accomplished with external "over the shoulder" lipstick cameras or may be directly recorded from the bus. Most aircraft equipped with a heads-up display (HUD) will have a manufacturer installed recorder. The problems that we saw with record time for avionics data should not be a problem with video recorders, since pilots normally have direct access to the recording device and can change tapes during flight. Most test organizations use Hi-8 record systems, but they are slowly giving way to digital recording systems.

Since this is an avionics program, we know that we will need TSPI data. We may elect to use one of the differential GPS/INS systems discussed in chapter 2. If we elect to use radar or laser tracking, the range directorate will need to install a beacon or laser reflectors.

The last items on the instrumentation "wish list" are a missile simulate unit (MSU) and an inadvertent jettison monitoring system (IJMS). The MSU allows us to simulate many missile firings during the captive carry portion of the program. The SMS will not allow multiple releases/firings of an inert or test munition because after the release signal is sent, if the SMS detects a store still present, it will declare it "hung" and not allow any further access to that station. The MSU allows the SMS to reset that station so another simulated firing can be accomplished.

During the vehicle portion of the program, the T&E team will clear the HARM for captive carry and firing. One of the tests they will perform before every load-out is a stray voltage check at the station. This test ensures that the missile will not be inadvertently fired because of stray voltage at the station. This test will also be performed during the avionics integration portion of the program, but is not sufficient to guarantee protection from an inadvertent release. The SMS sends a digital release signal to the decoders and the decoders perform the digital-to-analog (D/A) conversion to voltage. How can we guarantee that there is no stray digital release signal on the weapons bus? The IJMS is a device that monitors the bus for this signal. Safety will not allow carriage of the HARM until the T&E community can guarantee freedom from an inadvertent release.

11.3.2 Test Planning Requirements

Only after the program has been defined can we set about to plan the testing. Things to be considered during this phase are

- Forming a test team
- Generating a test plan
- Asset availability
- Test conduct and responsibilities

The test team members are going to be dictated by the contract or SOW; it is the project leader's responsibility to ensure that the members are used in the most efficient way possible. Remember that we know we are going to get screwed due to late delivery of software, and program management is not going to allow us more time or budget to accomplish the flight test portion of the program. Since I am throwing around a lot of titles, let me delineate responsibilities. The following generic organizational chart is common for most T&E programs (Figure 11.5).

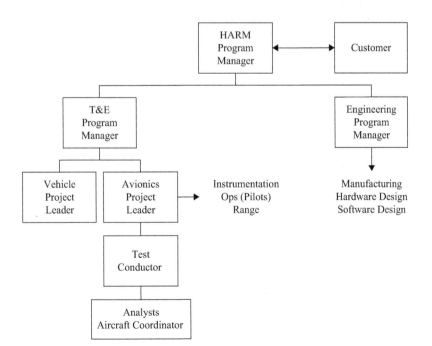

FIGURE 11.5 ■
Generic Program
Organizational Chart

The program manager is ultimately responsible for the success or failure of the program and is concerned with schedule and budget; he is the liaison with the customer. The engineering and T&E program managers report to the program manager and are responsible for their assigned phases of the program. Within T&E are an avionics and a vehicle project leader. These are usually two different individuals because of the diversity between vehicle and systems testing. The avionics project leader is the liaison with all of the support organizations for the project and supervises the flow of the test. The test conductor is responsible for the day-to-day operations of the test program. The responsibilities of the test conductor, analysts, and aircraft coordinator will be covered in a bit.

In large programs it may be wise to establish two or three working test groups under the avionics project leader. This would be the case with programs where major software releases are expected at specific intervals. The avionics project leader appoints a test conductor for each block of testing. She is responsible for the overall test planning and test conduct for the particular software release. The test conductor, as a rule, requires a mix of operational and technical expertise. The project leader assigns a mix of task team members to each block of testing and each team is responsible for the successful completion of their respective block/build of software. An example of three teams working independently would be:

- Team 1: Analysis and reporting of recent tests
- Team 2: Test conduct of ongoing tests
- Team 3: Preparation for upcoming tests

In most organizations the test plan format is governed by a set of operating instructions or corporate operating procedures. In general the test plan should contain an overview of all testable items, regardless of the demonstration requirements. It should contain system operation, method of test, and anticipated results. The system under test (SUT) can be described by test objectives, test conditions, test points, flight cards, or test information sheets. The objectives are driven by the top-level specifications; the test points are driven by the criteria used to satisfy the objectives (i.e., statistical population).

For example, a primary objective of our test plan may be to prove that bearing accuracy (polar coordinates) is properly displayed to within $\pm 2°$ throughout the operational field of view (FOV) of the HARM. Assume the FOV of the HARM is $20°$ (a notional number for the exercise). We may want to place a target at $10°$ left of boresight and measure the bearing accuracy. We repeat this test with the target at $5°$ left, at boresight, at $5°$ right, and at $10°$ right. These are the test conditions to prove bearing accuracy, but not the number of test points required. Say that for a 90% confidence in the data we calculate that we need 8 data points. For this exercise we would require 8 data points times 5 test conditions, or 40 total test points. If we had more than one station to evaluate, these would be multiplication factors in our test matrix. A test matrix for this objective is shown in Table 11.3.

This is a rather simple matrix, but it is important to understand which number you enter into the test plan (1 objective or 8 conditions or 80 test points). The U.S. Air Force offers a slight twist on matrix testing. In the design of experiments (DOE), the same matrix is built, but not every point in the matrix is evaluated. The rationale is that if some points are taken throughout the matrix and they all pass, then the probability of the other points passing is high and need not be accomplished. The net result is that less

TABLE 11.3 ■ HARM Bearing Accuracy Test Matrix

	System Under Test	Azimuth Accuracy	Stations to be Evaluated	Total Test Points for 90% Confidence
	HARM	$-10°$ to $+10°$	Sta 3 and Sta 6	8
Total Test Conditions	1	5	2	8
Total Test Points Required				80

flight testing is required. For information on DOE, the reader is should do an online search of "USAF Design of Experiments."

The completion of the test plan highlights which specific assets are required for the test program. For most programs there will be four major classes of assets required: targets, ranges, TSPI, and electronic warfare (EW) targets. There are many different classes and capabilities of targets that may be required. Some of the attributes that should be considered include

- Number of targets (resolution cells, raid count, track while scan [TWS], etc.)
- Size of targets (radar cross section [RCS], ability to adjust target size)
- Performance of targets (speed, altitude, turn performance)
- Availability and time on station (unique targets require long lead times, refueling support)
- Manned versus unmanned (missiles, target drones for live fire exercises)
- Cost (sometimes the ideal target is too costly)

TSPI has been covered in just about every section of this text, so I will not belabor the point. Just remember that whatever system is chosen should be at least four times as accurate as the system under test. EW assets, as for our HARM program, are specialized and only available at a few ranges around the globe. The types of EW assets required are covered in chapter 9 of this text. Some of the key considerations are the number and types of emitters required, the range and coverage of those emitters, the ability to change the emitter characteristics, and the ability to generate threat signals from multiple locations.

There is always going to be a contention for assets, as there are always multiple programs that require the same assets as you do. There could be problems with a data reduction and analysis backlog with multiple programs. It is always a good idea to maintain a close liaison with the scheduling office and be prepared to generate a sortie when an opportunity presents itself. As corny as it sounds, it is true that flexibility is the key to good flight testing.

11.4 | RESPONSIBILITIES OF THE TEST TEAM

This section covers the roles and responsibilities of the flight test team, which includes the flight crew, team and flight briefs, test conduct, and debriefs. The responsibilities assigned to individuals listed here are by no means an industry standard, but they have worked well in the past.

The flight crew is an integral part of the flight test team and not just the bus driver as we tend to think of them sometimes. They are the operational expertise of the joint test team and hopefully are qualified in the aircraft with some recent fleet or squadron experience. This may or may not be true for some test organizations in the United States. Typically, after test pilot school the U.S. Navy sends the graduate to an operational assignment. The U.S. Air Force, on the other hand, posts the graduate to a test organization. The longer a pilot is away from an operational squadron the less relevant they become in an avionics and weapons setting. In addition, the pilot should be knowledgeable of the system under test.

Ideally the flight crew is identified early in the program so they can follow the development of the system and become expert in the system's integration. One of the problems with military test crews is that they tend to be rotated to a different posting about every 3 years, which usually coincides with the first development flight of the program. Now we have a situation where the flying will start to become heavy using a new flight crew unfamiliar with the system.

The pilot is the safety of flight coordinator on the test team who identifies unrealistic/unsafe flight tests and is the key in anomaly tracking and resolution. He is the human factors man in the loop and the link to the control room during the tests. He is a member of the test planning group and one of the main liaisons with design engineering and software development, especially during controls and displays mock-up evaluations. The flight crew is instrumental in lab development and in troubleshooting and problem resolution.

Two hours prior to the flight is not the time to be discussing the merits of the flight cards. For this reason, the flight brief should actually be broken down into two briefings: a technical team brief and the actual flight brief. The technical team brief is held 24 hr prior to the flight. Flight cards are reviewed for correctness and merit and the flight is analyzed with respect to program goals (i.e., Are we on the right track?). Results from the analyses of previous flights are introduced and discussed as to the merits of repeating/rejecting a proposed test. The meeting is concluded only after all technical discussions have been resolved and the test cards have been rewritten to everyone's satisfaction. Test cards will be highlighted for any special actions required by maintenance, the flight crew, or the test team. Flight cards are firm (sealed in concrete) at the conclusion of the brief, and depending on your organization must be submitted to the flight card approval authority (normally safety or the project leader).

Flight briefs are normally held 2 hr prior to the flight. Flight briefs must be attended by

- Aircrew
- Test conductor
- Aircraft coordinator
- Manufacturing/aircraft engineering
- Test team

The test conductor is responsible for the flight brief and will follow a standardized test conductor's briefing guide or some facsimile thereof. The first item in the guide should be aircraft-specific items. In many organizations a technician is assigned the responsibility of aircraft coordinator, sometimes called a plane follower. It is her responsibility to basically live with the test aircraft; nothing happens on the airplane that she does not know about. At the beginning of the brief, the aircraft coordinator provides a brief of the latest configuration and status of the airplane, including

- Previous aircraft write-ups, gripes, snags (or whatever your organization calls them) and their resolution
- Aircraft configuration, both hardware and software (Has anything been modified since the last flight?)
- Weight and center of gravity (CG) (Test organizations have a tendency to add and remove items from the aircraft, such as instrumentation, which changes the weight and balance of the aircraft.)
- Limitations and flight operating limitations (All test aircraft will have some limitations that are over and above the basic airplane. This is usually due to development hardware and software. It could be speed or altitude requirements that are lower than the basic aircraft envelope or a function/mode that should not be engaged.)
- Special requirements (These may be maintenance or engineering adjustments or variations in the aircraft normal procedures.)
- Radio frequencies (frequencies to be used for the flight)

You will notice that all of the items covered are important to the test team and need to be briefed at some point in the flight brief. Some organizations assign these tasks to many individuals on the team, but by having one individual responsible for them makes life so much easier.

After the aircraft coordinator brief, the aircrew is called upon to brief some standard mission items. These include the weather forecast, flight time, takeoff time, and in some organizations, the emergency of the day. An emergency of the day is a hypothetical problem with the airplane, and the aircrew is polled to see what they would do. In some emergency situations there is what is known as "bold face," which are steps that must be memorized and followed verbatim.

The test conductor is responsible for briefing the specific test mission. Some of the items that are covered include

- Flight cards, objectives, test points
- Information from the technical team brief
- Instrumentation
- Station operations and setup
- Debrief time

Notice that I have added the item of debrief time to required briefing items; this takes away the excuse of not knowing about it. One other thing about debriefings: Always strive to conduct the debrief as soon as possible after the flight. People tend to forget things as time passes—usually about 1 nsec.

The aircrew is responsible for concluding the briefing with aircrew coordination items, a chase or target brief, if required, en route and recovery, and Joker and Bingo fuel. Bingo fuel is the minimum fuel needed to return to base, make an approach, and fly to an alternate field. All test activities will cease at a call of Bingo fuel. Joker fuel is a Bingo fuel plus some cushion, and is called as a warning to the test team that it is getting close to go-home time.

TABLE 11.4 ■ Partial Radio Log

Time	Event
11:02	Aircrew arrives at aircraft
11:12	Left engine start
11:14	Right engine start
11:16	INS to align
11:19	Radar BIT complete
11:22	INS to navigate
11:25	Taxi
11:28	Predeparture checks complete
11:30	Aircraft takes the active
11:32	Brake release
11:34	Bird strike right engine

The test conductor is responsible for the entire test operation. With the exception of flight following (air traffic control) and range control, the test conductor should be the only voice on the radio. Analysts in the control room should have assigned seating; overflow should be in a viewing room. An overflow room is usually set up to the rear of the control room where spectators can view the test operation. They can hear and see everything that is going on, but cannot intrude on the test team. Sometimes we call this room a VIP or wannabe room. It is amazing how many people show up at the first flight of a program just to say that they were there. All analysts should have access to Hot Mike, UHF/VHF, and separate integrated communication system (ICS) nets; however, access to UHF/VHF should be restricted to those with safety of flight responsibilities.

An individual should be assigned to keep a radio log. This is a rather inexpensive method of keeping track of what happens during a flight, and can prove invaluable when searching the data tapes for specific events. The individual writes down the times that events happen during the test. A partial example is shown in Table 11.4.

Later, in the debrief, an analyst may ask the pilot, "You said the velocity vector on the HUD looked a little funny. When was that?" The pilot replies, "Oh, I don't know the exact time, but it was just after that bird strike." By looking at the radio log we can see the approximate time where we should search for the data. This is a real time saver, as the analyst does not have to search the entire tape for the time slice he requires.

Contact with the TSPI site, instrumentation, and data operations should be maintained by the test conductor. Instrumentation start and stop times (such as starting a new tape when the Zulu time rolls to the next day) are the responsibility of the test conductor. Start and completion of each step of the flight cards is initiated by the test conductor. Any refly of an event is initiated by the test conductor after consultation with the test team. Overall security is also the responsibility of the test conductor. You can see that the bulk of the responsibility lies with the test conductor, which is where we would like it.

As mentioned previously, debriefs should be held as soon as possible after landing, even if there is more than one flight, because problems and anomalies are easily forgotten over a period of time. Attendance should include design engineering, maintenance, and the program/project manager. The aircrew first addresses aircraft crabs (gripes, squawks, write-ups) and then the mission anomalies.

The test conductor reviews the flight as seen by the control room; radio logs should be handed out at this time. Time slices for postflight processing should be determined.

TABLE 11.5 ■ Test Conductor's Summary Sheet

Test Conductor's Summary Sheet	
Aircraft tail no.	207
Flight no.	16
Software load	A2.3
Date	January 16, 2007
Takeoff time	11:32
Land time	11:47
Reason for delay	Weather
Test points accomplished/planned	0/125
Objectives and results	Objective was to test the navigational accuracy and drift rate of the AN/ASN XXX enhanced GPS/INS. Test not accomplished due to in-flight emergency.
Flight discrepancies	AN/ASN XXX failed in flight
Significant occurrences	Bird strike right engine at 11:34
Purpose of next flight	Functional check flight (FCF) after engine replacement
Action required prior to next flight	Engine ground run at thrust stand
Comments	This was the worst run test that I have ever been involved with. I do not think that I want to do this anymore. I just might transfer to instrumentation.
Distribution	The world

The debrief should conclude with a discussion of overall flight results, success rate, and probable goals for the next flight. The test conductor should then complete a test conductor's summary sheet or some record of the flight events.

The test conductor's summary sheet is a concise record of the events of a test flight that covers the key aspects of the flight. It is also designed to answer questions that may pop up about the flight days, months, or even years after the flight was performed. A sample of a test conductor's summary sheet is shown in Table 11.5.

Most of the initial information, such as tail number and flight number, date, takeoff time, and land time are pretty much self-explanatory. The listing of the software load provides us with some configuration control, especially if we have to go backward to review data. Whenever a scheduled flight is late, everyone wants to know the reason for the delay. I have found there are three major reasons for delays: weather, maintenance, and instrumentation; since weather cannot complain, I just put weather in this block. The test points accomplished/planned block provides the reader with a measure of performance or success; a good metric for program management (even though it is only an estimate by the test conductor, as data analysis has not yet been performed).

The objectives of the flight and their respective results are listed as the next entry and are the test conductor's best guess once again. Any unusual events are annotated as shown, and based on the results of this flight, the potential purpose of the next flight is predicted. Actions required before the next flight may include hardware replacement, software or instrumentation upgrade, or any other maintenance actions.

The comments section is left entirely up to the test conductor, and I used to urge them to be as truthful and verbose as they wanted. Occasionally we would receive comments like the ones in Table 11.5. The distribution is self-explanatory, but do not be surprised if it expands as the test program progresses, as everyone wants to know what happened on the last flight.

██ 11.5 █ | ANALYSIS AND REPORTING

The first place an analyst can look to gain an overview of the flight is the video review. If we were smart up front, we will have a complete record of the flight with time and pilot's voice. This is one of the more important tools in understanding what went on during the flight. Video can be obtained in many ways:

- Production HUD camera
- Production camera system
- Instrumentation cameras
- Video downlink

The aircrew will use the video to assist them in writing their daily flight reports, and analysts use it for assistance in troubleshooting anomalies, locating the proper time for batch processing, and determining specification compliance.

There are two basic types of flight analysis: specification and performance validation, and troubleshooting and problem resolution. The analysis can be performed in real time or postflight, but the analyst needs to be aware of some differences. Real-time data uses extrapolation to keep up with real time and is not as accurate as postflight batch processing, which uses interpolation. In real time, the rates (samples per second) are limited by the PCM frame setup and frame rate. Postflight batch processing allows the user to access all of the data at rates up to the limits of the data bus.

After the data are compiled they must be analyzed by approved software routines. These routines may be either government or contractor approved programs. All of the data we review as analysts are going to be sorted through some type of statistical analysis. These calculations include

- Circular error probable (CEP) and circular error average (CEA) about the mean and about the target
- Mean and standard deviation
- Dispersion
- Student's *t* distributions
- Normal distributions
- Root mean square (rms)
- Ensemble packages

After all of the hard work is done the reports can be written (although some engineers and most pilots will argue that the report is the hard work). As with the test plan, most organizations will dictate how the reports are to be written via an operational instruction (OI). Contractors also have their own formats, and most customers will allow the contractor to submit reports in a contractor format. Most test pilot schools state that the final report should stand alone; that is, a test can be duplicated by just using the final report. This concept assumes that the reader has no access to the approved test plan, and it puts quite a burden on the analyst since the plan has to be regurgitated in the report. Of course, this makes the final report a rather large document that will probably be classified if no references are allowed. But then again, in military circles the saying is, "The job is not complete until the paperwork equals the weight of the airplane."

I have always been of the opinion that shorter is better, and that test methodology can be referenced back to the approved test plan in the final report: "The test was accomplished IAW the approved test plan, reference X, section X.XX." No matter how it is done, there are a few items that must be included:

- Executive summary. In one page, summarize the key aspects of the test and state whether the system under test was found to be acceptable, marginally acceptable, or unacceptable. If unacceptable, the reasons are normally provided in the executive summary. Write the executive summary for a general officer or company director who does not have time to wade through the entire report to figure out what happened.

- Background. Why was this test performed? Who asked for it? Reference to a specific tasking.

- General objectives and specific objectives. These need to match the test plan, as unapproved tests are not allowed unless vetted through the test review board (TRB) and safety review board (SRB) process.

- Instrumentation. Which avionics and TSPI data were collected during the test to assist you in coming to your conclusions? The parameters can be shown in an appendix or referenced to the test plan, but should never be included in the body of the report.

- Test method. As mentioned previously, it can be regurgitated here or referenced to the approved test plan.

- Results. This is the body of the report. It is good to include graphs and tables in this section if it aids the reader in visualizing the results. Tabular listings and spreadsheets should be included as an appendix. The results should be identified as acceptable, marginally acceptable, or unacceptable.

- Conclusions and recommendations. Every unacceptable result needs a *must* recommendation and every marginally acceptable result needs a *should* recommendation. Recommendations are listed in order of priority: *musts* first and *shoulds* second.

- Supporting data. All supporting data should be included in the appendix.

- Witnessing sheets. If a customer representative was onsite and witnessed the tests, it is a good idea to obtain his/her signature on a document that states this fact. If the report is submitted and the customer has a question about the test, he will more often than not contact that representative rather than you. It makes the submittal process much easier.

The hardest part of the report writing is the signoff, because it seems that everyone loves to wordsmith your work. The signoff cycle must be in accordance with your procedures, and there is really nothing you can do about it other than to grin and bear it. Test reports are contract data requirements list (CDRL) items (deliverables) and are submitted to the customer under the terms and conditions of the data item description (DID). Military development test and evaluation (DT&E) will also be required to submit a final report under their SOW obligations.

11.6 | SELECTED QUESTIONS FOR CHAPTER 11

There are no selected questions for this chapter. However, there is one final question for you to answer. It is a question that I have asked all of my students in every course I have ever had the privilege to instruct. The question is: "What, if anything, have you learned from this text?" Hopefully everyone that picks up this book will be able to say that they learned at least one thing new, and that always makes my day.

There are plenty of flight test engineers out there who have followed these procedures but have never understood why it is done that way. They have always been given the excuse that "It has always been done that way." Now you know why.

One other thing. The procedures and techniques that have been discussed are by no means the only way to perform an avionics or weapons evaluation, but they are all time tested and they do work. As the Federal Aviation Administration (FAA) says in all of their advisory circulars: "Presented is a means, but not the only means, of showing compliance with the applicable regulations (specifications)." Good luck.

Unmanned Aerial Vehicles (UAV)

Chapter Outline

12.0	Introduction	851
12.1	UAV Types	852
12.2	Interoperability	856
12.3	The Airworthiness Certificate	857
12.4	UAS Communications Architecture	860
12.5	Navigation	886
12.6	Autopilots	889
12.7	Sense and Avoid Systems	892
12.8	Payload	899
12.9	Optionally Piloted Aircraft (OPA)	901
12.10	Summary	902
12.11	Selected Questions for Chapter 12	902

12.0 | INTRODUCTION

The term UAV is a misnomer since the UAV is a part of the UAS, or unmanned aerial system, which may contain a ground control element (GCE), communications system, the UAV, and perhaps a shared data system. Other terminology includes unmanned combat air system (UCAS), uninhabited air vehicle, unmanned aircraft vehicle, and unmanned aircraft system, depending on the documentation. For this chapter I will use the generic UAV and UAS to denote aircraft and system. The emphasis has been on military systems, but this technology has migrated into the civilian world with applications in law enforcement, fire detection, border enforcement, and traffic management. The benefits are easily recognizable: a pilot is not required to be in the air (hazardous operations will not endanger a human), machines are not subject to fatigue, systems can be built smaller which require less fuel and provide longer endurance, and the overall cost is much smaller.

Since a UAV is unmanned, there are restrictions on where it can fly. Currently, there is no fail-safe guarantee that a UAV can assure safe separation from other airborne traffic. The FAA has strict guidelines on the operation of unmanned aircraft. To mitigate this problem and to aid in the development of UAVs and UAV systems, some organizations have turned to surrogate UAVs (SUAV), sometimes called piloted UAVs. There are regulations on these systems as well, and they still need to go through an airworthiness certification process; the reader is referred to FAA Order 8130.34B, "Airworthiness Certification of Unmanned Aircraft Systems and Optionally Piloted Aircraft," November 28, 2011. The FAA uses aircraft instead of aerial and formalizes the term optionally piloted aircraft (OPA).

For an overview of UAV aircraft, history, missions, and applications, two texts will aid in the evaluator's understanding of the subject: *Introduction to Unmanned Aircraft Systems* (Richard K. Barnhart, Stephen B. Hottman, Douglas M. Marshall, and Eric Shappee, CRC Press, 2011) and *Unmanned Aircraft Systems: UAVS Design, Development and Deployment* (Reg Austin, John Wiley and Sons, 2010).

The following publications may also offer some insight into UAV systems: *Introduction to Unmanned Systems: Air, Ground, Sea & Space; Technologies and Commercial Applications* (Jeremy LeMieux, Unmanned Vehicle University, 2013) and *Unmanned Aerial Vehicle End to End Support Considerations* (John G. Drew, Rand Corporation, 2005).

As with the previous chapters in this text, we are concerned with evaluating UAV systems and will not delve into the performance and flying qualities of these systems. The previous chapters provide the foundation for these evaluations, and this chapter will discuss the nuances of testing when dealing with a UAS.

12.1 | UAV TYPES

UAVs run the gamut from small to large, high and low altitude, reconnaissance to interdiction, and military to civilian. The key to the design and what systems are required depends entirely on its designated mission. Troops on the battlefield desiring intelligence of the surrounding area or battle damage assessment may have a need for a system that is short range, compact, and easily deployable. An example of such a system may be the WASP, which is one of the smallest drones in operational use (Figure 12.1).

FIGURE 12.1 ■
WASP

For theatre reconnaissance, the mission may dictate large loiter times over a vast area of interest. The mission may also dictate that high-resolution sensors be used to produce images (either in near real time or analyzed offline) to be read by ground supporting elements. This may force the development of a larger aircraft (to carry the payload and the required fuel) capable of high altitude flight (for area coverage). The RQ-4A Global Hawk (35 hr endurance and a ceiling of 65,000 ft), used extensively by the military and now NASA, would meet these requirements (Figure 12.2).

There may be a requirement for a hunter–killer system operating beyond line of sight (BLOS) capable of detecting, recognizing, and identifying hostile targets at a safe distance and employing sufficient firepower to destroy these targets. In surveying the inventory, a possible candidate exists that can detect and identify hostile targets but lacks sufficient firepower to deal with them. An upgrade to this existing platform may be possible to carry out this mission. The MQ-9 Reaper is a scaled-up version of the MQ-1 Predator (which itself is derived from the Gnat-750) and is capable of carrying Hellfire missiles and LGBs (Figure 12.3).

FIGURE 12.2 ■
RQ-4A Global Hawk

FIGURE 12.3 ■
MQ-9 Reaper

UAVs don't have to be fixed wing aircraft; the MQ-8 Fire Scout is a rotary wing aircraft made by Northrop Grumman and operated by the U.S. Navy. It is capable of automated landing on a moving aircraft carrier. Typical missions include surveillance, locating targets, and directing friendly fire. There have also been weapons tests with a Fire Scout armed with 2.75 in. rockets; the U.S. Army has now shown interest in having its own version. The Battlehog 150 is intended to meet the Marine Corps requirement for a vertical takeoff drone capable of operating from aircraft carriers. It can fly at over 300 mph with a payload of 500 pounds, with armaments likely to include Hellfire missiles, rocket pods, and 7.62 mm mini-guns. The Battlehog series is designed to be as robust as possible, being able to withstand small-arms fire from close range. The U.S. Army has ordered three long-endurance multi-intelligence vehicle (LEMV) UAVs from Northrop Grumman; the LEMVs were to be delivered by 2013. LEMV is based on the existing (and tested) hybrid air vehicle (HAV), which was an aerodynamic blimp built to transport cargo. HAV looks like a flattened blimp, a wide airship with much better handling qualities. LEMV is an unmanned blimp that can carry 1.1 tons of sensors, stay aloft for 21 days at a time, supply 16 kW of power, and move at up to 148 km/hr at 6400 m (20,000 ft) altitude. The LEMV made its maiden flight on August 9, 2012, in New Jersey (Figure 12.4). (which was its last flight as the program was canceled).

The U.S. military tends to classify UAVs based on four categories: size/weight, range, endurance/altitude, and mission. Table 12.1 delineates these categories.

The pilot can operate a UAS in either internal or external mode. While operating in internal mode the pilot is located in a GCS viewing real-time video or instruments to control the UAS. While operating in external mode the operator is directly looking at the air vehicle outside of a GCS.

The two basic operating modes of a UAV are remotely piloted vehicle (RPV) or command directed. In RPV mode, a pilot in a control station remotely flies the A/C through a set of controls. The autopilot controls the A/C in the command direct mode, but the pilot can make changes to the parameters of heading, airspeed, altitude, and next waypoint via inputs over the command and control link, similar to autopilot inputs by the aircrew in a Part 25 aircraft.

There are two types of RPV control: rate and vector control. An RPV utilizing rate control is the most basic of control modes and provides a direct link between the input device and the position of the flight control surfaces. Rate control delivers the most resemblance to manned aircraft control inputs, stability, and handling qualities. The lack of proportional stick force for pilot feedback degrades efficacy of rate control. Rate control requires the highest level of pilot motor skills and training.

An RPV utilizing vector control can be highly augmented much like a sophisticated modern manned aircraft's autopilot. Using vector control, the operator makes discrete inputs to the autopilot outer loop (heading, altitude, or airspeed desired), and the autopilot or stability augmentation system (SAS) manipulates the flight control surfaces to achieve the desired condition. Operator inputs can be made by stick position, control knob position, or increasingly via computer interface selections.

In a command directed vehicle (CDV), the aircraft is essentially autonomous. The CDV takes off and lands automatically and executes all flight maneuvers based on a preprogrammed set of instructions, or flight plan, stored in the ground control station prior to flight. A CDV must have an intelligent and reliable fault detection and response system that ensures predictable autonomous action. The GCE can alter flight plan in flight by loading a new set of instructions through inputs to the ground.

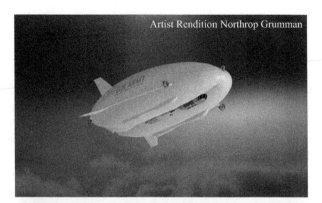

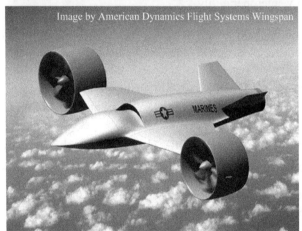

FIGURE 12.4 ▪
(from the top) LEMV,
Battlehog, and Fire
Scout

TABLE 12.1 ▪ UAV Categories

		Categories	
Size/Weight	Range	Endurance/Altitude	Mission
Micro Small	Short Medium	Medium Altitude and Endurance (MALE)	Reconnaissance
Medium Large	Long	High Altitude and Endurance (HALE)	Combat

■■■ 12.2 | INTEROPERABILITY

With the plethora of UAVs being fielded, it was necessary to create and use some sort of standardization in the data architecture of these systems. One military and one civilian standard are in use today. Interoperability between allied forces greatly increases efficiency and capability in a joint/combined service environment. This can be accomplished through the sharing of assets and the use of common information generated by UAV systems. By enabling the interoperability of multiple vehicles from a common STANAG (NATO Standardization Agreement) 4586 GCE, NATO operators can operate multiple UAVs of various types with widely different performance characteristics and features.

STANAG 4586, formally ratified by NATO in 2002 and now at edition 3, defines five levels of UAV interoperability:

- Level 1: Indirect receipt/transmission of UAV-related payload data
- Level 2: Direct receipt of intelligence, surveillance, and reconnaissance (ISR) data where "direct" covers reception of the UAV payload data by the unmanned control system when it has direct communication with the UAV
- Level 3: Control and monitoring of the UAV payload in addition to direct receipt of ISR and other data
- Level 4: Control and monitoring of the UAV, less launch and recovery
- Level 5: Control and monitoring of the UAV, plus launch and recovery

The Joint Architecture for Unmanned Systems (JAUS) (the full name of the standard is SAE-AS-4/JAUS) is an international civilian standard that defines communication protocols for unmanned vehicle systems (can also be ground vehicles or any other robot), some of their internal components, and their interaction with operator control stations. JAUS employs a service-oriented architecture (SOA) approach to enable distributed command and control of these systems. The standard defines message formatting for transport between system services and sets of standard services, which describe specific functional components for various unmanned system capabilities. The JAUS standards are owned and developed by SAE under the Aerospace Standards Unmanned Systems Steering Committee (AS-4).

The main goal of JAUS is to structure communication and interoperation of unmanned systems within a network. All information in a JAUS-compliant system is communicated in the form of messages, which are variable length sequences of bytes. Information on how to decode the messages is contained in a message header. The architecture is similar to the Link-16 TADIL-J Standard Messages and the terminal TIMs and TOMs (section 5.9.2.1). These messages were standardized so any JTIDS/MIDS terminal could decode them. There are seven message classes within JAUS: command, query, inform, event setup, event notification, node management, and experimental. The 16-byte header for these messages is in a format we saw in chapter 3. Each header contains the source and destination address, a command code that provides the decode information, and a sequence number to ensure that messages are not missed or received out of order.

Just as the TADIL-J messages are grouped according to functional areas such as network management, surveillance, and EW, so too are JAUS messages. Table 12.2 describes the functional areas (called service names in the document), functional purpose, and examples of the messages within that function. It should be noted that there

TABLE 12.2 ▪ JAUS Functionality

Service Name	Purpose	Example Messages
Transport	Acts as a common gateway for all messages entering and leaving the component.	None
Events	Enables other components to request messages from services that inherit from the events service on a fixed periodic or an on-change basis.	Create Event, Update Event, Cancel Event, Query Events, Reject Event Request, Event
Access Control	Allows services that inherit from the access control service to be exclusively controlled by a single source component. In turn, the controlled services will only accept commands from their controlling component. This is how JAUS components implement mutual exclusion.	Request Control, Release Control, Query Control, Query Authority, Report Control, Confirm Control
Management	Allows client components to control and access information about the internal state of another components function. For example, this allows client components to reset or shut down a server component.	Shutdown, Standby, Resume, Reset, Set Emergency, Clear Emergency, Query Status, Report Status
Time	Provides an interface for reporting a component's internal system time to other components.	Query Time, Report Time
Liveness	Allows client components to establish whether another component is online and responding to message communication. Similar to the concept of a network ping.	Query Heartbeat Pulse, Report Heartbeat Pulse
Discovery	Enables components to discover each other's presence and allows them to exchange information about which services they implement and support.	Register Services, Query Identification, Query Configuration, Report Identification, Report Configuration, Report Service List

are more than 150 standard messages in the current documentation. If users find that they require a message not included in the standard catalog they may develop experimental messages, but their incorporation will negate a JAUS compliance of their system.

The USN has declared that all UAV acquisitions will be STANAG 4586 compliant and all other systems will be JAUS compliant. The USN (PMA-263) (Program Management Activity) is the custodian for STANAG 4586.

▉ 12.3 | THE AIRWORTHINESS CERTIFICATE

A special airworthiness certificate in the experimental category, issued by the FAA, is required for all unmanned aircraft systems and optionally piloted aircraft to be flown in national airspace (NAS). As noted earlier in the chapter, the reader is referred to FAA Order 8130.34B, "Airworthiness Certification of Unmanned Aircraft Systems and Optionally Piloted Aircraft," November 28, 2011, for guidance on how to obtain this certificate. An important note for those seeking airworthiness certificates in this category is that the FAA considers the entire unmanned aircraft system, not just the UAV. The request will be evaluated by considering the UAV, data links, and the GCE as the complete system. This topic is addressed here because testing cannot commence until this certificate is granted. In addition, minimum aviation system performance standards (MASPS) are still in development at the time of this writing (hence an order and not an AC) and should be finalized by December 2014; further discussion on this subject is found in section 12.7.1.

This special airworthiness certificate will be issued only for research and development to nonpublic entities. Public or military organizations must obtain a certificate of authorization (COA). It is currently not possible to obtain a certificate for commercial operations. The certification restricts operation of the vehicle outside of a restricted area unless special permission is given requiring a manned chase aircraft or ground-based spotters providing visual separation from all other aircraft.

Additional information on keeping the vehicle within specified boundaries may be found in Range Commanders Council Document 323-99, "Range Safety Criteria for Unmanned Air Vehicles" (http://www.wsmr.army.mil/RCCsite/Documents/323-99_Range %20Safety%20Criteria%20for%20Unmanned%20Air%20Vehicles,%20Rationale%20and %20Methodology%20Supplement%20(Supplement)/323-99Sup.pdf).

Chapter 3, section 1, of the order describes the procedural requirements that must be followed in pursuit of the airworthiness certificate. In no case may any UAS or OPA be operated as civil aircraft unless a valid certificate has been issued. The process begins with a program letter that conveys to the authority the purpose of the certificate, the area over which the requested operations are to be conducted, the duration of the program, and other required information. Specific areas of concern that must be addressed in this letter are containment, lost link, and flight termination.

The letter must show the ability of the aircraft to be contained within the boundaries of the proposed flight area. The applicant's ability to provide information that satisfies this requirement will help define the operational area. The details of the lost link procedures must be included as part of a safety checklist. In the event that the command and control cannot be recovered, an independent means to safely terminate the flight must be provided.

A safety checklist is required, and the order goes into great detail of what is expected in this document. The applicant must provide a presentation consisting of detailed system descriptions augmented by block diagrams, wiring schematics, and S/W architecture. It is the FAA's responsibility to determine that the system is safe to operate in the NAS based on operational risk and safety assessments. Supporting documentation that must also be submitted includes the following:

- The proposed operating area plotted on an aeronautical chart with boundaries identified by latitude and longitude. The length of the boundary legs must be annotated as well as the proposed altitudes of operation.

- All appropriate operating manuals including operating limitations and checklists (normal, abnormal, and emergency).

- An appropriate training program for pilots, observers, chase, and ground personnel as well as documentation that all personnel have successfully completed such training.

- Evidence of an FAA pilot's license and current medical (private pilot's license and Class II medical); personnel not requiring a certificate but required to have successfully completed an FAA-accepted pilot ground school must document results of the written exam.

- Before conducting operations, the frequency spectrum used for operation and control must be approved by the Federal Communications Commission (FCC) or other appropriate government oversight agency. Depending on the frequency and power of the data link, a license for use may be required.

Chapter 3, section 3, covers the procedures for an OPA. The requirements are much the same as defined for the UAS with the exception of a statement allowing the applicant to opt out of the safety evaluation. Of course this comes with restrictions: there must always be a pilot onboard, and remote control equipment (RPV) must be removed. The only exception is within restricted airspace by permission of the controlling agency.

The appendices to the order present the applicant with examples of operating limitations for UAS and OPA systems and a sample program letter previously described. Appendix D provides the applicant with the safety checklist, which is quite lengthy (some 15 pages). It is imperative that organizations proposing civil applications or military applications operating in the NAS initiate work on this certificate early on in the program as ground and flight test operations will not be able to proceed without it.

A synopsis of the differences between the private industry and U.S. government operations.

Civil Operation: Private Industry

For civil operation, applicants may obtain a special airworthiness certificate, experimental category, by demonstrating that their unmanned aircraft system can operate safely within an assigned flight test area and cause no harm to the public. Applicants must be able to describe how their system is designed, constructed, and manufactured, including engineering processes, software development and control, configuration management, and quality assurance procedures used, along with how and where they intend to fly. If the FAA determines the project does not present an unreasonable safety risk, the local FAA Manufacturing Inspection District Office will issue a special airworthiness certificate in the experimental category with operating limitations applicable to the particular UAS.

Public Operation: U.S. Government Organizations

For public operation, the FAA issues a certificate of authorization or waiver (COA) that permits public agencies and organizations to operate a particular UA, for a particular purpose, in a particular area. The FAA works with these organizations to develop conditions and limitations for UA operations to ensure they do not jeopardize the safety of other aviation operations. The objective is to issue a COA with terms that ensure an equivalent level of safety as manned aircraft. Usually, this entails making sure that the UA does not operate in a populated area and that the aircraft is observed, either by someone in a manned aircraft or someone on the ground.

Additional information on airworthiness and COAs for UAV operations can be obtained from the FAA website (http://www.faa.gov/about/initiatives/uas/cert/).

12.3.1 Military Airworthiness Requirements

STANAG 4671, "Unmanned Aerial Vehicle Systems Airworthiness Requirements (USAR)," has established a baseline set of airworthiness standards in relation to the design and construction of military UAVs. When in compliance with this standard, a UAV shall be permitted to operate, or have a streamlined approval, to fly in the airspace of other NATO countries that have ratified the agreement. The airworthiness standards apply to fixed wing military UAVs with maximum gross weights of 150–20,000 kg (330–44,092 lb.)

TABLE 12.3 ▪ USAR Subparts

			UAV System			
		UAV	Command and Control Data Link	Communications System	UAV Control Station	Other Ancillary Elements
A	General	X	X	X	X	X
B	UAV Flight	X				
C	UAV Structure	X				X
D	UAV Design and Construction	X				X
E	UAV Powerplant	X				
F	Equipment	X				
G	Operating Limitations and Information	X			X	X
H	Command and Control Data Link		X	X		
I	UAV Control Station		X	X	X	

and have been modeled after EASA CS 23 (FAA Part 23) requirements. The agreement also states that compliance meets the minimum requirements of CS 23.1309 and the AMC (14 CFR 23.1309 and AC 23.1309). The agreement recognizes that certain unique features of UAS require particular additional requirements or subparts. The USAR does not cover "sense and avoid" but will include them as requirements are codified.

The USAR is presented like the current EASA regulations; Book 1 contains the applicable CS codes (FAA Part Paragraph Number) applicable to the UAV, and Book 2 contains the Acceptable Method of Compliance (AMC in EASA terms, Advisory Circular in FAA terms). Each of the books is divided into subparts (A through I) as shown in Table 12.3.

As an example, Subpart F, Equipment, starts with paragraph 1301 (which is very familiar) and is applicable as written in CS-23. Paragraph 1307 in the same subpart addresses ECS systems but only as applicable to the UAV; this paragraph is labeled U1307 to tell the reader that this paragraph is unique to USAR. When formulating a Test Plan for UAVs it is recommended that you consult USAR and incorporate the required testing called out to show airworthiness compliance; it is also an excellent method of ensuring that all testing is accounted for in the test plan. Book 2 will provide the applicant with an AMC to show airworthiness. The AMC for paragraph 1309 is to perform a safety assessment (which is exactly the same for manned aircraft but addresses UAV unique problems).

12.4 | UAS COMMUNICATIONS ARCHITECTURE

The communications network associated with a UAV or UAS is extensive and must be reliable. Four basic communications groups must be considered when performing an evaluation:

- In-vehicle communications
- Data link Communications

- Ground control element (GCE) communications
- Net-centric backbone communications

In-vehicle communications involve the movement of data and data commands within the vehicle. This is accomplished using one or many of the data busses previously covered. Data link communications involve the transfer of data and commands between the vehicle and the GCE. Uplink commands may control the autopilot, sensor operation, or requests for data. Downlink replies may consist of acknowledgments (ACK) or nonacknowledgments (NACK) (missed data), health and welfare status, or sensor data. Additional data link streams may also be used as in the case of Link-16 or TTNT participation. The GCE communications mostly involve the receipt and display of vehicle data to pilots, sensor operators, and analysts. Net-centric backbone communications forward UAV data (e.g., position, velocity, store status, sensor information) to other users such as command and control or network participants.

The in-vehicle system will be limited by the data bus architecture processing speed and memory. You may recall that if Mil-Std-1553 or AS15531 is used the throughput will be limited but deterministic. Data links have other concerns and these concerns are dependent on the types of data that are being exchanged. Autopilot commands and weapons arming and firing are critical commands that must be timely, correct, and interpreted correctly by the vehicle. Audio and video may be streaming or if not critical, be reassembled and viewed postflight or near real time. So within the data link, we may have the following:

- Multiple data streams
  - Sensors
  - Command and control
  - Status and health and welfare
- Each with different requirements
  - Urgency and priority
  - Reliability

The link itself can also encounter specific problems such as throughput, bandwidth, latency, dropouts, and what some call *lossy links* (which is really lost or missing packets of data and usually referring to transmission control protocol, or TCP). Requirements may also change based on whether the transmission is an uplink or downlink. For military applications, and in some cases civilian applications, the links are encrypted for obvious reasons. This adds another layer of difficulty to the operation and may add to latency issues. A generalized requirements list based on link type can be found in Table 12.4.

The GCE will receive multiple data streams of asynchronous and isochronous data; these data need to be time-aligned and merged to present this information to the UAV pilot, mission manager, or sensor operator/analysts. This can be a daunting task since all data may arrive at slightly different times. For this reason, it is imperative that all data be time-tagged when it was sensed for this is the time that will be used by the GCE to align and reassemble data.

As with the aircraft systems, there must be a contingency plan for cases where controls and displays within the GCE fail during an operation. What is the amount of redundancy required? Other concerns often overlooked in the development process are

TABLE 12.4 ▪ UAV Data Link Uses and Requirements

Data Link Types	High Reliability Data Link (HRDL)	High Capacity Data Link (HCDL)	Beyond Line of Sight (BLOS)	Redundant/Backup Data Link
Primary Use	Command and Control Own-ship Position and Status	Sensor Data (EO/IR/RADAR/ELINT)	Relay Transfer High Altitude	Command and Control and Status Data
Requirements	Low-to-Moderate Throughput with High Availability and Integrity	High Throughput Streaming Audio and Video	Combined Performance of High Reliability and High Capacity	Low Throughput with High Reliability and Integrity

FIGURE 12.5 ▪
Predator
Communications
Concept

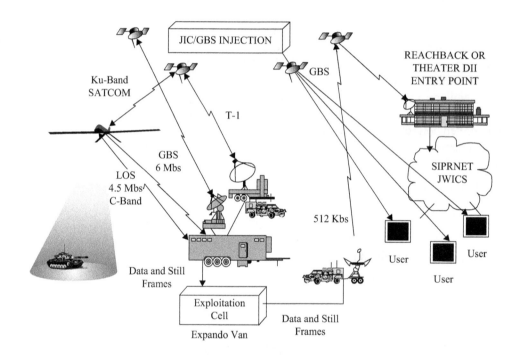

how much data are expected, how will they be captured, how will they be stored, how will they be retrieved, and what is their classification?

Shared data with other users are net-centric communications. The problems previously addressed with the GCE also apply to net-centric users; data must be in a readable (compatible) format. In large, multiparticipant scenarios, it would not be unreasonable to suggest that data would have to be distributed over multiple links (Links 4, 11, and 16 and TTNT) and in a timely manner.

12.4.1 Data Link Examples

Figure 12.5 is taken from the U.S. Air Combat Command (ACC), "Concept of Operations for Endurance Unmanned Aerial Vehicles," Version 2, December 3, 1996, the communications concept for the Predator UAV. The direct link to the GCE (called a theater exploitation system in the document) operates in C-band and can deliver a data rate of 4.5 Mbps. For the Predator, this link provides a full duplex command and control uplink and product (sensor) downlink out to approximately 100 nm LOS. The link is used to launch and recover the aircraft and provide EO full color and IR imagery at 30 frames per second; it supports EO/IR and SAR at 7 frames per second.

The Predator uses two satellite links: UHF SATCOM and Ku-Band SATCOM. Both provide over-the-horizon (OTH) command and control uplink and frame imagery (not including SAR) downlink. The UHF link supports a bandwidth of 16 Kbs, and command and control capabilities are limited to ingress and egress due to satellite control-response delays. The Ku-band commercial link has the same latency problems but can support a bandwidth of 1.544 Mbs (noted in the figure as a T-1 link), which allows downlinking high-resolution SAR frames and EO/IR black-and-white or color video.

The JIC/GBS satellites noted in the figure are part of the Joint Intelligence Center/ Global Broadcast Service. GBS operates as a one-way, wideband transmission service capable of supporting timely delivery of classified and unclassified data and video products for mission support and theater information transfer. GBS uses commercial satellite broadcast technology to deliver large imagery and data files that would overload typical tactical network capacity. GBS disseminates IP-based, real-time video and large data files (up to 4 GB in size) over-the-air (up to 45 Mbps) to combat forces using net-centric prioritized delivery based on unit mission reception priority profiles. More information on GBS can be obtained from the Los Angeles Air Force Base website (http://www.losangeles.af.mil/library/factsheets/factsheet.asp?id=7853).

Before continuing, a little clarity on nomenclature is required. You will notice in the previous paragraph that two terms are used to describe data size and data rate. These two terms are "4 GB in size" and "45 Mbps". When a capital B is used, the term is bytes; when a lowercase b is used, the term is bits. There are 8 bits to a byte, so a mistake in the understanding of the term will have you off by a factor of eight. Table 12.5 will hopefully eliminate any confusion.

Similar to the Range Commanders Council standardization of instrumentation and telemetry formats, the format and standards for exchange of command, control, communications, computers, intelligence, surveillance, and reconnaissance (C4ISR) architectures is governed by the Joint Technical Architecture (JTA) standards for interoperability. Per DoD Instruction 5000.2, December 8, 2008, highly sensitive, classified, cryptologic, and intelligence projects and programs shall follow this instruction to the extent practicable. Note that the DoD IT Standards Registry (DISR) is an online repository of IT standards formerly captured in the JTA, Version 6.0; DISR replaces JTA.

SIPRNET noted on the right side of the figure is the Secret Internet Protocol Router Network (SIPRNet), which is a system of interconnected computer networks used by the

TABLE 12.5 ■ Data Size Nomenclature

Meaning	Term
1 byte	8 bits
1024 bytes	1 KB (kilobyte)
1024 KB	1 MB (megabyte)
1024 MB	1 GB (gigabyte)
1024 GB	1 TB (terabyte)
1024 bits	1 Kb (kilobit)
1024 Kb	1 Mb (megabit)
1024 Mb	1 Gb (gigabit)
1024 Gb	1 Tb (terabit)

U.S. Department of Defense and the U.S. Department of State to transmit classified information (up to and including information classified SECRET) by packet switching over the TCP/IP protocols in a secure environment. The Joint Worldwide Intelligence Communications System (JWICS) is also noted on the right side of Figure 12.5 and is a 24 hr a day network designed to meet the requirements for secure multimedia intelligence communications worldwide. JWICS provides users with a high-speed multimedia network using high-capacity communications to handle data, voice, imagery, and graphics. Some sites have video and data capability on T1 lines, and some have strictly data capability (64 kbps lines).

The T1 line just mentioned may need a little more explanation. In the T&E programs performed in the 1980s, some organizations linked information over the satellite for remote operations and the organization paid for the transponder time when being used. The amount of data that are needed to be transmitted determines which link would be required. The nomenclature T1 and higher is the terminology for digital, two-way transmission of voice, data, or video over a single high-speed circuit. The transmission rate is based on the bandwidth for one voice channel in digital form. This channel is called DS-0 and consists of 64 kbps of bandwidth. Using time division multiplexing 24 of these DS-0 channels, the T1 link is formed. To separate the different channels a framing bit is used. For framing (see the discussion on instrumentation frame rates in section 3.19 of the text), 8000 bps are used. T1 therefore gives you 24 analog voice channels plus the framing rate. This defines the T1 speed as $24 \times 64,000 + 8000 = 1.544$ Mbps. Other T-versions use the same equation adding additional voice channels; the results are shown in Table 12.6.

Another family of data links in use with UAS is the common data link (CDL). This program is designed to achieve data link interoperability and provide seamless communications between multiple ISR collection systems operated by armed services and government agencies. CDL provides full-duplex, jam-resistant, digital microwave communications between the ISR sensor, sensor platform, and surface terminals. The CDL program establishes data link standards and specifications identifying compatibility and interoperability requirements between collection platforms and surface terminals across user organizations.

CDL is a full-duplex, jam-resistant spread spectrum, point-to-point digital link. The uplink operates at 200 kbps up to 45 Mbps. The downlink can operate at 10.71–45 Mbps, 137 Mbps, or 234 Mbps; rates of 548 Mbps and 1096 Mbps will be supported. The CDL family has five classes of links:

- Class I: Ground-based applications with airborne platforms operating at speeds up to Mach 2.3 and altitudes up to 80,000 ft
- Class II: Speeds up to Mach 5 and altitudes up to 150,000 ft

TABLE 12.6 ■ T(n) Link Capabilities

Carrier	Signal Level	# of T1 signals	# of Voice Channels	Speed
T1	DS-1	1	24	1544 kbps
T1c	DS-1c	2	48	3152 kbps
T2	DS-2	4	96	6312 kbps
T3	DS-3	28	672	44736 kbps
T4	DS-4	168	4032	274760 kbps

- Class III: Speeds up to Mach 5 and altitudes up to 500,000 ft
- Class IV: Terminals in satellites orbiting at 750 nm
- Class V: Terminals in relay satellites operating at greater altitudes

The CDL has the potential to utilize military satellites and carrier signals as well as commercial SATCOM. There are two military satellite CDL systems in use: Senior Span and Senior Spur. Senior Span operates in the I-band, whereas Senior Spur operates in the Ku-band to gain bandwidth. Other versions of CDL include the tactical common data link (TCDL) and the U.S. Navy sea-based common data link-Navy (CDL-N). The TCDL program provides a family of interoperable, secure, digital data links for use with both manned and unmanned airborne reconnaissance platforms. Possible platforms include UAVs, P-3 Orion, Guardrail, and JSTARS. It will transmit and receive data at rates from 1.544 Mbps to at least 10.7 Mbps over ranges of 200 km. TCDL should soon support the required higher CDL rates of 45, 137, and 274 Mbps. Predator was to have started TCDL testing in 2012.

12.4.2 Latency

Earlier in the text, we examined the concept of data latency with respect to the amount of time it takes for data to pass through the processing routine (senescence) and the delay, or transport time, to reach the receiver. We identified the total latency as the time from when a parameter is sensed to the time when it is received (e.g., telemetry room, another aircraft, the pilot's display). The Civil Certification of Automatic Dependent Surveillance Broadcast (ADS-B), also a data link, specifies that a latency study must be conducted (chapter 6). The reason given is that the information transmitted is being used by other aircraft and controlling agencies to mitigate traffic incursions. Latencies of over 2 seconds are deemed unacceptable for safety reasons. When evaluating UAS, the problem of latency should be obvious; critical functions which require as near real time as possible cannot be compromised based on "old" data.

LOS links will usually not have a problem with latency; since we are dealing with radio waves, the transmission time will be measured in microseconds. The largest component of the total latency will be that contribution of senescence, which is totally dependent on the system architecture of the in-vehicle communications.

BLOS communications via satellite is a completely different story as the largest contributor to the total latency will be the transport time. We can take an example of communicating from an aircraft to GCE over a geostationary satellite. For the satellite to be geostationary its orbit altitude above the earth must be 19,000 nm. If all other attenuation factors are removed, it would take a signal traveling at the speed of light about 250 ms to travel to the satellite and back to the ground. For an Internet packet this delay is doubled before a reply is received, which provides a typical latency of 500 ms. If an uplink and a downlink is required (i.e., command and control) this number would be doubled again to 1000 ms, which would be the theoretical minimum. In reality, the total latency across the satellite using TCP/IP would be somewhere between 1000 and 1400 ms (1–1.5 sec). Adding layers of security or encryption will increase the total transport delay. Some acceleration features can be incorporated in TCP/IP, but the latency will never be completely eliminated.

Medium and low earth orbiting (MEO/LEO) satellites do not have the severe latency problems associated with the geostationary satellites. LEO constellations have round trip delays of less than 40 ms, but some are severely limited in throughput of about 64 kbps. The proposed O3b Networks MEO is scheduled for deployment in 2013, will orbit at an altitude of 4350 nm, and claims round trip latency of 125 ms, and a throughput of 2.1 Mbps using TCP Reno (http://www.o3bnetworks.com/media/45606/latency%20matters.pdf).

Note: TCP Reno, like TCP Tahoe and TCP Vegas, is a congestion avoidance algorithm used in TCP, which is designed to limit the amount of lost packets of data. It accomplishes this by increasing a delay time for a new transmission and limiting the amount of data sent. More on this later.

MSCI's LEO COMMStellation will orbit the earth at a height of 540 nm in a polar orientation. The COMMStellation will be composed of 78 microsatellites in six orbital planes with an additional six redundant microsatellites (one per orbital plane). In its current configuration, COMMStellation will provide 100% global coverage with up to 15 times the speed and 10 times the total bandwidth capacity of a MEO constellation of comparable satellites (http://www.commstellation.com/constellation/index.html).

Congestion on TCP links will also add to latency. TCP ensures the error-free sequenced delivery of data transmitted from a source to a destination. Data are sent by the transmitter to the receiver in fragments called TCP segments or packets; a maximum size is defined for the segments. The receiver will confirm the correct reception of sent TCP segments by acknowledging them (sending the transmitter a response "ACK message" with a correct sequence number). This process can be seen in Figure 12.6. If an ACK message is not received or if it is not correct, the transmitter will return the message ensuring a reliable transfer. At the end of the session, it is correctly closed by exchange of FIN (finished) messages. The key factor in this context is that TCP will send only a limited amount of data before it needs an acknowledgment; in Figure 12.6 this number is four. The amount of data is governed by the TCP buffer size; Windows 7, for example, can operate at 64 KB. If 64 KB of data have been sent, but the acknowledgments have not yet been received, TCP will wait for an acknowledgment before it sends another packet. This acknowledgment mechanism is what limits the transmission rate over high latency links.

For data to be declared reliable the following protocol is followed:

– Each message is identified with a session identification and a sequence number.

– Periodic polling is accomplished to tell the receiver which sequence numbers should have been received.

FIGURE 12.6 ■
TCP Confirmation
(No Data Loss)

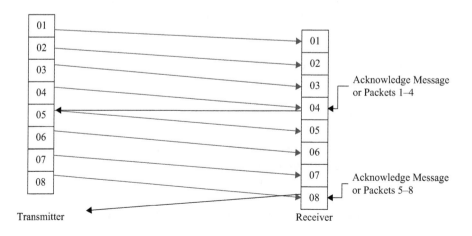

– Acknowledgments accepted at the transmitter, which clears the send buffer.

– Nonacknowledgments accepted at the transmitter, which then sends the repairs.

In the case of missing, lost, or error detected, the receiver will send a not-acknowledged (NACK) reply to the transmitter identifying the missing packets. This situation is shown in Figure 12.7. We can see that delays will increase as data need to be resent and then validated that it was received properly. A lost or missing packet will block all future traffic until that packet is repaired (because TCP sends all traffic over one stream). Similarly, a large message will block all future traffic until it is received. When this situation occurs the TCP will mistakenly attribute the packet loss to congestion. One could also assume that the larger the message, the higher the probability that one of the packets will be lost. If congestion is detected or anticipated, the congestion algorithms will be employed. One of the first things the algorithms will do is reducing the send window size (the amount of data allowed to be unacknowledged/outstanding in the network). For example, if you assume that the normal (uncongested send window) is 32 KB and congestion is detected the window size will be reduced to 16 KB; your 32 KB message will take twice as long. The window size will be adjusted upward when the congestion is eased. A new large window TCP extension (TCP-LW) allows windows up to 232 KB.

Other inherent problems with TCP contribute to latency. One of these items is slow start, which occurs when a TCP connection is first started or has been idle for a long time. The connection needs to determine the available bandwidth, and it does this by starting out with a window size of one packet (maybe 512 bytes) and ever increasing the window size until packet loss occurs: hence, an idea of bandwidth available (data in transit ~ bandwidth × delay). If we solve the window size problem noted in the previous paragraph the slow start problem becomes worse.

A final problem with TCP is the lack of prioritization as previously discussed in chapter 3 of this text. The reader will recall that for many data bus architectures the carrier sense multiple access (CSMA) can be handled in three ways: collision avoidance (CA), collision resolution (CR), and collision destruction (CD). Access can be handled

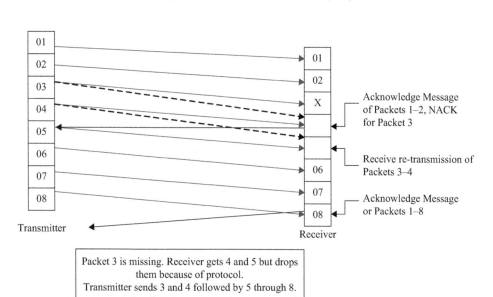

FIGURE 12.7 ■
TCP with Packet
Loss

FIGURE 12.8 ▪
TCP Data Buffers

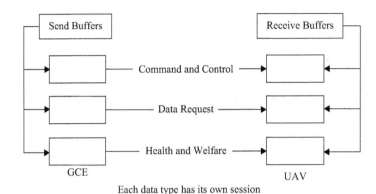

Each data type has its own session

by timing, prioritization, or a free-for-all; TCP uses the latter. Transmissions on TCP may have to be sent multiple times before they are acknowledged.

To recap the problems discussed:

– Blocking, where a byte cannot be delivered until all previous bytes have been received

– Insufficient buffer size

– Congestion control

– Slow start-up times

– Lack of prioritization

TCP has been addressed in detail here as the relevant discussion was deferred in chapter 3. In the discussion of military data links (Section 4.11.16 of the text) problems were noted with the excessive time devoted to complex participant net assignments, relatively low data throughput, and the inability to drop, add, or replace participants at will. It was suggested that TCP/IP could be extremely beneficial due to its peer-to-peer and plug-and-play attributes, which would allow a rapid entry into the net without complex assignments as the participant would be assigned an address at start-up.

The problems noted with TCP can be solved to some degree with modifications to the architecture. To solve the reliability, flow control, and disconnections, one would require data buffers at the transmit and receive side where data are sent from the transmit buffer to the receive buffer (similar to what was seen in JTIDS/MIDS systems). If multiple buffers were employed at both ends, it would be possible to exchange multiple sets of data between a single transmitter and receiver. This process is shown in Figure 12.8.

For this scenario to work properly, it must be reliable and must have some redundancy. This protocol must allow different buffer sizes, must keep the link active to prevent start-up times, must allow the buffers and links to survive link disconnection, and must support prioritization among the links. It should support duplicate packets (and know what to do with them) and identify (message or session ID) each message set. The protocol should batch small messages and parse and reassemble large messages.

Another aid to TCP is the implementation of a network middleware, which is a library between the operating system and the application, thereby insulating the application from the raw network; its implementation will increase reliability and allow for

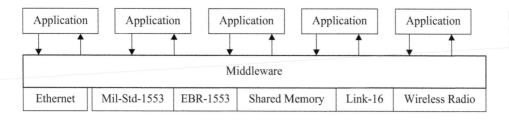

FIGURE 12.9 ■
Middleware
Implementation

buffering, supplying data when the applications need it. The network middleware will perform interface operations for all sources of data, not only TCP. Middleware has been mandated for data distribution within the U.S. DoD by DISR. Figure 12.9 provides a diagram of this concept.

12.4.3 UAV Communications Evaluations

The basic testing of the communications package within the UAV will be much the same as covered in chapter 4 of the text. The following evaluations must be planned:

– Engineering analysis

– Antenna patterns

– EMI/EMC

– Ground coverage

– Human factors and controls and displays

– Airborne coverage

– Latency

– USAR considerations

12.4.3.1 Engineering Analysis

The analysis listed here may be beyond the purview of the flight tester; it really depends on how your organization is structured. Either way, it is important for the evaluator to understand what analysis is required prior to formulating a test plan. Just as any other aircraft system, we must strive to understand as much as possible as to how the system under test works. The engineering analysis will be used as a predictor of in-flight performance. The analysis should be able to predict possible blockage areas based on aspect angle, possible interference, and effective maximum range of the communications systems.

Blockage areas can be surmised by examining the locations of the communications antennas on the aircraft for the primary and redundant systems. Because the size of the UAV may limit the amount of real estate, some blockage areas may not be avoided. This problem may be exacerbated by the requirements of separation of systems to avoid signal interference. Knowing this information ahead of time can direct the evaluator to verify the analysis as well as structure UAV operations to minimize the impact of dropouts.

We will plan to test the maximum LOS coverage of the communications systems, but what is the predicted maximum range? Similar to radar we can calculate the theoretical maximum range if we know the pertinent parameters which drive this equation. In some of the documentation, the maximum range can be calculated using a formula called the link equation, which calculates the received signal power as a function of

FIGURE 12.10 ▪
Link Equation
Contributors

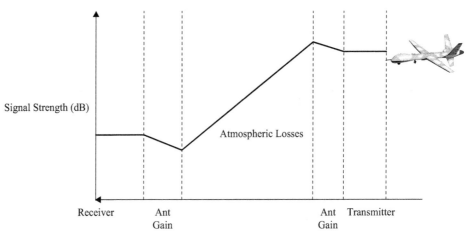

various link parameters. It is different from the radar range equation, which uses a round-trip propagation, whereas the link equation is a one-way propagation. The inputs to the calculation will consist of such variables as the transmitted power, the receiver sensitivity, transmitter and receiver gain, and the effective range of the transmitter. A graphic description of the data link signal path is shown in Figure 12.10.

If we were to sum up the variables in Figure 12.10:

$$\text{Received Power (dBm)} = \text{Transmitter Power(dBm)} + \text{Transmitter Antenna Gain (dB)}$$
$$- \text{Atmospheric Losses (dB)} + \text{Receiver Antenna Gain (dBm)}$$

$$(12.1)$$

where dBm is the normalized dB per milliwatt.

The first two terms of transmitter power and transmitter antenna gain, when combined, form the familiar effective radiated power (ERP), which is the signal power radiated from the transmitting antenna in the direction of the receiving antenna. ERP is in dBm as is the transmitter power (1 W ~ 30 dBm):

$$ERP \ (\text{dBm}) = P_T + G_T \qquad (12.2)$$

where P_T is transmitter power (dBm), and G_T is transmitter antenna gain (dB).

The conversion between watts to dBm and dBm to watts can be accomplished easily with online conversion tools, or one can remember some of the more common values. Table 12.7 provides some of these common values.

The term "atmospheric losses" shown in Figure 12.10 and equation (12.1) is really a combination of two losses: spreading (or diffusion) and attenuation. Spreading is akin to the inverse square law that was seen in chapter 7 for thermal energy and chapter 8 for radar energy. Remember that the power available at a receiver in a nonatmospheric case (i.e., transmission is not affected by the atmospheric medium) will vary inversely as the square of the distance to the source. The amount of energy captured will therefore depend on the distance from the transmitter and the size of the receiving antenna.

TABLE 12.7 ■ Watt/dBm Conversion

Power (W)	Power (dBm)
−30 dBm	0.0000010 W
−20 dBm	0.0000100 W
−10 dBm	0.0001000 W
0 dBm	0.0010000 W
1 dBm	0.0012589 W
2 dBm	0.0015849 W
3 dBm	0.0019953 W
4 dBm	0.0025119 W
5 dBm	0.0031628 W
6 dBm	0.0039811 W
7 dBm	0.0050119 W
8 dBm	0.0063096 W
9 dBm	0.0079433 W
10 dBm	0.0100000 W
20 dBm	0.1000000 W
30 dBm	1.0000000 W
40 dBm	10.0000000 W
50 dBm	100.0000000 W

Power (dBm)	Power (W)
0 W	not defined
0^+ W	$-\infty$ dBm
0.00001 W	−20.0000 dBm
0.0001 W	−10.0000 dBm
0.001 W	0.0000 dBm
0.01 W	10.0000 dBm
0.1 W	20.0000 dBm
1 W	30.0000 dBm
10 W	40.0000 dBm
100 W	50.0000 dBm
1000 W	60.0000 dBm
10000 W	70.0000 dBm
100000 W	80.0000 dBm
1000000 W	90.0000 dBm

For two isotropic antennas (as it turns out, the equation applies as well for transmissions between any two antennas), the spreading loss will be equal to a constant times the frequency squared times the distance squared:

$$L_S = K \times f^2 \times d^2$$

Or to put it in the form of dB (since it's easier to work with):

$$L_S = K + 20 \operatorname{Log}(f) + 20 \operatorname{Log}(d) \tag{12.3}$$

where L_S is spreading loss (dB), K is constant (32), f is frequency (MHz), and d is distance (km).

Nomographs are available in many antenna textbooks and can save you some of the calculations. Figure 12.11 is an example of such a nomograph and can help you calculate

FIGURE 12.11 ▪
Spreading Loss for
Frequency and
Range

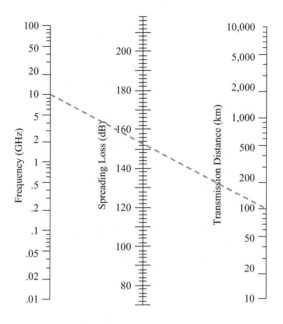

FIGURE 12.12 ▪
Losses due to
Absorption in dB

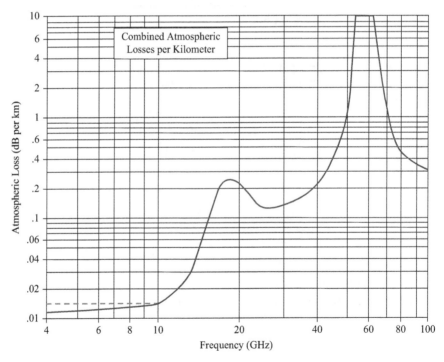

the loss with a known transmitting frequency and distance between the two antennas. Figure 12.11 shows a loss calculation for a 10 GHz signal over 100 km (~60 mi) is approximately equal to 153 dB; or you may expect a loss of 153 dB in this situation.

We have seen attenuation losses in radio waves when addressing radar in chapter 8 (Figure 8.1). Attenuation comprises absorption and scattering; losses are small for low frequencies and increase significantly above 10 GHz. Since Figure 8.1 is in terms of transmissivity, it is converted to terms of loss in dB in Figure 12.12. If we use our previous example of a 10 GHz signal traveling 100 km, Figure 12.12 shows a 0.015 dB loss per km, or 1.5 dB over the entire path length. By using the graphs in Figure 8.1 (for rain and fog) and Figure 12.12 for absorption, one can estimate the total attenuation for the link due to atmospherics. For frequencies less than 1 GHz it is usually acceptable to ignore atmospheric attenuation.

The signal arriving at the receiving antenna will be the ERP minus the atmospheric losses (spreading and attenuation) plus the receiving antenna gain. By putting all we know into equation (12.1) it now becomes

$$P_R = ERP - 32 - 20 \, \text{Log} \, (f) - 20 \, \text{Log} \, (d) - L_{atm} + G_R \tag{12.4}$$

where P_R is signal strength at receiving antenna (dBm), ERP is effective radiated power (dBm), f is frequency (MHz), d is distance (km), L_{atm} is atmospheric loss (dB), and G_R is receiving antenna gain (dB).

Let us suppose that in our example we generate an ERP of 70 dBm and utilize a receiving antenna with a 3 dB gain; what would be the received power? Using equation (12.4):

$$P_R = ERP - 32 - 20 \, \text{Log} \, (f) - 20 \, \text{Log} \, (d) - L_{atm} + G_R$$
$$P_R = 70 \, \text{dBm} - 32 - 20 \, \text{Log} \, (10000) \, \text{dB} - 20 \, \text{Log} \, (100) \, \text{dB} - 1.5 \, \text{dB} + 3 \, \text{dB}$$
$$P_R = 70 \, \text{dBm} - 32 \, \text{dB} - 80 \, \text{dB} - 40 \, \text{dB} - 1.5 \, \text{dB} + 3 \, \text{dB}$$
$$P_R = -80.5 \, \text{dBm}$$

Of course, this would be the answer in the perfect world; like the radar equation, other unknowns may adversely affect the received power. We could have rain, unwanted noise, interference, or antenna misalignment, which will affect what we receive. In radar we used S_{min} or minimum energy required to achieve detection. With data links, we have a minimum acceptable signal level at the receiving antenna required to ensure operational success; this is called a link margin and is defined as

$$M = P_R - S \tag{12.5}$$

where M is link margin (dB), P_R is power at the receiving antenna (dBm), S is minimum signal level that must be seen at the receiver for operational success (dBm).

The reader will remember that in radar we determined a minimum signal-to-noise (S/N) ratio required to enable detections and reduce false alarms; the S/N required was always above one (a margin). The link margin will be calculated based on operational need as a function of dropout time; the margin equal to the inverse of the average dropout time that can be tolerated (once again converted to dB):

- 10% dropout time equates to a 10 dB margin: $1/0.1 = 10$, $10 \, \text{Log} \, 10 = 10 \, \text{dB}$
- 1% dropout time equates to a 20 dB margin: $1/0.01 = 100$, $10 \, \text{Log} \, 100 = 20 \, \text{dB}$
- 0.1% dropout time equates to a 30 dB margin: $1/0.001 = 1000$, $10 \, \text{Log} \, 1000 = 30 \, \text{dB}$

Equation (12.4) can be rewritten as signal required with the added term of link margin and becomes

$$S_{REQD} = ERP - 32 - 20 \, \text{Log} \, (f) - 20 \, \text{Log} \, (d) - L_{Atm} + G_R - M \tag{12.6}$$

where S_{REQD} is signal strength at the receiving antenna (dBm), *ERP* is effective radiated power (dBm), *f* is frequency (MHz), *d* is distance (km), L_{atm} is atmospheric loss (dB), G_R is receiving antenna gain (dB), and *M* is required link margin (dB).

With this equation, we can solve for any one of the parameters in terms of the other five. For example, if we want to know the effective range of the link, we would rearrange the terms and solve for range.

12.4.3.2 Antenna Patterns

Antenna pattern testing is as described previously in section 4.10 of this text. In addition to communications antennas, self-protection antennas such as MAWS and RWR (if installed) and sensor coverage should be evaluated. Since UAVs are small, and unmanned, the logical place for the Engineering evaluation is first in modeling and simulation, followed by a full mock-up in an anechoic facility. Flight against an antenna farm would be inappropriate as we need to know the potential problem areas before flight is ever attempted. The communications coverage will be tested in flight but not before we have an idea of the predicted performance.

12.4.3.3 EMI/EMC

EMI/EMC will be evaluated in accordance with the procedures in section 4.14 of the text and, for the military, MIL-STD-461F and MIL-STD-464A. If the UAV is slated for shipboard operations, the evaluator should also consult MIL-STD-1605A (SH). Of particular concern for the evaluator will be the loss, degradation, or dropouts of data link communications due to interference. This evaluation may be thought of as susceptibility to RF interference, or as discussed previously, noise jamming.

In the electronic warfare section we introduced a term called the jamming-to-signal (J/S) ratio, which would determine when a target would burn through self-protection noise, or conversely, the effectiveness of the jamming signal. The major variables in the J/S ratio were the ERP of the jammer and the ERP of the radar. There are similar considerations with the evaluation of interfering signals on the data link. We would have to evaluate the ratio of a radiated interference signal to the desired signal as seen by the receiver. Logic would tell us that power, distance, and aspect would have serious considerations. In the simplest case, our receiver is pointed at the transmitter and tuned to the specific frequency and the interfering signal is located at some different range and direction of arrival (Figure 12.13). One can see that if the UAV had a high gain antenna (directivity) or if the interfering signal was out of band or of a very low power there may not be a problem.

In this geometry, rather than a J/S ratio, we have an interference-to-signal ratio (I/S), which is defined as the power ratio of interfering signal to the desired signal sensed at the receiver. In addition to ERP of the transmitters, it is also a function of the angular isolation (I_A) and the frequency isolation (I_F) of the signal. I_A is the bore sight gain of the receiving antenna referenced to the direction of the interference signal; I_F is the amount of filter attenuation in the receiving system that applies to the interfering signal. The I/S is given by

$$I/S = ERP_I - ERP_D - L_I + L_D - I_A - I_F \tag{12.7}$$

where *I/S* is interference-to-signal ratio (dB), $ERP_I =$ ERP of the interfering signal (dBm), ERP_D is ERP of the desired signal (dBm), L_I is link losses of the interfering signal (dB), L_D is link losses of the desired signal (dB), I_A is antenna isolation (dB), and I_F is frequency isolation (dB).

FIGURE 12.13 ■
Interference
Geometry

We know the link losses from equation (12.3) and can substitute in equation (12.7), which becomes

$$I/S = ERP_I - ERP_D - 32 - 20 \text{ Log } (F_I) - 20 \text{ Log } (D_I) + 32 + 20 \text{ Log } (F_D)$$

$$+ 20 \text{ Log } (D_D) - I_A - I_F$$

- The +32 and the −32 cancel.
- If the two frequencies are close to each other, the 20 Log (F) terms cancel.
- If they are at the same frequency, there is no frequency isolation so I_F drops out.
- Equation (12.7) then becomes the equation for interference within the band:

$$I/S = ERP_I - ERP_D - 20 \text{ Log } (D_I) + 20 \text{ Log } (D_D) - I_A \qquad (12.8)$$

It is readily seen in equation (12.8) that interference is a function of power, distance, and the aspect or off angle from boresight of the transmitters. The I_A term in the equation relates to the off boresight angle. If we assume that the UAV's antenna boresight in Figure 12.13 points directly at the GCE, then the angle to the interfering signal is something off boresight. If the gain of the antenna is a maximum at boresight (let's assume 10 dB), then the gain in the direction of the interfering signal will be less (let's assume 2 dB). It may be easier to visualize if we remember what a radar beam looks like: the main lobe would be equivalent to the boresight, whereas the off angle direction would be equivalent to a sidelobe. If I_A is equal to the difference between the receiving antenna's boresight gain and its gain in the direction of the interfering signal, our example would yield a value of 8 dB (10 dB – 2 dB). What would happen if the antenna on the UAV was an omni? Since the gain is equal about the sphere, I_A would disappear, making the link more susceptible to interference.

Suppose the GCE uplink operates at 2 GHz, ERP of +30 dBm at a range of 40 km; the interfering signal is operating at 1.8 GHz, an ERP of +30 dBm, at a range of 100 km with the same I_A as in the previous example:

$$I/S = ERP_I - ERP_D - 20 \text{ Log } (D_I) + 20 \text{ Log } (D_D) - I_A$$
$$I/S = 30 - 30 - 20 \text{ Log } (100) + 20 \text{ Log } (40) - 8$$
$$I/S = 30 - 30 - 40 + 32 - 8$$
$$I/S = -16 \text{ dB}$$

At 16 dB below the desired signal you are probably going to be all right; it may be seen but not to the degree of degrading the link. You can see from the equation that as the interfering signal gets closer in range or closer to the boresight gain, the I/S value will increase, thus creating problems. If we take the GPS satellite signal and assume losses were negligible (which they are not), how large of a jammer at 10 km away would produce detrimental effects? The GPS ERP is approximately 170 dB and orbits at 20,000 km; a 1 W jammer is converted to 30 dBm. When placing these parameters into equation (12.7):

$$I/S = ERP_I - ERP_D - 20 \, \text{Log} \, (D_I) + 20 \, \text{Log} \, (D_D) - I_A$$
$$I/S = 30 - 170 - 20 \, \text{Log} \, (10) + 20 \, \text{Log} \, (20{,}000) (\text{assume } I_A \text{ is negligible})$$
$$I/S = -140 - 20 + 80$$
$$I/S = -80 \, \text{dB}$$

It would appear that we are safe from this jammer, but what would be the result if the real ERP were half of the stated 170 dB?

This discussion tells us that a survey of the frequencies in use over our proposed operational/test area needs to be accomplished prior to executing our flight evaluation. Small, civil UAS may utilize link frequencies of 900 MHz, 2.4 GHz, or 5.8 GHz. The 900 MHz band is an industrial, scientific, and medical (ISM) radio band; it does not require a license below 1 W. Garage door openers, older wireless phones, some LAN systems, and RC hobbyists clobber this frequency. They use spread spectrum, frequency hopping to avoid interference, but at 1 W without an amplifier it can be a problem. The other available frequency spectrums for use without licenses have similar problems: 2.4 GHz is used by Wi-Fi, and newer phones and is heavily congested; 5.8 GHz is the least crowded of the three, but its use is rapidly expanding for wireless LANs and cordless phones.

12.4.3.4 Communications Ground Tests

As is the case with all systems evaluations, we initially start with a static test. This will involve external power to the aircraft, and exercising each of the links may be first to an external test cart and then to the GCE. If the system is employing multiple links, it would be advisable to test each link individually at first and then in combinations. We are looking for satisfactory communications between the UAV and the GCE. When we evaluated radios we used either a rhyme test or a digital STIDAS system. Since the UAS exchanges data, the exit criteria for this evaluation will be a satisfactory bit error rate (BER).

The BER testing involves the system's ability to detect and in some cases correct message errors. A parity bit, as in the case of 1553, cannot correct errors, and can detect only an odd number of errors. It would then be advisable to have some other error detection code to ensure that what was sent is indeed what was received. There are a number of data encoding schemes designed to meet this requirement. You may recall that Reed-Solomon encoding is utilized in 1760. Hamming codes, which are used in telecommunications, can detect up to two and correct up to one bit errors and is probably a system that you may have to deal with.

BER testing will be accomplished on the ground and in the air utilizing various link geometries, data rates, and data types comparing transmitted bits to received bits. A specification will govern the allowable BER; a typical value will be somewhere between 10^{-6} (one error per million bits sent) and 10^{-8}. When encryption or communications security (COMSEC) routines are used, the BER is usually relaxed. Remember that the link is tested on both sides: the UAV as well receipt by the GCE. If satellite link or SATCOM is to be employed, they need to be tested as well; this will be the first opportunity to evaluate the latency within the system.

The evaluation of video on the link is for the most part subjective; is the image usable to accomplish mission objectives? Things to look for that may be indicative of fading or lost signal include snow, pixilation, and frozen images. The picture quality will be a function of the resolution and frame rate. A high resolution and a high frame rate will require a large amount of bandwidth; a lower frame rate may cause a loss of resolution with dynamic targets. A Predator test team performing ground tests on link interference used a blowing flag in the camera field of view to provide a continuously moving target, which was an excellent idea. The report is titled "Predator UAV Line-of-Sight Datalink Terminal Radio Frequency Test Report," Report # JSC-CR-04-066, Department of Defense Joint Spectrum Center, Annapolis, Maryland 21402, and can be viewed online.

The direct links are LOS and will be affected by blockage areas such as buildings and other antennas and may be susceptible to multipath or refraction. One way to evaluate the possible ground coverage would be to tow the UAV around the airport (taxiways, holding areas, runways) and note link loss and duration of the loss. Operational procedures may have to be adjusted based on this testing; it may be advisable to tow the UAV to the runway rather than taxiing due to dropouts.

How do we track the UAV to ensure the highest gain for the antennas? The UAV can be tracked in many ways: visually, by signal strength, or some version of an auto tracker. The communications antennas are slaved to the tracker to ensure optimum reception. A visual track has obvious limitations: it will be affected by blockage areas, precipitation, lighting conditions and range. Track by signal strength can be accomplished in the same way ADF is implemented in an aircraft; but has the same limitations with azimuth accuracy. A high gain receiving antenna can be rendered useless with a tracker accurate to only 10 degrees. Early auto trackers utilized the aircraft's IFF transponder. The ARTCC system offers excellent azimuth information but requires an altitude reply from the aircraft due to poor elevation accuracy. The auto tracker uses the IFF azimuth and calculates an elevation from the altitude reply to point the antennas. This type of tracker will be affected by inoperative transponders or errors in the altitude encoder. A more common method is to use the aircraft's present position (as sent to the GCE) and the receiving antenna's present position, perform a coordinate conversion to range, azimuth and elevation, and drive the antennas with this information. This type of tracker requires excellent position information for the UAV, which is accomplished by using either GNSS or SBAS systems. Our static test and ground evaluation should include tracking accuracy of the auto tracker as part of the test. We will be looking for pointing accuracy as well as smoothness of the track. The optimum scenario would include an auto tracking antenna coupled with a spectrum analyzer to determine the actual strength of the received signal; the antenna pointing angle can be refined manually to obtain the maximum strength.

12.4.3.5 Human Factors and Controls and Displays

If the test and evaluation master plan was implemented correctly, the test/operational team will have had numerous hours of training and familiarization on GCE mockups and simulations. Many of the human factors concerns should have been addressed during this period. The ground evaluations, however, will afford the team the first real look at the installed system using actual links to control the UAV and receive information. It will offer the first chance to observe link connectivity, latency, and quality of the video.

Even though the aircraft operation is being viewed and controlled remotely, the human factors principles, certification, and problem issues covered in section 6.8 of the text still hold true. Crew station evaluation and workload will be concerned with the GCE operations: mission manager, UAV pilot, and sensor operator. There is little difference between this team and the team in a multiplace A/C: an AC-130 gunship or P-3 surveillance platform, for example. There are, however, unique concerns when evaluating workload and performance during long duration UAV missions. The need to keep the crew involved in the operation and focused on the mission during 14 hr missions at remote locations takes on an added meaning.

The human factors test plan should identify any critical tasks that must be completed during the mission: timeline, priority, and information required. The tasks should also state when and if operator intervention is required. For example, takeoff and landing may require pilot intervention and target identification, recognition, and designation may require sensor operator intervention. The plan should state the level of automation that is used and which information is provided, and in what format, to the aircrew. How are operators apprised of correct operations (e.g., valid receipt of a command) as well as failures (loss of navigation function)? At a minimum, each task should be described and potential problems discussed:

− How critical is the task?

− If errors are involved, what is the potential impact (safety and mission impact)?

− What information or lack of information could cause erroneous actions?

− Is there a limited amount of time for the task?

From the HMI perspective, what information is required by the operators and which actions must be taken?

− Observation of video or data displays

− CAWs, colors, or auditory alerts

− Execution of normal and abnormal procedures

− Ergonomic and anthropometric requirements

Finally, conduct a workload analysis and impact on operator performance:

− Bedford Rating Scale

− Subjective Workload Analysis

− Modified Cooper Harper Rating Scales

A modified form of the Cooper Harper Scale for use in sensor/UAV evaluation was proposed by Christopher Cotting as part of his dissertation at Virginia Tech (*Modified Cooper Harper Evaluation Tool for Unmanned Vehicle Displays,* with M.L. Cummings, Kevin Meyers, and Stacey D. Scott, available at http://web.mit.edu/aeroastro/labs/halab/ papers/Cummings2006_uvs_FINAL.pdf). The Cooper Harper Scale was modified to become a task generic scale for UAVs. Just as in piloted evaluation, the task description and comments by the UAV operator will be key in using the scale to properly evaluate the sensor/aircraft system. Using this new rating scale the UAV and the sensor system can be evaluated as an integrated system against mission performance. This rating scale also includes the ability to check an autonomous UAV against disturbances so that the system can be evaluated in a relevant operating environment. A copy of the Modified Cooper Harper Scale is shown in Figure 12.14. Specific guidelines for each of the ratings are described in the subject paper.

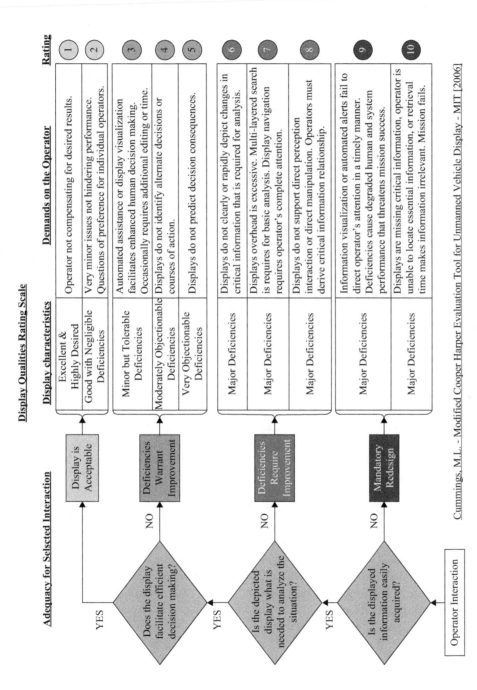

Cummings, M.L. - Modified Cooper Harper Evaluation Tool for Unmanned Vehicle Display - MIT [2006]

FIGURE 12.14 ■ Modified Cooper Harper Scale

Each of the following sections will incorporate human factors considerations for test. For communications, the most important aspect is clear, intelligible, and uninterrupted data transfer. How is the GCE made aware of a potential loss of communications, and what actions, either automatic or operator initiated, are taken to mitigate data dropouts? Signal strength was addressed earlier, and a method of displaying the strength of the data link signal should be displayed to the mission manager at all times. Colors make it easier for the operator to include this parameter in the scan: green for a strong signal, yellow for approaching the margin, and red for loss, or corrective action required. The actual numbers used will be determined by the test team and adjusted as dictated by the flight test results. A second parameter that should be displayed is the clear LOS region, dictated by altitude of the UAV and the GCE antennas and range to the UAV. Remembering the LOS equation from the previous chapters:

$$R_{Max} \sim .86 \left(\sqrt{2h_{UAV}} + \sqrt{2h_{GCE}} \right) \qquad (12.9)$$

where R_{Max} is maximum LOS range in nm, h_{UAV} is height of UAV in ft, and h_{GCE} is height of GCE antenna in ft.

This information can be displayed either on a vertical profile clearly defining the maximum LOS or on a moving map color coding or outlining the borders defined by the maximum LOS (an example is shown in Figure 12.15).

At a minimum, a nomograph for LOS range can be provided to the mission manager. If the GCE software can automatically vary the display presentations based on changes in UAV height, this function will need to be tested as well. Signal strength and LOS displays are easily testable during the ground evaluations while towing the aircraft. Each of these parameters should be superimposed on a moving map of sufficient detail to aid in situational awareness. Additional items on the moving map may be proposed: navigational routing, threat areas, targets, loiter areas, safe areas, and lost link procedures.

Cockpit displays for the UAV pilot are evaluated in the same fashion as in a manned aircraft: Are the displays readable and understandable? Is there enough information to make correct decisions? Is the aircraft controllable with the information provided? Is the information timely? Do the visuals provide adequate situational awareness? Are alerts sufficient in color, proximity, or loud enough? Is corrective action easily understood? There may be an added level of difficulty for the pilot when accounting for latency of the data or visuals. For RPV functions, pilot inputs may not be seen immediately due to the latency of the data. This may cause a pilot-induced oscillation (PIO) as the pilot holds in a command too long waiting for an aircraft response and then must apply an opposite command when the aircraft moves more than required. Sensor operators may experience a similar problem when attempting to slew a sensor to a desired position only to find out that the sensor overshoots the desired position and a reverse command is then required to correct for the overshoot. Operators may need to predict a position based on previous history. Another problem may arise if the out of the window video is out of synch with the cockpit display data: what the pilot is viewing is not what is displayed.

Although much of the groundwork for HMI and controls and displays testing was accomplished during modeling, mockups, and hardware-in-the-loop simulations, there is no better test than an actual operation. Aircrew workload will be significantly higher during an operation than during a simulation, and tasks that seemed easily manageable in simulation may overwhelm an aircrew during an actual operation.

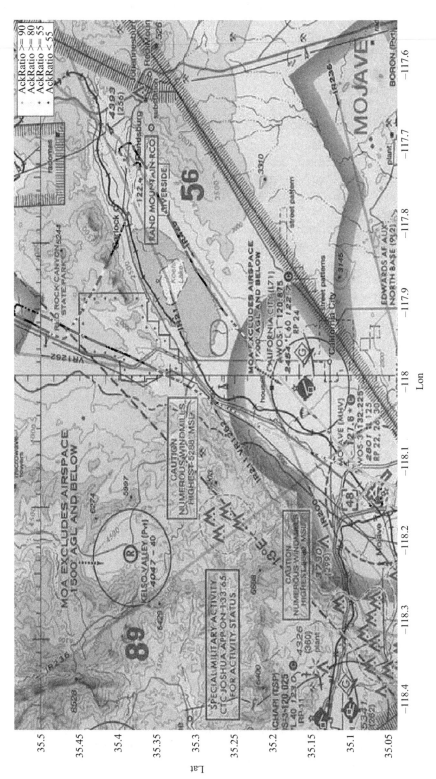

FIGURE 12.15 ■ Maximum LOS/Antenna Coverage

Other areas of interest that need to be addressed are the cockpit environment, which include items such as noise level, temperature and lighting, ergonomics, and GCE internal communications (intercom, radios, and SATCOM). The goal is to ensure that the mission can be accomplished in the most efficient and reliable manner. The end result of all of these ground evaluations will also provide an early indicator of what workload is expected.

12.4.3.6 USAR Human Factors and Controls and Displays Considerations

Subpart I, UAV Control Station, is dedicated in large part to human factors considerations. Many of the requirements and considerations can be found in GAMA Publication #10, "Recommended Practices and Guidelines for Part 23 Cockpit/Flight Deck Design," GAMA Publication #12, "Recommended Practices and Guidelines for an Integrated Cockpit/Flight Deck (Part 23)," and Public Statement PS-ACE100-2001-004, "Certification Plan for Human Factors (Part 23)," which were addressed in chapter 6 of this text. The subpart addresses HMI, controls and displays, and indicators and warnings.

HMI considerations involve the work environment and the evaluation of aircrew workload. The physical parameters of the GCE (e.g., size, temperature, maximum capacity) that are required to perform the mission must be stated in the UAV systems flight manual; the workplace must allow the crew to perform their duties without unreasonable concentration or fatigue. The workplace conditions such as temperature, humidity, noise, and vibration must not hamper the safe execution of the flight. The minimum flight crew must be established so that it is sufficient for safe operations; minimum flight crew is established via a workload analysis. The USAR states the following tasks should be part of the workload analysis:

– Operation and monitoring of all essential UAV system elements
– Navigation
– Flight path control
– Communications
– Compliance with airspace, air traffic, and air traffic control requirements
– Command decisions including crew resource management (CRM)

Workload should take into account the ease and accessibility of the necessary controls and under all lighting conditions. Crew members should be able to converse without difficulty and be able to recognize aural alerts under all foreseeable noise conditions. (Note: The USAR requires that a voice recorder be installed in the GCE to record and time-tag all radio and ICS communications)

Controls and displays evaluations are much the same as previously discussed; they must be clearly arranged and visible to the crew, grouped appropriately, and readable under all lighting conditions. Each control must be located and identified to provide convenient operation and to prevent confusion and inadvertent operation. Controls must be arranged so crewmembers have unrestricted access and ease of movement without restrictions in all flight conditions. The design, location, and accessibility of safety critical controls must be compatible with a rapid reaction by the UAV crew. The USAR notes that if pull-down menus are used, then controls that require a prompt reaction (as in an emergency or abnormal conditions) must be accessible on the first level; if not, then there must be dedicated controls.

When conventional flight controls and indicators are used, the form, location, and layout must ensure safe operation. When markings are on the cover glass of the indicator, there must be a means to maintain the correct alignment of the glass cover with the face of the dial. Each arc and line must be wide enough and located to be clearly visible to the crew, and all indicators must be calibrated in compatible units. The controls must be designed so that they operate intuitively, and are similar to those found in manned aircraft. The design of the controls must allow the crew to rapidly and easily change the UAV heading or track, altitude, and airspeed.

Cautions, advisories, and warnings must follow the color scheme defined in AC 23.1311-1C. The GCE must include an automatic diagnostic and monitoring capability for the status of the UAV and provide the crew with appropriate warning indications. The GCE must also be configured to warn the crew of any abnormal or emergency mode, including cases where an automatic corrective action has occurred. Warnings must be announced when an excessive deviation from the pre-planned flight path occurs or if an unsafe status exists within the UAV to alert recovery teams. There are other alerts required by the USAR pertaining to vehicle parameters (electrics, fuel, hydraulics, etc.) which are not discussed here.

Every control, switch, knob, or lever in the GCE must be plainly marked as to its function and method of operation; each remote control must also be suitably marked. Each emergency control must be red and marked as to method of operation; no control, other than an emergency control may be colored red. There must be an indication to the crew that the UAV must be operated IAW the operating limitations and permitted operations in the UAV systems manual.

There are other requirements if multiple UAVs are operated or if control of the UAV can be handed off between multiple GCEs:

– The in-control GCE must be clearly identified.

– Positive control must be maintained during handover.

– Handover cannot lead to an unsafe condition.

– The minimum crew must be established for control of multiple UAVs.

– All data, warnings, advisories and cautions previously discussed must be available to the controlling GCE.

12.4.3.7 Airborne Coverage

Link communications coverage will be evaluated in much the same way as radio communications testing described in chapter 4 of the text, with the exception of calculating BER rather than determining intelligibility. The geometry that will be used is the same as described in chapter 3: incremental range steps and executing orbits or cloverleaf turns to observe communications with changes in azimuth and elevation. Evaluators should resist the temptation to fly out to the maximum predicted range and start maneuvering; a lost link at the maximum LOS range will be much more difficult to recover than an A/C at a moderate range where a change in bank or heading should rapidly recover the link. A good understanding of the predicted range and possible blockage areas must be well understood before attempting any flight test. One would not like to lose the control link on the first test flight of the program.

Many organizations will test the link package on flying test beds or surrogate UAVs. Some testing has been done by slinging the package under a helicopter or tethering it to a large mast that can be rotated to test azimuth coverage. All of these test setups will enhance the evaluator's knowledge of predicted link performance and enhance safety when performing initial flight test.

As the A/C is flying, a real-time display of the signal/BER/ACK ratio should be available in the GCE. As suggested in the controls and displays section, we should color code the signal. Let's assume that we look at the acknowledgment ratio (packets acknowledged/packets sent) of the command signal and we color code this ratio. A typical scheme may be green for ≥90%, yellow for ≥80%, red for ≥55%, and black for <55%. The A/C will fly to a range and altitude away from the GCE and execute an orbit/cloverleaf and the acknowledgment ratio will be viewed and recorded. Any degradation in the ratio will be noted and correlated postflight: degradations at only one aspect could indicate an antenna blockage; degradation for all aspects in a portion of the operating area could indicate interference or multipath. As with all systems testing, the optimum condition should be tested first: clear LOS and short range. The envelope will be expanded from here. Problems that occur at the optimum condition will not magically disappear if we fly further away: diagnose the problem before continuing with the expansion.

The backup link and BLOS link (if installed) also needs to be tested and these are accomplished in the same manner. Remember that a comparison between the LOS link and the BLOS link will have to account for latency. Tests which can't be scheduled and still need to be evaluated will be link performance in weather, or the effects of pre-cipitation on communications. Once these tests are accomplished, a realistic operating area can now be defined for operations. Performance that is slightly less than predicted is expected, but if the performance is significantly less than it is necessary to trouble-shoot. To recap, the possible problems could be (in no particular order):

– Low power/no power at the transmitting antenna; possible bad transmitter or ampli-fier, transmitter/amplifier mismatch, broken lines, or bad connections

– Low power/no power received at the receiving antenna; broken lines or bad con-nections, UAV not in the beamwidth caused by wrong auto-tracker positioning or antenna misalignment, interference from another transmitter, multipath, refraction, attenuation/scattering due to atmospherics, blockage on the UAV, satellite transpon-der problems, crypto key problems, or frequency mismatch

12.4.3.8 Data Link Latency Evaluations

There will be latency in the system due to the problems previously described; some contributors will be fixed, such as senescence, whereas some will vary such as with TCP. Timing studies should be undertaken to quantify the latency. Earlier discussions in the text addressed this evaluation; we are interested in the delta time from when a parameter is sensed on the A/C to the time it is viewed in the GCE. In the testing of systems on a manned aircraft, we stressed the importance of time-aligning the data prior to presenting it to the analysts. By performing a time-alignment, all sources of asyn-chronous data (e.g., aircraft TM, TSPI, target data) could be merged and near real-time analysis could be performed. We also described the methods of time-alignment: inter-polation and extrapolation. Extrapolation is used in telemetry systems to stay as close as possible to real time and interpolation is used postflight to obtain the most accurate data.

Time-alignment of data for UAV missions becomes a little more complicated as streaming video from out-of-the-window TV and sensor data is also arriving with autopilot and health and welfare systems data; in some cases LOS data will also be arriving with BLOS data. The test team will have to determine the actual latency of all of these streams and then decide on how the data should be presented: time-aligned or throughput without adjustments. Additionally, we have the added problem of determining the delta time from a commanded input to the autopilot to the time it is executed. This is the cause of the PIO (for sensors or aircraft) discussed previously.

In most cases, data will be even further delayed by employing interpolation algorithms; we will want to stay as close as possible to real time. Video data will be throughput without modifications so the delay seen will be the link delay. Screen data, or data displayed on the pilots' MFDs may or may not be from multisources and hence the delay may be somewhat greater than the delay in video. The delta time between the video and the screen data must be known and understood by the pilot in order to safely conduct RPV portions of the mission. Time should be dedicated during the initial link testing to evaluate the workload tasking on the pilot using delayed data to control the UAV.

12.4.3.9 USAR Data Link Considerations

Subpart H of STANAG 4671 (USAR) states the airworthiness requirements for the command and control data link: Book 1 delineates the requirements, and Book 2 provides an acceptable means of compliance. This section will detail only the requirements set forth in the USAR. For detailed compliance statements, the reader is referred to the document. Requirements that must be considered by the evaluation team are as follows:

– The UAS must include a command and control data link (the USAR only addresses the command and control data link; ATC communications and sensor data links are deferred to the operations manuals).

– No single failure can lead to a hazardous or more severe event.

– Operation of data link equipment cannot adversely affect other equipment.

– EMI/EMC testing must be accomplished.

– The effective maximum range of the link must be stated in the UAV flight manual and must be displayed in the GCE.

– The integrity of the link must be continuously monitored; LOS information must be continuously displayed in the GCE.

– Latency must be specified in the UAV flight manual; latency must not lead to an unsafe condition.

– A lost link strategy must be established, approved and presented in the UAV flight manual and an aural and visual alert must be provided to the UAV crew.

– The lost link shall incorporate an automatic, autonomous procedure to reestablish communications in the shortest time possible.

– Masking must be stated in the UAV flight manual; warning cues shall be provided to the crew when approaching masking attitudes.

– Switchover (changing control from one channel/frequency to another) shall not lead to an unsafe condition; positive control cannot be lost during switchover.

▌ 12.5 │ NAVIGATION

The primary navigational system on the UAV will be an integrated GPS/INS and in most cases equipped to handle differential corrections. The navigation system may also incorporate VOR/TACAN/DME as a backup to short-range navigation, a magnetometer for emergency/backup magnetic heading, barometric altimeter, and a radar/laser altimeter.

12.5.1 UAV Navigation Evaluations

Accuracies for these systems are tested by the same means as described in chapter 5 of this text and will not be replicated here. Mission requirements may dictate a very high accuracy, and additional redundant systems such as a secondary GPS may be required. It will be extremely important to test the navigational hierarchy (also previously described in chapter 5) to ensure that the navigational system degrades gracefully and the system has a high probability of returning safely with a navigational failure.

Tests involving the navigational system unique to the UAV mission that should be considered are as follows:

– Ability to fly a preprogrammed route

– Ability to steer to a waypoint

– Ability to designate a point

– Ability to initiate an orbit about a designated point

– Navigation in the vertical plane

– Takeoff

– Landing

– Loss of navigation function

– Execution of lost link procedures

With the exception of takeoff and landing, all of these functions are handled through the autopilot (command directed), and for the most part we are evaluating the navigational system's ability to provide accuracy sufficient to accomplish these tasks. It should be noted that highly autonomous platforms do not always have the capability for the crew to rapidly and easily change the UAVs heading or track, altitude, and airspeed. Other sensors on the A/C will use commands based on navigational information; the ability to see a target will be dependent on how accurate the sensor is pointed. As pointed out earlier, some sensors have relatively narrow FOVs; an inaccurate pointing angle may place the target outside of the sensor FOV. The positional information on a designated point is dependent on the navigational system as well; correct bearing information from a sensor applied to an inaccurate ownship position will result in an inaccurate target position. It is incumbent upon the evaluator to understand the degree of accuracy required for each phase of the mission and then develop the test plans to verify that accuracy.

Accuracies should be tested in the primary and degraded modes of operation; it may be determined that certain portions of a planned mission cannot be accomplished in a degraded mode. It will, however, be a requirement to safely return in a degraded navigation mode. Ideally, within the NAS, the system will utilize a WAAS enabled GPS system which will provide differential corrections to the receiver to enhance positional accuracy. Regions outside of the NAS may be able to take advantage of other SBAS systems or require uplinked corrections from a portable system via a VHF data link. BLOS systems may require corrections be sent over a satellite link and then latency will become a problem especially in a dynamic environment.

Auto-takeoff and auto-land systems will require an even higher degree of accuracy. The reader is referred to section 6.14.5 of this text for "Airworthiness of Systems for Takeoff in Low Visibility and Airworthiness of Systems for Landing and Rollout in Low Visibility," which are the Category III systems requirements. From a safety point of view, the system will have to demonstrate very tight accuracy requirements for holding the runway centerline during many hundreds of takeoffs and landings. It is not realistic to incorporate an ILS system in a UAV, but it is quite probable that the system will incorporate a GBAS or JPALS-like takeoff/landing system. For RPV takeoffs and landings, it must be demonstrated that the navigation system can position the A/C at a point in space where the pilot can safely execute the takeoff/landing without excessive workload or piloting skill.

What will the UAV do if the GPS signal is lost? What happens if the differential corrections are lost on final approach? Will the UAV maintain heading and altitude for some period of time in hopes of reacquiring navigational information? If a coasting period times out, what is the desired maneuver of the UAV? Will the UAV plot a course for home or execute a turn to a safe heading? Questions like these will drive the flight tester's plan for a loss of navigation function. We know how the system is supposed to work by analyzing the Interface Design Specification and working closely with Engineering. Our job is to demonstrate that what was envisioned is actually coded correctly in the S/W.

A similar problem, although more severe, is when the communications link is lost. When the link is lost, the UAV will execute a series of lost link procedures. The UAV may initially hold heading, altitude, and airspeed for a period of time (coasting) or initiate an orbit about the point where the communications were lost. If the link is regained the UAV may continue on its mission. After a defined period of time (coasting timed out) the UAV may next execute a turn to home or a known safe heading. If the link is regained, there will be intervention of the GCE to inform the UAV on the course of action. It may be to continue the mission, re-plan the route or RTB. This test is easily accomplished by shutting down the transmitter/receiver at the GCE and then reinitiating the transmissions. Remember to evaluate this functional test in an area where communications are assured—not at the edge of the envelope.

12.5.2 USAR Navigation Considerations

USAR states that the following information must be displayed in the GCE *at all times* with update rates consistent with safe operation:

- Indicated airspeed

- Pressure altitude and associated altimeter setting

- Heading or track (when magnetic heading or track is used in the GCE, it must be automatically compensated for deviation)

- UAV position displayed on a map at a scale selectable by the UAV crew and at a level of detail to ensure safe operations
- When semi-automatic flight modes are activated, the commanded flight or navigation parameters sent to the UAV must be displayed in the GCE

The following are the minimum required flight and navigational data to be made available *when selected* by the UAV crew:

- Airspeed limitations
- Sideslip angle
- Free air temperature
- A speed warning device for turbine powered UAVs and for others that have an established V_{MO}/M_{MO} and V_D/M_D and V_{MO}/M_{MO} is greater than 0.8 V_D/M_D
- The speed warning device must give an effective aural warning and the setting should attempt to minimize nuisance trips
- Position of the UAV in terms of range and bearing from the GCE; deviation between actual track and planned track
- UAV attitude in terms of roll and pitch
- Vertical speed
- Time and navigation system status
- UAV identification when multiple UAVs are being operated
- Wind direction and speed at the UAV altitude when only track data is displayed to the crew
- For automatic takeoff and/or landing systems, the UAV flight path, deviation between UAV flight path and planned flight path including glideslope

Note: Many other parameters that address engine, oil, fuel, hydraulics, and electrical system are called out in the USAR but are not addressed here.

For systems equipped with auto-takeoff and/or auto-land, once the function is engaged the UAV crew monitors the progress from the GCE, but is not required to perform any manual intervention except for a manual abort. The automatic function will reside in the UAV control laws and will utilize navigation and flight path tracking inputs in such a manner as to not degrade redundancy or overall safety of the FGS. When off-board sensors are used via data link (e.g., differential corrections), the continued safe flight of the vehicle must be ensured in the event of the loss of the link. The automatic system may not cause unsafe sustained oscillations or undue altitude changes or control activity as a result of configuration or power changes. The status of the automatic takeoff and landing systems must be displayed in the GCE and designed to minimize errors.

Once the automatic takeoff mode has been engaged, the brake release, takeoff run, and rotation are automatic. The runway flight path, steering, speed, configuration, engine settings, and flight path after liftoff are all automatically controlled by the auto-takeoff system. In the event of a failure that could adversely affect safe flight or exceedance of limits at speeds up to rotation or refusal speed, an automatic abort function shall be provided to stop the UAV on the runway.

Once the automatic landing system is engaged, the approach, landing, and rollout are fully automatic until the UAV reaches a full stop, or a safe taxiing speed is reached and the crew switches to manual control. The UAV flight path, speed, configuration, engine settings, runway steering, and braking after touchdown are all controlled by the auto-land system. In the event of a failure that could adversely affect safe flight or exceedance of limits during the approach, an automatic go-around function shall be provided above the Decision Point (DP) at which a safe go-around may be safely performed.

For automatic systems, the GCE must continuously present to the crew:

– UAV flight path

– Deviation between the UAV flight path and the planned flight path

– Glideslope

If the UAV is designed for conventional takeoff and landing on a runway, it must provide a manual abort and go-around procedure. These functions:

– Must be easily accessible to the crew

– Must be able to stop the UAV on the runway up to rotation or refusal speed

– Must be able to initiate a go-around at every height down to the DP

■ 12.6 | AUTOPILOTS

At the heart of the command directed function of the UAV is the autopilot. The basic autopilot functions such as altitude hold, heading hold, GPS/Navigation guidance, etc., are the same as we have seen in chapter 6 when discussing Civil Certifications. A starting point for the UAV autopilot evaluation should be applicable Advisory Circulars, but there are a few differences between the Part 23/25 requirements and those required for the UAV mission. For example, during UAV operations, the default is usually to have the autopilot engaged, and is used to exercise full control of the vehicle. The exception is when the GCE operator may have to control the autopilot modes similar to a manned vehicle (e.g., heading change, altitude override (capture), Direct-To, etc.) Unlike manned vehicle operations, the UAV autopilot usually has full control authority; additionally, the UAV autopilot has no concern for passenger comfort, i.e., it may be "bang-bang" control. Limits are often built into the control laws of the Flight Control System or the autopilot; they may be mission or safety specific.

The autopilot integrated with the UAV may be controlled in one of two ways. It may be an autopilot controlled from the ground by a pilot (called semi-automatic in USAR), or it may be an autopilot controlled by an onboard computer that can be updated by an operator on the ground (called automatic in USAR). Some UAVs have a mode of operation that has separate (usually more simple) logic that requires constant pilot input. This method of integration has its pros and cons. The obvious good point is that the system is redundant (pilot and autopilot); the drawback is that it requires a lot of attention and the pilot is in a separate location.

12.6.1 UAV Autopilot Evaluations

Some of the autopilot evaluations that were discussed in chapter 6 may not be applicable to the UAS. Many of the autopilot functions seen previously in manned aircraft are replicated in the GCE. Tiles, knobs, and rotary controls may appear as exact duplicates

FIGURE 12.16 ■
Autopilot Submenu

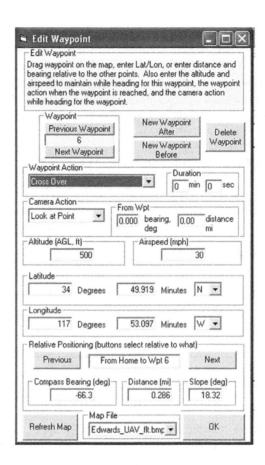

in the GCE, or as an alternate means, the method of selection may look different (point and click). Figure 12.16 provides an example of a point and click autopilot submenu.

The RPV or semi-automatic mode may replicate the displays in the GCE, whereas the command directed or automatic mode is more likely to use the mouse and keyboard. Autopilot disconnects are slightly different for the UAV as well. Usually, the autopilot is on and in control unless an alternate mode is selected by the operator.

There is no manual override by overcoming the autopilot by force; there may be a quick disconnect on the stick/yolk, similar to manned aircraft implementations. As demonstrated with manned aircraft, the controls necessary for the UAV crew to safely operate the autopilot must be clearly identified, visible in all lighting conditions and easily accessible. The basic modes of the autopilot (heading, airspeed, altitude hold, altitude capture, integration with the navigation system, orbits, VNAV, and LNAV) need to be demonstrated as functioning correctly. These tests can routinely be accomplished in concert with navigation testing. As mentioned previously, a good starting point for these evaluations is AC 23.1329, which contains, similar to the EASA certification specifications, a Book 1 requirements and a Book 2, which provides a test and evaluation matrix.

Malfunction testing is required; however, the reaction times will need to be adjusted. Since the crew is in the GCE, a malfunction first has to be announced, then recognized, and then acted upon (if necessary). The actual reaction time may be quite longer than seen in manned aircraft, and to add additional time may be unreasonable.

If there is an automatic contingency path following a detected malfunction, that contingency path must be tested as well.

Auto throttles, if installed, need to be evaluated for takeoff, landing, go-around, and VNAV. Of particular interest will be the low-speed and high-speed protection logic; is the airspeed controlled via auto throttles or pitch/angle of attack to prevent an overspeed or stall? Limiters in the control laws or autopilot need to be evaluated as functioning correctly; for example, can the UAV be commanded to an altitude and speed combination that places the aircraft in the buffet boundary with minimum excessive available thrust?

In manned aircraft, the minimum use height of the autopilot was determined and entered into the aircraft flight manual. In the event of a loss of GPS or RADAR altimeter on an approach, the UAV may execute a go-around, but what is the minimum altitude that this procedure can be executed safely? These functions need to be addressed by the test team and evaluated accordingly.

12.6.2 USAR Autopilot Considerations

The USAR references AMC 1329 whose U.S. equivalent is AC 23-17B and AC 23.1329; each of these ACs is addressed in chapter 6 of the text. It notes that the autopilot must be designed so that a UAV crew of average skill can operate the UAV system with acceptable workload (similar to manned systems where the AC notes that the autopilot should not cause the crew to bust a checkride). The flight director is not applicable in the USAR.

The UAV must have both an automatic and a semi-automatic mode selectable at any time by the crew except as noted as follows. The autopilot must apply limits to maneuvers to ensure the UAV is within flight envelope protection. There are cases where the UAV crew may be restricted from intervention (i.e., manual flight):

- During emergency situations such as total loss of the data link
- During launch phase, prior to achieving the minimum safe flight parameters
- During landing phase after reaching the decision point
- For UAVs recovered by parachute during landing phase under parachute
- For rocket and catapult-assisted launches, prior to reaching minimum safe flight parameters

The balance of the required testing comes directly from the civil airworthiness requirements:

- The FGS must be designed so that the range of adjustment by the crew cannot produce a hazardous situation.
- A single malfunction cannot produce a hard-over condition in more than one axis.
- If the FGS integrates signals from auxiliary controls or provides signals to other equipment, positive interlocks and sequencing of engagement to prevent improper operation are required.
- There must be protection from malfunctions of integrated equipment.
- The FGS must have a comprehensive self-test available and operating during all phases of flight including preflight.

– With the automatic flight mode engaged a warning must be displayed when excessive deviation from the preprogrammed flight path occurs (acceptable deviation is determined by the authority).

There must be a clear and distinctive low speed warning in the GCE with the flaps and landing gear in any normal position in straight and level and turning flight IAW the following:

– It should not be possible to command speed values lower than the minimum steady flight speed (except takeoff and landing) allowed by the flight envelope protection maintained by the flight control system.

– Adequate low speed cues and warning should be provided in the GCE when approaching stall speed or $V_{min\ DEMO}$ (if stalling is not demonstrated).

– The warning must begin by a margin of not less than 5 kts and must continue as long as the condition exists.

There is no mention of high-speed protection in the USAR.

12.7 | SENSE AND AVOID SYSTEMS

Manned aircraft have a number of systems at their disposal to enhance flight safety; these systems were identified in chapter 6 as proximity warning systems, such as RADAR, EO, TCAS, TAS, TAWS, GPWS, EVS, and ADS-B. TCAS, TAS, and ADS-B interrogate other transponders or receive aircraft position information and process these replies from other aircraft enabling them to show these aircraft on the PFD, MFD, or a stand-alone VSI indicator. TAWS and GPWS alert the aircrew of conflicts with the terrain, issue aural and visual warnings, and are installed to help prevent CFIT. RADAR and EO have obvious advantages of detecting and displaying ground and airborne targets and are used in EVS along with digital terrain to provide an aircrew with situational awareness of his surroundings. As the name implies, all of these systems are *warning* systems and are not coupled to the autopilot to provide active *avoidance*. In manned aircraft operations, pilots are schooled in the *see-and-avoid* concept: taking evasive action when a hazardous condition exists. Since the UAV is unmanned, most documentation addresses the need for sense and avoid, or in other documentation *detect, sense, and avoid*; the autopilot (or the GCE crew) takes evasive action when a hazardous condition exists. As far as the FAA (or the national authority) is concerned, flight of UAVs in the National Air Space (NAS) requires an equivalent level of safety to manned aircraft flight. An equivalent level of safety currently requires chase aircraft and five days prior notice to ARTCC and constant communication between the GCE pilot and ATC.

The FAA has not specifically approved any technology or suite of technologies as being sufficient to provide acceptable sense and avoid capabilities. Part of the challenge is that existing technologies do not assure avoidance capabilities under all operational conditions, including autonomous UAV operations or in situations when UAVs lose their command guidance links with ground control facilities. Standardized procedures for responding when UAV guidance has been lost are currently lacking, but will be needed to ensure that air traffic controllers and airspace managers can redirect nearby traffic and mitigate collision risks.

The USAF has been testing and developing active collision and avoidance systems (ACAS) for some years. The initial system was developed to aid in the saving of an aircraft and crew after a crewmember became incapacitated for some reason (e.g., hypoxia, g-lock). Aircraft parameters such as altitude, angle of attack, airspeed, and yaw rate were monitored by a processor, and when the processor determined that an unsafe situation was present, with a possibility of an incapacitated crew member, the autopilot took control of the aircraft and flew it to safe conditions. A system of this type installed in highly dynamic aircraft (such as fighters) may be susceptible to high false alarm rates.

The national aviation authorities are very concerned with the mixed use of airspace by manned and unmanned aircraft; hence the regulations restricting UAV operations. These concerns are justified as the current systems are not accurate enough to mitigate the possibility of a mid-air to acceptable safety assurance levels.

The most common proximity warning system for aircraft-to-aircraft collision avoidance is the TCAS II system; however, this system suffers from some problems that would make an automatic response less than optimum. Remembering our discussion from chapter 6, the current TCAS system suffers from false alarm rates and the inability to provide resolution advisories in the horizontal plane. There is an interesting paper on the subject titled "Safety Analysis of TCAS on Global Hawk using Airspace Encounter Models" (Thomas Billingsley, under an USAF contract to MIT, June 2006, http://dspace.mit.edu/bitstream/handle/1721.1/35294/74468299.pdf?seque). In a nutshell, a TCAS-equipped Global Hawk would enhance safety when compared with a transponder-only equipped Global Hawk; however, if the maneuver was controlled through the GCE, safety enhancement will drop with increased latency. Optimally, the RA maneuver should be automatic.

ADS-B, also addressed in chapter 6, provides a more accurate picture of the aircraft and its relation to other traffic in the area, but the system is not yet fully developed or fully deployed. An ADS-B system used for sense-and-avoid currently would have to be a hybridized ADS-B and TCAS implementation to capture all traffic (using either ADS-B or IFF codes) in the vicinity of the UAV. As a side note, all aircraft in the United States must be ADS-B compatible by 2020.

TAWS-based systems are in use on today's UAVs but like the airborne warning systems are not automatic. Many of the mission planners utilize terrain databases and TAWS logic to pre-plan the UAV missions to avoid any possibility of impact with the terrain. The mission designer interface is based on the FAA TAWS Standards TSO-C151c and provides terrain awareness for all waypoints of the mission route and real-time display of the terrain while enroute.

12.7.1 Airworthiness of Sense and Avoid Systems

Similar to the minimum operational performance standards (MOPS) we have seen developed for airborne systems, avionics equipment to be installed or integrated into UAVs must meet minimum aviation system performance standards (MASPS). MOPS, as a rule, are developed by RTCA. The SC-203 Committee of the RTCA is developing UAV standards related to MASPS for unmanned aircraft systems. MASPS will be released in two documents: "Guidance Material and Considerations for Unmanned Aircraft Systems," RTCA/DO-304, March 2007; and "Minimum Aviation System Performance Standards (MASPS) for Unmanned Aircraft Systems," which was due to be completed in December 2012 has been delayed. A third document, "Minimum Aviation

System Performance Standards (MASPS) for Sense and Avoid Systems for Unmanned Aircraft Systems," was scheduled for release in December 2013, and is also delayed.

The FAA will use these MASPS sections as support regulatory guidance recommendations, along with input from ASTM standards and specifications to develop regulations for UAVs for use in the U.S. National Airspace.

ASTM International, formerly known as the American Society for Testing and Materials, is a globally recognized leader in the development and delivery of international voluntary consensus standards. Some 12,000 ASTM standards are used around the world to improve product quality and enhance safety.

The ASTM committee F-38 on UAVs was established at the request of the UAV industry with participation from FAA on the subcommittees in the developing standards. According to its website (http://www.astm.org/) the mission of Committee F-38 is to produce cost-effective consensus standards that, when applied, will enhance the safe design, manufacture, maintenance, and operation of unmanned aircraft systems (UAS). This will be accomplished in the following steps:

- Define terms and scope of UAS standards
- Adopt current, safe practices and guidance as formal UAS standards
- Develop additional UAS standards as needed
- Maintain currency and relevancy of standards

The ASTM F-38 committee has issued and published a standard for DSA collision avoidance (F2411-04 DSA Collision Avoidance) which requires a UAV to be able to detect and avoid another airborne object within a range of $+/-15$ degrees in elevation and $+/-110$ degrees in azimuth and to be able to respond so that a collision is avoided by at least 500 ft. The 500 ft safety bubble is derived from the commonly accepted definition of what constitutes a near mid air collision (NMAC). This gives avionic electronics manufacturers a target for certification. It is likely that the ASTM standard will be incorporated by reference in eventual FAA certification requirements (MOPS, TSO and AC). This specification establishes the design, construction, and performance requirements necessary for the technical reliability of airborne sense and avoid systems that support the detection of, and safe separation from, airborne objects such as manned or unmanned aircrafts and air vehicles. It specifically applies to the manufacturer or component supplier of such systems seeking civil aviation authority approval, in the form of flight certificates, flight permits, or other like documentation, as providing an equivalent level of safety to the see-and-avoid capability of a human pilot. The reader will note that these words mimic the current verbiage of FAA Order 8130.34B. This specification does not cover transponder or broadcast-based cooperative sense and avoid systems as well as appliances onboard one or more airborne objects flying in formation flight. In particular, the specification:

- Covers requirements for the design and performance of airborne sense and avoid systems
- Includes requirements to support detection of, and safe separation from, airborne objects such as manned or unmanned aircraft and air vehicles
- Applies to the manufacturer of an appliance seeking civil aviation authority approval, in the form of flight certificates, flight permits, or other like documentation, as providing an equivalent level of safety to the see-and-avoid capability of a human pilot

- Is not intended to apply to the design and performance of cooperative sense and avoid systems

- Does not apply to appliances onboard one or more airborne objects flying in formation flight

- Does not purport to address all of the safety concerns, if any, associated with its use. It is the responsibility of the user of this standard to establish appropriate safety and health practices and determine the applicability of regulatory limitations prior to use

Existing standards and guidance should be referenced for specifications describing transponder or broadcast-based systems (examples of existing guidance and standards for cooperative proximity warning systems include FAA 20-131A, RTCA DO-289, and TSO-C119B; see chapter 6).

12.7.2 Developments in Sense and Avoid Systems

The proposed standards dictate a field-of-view ($+/-110°$ azimuth and $+/-15°$ elevation) and a safety bubble (500 ft) around the aircraft. Operators would have to translate these requirements to detection range requirements that would be influenced by closure and reaction time (remember that the closest point of approach allowable in TCAS systems is determined by range divided by range rate). If I want to establish a 500 ft safety bubble and I know the radial closure on an intruder and the performance of my aircraft, I can determine a minimum detection range to ensure safety.

These ranges were calculated for the Global Hawk and the Predator B by Northrop-Grumman and General Atomics-ASI, respectively, and presented by Dr. John F. McCalmont, USAF Research Laboratory, in "Small Sense and Avoid System" at the UAV 2007 Conference in Paris. The slides used to present the detection range requirements for the Global Hawk and Predator B are shown in Figure 12.17 and Figure 12.18.

Detection Range Requirements (NM)				
Scenario (Altitude)	"Manned" RQ-4	Actual RQ-4		
	Nominal Pilot[1]	AUTO (2, 6, 7)	LOS (3, 5, 6, 7)	BLOS (4, 5, 6, 7)
Low	2.4	0.9	1.7	2.1
Med	3.1	2	3	3.6
High	5.5	2.6	4.1	4.9

FIGURE 12.17 ▪ Global Hawk Minimum Detection Requirements

Low Altitude Scenario RQ-4-175 KTAS[8] Nose-on Intruder-300 KTAS	Medium Altitude Scenario RQ-4-230 KTAS[8] Nose-on Intruder-550 KTAS	High Altitude Scenario RQ-4-340 KTAS[8] Nose-on Intruder-550 KTAS

Notes
(1) Per U.S. Federal Aviation Administration Advisory Circular 90-48C
(2) Maneuvers is autonomous, so communications latency is 0.0
(3) Maneuvers is approved by human via Line-Of-Sight (LOS) data link
(4) Maneuvers is approved by human via beyond Line-Of-Sight (BLOS) link
(5) Includes 5 sec for human reponse, per U.S. air force research lab
(6) Guarantees minimum miss distance of 500 feet
(7) Climb/dive rates and turn rate are a function of a altitude
(8) Data provided by northrup-grumman

Communication Delays
Autonomous - 0.0 sec
Line-of-sight (LOS)-1.0 sec
Beyond-LOS (BLOS)-4.0 sec

FIGURE 12.18 ■
Predator Minimum
Detection Range
Requirements

Detection Range Requirements (NM)				
Scenario (Altitude)	"Manned" MQ-9	Actual MQ-9		
	Nominal Pilot[1]	AUTO (2, 6, 7)	LOS (3, 5, 6, 7)	BLOS (4, 5, 6, 7)
Low	2.1	0.7	1.4	1.5
Med	3.8	1	2.6	2.8
High	4.1	1	2.8	3

Low Altitude Scenario MQ-9-75 KTAS[8] Nose-on Intruder-300 KTAS	Medium Altitude Scenario MQ-9-150 KTAS[8] Nose-on Intruder-550 KTAS	High Altitude Scenario MQ-9-210 KTAS[8] Nose-on Intruder-550 KTAS

Notes
(1) Per U.S. Federal Aviation Administration Advisory Circular 90-48C
(2) Maneuvers is autonomous, so communications latency is 0.0
(3) Maneuvers is approved by human via Line-Of-Sight (LOS) data link
(4) Maneuvers is approved by human via Beyond Line-Of-Sight (BLOS) link
(5) Includes 5 sec for human reponse, per U.S. Air Force Research Lab
(6) Guarantees minimum miss distance of 500 feet
(7) Climb/dive rates and turn rate are a function of a altitude
(8) Data provided by General Atomics-ASI

Communication Delays[8]
Autonomous - 0.0 sec
Line-of-Sight (LOS)-0.5 sec
Beyond-LOS (BLOS)-1.3 sec

The minimum detection ranges to ensure a 500 ft safety bubble allow for three altitude scenarios (low, medium, and high), which affects closure speed and UAV performance. For the closure speeds in each of the scenarios, the first column provides the minimum detection range of the pilot's unaided eye and is based on appendix 1 of AC 90-48C; this assumes a 12.5 sec pilot reaction from the time an intruder is seen to the time that the aircraft responds to a pilot input. The three columns on the right indicate the minimum detection ranges required for each of the altitude scenarios and are dependent on an automatic maneuver upon detection, a maneuver commanded by the UAV pilot for LOS data link, and a maneuver commanded by the UAV pilot for a BLOS data link. Automatic reaction times are 0 and 5 sec for UAV pilot commands; communications delays are provided in the box at the bottom right of the slides.

Since detections are not guaranteed by TCAS II or ADS-B (e.g., no transponder, faulty altitude encoder, incompatible systems), it would seem that a system based on sensor detection (active or passive) would be a logical choice. In this case, TV, IR, or radar would be needed to scan the required volume and report intruders to either the autopilot or UAV crew. At the present time, automobile manufacturers have developed 77/94 GHz anti-collision systems that provide automatic braking when a potential collision is detected. There is no reason that this technology cannot be adapted for UAV sense and avoid systems. At 77/94 GHz, the components are quite small; monolithic microwave integrated circuit (MMIC) and radio frequency integrated circuit (RFIC) technology has led to a price drop in the production of RF transceivers.

NASA has done some research on MMW detect, sense, and avoid systems using modeling and simulation. Their conclusion is that MMW RADAR is a viable solution to the sense and avoid problem in UAVs. (NASA Technical Report, *Modeling and Simulation of an UAS Collision Avoidance Systems*, Edgardo V. Oliveros and A. Jennifer Murray, 2010, http://ntrs.nasa.gov/archive/nasa/casi.ntrs.nasa.gov/20100042541_2010046831.pdf).

Northrop-Grumman is in the final development stages of its airborne sense and avoid (ABSAA) system enabling the RQ-4 Global Hawk to fly in civil airspace.

 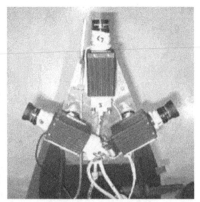

FIGURE 12.19 ■
Northrop-Grumman
ABSAA in a Calspan
Surrogate

The multi-sensor system has been in development at the U.S. Air Force Research Laboratory (AFRL) since 2008 under the multiple intruder autonomous avoidance (MIAA) science and technology program. The development program used a Calspan-operated Learjet equipped with the MIAA sensors and algorithms and acted as a surrogate for the unmanned Global Hawk. A system based on MIAA technology is expected to fly on the Global Hawk in 2015, with initial operating capability planned for 2017. The Air Force is leading development, but the Navy and Army are both partners in the ABSAA effort. The Navy's desire is to put the system on their aircraft (the MQ-4C Triton, commonly referred to as the broad area maritime surveillance, or BAMS), while the Army's interest is in controls and displays commonality with its ground-based sense and avoid system. The system uses electro-optical sensors and radar to detect and track noncooperative targets; the traffic collision avoidance system and ADS-B are used to detect and track cooperative commercial aircraft. The camera sensor setup and location in the test bed is shown in Figure 12.19.

The system can be mounted in the nose of the UAV or in aircraft with a space problem or, for rotary wing aircraft, in a dome above the aircraft or main rotor. The RQ-5A Hunter currently uses such a setup for its GCE to UAV data link. This design is shown in Figure 12.20.

The NASA Ikhana was issued a COA to operate within Class A airspace in the NAS. Team members worked with the FAA to safely and efficiently integrate the unmanned aircraft system into the national airspace. NASA pilots flew the Ikhana in close coordination with FAA air traffic controllers, allowing the aircraft to maintain safe separation from other aircraft. Additional information on this project may be found at http://history.nasa.gov/monograph44.pdf.

New Mexico State University (NMSU) has teamed with the FAA and has established the Unmanned Aircraft Systems Flight Test Center (FTC) and the Center of Excellence for General Aviation Research. University faculty and staff research, develop, test, and evaluate unmanned aircraft systems, which are capable of a wide range of military and civilian functions, including combat, search and rescue, tracking marine mammals, assessing forest fires, and gathering climate data. They have tested sense and avoid system technology aboard the NMSU-leased Aerostar UAV. They have evaluated The Defense Research Associates (DRA) sense and avoid system which uses optical technology; the full system uses three sensors and has a field of view of about 210 degrees. The system's processors select pixels moving differently from the

FIGURE 12.20 ■
Hunter with Data
Link Enclosure

Northrop-Grumman Photo

background and issue an alert. Since technology is better than human vision, the FAA hopes to migrate this innovation from unmanned to manned aircraft operations.

An excellent summary of detect, sense, and avoid technology can be found in the document DOT/FAA/AR-08/41, "Literature Review on Detect, Sense, and Avoid Technology for Unmanned Aircraft Systems," September 2009, published by the FAA in concert with NMSU, http://www.tc.faa.gov/its/worldpac/techrpt/ar0841.pdf). The reports provide a history of sense and avoid systems, the FAA's involvement with manned aircraft proximity systems, and a comparison of technology-based systems versus manned aircraft see and avoid. The report also provides a synopsis of the cooperative and noncooperative aircraft and ground based systems that are currently in use or in development:

Cooperative

– TCAS

– ADS-B

– TAS

Noncooperative

– RADAR

– LASER

– SONAR

– Motion detection

– EO

– IR

– Acoustic

– Passive ranging

– Electromagnetic visibility enhancement (similar to adding radar reflectors)

- Visibility enhancements (paint schemes and enhanced aircraft lighting)
- Ground-based radar

Appendix A of the report provides an excellent bibliography, and appendix C contains a capabilities matrix of technologies, in-service and in development, complete with notes relating to published articles on these technologies.

12.7.3 Test and Evaluation of Sense and Avoid Technologies

Cooperative systems (proximity warning) have been addressed in chapter 6 of this text. For tests of highly accurate sense and avoid systems, highly accurate TSPI systems must also be employed. For radar and EO sensors, the accuracies will be evaluated as described in the respective chapters of the text. A latency analysis will be required, and a nominal 2 sec total latency (as required by AC 20-165 and AC 20-172A for ADS-B systems) should be the maximum allowed.

12.8 | PAYLOAD

Payload entails sensors that are embedded or carried by the UAV required to successfully accomplish its mission. These sensors may be RADAR (SAR/ISAR), FLIR, TV, or LASER and may provide real-time imaging or archiving for post-flight processing. Probably the largest problem the evaluator will encounter is how to get large amounts of data generated by high-resolution sensors down to the GCE in an efficient and timely manner.

Latency in video distribution systems is the time it takes from when the video image is captured at the sensor until it is received on the sensor operator's screen in the GCE. The contributing factors to latency are the data transmission to the mission processors (senescence), data capture and compression, stream encoding, network transmission, stream reception, de-compression, and data display. Video capture and compression equipment will affect the front-end contributors to latency up to data transmission, and the back-end contributors such as stream reception and decompression are dependent on the software and hardware installed at the destination. As previously described in the data link section, the video stream transmission time can be the most difficult to predict or control. As a rule of thumb used by most organizations, 100 ms is the effective maximum latency for video compression and streaming systems that must be used in real-time analysis. The time limit ensures that the sensor information being acted on in the GCE is relevant and correct real time. As latency increases, relevance decreases and action taken may be too late as the situation may have changed. Latency can be minimized in a video handling system by an efficient video capture and video compression algorithms.

First, the data must be captured and put into a format which can be transmitted. In the case of multiple sensors, this operation may be carried out in the middleware described in section 12.4.2. Video compression can be accomplished by using constant bit rate (CBR), variable bit rate (VBR), or a hybrid known as capped variable bit rate (capped VBR).

CBR settings provide a video stream with a specified bits-per-second data rate. This scheme allows predictability but suffers in dynamic environments; significant scene changes or motion in the video stream will result in video quality degradation.

VBR settings ensure that the video will be compressed and streamed with a constant quality; i.e., the exact bits-per-second necessary to transmit the video stream will be dependent on the contents of the video data. If the bit-rate necessary to maintain quality exceeds the available bandwidth, frames will be dropped and the decompressed video being displayed at the GCE will experience ratcheting or skipping of frames. Again, this occurs with dynamic scene motion and critical information may be dropped.

Capped VBR is a hybrid option between the CBR and VBR approach and attempts to allow for variability in the bit-rate, but also considers the issue of maximum bandwidth available. It sets a bit-rate limit at which point the variable bit-rate approach is replaced with a constant bit-rate approach; when the bit-rate again falls below the set limit, VBR again is employed. The approach allows higher quality video transmission without maxing out the bandwidth.

Additional configuration parameters include the video quality setting (when in VBR mode), the video frame size and rate, the input video format, and the inclusion of audio and headers. Some of these settings are present all of the time, such as video quality settings, while others may be used only during initiation.

The delivery of this video (and audio) most likely will be encapsulated into a motion pictures experts group (MPEG) format. MPEG is an industry standard for video compression and transmission; if you have downloaded video files you may see that many of them are in an MPEG format. MPEG-1 and MPEG-2 combine audio and video into a digital format according to the MPEG standards. By knowing that the stream is formatted to this standard, the receiving device can decode the stream and provide the information to the user. HDTV uses MPEG-2 standards. Another version is MPEG-4 which specifies sending the video in other existing standards such as IP, RTP, and RTSP; MPEG-4 is used in 3-D broadcasts.

MPEG-1 and MPEG-2 are self-sufficient; however, they do not specify how they might be transmitted over the IP. You will remember during our discussions of the data link that we looked TCP/IP and how packets would be resent in the event of a loss or non-receipt of a packet. You will also recall that this lossiness contributes to latency. Streaming video and/or audio needs to be isochronous (remember, arriving in an orderly fashion) as opposed to bulk data transfer which is asynchronous. If one data packet is missing in a video transmission out of 1000 that were sent, it may not be that noticeable; but if we requested the missing packet be resent (as discussed earlier), the latency may make the information out of date.

User Datagram Protocol (UDP) does not have a dialog between sender and receiver; if a packet is lost or not received, the sender is never advised. This situation is analogous to an ARINC 429 data bus where there is no handshake between the source and the sink. UDP is better than TCP when latency is a concern and data needn't be acknowledged. Control and synchronization of streaming video is enhanced by using Real-time Transport Protocol (RTP), and it is usually sent via UDP. MPEG-1 and 2 provide their own synchronization, whereas MPEG-4 will almost always use RTP. What is normally used in payload streaming video applications is an MPEG-2 transport stream via raw UDP or via RTP over UDP.

ITU-T H.264 AVC is the current and most popular industry standard compression algorithm for MISB (Motion Imagery Standards Board)–compliant systems. H.264 compression, as it is called, provides very high compression ratios, low lossiness, convenient parameter controls, and can be encapsulated into an MPEG2 transport stream format for distribution over Ethernet using UDP/RTP packet protocol.

A second major consideration with payloads is the amount of available space and the associated power, video encoder equipment and transmitters required for the mission. One can see that for some applications, it may be physically or electrically impossible to carry all of the desired sensors at once. Carriage of sensors may have to be designed around specific missions (e.g., high-resolution video versus infrared for day or night missions). It may be decided, based on the available bandwidth, that only high value video will be sent in real time and all other video will be archived for postflight processing and analysis. Archiving will require a storage system on board and a compatible system on the ground for retrieving and analyzing the video. Just as with controls and displays evaluations addressed throughout the text, the information must be timely, understandable, and correct.

12.8.1 Test and Evaluation of Payload Sensors

The Test and Evaluation of sensors are described in chapters 7 and 8 of the text and will not be repeated here. STANAG 4671 makes no reference to payload, other than obligatory adherence to paragraphs 1301 and 1309.

As previously mentioned, the primary concerns of the evaluator will be: the usability of the data (can the scene be acted upon in a timely manner), the ability of the operator to detect, recognize, and identify objects, the ability of the sensor to track objects, and the ability to control, and store and retrieve sensor imagery. Identification and tracking will be accomplished IAW the techniques listed previously. Workload evaluations and SRS should be conducted over a number of scenarios with operators of various experience (low to high) to ensure that the system will be able to accomplish the mission. Data formatting and transmission was addressed earlier in this chapter.

12.9 | OPTIONALLY PILOTED AIRCRAFT (OPA)

The OPA is really a hybrid UAV where all of the capabilities of the UAS addressed in the previous sections are the same, but with the added benefit of a pilot onboard the aircraft. The FAA still considers the aircraft, pilot, GCE, and crew as a UAS and is still governed by the Airworthiness criteria (section 12.3). The requirements are somewhat lessened if the RPV capability is removed from the aircraft and a pilot is always on board.

The National Test Pilot School (NTPS) located in Mojave, CA has a Special Airworthiness Certificate for a C-150 Experimental Optionally Piloted Aircraft. The NTPS system is in accordance with FAA Order 8130.34B, defining an OPA as "an aircraft that is integrated with UAS technology and still retains the capability of being flown by an onboard pilot using conventional control methods."

The C-150 OPA can be operated in both CDV and RPV modes. The system is completely integrated and can also fly the UAV in the optimum orbit based on a target designated with the sensor. Figure 12.21 shows the National Test Pilot School's Cessna OPA and associated sensors.

The benefits of an OPA are obvious: in-flight emergencies such as aircraft and lost link can be handled by the pilot onboard, data link capabilities can be tested without concern of losing the link or the aircraft, automatic takeoff and landing development incorporates an extra layer of safety, remote control of sensors can be evaluated with or without the UAV under command direction, and new or upgraded payload or

FIGURE 12.21 ■
NTPS Cessna UAV
with EO/IR Sensor

National Test Pilot School Photo

communication packages can be tested more efficiently. The cost savings in evaluating S/W upgrades alone can be phenomenal. Because the OPA is, in effect a UAV, the development testing covered in this chapter applies to the OPA as well, with the added benefit of having the pilot in the loop.

12.10 | SUMMARY

It is envisioned by the FAA that by 2016 UAVs will be flying in the NAS in fairly large numbers. There will be an increased need for T&E personnel to validate the designs and performance of these systems. The test and evaluation of UAV aircraft will encompass all of the testing previously covered for manned aircraft, but presents some unique problems not seen with manned aircraft. Safety will play a major role in all UAV testing, even more so than with the development of manned aircraft; potential hazards cannot be easily mitigated by aircrew action. Human factors and workload must be evaluated with respect to the envisioned mission, which may require long missions at remote sites. Communications are critical and the evaluation team must be cognizant of potential latencies, lost link, data rates, and volume. Changes to S/W must be approached even more carefully than with manned aircraft as there is really no second chance.

12.11 | SELECTED QUESTIONS FOR CHAPTER 12

1. What are MALE and HALE systems?
2. Can a UAV currently operate in civilian U.S. airspace?

3. Name three considerations for UAS data link characteristics.

4. When considering safety, what is the maximum allowable latency in a data link system?

5. What is the difference between internal and external mode of UAS operation?

6. Which three frequency bands are available for UAV command and control links in the U.S.? Is a license required?

7. Where would you find a common data link?

8. How many bits are in a byte? Or vice versa?

9. What is the difference between command directed and RPV modes of operation?

10. What are the three components of a UAS?

11. What is an OPA? What are its benefits?

12. Name two applications of data links in UAS systems.

13. What is the difference between rate control and vector control?

14. Which documentation would you consult for UAS airworthiness in the U.S.? How about in Europe?

15. What are MASPS?

16. Should ACAS systems in a UAV be automatic or GCE controlled?

17. What is meant by slow start in a TCP connection?

18. Why are losses and gains calculated using decibels?

19. What is latency in data link transmissions? What is the total latency comprised of?

20. What is STANAG 4671, "Unmanned Aerial Vehicle Systems Airworthiness Requirements (USAR)" based on? Is it applicable to U.S. military operations?

21. What are ACK and NACKS?
    What is effective radiated power? What is it used for?

22. Where would you find a need for high capacity UAS data links?

23. Name two contributors to atmospheric losses in data link transmissions.

24. If you were to receive data across a satellite link, what would you expect for latency?

25. All other parameters remaining constant, which link would provide a greater range: 900 MHz or 10 GHz? Which would provide greater data throughput?

26. What is TCP? What is TCP/IP?

27. What are the major human factors concerns for the GCE during UAS operations?

28. What is the difference between 30 KB and 30 Kb?

29. What is a T link? Is there much difference between a T1 and a T3 link?

30. What is senescence?

31. How does a modified Cooper Harper scale aid in HMI testing of GCE controls and displays?

32. Name the important parameters that contribute to the link equation.

33. How is a UAV target tracked?

34. What is an antenna pattern? How is it measured for UAVs?

35. What is network middleware?

36. How does TCP handle congestion?

37. What is the difference between constant bit rate and variable bit rate in video links?

38. Has the FAA approved any type of sense and avoid systems for UAVs? What are they?

39. How does encryption affect the latency in a data link?

40. How is data sent on a TCP network?

41. What is meant by dBm? dBi? How would 1 W of power be expressed in dBm?

42. What is a TCP buffer size? How is it allocated?

43. How is the minimum flight crew determined for UAS operations?

44. What happens in the case of missing, lost or error detected by the receiver in the TCP network?

45. Which signals may possibly interfere with civilian UAV data links?

46. State three ways that carrier sense multiple access (CSMA) can be handled. What happens on Internet connections?

47. Will time-alignment of data be a problem with UAS operations?

48. What is meant by peer-to-peer and plug-and-play?

49. What are ACK ratios and received signal strength used for in the GCE?

50. How does the congestion algorithm work?

51. What is the industry standard for video compression and transmission?

52. What is a link equation used for?

53. Aside from a constant, which two parameters contribute to the calculation of spreading loss?

54. What is the spreading loss of a 5.8 GHz signal over 20 nm? How about a .9 GHz signal for the same conditions?

55. What is a link margin? What is it used for?

56. What are some of the contributors to latency in video streams?

57. Your effective range is much less than you calculated. How would you troubleshoot this anomaly?

58. What is the assumed reaction time from the time an intruder is seen to the time that the aircraft responds to a pilot input?

59. Are current proximity warning systems good enough to provide a sense and avoid capability for a UAV?

60. Does the USAR require FGS speed protection for the UAV?

61. Name a possible way of evaluating blockage areas during UAS ground tests.

62. How is jamming-to-signal ratio and interference-to-signal ratio related?

63. Name the three contributors to the I/S ratio.

64. What is the effective maximum latency for video compression and streaming systems that must be used in real-time analysis?

65. What is a BER? How is it employed in UAS testing?

66. Describe a possible ground test for UAV data links? What would be your primary concerns?

67. Are detections of intruders guaranteed with the installation of TCASII or ADS-B?

68. How can you avoid start-up time latency on a TCP link?

69. What are the proposed sense and avoid standards for field-of-view and safety bubble around the aircraft?

70. What is the normal default setting of the autopilot within the UAV?

71. What is BLOS? How is it calculated?

72. What problems may be encountered if the boresight of the GCE command and control antenna is not correctly aligned to true north?

73. What systems, aircraft or otherwise, could possibly contribute to interference on a 900 MHz command and control signal? How about a 2.4 GHz signal?

74. What problems may you encounter when employing a high gain directional antenna to control or receive data from a UAV?

75. What could possibly cause a PIO when operating in an RPV mode?

76. What is the difference between asynchronous and isochronous data?

77. The USAR references CS 23.1301 and CS 23.1309. What are they?

78. What is meant by the term navigational hierarchy? Why is it important?

79. What is a lost link procedure? What would you expect to see with a loss of GPS within the UAV?

80. Is there any benefit to having the GCE pilot constantly provide input to the UAV autopilot?

81. Are there any disadvantages to employing autopilot point-and-click technology in the GCE?

82. What reaction time is used in the GCE when evaluating autopilot malfunctions?

83. What are TCP Reno, TCP Tahoe, and TCP Vegas?

84. What is a reasonable TCP buffer/window size?

85. What is meant by CSMA/CD? Provide an example.

86. Name four types of possible proximity systems for a UAV.

87. What is lossiness?

88. Will airlines ever transition to unmanned vehicles?

Night Vision Imaging Systems (NVIS) and Helmet Mounted Displays (HMD)

Chapter Outline

	Acknowledgment	907
13.0	Introduction	907
13.1	Overview	908
13.2	Image Intensification (I^2) Technology	909
13.3	NVG Human Factors Issues	918
13.4	Lighting Specifications	920
13.5	Interior NVIS Lighting Methods	924
13.6	Exterior Lighting Methods	929
13.7	Test and Evaluation of NVIS Equipment	931
13.8	Some Final Considerations	938
13.9	Helmet Mounted Display Systems (HMD)	939
13.10	HMD Components	942
13.11	Test and Evaluation of HMD Equipment	949
13.12	Selected Questions for Chapter 13	954

ACKNOWLEDGMENT

The author would like to personally thank Chuck Antonio, aerospace medicine/night vision systems pilot instructor, National Test Pilot School (MD, Medical University of South Carolina, Charleston; graduate, US Navy Flight Training and Flight Surgeon School, NAS, Pensacola, Florida) and Edward "Dick" L. Downs, night vision systems pilot instructor, National Test Pilot School (BSc aeronautical engineering, Imperial College London), without whose help this chapter would not be possible.

13.0 | INTRODUCTION

The desire to improvement nighttime vision led to development of both FLIR (which has been discussed in chapter 7 of this text) as well as image intensification (I^2). Night vision imaging systems (NVIS) do not exploit thermal energy but rather attempt to intensify reflected light just outside of the visible light spectrum; it is referred to as invisible light in some of the documentation. Aviation operational capabilities with use of I^2 have grown with experience and with technological advances, and significant improvements in nighttime capabilities now exist. When applied to aircraft operations, the NVIS systems can simply be referred to as night vision goggles (NVG), which comes from the binocular design of most pilot systems. Helmet mounted display systems (HMD) are a natural extension of the NVG and can combine elements of FLIR, NVG, weapons, targeting cues, and HUD projected onto a visor patch, a combiner, or an eyepiece of the pilot's binocular system. This allows the pilot to view all tactical information no matter where his vision may be directed.

To understand the history of image intensifiers, the reader is referred to a wealth of documentation on the subject. The first is *NVG Fundamentals/Techniques, Night Vision Goggle Compatible Aircraft Lighting* (George W. Godfrey, Aerospace Lighting Institute [ALI], Clearwater, FL). In addition to lighting requirements and lighting compatibility, the author covers the history of NVGs and the basic physics behind their operation. SPIE is probably the authority on imaging systems. It was founded as the Society of Photographic Instrumentation Engineers in California to specialize in the application of photographic instrumentation. In 1964 the society formally changed its name to the Society of Photo-Optical Instrumentation Engineers and in 1981 began doing business as SPIE: The International Society for Optical Engineering. As of 2007, it is referred to simply as SPIE. An excellent compilation of papers is found in *Selected Papers on Night Vision Technology* (SPIE Optical Engineering Press, 2001), which contains 104 technical papers on the subject of night vision. The US military and the FAA have issued numerous standards, orders, and handbooks; these will be called out in the appropriate sections of this chapter.

13.1 | OVERVIEW

The human eye can see images that reflect light from violet to red in the electromagnetic visible light spectrum. Beyond the visible spectrum for humans is the light known as near infrared (NIR); the human eye is unable to see that light. The two methods of enhancing night vision are thermal imaging and image enhancing. Both are used regularly and both have their own advantages and drawbacks. Capabilities of NVGs include an improved situational awareness that in turn leads to enhancement of tactical operations. They aid in maneuverability, navigation, and threat detection. The image quality and resolution (the ability to identify objects) is better than that of a thermal imager, but the range of the NVG is much shorter. The thermal imager performs better in less than optimum atmospheric conditions (e.g., smoke, haze, precipitation) and can be used day and night. The best NVG visual performance is still worse than standard daytime uncorrected vision of 20/20; some of the better NVGs offering an equivalent 20/25 under optimal conditions, 20/40 under starlight conditions, and even worse at lower illumination levels. These numbers represent laboratory testing. The performance achieved in operational use will be less depending on factors discussed later in the chapter. The NVG FOV is smaller than that of a thermal imager and does not employ FOV changes or pan/zoom operations. The thermal imager is commonly used for targeting, whereas NVGs are not. NVGs add both a physical and mental workload and contribute to fatigue.

Thermal imaging systems use photon detectors or thermal detectors; from chapter 7, the reader will recall that photon detectors may be photovoltaic, photoelectric, photoconductive, or semiconductors. Thermal detectors are classified as thermocouples, bolometers, calorimeters, fluid expansion, or evaporographs. Image enhancing systems also use photon detectors, normally silicon semi conductors, and employ CCD or CMOS technology; these sensors were also previously described in chapter 7 under digital TV systems. The difference between the two technologies may be demonstrated by examining avoidance systems currently employed in the automobile market. Cadillac, BMW, and Mercedes all offer options to drivers for systems that help to identify objects at night. Cadillac and BMW offer an IR imaging system, whereas Mercedes offers an image enhancing system.

The Cadillac night vision infrared imaging system works with a staring focal planar array as the sensor and ties into an active matrix LCD for the HUD. The detector consists of a barium strontium titanate element bonded to an ROIC. The detector temperatures are stabilized with thermoelectric coolers for peak performance, but there is no attempt to achieve cryogenic temperatures; it incorporates a 240×320 FPA. A chopper disc (similar to an AIM-9 reticle) rotates in front of the detector to modulate the scene's energy by allowing the pixels to constantly blank on and off. The disc rotates in phase with the detector ROIC timing. The circuit under each element samples capacitance on a regular basis, and these readings are converted into a monochromatic video signal that is imaged on the driver HUD. The detectors operate in two colors, 3–5 μm and 8–12 μm, and are mounted in the grill. Raytheon is the system integrator.

In contrast, the Mercedes system uses NIR technology and produces an even, clear picture in the dark. This system is similar to NVGs; the NIR system in the Mercedes illuminates everything as if it were in the high beams of the vehicle. By utilizing a series of projection bulbs and cameras, the Mercedes' active nightvision system picks up the faintest traces of light and transforms it into a clear picture. The advantage is that the Mercedes' system can see warmer living things just as clear as it can spot colder objects (i.e., dead animals or inanimate objects). The drawback to the Mercedes' system is its range: maximum effective range is less than 600 ft. Mercedes' NIR system doesn't handle fog well, while the Cadillac night vision system can see through the dense conditions. Both systems incur an extra $2,000 to $2,500 added to the base price.

13.2 | IMAGE INTENSIFICATION (I^2) TECHNOLOGY

We will remember from our discussions about the visible light spectrum in chapter 7 that the human eye responds to frequencies between .39 and .72 μm; this band is called the visible light spectrum. Within this spectrum there is also a peak response for the eye. Figure 13.1 depicts the spectral response of the eye. The x-axis at the top depicts the energy of a photon at the corresponding wavelength located on the lower x-axis. *Photopic* in the upper right hand of the figure refers to lighting conditions (10^0 through 10^5 Lux) that allow good visual acuity and the ability to see color (below 1 Lux, human vision is said to be scotopic: poor acuity and a loss of color). The peak spectral response of the eye is at .555 μm with photon energy of 2.25 eV. This frequency/wavelength corresponds to the color green, which is why HUDs and NVGs are also in green.

NVG sensitivity ranges from .45 to .60 μm (depending on photocathode material) at the short end up to approximately .930 μm at the long end. This bandwidth extends into the visible light spectrum and overlays a portion of that bandwidth. Figure 13.2 depicts the spectral responses of the human eye and the NVG photocathode.

FIGURE 13.1 ▪
Human Eye Spectral
Response

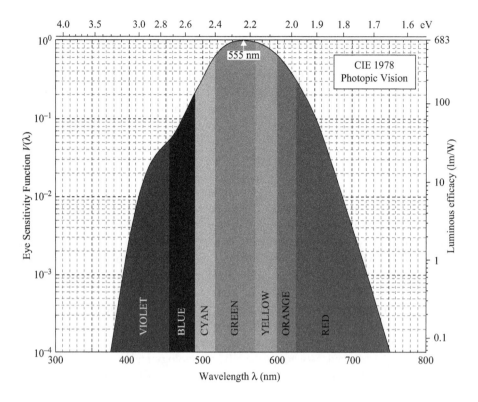

FIGURE 13.2 ▪
NVG versus Human
Spectral Response

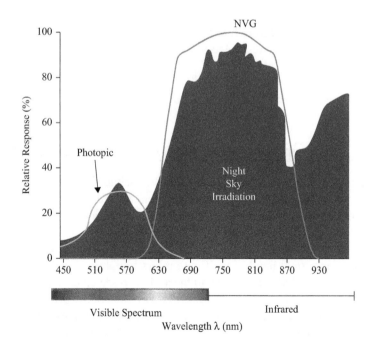

**Note: The graph assumes integration of a blue-green filter, hence the sharp drop-
off in response.**

Night sky irradiation is made up of moonlight, starlight, and cultural lighting. An overlap of the human eye and the NVG sensitivity spectrum includes the orange and red portions of the visual spectrum. The key to NVG compatibility is filtering the output of cockpit displays and exterior lights as well as visible spectrum light so that they don't enter the NVG.

Some definitions need to be covered prior to continuing the discussion. An I^2 tube is an electro-optic device used to detect and intensify reflected energy in the visible and near infrared region of the electromagnetic spectrum. Illuminance is the amount of light that strikes a surface or an object measured in either lux or foot-candle. Luminance is the intensity of reflected light measured in either foot-lambert or candela per square meter. Albedo is the ratio of the amount of reflected light to amount of light striking an object; varying Albedo in the scene is important to add contrast to the NVG image.

The evolution of NVGs was previously covered in section 7.1.1 of the text.

13.2.1 NVG Components

The basic components of NVGs are as follows:

- Objective lens
- Minus blue filter
- Photocathode
- Ion barrier film
- Micro-channel plate
- Phosphor screen
- Image inverter
- Internal power supply
- Eyepiece lens

The photocathode through the internal power supply are part of the I^2 tube, where the others are optics and filters. This grouping is shown in Figure 13.3.

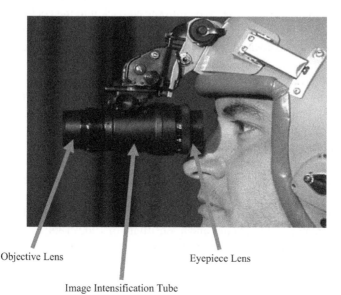

FIGURE 13.3 ■
NVG Components

Objective Lens

Eyepiece Lens

Image Intensification Tube

The objective lens is a combination of optical elements which are used to focus incoming energy (photons) onto the photocathode. The back side of lens is coated with a "minus blue" low pass (allows lower frequency/longer wavelength) filter that allows for compatible cockpit lighting. The objective lens may focus from several inches to infinity depending on NVG design. The space between the eye and the eyepiece lens (eye relief) allows the operator to use the NVG for viewing outside the cockpit or look beneath or around NVG for viewing inside the cockpit. A single optical element result in an inverted image being focused onto the photocathode. The image must be inverted again so that it is "the right way up" when viewed by the user; which is accomplished by incorporating a fiber optic twist between the phosphor and eyepiece lens. The purpose of the minus blue filter is to reject certain wavelengths of visible light and prevent these wavelengths from entering the image intensifier (refer to Figure 13.2 and review the overlap area). It allows for use of modified cockpit lighting that does not adversely affect NVG image performance. The minus blue filters are grouped by classes:

- Class A: blocks below 625 nm (blue/green)
- Class B: blocks below 665 nm (blue/green/some yellow/red which allows the use of color displays)
- Class C: same as Class B but with a notch or passband cutout added to permit direct viewing of aircraft HUD information (sometimes referred to as Modified Class B in the literature)
- Some European manufactured tubes block responses below .645 μm and are also referred to as Class C

The use of NVGs in aircraft with a fixed HUD will require a notch (Class C) of 1–2% around .555 μm to allow the green symbology or video of the HUD to be passed to the eye. Early photocathodes were not that sensitive, so they had to be exposed to a wide band of light, which included some red and yellow. As a result, early cockpits could not use red and yellow and had all green lighting schemes. When gallium arsenide (GaAs) photocathodes were introduced NVGs had a much higher sensitivity; as a result, the system could be exposed to a narrower bandwidth of light and still provide a useful image. This allowed some red and yellow to be introduced back into the cockpit lighting scheme.

The Class C filter is susceptible to a phenomenon called leaky green goggle glare. This is caused by green cockpit lighting entering the objective lens from at least 45° off-axis (an effect similar to thermal energy [noise] entering the FLIR external to the field of view). The notch filter at .545 μm allows the green light (energy) to impact the photo-cathode resulting in glare. The problem is most noticeable when the illumination is low and goggle gain is high. This problem mitigated by physically blocking off-axis energy with a hood at the objective lens without reducing NVG field of view (Figure 13.4).

The photocathode, like the detector element in IR systems, converts light energy to electrical energy (the reader is referred to IR Detectors, section 7.7.2 of the text). The inner surface of the objective lens is coated with photosensitive material, two of which are in common use. The first is multialkali antimonide (S-20/S-25) photocathodes, which have a response of .45–.90 μm and a sensitivity of 600–750 μA/lm. The second is GaAs, which has a response of .60–.90 μm and a sensitivity of 1000–1800 μA/lm. The sensitivity of an image intensifier tube is measured in microamperes per lumen (μA/lm). It defines how many electrons are produced per quantity of light that falls on the photocathode. Like the D* measurement (Section 7.7.2 of the text) for IR photon detectors, the measurement

FIGURE 13.4 ▪
Effects of the Lens
Hood Extension

Lens Hood
Extension (LHE)

NVG Image Without
an LHE

NVG Image with
an LHE

should be made at a specific color temperature, such as "at a color temperature of 2854 K." Typically, the higher the value, the more sensitive the tube is to light.

S-25 is still used in all European NVGs. It is argued that S-25 has the same area under the curve (Figure 13.2) as GaAs. This is a true statement until you filter it for aviation purposes, after which the area under the curve is greater for GaAs. S-25 is not as sensitive to damage from positive ions (to be discussed momentarily), so there is no ion barrier. Because no ion barrier is required there is a much smaller gap from the photo-cathode to the micro-channel plate (MCP), so the halo (also addressed momentarily) is much smaller. GaAs is used by ITT Exelis in Gen III NVGs, which represents all NVGs produced by them for aviation use since the late 1970s. GaAs is more sensitive than S-20/S-25, which implies better performance in low light, and a narrower band of light (caused by .665 μm cutoff of Class B filters) still results in a usable image. The life of I^2 tube depends on the life of the photocathode. Rapid loss of electrons, as in the case of exposure to bright light, reduces the tube life.

Positive ions are also given off by the MCP and are attracted to the negatively charged photocathode; these ions damage the PC and reduce its life (Figure 13.5). GaAs tubes have an aluminum oxide ion barrier film coated onto the MCP; this extends the tube life by trapping positive ions and keeping them from contaminating the photo-cathode. This barrier reduces system performance by degrading the signal-to-noise ratio (SNR). The barrier also requires an increased photocathode–MCP gap to prevent arcing between the MCP and the photocathode (i.e., arcing between the gallium arsenide and aluminimum oxide). As metioned previously, S-25 multi alkali tubes do not require ion barrier film.

Another phenomenom that occurs is the halo effect also shown in Figure 13.5. The electrons fan out in a cone as they depart the photocathode. The electrons are negatively charged, so if there is a bright light source in the image this results in electrons being liberated very close to each other on the photocathode. As these two adjacent electrons make the journey to the MCP, they are constantly repelling each other and move farther apart. This fanning out in a cone results in the circular halo seen in the NVG image. This is not as obvious in the S-25 photocathode NVG as the photocathode to MCP gap is much smaller due to the missing ion barrier.

The MCP is thin wafer containing millions of glass tubules or channels for electrons from photocathode to enter. The tube walls coated with a gallium-aluminum-arsenic compound which causes release of more electrons resulting in a cascading effect.

FIGURE 13.5 ■ Ion Barrier Film

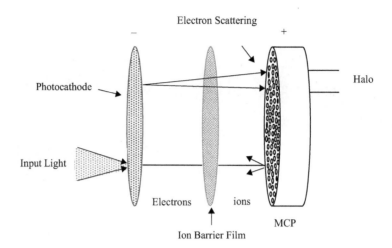

To ensure electrons will impact the tube wall, the tubes are tilted approximately 5°. The number of tubes is analogous to the number of pixels in a digital camera; more tubes/pixels result in a more detailed image. Not every electron that hits the MCP will fall into a tube; these losses along with the positive ions result in a reduced SNR.

The thin layer of phosphor at the output of the MCP is called the phosphor screen and is analogous to the analog TV systems described in section 7.8.3.1 of the text. Simply put, phosphor emits light when struck by electrons; the light emitted by phosphor creates a visible image. The NVG image is green due to the phosphor that is used. There are over 100 different types of phosphor, and each has its own chemistry, sensitivity, persistence, and manufacturing qualities. Phosphor selection will be based on persistence characteristics and manufacturing considerations. Persistence is the after image or tracking caused by bright lights on the phosphor; ideally a short persistence is better. Some examples of phosphor used in NVG applications and the colors they provide are P-20/22, which is green on green; P-43, which is green with a yellow tint; and P-45 (Y202S:Tb), which is used in hospital x-ray machines and provides a black-and-white contrast. A P-45 phosphor is used in the ONYX HyperGen IITs and shows a black-and-white image as opposed to a black-and-green image incorporated on most other NVGs. Use of a phosphor that allows production of a black-and-white image can arguably provide better scene interpretation.

The high-voltage power supply is built into image intensifier casing; it receives its power from source external to the tube (i.e., battery pack). The purpose of the high-voltage power supply is to amplify the relatively small electrical output of the MCP to a voltage high enough to produce a scene picture (roughly equivalent to the ROIC on an IR FPA). The supply is also responsible for providing an AGC function via two independent circuits: an auto brightness control (ABC) and a bright source protection (BSP).

The ABC provides for consistent image brightness over wide range of illumination levels by control of the MCP voltage (it controls number of electrons leaving the MCP). It protects the aircrew by limiting the effect of sudden bright flashes (such as a missile plume) and helps avoid eye fatigue by attempting to keep the output brightness constant regardless of scene brightness. High input light levels (e.g., an incompatible light source) increase the loss of electrons from photocathode. This results in creation of more ions, which can contaminate the photocathode and reduce its life. The BSP limits the number of electrons leaving the photocathode by automatically reducing the voltage between the photocathode and MCP at high light levels, which prolongs the tube life. However, this results in darkening of the scene outside the cone of light provided by a bright light source such as IR spotlight (an IR spotlight illuminates an area with energy at a frequency compatible with the NVG). Current NVGs have only a single gain value; a very bright source of light in the middle of a dark scene will cause the gain to accommodate the bright source. The image will be bright in the middle but dark in the edges. This is a danger when using a searchlight close to the ground as the horizon is often lost due to the BSP.

The ABC and BSP work together to form the AGC; as scene brightness decreases the gain increases to keep the image brightness constant. As the gain reaches 100%, any further reduction in scene brightness will result in a reduction of image brightness. As the gain increases both the signal (ambient light) and the noise (randomly liberated electrons) are magnified. The noise level remains constant, but as the scene gets darker the signal decreases and the noise begins to dominate the image; in total darkness only the noise remains and produces scintillation, or a white sparkling effect. The inference being that an NVG needs some energy to produce an image and in total darkness there will be no image.

Gain is the ratio of output brightness to input brightness for the total system. Typical values of gain are 2500–3000 for early Gen III NVGs and 5,000–10,000 for newer Gen III NVGs. NVG system gain can be measured by using an ANV-126 maintenance test set or other specialized lab equipment.

The final optical component of the NVG is the eyepiece lens. The lens allows for focusing of the visible image on the retina by moving the image plane forward and backward relative to the eye. It does permit some correction for vision variation within individuals, but corrective lens must still be worn by user and it does not correct for astigmatism. The size of lens impacts the degree of eye relief, which is important for two reasons: it must be compatible with equipment that must fit between the NVG and the eye (glasses or gas masks), and it must allow the user to look under and around the NVG to view cockpit lighting.

13.2.2 NVG System Characteristics

The evolution of NVIS technologies has been previously discussed in chapter 7 of the text. However, certain terminologies need to be understood when evaluating these systems. In addition to the Generation I, II, and III technologies, another US grouping called OMNIBUS is a US Army contract vehicle that incorporates minor technology and manufacturing improvements when available; systems may carry the designation of OMNIBUS II through OMNIBUS VI. There have been significant performance enhancements with the latest iterations.

The common designation used for NVGs is the aviators night vision imaging system (ANVIS); an example of a particular product would be AN/AVS-9, model item F4949R. This system is a Generation III system and the ninth version of an ANVIS system. Systems within a particular Generation may have vastly different performance characteristics; the generation classification is based on the technology being used. For example, an AN/AVS-6 is a Generation III system but does not have the performance of the AN/AVS-9 system, which is also Generation III. OMNIBUS IV outperforms them both but is still classified as Generation III. One must also be careful with the item nomenclature; an AN/AVS-9 is also called an ANVIS 9 or a F4949G (model) or a TACAIR NVG (used by the US Navy tactical air community). You may also put different I^2 tubes into different chassis; a new I^2 tube in an ANVIS-6 chassis will likely produce a better image than an older generation tube in an ANVIS-9 chassis.

The United States tends to favor the Generation III, whereas the European Union tends to favor the Generation II. The performance is very similar under high illumination, and the limiting resolutions are similar (~64 lp/mm for current IITs); 128 lines per millimeter means 64 dark lines alternating with 64 light lines, or 64 line pairs per millimeter. The performance differs under low light/low contrast: the Gen III SNR slightly better and the Pinnacle (ITT's OMNIBUS VI system) demonstrably better. Considering the differences between second- and third-generation NVGs, it was found that the difference in aided visual acuity (VA) between the two generations widens under two conditions: (1) when target contrast is constant and night sky irradiance decreases and (2) when night sky irradiance is constant and target contrast decreases. The halo size of Generation 3 tubes is approximately 1.0–2.0 mm for a point source (less for the Pinnacle at 0.7–1.25 mm) versus 0.6–0.8 mm for Super Generation 2 and HyperGen (Photonis). HyperGen is rated at 60 lp/mm, 700 μA/lm and an SNR of 20. In the overall scheme of things, current Generation II and Generation III system performance is comparable.

SNR is the ratio of signal power to noise power; the primary NVG noise source is the photocathode and the MCP. The signal portion is generated by photons of light, and the noise is a result of an imperfect photocathode that liberates electrons without an incident photon. The modulation transfer function (MTF) is identical to that previously described for thermal imager systems. The MTF is the system's ability to replicate a frequency/contrast to the observer. Image contrast is a function of spatial frequency; high contrast at the highest possible frequency is desirable. Figure 13.6 depicts resolution versus MTF for two NVG systems. NVG A and NVG B have the same limiting resolution of 64 lp/mm. System B has better performance because it provides greater contrast at high spatial frequencies (you can interchange resolution with spatial frequency),

FIGURE 13.6 ▪
Resolution versus
MTF

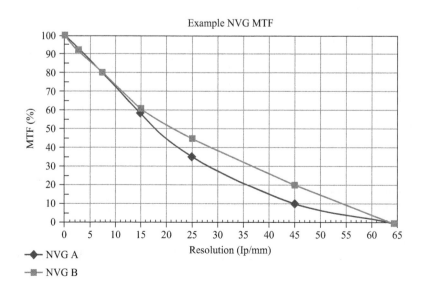

which results in a better image; the area under the curve becomes just as important as limiting resolution.

The U.S. State Department assigns a figure of merit (FOM) as a measure used for export control. The FOM is equal to the resolution (lp/mm) × the SNR. The product determines the level of control. US Domestic Emergency Medical Services (EMS) are authorized a product of 1400–1600 with auto-gating, whereas US Domestic Paramilitary (SWAT) are authorized a product of 1600–1800 with auto-gating. US military are authorized products in excess of 1800 with auto-gating. Auto-gating (ATG) is similar to a fast AGC; it was designed to improve the BSP feature to be faster and to keep the best resolution and contrast at all times. It is particularly suitable for ANVIS goggles, operations in urban areas or for special operations. ATG is a unique feature that operates constantly, electronically reducing the "duty cycle" of the photocathode voltage by very rapidly switching the voltage on and off. This maintains the optimum performance of the I^2 tube, continuously revealing mission critical details, safeguarding the I^2 tube from additional damage, and protecting the user from temporary blindness. It is particularly useful in the ITT Pinnacle tube by helping offset the loss in tube life created when thinning the ion barrier film to allow narrowing of the photocathode–MCP gap.

A comparison of current Generation II and Generation III technologies are shown in Table 13.1.

Most important to the user will be the image characteristics of the system. The image characteristics will be determined by the design of the optics and the image intensifier. Some of the parameters of concern will be the resolution (or image quality), FOV, chromaticity, halo size, blooming, scintillation, and goggle glare.

System resolution is accomplished in the same manner as spatial frequency for IR and EO devices. Since NVGs are an optical system, resolution will be measured by noting the sensitivity in separating shades of contrast (black/white, shades of gray). Resolution of an NVG can be expressed in line pairs per millimeter (as discussed previously), cycles per milliradian (spatial frequency; see chapter 7), or even Snellen acuity (e.g., 20/20, 20/40). The perceived image quality will be from a combination of system resolution, environmental conditions, and aircraft design constraints (see aero-mechanical and aero-optical effects in chapter 7). The FOV is determined by the optical design; a 40° circular FOV is common for most systems. Studies have shown that pilots prefer good resolution and limited FOV to poor resolution over a larger FOV (as the two are directly related). The chromaticity is determined by the type of phosphor used;

TABLE 13.1 ■ Generation II/III Comparison

	Euro Gen II Photonis, XR-5, S-25	US Gen III ITT, F9800VG, GaAs
Limiting Resolution	64 lp/mm	64 lp/mm
SNR	23	28
Figure of Merit	1472	1792
Halo	0.8 mm	0.7 mm
MTF		
@2.5 lp/mm	92%	92%
@7.5 lp/mm	80%	80%
@15.0 lp/mm	58%	61%
@25.0 lp/mm	45%	38%

FIGURE 13.7 ▪
Similar Scenes with
Generation III (GaAs,
left) and Generation II
(S-25, right)
Depicting the
Severity of the Halo

Photonis USA

a saturated green or green with a yellow tint is used for most systems. Green is used because it has the highest response of the human eye and lessens eye fatigue during prolonged use.

A halo, as previously described is a circle of opaque brightness around point light sources; halos will coalesce to form a larger area of coverage when many light sources are present. The size of the halo depends on the spacing between the photocathode and MCP, where Generation III systems will produce a larger halo than Generation II systems. The intensity of halo depends upon the strength, color, distance and aspect of the light source, and the gain setting of the NVG. An example of halos created in Generation II and Generation III systems is shown in Figure 13.7.

Blooming is a common term used to denote the washing out of all or part of the NVG image due to reduced gain when a bright light source is in or near the field of view. This can be a result of BSP or ATG activation. Scintillation is electronic noise created at high gain levels (e.g., during low illumination conditions) and appears as a sparkling effect (like snow on a television), or pixelating, throughout the image.

There are other system characteristics such as helmet mounting, backup power sources, low battery indicator, weight and CG, but they are really out of scope for this discussion and therefore are not covered.

13.3 | NVG HUMAN FACTORS ISSUES

As emphasized throughout the text, a system's compatibility with the aircrew's flight equipment and the cockpit environment is an essential test. Some of the items that need to be investigated are probably obvious. Since the NVG is mounted on the helmet, is other essential equipment adversely impacted or rendered unusable by the installation? Compatibility issues include helmet visors, laser eye protection, eye glasses, and chemical/biological/radiological (CBR) equipment (previously known as nuclear, biological, chemical [NBC]). Another issue will be the snugness of the fit of the helmet; a helmet that is allowed to shift on the head will provide an erroneous view of the world through the goggles.

Integration issues within the cockpit environment include the following:

- Fixed HUD
- Movable HUD

– NVG HUD

– Transparencies

– Crew station design

– NVG stowage

– Ejection seat concerns

A fixed HUD requires the use of a Class C objective filter to see the HUD symbology or the rastered imagery. A movable HUD (e.g., a flip-down installation) poses potential design issues relative to use of NVGs. There may be blockage of projected information (as in the case of a rear projector) or physical interference of the NVG with HUD, or the HUD with the NVG. The shift of eye position relative to HUD location may require the user to move his seat to obtain the design eye position. Several NVG HUD systems are available, but design differences result in varying concerns; in particular, how does one attach the HUD to the NVG? If it is attached to the eyepiece, it may affect diopter (a unit of measurement of the refractive power of lenses equal to the reciprocal of the focal length measured in meters; or in layman's terms: focus) control. If it is attached to the objective lens, it may cause cockpit obstruction problems, an increase in head-borne weight, and movement of CG forward. Additional concerns include the size, location, and length of cables from the control box to the NVG and the ability to control the brightness of the HUD symbology. Lead was used for bird-strike resistance in older transparencies (e.g., windows, windscreens); unfortunately, lead partially blocks transmission of near-IR energy degrading NVG image quality. Newer transparencies have good near-IR transmissivity as well as good bird-strike protection. Applied coatings on transparencies may also adversely affect transmission of near-IR energy. Transparencies may also have other deleterious effects; scratched surfaces, pitting, crazing and/or dirt may affect the user's ability to see through the transparency. The transparency can cause a refraction of incompatible light, a reflection of compatible light, or a reflection of interior lighting. A sharp curvature can create distortion and air vents and sliding windows (as used in helicopters) can result in a large reduction in image quality.

Crew station evaluations are performed as previously described with emphasis on the proximity and location of aircraft structures such as the canopy bow or window frame and their inter-relationship to the mounted NVG. Blockage areas caused by the NVG such as obstructions of the outside view, overhead panels, wires, and lights as well as any potential for interference with the NVG needs to be investigated—for example, when the NVG is in the operating position (i.e., down and in front of the eyes) or in the stowed position (i.e., still on the helmet but moved up and away from the eyes). NVGs must be shown to work as intended without adversely affecting the operation of other aircraft systems.

When NVGs are stowed in the aircraft, it must be shown that they remain secure during tactical maneuvering and, in USN applications, during carrier catapults and arrestments. When stowed, the NVGs should also be accessible, minimize heads-down time, and reduce the chance of spatial disorientation when donning and doffing the equipment.

Ejecting from an aircraft while wearing NVGs present other concerns; loss of the system during an ejection could cause a serious injury where none existed without the goggles. Whether the NVGs are on or off the helmet at ejection is a function of the mount design. If the goggles must be removed before ejection, it will require additional time.

As of this writing there have been automatic removal systems that have been developed and fielded. Two other concerns during the ejection sequence are an ejection through the canopy versus after the canopy deploys, and the location of the ejection handles. The NVG's physical relationship to canopy must be examined to determine if it will create an unsafe situation during the sequence. Ejection using a face curtain may be restricted based on interference with the NVG system. If the goggles must be removed and stowed prior to ejection, a modification to the minimum safe ejection altitudes must be made in the aircraft flight manual.

■ 13.4 ■ | LIGHTING SPECIFICATIONS

13.4.1 MIL-L-85762 and MIL-L-85762A

In the early 1980s, US government procurement documents began to impose NVG (NVIS) compatible lighting as requirement. Initially, green lighting was to be used and approval was sought from the surgeon general. Following this guidance, companies began to design and manufacture lighting using their own criteria. The systems were evaluated under lab and field conditions and the findings were mixed; some were good, whereas others were not. As a result, a NVIS compatible lighting requirements document was devised to ensure devices provided the same degree of compatibility. Work began in earnest in 1983 by a Joint Aeronautical Commanders Tri-Service committee composed of Naval Air Development Center and Industry representatives. The committee's objectives were to determine compatible lighting, how it would be measured, and how it would be implemented. MIL-L-85762 (a USN document) was released by the committee in January of 1986 and was strictly a lighting specification for NVIS compatibility. With the development of newer NVIS systems the specification was revised in August of 1988 (MIL-L-85762A). The revision added the requirement for lighting compatible with other types of night vision imaging systems such as Class B to allow an increased use of color (e.g., annunciators, displays). Use of the specification ensures cockpit lighting will not adversely affect NVG performance assuming there is proper matching of objective filter with class of lighting (e.g., Class A lighting with NVG using .625 µm filter).

The specification addresses most cockpit lighting methods to ensure all components can be modified for NVG use and allows for complete cockpit modification. The specification mandates that after modification there will be no change in the cockpit scan habit pattern and that operators can quickly and accurately gather heads-down information when using NVGs. It allows for smooth introduction of new NVIS equipment into an existing NVIS modified cockpit. The advantages of the specification are that it provides guidance for manufacturers to use when designing and producing NVIS lighting products and the same for integrators when developing plans for cockpit lighting modifications. It provides guidance for testers when conducting specification-compliance testing and standardizes terminology for NVIS types and classes (it matches the NVG objective filter to the filters required for NVIS lighting).

The specification does have limitations; it does not provide guidance for NVIS exterior lighting, nor does it provide guidance for the latest color display technology (it only addresses CRTs). It doesn't provide guidance for conducting qualitative evaluations of NVIS modifications and it may be more restrictive than necessary for a few current NVG designs.

There are four types of lighting called out in the specification:

- Type I: Lighting compatible with any direct view NVIS using Gen 3 image intensification (I^2) tubes

- Type II: Lighting compatible with any indirect view (projected image) NVIS using Gen 3 I^2 tubes

- Class A: Lighting compatible with any NVIS utilizing a .625 μm minus blue objective lens filter. The use of this filter allows energy from portions of the red spectrum to enter the I^2 tube

- Class B: Lighting compatible with any NVIS utilizing a .665 μm minus blue objective lens filter. This filter lets less of the visible spectrum enter the NVG and allows for the use of more color in cockpit lighting (e.g., annunciators, displays)

A light source that emits energy in the .60–.93 μm region can cause interference with the NVIS to the point where the AGC obscures the outside scene. How much interference is acceptable before the light source can be considered not compatible? Many methods have been developed to test the NVIS for lighting compatibility but with poor results.

The acceptable solution was to develop a new term called NVIS radiance (NR), which is defined as the amount of energy emitted by a light source that is visible through the NVIS. NR is the integral of the curve generated by multiplying the spectral radiance of a light source by the relative spectral response of the NVIS. The spectral radiance of a light source is a known quantity, a "standard" NVIS spectral response is not, and therefore must be defined. The equation for NR, which was developed empirically, is stated as follows:

$$\text{NVIS radiance} = G(\lambda)_{\max} \int_{.45}^{93} G(\lambda) i N(\lambda) d(\lambda) \qquad (13.1)$$

where:

$G(\lambda) = 1$ mA/W (for correct units)

$G(\lambda)i$ = Relative spectral response of the NVIS, where $i = A$ for Class A NVIS and $i = B$ for Class B NVIS

$N(\lambda)$ = Spectral radiance of the light source (W/cm^2)

$d(\lambda)$ = Wavelength increment (.005 μm)

Simply stated, the NVIS radiance is the product of the sensitivity of an NVG and the light source integrated from .45–.93 μm. As a baseline it was decided that compatibility could be achieved if the image of the cockpit lighting, when viewed through the NVIS, was no brighter than the outside scene. Operational experience had shown that because of its low reflectivity, a defoliated tree is the terrain feature that is the most difficult and important to see at night. The NVIS radiance of a defoliated tree was determined by multiplying the spectral radiance of starlight by the reflectivity of tree bark and inserting the resulting curve as $N(\lambda)$ in the NVIS radiance equation, which resulted in values of 1.7×10^{-10} NR and 1.6×10^{-10} NR$_B$. (The difference is due to the Class A and B filters.)

With this information, the illumination levels for cockpit lighting can be determined to ensure compatibility. Primary cockpit lighting is illuminated to 0.1 fL (foot-Lambert); this value is based on USAF tests of operationally representative lighting settings. Monochrome displays, when required to show gray scale (i.e., FLIR), are illuminated to 0.5 fL; higher illumination levels are required to make the image usable. Multi-color displays are not applicable to Class A NVIS. It was assumed that most color displays would be located so they would not be in the NVIS FOV when viewing outside the cockpit. It was also believed that the red portions of the display would occupy a very small percentage of the cockpit area. In reality, the backlighting of modern displays creates more issues than the red elements of the display. Even when fully dimmed, modern displays put out a significant amount of near IR energy (i.e., heat); therefore, when the display looks black to the unaided viewer it can still be clearly seen through the NVG. Modern multi-color displays are not adequately addressed in 85762A, so experience and empirical values are used. Warning, master caution, and emergency exit lights are illuminated to 15 fL as these lights must be clearly visible through Type I NVIS; the HUD is illuminated to 0.5 fL. Table 13.2 summarizes the Type I NVIS radiance limits.

MIL-L-85762A has additional information, which may be found in the Appendices. Appendix 5 contains information and descriptions on NVG Class A and Class B objective filters, including a discussion on the 50% cutoff and slope. Appendix 6 contains the rationale and justification behind the requirements in the specification. Appendix 7 is a PowerPoint overview of how the term NVIS radiance was derived.

TABLE 13.2 ▪ Type I NVIS Radiance Limits

Lighting Component	Paragraph 85762A	Class A		Class B	
Primary	3.10.9.1	0.1	1.7×10^{-10}	0.1	1.7×10^{-10}
Secondary	3.10.9.2	0.1	1.7×10^{-10}	0.1	1.7×10^{-10}
Illuminated Controls	3.10.9.3	0.1	1.7×10^{-10}	0.1	1.7×10^{-10}
Compartment	3.10.9.4	0.1	1.7×10^{-10}	0.1	1.7×10^{-10}
Caution/Advisory	3.10.9.6	0.1	1.7×10^{-10}	0.1	1.7×10^{-10}
Jump Lights	3.10.9.7	5.0	1.7×10^{-8} to 5×10^{-8}	5.0	1.6×10^{-8} to 4.7×10^{-8}
Warning Signal	3.10.9.8	15.0	5×10^{-8} to 1.5×10^{-7}	15.0	4.7×10^{-8} to 1.4×10^{-7}
Master Caution Signal	3.10.9.8	15.0	5×10^{-8} to 1.5×10^{-7}	15.0	4.7×10^{-8} to 1.4×10^{-7}
Emergency Exit	3.10.9.8	15.0	5×10^{-8} to 1.5×10^{-7}	15.0	4.7×10^{-8} to 1.4×10^{-7}
HUD	3.10.9.10	5.0	1.7×10^{-9} to 5.1×10^{-9}	5.0	1.6×10^{-9} to 4.7×10^{-9}
Monochromatic Displays	3.10.9.9.1	0.5	1.7×10^{-10}	0.5	1.6×10^{-10}
Multi-Color Display White Max	3.10.9.9.2	0.5	2.3×10^{-9} 1.2×10^{-8}	0.5	2.3×10^{-9} 1.2×10^{-8}

The US Navy has determined that for modern multi-color displays if the background NVIS radiance is $<5 \times 10^{-11}$ NR$_B$ when the white part of the display is set to 0.5 fL the display is likely to be satisfactory.

13.4.2 Other Lighting Specifications and Guidance

MIL-STD-3009 is a USAF document derived from MIL-L-85762A (USN document); it did not supersede or replace 85762A, but most manufacturers now refer to 3009. The USN, of course, still references 85762A. The major differences between 3009 and 85762A are that 3009 added Class C as terminology for Modified Class B (leaky green) filters. As mentioned earlier, Class C is also used by others, particularly in Europe, for the .645 μm filter. MIL-STD-3009 does not provide daylight readability information although some external lighting guidance is provided.

Since the exterior lighting requirements were derived from a system designed and tested on a single aircraft type for specific operational requirements, they cannot be universally applied. NVIS radiance requirements are the same as those called out in 85762A.

Several SAE documents concern NVIS use but are Aerospace Recommended Practices rather than specifications. SAE ARP4392, "Lighting, Aircraft Exterior, NVIS Compatible," describes the recommended performance levels for equipment located on the aircraft exterior that will produce radiant energy and thus provide desired information when viewed with NVIS goggles. SAE ARP5825, "Design Requirements and Test Procedures for Dual Mode Exterior Lights," contains the general requirements and test procedures for dual mode (NVIS friendly visible and covert) exterior lighting for most rotorcraft and fixed wing aircraft and could be applicable to ground vehicles in which it is desirable to have a dual mode lighting system. SAE ARP4168, "NVG Compatible Light Sources," recommends certain basic considerations that the design engineer should observe when designing NVG compatible lighting. Night vision goggle compatible light is the condition in which the spectral wavelengths, luminance level, and uniformity of the cockpit lighting do not interfere with the operation of night vision goggles. SAE ARP4169, "NVG Filters," describes the functions and characteristics of NVG filters used in NVG compatible lighting. (For more information, see http://www.sae.org/search?searchfield=nvis&typ=std.)

The RTCA Steering Committee 196 was established in December 1999 to provide the FAA with guidelines for developing rules and regulations regarding civilian use of NVGs. One such document, RTCA/DO 268, "Concept of Operations, Night Vision Imaging System for Civil Operators," describes the concept of operations supporting the implementation of aviation night vision imaging system (NVIS) technology into the National Airspace System by civilian aviation operators. Terminology, capabilities, limitations, and operations for NVIS are discussed as well as training and supporting agencies. The focus of the document is the safe and efficient implementation of NVIS during various phases of flight. Another document from this committee, RTCA/DO 275, "Minimum Operational Performance Standards for Integrated Night Vision Imaging System Equipment," contains the MOPS for the ANVIS used to supplement night VFR operations. NVIS consists of the night vision goggle, interior and exterior lighting, cockpit transparencies and crew station design and components. Performance and test procedures are provided for the night vision goggle and lighting. A section on continued airworthiness contains guidance to ensure the integrated NVIS equipment installation continues to meet the minimum performance standard once in operational use. Also, RTCA/DO 295, "Civil Operators' Training Guidelines for Integrated Night Vision Imaging System Equipment," presents training guidance that has been generated from lessons learned by agencies having many years of experience in the training and operational application of night vision imaging systems.

By tapping this experience base it is hoped that civil aircrew, through appropriate ground and flight training, will learn how to properly use night vision imaging systems, thus enhancing the effectiveness of night operations while mitigating the potential for mishaps relating to the use of these systems. Further information can be obtained from the RTCA website (http://www.rtca.org).

The FAA also has guidance and regulations regarding the civilian use of NVGs. This guidance is provided via orders, notices, changes to the operating procedures (Parts 61, 91, and 141), and advisory circulars. Notice N 8000.349 provides guidance for evaluating the operator's formal application for an NVIS supplemental type certification (STC) (remember that notices are internal policy procedures for FAA employees). Notice N 8900.100 provides guidance for NVG training and training plans and refers to Part 61 of the operating instructions. Amendments to 14 CFR Parts 61, 91, and 141 include changes to paragraphs 61.31(k), 61.57(f), 61.57(g), and 61.195(k), which refers to NVG training, currency, proficiency, and flight instructor requirements; 91.205 dictates the minimum equipment required for NVG and requires the installation of a radar altimeter.

Paragraph 61.31(k) states that a pilot in command's (PIC) use of NVGs requires completion of ground and flight training from an authorized instructor with an endorsement of such training in her logbook or training record. The paragraph also delineates the required syllabus for the training; it does provide exceptions for pilots with previous experience. The pilot must also retain proficiency in NVG usage. Paragraph 61.57(f) lists the currency requirements for flight with NVGs. An operator is current if:

– A flight with NVGs within 2 months preceding the proposed flight (with passengers)
– A flight with NVGs within 4 months preceding the proposed flight (without passengers)

To obtain a currency (if currency was lost) an operator must perform:

– 3 takeoff and landings (climb, cruise, descent, approach)
– 3 hovering tasks (helicopters or powered lifts only)
– 3 area departure and arrival
– 3 transitions from aided to unaided to aided
– 3 NVG operations (6 for helicopters or powered lifts)

Paragraph 61.57(g) identifies the requirements for obtaining an initial NVG currency. An NVG proficiency check is required if the operator does not meet the requirements in 61.57(f). The training requirements in 61.31(k) must be completed and a check ride performed by a qualified person. Paragraph 61.195(k) defines the NVG flight instructor requirements. In general, the candidate must be a CFI with applicable category and class rating and a type rating if appropriate. Eligibility requires 100 NVG operations as sole manipulator at controls and 20 NVG operations as sole manipulator at controls in category and class aircraft if appropriate; a logbook endorsement is required.

13.5 | INTERIOR NVIS LIGHTING METHODS

The AGC circuit is activated by energy to which the NVG is sensitive, and this results in reduced contrast, especially in darker areas of the image. This energy is emitted from cockpit lighting, exterior lighting, and environmental illumination, and it does not take much energy for AGC to reduce the gain. To mitigate this problem, NVIS lighting is necessary for NVG compatibility; even a minor incompatibility may cause imperceptible but real loss of NVG image quality. The human operator can lose as much as 50% of aided acuity without noticing it, especially in medium/low contrast environments. To put this in context, the operator would have to be twice as close to see the same image as before the loss.

13.5.1 Interior NVIS Lighting Modifications

Filters are used to make interior lighting compatible with NVGs. Filters reduce emission of near-IR energy as well as visible wavelengths within NVG sensitivity range (based on type of NVG filter—Class A or Class B). The filters will notch out red and near-red colors while allowing good color transmission in displays. This must be done judiciously so as to minimize color shifts and avoid adversely affecting daylight readability. Compatibility will use two filters: the display filter is used to stop IR wavelengths from exiting the display, and the NVG filter is used to block visible wavelengths from entering the NVG. The display filters and the NVG filter should match and be the same class. Remember that the filters are not perfect and that at the rated wavelength they are still passing 50% of the energy. A visual presentation of the filter setup is shown in Figure 13.8.

There are many NVIS lighting modification methods in use, and choosing one depends somewhat on the age of the aircraft. For older aircraft it may be necessary to modify analog instruments, displays, and annunciators that use incandescent technology, whereas for newer aircraft the use of digital instrumentation and light emitting diodes (LEDs) may predominate. Also, older aircraft will usually be modified postproduction, whereas newer aircraft will be built with NVIS lighting installed as part of production. Methods used for NVIS modifications include the following:

– Internal modification

– NVIS bezels

– Post lights

– Bridge lights

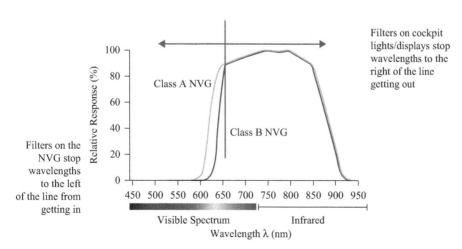

FIGURE 13.8 ■
NVIS Filters

- Annunciators
- Cockpit displays
- Panel lighting
- Flood lighting

Internal or integral modification may provide best display illumination and depends on age and design of instrument. The modification involves a complete disassembly of the instrument and filtering all the internal lights with NVIS compatible filters. This modification is usually more expensive, is technically difficult, and may revoke the original equipment manufacturer (OEM) warranty. After the modification, a recertification will be required and the modified and non-modified parts should have different part numbers. This is a built-in trap as it is easy to accidentally replace modified components with unmodified components during maintenance operations.

An NVIS bezel is an open ring design with lights embedded in bezel hardware. Older designs incorporated glass cover that transilluminated light across instrument face. The two layers of glass with associated dirt and scratches was very thick causing blockage of information. The current bezel matches the size and contour of existing instrument bezel that holds the original glass cover in place. The control knobs are incorporated on existing bezel with cutouts for the NVIS bezel. Power to the bezel may present some problems, and the most effective implementation is a design that allows an easy removal and replacement. Advantages of the bezel modification are that it provides a balanced illumination across instrument face and can be designed to fit unique instrument shapes. They can be cut out to provide parallax readability. If designed correctly, a bezel panel can cover a cluster of smaller instruments. With light redundancy, the system can lose some lights and maintain readability. The NVIS bezel reduces the use of post lights and provides superior readability. There are disadvantages as well; bezels are generally more expensive than most other methods and may block information on instrument face depending on design and viewing angle. A luminance balance may be difficult to accomplish if varying bezel designs are incorporated in a display suite.

A post light is essentially a small floodlight. The design consists of a post and a hood; the post used for attaching the light to surface and a hood is fitted to the top of the post. The post height is designed to shine light from above the instrument face. In some applications, you may be able to change the hood without removing the post, or you may be able to rotate the hood on the post to change the area that is illuminated. A window in the hood, usually glass, contains the filter material—either NVIS Green A or NVIS Green B. The largest advantage of the post light is that they are inexpensive and may provide adequate illumination on smaller instruments or small areas. They may be difficult to place (especially in multiplace aircraft) because the height of the post may block some information. Since the light is reflected downward, it may cause reflections in transparencies or off the instrument face. It is difficult to obtain adequate balance when used in large quantities and many have reliability and maintainability problems. Hoods may become loose and create light leaks or get kicked off when entering or exiting the aircraft.

Bridge lights are essentially two post lights on a bracket and may be used when a single post light does not provide enough illumination. They are not adequate for larger instruments as they lack even illumination across the instrument face. A bridge light may be needed due to space limitations where there is no room for a bezel or other method. As with the post light, there is an increased risk of blocking information from the instrument face due to viewing angle.

Annunciators must be recognized in all lighting conditions, must be attention grabbing and need to be unambiguous. If warning (red) annunciators are filtered, a color shift will be observed, the red will be less saturated, and the warning will appear orange-like. This can lead to confusion between red and amber (cautions) annunciations. There can also be a problem with normal daylight readability due to the reduction of intensity. Dead-face filters (inverse video) reduce the total amount of red and yellow energy emitted from the annunciator, which will reduce the possibility of a false on indication. The emitted energy however, may not be sufficient for some annunciators depending on location and purpose (e.g., fire warning); the filter may also negatively impact daylight readability.

It should be obvious that the control of colors will be critical in the NVIS-compatible cockpit. One needs to apply judicious use of reds and yellows, and limit their usage to emergency and caution annunciations. The use of NVIS Green B for annunciators may be a possible alternative to filtering incandescent light. Warnings and cautions may also be supplemented by aural tones, or placing the annunciation in other locations (e.g., HUD, HMD). Marginal findings or test failures during evaluations of filtered incandescent annunciators may be acceptable in certain situations, but the user will find it difficult indeed to meet the airworthiness requirements of AC 25.1322-1 and AC 25-11A for civilian applications. In all installations, it will probably necessitate the capability to increase the brightness of the annunciator for daytime use and allow them to be dimmable for night operations.

The cockpit displays must also be compatible with NVIS. A glass cockpit contains multiple displays such as EFIS and MFDs; there are a reduced number of standard gauges and annunciators. The displays may employ various technologies such as CRT, LCD, and LED; the displays may be color or monochromatic (although most new cockpits use color). The displays may or may not be NVIS compatible even if internal filters are incorporated. Because of its age, the MILSPEC addresses CRT displays but not the newer technologies.

For displays that are poorly modified or do not include an internal NVIS filter, a COTS external, removable filter may be required. As noted with annunciators, filters create daylight readability problems and a color shift. If the filters are removable (for daytime or unaided night operations) additional problems may be created:

- Aircraft stowage issues
- Location and accessibility
- Lost filters
- Handling problems
- Breakage
- Fingerprints

The use of external filters may require trade-off analysis evaluating the impact on NVG performance versus color shift and the cost of the units.

Other displays such as the primary flight instruments may also cause unique problems for NVIS performance. Reflection issues caused by sunlight and nighttime illumination on the canopy will cause noise to enter the processor. Antireflective coatings may help but may also reduce transmission of colors or further reduce luminance. If the same dimming circuit is used for all displays regardless of their modification methods there may tend to be a luminance imbalance where some information is barely visible, whereas other information produces glare.

Panel lighting is grouped into three classes: Types 3, 4, and 5. Type 3 lights have replaceable bulbs that internally light the Plexiglas to provide the legend lighting. There is normally a screw-off cap to access and cover the bulb with a filter. To make this type of panel lighting compatible, only panel filters are required. In Type 4 panel lighting, the bulbs are imbedded in the Plexiglas and therefore may only require donut filters over the lights in the circuit board, but at the trouble and expense of first digging them out from the Plexiglas. In Type 5 lighting, the bulbs are imbedded in the circuit board, and will require reworking the circuit boards to obtain compatibility. For all types of panel lighting, a generalized cleanup effort will need to be undertaken: old panels may need to be refurbished, the lettering will need to be cleaned, and scratches will need to be painted over to prevent leakages. A combination of flood lighting, edge lit, and panel lighting may be needed to see switch position and the windows in the panels may need to have separate filters installed. Edge lit lighting provides uniform illumination over flat instruments and panels such as switches and circuit breakers but is not suitable for contoured displays or displays with multiple types of information.

Flood lighting may be incandescent, LED, or UV light or Saturn yellow (the planet Saturn is yellow due to reflected sunlight passing through ammonia; in this application UV light is used to illuminate Saturn yellow labels). These lights may be suitable for general lighting, determining switch positions, and for reading placards. Flood lighting may be adequate as backup or emergency lighting depending on the power source, such as a battery. They are inexpensive but do present potential problem areas: extensive use or incorrect positioning leads to increased reflections (canopy, instrument face), shadowing can reduce readability, it is difficult to achieve adequate balance, adequate locations can be difficult to achieve, and the physical size can obscure part of the instruments. The aircraft utility light may be selected as a backup source but should not be used as the primary lighting system. Of course, if the utility light is used for NVIS lighting then its primary use is lost. Utility lights are not fixed, so the light source is not stable and they are often awkward to use.

As with everything else today, LED annunciators are being used more often. LEDs offer low power consumption, low IR emissions and have a long life (>10,000 hours). LEDs are not easily mixed with incandescent lights. LEDs use a duty cycle to vary brightness (i.e., pulse width modulation), whereas incandescent bulbs use voltage to vary brightness (they dim as the voltage drops) which will present a problem for automatic dimming circuits (day/night setting). Brightness can be a problem for older LEDs; they can be hard to read in daylight conditions, although modern LEDs are brighter. A multi-LED array is normally used to ensure coverage angles and brightness.

As a recap, standard lighting cannot be turned down enough to be NVG compatible. The designer/evaluator must ensure that the initial NVIS lighting design includes the type of NVG filter required. The same design rationale for standard lighting applies to NVIS lighting, i.e., a poor installation can negate a good modification design. Unaided readability is just as important as compatibility; a loss of NVIS capability will force a reversion to unaided flight.

■ 13.6 | EXTERIOR LIGHTING METHODS

Modification of interior lighting makes it possible to use NVGs, and the modification of exterior lighting makes it possible to use NVGs more effectively. For interior lighting, this requires meeting a formal specification and results in most interior lighting being invisible when viewed with an NVG (except for some warning/caution lights). For exterior lighting, there is no formal specification but the intent is to reduce the adverse effects on NVGs while being able to see the lights in the image. As opposed to NVIS compatible, NVIS friendly is a term used to describe NVIS compatibility for exterior lights. Covert lighting is a term initially used to describe IR lighting since it could not be seen with unaided vision. The proliferation of night vision technology worldwide has reduced the effectiveness of IR lighting so now the term is used to describe lighting that will reduce probability of detection.

The aircraft exterior lighting system must meet civil regulations and military requirements. Civil regulations dictate position lights and anti-collision/strobe lights, the intent of which are to see and avoid other aircraft in civil airspace. Military requirements are designed to ensure that aircraft can successfully meet the tactical goals for training and operations. The two organizations dictate potentially conflicting requirements; a military tactical requirement may result in diminished visual detection range that could result in noncompliance with civil regulations.

The original concept was to have two separate lighting systems and/or turn off exterior lighting off during training; this was not acceptable when military training was conducted in civil airspace as the aircraft would be noncompliant. It was generally decided to have one system that met both constraints (civil regulations and tactical requirements); this, in turn, resulted in the development of NVIS friendly exterior lighting. The attributes of NVIS friendly exterior lighting include:

– A reduced IR emission from existing or new exterior lights

– Less energy that will adversely affect NVG performance

– Lights are visible without NVGs in colors, distances, and aspects that meet FAA regulations

– Lights are visible with NVGs

– There is a potential for adverse effects if viewing at closer ranges depending on modification methods used

– Most needs can be met by the use of LEDs or filtration

All external lights and their potential effects on NVG use should be examined:

– Position lights

– Anti-collision and strobe lights

– Formation lights

– Taxi lights

– Landing lights

– In-flight refueling lights

– Probe lights

– Tanker lights

- Spotlights
- Other lights
- Wind indicator
- Angle of attack indicators

During this examination, it is helpful to ask the following questions:

- Does location of exterior lights cause problems with NVGs?
  - Light intruding in the cockpit
  - Reflections from props or rotor blades
- How much control should the aircrew have?
  - Selection of flash patterns
  - Brightness control
  - On/off control

When evaluating lighting controls, it is important to determine which lights must be on and at what times during the flight must they be on. Pilot control of each set of lights provides more flexibility when given tactical requirements and knowing environmental conditions. The optical profile (aided or unaided) can be managed by controlling the brightness. For example, for a constant brightness, the detection range increases when using NVGs on a dark night due to the increased gain. Providing flash patterns for various sets of lights enhances tactical flexibility; it is easier to identify an aircraft against star or cultural backgrounds. As previously discussed, the location and tactile feel of cockpit controls will limit heads-down time and reduce aircrew workload.

13.6.1 Exterior NVIS Lighting Modifications

The use of LEDs for exterior lighting is currently used on most new aircraft and some legacy aircraft. LEDs can be compliant with civil regulations and can meet tactical requirements if clearly stated early in the design phase. They allow a good control of color and improve brightness. The design must use an array of LEDs to gain proper coverage; viewing a single LED light off-axis may result in color shift. As mentioned in the interior lighting discussion, the design cannot mix LEDs with incandescent lights nor use incandescent dimming controls. The filtration of incandescent lights is most commonly used on legacy aircraft.

The two types of filtration used with incandescent lighting are absorption filters and interference filters. The absorption filter involves the conversion of the transmitted energy into another form (usually thermal). The conversion takes place as a result of interaction between the incident energy and the material medium at the molecular or atomic level. An interference filter consists of multiple thin layers of dielectric material having different refractive indices; there also may be metallic layers. Interference filters are wavelength-selective by virtue of the interference effects that take place between the incident and reflected waves at the thin-film boundaries. The single biggest problem with filtration is heat. Heat may be trapped within the light unit, and if not vented properly, or transferred in some way, it will reduce the life of the incandescent light or worse, pose a fire hazard. It also may be difficult to obtain uniform brightness from different light sources (i.e., red and green position lights).

A significant aviation safety problem that has resulted secondary to the use of red LEDs for NVIS friendly exterior lighting designs involves their use as red obstruction lights. Commonly used red LED obstruction lights (.645 μm) are visible unaided but not as well with NVGs (depending on which objective filter is incorporated). Remember that the NVG objective filter blocks wavelengths from the intensification process, which allows for cockpit lighting that will not adversely affect the NVG image. This has resulted in red LED obstruction lights being much less visible to pilots using NVGs. The problem has been identified by numerous agencies and noted in a Flight Safety Flash in 2008 by Canadian Air Force's Directorate of Flight Safety. An FAA SAFO (Safety Alerts for Operators) #09007, dated 3/6/09, and an FAA InFO (Information for Operators) #11004, dated 2/15/11, also address this problem.

13.7 | TEST AND EVALUATION OF NVIS EQUIPMENT

The evaluation will consist of a series of NVG ground tests, aircraft ground tests, and aircraft flight tests. The ground evaluation will examine human factors and MMI as well as qualitative and quantitative performance of the devices. Aircraft ground tests will incorporate daylight readability, unaided nighttime readability and aided visual acuity. Aircraft flight tests will ensure safety of flight and mission suitability.

13.7.1 NVG Ground Tests

The scope of the ground evaluation will be determined by maturity of the system. The level of testing will be affected if the system is an initial capability or an existing system, whether it is a newly developed system or an upgrade or replacement of an existing system. The testing will contain a mix of quantitative and qualitative testing; component level testing versus system level testing. Normally, qualitative testing is done by flight testers, whereas quantitative testing is done by special labs.

The preparation for the ground evaluation will be critical as improper preparation will adversely affect the results. The test area should be in a location where the light level can be adequately controlled; an ideal setup would be an interior room with no windows. The test area should have controlled access in order to avoid inadvertent light intrusion; fluorescent lighting is to be avoided. The size of the area should allow for inclusion of all test equipment and for the conduct of NVG visual acuity and Human Factors evaluations. Care should be taken to ensure that paint and flooring (e.g., carpet, vinyl) do not reflect IR energy, which may interfere with the tests.

Human Factors is evaluated with respect to the helmet, mount and NVG system. Tests will include the ease of donning and doffing the helmet, its stability and comfort, the nape and chin strap supports, and the system weight and center of gravity. As always, the evaluation should be done with the configuration used for flight and with the normal flight clothing (don't forget special equipment that may be used on a particular mission, such as NBC, body armor, weapons, floatation vest, spectacles, and laser eye protection). Test for the compatibility with the oxygen mask or boom mike, battery pack, and lip light. The mounting mechanism should be evaluated for ease of attaching and removing the binocular assembly, stowed versus operating positions, and the position of the mount on the helmet. NVG items of interest include adjustments and image positioning, eye relief and tilt, interpupillary distance, and focusing. The location and tactile feel of the

FIGURE 13.9 ▪
NVG Visual Acuity

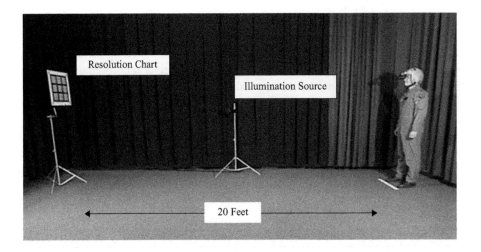

controls should be evaluated with one hand operation while wearing gloves. The power supply should be tested for accessibility, redundancy, and proper operation of the low power warning. NVG performance includes aided visual acuity, image defects, and cosmetic blemishes. Interoperability issues may include restrictions to reach and manipulation, restrictions in head and neck mobility and overall creature comfort.

Visual acuity is tested much the same as previously described in spatial frequency testing for EO systems. The NVG community has its own description of the test (called an eye lane) and it is really an eye test while wearing NVGs. As with the Snellen eye exam, the subject should be at a viewing distance of 20 ft to charts, which helps reduce large visual acuity differences caused by small head movements (Figure 13.9).

The test needs to be accomplished in an area that can be adequately darkened, with a control of illumination levels for resolution and contrast charts. The tests should be run against low, medium, and high contrast resolution charts (20%, 70%, and 100%); this provides for a more critical evaluation of visual performance. Other than the USAF tri-bar charts used in EO spatial frequency tests, NVG evaluations also use square wave grating, Landolt C, Open E, and Snellen. Examples of these charts are found in Figure 13.10. Square wave grating charts can be made with a chart-maker program, and the others may be obtained from numerous online sources. The illumination source should provide for various illumination conditions (e.g., full moon, starlight, and overcast starlight). The method of test is identical to the spatial frequency ground tests described in chapter 7 of the text: Adjust the NVG to obtain the maximum performance, use a high contrast chart with full moon (high) illumination and note the smallest resolution element seen on each chart for all illumination conditions. The cutoff (best acuity) is where the resolution lines are clearly vertical or horizontal and distinctly separated. The test is repeated for each test subject and each NVG. If performance of the NVG is the sole object of the test (and not human factors or MMI), then it is best to use trained test subjects that will participate in all of the testing. If novice test subjects are used, then we must increase the number of subjects considerably.

The complexity of illumination device is determined by the test requirements: R&D of a newly developed system will require more effort than the evaluation of an off-the-shelf system. If precise illumination settings are required, the evaluation will require test equipment such as a color-controlled illumination source. One such system

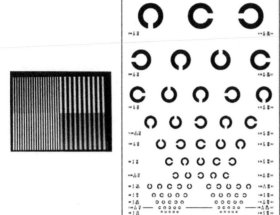 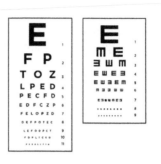

FIGURE 13.10 ▪
Visual Acuity Charts, from Left: Square Wave Grating, Landolt C, Open E, and Snellen

is the Hoffman LM-33-80 Starlight Projector (http://www.hoffmanengineering.com/products_detail.php?item_id=137&cat_id=). If measurement of low illumination levels is a requirement, then a spot photometer with a Class A or B filter will be used. A typical system in current use is the Pritchard 1530-AR (http://pdf.directindustry.com/pdf/photo-research-inc/pr-1530ar-nvispof-night-vision-radiometer-photometer-colorimeter/32447-184616.html). In order to verify compatibility with internal and external lighting, a quantitative analysis can be obtained with the use of an inspection scope. Inspection scopes are precision-calibrated, certified reference tools designed for the purpose of evaluating cockpit lighting compatibility and external aircraft covert lighting. These inspection scopes feature up to four calibrated gain settings that enable them to evaluate a broad range of lighted devices. A typical employable device is the Hoffman NVG-102B or NVG-103 (http://www.transaeroinc.com/Life%20Support/man/Hoffman%20Engineering/NVG-102%20&%20103%20NVIS%20Inspection%20Scope). Less precise settings or self-made illumination devices require a thorough knowledge of NVG design and performance. During the test, the control of illumination to effect perceived change in image quality will yield qualitative results.

During the ground evaluation, the operator should note any operational defects or cosmetic blemishes. If operational defects are noted then the NVGs should not be used; cosmetic blemishes, although distracting, may be acceptable if still within specifications. Operational defects are classified as:

- Shading
- Edge glow
- Flashing
- Flicker

Shading is indicative of a dying photocathode. It appears as a dark high contrast area along the outer edge of the image that will migrate inwards. *Edge glow* appears as a bright area along the outer edge of the image. It is caused by a light source outside the image FOV or from a defective phosphor screen that allows light feedback to the photocathode. If cupping a hand over the lens does not eliminate the edge glow then the problem is within the image intensifier. *Flashing* or *flickering* may be intermittent or may occur at different rates in one or both monocular and for a number of reasons, including impending tube failure, electromagnetic interference, faulty wiring, and battery problems. If it cannot be repaired the problem is within the image intensifier.

Cosmetic blemishes are characterized as:

– Bright spots

– Emission points

– Dark (black) spots

– Fixed pattern noise (honeycomb)

– Chicken wire

– Image disparity

– Image distortion

– Output brightness variation

A *bright spot* is a small, nonuniform bright spot in the image resulting from a flaw on the film coated onto the MCP. An *emission point* is a steady or fluctuating pinpoint of bright light that does not go away when the lens is covered. A *dark (black) spot* is seen as a dark speck in the image that results from dirt or debris between the lenses or from the image intensifier manufacturing process. Dark spots may become defects depending on their size, number and location within the intensified image. A *honeycomb* pattern in the image results from the fiber optic inverter. It is most often seen against a uniform background during high illumination conditions. *Chicken wire* is an irregular pattern of dark thin lines throughout or in parts of the image. It is a result of defective fibers in the fiber optic inverter. *Image disparity* is a difference in brightness between the two monocular images. Various types of *distortion* result from problems with the fiber optic inverter (e.g., shear, wave or bending). *Output brightness variation* is evidenced by areas of varied brightness in or across the image area. Examples of image defects are shown in Figure 13.11.

FIGURE 13.11 ■
NVG Image Defects

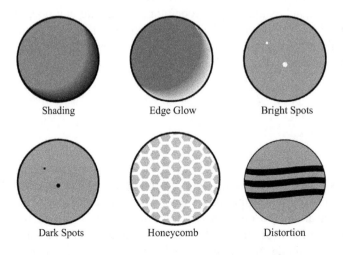

| Shading | Edge Glow | Bright Spots |
| Dark Spots | Honeycomb | Distortion |

13.7.2 Aircraft NVIS Lighting Ground Tests

NVIS lighting standards and associated documentation were covered in section 13.4 and will not be addressed again here except to recap the section:

- The basic standard for interior lighting is MIL-L-85762A.
- Data were taken from this specification to develop other NVIS lighting specs (e.g., MIL-STD-3009, NATO STANAG).
- There are limitations resident in each standard, so it is important to know which will be used.
- Contractors use information in the specification to develop and test lighting components.
- Testers use information in the specification for developing test methods and acquiring test equipment.
- There are currently no universally accepted specifications for NVIS exterior lighting.

A NVIS interior lighting modification is considered compatible (suitable) if it meets the following objectives:

- Does not adversely affect perceived NVG performance
- Provides adequate readability and use of instruments, displays, warnings, and controls during daytime and nighttime
- Does not cause an increase in aircrew workload (although NVG use inherently increases workload)
- Compliance with the relevant specifications

The data obtained during the aircraft NVIS lighting Ground Tests can be qualitative as well as quantitative; the tests can be structured in the following way:

- Daylight readability (qualitative)
- Unaided nighttime readability (qualitative)
- Aided visual acuity (qualitative)
- Lighting controls (qualitative)
- Other devices carried aboard the aircraft (qualitative and quantitative)
- Spectra-radiometric data (quantitative)

The testing labeled "other devices carried aboard the aircraft" applies specifically to items that contain lights or that may cause reflective problems. These items may be additional cockpit lighting such as flashlights, finger lights, or lip lights. Other lighted equipment may include handheld laser pointers, emergency radios, laptop computers, and iPads.

When evaluating daylight readability, the filtered displays are of the most concern. The lights will include the warning, caution, and advisory annunciators and the filtered tactical displays (CRT, LCD and EFIS). The evaluator should determine if the filters should be removable for day operations. The daylight readability can be accomplished at various times and at various locations, day or night, in a test facility or on the ramp, but a method of controlling the lighting and measuring the illuminance must be available.

TABLE 13.3 ■ Nighttime Readability Test Matrix

Nighttime Readability		
For each display evaluate each test point and specific conditions; note discrepancies or enhancing features in the notes column		
Test Point	**Specifics**	**Notes**
Blockage of display data by modification design and/or location	Bezel edges	
	Post light or bridge light location	
Cross-cockpit readability		
Display readability	Color discrimination	
	Range of brightness and contrast	
	Readability at various head positions	
Chromaticity of annunciators and other instruments	Discrimination and uniformity	
Illumination uniformity	Across each instrument face	
	All illuminated items on a single dimming control (lighting balance)	
	All illuminated items on all dimming controls (system uniformity)	
Reflections on instrument faces	Filtered displays	
	Floodlights and post lights	
Reflections on transparencies	Note locations and offending sources	
Degraded modes/secondary lighting	Readability of critical displays	
Lighting controls	Brightness range adequate for all conditions	
	Control linearity	
	Increases and decreases in a linear fashion	
	Grouping logic	
	Control location	
	Easily identifiable	
	Visually and by touch	
	Easily reachable	

The method of test will require the evaluator to hold an illuminance meter next to display being evaluated and adjust the distance of a sun gun to obtain relevant energy level. The standard calls for 10,000 fc for diffuse reflections and 2,000 fc for specular reflections, but these numbers can only be accurately measured in the laboratory. Measure all of the relevant displays and note the location of the displays; some displays may not be susceptible to bright conditions. If the component is not readable at 10,000 fc, the sun gun distance should be increased until it is readable, and then note the illuminance.

Nighttime readability is best evaluated by asking a series of questions similar to an SRS and then noting specific problem areas. Table 13.3 provides a sample test matrix for evaluating the cockpit nighttime readability. For each display, evaluate each test point and specific conditions and note any discrepancies or enhancing features in the notes column.

NVG aided visual acuity will be accomplished by viewing the charts described in section 13.7 under measured lighting conditions. The test setup, data collection, and analysis are the same as described for the spatial frequency ground tests described in chapter 7 of the text. The results are not specific enough for it to be used as a standalone test method, but do provide a gross evaluation of NVG compatibility and a sanity check of NVIS lighting system performance. The evaluation will help identify problem areas and is a balance against spectra-radiometric results. The testing will also determine the impact of unaided field of view on instrument readability.

The visual acuity will be tested under three conditions: charts viewed directly (outside of the aircraft), charts viewed through the transparencies with the cockpit lighting off, and when viewed with the cockpit lighting set at varying configurations. By performing these steps, a baseline for NVG performance and a baseline for cockpit lighting effects can be obtained. The cockpit lighting configuration will be operational brightness setting with no displays, cockpit lighting and displays at operational brightness setting, cockpit lighting and displays at operational brightness setting with all warning, caution, and advisory (WCA) lights on, and finally, cockpit lighting and displays at operational brightness setting with all WCAs and fire lights on.

As mentioned the procedure will be the same as spatial frequency measurements:

- Measure and record distances to all charts from the eye position of each station from which NVG aided visual acuity is being acquired
- Adjust illumination source for resolution chart such that NVIS radiance on the white portion of the chart equals:

  1.7×10^{-10} NRA for Class A requirements

  1.6×10^{-10} NRB for Class B requirements

- This illumination level (near starlight) allows for the NVG gain to be maximized which increases its sensitivity to incompatible light sources
- Measure NVIS radiance with the spot photometer
- Use an inspection scope, such as the NVG-103, to verify results

If degradation is noted, isolate lighting sources to identify problem area. Turn various components or dimming controls off one at a time, and ensure all lighting within the aircraft that may intrude into the cockpit is evaluated (e.g., from rear of aircraft). Assess reflections on transparencies, which are usually caused by displays and floodlights. Determine the aided and unaided effects at varying brightness levels and consider moving the acuity charts so that they are behind area of reflection. The object of the test is to compare the visual acuity at different brightness levels with the baseline; any potential impact on operations will require flight test for final determination.

In NVG analysis, the visual acuity is presented in terms of the Snellen test—that is, 20/20, or what a person with "normal vision" is able to see. Spatial frequency as previously used was in cycles per milliradian. Conversion between the two can be accomplished easily. We know that

$$\text{NVG Resolution} = \text{Spatial Frequency} = \gamma = R \times D \qquad (13.2)$$

where γ is NVG resolution (cy/mr), R is chart resolution (lp/mm), and D is distance from NVG to chart (m).

$$\text{Aircrew Visual Acuity } (20/xx); \; xx = \frac{34.384}{\gamma} \qquad (13.3)$$

where xx is denominator of the visual acuity term, and γ is NVG resolution (cy/mr).

NVIS exterior lighting ground testing considerations are limited to an evaluation of the functionality due to reflection of bright lights in a closed space (test facility). A human factors evaluation of the controls (accessibility, tactile feel, and discrimination) and control linearity if applicable can be assessed. A functional test to ascertain whether the appropriate lights illuminate when selected and that they perform as required (flash) can also be tested. An extended evaluation may be accomplished by positioning aircraft at night on a runway or taxiway and viewing aided and unaided at longer distances (locate an area with limited cultural lighting). The true usability of NVIS external lighting will require flight testing.

13.7.3 Aircraft NVIS Flight Tests

Flight Test will evaluate issues discovered during ground test and evaluate anomalies discovered during flight that were not noted during ground test. Throughout the course of the test, the evaluator should strive to develop techniques or design recommendations that help reduce pilot workload. The exterior lighting evaluation may involve the bulk of the flight tests. For NVG-only evaluations the test must be conducted in the relevant aircraft because this is important for aircraft interface issues such as lighting, obstructions to scan, and transparency effects.

The phases of flight test mimic the ground evaluations. Daylight readability flight during daytime should be conducted first in order to note any discrepancies that may affect safety of flight; maneuver the aircraft to change sun angles relative to the cockpit during flight. Unaided nighttime readability is conducted in the same manner as the ground evaluations. Effects on NVG performance are subjective; use the data from the ground tests to identify potential problems such as reflections on instrument faces or transparencies, viewing over displays and effects due to WCAs.

The exterior lighting evaluation can be complex, and flight safety is most important. Test objectives which are required include: the maximum detection ranges of all exterior lighting from various altitudes and aspects, the suitability of flash patterns for acquisition against various backgrounds (e.g., stars, cultural lights), the suitability for formation and rendezvous, and suitability for civil operations. Tactical suitability is normally accomplished during operational test and evaluation.

If tasked with performing a tactical suitability evaluation, remember to start with the most benign condition before advancing to real-world operational scenarios. Initial testing should be conducted using existing night mission profiles with a logical approach to mission expansion. Conduct evaluation during varying environmental conditions, over various terrains, and in differing cultural areas.

13.8 | SOME FINAL CONSIDERATIONS

The use of NVGs increases workload and adds focal tasking; you are essentially asking the pilot to perform an integration of four scans simultaneously: NVG outside, beneath NVG inside (instrument scan), beneath NVG outside, and the NVG image itself. There have been mishaps related to NVG use, and the contributing factors are noteworthy: task saturation, breakdown of crew coordination, poor judgment, complacency, fatigue, overconfidence, training deficiencies, and inexperience. This is an important point for testers when evaluating a new aircraft system while wearing NVGs as the workload will be even higher. NVG training and currency is key; try to avoid learning or relearning how to use NVGs while testing other systems.

If test aircrew do not have NVG experience, it is necessary to begin training well ahead of the scheduled test dates and through lessons from past experience will require at least six months of lead time. The team should allow for an NVG build-up in the test aircraft once the training has been completed. You may want to consider the use of NVG experienced operational aircrew to augment test aircrew that may have limited experience.

On a final note, don't reinvent the wheel for each system evaluation. Some of the techniques which have been tried in the past and proven unsuccessful have been: pilot in a chase aircraft without NVGs, a safety pilot on board without NVGs, and a pilot at the controls without NVGs.

13.9 | HELMET MOUNTED DISPLAY SYSTEMS (HMD)

The HMD is really the next step in aiding vision and increasing situation awareness for the pilot. HMD refers to any system in which a display is helmet mounted. Various types of information may be presented on a display, such as flight and tactical symbology, sensor imagery, or any combination thereof. There are two basic types of HMDs: an aircraft-mounted head-up weapons sighting system, which allows heads-up viewing of target information but does not portray flight following information; and the aircraft-mounted head-up-display, which provides flight following and tactical information. In both cases, it is necessary for the pilot to point the nose of the aircraft at the target. HMDs have evolved with advances in technology and miniaturization in similar fashion with FLIR, NVG, and EO systems. As a reference, the reader is referred to two texts on the subject: *Helmet-Mounted Displays: Design Issues for Rotary-Wing Aircraft* (Clarence E. Rash, SPIE Press Book, 2001) and *Head-Mounted Displays: Designing for the User* (James E. Melzer and Kirk Wayne Moffitt, McGraw-Hill, 1997).

13.9.1 HMD Overview

The HMD idea is not new; a patent was issued in 1909 for a helmet sighting device, but it was never developed. The first functional HMD was a helmet mounted sight (HMS) deployed in the 1970s for the US Navy F-4 aircraft. It was called the visual target acquisition system (VTAS), and it incorporated an IR optical head tracking system that allowed the pilot to look off-boresight and slave the radar or AIM-9G Sidewinder missile to pilot's line of sight. NVGs were initially deployed by the US Army in the late 1970s and later as the first successful HMD, which was eventually adopted for all services. The HMD was originally developed to provide a visual reference during nighttime. Flight following via a clip-on NVG HUD was developed later. Advanced capabilities such as an integrated display, head tracking, and wide field of view are continuously being developed and improved. Various HMD designs are currently in use by most services worldwide.

There exists an increasing requirement by many countries to incorporate helmet mounted off-boresight targeting, flight following and visual imagery into their aircraft.

These advanced capabilities require the integration of a head tracking system. There are many different HMD design approaches to accomplish some or all of these requirements, the most common of which are as follows:

– Single component: symbology only; day and unaided night capable
– Single component: visual imagery only; aided night capable
– Two components: symbology and visual imagery; aided night capable

  – First component: visual imagery only
  – Second component: symbology only

– Two components: symbology and visual imagery; day, unaided night and aided night capable

  – Approach 1
  – First component: symbology only; day and unaided night capable
  – Second component: symbology and night vision imagery; aided night capable
  – Components must be exchanged for day/night operations
  – Approach 2
  – First component: symbology only; day, unaided night and aided night capable
  – Second component: visual imagery only; aided night capable
  – First component remains on for all operations and second component added at night

– Fully integrated single component: symbology and visual imagery; day, unaided night and aided night capable

Note: Head tracking may be integrated in most HMD designs.

Single component systems (Figure 13.12) provide a capability to more effectively employ a weapon system by the use of off-boresight targeting or provide the user with night vision imagery. Off-boresight systems require a head tracking system and may employ a single combiner positioned over the right eye or may be binocular. Night vision imagery systems are mostly NVGs, but some HMD systems have the capability to provide imagery from other sensors (e.g., FLIR).

FIGURE 13.12 ■
Single Component
Systems; Topsight
(left) and
USN/USMC F4949G
NVG System (right)

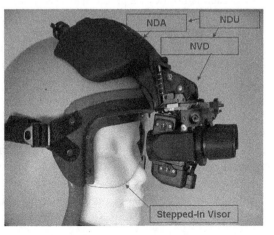

The purpose of one of the two component system designs (symbology and visual imagery; aided night capable only) is to provide the user with night vision imagery and symbology. This is accomplished by inserting symbology onto an NVG image through one of two techniques: an NVG HUD clips onto either the objective or eyepiece of an NVG, or the use of a display integrated into NVG optics. Approach 1 for a two component design (symbology and visual imagery; day, unaided night and aided night capable) provides the user with symbology and night vision imagery (Figure 13.13). As the name implies, there are two components: a visor module provides symbology for day and unaided night operations, and a night vision module provides symbology and imagery for aided night operations. The drawback to this approach is that the user must change components when transitioning from day to night or from night to day.

Approach 2 differs in that the combiner for day and unaided night operations remains on for all operations and NVG is added for aided night operations (Figure 13.14).

The fully integrated single component (symbology and visual imagery; day, unaided night and aided night capable) system provides symbology and visual imagery in a single helmet design for all operations (Figure 13.15). There is no need to change systems or add systems when transitioning from day to night or from night to day. Night vision sensors are integrated into the helmet shell, and symbology is inserted in an optical train; these systems usually include head tracking capability.

FIGURE 13.13 ■ Approach 1, Joint Helmet Mounted Cueing System (JHMCS) Day and Unaided Night Operations (left) and Aided Night Operations (right)

FIGURE 13.14 ■ Approach 2, Q Sight (left) and Q Sight with NVG (right)

FIGURE 13.15 ■
Fully Integrated
Systems, Typhoon
HMD (left) and Top
Owl (right)

FIGURE 13.15 ■ Fully Integrated Systems, Typhoon HMD (left) and Top Owl (right)

13.10 | HMD COMPONENTS

As with all avionic systems, the design of the HMD starts with the operational requirements; what information needs to be presented to the eye. Like the HUD design, some care must be given as to not provide too much information which renders the display useless. There may be a need to add a declutter mode. Information which may be required (based on mission type) include: navigation, situation awareness, weapon aiming, takeoff and landing, primary flight displays, or simply a HUD replication. The basic components of the system include: the display device, the drive electronics, adjustments, viewing optics, and perhaps a helmet tracking system.

The display device can be one of three basic systems: it can be emissive, such as a cathode ray tube (CRT) or field emission display (FED); it can be transmissive, such as a liquid crystal display (LCD); or it can be reflective, which may be liquid crystal on silicon (LCoS) or similar to projection televisions. The drive electronics primary function is to convert the video signal into the proper display format (PAL/NTSC/VGA). They also need to be small, lightweight, and at a low cost. The mechanical components provide support, image adjustments, alignment and focus.

Because of the 25 mm minimum distance from the eye (to allow for eyeglasses, laser protection, etc.), the HMD requires viewing optics; these optics must provide suitable performance and be matched to the display. The FOV of the human eye is approximately 200°. If the HMD provides a WFOV (200°), then it should account for the foveal vision versus the peripheral vision. (Foveal vision is literally vision with the fovea; it is the normal photopic vision in daylight.) Resolution also needs to be maintained: the human vision resolution is approximately 1.7–1.9 cycles/milliradian whereas Generation III NVIS is approximately 1.3 cycles/milliradian. Since we are adding another optical path, a new MTF will be generated and the resolutions noted can only degrade.

It is not possible to simply state that we should optimize all of the components since HMD parameters are mutually interdependent. As an example:

- Wider field of view will dictate a higher weight.

- Wider field of view will provide increased capability.

- Increased capability also has a penalty of heavier weight.
- Higher protection levels also implies heavier weight.
- A binocular system provides a better capability at higher weight.
- Higher weight equals a loss of overall capability.
- Improved capability will come at higher cost.

The HMD must provide all of the traditional protections of an aircrew helmet, so in addition to HMD component design, the developer must also take into account the biodynamics of the design. The concept of weight and CG was addressed previously in NVIS design, and is included in biodynamics. Simply put, biodynamics is the study of the effects of dynamic processes, such as motion or acceleration, on living organisms. In our case, the living organism is the pilot. The basic helmet provides physical protection (impact, ejection), eye protection (blast visor), and aural protection (environmental noise). In addition, the HMD must be operationally viable by allowing the pilot to perform his mission effectively. The HMD must be stable and have a comfortable fit, especially under high G-loadings. Like the enhanced flight vision systems (EFVS) addressed in the Civil Certification section, it is imperative that display of flight/tactical information and NVIS or IR/EO sensor video be harmonized with the outside world.

The packaging of display or on-helmet sensor elements will adversely affect total head-borne mass (compared to basic helmet) and may adversely affect the CG in some or all of the X, Y, and Z axes as seen previously with NVIS systems. In the United States, as part of initial research into HMDs in Army helicopters, studies sought to actually define a safe limit on flight helmet mass. In 1982, the US Army Advanced Research Laboratory (USAARL) proposed a limit of 1.8 kg during the development of the AH-64 Apache Integrated Helmet and Display Sighting System (IHADSS) HMD. The helmet system subsequently developed met this mass limitation while providing the desired platform for the monocular HMD and the required acoustic and impact protection. Unfortunately, the legacy SPH-4 plus AN/AVS-6/9 combination used for night operations in all other Army helicopters exceeded the proposed 1.8 kg limit by more than 1 kg. Additional scientific studies have been conducted since, trying to define the maximum CG (or in some studies, center of mass) in the vertical and longitudinal axes of the human skull. The information obtained is important as it is used to prevent injuries, especially during ejection, and to reduce fatigue. It is also important to note that the initial studies were done by the US Army to support helicopter operations, which is somewhat benign when compared to fighter operations under sustained G-loadings. Studies have shown that transport aircraft and helicopters are similar enough to use the same guidance, however allowances must be made for the cockpit environment.

Fighter HMD implementation has significant differences from the helicopter/fixed-wing transport cases. Aircrew are subjected to higher snap and sustained-g capability which will certainly contribute to neck fatigue. The added head-borne mass, cockpit interference and potential of high speeds in an ejection scenario is more than likely a far worse scenario than the worst case of a crash landing in helicopters or transport category aircraft. The HMD equipment may create adverse aerodynamic loads during separation that could be catastropic; the decision will have to be made as to whether the display module is discarded, either manually or automatically, or retained during an ejection (Figure 13.16). The capability for high speed ejection (aircraft velocity approaching 600 kts) may have to be adjusted downward with HMD equipment for safety reasons.

FIGURE 13.16 ■
Unacceptable HMD
Design

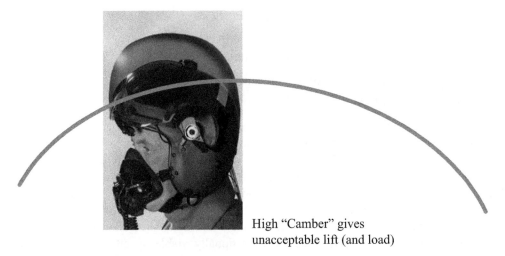

High "Camber" gives
unacceptable lift (and load)

This text gives only a cursory examination of the biodynamics in HMD design. For further information, the reader is referred to a number of studies in this area:

- "Cervical Range of Motion and Dynamic Response and Strength of Cervical Muscles," D. Foust, D. Chaffin, R. Snyder, and J. Baum, SAE Technical Paper 730975

- "Neck Muscle Endurance and Fatigue as a Function of Helmet Loading: The Definitive Mathematical Model," Chandler Allen Phillips and P.E. Jerrold Scott Petrofsky. U.S. Army Medical Research and Development Command Fort Detrick, Frederick, MD

- *Mechanics of Skeletal and Cardiac Muscle*, Chandler A. Phillips and Jerrold Scott Petrofsky, Thomas, 1983

- "Simulation of Head/Neck Response to + Gz Imoact Acceleration due to Additional Head Mass," Amit Lal Patra and Christina Estep, Virginia Polytechnic and State University, 1991 Graduate Research Report, Armstrong Laboratory, Wilford Hall Medical Center

- "Vertical Impact Tests of the X-31 Helmet-Mounted Visual and Audio Display (HMVAD) System," Chris E. Perry, USAF Armstrong Laboratory, October 1993

13.10.1 HMD Optical Design

An aircraft sensor (EO, IR, and in some cases radar) renders the outside scene within its FOV into an electrical signal for relay via databus to a suitable display source. The display source is composed of an image generator (which converts the electrical signal into a visual image), a projector (which outputs the image) and a screen (which displays the resultant image), such as a color MFD. The concept of an HMD is almost exactly the same; the sensor still outputs the scene within its FOV as an electrical signal, although in this case, the image generator, projector and screen are integrated into the pilot's helmet. Since the scene capture and conversion to an electrical output all takes place within the aircraft, the HMD is only concerned with the image signal arriving at the helmet. By convention in HMD documentation, the combination of image generator and projector is termed the image source.

FIGURE 13.17 ■ Miniature CRTs, (from left) Elbit Systems of America Integrated Helmet and Display Sighting System (IHADSS), Kaiser Strike Eye, GEC Marconi Dual Combiner HMD

HMDs have two basic optical requirements: the display itself and the optical train. The display requires a method (such as a flat combiner or projection) and a presentation (monocular, biocular, or binocular). The optical train is concerned with optical characteristics, FOV versus resolution, eye relief, luminance, and focus.

In addition to the concerns of size, weight, cost, and power consumption, a determination must be made as to the type of display to be implemented. As noted earlier, displays may be emissive, which is a direct-view device with combined light and image generation, or they may be transmissive/reflective where a projection is used with separate light and image generation. For all types, the key performance parameters will be luminance, contrast, and resolution. All types have their advantages and disadvantages; the key is to match the capabilities to mission requirements.

The family of emissive display image sources includes miniature CRTs, phosphors, electroluminescent (EL), FEDs, lasers, LEDs, and organic LEDs (OLED). Transmissive/reflective display image sources include the different types of LCDs. Miniature CRTs (Figure 13.17) are efficient in light conversion and have a high resolution performance; on the other hand, they are large and heavy and require high-voltage power.

Electroluminescence or EL is the nonthermal conversion of electrical energy into light energy. This phenomenon is used in EL lamps, LEDs, and OLEDs. EL lamps or sometimes called high field electroluminescent lamps use electric current directly through a phosphor (such as $ZnS:Mn$ or $GaAs$) to produce light. They can be shaped to be extremely flat, or in narrow wire-like shapes. They are different from LEDs/OLEDs, which use a p/n junction (two semiconductive materials where electrons and holes combine on the boundary). In EL there is a layer called an activator in which the whole layer is emitting light, not just the boundary. EL displays have the advantages of low power consumption, small size, long life, low weight and can be manufactured into flat flexible panels, narrow strings, and other small shapes. They are not directional like LCDs when used as a display, they are viewable at all angles. They are not practical for general lighting of large areas due to the low lumen output of the phosphors; typical brightness values are monochrome luminance of 2000 fL and color lumiance of 200 fL.

Phosphors, and particulate phosphors, are a variation on miniature CRTs and use the same manufacturing techniques. They optimize resolution, luminous efficiency and contrast and have good adaptation to glass or fiber-optic faceplates. OLEDs have a layer of organic material sandwiched between two conductors (anode/cathode); the glass top plate is the seal, and the glass bottom plate is the substrate. When current is applied to the conductors, electro-luminescent light is produced directly from the organic material.

FDs contain millions of micro electron guns and employ high energy efficiency, good brightness, high video speed, high contrast, wide viewing angle, and low weight.

A scanning laser with associated optics may also be used and have the attributes of high brightness capability and low power. A concern, of course, would be retinal safety.

Nonemissive image sources include the family of LCDs which were previously descibed in section 7.7.6.2 of this text. An active-matrix LCD (AMLCD) display uses a grid of transistors and capacitors with the ability to hold a charge for a limited period of time. Because of the switching action of transistors, only the desired pixel receives a charge, and the pixel acts as a capacitor to hold the charge until the next refresh cycle, improving image quality over a passive matrix. The F-35 HMD system employs a flat panel AMLCD, combined with a high-intensity back light, as its image source. The fully overlapped display provides up to a binocular $40° \times 30°$ image, depending on which sensor is displayed. The digital image source provides both symbol writing and video capability with sensor overlap from the distributed aperture system (DAS) (Figure 13.18). Whatever optics are employed, it should be remembered that the addition of another transparency, especially in fighter applications, will further degrade the pilot's ability to see the real world.

The presentation may be accomplished by using one of three methods: monocular, biocular or binocular. As with the display types, each has its advantages and disadvantages. The monocular presentatation is lightweight (compared with biocular/binocular systems) and provides aided information to only one eye. This may cause physiological issues such as eye dominance and binocular rivalry issues. Biocular systems utilize a single projection system and this single image is presented to both eyes. This method is lighter than the binocular system, but is costly when compared to monocular systems. Binocular systems utilize a dual image projection system, and individual images with up to 100% overlap are presented to each eye. The image quality and resolution are significantly better than monocular systems, but are costly and are more problematic in the areas of weight and CG. Biocular resolving performance is equivalent only to that of monocular because both displayed features and resolution-limiting noise are spatiotemporally correlated between the left and right eye images. Only discrete two-channel binocular HMD systems offer better resolving performance

FIGURE 13.18 ■
F-35 AMLCD HMD

than monocular; however the size, weight, and design complexities of visually coupled binocular imaging sensors are far greater than monocular systems. Compared with the monocular approach, no biocular range resolution performance benefit has been in evidence, while the binocular sensor system with its potential performance benefit comes at the price of significantly greater design complexity and cost.

The last two parameters of note within the optical system are the resolution and the FOV. The relationship of these two parameters was discussed in the EO Chapter, section 7.7.4. Two definitions not used in chapter 7, but common within HMD documentation are eye relief and exit pupil/exit pupil diameter. Figure 13.19 provides an optical path with the HMD nomenclature. In our previous discussions, the exit pupil was identified as the focal point and the eye relief was defined as the focal length. The exit pupil diameter is the size of the shaft of light transmitted to the eye. The larger the exit pupil, the brighter the image will appear. The f-stop, or focal ratio was defined as the focal length divided by the effective aperture diameter. By increasing the focal length, we decrease the FOV and increase resolution. We can see that if we limit the eye relief to no less than 25 mm we limit the FOV unless we adjust the effective aperture. If we again limit the aperture as in HMD design (most HMD design specifications call for an approximately 18 mm exit pupil diameter, to allow for helmet movement under high G) the FOV will be limited by physics. The only other way to increase the FOV would be to have 2 separate optical systems and overlap, or fuze them in some way. By only partially overlapping the individual FOVs for each eye, an expanded azimuth total FOV is obtained.

In our discussions of EO/IR technology, we stated that if we increased the resolution of the system (e.g., more detector elements) and failed to increase the resolution of the display, then the net result would be a decreased resolution. The same applies in HMDs with an overlapped FOV. If we have stretched the horizontal FOV utilizing a partial overlap in order to maintain system resolution, extra pixels will be required. If we wanted to maintain a 20/20 resolution, we would require 60 pixels per degree (60 pixels per degree = 30 cycles per degree = 1.7 cycles per milliradian = acuity of 20/20, Equation (13.3)). Therefore, for any FOV, we can determine the minimum number of pixels required to maintain optimum aircrew resolution.

Partial binocular overlap can provide a larger horizontal total field of view without the penalty of decreased resolution or increased weight from the larger collimating optics. It requires two separate video channels, high quality optical correction, distortion mapping, canting correction, and more attention to the details of optical alignment. There are concerns by some researchers about the minimum amount of overlap required as well as the possible effects of eyestrain and image fragmentation.

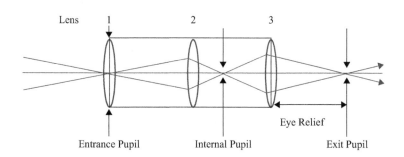

FIGURE 13.19 ■
HMD Optical Path

13.10.2 Helmet Tracking Systems

There are two types of helmet tracking systems: a visually coupled system (VCS) and the helmet/head tracking systems (HTS). A VCS is a special subsystem that integrates the natural visual and motor skills of an operator with the machine he is controlling. An operator visually searches for, finds, and tracks an object of interest. His line of sight is measured and used to aim sensors and/or weapons toward the object. Information related to his visual/motor task from sensors, weapons, or central data sources is fed back directly to his vision by special displays so as to enhance his task performance.

Two functions are performed: a LoS sensing/control function and a display feedback function. Although each may be used separately, a fully visually coupled system includes both. Thus, it is a unique control/display subsystem in which LoS is measured and used for control, and visual information is fed back directly to his eyes for his attention and use.

The Defence Evaluation Research Agency (DERA) carried out an airborne demonstration and evaluation of a fast-jet VCS installed in ZD902, the Tornado Integrated Avionics Research Aircraft (TIARA) for the UK Ministry of Defence. The installed VCS used a head steered forward looking infra-red (HSFLIR) sensor and an HTS to provide the pilot with an image of the outside world projected onto a binocular helmet mounted display (BHMD). In addition to the sensor image, information such as aircraft altitude, attitude, and airspeed were also presented to the pilot through the HMD to eliminate the need to look inside the cockpit for critical flight data. The aim of the trial was to demonstrate the day and night benefits of a fast-jet integrated with HSFLIR and HMD as an aid to low-level flight, navigation, target acquisition, take-off, and landing. The outcome of this flight test program was very encouraging and, although testing has identified that improvements are necessary, in particular to HSFLIR image quality, AGC performance, helmet fit and symbology design, test aircrew endorsed the acceptability of a VCS.

In order to achieve a VCS, it must be able to calculate the position and LOS of the wearer's eyeline. By utilizing Type I NVG systems, this was simply a matter of defining the helmet position and orientation; it was assumed that the helmet/head relationship is constant. The setup allowed stability and comfort, and the relatively small FOV and Type I on-axis requirement made the calculations relatively easy.

The HMT (head/helmet mounted tracker) is a type of HTS. The system requires a position and orientation tracker (POT) and head orientation and position (HOP) information. These data can be obtain in a multitude of ways: transducer pairing using transmitters and receivers or ultrasonic, electro-magnetic or electro-optic sensors; the Thales Scorpion uses inertial combined with optics. The information may also be obtained from a platform stabilization system much like a FLIR or radar system. No matter which type of system is selected, it must meet certain performance requirements, including the following, which have been determined through operational use:

- Resolution (linear < 1 mm and angular < 1 milliradian)
- Accuracy (at the design eye ~ 1 milliradian, and elswhere in the head motion box < 6 milliradians) (accuracy requirements may change based on the FOV of the weapon system employed; the target must be in the system's FOV)
- Update rate > 100 Hz
- Latency < 10 ms (with rapid head movement)

As mentioned multiple times within the text, as more and more processing of data is accomplished, the probability exists that there will be either senescence or latency when reported to the pilot. To decrease the total latency, it may be necessary to relax some of the other performance requirements. The more complicated the system, the higher probability of latency; managing this latency is extremely important as it can render even the most advanced system unusable.

13.11 | TEST AND EVALUATION OF HMD EQUIPMENT

As seen with NVIS equipment, the T&E of HMD equipment will involve lab, ground, and flight test. The quantitative testing will, for the most part, be accomplished in the lab, whereas most of the on-aircraft evaluation will be qualitative, concentrating on the user's ability to increase performance and capability.

13.11.1 HMD Quantitative Testing

Due to the complex nature of HMD designs, several subsystems can contain varying technologies, and each technology will require slightly different evaluation techniques. Consequently, it is difficult to develop a matrix that covers all possible combinations (e.g., CRT versus various flat panel displays, I^2 versus coupled designs, magnetic versus optical trackers, analog versus digital designs). The degree of test complexity is determined by the system design. Laboratory testing is important and necessary to accomplish but alone cannot fully evaluate the performance of an HMD system. To achieve this goal requires qualitative ground and flight evaluations under actual operating conditions.

The flow of the lab testing looks very much the same as other avionics systems covered previously. System level testing investigates integration issues, whereas the subsystem level looks at the contributing pieces such as the helmet, image source, display optics, acoustics, tracker electronics, and software. The CRT display performance measures are legion:

Geometric	Electronic	Photometric
Viewing distance	Bandwidth	Luminance
Display size	Dynamic range	Luminance uniformity
Aspect ratio	Signal/noise ratio	Grey shades
Number of scan lines	Frame rate	Contrast ratio
Interlace ratio	Field rate	Halation
Scan line spacing		Ambient illuminance
Linearity		Color
		Resolution
		Spot size and shape
		MTF
		Gamma

Likewise, the optical test parameters:

<table>
<tr><td align="center">**System**</td><td align="center">**Display**</td></tr>
<tr><td valign="top">

Visual field
Spectral transmittance
Physical eye relief
Interpupillary distance range
Luminous transmittance chromaticity
Neutrality
Prismatic deviation
Refractive power
Cockpit display emission transmittance

</td><td valign="top">

Field of view
Image overlap
Resolution (visual acuity)
Extraneous reflections
Luminance range
Grey levels
Chromatic aberration
Contrast ratio
Exit pupil size
Focus range
Spherical/astigmatic aberration
Image rotation
Image luminance disparity
Vertical/horizontal alignment
Distortion
Luminance uniformity
Static/dynamic uniformity

</td></tr>
</table>

As well as the biodynamics:

<table>
<tr><td align="center">**System**</td><td align="center">**Helmet**</td><td align="center">**Head Tracker**</td></tr>
<tr><td valign="top">

Mass properties
Impact attenuation
Stability
Dynamic retention
Anthropometric
  fit/comfort
Ballistic protection
HMD breakaway force

</td><td valign="top">

Shell tear resistance
Chin strap assembly
  integrity

</td><td valign="top">

Motion box size
Update rate (latency)
Jitter
Pointing angle
  accuracy
Pointing angle
  resolution

</td></tr>
</table>

And finally, the acoustics:

<table>
<tr><td align="center">**System**</td><td align="center">**Earphones**</td></tr>
<tr><td valign="top">

Real-ear attenuation
Physical-ear attenuation
Speech intelligibility

</td><td valign="top">

Sensitivity
Distortion
Frequency response

</td></tr>
</table>

The test equipment required for conducting most HMD subsystem laboratory testing is just as complicated as the subsystem itself, requiring technical training and expertise, proper facilities, and the infamous necessary funding. Much of the specialized sub-system testing may be conducted by other laboratories within or outside the responsible organization. These tests may include RTCA/DO-160G environmental testing, helmet retention during windblast, and ejection sled testing. Some laboratory testing may be required before safety of flight approval can be obtained.

13.11.2 HMD Qualitative Ground Testing

The ground testing can be set up in a similar fashion to previously discussed with NVIS systems. The test will encompass a series of questions for the pilot to answer while utilizing the HMD equipment. It is easiest to set up a subjective rating scale or Bedford rating scale to answer pertinent questions about the system. Table 13.4 contains the questions that should be addressed for a generic HMD system.

TABLE 13.4 ▪ HMD Subjective Rating Scale Questionnaire and Functional Evaluation

HMD Subjective Rating Scale Questionnaire and Functional Evaluation		
Subject	Question	Rating/Comments
Integrated Sensors	What image controls are available and how do they function?	
	Are I^2 modules removable?	
	One hand, gloved hand?	
	Can system be focused using existing preflight focusing equipment?	
	Is there a new system for preflight focusing?	
Integration with NVGs	Does the HMD integration interfere with NVG controls?	
	Does the integration allow for normal positioning of the NVG image?	
	Vertical, horizontal and eye relief	
Visor Patch Considerations		
(1) Monocular system	Can imagery or symbology be correctly aligned with the eye?	
	Can imagery and symbology be selected independently?	
(2) Binocular system	Can imagery from each input be correctly registered and aligned with the eyes?	
	Is imagery 100% or partially overlapped?	
	Can symbology from each input be correctly registered and aligned with the eyes?	
	Is presentation of symbology selectable (binocular, right eye or left eye)?	
	Can imagery and symbology be selected independently?	
Monocular and Binocular Systems	What adjustments are available to the pilot for positioning imagery or symbology?	
	Can imagery or symbology be moved independently in the area of the visor patch?	
	Ease of reach, identification and control	
	What adjustments are hard-mounted (can be set on the ground but not available to the pilot)?	
	How is fitting accomplished for each pilot?	
	If no adjustments are provided, what design technique is used to allow for variances in eye position (e.g., various sized visors, various sized modular attachments, helmet suspension system, helmet liner molding)?	
Combiner Considerations	Is the combiner used for symbology only, for both imagery and symbology, or used in conjunction with a NVG?	
(1) Monocular system	Can imagery or symbology be correctly aligned with the eye?	
	Can the pilot adjust the position of the combiner?	
	On the ground or in the air?	
	Can imagery or symbology be moved within the combiner?	
	Can imagery or symbology be selected independently?	

(Continued)

TABLE 13.4 ■ *(Continued)*

HMD Subjective Rating Scale Questionnaire and Functional Evaluation		
Subject	**Question**	**Rating/Comments**
(2) Binocular system	Can imagery from each combiner be correctly registered and aligned with the eyes?	
	Can the pilot adjust the position of the combiner?	
	On the ground or in the air?	
	Can symbology from each combiner be correctly registered and aligned with the eyes?	
	Is presentation of symbology selectable (binocular, right eye or left eye)?	
(3) Integration with NVGs	With the NVG image in the proper location relative to the pilot's eyes, can the combiner be adjusted to correctly position the symbology?	
	Can symbology be moved within the combiner?	
Compatibility with Aircrew Equipment	Compatible with aircrew spectacles?	
	Compatible with laser eye protection	
	Either spectacles or visors	
	Comfort?	
	Ease of donning and doffing?	
	Storage?	
	Compatible with existing equipment or designed specifically for HMD Comfort?	
	Restriction of head motion?	
	Restriction of visual fields?	
	Additional physical obstruction with cockpit components	
	Effects on head movement (e.g., scan, targeting)?	
	Stowage and accessibility of equipment?	
Aircraft Integration	Entering and exiting the aircraft	
	Normal entrance and egress?	
	Emergency egress?	
	Connecting HMD to aircraft	
	Location of attachment points and how easy to connect?	
	Quick release points (emergency egress)?	
	Effects of HMD and related cables and wires	
	Restrictions to head movement?	
	Interference with cockpit structures?	
	Physical interference of helmet, display module or visor	
	Contact with canopy and windows?	
	Contact with overhead panels and wires?	
	Effects due to location of head tracking components	
	Physical interference with HMD cables?	
	Visual blockage?	
	Changing and stowing HMD modules	
	Location and accessibility of HMD components (e.g., two-part design)?	
	Ease of donning and doffing components?	
	NVIS compatible lighting testing if required	

TABLE 13.4 ■ (Continued)

HMD Subjective Rating Scale Questionnaire and Functional Evaluation		
Subject	Question	Rating/Comments
	Effects on the visual field	
	Does the helmet or combiner hardware block peripheral vision?	
	Up, down, or lateral directions?	
	Is there blockage only in one direction?	
	Combiner attached to the right side or along the upper brow of the helmet	
	Is there blockage caused within or outside the aircraft?	
	Is information on head down displays or overhead consoles blocked most of the time or only when viewing in specific directions?	
	Are areas of the outside scene blocked most of the time or only when viewing in specific directions?	
	Will either or both of these require changes in scan patterns or CRM procedures relative to current applications?	
	Are there different considerations based on seat position (e.g., side by side or tandem)?	
	Pilot in right seat cannot see lower right terrain when viewing straight ahead?	
	Pilot in left seat cannot see tactical display in center console when viewing straight ahead?	
Other human factors considerations	Consider any other human factors considerations based on the aircraft unique system to include integration with weapon systems	

13.11.3 Flight Test of HMD Systems

Flight test of HMD systems is complex and, in addition to myriad human factors issues, includes the integration of other aircraft avionics and tactical suites. What may appear as a good system when tested in a lab may prove to have many problems when tested in the aircraft, partially depending on the design of the HMD and the interface with cockpit geometry. Ensuring test pilots have the necessary experience to test HMD systems may require preparation well in advance of the start of the test program.

The initial portion of the flight test program will be retest of the ground test matrix, only now in a flight environment. Table 13.4 may be reused in-flight to complete the human factors and functionality test requirements. In-flight testing will really provide the first opportunity to judge symbology, imagery and latency. How is the ease of visual focus in binocular and monocular modes; what is the focal point of imagery and symbology. Is it possible to easily focus on imagery and symbology concurrently, and what is the range of brightness in day and night conditions? The tests should evaluate the readability and clarity of the presentation, access and ease of controls, declutter modes, and the ability to switching between tactical and navigation modes. Symbology latency can be evaluated only in dynamic conditions. The evaluator should note any basic latency during benign conditions and then compare to the latency experienced during dynamic conditions and head movements. The evaluator should consider the effects during various flight profiles (e.g., landing, formation flying).

Integration of the HMD with the sensor and weapons systems will need to be evaluated in-flight. Of particular concern will be the system boresight, the accuracy of designation, navigation and targeting, drift, tracking accuracy, the ease of access, and use of the weapon and sensor interface controls. The bottom line is to provide the aircrew with improved capability, reduce workload, and not lose any system capability.

13.12 | SELECTED QUESTIONS FOR CHAPTER 13

1. What is meant by the term image intensification?
2. Where might I find currency requirements for flight with NVGs?
3. What causes leaky green goggle glare? Can it be rectified?
4. What is blooming? What is it caused by?
5. Name three areas of evaluation for NVG human factors.
6. Which modification may provide the best display illumination for NVG use?
7. Are multicolor displays applicable to Class A NVIS?
8. Are there any advantages to using NVGs? If so, what are they?
9. Which spectrum of energy is visible by the human eye?
10. What causes a halo effect?
11. What other spectrum (other than visible light) may be used in NVIS systems?
12. Can NVGs improve upon the standard uncorrected vision of 20/20?
13. What is the best visual performance an operator can expect while wearing NVGs?
14. What technology is employed in NVGs?
15. Briefly describe a single component HMD system.
16. What is the approximate human vision resolution (cycles/mrad)?
17. Why is the NVG image normally green and black?
18. What is an MTF?
19. How is system resolution determined for NVG devices?
20. What is the peak spectral response of the human eye? At what color?
21. What is meant by lighting compatibility?
22. What is the spectrum of energy used in NVGs?
23. What is the difference between absorption filters and interference filters?
24. What is the purpose of an ion barrier in GaAs systems?
25. Provide three sources of night sky irradiation.
26. What is an NVIS bezel?
27. What problems arise from the use of red LED obstruction lights?
28. What is the difference between civil and military external lighting requirements?

29. How can HUD symbology be viewed while wearing NVGs?

30. What must the eye relief account for?

31. How is the electrical output of the MCP amplified to create a usable picture?

32. What is the difference between luminance and illuminance?

33. What is the difference between a display filter and an NVG filter?

34. Name the two types of Helmet Tracking Systems.

35. What is albedo? Does varying albedo in the NVG scene enhance performance?

36. What is an objective lens and what is its function in an NVG system?

37. What is a minus blue filter? What is its purpose?

38. What is the FOV of the human eye?

39. What problems may occur if warning (red) annunciators are filtered?

40. What is meant by the term Electroluminescence or EL?

41. How are aircrew protected from the effects of a sudden flash of light?

42. What are the two circuits that contribute to the NVG AGC function?

43. Briefly describe a two component HMD system.

44. What is eye relief? Why is its purpose?

45. What concerns should be addressed when integrating HMDs with weapon systems?

46. Describe a visual acuity test for NVGs.

47. Where can I find documentation on illumination levels for cockpit lighting?

48. What is Type I lighting?

49. What is the minimum distance from the eye to the HMD?

50. What is a modified Class B minus blue filter?

51. Is it possible to optimize all HMD components?

52. Does NVG use contribute to a decrease in workload?

53. Name three compatibility issues that should be addressed when evaluating NVGs.

54. What does the term NVIS friendly refer to?

55. What filter is required in the NVG if the aircraft has a fixed HUD?

56. What is scintillation in an NVG scene? What is it caused by?

57. What is a post light? Are there advantages to using this modification?

58. What would you think a reasonable value of gain would be for a Gen III NVG?

59. How can I focus an NVG?

60. What are the major differences between Gen II and Gen III systems?

61. Name three causes of degradation which may be caused by transparencies.

62. Is latency a problem in HMD systems?

63. Name three problematic areas associated with ejection and the use of NVGs.

64. What is meant by NVIS radiance (NR)?

65. What are the three conditions for evaluating on-aircraft NVG visual acuity?

66. Can exterior lighting requirements be universally applied?

67. Is there a civil standard for NVIS use in the NAS?

68. Name three NVIS lighting modification methods currently in use.

69. What are bridge lights? When are they used?

70. Are LEDs easily mixed with incandescent lights? Why, or why not?

71. Name three attributes of NVIS friendly exterior lighting.

72. Is the flight test of NVIS usually quantitative or qualitative?

73. What is biodynamics?

74. Give four examples of cosmetic blemishes.

75. What is the relationship between visual acuity and NVG resolution?

76. Name the two basic types of HMDs.

77. Is there any possibility that an engineer designed HMD would provide too much information?

78. Name the basic components of the HMD system.

79. What are the biodynamics concerns in the development of an HMD?

80. What is the difference between monocular, biocular, and binocular?

Acquisition, Test Management, and Operational Test and Evaluation

Chapter Outline

14.0 Overview 957
14.1 Applicable Documentation 958
14.2 The Acquisition Process. 958
14.3 The Operational Requirements Document (ORD) 965
14.4 The Test and Evaluation Master Plan (TEMP) 966
14.5 Operational Test and Evaluation. 967
14.6 OT&E Test Plan Structure. 970
14.7 Reliability, Maintainability, Logistics Supportability, and Availability (RML&A) . . 978
14.8 Summary 981
14.9 Selected Questions for Chapter 14. 981

14.0 | OVERVIEW

Until now the testing that has been described has been applicable to development test and evaluation (DT&E) and specification compliance. The final area of testing is operational test and evaluation (OT&E), where it is the tester's responsibility to determine if the system under test can perform the mission. The charter of the OT&E evaluator is best summed up in the Air Force Operational Test and Evaluation Center (AFOTEC) Test Director's Toolkit, April 12, 2010: "When our Airmen go to war they require the weapon system capabilities their commanders requested to defeat an enemy. Therefore, we will ensure they receive those capabilities at the right time and with confidence in their ability to accomplish the mission."

An overview of the test and evaluation process was covered in chapter 1, and the reader may want to review that chapter before proceeding. The examples used here are based on the U.S. military OT&E practices, in particular the U.S. Air Force (AFOTEC) and the U.S. Navy Operational Test and Evaluation Force (OPTEVFOR). It is recognized that most other test organizations around the globe do not have dedicated organizations such as these in place; rather, they use operational units who are trained as OT&E evaluators. The procedures and policies described here may be used as guidance in any organization whose goal is to ensure mission effectiveness and suitability.

▉ 14.1 ▉ | APPLICABLE DOCUMENTATION

The documentation that covers the acquisition, management, and test of a systems program within the United States is burdensome and can be equated to the plethora of documentation previously described in the civil certification chapter. The overall guidance for any acquisition program is the Department of Defense Directive (DoDD) 5000.01, "The Defense Acquisition System" and the Department of Defense Instruction (DoDI) 5000.02, "Operation of the Defense Acquisition System," collectively called the DoD 5000 series. These documents have been revised numerous times as acquisition has evolved from specification compliant to capabilities based. The most current versions are 2003 and 2008, respectively.

The services, in turn, provide operating instructions on how to implement the DoD 5000 series. The USAF has three instructions that are classified as mandatory compliance: Air Force Instruction (AFI) 99-103, "Capabilities-Based Test and Evaluation," March 20, 2009; Air Force Instruction (AFI) 10-601, "Operational Capability Requirements Development," July 12, 2010; and Air Force Instruction (AFI) 63-101, "Operation of the Capabilities Based Acquisition System." The USN uses Secretary of the Navy Instruction (SECNAVINST) 5000.2D, "Implementation and Operation of the Defense Acquisition System and the Joint Capabilities Integration and Development System," October 16, 2008. AFOTEC uses further guidance contained in Air Force Operational Test and Evaluation Center Instruction (AFOTECI) 99-103, "Conduct of Operational Test and Evaluation," November 24, 2004.

For the evaluators, the USAF has issued the HQ USAF/TE, "Air Force Test and Evaluation Guidebook," December 2004. AFOTEC provides a checklist of operational test and evaluation activities and events for program acquisition processes. This checklist is contained in the AFOTEC "Test Director's Toolkit," April 12, 2010; AFOTEC also publishes AFOTEC OT&E guide, which is now in its sixth edition, February 12, 2009.

A textbook that may be useful as a reference, although somewhat dated, is *Operations Research Analysis in Test and Evaluation* (Donald l. Giadrosich, AIAA Education Series, 1995).

▉ 14.2 ▉ | THE ACQUISITION PROCESS

As noted previously, the overall guidance for acquisition programs within the United States is contained in the DoD 5000 series. Deputy secretary of defense Packard signed the first DoD Directive 5000.1 in 1971. This document was only eight pages long and created the Defense Systems Acquisition Review Council (DSARC), requiring three major decision points (milestones) with one supporting document. The document was applicable for all major programs entailing costs of $50M RDT&E or $200M procurement. Over the next 37 years this DoD 5000 process was revised 14 times. The acquisition milestones and the relative lengths of the phases for the original directive are shown in Figure 14.1.

The conceptual phase begins with program initiation and ends with full-scale development decision. The second phase concludes with a production go-ahead decision.

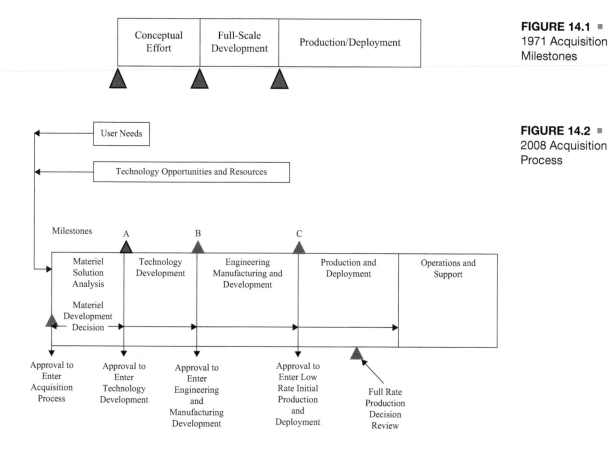

FIGURE 14.1 ■
1971 Acquisition
Milestones

FIGURE 14.2 ■
2008 Acquisition
Process

The 5000 process has evolved into five phases with three critical milestones. Decisions are made prior to entering each new phase and are called materiel development decisions; they may be thought of as a successful exit criteria from the preceeding phase and entrance criteria for the new phase. The process is now applicable for all major programs entailing costs of $365M RDT&E or $2.19B procurement (in FY 2000 dollars). The updated flow for the 2008 acquisition process is shown in Figure 14.2.

The five phases of the acquisition process are as follows:

– Materiel solution analysis (also called the concept refinement phase)

– Technology development

– Engineering manufacturing and development

– Production and deployment

– Operation and support

In addition to the five phases there are three critical milestones:

– Milestone A, which is concept and technology approval

– Milestone B, which is system development and demonstration approval

– Milestone C, which authorizes production or limited deployment approval

The two items that will kick off the acquisition process are user needs and technology opportunities and resources; these are shown as the initial entries in Figure 12.2.

User needs are generated by field commanders, military departments, or even the operational users themselves. A user need could be a suggestion (unsolicited) by a commercial enterprise. A user need is the one that seeks to rectify a deficiency in a capability or one that may enhance the efficiency or capabilities of the user. A mission need analysis will validate the need which in turn will be documented as a mission need or capability. The document is further assessed with respect to other available options and the final outcome is an operational requirements document (ORD). The structure of the ORD may be found in section 14.3.

In today's strategy of capabilities based acquisitions in the United States, a capability-based assessment (CBA) is performed, and a capability development document (CDD) will be produced. The CBA is the joint capabilities integration and development system analysis process. It answers several key questions for the validation authority prior to their approval. It defines the mission, identifies the capabilities required, determines the attributes and standards of the capabilities, identifies gaps and shortfalls, identifies and assesses potential non-materiel solutions, and provides recommendations for addressing the gaps and shortfalls. A non-materiel solution may be thought of as a procedural change to satisfy a need. Changes may be in doctrine, organization, training, leadership and education, personnel, facilities, or policy to satisfy identified functional capabilities.

The CDD captures the information necessary to develop a proposed program normally using evolutionary strategy (which will be addressed later). It outlines an affordable increment of militarily useful, logistically supportable, and technically mature capability. The CDD may define multiple increments if there is sufficient definition of the performance attributes (key performance parameters (KPP) and key system attributes (KSA) to allow approval of multiple increments. A CDD requirements correlation table (CRT) is a summary of all required capability characteristics listed as threshold or objective values within the CDD text. The CRT provides a concise summary to assist requirements traceability. The CRT follows a set format that may be found in Air Force Instruction 10-601, "Operational Capability Requirements Development," July 12, 2010. An example of a CRT is shown in Table 14.1.

A National Test Pilot School example of a possible completed RCT for a fictitious aircraft and associated systems may be found in Table 14.2.

TABLE 14.1 ■ Requirements Correlation Table

Paragraph # (1)	JCA Tier ½ (2)	KPP or KSA (3)	Development Threshold (4)	Development Objective (5)	Rationale and Analytical References (6)
		KPP/KSA 1	Value	Value	
		KPP/KSA 2	Value	Value	
		KPP/KSA 3	Value	Value	

Where:

(1) The CDD Paragraph #
(2) Joint Capability Area is a collection of like DoD capabilities functionally grouped (per Joint Instruction)
(3) Insert the Operational Performance characteristic
(4) The minimum acceptable threshold value below which utility of the subject characteristic becomes questionable
(5) The desired operational value above the threshold value; the need for improvement over the threshold must be justified
(6) A summarization of the rationale for the KPP or KSA parameter identified within the CDD

TABLE 14.2 ▪ Completed RCT for a Fictitious Aircraft and Associated Systems

System Capabilities and Characteristics	ORCD I		ORCD II		ORCD III	
	Threshold	Objective	Threshold	Objective	Threshold	Objective
1. Non Afterburner Supersonic Cruise[1]	Yes					
a. Sustained	TBD	1.5 M	1.5–1.7 M	2.0 M	1.5 M	2.0 M
b. Dash	TBD	>1.5 M	TBD	2.4 M	2.1 M	2.4 M
2. System Availability	YES					
a. MCP	TBD	90%	85%	90%	85%	90%
b. Break	TBD	TBD	TBD	5–10%	10%	5%
c. Fix	TBD	TBD	TBD	80–90%	82%	83%
3. Radar	Yes					
a. Search (tgts)	TBD	12	TBD	16	8	6
b. Track (tgts)	TBD	6	4	8	4	8
c. Search range	TBD	250	150	200	150	200
d. Track range	TBD	TBD	TBD	TBD	TBD	TBD
4. Terrain Following	YES					
a. Min Altitude	TBD	TBD	TBD	100 all WX	200 all WX	100 all WX

Note 1: Values of system capabilities, system characteristics and critical system characteristics may change as the system matures.

In much of the documentation, an ORD, CDD, or an initial capabilities document (ICD) is covered by the universal term of operational capability requirements document (OCRD).

The user need evolution can be equated to a new product development in the commercial market. The vendor surveys the marketplace for market demand. If a need is found, technology is assessed to see if a product or service can be developed to meet the need. An analysis of the demand versus the potential cost is performed to see if the venture has potential profit. If the venture appears promising, a directive will be issued to forge ahead with the project. User needs are evaluated in the same way, substituting mission capability for profit.

Technology opportunities and resources provide the user with increased or enhanced capability by exploiting advances in science and technology. Advances may be accomplished in the form of: basic research as in the case of the NAVSTAR/GPS program, applied research as in the case of differential navigation, or in advanced research such CEP enhancement for "smart" weapons. In all cases, emerging technology allows enhancements in capabilities not previously achievable. The basic question should be, "What will this technology allow us to do that were up till now unfeasible?" Science and technology should not be limited in scope to government initiatives; Commercial off the shelf (COTS) also may provide enhanced capabilities.

The first phase to be entered is the materiel solution analysis, sometimes called the concept refinement phase. As noted, prior to entering this phase or any other subsequent phase, a materiel development decision (MDD) must be made. The MDD review is the formal entry point into the acquisition management system and is mandatory for all programs. At the MDD the milestone decision authority (approval authority) approves the analysis of alternatives study guidance, determines the acquisition phase of entry, identifies the initial review milestone and designates the lead DoD components. The purpose of this first phase is to refine the initial concept and develop a technology strategy. A secondary task is to analyze possible alternatives.

The steps in the concept refinement phase should, at a minimum, cover the following topics:

– Assess the critical technologies required for the success of the system. Hardware and software should be evaluated for maturity technical and financial risk.

– Evaluators should consider innovation on the original idea, competition with other systems, and COTS.

– Evaluators need to verify that the concept will meet the user's needs; the system needs to be interoperable, survivable, and able to be maintained.

– The system needs to be evaluated against program metrics of cost, schedule, and performance.

– Will the acquisition be evolutionary in nature (incremental), or will it be accomplished in a single step?

– It is advisable to consult industry in performing this analysis.

Specific results must be obtained within each phase before the program is allowed to progress to the next phase. The specific results come under two headings: affordability and performance. Affordability contains the program metrics of cost and schedule; performance is the validation that the concept meets user needs and that the technology is mature enough to adequately support the concept. Affordability and performance measurements must be agreed on by the acquisition authority and the user community before satisfying the entrance criteria to the new phase. These results are documented in a decision memorandum, which

– Documents the milestone decision

– Establishes the entrance criteria

– Provides direction for the next phase

– Provides an audit trail for the program

The content of the decision memorandum will vary between milestones and programs. A flow depicting the gates that must be satisfied prior to progressing to the next phase is shown in Figure 14.3.

Technology development is entered when Milestone A is satisfied. A prerequisite for this approval is an updated mission need statement (based on the analysis of the previous phase), a draft ORD or CDD and an approved technology development strategy (TDS). The initial portion of this phase is devoted toward concept exploration, most of which has been described previously: Does the system meet user needs? What is the cost and performance? How long will it take? Will it integrate with other systems already fielded? Can it survive in the operational environment? Can it be maintained? As the concept exploration progresses, it may be noted that additional components may be necessary for the system to be optimized. A decision review will deem whether or not component development is necessary.

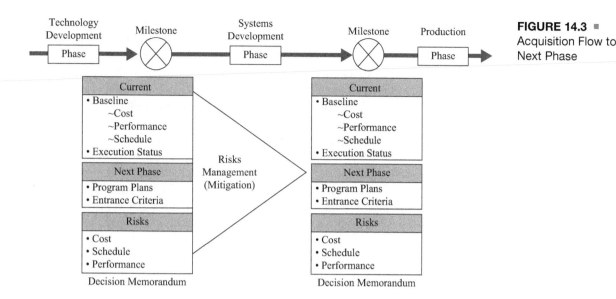

FIGURE 14.3 ■
Acquisition Flow to
Next Phase

If the decision is made to develop additional components, the phase continues with work intended to reduce the overall risk. The tasks may be:

— Develop components

— Develop subsystems

— Develop system architecture

— Build prototypes

The true intent of this phase is to demonstrate that the technology will work.

The outputs of the technology development phase are a final ORD and a draft program performance specification. In order to progress to the next phase, the system must be affordable, have demonstrated capability and the ability to be produced.

Milestone B clears the way for entrance into Phase 3, system development and demonstration. It approves the start of development and initiates the acquisition program. The primary goal of this phase is to develop a system and reduce integration and manufacturing risk. The prerequisites for entering this phase are to have a concept designed, technology mature enough to mitigate risk, validated operational requirements and capabilities and an approved test and evaluation master plan (TEMP) (section 14.4).

The validated operational requirements and capabilities will have been accomplished in order to finalize the ORD/CDD/OCRD. An operational capability, when discussed within the framework of government acquisitions, is the ability to achieve a desired effect under specified standards and conditions through combinations of means and ways across the doctrine, organization, training, materiel, leadership and education, personnel and facilities (DOTMLPF) to perform a set of tasks to execute a specified course of action. It is defined by an operational sponsor and expressed in broad operational terms in the format of an operational capabilities document or a Joint DOTMLPF change recommendation. In the case of materiel proposals/documents, the definition will progressively evolve to DOTMLPF performance attributes identified in the capability development document and the capability production document.

The TEMP documents the overall structure and objectives of the T&E program. It provides a framework to generate detailed T&E plans and it documents schedule and resource implications associated with the T&E program. The TEMP identifies the necessary developmental, operational and live-fire test activities. It relates program schedule, test management strategy and structure, and required resources to: critical operational issues, critical technical parameters, objectives and thresholds in the requirements document and milestone decision points. The TEMP is written in terms of measures of assessment and performance. The structure of the TEMP is described in section 14.4.

The system development and demonstration phase contains all of the elements addressed previously in chapter 1 of the text: systems integration/development, system demonstration/qualification, regression, and OT&E. The DT&E portion of the phase encompasses development, demonstration, and regression testing. The primary purpose of development is to ensure proper integration and functionality, proper operation in its operational environment, proper S/W interfaces with other systems, freedom from interference from those systems, and validation of modeling and simulation results. The purpose of system demonstration is to collect data and statistically prove that the system performs to some basic level of performance such as specification compliance. In incremental programs, regression testing (proving the system performs at the same level with changes of S/W) needs to be planned and budgeted. OT&E evaluates the system against the operational requirements in environments that replicate real-world scenarios.

Milestone C provides the go-ahead for low rate initial production; it may allow limited deployment or fielding of the system. It is the precursor for entry into the production and deployment phase. Low rate initial production is defined as production of the system in the minimum quantity to provide production configured or representative articles for operational test, establish an initial production base for the system, and permit an orderly increase in the production rate for the system sufficient to lead to full rate production upon the successful completion of operational testing. The entry criteria for this phase are an updated TEMP (to include OT&E and live fire testing), a production plan, mature technology, and a system that is interoperable, supportable, and affordable. The output of this phase is full rate production and deployment of an operationally effective and suitable system with full operational capability.

The final phase of the acquisition cycle is operations and support, which is the basic care and feeding of the system. The operational support of the system includes:

- Supply
- Maintenance
- Transportation
- Sustained engineering
- Data management
- Configuration management
- Manpower
- Training
- Safety

Throughout the life cycle of the system there will be occasions of deficiency corrections, product improvement, and life extension. In many of these cases, a follow-on OT&E

program may be required to ensure effectiveness, survivability, sustainability, and interoperability. When the system is phased out, it will be demilitarized and disposed of.

14.3 | THE OPERATIONAL REQUIREMENTS DOCUMENT (ORD)

If we remember from our previous discussion of the acquisition process, a mission need analysis will validate the need that in turn will be documented as a mission need or capability. The document is further assessed with respect to other available options and the final outcome is an operational requirements document (ORD). We will also recall that in much of the documentation, an ORD, CDD, or an initial capabilities document (ICD) is covered by the universal term of operational capability requirements document (OCRD). In this section I will use the term ORD as I think most readers are familiar with this nomenclature.

The ORD translates broad capabilities into specific operational performance parameters and minimum acceptable operational performance requirements. It is important to remember that the ORD represents specific user requirements and capabilities that are solution specific. It is also important to remember that the ORD is going to be the source of all testing requirements and that testing will need to show traceability back to the ORD.

The ORD should be written to describe the system concept; it should relate the shortcomings of existing systems and capabilities. For example, collateral damage to the civilian population can occur when trying to eradicate terrorists from urban areas. The basic problem may be that current EO sensors do not have the resolution necessary to correctly identify friend from foe. A sensor capable of xx.x mrad at a range of xx.x ft at an altitude of xxx.x ft in a defined urban environment is required. So, the ORD should describe the required performance capabilities. Additionally, the ORD should address infrastructure, interoperability, and integrated logistics support. The document should identify the proposed buy (quantities and spares) and when the system is needed, that is, schedule.

I tell all of my students that if you are writing specifications for anything, pretend that you are writing them for the dumbest person in the world; don't take for granted that the reader will be as familiar with the concept as you. It's very rarely the case. Also, if you want something to operate in a specific way write it as a requirement, or else it will be implemented the way the designer thinks is best. The ORD should be concise, noting the performance and related operational parameters for the proposed system; operational effectiveness and suitability issues should be included. It should be prepared and updated by the user and not by program or management. We expect the ORD to be done as early as possible in the program, remembering that it will drive the requirements for test and evaluation. The ORD is a living document and will be updated and perhaps expanded as the program progresses and more information is gained on technology and viable alternatives.

The ORD, as one might expect, follows a standardized format. A template that may be used for constructing an ORD may be found at http://www.dhs.gov/xlibrary/assets/ORD_Template.pdf. An example of a completed ORD may be found at http://www.fas.org/nuke/guide/usa/c3i/VLFORD.html. The required operational effectiveness and suitability capabilities must be identified as well as prioritized. Each requirement must be described in terms of a threshold, which is in reality a minimum acceptable level of performance and an objective, which is the desired level of performance. The ORD will identify the key system attributes, key system parameters, and critical operational issues. These will be covered in detail later, but suffice it to say that these parameters are so critical that a failure to meet the threshold on any of them would necessitate a reevaluation of the solution or termination of the program. Last, the ORD becomes the basis for developing the contractual system specification.

Since the ORD will be the basis for the contractual specification, the writer needs to pay attention to some items when formulating an ORD. For example:

- Each requirement should stand alone; do not combine requirements with *and, also,* and *with.*

- Do not write open-ended requirements that may be subject to interpretation; that is, don't use *etc.* within the requirement.

- Similarly, don't use the terms *except, if, when, unless,* or *but* as they provide ambiguity.

- The terms *shall, will,* and *must* make the requirements mandatory; the terms *may, might, could, perhaps,* and *probably* do not.

- Ambiguities and interpretation also come into play when using terms such as the *greatest extent possible, maximum* and *minimum, state-of-the-art, user-friendly, flexible, efficient, several, improved, adequate, adaptable,* and *simple.*

- The system should not be engineered; only the requirements should be stated.

The ORD should be concise, but it should be specific.

14.4 | THE TEST AND EVALUATION MASTER PLAN (TEMP)

The TEMP takes the requirements from the ORD, along with inputs from the program plan (cost, schedule, performance), and generates the test and evaluation plans and procedures. The TEMP, like the ORD, follows a specified format, and is a living document being updated as T&E progresses. A template for the TEMP may be found at http://www.scribd.com/doc/507305/Test-Evaluation-Master-Plan-Template. A good set of instructions for formulating a TEMP is found in the Department of the Army, Pamphlet 73-2, "Test and Evaluation Master Plan Procedures and Guidance," a copy of which may be found at http://www.if.uidaho.edu/.../Test%20and%20Evaluation%20Master%20Plan.pdf. An example of an FAA completed TEMP regarding next-generation systems can be seen at http://www.faa.gov/nextgen/portfolio/trans.../SE%20RNAV%20Temp.pdf.

The structure of the TEMP will follow the format:

- Objectives
- Evaluation criteria
- Data analysis
- Resources
- Responsibilities

– Decision milestones

– Schedules

Copies of the mission need statement, system threat assessment, and ORD are normally attached to the TEMP.

The scope of the TEMP provides a framework for all testing on the program, delineating both DT&E and OT&E with enough detail to generate test plans. The TEMP will document the schedule and resource implications associated with the test activities. For example, a word grabber or bus editor must be available and tested prior to first flight, or test team training must be accomplished 30 days prior to first flight. The TEMP will cover all test activities, starting with ground evaluations, modeling and simulation and progressing through combined contractor/DT and operational test. If it is done correctly, the TEMP will show traceability from the ORD to critical operational issues to critical technical parameters to the minimum operational performance requirements and finally the evaluation criteria. Successful completion of the chain will lead to the milestone decision points.

Just as we have seen previously, a good plan will identify all test objectives and the logistical support required to achieve success. Logistical support can include:

– Hardware

– Software

– Maintenance

– Training

– Instrumentation

– TSPI

– Analysis software

Testing will always require resources, some of which may be:

– Test articles with proven functionality

– Ranges (e.g., open air, anechoic chamber, live fire)

– Threat systems (real or simulated)

– Targets (e.g., air, ground, augmented, jamming)

– Simulators, test beds

– A trained test team

– A budget

Where we considered test objectives, test points and exit criteria in the DT&E phase, the OT&E phase will be structured with critical operation issues (COI), measures of effectiveness (MOE), and measures of performance (MOP).

▮ 14.5 | OPERATIONAL TEST AND EVALUATION

In chapter 1 of the text, we explained the basic differences between DT&E and OT&E. As a generalized statement we said that the purpose of DT&E was to ensure specification compliance and, from the contractor's standpoint, meet the terms of the contract.

OT&E ensures that the system can meet mission requirements and that the procuring agency gets what it paid for. We also stated that as a rule, the military DT&E group wears two hats: one to verify contractor compliance and another to ensure that the system is ready for OT&E.

The Defense Acquisition Guidebook defines DT&E as:

> "Test and evaluation conducted to evaluate design approaches, validate analytical models, quantify contract technical performance and manufacturing quality, measure progress in system engineering design and development, minimize design risks, predict integrated system operational performance (effectiveness and suitability) in the intended environment, and identify system problems (or deficiencies) to allow for early and timely resolution. DT&E includes contractor testing and is conducted over the life of the system to support acquisition and sustainment efforts."

Air Force Instruction 99-103 defines OT&E in two ways:

1. "The field test, under realistic combat conditions, of any item of (or key component of) weapons, equipment, or munitions for the purpose of determining the effectiveness and suitability of the weapons, equipment, or munitions for use in combat by typical military users; and the evaluation of the results of such test."

2. "Testing and evaluation conducted in as realistic environment as possible to estimate the prospective system's operational effectiveness, suitability, and operational capabilities. In addition, OT&E provides information on organization, personnel requirements, doctrine and tactics. It may also provide data to support or verify material in operating instructions, publications and handbooks."

Table 14.3 breaks down the differences between the two test organizations.

OT&E will normally not use contractor test personnel during their evaluation. In some cases, the OT&E team has designated some contractor personnel as *trusted agents* (the individuals will not divulge information on the tests performed or the results obtained). Many times, these trusted agents assist the team in Instrumentation and data collection or, in some cases, test setups and range control.

With the emphasis now on capabilities-based acquisition, it is even more important that the OT&E evaluation team assess system impacts to combat and peacetime

TABLE 14.3 ■ DT&E and OT&E Charters

DT&E	OT&E
Specification Compliance	Mission Performance
	System Suitability
Contractor Driven	User Driven
Technical Personnel	Operationally Current Personnel
Heavily Instrumented	Not required; but used in some cases
Development Intensive	Determines Capability Using Production Representative Systems Against Threat Representative Systems
Reports to Program Management	Reports to the Operational Authority
Works with OT&E	Relies on DT&E Findings

operations. OT&E is also tasked with determining the impacts of fielding or employing a system across a full range of operations in many diverse theatres. OT&E is tasked with determining the reliability, maintainability, and support for these systems as well as the ability of the maintainers to operate and work on these systems.

14.5.1 Types of OT&E

The documentation makes note of the differences between evaluations and assessments. Evaluations collect, analyze, and report data against stated criteria with a high degree of analytical rigor and are used to support full rate production (FRP) or fielding decisions. This wording is quite similar to the demonstration testing discussed in chapter 1 of this text. Assessments usually collect and analyze data with less analytical rigor, need not report against stated criteria, and cannot be the sole source of T&E data to support FRP or fielding decisions. These words are used in chapter 1 to describe OT&E. The type and level of OT&E testing varies from program to program and is decided by a consensus of the OT&E organization, user, and the responsible test organization (RTO). Operational testing is based on the requirement documents for the capabilities being fielded and can include COTS, nondevelopmental items (NDI), or GFE. The types of OT&E are many and varied:

- Initial OT&E (IOT&E): Within the USAF, this evaluation is conducted only by AFOTEC and is used to determine the operational effectiveness and suitability of the items under test using production representative hardware and software, stabilized performance, and operationally representative personnel. The tests are conducted under real-world conditions, including simulated combat conditions; it determines if the critical operational issues (COI) have been met (COI will be addressed in detail).

- Qualification OT&E (QOT&E): Like IOT&E, this evaluation is conducted only by AFOTEC within the USAF. It is used to evaluate the military applications of COTS, NDI, and GFE when little or no development takes place. The reader will remember from previous discussions that this is a trap as all programs will require development and integration. History is littered with programs that were qualification-only, which proved not to be the case.

- Follow-on OT&E (FOT&E): As the name implies, FOT&E is the continuation of IOT&E or QOT&E to answer questions of unresolved COIs or previous deficiencies; it is also conducted only by AFOTEC within the USAF. It may also complete T&E if those areas not finished during OT&E.

- Multi-service OT&E (MSOT&E): This evaluation can be I, Q, or FOT&E when two or more military services are involved. It may be called a force development evaluation (FDE) if it is performed and managed by a major command (MAJCOM). The FDE is used for multiple purposes:

  ~ Evaluation to verify the resolution of deficiencies or shortfalls

  ~ Evaluation of routine software modifications such as operational flight programs (OFP), or incremental software to support upgrades or improvements

  ~ Evaluation of shortfalls discovered after fielding the system

  ~ Evaluation of systems against foreign equipment or new or modified threats

  ~ Evaluation of COTS, NDI, or GFE for military applications

– Tactics development and evaluation (TD&E): TD&E is a specialized form of FDE conducted by the MAJCOMs to refine doctrine and tactics.

– Weapons system evaluation program (WSEP): WSEP is another specialized FDE conducted to provide an end-to-end validation of fielded weapons systems and their support systems using realistic combat scenarios. WSEP is also used to conduct investigative live firings to revalidate capabilities or better understand malfunctions.

– Operational utility evaluation (OUE): OUE is performed whenever an operational assessment is required, but a full scope evaluation is not appropriate.

14.5.2 Operational Assessments (OA)

The OA is conducted by AFOTEC or the MAJCOM in preparation for a dedicated OT&E period; the results are normally used to support the go-ahead for LRIP. They are supposed to be progress reports and not an evaluation of the overall mission capability of the system. An OA provides early operational data and feedback to developers, operators, and decision makers. The objectives of an OA are as follows:

– Assess and report on a system's maturity and the potential to meet operational requirements

– Support the decision for LRIP or increments of evolutionary acquisition programs

– Identify design problems and deficiencies that may prevent successful demonstration of overall mission capability

– Uncover potential system changes needed to update operational requirements

– Support the demonstration of prototypes, new technologies, or new applications

– Support proof of concept initiatives

It is important for operational testers to realize that these assessments are progress reports and not a declaration of mission capabilities. The author has seen on previous systems programs, during early operational assessments (OSA), a tendency to report on functions and capabilities that are not yet present in the test article.

▌ 14.6 ▐ OT&E TEST PLAN STRUCTURE

The OT&E test plan can be summarized by critical operational issues, critical technical parameters, measures of effectiveness and suitability, and measures of assessment and performance. These parameters provide the structure for all OT&E evaluations:

Critical operational issues (COIs): These are the top-level issues that must be examined in OT&E to determine the system's capability to perform its mission. COIs are included in the TEMP and derived from the ORD. There are two categories of COIs: effectiveness and suitability. COIs are typically phrased as a question:

– Effectiveness: "Will the system detect the threat in a combat environment at adequate range to allow successful engagement?"

– Suitability: "Will the system be safe to operate in a combat environment?"

The CDD is the primary source for developing COIs.

There is quite a difference between effectiveness and suitability; if a system performs its function perfectly it is said to be effective. If the system cannot be maintained, it will not be suitable. A classic example is the emergency power system selected for the F-16. A backup system was required to provide hydraulic/electrical power when the engine failed. The lightest and most compact solution was a hydrazine-powered gas generator. Hydrazine is a highly toxic hazardous material and difficult to handle at home station and becomes a huge challenge when deployed off-site. The decision to proceed with the hydrazine system was good for effectiveness but not for suitability.

Operational suitability can be described as the degree to which a system can be satisfactorily placed in the field. Issues that should be considered when evaluating a system's suitability are as follows:

- Availability
- Compatibility
- Transportability
- Interoperability
- Reliability
- Normal and surge usage rates
- Maintainability
- Safety
- Logistics supportability
- Manpower requirements
- Spares
- Documentation
- Training
- Sustainment
- Simulators
- Software
- Human factors

Although the meaning of most of these parameters is obvious, some require some discussion when addressing an operational suitability requirement. *Reliability* is the ability of a system to perform its mission without failure, degradation, or excessive demand on the support system. *Operational reliability* can be thought of as the probability that an operationally ready system will work as required to accomplish its intended mission or function as planned. *Maintainability* is the ability of an item to be retained in or restored to specified condition when personnel with specified skill levels maintain it using prescribed procedures and resources, at each prescribed level of maintenance and repair. *Availability* is the ability of a system to perform its mission when needed. When dealing with hardware, availability is a function of MTBF, MTTR (mean time to repair), MTTF (mean time to failure), and FIT (failure in time). *Compatibility* is the capability of two or more items or components of equipment or material to exist or function in the same system or environment without mutual interference. *Interoperability* is the ability of systems, units, or forces to provide services to and accept services from other systems, units, or forces and to use these services to help them operate effectively together. *Supportability* is the degree to which system design characteristics and planned logistics resources, including manpower, meet system peacetime readiness and wartime utilization requirements. *Logistics Supportability* is the degree to which planned logistics support allows the system to meet its availability and surge (a period of higher than normal utilization or operation, normally over a relatively short period of time) usage requirements. *Planned logistics support* would include: test, measurement, and diagnostic equipment, spare and repair parts, technical data, support facilities, transportation requirements, training, manpower, and software. *Sustainment* is the ability to maintain a necessary level and duration of operational activity to achieve user objectives in both peacetime (readiness) and wartime (sustainability).

Critical technical parameters (CTPs): These are the engineering design factors that a system must meet or exceed to ensure that established performance thresholds are achieved. CTPs are derived from the CDD, critical system characteristics, and systems engineering documents. CTPs are listed in a matrix, along with the performance objectives and thresholds in Part I of the TEMP.

An example of a CTP may be the weight of the system. The CTP is stated as a specific numerical value and may change as the system matures during development. The system should show improvement in the performance parameter over time.

Measures of effectiveness and suitability (MOEs & MOSs) - Part III of the TEMP lists the measures of effectiveness (MOE) and measures of suitability (MOS). They are key in assessing the performance capabilities and characteristics identified in the CDD. These measures (MOEs and MOSs) are used to determine the attainment of the top-level performance issues.

Examples of an MOE may be maximum effective range or land mobility speed. Examples of MOS may be interoperability or mean time between operational mission failures.

Measures of assessment and measures of performance (MOAs and MOPs): These are the data collected to show suitability or compliance with the MOE/MOS.

The best description of the interrelationship of the COI/CTP/MOE/MOS/MOA/MOP is found in chapter 4 of the *AFOTEC OT&E Guide,* 6th ed., February 12, 2009. The guide provides readers with examples of each parameter and how they show traceability back to the ORD. A companion to the OT&E guide is the "AFOTEC Test Director's Toolkit," April 12, 2010, which provides a checklist of OT&E activities and events in an acquisition program.

Coming from the development test world, the OT&E flow might be more understandable if related to a typical DT&E flow previously described. Figure 14.4 compares the OT and DT test process.

14.6.1 Critical Operational Issues

COIs describe key operational effectiveness or suitability issues to determine a system's capability to perform its mission. They are selected from list of system operational issues and usually are stated as a question. Should the evaluator fail to answer the question in the positive the system is not worthy of procurement. For this reason, COIs must be written with care and specificity as the outcome of the program depends on them.

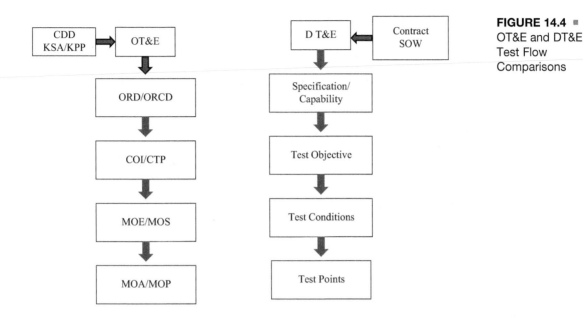

FIGURE 14.4 ■
OT&E and DT&E
Test Flow
Comparisons

Let's assume that a requirement exists for an ILS Cat 1 minimums landing system to be used by transport aircraft worldwide. The basic COI could read: Can the proposed system provide landing guidance down to ILS Cat 1 minima? A second COI may read: Is the proposed system available worldwide. A COTS WAAS-enabled GNSS would satisfactorily meet the requirements of COI#1, but would fail the requirements of COI#2. The way the COIs are written automatically discards the WAAS system as a candidate (and, in fact, any other known system available today). COI#2 could have been written to ask if it were feasible to develop a portable landing system based on the WAAS system to be used on remote, unimproved airstrips.

Some organizations relate COI development to a series of questions containing key elements of concern:

- Do the COIs reflect the user's priorities?
- Can all COIs be answered by a yes or no?
- Are all COIs supported by MOE/MOS and MOAs?
- Are COIs consistent with the strategy-to-task methodology?
- Are COIs specific to the mission tasks the system must perform?
- Are all COIs consistent with the primary purpose of the OT&E?
- Can the COIs, as stated, be answered through OT&E?
- Do any of the COIs overlap?
- Are the COIs consistent with the existing or planned operations and maintenance concepts, policies, procedures, and techniques?
- Are new policies and procedures necessary for the test and are they properly factored into the COIs?
- Do COIs address areas of interface with other end user systems?

Strategy-to-task methodology as well as MOE and MOA will be addressed shortly.

14.6.2 Measure of Effectiveness/Measure of Suitability

The Measure of Effectiveness (MOE) is a qualitative or quantitative measure of a system's performance or a characteristic that indicates the degree to which it performs the task or meets a requirement under specified conditions. MOEs should be constructed to measure the system's capability to produce or accomplish the desired result. The Measure of Suitability (MOS) is similar to the MOE in that it is a qualitative or quantitative measure of a system's capability to support mission or task accomplishment with respect to reliability, availability, maintainability, transportability, supportability, training, and other suitability considerations. In our previous example, the WAAS-enabled GNSS may provide a 95% success rate in CAT 1 landings (MOE), but the system is not transportable (MOS).

The MOE/MOS can be thought of as a measure of a system's intended task or mission accomplishment in terms of representative operational scenarios. These scenarios are usually at the engagement or battle outcome level (CAT 1 Landings at an unimproved airstrip). The evaluators should ask the basic question: "How well did we achieve what we wanted?" The MOE/MOS must be quantifiable, relevant to COIs, and measurable/testable.

14.6.3 Measure of Assessment/Measure of Performance

The Measure of Assessment (MOA)/Measure of Performance (MOP) is a quantitative measure of a system's capability to accomplish a task. These measures are typically in the area of physical performance (e.g., range, velocity, memory, throughput). They are a determination of the system's actual capabilities or characteristics in real world operations. They are usually quantitative but can be qualitative and measure the degree to which a system performs or meets a requirement or specification. The MOA/MOP must show traceability to a MOE/MOS and then back to a COI. As opposed to "How well did we achieve what we wanted?" we now would ask "What was achieved?"

The way the MOA/MOPs are written will have a bearing on which testing is required. If the term *demonstrate* is used it implies the test must show operation/function under operational conditions. If the terms *determine* or *measure* are used, actual system measurement in operational conditions is required. Conversely, if the terms *assess* or *evaluate* are used, a subjective estimate is expected. *Verify* requires quantitative data against the ORD under operational conditions, whereas *validate* produces quantitative data against a prediction under operational conditions.

As noted previously for COIs, organizations may relate MOE and MOA development to a series of questions containing key elements of concern:

– Are all MOE related to the COIs?

– Are the MOA linked to the MOEs? To the COIs?

– Are MOE/MOA quantitative where practicable?

– Do the MOE/MOA address the user requirements?

– Are MOE/MOA feasible and executable in terms of time, cost, and resources?

– Can any of the MOE/MOA be satisfied by answering other COI or through add-on data collection performed on a noninterference basis?

– Which MOE/MOA can be satisfied by field testing, and which must be addressed by modeling or a combination of field testing and modeling?

– Do MOE/MOA address the information and data requirements in controversial areas?

– Can any of the tests be streamlined and still satisfy data and information requirements?

– What shortfalls and limitations exist in addressing the requirements? What are the impacts? Are there any workarounds?

– Do MOEs/MOAs call for evaluation of maintenance and other logistics requirements?

– Are there unique support aspects related to the system that should be integrated into the MOEs/MOAs?

– Are RML&A issues properly considered?

– Is software supportability and maturity sufficiently addressed?

– Are databases current and accurate for the users needs?

– Are support issues such as training (both maintenance and flight crew), simulators, mission planning system, technical data, human factors, safety, and integrated diagnostics sufficiently addressed?

– Are the methods of determining the impacts of support issues on system effectiveness and suitability adequately described?

– Are compatibility aspects properly considered particularly the conditions between other systems and equipment involved?

– Are environmental compliance aspects properly factored into the OT&E?

The nomenclature differs from organization to organization around the world, but the basic tenets are the same. Table 14.4 compares the terminology in use by three military organizations.

14.6.4 Developing the Test Plan

The test plan will be composed of the COI and the measurements discussed in the previous section. The most common way of developing the plan is the strategy-to-task

TABLE 14.4 ■ Comparison of OT&E Terminology

The United States	Canada	Australia
Initial Capabilities Document (ICD)	Statement of Intent (SOI)	Mission Need Statement (MNS)
Capability Development Document (CDD)	Statement of Requirements (SOR)	Statement of Requirements (SOR)
Attributes	Operational Issues (OI)	Operational Issues (OI)
Key System Attributes (KSA)		
Key Performance Parameters (KPP)		
Critical Operational Issues (COI)	Critical Operational Issues (COI)	Critical Operational Issues (COI)
Measures of Effectiveness (MOE)	Measures of Effectiveness (MOE)	Measures of Effectiveness (MOE)
Measures of Suitability (MOS)	Measures of Suitability (MOS)	Measures of Suitability (MOS)
Measures of Assessment (MOA)	Measures of Assessment (MOA)	Measures of Assessment (MOA)
Measures of Performance (MOP)		Measures of Performance (MOP)

method. In this process the task definition is related to OT&E (i.e., first define the task and then go about the methods for accomplishing the task). This method is sometimes called the top-down method as it is initiated with national doctrine or strategy. The tasks provide the critical linkage between the need statement (top), the operational requirements document (ORD), the COI, MOE, MOS, and MOA, and finally OT&E (bottom).

If we take the time frame of Desert Storm and propose a hypothetical need, the process can be illustrated. First, the task is derived from the national strategy; the national security strategy may have been to support our Middle East allies. The national military strategy may have been to free Kuwait. Within the theatre, the regional military strategy may have been to neutralize/isolate enemy ground forces. The operational objective would now be to disrupt enemy command and control; this, in turn would become COI #1.

The task (MOE) for this COI may have been to destroy enemy command and control centers and communication links. The assessment (MOA) of how well we could achieve this objective could dependent on many things:

– System functions/characteristics

– Performance

– Endurance

– Payload

– Navigation

– Target identification

– Warhead effects

The outcome of this exercise would then become the test matrices, showing traceability between the need and the assessment. Figure 14.5 shows an example of a completed matrix, which is taken from the AFOTEC OT&E Guide.

A second method for developing the structure of the plan is to use the systems engineering approach, sometimes called the bottom-up approach. In this method, the evaluators concentrate on the subsystem engineering performance characteristics; for example, the performance of a litening target designator. Resolution, range, discrimination, and performance in weather may be evaluated and then applied to the broader requirement, which may be identification and eradication of hostiles in an urban area. Engineers tend to prefer this easier method, but it's possible to lose sight of a user's initial need statement.

14.6.5 Executing the Plan

Depending on the complexity of the test scenarios, executing the plan may be more difficult than plans developed for DT&E. One of the key tenets of OT&E is that capabilities are evaluated in real-world, operational conditions. The choice of where the testing will be done can be a daunting task. A budget and a timeline for task completion must still be adhered to. We still can't schedule weather, and the system will still need to be supported wherever the testing is scheduled. The points made in chapter 1 of this text still hold true for OT&E.

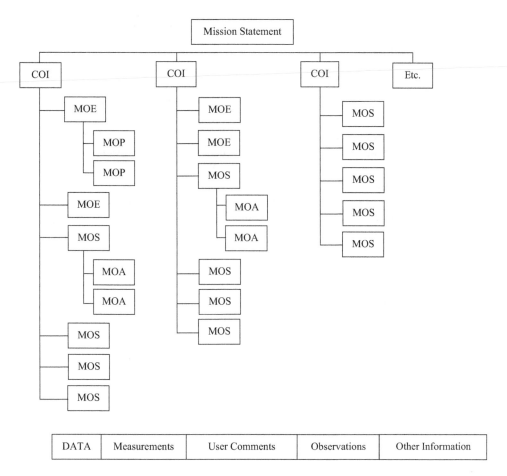

FIGURE 14.5 ▪
OT&E Evaluation
Structure

Data are identified within the measure or in the measurement criteria (Figure 14.5). When reviewing the OT&E structure ensure that the data required for evaluation can indeed be collected. As we have seen with DT&E, instrumenting aircraft to obtain data can be a costly and, more important, a lengthy process. If rating scales are to be used, evaluators will have to be trained in their use. If contractors are used to assist in data collection, they will need to be identified and approved.

Ranges will also have to be coordinated and scheduled; there is a finite amount of time as well as assets. At a minimum, ranges will need some sort of support agreement documentation as explained in chapter 2 of this text. The reader will recall that scheduling of specific assets (e.g., anechoic chamber, ground and airborne targets, emitters, clearance for chaff and flare deployment, RF radiation) must be accomplished well in advance of their intended use. Six months is probably not unrealistic for most events, although live fires may be significantly more coordination time. Test review boards (TRB), flight test readiness reviews (FTRR), and safety review boards (SRB) will all be required by the support facility.

Chapter 4 of the AFOTEC OT&E Guide provides the reader with test support considerations for OT&E evaluations as well as assessments. It should be noted that all of the cited documentation encourages joint or combined test teams during any program. The author has stated some of the pitfalls to these arrangements in chapter 1 of the text.

■ 14.7 | RELIABILITY, MAINTAINABILITY, LOGISTICS SUPPORTABILITY, AND AVAILABILITY (RML&A)

For a system to effectively accomplish its mission tasks, it must be in an operable condition when needed. Operational effectiveness determines if the system can accomplish the mission whereas operational suitability determines the ability to field and support the system. Complex interrelationships exist between reliability, maintainability, and logistics supportability, which combine to produce a system's operational availability (which in reality is suitability). The best system in the world cannot be effective if its availability rate is 50%.

For information, the reader is referred to AGARD-RTO-Vol-300, chapter 22, "Reliability and Maintainability," and chapter 23, "Logistics Test and Evaluation." Chapter 22 is from the perspective of the USAF at Edwards AFB, and chapter 23 is written from the USN perspective at Patuxent River.

RML&A measurements can be difficult and require a great deal of data. It's not unusual for the suitability portion of the OT&E program to account for over 75% of the testing. A correct assessment can identify system deficiencies, correct the deficiencies in a timely fashion, enhance system capabilities, determine how to best support the system under operational conditions, and help determine the number of spares required, manpower to adequately support the system, and training required for the maintainers. In the perfect world, the system under test would have a high MTBF, is easily and quickly repaired, requires unskilled labor to fix it, and is easy to deploy with limited mobility needs. This section will not present an in-depth explanation of the types of analysis, modeling and simulation used in performing RML&A evaluations. It will attempt to highlight the areas where testers should be knowledgeable enough to have an intelligent conversation with their maintenance counterparts.

The operational user should state the RML&A system requirements in terms that can be measured during test and evaluation (e.g., downtime, weapons upload time, surge). These measures are going to be dependent on the maintenance concept selected for the system. For example, is the system expected to be contractor maintained, organic, or a mix of the two? What will be the operational maintenance environment (basing concept, climate, weather, and deployment scenarios)? What, if any, are the maintenance constraints (electronic test equipment, occupational health and safety, nuclear hardness, NBC gear)? The concept is the same as described in DT&E, where it was stressed that the evaluator needs to perform the test with the same type of equipment that the operational pilot would use (i.e., wearing gloves when conducting a cockpit evaluation).

Other considerations when developing the MOS are as follows:

– Spares, technical data, and support equipment
– Special facilities, special training, special tools
– Types and number of maintenance personnel
– Required skill levels
– Mockups, training aids, and simulators (ops & maintenance)
– Automated test equipment
– Software maturity

OT&E, and in particular IOT&E, face significant limitations when conducting the RML&A evaluation. The primary reason for this is that the system being evaluated is not mature. For the most part, IOT&E will be conducted on an asset that is still in the throes of a development program. The test environment is different from intended operational environment and is still dependent on contractor support. There will probably be a lack of technical data or immature technical data. There will be limited time to conduct the assessment, and there will be a limited number of test articles. One of the more unfortunate outcomes can be an unfavorable rating of the system by evaluators who do not understand the functional capabilities of the system in its present state.

As with the previous sections, RML&A uses specific definitions when defining suitability. Giadrosich's text covers the subject very well, and he has culled out the specific terminology that evaluators should be familiar with. All of the terms have commonality in that four characteristics are a common theme: probability, satisfactory system performance, time, and conditions of use.

Reliability is a measure of a system's capability to satisfactorily perform required functions under stated conditions for a specific period of time. It entails a measure of human and machine performance and the probability of successful performance. Critical parameters that determine reliability are mean time between critical failure and the probability of a loss of a mission essential function. As with CFR 14, 1301, studied during the civil certification section, a failure is the loss of an intended function. Because reliability cannot be measured accurately during DT&E and OT&E due to system maturity and time constraints, evaluators often depend on a term called *reliability growth,* which is projected by the use of modeling and simulation. The results are only as good as model effort and how well the modelers understand how the system will be used. The input to the model is the reliability data gathered during test.

There are three types of reliability projection estimates:

– Demonstrated/observed reliability

– End-of-test reliability

– Projected (mature system) reliability

Demonstrated/observed reliability is system reliability information directly obtained during the DT&E and OT&E test programs. The results obtained during OT&E will be different from those obtained from DT&E. The reasons for this discrepancy are that OT&E counts all failures equally, even if a particular failure mode has been identified and corrected, and the failure is not expected to occur again. *End-of-test reliability* is also called current or instantaneous reliability. It is the demonstrated reliability as adjusted by agreed upon growth factors or margins. *Projected (mature system) reliability* uses agreed upon models and growth factors that extend or project end-of-test reliability forward to a mature system. It assumes a continuation of the system design corrections and improvements to correct identified system deficiencies. It also assumes that the implemented fixes will work; FOT&E data are used to validate fixes.

Maintainability of a system is the ability to return or restore the system to an operable state. It is really the cost of care and feeding for the system. Maintainability looks at manpower and skill levels required to repair the system and can be a significant cost when calculating life-cycle costs. Terms associated with maintainability are mean down time (MDT) and mean time to repair (MTTR). Reliability and maintainability are interrelated insofar as their effect on operational availability and system effectiveness.

Cost-benefit analysis of the system looks at both acquisition cost and maintenance cost; a system may be very reliable but with a high acquisition cost, a competing system may cost less but is less reliable system or has higher maintenance costs.

Logistics supportability is the degree to which the planned logistics and manpower meet system availability and usage needs. It is measured during required maintenance actions using trained user operators and maintenance personnel. Actions requiring an inordinate amount of time will obviously have a direct impact on system availability.

Availability is the degree to which a system is in an operable state. It is different from reliability in that reliability depends on a time interval and availability is at an instant of time (i.e., is the system operable when needed). Availability will depend upon how often the system is inoperable and how quickly it can be repaired. Availability is expressed in operational terms:

– Uptime ratio, availability = uptime/total time
– Inherent availability
   Ai = Successful operate time/(successful operate time + active repair time)
   $Ai = MTBF/(MBTF + MTTR)$
– Operational availability, all segments of time the system is intended to be operational.
   $Ao = MTBM/(MTBM + MDT)$

where *MTBF* is mean time between failure, *MTTR* is mean time to repair, *MTBM* is mean time between maintenance, and *MDT* is mean down time.

Other measures are used by the RML&A community to help describe availability and sustainability and are mentioned here for the sake of completeness:

– Mission Capable (MC) Rate – The sum of: Fully Mission Capable (FMC) (that is the system can perform all aspects of the mission) and Partially Mission Capable (PMC) (that is the system can perform at least one assigned wartime mission)

$$MC = FMC + PMC$$

– Utilization Rate (UR) – Defined as flight hours or sorties per aircraft per relevant period of time, normally calculated for peacetime and wartime utilization rates

$$UR = \text{number sorties per day/number aircraft}$$

– Break Rate (BR) - Measures the percentage of sorties from which an aircraft returns with an inoperable mission essential system; break rate includes both air and ground aborts

$$BR(\%) = \text{Number of breaks/Number of sorties flown}$$

– Combat Rate (CR) - Measures the number of consecutively scheduled missions flown before an aircraft experiences a critical failure

$$CR = \text{Number of successful sorties/(scheduled sorties − ground aborts − air aborts)}$$

– Mean Time Between Critical Failure (MTBCF) – Measures the average time between failures of mission essential system functions (can be hardware or software)

$$MTBCF = \text{Number of operating hours/Number of Critical Failures}$$

▌ 14.8 ▌ | SUMMARY

In DT&E, the evaluators are tasked to determine if the system meets the requirements of the specification, the stated capability or the terms of the contract. In fact, the majority of the time spent in DT&E is to get the system to perform its intended function. DT&E progresses from the development to demonstration phases of test until enough data are collected to statistically prove that the requirements have been met. OT&E ensures that the system is effective and suitable in real-world scenarios under operational conditions. The test methodology is different, and more importantly the evaluators need to know how the system will be utilized operationally as opposed to operations in a sanitized test world.

▌ 14.9 ▌ | SELECTED QUESTIONS FOR CHAPTER 14

As chapter 14 is really a study of a process and can vary from organization to organization around the world, there are no questions for this chapter. For generalized Test and Evaluation fundamentals, the reader is referred back to chapter 1.

Index

A

A380 121
A-32767 244
AAA: *see* antiaircraft artillery (AAA)
ABC: *see* auto brightness control (ABC)
ABF: *see* annular blast fragmentation (ABF)
above ground level (AGL) 52
    altitude 173
ABSAA system: *see* airborne sense and avoid (ABSAA) system
absorption filters 930
AC: *see* advisory circular (AC); alternating current (AC)
AC 20-131A 238
AC 20-138 330, 333
AC 20-138A 330
AC 20-138C 330
AC 20-158 157
AC 20-165 377, 378
AC 20-168 323
AC 20-172A 377
AC 20-182 323
AC 23-8B 329
AC 23-8C 323, 324, 329
AC 23-18 382, 391, 395
AC 25-7C 323, 324, 329
    weather radar tests 324–9
AC 25-23 383, 391, 395
AC 25.899-1 157
AC 27-1B 324, 383
AC 29-2C 324, 383
AC 90-105 330
ACAS: *see* active collision and avoidance systems (ACAS)
acceleration 46
    and velocity accuracy 205–6
accelerometer 191
    types of 191
Acceptable Means of Compliance (AMC) 300
accuracies
    GPS 223–4
    positional 223
accuracy tests, air-to-ground 803
AC 20-138D 330
ACJ: *see* Advisory Circular Joint (ACJ)

ACK field: *see* Acknowledge (ACK) field
acknowledge (ACK) field 86
ACLS: *see* automatic carrier landing system (ACLS)
ACM: *see* air combat maneuvering (ACM)
ACO: *see* Aircraft Certification Office (ACO)
acquisition programs 958–65
    affordability and performance measurements 962
    operations and support 964
    phases of 959
    technology development 962–3
    user needs 960, 961
active collision and avoidance systems (ACAS) 893
active detection 729
active EO systems
    boresight accuracy 614
    laser 614
        beam divergence 616–17
        output power 617
        pulse amplitude 617–18
        pulse repetition interval 618
        pulse width 618
        ranging accuracy 615–16
        see also Laser system
    line-of-sight (LOS)
        drift rate 615
        jitter 615
active-matrix LCD (AMLCD) 946
AD: *see* airworthiness directives (AD)
Ada 75
ADC: *see* air data computer (ADC)
ADF: *see* automatic direction finder (ADF)
ADL: *see* avionics development lab (ADL)
ADS-B 893
advanced medium-range air-to-air missile (AMRAAM) 764, 772
    operations 163–4
advanced range data system (ARDS) 224
Advanced Safe Escape Program (ASEP) 816
Advanced Tactical Airborne Reconnaissance System (ATARS) 582

advanced targeting FLIR (AT-FLIR) 544
advanced targeting pod (ATP) 544
advisory circular (AC) 238, 300
Advisory Circular Joint (ACJ) 300
Advisory Group for Aerospace Research and Development (AGARD) 536, 625, 832
    AGARD-AG-300, volume 10 803–4
    flight test techniques series 820
    separation effects testing 820
AEEC: *see* Airlines Electronic Engineering Committee (AEEC)
AEG: *see* Aircraft Evaluation Group (AEG)
AEIS: *see* aircraft/store electrical interconnection system (AEIS)
Aeronautical Information Manual (AIM) 384
aerotriangulation 42
    by arcs of range 43
    cinetheodolites 42
AFCSS: *see* armament and fire control system survey (AFCSS)
AFDX: *see* avionics full-duplex switched Ethernet (AFDX)
AGARD: *see* Advisory Group for Aerospace Research and Development (AGARD)
AGC: *see* automatic gain control (AGC)
AGL: *see* above ground level (AGL)
AHRS: *see* attitude heading reference system (AHRS)
AIF: *see* avionics integration facility (AIF)
AIL: *see* avionics integration lab (AIL)
AIM-9 801
AIM-120 801
AIM: *see* Aeronautical Information Manual (AIM)
AIM-54 A/C 787
AIM-120 AMRAAM 775, 778, 779
AIM-54C Phoenix 163–4, 801
AIM-9 Sidewinder 764, 772
AIM-7 Sparrow 163, 764, 771, 772, 774–5, 776, 777
AI radars: *see* airborne intercept (AI) radars

airborne early warning radar parameters 649
airborne intercept (AI) radars 726–7
airborne radar display processor (ARDP) 654
airborne radar signal processor (ARSP) 654
airborne sense and avoid (ABSAA) system 896–7
airborne targets 566, 567
airborne warning and control system (AWACS) 729
Airbus 121
air combat maneuvering (ACM) 689
    modes 651, 691, 692
Aircraft Certification Office (ACO) 275
Aircraft Evaluation Group (AEG) 277
aircraft(s)
    attitude 45
    clocks 53
    communications equipment 142–4
    designers 640
    exterior lighting 929
    glinting and HARM 837–8
    NVIS flight tests 938
    NVIS lighting ground tests 935–8
aircraft/store electrical interconnection system (AEIS) 100
Aircraft Stores Interface Manual (ASIM) 813
aircrews, responsibilities of 845
air data computer (ADC) 74
Air Force Operational Test and Evaluation Center (AFOTEC) Test Director's Toolkit 957
airframe types, for air-to-air missiles 786
air-launched antiradiation missiles (ALARMs) 721
Airline Deregulation Act of 1978 270
Airlines Electronic Engineering Committee (AEEC) 87
Airport Facility Directory 187
airport surveillance radars (ASRs) 728
airspeed 7, 172–3
air-to-air missiles
    airframe types 786
    basic configuration 785–9
    coordinate systems 786
    lab and ground tests
        fuse tests 792
        launch acceptability region determination 793
        radar cross section measurements 792–3
        seeker tests 791–2
        trajectory simulations 791–2
        warhead tests 792
    launcher qualification tests 793–6
    missile coordinate system 787

modes of 787
navigation 787
propulsion section profiles 787
reliability and maintainability 796–7
systems test 797–803
test and evaluation 789–91
warhead section 787
air-to-ground accuracy tests 803–4
    see also Air-to-ground weapons
air-to-ground range resolution test 695–6
air-to-ground ranging radar 694–5
    air-to-ground radar test 695–6
    ground moving target indicator (GMTI) mode 696–7
air-to-ground release geometry 806–12
    continuous computing impact point (CCIP) 808
    manual weapons delivery 806–9
    modern weapons delivery 807–8
air-to-ground weapons 803
    accuracy tests 803
    horizontal plane bomb trajectory 805
    minimum release altitude 811–12
    operational flight program weapons accuracy testing 820–1
    sources of errors 809–11
    vertical plane bomb trajectory 804
air traffic control (ATC) mode 235
Air Traffic Control Radar Beacon System (ATCRBS) 234–5
air traffic management (ATM) 330
airways, defined 175
airworthiness
    enhanced flight vision systems (EFVS) 523
    flight guidance systems (Part 23) 412–18
        cockpit control 413–14
        compliance, alternate means of 417–18
        flight control recovery 418
    flight guidance systems (Part 25) 418
    flight management systems (FMS) 506–7, 507–8
        AC 20-167 523
    landing systems
        category I systems 471–9
        category II systems 479–88
        category III systems 488
        for landing and rollout in low visibility 495–503
        for takeoff in low visibility 489–95
    malfunction evaluations 414, 416
    maneuvering and approach malfunction 416–17
    normal flight malfunctions 415

    performance evaluation 418
    reduced vertical separation minimums (RVSM) 352–4
    single-engine approach 418
airworthiness, positioning systems: see positioning systems airworthiness
airworthiness certificate 857–8
    military airworthiness requirements 859–60
airworthiness directives (AD) 273
airworthiness release (AWR) 825
ALARMs: see air-launched antiradiation missiles (ALARMs)
alignment types, for INS 200–2
Allied Signal Bendix/King KRA-10A Radar Altimeter 174
all-up-round test (AURT) 797
ALQ-167 countermeasures set 750
alternating current (AC), defined 155
altitude 32, 52, 173–4
    above ground level (AGL) 173
    flow chart 80
    mean sea level (MSL) 173
    measurement 173–4
    relationships 52
    return 672, 673
AM: see amplitude modulation (AM)
AMC: see Acceptable Means of Compliance (AMC)
American National Standards Institute (ANSI) 108
American Radio Inc. (ARINC) 62, 87
American Standards Association (ASA) 152
AMLCD: see active-matrix LCD (AMLCD)
amplitude modulation (AM) 143
AMRAAM: see advanced medium-range air-to-air missile (AMRAAM)
analog data 839
analog-to-digital (A/D) converter 64, 77
analog TV systems 583–5
    camera tubes 583
        orthicon 584–5
        vidicon 583–4
    CRT 583
analysis and reporting, of flight test program 848–9
    in real time 848
    software routines 848
    statistical analysis 848
    video obtaining ways 848
    writing report 848
    see also Report writing
AN/APX-109 238
AN/AVS-6 915
AN/AVS-9 915
AN/DLQ-3C 750
anechoic chamber 50–1

angle deception 739, 741
angle discrimination 747
angle-on-jam (AOJ) 727, 746
ANL: *see* automatic noise leveling (ANL)
annular blast fragmentation (ABF) 788
annunciators 927
    LEDs 928
ANSI: *see* American National Standards
        Institute (ANSI)
antenna beamwidth *versus* resolution 666
antenna manual positioning 747
antenna pattern testing 149–50
antennas 142
antenna switching, radio aids testing and
        188–90
antiaircraft artillery (AAA) 720
antiradiation missile (ARM) 721
AN/ULQ-21(V) 750
ANVIS: *see* aviators night vision imaging
        system (ANVIS)
AOJ: *see* angle-on-jam (AOJ)
aperiodic synchronization gap (ASG) 92
APG-XXX radar 655
ARAC: *see* Aviation Rulemaking
        Advisory Committee (ARAC)
arbiter 53
arbitrated loop topology 109–10, 113
ARC: *see* automatic radio compass
        (ARC)
ARC-XXX 144
ARDP: *see* airborne radar display
        processor (ARDP)
ARDS: *see* advanced range data system
        (ARDS)
area navigation (RNAV) 336
    B-RNAV 329
    defined 329
    P-RNAV 329–30
ARINC 429 86–9
    bus-drop topology 87
    multiple bus topology 88
    RTZ tristate modulation 87–8
    star topology 87
    word format 88
ARINC 629 89–93
    basic protocol aperiodic mode 92
    basic protocol periodic mode 91
    combined protocol 91–2
    message structure 90
    word string 90
ARINC 629-2 90
ARINC 629-3 90
ARINC: *see* American Radio Inc. (ARINC)
ARM: *see* antiradiation missile (ARM)
armament and fire control system survey
        (AFCSS) 825
ARN-118 144
ARSP: *see* airborne radar signal processor
        (ARSP)

ARTCC system 877
artificial aperture 728
ASA: *see* American Standards
        Association (ASA)
A scope 651
ASEP: *see* Advanced Safe Escape
        Program (ASEP)
ASG: *see* aperiodic synchronization gap
        (ASG)
ASIM: *see* Aircraft Stores Interface
        Manual (ASIM)
ASN-130 144
ASRs: *see* airport surveillance radars
        (ASRs)
Astec Z-12 48
ASTM International 894
ATARS: *see* Advanced Tactical Airborne
        Reconnaissance System
        (ATARS)
AT-FLIR: *see* advanced targeting FLIR
        (AT-FLIR)
ATC mode: *see* air traffic control (ATC)
        mode
ATCRBS: *see* Air Traffic Control Radar
        Beacon System (ATCRBS)
ATM: *see* air traffic management
        (ATM); automatic teller machine
        (ATM)
atmospheric absorption *versus*
        frequency 629
atmospheric effects
    radio aids testing and 190
atmospheric noise 142
atmospheric propagation of radiation
        556–62
    aerosols 557
    carbon dioxide 557, 558
    extinction coefficient 556
        absorption and scattering 556
        Beer–Lambert law for
            transmittance 556
    gases 557
    long-wave infrared (LWIR) 558, 562
    LOWTRAN model 561
    mid-wave infrared (MWIR) 558, 562
    modifying phenomena in 556
    path radiance 557
    refractive index fluctuations 557
    transmittance 558–60
atomic frequency standard (AFS)
        anomaly on satellite 23 217–18
ATP: *see* advanced targeting pod
        (ATP)
attitude heading reference system
        (AHRS) 705
AURT: *see* all-up-round test (AURT)
Australian ground-based regional
        augmentation system (GRAS)
        229, 332

autoacquisition mode 651
auto brightness control (ABC) 914
autoguns 651
automatic carrier landing system
        (ACLS) 239
Automatic Dependent Surveillance
        Broadcast (ADS-B) 374–382
    ADS-B Certification 377
    ADS-B Out Airworthiness 377–379
    ADS-B In Airworthiness 379–382
automatic direction finder (ADF)
    defined 183
    loop antenna 183
    operation of 183
automatic gain control (AGC)
        213, 746
automatic noise leveling (ANL) 746
automatic radio compass (ARC) 183
automatic teller machine (ATM) 580
autopilots, UAV 889
    auto throttles 891
    evaluations 889–91
    malfunction testing 890
    submenu 890
    USAR considerations 891–2
auto-takeoff and auto-land systems 887
availability 971, 980
average power 638
average power *versus* peak 638
Aviation Rulemaking Advisory
        Committee (ARAC) 272
aviators night vision imaging system
        (ANVIS) 915
avionics development lab (ADL) 14
avionics flight tester 21
avionics full-duplex switched Ethernet
        (AFDX) 121
avionics integration facility (AIF) 14
avionics integration lab (AIL) 14
Avionics Navigation Systems, 2nd
        edition 167
avionics test 10
    programs 8–9
    requirements 833–5
AWACS: *see* airborne warning and
        control system (AWACS)
AWACS Extend Sentry, B-1 115
awareness and warning system (TAWS)
    FLTA 383
    functions 383
    PDA 383
AWR: *see* airworthiness release (AWR)
azimuth
    accuracies 37, 666–7, 688
    defined 32
    resolution 665, 695
        geometry 667
        sample data card 668
        tests 667

B

background noise 640
backscatter 686
backscattering coefficients 694, 701
backup, PSSA 293
ballistics accuracy decision tree 819
bank error 184
barometric altimeter 174, 333
barometric altitude 45
baro-VNAV system 340, 342, 347
    flight test 347
    ground test 347
barrage jamming 736
basic area navigation (B-RNAV) 329
batch processing 135
Battlehog series 854, 855
BC: see bus controller (BC)
beacon mode 693
    flight testing 693
beam shaping 630
beamwidth 142, 646, 667
bearing accuracy 611
    data card 186
    geometry 186
    HARM test matrix 842
    NDB 185
    radar 688
    VOR 185
Beidou 227
Bell Laboratories 81
BER: see bit error rate (BER)
best estimate trajectory (BET) 48
BET: see best estimate trajectory (BET)
beyond line of sight (BLOS)
        communications 865
beyond visual range (BVR) 646
BHMD: see binocular helmet mounted
        display (BHMD)
Biezad, Daniel J., 167
Bingo fuel 845
binocular helmet mounted display
        (BHMD) 948
binocular systems 946
BIT: see built-in test (BIT)
bit error rate (BER) 154, 876
bits 65–7
16-bit word 66
blackbody
    defined 547
    emissivity factor of 547
    Planck's law 549–50
    spectral radiation 549
    Wien's displacement law 549
Black-McKeller Act of 1934 269
Block IIR-M constellation 215
BLOS communications: see beyond line
        of sight (BLOS) communications
BLU-107 Durandal Runway Buster 782
B-52 missile interface unit test 156

boarding steps 784
Boeing 777 90
Boeing Phantom Works
    iNET by 130
bolometer 575
bomb drag parameter (KD) 805
bomb drop and scoring 50
bomb release unit (BRU) 778
Booker, Graham H., 535
boredom, communications flight testing
        and 151
boresight 608–9, 614
    accuracy 614
    errors 809
BR: see break rate (BR)
break rate (BR) 980
bright source protection (BSP) 914
bright spot 934
British Aerospace 91
B-RNAV: see basic area navigation
        (B-RNAV)
BRU: see bomb release unit (BRU)
B scope 647
BSP: see bright source protection (BSP)
built-in test (BIT) 185, 293, 772
    pulse radar 653
bulk transfers 98
bus
    analyzer 63
    editors 64
    monitor
        defined 63
        types of 63–4
bus controller (BC) 63
    design 75–6
BVR: see beyond visual range (BVR)

C

C-32767 244
CAA: see Civil Aeronautics
        Administration (CAA); Civil
        Aeronautics Authority (CAA)
CAB: see Civil Aeronautics Board (CAB)
CAGC: see clutter automatic gain control
        (CAGC)
calibrated airspeed (CAS) 172
calibration curve, for thermocouple 126
calorimeter 575
CAM: see Civil Aeronautics Manuals
        (CAM)
CAN 1.0 85
CAN 2.0 85
Canadian CF-18s 732
CAN bus: see controller area network
        (CAN) bus
capability-based assessment (CBA) 960
capability development document
        (CDD) 960, 970
captive carry tests 798–9

CAR: see Civil Aeronautics Regulation
        (CAR)
carrier (CV) alignment 201
carrier sense multiple access/collision
        detection (CSMA/CD) media
        access 115
carrier sense multiple access (CSMA) 90
CAS: see calibrated airspeed (CAS)
"catcher's mitt" test 795
cathode ray tube (CRT) 942
    miniature 945
cathode ray tube (CRT) displays 577–9
cats and traps 50
CBA: see capability-based assessment
        (CBA)
CBIT: see continuous BIT (CBIT)
CBR: see constant bit rate (CBR)
CCA: see common cause analysis (CCA)
CCIP: see continuous computing impact
        point (CCIP)
CDD: see capability development
        document (CDD)
CDI: see course deviation indicator (CDI)
CDL: see common data link (CDL)
CDRL: see contract data requirement list
        (CDRL)
CDV: see command directed vehicle
        (CDV)
CE: see concatenation event (CE);
        conducted emission (CE)
CE circuit: see clutter elimination (CE)
        circuit
Center of Excellence for General
        Aviation Research 897
center of gravity (CG) 50
centripetal accelerations 194
certificate of authorization (COA) 858,
        859
certification 24
    highly integrated system 285–98
    history 268–70
    process 274–82
    of radio aids 190–1
    Type Certification Board 277–82
    see also airworthiness
CFAR detectors: see constant false alarm
        rate (CFAR) detectors
CFE: see contractor furnished equipment
        (CFE)
CFIT: see controlled flight into terrain
        (CFIT)
CFR: see Code of Federal Regulations
        (CFR)
CG: see center of gravity (CG)
CGS: see chirp gate stealer (CGS)
chaff 33, 50, 731
chaff employment 742–3
chemical/biological/radiological (CBR)
        equipment 918

chicken wire 934

Chinese satellite navigation augmentation system (SNAS) 229

chirp gate stealer (CGS) 740

C4I
advances in 258–9

cinetheodolites 41–3
aerotriangulation 42
film-read 41
tracking system 39

circular error probable (CEP) and drift rate 202–5

CIU: *see* converter interface unit (CIU)

Civil Aeronautics Act of 1938 269

Civil Aeronautics Administration (CAA) 269

Civil Aeronautics Authority (CAA) 269

Civil Aeronautics Board (CAB) 269

Civil Aeronautics Manuals (CAM) 270

Civil Aeronautics Regulation (CAR) 269–70
recodification of 270

Civil Certification of Automatic Dependent Surveillance Broadcast (ADS-B) 865

Class Beta 331

Class Beta-1 equipment 331

Class Beta-2 equipment 332

Class Beta-3 equipment 332

Class 4 Gamma and Delta equipment 332

clean emitter 754

clear to send (CTS) 83

clutter automatic gain control (CAGC) 689

clutter elimination (CE) circuit 747

CMDS: *see* countermeasure dispenser system (CMDS)

CMOS: *see* complementary metal oxide semiconductor (CMOS)

COA: *see* certificate of authorization (COA)

cockpits 301–22
controls and displays
design principles 304–9
standardization 321–2
tests and evaluation 314–21
FAA identified problems 303–4
crew interface 303
disengagement behavior 303
mode awareness 303
mode behavior 303
speed and altitude protection 303

Cockpit Display of Traffic Information (CDTI) 375

Code of Federal Regulations (CFR) 270
Federal Aviation Administration (FAA) in 270–3

coefficient and ballistic dispersion (CEP) 815

COIs: *see* critical operational issues (COIs)

combat rate (CR) 980

Combined Test Force (CTF) 3, 11

command, control, and communications intelligence (C3I) 258

command and control processor (C2P) 242

command and control system 729

command directed vehicle (CDV) 854

command words
1553 systems 68–70
anomalous conditions 70–1
subaddress fields MIL-STD-1760 102
synchronization pattern 67–8

commercial off the shelf (COTS) 961

common cause analysis (CCA) 294
common mode analysis 294
particular risks analysis 293–4
zonal safety analysis 294

common data link (CDL) 864–5

common mode analysis, CCA 294

communications flight test
aircraft communications equipment 142–4
basics 140–2
boredom 151
effects of stores, landing gear, and flaps 149
EMI/EMC 154–8
elimination 161–4
testing 158–9, 161
executing the matrix 148–9
logistics 150
overview 139
requirements 144
speech intelligibility 151–4
test plan matrix 146–8
three basic steps
compliance tests 145
static test 144–5
"warm fuzzy" test 145
weather and 149–50

communications security (COMSEC) custodian 150

COMPASS BeiDou 213

compatibility, defined 971

complementary metal oxide semiconductor (CMOS) 545

complex system
defined 285

compliance tests 145

COMSEC custodian: *see* communications security (COMSEC) custodian

concatenation event (CE) 92

conducted emission (CE) 155

conducted susceptibility (CS) 155

cone of confusion, radio aids testing and 190

conical scan modulated jamming 738

conical scan radars 726

conical scan receive only (COSRO) 663

consent switch 784

constant bit rate (CBR) 899

constant false alarm rate (CFAR) detectors 644

continuous BIT (CBIT) 755

continuous computing impact point (CCIP) 808

continuous wave (CW) 210

contract data requirement list (CDRL) 9–11, 849

contractor and subtier specifications 20–1

contractor flight testing 2

"contractor format," 11

contractor furnished equipment (CFE) 839

contractual requirements 9–11

controlled flight into terrain (CFIT) 382

controller area network (CAN) bus 84–6
CAN 2.0A frame format 85
CAN 2.0B format 85–6

controller pilot data link communications (CPDLC) 239

control room 51

control transfers 98

converter interface unit (CIU) 64
arrangement 77

C-150 OPA 901

Coriolis effect 195

Coriolis forces 194

correlation table (CRT) 960

cosmetic blemishes 934

COSRO: *see* conical scan receive only (COSRO)

cost-benefit analysis 980

COTS: *see* commercial off the shelf (COTS)

countermeasure dispenser system (CMDS) 756

course deviation indicator (CDI) 178

cover pulse jamming 736

covert lighting 929

CPDLC: *see* controller pilot data link communications (CPDLC)

CR: *see* combat rate (CR)

CRC: *see* cyclical redundancy check (CRC)

critical operational issues (COIs) 970, 972–3

critical technical parameters (CTPs) 972

cross-eye (X-EYE) jammer 741–2

cross-polarization (X-POLE) 740, 741

CRT: *see* cathode ray tube (CRT); correlation table (CRT)

crypto keys 150

CS: *see* conducted susceptibility (CS)

C scope 651
CSMA: *see* carrier sense multiple access (CSMA)
CTF: *see* Combined Test Force (CTF)
CTPs: *see* critical technical parameters (CTPs)
CTS: *see* clear to send (CTS)
cumulative probability of detection (Pc) 645
customer expectations 21–2
CW: *see* continuous wave (CW)
cyclical redundancy check (CRC) 86
   MIL-STD-1760 104

D
DA: *see* destination address (DA)
dark (black) spot 934
DAS: *see* detector angular subtense (DAS); distributed aperture system (DAS)
DASS: *see* distress alerting satellite system (DASS)
data, processing 51
data acquisition, reduction and analysis instrumentation
   calibration curve for thermocouple 126
   counts 126
   for forced-air cooling 126
   PCM system 124–30
   real-time telemetry 130–1
   postflight data reduction 135
   real-time data 132–4
data acquisition routines 17
data arrival, asynchronous 54
data bus editors 25
data busses 86–9, 89–93, 93–6, 99–100, 118–21
   AFDX 121
   CAN bus 84–6
   Ethernet 115–18
   fiber channel (FC) 108–15
   MIL-STD-1760D 100–8
   problems 122–3
   RS-232C 81–4
   USB 96–9
data carrier detect (DCD) 83
data communications equipment (DCE) 81–3
data encoding
   1553 systems 67–8
data item description (DID) 10, 849
data link latency evaluations 884–5
data link systems, navigation systems 238
   C4I, advances in 258
   comparisons 256–7
   flight testing of 257–8
   Link 16 operation 240
   tactical background 239–40

data reduction and processing facilities 24–5
data set ready (DSR) 82
data terminal ready (DTR) 82
data transmission equipment (DTE) 81–3
data words
   1553 systems 73–4
DBS: *see* Doppler beam sharpening (DBS)
DCD: *see* data carrier detect (DCD)
DCE: *see* data communications equipment (DCE)
DE: *see* directed energy (DE)
dead-face filters 927
deception 721
deception jammer 737
deception jamming systems 738–9
   types of 739–42
decibel 626
   negative 627
decoders (type 1, 2) 768
decoy deployment 50
Defence Evaluation Research Agency (DERA) 948
Defense Research Associates (DRA) sense and avoid system 897–8
Defense Systems Acquisition Review Council (DSARC) 958
Defense Test and Evaluation Professional Institute 30
delta azimuth (ΔAz) 41
delta elevation (ΔEl) 41
demonstration program 18
DEP: *see* design eye point (DEP)
Department of Defense Directive (DoDD) 958
Department of Defense (DOD) 748
depression angle
   DNS 208
DERA: *see* Defence Evaluation Research Agency (DERA)
design eye point (DEP) 809
destination address (DA) 116
detection of targets at sea 684–7
detector angular subtense (DAS) 594
detectors 545
   cooling 575
   IR 569–70
determinism
   Link 16 systems 256
development test and evaluation (DT&E) 2, 3, 10, 957, 964, 967
   definition 968
   versus OT&E 968
   TEMP documents 964
DFCS: *see* digital flight control system (DFCS)
DF mode: *see* direction finding (DF) mode

DID: *see* data item description (DID)
differentially corrected GPS 48
differentially corrected positioning service (DCPS) 335
diffraction 140
digital converter unit 64
digital data 839
digital flight control system (DFCS) 155
digital subscriber line (DSL) connection 99
digital terrain elevations (DTEs) 700
digital TV systems 585–6
   CCD sensors 585
      advantages and disadvantages 585
   CMOS 585
      advantages and disadvantages 585
   sensors 585–6
   SMPD 585, 586
dilution of precision (DOP) 39, 44–5, 221
   effect of satellite removal on 221
directed energy (DE) 721
direction finding (DF) mode 144
direct threat radar systems 726
dirty emitter 754
displays
   CRT 577–9
   LCD 579–81
   overview 577
dissimilarity, PSSA 293
distance measuring equipment (DME) 179–80
distress alerting satellite system (DASS) 215
distributed aperture system (DAS) 946
diurnal cycle 564
DLQ-38 791, 792
DLZ: *see* dynamic launch zone (DLZ)
DME: *see* distance measuring equipment (DME)
DNS: *see* Doppler navigation system (DNS)
doctrine, organization, training, materiel, leadership and education, personnel and facilities (DOTMLPF) 963
documentation 298–9, 958
   EASA and FAA, 300–1
DOD: *see* Department of Defense (DOD)
DoDD: *see* Department of Defense Directive (DoDD)
DoD 5000 series 958
DOP: *see* dilution of precision (DOP)
Doppler, PRF *versus* closure for unambiguous 679
Doppler ambiguities 676–80
Doppler beam sharpening (DBS) 649, 706–10
Doppler characteristics 668–71
Doppler clutter regions 671–6

Doppler clutter spectrum 674, 677
Doppler discrimination 747
Doppler effect 207
  elimination 682
Doppler filters 644
Doppler frequency 668, 670, 671
  accounted for with a constant
    frequency segment 683
  computation 669
  computation of the true 671
Doppler GPS 45
  see also Global positioning system
    (GPS)
Doppler navigation system (DNS) 206,
  207, 684
  advantages and disadvantages 211–12
  aircraft velocity components 210
  beam arrangements 209
  depression angle 208
  errors 211
  flight testing of 212–13
Doppler noise jammer 740
Doppler profile 678
Doppler proximity fusing 765
Doppler radar 727
Doppler resolution 671
Doppler return
  zero, as a function of dive angle 674
Doppler sensitivity 630
Doppler shift 45, 207–11, 630, 672, 772
Doppler-shifted carrier frequency 678
drift rate
  CEP and 202–5
  EO system 611
  of laser designator 615
DSARC: *see* Defense Systems Acquisition
    Review Council (DSARC)
DS-0 channels 864
DSL connection: *see* digital subscriber
    line (DSL) connection
DSR: *see* data set ready (DSR)
DTE: *see* data transmission equipment
    (DTE)
DT&E: *see* development test and
    evaluation (DT&E)
DTEs: *see* digital terrain elevations
    (DTEs)
DT&E testers 12–13, 14–15
DTR: *see* data terminal ready (DTR)
Duty cycle 638
Duty factor 638
Dynamic launch zone (DLZ) 792

E

EA: *see* electronic attack (EA)
EA-6B Prowlers 733
early warning radars (EWRs) 724
earth models 39–41
  differences 40

earth rotation rate compensation 193
ECCM: *see* electronic counter-
    countermeasures (ECCM)
ECM threat simulation pod 750
ECP: *see* engineering change proposal
    (ECP)
edge glow 933–4
EDR: *see* expanded data request (EDR)
Edwards Air Force Base 23
EEC: *see* essential employment capability
    (EEC)
EFAbus (Eurofighter bus) 93, 94
effective radiated power (ERP) 736
EFVS: *see* enhanced flight vision systems
    (EFVS)
EGNOS: *see* European geostationary
    navigation overlay service
    (EGNOS)
EGPWS: *see* enhanced ground proximity
    warning system (EGPWS)
EHF bands: *see* extra-high-frequency
    (EHF) bands
EIA 422 82
EIA 423 82
EIA: *see* Electronic Industries Alliance
    (EIA)
EIA 232E 82
EIA (RS) 232C signals 84
Einstein, Albert 588
ejection launcher 794
  test requirements 796
EL: *see* electroluminescent (EL)
electrical noise, defined 141
electric noise 630
electroluminescent (EL) 945
electromagnetic (EM) fields 154–5
electromagnetic interference/
    electromagnetic compatibility
    (EMI/EMC) 139–40, 154–8,
    874–6
  elimination 161–4
  testing 158–61
    four phases of 158–9
    issues 161
    method of 160–1
electromagnetics 154, 155
electronic attack (EA) 721, 730
electronic counter-countermeasures
    (ECCM) 720
electronic countermeasures 729
electronic imaging system 536–7
Electronic Industries Alliance (EIA) 81
electronic protection (EP) measures 745–7
electronic warfare (EW)
  air defense systems 728–9
  chaff employment 742–3
  deception jamming 738–42
  electronic attack 729–33
  electronic protection measures 745–7

electronic warfare systems test and
    evaluation 748–60
  flare employment 743–5
  noise jamming 734–8
  terminology 721–3
  threat 724–8
Electronic Warfare Integrated Systems
    Test Laboratory (EWISTL) 838
electronic warfare systems test and
    evaluation 748–60
  architecture of 749
  test range parameters 753
electro-optical (EO) systems 810–11
  components and performance
    CRT displays 577–9
    detector cooling 575
    displays 577
    field of view 576–7
    LCD 579–81
    optics 576
    scanning techniques 568–9
  flight test evaluations
    basic descriptions 594
    bearing accuracy 611
    boresight 608–9
    drift rate 610
    equivalent bar pattern 596, 597, 598
    field of regard (FOR) 608
    field of view (FOV) 608
    jitter 611
    minimum resolvable temperature
      differential 613
    MTF 596
    noise equivalent temperature
      differential 613–14
    other tests 612–13
    sensor slew rate 609–10
    target discrimination 597, 598
    target tracking 612
  passive devices
    FLIR 582
    IRLS 581–2
    IRST 586–7
    television systems 583–7
Electro-Optical Imaging Performance,
    5th edition 535
electro-optical sensor 64
elevation, defined 32
elevation accuracies 37
EMCON: *see* emission control (EMCON)
EM fields: *see* electromagnetic (EM) fields
emission control (EMCON) 721
emission point 934
emissions reduction 731
emissivity, IR systems 547–9
  of blackbody 547
  Kirchhoff's law 547
  Stefan–Boltzmann law 548
  of surface 547

encryption 24
    keys 143
end of frame (EOF) field 86
end of transmission (EOT) word 89
engineering change proposal (ECP) 12
engineering unit (EU) converted data
    63
Engineer's Manual, The 154
enhanced bit rate (EBR) 1553, 99–100
    topology 99
enhanced flight vision systems (EFVS)
    airworthiness certification 523
    with HUD 526
enhanced flight vision systems (EFVS)
    522–3, 943
    general and specific requirements
        525–6
enhanced ground proximity warning
        system (EGPWS) 382
    see also Terrain awareness and
        warning system (TAWS)
enhanced vision systems (EVS) 520
    airworthiness 523–7
        general requirements 524
        installation requirements 527
        intended function 523
        performance evaluation 527
        specific requirements 524–5
    enhanced flight vision systems
        (EFVS) 522–3
EOF field: *see* end of frame (EOF)
    field
EO systems: *see* electro-optical (EO)
    systems
EOT word: *see* end of transmission
    (EOT) word
EP: *see* electronic protection (EP)
EPE: *see* expected positional error
    (EPE)
E Port: *see* expansion port (E Port)
ERP: *see* effective radiated power (ERP)
errors
    air-to-ground sources of 809–11
    DNS 211
    GPS 217–21
ESA: *see* European Space Agency
    (ESA)
escort jamming 733
ESD-TR-86-278 322
essential employment capability (EEC)
    10, 18
ethernet 115–18
    fields and number of bytes per field
        116
    star topology 117
ETSO: *see* European technical standard
    order (ETSO)
EU converted data: *see* engineering unit
    (EU) converted data

European Aviation Safety Administration
        (EASA) 267
    EASA and FAA documents,
        differences between 300–1
European Aviation Safety Agency 300
European geostationary navigation
        overlay service (EGNOS) 225,
        229, 231–2
European Geostationary Navigation
        Overlay Service (EGNOS)
        (ESA) 225, 229
European Space Agency (ESA) 226
European technical standard order
        (ETSO) 300
evaluators, notes for
    concerning documentation 298–9
evaporograph 575
EVS: *see* enhanced vision systems (EVS)
EW: *see* electronic warfare (EW)
EWRs: *see* early warning radars (EWRs)
expanded data request (EDR) 690
expansion port (E Port) 113
expected positional error (EPE) 221–2
explosive train 765
exterior NVIS lighting methods 929–30
    modifications 930–1
external noise 142
extra-high-frequency (EHF) bands 141
extrapolation 54

F
F-14A/D 646
F-15 76
    system boresighting and 810
F-16 76, 156
F-18 76
F-117 156, 641
F/A-18 115
F/A-18C 23
FA: *see* false alarm (FA)
FAA: *see* Federal Aviation
    Administration (FAA)
FAA data tapes 654
FAA Form 337 274
FAA Part 25 76
fabric port (F Port) 113
false alarm (FA)
    rate 643
    threshold 643
    time 643
false replies uncorrelated in time
        (FRUIT) 237
false target creation 731
false target generation 738, 739
F-35 AMLCD HMD 946, 947
FAR 21.21 279
FAR 21.33 281
fast time constant (FTC) 747
fault detection and exclusion (FDE) 224

FC: *see* fiber channel (FC)
FDE: *see* fault detection and exclusion
    (FDE)
F-15E 23
FEATS: *see* Future European Air
    Traffic Management System
    (FEATS)
FED: *see* field emission display (FED)
Federal Aviation Act of 1958 270
Federal Aviation Administration (FAA)
    48, 152, 267
    in Code of Federal Regulations (CFR)
        270–3
    EASA and FAA documents,
        differences between 300–1
Federal Republic of Yugoslavia (FRV)
    720
FFAR: *see* folding fin aircraft rocket
    (FFAR)
FHA: *see* functional hazard assessment
    (FHA)
fiber channel (FC) 108–15
    frame, sequence, and exchange
        structure 114
    port types 113
    topologies 109, 113
fiber-optic gyro (FOG) 199
field emission display (FED) 942
field of regard (FOR) 608, 764
field of view (FOV) 607, 608, 651, 764
    of HARM 842
fields, defined 154
figure of merit (FOM) 917
fire Scout 855
fiscal accountability 21
fixed-format messages 248–50
fixed target track (FTT) 697
flaps, communications flight testing and
    149
flare employment 743–4
    infrared protective measures 744–5
flare rejection capability 744
flashing 934
flickering 934
flight briefs 844
flight crew 843–4
flight guidance systems 410–59
    airworthiness certification (Part 23)
        412–18
        cockpit control 413–14
        compliance, alternate means of
            417–18
        flight control recovery 418
        malfunction evaluations 414, 416
        maneuvering and approach
            malfunction 416–17
        normal flight malfunctions 415
        performance evaluation 418
        single-engine approach 418

airworthiness certification (Part 25) 418
  autopilot MUH 440–1
  compliance 437–40
  control, indications and alerts 422–3
  disengagement failure 419–20
  disengagements, under rare and
      nonnormal conditions 420–2
  hazardous loads and deviations,
      protection from 423–6
  HUD considerations 429–31
  mode characteristics 431–6
  quick disengagement 418–19
  speed protection 426–9
  transients, by disengagements 420
  transients, by engagement 420
  components 412
  documentation for 411
Flight Information Services-Broadcast
    (FIS-B) 375, 379
flight management systems (FMS)
  airworthiness certification 506–7
  airworthiness requirements 507–8
    AC 20-167 523
  flight tests 509–10
    multisensor 510
  multisensor flight tests 510
flight test
  of data link systems 257–8
  of DNS 212–13
  of GPS 233
  of IFF systems 238
  inertial navigation system (INS)
    acceleration and velocity accuracy
        205–6
    alignment types 200–2
    circular error probable and drift
        rate 202–5
    navigation updates 206–7
    time to align 199–200
  of INS
    acceleration and velocity accuracy
        205–6
    alignment types 200–2
    circular error probable and drift
        rate 202–5
    navigation updates 206–7
    time to align 199–200
  need for 6–7
  objective 5–6
  organization 4–5
  SAR 710
  TAWS 391–401
    advisory callout 400
    excessive closure rate to terrain 396–7
    excessive downward deviation from
        an ILS glideslope 400
    excessive rate of descent 396
    flight into terrain when not in
        landing configuration 398
    for FLTA 392–4
    installation types and 391–2
    negative climb rate or altitude loss
        after takeoff 397–8
    for PDA function 394, 395
    wind shear warnings 401
  TCAS 365
  terrain avoidance (TA) 702–5
  terrain following (TF) 702–5
flight tester's task
  1553 systems 79–80
*Flight Testing of Radio Navigation
    Systems* 167
flight test readiness reviews (FTRR) 977
flight test team, responsibilities of 843,
    844
flight test techniques series
  AGARD 820
FLIR: *see* forward-looking infrared
    (FLIR)
Flood lighting 928
FLTA: *see* forward-looking terrain
    avoidance (FLTA)
Fluid-expansion thermometer 575
Fly-fix-fly 15–16, 23
FM: *see* frequency modulation (FM)
FM-CW Doppler radar 727
FM ranging 681
  flight testing of radar altimeters 683–4
  radar altimeters 681–3
FMR methods: *see* frequency modulation
    ranging (FMR) methods
FOC: *see* full operational capability
    (FOC)
focal plane array (FPA) 545
FOG: *see* fiber-optic gyro (FOG)
folding fin aircraft rocket (FFAR) 827
follow-on OT&E (FOT&E) 969
FOM: *see* figure of merit (FOM)
FOR: *see* field of regard (FOR)
forced-air cooling, instrumentation for
    126
forward-looking infrared (FLIR) 62, 382,
    582
  categories of 582
  line-of-sight (LOS) 587
  long-wave infrared (LWIR) 582
  mid-wave infrared (MWIR) 582
  purpose of 582
  staring systems 582
  thermal imaging systems 540, 541, 542
forward-looking terrain avoidance
    (FLTA) 383
FOT&E: *see* follow-on OT&E (FOT&E)
FOV: *see* field of view (FOV)
FPA: *see* focal plane array (FPA)
FRD: *see* functional requirements
    document (FRD)
free-stream ballistics 815

free-text messages 248–9
frequency Allocation Council 51
frequency allocations 51
frequency-hopping mode 143
frequency modulation (FM) 143, 646
frequency modulation ranging (FMR)
    methods 739
Fried, Walter R., 167, 168
FRP: *see* full rate production (FRP)
FRUIT: *see* false replies uncorrelated in
    time (FRUIT)
FSD program: *see* full-scale development
    (FSD) program
FSPI: *see* full-speed peripheral interface
    (FSPI)
FTC: *see* fast time constant (FTC)
FTJU: *see* fuel tank jettison unit (FTJU)
FTRR: *see* flight test readiness reviews
    (FTRR)
FTT: *see* fixed target track (FTT)
fuel lab 50
fuel tank jettison unit (FTJU) 768
full operational capability (FOC) 227
full rate production (FRP) 969
full-scale development (FSD) program 8
full-speed peripheral interface (FSPI) 97
functional hazard assessment (FHA)
    290–2
functional requirements document
    (FRD) 14, 79, 835
fuse tests, for air-to-air missiles 792
fusing mechanisms weapons integration
    764–5
Future European Air Traffic Management
    System (FEATS) 329

G
GAGAN: *see* GPS-Aided Geo
    Augmented Navigation
    (GAGAN)
GAI systems: *see* GPS-aided INS (GAI)
    systems
galactic noise 142, 630
Galileo 213, 226, 226–7
GBAS: *see* ground-based augmentation
    system (GBAS)
GBS 863
GBU-15 764
GCE communications: *see* ground control
    element (GCE) communications
GCI: *see* ground controlled intercept
    (GCI)
GCI-type display 751
GCU: *see* gun control unit (GCU)
GDOP: *see* geometric dilution of
    precision (GDOP)
gear handle, position of 784
geodesy for the Layman 39–40
geoid 40

geometric dilution of precision (GDOP) 38–9, 221
  cinetheodolite 39
GFE: *see* government furnished equipment (GFE)
Gigabit Ethernet 116
GIVE: *see* grid ionospheric vertical error (GIVE)
glinting/quadrantal effect 184
global navigation satellite system (GNSS) 64, 213, 333
  GPS errors 217–21
  GPS inaccuracy, contributors to 221–4
  Pseudoranging 216
  satellite augmentation systems 229–32
  satellite systems 225–8
global positioning system (GPS) 30, 44, 839
  accuracies 223–4
  certifications
    cockpit display and 318
    equipment classes 331
    MOPS 331, 333
    TSO 331
  degradations 39
  differentially corrected 48
  errors 217–21
  flight test objectives 233
  inaccuracy, contributors to 221–4
  as navigational aid 224–5
  nine-state vector 216
  operational testing 234
  operation of 214
  pseudoranging 216–17
  recap 233
  receivers, stand-alone 44–6
  satellite augmentation systems 225–8
  satellite systems 225–8
  total electron content map 219
  user-defined outage levels 220
  weapons integration testing 233–4
GLONASS operational system 225–6
GMTI mode: *see* ground moving target indicator (GMTI) mode
GMTT: *see* groundmoving target track (GMTT)
GNS landing systems 468–70
GNSS: *see* global navigation satellite system (GNSS)
GNSS/SBAS equipment 335
government furnished equipment/contractor furnished equipment (GFE/CFE) 14
government furnished equipment (GFE) 8, 833, 839
GPS: *see* global positioning system (GPS)
GPS-Aided Geo Augmented Navigation (GAGAN) 229, 232
GPS-aided INS (GAI) systems 46

GPWS: *see* ground proximity warning system (GPWS)
GRAS: *see* ground-based regional augmentation system (GRAS)
grazing angle 672, 684–5
  scattering regions defined by 686
grid ionospheric vertical error (GIVE) 220
ground-based augmentation system (GBAS) 225, 229, 330
ground-based regional augmentation system (GRAS) 229, 232, 332
ground-based tracking radars 48
ground control element (GCE)
  communications 861, 889
ground controlled intercept (GCI) 725
ground mapping radar 648
  parameters 649
ground moving target indicator (GMTI) mode 696–7
  GMTI and GMTT flight testing 697–8
groundmoving target track (GMTT) 696–7
ground proximity warning system (GPWS) 382
  see also Terrain awareness and warning system (TAWS)
ground tests 748
  NVG 931–4
ground wave 141
guided missile 827–8
gun control unit (GCU) 768
guns 721, 776, 781, 826–7
gyrocompass alignment 195, 196
gyro-stabilized platforms INS 192–7

H

HARM: *see* high-speed antiradiation missile (HARM)
HAT: *see* height above terrain (HAT)
Have Quick/Have Quick II 143
HDDR: *see* high-density digital recorders (HDDR)
HDOP: *see* horizontal dilution of precision (HDOP)
header word format
  MIL-STD-1760 104
heading 170–2
  measurement of 170–2
heading scale 171
head orientation and position (HOP) 948
head steered forward looking infra-red (HSFLIR) sensor 948
heads-up display (HUD) 177
  mode 651, 692
heat flux 554
height above terrain (HAT) 698
height finder radars (HFRs) 725
HEL: *see* high-energy lasers (HEL)

helicopters
  delivery considerations for 824–5
  guided missile 827–8
  guns 826–7
  rockets 827
  specific weapons types 826
helicopter TAWS (HTAWS) 382, 409–10
helmet mounted display systems (HMD) 908, 939
  components 942–4
  equipment, test and evaluation of 949
  flight test of 953–4
  helmet tracking systems (HTS) 948–9
  optical design 944–7
  overview 939–42
  qualitative ground testing 950–3
  quantitative testing 949–50
  requirements 940
  unacceptable design 944
helmet tracking systems (HTS) 948–9
Hershel, Sir William 536
HF bands: *see* high-frequency (HF) bands
HFRs: *see* height finder radars (HFRs)
high-density digital recorders (HDDR) 131
high-energy lasers (HEL) 566
high-frequency (HF) bands 141
high-intensity radiated fields (HIRF) 156, 279
highly integrated system, certification considerations for 285–98
  compliance 286
  defined 285
  references 285
  safety assessment 290–6
    CCA 294
    certification plan 296–8
    FHA 290–2
    PSSA 293, 295
    SSA 295
high-speed antiradiation missile (HARM)
  accuracy test matrix 843
  aircraft glinting and 837–8
  carriage, allowable flight envelope for 832
  field of view (FOV) of 842
  integration flight test program
    analysis and reporting 848–9
    avionics test requirements 833–5
    responsibilities of test team 843–7
    test planning 835–43
    vehicle test requirements 831–3
  operations 163
  responsible test organizations 835
high-speed antiradiation missiles (HARMs) 721
high-speed peripheral interface (HSPI) 97
Hi-8 record systems 845

HIRF: *see* high-intensity radiated fields (HIRF)
HMD: *see* helmet mounted display systems (HMD)
HMI: *see* human machine interface (HMI)
HOJ: *see* home-on-jam (HOJ)
Holst, Gerald C., 535
home-on-jam (HOJ) 727, 746
honeycomb pattern 934
hood extension, effects of lens 913
HOP: *see* head orientation and position (HOP)
horizontal dilution of precision (HDOP) 39
horizontal DOP (HDOP) 221
host controller 97
Hot Mike 128, 134
HQ USAF/TE 958
HSFLIR sensor: *see* head steered forward looking infra-red (HSFLIR) sensor
HSPI: *see* high-speed peripheral interface (HSPI)
HTAWS: *see* helicopter TAWS (HTAWS)
HTS: *see* helmet tracking systems (HTS)
hubbed loop topology 110, 113
hubs 97
HUD: *see* heads-up display (HUD)
Hudson, R. D., 536
Hudson, Ralph G., 154
human factors 303–4
human factors issues
    NVG 918–20
human machine interface (HMI) 4

I

IADS: *see* Integrated Air Defense System (IADS)
IAG: *see* inertially aided GPS (IAG)
IAS: *see* indicated airspeed (IAS)
ICAO: *see* International Civil Aviation Organization (ICAO)
ICD: *see* interface control document (ICD)
identification friend or foe (IFF) system
    airport surveillance radar 234–5
    flight testing of 238
    interrogation timing 236
    interrogators 237–8
    mode S 237
    modes used in 65, 235–7
IDS: *see* integration design specification (IDS); interface design specification (IDS)
IEEE 802 115
IEEE 1394 118–21
    asynchronous subaction 120
    cable configuration completion 119
    isochronous subaction 121

network after bus initialize 119
packet frame 121
topology 118
IFAST: *see* Integrated Facility for Avionic Systems Testing (IFAST)
IFF system: *see* identification friend or foe (IFF) system
IFF tracks 654
IFR: *see* instrument flight rules (IFR)
IFR high-altitude chart legend 176
IGI: *see* integrated GPS/INS (IGI)
IHADSS: *see* Integrated Helmet and Display Sighting System (IHADSS)
IJMS: *see* inadvertent jettison monitoring system (IJMS)
ILC Data Device Corp., 62
ILS: *see* instrument landing systems (ILS); integrated logistics support (ILS)
image disparity 934
image distortion 934
image intensification (I2) technology 909–11
imminent destruction 731
inadvertent jettison monitoring system (IJMS) 840
incandescent lights 930
    absorption filters 930
    interference filters 930
INCITS: *see* International Committee for Information Technology Standards (INCITS)
"index of refraction" 34
Indian GPS-aided geo augmented navigation (GAGAN) 229
Indian regional navigation satellite system (IRNSS) 214, 228, 229
Indian Space Research Organization (ISRO) 228
indicated airspeed (IAS) 172
inertial-aided GPS receivers 46–7
inertially aided GPS (IAG) 46
inertial navigation system (INS) 5, 38, 46, 64, 78, 80, 728
    accelerometer 191
    flight testing
        acceleration and velocity accuracy 205–6
        alignment types 200–2
        circular error probable and drift rate 202–5
        navigation updates 206–7
        time to align 199–200
    gyro-stabilized platforms 192–7
    platform structure 192
    strap-down systems 197–9
    test 38
    theory 46

iNET: *see* integrated network-enhanced telemetry (iNET)
in-flight alignment 201
infrared (IR) missiles, characteristics of 744–5
infrared (IR) signature 7
infrared (IR) systems 536
    radiation fundamentals 547–51
        emissivity 547–9
        Lambert's law of cosines 551
        Planck's law 549, 550
        spectral distribution of energy 549
        Stefan–Boltzmann law 548
        transmission properties 550, 551
        Wien's displacement law 549
    terminology 546–7
    see also specific Infrared (IR)
infrared search and track (IRST) 586–7
    AN/AAS-42 587
    concept 586
    DAS-IRST 587
    EO-DAS 587
    EOTS 587
    IR-OTIS 587
    PIRATE 587
    range calculation 586–7
    technologies used in 586–7
initial OT&E (IOT&E) 969, 979
Innovative Solutions and Support (IS&S) 279
INS: *see* inertial navigation system (INS)
INS/GPS program 48
INS/IRU 336
instrumentation
    calibration curve for thermocouple 126
    counts 126
    for forced-air cooling 126
    PCM system 124–30
    real-time telemetry 130–1
instrument flight rules (IFR) 703
instrument landing systems (ILS) 62, 459–67
    installations 183
Integrated Air Defense System (IADS) 728
Integrated Facility for Avionic Systems Testing (IFAST) 838
integrated GPS/INS (IGI) 46
Integrated Helmet and Display Sighting System (IHADSS) 943
integrated logistics support (ILS) 5
Integrated Navigation and Guidance Systems 167
integrated network-enhanced telemetry (iNET) 130
    system interfaces 130–1
integration design specification (IDS) 13
integration flight test program, for HARM analysis and reporting 848–9

avionics test requirements 833–5
responsibilities of test team 843–7
test planning 835–43
vehicle test requirements 831–3
integrity
defined 224
types of 224
interactive batch processing 135
interface control document (ICD) 73, 74,
835
interface design specification (IDS) 73,
79, 835
interference blanking 759
interference filters 930
interim operational flight program
(IOFP) 16
interior NVIS lighting methods 924
modifications 925–8
interleave function 705
interleaving 692
interlocks 785
intermission (Int) field 86
internal inertial reference system (IRS) 46
internal noise 142
International Civil Aviation Organization
(ICAO) 329
International Committee for Information
Technology Standards (INCITS)
108
Interoperability, defined 971–2
interpolation 54
Interrange Instrumentation Group (IRIG)
Standard 106-07 840
interrogators, IFF 237–8
interrupted alignments
power interrupt 202
taxi interrupt 202
interrupt transfers 98
In-Trail Procedure (ITP) application 380
*Introduction to Airborne Early Warning
Radar Flight Test* 625
Introduction to Airborne Radar, 3rd
edition 625
Introduction to Avionics Flight Test 626
*Introduction to Infrared and Electro-
Optical Systems* 535
Introduction to Radar Systems 684
3rd edition 625
*Introduction to Sensors for Ranging and
Imaging* 535
Invalidity word format
MIL-STD-1760 106
Inverse SAR (ISAR) 706–10
see also Synthetic aperture radar (SAR)
IOFP: *see* interim operational flight
program (IOFP)
Ion barrier film 914
ionosphere 140–1
effects on wave frequencies 140–1

IPT-based flight test program 12–13
IR detectors
cooling 575
photon detectors
photoconductive radiation 573
photoemissive radiation 572
photovoltaic radiation 573–4
semiconductors 574–5
thermal detectors
bolometer 575
calorimeter 575
evaporograph 575
fluid-expansion thermometer 575
thermocouple 574
IR line scanner (IRLS)
analog 581, 582
defined 581
digital 582
framing sensors 582
push broom sensors 582
scan rate 581
velocity-to-height (V:H) ratio 581
whisk broom sensors 582
IRLS: *see* IR line scanner (IRLS)
IRNSS: *see* Indian regional navigation
satellite system (IRNSS)
IRS: *see* internal inertial reference system
(IRS)
IR sources
controlled sources 552
natural sources 552–4
celestial sources 552–3
defined 552
moon 552
reflectance/emittance of 552
sun 552, 553–4
terrestrial sources 552
IRST: *see* infrared search and track (IRST)
IR systems: *see* infrared (IR) systems
isochronous transfers 98
ISRO: *see* Indian Space Research
Organization (ISRO)
IS&S: *see* Innovative Solutions and
Support (IS&S)
iteration rate 45, 46
ITU-T H.264 AVC 900

J
jamming 721
Janus Lambda system 209
JAUS: *see* Joint Architecture for
Unmanned Systems (JAUS)
JDAM: *see* joint direct attack munition
(JDAM)
JETDS: *see* Joint Electronics Type
Designation System (JETDS)
jet engine modulation (JEM) 687, 693
effects 688
lines on a velocity search display 689

Johnson, John 597, 598
Joint Architecture for Unmanned Systems
(JAUS) 856
functionality 857
joint direct attack munition (JDAM) 721
Joint ElectronicsType Designation
System (JETDS) 144
Joint Strike Fighter 115, 123
Joint Tactical Information Distribution
System (JTIDS) 240
architecture 244–8
communications modes 250–1
RF signal 250–2
Joint Tactical Radio System (JTRS) 258
Joint Technical Architecture (JTA)
standards 863
Joint Test Team (JTT) 3, 12
Joint Worldwide Intelligence
Communications System
(JWICS) 864
Joker fuel 845
J series messages
of Link 16 system 240
JSTARS 729
JTA standards: *see* Joint Technical
Architecture (JTA) standards
JTIDS: *see* Joint Tactical Information
Distribution System (JTIDS)
JTIDS TDMA 244–5
JTRS: *see* Joint Tactical Radio System
(JTRS)
JTT: *see* Joint Test Team (JTT)
JWICS: *see* Joint Worldwide Intelligence
Communications System
(JWICS)

K
Kalman filter 46–7, 697
Kayton, Myron 167, 168
KEAS: *see* knots equivalent airspeed
(KEAS)
KG-84 150
Kinemetrics 53
Kirchhoff, Gustav 536
Kirchhoff's law 547
Knots equivalent airspeed (KEAS) 172
Koni 226
Ku-band radar 648
KY-28s 143, 150
KY-58s 143, 150

L
LAAS: *see* Local Area Augmentation
System (LAAS)
labs, major drawbacks to 6
Lambert's law of cosines 551
landing gear
communications flight testing and
149

landing systems
    airworthiness certification
        category I systems 471–9
        category II systems 479–88
        category III systems 488
        for landing and rollout in low
            visibility 495–503
        for takeoff in low visibility 489–95
    GNS 468–70
    ILS 459–67
    MLS 467–8
LANDO 175
land remote sensing satellite
    (LANDSAT) 582
LANDSAT: *see* land remote sensing
    satellite (LANDSAT)
LANTIRN: *see* low-altitude navigation
    and targeting infrared for night
    (LANTIRN)
LAR: *see* launch acceptability region
    (LAR)
laser
    beam divergence 616–17
    designators 62
    output power 617
    pulse amplitude 617–18
    pulse repetition interval 618
    pulse width 618
    rangefinder 44
    ranging accuracy 615–16
    trackers 43
laser systems 588
    collimation 591
    concept 588
    eyes
        damage 592
        protection 592–3
    high spectral radiance 591, 592
    military application 593
        communication 593
        direct fire simulators 594
        HEL 594
        optical radar 594
        range finding 593
        target marking 593
        thermonuclear fission 594
    monochromaticity 591
    neodymium: yttrium aluminum garnet
        (Nd:YAG) 592
    output power of 592
    process 588
    Raman shifting 593
    short-pulsed lasers 592
latitude 32, 168, 169
launch acceptability region (LAR) 792
launcher qualification tests for air-to-air
    missile 793–6
launch-to-eject (LTE) cycle 163, 772
L-band spectrum 214

LCD: *see* liquid crystal display (LCD)
LCoS: *see* liquid crystal on silicon
    (LCoS)
LDU: *see* link data units (LDU)
least significant bit (LSB) 89
LEDs: *see* light emitting diodes (LEDs)
lens hood extension 913
L1 frequency 214
L2 frequency 214
light emitting diodes (LEDs) 925
    annunciators 928
lighting specifications 920
    Class A 921
    Class B 921
    and guidance 923–4
    MIL-L-85762 920–2
    MIL-L-85762A 920–2
    Type I 921
    Type II 921
line-of-sight (LOS) 33, 141, 727
    closure 688
    drift rate 610, 615
    equation 880
    in FLIR 587
    jitter 611, 615
    limitations 48
    sensor reference 610
    sensor slew rates 610
    stabilization for IRLS 582
line replaceable units (LRU) 6, 21
Link 4A 239
Link 4C 239
link data units (LDU) 89
Link 16 system
    determinism 256
    J series messages 240–4
    JTIDS architecture 244–8
    JTIDS RF signal 250–2
    message types 248–50
        fixed-format 248–50
        free-text 248–9
        round-trip timing (RTT) 249
        variable-format 249
    participants 252–4
    ports of 243
    PPLI and RelNav 254–6
liquid crystal displays (LCD) 579–81,
    942
    active matrix addressing 580
    color filter 580
    panel 579
liquid crystal on silicon (LCoS) 942
"Live fire," 800–1
lobe-on receive only (LORO) 663
Local Area Augmentation System
    (LAAS) 231
localizer performance with vertical
    guidance (LPV) approach
    WAAS 230–1

logical link control (LLC) standard 115
logistics
    communications flight testing and 150
Logistics Supportability 972, 980
long-endurance multi-intelligence vehicle
    (LEMV) UAVs 854, 855
longitude 32, 168, 169
long-range radars (LRRs) 724
"Long words," 74
loop port (NL Port) 113
LORO: *see* lobe-on receive only (LORO)
LOS: *see* line of sight (LOS)
low-altitude navigation and targeting
    infrared for night (LANTIRN)
    543
low probability of intercept (LPI) 630,
    699, 700, 731
low-speed peripheral interface (LSPI) 97
LPI: *see* low probability of intercept (LPI)
LRRs: *see* long-range radars (LRRs)
LRU: *see* line replaceable units (LRU)
LSB: *see* least significant bit (LSB)
LSPI: *see* low-speed peripheral interface
    (LSPI)
LTE cycle: *see* launch-to-eject (LTE)
    cycle

M
MAC standards: *see* media access control
    (MAC) standards
magnetic variation
    radio aids testing and 190
main lobe clutter (MLC) 672, 674, 743
    with radar look angle 672
maintainability 971, 979
maintenance fault list (MFL) 755
MALD: *see* miniature air-launched decoy
    (MALD)
Manchester II biphase encoding 67–8
man-made noise 142
man-portable air defense (MPAD) 720
Manufacturing Inspection District Office
    (MIDO) 276
MASPS: *see* Minimum Aviation System
    Performance Standards (MASPS)
master arm 784
master control centers (MCC) 232
master equipment list (MEL) 277
materiel development decision (MDD)
    961–2
mathematical models 7, 644
MAWS: *see* missile approach warning
    systems (MAWS); missile attack
    warning systems (MAWS)
maximum detection range 652, 655, 656,
    690, 693
maximum unambiguous Doppler 678
MCC: *see* master control centers (MCC)
MC rate: *see* mission capable (MC) rate

MCT: *see* mercury cadmium telluride (MCT)

MDB: *see* measurement descriptor blocks (MDB)

MDD: *see* materiel development decision (MDD)

MDT: *see* mean down time (MDT)

MDTCI: *see* MUX data transfer complete interrupt (MDTCI)

mean down time (MDT) 979

mean impact point (MIP) 205

mean sea level (MSL) 52
    altitude 173

mean time between critical failure (MTBCF) 980

mean time between failure (MTBF) 8, 796–8

mean time to repair (MTTR) 979

measurement accuracy 661
    geometry 661

measurement descriptor blocks (MDB) 13

Measure of Assessment (MOA)/Measure of Performance (MOP) 974–5

Measure of Suitability (MOS) 974

measures of effectiveness and suitability (MOEs & MOSs) 972, 974

measures of effectiveness (MOE) 972

measures of suitability (MOS) 972, 978

media access control (MAC) standards 115

media interface unit (MIU) 95

medium and low earth orbiting (MEO/LEO) satellites 865–6

MEL: *see* master equipment list (MEL)

MEO/LEO satellites: *see* medium and low earth orbiting (MEO/LEO) satellites

mercury cadmium telluride (MCT) 544

message contents 74–5

message formats
    1553 systems 74–5
    3910 systems 94

message header 249

messages
    defined 74
    types of 74
        Link 16 systems 248

message words, format of 249

metric radars 628

MFD: *see* multifunction displays (MFD)

MFL: *see* maintenance fault list (MFL)

MFR: *see* multiple frequency repeater (MFR)

MG-18 409

microwave frequencies 141

microwave landing system (MLS) 225, 467–8

MIDO: *see* Manufacturing Inspection District Office (MIDO)

MIDS: *see* Multifunctional Information Distribution System (MIDS)

Mil 36

MIL-D-8708B(AS) 18–19

MIL-HDBK-1763 831, 832
    aerodynamic analysis 814
    aeroelasticity 814
    air-to-ground accuracy tests 803–4
    detailed test requirements 816–20
    flight tests 816
    general test requirements 813–16
    ground tests 816
    physical compatibility review 813–14
    test and evaluation of air-to-air missiles in 789–91
    test requirements 812–13

MIL-HDBK-237C 157

military airworthiness requirements 859–60

MIL-L-85762 920–2

MIL-L-85762A 920–2

Miller, Gary M., 140

millimetric radar 630

millimetric wave (MMW)
    frequencies 141
    radar 710–11

milliradian 36

MIL-STD-461D 157

MIL-STD-462D 157

MIL-STD-464 157

MIL-STD-1472D 322

MIL-STD-1553 bus
    bits 65–7
    bus controller (BC) 63
        design 75–6
    bus monitor 63–4
    command words 68–70
        anomalous conditions 70–1
    data encoding 67–8
    data words 73–4
    flight tester's task 79–80
    high-speed data bus implementation 123
    historical overview 62
    message formats 74–5
    mode codes 101
    problems 122–3
    remote terminals (RTs) 64
    simple bus configuration 76–9, 122
    stand-alone remote terminal 64
    status words 71–3, 101
    system architecture 62–4
    twisted pair shielded cable 64
    word formats 68
    word types 67

MIL-STD-1553 compatible tactical air navigation (TACAN) system 18

MIL-STD-1553 Designer's Guide 62

MIL-STD-1605 (SHIPS) 157

MIL-STD-1760 62
    command words subaddress fields 102
    critical authority format 108
    critical control 2 format 104, 107
    cyclical redundancy check (CRC) 104
    header word format 104
    invalidity word format 106
    MIL-STD-1760D 100–8
        primary verifications 103
        stores interfaces 100–1
    store control message 109
    store monitor message 110
    transfer control message
        format 110
        instruction word 111
    transfer monitor message
        format 112
        transfer status word 112
    vector word format 103

MIL-STD-1773 62
    data encoding 68

MIL-STD specifications 4

miniature air-launched decoy (MALD) 50, 731

miniature CRTs 945

miniature munitions/stores interface (MMSI) standard 100

Minimum Aviation System Performance Standards (MASPS) 893–4

minimum navigation performance standards (MNPS) 329

minimum operational performance standards (MOPS) 238, 282
    GPS 331, 333

minimum release altitude (MRA) 811–12

minimum resolvable temperature difference (MRΔT) 582, 613

minimum system performance standards (MSPS) 329

Minor, John 536

MIP: *see* mean impact point (MIP)

MISB (Motion Imagery Standards Board)–compliant systems 90

missile approach warning systems (MAWS) 756

missile attack warning systems (MAWS) 744

missile designations 766

missile designations codes 766

missile firings 50

missile launch
    data analysis of 800–3

missile launch envelope (MLE) 792

missile on aircraft test (MOAT) 770–1
    commands 771

missile power relay unit (MPRU) 768

missile safety footprint 801

missile simulate unit (MSU) 840

missile system errors 815

mission capable (MC) rate 980
MIU: *see* media interface unit (MIU)
MJU-7 flare 743–4
Mk 82 782
    low-drag deliveries, error sources for
        808
MLC: *see* main lobe clutter (MLC)
MLE: *see* missile launch envelope (MLE)
MLS: *see* microwave landing system
        (MLS)
MMSI standard: *see* miniature munitions/
        stores interface (MMSI) standard
MMW: *see* millimetric wave (MMW)
MNPS: *see* minimum navigation
        performance standards (MNPS)
MOAT: *see* missile on aircraft test
        (MOAT)
Mode 1, of IFF system 235
Mode 2, of IFF system 235
Mode 3/A, of IFF system 235–6
Mode 4, of IFF system 237
Modern Electronic Communication 140
Mode S, of IFF system 237
modulation transfer function (MTF) 596,
        916
    resolution *versus* 916
MOE: *see* measures of effectiveness
        (MOE)
monitoring, PSSA 293
monopulse 663
    tracking 664
monopulse radars 727
MOPS: *see* minimum operational
        performance standards (MOPS)
MOS: *see* measures of suitability (MOS)
most significant bit (MSB) 89
motion pictures experts group (MPEG)
        format 900
mountain effect 184
moving mass accelerometer 191
moving target rejection (MTR) 687
MPAD: *see* man-portable air defense
        (MPAD)
MPEG format: *see* motion pictures
        experts group (MPEG) format
MPRU: *see* missile power relay unit
        (MPRU)
MQ-8 Fire Scout 854
MQ-9 Reaper 853
MRA: *see* minimum release altitude
        (MRA)
mrad 36
MSAS: *see* MTSAT satellite-based
        augmentation system (MSAS)
MSB: *see* most significant bit (MSB)
MSL: *see* mean sea level (MSL)
MSPS: *see* minimum system performance
        standards (MSPS)
MSU: *see* missile simulate unit (MSU)

MTBCF: *see* mean time between critical
        failure (MTBCF)
MTBF: *see* mean time between failure
        (MTBF)
MTF: *see* modulation transfer function
        (MTF)
MTR: *see* moving target rejection (MTR)
MTSAT: *see* multifunction transport
        satellite (MTSAT)
MTSAT satellite-based augmentation
        system (MSAS) 225, 229, 232
MTTR: *see* mean time to repair (MTTR)
multibus architecture 78
multifaceted mirrors 43
Multifunctional Information Distribution
        System (MIDS) 240
multifunction displays (MFD) 81
multifunction transport satellite
        (MTSAT) 225
multipath, effects of, on radar reception 34
multiple frequency repeater (MFR) 739
multiple intruder autonomous avoidance
        (MIAA) science and technology
        program 897
multiple track accuracy (TWS) 693
multiplex 62
multisensor flight tests 510
multi-service OT&E (MSOT&E) 969
munitions designations 766
    codes 767
MUX data transfer complete interrupt
        (MDTCI) 242–3

N
NAD-27 40–1
NAD: *see* North American Datum (NAD)
NAGU: *see* notice advisories to
        GLONASS users (NAGU)
NANU system 218
narrowband repeater noise jammer
        (NBRN) 739
NAS: *see* National Air Space (NAS)
NASA FTB 3000 (Rev A) 322
NASA Ikhana 897
National Air Space (NAS) 887, 892
National Oceanic and Atmospheric
        Administration (NOAA)
        satellite 157
National Test Pilot School (NTPS) 901,
        902
NATO standard: *see* North Atlantic Treaty
        Organization (NATO) standard
NATO threat 751
Naval Air Instruction (NAVAIRINST)
        3960.4B 11
navigation systems 167
    basic concepts
        airspeed 172–3
        altitude 173–4

        heading 170–2
        latitude 168, 169
        longitude 168, 169
        position 168–9
    data links 238
        C4I, advances in 258
        comparisons 256–7
        flight testing of 257–8
        Link 16 operation 240
        tactical background 239–40
    definition of 168
    historical background 168
    radio aids testing
        aircraft configurations and 188–9
        antenna switching and 189–90
        atmospheric effects and 190
        civil certification 190–1
        cone of confusion and 190
        magnetic variation and 190
        navigation flight testing 185–8
        navigation ground testing 184–5
        propeller modulation and 190
        statistical accuracy and 190
    radio aids to navigation 175
        DME 179–80
        NDB and ADF 181–3
        TACAN system 180–1
        VOR 177–9
NAVSTAR/GPS program 961
NBRN: *see* narrowband repeater noise
        jammer (NBRN)
NCTR: *see* noncooperative target
        recognition (NCTR)
NDB: *see* nondirectional beacons
        (NDB)
near infrared (NIR) 908
near mid air collision (NMAC) 894
NETD: *see* noise equivalent temperature
        difference (NETD)
net time reference (NTR) 247
network participation groups (NPG)
        assignments and functions 252
newer-generation aircraft 749
New Mexico State University (NMSU)
        897
night effect 184
Nighttime Readability Test Matrix 936
night vision goggles (NVG) 537, 908
    components 911–15
    generation I (Gen I) 538–9
    generation II (Gen II) 539
    generation III (Gen III) 539
    ground tests 931–4
        Operational defects 933
    human factors issues 918–20
    image defects 934
    pilot in command's (PIC) 924
    system characteristics 915–16
    Visual Acuity 932

night vision imaging systems (NVIS) 907
  aircraft flight tests 938
  aircraft lighting ground tests 935–8
  equipment, test and evaluation of 931
  exterior lighting methods 929–30
    modifications 930–1
  filters 925
  image intensification (I2) technology 909–11
  interior lighting methods 924
    modifications 925–8
  lighting specifications 920
    and guidance 923–4
    MIL-L-85762 920–2
    MIL-L-85762A 920–2
  original equipment manufacturer (OEM) 926
  supplemental type certification (STC) 924
nine-state vector 44, 47, 192
NIR: *see* near infrared (NIR)
NMAC: *see* near mid air collision (NMAC)
NMSU: *see* New Mexico State University (NMSU)
NOAA satellite: *see* National Oceanic and Atmospheric Administration (NOAA) satellite
node port (N Port) 113
noise, categories of 142
noise equivalent temperature difference (NETD) 582, 613–14
noise jammers 746
noise jamming 733, 734–8
  signal 735
noise swept amplitude modulation (NSAM) 19
noncooperative target recognition (NCTR) 691, 693
nondirectional beacons (NDB)
  and ADF 181–3
  classifications 183
non-return-to-zero invert (NRZI) encoding 96
non-return-to-zero (NRZ) encoding 67
normal alignment 200
North American Datum (NAD) 40
North Atlantic Treaty Organization (NATO) standard 33
Northrop-Grumman 756
Northrop-Grumman ABSAA 896–7
nose-on, clear region 675
NOTAM system: *see* notices to airmen (NOTAM) system
notice advisories to GLONASS users (NAGU) 226

notices to airmen (NOTAM) system 218
NRZ encoding: *see* non-return-to-zero (NRZ) encoding
NRZI encoding: *see* non-return-to-zero invert (NRZI) encoding
NSAM: *see* noise swept amplitude modulation (NSAM)
NTPS: *see* National Test Pilot School (NTPS)
NTR: *see* net time reference (NTR)
NVG: *see* night vision goggles (NVG)
NVIS: *see* night vision imaging systems (NVIS)
NVIS bezel 926

O

oblate earth effect 194, 195
OBS knob: *see* omnibearing selector (OBS) knob
OCRD: *see* operational capability requirements document (OCRD)
offset correction method 809
OFP: *see* operational flight program (OFP)
OLED: *see* organic LEDs (OLED)
omnibearing selector (OBS) knob 178
OMNIBUS 915
OMNIBUS II 915
OMNIBUS VI 915
OPA: *see* optionally piloted aircraft (OPA)
operational assessments (OA) 970
operational capability requirements document (OCRD) 961
operational flight program (OFP) 16, 819
  ballistics evaluation test 803
  weapons accuracy testing 820–1
operational reliability 971
operational requirements document (ORD) 960, 965–6
  do's and don'ts 966
  template for 965
  writing method 965
operational suitability 971
operational test and evaluation (OT&E) 2, 3–4, 957, 967–9
  AFOTEC 958
  definition 968
  versus DT&E 968
  evaluation structure 977
  follow-on OT&E (FOT&E) 969
  initial OT&E (IOT&E) 969
  multi-service OT&E (MSOT&E) 969
  operational assessments (OA) 970
  operational utility evaluation (OUE) 970
  organizations 10
  qualification OT&E (QOT&E) 969
  tactics development and evaluation (TD&E) 970

terminology 975
testers 12–13, 14–15
test plan structure 970–2
  critical operational issues 972–3
  developing 975–6
  executing 976–7
  Measure of Assessment (MOA)/ Measure of Performance (MOP) 974–5
  Measure of Effectiveness (MOE)/ Measure of Suitability (MOS) 974
  types of 969–70
  weapons system evaluation program (WSEP) 970
operational test plan, for GPS 234
operational utility evaluation (OUE) 970
operations security (OPSEC) 722
operator initiated BIT (IBIT) 755
OPSEC: *see* operations security (OPSEC)
optical systems, target signatures
  aeromechanical effects on 565
  aero-optical effects on 565
optical trackers 42
optical turbulence 565
optics 576
optionally piloted aircraft (OPA) 901–2
ORD: *see* operational requirements document (ORD)
organic LEDs (OLED) 945
organizationally unique identifiers (OUI) 116
OT&E: *see* operational test and evaluation (OT&E)
OUE: *see* operational utility evaluation (OUE)
OUI: *see* organizationally unique identifiers (OUI)
output brightness variation 934

P

packed-2 (P2) format, of message words 249
packed-4 (P4) format, of message words 249
packet identifier (PID) USB 98
PAL modes: *see* pilot assist lock-on (PAL) modes
panel lighting 928
partial binocular 947
participant location and identification (PPLI) 254–6
particular risks analysis, PCA 294
partitioning, PSSA 292
passive detection 729
passive techniques 730
patch map 651
  display 652

PATH manuals: *see* Probability of Aircraft Tactical Hazard (PATH) manuals
payload 899–901
  test and evaluation of 901
PBIT: *see* power-up BIT (PBIT)
PCM: *see* pulse code modulation (PCM)
PDA: *see* premature descent alert (PDA)
peak *versus* average power 638
pendulum 195
  device accelerometer 191
periodic synchronization gap (PSG) 92
peripheral devices 97
photoconductive radiation detectors 573
  uses 574
photoemissive radiation detectors 572
photon detectors
  low temperatures 571
  photoconductive radiation 573, 574
  photoemissive radiation 572
  photovoltaic radiation 573–4
  process 569–70
  semiconductors 574–5
photovoltaic radiation detectors 573–4
  uses 574
piezo-electric effect 199
pilot assist lock-on (PAL) modes 689
piloted UAVs 851
"Pipper," 36
pitot-static system 73
"Pit" test 795
planar array mechanism 663
Planck's law 549, 550
plan position indicator 648
plasma 141
POC: *see* point of contact (POC)
pods 544
point of contact (POC) 14
point-to-point topology 109, 113, 118
polarization 626
  signal losses due to differences in 626
ports, of Link 16 system 243
positional accuracies 223
positional DOP (PDOP) 221
positional velocity 698
position and orientation tracker (POT) 948
positioning systems airworthiness 333
  Baro-VNAV system
    flight test 347
    ground test 347
  equipment performance 333
    advisory vertical guidance 333
    Baro-VNAV system 340–2
    GNSS equipment 333
    RNAV multisensor 335
    RNP approach 337–8
    RNP general 337
    RNP terminal 338–40

general installation considerations 342–4
installed performance-test 344, flight test
  aircraft flight manual 349
  ground test 344–5
  VFR installation of GNSS equipment 348
  RNAV multisensor 346–7
positioning systems TSOA 331–2
postflight data reduction 135
postflight systems 54
POT: *see* position and orientation tracker (POT)
power loss factor 626
power ratios 626–7, 627
power-up BIT (PBIT) 755
PPLI: *see* participant location and identification (PPLI)
PPS: *see* program performance specification (PPS)
PR: *see* pulse repeater (PR)
preamble 143
precision area navigation (P-RNAV) 329–30, 329
predator communications concept 862
predator minimum detection range requirements 896
predetection integration 643
preliminary system safety assessment (PSSA) 293, 295
  backup 295
  built-in test (BIT) 293
  dissimilarity 293
  monitoring 293
  partitioning 292
  redundancy 293
Premature descent alert (PDA) 383, 385, 394
  test conditions 395, 403
Present position indicator 648
PRF: *see* pulse repetition frequency (PRF)
PRN: *see* pseudorandom noise jammer (PRN)
Probability of Aircraft Tactical Hazard (PATH) manuals 816
probability of unsafe flight (PUF) 295
program formulation 22–5
program performance specification (PPS) 13–14, 835
propeller modulation, radio aids testing and 190
proximity warning systems
  TAS 372–4
  TAWS 382–410
pseudorandom noise jammer (PRN) 740
pseudoranging 216, 216–17
  calculation 217

PSG: *see* periodic synchronization gap (PSG)
PSSA: *see* preliminary system safety assessment (PSSA)
PSTT: *see* pulse single-target track (PSTT)
PSTT Conical track 663
PSTT Lobing 662
PUF: *see* probability of unsafe flight (PUF)
pulse code modulation (PCM) 10, instrumentation system
  instrumentation system 124–30, 839, 840
    frame format 127
    process 124
    supercommutation 128
pulse compression (Chirp) 639
pulsed Doppler radar 727
pulse Doppler search and track 690
  RWS, TWS, and STT flight testing 690–3
pulse radar 652
  built-in test 653
  maximum range detections 655–9
  measurement accuracy 660–2
  minimum detection range 659–60
  observations and detections 654–5
  resolutions 665–8
  track accuracy 664
  track initiation range 662–4
  see also Radar
pulse repeater (PR) 740
pulse repetition frequency (PRF) 210, 638
  versus closure for a 1 cm wavelength 680
  versus closure for unambiguous Doppler 679
  lines 678
  with sideband frequencies 669–70
  switching 668, 670
    change in observed frequency with 670
pulse search measurement accuracy test matrix 660
pulse search minimum detection range geometry 659
pulse search R_{90} detections data card 658
pulse single-target track (PSTT) 662, 663, 664
pulse width discriminator (PWD) 747
pulse width (PW) 638
PVU enhancement 697
PW: *see* pulse width (PW)
PWD: *see* pulse width discriminator (PWD)
PW variations with range scale selection 665
PZ-90 226

Q

Q factor, INS and 47, 200
QOT&E: *see* qualification OT&E
    (QOT&E)
Q routes 333
quadrantal error 184
qualification OT&E (QOT&E) 969
quasi-zenith satellite system (QZSS) 214,
    228, 229
quick disengagement 418–19
quick look 134
QZSS: *see* quasi-zenith satellite system
    (QZSS)

R

R_{90} 655, 690
  testing 658–9, 692, 693
RA: *see* raid assessment (RA)
radar 746–7
  altimeter 683–4
  defined 626
  display processor 64
  error budget 38
  frequency designations 628
  ground array 695
  lobes 662
  lobing 662
  operations 163
  performance considerations 629–30
  position velocity update (PVU) 206
  processor 664
  raster scan 646
  reception, effects of multipath on 34
  resolution cell 665
  resolutions 665–8
  sample applications 645–51
  size as a function of frequency 631
  testing 712–13
radar absorption material (RAM) 641
radar accuracies 35–8
  degradation 38–9
  sample range handbook 36
radar band designations 34
radar center frequencies 629
radar cross-section (RCS) 33, 640,
    656–7, 684, 731
  for air-to-air missiles 792–3
  of chaff bundle 742
  effect on reflectivity of radio waves 33
  meter 50
radar detections 636–41, 641
  blip-scan ratio (BSR) 641–2
  parameters 528
  prediction of 642–5
  range, defined 630–1
  target detection 630
radar homing and warning (RHAW)
    systems 751
radar jamming 733

radar noise jamming system 735
radar receiver protection 746
Radar Roof House facility 713
radar systems 726
radar track accuracy 664
radar track initiation range 662–4
radar tracks, versions of 662
radar warning receivers (RWR) 62, 751
  frequency bands 755
  generic approach to 752
  operations 163
  self-test band 1 756
radiated emission (RE) 155
radiated susceptibility (RS) 155
radiation
  atmospheric propagation of 556–62
    absorption and scattering 556
    aerosols 557
    Beer–Lambert law for
        transmittance 556
    carbon dioxide 557, 558
    extinction coefficient 556
    gases 557
    long-wave infrared (LWIR) 558,
        562
    LOWTRAN model 561
    mid-wave infrared (MWIR) 558,
        562
    modifying phenomena in 556
    path radiance 557
    refractive index fluctuations 557
    transmittance 558–60
  into free space 51
radio aids testing
  aircraft configuration and 188–9
  antenna switching and 189–90
  atmospheric effects and 190
  civil certification 190–1
  cone of confusion and 190
  magnetic variation and 190
  navigation flight testing 185–8
  navigation ground testing 184–5
  propeller modulation and 190
  range/bearing accuracy data card 186
  statistical accuracy and 190
radio aids to navigation 175
  DME 179–80
  NDBs and ADF 181–3
  TACAN system 180–1
  VOR 177–9
radio frequency (RF)
  band designation 141
  energy 139
  slots 664
radio magnetic indicator (RMI)
  VOR display and 178
radio(s)
  communications flight testing and 139
  energy 31–2

log 846
modes of operation 143
noise 142
waves
  attenuation 629
  properties of 33–4
radome errors 713
radome test fixture 712
raid assessment (RA) 690
raid count request (RCR) 690
rail launcher 795
RAIMs: *see* receiver autonomous
    integrity monitors (RAIMs)
rain backscatter 710
RAM: *see* radar absorption material
    (RAM)
ram air turbine (RAT) 765
random Doppler (RD) 740
Range Commanders Council Telemetry
    Group 124, 130, 840
range deception jammers 739
range gate pull-off (RGPO) 737
range interface memorandum (RIM) 52–3
range predictions 603–7
  adjustments based on atmospherics
      606–7
  calculating specific targets 604
  critical dimension and 603
  critical distance and 604
  operational considerations and 607
  spatial frequency and 603
range(s) 21
  ambiguity 654
  assets 21
  differentiation 668
  gates 644
  interfacing with 51–3
  rate resolution setup 691
  resolution 638, 646
    geometry 665
    sample data card 666
range-while-search (RWS) 690
RAT: *see* ram air turbine (RAT)
RCR: *see* raid count request (RCR)
RCS: *see* radar cross section (RCS)
RCT
  for fictitious aircraft and associated
      systems 961
RD: *see* random Doppler (RD)
RE: *see* radiated emission (RE)
readout integrated circuit (ROIC) 545
real-time data 24
  analysis and 848
  defined 132
  importance of 132
  reduction and analysis 132–4
real-time telemetry 54, 130–1
  see also Telemetry
Real-time Transport Protocol (RTP) 900

received data (RXD) 82
receiver autonomous integrity monitors (RAIM) 45, 221
Receiving antenna 626
Recurrence rate number (RRN) 244–6
reduced vertical separation minimums (RVSM) 329, 348–56
  airworthiness approval 352–4
  errors 355–6
  loss of approval 355–6
  minimum equipment requirements for 350
  operational approval 354–5
  system performance 350
redundancy, PSSA 293
Reed-Solomon (R-S) encoding 248
reflection 140, 927–8
refraction 140
regression testing 17
relative navigation (RelNav) 254–6
release geometry
  air-to-ground 806–12
reliability 971, 979
  demonstrated/observed reliability 979
  end-of-test reliability 979
  projected (mature system) reliability 979
reliability, maintainability, logistics supportability, and availability (RML&A) 978–81
reliability ground tests, for air-to-air missile 796–7
remotely piloted vehicle (RPV) 854
Remote terminals (RT) 64
  arrangement 77
Remote transmission request (RTR) bit 85
repeater swept amplitude modulation (RSAM) 740
report writing
  background 848
  conclusions 849
  executive summary 849
  instrumentation 849
  items in 849
  objectives 849
  recommendations 849
  results 849
  signoff 849
  supporting data 849
  test method 849
  via operational instruction 848
  witnessing sheet 849
request to send (RTS) 83
required obstacle (terrain) clearance (ROC) 384
reticle matching 36
Return-to-Zero (RTZ) tristate modulation ARINC 429 87–8

RF: see radio frequency (RF)
RGPO: see range gate pull-off (RGPO)
RHAW systems: see radar homing and warning (RHAW) systems
rhyme list
  for speech intelligibility test 153
RIM: see range interface memorandum (RIM)
ring laser gyro (RLG) INS 197–8
RIVET JOINT 239
RMI: see radio magnetic indicator (RMI)
RNAV multisensor 335
RNP Approach 337–338
RNP General 337
RNP Terminal 338–340
Robert Bosch GmbH 84–5
ROC: see required obstacle (terrain) clearance (ROC)
rockets 827
Rockwell Collins 121
ROIC: see readout integrated circuit (ROIC)
round-trip timing (RTT) message 249
RPV: see remotely piloted vehicle (RPV)
RQ-4A Global Hawk 853
RR-180 chaff cartridge 742, 742
RRN: see recurrence rate number (RRN)
RS-422 84
RS-423 84
RS-485 84
RS: see radiated susceptibility (RS)
RSAM: see repeater swept amplitude modulation (RSAM)
RS-232C 81–4
  basic configuration of 82, 83
  null modem configuration 83
  signal levels and character frame 82
R-S encoding: see Reed-Solomon (R-S) encoding
RT: see remote terminals (RT)
RTCA DO-178 295
RTCA/DO-309 383
RTCA/DO-161A 382
RTP: see Real-time Transport Protocol (RTP)
RTR bit: see remote transmission request (RTR) bit
RTS: see request to send (RTS)
RTT message: see round-trip timing (RTT) message
Ruggedization of equipment 75
RVSM: see reduced vertical separation minimums (RVSM)
RWR: see radar warning receiver (RWR)
RWS: see range-while-search (RWS)
RWS measurement accuracy 691
RXD: see received data (RXD)

S
SA: see source address (SA)
SA-2 Fan Song 727
SAE Aerospace Recommended Practice ARP 5583 157
SAE AIR 1093 322
SAE ARP 1874 322
SAE ARP 4102 322
SAE ARP 4107 322
SAE ARP 571C 322
SAEAS 8034 322
safety assessment, for highly integrated or complex systems 290–6
  CCA 294
  certification plan 296–8
    configuration index 296
  FHA 290–2
  PSSA 293, 295
  requirements 297
  SSA 295
  validation and verification (V&V) methods 296–7
safety-of-flight (SOF) analysis 825
safety review boards (SRB) 977
Sagnac, G., 197
SAL: see system address label (SAL)
SAM: see situation awareness mode (SAM); surface-to-air missile (SAM)
SAR: see synthetic aperture radar (SAR)
SAR helicopter: see search and rescue (SAR) helicopter
satellite-based augmentation systems (SBAS) 225–8
Satellite Navigation Augmentation System (SNAS) (China) 229
satellite systems
  Beidou 227
  Galileo 225–8
  GLONASS 225–6
  Indian Regional Navigation Satellite System (IRNSS) 228
  message content 215
  Quasi-Zenith Satellite System (QZSS) 228
saturation 731
SAW: see selectively aimable warhead (SAW)
S band 630, 631
SBAS: see satellite-based augmentation systems (SBAS)
scanning systems 568–9
  direct imaging 568
  optical principles 568
scattering 140
scheduling 50–1
Schuler cycle 147
Schuler effect, true error and 204
Schuler-tuned platforms 204

Schuler tuning 194
SDU: *see* secure data unit (SDU)
SEAD: *see* suppression of enemy air
    defenses (SEAD)
SEAM: *see* sidewinder expanded
    acquisition mode (SEAM)
search and rescue (SAR) helicopter 234
Secret Internet Protocol Router Network
    (SIPRNet) 863
sectional aeronautical charts 170
sector attenuation response 745
secure data unit (SDU) 249
seeker head 764
seeker push-ahead response 745
seeker push-pull 745
seeker tests, for air-to-air missiles 791–2
selectively aimable warhead (SAW)
    788–9
self-protection flares 743–5
self-protection jammers 758
self-protection jamming systems 733
self-protect jammer operations 163–4
semiactive homing missile
    AIM-7 Sparrow 774, 776, 777
    speed gates 775
semiconductors 574–5
    extrinsic 574
        characteristics 575
    intrinsic 574
sense and avoid systems 892
    airworthiness of 893–5
    developments in 895–9
    test and evaluation of 899
sensitivity time constant (STC) 712
sensitivity time control (STC) 746
sensor fusion 48–9
sensor slew rate 609–10
separation effects testing 803, 820
service failures, for air-to-air missile 797
SFAR: *see* Special Federal Aviation
    Regulation (SFAR)
SFD: *see* start frame delimiter (SFD)
SG: *see* synchronization gap (SG)
shading 933–4
shoreline effect 184
short pulse detection 659
SICP: *see* subscriber interface control
    program (SICP)
sidelobe blanking (SLB) 747
sidelobe cancellation (SLC) 747
sidelobe clutter (SLC) 672
    spectrum 673
sidelobes 672
    blanking or cancellation 712
side lobe suppression (SLS) 236, 747
side-looking radars (SLARs) 728
sidewinder expanded acquisition mode
    (SEAM) 780, 782, 784
signal interface unit 64

signal losses, differences in polarization
    626
signal-to-noise ratio (SNR) 637–8, 913,
    916
SIL: *see* software integration lab (SIL)
simple memory response 745
simulations 7, 748
simulators, major drawbacks to 6
single-engine approach 418
single-look probability 655
single-pilot cockpit 749
single-scan probability 655
single target track (STT) 689, 690
    flight testing 691, 693
SINPO rating scale 152
SIPRNet: *see* Secret Internet Protocol
    Router Network (SIPRNet)
situation awareness mode (SAM) 690
Skyflash, the 786–7
sky wave 141
SLAM: *see* stand-off land attack missile
    (SLAM)
slant range 673
SLARs: *see* side-looking radars (SLARs)
SLB: *see* sidelobe blanking (SLB)
SLS: *see* side lobe suppression (SLS)
Smiths Industries 91
SMP: *see* stores management processor
    (SMP)
SMS: *see* stores management system
    (SMS)
SNAS: *see* Satellite Navigation
    Augmentation System (SNAS)
SNR: *see* signal-to-noise ratio (SNR)
SOF analysis: *see* safety-of-flight (SOF)
    analysis
SOF field: *see* start of frame (SOF) field
software build contents 17
software integration lab (SIL) 14
software support and integration (SSI) 14
SOJ: *see* stand-off jamming (SOJ)
solar noise 142
sole-means accuracy, performance
    requirements for 224
sole-means system accuracy 224
SOT word: *see* start of transmission
    (SOT) word
source address (SA) 116
Soviet Azovsky MAK-UT MAWS 756
Space Debris and Tracking System
    (SPADATS) 728
space noise 142
space surveillance radars 728
SPADATS: *see* Space Debris and
    Tracking System (SPADATS)
spatial frequency 598
    airborne test 602
    angular resolution and 611
    board

light intensity at 601
    placement 601, 602
    temperature measurements 601
    ground test 601, 602
    plotting 602
    scientific method for 599–600
    vertical or horizontal 603
Special Federal Aviation Regulation
    (SFAR) 273
speech intelligibility 151–4
speech transmission index (STI) 152
speed protection 414–16
Sperry, Lawrence 410
spot jammers, advantage of 736
SRB: *see* safety review boards (SRB)
SRR bit: *see* substitute remote request
    (SRR) bit
SSA: *see* system safety assessment (SSA)
SSI: *see* software support and integration
    (SSI)
SSV: *see* standard service volume (SSV)
stabilization 211, 705
    aircraft 635, 647
    antenna 327
    DNS 209
    earth 635, 647
    gyro system 195
    line-of-sight (LOS) 582
    radar 653–4
stadiametric ranging 36
STANAG 3910 93–6
    high-speed action and status words 94
    high-speed message frame format 96
    message formats 94
    type A network 93
    types of 93
STANAG 4586 856
stand-alone GPS receivers 44–6
stand-alone remote terminal 64
standard lighting 928
standard service volume (SSV) 178
stand-off jamming (SOJ) 733
stand-off land attack missile (SLAM)
    764
start frame delimiter (SFD) 116
start of frame (SOF) field 85
start of transmission (SOT) word 89
static test 144–5
statistical accuracy
    radio aids testing and 190
status bits 72
status words
    synchronization pattern 67–8
    1553 systems 71–3, 101
    3910 systems 94
STC: *see* sensitivity time constant (STC);
    sensitivity time control (STC);
    supplemental type certificate
    (STC)

"Stealth" aircraft 33
Stealth Sea Shadow 685
Stefan–Boltzmann law 548
STI: *see* speech transmission index (STI)
stored heading alignment 200
stores, communications flight testing and 149
stores management processor (SMP) 767, 768
stores management system (SMS) 63
  architecture 769
  combinations of weapons 771
  documentation 835
  interfaces 780, 781, 784
  inventory 770
  key interlocks 784
  processes 769–72
    cold start 770
    warm start 769
  1553 relationship 768
  requirements for 767
  test and evaluation 785
  weapons interfaces
    AIM-120 AMRAAM 775, 779
    AIM-9 Sidewinder 772, 773, 774
    AIM-7 Sparrow 774, 777
    bombs 778–80, 783
    guns 776, 778, 781
strap-down inertial systems 197–9
STT: *see* single target track (STT)
student's t variable 656
subscriber interface control program (SICP) 242
substitute remote request (SRR) bit 86
subsystem/system development 8–9
supercommutation 128
supplemental type certificate (STC) 274, 300
  IS&S for 279
  process for obtaining 274
supportability, defined 972
suppression of enemy air defenses (SEAD) 720–1, 731
suppression of IR radiation 731
surface-to-air missile (SAM) 651, 720
surrogate UAVs (SUAV) 851
sustainment, defined 972
SUT: *see* system under test (SUT)
swerling cases 645
swerling curves 645
switched fabric topology 110, 113
synchronization gap (SG) 90
synthetic aperture radar (SAR) 651, 706–10
  flight testing of 710
system address label (SAL) 89
system response time 751, 760

system(s)
  boresighting 810
  integration 8
    air-to-air missile 797–803
system design assurance (SDA) parameter 377
system safety assessment (SSA) 295, 378
systems interface unit 64
system under test (SUT) 842

T
TA: *see* terrain avoidance (TA)
TACAN system: *see* tactical air navigation (TACAN) system
tactical air navigation (TACAN) system 18, 143, 180–1
  frequency pairings 182
  ground station 180
  signal pattern 181
tactical air-to-air radar parameters 646
tactical air-to-ground radar parameters 647
tactical digital information link (TADIL) 239
tactical targeting network technology (TTNT) 131, 259
tactics development and evaluation (TD&E) 970
TAD: *see* towed air decoy (TAD)
TADIL: *see* tactical digital information link (TADIL)
TADIL-A 239
TADIL-B 239
TADIL-C 239
TADIL-J 240
  message catalog 240
takeoff and go-around (TOGA) 304
target aspect angle 52
targets and target availability 50
target signatures 562–7
  airborne targets 566, 567
  with area-weighted target temperature 563
  attributes of 566
  diurnal cycle 564
  high-energy lasers (HEL) 566
  optical systems
    aeromechanical effects on 566
    aero-optical effects on 565
  optical turbulence 565
  surface targets 566
  thermal lens 566
target tracking 612
target velocity, effects of opening 676
TAS: *see* traffic advisory system (TAS); true airspeed (TAS)
TAWS: *see* terrain awareness and warning system (TAWS)
TAWS-based systems 893
TC: *see* type certificate (TC)

TCAS: *see* traffic alert and collision avoidance systems (TCAS)
TCP Reno 866
TD&E: *see* tactics development and evaluation (TD&E)
TDMA: *see* time division multiple access (TDMA)
team member responsibilities 13–15
technical standard order (TSO) 238, 282–4, 299
  component identification in 284
  GPS certifications 331
  interchangeability in 284
  mixing of TSO and non-TSO functions in 284
TEC map: *see* total electron content (TEC) map
telemetry 51, 54, 124
telemetry network system (TmNS) 130
telemetry station clocks 53
television (TV) 62
  analog TV 583–5
  digital TV 585–6
TEMP: *see* test and evaluation master plan (TEMP)
TEMP documents 964
terminal gap (TG) 90, 92
terminal input messages (TIMs) 240
  functions 242
terminal output messages (TOMs) 240
  functions 242
terrain avoidance (TA) 692, 698–702
  with digital terrain database 700
  flight test evaluations 702–5
  with radar sensor 701
terrain awareness and warning system (TAWS) 299, 382
  alerts
    class C equipments 401–3
    internal priority scheme 388
    prioritization scheme 388
    standard for visual and aural 385–7, 389
  aural alert, standard for 385–7, 389
    class C 401, 402
  class C equipment
    cruise 401
    landing 401
    MSL altitude 401
    phases of flight 401
    RTC 401
    takeoff 401
  class C evaluations 401–8
  documentation 382
  evaluations 390
    autoranging 390
    installation and flight deck integration 390
    pop-up feature 390

failure monitor function 389
flight test 391–401
  advisory callout 400
  excessive closure rate to terrain
    396–7
  excessive downward deviation from
    an ILS glideslope 400
  excessive rate of descent 396
  flight into terrain when not in
    landing configuration 398
  for FLTA 392–4
  installation types and 391–2
  negative climb rate or altitude loss
    after takeoff 397–8
  for PDA function 394, 395
  wind shear warnings 401
  GPS internal 385
  ground test 391
  HTAWS 409–10
  map itself 385
  onboard navigation system 385
  phases of flight for 389
    approach 389
    departure 389
    enroute 389
    terminal 389
  visual alert, standard for 385–7, 389
    class C 401
terrain following processor (TFP) 699,
  701, 704
terrain following (TF) 648, 698–702
  flight test evaluations 702–5
  step-down logic 702
  transverse slope 704
terrain masking 731
test and evaluation master plan (TEMP)
  966–7
  format 966–7
  logistical support 967
  scope of 967
test conductor's summary sheet 847
test information sheets (TIS) 835
test matrix
  communications systems 146–8
  executing 148–9
test planning, for HARM integration
  program
  data requirements 838–41
  generic organizational chart 841
  requirements 841–3
  test matrix 843
  test summary sheets 835–7
test planning working groups (TPWG)
  53
test review boards (TRB) 977
test summary sheets (TSS) 835–7
test team composition 11–13
tetrahedral inertial platforms (TIP) 199

TF: see terrain following (TF)
TFP: see terrain following processor
  (TFP)
TFT: see thin-film transistor (TFT)
TF/TA tower detection and classification
  sequence 704
TG: see terminal gap (TG)
theodolite
  instrumentation 42
  limitations 44
thermal detectors
  bolometer 575
  calorimeter 575
  evaporograph 575
  fluid-expansion thermometer 575
  thermocouple 574
thermal emission 554, 555
thermal imaging systems 536, 908–9
  electrical signal processor 542
  EMUX FLIR 542, 543
  first-generation systems 542
  FLIR 540, 541, 542, 544
  scanning systems 542–3
  second generation system 542
  third generation system 545
thermal inertia 555, 564
thermal lens 566
thermocouple 574
thin-film transistor (TFT) 580
1553 word types 67
three-plane captive carry tests 799
thunderstorm effect 184
TI: see transmit interval (TI)
TIA: see type inspection authorization
  (TIA)
TIARA: see Tornado Integrated Avionics
  Research Aircraft (TIARA)
time 764
time alignment 53–4, 132–3
time division multiple access (TDMA)
  91
  architecture 240
time on target term 640
time slot block (TSB) 246
  defined 253
time slot duty factor (TSDF) 251
time space position information (TSPI)
  10, 29–30, 714
  assets 21, 47
  clocks 53
  data 661
  devices 45
  nine-state vector 47
  sources 38–9
    summary of 30
  system 35–6, 44
time stamping of data 133
time to align, for INS 199–200

TIMs: see terminal input messages (TIMs)
TIP: see tetrahedral inertial platforms
  (TIP)
TIS: see test information sheets (TIS)
TmNS: see telemetry network system
  (TmNS)
TOGA: see takeoff and go-around
  (TOGA)
TOMs: see terminal output messages
  (TOMs)
Tornado Integrated Avionics Research
  Aircraft (TIARA) 948
total electron content (TEC) map 219
Total latency 378
Total System Error (TSE) 337
towed air decoy (TAD) 50
towed decoys 731
TPIU: see transmission protocol interface
  unit (TPIU)
TPWG: see test planning working groups
  (TPWG)
track breaker 738, 739
tracking radars, on same target 48, 49
track initiation range 664
track-while-scan (TWS) 689, 690, 727
traffic advisory system (TAS) 368, 372–4
  class A equipment 372–3
    defined 372
  class B equipment 373–4
    defined 372
traffic alert and collision avoidance
  system (TCAS) 237, 356, 895
  advisories 356
  Airworthiness Versions 7.0 and 7.1
    367–368
  display 359–60
  ground test 364–5
  inhibits 360, 362
  minimum reaction time 357
  resolution advisories (RA) 356, 359
    alert thresholds 359
    aural alerts 360
  TCAS I 357
  TCAS II 357–8, 893
    certification 361–4
    mode S transponders 357, 363–4
  tests and evaluation 364
    flight tests 365
    ground tests 364–5
    knock it off 365
    planned encounter flight tests 365–6
    surveillance flight tests 366–7
  traffic advisories (TA) 358
    alert thresholds 359
    aural alerts 360
  versions 357–8
Traffic Information Services-Broadcast
  (TIS-B) 375

trajectory simulations, air-to-air missiles 791–2
transmission control protocol (TCP) 861, 866, 867–9
  data buffers 868
transmission properties, of IR radiation 550
transmission protocol interface unit (TPIU) 95
transmit interval (TI) 90, 92
transmitted data (TXD) 82
transmitter frequency 682
transverse flight 704
TRB: *see* test review boards (TRB)
troposphere 140–1
  effects on wave frequencies 141
T routes 333
true airspeed (TAS) 172
true error, Schuler effect and 204
trusted agents 968
truth data 839
TSB: *see* time slot block (TSB)
TSB A-2-11 246
TSDF: *see* time slot duty factor (TSDF)
TSE: *see* Total System Error (TSE)
TSO: *see* technical standard order (TSO)
TSOA 331–2
TSO C-112 238
TSO C-129 sensor 333
TSO C-161a 332
TSO C-162a 332
TSO-C194 383
TSO C-196 sensor 333
TSO-C47c 238
TSO-C151c 382
TSO C-145c and 146c equipment 331
TSPI: *see* time space position information (TSPI)
TSS: *see* test summary sheets (TSS)
TTNT: *see* tactical targeting network technology (TTNT)
TV: *see* television (TV)
twisted pair shielded cable 64
two-color switch sample IR 745
TWS: *see* track-while-scan (TWS)
TXD: *see* transmitted data (TXD)
type certificate (TC) 274, 300
Type Certification Board 277–82
  primary functions 277
type inspection authorization (TIA) 277

U
UAS communications architecture 860
  data link examples 862–5
  latency 865–9
  UAV communications evaluations 869–85
UAV: *see* unmanned aerial vehicles (UAV)

UDP: *see* User Datagram Protocol (UDP)
UH-60 Blackhawk helicopters 156
UHF bands: *see* ultra-high-frequency (UHF) bands
ULQ-21 countermeasures set 750
ultra-high-frequency (UHF) bands 141
universal serial bus (USB) 96–9
  packet formats 98
  tiered star topology 97
  transfer types 98
universal time code (UTC) 226
unmanned aerial vehicles (UAV) 851
  airworthiness certificate 857–8
    military airworthiness requirements 859–60
  autopilots 889
    considerations 891–2
    evaluations 889–91
  BER testing 876
  categories 855
  civil operation *versus* public operation 859
  command directed vehicle (CDV) 854
  communications evaluations 869
    airborne coverage 883–4
    antenna patterns 874
    communications ground tests 876–7
    data link latency evaluations 884–5
    EMI/EMC 874–6
    engineering analysis 869–74
    human factors and controls and displays 877–82
    USAR data link considerations 885
    USAR human factors and controls and displays considerations 882–3
  data link uses and requirements 862
  interoperability 856–7
  link equation contributors 870
  long-endurance multi-intelligence vehicle (LEMV) 854
  maximum los/antenna coverage 881
  modified Cooper Harper scale 879
  navigation 886
    considerations 887–9
    evaluations 886
  operating modes of 854
  optionally piloted aircraft (OPA) 901–2
  payload 899–901
    test and evaluation of 901
  pilot-induced oscillation (PIO) 880, 885
  remotely piloted vehicle (RPV) 854
  sense and avoid systems 892
    airworthiness of 893–5
    developments in 895–9
    test and evaluation of 899
  types 852–5

UAS communications architecture 860–2
  data link examples 862–5
  latency 865–9
  warning systems 862
Unmanned Aerial Vehicle Systems Airworthiness Requirements (USAR) 859–60
  data link considerations 885
  human factors and controls and displays considerations 882–3
  navigation considerations 887–9
Unmanned Aircraft Systems Flight Test Center (FTC) 897
updates, navigation 206–7
  via onboard sensor 206
UR: *see* utilization rate (UR)
U.S. Air Force (USAF) 11
U.S. Army Nike Hercules Air Defense Systems 30–1
U.S. Department of Defense (DOD) 72, 130, 239
U.S. Geological Survey (USGS) 170
U.S. Navy (USN) 11
U.S. Space Command 217
U.S. Standard for Terminal Instrument Procedures (TERPS) 384
USAARL: *see* US Army Advanced Research Laboratory (USAARL)
USAR: *see* Unmanned Aerial Vehicle Systems Airworthiness Requirements (USAR)
US Army Advanced Research Laboratory (USAARL) 943
USB: *see* universal serial bus (USB)
User Datagram Protocol (UDP) 900
user needs 960
  evaluation 961
USGS: *see* U.S. Geological Survey (USGS)
UTC: *see* universal time code (UTC)
utilization rate (UR) 980

V
variable bit rate (VBR) 899, 900
variable-format messages 249
V-beam radar 726
VBR: *see* variable bit rate (VBR)
VCS: *see* visually coupled system (VCS)
VDB antenna 343
VDOP: *see* vertical dilution of precision (VDOP)
vector word format
  MIL-STD-1760 103
vehicle programs 8
velocity accuracy, acceleration and 205–6
velocity gate pull-off (VGPO) 737
velocity search 687–8
  test of MTR feature 688

vertical dilution of precision (VDOP) 39
vertical DOP (VDOP) 221
very-high-frequency (VHF) bands 141
VGPO: *see* velocity gate pull-off (VGPO)
VHF bands: *see* very-high-frequency (VHF) bands
VHF omnidirectional range (VOR) 177–9
  display, RMI and 178
  radials emanating from 177
video, flight analysis and reporting 848
visual acuity 932, 937
  charts 933
visually coupled system (VCS) 948
VOR: *see* VHF omnidirectional range (VOR)
VOR test facility (VOT) 184
Vulcan cannon 780

W
WAAS: *see* wide area augmentation system (WAAS)
wander angle 196–7
warhead section, of air-to-air missiles 788–9
warhead tests, for air-to-air missiles 792
"Warm fuzzy" test 145, 685
warning systems 862
warning system (TAWS)
  functions 383
WASP 852
waves
  interaction of, with obstructions 140
  propagation of 140–1
weapon(s)
  boresighting 809

delivery considerations for helicopters 824–5
five basic classes of 765–6
free-stream ballistics test 803
integration, fusing mechanisms 764–5
integration testing, GPS 233–4
inventory 785
prep 785
release sequence 63–4
sequencing 785
weapons inhibits, limits, and interrupts (WILIs) 825
weapons system accuracy report (WSAR) 10
weapons system evaluation program (WSEP) 970
weather, communications flight testing and 149–50
weather radar
  certification 322–9
    antenna stability 327
    beam tilting and structural clearances 325–6
    contour display (iso echo) 326
    display 325
    electromagnetic compatibility 327–9
    ground mapping 327
    mutual interference 327
    range capabilities 325
    specific tests 324–9
    stability 327
    warm-up period 325
  parameters 550
  precipitation 313–14
weight-on-wheels (WOW) switch 52, 784
WGS-72 40

WGS-84 40–1, 52, 101
WGS: *see* World Geodetic System (WGS)
wide area augmentation system (WAAS) 45, 48, 225, 229, 330
  LPV approach 230–1
  reference and master stations 230
  simplified architecture 229
Wien's displacement law 549
Williams, James H., 282
wind effects, on heading and airspeed 173
word formats 1553 systems 68
word grabbers 25, 64
World Geodetic System (WGS) 40
World Meteorological Organization Sea States 686
WOW switch: *see* weight-on-wheels (WOW) switch
WSAR: *see* weapons system accuracy report (WSAR)
WSEP: *see* weapons system evaluation program (WSEP)

X
X-band
  beamwidth rule of thumb 706
  radar 630, 631
  system 647
X-EYE jammer: *see* cross-eye (X-EYE) jammer
X-POLE: *see* cross-polarization (X-POLE)

Y
Yom Kippur War 720

Z
zonal safety analysis, CCA 294
Zulu time 143, 237